Artificial Intelligence (AI) and Machine Learning (ML) in Human Health and Healthcare

Artificial Intelligence (AI) and Machine Learning (ML) in Human Health and Healthcare

Editor

Mahmudur Rahman

MDPI • Basel • Beijing • Wuhan • Barcelona • Belgrade • Manchester • Tokyo • Cluj • Tianjin

Editor
Mahmudur Rahman
Computer Science
Morgan State University
Baltimore
United States

Editorial Office
MDPI
St. Alban-Anlage 66
4052 Basel, Switzerland

This is a reprint of articles from the Special Issue published online in the open access journal *Healthcare* (ISSN 2227-9032) (available at: www.mdpi.com/journal/healthcare/special_issues/AIML_Healthcare).

For citation purposes, cite each article independently as indicated on the article page online and as indicated below:

LastName, A.A.; LastName, B.B.; LastName, C.C. Article Title. *Journal Name* **Year**, *Volume Number*, Page Range.

ISBN 978-3-0365-3742-9 (Hbk)
ISBN 978-3-0365-3741-2 (PDF)

Contents

About the Editor

Mahmudur Rahman

Dr. Mahmudur Rahman is currently an Associate Professor and Program Director of BS for the Cloud Computing program at Morgan State University, Baltimore, Maryland, USA. He received his PhD (2008) in Computer Science from Concordia University, Montreal, Canada, with an focus on medical informatics and image retrieval. Prior to joining as an Assistant Professor at Morgan State University in 2014, Dr. Rahman extensively conducted research at the National Institutes of Health (NIH), USA, for almost six years as a Research Scientist. He has good expertise in the fields of data science, AI and MaFchine learning, image processing and computer vision, database and information retrieval, and their applications to classification, annotation, and the retrieval of biomedical images from large collections. Dr. Rahman's is currently researching crowdsourcing and deep learning techniques and their application in the medical field.

healthcare

MDPI

Article

Intelligent IoT (IIoT) Device to Identifying Suspected COVID-19 Infections Using Sensor Fusion Algorithm and Real-Time Mask Detection Based on the Enhanced MobileNetV2 Model

Rupali Kiran Shinde [1], Md. Shahinur Alam [1], Seong Gyoon Park [2], Sang Myeong Park [1] and Nam Kim [1,*]

1 Department of Information and Communication Engineering, Chungbuk National University, Cheongju 28644, Korea; rups@chungbuk.ac.kr (R.K.S.); shahinur@chungbuk.ac.kr (M.S.A.); smpark11@korea.kr (S.M.P.)

2 Department of Smart Information Technology Engineering, Kongju National University, Gongju 32588, Korea; psk@kongju.ac.kr

* Correspondence: namkim@chungbuk.ac.kr; Tel.: +82-43-261-2482

Abstract: This paper employs a unique sensor fusion (SF) approach to detect a COVID-19 suspect and the enhanced MobileNetV2 model is used for face mask detection on an Internet-of-Things (IoT) platform. The SF algorithm avoids incorrect predictions of the suspect. Health data are continuously monitored and recorded on the ThingSpeak cloud server. When a COVID-19 suspect is detected, an emergency email is sent to healthcare personnel with the GPS position of the suspect. A lightweight and fast deep learning model is used to recognize appropriate mask positioning; this restricts virus transmission. When tested with the real-world masked face dataset (RMFD) dataset, the enhanced MobileNetV2 neural network is optimal for Raspberry Pi. Our IoT device and deep learning model are 98.50% (compared to commercial devices) and 99.26% accurate, respectively, and the time required for face mask evaluation is 31.1 milliseconds. The proposed device is useful for remote monitoring of covid patients. Thus, the method will find medical application in the detection of COVID-19-positive patients. The device is also wearable.

Keywords: COVID-19; enhanced MobileNetV2; IoT device sensor fusion; suspect detection and s tracking; face mask detection; remote monitoring

Citation: Shinde, R.K.; Alam, M.S.; Park, S.G.; Park, S.M.; Kim, N. Intelligent IoT (IIoT) Device to Identifying Suspected COVID-19 Infections Using Sensor Fusion Algorithm and Real-Time Mask Detection Based on the Enhanced MobileNetV2 Model. *Healthcare* **2022**, *10*, 454. https://doi.org/10.3390/healthcare10030454

Academic Editors: Mahmudur Rahman and Joaquim Carreras

Received: 31 December 2021
Accepted: 24 February 2022
Published: 28 February 2022

Publisher's Note: MDPI stays neutral with regard to jurisdictional claims in published maps and institutional affiliations.

1. Introduction

In December 2019, a pneumonia-like disease began to spread worldwide, accompanied by fever and cold-like symptoms [1,2], caused by the COVID-19 (Coronavirus disease of 2019) virus [3,4]. The World Health Organization (WHO) declared COVID-19 a Public Health Emergency of International Concern on 30 January followed by declaration of pandemic on 11 March 2020. pandemic affects people's mental and physical health. To date, 401 million COVID-19 cases have been detected, with 5.76 million deaths confirmed. The increasing number of COVID-19 cases and deaths have led to worldwide lockdowns, quarantines, and restrictions on human movements. Abdulkadir Atalan mentioned that lockdowns could suppress the spread of the virus. Reference [4] also mentioned the effects of lockdowns on psychology, the environment, and the economy. Various studies have shown the effects of lockdowns on economics, domestic abuse, mental health, and social health [5].

Even though many types of vaccines are in the market, but there are new virus strains coming due to mutations. Vaccinating the entire world population is an ideal way to stop pandemics, but many countries are poor, and their healthcare systems are not advanced enough to provide vaccine for all population. Moreover, H.C Hsu presented the effects of COVID-19 on healthcare workers; for example, nurses are overworking and are under pressure; thus, it will take a long time to reach an ideal situation [6]. In the era of globalization, it has been difficult to travel during pandemic conditions. At present, the omicron strain is a major concern worldwide and many countries have announced restrictions on gathering and traveling, causing harm to economies and social welfare. Early testing and tracing are used to control the number of cases and outbreaks.

Here, we present an Internet-of-Things (IoT)-based device for early detection of infected subjects and to control spread via face mask detection. IoT devices collect and share data (with minimal human interaction) using various transfer protocols [7]. IoT applications are used in healthcare and smart factories, homes, and education. A fitness band is an IoT-based wearable device that monitors user activities and health. Lockdowns create economic and mental health difficulties [8]. This paper presents a wearable device that detects suspected COVID-19-infected individuals.

Dong et al. [9] developed a wearable device for continuous blood pressure monitoring [10]; this device did not store health data for future analysis. Aadil et al. [11] described a wireless body area network (WBAN) that used the IoT for remote health monitoring. A ZigBee network was implemented by Li et al. [12] to connect devices to a base station. Fu et al. [13] utilized a wireless sensor network and a Wi-Fi transmission protocol to measure blood oxygen levels in athletes but this paper focuses only on one health parameter, which makes overall health-checking difficult. The literature indicates that Wi-Fi protocols are appropriate and cost-effective for wearable devices. Artificial intelligence (AI) has played an important role during the pandemic. AI algorithms have been used to identify COVID-19 infections using features extracted from electrocardiograms or chest x-rays. Machine-learning algorithms that rapidly analyzed blood samples were 90% accurate when used to estimate the survival of COVID-19-infected patients [14,15]. M. Phan et.al. proposed a patent to detect COVID-19 using breathing data trained on IoT devices, but the sample size of the data was small [16]. Several authors have used deep learning techniques for face mask detection. The databases includes Kaggle, the face mask label dataset (FMLD), the masked face analysis (MAFA) dataset, and the real-world masked face recognition dataset (RMFRD) [17,18]. The YOLOv2, YOLOv3, SSDMNV2, MobileNetV2, and ResNet50 deep learning models for face mask detection are over 95% accurate. Some models are compatible with IoT platforms; others require high-performance graphic processing units (GPUs) [19–21].

This paper presents a preventive approach to avoid virus outbreaks and control the pandemic. The major contribution of this work is the application of the sensor fusion method for covid detection automatically using artificial intelligence. The proposed device takes percussion to avoid false-positive alerts. False-positives will create trouble for the healthcare system instead of helping it. The enhanced MobileNetV2 model is the optimal solution for IoT platforms due to the small model size, higher accuracy, and lower detection time.

Here, artificial intelligence (AI) is used to aid healthcare systems. This work detects and traces infected persons in real-time; this limits viral spread and outbreak. Automatic and correct locations of masks detected that control spread. This method is preventative and rapid. This paper is divided into five sections. Section 2 focuses on the proposed method, Section 3 presents the experimental setup, the results and discussion are presented in Section 4, and the conclusion and future scope are presented in Section 5.

2. The Methodology

The proposed method uses a sensor fusion (SF) algorithm to detect infected suspects in the early stage of infection and detect face masks. We implemented a deep learning model on an IoT platform. The decision-making intelligence was provided by the SF algorithm and the deep learning model. Section 2.1 explains the SF algorithm and Section 2.2—face mask detection. The overall architecture of intelligent IoT (IIoT) devices is shown in Figure 1, with separate layers and the functionality of each layer. The data flow is shown in Figure 2, as well as the feature data collection and processing by the SF and deep neural network

(DNN) algorithms, along with hardware and software components used in the system. SF merges sensory inputs from various channels to improve the information (compared to that available if the sources is used separately) [22]. SF finds applications in autonomous cars [23], robotics [24], and biomedical appliances [25]. To the best of our knowledge, this is the first work to use SF for COVID-19 disease prediction. The SF algorithm fuses inputs from blood oxygen, body temperature, and heart rate sensors. Low oxygen levels and fevers are the most common symptoms in COVID-19 patients; these are often misunderstood as normal colds in the early stages of the disease. Our method focuses on these three factors. Even if only one symptom is apparent, the AI algorithm sends an android alert of the unusual reading. The subject can now consider self-isolation and a possible need for medical care. The proposed approach does not detect asymptomatic people. This method does not confirm infection but, rather, anticipates who might be infected with COVID; this assists in early testing and tracing.

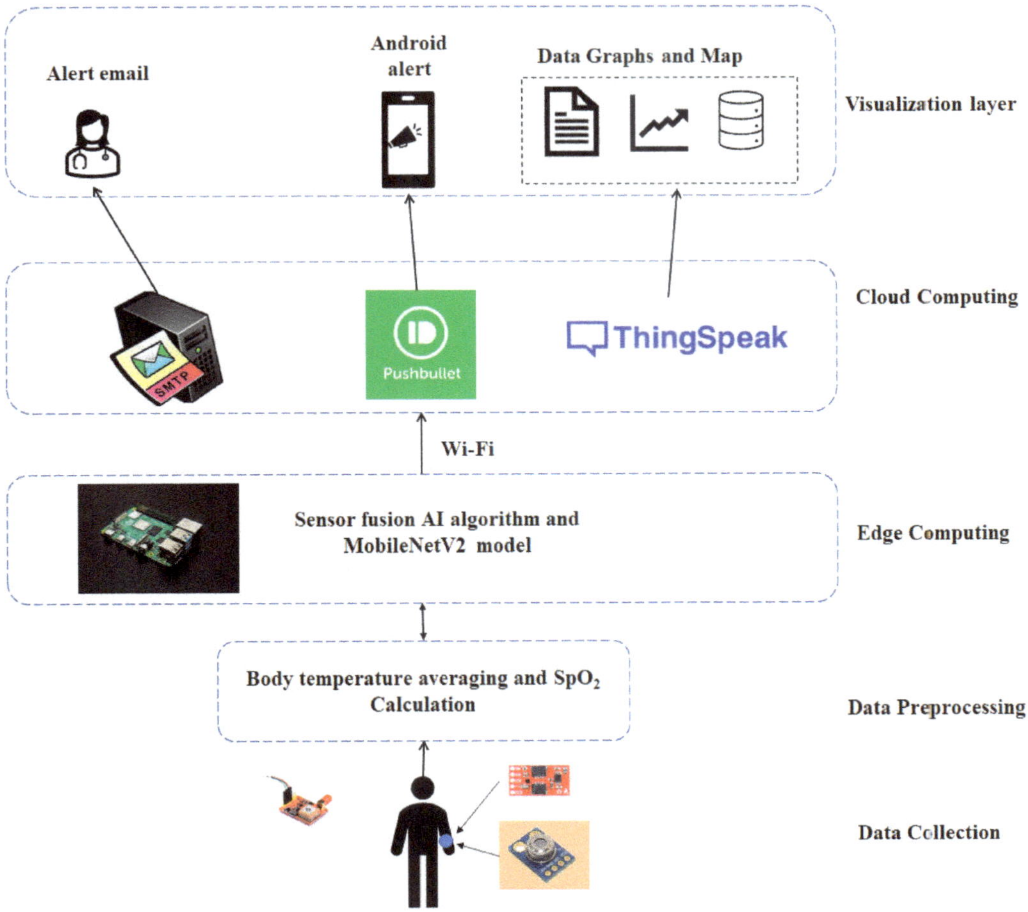

Figure 1. The overall architecture of the proposed IIOT device.

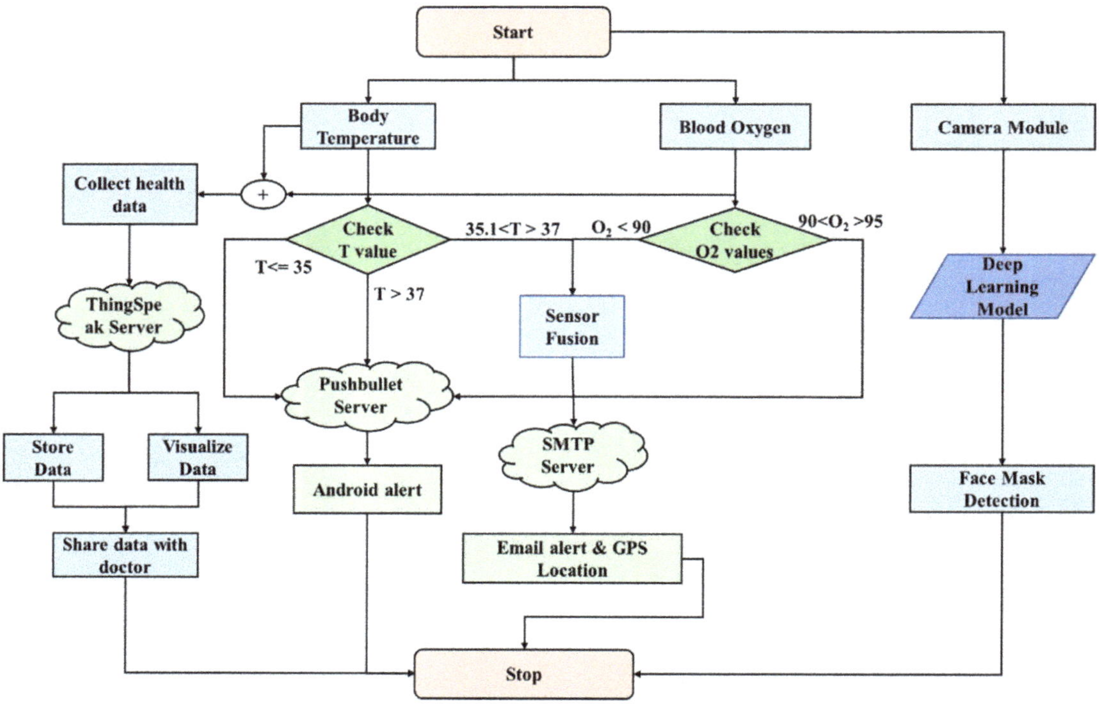

Figure 2. The data flow.

In this method, three different cloud servers are implemented for the respective functionality, as shown in Figure 1. ThingSpeak [26] is a cloud-based IoT platform that aggregates, visualizes, and analyzes real data streams. A private channel is created; the cloud provides a write API key used to save data, and a read API key to receive saved data in JSON, XML, or in text format. We installed the simple mail transfer protocol (SMTP) on the Raspberry Pi [27]. The SMTP server sends an alert email with crucial health data and the GPS position of a suspect to a healthcare provider. The Pushbullet server [28] is used to transfer links, text, and files between devices. This server sends android alerts that are not urgent but that require attention soon. After registering a device using its ID, the Pushbullet server delivers messages and notifications. Data collection and cloud storage are shown in Figure 2. The edge device features SF and notification servers. Real-time face detection (using a spy camera) predicts an output with the aid of the trained deep learning model (Figure 1).

2.1. Sensor Fusion (SF)

The sensor fusion (SF) approach is used to identify COVID-19 suspects. A body temperature of 35–37 °C is normal; an alarm is sent if the temperature exceeds this range. The normal blood oxygen level is 95–100%; anything below that range is considered serious. To generate emergency alerts, the data from the two sensors are fused and the threshold values evaluated. The SF algorithm and its implementation are shown in Algorithm 1.

Algorithm 1. Pseudo-code: COVID suspect prediction

1. Save the input from the temperature sensor;
2. If a finger is on the sensor, go to step 3; otherwise stop;
3. If sensor reading confidence level is above 90% collect data;
4. Save the input from the oximeter;
5. If $90 < O_2 < 95$:
Send an android alert message via the Pushbullet server;
6. If $O_2 \leq 90$:
If fever > 37.5:
Assign Array [] and store the values for 30 min;
If max of array [] < 90:
Send an email stating that a suspect has been detected; include the GPS location;
Otherwise, clear the array;
else, send "low O_2 need attention" alert to user;
else, collect and save data in real-time.

SF algorithm features:

- The SF algorithm receives input data from fever, oximeter sensors, and heart rate, all of which are calibrated to commercial-level precision.
- To eliminate errors, the oximeter sensor accepts readings only when the sensor is in contact with human skin and the sensor's confidence level is above 90%.
- When the oximeter indicates a low oxygen level, this might be transient (caused by exercise or stress). To avoid false positives, the SF system waits and examines additional health metrics.
- When the oxygen level drops, the system seeks information from the body temperature sensor.
- If both sensors produce anomalous results, the SF algorithm records all inputs for 30 min in an array and saves them for future study.
- If all values are below the usual levels for an extended period, only then does the SF algorithm send an email alert with a GPS position. If the values are not anomalous over an extended period, the algorithm concludes that no emergency exists, wipes all data from the array, and sends a simple notice to an Android smartphone.

2.2. *Face Mask Detection Using Deep Learning on an IoT Platform*

Deep learning is a form of image processing for AI that employs feature extraction algorithms. This requires a powerful GPU, but IoT devices lack a powerful GPU, which makes rendering deep learning difficult. Image processing employs the OpenCV and TensorFlow platforms. Raspberry Pi 4 includes support for image processing systems, such as Keras. MobileNetV2 [29] is an efficient neural network for IoT devices featuring an inverted residual structure with connections between the bottleneck levels, so we used this as a backbone network.

We used the RFMD dataset (which includes 2165 pictures with masks and 1930 without masks) for testing and training. Sample pictures are shown in Figure 3, along with pictures from the Bing search API and the Kaggle datasets. The manually morphed pictures are not included in the dataset; corrupt and duplicate pictures are removed. Clearing, detection, and correction improved prediction. The dataset was divided into 80% for training and 20% for testing subsets before pre-processing. A function was implemented that accepted dataset folders as inputs, loaded all files, and resized the pictures. The list was then sorted alphabetically, and the pictures were transformed into tensors. The list was then transformed to a NumPy array (to accelerate computation).

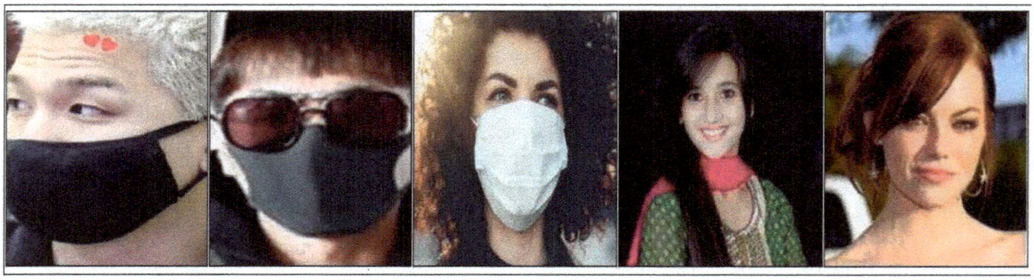

Figure 3. Sample images used for neural network training.

The OpenCV library was used to recognize human faces rapidly before training. To eliminate recursive scan latency, several faces could be identified in a single shot; only one image was required to identify numerous objects. This determined the region of interest for MobileNetV2 feature extraction. Figure 3 presents sample images used to train the model. We had a diversified dataset with different nationalities, age groups, sexes, ethnicities, and types of masks for better accuracy.

MobileNetV2 is a lightweight, deep learning neural network for picture classification. The standard MobileNetV2 model is in this work base model; the head model is added to enhance to base model output. The head model enhances the accuracy and it includes an averaging pooling layer followed by flattening operations. There were five dense layers added before the output layer. Whereas in the base model, TensorFlow was used to load the pre-trained weights. Then, to allow feature extraction, additional layers were added to (and trained on) the database. The model was then fine-tuned, and the weights were saved on the layers. Transfer learning saves time; existing biased weights were used without sacrificing previously learned features. MobileNetV2 features a core convolutional neural network layer. A pooling layer accelerates calculations by decreasing the size of the input matrix without changing its features. The dropout layer prevents overfitting during model training. The non-linear functions include several types of rectified linear units (ReLUs). The fully connected layers are linked to the activation layers. If connections are skipped, network execution may suffer. Thus, a linear bottleneck was added. Figure 4 shows the detailed architecture of the model. The method precisely identifies mask location. If a person is not wearing a mask, the model draws a red box around the face. The model can detect several faces in the same frame at the same time. This model can employ a basic picture as an input, or a real-time video stream from the Raspberry Pi camera. Figure 5 shows face mask detection and the percentage accuracies (red or green boxes). For critical analysis, images were taken from a side view and multiple faces on the same image to test the model. Figure A1a,b shows that face mask identification was 99.26% accurate; the loss and accuracy were plotted by the epoch, respectively. The Figure A1a,b shows that, after the 20th epoch, accuracy was close to 99.26%, and the "after loss" per epoch, was also minimum, which satisfied the well-fitted model condition. The time required to train the model on Raspberry Pi was almost twice that required when a PC equipped with a GeForce GTX 750 GPU, an Intel Core i5 processor, and 8 GB of RAM, were employed. After training, the real-time mask detection speeds on a PC and the IoT devices were identical. The model was tested by placing different objects on faces, altering the mask positions, and capturing faces from the side. Even in such unusual circumstances, model performance was unaffected.

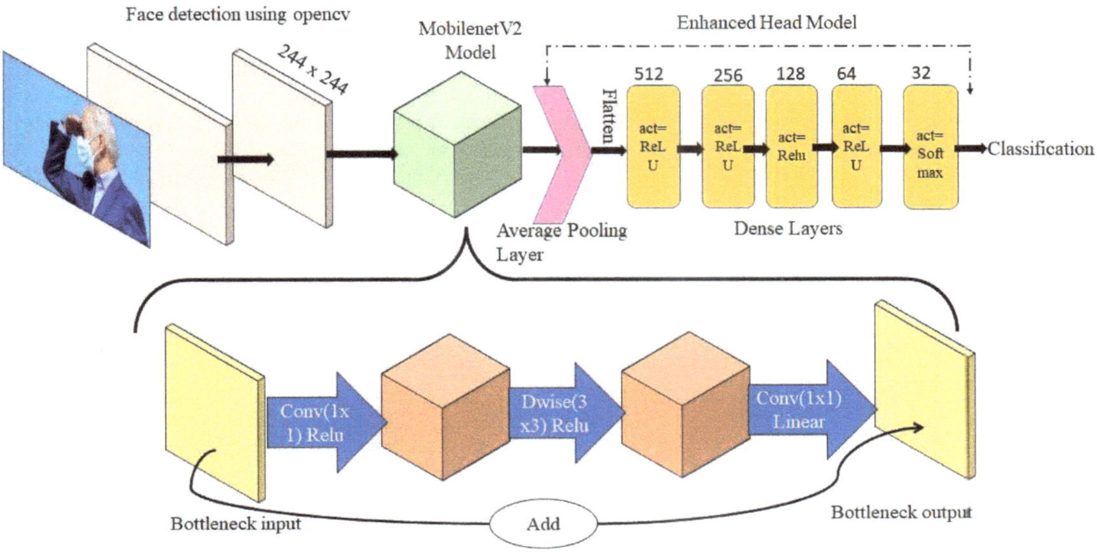

Figure 4. Face mask detection.

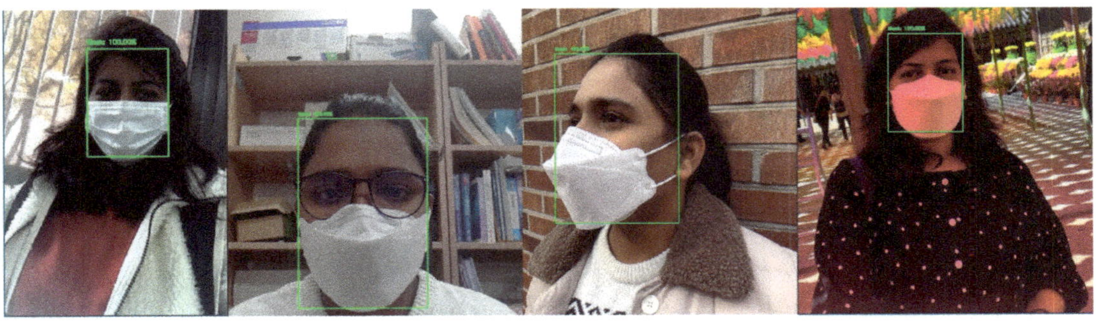

(**a**)

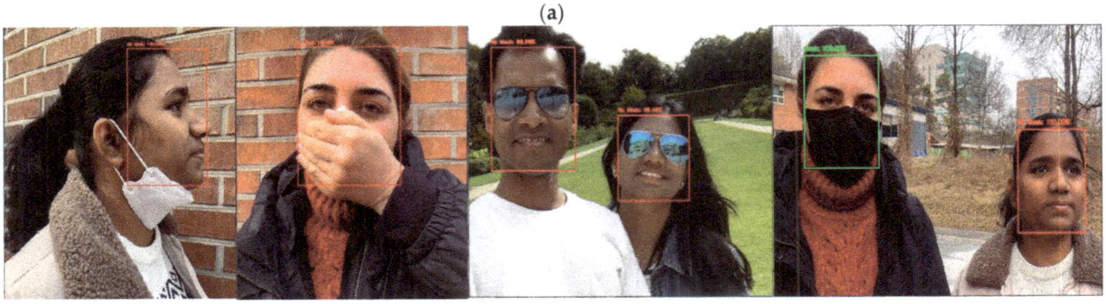

(**b**)

Figure 5. (**a**) Real-time face mask detection from different viewpoints. (**b**) Real-time face detection without a mask, capable of detecting the incorrect position of the mask and identifying it as "without mask".

3. Experimental

In a serial communication system, Raspberry Pi 4 plays the role of a host and an Arduino the role of a slave. The MLX 90614 sensor detects body temperature; the SparkFun sensor detects the blood oxygen level and heartbeat [30–33]. The GPS signal is detected by an LM80 sensor connected to a USB port. The MLX 90614 and SparkFun biosensors are integrated into the Raspberry Pi and the Arduino, respectively. The I2C protocol is used to link the biometric sensors. The spy camera is installed on the Raspberry Pi camera slot for real-time video-streaming and face mask recognition [34]. As we propose, this device for wearable purposes, a small size camera is necessary. The detailed pin connections with Raspberry Pi 4 and Arduino Uno are explained in Table A1 (Appendix A) and Table A2 (Appendix A) respectively.

Figure 6 shows the experimental setup. The Raspberry Pi 4 microprocessor is optimal for the TensorFlow platform. The analog sensor is powered by an Arduino Uno. To allow for future expansion, we used an Arduino rather than an analog-to-digital converter (ADC). During implementation, the multithreading feature of the Python language was used to effectively run the multiple sensors concurrently. There was a dedicated python thread; running concurrently for each sensor, Pi camera, and GUI data update featured.

Figure 6. The experimental testbed.

Temperature sensor: the temperature sensor determines whether a person has a fever. Five hundred continuous inputs from the sensor are averaged in real-time before display to the user; the processing time is less than 1 s. A few milliseconds are required to provide the results, but health data are enormous; a short delay is acceptable. The enhancement algorithm is based on Equation (1):

$$\text{Output temperature} = \sum_{1}^{n} \frac{\text{Temp}}{\text{n}} \quad (1)$$

where temp = current temperature in Celsius and n = number of inputs.

The SparkFun sensor: the SparkFun sensor works as a pulse oximeter and the heart rate sensor is an I2C-based biometric sensor that features two Maxim Integrated chips; the MAX32664 sensor analyzes data collected by the MAX30101 sensor and the photoplethysmogram (PPG).

4. Results and Discussion

4.1. Device Performance

The accuracies of sensor data and face mask identification were evaluated. The MLX 90614 sensor was tested on the same individual; readings were obtained at 10-min intervals and compared to those of a commercial thermometer (Figure 7). All temperature measurements are in Celsius. The MLX 90610 sensor error was about 0.1 °C; the accuracy was thus about 98%. The temperature sensor gave the best accuracy when the user and sensor were stable.

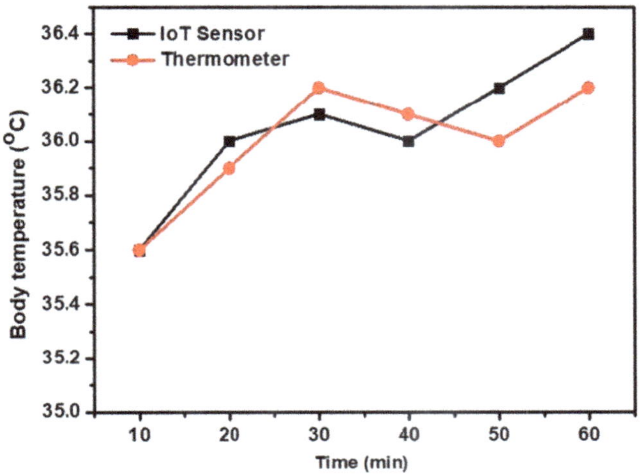

Figure 7. Comparison of the MLX 90614 sensor and the thermometer.

The SparkFun sensor is a pulse oximeter. The values obtained are plotted against those of the commercial Britz band (Figure 8). The picture of the commercial health band is shown in Figure A2 (Appendix A). The values were near-identical. The percentage accuracies at each time were averaged to yield an overall accuracy. Equation (2) shows the accuracy percentages at specific times; the average accuracy was then determined.

$$\frac{\text{IoT value}}{\text{Commercial device value}} \times 100 = \text{Accuracy percentage} \quad (2)$$

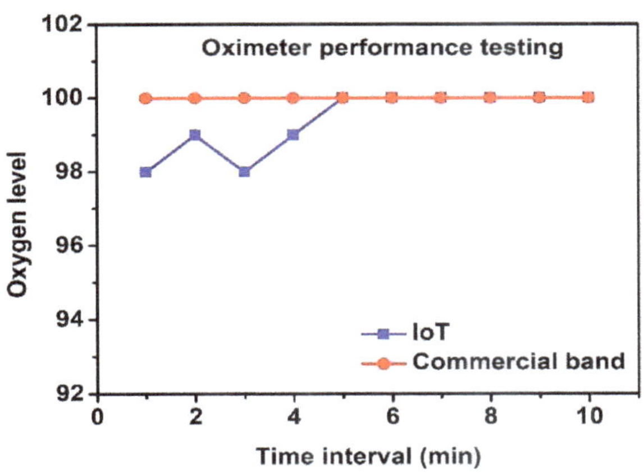

Figure 8. SparkFun sensor accuracy.

The average accuracy was 99.1%. The sensor also yielded the heart rate and raw data. Heart rate monitoring is critical in COVID-19-infected and cardiac patients because, according to Dr. Nisha Parekh, "There are numerous ways COVID-19 can damage the heart during the first period when someone has the infection, particularly in the first few weeks. These side effects might include new or worsening difficulties with blood pumping, inflammation of the heart muscle, and inflammation of the membrane around the heart. It should be emphasized that other infections can potentially cause the same symptoms." [35]. Heart rate data were collected on the IoT server; however, the it was not included in suspected detection conditions.

An android message from the Pushbullet server is shown in Figure A3 (Appendix A). The android alert is issued only when the temperature falls below 30 °C or rises above 37 °C. Regarding the ThingSpeak channel connectivity and real-time data visualization is in MATLAB and each sensor value is represented as a single field and implementation output is provided in Figure A4 (Appendix A). The geographical position and the temperature are shown in Figure A5 (Appendix A). Heartbeat data were saved in field 3 of the ThingSpeak channel and values are plotted as shown in Figure A6 (Aappendix A). This shows our device is collecting data after every 15 min and saves over the cloud server. Along with data collection, data analysis is also performed over edge servers in real-time.

It is difficult to test the device on actual COVID patients due to social distancing rules; validation of the device was performed by Dr. Anuja Padwal, a practicing medical student at the Maharashtra University of Health Sciences (MUHS). According to Padwal, "The proposed method is beneficial for COVID perspective and automatic precautions for false positive is worth noting in the study. This method is beneficial and practical to control pandemics in developing countries because of the low manufacturing cost".

The comparison of the our device with the available market devices are shown in Table 1, considering the various factors such as heart rate, body temperature, cost of the device, etc.

Table 1. Comparative study of the commercial device and proposed device.

Device Features	Apple Watch Series 6	Apple Watch Series 5	Proposed Device
Heart rate	✓	✓	✓
Body temperature	×	×	✓
Oximeter	✓	×	✓
Charging	Wireless	Wireless	USB
Database	Apple app	Apple app	IoT cloud
Data visualization	×	×	✓
Data sharing	×	×	✓
Alert and notification	×	×	✓
Sensor fusion for AI	×	×	✓
Covid suspect tracking	×	×	✓
Price (USD)	400+	400+	100

4.2. Training and Testing of the Deep Learning Model

For accuracy testing, we performed several tests of system performance, in terms of finding masked faces. For training purposes, the Adam optimizer with 30 epochs and a batch size of 32 was used. Loey et al. [26] evaluated training using Adam and SGDM and concluded that Adam outperformed SGDM in terms of a mini-batch root mean square error and loss. The Adam training is shown in Table 2; any loss was minor. Model performance was quantitatively compared to those of the InceptionV3 and ResNet50 architectures (using the RMFD dataset); the values are listed in Table 3 and plotted in Figure 9. The sizes of the deep learning model, the detection times, and the accuracies, were computed. Figure 9 shows that the ResNet50 architecture afforded the highest accuracy; however, this model includes more parameters than MobileNetV2, rendering it larger and slower. Figure 9c shows that the MobileNetV2 architecture is lightweight, with a size of 11.3 MB and a detection speed nearly half that of the ResNet50 model.

Table 2. Training and validation of the enhanced MobileNetV2 model on Adam.

Epoch	Iteration	Training Time (s)	Batch Loss	Accuracy (%)	F1 Score
5	120	569.19	0.0711	98.29	0.98
10	240	1164.06	0.0420	98.46	0.99
15	360	1709.57	0.0336	98.90	0.99
20	480	2165.29	0.0305	99.15	0.99
25	600	2538.23	0.029	99.20	0.99
30	720	3248.43	0.025	99.26	0.99

Table 3. Comparison of model sizes and detection times.

Model	Model Size (MB)	Detection Time (ms)	Accuracy (%)	Raspberry Pi Support
MobileNetV2	11.3	31.3	99.11	✓
InceptionV3	478.08	58.8	96.00	×
ResNet50	1296.62	74.9	99.51	✓
Enhanced MobileNetV2	11	31.3	99.26	✓

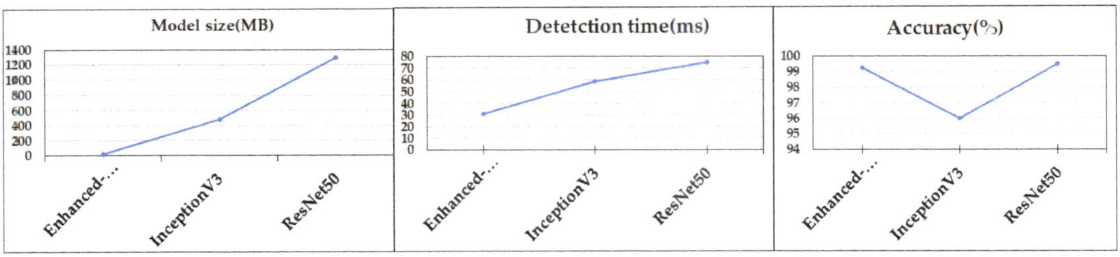

(a) (b) (c)

Figure 9. A comparison of the proposed model (enhanced MobileNetV2) with InceptionV3 and ResNet50 in terms of (a) size; (b) detection time; and (c) accuracy when evaluating the RMFD dataset.

The training and validation loss curve is shown in Figure A1b. We observed that our model neither overfits nor underfits. Generally, the cost function is a way to compute error and to quantify how good or bad the model is performing. The less the loss, the more accurate the model is. From Figure A1b and Table 2, it could be concluded that the model is fine-tuned with minimal loss. In this experiment, the binary cross entropy function was used to optimize the model; the formula of the function is as given in Equation (3).

$$\text{Log loss} = \frac{1}{N}\sum_{i=1}^{N} -(Yi * \log(pi) + (1 + Yi) * \log(1 - pi)) \tag{3}$$

Here, *pi* is the probability of class with mask and (1 − *pi*) is the probability of class without a mask.

The model was further evaluated using the properly wearing masked face detection (PWMFD) dataset and compared with the results of Loey et al. [21]. Table 4 shows that the MobileNetV2 model size was the smallest and that our improvements reduced the detection time. The model accuracy using the RMFD dataset, PWMFD dataset, and combined dataset was only 99.11%, 89.00%, 90.14%, respectively, but when tested against the enhanced model, the accuracy was 99.26%, 99.15%, and 92.51%, respectively. We conclude that the enhanced model gives better accuracy with both datasets. The RMFD dataset performed better than PWMFD in all instances because many PWMFD pictures were blurred, rendering single-shot face identification difficult. Table 4 compares our system to that of Loey et al. [21].

Table 4. Real-time face mask detection model comparative study.

Method	Backbone	Input Image Size	Detection Time (ms)	Accuracy %	Raspberry Pi Support
RetinaNet	ResNet-50	800	76.8	94.9	✓
EfficientDet-D0	EfficientDet-Bo	512	99.3	84.5	×
EfficientDet-D1	EfficientDet-B1	608	122.0	85.1	×
SSD	VGG-16	512	34.5	92.7	×
YOLOv3	Darknet53	608	61.5	95.3	✓
SE-YOLOv3	SE-Darknet53	512	49.2	96.2	×
MobileNet	MobileNetV2	512	31.9	90.1	✓
Enhanced-Mobile net	MobileNetV2	512	31.9	95.	✓

In Table 4, we compare our model with other papers to show that the proposed model outperforms previously reported models. Whereas in Table 5, we combine RMFD and PWMFD datasets to compare the results of using the proposed model. In all instances, enhanced MobileNetV2 performs better than any other model. In [34], the authors presented face mask detection using SSD-MobileNetV2 and had 92.64% accuracy, whereas the presented model had 99.26% accuracy; hence, we can conclude that our model is accurate and lightweight compared to the other proposed models, which makes it suitable for IoT devices.

Table 5. Enhanced MobileNetV2 compared with MobileNetV2 against different datasets.

Model Name	Accuracy (%)		
	RMFD Dataset	PWMFD Dataset	RMFD + PEMFD Combine Dataset
MobileNetV2	99.11	89.00	90.14
Enhanced MobileNetV2	99.26	91.15	92.51

To further evaluate the model, we calculated true positive (TP), true negative (TN), false positive (FP), and false negative (FN) on 30 random images with 38 random faces. The confusion matrix is shown in Figure 10. The experiment results show that 15 TP, 19 TN, 2 FP, and 2 FN were detected. Additionally, the precision and recall were calculated based on Equations (4) and (5). The values of the precision and recall were 0.88 and 0.88, respectively.

$$\text{Precision} = \frac{\text{TP}}{\text{TP} + \text{FP}} = 0.88 \tag{4}$$

$$\text{Recall} = \frac{\text{TP}}{\text{TP} + \text{FN}} = 0.88 \tag{5}$$

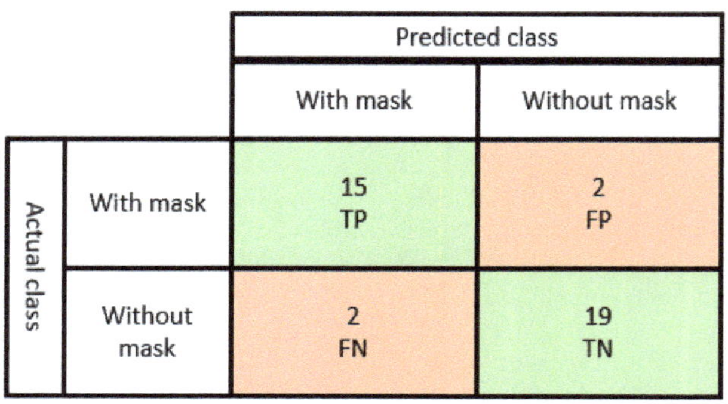

Figure 10. Confusion matrix for face mask detection model.

Here, FP and FN values are low, meaning we could predict that the algorithm is precise and accurate, with a large-sized dataset; we are expecting higher TP and TN values.

5. Conclusions

We present a novel SF technique embedded in a device with a deep neural network; this method "seeks" ways to help control a pandemic. The accuracy of the device is 98–99%, compared to commercial devices. To avoid false-positive alerts, precautionary measures were automatically taken by the SF algorithm without human interference (key features of this paper). The proposed method identifies suspected COVID-infected individuals in real-time, and facilitates tracing and tracking using a GPS sensor. The presented method is economical, practical, scalable, easy to use, and pandemic-focused. To the best of our knowledge, this method is the first to implement SF technology in a wearable device for pandemic control. The proposed device mainly has application in two major categories—wearable gadgets and devices for public areas. Wearable devices can be used by COVID-19 patients or those with other critical conditions who require continuous real-time data monitoring in the absence of a doctor. If the device is used in public places (e.g., schools, malls, train and bus stations, airports, tourist places), face mask detection would ensure that people wear their masks correctly. The device is scalable, inexpensive, simple to deploy, user-friendly, and securely saves health data. Remote monitoring (without face-to-face medical consultation) is possible; continuously recorded data are shared. The read data API key allows a user to control the data completely; anyone else needs specific permission to view the data.

In the future, we will enhance device accuracy and attempt to reduce the size of the wearable device to make it more user-friendly. Furthermore, we plan to include additional sensors with microprocessors for other types of diseases, such as diabetics and cardiac arrest. IoT devices are vulnerable to cyber-attacks. Thus, data flowing from the device to the cloud must be encrypted and, therefore, security measures need to be added to prevent cyber-attacks. Health data are "big data"; data storage and access are challenging and researchers aim to address these issues.

Author Contributions: Conceptualization, R.K.S.; methodology, R.K.S., M.S.A.; software, R.K.S., M.S.A.; validation, R.K.S., M.S.A.; formal analysis, S.G.P., S.M.P.; Investigation, R.K.S. and N.K.; data curation, R.K.S., M.S.A.; writing—original draft preparation, R.K.S.; writing—review and editing, R.K.S., M.S.A., S.G.P., S.M.P., N.K.; visualization, R.K.S.; supervision, N.K.; funding acquisition, N.K. All authors have read and agreed to the published version of the manuscript.

Funding: This research was funded by the Ministry of Science and ICT (MSIT), Korea, under the Information Technology Research Center (ITRC) support program (IITP-2020-0-01846) supervised by the Institute for Information & Communications Technology Planning & Evaluation (IITP), the National Research Foundation of Korea (NRF) grant funded by the Korea government (NRF 2018R1D1A3B07044041 and NRF 2020R1A2C1101258).

Institutional Review Board Statement: Not applicable.

Informed Consent Statement: Not applicable.

Data Availability Statement: The dataset and python code are available at https://github.com/shahinur-alam/Covid-Project. The ThingSpeak cloud channel is available at https://thingspeak.com/channels/1423804/private_show (Last accessed on 20 February 2022).

Acknowledgments: We thank Anuja Padwal for validating the proposed method from a medical perspective.

Conflicts of Interest: The authors declare no conflict of interest.

Appendix A

Raspberry Pi 4 connections with sensors are connected via a I2C protocol using GPIO pins.

Table A1. Pin connections of Raspberry Pi 4.

Raspberry Pi 4	IoT Device
Body temperature sensor	
5V	VCC
Pin 6	GND
GPIO 2 Serial Data	SDA
GPIO 3 Serial Clock	SCL
LCD screen	
5V	VCC
Pin 7	GND
GPIO 17	SDA
GPIO 27	SCL
Alert buzzer	
GPIO	Positive
Pin 39	GND

Arduino Uno works as an analog sensor data collection and preprocessing analog data.

Table A2. Arduino Uno pin connection.

Arduino Uno	Oximeter Sensor
3.3 V	VCC
GND	GND
Analog (A4)	SDA
Analog (A5)	SCL

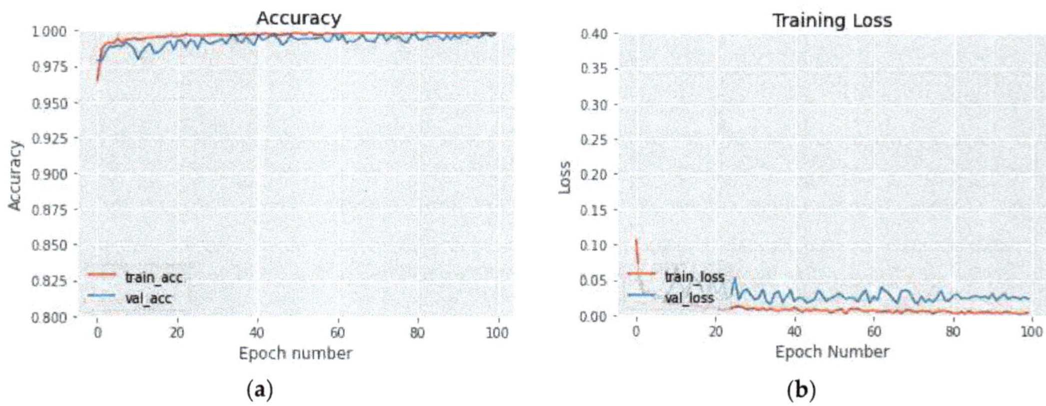

Figure A1. Accuracy (**a**) and loss (**b**) of the proposed model per epoch.

Figure A2. The Britz health band.

Human health-related vital data sometimes show abnormal readings, e.g., concerning abnormal condition emergency alerts, which are sent to users and relatives. These alerts will be useful for healthcare workers and in remote monitoring.

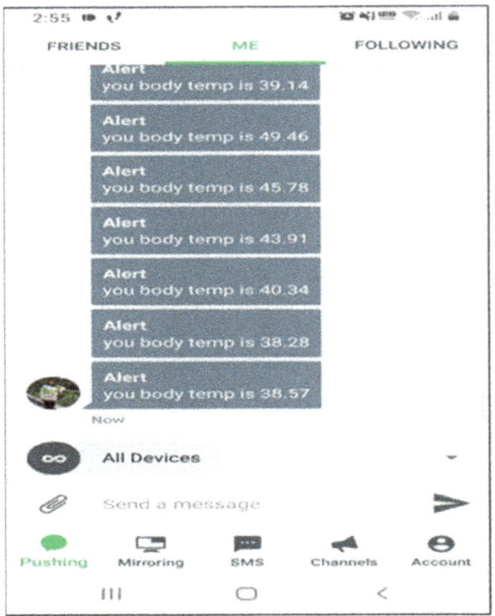

Figure A3. Android alert message.

Fever and low oxygen are common signs of COVID-19; when both conditions occur at the same time, emergency tracing and testing is needed. To provide emergency services, location and data history are provided to healthcare workers through a read API key of the IoT cloud server.

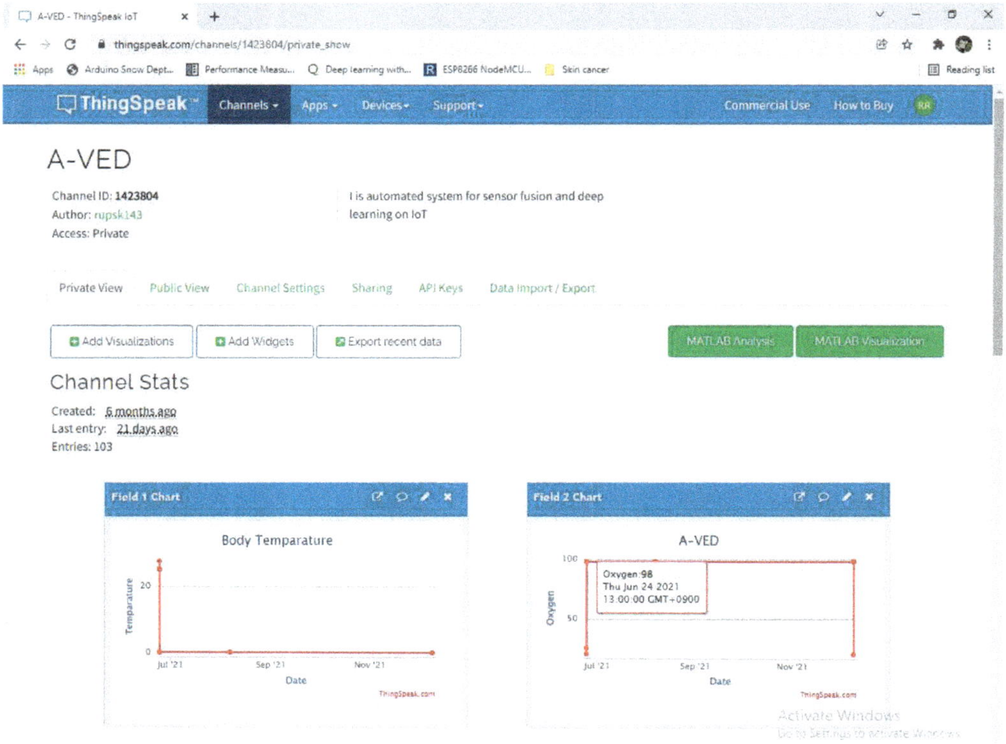

Figure A4. IoT cloud server ThingSpeak channel.

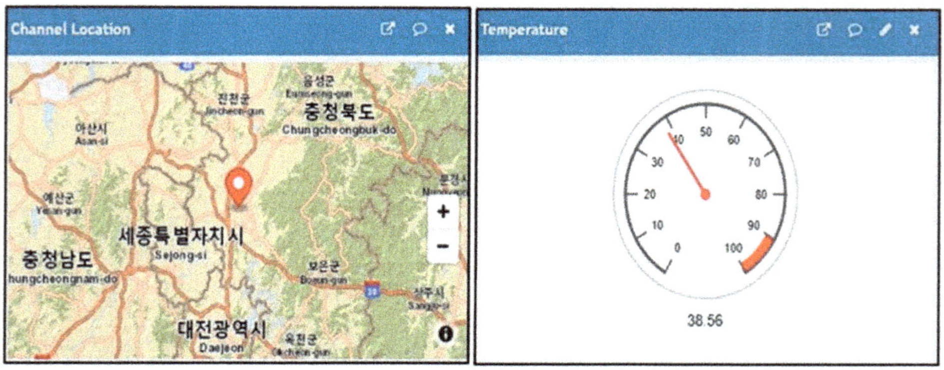

Figure A5. GPS and body temperature sensor data visualization on ThingSpeak cloud.

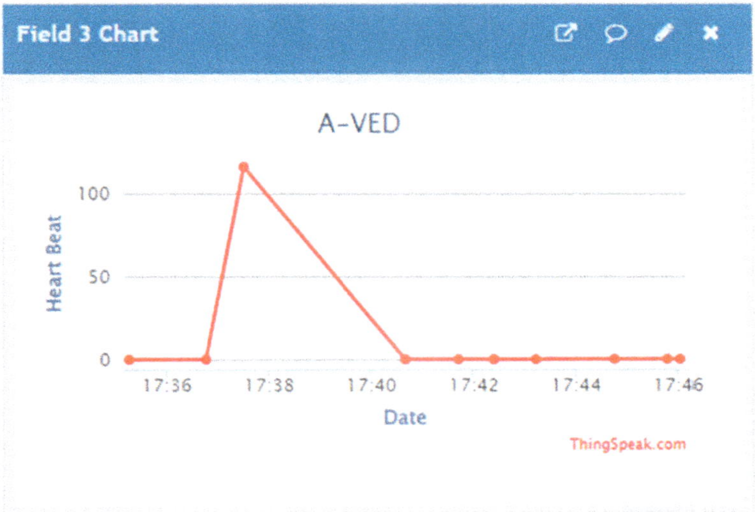

Figure A6. Heartbeat graph on IoT cloud.

References

1. Tobías, A.; Carnerero, C.; Reche, C.; Massagué, J.; Via, M. Changes in air quality during the lockdown in Barcelona (Spain) one month into the SARS-CoV-2 epidemic. *Sci. Total. Environ.* **2020**, *726*, 138540. [CrossRef] [PubMed]
2. Saglietto, A.; D'Ascenzo, F.; Zoccai, G.B.; Ferrari, G.M.D. COVID-19 in Europe: The Italian lesson. *Lancet* **2020**, *359*, 1110–1111. [CrossRef]
3. Republic of Turkey Ministry of Health. Available online: https://covid19.saglik.gov.tr/ (accessed on 25 August 2021).
4. Atalan, A. Is the lockdown important to prevent the COVID-19 pandemic? Effects on psychology, environment and economy perspective. *Ann. Med. Surg.* **2020**, *56*, 38–42. [CrossRef] [PubMed]
5. Rymer-Diez, A.; Roca-Millan, E.; Estrugo-Devesa, A.; González-Navarro, B.; López-López, J. Confinement by COVID-19 and Degree of Mental Health of a Sample of Students of Health Sciences. *Healthcare* **2021**, *9*, 1756. [CrossRef] [PubMed]
6. Hsu, H.-C.; Chou, H.-J.; Tseng, K.-Y. A Qualitative Study on the Care Experience of Emergency Department Nurses during the COVID-19 Pandemic. *Healthcare* **2021**, *9*, 1759. [CrossRef]
7. Patel, K.; Patel, S.M. Internet of Things-IOT: Definition, characteristics, architecture, enabling Technologies, application & future challenges. *Int. J. Eng. Sci. Comput.* **2016**, *6*, 6122–6131.
8. Bostan, S.; Erdem, R.; Ozturk, Y.E.; Kilic, T.; Yilmaz, A. The Effect of COVID-19 Pandemic on the Turkish Society. *Electr. J. Gen. Med.* **2020**, *237*, em237.
9. Ferretto, L.R.; Bellei, E.; Biduski, D.; Bin, L.C.; Moro, M.M. A Physical Activity Recommender System for Patients with Arterial Hypertension. *Access IEEE* **2020**, *8*, 61656–61664. [CrossRef]
10. Pradhan, B.; Bhattacharyya, S.; Pal, K. IoT-Based Applications in Healthcare Devices. *J. Healthc. Eng.* **2021**, *2*, 6632599. [CrossRef]
11. Aadil, F.; Mehmood, B.; Hasan, N.; Lim, S.; Ejaz, S. Remote Health Monitoring Using IoT-Based Smart Wireless Body Area Network. *Comput. Mater. Contin.* **2021**, *68*, 2499–2513. [CrossRef]
12. Sheikh, F.; Li, X. Wireless sensor network system design using Raspberry Pi and Arduino for environment monitoring applications. *Procedia Comput. Sci.* **2014**, *34*, 103–110.
13. Fu, Y.; Liu, X. System Design for Wearable Blood Oxygen Saturation and Pulse Measurement Device. *Procedia Manuf.* **2015**, *3*, 1187–1194. [CrossRef]
14. Dananjayan, S.; Raj, G. Artificial Intelligence during a pandemic: The COVID-19 example. *Int. J. Health Plan. Manag.* **2020**, *35*, 1260–1262. [CrossRef] [PubMed]
15. Wallis, C. How artificial intelligence will change medicine. *Nat. Artic.* **2019**, *567*, 48. [CrossRef]
16. Phan, M.H.; Hwang, K.Y.; Jimenez, V.O.; Muchharla, B. Real Time Monitoring of COVID-19 Progress Using Magneti Sensing and Machine Learning. U.S. Patent 0369137 A1, 2 December 2021.
17. Batageli, B.; Peer, P.; Stuc, V.; Dobrisek, S. How to correctly detect face mask for COVID-19 from visual information? *Appl. Sci.* **2021**, *11*, 2070. [CrossRef]
18. Larxel. Face Mask Detection Dataset. Available online: https://www.kaggle.com/andrewmvd/face-mask-detection (accessed on 5 May 2021).

19. Nagrath, P.; Jain, R.; Madan, A.; Arora, R.; Kataria, P.; Hemanth, J. SSDMNV2: A real time DNN-based face mask detection system using single shot multibox detector and MobileNetV2. *Sustain. Cities Soc.* **2021**, *66*, 102692–102710. [CrossRef]
20. Jiang, X.; Gao, T.; Zhu, Z.; Zhao, Y. Real-Time Face Mask Detection Method Based on YOLOv3. *Electronics* **2021**, *10*, 837. [CrossRef]
21. Loey, M.; Manogaran, G.; Taha, M.H.; Khalifa, N.E. Fighting against COVID-19: A novel deep learning model based on YOLO-v2 with ResNet-50 for medical face mask detection. *Sustain. Cities Soc.* **2020**, *65*, 102600. [CrossRef]
22. Elmenreich, W. An introduction to sensor fusion. *Vienna Univ. Technol. Austria* **2002**, *502*, 1–28.
23. Wang, R.; Shen, M.; Li, T.; Gomes, S. Multi-task joint sparse representation classification based on fisher discrimination dictionary learning. *CMC Comput. Mater. Contin.* **2018**, *57*, 25–48. [CrossRef]
24. Alatise, M.; Hancke, G. A Review on Challenges of Autonomous Mobile Robot and Sensor Fusion Methods. *IEEE Access* **2020**, *8*, 39830–39846. [CrossRef]
25. Kelechi, A.; Alsharif, M.; Agbaetuo, C.; Ubadike, O.; Aligbe, A. Design of a low-cost air quality monitoring system using Arduino and ThigSpeak. *Comput. Mater. Contin.* **2021**, *70*, 151–169. [CrossRef]
26. Enea, C.; Charlotte, F.; Bam, S.; Gemma, T.; Lyes, K. Hand-gesture recognition based on EMG and event-based camera sensor fusion: A benchmark in Neuromorphic Computing. *Front. Neurosci.* **2020**, *14*, 637.
27. Lee, D.; Kang, J.; Dahouda, M.K.; Joe, I.; Lee, K. DNT-SMTP: A novel mail transfer protocol with minimized interactions for space internet. In Proceedings of the 20th International Computational Science and Its Applications, Cagliari, Italy, 1–4 July 2020.
28. Razali, R.A.B.; Hashim, I.B.; Mohamed, R.B.; Raj, M.A. A development of smart aquarium prototype: Water temperature system for shrimp. *Adv. Sci. Lett.* **2018**, *24*, 773–776.
29. Sandler, M.; Howard, A.; Zhu, M.; Chen, L. MoboleNetV2: Inverted residuals and linear bottlenecks. In Proceedings of the IEEE Conference on Computer Vision and Pattern Recognition, Salt Lake City, UT, USA, 18–22 June 2018.
30. Maques, G.; Pitarma, R. Non-contact Infrared Temperature Acquisition System based on Internet of Things for Laboratory Activities Monitoring. *Procedia Comput. Sci.* **2019**, *155*, 487–494. [CrossRef]
31. Bassam, N.; Hussain, S.A.; Qaraghuli, A.; Khan, J.; Sumesh, E.P.; Lavanya, V. IoT based wearable device to monitor the signs of quarantined remote patients of COVID-19. *Inform. Med. Unlocked* **2021**, *24*, 100588. [CrossRef] [PubMed]
32. Shinde, R.; Choi, M. An experimental of health monitoring system using wearable devices and IoT. In Proceedings of the 2021 Winter Comprehensive Conference of the Korea Telecommunications Society, Pyeongchang, Korea, 3–5 February 2021.
33. Shinde, R.; Alam, M.S.; Choi, M.; Kim, N. Economical and wearable pulse oximeter using IoT. In Proceedings of the 16th International Conference on Computer Science & Education (ICCSE), Lancaster, UK, 17–21 August 2021.
34. Saha, S.; Singh, A.; Bera, P.; Kamal, M.; Dutta, S. GPS based smart spy surveillance robotic system using Raspberry Pi for security application and remote sensing. In Proceedings of the 8th IEEE Annual Information Technology, Electronics and Mobile Communication Conference, Vancouver, BC, Canada, 3–5 October 2017.
35. University of California San Francisco Magazine. Available online: https://www.ucsf.edu/magazine/covid-hearts (accessed on 29 July 2021).

healthcare

Article

Intelligent Diagnostic Prediction and Classification Models for Detection of Kidney Disease

Ramesh Chandra Poonia [1], Mukesh Kumar Gupta [2], Ibrahim Abunadi [3], Amani Abdulrahman Albraikan [4], Fahd N. Al-Wesabi [5,*], Manar Ahmed Hamza [6] and Tulasi B [1]

1 Department of Computer Science, CHRIST (Deemed to be University), Bangalore 560029, India; rameshcpoonia@gmail.com (R.C.P.); tulasi.b@christuniversity.in (T.B.)
2 Department of Computer Science & Engineering, Swami Keshvanand Institute of Technology, Management & Gramothan (SKIT), Jaipur 302017, India; mukeshgupta@skit.ac.in
3 Department of Information Systems, Prince Sultan University, P.O. Box No. 66833 Rafha Street, Riyadh 11586, Saudi Arabia; i.abunadi@psu.edu.sa
4 Department of Computer Sciences, College of Computer and Information Sciences, Princess Nourah bint Abdulrahman University, P.O. Box 84428, Riyadh 11671, Saudi Arabia; aalbraikan@pnu.edu.sa
5 Department of Computer Science, College of Science & Art at Mahayil, King Khalid University, Abha 61421, Saudi Arabia
6 Department of Computer and Self Development, Preparatory Year Deanship, Prince Sattam bin Abdulaziz University, Al-Kharj 16273, Saudi Arabia; mahamza@psau.edu.sa
* Correspondence: falwesabi@kku.edu.sa; Tel.: +966-534227096

Abstract: Kidney disease is a major public health concern that has only recently emerged. Toxins are removed from the body by the kidneys through urine. In the early stages of the condition, the patient has no problems, but recovery is difficult in the later stages. Doctors must be able to recognize this condition early in order to save the lives of their patients. To detect this illness early on, researchers have used a variety of methods. Prediction analysis based on machine learning has been shown to be more accurate than other methodologies. This research can help us to better understand global disparities in kidney disease, as well as what we can do to address them and coordinate our efforts to achieve global kidney health equity. This study provides an excellent feature-based prediction model for detecting kidney disease. Various machine learning algorithms, including k-nearest neighbors algorithm (KNN), artificial neural networks (ANN), support vector machines (SVM), naive bayes (NB), and others, as well as Re-cursive Feature Elimination (RFE) and Chi-Square test feature-selection techniques, were used to build and analyze various prediction models on a publicly available dataset of healthy and kidney disease patients. The studies found that a logistic regression-based prediction model with optimal features chosen using the Chi-Square technique had the highest accuracy of 98.75 percent. White Blood Cell Count (Wbcc), Blood Glucose Random (bgr), Blood Urea (Bu), Serum Creatinine (Sc), Packed Cell Volume (Pcv), Albumin (Al), Hemoglobin (Hemo), Age, Sugar (Su), Hypertension (Htn), Diabetes Mellitus (Dm), and Blood Pressure (Bp) are examples of these traits.

Keywords: usability score artificial intelligence; medical information systems; image matching; machine learning algorithms; morphological operations

Citation: Poonia, R.C.; Gupta, M.K.; Abunadi, I.; Albraikan, A.A.; Al-Wesabi, F.N.; Hamza, M.A.; B, T. Intelligent Diagnostic Prediction and Classification Models for Detection of Kidney Disease. *Healthcare* **2022**, *10*, 371. https://doi.org/10.3390/healthcare10020371

Academic Editor: Mahmudur Rahman

Received: 7 December 2021
Accepted: 3 February 2022
Published: 14 February 2022

1. Introduction

Kidney disease affects over 750 million people worldwide, a figure that is growing. Kidney disease is a condition that affects people all over the world, but the disease's prevalence, identification, and treatment are all very different. Renal failure is the leading cause of death among people living in modern society. Cigarette smoking, excessive alcohol consumption, high cholesterol, and a variety of other risk factors all play a role in the disease. The kidney is a vital organ in the human body, performing a variety of vital functions. Despite the fact that kidney disease is better understood in developed countries, new research indicates that the condition is more prevalent in developing countries. The primary function is to collect waste and excess fluid from the circulatory system and excrete it via the kidneys via urine. If the function of this organ is compromised, the amount of harmful liquids and wastes in our systems may have disastrous consequences [1]. It is critical to emphasize that there are two kinds of kidney disease: acute kidney disease and chronic (long-term) kidney disease [2]. The most common type of kidney illness is acute renal disease. Chronic kidney failure is characterized by a progressive decline in kidney function over time (usually years). When the kidney's blood supply is cut off, the flow of urine is hampered by an enlarged prostate, or the kidney itself is injured and becomes ineffective, this type of kidney failure occurs. As a result of a chronic renal condition, kidney failure does not occur overnight. In the early stages of the disease, the patient exhibits no signs or symptoms of the illness. Patients who have had diabetes and high blood pressure for a long time are more likely to develop this syndrome. Patients who have been exposed for an extended period of time to lead-based medications and poisons are at risk of developing this disease. According to a poll, this condition affects a large number of people in our country, and thousands of people die from it each year. Only the most affluent countries have access to renal failure treatment. According to the World Health Organization, only 11% of the world's population receive adequate treatment for renal failure. Because they cannot afford dialysis or a kidney transplant, low-income patients die of renal failure. Patients who are identified and treated early on have a better chance of avoiding renal failure entirely. Scientists have developed a number of methods for detecting kidney disease at an early stage [3,4]. Patients' doctors may inform them ahead of time. Taking preventative measures before things get out of hand is a viable option.

Chronic Kidney Disease

Humans have two kidneys that are roughly the size of a fist. Their primary purpose is to filter blood. They remove waste and excess water, which turn into urine. They also help to keep the body's chemical balance, control blood pressure, and produce hormones. Chronic kidney disease means that the kidneys are damaged and are unable to filter blood as effectively as they should. This damage can cause waste to accumulate in the body and cause other issues that can be harmful to health. The most common causes of chronic kidney disease are diabetes and high blood pressure. Kidney damage occurs gradually over a long period of time. Many people have no symptoms until their kidney disease is advanced. Only blood and urine tests can inform you if you have kidney disease. Treatments cannot cure kidney disease, but they can help to slow its progression. They include blood pressure medications, blood sugar control medications, and cholesterol-lowering medications. Chronic kidney disease can worsen over time. It can occasionally result in kidney failure. Dialysis or a kidney transplant will be required if your kidneys fail. Based on population studies from developed countries, a systematic review found a mean prevalence of 7.2% in individuals older than 30 years. According to WHO data, it affects approximately 10% of the adult population and more than 20% of those over the age of 60, and it is undoubtedly underdiagnosed. The prevalence of CKD can reach 35–40% in patients followed up in primary care for diseases as common as high blood pressure (HBP) or diabetes mellitus (DM). The magnitude of the problem is magnified by the increase in morbidity and mortality, particularly cardiovascular mortality, caused by renal deterioration. CKD is thought to be the common final destination of a group of pathologies that affect the kidney in a chronic and irreversible manner. Once the diagnostic and therapeutic options for primary kidney disease have been exhausted, CKD necessitates common protocols of action that are, in general, independent of it. The most common causes of ACKD are described below, along with links to further information. More than one cause frequently coexists and worsens kidney damage.

In this work, the primary objective is to identify the best early-stage prediction model [5] for renal disease based on the most optimal attributes possible [6]. The following sub-goals are included:

- Review the existing approaches for the detection of kidney disease.
- Determine the best feature by applying various feature selection techniques.
- Build various prediction models on a kidney dataset using different machine learning algorithms and analyze their accuracy in the detection of kidney disease.

The rest of the article is organized as follows: Section 2 provides a review of the literature on the detection of kidney disease. Section 3 proposes a method for detecting kidney disease that makes use of machine learning and feature extraction. Section 4 discusses the kidney dataset, experimental results, and comparisons with existing methods. Section 5 discusses the conclusion and future work.

2. Related Works

The diagnosis of kidney illness using machine learning algorithms is an emerging subject of computer vision in healthcare. Because of their great accuracy in identifying illnesses, these procedures are gaining prominence. Using machine learning algorithms, such as decision trees, J48, Support Vector Machine (SVM), and others, researchers have developed several methods for identifying kidney illness. This section describes previous research ideas proposed by a variety of scholars.

Boukenze, B. et al. [6] suggested a machine learning-based method for identifying renal disorders. They employed the k-nearest neighbors algorithm (KNN), support vector machine (SVM), decision tree, and artificial neural network (ANN) machine learning algorithms. They used a number of performance measures to evaluate the accuracy of prediction models. They observed that the decision tree-based model outperformed all other models in diagnosing chronic failure, with an accuracy of 63 percent.

A. Salekin and colleagues employed SVM, KNN, and random forest techniques to build prediction models. They based their findings on a dataset of 400 cases. There were 24 properties in each record. Different machine learning algorithm-based models produced variable degrees of accuracy, it was revealed. The accuracy of the decision tree-based model was 98 percent, which was greater than that of earlier models.

H. Polat et al. [7] predicted renal disease using the SVM machine learning technique. They had a 97.5 percent accuracy rate. In order to enhance the accuracy, they applied a variety of feature selection methodologies. They improved the accuracy by 1% by employing feature selection.

Panwong, P. et al. [8] proposed an approach using KNN, NB, and decision tree classifiers. They also reduced the number of features by using the wrapper technique. Using the decision tree technique, they attained a maximum accuracy of 85 percent.

Dulhare, U. N. et al. [9] suggested a technique for diagnosing kidney illness using the naive Bayes machine learning algorithm in combination with the R attribute selector. They were 97.5% accurate in diagnosing renal illness.

Vasquez-Morales et al. [10] developed a neural network classifier based on massive quantities of CKD data, and the model proved to be 95 percent accurate in its predictions. To predict the advancement of diabetic kidney disease, Makino et al. [11] collected patient diagnoses and treatment information from textual data in an attempt to predict the progression of diabetic kidney disease.

According to Ren et al. [12], they developed a prediction model for diagnosing chronic kidney disease (CKD) using data from electronic health records (EHR). Based on a neural network architecture, the proposed model encoded and decoded textual and numerical data from electronic health records (EHR). A deep neural network model for identifying chronic renal disease was developed by Ma F. et al. [13]. Comparing the supplied model with ANN and SVM, the accuracy of the given model was the highest.

Almansour and colleagues [14] utilized machine learning to develop a technique for preventing chronic kidney disease. Researchers used machine learning classification methods, such as SVM and ANN, to make their findings. The experiments revealed that ANN outperformed SVM in terms of accuracy, with a 99.75% accuracy rate.

J. Qin and colleagues [15] presented a machine learning strategy for diagnosing chronic kidney disease (CKD) in its early stages. In order to construct their models, they used logistic regression, random forest, SVM, naive Bayes classifier, KNN, and the feedforward neural network as techniques. With an accuracy rating of 99.75%, the random forest classification model was shown to be the most accurate.

Z. Segal and colleagues [16] developed an ensemble tree-based machine learning algorithm (XGBoost) for the diagnosis of kidney disease in its early stages. Models such as random forest, CatBoost, and regression with regularization were used to compare the results of the stated model. All matrices were improved by using the proposed model, which had c-statistics of 0.93, sensitivity of 0.715%, and specificity of 0.958, among other improvements.

Khamparia et al. [17] developed a deep learning model for the early identification of chronic kidney disease (CKD) that employed a stacked autoencoder model to extract features from multimedia data and was published in Nature Communications. The authors used a SoftMax classifier to predict the final class, which they found to be accurate. Using the UC Irvine Machine Learning Repository (UCI) chronic kidney disease (CKD) dataset [18], it was revealed that the recommended model outperformed standard classification algorithms when compared to the data set in question.

Ebiaredoh Mienye Sarah A. et al. [19] developed a robust model for predicting chronic kidney disease (CKD) by combining an enhanced sparse autoencoder (SAE) with Softmax regression. The autoencoders in our proposed model achieved sparsity by penalizing the weights, as previously stated. Because the SoftMax regression model was specifically tailored for the classification task, the proposed model performed wonderfully in the testing environment. On the chronic kidney disease (CKD) data set, the proposed model had a precision of 98 percent, according to the researchers. When it came to performance, the proposed model outperformed other already available strategies.

According to Zhiyong Pang et al. [20], a fully automated computer-aided diagnostic approach that employed breast magnetic resonance imaging to differentiate between malignant and benign masses was proposed.

Using a combination of the support vector machine and the ReliefF feature selection approaches, the texture features were selected for use. It was found that this method was 92.3% accurate.

Chen, G. et al. [21] developed a model for identifying Hepatitis C virus infection that used the Fisher discriminating analysis method with an SVM classifier to obtain a more accurate diagnosis. The comparison of the proposed methodology to current methods showed that the hybrid method outperformed all other methods, reaching the highest classification accuracy of 96.77%. The authors of this paper developed a breast cancer diagnosis model [22]. Artificial neural networks are used to classify breast cancer based on qualities that have been selected using sequential forward and backward selection processes. SBSP obtained the highest level of accuracy, with a score of 98.75%.

Table 1 outlines prior studies by different researchers. According to the table, researchers employed multiple machine learning algorithm-based prediction models to predict renal disorders. The accuracy of these models varied and was inadequate. We noticed that many researchers did not pre-process their data and used no feature selection strategy.

Table 1. Summary of related work.

Sr. No.	Author	Year	Machine Learning Algorithms and Accuracy (%)
1.	A. J. Aljaaf et al. [1]	2018	Naïve Bayes: 83.4%, J48: 86.23%
2.	N. Borisagar, D. Barad, and P. Raval [5]	2017	ANN: 99.5
3.	B. Boukenze, A. Haqiq, and H. Mousannif [6]	2018	SVM: 63.5%, LR: 64.0, C4.5: 63%, KNN: 55.15%
4.	H. Polat, H. D. Mehr and A. Cetin [7]	2019	SVM: 97.5%
5.	P. Panwong and N. Iam-On [8]	2016	KNN: 86.32%, naïve Bayes: 60.46%, ANN: 83.24%, RF: 86.60%, J48: 79.52%
6.	Makino et al. [11]	2019	KNN, Naïve Bayes + LDA + random subspace + Tree-based decision: 94%
7.	Ren et al. [12]	2019	SVM + ReliefF: 92.7%
8.	Ma F. et al. [13]	2019	Fisher discriminatory analysis and SVM: 96.7%
9.	Almansour and colleagues [14]	2020	KNN and SVM: 99%
10.	J. Qin and colleagues [15]	2019	SVM, KNN, and naïve Bayes decision tree: 99.7%
11.	Z. Segal and colleagues [16]	2019	SVM, KNN, and decision tree: 99.1%
12.	Khamparia et al. [17]	2020	Logistic regression, KNN, SVM, random forest, naive Bayes, and ANN: 99.7%
13.	Ebiaredoh-Mienye Sarah A. et al. [18]	2017	SVM 98.5%
14.	Zhiyong Pang et al. [19]	2020	Softmax regression 98%
15.	Tabassum, Mamatha et al. [23]	2017	DT: 85%, RF: 85%
16.	K. R. A. Padmanaban and G. Parthiban [24]	2016	DT: 91%, naïve Bayes: 86%
17.	Sahil Sharma, Vinod Sharma, and Atul Sharma [25]	2018	ANN: 80.4%, RF: 78.6%
18.	Pratibha Devishri [26]	2019	ANN: 86.40%, SVM: 77.12%
19.	Sujata Drall, G. Singh Drall, S. Singh, Bharat Naib [27]	2018	Naïve Bayes: 94.8%, KNN: 93.75%, SVM: 96.55%

LR: Logistic Regression; KNN: k-Nearest Neighbors; SVM: Support Vector Machines; CART: Classification and Regression Trees; ANN: Artificial Neural Networks; LDA: Linear Discriminant Analysis; DT: Decision Tree; RF: Random Forest.

3. Support Vector Machine

The first concepts and foundational principles of SVM were provided by the statistical learning theory (structural risk minimization). It can be used in classification and nonlinear regression. This broad classification of SVM can be further subdivided into two subcategories: linear SVM (linear SVM) and nonlinear SVM (nonlinear SVM) [28].

L-SVM [29] training data of different types are classified using linear SVM, which classifies training data by giving Class 1 to the "+1" and Class 2 to the "−1" symbols, then uses the mathematical notation

$$\begin{gathered}\{\{x_i, y_i\}_{i-1}^{T}, x_i \in R^m, y_i \in \{-1, +1\}\} \\ w \cdot x + b = 0\end{gathered} \quad (1)$$

here w is the weight vector, x is the input dataset, and b is a bias in the hyper plane, which is referred to as a displacement. Bias is used to make sure that the hyper plane [11] is positioned correctly following movement in a horizontal plane. Thus, prejudice is affected by training with bias. A hyper plane has its parameters, which are w and b. A decision surface G. Chen et al. (2020) [29] is considered to be a function when SVM is used for classification.

$$f(x) = sign(w \cdot x + b) \quad (2)$$

SVM generally serves to increase the marginal distance of the data set and therefore enhance the distinguishing function, allowing better categorization. Improving the hyperplane's distinguishing function is a quantic programming issue.

$$minimize\ L_p = \frac{1}{2} \parallel w \parallel^2\ subject\ to\ y_i(x_i \cdot w + b) - 1 \geq 0,\ i = 1, \ldots, l \tag{3}$$

To solve the initial minimization issue, we apply the Lagrange theory:

$$\begin{aligned} L_D(\alpha) &= \sum_{i=1}^{l} \alpha_i - \frac{1}{2} \sum_{i=1}^{l} \sum_{j=1}^{l} \alpha_i \alpha_j y_i y_{j(x_i x_j)} \\ \text{subject to } & \sum_{i=1}^{l} \alpha_i y_i = 0, i = 1, \ldots, l \\ & \alpha_i \geq 0, i = 1, \ldots, l \end{aligned} \tag{4}$$

In the end, the linear divisive decision-making function has been completed.

$$f(x) = \text{sign}\left(\sum_{i=1}^{n} y_i \alpha_i^* (x \cdot x_i) + b^* \right) \tag{5}$$

To sum up, when $f(x) > 0$, it indicates that the sample is marked +1 and is in the same category as samples marked with "+1"; otherwise, it indicates that the sample is marked -1 and is in the same category as samples marked with "-1". Linear hyper planes [30] cannot properly identify data points when training data include noise. Slack variables ξ_i are introduced to the constraint, resulting in a modification of the original (3):

$$\begin{aligned} \text{minimize } & \tfrac{1}{2} \parallel w \parallel^2 + C\left(\sum_{i=1}^{l} \xi_i \right) \\ \text{subject to } & y_i(x_i \cdot w + b) - 1 + \xi_i \geq 0,\ i = 1, \ldots, l \\ & \xi_i \geq 0,\ i = 1, \ldots, l \end{aligned} \tag{6}$$

The position of the border and the classification point are separated by a distance of ξ_i; in this case, C represents the cost of the training data classification mistake, as specified by the user. A lower C value means that the margin will be narrower, suggesting that fault tolerance has a lower chance of working in the event of a problem [31,32]. The fault tolerance rate will be larger if C is lower. The linear inseparable issue (also known as the infinitely large linear problem) will degenerate into a linear separable problem as $C \rightarrow \infty$. In this instance, the parameters and the optimal solution of the target function may be found by using the Lagrangian coefficient [33,34] in order to solve the linear inseparable dual optimization issue; hence, the solution of the linear inseparable dual optimization problem is as follows:

$$\begin{aligned} \text{Max} \quad L_D(\alpha) &= \sum_{i=1}^{l} \alpha_i - \tfrac{1}{2} \sum_{i=1}^{l} \sum_{j=1}^{l} \alpha_i \alpha_j y_i y_j (x_i x_j) \\ \text{subject to } & \sum_{i=1}^{l} \alpha_i y_i = 0, i = 1, \ldots, l \\ & 0 \leq \alpha_i \leq C, i = 1, \ldots, l \end{aligned} \tag{7}$$

Finally, the linear decision-making function is

$$f(x) = \text{sign}\left(\sum_{i=1}^{n} y_i \alpha_i^* (x \cdot x_i) + b^* \right), \tag{8}$$

a support vector machine whose operation can include nonlinear inputs (nonlinear SVM). In the case where we cannot separate training samples using linear SVM, we may apply feature transformation, such as the function φ, to convert original 2-D data into a new, high-dimensional feature space that allows us to solve linear separable problems. SVM can use the kernel technique to effectively conduct nonlinear classification utilizing an approach known as the kernel trick. For the time being, there are many diverse foundational components being put forward. Differentiating distinct data characteristics with respect to different core functions allows for more efficient computation with SVMs [7]. Of the very common fundamental functions, these four functions have something in common:

Linear kernel function:

$$K(x_i, y_i) = x_i^t \cdot y_j \tag{9}$$

Polynomial kernel function:

$$K(x_i, y_j) = \left(\gamma x_i^t x_j + r\right)^m, \ \gamma > 0 \tag{10}$$

Radial basis kernel function:

$$K(x_i, y_j) = exp\left(\frac{- \parallel x_i - y_j \parallel^2}{2\sigma^2}\right), \ \gamma > 0 \tag{11}$$

Sigmoid kernel function:

$$K(x_i, y_j) = tanh\left(\gamma x_i^t \cdot y_j + r\right) \tag{12}$$

This study utilizes the emissive core function, because settings such as γ and C can increase computation efficiency and lower SVM complexity.

4. Materials and Methods

The proposed strategy is based on data mining framework as shown in Figure 1. Data mining employs computational approaches at the intersection of artificial intelligence, machine learning, statistics, and database systems [35]. Data mining is predicated on the idea that data can be analyzed from a variety of perspectives. The "Knowledge Discovery in Databases" (KDD) process is employed in this study to extract unknown patterns from web data [36]. This section describes the suggested method for detecting kidney disease. The availability of kidney disease care is directly affected by each country's public policies and financial situation. A lower dialysis–to–transplant ratio, for example, suggests that more affluent countries have a higher rate of kidney transplantation.

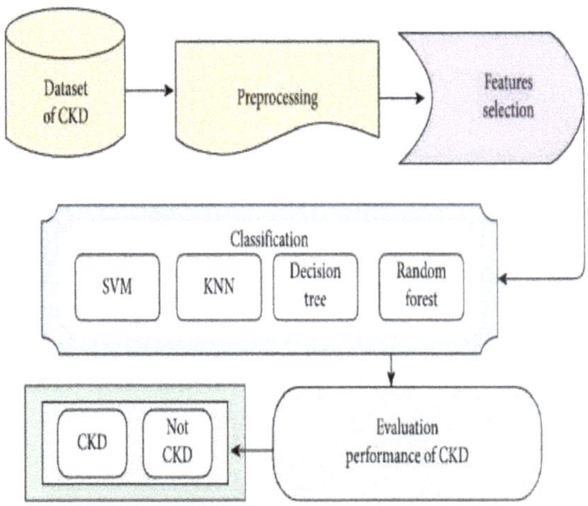

Figure 1. Detection of chronic kidney disease using recursive feature elimination and classification algorithms. CKD: Chronic Kidney Disease; SVM: Support Vector Machine; KNN: K-Nearest Neighbors.

4.1. Kidney Disease Dataset

In this work, we used a dataset of 400 patients, each with 24 attributes [18,37]. The dataset had 250 records of patients who were suffering from kidney disease and 150 medical records for completely healthy people. This dataset has medical data for different age groups. It has 50 records of people less than 30 years old and 55 records of people greater than 70 years old. The remaining records belong to people aged 31–69. From the various studies, it was found that people of any age group may suffer from kidney disease. Therefore, there is no risk of bias in evaluating the performance of prediction models. Table 2 shows the details of the various kidney disease-related attributes.

Table 2. Details of the various kidney disease-related attributes.

Name	Feature	Description
Age	Age	Patient's age
Blood pressure	Bp	Blood pressure of the patient
Sugar level	Su	Sugar level of the patient
Bacteria	Ba	Presence of bacteria in the blood
Ratio of the density of urine	Sg	Ratio of the density of urine
Albumin level in the blood	Al	Ratio of the albumin level in the blood
Pedal edema	Pe	Does the patient have pedal edema or not
Red blood cells	Rbc	Patients' red blood cell counts
Patient class	Class	Does the patient have kidney disease or not
Pus cell clumps	Pcc	Presence of pus cell clumps in the blood
Anemia	Ane	Does the patient have anemia or not
Red blood cell count	Rc	Red blood cell count of the patient
Hypertension	Htn	Does the patient have hypertension on not
Serum creatinine	Sc	Serum creatinine level in the blood
Diabetes mellitus	Dm	Does the patient have diabetes or not
Blood urea	Bu	Blood urea level of the patient
Blood glucose	Bgr	Blood glucose random count
Sodium	Sod	Sodium level in the blood
White blood cell count	Wc	White blood cell count of the patient
Hemoglobin	Hemo	Hemoglobin level in the blood
Packed cell volume	Pcv	Packed cell volume in the blood
Pus cell	Pc	pus cell count of patient
Potassium	Pot	Potassium level in the blood
Appetite	Appet	Patient's appetite
Coronary artery disease	Cad	Does the patient have coronary artery disease or not

To explain the proposed approach in an easy and efficient manner, a flow chart of the whole procedure is given in Figure 2 and the steps are explained one–by–one as follows:

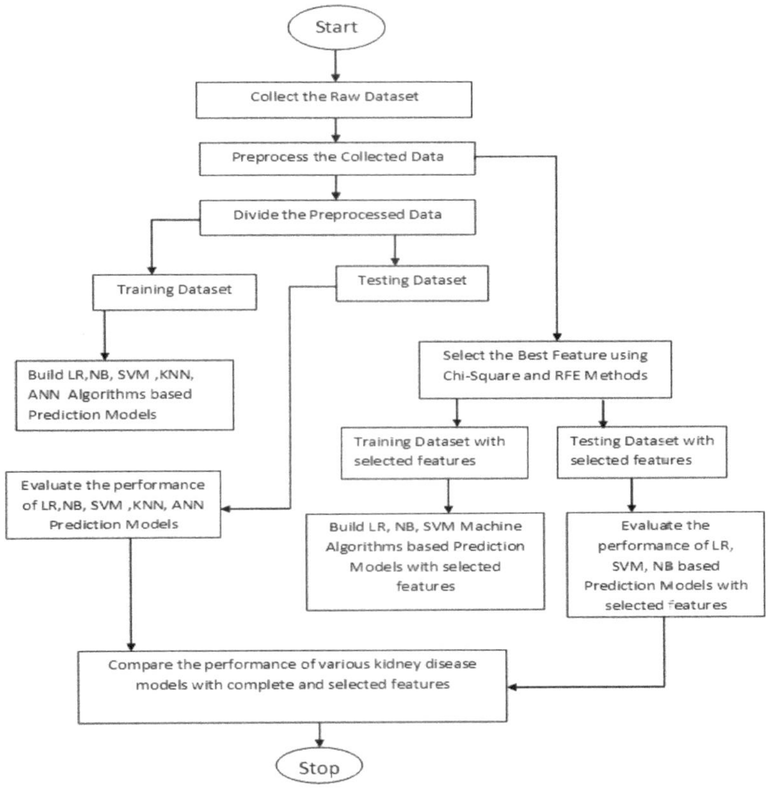

Figure 2. Flow chart of the proposed model. LR: Logistic Regression; NB: Naïve Bayes; SVM: Support Vector Machine; KNN: Nearest Neighbors; ANN: Artificial Neural Network; RFE: Recursive Feature Elimination.

4.2. *Proposed Algorithm*

Procedure: The proposed approach for the detection of kidney disease

Input: Dataset of kidney disease records

Output: Performance of the prediction models in detecting kidney disease.

It has the following steps:

Step 1: The Glomerular Filtration Rate (GFR) is the most often utilized measure of kidney health function in CKD medical therapy. In order to calculate which, the formula uses information such as the patient's blood creatinine, age, race, gender, and other variables. As is widely accepted, the standard formula for renal disease modification of diet (MDRD).

$$GfR = 186 \times (create\,)^{-1.154} \times (age\,)^{-0.203} \times \left(\frac{\frac{\text{mL}}{\text{min}}}{173\ \text{m}^2}\right) \tag{13}$$

Then preprocess the collected data: In this step, we preprocess the collected kidney disease dataset. In the original dataset, the 'rbc' and 'pc' columns have normal, abnormal, and empty values. The 'rbc' and 'pc' columns have 150 and 120 entries without any values, respectively. In this dataset, the 'pcc' and 'ba' columns have 'present' and 'not present' values. The 'cad', 'pe', 'htn', 'dm', and 'ane' columns have the values 'yes' and 'no'. Also, in this dataset, 'appet' has the values 'poor' and 'good'. Therefore, preprocessing of this dataset is a mandatory task for correct results. In this step, the empty values are replaced by NaN. We converted nominal values to binary values as follows:

1. In the 'rbc' and 'pc' columns, 'normal' and 'abnormal' nominal values are replaced with 1 and 0, respectively.
2. In the 'pcc' and 'ba' columns, the 'present' and 'nonpresent' values are replaced with 1 and 0, respectively.
3. In the 'htn', 'pe', 'ane', 'dm', and 'cad' columns, the values 'yes' and 'no' are replaced with 1 and 0, respectively.
4. Finally, in the 'appet' column, 'good' and 'poor' are replaced with 1 and 0, respectively.

In the next step, null values are replaced by the average value of that particular column's values.

Step 2. Observe the relationship between different features. In this step, we find the relationship between input and target features. We found that 'pot' and 'ba' are weakly related to the target feature.

Step 3. Divide the dataset. In this step, we divide the dataset into training and testing datasets using an 80:20 ratio. It means that 80% of data are used for training and 20% of data are used for testing purposes.

Step 4. Set the parameters of the machine learning algorithms. In this step, the kidney disease dataset's processed features are used with machine learning algorithms to build prediction models. We used Logistic Regression, Naive Bayes, Support Vector Machine, K-Nearest Neighbors (KNN), and Artificial Neural Network (ANN) machine learning algorithms. We applied a 10-fold cross validation for building the prediction models.

Let $\phi(x)$ be a ridge basis function, nonconstant, limited, and monotonically growing. If K is a compact subset on R^n, and $f(x_1,\ldots,x_n)$ is a real-valued continuous function on K, then K may be represented as a subset of $R\hat{}n$, where f is a collection of real numbers. Given an arbitrary positive parameter, there are integer N and real parameters

$$\begin{gathered} v_j, \theta_j, w_{ij} \text{ for } i = 1,\ldots,n \text{ and } j = 1,\ldots,m. \\ f^{\sim}(x_1,\ldots,x_n) = \textstyle\sum_{j=1}^{m} v_j \phi_j \left(\sum_{i=1}^{n} w_{ij} x_i + \theta_j\right) + d \end{gathered} \tag{14}$$

it satisfies the condition

$$max_{X\in K}|f^{\sim}(X) - f(X)| < \varepsilon \tag{15}$$

We are saying that, for every given $\varepsilon > 0$, there exists a three-layer network, where the hidden layer represented by the ridge basis function $\phi(x)$ and whose input–output function is $f^{\sim}(x_1,\ldots,x_n)$, which has a $max_{X\in K}|f^{\sim}(X) - f(X)| < \varepsilon$ mapping function $f^{\sim}(x_1,\ldots,x_n)$ that results in $f(x_1,\ldots,x_n)$ being greater than or equal to ε.

Step 5. Feature selection. In this step, we select the best features using the Recursive Feature Selection (RFE) and Chi-Square feature selection methods. As our kidney disease data set was a labeled dataset, we used the wrapper and filter technique that is the supervised feature selection technique. As we discussed earlier, the supervised feature selection techniques were divided into three categories, which had different methods in each category.

For feature selection, we used $S = (U, C \cup D)$ and $B \subseteq C$, where S is the set of attributes of feature and attribute set D with respect to the conditional attribute subset B, then the evaluation function for feature selection is defined by

$$\sigma(B,D) = \frac{1}{N}(\sigma_B(D_1) + \sigma_B(D_2) + \cdots + \sigma_B(D_N)) \tag{16}$$

In this case, N is the number of decision classes generated by the decision attribute set D, and is equal to $\sigma_B(D_i), i = 1,2,\ldots,N$, reflecting the uncertainty measure of each decision class, and $\sigma(B,D)$ describes the integrated uncertainty degree of blocks $D_1, D_2, \ldots, D_N$.

Recursive Feature Elimination (RFE) is a feature selection algorithm of the wrapper type. Internally, it employs filter-based techniques that are distinct from the filter approach. It has two important configuration options: a. it specifies the number of features to be selected, and b. it specifies the machine learning algorithm used in feature selection. In the first case, it searches for a subset of features by considering all of the features in the training dataset and removing them until the required number of features remains. In the second case, it employs a machine learning algorithm that ranks features [38] based on their importance. It removes the least important features and then repeats the model fitting process. The process is repeated until the specified number of features remain.

The Chi-Squared feature selection method investigates the relationship between the input features and the target class. In this test, the Chi-Square value is calculated for each input feature and the target class. It has the required number of features, as well as the highest Chi-Square scores. We used the formula below to calculate the chi-square metric (Xc2) between each target class feature and each input feature. It chooses only the input features with the highest Chi-Squared values.

Chi-Square feature selection in data with m attribute values and k class labels as output. Then, the value of χ^2 is

$$\chi^2 = \sum_{i=1}^{m}\sum_{j=1}^{k} \frac{\left(O_{ij} - E_{ij}\right)^2}{E_{ij}} \tag{17}$$

where O_{ij} is the observed frequency.

Step 6. Build the prediction model using the selected features. In this step, again, we applied 10-fold cross validation with the selected features and various machine learning algorithms to build different prediction models.

Step 7. Finally, the performance of prediction models with all features and selected features are compared.

5. Results and Analysis

To assess the performance of machine learning approaches, researchers use a variety of performance metrics. To evaluate and compare the performance of proposed prediction models, we used the precision, recall, F-measure, and accuracy performance measures.

5.1. Performance Measures

Accuracy is calculated by dividing the number of test records by the number of successfully classified records. The percentage of True Positive (TP) records to the total number of True Positive (TP) records in a certain class is called precision. There are two types of recall: true positives and false negatives. The total number of records properly categorized to the total number of records in a class is known as the recall ratio (FN). The precision, recall, F-measure, and accuracy were calculated using the following formulas:

$$Precision_i = \frac{TP_i}{TP_i + FP_i} \tag{18}$$

$$Recall_i = \frac{TP_i}{TP_i + FN_i} \tag{19}$$

$$F_\beta = \frac{\left(1+\beta^2\right)\ precision * recall}{\beta^2 * precision + recall} \tag{20}$$

where β is a parameter that can be used to give the importance to any one precision or recall.

Accuracy is commonly used as a measure for categorization techniques.

$$Accuracy_i = \frac{TP_i + TN_i}{TP_i + FP_i + FN_i + TN_i} \tag{21}$$

where TP_i is the number of records correctly classified as belonging to the kidney disease class, FP_i is the number of records incorrectly classified as having kidney disease, FN_i is the number of records that were not classified as having a kidney disease, and TN_i is the number of images that were not assigned to the correct kidney disease class.

Precision (P) is a metric that quantifies the proportion of correct positive outcomes among all possible outcomes. It is computed as follows:

$$P = TP/(TP + FP) \tag{22}$$

Specificity: The system's ability to accurately recognize the absence of impurities in the ghee picture is measured in this category. To obtain it, the number of true negatives recognized in the photographs must be counted and divided by the amount of pure milk included in the images. It was utilized to determine the specificity of the data.

$$Specificity\ (SP) = (TN)/(TN + FP) \tag{23}$$

Mean: Means are a straightforward approach commonly used in pure mathematics, as well as in analysis and computing; a wide variety of means have been invented to perform these duties. During an image processing competition, the technique of filtering by the mean is evaluated as abstraction filtering and is utilized for noise reduction.

$$X^{-} = \frac{\sum_{i=0}^{n} X_i}{n} \tag{24}$$

A measure of variability or diversity in statistics, the standard deviation is the most widely used measure available to researchers. In the context of image processing, it indicates what fraction of variance or dispersion occurs between the predicted value and the observed value. An extremely low standard deviation suggests that the data points have a strong tendency to be extremely near to one another. A large standard deviation, on the other hand, shows that the data points are evenly distributed throughout a wide range of values.

$$X^{-}{}_{rms} = \sqrt{\frac{\sum_{X_{i=1}}^{n} (X_i - X^{-})^2}{(n-1)}} \tag{25}$$

We used Anaconda, an enterprise-ready, secure, and scalable data science platform, and Spyder to build and analyze the prediction approaches (Python 3.6). To evaluate the proposed method's performance, we downloaded a kidney disease dataset containing 400 patient records. We pre-processed the data to remove null values and for other purposes. The data set was divided into two parts: training and testing, with 80 percent of the records in training and 20% in testing. Using machine learning algorithms, such as Logistic Regression, NB, SVM, K-Nearest Neighbors (KNN), and Artificial Neural Network, we developed a variety of prediction models (ANN).

The data correlation matrix was represented using Heatmap [6]. It shows how different features interact with one another. It is a useful visualization technique for comparing the values of any two features. A positive correlation indicates that, as the value of a feature increases, so does the value of the target variable. It could be negative, implying that increasing the value of a feature decreases the value of the target variable. The heatmap was created with the help of the seaborn library. It visually displays which features are closely related to the target variable. By simply looking at the different color tones used, it can be determined which value is higher, lower, and so on. A heatmap correlation matrix of kidney disease data was displayed. It showed that the Ane, Bgr, Bu, Sc, Pcv, Al, Hemo, Age, Su, Htn, Dm, and Bp characteristics were highly related to the target variable (represented in green color). This means that raising these parameter values raises the risk of kidney disease.

5.2. Prediction Models with All Features

Table 3 and Figure 3 show the performance of the prediction models by considering all features or, in other words, without applying any feature selection technique. From the table and graph, we can see that the accuracies of the Logistic Regression, Naïve Bayes,

SVM, KNN, and ANN-based prediction models with all features were 97.5%, 95%, 97.5%, 66.25%, and 65% respectively.

Table 3. Results of the prediction models with all features.

Machine Learning	Precision	Recall	F-Measure	Accuracy
Algorithms	(%)	(%)	(%)	(%)
Logistic regression	98	97	98	97.5
Naïve Bayes	95	95	95	95
Support Vector Machines	98	97	98	97.5
k-Nearest Neighbors	76	66	66	66.25
Artificial Neural Networks	42	65	51	65

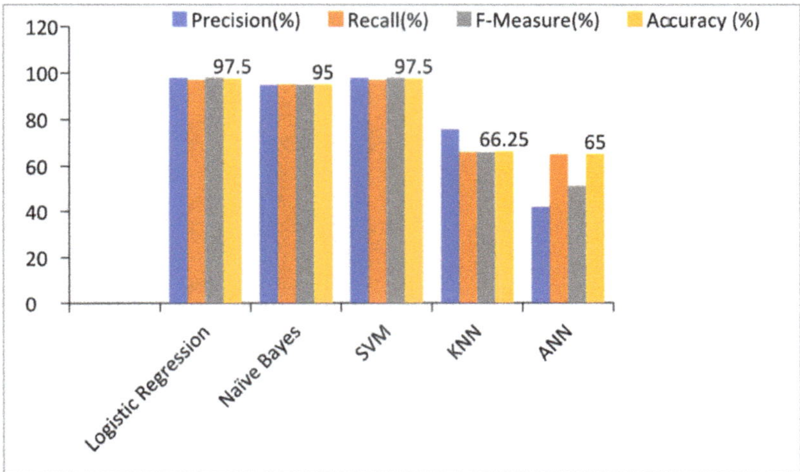

Figure 3. Results of the prediction models with all features. SVM: Support Vector Machine; KNN: K-Nearest Neighbors; ANN: Artificial Neural Network.

These also show that the accuracy of the Logistic Regression and SVM algorithm-based prediction models were highest i.e., 97.5%. The ANN-based prediction model achieved the lowest accuracy in the detection of kidney diseases. The performances of Logistic Regression and SVM were the same and can be used interchangeably for the detection of kidney diseases in the early stage. We can also see that the precision, recall, and F-measure values were the highest for the Logistic Regression and SVM-based prediction models.

5.3. Prediction Models with RFE Feature Selection Technique

Recursive Feature Elimination (RFE) is a feature selection algorithm of the wrapper type. It internally uses filter-based techniques; however, it is different to the filter approach. It has two important configuration options: a. it specifies the number of features to be selected, and b. it sets the machine learning algorithm in choosing the features. In the first case, it searches a subset of features by considering all features present in the training dataset and removes the features until the required number of features remains. In the second case, it uses a machine learning algorithm and ranks the features by their importance. It discards the least important features and repeats the model fitting process. The whole process is repeated until the mentioned number of features remains.

Table 4 and Figure 4 show the results of the prediction models built with basic logistic regression and with the RFE feature selection technique. From the table and graph, we can see that it achieved 97.5% accuracy without feature selection and 91.25% accuracy with RFE feature selection. It was also observed that the values of precision, recall, and F-measure were better without the RFE feature selection technique. Therefore, we conclude that the accuracy of the basic logistic model is higher than that with the RFE feature selection technique. Table 5 shows the results of the prediction models built with basic SVM and with the RFE feature selection technique.

Table 4. Results of the LR model with RFE feature selection technique.

Performance Measure	Basic Logistic Regression	Logistic Regression with RFE Feature Selection
Precision (%)	98	92
Recall (%)	97	94
F-Measure (%)	98	93
Accuracy (%)	97.5	91.25

RFE: Recursive Feature Selection.

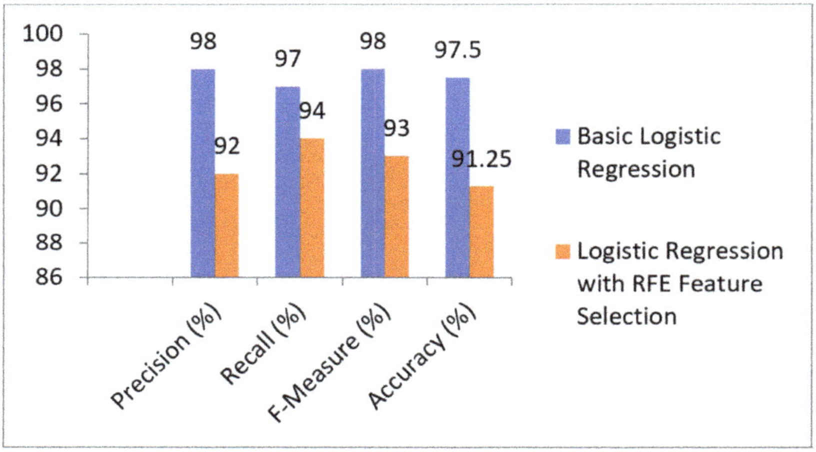

Figure 4. Comparison of LR Models with and without RFE feature selection. RFE: Recursive Feature Selection.

Table 5. Results of the SVM model with the RFE feature selection technique.

Performance Measure	Basic SVM	SVM with RFE Feature Selection
Precision (%)	98	98
Recall (%)	97	96
F-Measure (%)	98	97
Accuracy (%)	97.5	96.25

SVM: Support Vector Machine; RFE: Recursive Feature Elimination.

From this, we can see that it achieved 97.5% accuracy without feature selection and 96.25% accuracy with RFE feature selection. It was also observed that the values of precision, recall, and F-measure were also better without the RFE feature selection technique. Therefore, we conclude that the accuracy of the basic SVM model is higher than that with the RFE feature selection technique.

5.4. Performance of Prediction Models with Chi-Square Feature Selection

In this subsection, from Table 3, we found that the accuracy of the Logistic Regression-based model was highest among the other built models in the detection of kidney disease. As we know, the feature selection technique may improve the performance of the model. In this section, we applied the Chi-Squared (chi2) statistical test to select the K-best features from the kidney disease-prediction dataset.

The Chi-Square feature selection method checks the relationship between input features and the target class. In this test, Chi-Square is determined among each input feature and the target class. It provides the required number of features with the best Chi-Square scores. It selects only those input features that have the maximum Chi-Square values. The scikit-learn library provides the SelectKBest class that is used to select a specific number of features in a suite of different statistical tests. Table 6 shows the scores of various features. It shows that the Wbcc, Bgr, Bu, Sc, Pcv, Al, Hemo, Age, Su, Htn, Dm, and Bp features have high scores in comparison with the other features.

Table 6. Features and their scores by the Chi-Square test.

Features	Score
Wbcc	12,733.73
Bgr	2428.328
Bu	2336.005
Sc	354.4105
Pcv	324.7065
Al	228.1047
Hemo	125.0657
Age	113.4602
Su	100.95
Htn	86.29181
Dm	82.2
Bp	80.02432
Pe	45.10802
Ane	35.6116
Sod	28.7933
Pcc	24.07546
Rbcc	20.848
Cad	19.93604
Pc	14.16913
Ba	12.58705
Appet	12.58703
Rbc	9.416036
Pot	4.071145
Sg	0.005035

Wbcc: White Blood Cell Count; brg: Blood Glucose Random; Bu: Blood Urea; Sc: Serum Creatinine; Pcv: Packed Cell Volume; Al: Albumin; Hemo: Hemoglobin; Su: Sugar; Htn: Hypertension; Dm: Diabetes Mellitus; Bp: Blood Pressure; Pe: Pedal edema; Ane: Anemia; Sod: Sodium; Pcc: Pus cell clumps; Rbcc: Red blood cells count; Cad: Coronary artery disease; Pc: Pus cell; Ba: Bacteria; Appet: Appetite; Rbc: Red blood cells; Pot: Potassium; Sg: Ratio of the density of urine.

Table 7 and Figure 5 show the performance of the LR prediction model with the Chi-Square feature-selection technique.

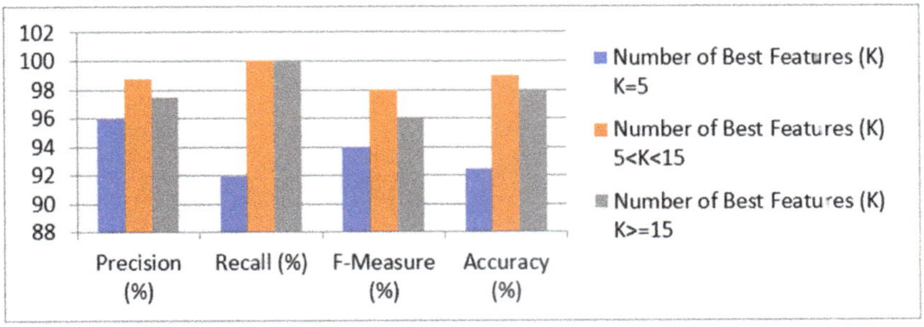

Figure 5. Results of the LR prediction model with Chi-Square feature selection.

Table 7. Results of the LR prediction model with Chi-Square feature selection.

Performance Measure	Number of Features (K) K = 5 5 < K < 15	Best K > = 15
Precision (%)	96 100	100
Recall (%)	92 98	96
F-Measure (%)	94 99	98
Accuracy (%)	92.5 98.75	97.5

We evaluated the technique using a variety of best features. It was discovered that, when the k-values ranged from 6 to 14, the model provided the best precision, recall, f-measure, and accuracy, i.e., 100 percent, 98 percent, 99 percent, and 98.75 percent, respectively. When k = 5 or fewer features are used, the model had the lowest accuracy. The table also shows that, when more than 15 features were used, the model's performance suffered. As a result, it can be concluded that the model with more than 5 and less than 15 features provided the highest accuracy in detecting kidney disease. The performance of the SVM prediction model with the Chi-Square feature-selection technique is shown in Table 8.

Table 8. Comparative analysis of existing models on a dataset of 400 patients each with 24 attributes [2,27].

Method	Accuracy	Recall	Precision	F-Measure
Logistic regression [28]	91.8	1	0.98	0.98
KNN [29]	92.7	0.88	0.98	0.92
Naïve Bayes [30]	95.21%	0.92	1.00	0.94
SVM [31]	92.32	0.87	0.96	0.93
Decision tree [32]	93.45	0.95	1.00	0.96
Proposed method [33]	97.54	0.99	1.00	1.0

KNN: k-nearest neighbors algorithm; SVM: support vector machines.

We evaluated the technique with different numbers of best features. It was found that the model achieved the best precision, recall, f-measure, and accuracy when the k-values were greater than 15, i.e., 100%, 96%, 98%, and 97.5%, respectively. The model gave the lowest accuracy when K = 5 or a smaller number of features was taken. From the table, we also see that the performance of the model decreased whenever fewer than 15 features were taken. Therefore, it can be concluded that the accuracy of the SVM model did not increase by applying the Chi-Square test.

5.5. Comparison of Models with and without Feature Selection Technique

From all of the results, it can be seen that the accuracy of the Logistic Regression model with the Chi-Square feature selection techniques was the best in the detection of kidney disease. This result was the best among the other approaches in the detection of kidney disease. Table 9 shows the results of various combinations of LR models and Figure 6 graphically compares the accuracy of the different models.

Table 9. Prediction models with and without various feature-selection techniques.

Prediction Model	Accuracy (%)
Basic LR model	91.25
LR model + RFE feature selection	97.5
LR model + Chi-Square feature selection (K = 5)	92.5
LR model + Chi-Square feature selection (5 < K < 14)	98.75
LR model + Chi-Square feature selection (K > 14)	97.5

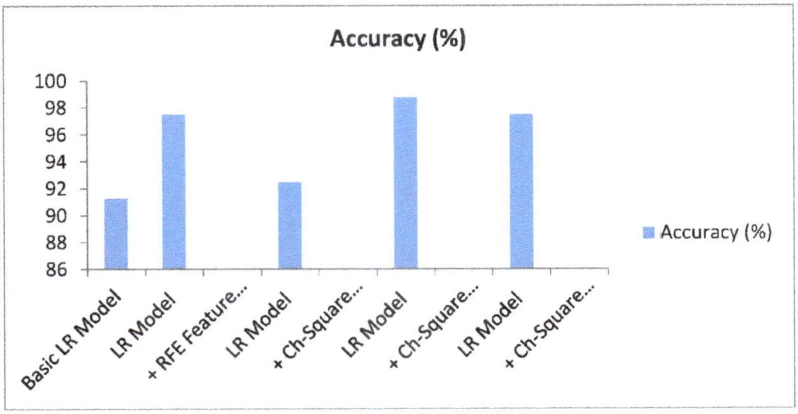

Figure 6. Results of the models with and without feature selection. LR: Logistic Regression; REF: Recursive Feature Elimination.

The accuracies of the basic LR model, LR model with RFE feature selection, LR model with Chi-Square feature selection (K = 5), LR model with Chi-Square feature selection (5K14), and LR model with Chi-Square feature selection (K > 14) were 91.25 percent, 97.5 percent, 92.5 percent, 98.75 percent, and 97.5 percent, respectively, as shown in Table 9. This demonstrates that the Chi-Square method outperformed the RFE feature method in terms of accuracy. It is also worth noting that the model produced good results, with 5 to 15 of the best features out of a total of 24. In summary, we achieved 98.5 percent accuracy in detecting kidney disease. In comparison to existing approaches, this has the highest accuracy.

As a result of the random forest algorithm, 250 positive samples (TP) and 150 negative samples (TN) were correctly identified as positive. Positive (TP) samples were scored at 94.74 percent by the SVM, KNN, and Decision Tree algorithms with an error (TN) of 5.26 percent each, and 97.37 percent by the SVM, KNN, and Decision Tree algorithms with an error (TN) of 1.32 percent each. Table 6 shows the results of the four classifiers that were used. The random forest method outperformed the other classifiers on all metrics, including accuracy, precision, recall, and F1-score. The decision tree algorithm came in second, with accuracy, precision, recall, and F1-score values of 99.17 percent, 100 percent, 98.68 percent, and 99.34 percent, respectively. As a result, the KNN algorithm achieved 98.33 percent accuracy, precision, recall, and an F1-score of 98.67 percent. The final SVM accuracy, precision, recall, and F1-score were 96 percent, 92 percent, 93 percent, and 97 percent, respectively.

6. Conclusions and Future Work

In this paper, we developed many prediction models by using different machine learning algorithms and feature-selection techniques. We used a dataset that contained a large set of healthy and unhealthy patients with kidney disease. We used LR, SVM, and many other classifiers to develop various prediction models. We exercised the prediction models with Recursive Feature Elimination (RFE) and Chi-Square test feature selection techniques. From the results, it was shown that the accuracy of the Logistic Regression model with the Chi-Square feature selection technique achieved the best result in the detection of kidney disease. This result was the best among other approaches in the detection of kidney disease. It was also observed that the model achieved good results with 5 to 15 best features among 24 features. It was also found that the Wbcc, Bgr, Bu, Sc, Pcv, Al, Hemo, Age, Su, Htn, Dm, and Bp features had more significance in the detection of kidney diseases. In the future, we will develop a hybrid approach for improving disease detection accuracy before actual disease arises in humans.

Author Contributions: Conceptualization, methodology; validation and writing—review and editing, R.C.P. and T.B.; formal analysis, M.K.G.; investigation, I.A.; resources, data curation, A.A.A.; visualization, M.A.H.; funding acquisition, F.N.A.-W. All authors have read and agreed to the published version of the manuscript.

Funding: This work was supported by the Deanship of Scientific Research, King Khalid University, Saudi Arabia under grant number (RGP 1/190/43).

Institutional Review Board Statement: Not applicable.

Informed Consent Statement: Informed consent was obtained from all participants involved in the study.

Data Availability Statement: Data is available on reasonable request.

Acknowledgments: The authors extend their appreciation to the Deanship of Scientific Research at King Khalid University for funding this work under grant number (RGP 1/190/43), Princess Nourah bint Ab-dulrahman University Researchers Supporting Project number (PNURSP2022R191), and Princess Nourah bint Abdulrahman University, Riyadh, Saudi Arabia. The authors would like to acknowledge the support of Prince Sultan University for paying the Article Processing Charges (APC) of this publication.

Conflicts of Interest: The authors declare no conflict of interest.

References

1. Aljaaf, A.J. Early Prediction of Chronic Kidney Disease Using Machine Learning Supported by Predictive Analytics. In Proceedings of the IEEE Congress on Evolutionary Computation (CEC), Wellington, New Zealand, 8–13 July 2018.
2. Nishanth, A.; Thiruvaran, T. Identifying Important Attributes for Early Detection of Chronic Kidney Disease. *IEEE Rev. Biomed. Eng.* **2018**, *11*, 208–216. [CrossRef] [PubMed]
3. Ogunleye, A.; Wang, Q.-G. XGBoost Model for Chronic Kidney Disease Diagnosis. *IEEE/ACM Trans. Comput. Biol. Bioinform.* **2020**, *17*, 2131–2140. [CrossRef] [PubMed]
4. Aqlan, F.; Markle, R.; Shamsan, A. Data Mining for Chronic Kidney Disease Prediction. In Proceedings of the 67th Annual Conference and Expo of the Institute of Industrial Engineers, Pittsburgh, PA, USA, 20–23 May 2017.
5. Borisagar, N.; Barad, D.; Raval, P. Chronic Kidney Disease Prediction Using Back Propagation Neural Network Algorithm. In Proceedings of the International Conference on Communication and Networks, Ahmedabad, India, 19–20 February 2017; pp. 295–303.
6. Boukenze, B.; Haqiq, A.; Mousannif, H. Predicting Chronic Kidney Failure Disease Using Data Mining Techniques. In *Advances in Ubiquitous Networking*; El-Azouzi, R., Menasche, D.S., Sabir, E., De Pellegrini, F., Benjillali, M., Eds.; Springer: New York, NY, USA, 2018; Volume 2, pp. 701–712.
7. Polat, H.; Mehr, H.D.; Cetin, A. Diagnosis of chronic kidney disease based on support vector machine by feature selection methods. *J. Med. Syst.* **2017**, *41*, 55. [CrossRef] [PubMed]
8. Panwong, P.; Iam-On, N. Predicting transitional interval of kidney disease stages 3 to 5 using data mining method. In Proceedings of the 2016 Second Asian Conference on Defence Technology (ACDT), Chiang Mai, Thailand, 21–23 January 2016; pp. 145–150. [CrossRef]
9. Dulhare, U.N.; Ayesha, M. Extraction of action rules for chronic kidney disease using Naïve bayes classifier. In Proceedings of the 2016 IEEE International Conference on Computational Intelligence and Computing Research (ICCIC), Chennai, India, 15–17 December 2016; pp. 1–5. [CrossRef]
10. Vasquez-Morales, G.R.; Martinez-Monterrubio, S.M.; Moreno-Ger, P.; Recio-Garcia, J.A. Explainable Prediction of Chronic Renal Disease in the Colombian Population Using Neural Networks and Case-Based Reasoning. *IEEE Access* **2019**, *7*, 152900–152910. [CrossRef]
11. Makino, M.; Yoshimoto, R.; Ono, M.; Itoko, T.; Katsuki, T.; Koseki, A.; Kudo, M.; Haida, K.; Kuroda, J.; Yanagiya, R.; et al. Artificial intelligence predicts the progression of diabetic kidney disease using big data machine learning. *Sci. Rep.* **2019**, *9*, 11862. [CrossRef]
12. Ren, Y.; Fei, H.; Liang, X.; Ji, D.; Cheng, M. A hybrid neural network model for predicting kidney disease in hypertension patients based on electronic health records. *BMC Med. Inf. Decis. Mak.* **2019**, *19*, 131–138. [CrossRef]
13. Ma, F.; Sun, T.; Liu, L.; Jing, H. Detection and diagnosis of chronic kidney disease using deep learning-based heterogeneous modified artificial neural network. *Future Gener. Comput. Syst.* **2020**, *111*, 17–26. [CrossRef]
14. Almansour, N.; Syed, H.F.; Khayat, N.R.; Altheeb, R.K.; Juri, R.E.; Alhiyafi, J.; Alrashed, S.; Olatunji, S.O. Neural network and support vector machine for the prediction of chronic kidney disease: A comparative study. *Comput. Biol. Med.* **2019**, *109*, 101–111. [CrossRef]
15. Qin, J.; Chen, L.; Liu, Y.; Liu, C.; Feng, C.; Chen, B. A Machine Learning Methodology for Diagnosing Chronic Kidney Disease. *IEEE Access* **2019**, *8*, 20991–21002. [CrossRef]

16. Segal, Z.; Kalifa, D.; Radinsky, K.; Ehrenberg, B.; Elad, G.; Maor, G.; Lewis, M.; Tibi, M.; Korn, L.; Koren, G. Machine learning algorithm for early detection of end-stage renal disease. *BMC Nephrol.* **2020**, *21*, 518. [CrossRef]
17. Khamparia, A.; Saini, G.; Pandey, B.; Tiwari, S.; Gupta, D.; Khanna, A. KDSAE: Chronic kidney disease classification with multimedia data learning using deep stacked autoencoder network. *Multimed. Tools Appl.* **2020**, *79*, 35425–35440. [CrossRef]
18. Dua, D.; Graff, C. UCI Machine Learning Repository. 2019. Available online: http://archive.ics.uci.edu/ml (accessed on 6 December 2021).
19. Ebiaredoh-Mienye, S.A.; Esenogho, E.; Swart, T.G. Integrating Enhanced Sparse Autoencoder-Based Artificial Neural Network Technique and Softmax Regression for Medical Diagnosis. *Electronics* **2020**, *9*, 1963. [CrossRef]
20. Pang, Z.; Zhu, D.; Chen, D.; Li, L.; Shao, Y. A Computer-Aided Diagnosis System for Dynamic Contrast-Enhanced MR Images Based on Level Set Segmentation and ReliefF Feature Selection. *Comput. Math. Methods Med.* **2015**, *2015*, 450531. [CrossRef] [PubMed]
21. Chen, G.; Ding, C.; Li, Y.; Hu, X.; Li, X.; Ren, L.; Ding, X.; Tian, P.; Xue, W. Prediction of Chronic Kidney Disease Using Adaptive Hybridized Deep Convolutional Neural Network on the Internet of Medical Things Platform. *IEEE Access* **2020**, *8*, 100497–100508. [CrossRef]
22. Khan, B.; Naseem, R.; Muhammad, F.; Abbas, G.; Kim, S. An Empirical Evaluation of Machine Learning Techniques for Chronic Kidney Disease Prophecy. *IEEE Access* **2020**, *8*, 55012–55022. [CrossRef]
23. Tabassum, M.; Bai, B.G.; Majumdar, J. Analysis and Prediction of Chronic Kidney Disease using Data Mining Techniques. *Int. J. Eng. Res. Comput. Sci. Eng.* **2018**, *4*, 25–31.
24. Padmanaban, K.R.A.; Parthiban, G. Applying Machine Learning Techniques for Predicting the Risk of Chronic Kidney Disease. *Indian J. Sci. Technol.* **2016**, *9*. [CrossRef]
25. Sharma, S.; Sharma, V.; Sharma, A. Performance Based Evaluation of Various Machine Learning Classification Techniques for Chronic Kidney Disease Diagnosis. *Int. J. Mod. Comput. Sci.* **2018**, *4*, 11–15.
26. Devishri, P.; Ragin, S.; Anisha, O.R. Comparative Study of Classification Algorithms in Chronic Kidney Disease. *Int. J. Recent Technol. Eng.* **2019**, *8*, 180–184.
27. Drall, S.; Drall, G.S.; Singh, S. Chronic Kidney Disease Prediction Using Machine Learning: A New Approach. *Int. J. Manag.* **2018**, *8*, 278.
28. Dang, B.V.; Taylor, R.A.; Charlton, A.J.; Le-Clech, P.; Barber, T.J. Toward Portable Artificial Kidneys: The Role of Advanced Microfluidics and Membrane Technologies in Implantable Systems. *IEEE Rev. Biomed. Eng.* **2020**, *13*, 261–279. [CrossRef] [PubMed]
29. Cheng, L.C.; Hu, Y.H.; Chiou, S.H. Applying the Temporal Abstraction Technique to the Prediction of Chronic Kidney Disease Progression. *J. Med. Syst.* **2019**, *41*, 85. [CrossRef] [PubMed]
30. Hodneland, E. In Vivo Detection of Chronic Kidney Disease Using Tissue Deformation Fields from Dynamic MR Imaging. *IEEE Trans. Biomed. Eng.* **2019**, *66*, 1779–1790. [CrossRef] [PubMed]
31. Marsh, J.N.; Matlock, M.K.; Kudose, S.; Liu, T.-C.; Stappenbeck, T.S.; Gaut, J.P.; Swamidass, S.J. Deep Learning Global Glomerulosclerosis in Transplant Kidney Frozen Sections. *IEEE Trans. Med. Imaging* **2018**, *37*, 2718–2728. [CrossRef] [PubMed]
32. Antony, L.; Azam, S.; Ignatious, E.; Quadir, R.; Beeravolu, A.R.; Jonkman, M.; De Boer, F. A Comprehensive Unsupervised Framework for Chronic Kidney Disease Prediction. *IEEE Access* **2021**, *9*, 126481–126501. [CrossRef]
33. Hossain, M.; Detwiler, R.K.; Chang, E.H.; Caughey, M.C.; Fisher, M.W.; Nichols, T.C.; Merricks, E.P.; Raymer, R.A.; Whitford, M.; Bellinger, D.A.; et al. Mechanical Anisotropy Assessment in Kidney Cortex Using ARFI Peak Displacement: Preclinical Validation and Pilot In Vivo Clinical Results in Kidney Allografts. *IEEE Trans. Ultrason. Ferroelectr. Freq. Control* **2018**, *66*, 551–562. [CrossRef] [PubMed]
34. Hussain, M.A.; Hamarneh, G.; Garbi, R. Cascaded Localization Regression Neural Nets for Kidney Localization and Segmentation-free Volume Estimation. *IEEE Trans. Med. Imaging* **2021**, *40*, 1555–1567. [CrossRef]
35. Shehata, M.; Khalifa, F.; Soliman, A.; Ghazal, M.; Taher, F.; El-Ghar, M.A.; Dwyer, A.C.; Gimel'Farb, G.; Keynton, R.S.; El-Baz, A. Computer-Aided Diagnostic System for Early Detection of Acute Renal Transplant Rejection Using Diffusion-Weighted MRI. *IEEE Trans. Biomed. Eng.* **2018**, *66*, 539–552. [CrossRef]
36. Bhaskar, N.; Manikandan, S.M.S. A Deep-Learning-Based System for Automated Sensing of Chronic Kidney Disease. *IEEE Sens. Lett.* **2019**, *3*, 1–4. [CrossRef]
37. Chittora, P.; Chaurasia, S.; Chakrabarti, P.; Kumawat, G.; Chakrabarti, T.; Leonowicz, Z.; Jasinski, M.; Jasinski, L.; Gono, R.; Jasinska, E.; et al. Prediction of Chronic Kidney Disease—A Machine Learning Perspective. *IEEE Access* **2021**, *9*, 17312–17334. [CrossRef]
38. Zollner, F.G.; Kocinski, M.; Hansen, L.; Golla, A.-K.; Trbalic, A.S.; Lundervold, A.; Materka, A.; Rogelj, P. Kidney Segmentation in Renal Magnetic Resonance Imaging—Current Status and Prospects. *IEEE Access* **2021**, *9*, 71577–71605. [CrossRef]

healthcare MDPI

Review

mHealth Apps for Self-Management of Cardiovascular Diseases: A Scoping Review

Nancy Aracely Cruz-Ramos [1], Giner Alor-Hernández [1,*], Luis Omar Colombo-Mendoza [2], José Luis Sánchez-Cervantes [3], Lisbeth Rodríguez-Mazahua [1] and Luis Rolando Guarneros-Nolasco [1]

1 Tecnológico Nacional de México/I. T. Orizaba, Av. Oriente 9, No. 852, Col. Emiliano Zapata, Orizaba 94320, Mexico; dci.ncruz@ito-depi.edu.mx (N.A.C.-R.); lrodriguezm@ito-depi.edu.mx (L.R.-M.); luisguarneros@gmail.com (L.R.G.-N.)

2 Tecnológico Nacional de México/Instituto Tecnológico Superior de Teziutlán, Fracción l y ll, Teziutlán 73960, Mexico; luis.cm@teziutlan.tecnm.mx

3 CONACYT-Tecnológico Nacional de México/I. T. Orizaba, Av. Oriente 9, No. 852, Col. Emiliano Zapata, Orizaba 94320, Mexico; jlsanchez@conacyt.mx

* Correspondence: giner.ah@orizaba.tecnm.mx; Tel.: +52-272-725-7056

Abstract: The use of mHealth apps for the self-management of cardiovascular diseases (CVDs) is an increasing trend in patient-centered care. In this research, we conduct a scoping review of mHealth apps for CVD self-management within the period 2014 to 2021. Our review revolves around six main aspects of the current status of mHealth apps for CVD self-management: main CVDs managed, main app functionalities, disease stages managed, common approaches used for data extraction, analysis, management, common wearables used for CVD detection, monitoring and/or identification, and major challenges to overcome and future work remarks. Our review is based on Arksey and O'Malley's methodological framework for conducting studies. Similarly, we adopted the PRISMA model for reporting systematic reviews and meta-analyses. Of the 442 works initially retrieved, the review comprised 38 primary studies. According to our results, the most common CVDs include arrhythmia (34%), heart failure (32%), and coronary heart disease (18%). Additionally, we found that the majority mHealth apps for CVD self-management can provide medical recommendations, medical appointments, reminders, and notifications for CVD monitoring. Main challenges in the use of mHealth apps for CVD self-management include overcoming patient reluctance to use the technology and achieving the interoperability of mHealth applications with other systems.

Keywords: cardiovascular diseases; mHealth; self-management

Citation: Cruz-Ramos, N.A.; Alor-Hernández, G.; Colombo-Mendoza, L.O.; Sánchez-Cervantes, J.L.; Rodríguez-Mazahua, L.; Guarneros-Nolasco, L.R. mHealth Apps for Self-Management of Cardiovascular Diseases: A Scoping Review. *Healthcare* **2022**, *10*, 322. https://doi.org/10.3390/healthcare10020322

Academic Editor: Mahmudur Rahman

Received: 29 December 2021
Accepted: 7 February 2022
Published: 8 February 2022

1. Introduction

According to the World Health Organization (WHO), cardiovascular diseases (CVDs) are a group of disorders of the heart and blood vessels. They affect the normal behavior of the organism and have an adverse impact on a patient's emotional wellbeing, as well as on their work, family, social, and economic environments. On a much larger scale, CVDs are a public health concern due to their high prevalence, mortality, vulnerability, and the high public costs implied in their management. According to the WHO, CVDs are and will remain the number one cause of death globally at least for the following eight years. In fact, by 2030 almost 23.6 million people are estimated to die from some form of CVD [1]. In 2017 alone, approximately 17.9 million people died from CVDs, representing 31% of all global deaths. Overall, 85% of CVD-related deaths are due to either heart attacks or strokes. People suffering from CVDs or those at greater risk of developing them need to rely on effective means, such as counseling and medicines, for early CVD detection and management.

Mobile health (mHealth) is a medical and public health practice supported by mobile devices, such as mobile phones, portable monitoring devices, and personal digital assistants.

It involves using strategies such as smartphone apps, global positioning systems (GPS), and Bluetooth technologies. Approximately 500 million patients use mobile health (mHealth) applications to support their self-healthcare activities [2]. In this sense, cardiovascular mHealth is the most used in the mHealth domain through innovation, research, and implementation in the areas of CVD prevention, cardiac rehabilitation, and education [3]. Additionally, the most promising domains of mHealth use have to do with blood pressure monitoring, cardiac rehabilitation, arrhythmia monitoring, medication management, and social support.

mHealth apps hold promise for delivering health information and services to patients, especially for chronic diseases such as CVDs, which require extensive self-management. Self-management is key to person-centered care, but its support requires an understanding of individual preferences for different types of health information and decision-making autonomy. The self-management of chronic conditions requires the ability to manage the symptoms, treatment, physical, and psychosocial consequences and lifestyle changes inherent to living with a chronic condition. Additionally, self-management is inherent to person-centered care that promotes a balanced consideration of the values, needs, expectations, preferences, capacities, health, and wellbeing of all the constituents and stakeholders of the healthcare system. Effective self-management and person-centered care require full accommodation of people's needs and preferences for different types and amounts of information and other care services, a degree of autonomy in health-related decision-making, and support from their healthcare professionals and family members [4].

Current studies investigating mHealth interventions for patients with CVDs have returned mixed findings. Hence, more effort and work are needed to create engaging mHealth platforms that provide the necessary level of support to make sustained behavioral change. Similarly, addressing specific motivational, physical, and cognitive barriers to mHealth adoption among patients might increase the utilization of future interventions. It is also important to adopt new approaches that minimize the weaknesses of commercially available mobile apps [5].

We found related reviews that are focused on mobile apps for CVDs self-management using different technologies. These reviews studied the impact of incorporating mobile applications for symptom tracking, medication reminding, self-care support, and physiological state monitoring on the self-management of CVD patients' health. In addition, we identified that most of these proposals addressed the prevention and treatment of heart failure, arrhythmias, and coronary disease. Searcy et al. [5] documented the use domains of mHealth in CVD management, the barriers to mHealth adoption in older adults, and future directions for mHealth to increase engagement in this population. Furthermore, other studies [6–11] have explored the effectiveness, acceptability, and usefulness of mobile applications for CVD self-management and risk factor control using a variety of performance metrics. These studies have identified the most attractive features of the applications, such as the monitoring of healthy behaviors and the personalization of content. In addition, they have concluded that cardiovascular disease risk factors and behaviors are modifiable in the short term. Other authors [3,4,12] studied the mHealth apps for CVD prevention and management. Likewise, Cruz-Martínez et al. [13] identified interventions of self-management through the use of remote monitoring technologies. Other studies have rather focused on the self-management of specific CVDs. For instance, Refs. [14,15] analyzed the effect of the use of wearables and apps for cardiac rehabilitation of arrhythmia patients, whereas [16–18] studied a series of prevention and treatment programs for heart failure management through mHealth. Additionally, in [2,19–22] the functionalities of mHealth apps for heart failure self-management were evaluated.

The main difference between our scoping review and similar state-of-the-art reviews is that ours addresses more CVDs than those more frequently addressed in other reviews. Our scoping review aims at describing the current state of mHealth apps for CVD self-management by analyzing six aspects: (1) main CVDs managed, (2) main app functionalities, (3) common wearables used with these apps, (4) disease stages managed, (5) common approaches used for data extraction, analysis, and management, and (6) challenges and future work remarks. The review comprises a body of scientific literature issued from 2014 to mid-2021. The remainder of this paper is organized as follows: Section 2 introduces the materials and methods used to conduct the review. In Section 3, we present our results with respect to the research questions, whereas in Section 4, we discuss such findings. Finally, our conclusions are summarized in Section 5.

2. Materials and Methods

Our review is based on Arksey and O'Malley's [23] methodological framework for conducting studies as well as on the recommendations of Levac regarding such a framework [24]. Similarly, we adopted the PRISMA model proposed by Moher et al. [25] for reporting systematic reviews and meta-analyses and the PRISMA-ScR model extension. Next, we relied on the work of Tricco et al. [26] to determine how to organize and present the scoping review findings. The scoping review comprises five development phases: (1) identify research questions, (2) identify relevant studies, (3) select relevant studies, (4) chart the data, and (5) collate, summarize, and report findings.

2.1. Research Questions

We formulated seven research questions that framed our scoping review, helped us meet our research goals, and guided us throughout the reviewing process.

- RQ1. Which CVDs are most commonly managed by mHealth apps?
- RQ2. Which mHealth apps for CVD self-management are reported in the literature?
- RQ3. What are the main functionalities of mHealth apps for CVD self-management?
- RQ4. What are the major remarks for future work and challenges to be overcome by mHealth apps for CVD self-management?
- RQ5. Which approaches to data extraction, analysis, and management are commonly implemented in mHealth apps for CVD self-management?
- RQ6. Which wearables are commonly used to detect, monitor, and/or identify CVDs?
- RQ7. Which CVD stages are commonly managed by mHealth apps?

2.2. Inclusion and Exclusion Criteria

At the first stage of the search strategy, we defined the repositories in which we would search for the primary studies. These repositories included IEEE Xplore Digital Library, PubMed, ScienceDirect (Elsevier), SpringerLink, and Wiley Online Library. According to our preliminary search, these digital libraries hosted a greater amount of related literature when compared to other repositories, such as ACM (Association for Computing Machinery) Digital Library and Web of Science. Additionally, we relied on Google Scholar to expand our search. At the second stage of the search strategy, we performed a keyword search for primary studies issued within the 2014–2021 period. Table 1 lists such keywords, which were used both individually and combined using the conjunctions "and" and "or" to broaden our results.

The following queries were built to search for primary studies in each selected repository.

1. 'Cardiovascular disease' AND ('Self-management' OR 'Self-care' OR 'Self-monitoring') AND ('mHealth' OR 'mobile application' OR 'smart application' OR 'wearable' OR 'smartwatch' OR 'app'). The analysis of the preliminary results of this query revealed relevant search terms related to different cardiovascular disease types. Query 2 includes these search terms to expand on the relationship identified.
2. ('Heart disease' OR 'Cardiac issues' OR 'Heart failure' OR 'Arrhythmia' OR 'Coronary heart disease' OR 'Atrial Fibrillation' OR 'Hypertension' OR 'Cardiac arrest' OR 'Peripheral artery disease') AND ('Self-management' OR 'Self-care' OR 'Self-monitoring') AND ('mHealth' OR 'mobile application' OR 'smart application' OR 'wearable' OR 'smartwatch' OR 'app').

Finally, we used the PRISMA model as a guide to organize and report our results.

Table 1. Keywords and related concepts.

Area	Keywords	Related Concepts
Cardiovascular disease	Self-management Self-care Self-monitoring Heart disease Cardiac issues Heart failure Arrhythmia Coronary heart disease Atrial Fibrillation (AF) Hypertension Cardiac arrest Peripheral artery disease	mHealth mobile application smart application wearable smartwatch app

2.3. *Study Selection and Eligibility*

At the end of the search process, we found 442 relevant results: 33 from IEEE Xplore Digital Library, 105 from PubMed, 96 from ScienceDirect (Elsevier), 57 from SpringerLink, 16 from Wiley Online Library, and 135 from Google Scholar. Then, after removing duplicates, we relied on 159 articles for the first analysis, which necessitated classifying these papers by title and abstract. We performed a full-text reading of 84 of these articles, 38 of which were finally used in the scoping review (see Figure 1).

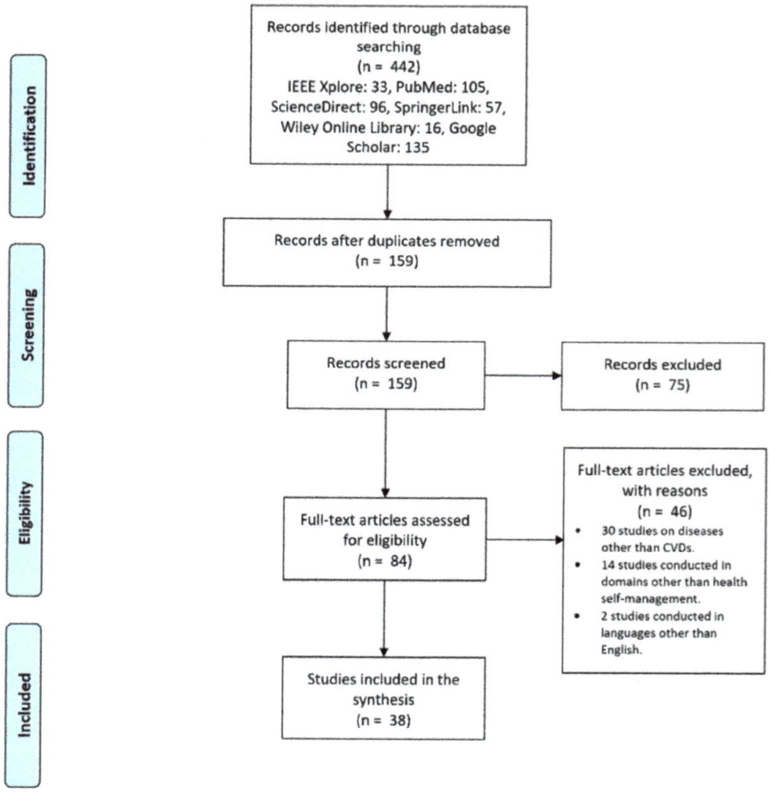

Figure 1. Study selection process—PRISMA diagram flow.

Once we gathered the initial 442 studies, we selected those containing at least two of the keywords listed in Table 1 in their abstract. Then, we removed those papers that were not directly related to CVD self-management. Following this step, we kept just 159 studies: 16 from IEEE Xplore Digital Library, 35 from PubMed, 27 from ScienceDirect (Elsevier), 12 from SpringerLink, 10 from Wiley Online Library, and 59 from Google Scholar. Next, we analyzed these papers with respect to our set of established exclusion criteria:

1. Studies on diseases other than CVDs;
2. Studies conducted in domains other than health self-management;
3. Studies written in languages other than English.

The remaining 38 primary studies were those comprising the scoping review. We downloaded the entire file of each study to ensure its proper analysis. As depicted in Figure 2, the majority of the studies (92.1%) were published in journals, 2.6% were issued as book chapters, and 5.3% were published in conference proceedings. Moreover, most of the studies were published between 2017 and 2018.

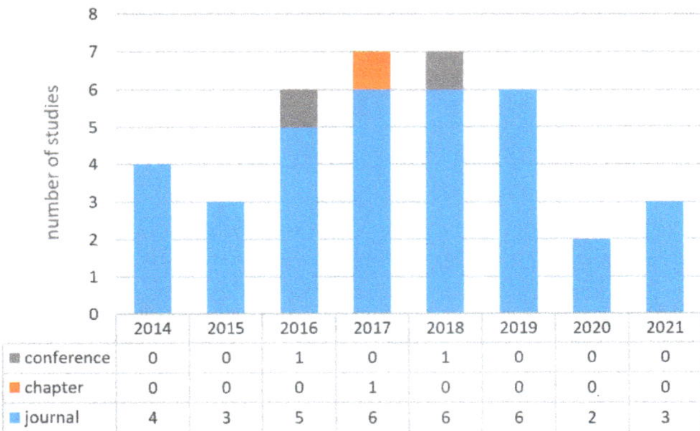

Figure 2. Type of publication from 2014 to 2021.

Figure 3 illustrates the geographical distribution of our primary studies. As can be observed, most of them were conducted in the United States.

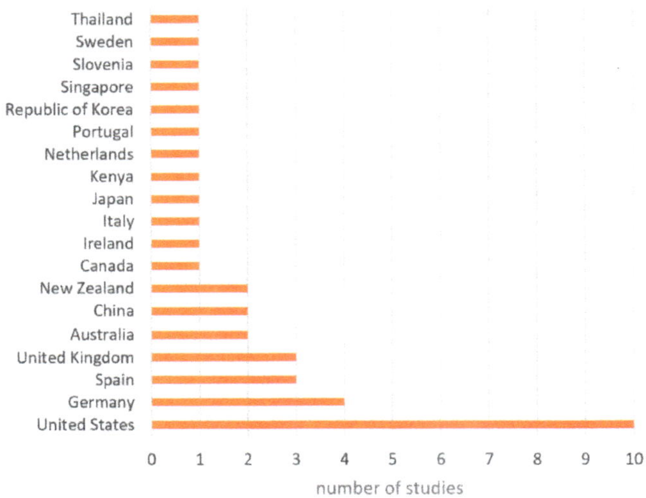

Figure 3. Geographical distribution of primary studies.

As regards allocation (see Figure 4), the majority of the primary studies were collected from PubMed, followed by Google Scholar, and then ScienceDirect. Research articles retrieved from SpringerLink, Wiley Online Library, and IEEE were less frequent.

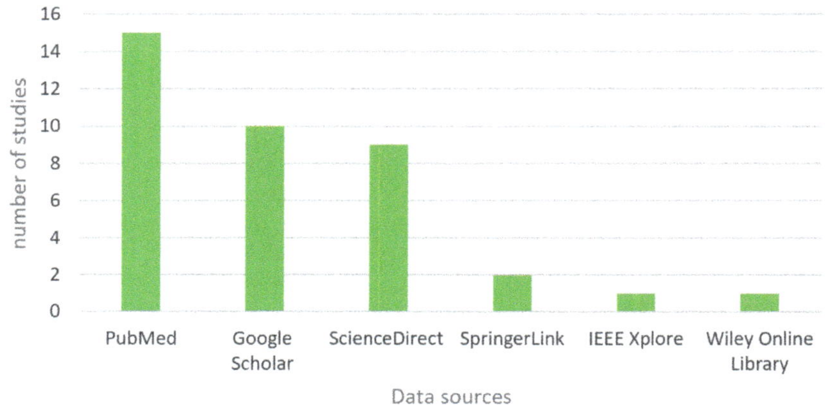

Figure 4. Primary studies by digital libraries.

2.4. Data Collection and Analysis

Once we defined the primary studies to be used in the review, we retrieved their bibliographic data and content data. The former included research title, authors, research goals, and database of provenance. The latter refer to the information contained in each study helping us answer our research questions (see Section 2.1).

3. Results

We reviewed the studies with respect to six aspects (see Table 2) aligned with our research questions. These aspects are listed and briefly explained below:

1. Type of CVD that is managed by each mHealth app.
2. Main app functionalities. Central capabilities of mHealth apps for CVD self-management, including (a) medical recommendations for patient follow-up, (b) real-time alerts before vital sign alterations, (c) medication management, (d) report of monitored parameters, (e) reminders for patient adherence to medication, physical activity, and/or dietary plans, (f) patient–physician communication via text messages, and (g) atrial fibrillation (AF) detection.
3. Challenges and/or future work remarks (when applicable). Main challenges to overcome and/or suggestions for future work for mHealth apps used in CVD self-management.
4. Approaches to data analysis, extraction, and management. The approaches were identified such as (a) machine learning techniques, (b) machine learning tasks, (c) big data types, and (d) device/sensor types. We identified mHealth apps relying on large datasets and big data analysis techniques. Additionally, there are apps relying on machine learning algorithms (MLAs) or techniques. Finally, we detected mHealth apps relying on sensors/wearables to obtain patient data (e.g., vital signs).
5. Device and apps. Information on the wearables and web and mobile apps—either commercially available or purposefully developed in the study itself—used by each mHealth app to retrieve patient data and biomedical variables.
6. CVD phase or set of phases managed by each mHealth app reviewed. The main CVD phases identified were diagnosis, prevention, monitoring, and treatment.

Table 2. Comparison of the main characteristics of mHealth apps.

Study Reference	CVD	Main App Functionalities	Challenges and/or Future Work Remarks	Approaches	Device or Web/Mobile Application	CVD Phase
Zisis et al. [27]	Heart failure	Medical recommendations, reminders, weight control	Computer skills of the patient, hearing problems, impaired vision, and cognitive impairment	Supervised machine learning (classification)	Smartphone or Tablet, Heart Failure app	Monitoring, treatment
Bohanec et al. [28]	Heart failure	Nutrition management, managing medication intake, psychological support, daily Exercise management, monitoring biomedical variables, medical recommendations	Increased adaptation to the patients' lifestyle, add methods for recognizing patients' activities, and integrating the optimization module in a smart-home environment	Supervised machine learning (random forest algorithm), classic differential evolution algorithm, and IoT device (heart rate, blood pressure)	Wristband, Blood pressure monitor, HeartMan Web app	Monitoring, treatment
Heiney et al. [29]	Heart failure	Text messages for communication between patients and physicians, weight and symptoms control, medical recommendations, medication management	Disparate population with low literacy, low health literacy, and limited smartphone use	IoT device (heart rate)	Smartphone, Healthy Heart app	Monitoring, treatment
Koirala et al. [30]	Heart failure	Medical recommendations	Implement the app in a real environment	Big data type (unstructured data), Supervised machine learning	Smartphone	Prevention, diagnosis
Gonzalez-Sanchez et al. [31]	Heart failure	Medical recommendations	Overcome patient resistance behavior toward using technology Add more functionality to the mobile app	Unsupervised machine learning	Smartphone, Evident II app	Prevention
Barret et al. [32]	Heart failure	Medical recommendations	Measure patient variables Greater focus on CVD asymptomatic patients	Unsupervised machine learning	Smartphone, Abby Web app	Prevention, treatment
Silva et al. [33]	Heart failure	Medical recommendations	Ensure interoperability of mHealth apps for remote monitoring, Heart rate measurement automation	Unsupervised machine learning	Smartphone, MOVIDA.eros app	Monitoring, treatment
Foster [34]	Heart failure	Medical recommendations, alerts	Implement the app in a real environment	Unsupervised machine learning	Smartphone, HF mobile app	Monitoring, treatment
Sakakibara et al. [35]	Heart failure	Medical recommendations, alerts, medication management	Implement the app in a real environment	Big data type (unstructured data)	Smartphone, mobile app	Prevention, treatment

Table 2. *Cont.*

Study Reference	CVD	Main App Functionalities	Challenges and/or Future Work Remarks	Approaches	Device or Web/Mobile Application	CVD Phase
De la Torre-Diez et al. [36]	Heart failure	Medical recommendations, alerts	Integrate the app system with EMR systems, Improve the usability of the mobile app, Add serious games to the app	Unsupervised machine learning	Smartphone, Heartkeeper app	Treatment
K. Rahimi et al. [37]	Heart failure	Medical recommendations, alerts, medication management	Integrate the app system with EMR systems, Increase wearable precision	Unsupervised machine learning, IoT device (heart rate, sensor Sp02)	Smartphone, SUPPORT-HF app, Oximeter	Monitoring, treatment
Bartlett et al. [38]	Heart failure	Step count calculation, weight control, blood pressure control	Overcome technological problems	IoT device (heart rate, blood pressure)	SMART Personalized Self-Management System (PSMS), HTC HD2 phone, MiFi device, mobile app	Monitoring, treatment
Turchioe et al. [39]	Arrhythmia	Medical recommendations	Overcome patient resistance to technology	Unsupervised machine learning	Smartphone	Prevention, monitoring
Pierleoni et al. [40]	Arrhythmia	Medical recommendations, alerts	Implement application in a real environment	Big data type (unstructured data), Unsupervised machine learning	Smartphone	Monitoring, treatment
Reverberi et al. [41]	Arrhythmia	AF detection	Implement algorithm for AF detection	IoT device (heart rate, ECG), Supervised machine learning (classification)	HR monitor of the chest-strap type, RITMIA app	Prevention
Fukuma et al. [42]	Arrhythmia	AF detection	Increase patient monitoring time	IoT device (heart rate, ECG)	T-Shirt-type wearable, ECG monitor, Hitoe Transmitter 01, smartphone	Prevention, treatment
Bumgarner et al. [43]	Arrhythmia	AF detection	Increase sample size, Increase the performance of the KB smartwatch algorithm, Review the real-time display of the ECG recording	IoT device (heart rate, blood pressure), Unsupervised machine learning	Kardia Band, Apple Watch, KB app	Prevention, monitoring
Krivoshei et al. [44]	Arrhythmia	AF detection, monitoring of heart rate, pulse wave analysis	Test the algorithm on a smartwatch	Unsupervised machine learning	Smartphone, iPhone 4S	Prevention
Guo et al. [45]	Arrhythmia	Medical recommendations, medication management, alerts, medical record	Overcome patient resistance to using technology	Supervised machine learning	Smartphone, mAF app	Treatment

Table 2. *Cont.*

Study Reference	CVD	Main App Functionalities	Challenges and/or Future Work Remarks	Approaches	Device or Web/Mobile Application	CVD Phase
Evans et al. [46]	Arrhythmia	AF detection	Extend study to other hospitals serving low-resource areas, Ensure interoperability with further systems	IoT device (heart rate, blood pressure), Supervised machine learning (classification)	AliveCor Kardia mobile ECG device, iPhone and iPad	Diagnosis, monitoring
Halcox et al. [47]	Arrhythmia	AF detection	The relatively high false-positive rate in the minor proportion of those reported as AF by the device	IoT device (heart rate, blood pressure), Supervised machine learning (classification)	AliveCor Kardia device, iPad	Diagnosis, monitoring
Lowres et al. [48]	Arrhythmia	iPhone handheld electrocardiogram (iECG)	Using iECG self-monitoring among other patient groups	Supervised machine learning	iPhone and AliveCor Heart monitor (iECG)	Monitoring
Hickey et al. [49]	Arrhythmia	AF detection	Implement the application in a real environment	IoT device (heart rate, blood pressure), Supervised machine learning (classification)	AliveCor Kardia mobile ECG device, iPhone	Diagnosis, monitoring
McManus et al. [50]	Arrhythmia	AF detection	Improve pulse recording and app performance	IoT device (heart rate), Supervised machine learning (classification)	PULSE-SMART app, iPhone 4S	Diagnosis, monitoring
Kakria et al. [51]	Arrhythmia	Alerts, monitoring of heart rate, blood pressure, and temperature	Solve the problem of delayed alarms in remote areas	IoT device (heart rate, blood pressure, stress level)	Smartphone, Zephyr BT system, G plus sensor, the Omron Wireless Upper Arm blood pressure monitor	Diagnosis, monitoring
Brouwers et al. [52]	Coronary heart disease	Medical recommendations, alerts	Sedentary patients	IoT device (heart rate)	Patient-centered web app, accelerometer, heart rate monitor	Monitoring, treatment
Zhang et al. [53]	Coronary heart disease	Medical recommendations	Ensure interoperability of applications for remote monitoring	Big data type (unstructured data), Unsupervised machine learning	Smartphone, Care4Heart app	Prevention
Athilingam [54]	Coronary heart disease	Medical recommendations, alerts, medication management	Overcome patient resistance to using technology Replace current sensor with handheld sensor	IoT device (heart rate), Supervised machine learning	Smartphone, HeartMapp, BioHarness Bluetooth sensor	Monitoring, treatment
Dale et al. [55]	Coronary heart disease	Text messages for communication of patients and physicians	Implement the app in a real environment	Big data type (structured data)	Smartphone	Treatment

Table 2. *Cont.*

Study Reference	CVD	Main App Functionalities	Challenges and/or Future Work Remarks	Approaches	Device or Web/Mobile Application	CVD Phase
Skobel et al. [56]	Coronary heart disease	Exercise module, activity level monitoring	Automatic arrhythmia detection	IoT device (heart rate, ECG, respiration, activity), Supervised machine learning	HeartCycle's guided exercise (GEX) system, tablet or laptop, portable PDA for ECG display, shirt with sensors	Diagnosis, monitoring
AM et al. [57]	Coronary heart disease	Educational material, medication reminders, and activity level monitoring	Train medical personnel and patients	IoT device (heart rate)	Smartphone	Monitoring, treatment
Dale et al. [58]	Coronary heart disease	Text messages for communication of patients and physicians, medical recommendations, weight control	Implement app in a real environment	IoT device (heart rate)	Smartphone, web app Text4Heart	Treatment
Jiang et al. [59]	Several (coronary heart disease and hypertension)	Alerts, medication management	Achieve acceptance of mHealth solutions among older patient populations, Improve app design	Supervised machine learning (Regression)	Smartphone, mobile app	Treatment
Baek et al. [60]	Several (atrial fibrillation, hypertension, chest pain, vasovagal syncope, variant angina, and dyspnea on exertion)	Medical recommendations, alerts, diary, weight control	Improve app usability, Integrate app system with EMR (Electronic Medical Record) systems	IoT device (heart rate)	Smartphone	Treatment, monitoring
Supervía & López-Jimenez [61]	Several (heart failure, coronary heart disease, tachycardias, arrhythmia, and hypertension)	Medical recommendations	Guarantee patient data protection and confidentiality	Unsupervised machine learning	Smartphone	Treatment
Tinsel et al. [62]	Several (heart failure, Coronary heart disease, tachycardias, arrhythmia, and hypertension)	Medical recommendations, alerts	Overcome patient resistance to using technology	IoT device (heart rate)	Mobile app	Prevention, treatment
Martorella et al. [63]	Several (heart failure, coronary heart disease, tachycardias, arrhythmia and hypertension)	Medical recommendations, medication management	Screen questionnaire to tailor content according to chronic postsurgical pain (CPSP) risk factors	Not specified	Web app	Monitoring, treatment
Johnston et al. [64]	Several (myocardial infarction, angina pectoris, heart failure, atrial fibrillation, embolic stroke, peripheral artery disease, hypertension)	Medication management, text messaging, reminders, e-diary, exercise module, BMI module, and blood pressure module	Improve patient self-reported drug adherence	IoT device (heart rate)	Smartphone, web-based app	Treatment

4. Discussion

4.1. RQ1. Which CVDs Are Most Commonly Managed by mHealth Apps?

The CVDs most commonly managed by mHealth apps in the literature are arrhythmias, heart failure, and coronary heart disease. Arrhythmia self-management is present in 34% of the reviewed studies [39–51], whereas heart failure self-management exhibited a frequency of 32% [27–38]. In turn, coronary heart disease self-management is present in 18% [52–58]. Additionally, 16% of the papers explore the self-management of several CVDs simultaneously [59–64].

We attribute this to the fact that these diseases have a high prevalence and mortality worldwide. In this regard, we believe that researchers are mainly focusing on solutions for the management of arrhythmias, specifically atrial fibrillation, because it affects 25% of the population aged over 40 years.

Some studies analyzed mHealth apps that address more than one cardiovascular disease at a time; we believe that this is due to the patients' risk factors, which can cause comorbidities, i.e., one disease can develop from another. However, our recommendation is that mHealth applications should focus only on one particular disease to provide more accurate forecasts, as each disease has its own characteristics.

4.2. RQ2. Which mHealth Apps for CVD Self-Management Are Reported in the Literature?

As regards arrhythmia self-management, mHealth apps include the RITMIA smartphone app [41], the KB app [43], the mAF app [45], and PULSE-SMART [50]. Generally speaking, these apps issue medical recommendations and allow for the early detection of AF, the most common type of arrhythmia. mHealth apps for arrhythmia self-management generally focus on arrhythmia prevention, diagnosis, and monitoring. The monitoring devices that can be connected to these apps are the T-shirt-type wearable ECG monitor and the AliveCor Kardia Mobile ECG.

mHealth apps for heart failure self-management include the Heart Failure app [27], HeartMan [28], Healthy Heart [29], Evident II [31], Abby [32], MOVIDA.eros [33], HeartKeeper [36], and SUPPORT-HF [37]. The majority of them focus on heart failure monitoring and treatment through issuing medical recommendations and medication management. Oximeters and sensors for blood pressure measurement are the most common devices connected to these apps.

The Care4Heart [53], HeartMapp [54], and Text4Heart [58] apps support coronary heart disease self-management. They primarily issue medical recommendations, reminders, and alerts and offer medication management. Most of these apps focus only on coronary heart disease monitoring. Devices and wearables such as heart rate monitors, the BioHarness Bluetooth sensor, portable ECG monitors, and T-shirts with sensors are usually connected to these applications.

mHealth apps such as those reported in [59–64] aim at supporting the self-management of multiple CVDs. These mHealth apps mainly provide medical follow-up recommendations for physical activity or dietary plans. Likewise, they issue medication reminders and real-time warnings before potential vital sign alterations. We found that 63.6% of the mHealth apps are compatible with the Android operating system, whereas 13.6% support iOS, and 22.8% support both (see Table 3).

It is reasonable to believe that there are more mHealth applications for the Android operating system because it is the most popular operating system in the world. However, we found in this research that the most complete mHealth application is the Kardia app, which is available for the iOS operating system only. Therefore, we believe that it is important to develop cross-platform mHealth applications; in this regard, an alternative would be the use of PWA (progressive web app) development technologies.

Another remarkable finding is that only 2 of the 16 mobile applications analyzed are available through digital distribution platforms: (1) MOVIDA.eros and (2) the Kardia app. Moreover, some of the applications analyzed were subjected to user acceptance tests with small groups of patients; in addition, some of them are not widely available because they are still in development. In this regard, we believe there is an opportunity to release free trial versions of the mHealth applications to test them with larger patient samples.

Table 3. mHealth applications for CVD self-management.

CVD	Study	Mobile App Name	Android	iOS
Heart failure	Zisis et al. [27]	Heart Failure app	✓	
	Bohanec et al. [28]	HeartMan	✓	
	Heiney et al. [29]	Healthy Heart	✓	
	Gonzalez-Sanchez et al. [31]	Evident II	✓	
	Barret et al. [32]	Abby	✓	
	Silva et al. [33]	MOVIDA.eros	✓	✓
	Foster [34]	HF mobile app	✓	✓
	Sakakibara et al. [35] Bartlett et al. [38]	Not specified	✓	
	De la Torre-Diez et al. [36]	HeartKeeper	✓	
	K. Rahimi et al. [37]	SUPPORT-HF	✓	
Arrhythmia	Reverberi et al. [41]	RITMIA	✓	
	Bumgarner et al. [43] Evans et al. [46] Halcox et al. [47] Lowres et al. [48] Hickey et al. [49]	Kardia app		✓
	Krivoshei et al. [44]	Unstated		✓
	Guo et al. [45]	mAF app	✓	✓
	McManus et al. [50]	PULSE-SMART		✓
	Kakria et al. [51]	Not specified	✓	
Coronary heart disease	Zhang et al. [53]	Care4Heart	✓	✓
	Athilingam [54]	HeartMapp	✓	
	AM et al. [57]	Not specified	✓	
	Dale et al. [58]	Text4Heart	✓	
Other CVDs	Jiang et al. [59]	Not specified	✓	
	Supervía & López-Jimenez [61] Tinsel et al. [62]	Not specified	✓	✓

4.3. *RQ3. What Are the Main Functionalities of mHealth Apps for CVD Self-Management?*

We identified six main functionalities of mHealth apps for CVD self-management (see Table 4):

- Recommendations (F1). Medical recommendations issued for patient follow-up in terms of dietary plans, physical activity, and overall health status.
- Alerts/reminders/text messages (F2). (a) Early, real-time warnings issued before potential vital signal alterations, (b) medication, physical activity, and/or dietary reminders, and (c) text messages communication between patients and physicians.
- Parameter monitoring (F3). Reports of monitored patient parameters, such as active minutes, burned calories, weight, step count, traveled distance, heart rate, blood pressure, body temperature, and physical activity.
- Medication management (F4). Control and follow-up of patient medication.

- Patient medical history (F5). Electronic health records (EHRs) including clinical data, medical history, diagnoses, medications, treatment plans, allergy test records, and laboratory and test results.
- AF detection (F6). Early detection of AF using heart rate monitoring and ECG results.

To summarize our findings on the main functionalities of mHealth apps for CVD self-management, 57.9% of these apps issue medical recommendations to patients [27–37,39,40,45,52–54,58,60–63], whereas 47.3% can generate reminding notifications or alerts for medical appointments [27,29,34–37,40,45,51,52,54,55,57–60,62,64]. Additionally, 34.2% of these apps monitor patient parameters, such as physical activity, step count, weight control, blood pressure, heart rate, pulse wave, and body temperature [27–29,38,44,48,51,56–58,60,64], while 21% allow for AF detection [41–44,46,47,49,50]. Finally, 21% allow for medication management [27,29,35,37,45,54,59,63,64]. and 10.5% allow patients to access their electronic health records [28,45,60,64].

Table 4. Main functionalities of mHealth apps for CVD self-management.

CVD	Study	F1	F2	F3	F4	F5	F6
Heart failure	Zisis et al. [27]	✓	✓	✓	✓		
	Bohanec et al. [28]	✓		✓		✓	
	Heiney et al. [29]	✓	✓	✓	✓		
	Koirala et al. [30]	✓					
	Gonzalez-Sanchez et al. [31]	✓					
	Barret et al. [32]	✓					
	Silva et al. [33]	✓					
	Foster [34]	✓	✓				
	Sakakibara et al. [35]	✓	✓		✓		
	De la Torre-Diez et al. [36]	✓	✓				
	K. Rahimi et al. [37]	✓	✓		✓		
	Bartlett et al. [38]			✓			
Arrhythmia	Turchioe et al. [39]	✓					
	Pierleoni et al. [40]	✓	✓				
	Reverberi et al. [41]						✓
	Fukuma et al. [42]						✓
	Bumgarner et al. [43]						✓
	Krivoshei et al. [44]			✓			✓
	Guo et al. [45]	✓	✓		✓	✓	
	Evans et al. [46]						✓
	Halcox et al. [47]						✓
	Lowres et al. [48]			✓			
	Hickey et al. [49]						✓
	McManus et al. [50]						✓
	Kakria et al. [51]		✓	✓			
Coronary heart disease	Brouwers et al. [52]	✓	✓				
	Zhang et al. [53]	✓					
	Athilingam [54]	✓	✓		✓		
	Dale et al. [55]		✓				
	Skobel et al. [56]			✓			
	AM et al. [57]		✓	✓			
	Dale et al. [58]	✓	✓	✓			
Several	Jiang et al. [59]		✓	✓			
	Baek et al. [60]	✓	✓	✓		✓	
	Supervía & López-Jimenez [61]	✓					
	Tinsel et al. [62]	✓	✓				
	Martorella et al. [63]	✓			✓		
	Johnston et al. [64]		✓	✓	✓	✓	

Most of the mHealth applications studied in this work have been demonstrated to be useful in the self-management of CVDs. There is evidence that these applications have changed the behavior of CVDS patients. This can be attributed to the self-alignment of patients to healthier lifestyles and to the constant monitoring of their vital signs. In addition, in the event of any change in patients' health status, these applications allow relatives and doctors to be notified to provide immediate care and avoid any health complications.

As part of the findings of this research, we identified six main features of the analyzed applications: (1) simplicity of user interface, (2) professional medical assistance, (3) connection with other services, (4) management of medical record, (5) reliable information, and (6) real-time biometric data tracking. In addition, we identified characteristics that are currently not considered in the development of applications to prevent and detect cardiovascular diseases: management of psychological health and family participation. Additionally, we suggest incorporating the following features: virtual rewards/gaming features, social media integration, and data privacy, since they are characteristics commonly sought by users.

The results of the usability tests performed for the mHealth applications have shown that the age factor influences the importance that users give to the applications' characteristics. Therefore, for children and adolescents, we recommend applications with simple user interfaces, which include social media integration and are oriented towards virtual rewards/gaming. We recommend, however, fully customizable applications with features such as psychological health management and family integration for adult patients.

4.4. RQ4. What Are the Major Remarks for Future Work and Challenges to Be Overcome by mHealth Apps for CVD Self-Management?

Since CVD self-management implies dealing with and managing a significant number of data, a lack of comprehensive information may hinder the correct functioning of mHealth apps for CVD self-management. To overcome this problem, many studies recommend implementing scalable app designs and ensuring the interoperability of these apps with other systems. In this sense, we found that only 4 of the 38 applications reviewed allow patients to access their electronic health records, yet this information is crucial both for patients and for CVD self-management.

Additionally, over 60% of the reviewed apps request access to patient personal information without a clear indication of how such information would be stored or used. In this sense, since privacy concerns might affect app usage, application developers should integrate privacy protection measures into their future designs. Other challenges to overcome include improving user satisfaction with respect to app functionalities and supporting patients in their learning of how to use the applications correctly. It is also important that future mHealth apps for CVD self-management address patient psychological health in their design [4]. We also found that none of the reviewed applications possess all the six functionalities for CVD self-management listed in Section 4.3. Hence, we conclude that the apps lack sufficient functions to support patients in effectively self-managing their CVD. Finally, functionalities for patient family involvement have not been sufficiently implemented in these apps.

4.5. RQ5. Which Approaches to Data Extraction, Analysis, and Management Are Commonly Implemented in mHealth Apps for CVD Self-Management?

The approaches to data extraction, analysis, and management used by mHealth apps for CVD self-management include machine learning techniques (supervised and unsupervised approaches), machine learning tasks (classification, clustering, regression), big data (structured and unstructured data), and IoT devices/sensors (see Table 5).

Big data make it possible to take advantage of the large amount of information that results from patients accessing health services. These data include, for instance, personal information, electronic medical records, social media data, telehealth data, clinical trials, and even biometric data from wearables [65–67]. In this context, we also found that mHealth apps may equally rely on data mining and sentiment analysis techniques. As for association rules and neural networks, they allow mHealth apps to create solutions for better decision making based on real data, thus improving CVD diagnosis, proposing customized treatment plans, reducing medical errors, increasing the effectiveness of CVD prevention measures, and promoting better CVD self-management.

Table 5. Main approaches to data extraction and analysis in mHealth apps for CVD self-management.

CVD	Study	Machine Learning Techniques and Tasks	Big Data Types	IoT Devices/Sensors
Heart failure	Zisis et al. [27]	✓		✓
	Bohanec et al. [28]	✓		✓
	Heiney et al. [29]			✓
	Koirala et al. [30]	✓	✓	
	Gonzalez-Sanchez et al. [31]	✓		
	Barret et al. [32]	✓		
	Silva et al. [33]	✓		
	Foster [34]	✓		
	Sakakibara et al. [35]		✓	
	De la Torre-Diez et al. [36]	✓		
	K. Rahimi et al. [37]	✓		✓
	Bartlett et al. [38]			✓
Arrhythmia	Turchioe et al. [39]	✓		
	Pierleoni et al. [40]	✓	✓	
	Reverberi et al. [41]	✓		
	Fukuma et al. [42]			✓
	Bumgarner et al. [43]	✓		✓
	Krivoshei et al. [44]	✓		
	Guo et al. [45]	✓		
	Evans et al. [46]	✓		✓
	Halcox et al. [47]	✓		✓
	Lowres et al. [48]			✓
	Hickey et al. [49]	✓		✓
	McManus et al. [50]	✓		✓
	Kakria et al. [51]			✓
Coronary heart disease	Brouwers et al. [52]			✓
	Zhang et al. [53]	✓	✓	
	Athilingam [54]	✓		
	Skobel et al. [56]	✓		✓
	AM et al. [57]			✓
	Dale et al. [58]			✓
Several	Jiang et al. [59]	✓		
	Baek et al. [60]			✓
	Supervía & López-Jimenez [61]	✓		
	Tinsel et al. [62]			✓

In regard to the IoT devices/sensors, this approach allows mHealth apps to retrieve real-time data on patient biometric variables, such as body temperature, heart rate, and blood pressure, through wearables, which in turn allow physicians to monitor patients remotely [27,28,68–70]. Additionally, wearables provide apps with real-time data that facilitate risk factor tracking and prevent CVD events [71]. In this regard, even though IoT platforms can integrate data from medical devices, wearables, and apps, defining data privacy parameters seems to be a considerable challenge to overcome; nevertheless, wearables have been shown to enable effective CVD detection outside of clinics [72].

Finally, mHealth apps for CVD self-management may also resort to machine learning techniques to mainly create predictive models that support—for example—medical diagnosis and treatment plans and predict the evolution of CVDs and their potential complications [27,28,73–77].

4.6. RQ6. Which Wearables Are Commonly Used to Detect, Monitor, and/or Identify CVDs?

According to our findings, 85% of the reviewed mHealth apps for CVD self-management rely on smartphones, whereas the remaining 15% use some type of wearable. We identified the five wearable devices most commonly connected to mHealth apps for CVD self-management: chest strap (W1), heart rate monitors (W2), T-shirt-type wearable ECG monitor (W3), the portable ECG monitor (W4), and the smartwatch/smartbands (W5). Table 6 below summarizes such findings. On the other hand, less common devices include pulse oximeters (Sp02 sensors), MiFi devices, and the Hitoe Transmitter 01 device.

Table 6. Main Wearables for CVD Monitoring.

CVD	Study	W1	W2	W3	W4	W5
Heart failure	Bohanec et al. [28]		✓			✓
	Bartlett et al. [38]		✓			
Arrhythmia	Reverberi et al. [41]		✓			
	Fukuma et al. [42]			✓		
	Bumgarner et al. [43]				✓	✓
	Evans et al. [46]				✓	✓
	Halcox et al. [47]				✓	✓
	Lowres et al. [48]				✓	✓
	Hickey et al. [49]				✓	✓
	Kakria et al. [51]	✓	✓			
Coronary heart disease	Brouwers et al. [52]		✓			
	Athilingam [54]	✓	✓			
	Skobel et al. [56]			✓	✓	

We found that it is essential to consider the use of wearables and other types of devices for monitoring biomedical variables automatically. Wearables such as smartwatches and smartbands can successfully assist in CVD detection and prevention. In addition, in most cases, these devices can be synchronized with cloud platforms such as Google Fit, thus storing all the data generated in the cloud. These platforms also allow synchronized data to be retrieved and integrated into mHealth applications.

The Xiaomi Mi Band is one of the most successful families of sport bracelets on the market, whose success could be due to its low price. It works, however, with another mobile application called Mi Fit, which can also be synchronized with Google Fit. We recommend this smartband as a great option for monitoring blood pressure and heart rate with high precision.

4.7. RQ7. Which CVD Stages Are Commonly Managed by mHealth Apps?

Many mHealth apps for CVD self-management can support patients throughout multiple stages of a CVD. As can be observed from Table 7, 63.2% of the mHealth apps can manage CVD treatment, 57.9% cover CVD monitoring, 28.9% focus on CVD prevention, and 18.4% allow for CVD diagnosis.

Table 7. Disease stages managed by mHealth apps for CVD self-management.

CVD	Study	Prevention	Diagnosis	Monitoring	Treatment
Heart failure	Zisis et al. [27]			✓	✓
	Bohanec et al. [28]			✓	✓
	Heiney et al. [29]			✓	✓
	Koirala et al. [30]	✓	✓		
	Gonzalez-Sanchez et al. [31]	✓			
	Barret et al. [32]	✓			✓
	Silva et al. [33]			✓	✓
	Foster [34]			✓	✓
	Sakakibara et al. [35]	✓			✓
	De la Torre-Diez et al. [36]				✓
	K. Rahimi et al. [37]			✓	✓
	Bartlett et al. [38]			✓	✓
Arrhythmia	Turchioe et al. [39]	✓		✓	
	Pierleoni et al. [40]			✓	✓
	Reverberi et al. [41]	✓			
	Fukuma et al. [42]	✓			✓
	Bumgarner et al. [43]	✓		✓	
	Krivoshei et al. [44]	✓			
	Guo et al. [45]				✓
	Evans et al. [46]		✓	✓	
	Halcox et al. [47]		✓	✓	
	Lowres et al. [48]			✓	
	Hickey et al. [49]		✓	✓	
	McManus et al. [50]		✓	✓	
	Kakria et al. [51]		✓	✓	
Coronary heart disease	Brouwers et al. [52]			✓	✓
	Zhang et al. [53]	✓			
	Athilingam [54]			✓	✓
	Dale et al. [55]				✓
	Skobel et al. [56]		✓	✓	
	AM et al. [57]			✓	✓
	Dale et al. [58]				✓
Several	Jiang et al. [59]				✓
	Baek et al. [60]			✓	✓
	Supervía & López-Jimenez [61]				✓
	Tinsel et al. [62]	✓			✓
	Martorella et al. [63]			✓	✓
	Johnston et al. [64]				✓

Most of the analyzed applications focused on the treatment of CVDs. These apps were tested by patients diagnosed with a heart disease, showing positive results. We suggest that new mHealth apps focus on the early stages of CVD management, specifically on detection, to allow doctors and patients to prevent medical complications.

5. Conclusions

The goal of this scoping review was to describe the current state of mHealth apps for CVD self-management through our analysis of six aspects: (1) CVDs commonly addressed, (2) main functionalities of mHealth apps for CVD self-management, (3) wearables used for CVD detection, monitoring, and identification, (4) disease stages managed by mHealth apps, (5) current approaches to data extraction, analysis, and management, and (6) current challenges to overcome and future work remarks for mHealth apps used in CVD self-management. The scoping review was performed on 38 primary studies, from which we propose the following conclusions: First, arrhythmia is the most common CVD addressed by mHealth apps, with a frequency of 34% (RQ1). Additionally, 63.6% of the mobile applications used by these mHealth apps are compatible with the Android operating system, whereas 13.6% support iOS, and 22.8% support both (RQ2). Additionally, the majority of the reviewed mHealth apps can provide patients medical recommendations, issue medical appointment reminders, and generate notifications for CVD monitoring (RQ3). The two major challenges these applications must overcome are patient resistance to using the technology and the lack of interoperability between mHealth apps and other systems (RQ4). In regard to the approaches for data extraction, analysis, and management, we found that the majority of the mHealth apps for CVD management rely on big data (structured and unstructured data), IoT devices/sensors and machine learning techniques (supervised and unsupervised approaches), and implementing classification, clustering, and regression algorithms (RQ5). Finally, smartphones—specifically Android smartphones—are commonly connected to mHealth apps for CVD self-management, even though wearables are becoming increasingly used (RQ6). Finally, the great majority of mHealth apps for CVD self-management focus on CVD treatment rather than on any other disease phase (RQ7). As regards our suggestions for future work, we first recommend conducting a systematic review of diseases that are correlated with CVD, such as diabetes and hypertension. Likewise, new research efforts should concentrate on exploring the implications of the increasing use of wearables for managing CVDs such as arrhythmia, heart failure, coronary heart disease, and cardiopathies.

Author Contributions: Conceptualization, N.A.C.-R., G.A.-H. and L.O.C.-M.; Data curation, N.A.C.-R.; Formal analysis, N.A.C.-R. and G.A.-H.; Investigation, L.O.C.-M.; Methodology, N.A.C.-R. and L.O.C.-M.; Supervision, J.L.S.-C. and L.R.-M.; Validation, J.L.S.-C., G.A.-H. and L.R.-M.; Visualization, N.A.C.-R. and L.R.G.-N.; Writing—original draft, N.A.C.-R.; Writing—review and editing, N.A.C.-R. and G.A.-H. All authors have read and agreed to the published version of the manuscript.

Funding: This research was funded by Mexico's National Council of Science and Technology (CONA-CYT), the Public Secretariat of Education (SEP) through the Sectorial Fund of Research for Education, grant number A1-S-51808, the Council for Scientific Research and Technological Development in Veracruz (COVEICYDET), grant number 12-1806, and the project 52–2016: "Application of Big Data and Semantic Web Techniques to Develop Intelligent Systems", a postdoctoral grant, and a doctoral grant.

Institutional Review Board Statement: Not applicable.

Informed Consent Statement: Informed consent was obtained from all subjects involved in the study.

Data Availability Statement: Not applicable.

Acknowledgments: This work was supported by Mexico's National Technological Institute (TecNM) and sponsored by both Mexico's National Council of Science and Technology (CONA-CYT) and the Secretariat of Public Education (SEP) through the PRODEP project (Programa para el Desarrollo Profesional Docente).

Conflicts of Interest: The authors declare no potential conflict of interest with respect to the publication of this research.

References

1. WHO, "Noncommunicable Diseases", World Health Organization (WHO), 13 April 2017. Available online: https://www.who.int/en/news-room/fact-sheets/detail/noncommunicable-diseases (accessed on 7 February 2021).
2. Athilingam, P.; Jenkins, B. Mobile Phone Apps to Support Heart Failure Self-Care Management: Integrative Review. *JMIR Cardio* **2018**, *2*, e10057. [CrossRef]
3. Chow, C.K.; Ariyarathna, N.; Islam, S.M.S.; Thiagalingam, A.; Redfern, J. mHealth in Cardiovascular Health Care. *Heart Lung Circ.* **2016**, *25*, 802–807. [CrossRef]
4. Xie, B.; Su, Z.; Zhang, W.; Cai, R. Chinese Cardiovascular Disease Mobile Apps' Information Types, Information Quality, and Interactive Functions for Self-Management: Systematic Review. *JMIR mHealth uHealth* **2017**, *5*, e195. [CrossRef] [PubMed]
5. Searcy, R.P.; Summapund, J.; Estrin, D.; Pollak, J.P.; Schoenthaler, A.; Troxel, A.B.; Dodson, J.A. Mobile Health Technologies for Older Adults with Cardiovascular Disease: Current Evidence and Future Directions. *Curr. Geriatr. Rep.* **2019**, *8*, 31–42. [CrossRef]
6. Coorey, G.M.; Neubeck, L.; Mulley, J.; Redfern, J. Effectiveness, acceptability and usefulness of mobile applications for cardiovascular disease self-management: Systematic review with meta-synthesis of quantitative and qualitative data. *Eur. J. Prev. Cardiol.* **2018**, *25*, 505–521. [CrossRef]

7. Dale, L.P.; Dobson, R.; Whittaker, R.; Maddison, R. The effectiveness of mobile-health behaviour change interventions for cardiovascular disease self-management: A systematic review. *Eur. J. Prev. Cardiol.* **2016**, *23*, 801–817. [CrossRef] [PubMed]
8. Whitehead, L.; Seaton, P. The Effectiveness of Self-Management Mobile Phone and Tablet Apps in Long-term Condition Management: A Systematic Review. *J. Med. Internet Res.* **2016**, *18*, e97. [CrossRef]
9. Gandhi, S.; Chen, S.; Hong, L.; Sun, K.; Gong, E.; Li, C.; Yan, L.L.; Schwalm, J.-D. Effect of Mobile Health Interventions on the Secondary Prevention of Cardiovascular Disease: Systematic Review and Meta-analysis. *Can. J. Cardiol* **2017**, *33*, 219–231. [CrossRef]
10. Pearsons, A.; Hanson, C.L.; Gallagher, R.; O'Carroll, R.E.; Khonsari, S.; Hanley, J.; Strachan, F.E.; Mills, N.L.; Quinn, T.J.; McKinstry, B.; et al. Atrial fibrillation self-management: A mobile telephone app scoping review and content analysis. *Eur. J. Cardiovasc. Nurs.* **2021**, *20*, 305–314. [CrossRef]
11. Villarreal, V.; Alvarez, A. Evaluation of mHealth Applications Related to Cardiovascular Diseases: A Systematic Review. *Acta Inform. Med.* **2020**, *28*, 130–137. [CrossRef]
12. Neubeck, L.; Lowres, N.; Benjamin, E.; Freedman, B.; Coorey, G.; Redfern, J. The mobile revolution—using smartphone apps to prevent cardiovascular disease. *Nat. Rev. Cardiol.* **2015**, *12*, 350–360. [CrossRef] [PubMed]
13. Cruz-Martínez, R.R.; Wentzel, J.; Asbjørnsen, R.A.; Noort, P.D.; van Niekerk, J.M.; Sanderman, R.; van Gemert-Pijnen, J.E. Supporting Self-Management of Cardiovascular Diseases Through Remote Monitoring Technologies: Metaethnography Review of Frameworks, Models, and Theories Used in Research and Development. *J. Med. Internet Res.* **2020**, *22*, e16157. [CrossRef]
14. Hannan, A.L.; Harders, M.P.; Hing, W.; Climstein, M.; Coombes, J.S.; Furness, J. Impact of wearable physical activity monitoring devices with exercise prescription or advice in the maintenance phase of cardiac rehabilitation: Systematic review and meta-analysis. *BMC Sports Sci. Med. Rehabil.* **2019**, *11*, 1–21. [CrossRef] [PubMed]
15. Marston, H.R.; Hadley, R.; Banks, D.; Duro, M.D.C.M. Mobile Self-Monitoring ECG Devices to Diagnose Arrhythmia that Coincide with Palpitations: A Scoping Review. *Healthcare* **2019**, *7*, 96. [CrossRef]
16. Brørs, G.; Pettersen, T.R.; Hansen, T.B.; Fridlund, B.; Hølvold, L.B.; Lund, H.; Norekvål, T.M. Modes of e-Health delivery in secondary prevention programmes for patients with coronary artery disease: A systematic review. *BMC Health Serv. Res.* **2019**, *19*, 364. [CrossRef]
17. Villarreal, V.; Castillo-Sanchez, G.; Hamrioui, S.; Alvarez, A.B.; Díez, I.D.L.T.; Lorenz, P. A Systematic Review of mHealth apps Evaluations for Cardiac Issues. *Multidiscip. Digit. Publ. Inst. Proc.* **2018**, *2*, 481. [CrossRef]
18. Hamilton, S.J.; Mills, B.; Birch, E.M.; Thompson, S.C. Smartphones in the secondary prevention of cardiovascular disease: A systematic review. *BMC Cardiovasc. Disord.* **2018**, *18*, 25. [CrossRef]
19. Bochicchio, M.A.; Vaira, L.; Mortara, A.; De Maria, R. A preliminar analysis and comparison of international projects on mobile devices and mHealth Apps for heart failure. In Proceedings of the 2019 5th Experiment International Conference (exp.at'19), Madeira Island, Portugal, 12–14 June 2019; pp. 280–285.
20. Allida, S.; Du, H.; Xu, X.; Prichard, R.; Chang, S.; Hickman, L.D.; Davidson, P.M.; Inglis, S.C. mHealth education interventions in heart failure. *Cochrane Database Syst. Rev.* **2020**, *2020*, CD011845.
21. Schmaderer, M.S.; Struwe, L.; Loecker, C.; Lier, L.; Lundgren, S.W.; Wichman, C.; Pozehl, B.; Zimmerman, L. Mobile Health Self-management Interventions for Patients With Heart Failure. *J. Cardiovasc. Nurs.* **2021**, 1–11. [CrossRef]
22. Creber, R.M.M.; Maurer, M.S.; Reading, M.; Hiraldo, G.; Hickey, K.T.; Iribarren, S. Review and Analysis of Existing Mobile Phone Apps to Support Heart Failure Symptom Monitoring and Self-Care Management Using the Mobile Application Rating Scale (MARS). *JMIR mHealth uHealth* **2016**, *4*, e74. [CrossRef]
23. Arksey, H.; O'Malley, L. Scoping studies: Towards a methodological framework. *Int. J. Soc. Res. Methodol.* **2005**, *8*, 19–32. [CrossRef]
24. Levac, D.; Colquhoun, H.; O'Brien, K.K. Scoping studies: Advancing the methodology. *Implement. Sci.* **2010**, *5*, 1–9. [CrossRef] [PubMed]
25. Moher, D.; Liberati, A.; Tetzlaff, J.; Altman, D.G.; PRISMA Group. Preferred reporting items for systematic reviews and meta-analyses: The PRISMA statement. *PLoS Med.* **2009**, *6*, e1000097. [CrossRef] [PubMed]
26. Tricco, A.C.; Lillie, E.; Zarin, W.; O'Brien, K.K.; Colquhoun, H.; Levac, D.; Moher, D.; Peters, M.D.J.; Horsley, T.; Weeks, L.; et al. PRISMA extension for scoping reviews (PRISMA-ScR): Checklist and explanation. *Ann. Intern. Med.* **2018**, *169*, 467–473. [CrossRef]
27. Zisis, G.; Carrington, M.J.; Oldenburg, B.; Whitmore, K.; Lay, M.; Huynh, Q.; Neil, C.; Ball, J.; Marwick, T.H. An m-Health intervention to improve education, self-management, and outcomes in patients admitted for acute decompensated heart failure: Barriers to effective implementation. *Eur. Heart J. Digit. Health* **2021**, *2*, 649–657. [CrossRef]
28. Bohanec, M.; Tartarisco, G.; Marino, F.; Pioggia, G.; Puddu, P.E.; Schiariti, M.S.; Baert, A.; Pardaens, S.; Clays, E.; Vodopija, A.; et al. HeartMan DSS: A decision support system for self-management of congestive heart failure. *Expert Syst. Appl.* **2021**, *186*, 115688. [CrossRef]
29. Heiney, S.P.; Donevant, S.B.; Adams, S.A.; Parker, P.D.; Chen, H.; Levkoff, S. A Smartphone App for Self-Management of Heart Failure in Older African Americans: Feasibility and Usability Study. *JMIR Aging* **2020**, *3*, e17142. [CrossRef]
30. Koirala, B.; Himmelfarb, C.R.D.; Budhathoki, C.; Davidson, P.M. Heart failure self-care, factors influencing self-care and the relationship with health-related quality of life: A cross-sectional observational study. *Heliyon* **2020**, *6*, e03412. [CrossRef]

31. Gonzalez-Sanchez, J.; Recio-Rodriguez, J.I.; Fernandez-Delrio, A.; Sanchez-Perez, A.; Magdalena-Belio, J.F.; Gomez-Marcos, M.A.; Garcia-Ortiz, L. Using a smartphone app in changing cardiovascular risk factors: A randomized controlled trial (EVIDENT II study). *Int. J. Med. Inform.* **2019**, *125*, 13–21. [CrossRef]
32. Barrett, M.; Boyne, J.; Brandts, J.; Rocca, H.-P.B.-L.; De Maesschalck, L.; De Wit, K.; Dixon, L.; Eurlings, C.; Fitzsimons, D.; Golubnitschaja, O.; et al. Artificial intelligence supported patient self-care in chronic heart failure: A paradigm shift from reactive to predictive, preventive and personalised care. *EPMA J.* **2019**, *10*, 445–464. [CrossRef]
33. Silva, E.; Rijo, R.; Martinho, R.; Assuncao, P.; Seco, A.; Fonseca-Pinto, R. A Cardiac Rehabilitation Program Supported by mHealth Technology: The MOVIDA.eros Platform. *Procedia Comput. Sci.* **2018**, *138*, 119–124. [CrossRef]
34. Foster, M. A Mobile Application for Patients with Heart Failure: Theory—and Evidence-Based Design and Testing. *CIN Comput. Inform. Nurs.* **2018**, *36*, 540–549. [CrossRef] [PubMed]
35. Sakakibara, B.M.; Ross, E.; Arthur, G.; Brown-Ganzert, L.; Petrin, S.; Sedlak, T.; Lear, S.A. Using Mobile-Health to Connect Women with Cardiovascular Disease and Improve Self-Management. *Telemed. e-Health* **2017**, *23*, 233–239. [CrossRef]
36. De la Torre-Diez, I.; Martinez-Perez, B.; Lopez-Coronado, M.; Rodrigues, J.J.P.C.; Arambarri, J. Development and validation of a mobile health app for the self-management and education of cardiac patients. In Proceedings of the 2016 11th Iberian Conference on Information Systems and Technologies (CISTI), Gran Canaria, Spain, 15–18 June 2016; pp. 1–5.
37. Rahimi, K.; Velardo, C.; Triantafyllidis, A.; Conrad, N.; Shah, S.A.; Chantler, T.; Mohseni, H.; Stoppani, E.; Moore, F.; Paton, C.; et al. A user-centred home monitoring and self-management system for patients with heart failure: A multicentre cohort study. *Eur. Heart J. Qual. Care Clin. Outcomes* **2015**, *1*, 66–71. [CrossRef] [PubMed]
38. Bartlett, Y.K.; Haywood, A.; Bentley, C.L.; Parker, J.; Hawley, M.S.; Mountain, G.A.; Mawson, S. The SMART personalised self-management system for congestive heart failure: Results of a realist evaluation. *BMC Med. Inform. Decis. Mak.* **2014**, *14*, 1–13. [CrossRef]
39. Turchioe, M.R.; Jimenez, V.; Isaac, S.; Alshalabi, M.; Slotwiner, D.; Creber, R.M. Review of mobile applications for the detection and management of atrial fibrillation. *Heart Rhythm O2* **2020**, *1*, 35–43. [CrossRef] [PubMed]
40. Pierleoni, P.; Belli, A.; Gentili, A.; Incipini, L.; Palma, L.; Raggiunto, S.; Sbrollini, A.; Burattini, L. Real-time smart monitoring system for atrial fibrillation pathology. *J. Ambient Intell. Humaniz. Comput.* **2021**, *12*, 4461–4469. [CrossRef]
41. Reverberi, C.; Rabia, G.; De Rosa, F.; Bosi, D.; Botti, A.; Benatti, G. The RITMIATM Smartphone App for Automated Detection of Atrial Fibrillation: Accuracy in Consecutive Patients Undergoing Elective Electrical Cardioversion. *BioMed Res. Int.* **2019**, *2019*, 4861951. [CrossRef]
42. Fukuma, N.; Hasumi, E.; Fujiu, K.; Waki, K.; Toyooka, T.; Komuro, I.; Ohe, K. Feasibility of a T-Shirt-Type Wearable Electrocardiography Monitor for Detection of Covert Atrial Fibrillation in Young Healthy Adults. *Sci. Rep.* **2019**, *9*, 1–6. [CrossRef]
43. Bumgarner, J.M.; Lambert, C.T.; Hussein, A.A.; Cantillon, D.J.; Baranowski, B.; Wolski, K.; Lindsay, B.D.; Wazni, O.M.; Tarakji, K.G. Smartwatch Algorithm for Automated Detection of Atrial Fibrillation. *J. Am. Coll. Cardiol.* **2018**, *71*, 2381–2388. [CrossRef]
44. Krivoshei, L.; Weber, S.; Burkard, T.; Maseli, A.; Brasier, N.; Kühne, M.; Conen, D.; Huebner, T.; Seeck, A.; Eckstein, J. Smart detection of atrial fibrillation. *Europace* **2016**, *19*, 753–757. [CrossRef]
45. Guo, Y.; Chen, Y.; Lane, D.A.; Liu, L.; Wang, Y.; Lip, G.Y. Mobile Health Technology for Atrial Fibrillation Management Integrating Decision Support, Education, and Patient Involvement: mAF App Trial. *Am. J. Med.* **2017**, *130*, 1388–1396. [CrossRef]
46. Evans, G.F.; Shirk, A.; Muturi, P.; Soliman, E.Z. Feasibility of Using Mobile ECG Recording Technology to Detect Atrial Fibrillation in Low-Resource Settings. *Glob. Heart* **2017**, *12*, 285–289. [CrossRef] [PubMed]
47. Halcox, J.P.; Wareham, K.; Cardew, A.; Gilmore, M.; Barry, J.P.; Phillips, C.; Gravenor, M.B. Assessment of remote heart rhythm sampling using the AliveCor heart monitor to screen for atrial fibrillation the REHEARSE-AF study. *Circulation* **2017**, *136*, 1784–1794. [CrossRef]
48. Lowres, N.; Mulcahy, G.; Gallagher, R.; Ben Freedman, S.; Marshman, D.; Kirkness, A.; Orchard, J.; Neubeck, L. Self-monitoring for atrial fibrillation recurrence in the discharge period post-cardiac surgery using an iPhone electrocardiogram. *Eur. J. Cardio-Thorac. Surg.* **2016**, *50*, 44–51. [CrossRef]
49. Hickey, K.T.; Hauser, N.R.; Valente, L.E.; Riga, T.C.; Frulla, A.P.; Creber, R.M.; Whang, W.; Garan, H.; Jia, H.; Sciacca, R.R.; et al. A single-center randomized, controlled trial investigating the efficacy of a mHealth ECG technology intervention to improve the detection of atrial fibrillation: The iHEART study protocol. *BMC Cardiovasc. Disord.* **2016**, *16*, 152. [CrossRef]
50. McManus, D.D.; Chong, J.W.; Soni, A.; Saczynski, J.S.; Esa, N.; Napolitano, C.; Darling, C.E.; Boyer, E.; Rosen, R.K.; Floyd, K.C.; et al. PULSE-SMART: Pulse-Based Arrhythmia Discrimination Using a Novel Smartphone Application. *J. Cardiovasc. Electrophysiol.* **2015**, *27*, 51–57. [CrossRef]
51. Kakria, P.; Tripathi, N.K.; Kitipawang, P. A Real-Time Health Monitoring System for Remote Cardiac Patients Using Smartphone and Wearable Sensors. *Int. J. Telemed. Appl.* **2015**, *2015*, 373474. [CrossRef]
52. Brouwers, R.W.; Kraal, J.J.; Traa, S.C.; Spee, R.F.; Oostveen, L.M.; Kemps, H.M. Effects of cardiac telerehabilitation in patients with coronary artery disease using a personalised patient-centred web application: Protocol for the SmartCare-CAD randomised controlled trial. *BMC Cardiovasc. Disord.* **2017**, *17*, 46. [CrossRef]
53. Zhang, H.; Jiang, Y.; Nguyen, H.D.; Poo, D.C.C.; Wang, W. The effect of a smartphone-based coronary heart disease prevention (SBCHDP) programme on awareness and knowledge of CHD, stress, and cardiac-related lifestyle behaviours among the working population in Singapore: A pilot randomised controlled trial. *Health Qual. Life Outcomes* **2017**, *15*, 1–13. [CrossRef]

54. Athilingam, P.; Labrador, M.A.; Remo, E.F.J.; Mack, L.; San Juan, A.B.; Elliott, A.F. Features and usability assessment of a patient-centered mobile application (HeartMapp) for self-management of heart failure. *Appl. Nurs. Res.* **2016**, *32*, 156–163. [CrossRef] [PubMed]
55. Dale, L.P.; Whittaker, R.; Jiang, Y.; Stewart, R.; Rolleston, A.; Maddison, R. Text Message and Internet Support for Coronary Heart Disease Self-Management: Results From the Text4Heart Randomized Controlled Trial. *J. Med. Internet Res.* **2015**, *17*, e237. [CrossRef] [PubMed]
56. Skobel, E.; Martinez-Romero, A.; Scheibe, B.; Schauerte, P.; Marx, N.; Luprano, J.; Knackstedt, C. Evaluation of a newly designed shirt-based ECG and breathing sensor for home-based training as part of cardiac rehabilitation for coronary artery disease. *Eur. J. Prev. Cardiol.* **2014**, *21*, 1332–1340. [CrossRef]
57. Layton, A.M.; Whitworth, J.; Peacock, J.; Bartels, M.N.; Jellen, P.A.; Thomashow, B.M. Feasibility and Acceptability of Utilizing a Smartphone Based Application to Monitor Outpatient Discharge Instruction Compliance in Cardiac Disease Patients around Discharge from Hospitalization. *Int. J. Telemed. Appl.* **2014**, *2014*, 415868. [CrossRef]
58. Dale, L.P.; Whittaker, R.; Jiang, Y.; Stewart, R.; Rolleston, A.; Maddison, R. Improving coronary heart disease self-management using mobile technologies (Text4Heart): A randomised controlled trial protocol. *Trials* **2014**, *15*, 71. [CrossRef]
59. Jiang, J.; Zhu, Q.; Zheng, Y.; Zhu, Y.; Li, Y.; Huo, Y. Perceptions and Acceptance of mHealth in Patients With Cardiovascular Diseases: A Cross-Sectional Study. *JMIR mHealth uHealth* **2019**, *7*, e10117. [CrossRef]
60. Baek, H.; Suh, J.-W.; Kang, S.-H.; Kang, S.; Lim, T.H.; Hwang, H.; Yoo, S. Enhancing User Experience Through User Study: Design of an mHealth Tool for Self-Management and Care Engagement of Cardiovascular Disease Patients. *JMIR Cardio* **2018**, *2*, e3. [CrossRef]
61. Supervía, M.; López-Jimenez, F. mHealth and cardiovascular diseases self-management: There is still a long way ahead of us. *Eur. J. Prev. Cardiol.* **2018**, *25*, 974–975. [CrossRef]
62. Tinsel, I.; Siegel, A.; Schmoor, C.; Poguntke, I.; Maun, A.; Niebling, W. Encouraging Self-Management in Cardiovascular Disease Prevention. *Dtsch. Ärztebl. Int.* **2018**, *115*, 469–476. [CrossRef]
63. Martorella, G.; Graven, L.; Schluck, G.; Bérubé, M.; Gélinas, C. Nurses' Perception of a Tailored Web-Based Intervention for the Self-Management of Pain After Cardiac Surgery. *SAGE Open Nurs.* **2018**, *4*, 237796081880627. [CrossRef]
64. Johnston, N.; Bodegard, J.; Jerström, S.; Åkesson, J.; Brorsson, H.; Alfredsson, J.; Albertsson, P.A.; Karlsson, J.-E.; Varenhorst, C. Effects of interactive patient smartphone support app on drug adherence and lifestyle changes in myocardial infarction patients: A randomized study. *Am. Heart J.* **2016**, *178*, 85–94. [CrossRef]
65. Gökalp, M.O.; Kayabay, K.; Akyol, M.A.; Koçyiğit, A.; Eren, P.E. Big data in mHealth. In *Current and Emerging mHealth Technologies: Adoption, Implementation, and Use*; Springer International Publishing: Berlin/Heidelberg, Germany, 2018; pp. 241–256.
66. Khan, Z.F.; Alotaibi, S.R. Applications of Artificial Intelligence and Big Data Analytics in m-Health: A Healthcare System Perspective. *J. Healthc. Eng.* **2020**, *2020*, 8894694. [CrossRef] [PubMed]
67. Baladrón, C.; de Diego, J.J.G.; Amat-Santos, I.J. Big data and new information technology: What cardiologists need to know. *Rev. Esp. Cardiol.* **2021**, *74*, 81–89. [CrossRef] [PubMed]
68. Code, R. Wearable technology in healthcare. *Nat. Biotechnol.* **2019**, *37*, 376.
69. Singhal, A.; Cowie, M.R. The Role of Wearables in Heart Failure. *Curr. Heart Fail. Rep.* **2020**, *17*, 125–132. [CrossRef]
70. Dagher, L.; Shi, H.; Zhao, Y.; Marrouche, N.F. Wearables in cardiology: Here to stay. *Heart Rhythm* **2020**, *17*, 889–895. [CrossRef] [PubMed]
71. Kario, K. Management of Hypertension in the Digital Era: Small Wearable Monitoring Devices for Remote Blood Pressure Monitoring. *Hypertension* **2020**, *76*, 640–650. [CrossRef]
72. Dunn, J.; Runge, R.; Snyder, M. Wearables and the medical revolution. *Pers. Med.* **2018**, *15*, 429–448. [CrossRef]
73. Ambhore, S. Early Detection of Cardiovascular Diseases Using Deep Convolutional Neural Network & Fourier Wavelet Transform. Available online: https://www.sciencedirect.com/science/article/pii/S2214785320392324 (accessed on 20 January 2021).
74. Mohan, S.; Thirumalai, C.; Srivastava, G. Effective Heart Disease Prediction Using Hybrid Machine Learning Techniques. *IEEE Access* **2019**, *7*, 81542–81554. [CrossRef]
75. Khan, M.A. An IoT Framework for Heart Disease Prediction Based on MDCNN Classifier. *IEEE Access* **2020**, *8*, 34717–34727. [CrossRef]
76. Raj, S. An Efficient IoT-Based Platform for Remote Real-Time Cardiac Activity Monitoring. *IEEE Trans. Consum. Electron.* **2020**, *66*, 106–114. [CrossRef]
77. Khan, M.A.; Algarni, F. A Healthcare Monitoring System for the Diagnosis of Heart Disease in the IoMT Cloud Environment Using MSSO-ANFIS. *IEEE Access* **2020**, *8*, 122259–122269. [CrossRef]

healthcare

MDPI

Review

Detecting Depression Signs on Social Media: A Systematic Literature Review

Rafael Salas-Zárate [1], Giner Alor-Hernández [1,*], María del Pilar Salas-Zárate [2], Mario Andrés Paredes-Valverde [2], Maritza Bustos-López [3] and José Luis Sánchez-Cervantes [4]

1 Tecnológico Nacional de México/I. T. Orizaba, Av. Oriente 9 No. 852, Col. Emiliano Zapata, Orizaba 94320, Veracruz, Mexico; dci.rsalas@ito-depi.edu.mx

2 Tecnológico Nacional de México/I.T.S. Teziutlán, Fracción I y II S/N, Aire Libre, Teziutlán 73960, Puebla, Mexico; maria.sz@teziutlan.tecnm.mx (M.d.P.S.-Z.); mario.pv@teziutlan.tecnm.mx (M.A.P.-V.)

3 Centro de Investigación en Inteligencia Artificial/Universidad Veracruzana, Sebastián Camacho 5, Zona Centro, Centro, Xalapa-Enríquez 91000, Veracruz, Mexico; maritbustos@gmail.com

4 CONACYT-Tecnológico Nacional de México/I. T. Orizaba, Av. Oriente 9 No. 852, Col. Emiliano Zapata, Orizaba 94320, Veracruz, Mexico; jlsanchez@conacyt.mx

* Correspondence: giner.ah@orizaba.tecnm.mx; Tel.: +52-(272)-725-7056

Abstract: Among mental health diseases, depression is one of the most severe, as it often leads to suicide; due to this, it is important to identify and summarize existing evidence concerning depression sign detection research on social media using the data provided by users. This review examines aspects of primary studies exploring depression detection from social media submissions (from 2016 to mid-2021). The search for primary studies was conducted in five digital libraries: ACM Digital Library, IEEE Xplore Digital Library, SpringerLink, Science Direct, and PubMed, as well as on the search engine Google Scholar to broaden the results. Extracting and synthesizing the data from each paper was the main activity of this work. Thirty-four primary studies were analyzed and evaluated. Twitter was the most studied social media for depression sign detection. Word embedding was the most prominent linguistic feature extraction method. Support vector machine (SVM) was the most used machine-learning algorithm. Similarly, the most popular computing tool was from Python libraries. Finally, cross-validation (CV) was the most common statistical analysis method used to evaluate the results obtained. Using social media along with computing tools and classification methods contributes to current efforts in public healthcare to detect signs of depression from sources close to patients.

Keywords: depression; social media; sentiment analysis

Citation: Salas-Zárate, R.; Alor-Hernández, G.; Salas-Zárate, M.d.P.; Paredes-Valverde, M.A.; Bustos-López, M.; Sánchez-Cervantes, J.L. Detecting Depression Signs on Social Media: A Systematic Literature Review. *Healthcare* **2022**, *10*, 291. https://doi.org/10.3390/healthcare10020291

Academic Editor: Tin-Chih Toly Chen

Received: 28 December 2021
Accepted: 29 January 2022
Published: 1 February 2022

1. Introduction

Mental disorders are a worldwide health problem affecting a large number of people and causing numerous deaths every year. According to a World Health Organization (WHO) report, the most common major disorders in 2017 included anxiety (284 million sufferers), depression (264 million), bipolar disorder (46 million), schizophrenia (20 million), and eating disorders (16 million) [1].

According to the American Psychiatric Association (APA), depression is a serious and common medical condition that negatively affects how people feel and act and the way they think. Fortunately, major depression is also treatable. Depression is an important factor in suicide among both adolescents and the elderly, but those with a late onset of depression are at higher risk [2]. In fact, nearly 800,000 people die due to suicide every year, and suicide alone is the second leading cause of death among 15–29 year-old people (WHO). Depression can lead to physical and emotional problems and can affect a person's ability to work [3]. Furthermore, the stress factors of the COVID-19 crisis indicate that a great number of people in the world may be in the course of developing depression as a result of the new and unusual lifestyle caused by the pandemic. It is also common for the effects of a viral disease to affect people's moods, causing them to go into depressive states; moreover, the COVID-19 crisis has increased the chances of depression, which in turn will make recovery from the pandemic harder across a spectrum of needs [4]. According to Szmuda [5], during the current situation, telemedicine and social media allow patients to receive healthcare while still practicing social distancing, the principal anti-pandemic defense. Moreover, bots can be adjusted quickly based on the latest research findings and WHO recommendations on COVID-19. With triage being exclusively handled by bots, nurses and clinicians can devote more of their time to patient care. We can say that the focus of this research is valuable in the application of tools to detect the onset of depressive problems in people so that they can be used in healthcare institutions, as well as in the support of individuals, making those who suffer from mental problems more participatory in relation to their mental health. When the period of social isolation finishes, people suffering from depression will have a harder time returning to their common social activities and exercise, and when the virus infection abates, people with depression are more likely to suffer from immunological problems, making them more prone to other conditions [6].

During this time, it is crucial for psychiatrists to become familiar with screening and triage procedures and work closely with public health specialists and physicians to reduce the problems that their patients face [7].

The study of social media, particularly in the public health domain, is a rapidly growing research area. For instance, social media are commonly used to monitor outbreaks of infectious diseases [8–11] and understand trends in prescription medication usage [12]. Furthermore, several authors [13–16] claim that the value of social media in understanding mental health is of the utmost importance, since they provide access to the public accounts, behaviors, activities, thoughts, and feelings of users that may be indicative of their emotional wellbeing.

Since social media information is of great value for identifying people at risk of depression or with other mental disorders, many models and systems have been developed to detect the signs and symptoms of mental illnesses from social media data. For instance, Renara et al. [17] found that sentiment analysis on social media could help monitor the mood of a person, which is particularly important since people with depression symptoms experience similar feelings and have similar behavior, which are often expressed through what they post on their social media platforms. To perform sentiment analysis, the n-gram model, i.e., a set of n consecutive words, is commonly used. In fact, several authors [18–21] use the n-gram model for the specific case of n equals one ($n = 1$), which is also called unigram. According to De Choudhury and Gamon [13], the following unigrams are associated with depression signs or symptoms: retraction, psychosis, harsh, delusions, ADHD, imbalance, sleeplessness, suicidal, vertigo, retching, attacks, sleep, seizures, addictive, weaned, swings, dysfunction, appetite, fuzzy, irritability, episodes, headache, tiredness, edging, anxiety, burden, heaviness, and somnolent. On the other hand, investigations from these authors [22–25] have demonstrated the results obtained in this topic. From this perspective, it seems relevant for the scientific community to perform a systematic literature review to identify and become familiar with the social media sites and features of datasets, methods for linguistic feature extraction, machine-learning algorithms, computing tools, and statistical analysis methods currently employed to determine depression on social media.

The scope of this research is to identify and summarize the existing evidence concerning depression sign detection on social media via computing tools, methods for linguistic feature extraction, statistical analysis techniques, and machine-learning algorithms. The research follows the methodology proposed by Brereton et al. [26] to review relevant literature from the last five years (from 2016 to mid-2021), which were retrieved from major academic digital libraries. Then, we synthetize the results from our primary sources using strategies for reducing bias and random errors. Our findings highlight the social media sites, computing tools, methods for linguistic feature extraction, statistical analysis techniques, and machine-learning algorithms most used in depression sign detection research. We also analyze and discuss literature reviews similar to ours to emphasize the progress being made in terms of depression sign detection via innovative techniques. The review is focused on the research into depression sign detection and seeks to elucidate the different methods used for detecting depression on social media using sentiment analysis.

An Overview of Machine-Learning Techniques, Dataset Features, and Social Media

Sentiment analysis (SA) is a technique for analyzing consumer opinions and producing data that can depict these opinions as a whole [27]. SA is also known as opinion mining, a text analysis technique that analyzes the opinions of human emotions toward entities and the features that exist in these entities [28]. In the context of SA, a feature is an item that people talk about in relation to services, products, policies, events, organizations, or individuals. The combination of features and corresponding sentiment words can help produce accurate, meaningful, and high-quality sentiment analysis results [27].

Machine-learning (ML) techniques are applied in sentiment classification to organize text into positive, negative, or neutral categories. Training datasets and testing datasets are used in ML techniques. The training datasets are applied to learn the documents, while the testing datasets are used to validate the execution of ML techniques [29]. As Maetschke et al. [30] explain, machine-learning algorithms comprise supervised, unsupervised, and semisupervised methods. Unsupervised methods are applied on expression data but have a lower prediction capability than supervised methods. Supervised methods need data on known associates for training, and these are often scarce. Semisupervised methods can be trained with fewer interaction data but are generally less accurate predictors than supervised methods.

Social media allows researchers to obtain behavioral data relevant to a person's way of thinking, emotional state, communication, activities, and means of relating. The texts that are published on social networks allow the detection of feelings of uselessness, guilt, powerlessness, and self-aversion that determine the signs of depression. According to De Choudhury and Gamon [13], changes in social relationships, activity, and language can be applied to build statistical models that allow the detection and prediction of depression in a more precise way, including ways that can complement traditional diagnostic approaches.

The rest of this paper is organized as follows: Section 2 discusses the goal and justification of the research, while Section 3 explains the methods, which include our research questions, search strategy, selection process of primary studies, and data extraction process. The results of the review are included in Section 4, whereas in Section 5 we introduce a discussion of the results. At the end, in Section 6 we define the conclusions and suggestions for future work.

2. Research Goal and Need for Literature Review

This literature review seeks to identify and summarize existing evidence concerning depression sign detection research on social media using methods of linguistic feature extraction, machine-learning algorithms, computing tools, and statistical analysis methods. Currently, there are works that address a theme similar to that of this work. Table 1 lists research works similar to ours, for example, Guntuku et al. [31] focus on studies aimed at predicting mental illness using social media. First, they consider the methods used to predict depression, and then they consider four approaches that have been used in the literature: prediction based on survey responses, prediction based on self-declared mental health status, prediction based on forum membership, and prediction based on annotated posts. Wang et al. [32] examined relevant investigations with the Beck Depression Inventory-II for measuring depression in medical settings to provide guidelines for practicing clinicians. The Beck Depression Inventory-II showed high reliability and good correlation with the measures of depression and anxiety. Its threshold for detecting depression varied according to the type of patient, suggesting the need for adjusted cutoff points. The somatic and cognitive–affective dimension described the latent structure of the instrument. Gottlieb et al. [33] showed that contextual interventions for the prevention and treatment of depressive symptoms and psychological distress can be effective, though very limited data exist in this field. Policy implications include a greater emphasis on improving conditions to decrease the incidence of depression and other mental disorders.

Although the aforementioned works share some similarities with our research, none of them review sentiment-analysis-based initiatives. Moreover, only one of the works reviewed social media for predicting mental illnesses, but it did not specifically focus on depression sign detection. From this perspective, we conclude that the principal differences between our literature review and similar works are as follows: (1) we analyze the most recent relevant works; (2) we identify the social media sites most commonly studied and the features of the datasets retrieved; and we determine (3) the linguistic feature extraction methods, (4) machine-learning algorithms, (5) computing tools, and (6) mathematical analysis methods most commonly applied in depression sign detection from social media.

Table 1. Summary of related studies.

Study Reference	Approach	Year	Studies Reviewed	Years Covered
Guntuku et al. [31]	Predictive models	2017	12	2013–2017
Wang and Gorenstein [32]	Beck Depression Inventory-II	2013	70	1996–2012
Gottlieb et al. [33]	Social contexts	2011	30	1997–2008

3. Methods

This literature review examines quantitative and qualitative aspects of primary studies exploring depression detection from social media submissions via novel approaches and methods. We followed the three-stage methodology depicted in Figure 1, which was proposed by Brereton et al. [26] as a straightforward method for conducting systematic literature reviews. The planning stage of the methodology comprises three steps: (a) determine need for literature review, (b) state research questions, and (c) review the protocol. Next, the conducting stage of the methodology comprises four steps: (a) determine search strategy, (b) select primary studies, (c) extract data, and (d) synthesize data. In the end, the documenting stage involves three steps: (a) obtain results, (b) identify threats to validity, and (c) establish conclusions.

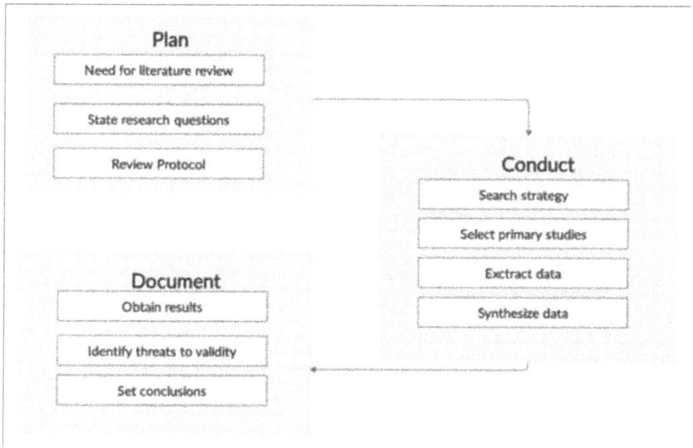

Figure 1. Literature review process.

3.1. Research Questions and Motivations

Five research questions were formulated that oriented the research and helped meet the objectives of the review. These questions are listed in Table 2.

Table 2. Research questions.

Research Question (RQ)	Question
RQ1	Which social media sites and features of datasets are mainly used in depression sign detection research?
RQ2	Which are the main linguistic feature extraction methods used for detecting depression signs on social media?
RQ3	Which are the main machine-learning algorithms used in depression sign detection from social media?
RQ4	Which are the main computing tools applied in detecting depression signs on social media?
RQ5	Which are the main statistical analysis methods used to validate results in detecting depression signs on social media?

3.2. Search Strategy

The search for primary studies was conducted in five digital libraries: ACM Digital Library, IEEE Xplore Digital Library, SpringerLink, Science Direct, and PubMed, as well as on the search engine Google Scholar to broaden our results. We selected the libraries based on their prestige and popularity in the scientific community, since they all provide access to a large proportion of digital literature, especially peer-reviewed articles, on a wide range of topics, including those related to our research. In a second step, we conducted a search based on keywords. To do this, we performed two tasks: we first identified a set of words or phrases in relation to our search topic (i.e., keywords); then, we identified related concepts. As for the search period, our review was intended to be not only accurate, but also up to date. To this end, the search covered the last six years—from 2016 to mid-2021. Finally, regarding the keyword search, Table 3 lists the set of keywords and related concepts used.

Table 3. Keywords and related concepts of the literature review.

Area	Keywords	Related Concepts
Mental health	Depression	Mental illness
Social media	Social media	Mental disorder
		Social networks
		Social web
		Microblogs
		Twitter
		Facebook
		Reddit
		Instagram
		Weibo
		NHANES

The search strings were formed by combining the keywords listed in Table 3 using connectors "AND" and "OR" as follows: ((Depression) OR (Mental Health) OR (Mental illness) OR (Mental disorder) AND (Social media OR Social networks OR Social web OR Microblogs OR Twitter OR Facebook OR Reddit OR Instagram OR Weibo OR NHANES)) Year: 2016–2021. As Figure 2 shows graphically, we found 482 relevant search results: 154 from IEEE Xplore Digital Library, 89 from SpringerLink, 78 from ACM Digital Library, 62 from Google Scholar, 62 from PubMed, and 37 from ScienceDirect.

According to Figure 2, the majority of the literature regarding depression detection on social media is produced by IEEE, followed by SpringerLink and ACM. Conversely,

Google Scholar and PubMed provide access to fewer research articles on the subject matter. Finally, we found lowest number of publications relevant to our search on Science Direct.

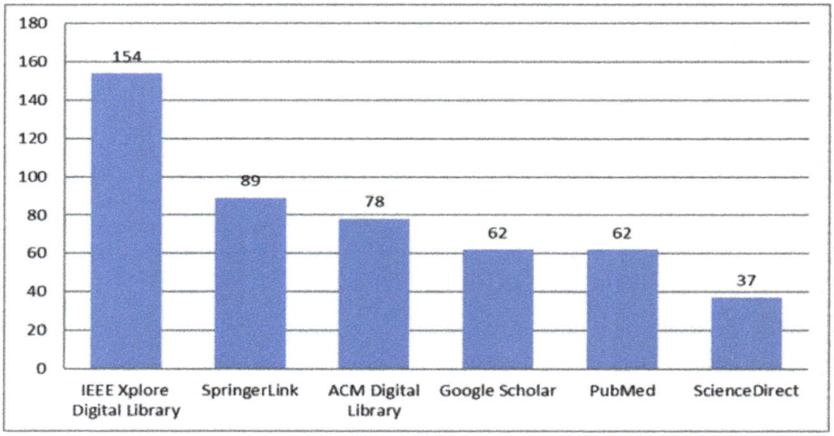

Figure 2. Research papers by digital libraries.

3.3. Selection of Primary Studies

We selected only studies including at least one of the keywords such as *Depression*, *Social Media*, and related concepts (see Table 3).

We identified 420 records through database searching; furthermore, we identified 62 additional records through other sources such as Google Scholar. After the duplicates were removed, we obtained 287 papers that determined the records screened. Once we had read the abstracts, were excluded 95 (57 master and doctoral dissertations and 38 papers not written in English). Then, we read the full articles assessed for eligibility and excluded 158 studies conducted in domains other than detecting depression signs on social media to obtain the studies included in the synthesis (192). Finally, we obtained 34 studies that constituted the studies included in the quantitative synthesis.

A PRISMA diagram [34] is shown in Figure 3 that represents the flow diagram of the papers searched and chosen for our review.

We retrieved and analyzed 192 full text articles assessed for eligibility but only considered 34 primary studies. As depicted in Figure 4, 59% of the retrieved publications were published in journals, 32% in conference proceedings, and 9% as book chapters. As regards the year of publication, 8 papers were issued in 2016 (journals); 26 papers were published in 2017 (7 in conference proceedings, 18 in journals, and 1 as a book chapter); 35 papers were published in 2018 (12 in conference proceedings, 20 in journals, and 3 as book chapters); 40 were issued in 2019 (14 in conference proceedings, 22 in journals, and 4 as book chapters); 49 papers were published in 2020 (18 in conference proceedings, 25 in journals, and 6 as book chapters); and finally, 34 papers were published in the first half of 2021 (10 in conference proceedings, 20 in journals, and 4 as book chapters).

Figure 5 graphically represents the geographical distribution of the retrieved publications. As can be seen, the majority of the research was conducted in the United States (29%), China (24%), India (12%), England (9%), Spain (5%), Taiwan (5%), Thailand (3%), Switzerland (3%), Germany (3%), Brazil (1%), Israel (1%), Saudi Arabia (1%), Argentina (1%), Canada (1%), Mexico (1%), Australia (1%), and Iran (1%).

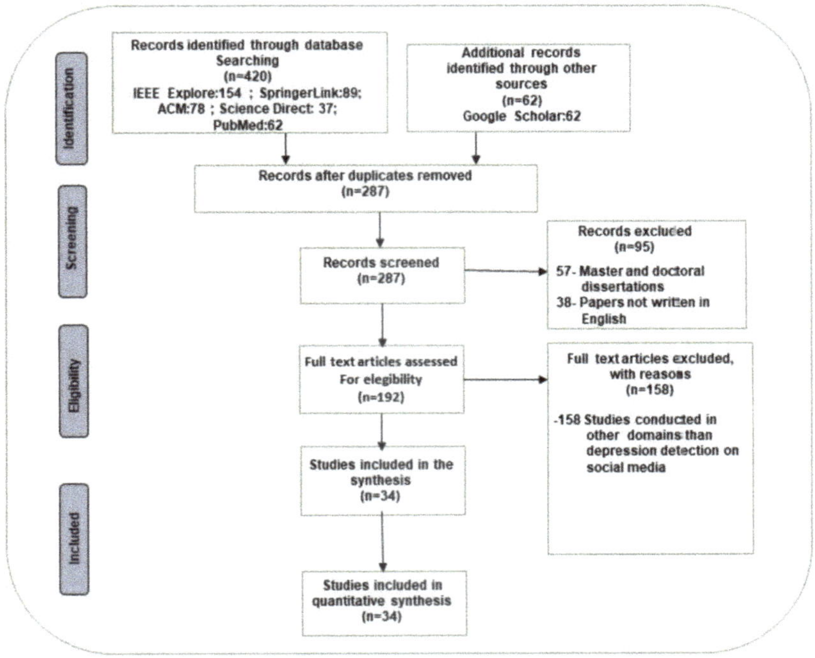

Figure 3. PRISMA flow diagram for the literature search.

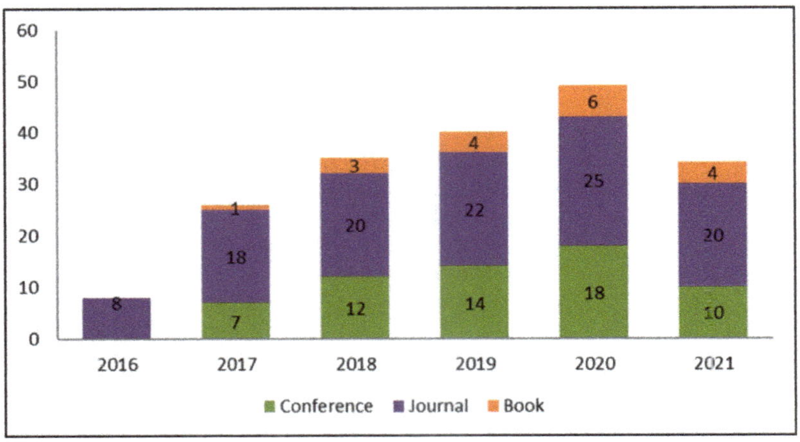

Figure 4. Type of publication from 2016 to mid-2021.

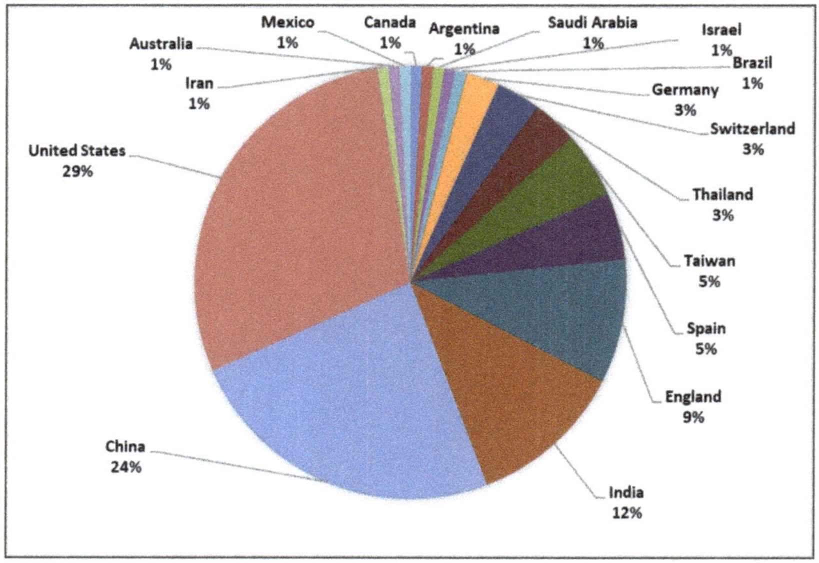

Figure 5. Geographical distribution.

3.4. *Data Extraction*

We retrieved two types of data from the papers: bibliographic data and content data. The former included information such as research title, author names, research goal, and research database; the latter concerned actual information on the research, namely, the studied social media sites and dataset features, along with the computing tools, linguistic feature extraction models, mathematical analysis methods, and machine-learning algorithms used for depression sign detection. The following section discusses our findings.

4. Results

As previously mentioned, we initially retrieved 192 relevant works but ultimately selected and reviewed 34 primary studies, which better described the researched topic. The findings of the review are discussed in the following five subsections, corresponding to our five research questions. The first subsection discusses the most common social media sites and corresponding features of datasets used for depression detection on social media. In the second subsection, we discuss linguistic feature extraction methods from sentiment analysis found in the literature. Then, in the third subsection, we discuss the machine-learning algorithms most commonly applied when trying to detect depression signs from social media data, whereas the fourth subsection identifies the most common computing tools used to process the data. Finally, the fifth subsection reviews the main statistical analysis methods used to validate the results of the classification algorithms applied.

4.1. *RQ1: Which Are the Main Social Media Sites and Dataset Features Used in Depression Detection?*

Table 4 lists the social media sites and features of datasets most commonly studied in depression detection research during the period of 2016 to mid-2021.

According to Table 4 and Figure 6, Twitter, Reddit, and Facebook—in that specific order—are the social media sites most commonly studied. In the case of Twitter, the study of Leis et al. [35] was applied to texts in Spanish and was developed in two steps. In the first step, the selection of users and the compilation of tweets were performed. A total of three datasets of tweets were created, a depressive users dataset (made up of the timeline of 90 users who explicitly mentioned that they suffer from depression), a depressive tweets dataset (a manual selection of tweets from the previous users, which included expressions indicative of depression), and a control dataset (made up of the timeline of 450 randomly selected users). In the second step, the comparison and analysis of the three datasets of tweets were carried out.

Table 4. Social media and corresponding features of datasets used in depression detection research.

Social Media	Study	Features of Dataset
Twitter	Leis et al. [35]	140,946 tweets
	Kr [36]	4000+ tweets
	Shen et al. [37]	36,993 depression-candidate dataset users
	Chen et al. [38]	585 and 6596 unique and valid users with their past tweets
	Arora and Arora [39]	3754 tweets
	Biradar and Totad [40]	60,400 tweets
	Ma et al. [41]	54 million tweets
	Nadeem [42]	1,253,594 documents (tweets) as control variables
	Yazdavar et al. [43]	8770 users, including 3981 depressed users and 4789 control subjects
	Titla-Tlatelpa et al. [44]	7999 users with Twitter submissions
	Chiong et al. [45]	22,191 records
	Safa et al. [46]	570 users from the control group of 16,623,164 tweets
Reddit	Leiva and Freire [47]	135 depressive users, 752 control-group users
	Rissola et al. [48]	1,076,582 submissions from 1707 unique users
	Sadeque et al. [49]	531,453 submissions from 892 unique users
	Tadesse et al. [50]	1293 depression-indicative posts, 548 standard posts
	Wolohan et al. [51]	Reddit posts from a sample of 12,106 users
	Burdisso et al. [52]	887 subjects with 531,394 submissions
	Trotzek et al. [53]	135 depressed users and a random control group of 752 users
	Titla-Tlatelpa et al. [44]	1707 users, Reddit eRisk 2018 task
	Martinez-Castaño et al. [54]	eRisk collections containing up to 1000 posts and 1000 comments
Facebook	Tai et al. [55]	3599 diaries
	Katchapakirin et al. [56]	35 Facebook users
	Wongkoblap et al. [57]	509 users in the final dataset
	Wu et al. [58]	1294 students with their data
	Yang, Mcewen, et al. [59]	22,043,394 status updates from 153,727 users
	Aldarwish and Ahmad [60]	2287 posts
	Ophir et al. [61]	190 Facebook status updates of at-risk adolescents
	Chiong et al. [45]	Facebook, Virahonda, 9178 records
Instagram	Ricard et al. [62]	data from 749 participants
	Reece and Danforth [63]	43,950 user photographs and data
	Mann et al. [64]	221 students, mean of 16.73 posts per student (60 days)
	Chun et al. [65]	520 users from Instagram through the data collection method
Weibo	Li et al. [66]	15,879 Weibo posts from 10,130 distinct Weibo users
	Lixia Yu et al. [67]	7,116,958 posts
NHANES, K-NHANES	Oh et al. [68]	dataset of 28,280 participants with 157 variables for NHANES and 4949 participants with 314 variables for K-NHANES

In the case of Reddit, Rissola et al. [48] introduced a methodology to automatically gather post samples in English of depression and nondepression and used the dataset to train models which are able to determine whether a post conveys evidence of depression.

Katchapakirin et al. [56] employed Natural Language Processing (NLP) techniques to develop a depression detection algorithm for the Thai language on Facebook, which people use as a tool for sharing opinions, feelings, and life events. Results from 35 Facebook users indicated that Facebook behaviors could predict depression level.

Instagram is less prominently researched form of social media, since the platform emphasizes photograph and video sharing rather than text sharing, although some researchers have focused on the alternative text descriptions from Instagram posts to develop their research. We also found a few social media sites that are distinctive to a particular region. For instance, Weibo was studied in China by Li et al. [66], and K-NHANES and NHANES in Korea and the US, respectively, by Oh et al. [68]. Some of these studies were designed to be applied among speakers of other languages, such as Chinese, Thai, Korean, Arabic, and Portuguese. Overall, our findings indicate a growing use of social networking services around the globe.

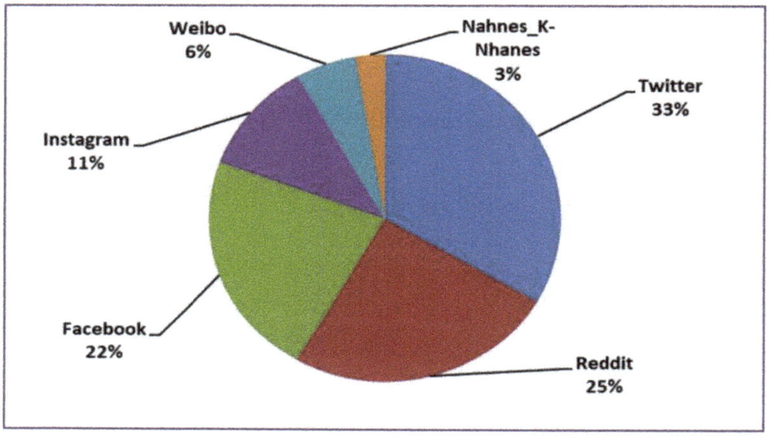

Figure 6. Social media sites explored in depression sign detection research.

4.2. RQ2: Which Are the Main Linguistic Feature Extraction Methods Used for Detecting Depression Signs on Social Media?

Table 5 lists our findings in response to the second research question.

Table 5. Linguistic feature extraction methods used for detecting depression signs on social media.

Model	Study
Word embedding	Rissola et al. [48] Wongkoblap et al. [57] Wu et al. [58] Ma et al. [41] Yazdavar et al. [43] Trotzek et al. [53] Mann et al. [64] Titla-Tlatelpa et al. [44] Yueh et al. [65]
N-grams	Wolohan et al. [51] Rissola et al. [48] Sadeque et al. [49] Arora and Arora [39] Wolohan et al. [51] Nadeem [42] Titla-Tlatelpa et al. [44] Chiong et al. [45] Safa et al. [46]
Tokenization	Tadesse et al. [50]

Table 5. *Cont.*

Model	Study
	Arora and Arora [39]
	Biradar and Totad [40]
	Aldarwish and Ahmad [60]
	Trotzek et al. [53]
	Titla-Tlatelpa et al. [44]
	Chiong et al. [45]
	Safa et al. [46]
Bag of words	Ricard et al. [62]
	Rissola et al. [48]
	Nadeem [42]
	Mann et al. [64]
	Titla-Tlatelpa et al. [44]
	Chiong et al. [45]
	Safa et al. [46]
Stemming	Tadesse et al. [50]
	Arora and Arora [39]
	Aldarwish and Ahmad [60]
Emotion analysis	Leis et al. [35]
	Shen et al. [37]
	Chen et al. [38]
Part-of-Speech (POS) tagging	Wu et al. [58]
	Leis et al. [35]
	Chiong et al. [45]
Behavior features	Wu et al. [58]
	Yang, McEwen, et al. [59]
Sentiment polarity	Leis et al. [35]
	Rissola et al. [48]

Methods for linguistic feature extraction are important since researchers need to use basic elements to determine whether a person shows or does not show depression symptoms. As can be observed from Table 5, word embedding is a prominent model used to detect depression from social media data. In word embedding, each word from a text is listed as a continuous, low dimensional, and real-valued vector [58], and researchers may combine word embedding with other methods for better results. For instance, Rissola et al. [48] combined word embedding with the bag-of-words model to build a depression-post classifier using depression-positive sample posts (D+); depression-negative sample posts (D−); unigrams; word count; and the polarity scores, sadness scores, and happiness scores of words.

The n-gram model is another effective tool in depression sign research. According to Damashek [69], in the n-gram model a document can be listed as a vector whose components are the relative frequencies of its distinct constituent n-grams. In their work, Wolohan et al. [51] found that the best performing model for depression sign identification mixes word-and-character n-grams with LIWC features. As for tokenization, another model for linguistic feature extraction, Arora and Arora [39] explain that it is a process of a giving a token to a sequence of characters that we want to treat as a group; treating text as a token enables the creation of counts of tokens, which can be used as features. In the work of Aldarwish [60], the tokenize operator splits the text of a document into a sequence of tokens. For instance, the research of Tadesse et al. [50] reports the use of tokenization for data preprocessing in order to divide social media posts into individual tokens. Next, all the URLs are divided by punctuation and stop words. Then, the researchers applied stemming to decrease the words to their root form and join similar words together. As for the bag-of-words model, Nadeem [42] describe it as an approach that uses the frequency of word occurrence to determine the content of a tweet. In the bag-of-words model used by Rissola et al. [48], each post is depicted with the raw frequency of the unigrams from the textual content of the posts.

According to Arora and Arora [39], the stemming model for linguistic feature extraction refers to the process of grouping words that are close in meaning. In the study of Arora and Arora [39], the goal was to remove the suffix of a word to retrieve its base form, thus reducing redundancy. In the process of feature extraction, stemming is regularly combined with tokenization. Emotion analysis, behavior feature extraction, polarity, and POS tagging are less frequently used to detect depression from social media. As Shen et al. [36] claim, an emotion analysis determines whether the emotional state of depressed users differs from that of common users. Authors Shen et al. [37] studied emotion-related words and extracted positive and negative word counts from recent tweets using LIWC. As for the behavior feature extraction model, its usefulness is related to the fact that depression sufferers are inclined to focus on themselves and detach from others; moreover, they rarely succeed at communicating with others. Researchers Ramirez-Esparza et al. [70] performed behavior feature extraction on social media posts to identify the behavior of depression sufferers. Additionally, Wu et al. [58] applied this model with POS tagging, UKW (unknown word), word embedding, content-based features, and living-environment features.

In the polarity model, emotions can be tied to the sentiment polarity of a message defined by the text. In their research, Liu and Liu [28] consider that the negative polarity of social media posts (i.e., a value below zero) is a good indicator of unhappiness or distress, especially when the posts come from users with depression. In their work, Rissola et al. [48] combined the polarity score, word count, happiness score, and sadness score of social media posts to build a depression predictor model. Finally, POS tagging is a form of syntactic analysis with countless applications in Natural Language Processing (NLP). According to Lovins [71], it is also one of the most basic parts of the linguistic pipeline.

4.3. *RQ3: Which Are the Main Machine-Learning Algorithms Used for Detecting Depression Signs on Social Media?*

To respond to this question, Table 6 lists our review of the machine-learning algorithms used in depression sign detection research.

Table 6. Machine-learning algorithms.

Machine-Learning Algorithm	Study
Support vector machine (SVM)	Leiva and Freire [47] Rissola et al. [48] Katchapakirin et al. [56] Sadeque et al. [49] Chen et al. [38] Tadesse et al. [50] Arora and Arora [39] Wolohan et al. [51] Yang, McEwen, et al. [59] Burdisso et al. [52] Li et al. [66] Nadeem [42] Yazdavar et al. [43] Oh et al. [68] Aldarwish and Ahmad [60] Mann et al. [64] Titla-Tlatelpa et al. [44] Chiong et al. [45] Safa et al. [46]
Logistic regression	Leiva and Freire [47] Rissola et al. [48] Chen et al. [38] Tadesse et al. [50] Reece and Danforth [63] Yang, McEwen, et al. [59]

Table 6. *Cont.*

Machine-Learning Algorithm	Study
	Burdisso et al. [52]
	Li et al. [66]
	Nadeem [42]
	Yazdavar et al. [43]
	Oh et al. [68]
	Trotzek et al. [53]
	Martinez-Cataño et al. [54]
	Chiong et al. [45]
	Safa et al. [46]
Neural networks	Kr [36]
	Sadeque et al. [49]
	Wongkoblap et al. [57]
	Wu et al. [58]
	Biradar and Totad [40]
	Yang, McEwen, et al. [59]
	Li et al. [66]
	Yazdavar et al. [43]
	Trotzek et al. [53]
	Mann et al. [64]
	Yueh et al. [65]
Random forests	Leiva and Freire [47]
	Katckapakirin et al. [56]
	Chen et al. [38]
	Tadesse et al. [50]
	Reece and Danforth [63]
	Yang, McEwen, et al. [59]
	Li et al. [66]
	Yazdavar et al. [43]
	Titla-Tlatelpa et al. [44]
	Chiong et al. [45]
	Safa et al. [46]
	Yueh et al. [65]
Bayesian statistics	Tai et al. [55]
	Chen et al. [38]
	Arora and Arora [39]
	Reece and Danforth [63]
	Yang, McEwen, et al. [59]
	Burdisso et al. [52]
	Nadeem [42]
Decision trees	Yang, McEwen, et al. [59]
	Nadeem [42]
	J Oh et al. [68]
	Titla-Tlatelpa et al. [44]
	Chiong et al. [45]
	Safa et al. [46]
K-Nearest Neighbor	Leiva and Freire [47]
	Yang, McEwen, et al. [59]
	Burdisso et al. [52]
	Oh et al. [68]
Linear regression	Leiva and Freire [47]
	Ricard et al. [62]
	Yu et al. [67]
Ensemble classifiers	Leiva and Freire [47]
	Oh et al. [68]
Multilayer Perceptron	Chiong et al. [45]
	Safa et al. [46]
Boosting	Tadeesse et al. [50]
K-Means	Ma et al. [41]

Machine-learning algorithms are powerful generalizers and predictors [72]. According to Baharudin et al. [73], many algorithms and techniques have been recently proposed for the classification and clustering of digital documents.

According to Batta [74], Support Vector Machines are supervised learning models with associated learning algorithms that analyze data used for classification and regression analysis. In addition to performing linear classification, SVMs can efficiently perform a nonlinear classification using what is called the kernel trick, implicitly mapping their inputs into high-dimensional feature space. Ray [75] explains that logistic regression is used to deal with classification problems. It gives a binomial outcome for the probability of whether or not an event will occur (in terms of 0 and 1), based on the values of input variables. For example, predicting whether a tumor is malignant or benign or an e-mail is classified as spam or not. Logistic regression deals with the prediction of target variables that are categorical. According to Batta [74], a neural network is a series of algorithms that endeavors to recognize underlying relationships in a set of data through a process that mimics the way the human brain operates. In this sense, neural networks refer to systems of neurons, either organic or artificial in nature. Neural networks can adapt to changing input; thus, the network generates the best possible result without needing to redesign the output criteria.

Related to our review, machine-learning algorithms increase the accuracy of predictions in multiple types of datasets. In some cases, several algorithms are used in a single research work. For example, Leiva and Freire [47] use support vector machine, logistic regression, random forest, k-nearest neighbor, linear regression, and ensemble classifiers; Rissola et al. [48] use support vector machine and logistic regression.

As can be observed from Figure 7, researchers generally rely on SVM, logistic regression, or neural networks to complete their diagnosis of depression from social media data. Other machine-learning algorithms less frequently employed include random forests (14%), Bayesian statistics (9%), decision trees (7%), k-nearest neighbor classifiers (6%), linear regression (4%), ensemble classifiers (2%), multilayer perceptron (2%), and boosting and k-means (1%).

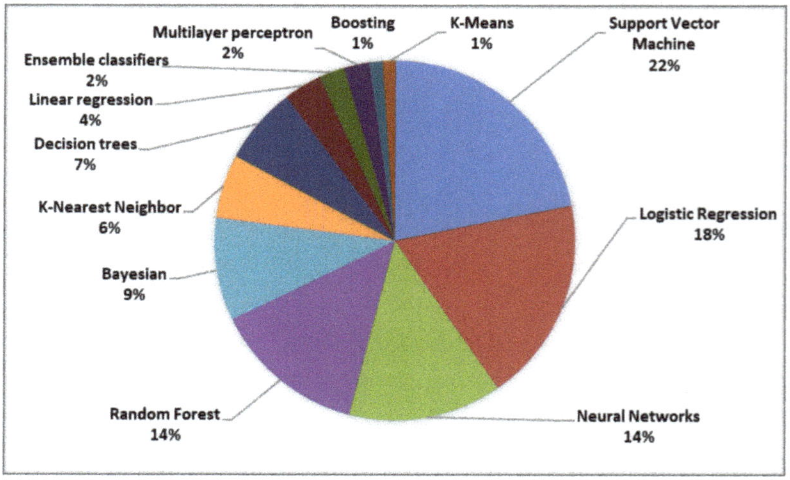

Figure 7. Machine-learning algorithms used for detecting depression signs on social media.

4.4. RQ4L: Which Are the Main Computing Tools Used for Detecting Depression Signs on Social Media?

To respond to this question, Table 7 shows the main computing tools used for detecting depression signs on social media.

Table 7. Computing tools used for detecting depression signs on social media.

Computing Tool	Study
Python libraries	Kr [36]
	Leiva and Freire [47]
	Rissola et al. [48]
	Katchapakirin et al. [56]
	Tadesse et al. [50]
	Wongkoblap et al. [57]
	Biradar and Totad [40]
	Ma et al. [41]
	Burdisso et al. [52]
	Nadeem [42]
	Yazdavar et al. [43]
	Trotzek et al. [53]
	Mann et al. [64]
	Martinez-Cataño et al. [54]
	Safa et al. [46]
	Lu et al. [67]
LIWC	Shen et al. [37]
	Chen et al. [38]
	Tadesse et al. [50]
	Wolohan et al. [51]
	Yang, McEwen, et al. [59]
	Li et al. [66]
	Yazdavar et al. [43]
	Trotzek et al. [53]
	Safa et al. [46]
Word2Vec	Shen et al. [37]
	Rissola et al. [48]
	Wu et al. [58]
	Ma et al. [41]
	Yueh et al. [65]
Twitter APIs	Chen et al. [38]
	Biradar and Totad [40]
	Leis et al. [35]
	Kr [36]
WordNet	Shen et al. [37]
	Arora and Arora [39]
FastText	Rissola et al. [48]
	Trotzek et al. [53]
Weka	Katchapakirin et al. [56]
	Li et al. [66]
RapidMiner	Katchapakirin et al. [56]
	Aldarwish and Ahmad [60]
Google Apps	Katchapakirin et al. [56]
	Wu et al. [58]
Microsoft Excel	Li et al. [66]
	Aldarwish and Ahmad [60]

Figure 8, below, introduces a graphic representation of the most common computing tools used for detecting depression signs from social media data. As can be observed, the authors use Python in first place; for example, Rissola et al. [48] use the TextBlob2 Python library to compute the polarity score of the posts in negative samples and sort them in ascending order. In the study of Leyva and Freire [47], the implementation of the learning algorithms and the vectorization were implemented with the scikit-learn library, version 0.18, for Python. In second place is LIWC (Linguistic Inquiry and Word Count). Tausczik and Pennebaker [76] explain that LIWC is a program for text analysis that counts words in psychologically meaningful categories. In their work, Shen et al. [37] extracted positive and negative word counts in recent tweets with LIWC, while Tadesse et al. [50] explored the users' linguistic usage in the posts, employing the LIWC dictionary. Word2vec and Twitter APIs are also popular but less commonly used, followed in the list by WordNet; FastText; Weka; RapidMiner; Google Apps (in this case, it is interesting to mention that this program was used as a language translator with the Google Cloud Translation API [56]); and Microsoft Excel [60]. In the case of Microsoft Excel, the supervised dataset used in the two classifiers were created using three columns: the first being the sentiment (depressed or not depressed); the second being the depression category, which consists of one of the nine depression categories; and the third containing the manually trained posts. Finally, much less prominent tools include SPSS, Clickworker (a crowdsourcing platform), Instagram Graph API, Java, Jade, Google Cloud Translation API, and MATLAB. All these are applied along with mathematical analysis methods and machine-learning algorithms for higher accuracy in the results. Herein lies the importance of knowing which computing tools can be applied in combination with other methods.

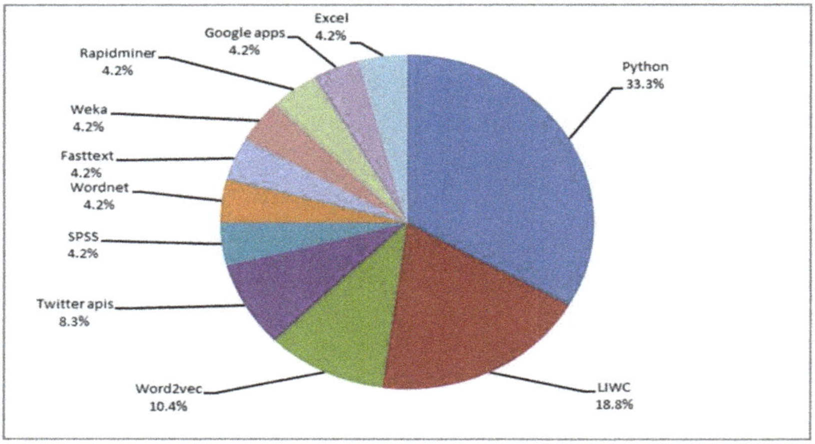

Figure 8. Computing tools used for detecting depression signs on social media.

4.5. RQ5: Which Are the Main Statistical Analysis Methods Used to Validate Results in Detecting Depression Signs on Social Media?

Our findings summarized in Table 8 respond to our fifth research question.

Statistical analysis is the use of mathematics to analyze data. According to our review, and as summarized in Table 8, the most common statistical analysis methods applied to validate results in depression detection research from social media include cross-validation (CV), term frequency/inverse document frequency (TF–IDF), and Cohen's kappa statistic. On the one hand, CV is remarkably versatile; it is applicable to a wide range of problems across multiple areas. For instance, CV has been used for smoothing parameters in non-parametric smoothing and for variable selection in regression. The idea behind this method is simply splitting the data into two parts, applying the first part to determine a prediction rule, and then assessing the quality of the prediction by matching its outputs with the rest of the data; hence, the name cross-validation [77]. In the work of Ricard et al. [62], the mean and SD of the text-based scores for the most recent k posts were utilized as features in their model training, with k as a hyperparameter tuned through cross-validation. Wongkoblap et al. [57] created a predictive model and used n-fold cross-validation to report the performance of the model. The results of the evaluation are presented with accuracy, precision, recall, and the f1-score achieved by the model after training and testing with five-fold cross-validation. Oh et al. [68] ran 10-fold cross-validation for all algorithms and datasets to validate the performance of each classifier and to avoid overfitting. On the other hand, TF–IDF is a statistic used to determine the relevance of a search query to a document in a collection of documents or the occurrences of a given query in a document. It is commonly used as a basic weighting factor for text retrieval [78]. In their work, Tadesse et al. [50] used the term frequency/inverse document frequency (TF–IDF) as a numeric statistic for n-gram modelling, where the importance of a word with respect to each document in the corpora is highlighted. The main goal of its usage is to scale down the impact of empirically less-informative tokens that occur frequently to provide space for the more informative words occurring in a smaller fraction.

Table 8. Statistical analysis methods used to validate results in detecting depression signs on social media.

Statistical Analysis Method	Study
Cross-validation	Ricard et al. [62]
	Wongkoblap et al. [57]
	Oh et al. [68]
	Tai et al. [55]
	Sadeque et al. [49]
	Burdisso et al. [52]
	Li et al. [66]
	Nadeem [42]
	Yazdavar et al. [43]
	Mann et al. [64]
	Titla-Tlatelpa et al. [44]
	Chiong et al. [45]
Term frequency/inverse document frequency (TF–IDF)	Leiva and Freire [47]
	Tadesse et al. [50]
	Wolohan et al. [51]
	Yang, McEwen, et al. [59]
	Aldarwish and Ahmad [60]
	Martinez-Cataño et al. [54]
	Titla-Tlatelpa et al. [44]
Cohen's kappa statistic	Rissola et al. [48]
	Li et al. [66]
	Yazdavar et al. [43]
	Yang, McEwen, et al. [59]
Mean/standard deviation	Chen et al. [38]
	Ricard et al. [62]
	Mann et al. [64]
Mann–Whitney	Ricard et al. [62]
	Ophir et al. [61]
Likert scale	Kr [36]
	Ophir et al. [61]
Softmax function	Wongkoblap et al. [57]
variance	Leis et al. [35]
Direction method of multipliers	Shen et al. [37]
Adam optimizer	Biradar and Totad [40]
Pixel-level averages	Reece and Danforth [63]

Finally, Cohen's kappa statistic is a measure for assessing the degree of agreement between evaluators for the absence or presence of a trait [79]. In the work of Yazdavar et al. [43], the dataset used provided the users' profile information, including screen name, profile description, follower/followee counts, profile image, and tweet content, which could express various depression-relevant characteristics and determine whether a user indicated any depressive behavior. They reported the inter-rater agreement as K = 0.74, based on Cohen's kappa statistics.

Other common mathematical analysis methods include mean/standard deviation, the Mann–Whitney U test, Likert scales, and SoftMax functions, which help improve the accuracy of the results. We also found evidence of the use of variance analysis, the alternating direction method of multipliers (ADMM), Adam optimization, and Pixel-level weighted averaging.

5. Discussion

Depression sign detection from social media data is a growing area of interest, as the literature confirms. Data sources may vary across studies (e.g., Twitter, Facebook, Reddit, Instagram, Weibo, and NHANES). Users tend to employ social media to write about how they feel according to their interest in doing so and the facility of the use of such social media; however, in our study, we could see that much of the research into this is based on the tools that are most commonly used worldwide and that the datasets examined range from a few tweets to millions of posts. As new social media services constantly emerge, their focus continues to vary. Nowadays, a growing number of social networking services focus more on photo and video sharing rather than text sharing, thus making mental disease prediction efforts more challenging. As internet tools become more user-friendly, an increasing number of people join the social media community every day. In our study, we could see that there have been many different methods applied by researchers to extract data from tweets or posts written by users. These tools can be combined to gain better results. Machine-learning algorithms allow for the classification and clustering of data. Such tools are helpful in the process of obtaining precise results. Some authors use several of these tools in combination to ascertain which is the best for the study in question. Computer tools are necessary to process the information obtained. They perform an essential task in the sense that they help to obtain natural language information and translate or process the data to be classified. Many authors use a wide range of mathematical analysis methods; in our study, we could see that these statistical tools are useful to validate results for the detection of depression from social media. All the studies explored in this review were written in English, which is considered as the language of global scientific understanding. However, some of these studies were designed to be applied among speakers of other languages, such as Chinese, Thai, Korean, Arabic, and Portuguese.

6. Conclusions and Future Work

The objective of this review work was to identify all the tools necessary to detect signs of depression via social media. Using social media along with computing tools and increasingly efficient classification methods contributes to current efforts to detect signs of depression or any other mental illness from sources close to patients. This is important because, with the advance in technology, more and more people are using new media to communicate and to share experiences in the treatment of mental illnesses. Some of the studies we considered were applied in real environments and demonstrated the benefit of the research's application in real life situations. Depression diagnosis from social media data is being widely explored around the world using a variety of networking sites, datasets, linguistic feature extraction methods, machine-learning algorithms, computing tools, and statistical analysis methods. The results obtained in most of the research works indicate that the use of new digital tools related to mental health is an incentive to continue investigating in this area. Finally, we believe that this work paves the way for further exploration of initiatives for diagnosing other mental illnesses, such as anxiety, in the sense that most of the symptoms presented in anxiety are also presented in depression. Additionally, researchers can go beyond by exploring current efforts in the monitoring and treatment of mental disorders using the Internet of Things.

Author Contributions: Conceptualization, R.S.-Z., G.A.-H. and M.d.P.S.-Z.; data curation, R.S.-Z.; formal analysis, R.S.-Z. and G.A.-H.; investigation, M.d.P.S.-Z. and M.A.P.-V.; methodology, R.S.-Z. and M.A.P.-V.; supervision, M.A.P.-V. and M.B.-L.; validation, M.A.P.-V., J.L.S.-C. and M.B.-L.; visualization, R.S.-Z. and M.B.-L.; writing—original draft preparation, R.S.-Z. and M.d.P.S.-Z.; writing—review and editing, R.S.-Z. and G.A.-H.; project administration, J.L.S.-C. All authors have read and agreed to the published version of the manuscript.

Funding: This research received no external funding.

Institutional Review Board Statement: This study did not require ethical approval.

Informed Consent Statement: Not applicable.

Data Availability Statement: Not applicable.

Acknowledgments: We are grateful to the Tecnológico Nacional de Mexico (TecNM, by its Spanish acronym) for supporting this work. This research was also sponsored by Mexico's National Council of Science and Technology (CONACYT) and Mexico's Secretariat of Public Education (SEP) through the PRODEP program.

Conflicts of Interest: The authors declare no conflict of interest.

References

1. James, S.L.; Abate, D.; Abate, K.H.; Abay, S.M.; Zucker, I.; Vos, T.; Murray, C.J.L. Global, regional, and national incidence, prevalence, and years lived with disability for 354 Diseases and Injuries for 195 countries and territories, 1990–2017: A systematic analysis for the Global Burden of Disease Study 2017. *Lancet* **2018**, *392*, 1789–1858. [CrossRef]
2. WHO. *Preventing Suicide a Resource for General Physician*; World Health Organization: Geneva, Switzerland, 2000; pp. 1–17.
3. American Psychiatric Association. Help With Depression (n.d.). Available online: https://www.psychiatry.org/patients-families/depression/what-is-depression (accessed on 20 January 2021).
4. All Documents (n.d.). Available online: https://theconversation.com/what-causes-depression-what-we-know-dont-know-and-suspect-81483 (accessed on 4 August 2020).
5. Szmuda, T.; Shan, A.; Pawel, S.; Nsurg4WL Group. Telemedicine in neurosurgery during the novel coronavirus (COVID-19) pandemic. *Pol. J. Neurol. Neurosurg.* **2020**, *54*, 207–208. [CrossRef]
6. Kanther, J.; Manbeck, K. No Title. Available online: https://theconversation.com/covid-19-could-lead-to-an-epidemic-of-clinical-depression-and-the-health-care-system-isnt-ready-for-that-either-134528 (accessed on 12 October 2020).
7. Zhu, N.; Zhang, D.; Wang, W.; Li, X.; Yang, B.; Song, J.; Zhao, X.; Huang, B.; Shi, W.; Lu, R.; et al. A novel coronavirus from patients with pneumonia in China, 2019. *N. Engl. J. Med.* **2000**, *382*, 727–733. [CrossRef] [PubMed]
8. Oh, S.H.; Lee, S.Y.; Han, C. The Effects of Social Media Use on Preventive Behaviors during Infectious Disease Outbreaks: The Mediating Role of Self-relevant Emotions and Public Risk Perception. *Health Commun.* **2020**, *36*, 972–981. [CrossRef]
9. Lazard, A.J.; Scheinfeld, E.; Bernhardt, J.M.; Wilcox, G.B.; Suran, M. Detecting themes of public concern: A text mining analysis of the Centers for Disease Control and Prevention's Ebola live Twitter chat. *Am. J. Infect. Control* **2015**, *43*, 1109–1111. [CrossRef]
10. Odlum, M.; Yoon, S. What can we learn about the Ebola outbreak from tweets? *Am. J. Infect. Control* **2015**, *43*, 563–571. [CrossRef]
11. Ahmed, W.; Bath, P.A.; Sbaffi, L.; Demartini, G. Novel insights into views towards H1N1 during the 2009 Pandemic: A thematic analysis of Twitter data. *Health Inf. Libr. J.* **2019**, *36*, 60–72. [CrossRef]
12. Sarker, A.; O'Connor, K.; Ginn, R.; Scotch, M.; Smith, K.; Malone, D.; Gonzalez, G. Social media mining for toxicovigilance: Automatic monitoring of prescription medication abuse from twitter. *Drug Saf.* **2016**, *39*, 231–240. [CrossRef]
13. Choudhury, M.D.; Gamon, M.; Counts, S.; Horvitz, E. Predicting depression via social media. In *Proceedings of the Seventh International AAAI Conference on Weblogs and Social Media*; Cambridge, MA, USA, 8–11 July 2013, IAAA Publisher: Palo Alto, CA, USA, 2013; Volume 2, pp. 128–137.
14. Tsugawa, S.; Kikuchi, Y.; Kishino, F.; Nakajima, K.; Itoh, Y.; Ohsaki, H. Recognizing depression from twitter activity. In Proceedings of the 33rd Annual ACM Conference on Human Factors in Computing Systems, Seoul, Korea, 18–23 April 2015; Association for Computing Machinery: New York, NY, USA, 2015; pp. 3187–3196. [CrossRef]
15. Hu, H.W.; Hsu, K.S.; Lee, C.; Hu, H.L.; Hsu, C.Y.; Yang, W.H.; Wang, L.; Chen, T.A. Keyword-Driven Depressive Tendency Model for Social Media Posts. In *Business Information Systems*; Lecture Notes in Business Information Processing; Springer: Berlin/Heidelberg, Germany, 2019; pp. 14–22. [CrossRef]
16. Calvo, R.A.; Milne, D.N.; Hussain, M.S.; Christesen, H. Natural language processing in mental health applications using non-clinical texts. *Nat. Lang. Eng.* **2017**, *23*, 649–685. [CrossRef]
17. Rosa, R.L.; Rodríguez, D.Z.; Schwartz, G.M.; de Campos Ribeiro, I.; Bressan, G. Monitoring System for Potential Users with Depression Using Sentiment Analysis. In Proceedings of the 2016 IEEE International Conference on Consumer Electronics (ICCE), Las Vegas, NV, USA, 7–11 January 2016; pp. 381–382.
18. Saif, H.; He, Y.; Alani, H. *Semantic Sentiment Analysis of Twitter*; Lecture Notes in Computer Science (LNCS); Including Subser. Lect. Notes Artif. Intell. Lect. Notes Bioinformatics; Springer: Berlin/Heidelberg, Germany, 2012; Volume 7649, pp. 508–524. [CrossRef]
19. Tayal, D.K.; Yadav, S.K. Sentiment analysis on social campaign "Swachh Bharat Abhiyan" using unigram method *AI Soc.* **2017**, *32*, 633–645. [CrossRef]
20. Venugopalan, M.; Gupta, D. Exploring sentiment analysis on twitter data. In Proceedings of the 2015 Eighth International Conference on Contemporary Computing (IC3), Noida, India, 20–22 August 2015; pp. 241–247. [CrossRef]
21. Altrabsheh, N.; Cocea, M.; Fallahkhair, S. Sentiment Analysis: Towards a Tool for Analysing Real-Time Students Feedback. In Proceedings of the 2014 IEEE 26th International Conference on Tools with Artificial Intelligence, Limassol, Cyprus, 10–12 November 2014; pp. 419–423. [CrossRef]

22. Naslund, J.A.; Aschbrenner, K.A.; McHugo, G.J.; Unützer, J.; Marsch, L.A.; Bartels, S.J. Exploring opportunities to support mental health care using social media: A survey of social media users with mental illness. *Early Interv. Psychiatry* **2019**, *13*, 405–413. [CrossRef] [PubMed]
23. Gkotsis, G.; Oellrich, A.; Hubbard, T.; Dobson, R.; Liakata, M.; Velupillai, S.; Dutta, R. The language of mental health problems in social media. In Proceedings of the Third Workshop on Computational Linguistics and Clinical Psychology, San Diego, CA, USA, 16 June 2016; pp. 63–73. [CrossRef]
24. Conway, M.; O'Connor, D. Social media, big data, and mental health: Current advances and ethical implications. *Curr. Opin. Psychol.* **2016**, *9*, 77–82. [CrossRef] [PubMed]
25. de Choudhury, M. Role of social media in tackling challenges in mental health. In *SAM'13: Proceedings of the 2nd International Workshop on Socially-Aware Multimediasam*; Co-Located with ACM Multimed; Association for Computing Machinery: New York, NY, USA, 2013; pp. 49–52. [CrossRef]
26. Brereton, P.; Kitchenham, B.A.; Budgen, D.; Turner, M.; Khalil, M. Lessons from applying the systematic literature review process within the software engineering domain. *J. Syst. Softw.* **2007**, *80*, 571–583. [CrossRef]
27. Ahmad, S.R.; Bakar, A.A.; Yaakub, M.R. A review of feature selection techniques in sentiment analysis. *Intell. Data Anal.* **2019**, *23*, 159–189. [CrossRef]
28. Liu, B.; Liu, B. The Problem of Sentiment Analysis. In *Sentiment Analysis*; Cambridge University Press: Cambridge, UK, 2015. [CrossRef]
29. Moralwar, S.B.; Deshmukh, S.N. Different Approaches of Sentiment Analysis. *Int. J. Comput. Sci. Eng.* **2015**, *3*, 160–165.
30. Maetschke, S.R.; Madhamshettiwar, P.B.; Davis, M.J.; Ragan, M.A. Supervised, semi-supervised and unsupervised inference of gene regulatory networks. *Brief. Bioinform.* **2014**, *15*, 195–211. [CrossRef] [PubMed]
31. Guntuku, S.C.; Yaden, D.B.; Kern, M.L.; Ungar, L.H.; Eichstaedt, J.C. Detecting depression and mental illness on social media: An integrative review. *Curr. Opin. Behav. Sci.* **2017**, *18*, 43–49. [CrossRef]
32. Wang, Y.P.; Gorenstein, C. Assessment of depression in medical patients: A systematic review of the utility of the Beck Depression Inventory-II. *Clinics* **2013**, *68*, 1274–1287. [CrossRef]
33. Gottlieb, L.; Waitzkin, H.; Miranda, J. Depressive symptoms and their social contexts: A qualitative systematic literature review of contextual interventions. *Int. J. Soc. Psychiatry* **2011**, *57*, 402–417. [CrossRef]
34. Moher, D.; Liberati, A.; Tetzlaff, J.; Altman, D.G.; Altman, D.; Antes, G.; Atkins, D.; Barbour, V.; Barrowman, N.; Berlin, J.A.; et al. Preferred reporting items for systematic reviews and meta-analyses: The PRISMA statement. *PLoS Med.* **2009**, *6*, e1000097. [CrossRef]
35. Leis, A.; Ronzano, F.; Mayer, M.A.; Furlong, L.I.; Sanz, F. Detecting Signs of Depression in Tweets in Spanish: Behavioral and Linguistic Analysis. *J. Med. Internet Res.* **2019**, *21*, e14199. [CrossRef] [PubMed]
36. Kr, P. Neural Network Based System to Detect Depression in Twitter Users via Sentiment Analysis. *IRJET* **2018**, *5*, 1449–1451.
37. Shen, G.; Jia, J.; Nie, L.; Feng, F.; Zhang, C.; Hu, T.; Chua, T.; Zhu, W. Depression Detection via Harvesting Social Media: A Multimodal Dictionary Learning Solution. In Proceedings of the Twenty-Sixth International Joint Conference on Artificial Intelligence (IJCAI-17), Melbourne, Australia, 19–25 August 2017; pp. 3838–3844.
38. Chen, X.; Sykora, M.D.; Jackson, T.W.; Elayan, S. What about Mood Swings. In *WWW '18: Companion Proceedings of the the Web Conference 2018*; Association for Computing Machinery (ACM): New York, NY, USA, 2018; pp. 1653–1660. [CrossRef]
39. Arora, P.; Arora, P. Mining Twitter Data for Depression Detection. In Proceedings of the 2019 International Conference on Signal Processing and Communication (ICSC), Noida, India, 7–9 March 2019; pp. 186–189. [CrossRef]
40. Biradar, A.; Totad, S.G. *Detecting Depression in Social Media Posts Using Machine Learning*; Springer: Singapore, 2019. [CrossRef]
41. Ma, L.; Wang, Z.; Zhang, Y. *Extracting Depression Symptoms from Social Networks and Web Blogs via Text Mining*; Lecture Notes in Computer Science; Including Subser. Lect. Notes Artif. Intell. Lect. Notes Bioinformatics, LNBI; Springer: Berlin/Heidelberg, Germany, 2017; Volume 10330. [CrossRef]
42. Nadeem, M. Identifying Depression on Twitter. *arXiv* **2016**, arXiv:1607.07384.
43. Yazdavar, A.H.; Mahdavinejad, M.S.; Bajaj, G.; Romine, W.; Sheth, A.; Monadjemi, A.H.; Thirunarayan, K.; Meddar, J.M.; Myers, A.; Pathak, J.; et al. Multimodal mental health analysis in social media. *PLoS ONE* **2020**, *15*, e0226248. [CrossRef] [PubMed]
44. Titla-Tlatelpa, J.D.; Ortega-Mendoza, R.M.; Montes-y-Gómez, M.; Villaseñor-Pineda, L. A profile-based sentiment-aware approach for depression detection in social media. *EPJ Data Sci.* **2021**, *10*, 54. [CrossRef]
45. Chiong, R.; Budhi, G.S.; Dhakal, S.; Chiong, F. A textual-based featuring approach for depression detection using machine learning classifiers and social media texts. *Comput. Biol. Med.* **2021**, *135*, 104499. [CrossRef] [PubMed]
46. Safa, R.; Bayat, P.; Moghtader, L. *Automatic detection of depression symptoms in twitter using multimodal analysis*; Springer: New York, NY, USA, 2021. [CrossRef]
47. Leiva, V.; Freire, A. Towards suicide prevention: Early detection of depression on social media. In *Lecture Notes in Computer Science*; Including Subser. Lect. Notes Artif. Intell. Lect. Notes Bioinformatics; Springer: Berlin/Heidelberg, Germany, 2017; pp. 428–436. [CrossRef]
48. Ríssola, E.A.; Bahrainian, S.A.; Crestani, F. Anticipating Depression Based on Online Social Media Behaviour. In *Lecture Notes in Computer Science*; Including Subser. Lect. Notes Artif. Intell. Lect. Notes Bioinformatics; Springer: Berlin/Heidelberg, Germany, 2019; pp. 278–290. [CrossRef]

49. Sadeque, F.; Xu, D.; Bethard, S. Measuring the latency of depression detection in social media. In *WSDM '18: Proceedings of the Eleventh ACM International Conference on Web Search and Data Mining*; Association for Computing Machinery, Inc.: New York, NY, USA, 2018; pp. 495–503. [CrossRef]
50. Tadesse, M.M.; Lin, H.; Xu, B.; Yang, L. Detection of depression-related posts in reddit social media forum. *IEEE Access* **2019**, *7*, 44883–44893. [CrossRef]
51. Wolohan, J.T.; Hiraga, M.; Mukherjee, A.; Sayyed, Z.A.; Millard, M. Detecting Linguistic Traces of Depression in Topic-Restricted Text: Attending to Self-Stigmatized Depression with {NLP}. In Proceedings of the First International Workshop on Language Cognition and Computational Models, Santa Fe, NM, USA, 20 August 2018; pp. 11–21. [CrossRef]
52. Burdisso, S.G.; Errecalde, M.; Montes-y-Gómez, M. A text classification framework for simple and effective early depression detection over social media streams. *Expert Syst. Appl.* **2019**, *133*, 182–197. [CrossRef]
53. Trotzek, M.; Koitka, S.; Friedrich, C.M. Utilizing Neural Networks and Linguistic Metadata for Early Detection of Depression Indications in Text Sequences. *IEEE Trans. Knowl. Data Eng.* **2020**, *32*, 588–601. [CrossRef]
54. Martínez-Castaño, R.; Pichel, J.C.; Losada, D.E. A big data platform for real time analysis of signs of depression in social media. *Int. J. Environ. Res. Public Health* **2020**, *17*, 4752. [CrossRef]
55. Tai, C.H.; Fang, Y.E.; Chang, Y.S. SOS-DR: A social warning system for detecting users at high risk of depression. *Pers. Ubiquitous Comput.* **2017**, *1*, 1–12. [CrossRef]
56. Katchapakirin, K.; Wongpatikaseree, K.; Yomaboot, P.; Kaewpitakkun, Y. Facebook Social Media for Depression Detection in the Thai Community. In Proceedings of the 2018 15th International Joint Conference on Computer Science and Software Engineering (JCSSE); Institute of Electrical and Electronics Engineers Inc, Piscataway, NJ, USA, 11–13 July 2018. [CrossRef]
57. Wongkoblap, A.; Vadillo, M.A.; Curcin, V. Predicting Social Network Users with Depression from Simulated Temporal Data. In Proceedings of the IEEE EUROCON 2019-18th International Conference on Smart Technologies; Institute of Electrical and Electronics Engineers Inc., Piscataway, NJ, USA, 1–4 July 2019; 2019. [CrossRef]
58. Wu, M.Y.; Shen, C.Y.; Wang, E.T.; Chen, A.L.P. A deep architecture for depression detection using posting, behavior, and living environment data. *J. Intell. Inf. Syst.* **2020**, *54*, 225–244. [CrossRef]
59. Yang, X.; Mcewen, R.; Robee, L.; Zihayat, M. International Journal of Information Management A big data analytics framework for detecting user-level depression from social networks. *Int. J. Inf. Manag.* **2020**, *54*, 102141. [CrossRef]
60. Aldarwish, M.M.; Ahmad, H.F. Predicting Depression Levels Using Social Media Posts. In Proceedings of the 2017 IEEE 13th International Symposium on Autonomous Decentralized System (ISADS), Bangkok, Thailand, 22–24 March 2017; pp. 277–280. [CrossRef]
61. Ophir, Y.; Asterhan, C.S.C.; Schwarz, B.B. Unfolding the notes from the walls: Adolescents' depression manifestations on Facebook. *Comput. Hum. Behav.* **2017**, *72*, 96–107. [CrossRef]
62. Ricard, B.J.; Marsch, L.A.; Crosier, B.; Hassanpour, S. Exploring the Utility of Community-Generated Social Media Content for Detecting Depression: An Analytical Study on Instagram. *J. Med. Internet Res.* **2018**, *20*, e11817. [CrossRef]
63. Reece, A.G.; Danforth, C.M. Instagram photos reveal predictive markers of depression. *EPJ Data Sci.* **2017**, *6*, 15. [CrossRef]
64. Mann, P.; Paes, A.; Matsushima, E.H. See and Read: Detecting Depression Symptoms in Higher Education Students Using Multimodal Social Media Data. *arXiv* **2020**, arXiv:1912.01131.
65. Yueh, C.; Hsien, C.; Lane, Y.; Ling, J.; Arbee, K. Available online: 10.1007/s10844-020-00599-5 (accessed on 20 May 2020).
66. Li, A.; Jiao, D.; Zhu, T. Detecting depression stigma on social media: A linguistic analysis. *J. Affect. Disord.* **2018**, *232*, 358–362. [CrossRef]
67. Yu, L.; Jiang, W.; Ren, Z.; Xu, S.; Zhang, L.; Hu, X. Detecting changes in attitudes toward depression on Chinese social media: A text analysis. *J. Affect. Disord.* **2021**, *280*, 354–363. [CrossRef]
68. Oh, J.; Yun, K.; Maoz, U.; Kim, T.S.; Chae, J.H. Identifying depression in the National Health and Nutrition Examination Survey data using a deep learning algorithm. *J. Affect. Disord.* **2019**, *257*, 623–631. [CrossRef]
69. Damashek, M. Gauging Similarity with Categorization of Text. *Data Min. Introd. Adv. Top.* **1994**, *23*, 843–848.
70. Ramirez-esparza, N.; Chung, C.K.; Kacewicz, E.; Pennebaker, J.W. *The Psychology of Word Use in Depression Forums in English and in Spanish: Testing Two Text Analytic Approaches*; Association for the Advancement of Artificial Intelligence: Menlo Park, CA, USA, 2008; pp. 102–108.
71. Lovins, B. Development of a Stemming Algorithm *. *Mech. Transl. Comput. Linguist.* **1968**, *11*, 22–31.
72. Burrell, J. How the machine 'thinks': Understanding opacity in machine learning algorithms. *Big Data Soc.* **2016**, *3*, 1–12. [CrossRef]
73. Baharudin, B.; Lee, L.H.; Khan, K. A Review of Machine Learning Algorithms for Text-Documents Classification. *J. Adv. Inf. Technol.* **2010**, *1*, 4–20. [CrossRef]
74. Batta, M. Machine Learning Algorithms—A Review. *Int. J. Sci. Res.* **2020**, *9*, 381–386. [CrossRef]
75. Ray, S. A Quick Review of Machine Learning Algorithms. In Proceedings of the 2019 International Conference on Machine Learning, Big Data, Cloud and Parallel Computing (COMITCon), Faridabad, India, 14–16 February 2019; pp. 35–39. [CrossRef]
76. Tausczik, Y.; Pennebaker, W. The Psychological Meaning of Words: LIWC and Computerized Text Analysis Methods. *J. Lang. Soc.* **2010**, *29*, 24–54. [CrossRef]
77. Zhang, P. Model selection via multifold cross validation. *Ann. Stat.* **1993**, *21*, 299–313. [CrossRef]

78. He, B.; Ounis, I. Term frequency normalisation tuning for BM25 and DFR models. In *Advances in Information Retrieval*; Lecture Notes in Computer Science; Springer: Berlin/Heidelberg, Germany, 2005; Volume 3408, pp. 200–214. [CrossRef]
79. Donner, A.; Eliasziw, M.; Klar, N. Testing the Homogeneity of Kappa Statistics. *Biometrics* **1996**, *52*, 176. [CrossRef]

healthcare

MDPI

Review

Current and Future Applications of Artificial Intelligence in Coronary Artery Disease

Nitesh Gautam [1], Prachi Saluja [1], Abdallah Malkawi [2], Mark G. Rabbat [3], Mouaz H. Al-Mallah [4], Gianluca Pontone [5], Yiye Zhang [6], Benjamin C. Lee [7] and Subhi J. Al'Aref [2,*]

1 Department of Internal Medicine, University of Arkansas for Medical Sciences, Little Rock, AR 72205, USA; ngautam@uams.edu (N.G.); psaluja@uams.edu (P.S.)
2 Department of Medicine, Division of Cardiology, University of Arkansas for Medical Sciences, Little Rock, AR 72205, USA; amalkawi@uams.edu
3 Division of Cardiology, Loyola University Medical Center, Chicago, IL 60153, USA; mrabbat@lumc.edu
4 Houston Methodist DeBakey Heart & Vascular Center, Houston, TX 77030, USA; mouaz74@gmail.com
5 Centro Cardiologico Monzino IRCCS, 20138 Milan, Italy; gianluca.pontone@ccfm.it
6 Department of Population Health Sciences, Weill Cornell Medicine, New York, NY 10021, USA; yiz2014@med.cornell.edu
7 Department of Radiology, Dalio Institute of Cardiovascular Imaging, New York-Presbyterian Hospital and Weill Cornell Medicine, New York, NY 10021, USA; bcl2004@med.cornell.edu
* Correspondence: sjalaref@uams.edu

Abstract: Cardiovascular diseases (CVDs) carry significant morbidity and mortality and are associated with substantial economic burden on healthcare systems around the world. Coronary artery disease, as one disease entity under the CVDs umbrella, had a prevalence of 7.2% among adults in the United States and incurred a financial burden of 360 billion US dollars in the years 2016–2017. The introduction of artificial intelligence (AI) and machine learning over the last two decades has unlocked new dimensions in the field of cardiovascular medicine. From automatic interpretations of heart rhythm disorders via smartwatches, to assisting in complex decision-making, AI has quickly expanded its realms in medicine and has demonstrated itself as a promising tool in helping clinicians guide treatment decisions. Understanding complex genetic interactions and developing clinical risk prediction models, advanced cardiac imaging, and improving mortality outcomes are just a few areas where AI has been applied in the domain of coronary artery disease. Through this review, we sought to summarize the advances in AI relating to coronary artery disease, current limitations, and future perspectives.

Keywords: artificial intelligence; coronary artery disease; major adverse cardiovascular events; fractional flow reserve; cardiac computed tomography

Citation: Gautam, N.; Saluja, P.; Malkawi, A.; Rabbat, M.G.; Al-Mallah, M.H.; Pontone, G.; Zhang, Y.; Lee, B.C.; Al'Aref, S.J. Current and Future Applications of Artificial Intelligence in Coronary Artery Disease. *Healthcare* **2022**, *10*, 232. https://doi.org/10.3390/healthcare10020232

Academic Editor: Mahmudur Rahman

Received: 14 December 2021
Accepted: 24 January 2022
Published: 26 January 2022

Publisher's Note: MDPI stays neutral with regard to jurisdictional claims in published maps and institutional affiliations.

1. Introduction

Clinically significant atherosclerosis of the coronary arteries, known as coronary artery disease (CAD), is an endemic condition that is associated with significant morbidity and mortality [1]. For instance, CAD is reported to have affected 20.1 million American adults between 2015 and 2018 [2]. Current societal guidelines emphasize the importance of early detection and risk stratification in the appropriate age and risk groups, with the goal of implementation of goal-directed medical therapies that can alter the natural trajectory of CAD to a less morbid course [3]. Traditional population-derived primary and secondary prevention cardiovascular risk assessment tools (e.g., Framingham risk score, ASCVD, TIMI score, GRACE score, etc.) have historically relied on patient-level data that are easily retrievable and practical to utilize in the clinical setting. Despite their importance, such tools are inherently limited by design due to relying on regression models that make many mathematical assumptions that often do not hold in a real-world setting, such as collinearity between variables and homogeneity of effects. The complex nature and multifactorial pathology of CAD make such regression-based tools less generalizable across different populations.

Recently, the digitization of health records has improved access to large repositories of clinical and imaging datasets for clinical care and research purposes. This is coupled with advances in diagnostic tools that are available for the detection and quantification of CAD. To that end, recent studies have highlighted the usefulness of these tools in enhancing risk assessment and decision making through incorporation of different yet complementary findings from these imaging modalities (e.g., quantitative and qualitative plaque features on computed tomographic imaging of the coronary circulation, coupled with functional and physiologic findings on stress-test imaging). In addition, there has been an increasing interest in using the plethora of data in electronic health records and genomic data for better risk assessment [4]. Such tools are being integrated in practice as complementary methods to traditional tools [5–7].

Yet, despite the ever-increasing amounts of data, risk-prediction methods have been historically limited by what was possible with traditional statistical tools. The concept of Artificial Intelligence (AI) was introduced to mankind as early as the 1950s, with its employment in medical sciences commencing in the 1970s [8]. AI has gained momentum recently, fueled by an improvement in computational power, accumulation of data, and cloud processing. With the attempt to transfer a significant portion of human intelligence to machines, there has been a concerted effort aimed at harnessing the power of AI for biomedical applications in the past two decades [9,10]. Machine learning (ML) is a subfield of AI that involves the creation of algorithms that analyze large datasets without prior assumptions and learn rules and patterns between variables to make predictions and classifications [11]. On the other hand, deep learning (DL) is a subset of ML geared towards image analysis and utilizes more intricate algorithms known as neural networks with multiple deep, hidden layers. Specifically, while ML usually relies on structured data with handcrafted features often in tabular form, DL algorithms can input both structured and raw, unstructured data (e.g., images, video, and text) and extract their own features.

ML algorithms can incorporate a larger number of variables from different modalities, including both patient-level clinical parameters as well as two- and three-dimensional imaging data that take into account the multidimensional nonlinear interactions between variables [11]. Implementing such techniques in healthcare mainly aims to improve the accuracy of risk prediction and customize clinical decisions to each individual, which is the overarching theme in the goal of achieving precision medicine. In this paper, we summarize the recent advances in ML and current attempts at improving predictive analytics with relevance to CAD. We also elucidate on the role of AI in genetics, the incremental role of AI in improving post-procedure risk prediction and long-term mortality. Lastly, we discuss the limitations and potential near-future applications of AI within cardiovascular medicine.

2. Integration of Genetics and AI in Cardiovascular Diseases

Over the last two decades, the emergence of technologies able to measure biological processes at a large scale have resulted in an enormous influx of data. For instance, the completion of the Human Genome Project has paved the way to design single-nucleotide polymorphism (SNP) and mRNA microarrays, which can broadly test for millions of genetic variants in a simple point-of-care test. This has paved the way for the emergence of modern data-driven sciences such as genomics and other "omics" [12]. Genome-wide association studies (GWASs) operate by simultaneous comparison of millions of SNPs between diseased individuals and disease-free controls to detect a statistically significant association between an SNP locus and a particular condition [12]. Erdmann et al. reported that up until the year 2018, GWASs have successfully identified 163 distinct genetic loci for SNPs that are associated with CAD [13]. The risk for expressing a complex trait like CAD can be represented by a mathematical model that assumes a normal distribution of a binary outcome (i.e., CAD or no CAD) and captures the aggregate influence of multiple genetic variants that are predisposed to disease. Such a model is referred to as a polygenic risk score (PRS). PRSs were proposed early on to improve risk stratification in CAD risk models, especially when combined with traditional cardiovascular risk factors. However, the complex genetic architecture along with the multifactorial nature of CAD have been major challenges in CAD risk prediction [14]. For instance, Kathiresan et al. built a genetic risk score to predict major adverse cardiovascular events based on nine different dyslipidemia-related SNPs previously identified in GWASs. Adding the genetic score to a Cox proportional hazard model along with traditional risk factors did not improve predictive accuracy as measured by the C statistic model; however, there was a significant improvement in the net reclassification index, which accounts for correct movement of categories (assigning high-risk for patients who developed the disease, and low-risk for those who were disease-free) [15]. Brautbar et al. also suggested a genetic risk score to predict coronary heart disease based on SNPs. Adding the genetic risk score to traditional risk factors in a Cox proportional hazard model only modestly improved the area under the curve (AUC) for prediction of coronary heart disease from 0.742 to 0.749 ($\Delta = 0.007$; 95% CI, 0.004–0.013) [16]. ML and particularly DL algorithms are inherently designed to extract patterns and associations from large-scale data, including clinical and genomic data. Given the complexity and multifaceted nature of cardiovascular diseases in general, and CAD in particular, an approach that integrates all these factors into a risk-stratification model would be expected to better predict incident events than existent models [17].

Multiple studies have emphasized the role of ML in identifying genetic variants and expression patterns associated with CAD from mRNA arrays using differential expression analysis and protein–protein interaction networks [18,19]. For example, Zhang et al. used ML to perform differential expression analysis on mRNA profiles from CAD patients and healthy controls to identify a set of differentially expressed genes between the two groups, then built a network representation of functional protein–protein interaction. The top 20 genes in the network were identified using a maximal clique centrality (MCC) algorithm. Finally, to test the performance, a logistic regression model was built using the top five predictor genes to classify individuals into the presence or absence of CAD. The model achieved an AUC of 0.9295 and 0.8674 in the training and internal validation sets respectively [20].

Dogan et al. built an ensemble model of eight random-forest (RF) classifiers to predict the risk of symptomatic CAD using genetic and epigenetic variables along with clinical risk factors. The model was trained on a cohort derived from the Framingham heart study ($n = 1545$) and utilized variables derived from genome-wide array chips to extract epigenetic (DNA methylation loci) and genetic (SNP) profiles. The initial number of available variables were 876,014 SNP and DNA methylation (CpG) loci, which required multiple reduction steps, ending up with 4 CpG and 2 SNP predictors fed into the model along with age and gender. The model predicted symptomatic CAD with an accuracy, sensitivity, and specificity of 0.78, 0.75, and 0.80, respectively, in the internal validation cohort ($n = 142$). For comparison, a similar ensemble model was built using clinical risk factors only as predictor variables and had an accuracy, sensitivity, and specificity of 0.65, 0.42, and 0.89, respectively [21]. Pattarabanjird et al. tested multiple ML models to predict anatomical CAD severity (extent of diameter stenosis) in a binary fashion using clinical variables along with SNP loci. Quantitative coronary angiography and the Gensini score, which is a summation score that quantifies the severity of CAD by accounting for the segment-based most severe stenosis and the location of the stenosis within the coronary arteries, were used to assess model performance. The best-performing model (Sequential Neural Network; training set $n = 325$ and internal validation set $n = 82$) accurately classified CAD severity with AUC of 0.84 in the validation set [22]. Similarly, Naushad et al. trained ML models to predict the presence of CAD and the percentage of coronary diameter stenosis using clinical and genetic variables. The best-performing model (an ensemble model; training set $n = 648$) accurately predicted CAD using 11 variables (clinical and genetic variants) with an AUC of 0.96 in the training set. The model also predicted the percentage of diameter stenosis with a correlation of 82.5% with the actual stenosis assessed using the gold-standard invasive angiography. However, these models were not internally nor externally validated [23].

Finally, the coronary artery calcium (CAC) score, calculated using the Agatston method on noncontrast ECG-gated cardiac computed tomography, is an established strong predictor of major adverse cardiovascular events in asymptomatic individuals. Genomic studies have previously focused on identifying genetic loci linked to CAC [24,25]. Oguz et al. suggested the use of ML algorithms to predict advanced CAC from SNP arrays and clinical variables. They identified a set of SNPs that ranked the highest in predictive importance and correlated with advanced CAC scores, defined as the 89th–99th percentile CAC scores in the derivation and replication cohorts, and trained different RF models to predict advanced CAC scores using clinical and genetic variables. Adding SNPs to clinical variables significantly improved AUC from 0.61 ± 0.02 to 0.83 ± 0.01; ($p < 0.001$) [26] for prediction of advanced CAC scores.

3. Risk Prediction Models and Imaging Modalities for Estimating Pretest Probability of CAD

Traditionally, stratifying patients presenting with stable chest pain using pretest probability (PTP) estimates of CAD has been commonly used to help with decision-making regarding downstream testing and the choice of an appropriate diagnostic modality. Historically, the Diamond–Forrester model—developed using age, sex, and chest pain characteristics—was used as a clinician's risk stratification tool in predicting the PTP of CAD [27]. However, numerous studies showed its limitation in overestimating PTP by approximately threefold, especially in women [28]. This led to the development of the updated Diamond–Forrester model (UDF) and the CAD consortium score [29–31]. These scores, incorporating demographic and clinical risk factors, have been proven to be better at predicting the risk of CAD. Therefore, improving the ability to predict CAD using more accurate risk-assessment modeling is imperative, given the potential to reduce downstream testing and associated costs. Using clinical and demographic features, ML models have been employed to estimate the PTP of CAD [32–34]. In a recent multicenter cross-sectional study, a deep neural network algorithm based on the facial profile of individuals was able to achieve a higher performance than traditional risk scores in predicting PTP of CAD (AUC for the ML model 0.730 vs. 0.623 for Diamond –Forrester and 0.652 for the CAD consortium, $p < 0.001$) [35]. Though the study is limited by the lack of external validity and low specificity (54%), such approaches can potentially lead to a paradigm change in CAD management by facilitating earlier detection and initiation of primary prevention using readily available parameters, such as an individual's facial profile.

When available, a CAC score has been shown to add to the PTP of CAD, with a CAC score of zero identifying low-risk patients who might not need additional testing [7,36]. ML models, combining clinical and imaging parameters, have been shown to have higher predictive power than traditional risk scores when predicting the PTP of obstructive CAD [37,38]. Al'Aref et al. included 25 clinical and demographic features to devise a ML model which, when combined with the Agatston CAC score, fared better than the ML model or CAD consortium score alone or in combination with the CAC score (AUC 0.881 for ML + CAC as compared to 0.866, 0.773, and 0.734 for the CAD consortium + CAC, ML model, and CAD consortium respectively, $p < 0.05$) [38]. As expected, CAC, age, and gender were the highest-ranked features in the model (Figure 1).

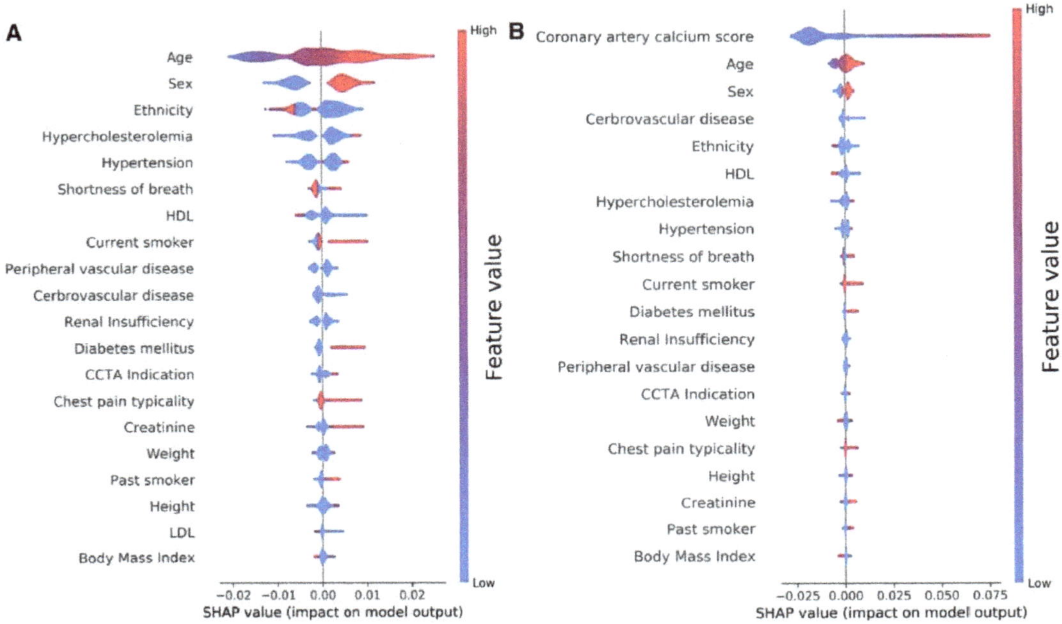

Figure 1. Feature ranking in the machine-learning model developed by Al'Aref et al. based on clinical and demographic factors (**A**) and when combined with the Agatston calcium score (**B**), for the prediction of the presence of obstructive CAD on coronary CT angiography. A more positive SHAP (Shapley additive explanation value) indicates higher importance of the variable in the machine-learning model. Adapted with permission from Al'Aref et al. [38], Oxford University Press.

Various ML algorithms based on stress imaging, particularly single-photon emission computed tomography (SPECT), have been devised to facilitate the prediction of CAD. These models combined the clinical and demographic characteristics with the quantitative variables, as evaluated via SPECT to better predict CAD compared with the visual interpretation or quantitative variables alone [39–44]. More details about the parameters used to develop these models have been provided in Section 4, and a summary of the study results is included in Table 1.

Cardiac phase-space analysis is a novel noninvasive diagnostic platform that combines advanced disciplines of mathematics and physics with ML [45]. Thoracic orthogonal voltage gradient (OVG) signals from a patient are evaluated by cardiac phase-space analysis to quantify physiological and mathematical features associated with CAD. The analysis is performed at the point of care without the need for a change in physiologic status or radiation. Initial multicenter results suggest that resting cardiac phase-space analysis may have comparable diagnostic utility to functional tests currently used to assess CAD [46].

Finally, the assessment of regional wall motion abnormalities (RWMAs) on echocardiography has been associated with the presence of obstructive CAD, and as such can be useful in helping clinicians with downstream decision-making [47]. Recently, a deep-learning model developed by Kusunose et al. achieved performance similar to that of experienced cardiologists in the assessment of RWMAs on echocardiography (AUC of 0.99 vs. 0.98, $p = 0.15$) [48]. Other than the assessment of obstructive CAD, machine learning has found its wide applicability in echocardiography to predict ventricular capacities, abnormal valvular function, as well as cardiac hemodynamics, the discussion of which is outside the scope of this review paper [49–51].

Table 1. Studies comparing ML models developed using SPECT variables with those using the qualitative or quantitative variables for prediction of CAD.

Study	Center/Sample Size	ML Technology	Brief Description and Outcomes	Result	Limitations
Guner et al. [41] 2010	Retrospective Single-center study 243 patients	Artificial neural networks	ML model trained from image data from stress and difference (devised from rest and stress maps) polar maps. Outcome: ML model vs. expert interpretation in the prediction of obstructive (>70% stenosis) CAD	AUC 0.74 and 0.84 for ML and expert read, no statistical difference found between ML-trained model and expert read.	1. Small sample size 2. Limited availability of software used.
Arsanjani et al. [44] 2013	Retrospective Single-center study 1181 patients	Boosted ensemble	ML model using quantitative variables (TPD, stress/rest perfusion change, TID) and clinical variables (age, sex, and post-ECG probability) created. Outcome: ML vs. visual analysis and TPD in prediction of obstructive CAD.	AUC: ML (quantitative + clinical – 0.94) > ML (quantitative, 0.90) > combined supine/prone TPD – 0.88. Also, better than experts (0.89 and 0.85 for two different experts).	1. Dual isotope imaging protocol used, leading to difficulty in comparing rest and stress images. 2. No information was given on localization of ischemia (didn't provide information about the culprit vessel).
Arsanjani et al. [39] 2013	Retrospective Single-center study 957 patients with no history of CAD.	Support vector machines	ML model using quantitative and functional variables derived from SPECT. Outcome: ML model vs. quantitative and visual analysis in prediction of obstructive CAD or LAD stenosis > 50%.	AUC: ML (0.92) > TPD (0.90) > Expert analysis (0.88 and 0.87 for two different experts)	1. Limited generalizability (patients with a history of CAD and valvular disease were excluded). 2. Stenosis on CAG determined qualitatively rather than quantitatively.
Betancur et al. [43] 2018	Retrospective Multicenter study 1638 patients	Convolutional neural networks	DL model developed from single-view polar maps; trained and compared with TPD for prediction of CAD. Outcome: ML model vs. TPD for prediction of obstructive CAD.	DL > TPD on per patient (AUC 0.80 vs. 0.78) and per vessel level (AUC 0.76 vs. 0.73) for prediction of obstructive CAD, $p < 0.01$.	1. Stenosis on CAG determined qualitatively rather than quantitatively. 2. Only stress static images used to train the algorithm.
Betancur et al. [40] 2018	Retrospective Multicenter study 1160 patients with no history of CAD	Convolutional neural networks	DL model developed to automatically combine upright and supine MPI polar maps. Outcome: ML model vs. TPD for prediction of obstructive CAD.	DL > TPD on per patient (AUC 0.81 vs. 0.78) and per vessel (AUC 0.77 vs. 0.73) for prediction of obstructive CAD, $p < 0.001$	1. Stenosis on CAG determined visually. 2. Only stress MPI images were taken.
Rahmani et al. [42] 2019	Retrospective Single-center study 93 patients	Artificial neural networks	ML model created using clinical, demographic, and polar-map data. Outcome: ML model vs. expert interpretation in prediction of obstructive CAD and abnormal angiographic results.	Accuracy for ML vs. visual interpretation for prediction of: Obstructive CAD:85.7% vs. 65.0% Abnormal angiographic results: 92.9 % vs. 81.7%	1. Small sample size 2. Patients with a high pretest probability included, hence possible over- and underestimation of sensitivity and specificity respectively.

CAG: coronary angiography; LAD: left anterior descending; MPI: myocardial perfusion imaging, TPD: total perfusion deficit, TID: transient ischemic dilation.

4. Artificial Intelligence in Management of CAD in the Emergency Department

Chest pain is a common emergency department presentation, and distinguishing cardiac from noncardiac pain causes is crucial for optimal management. Modalities such as electrocardiography (ECG) serve as a quick way to recognize patterns associated with unstable CAD, and in particular acute coronary syndromes (ACSs). Deep neural networks have shown a consistent performance in image recognition, and models have hence been devised to identify patterns related to CAD and myocardial infarction (MI) [52–54]. By reducing interobserver variability and providing accurate results efficiently, this approach holds the promise of improving workflow across healthcare systems, while helping patients in areas of limited medical infrastructure and specialized care.

Cardiac biomarkers, such as high-sensitivity troponin, have been well-validated as markers of myocardial ischemia and damage [55]. High-sensitivity troponin I (hs-cTnI) assay forms the core of the 'rule in and rule out' clinical decision pathway as per ESC 2020 chest pain guidelines and 2021 ACC/AHA chest pain guidelines [36,56]. For instance, a very low hs-cTnI at hospital admission or a negative one-hour delta troponin (in the background of a low hs-cTnI value at admission) has a high negative predictive value (>99%) for ACS [57–61]. On the other hand, a high admission hs-cTnI value or a significant increase in values in an hour portends a high positive-predictive value (70–75%), warranting additional downstream testing [56,62,63].

Using the strategy mentioned above, approximately one-third of the patients fall in the 'indeterminate' zone. Diagnosis and management of this group is challenging, necessitating an approach based on clinical history, pre-existing risk factors, serial hs-cTnI trends, and further imaging. A recent ML model based on three clinical (age, sex, and prior percutaneous coronary intervention) as well as levels of three biomarkers (hs-cTnI, KIM-1, and adiponectin) demonstrated excellence in predicting obstructive CAD in the validation cohort (AUC 0.86 for prediction of >50% diameter stenosis) [64]. Notably, the model performed remarkably well in patients in the 'indeterminate' zone, with AUC of 0.88 and a positive predictive value of 93%, hence identifying patients who will benefit from further testing.

The 2021 American College of Cardiology/American Heart Association (ACC/AHA) chest pain guidelines advocate for the use of coronary CT angiography (CCTA) in intermediate-risk patients presenting with acute chest pain who either have no known history or a history of nonobstructive CAD (defined as coronary artery disease with less than 50% diameter stenosis) [36]. Given the ability of CCTA to accurately define coronary anatomy and extent/distribution of atherosclerotic plaque, it has been consistently shown to be a useful noninvasive imaging modality for patient selection, particularly for those who might require further invasive evaluation. However, interpretation of CCTA scans requires expertise and is time-intensive. Therefore, automatic interpretation of CCTA, which can lead to a significant reduction in the processing times, is highly desirable. ML algorithms have recently been developed, achieving a 70–75% reduction in reading time compared to that required for human interpretation (2.3 min for AI vs. 7.6–9.6 min for human readers). Though the model described performed slightly lower than highly experienced readers in interpreting CCTA (AUC 0.93 vs. 0.90 for human vs. AI, $p < 0.05$), when combined with low-experience human readers, it augmented the reader's ability to correctly reclassify obstructive CAD (per-vessel net reclassification index (NRI) 0.07, $p < 0.001$) [65]. In addition, ML has been applied for various segmentation and classification tasks on cardiac CT imaging, from automatic segmentation of calcified and noncalcified plaque to automated calculation of the Agatston CAC score, and finally quantification of cardiac structures on CT imaging (Figure 2) [66–73]. Therefore, the application of ML could provide reliable results in real time, while bridging the dearth of experts in low-resource settings.

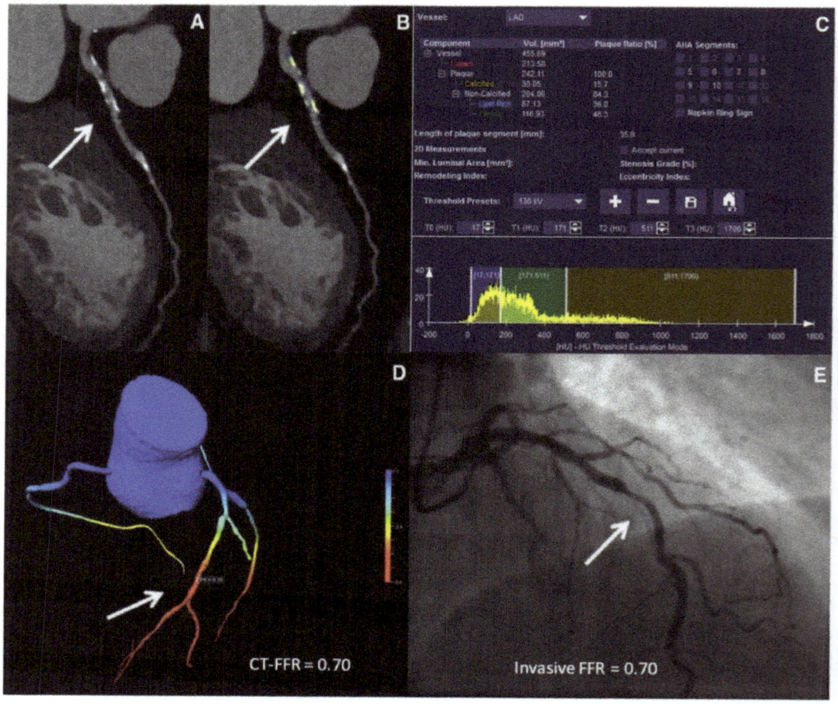

Figure 2. ML-based fractional flow reserve from cardiac CT (CT-FFR$_{ML}$). Machine-learning-based coronary plaque analysis quantifies atherosclerotic plaque into calcified and noncalcified components (**A**,**B**). This is further integrated with other quantitative parameters (**C**) and transformed into 3-D images of the vessels to give CT-FFR$_{ML}$ (**D**), which has been shown to have a good correlation with invasive fractional flow reserve (FFR—**E**). Adapted with permission from Von Knebel Doeberitz et al. [65], Elsevier.

Stress testing, which provides an estimate of myocardial perfusion and viability, has been recommended as an alternative to CCTA in intermediate-risk chest pain patients [36]. Myocardial perfusion imaging, particularly SPECT, has been employed to recognize patients who might need an invasive evaluation, with a diagnostic sensitivity of 75–88% and specificity of 60–79% [74–79]. SPECT can be evaluated qualitatively in terms of size, severity, location, and reversibility of perfusion defect, and quantitatively, in terms of total perfusion deficit (TPD), summed stress score (SSS), summed rest score (SRS), as well as stress and rest volumes [80]. Automatically generated polar maps (representing radiotracer distribution in a two-dimensional plane) after three-dimensional segmentation of the left ventricle (LV) have been used as raw data for quantitative analysis. After the LV polar map is divided into 17 segments, each of the segments is graded on a scale of 0–4 based on the severity of ischemia. The scores are then summated to generate SSS and SRS [81]. Polar maps also provide information about the overall extent and magnitude of ischemia, in terms of TPD [81,82]. These objective variables extracted from the quantitative analysis offer an increased degree of reproducibility and can be incorporated into risk scores to predict mortality [82,83]. The diagnostic accuracy of qualitative and quantitative approaches is comparable, as has been shown in numerous studies [84]. A deep convolutional neural network-based model derived from polar maps (Figure 3) had a superior performance compared to TPD in predicting obstructive coronary artery disease (the AUC for ML were 0.80 and 0.76 vs. 0.78 and 0.73 for TPD on a per-patient and per-vessel basis respectively, $p < 0.01$). In addition to diagnosis, models to predict early revascularization (<90 days from

SPECT) have been developed and have demonstrated better performance than individual SPECT variables on a per-patient and a per-vessel level [85,86].

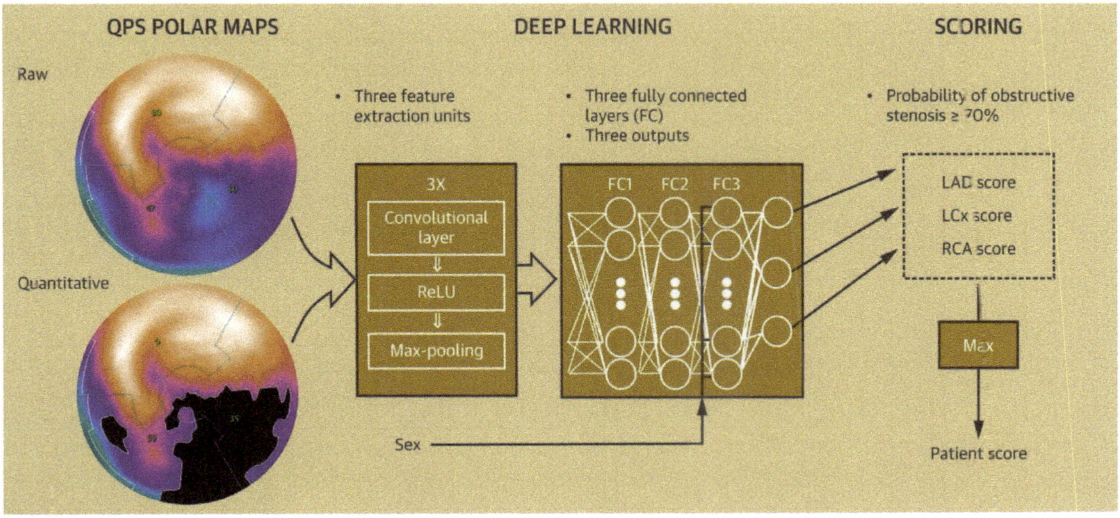

Figure 3. Deep-learning model to predict obstructive CAD from polar maps. Raw polar maps and extent polar maps (maps with abnormal pixels representing ischemia blackened out) are fed into deep neural networks, with the extracted data used to calculate scores for individual vessels to predict the probability of CAD. Adapted with permission from Betancur et al. [43], Elsevier.

5. Artificial Intelligence to Predict Functionally Obstructive CAD and Lesion-Specific Ischemia—As a Gatekeeper to the Catheterization Laboratory

One of the inherent limitations of CCTA is its limited ability to predict the functional significance of coronary stenosis. To overcome this shortcoming, CT-derived fractional flow reserve (FFR_{CT}) was developed based on the critical concept of computational fluid dynamics (CFD), with numerous trials demonstrating its strong correlation with invasive fractional flow reserve (FFR) as determined by invasive coronary angiography (ICA) [87–90]. Rabbat et al. demonstrated that FFR_{CT} added to CCTA safely deferred ICA in patients with CAD of indeterminate hemodynamic significance. In addition, a high proportion of those who underwent ICA were revascularized [91]. These studies and others led to FFR_{CT} being incorporated in the 2021 ACC/AHA chest pain guidelines in intermediate-risk patients to detect lesion-specific ischemia in proximal or middle segments of the coronary arteries and determined to have atherosclerotic plaque with 40% to 90% diameter stenosis [36]. Despite its excellent correlation, the off-site computation of FFR_{CT} hampers its use in real time, owing to the need for longer processing times [92]. To overcome this limitation and to allow for quick computation of a value for the functional significance of a particular lesion, novel ML approaches based on artery lumen segmentation [93], left ventricular myocardial segmentation [94,95], and artery centerline tracking [96], have been proposed.

5.1. ML-Based CT-FFR Estimation and Diagnostic Accuracy

Based on the concept of artery lumen segmentation, the ML-based FFR estimation (CT-FFR_{ML}) has generated significant interest in the past few years. The CT-FFR_{ML} model was trained on 12,000 synthetically generated coronary geometric datasets and used deep neural networks, allowing for automatic computation of FFR in real-time [93]. Coenen et al. performed a multicenter, prospective study to evaluate the diagnostic performance of CT-FFR_{ML} to predict lesion-specific ischemia, comparing it with traditional CCTA parameters, with invasive FFR being the gold standard [97]. They demonstrated an excellent correlation between CT-FFR_{ML} and FFR_{CT} (r = 0.997) and a superior performance of CT-FFR_{ML} over traditional CCTA in predicting lesion-specific ischemia (AUC: 0.84 vs. 0.69, $p < 0.001$ on a per-vessel level). Since then, multiple retrospective studies have been performed to evaluate the diagnostic accuracy of CT-FFR_{ML}, validated against the gold-standard invasive FFR. They have further demonstrated superior diagnostic performance of CT-FFR_{ML} over CTA stenosis severity and quantitative atherosclerotic plaque features derived from CCTA [70,93,97–107].

To further highlight the incremental diagnostic value of CT-FFR_{ML} over anatomic plaque features derived from CCTA in vessels with intermediate stenosis, several other studies have been performed [99,102,103,106]. Tang et al. evaluated the diagnostic value of CT-FFR_{ML} in predicting lesion-specific ischemia [103]. Based on a study sample of 122 vessels in 101 patients, CT-FFR_{ML} performed better than anatomic CCTA parameters (AUC 0.96 for CT-FFR_{ML} vs. 0.63 for CCTA on a per-vessel basis $p < 0.05$).

5.2. Impact of Calcification Burden on the Performance of CT-FFR_{ML}

The impact of coronary calcification on the diagnostic performance of CCTA has been well-established, with more extensive calcification limiting the ability of CCTA to evaluate for the presence of obstructive CAD [108–110]. Multiple indices have been devised to compute a CAC score, with the Agatston score, calcium volume, calcification remodeling index (CRI), and segmental arc calcification method being common examples [111]. The Agatston Score (AS) is the most widely validated approach, which summates the calcium score (function of peak density and area of the lesion) of the individual lesions across all coronary artery segments [112]. CRI provides a lesion-specific calcium estimate and is calculated as a ratio of the cross-sectional luminal area of the most severely calcified site to the proximal luminal area [113]. The segmental arc calcification method estimates lesion-specific calcium burden by measuring the greatest circumferential extent of coronary calcium, grading as nil (noncalcified), mild (0–90°), moderate (90–180°), and severe (>180°) calcification [110,114]. Recent studies have evaluated the performance of CT-FFR_{ML} with varying calcification burden as assessed by the parameters mentioned above [97–99,104]. Tesche et al. did a retrospective analysis using 482 vessels in 314 patients to evaluate the impact of calcifications on the performance of CT-FFR_{ML} [104]. They showed a statistically significant decrease in discriminatory power of CT-FFR_{ML}, measured in terms of AUC with increasing Agatston scores (AUC for CT-FFR_{ML} 0.85 and 0.81 in low–intermediate Agatston score (1–400) and high Agatston score (>400) ranges respectively, $p = 0.04$).

Di Jiang et al. [98] evaluated the impact of calcification arc and CRI on the performance of CT-FFR_{ML}. No statistically significant difference was found in the discriminatory power of CT-FFR_{ML} with increasing calcification burden. In the proportion of patients where the Agatston score was available, there was no difference in the diagnostic performance of CT-FFR_{ML} across severity of calcification. The difference from Tesche et al. can be explained by a lower mean Agatston score (288 vs. 492 and 138 vs. 187 at a per-patient and per-vessel level, respectively) and smaller sample size ($n = 150$) for whom the Agatston score was available, resulting in low power to detect a difference.

Furthermore, Koo et al. [99] carried out a similar study and found no impact of increasing Agatston score on the performance of CT-FFR_{ML}. Interestingly, a sizeable proportion of the sample had higher coronary calcification (mean Agatston score of 311 on a per-vessel basis). More research in this area is needed in order to further validate the diagnostic performance of CT-FFR_{ML} across varying degrees of coronary calcification.

5.3. CT-FFR_{ML} in Predicting Revascularization Events

CT-FFR_{ML} has been shown to be a better predictor than plaque features derived from CCTA for the determination of the presence of lesion-specific ischemia, but whether CT-FFR_{ML} influences the eventual treatment plan and outcomes (as guided by ICA-FFR) remains an active area of investigation [115–118]. Qiao et al. demonstrated the added benefit of CT-FFR_{ML} compared to relying on an anatomy-based strategy in patients with stable chest pain (reduction rate of ICA by 54.5% and 4.4% fewer revascularizations) [115]. Additionally, this study demonstrated that adding CT-FFR_{ML} to CCTA can decrease the rate of unnecessary ICA by 35.2% (thereby increasing the proportion of revascularizations when ICA is undertaken), truly acting as a gatekeeper to ICA. Furthermore, lower CT-FFR_{ML} was associated with higher major adverse cardiovascular event (MACE) risk when compared to diameter stenosis on CCTA (HR, 6.84 vs. 1.47) or ICA (HR, 6.84 vs. 1.84). Liu et al. found a similar rate of MACE (2.9%) after revascularization based on either combining CCTA stenosis $\geq$ 50% and CT-FFR_{ML} $\leq$ 0.8 or ICA stenosis $\geq$ 75% in a 2-year follow-up [116]. This study further highlighted the use of CT-FFR_{ML} as a gatekeeper to ICA with a positive impact on lower healthcare costs.

CT-FFR_{ML} comes with its own set of shortcomings. The diagnostic performance of the CT-FFR_{ML} model is lower, with the invasive FFR closely approaching the diagnostic threshold of 0.8 [97,99,119]. Traditional statistical and DL approaches have shown that stenosis severity; plaque characteristics, such as low-density, noncalcified plaque; and remodeling index are independent predictors of lesion-specific ischemia that are not related to CT-FFR_{ML} [120,121]. An integrated DL approach in the future that combines clinical features, anatomical plaque characteristics, vessel features, and functional assessment could potentially overcome this limitation.

6. Artificial Intelligence in the Field of Intracoronary Imaging

During ICA, intravascular ultrasound (IVUS) and optical coherence tomography (OCT) have been widely adopted for coronary luminal imaging, and some of the main applications involve assessment of plaque burden and optimization of stent placement [122]. IVUS uses ultrasound waves to generate cross-sectional images of coronary vessels with axial and lateral resolution ranging from 70–200 microns and 200–400 microns, respectively [122–124]. The penetration depth of IVUS is 10 mm, which allows for a complete cross-sectional analysis of the coronary vessel walls [124]. IVUS can help describe plaque characteristics, with high-risk plaques (plaques with large necrotic cores) appearing as areas of echo-attenuation [125]. On the other hand, calcifications in the IVUS frame indicate a calcified plaque, with heavily calcified plaque increasing the risk of stent underexpansion during percutaneous coronary intervention (PCI) [126,127]. Virtual histology IVUS (VH-IVUS) is another technique derived from radiofrequency data from IVUS, allowing for in vivo assessment of plaque composition [128]. By characterizing plaque features and vessel dimensions, IVUS has found its pre-procedural role in the quantitative and qualitative assessment of atherosclerotic plaque as well as interventional planning, ranging from vessel dimension assessment and evaluation of stent placement. Post-procedurally, IVUS can be employed to visualize stent expansion, identify stent edge dissection, stent mal-apposition, and confirm the presence of in-stent thrombosis in the right clinical context [129,130]. Given the benefits, IVUS has been shown to optimize stent implantation and improve outcomes, including revascularization, MACE, and mortality when used routinely in the cardiac catheterization laboratory [130–132].

On the other hand, OCT works on the principle of near-infrared light waves, generating cross-sectional images with a much higher axial and lateral resolution of 10 microns and 20–40 microns, respectively [133]. This allows for a detailed view of the lumen–plaque interface, providing accurate dimensions of the luminal area and better plaque characterization. The vulnerability of a plaque is a function of the thickness of its fibrous cap, the size of the necrotic core, and the presence of macrophages. A thin, fibrous cap; sizeable necrotic core; and increased macrophages increase the risk of plaque rupture and subsequent ACS [113]. Given the high resolution provided by OCT, it is considered a gold-standard invasive imaging modality for detecting thin-cap fibroatheroma (TCFA), which, pathologically, is a precursor of vulnerable plaque and clinically proven to be an independent predictor of MACE [134]. A significant drawback of OCT is its inherent low penetration depth (1–2 mm), which makes IVUS a better modality for a full-thickness analysis of vessel wall [130].

Though fascinating, IVUS and OCT have a low adoption rate in the US, being employed only at tertiary-care centers owing to cost, need for additional procedural time, and the associated technical complexities [135,136]. By using deep-learning algorithms to optimize the workflow associated with image acquisition and interpretation, ML has the potential to reduce procedural costs and time required, which are the two major hindrances to the widespread use of IVUS and OCT.

6.1. Artificial Intelligence to Optimize Peri-Intervention Workflow

To predict OCT-derived TCFA on IVUS images, Bae et al. created a ML model, enrolling 517 patients who underwent ICA [137]. A total of 40,908 IVUS-OCT co-registered sections in 517 coronary arteries were divided into training and testing sets in a ratio of 4:1. An artificial-neural-network-based model using 17 features achieved the highest performance with a sensitivity and specificity of 85 $\pm$ 4% and 79 $\pm$ 6%, respectively, and good discriminatory power (AUC of 0.80 $\pm$ 0.08). Larger plaque burden, minimal diameter, decreased lumen area, and increased lumen eccentricity were seen to be strongly associated with OCT-derived TCFAs. Min et al. utilized a deep learning algorithm (densely connected convolutional neural network) on 35,678 OCT frames to automatically detect TCFAs from OCT images [138]. After the frames were interpreted for the presence/absence of TCFA, data was fed into the algorithm to devise a deep-learning model. By achieving high sensitivity and specificity of 88.7 $\pm$ 3.4% and 91.8 $\pm$ 2.0% on the test data, such deep-learning models can significantly reduce processing times and allow for easy interpretation when it comes to identifying a vulnerable high-risk plaque.

As mentioned earlier, IVUS can help characterize high-risk plaques, which appear as areas of attenuation on IVUS frames due to the presence of a large necrotic core. Identifying such lesions becomes imperative to reduce the incidence of complications such as periprocedural MI. To accurately classify plaque characteristics and to facilitate detection of high-risk lesions, Cho et al. described a deep-learning algorithm to accurately differentiate IVUS segments as attenuated or calcified, or plaque without attenuation or calcification [139]. A total 598 vessels in 598 patients were evaluated, and a DL model with five-fold cross-validation was developed. The deep-learning model closely correlated with the expert read, and correlation coefficients for calcification, attenuation, and no attenuation or calcification were 0.79, 0.74, and 0.99, respectively (Figure 4).

Stent underexpansion is a frequently encountered entity that has been associated with an increased risk of in-stent restenosis. Studies have demonstrated the postprocedural minimum stent area (MSA) and IVUS-measured stent length to be independent predictors for in-stent restenosis [130,140–143]. Min et al. devised a deep-learning model to predict stent underexpansion based on pre-PCI IVUS frames [144]. They evaluated 618 coronary lesions from 618 patients undergoing pre- and postprocedural IVUS and divided them into training and testing sets in a 5:1 ratio. A convolutional neural network (CNN) model was used to predict the poststenting stent area. Features extracted from the CNN were combined with additional image-derived features via a boosted ensemble algorithm, which yielded sensitivity and specificity of 68% and 98%, respectively, and an AUC of 0.95 to predict stent underexpansion. The stent areas and volumes predicted via the CNN correlated well with poststenting IVUS (r for stent area and volume 0.832 and 0.958, respectively). The most important features predicting stent underexpansion were luminal area, external elastic membrane (EEM) area (both at the reference and the target), and plaque area of the region of interest.

6.2. Applications of Artificial Intelligence in Intra and Post-Intervention Workflow

Optimal stent expansion is vital to successful outcomes, with stent underexpansion predisposing to stent restenosis and a greater stent expansion exposing the procedure to a risk of stent edge dissection [130]. IVUS, by allowing direct visualization of vessel architecture, can help in the earlier identification and management of these complications. Nishi et al. developed a ML model to compute the luminal area and the vessel area accurately, as well as the stent area, which exhibited an excellent correlation between ML-derived and expert-derived dimensions while dramatically reducing the time required for segmentation of IVUS images (37 s) compared with expert analysis (30 h) [145].

Virtual histology IVUS (VH-IVUS) is a well-studied intracoronary imaging modality used for in vivo visualization of high-risk plaques [146–149]. Zhang et al. devised a deep-learning model to predict the location of high-risk plaques in nonculprit vessels in patients who underwent IVUS at baseline and after one year [150]. Though large-scale validation is required, the model predicted the occurrence of TCFAs, plaque burden >70%, and minimal luminal area $\leq$4 mm^2 reasonably well at a one-year follow-up on a per-lesion level.

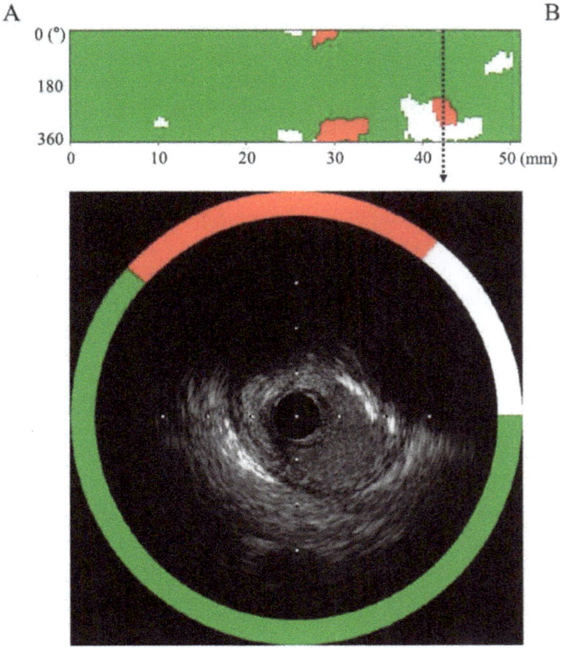

Figure 4. ML(**A**) vs. human (**B**) interpretations for plaque characterization for IVUS images. The upper panel shows representation of plaque features along the long axis of the vessel (x-axis represents the distance from ROI (region of interest) and y-axis represents the angular position (0–360°) of the plaque. The lower panel shows the plaque characterization on a cross-sectional view of the IVUS frame. Attenuation, calcification, and regions without attenuation or calcification are represented by red, white, and green respectively. Adapted with permission from Cho et al. [135], Elsevier.

7. Artificial Intelligence-Based Post-Procedure Risk Prediction Models

In addition to early detection and the institution of guideline-directed therapy in the appropriate risk strata, accurate prediction of unheralded adverse events forms the cornerstone for managing CAD. Identifying the high-risk target population can potentially provide a window for aggressive risk factor modulation, thereby reducing mortality and contributing towards better health at a population level. Multiple risk-prediction models have been developed to predict in-hospital mortality and the long-term risk of MACE in high-risk cohorts [151–158].

PCI is a relatively safe procedure, with a reported overall in-hospital mortality rate of 1–2% [159]. The risk of complications increases with increasing patient morbidity, with an incidence of technical difficulties and periprocedural complications 2.2 times higher than in the average population [160]. The Mayo clinic risk score (MCRS) and New York State risk score (NYSRS) were developed to predict in-hospital and 30-day mortality in patients undergoing PCI. Both scores performed equivalently well, showing an excellent discriminative ability to identify patients at a higher risk for in-hospital and 30-day mortality [161]. They employed regression-based models, assuming a linear interplay between patient variables and mortality outcomes. ML models have been recently developed to potentially uncover complex and nonlinear relationships between multiple factors, hence improving diagnostic accuracy over current models.

Zack et al. evaluated 11,709 patients to train two RF regression models—one using 52 demographic and clinical parameters to predict in-hospital mortality and the second model also incorporating 358 discharge variables in addition to the 52 admission parameters to predict 180-day cardiovascular mortality and 30-day heart failure rehospitalization [162]. They compared the model performances against logistic regression models trained using the same variables. No significant difference was found between the RF model and logistic regression in predicting in-hospital mortality (AUC 0.923 vs. 0.925, $p = 0.84$). The ML model performed significantly better than the logistic regression model for prediction of 30-day heart failure hospitalizations (AUC 0.899 vs. 0.846, $p = 0.003$) and 180-day cardiovascular death (AUC 0.881 vs. 0.812, $p = 0.02$).

Al'Aref et al. [163] developed a supervised machine learning approach to predict in-hospital mortality among patients undergoing PCI. Utilizing 479,804 patients from the New York state registry, they utilized 49 clinical, angiographic, and periprocedural event characteristics to create a ML model via adaptive boosting. It performed better than the logistic regression model (AUC 0.927 for ML vs. 0.908 for logistic regression, $p < 0.01$). Age and ejection fraction emerged as the most important variables predicting mortality.

Periprocedural bleeding is one of the most common complications of PCI and has been linked to adverse in-hospital outcomes [164,165]. Current risk scores such as the NCDR bleeding risk-prediction model and the simplified NCDR bleeding-risk score have performed modestly well in identifying patients at a high risk of periprocedural bleeding [166]. To improve the performance of the existing risk model, an ML-based model was developed on 3,316,465 patients enrolled in the CathPCI registry [167]. In addition to the 31 variables used in the existing model, 28 new variables were incorporated to devise an integrated model via the gradient-boosting approach. The blended model using ML had a higher discriminatory power than the existing model (C statistic 0.82 vs. 0.78, $p < 0.05$) and improved the positive predictive value to 26.6%, compared with 21.5% for the existent model.

One of the primary challenges faced in the PCI era is in-stent restenosis, which is linked to neointimal proliferation due to vascular wall damage [168]. The incidence of ISR has been estimated to be 20–40% for bare metallic stents and 10–15% for drug-eluting stents [168,169]. Smaller vessel size, increasing stent length, complex lesion morphology, diabetes mellitus, and prior bypass surgery are risk factors for stent restenosis [169]. These factors have been incorporated with other variables to devise risk models such as PRESTO 1, PRESTO 2, and EVENT scores to provide an estimated risk of ISR [170,171]. These models have a modest discriminatory power in predicting ISR, leaving room for improvement. A big-data approach incorporated 68 variables relating to clinical, demographic, and angiographic characteristics to devise a risk prediction model for ISR [172]. The ML model, when applied post-PCI, achieved a higher discriminatory power (AUC for the precision recall curve was 0.45 vs. 0.31, 0.27, and 0.18 for PRESTO-1, PRESTO-2, and EVENT, respectively, $p < 0.05$) to predict ISR at 12 months. Interestingly, post-PCI TIMI flow was one of the prominent predictors of ISR, alongside diabetes mellitus and the presence of $\geq$2 vessel CAD. Though the model requires external validation, given the small sample size of the population ($n = 263$), the study yet again underscores the merit of ML in identifying crucial parameters from a vast dataset to predict outcomes.

8. Artificial Intelligence-Based Long-Term Mortality and MACE Prediction Models

Prognostic modeling via ML has been validated with the use of electronic health records (EHRs) integrated with clinical scores and imaging modalities to predict MACE [173–175].

Utilizing the array of data available in EMR and identifying patterns based on clinical course, ML models have been used to create a personalized treatment algorithm (ML4CAD) for every patient, based on risk factors, past medical history, time present in the EMR system, and medications. The illustrated model makes clinical decisions for patients based on these factors and suggests a decision with an aim to increase prescription effectiveness, evaluated in the terms of time from initial diagnosis to the first potential adverse event (time to adverse event, TAE). The model had superior performance when compared to standard of care, increasing the time to adverse event (TAE) from 4.56 to 5.66 years (24.3% increase), hence furthering the idea of precision medicine [174,176].

Imaging findings, such as CAC score quantified from cardiac computed tomography, are an independent risk factor adding to the traditional clinical risk factors in predicting long-term risk of cardiovascular events [177–179]. Noncontrast CT imaging, other than providing information on the CAC score, provides valuable measures such as epicardial adipose tissue (EAT) volume, and EAT attenuation, all of which have been shown to provide additional information regarding the long-term risk of cardiovascular disease [180–182]. Extracting these pieces of data can be tedious and labor-intensive, and automated techniques can result in more standardized evaluations in a more time-efficient manner.

Multiple ML techniques have been proposed to automatically evaluate CAC score from dedicated cardiac and non-EKG gated chest CT scans [66,67,183–185]. ML techniques incorporating CAC score and other imaging parameters have been shown to be a better predictor than the traditional risk scores employed for cardiovascular disease risk stratification [181,186–189]. An ensemble-boosting model developed by Nakanishi et al. incorporating a total of 77 clinical and imaging variables had a superior discriminatory power for predicting coronary heart disease deaths than imaging and clinical data alone (AUC for ML model: 0.845 compared to 0.821 and 0.781 for clinical data and CAC respectively, $p < 0.001$) (Figure 5) [190].

Apart from CAC scoring and traditional CT metrics, the role of EAT volume and attenuation in the prediction of future cardiovascular risk has been an active area of research. Deep-learning approaches to automatically compute EAT volume and EAT attenuation from CT have been developed, significantly reducing generation time from 15 min to 2 s [186]. Eisenberg et al. demonstrated an independent association between deep-learning-derived EAT volume and attenuation with the risk of future MACE, defined as myocardial infarction, late (>180 days) revascularization, and cardiac death (HR:1.35, $p < 0.01$ and 0.83, $p = 0.01$, demonstrating a direct correlation with EAT volume and an inverse correlation with EAT attenuation respectively) [187]. Subsequently, these parameters have been combined with other physiologic and radiology variables to develop new deep-learning approaches, which have further been shown to have a higher predictive value than the traditional risk scores [186,189]. These have been summarized in Table 2.

Apart from its role in CAD diagnosis, CCTA has been shown to have an incremental prognostic value in terms of short- and long-term risk prediction. Results from the CONFIRM registry validated two CCTA parameters, namely the number of proximal segments with stenosis > 50% and the number of proximal segments with mixed or calcified plaque as important prognostic markers above the predictive value of the Framingham risk score (FRS) [191–193].

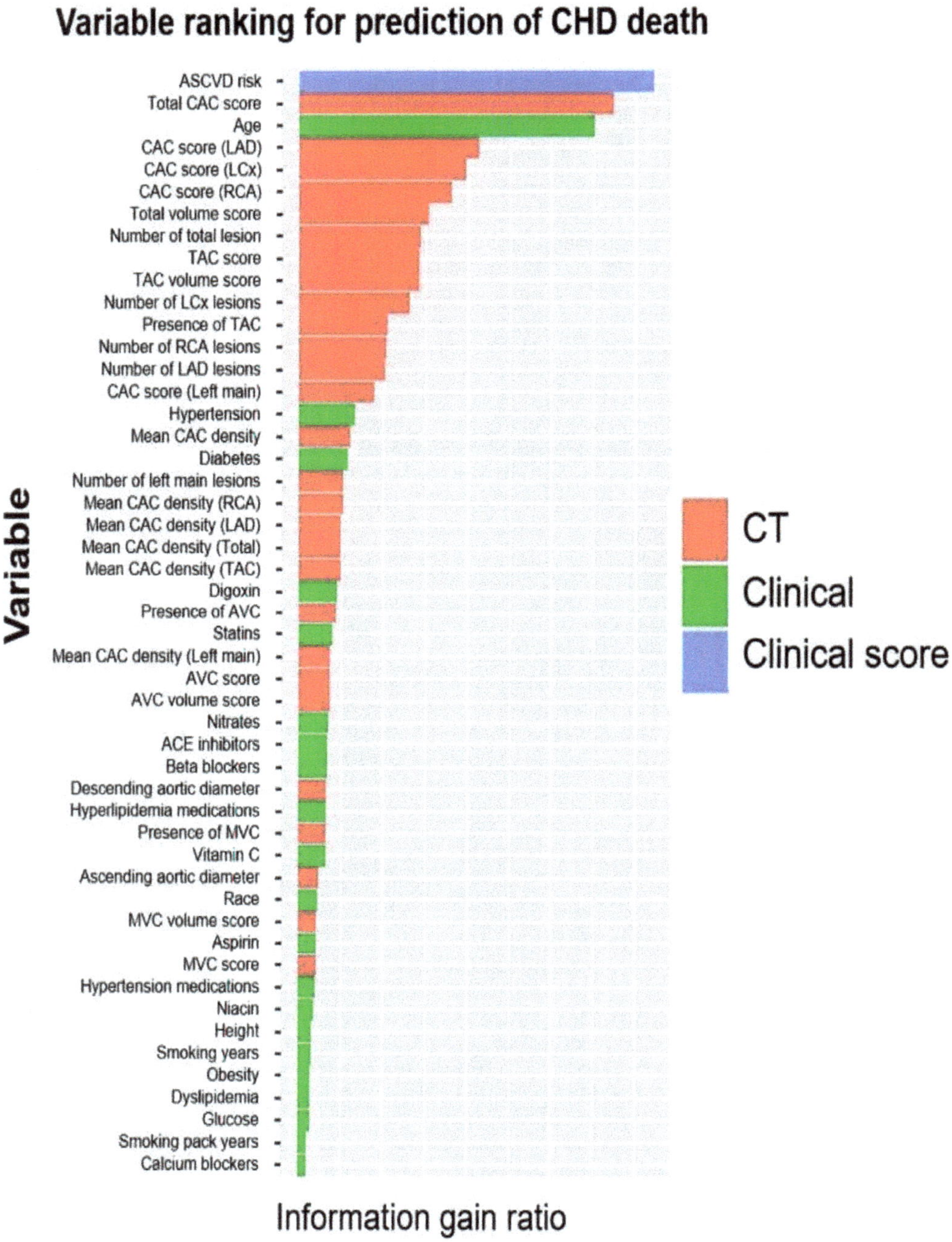

Figure 5. Variable importance as determined by the ML model for prediction of coronary heart disease deaths. Abbreviations: CAC: coronary artery calcium; TAC: thoracic aortic calcification; AVC: aortic valve calcification; MVC: mitral valve calcifications; LAD: left anterior descending; LCx: left circumflex RCA: right coronary artery. Adapted with permission from Nakanishi et al. [190], Elsevier.

Table 2. Studies evaluating the impact of coronary artery calcium score (CACS) among other variables in the prediction of mortality in patients with no history of coronary artery disease.

Study	Study Design/Sample Size	ML Model	Brief Description and Follow-Up	Results	Limitations
Eisenberg et al. [187] 2020	Prospective single-center study, 2068 asymptomatic patients	Convolutional neural network	To check for impact of EAT volume and EAT attenuation computed via deep learning in prediction of MACE, defined as defined as MI, late (>180 days) revascularization and cardiac death. Follow up: >14 years	Increased EAT volume (HR: 1.35) and decreased EAT attenuation (HR 0.83) independently associated with MACE in addition to CACS (HR 1.25) and ASCVD score (HR 1.03), $p < 0.01$ for all.	1. Study done on asymptomatic patients; external validation needed if applied on symptomatic patients. 2. Previous-generation CT scanners used (data collected from 1998–2005).
Han et al. [188] 2020	Retrospective multicenter study, 86,155 asymptomatic patients	Boosted ensemble	ML model with 35 clinical, 32 lab, and 3 CACS parameters (CACS, calcium volume, and calcium mass) in prediction of all-cause mortality Median follow up: 4.6 years	ML (0.82) > ASCVD score + CACS (0.74) > Framingham risk score + CACS (0.70)—reported as AUC in the test set. No statistical difference in the performance in the validation set.	1. Retrospective 2. All-cause mortality reported rather than specific cardiac endpoints.
Nakanishi et al. [190] 2021	Multicenter observational study, 66,636 asymptomatic patients	Boosted ensemble (Logitboost)	ML model incorporating 46 clinical and 31 CT variables—CAC score, extra coronary scores (not including EAT) in prediction of cardiovascular (CHD + stroke + CHF + other circulatory diseases), and coronary heart disease (CHD) deaths Follow up: 10 years	1. For cardiovascular deaths: AUC for ML (all) 0.845 > ASCVD (0.821) > CAC score (0.78). 2. For coronary heart disease deaths: AUC for ML (all) 0.860 > ASCVD (0.835) > CAC score (0.816).	1. Multiple CT variables, including EAT, were not available for some patients.
Commandeur et al. [186] 2020	Prospective single-center study, 1912 asymptomatic patients	Boosted ensemble (XgBoost)	ML model using clinical variables, plasma lipid panel measurements, CAC, aortic calcium, and automated EAT measures in prediction of MI and cardiac deaths. Median follow up: 14.5 years	1. ML model 0.82 > ASCVD risk score 0.77 ~ CAC 0.77. 2. Age, ASCVD risk score, and CACS were the three most important features seen in the model.	1. Overfitting; since small number of events (<4%). 2. Study done on asymptomatic patients; external validation needed if applied on symptomatic patients.
Tamarappoo et al. [189] 2021	Prospective single-center study, 1069 asymptomatic patients	Boosted ensemble (XgBoost)	ML model using 12 variables from ASCVD score, 5 CT parameters (including EAT volume and attenuation) and top 15 serum biomarkers) to predict cardiac events Mean follow up: 14.5 years	ML (0.81) > CAC (0.75) > ASCVD (0.74).	1. Single-center study 2. Overfitting; given the small number of cardiac events during follow up (~2%)

ASCVD: atherosclerotic cardiovascular disease; CHF: congestive heart failure; EAT: epicardial adipose tissue; HR: hazard ratio; MI: myocardial infarction.

A multitude of ML approaches have been described, combining imaging parameters with clinical and demographic parameters for better prognostication of cardiovascular outcomes [194–199] Including 10,030 patients with suspected CAD from the CONFIRM registry, Motwani et al. utilized a boosting ensemble algorithm using 25 clinical and 44 CCTA parameters [195]. The ML algorithm performed better in predicting 5-year all-cause mortality than CCTA segment stenosis score or FRS (AUC 0.79 for ML vs. 0.664 for segment stenosis score and 0.61 for FRS, respectively, $p < 0.001$). More recently, models incorporating high-risk plaque features with the traditional imaging and clinical parameters have performed better than either of the parameters in isolation [196,197]. A review of literature summarizing all the studies is presented in Table 3.

Although anatomical CT scores and plaque features provide useful diagnostic and prognostic data, the complex interplay of factors at the molecular level, in addition to patient-level characteristics leading to specific phenotypic manifestations in terms of plaque burden and features, is not well-elucidated and remains an area of active research. In particular, elucidating important factors that "drive" the process of atherosclerotic plaque formation and progression is not only vital from a therapeutic perspective, but it can also improve risk-assessment strategies. Recent studies have demonstrated that coronary artery inflammation inhibits lipid accumulation in the perivascular adipose tissue [200]. This results in a higher attenuation of the affected perivascular area, identified on CCTA as the fat attenuation index (FAI). FAI has been shown to be a sensitive marker of coronary inflammation, with higher FAI values (≥ -70.1 HU) independently predicting cardiovascular mortality [200,201]. A posthoc analysis of the CRISP-CT study showed an incremental value of adding FAI to high-risk plaque characteristics, pointing towards a more significant role of these precursor lesions in predicting patient outcomes [202]. A more recent ML approach created a pericoronary fat 'radiomic' profile (FRP), identifying radiomic variables predicting tissue inflammation, fibrosis, and vascularity on CCTA [203]. The incorporation of FRP significantly improved the MACE predictive ability of the traditional model (AUC for traditional + FRP 0.88 vs. 0.754 for the traditional model, $p < 0.001$). Using a cut-off of 0.63, individuals in the high FRP group were at a higher risk of MACE (HR = 10.84, $p < 0.001$). Importantly, Kaplan–Meir analysis showed an additional value of FRP over high-risk plaque (HRP) characteristics in predicting long-term survival (HR for the FRP-/HRP+ subgroup 5.97, $p = 0.03$ compared to 43.33 for the FRP+/HRP+ subgroup). Such 'radiotranscriptomic' approaches incorporating molecular biology and radiology and evaluating their interaction via artificial intelligence can help uncover deeper relationships between metabolic pathways and clinical outcomes, helping to better understand the pathophysiology and elements involved in the clinical progression of cardiovascular disease.

Table 3. Summary of literature regarding mortality outcomes using CCTA data.

Study	Study Design/Sample Size	ML	Brief Description and Outcomes	Results	Limitations
Motwani et al. [195] 2016	Multicenter prospective study, 10,030 patients with suspected CAD	Boosted ensemble (LogitBoost)	25 clinical and 44 CCTA parameter used to create ML model Outcome: Prediction of 5-year ACM; compared against clinical risk scores and CCTA parameters.	AUC: ML (0.79) > Segment stenosis score (SSS) (0.64) and FRS (0.61); $p < 0.001$.	1. Observational; concern for selection bias 2. Cardiac-specific endpoints were not defined, given the data unavailability.
Hoshino et al. [198] 2016	Multicenter retrospective study, 220 patients with intermediate LAD stenosis	Unsupervised hierarchical clustering	Two clusters (CS1 and CS2) using 42 variables created via ML. Outcome: 1. Relation between FAI and CCTA defined clusters, 2. Prognostic value of ML-derived clusters in combination with FAI.	1. Age, CS1 features (higher plaque volume, remodeling index, higher FAI amongst others), and FAI were independent predictors of MACE. 2. Improved NRI with (FRS + CS1 + FAI) as compared to FRS alone.	1. Retrospective, small size 2. Majority of vessels were LAD; hence the study was restricted to a specific population. 3. 40% cardiac events were non-LAD revascularization; hence the results were not generalizable.
Van Rosendael et al. [197] 2018	Multicenter prospective study, 8844 patients with suspected CAD	Boosted ensemble	35 variables (SS and plaque composition for 16 coronary segments and 3 additional variables) compared with traditional CT scores. Outcome: ML vs. traditional CT scores in predicting 5-year composite MI and death.	AUC for ML (0.77) > SSS (0.70)	1. No comparison with clinical risk scores 2. Retrospective study with risk of selection bias
Johnson et al. [194] 2019	Single-center retrospective study, 6892 patients	K nearest neighbors	ML model (64 vessel-related features) vs. CAD-RADS. Outcome: Prediction of ACM, CAD-related deaths. Also, decision to start statin.	1. AUC for all-cause mortality (0.77) > CAD-RADS (0.72); AUC for CAD-related deaths—ML (0.85) > CAD-RADS (0.79). 2. Significant increase in sensitivity with ML model.	1. Retrospective study with limited population diversity 2. Unblinded CCTA results that might have affected event incidence
Johnson et al. [199] 2020	Single-center retrospective study, 6892 patients		ML model developed via radiologist report. Outcome: Prediction of ACM and CAD-related mortality; compared against FRS. Also, decision to start statin.	1. ACM: AUC for ML (0.85) > FRS (0.79) CAD related deaths: AUC for ML (0.87) > FRS (0.82) 2. Using ML, equally high sensitivity but significant reduction in unnecessary statin prescription (AUC for ML 0.89 vs. FRS 0.75).	1. Retrospective study design 2. Concern for misclassification bias due to incomplete follow-up
Tesche et al. [196] 2021	Single-center retrospective study, 361 patients with suspected and confirmed CAD	Boosted ensemble (RUSBoost)	28 clinical, CCTA scores and adverse plaque characteristics included. Outcome: 5-year MACE prediction; compared against FRS, CCTA scores and adverse plaque features.	1. AUC for ML (0.96) > AS (0.84) > FRS (0.76). 2. Important imaging parameters: SSS, obstructive CAD of RCA. 3. Important clinical factors: age, FRS	1. Small sample size, retrospective study design 2. Follow-up using medical records 3. No external validation to test prognostic accuracy

ACM: all-cause mortality; AS: Agatston score; CAD-RADS: coronary artery disease reporting and data system; CS: cluster sample; FAI: fat attenuation index; FRS: Framingham risk score; RCA: right coronary artery; SSS: segment stenosis score.

9. Discussions

With significant developments occurring in the last decade in terms of data processing and analytics, AI can provide new and sophisticated tools that could help us to better understand disease processes, which ultimately should translate into better patient care and outcomes (Figure 6). Nevertheless, AI comes with its own set of limitations. ML models lack interpretability and suffer from the 'black box' problem [204]. ML models based on neural networks and ensemble methods are inherently complex and are derived from complicated mathematical algorithms. 'Explainable (interpretable) machine learning', whereby simple approximations of the model are devised to make it more understandable, is being developed to overcome the black box problem [205,206].

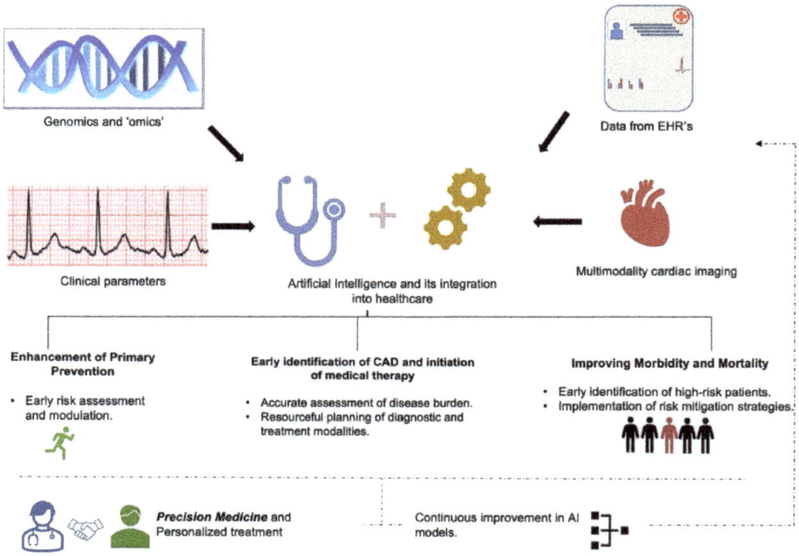

Figure 6. Current applicability and future directions for AI in coronary artery disease.

Another limitation of ML encountered at the model-development phase is sampling bias and lack of external validation [207,208]. ML learning models usually derive their weights from large datasets. Datasets, particularly those derived from EHRs, might be skewed and not representative of the entire population, leading to significant sampling bias and limited generalizability. A few models have tried to address this problem by stratifying the datasets at the model-development phase to ensure not to lose representation of any subgroup and preserve the model's generalizability. Nevertheless, randomized controlled trials are needed to potentially overcome this bias and establish the model performance against the standard clinical parameters. In addition, imputation methods such as MICE have been used to address the missing data issue [209].

Furthermore, the creation of bigger datasets by pooling data from multiple hospital systems has led to a lack of standardization of datasets, potentially compromising the quality of analysis. Datasets might internally differ from each other because of the different mechanisms used to generate them. For instance, one dataset might define the presence of diabetes mellitus through ICD-10 codes, while another dataset might define it using glycemic indices, such as the hemoglobin A1c. On a similar theme, ML models developed by using imaging modalities deserve a special mention. For instance, differences can exist at the level of image scanning (different scanner characteristics and vendors), image quality (radiation dose, motion artifacts), and image processing (reconstruction filters, post-processing) which can potentially lead to significant variability and differences of the assimilated data. A prerequisite to the development of any ML model is the centralization of data, which is tedious given the different image processing algorithms employed at various institutions. This lack of standardization needs to be addressed before AI can be fully integrated into clinical practice.

Overfitting is another concern encountered during ML model development, which occurs when the algorithm learns the data 'too well' and interprets the signal noise as concepts [210]. This usually happens with smaller datasets and can lead to a lack of external validity, despite high performance in the training and internal validation datasets. A definite solution is k-fold cross-validation, whereby data is randomly divided into an arbitrary k number of partitions. The model is trained using $k - 1$ number of data subsets and tested on the remaining subset. This process is repeated k total number of times, using different combinations of training and testing datasets to select the best model hyper-parameters to yield the final model. This can potentially reduce noise and lead to better generalizability of the model in the overall population.

Apart from the problems encountered at the model development and training phase, there are a few noteworthy practical limitations to the implementation of ML within health-care workflows. Firstly, unauthorized data access is an issue, as handling such large amounts of data also poses a risk of leaking sensitive patient information, thereby violating patient confidentiality and privacy [211]. Furthermore, comparisons between various machine-learning methods are difficult, given the different combinations of model parameters and different population characteristics used for in model development. Hence, it becomes difficult for physicians to compare and choose one model over the other. Prospective future trials, comparing these models on the same dataset, are needed to select the best algorithm fit for integration into routine clinical decision-making. Proper integration of AI can only be achieved once these models are embedded within EHRs. However, the full implementation and assimilation of developed AI models into EHRs can be a complex issue, as it depends on organizational resources and patient-privacy policies. Furthermore, available algorithms may be limited to off-the-shelf ML models, rather than more intricate and complex neural networks, which is easier to implement in a real clinical setting. Yet, a data-driven approach utilizing advanced analytic techniques can help clinicians and patients to make informed decisions, improve care, and optimize workflow efficiency.

10. Conclusions

In conclusion, AI provides an unprecedented potential to transform healthcare and enhance the current system's ability to serve populations at large, while providing tools to focus on individualized yet comprehensive and precise care.

Author Contributions: Conceptualization, S.J.A.; data curation, N.G. and P.S.; writing—original draft preparation, N.G., P.S. and A.M.; writing—review and editing, N.G., M.G.R., M.H.A.-M., G.P., Y.Z., B.C.L. and S.J.A. All authors have read and agreed to the published version of the manuscript.

Funding: This research received no external funding.

Data Availability Statement: Not applicable.

Conflicts of Interest: Subhi J. Al'Aref is supported by NIH 2R01 HL12766105 & 1R21 EB030654 and receives royalty fees from Elsevier. Gianluca Pontone receives honorarium and institutional research funding from GE Healthcare, Bracco, Boehringer Ingelheim, Bayer, and Heartflow. Mouaz Al-Mallah receives research support from SiEMENS and consulting fee/honorarium from Philips, Pfizer, and Draximage. All other authors have reported that they have no relationships relevant to the contents of this paper to disclose.

References

1. Virani, S.S.; Alonso, A.; Aparicio, H.J.; Benjamin, E.J.; Bittencourt, M.S.; Callaway, C.W.; Carson, A.P.; Chamberlain, A.M.; Cheng, S.; Delling, F.N.; et al. Heart Disease and Stroke Statistics—2021 Update. *Circulation* **2021**, *143*, e254–e743. [CrossRef]
2. Benjamin, E.J.; Muntner, P.; Alonso, A.; Bittencourt, M.S.; Callaway, C.W.; Carson, A.P.; Chamberlain, A.M.; Chang, A.R.; Cheng, S.; Das, S.R.; et al. Heart Disease and Stroke Statistics—2019 Update: A Report From the American Heart Association. *Circulation* **2019**, *139*, e56–e528. [CrossRef]
3. Arnett, D.K.; Blumenthal, R.S.; Albert, M.A.; Buroker, A.B.; Goldberger, Z.D.; Hahn, E.J.; Himmelfarb, C.D.; Khera, A.; Lloyd-Jones, D.; McEvoy, J.W.; et al. 2019 ACC/AHA Guideline on the Primary Prevention of Cardiovascular Disease: A Report of the American College of Cardiology/American Heart Association Task Force on Clinical Practice Guidelines. *Circulation* **2019**, *140*, e596–e646. [CrossRef]
4. Aragam, K.G.; Natarajan, P. Polygenic Scores to Assess Atherosclerotic Cardiovascular Disease Risk: Clinical Perspectives and Basic Implications. *Circ. Res.* **2020**, *126*, 1159–1177. [CrossRef]
5. Schaap, J.; de Groot, J.A.H.; Nieman, K.; Meijboom, W.B.; Boekholdt, S.M.; Kauling, R.M.; Post, M.C.; Van der Heyden, J.A.; de Kroon, T.L.; Rensing, B.J.W.M.; et al. Added value of hybrid myocardial perfusion SPECT and CT coronary angiography in the diagnosis of coronary artery disease. *Eur. Heart J.-Cardiovasc. Imaging* **2014**, *15*, 1281–1288. [CrossRef]
6. Andreini, D.; Magnoni, M.; Conte, E.; Masson, S.; Mushtaq, S.; Berti, S.; Canestrari, M.; Casolo, G.; Gabrielli, D.; Latini, R.; et al. Coronary Plaque Features on CTA Can Identify Patients at Increased Risk of Cardiovascular Events. *JACC Cardiovasc. Imaging* **2020**, *13*, 1704–1717. [CrossRef]
7. Budoff, M.J.; Mayrhofer, T.; Ferencik, M.; Bittner, D.; Lee, K.L.; Lu, M.T.; Coles, A.; Jang, J.; Krishnam, M.; Douglas, P.S.; et al. Prognostic Value of Coronary Artery Calcium in the PROMISE Study (Prospective Multicenter Imaging Study for Evaluation of Chest Pain). *Circulation* **2017**, *136*, 1993–2005. [CrossRef]
8. Patel, V.L.; Shortliffe, E.H.; Stefanelli, M.; Szolovits, P.; Berthold, M.R.; Bellazzi, R.; Abu-Hanna, A. The coming of age of artificial intelligence in medicine. *Artif. Intell. Med.* **2009**, *46*, 5–17. [CrossRef]
9. Ranka, S.; Reddy, M.; Noheria, A. Artificial intelligence in cardiovascular medicine. *Curr. Opin. Cardiol.* **2021**, *36*, 26–35. [CrossRef]
10. Dey, D.; Slomka, P.J.; Leeson, P.; Comaniciu, D.; Shrestha, S.; Sengupta, P.P.; Marwick, T.H. Artificial Intelligence in Cardiovascular Imaging: JACC State-of-the-Art Review. *J. Am. Coll. Cardiol.* **2019**, *73*, 1317–1335. [CrossRef]
11. Johnson, K.W.; Torres Soto, J.; Glicksberg, B.S.; Shameer, K.; Miotto, R.; Ali, M.; Ashley, E.; Dudley, J.T. Artificial Intelligence in Cardiology. *J. Am. Coll. Cardiol.* **2018**, *71*, 2668–2679. [CrossRef] [PubMed]
12. Sprangers, M.A.G.; Sloan, J.A.; Barsevick, A.; Chauhan, C.; Dueck, A.C.; Raat, H.; Shi, Q.; Van Noorden, C.J.F.; Consortium, G. Scientific imperatives, clinical implications, and theoretical underpinnings for the investigation of the relationship between genetic variables and patient-reported quality-of-life outcomes. *Qual. Life Res.* **2010**, *19*, 1395–1403. [CrossRef] [PubMed]
13. Erdmann, J.; Kessler, T.; Munoz Venegas, L.; Schunkert, H. A decade of genome-wide association studies for coronary artery disease: The challenges ahead. *Cardiovasc. Res.* **2018**, *114*, 1241–1257. [CrossRef] [PubMed]
14. Noll, D.R.; Ginsberg, T.; Elahi, A.; Cavalieri, T.A. Effective Patient-Physician Communication Based on Osteopathic Philosophy in Caring for Elderly Patients. *J. Osteopath. Med.* **2016**, *116*, 42–47. [CrossRef]
15. Kathiresan, S.; Melander, O.; Anevski, D.; Guiducci, C.; Burtt, N.P.; Roos, C.; Hirschhorn, J.N.; Berglund, G.; Hedblad, B.; Groop, L.; et al. Polymorphisms associated with cholesterol and risk of cardiovascular events. *N. Engl. J. Med.* **2008**, *358*, 1240–1249. [CrossRef]
16. Brautbar, A.; Pompeii, L.A.; Dehghan, A.; Ngwa, J.S.; Nambi, V.; Virani, S.S.; Rivadeneira, F.; Uitterlinden, A.G.; Hofman, A.; Witteman, J.C.; et al. A genetic risk score based on direct associations with coronary heart disease improves coronary heart disease risk prediction in the Atherosclerosis Risk in Communities (ARIC), but not in the Rotterdam and Framingham Offspring, Studies. *Atherosclerosis* **2012**, *223*, 421–426. [CrossRef]
17. Eraslan, G.; Avsec, Ž.; Gagneur, J.; Theis, F.J. Deep learning: New computational modelling techniques for genomics. *Nat. Rev. Genet.* **2019**, *20*, 389–403. [CrossRef]
18. Wang, Y.; Liu, T.; Liu, Y.; Chen, J.; Xin, B.; Wu, M.; Cui, W. Coronary artery disease associated specific modules and feature genes revealed by integrative methods of WGCNA, MetaDE and machine learning. *Gene* **2019**, *710*, 122–130. [CrossRef]
19. Balashanmugam, M.V.; Shivanandappa, T.B.; Nagarethinam, S.; Vastrad, B.; Vastrad, C. Analysis of Differentially Expressed Genes in Coronary Artery Disease by Integrated Microarray Analysis. *Biomolecules* **2019**, *10*, 35. [CrossRef]
20. Zhang, D.; Guan, L.; Li, X. Bioinformatics analysis identifies potential diagnostic signatures for coronary artery disease. *J. Int. Med. Res.* **2020**, *48*, 300060520979856. [CrossRef]
21. Dogan, M.V.; Grumbach, I.M.; Michaelson, J.J.; Philibert, R.A. Integrated genetic and epigenetic prediction of coronary heart disease in the Framingham Heart Study. *PLoS ONE* **2018**, *13*, e0190549. [CrossRef]
22. Pattarabanjird, T.; Cress, C.; Nguyen, A.; Taylor, A.; Bekiranov, S.; McNamara, C. A Machine Learning Model Utilizing a Novel SNP Shows Enhanced Prediction of Coronary Artery Disease Severity. *Genes* **2020**, *11*, 1446. [CrossRef]
23. Naushad, S.M.; Hussain, T.; Indumathi, B.; Samreen, K.; Alrokayan, S.A.; Kutala, V.K. Machine learning algorithm-based risk prediction model of coronary artery disease. *Mol. Biol. Rep.* **2018**, *45*, 901–910. [CrossRef]
24. Ferguson, J.F.; Matthews, G.J.; Townsend, R.R.; Raj, D.S.; Kanetsky, P.A.; Budoff, M.; Fischer, M.J.; Rosas, S.E.; Kanthety, R.; Rahman, M.; et al. Candidate gene association study of coronary artery calcification in chronic kidney disease: Findings from the CRIC study (Chronic Renal Insufficiency Cohort). *J. Am. Coll. Cardiol.* **2013**, *62*, 789–798. [CrossRef]

25. O'Donnell, C.J.; Kavousi, M.; Smith, A.V.; Kardia, S.L.; Feitosa, M.F.; Hwang, S.J.; Sun, Y.V.; Province, M.A.; Aspelund, T.; Dehghan, A.; et al. Genome-wide association study for coronary artery calcification with follow-up in myocardial infarction. *Circulation* **2011**, *124*, 2855–2864. [CrossRef]
26. Oguz, C.; Sen, S.K.; Davis, A.R.; Fu, Y.P.; O'Donnell, C.J.; Gibbons, G.H. Genotype-driven identification of a molecular network predictive of advanced coronary calcium in ClinSeq® and Framingham Heart Study cohorts. *BMC Syst. Biol.* **2017**, *11*, 99. [CrossRef]
27. Diamond, G.A.; Forrester, J.S. Analysis of probability as an aid in the clinical diagnosis of coronary-artery disease. *N. Engl. J. Med.* **1979**, *300*, 1350–1358. [CrossRef]
28. Foldyna, B.; Udelson, J.E.; Karády, J.; Banerji, D.; Lu, M.T.; Mayrhofer, T.; Bittner, D.O.; Meyersohn, N.M.; Emami, H.; Genders, T.S.S.; et al. Pretest probability for patients with suspected obstructive coronary artery disease: Re-evaluating Diamond-Forrester for the contemporary era and clinical implications: Insights from the PROMISE trial. *Eur. Heart J. Cardiovasc. Imaging* **2019**, *20*, 574–581. [CrossRef]
29. Genders, T.S.; Steyerberg, E.W.; Alkadhi, H.; Leschka, S.; Desbiolles, L.; Nieman, K.; Galema, T.W.; Meijboom, W.B.; Mollet, N.R.; de Feyter, P.J.; et al. A clinical prediction rule for the diagnosis of coronary artery disease: Validation, updating, and extension. *Eur. Heart J.* **2011**, *32*, 1316–1330. [CrossRef]
30. Genders, T.S.; Steyerberg, E.W.; Hunink, M.G.; Nieman, K.; Galema, T.W.; Mollet, N.R.; de Feyter, P.J.; Krestin, G.P.; Alkadhi, H.; Leschka, S.; et al. Prediction model to estimate presence of coronary artery disease: Retrospective pooled analysis of existing cohorts. *BMJ* **2012**, *344*, e3485. [CrossRef]
31. Bittencourt, M.S.; Hulten, E.; Polonsky, T.S.; Hoffman, U.; Nasir, K.; Abbara, S.; Di Carli, M.; Blankstein, R. European Society of Cardiology-Recommended Coronary Artery Disease Consortium Pretest Probability Scores More Accurately Predict Obstructive Coronary Disease and Cardiovascular Events Than the Diamond and Forrester Score: The Partners Registry. *Circulation* **2016**, *134*, 201–211. [CrossRef] [PubMed]
32. Li, D.; Xiong, G.; Zeng, H.; Zhou, Q.; Jiang, J.; Guo, X. Machine learning-aided risk stratification system for the prediction of coronary artery disease. *Int. J. Cardiol.* **2021**, *326*, 30–34. [CrossRef] [PubMed]
33. Velusamy, D.; Ramasamy, K. Ensemble of heterogeneous classifiers for diagnosis and prediction of coronary artery disease with reduced feature subset. *Comput. Methods Programs Biomed.* **2021**, *198*, 105770. [CrossRef] [PubMed]
34. Muhammad, L.J.; Al-Shourbaji, I.; Haruna, A.A.; Mohammed, I.A.; Ahmad, A.; Jibrin, M.B. Machine Learning Predictive Models for Coronary Artery Disease. *SN Comput. Sci.* **2021**, *2*, 350. [CrossRef]
35. Lin, S.; Li, Z.; Fu, B.; Chen, S.; Li, X.; Wang, Y.; Wang, X.; Lv, B.; Xu, B.; Song, X.; et al. Feasibility of using deep learning to detect coronary artery disease based on facial photo. *Eur. Heart J.* **2020**, *41*, 4400–4411. [CrossRef]
36. Gulati, M.; Levy, P.D.; Mukherjee, D.; Amsterdam, E.; Bhatt, D.L.; Birtcher, K.K.; Blankstein, R.; Boyd, J.; Bullock-Palmer, R.P.; Conejo, T.; et al. 2021 AHA/ACC/ASE/CHEST/SAEM/SCCT/SCMR Guideline for the Evaluation and Diagnosis of Chest Pain: A Report of the American College of Cardiology/American Heart Association Joint Committee on Clinical Practice Guidelines. *Circulation* **2021**, *78*, e187–e285. [CrossRef]
37. Baskaran, L.; Ying, X.; Xu, Z.; Al'Aref, S.J.; Lee, B.C.; Lee, S.E.; Danad, I.; Park, H.B.; Bathina, R.; Baggiano, A.; et al. Machine learning insight into the role of imaging and clinical variables for the prediction of obstructive coronary artery disease and revascularization: An exploratory analysis of the CONSERVE study. *PLoS ONE* **2020**, *15*, e0233791. [CrossRef]
38. Al'Aref, S.J.; Maliakal, G.; Singh, G.; van Rosendael, A.R.; Ma, X.; Xu, Z.; Alawamlh, O.A.H.; Lee, B.; Pandey, M.; Achenbach, S.; et al. Machine learning of clinical variables and coronary artery calcium scoring for the prediction of obstructive coronary artery disease on coronary computed tomography angiography: Analysis from the CONFIRM registry. *Eur. Heart J.* **2020**, *41*, 359–367. [CrossRef]
39. Arsanjani, R.; Xu, Y.; Dey, D.; Fish, M.; Dorbala, S.; Hayes, S.; Berman, D.; Germano, G.; Slomka, P. Improved accuracy of myocardial perfusion SPECT for the detection of coronary artery disease using a support vector machine algorithm. *J. Nucl. Med.* **2013**, *54*, 549–555. [CrossRef]
40. Betancur, J.; Hu, L.H.; Commandeur, F.; Sharir, T.; Einstein, A.J.; Fish, M.B.; Ruddy, T.D.; Kaufmann, P.A.; Sinusas, A.J.; Miller, E.J.; et al. Deep Learning Analysis of Upright-Supine High-Efficiency SPECT Myocardial Perfusion Imaging for Prediction of Obstructive Coronary Artery Disease: A Multicenter Study. *J. Nucl. Med.* **2019**, *60*, 664–670. [CrossRef]
41. Guner, L.A.; Karabacak, N.I.; Akdemir, O.U.; Karagoz, P.S.; Kocaman, S.A.; Cengel, A.; Unlu, M. An open-source framework of neural networks for diagnosis of coronary artery disease from myocardial perfusion SPECT. *J. Nucl. Cardiol.* **2010**, *17*, 405–413. [CrossRef]
42. Rahmani, R.; Niazi, P.; Naseri, M.; Neishabouri, M.; Farzanefar, S.; Eftekhari, M.; Derakhshan, F.; Mollazadeh, R.; Meysami, A.; Abbasi, M. Improved diagnostic accuracy for myocardial perfusion imaging using artificial neural networks on different input variables including clinical and quantification data. *Rev. Esp. Med. Nucl. E Imagen. Mol.* **2019**, *38*, 275–279. [CrossRef]
43. Betancur, J.; Commandeur, F.; Motlagh, M.; Sharir, T.; Einstein, A.J.; Bokhari, S.; Fish, M.B.; Ruddy, T.D.; Kaufmann, P.; Sinusas, A.J.; et al. Deep Learning for Prediction of Obstructive Disease From Fast Myocardial Perfusion SPECT: A Multicenter Study. *JACC Cardiovasc. Imaging* **2018**, *11*, 1654–1663. [CrossRef]
44. Arsanjani, R.; Xu, Y.; Dey, D.; Vahistha, V.; Shalev, A.; Nakanishi, R.; Hayes, S.; Fish, M.; Berman, D.; Germano, G.; et al. Improved accuracy of myocardial perfusion SPECT for detection of coronary artery disease by machine learning in a large population. *J. Nucl. Cardiol.* **2013**, *20*, 553–562. [CrossRef]

45. Rabbat, M.G.; Ramchandani, S.; Sanders, W.E., Jr. Cardiac Phase Space Analysis: Assessing Coronary Artery Disease Utilizing Artificial Intelligence. *Biomed. Res. Int.* **2021**, *2021*, 6637039. [CrossRef]
46. Stuckey, T.D.; Gammon, R.S.; Goswami, R.; Depta, J.P.; Steuter, J.A.; Meine, F.J., 3rd; Roberts, M.C.; Singh, N.; Ramchandani, S.; Burton, T.; et al. Cardiac Phase Space Tomography: A novel method of assessing coronary artery disease utilizing machine learning. *PLoS ONE* **2018**, *13*, e0198603. [CrossRef]
47. Medina, R.; Panidis, I.P.; Morganroth, J.; Kotler, M.N.; Mintz, G.S. The value of echocardiographic regional wall motion abnormalities in detecting coronary artery disease in patients with or without a dilated left ventricle. *Am. Heart J.* **1985**, *109*, 799–803. [CrossRef]
48. Kusunose, K.; Abe, T.; Haga, A.; Fukuda, D.; Yamada, H.; Harada, M.; Sata, M. A Deep Learning Approach for Assessment of Regional Wall Motion Abnormality From Echocardiographic Images. *JACC Cardiovasc. Imaging* **2020**, *13*, 374–381. [CrossRef]
49. Huang, M.-S.; Wang, C.-S.; Chiang, J.-H.; Liu, P.-Y.; Tsai, W.-C. Automated Recognition of Regional Wall Motion Abnormalities Through Deep Neural Network Interpretation of Transthoracic Echocardiography. *Circulation* **2020**, *142*, 1510–1520. [CrossRef]
50. Asch, F.M.; Poilvert, N.; Abraham, T.; Jankowski, M.; Cleve, J.; Adams, M.; Romano, N.; Hong, H.; Mor-Avi, V.; Martin, R.P.; et al. Automated Echocardiographic Quantification of Left Ventricular Ejection Fraction Without Volume Measurements Using a Machine Learning Algorithm Mimicking a Human Expert. *Circ. Cardiovasc. Imaging* **2019**, *12*, e009303. [CrossRef]
51. Kwon, J.M.; Lee, S.Y.; Jeon, K.H.; Lee, Y.; Kim, K.H.; Park, J.; Oh, B.H.; Lee, M.M. Deep Learning—Based Algorithm for Detecting Aortic Stenosis Using Electrocardiography. *J. Am. Heart Assoc.* **2020**, *9*, e014717. [CrossRef]
52. Acharya, U.R.; Fujita, H.; Sudarshan, V.K.; Oh, S.L.; Adam, M.; Koh, J.E.W.; Tan, J.H.; Ghista, D.N.; Martis, R.J.; Chua, C.K.; et al. Automated detection and localization of myocardial infarction using electrocardiogram: A comparative study of different leads. *Knowl.-Based Syst.* **2016**, *99*, 146–156. [CrossRef]
53. Han, C.; Shi, L. ML–ResNet: A novel network to detect and locate myocardial infarction using 12 leads ECG. *Comput. Methods Programs Biomed.* **2020**, *185*, 105138. [CrossRef]
54. Lih, O.S.; Jahmunah, V.; San, T.R.; Ciaccio, E.J.; Yamakawa, T.; Tanabe, M.; Kobayashi, M.; Faust, O.; Acharya, U.R. Comprehensive electrocardiographic diagnosis based on deep learning. *Artif. Intell. Med.* **2020**, *103*, 101789. [CrossRef]
55. Keller, T.; Zeller, T.; Ojeda, F.; Tzikas, S.; Lillpopp, L.; Sinning, C.; Wild, P.; Genth-Zotz, S.; Warnholtz, A.; Giannitsis, E. Serial changes in highly sensitive troponin I assay and early diagnosis of myocardial infarction. *JAMA* **2011**, *306*, 2684–2693. [CrossRef]
56. Collet, J.P.; Thiele, H.; Barbato, E.; Barthelemy, O.; Bauersachs, J.; Bhatt, D.L.; Dendale, P.; Dorobantu, M.; Edvardsen, T.; Folliguet, T.; et al. 2020 ESC Guidelines for the management of acute coronary syndromes in patients presenting without persistent ST-segment elevation. *Eur. Heart J.* **2021**, *42*, 1289–1367. [CrossRef]
57. Reichlin, T.; Schindler, C.; Drexler, B.; Twerenbold, R.; Reiter, M.; Zellweger, C.; Moehring, B.; Ziller, R.; Hoeller, R.; Rubini Gimenez, M.; et al. One-hour rule-out and rule-in of acute myocardial infarction using high-sensitivity cardiac troponin T. *Arch. Intern. Med.* **2012**, *172*, 1211–1218. [CrossRef]
58. Reichlin, T.; Twerenbold, R.; Wildi, K.; Gimenez, M.R.; Bergsma, N.; Haaf, P.; Druey, S.; Puelacher, C.; Moehring, B.; Freese, M.; et al. Prospective validation of a 1-hour algorithm to rule-out and rule-in acute myocardial infarction using a high-sensitivity cardiac troponin T assay. *Can. Med. Assoc. J.* **2015**, *187*, E243. [CrossRef]
59. Gimenez, M.R.; Twerenbold, R.; Jaeger, C.; Schindler, C.; Puelacher, C.; Wildi, K.; Reichlin, T.; Haaf, P.; Merk, S.; Honegger, U. One-hour rule-in and rule-out of acute myocardial infarction using high-sensitivity cardiac troponin I. *Am. J. Med.* **2015**, *128*, 861–870.e864. [CrossRef]
60. Druey, S.; Wildi, K.; Twerenbold, R.; Jaeger, C.; Reichlin, T.; Haaf, P.; Gimenez, M.R.; Puelacher, C.; Wagener, M.; Radosavac, M. Early rule-out and rule-in of myocardial infarction using sensitive cardiac Troponin I. *Int. J. Cardiol.* **2015**, *195*, 163–170. [CrossRef]
61. Neumann, J.T.; Sörensen, N.A.; Schwemer, T.; Ojeda, F.; Bourry, R.; Sciacca, V.; Schaefer, S.; Waldeyer, C.; Sinning, C.; Renné, T.; et al. Diagnosis of Myocardial Infarction Using a High-Sensitivity Troponin I 1-Hour Algorithm. *JAMA Cardiol.* **2016**, *1*, 397–404. [CrossRef]
62. Twerenbold, R.; Badertscher, P.; Boeddinghaus, J.; Nestelberger, T.; Wildi, K.; Puelacher, C.; Sabti, Z.; Gimenez, M.R.; Tschirky, S.; Lavallaz, J.d.F.d.; et al. 0/1-Hour Triage Algorithm for Myocardial Infarction in Patients with Renal Dysfunction. *Circulation* **2018**, *137*, 436–451. [CrossRef]
63. Boeddinghaus, J.; Nestelberger, T.; Twerenbold, R.; Neumann, J.T.; Lindahl, B.; Giannitsis, E.; Sörensen, N.A.; Badertscher, P.; Jann, J.E.; Wussler, D.; et al. Impact of age on the performance of the ESC 0/1h-algorithms for early diagnosis of myocardial infarction. *Eur. Heart J.* **2018**, *39*, 3780–3794. [CrossRef]
64. McCarthy, C.P.; Neumann, J.T.; Michelhaugh, S.A.; Ibrahim, N.E.; Gaggin, H.K.; Sorensen, N.A.; Schaefer, S.; Zeller, T.; Magaret, C.A.; Barnes, G.; et al. Derivation and External Validation of a High-Sensitivity Cardiac Troponin-Based Proteomic Model to Predict the Presence of Obstructive Coronary Artery Disease. *J. Am. Heart Assoc.* **2020**, *9*, e017221. [CrossRef]
65. Liu, C.Y.; Tang, C.X.; Zhang, X.L.; Chen, S.; Xie, Y.; Zhang, X.Y.; Qiao, H.Y.; Zhou, C.S.; Xu, P.P.; Lu, M.J.; et al. Deep learning powered coronary CT angiography for detecting obstructive coronary artery disease: The effect of reader experience, calcification and image quality. *Eur. J. Radiol.* **2021**, *142*, 109835. [CrossRef]
66. Lee, J.-G.; Kim, H.; Kang, H.; Koo, H.J.; Kang, J.-W.; Kim, Y.-H.; Yang, D.H. Fully Automatic Coronary Calcium Score Software Empowered by Artificial Intelligence Technology: Validation Study Using Three CT Cohorts. *Korean J. Radiol.* **2021**, *22*, 1764–1776. [CrossRef]

67. van Velzen, S.G.M.; Lessmann, N.; Velthuis, B.K.; Bank, I.E.M.; van den Bongard, D.; Leiner, T.; de Jong, P.A.; Veldhuis, W.B.; Correa, A.; Terry, J.G.; et al. Deep Learning for Automatic Calcium Scoring in CT: Validation Using Multiple Cardiac CT and Chest CT Protocols. *Radiology* **2020**, *295*, 66–79. [CrossRef]
68. Baskaran, L.; Maliakal, G.; Al'Aref, S.J.; Singh, G.; Xu, Z.; Michalak, K.; Dolan, K.; Gianni, U.; van Rosendael, A.; van den Hoogen, I.; et al. Identification and Quantification of Cardiovascular Structures From CCTA: An End-to-End, Rapid, Pixel-Wise, Deep-Learning Method. *JACC Cardiovasc. Imaging* **2020**, *13*, 1163–1171. [CrossRef]
69. Wang, W.; Wang, H.; Chen, Q.; Zhou, Z.; Wang, R.; Wang, H.; Zhang, N.; Chen, Y.; Sun, Z.; Xu, L. Coronary artery calcium score quantification using a deep-learning algorithm. *Clin. Radiol.* **2020**, *75*, 237.e11–237.e16. [CrossRef]
70. von Knebel Doeberitz, P.L.; De Cecco, C.N.; Schoepf, U.J.; Duguay, T.M.; Albrecht, M.H.; van Assen, M.; Bauer, M.J.; Savage, R.H.; Pannell, J.T.; De Santis, D.; et al. Coronary CT angiography-derived plaque quantification with artificial intelligence CT fractional flow reserve for the identification of lesion-specific ischemia. *Eur. Radiol.* **2019**, *29*, 2378–2387. [CrossRef]
71. Koo, H.J.; Lee, J.G.; Ko, J.Y.; Lee, G.; Kang, J.W.; Kim, Y.H.; Yang, D.H. Automated Segmentation of Left Ventricular Myocardium on Cardiac Computed Tomography Using Deep Learning. *Korean J. Radiol.* **2020**, *21*, 660–669. [CrossRef]
72. Morris, E.D.; Ghanem, A.I.; Dong, M.; Pantelic, M.V.; Walker, E.M.; Glide-Hurst, C.K. Cardiac substructure segmentation with deep learning for improved cardiac sparing. *Med. Phys.* **2020**, *47*, 576–586. [CrossRef]
73. Muscogiuri, G.; Chiesa, M.; Trotta, M.; Gatti, M.; Palmisano, V.; Dell'Aversana, S.; Baessato, F.; Cavaliere, A.; Cicala, G.; Loffreno, A.; et al. Performance of a deep learning algorithm for the evaluation of CAD-RADS classification with CCTA. *Atherosclerosis* **2020**, *294*, 25–32. [CrossRef]
74. Fihn, S.D.; Gardin, J.M.; Abrams, J.; Berra, K.; Blankenship, J.C.; Dallas, A.P.; Douglas, P.S.; Foody, J.M.; Gerber, T.C.; Hinderliter, A.L.; et al. 2012 ACCF/AHA/ACP/AATS/PCNA/SCAI/STS Guideline for the Diagnosis and Management of Patients With Stable Ischemic Heart Disease. *Circulation* **2012**, *126*, e354–e471. [CrossRef]
75. Biagini, E.; Shaw, L.J.; Poldermans, D.; Schinkel, A.F.; Rizzello, V.; Elhendy, A.; Rapezzi, C.; Bax, J.J. Accuracy of non-invasive techniques for diagnosis of coronary artery disease and prediction of cardiac events in patients with left bundle branch block: A meta-analysis. *Eur. J. Nucl. Med. Mol. Imaging* **2006**, *33*, 1442–1451. [CrossRef]
76. Mahajan, N.; Polavaram, L.; Vankayala, H.; Ference, B.; Wang, Y.; Ager, J.; Kovach, J.; Afonso, L. Diagnostic accuracy of myocardial perfusion imaging and stress echocardiography for the diagnosis of left main and triple vessel coronary artery disease: A comparative meta-analysis. *Heart* **2010**, *96*, 956–966. [CrossRef]
77. Jaarsma, C.; Leiner, T.; Bekkers Sebastiaan, C.; Crijns Harry, J.; Wildberger Joachim, E.; Nagel, E.; Nelemans Patricia, J.; Schalla, S. Diagnostic Performance of Noninvasive Myocardial Perfusion Imaging Using Single-Photon Emission Computed Tomography, Cardiac Magnetic Resonance, and Positron Emission Tomography Imaging for the Detection of Obstructive Coronary Artery Disease. *J. Am. Coll. Cardiol.* **2012**, *59*, 1719–1728. [CrossRef]
78. Takx, R.A.P.; Blomberg, B.A.; Aidi, H.E.; Habets, J.; de Jong, P.A.; Nagel, E.; Hoffmann, U.; Leiner, T. Diagnostic Accuracy of Stress Myocardial Perfusion Imaging Compared to Invasive Coronary Angiography With Fractional Flow Reserve Meta-Analysis. *Circ. Cardiovasc. Imaging* **2015**, *8*, e002666. [CrossRef]
79. Fleischmann, K.E.; Hunink, M.G.; Kuntz, K.M.; Douglas, P.S. Exercise echocardiography or exercise SPECT imaging? A meta-analysis of diagnostic test performance. *JAMA* **1998**, *280*, 913–920. [CrossRef]
80. Holder, L.; Lewis, S.; Abrames, E.; Wolin, E.A. Review of SPECT myocardial perfusion imaging. *J. Am. Osteopath. Coll. Radiol.* **2016**, *5*, 5–13.
81. Czaja, M.; Wygoda, Z.; Duszańska, A.; Szczerba, D.; Głowacki, J.; Gąsior, M.; Wasilewski, J.P. Interpreting myocardial perfusion scintigraphy using single-photon emission computed tomography. Part 1. *Kardiochir. Torakochirurgia Pol.* **2017**, *14*, 192–199. [CrossRef]
82. Slomka, P.; Xu, Y.; Berman, D.; Germano, G. Quantitative analysis of perfusion studies: Strengths and pitfalls. *J. Nucl. Cardiol. Off. Publ. Am. Soc. Nucl. Cardiol.* **2012**, *19*, 338–346. [CrossRef]
83. Hachamovitch, R.; Hayes, S.W.; Friedman, J.D.; Cohen, I.; Berman, D.S. A prognostic score for prediction of cardiac mortality risk after adenosine stress myocardial perfusion scintigraphy. *J. Am. Coll. Cardiol.* **2005**, *45*, 722–729. [CrossRef]
84. Arsanjani, R.; Xu, Y.; Hayes, S.W.; Fish, M.; Lemley, M., Jr.; Gerlach, J.; Dorbala, S.; Berman, D.S.; Germano, G.; Slomka, P. Comparison of fully automated computer analysis and visual scoring for detection of coronary artery disease from myocardial perfusion SPECT in a large population. *J. Nucl. Med.* **2013**, *54*, 221–228. [CrossRef]
85. Hu, L.H.; Betancur, J.; Sharir, T.; Einstein, A.J.; Bokhari, S.; Fish, M.B.; Ruddy, T.D.; Kaufmann, P.A.; Sinusas, A.J.; Miller, E.J.; et al. Machine learning predicts per-vessel early coronary revascularization after fast myocardial perfusion SPECT: Results from multicentre REFINE SPECT registry. *Eur. Heart J. Cardiovasc. Imaging* **2020**, *21*, 549–559. [CrossRef]
86. Arsanjani, R.; Dey, D.; Khachatryan, T.; Shalev, A.; Hayes, S.W.; Fish, M.; Nakanishi, R.; Germano, G.; Berman, D.S.; Slomka, P. Prediction of revascularization after myocardial perfusion SPECT by machine learning in a large population. *J. Nucl. Cardiol.* **2015**, *22*, 877–884. [CrossRef]
87. Koo, B.K.; Erglis, A.; Doh, J.H.; Daniels, D.V.; Jegere, S.; Kim, H.S.; Dunning, A.; DeFrance, T.; Lansky, A.; Leipsic, J.; et al. Diagnosis of ischemia-causing coronary stenoses by noninvasive fractional flow reserve computed from coronary computed tomographic angiograms. Results from the prospective multicenter DISCOVER-FLOW (Diagnosis of Ischemia-Causing Stenoses Obtained Via Noninvasive Fractional Flow Reserve) study. *J. Am. Coll. Cardiol.* **2011**, *58*, 1989–1997. [CrossRef]

88. Min, J.K.; Berman, D.S.; Budoff, M.J.; Jaffer, F.A.; Leipsic, J.; Leon, M.B.; Mancini, G.B.; Mauri, L.; Schwartz, R.S.; Shaw, L.J. Rationale and design of the DeFACTO (Determination of Fractional Flow Reserve by Anatomic Computed Tomographic AngiOgraphy) study. *J. Cardiovasc. Comput. Tomogr.* **2011**, *5*, 301–309. [CrossRef]
89. Nørgaard, B.L.; Leipsic, J.; Gaur, S.; Seneviratne, S.; Ko, B.S.; Ito, H.; Jensen, J.M.; Mauri, L.; Bruyne, B.D.; Bezerra, H.; et al. Diagnostic Performance of Noninvasive Fractional Flow Reserve Derived From Coronary Computed Tomography Angiography in Suspected Coronary Artery Disease. *J. Am. Coll. Cardiol.* **2014**, *63*, 1145–1155. [CrossRef]
90. Rabbat, M.G.; Berman, D.S.; Kern, M.; Raff, G.; Chinnaiyan, K.; Koweek, L.; Shaw, L.J.; Blanke, P.; Scherer, M.; Jensen, J.M.; et al. Interpreting results of coronary computed tomography angiography-derived fractional flow reserve in clinical practice. *J. Cardiovasc. Comput. Tomogr.* **2017**, *11*, 383–388. [CrossRef]
91. Rabbat, M.; Leipsic, J.; Bax, J.; Kauh, B.; Verma, R.; Doukas, D.; Allen, S.; Pontone, G.; Wilber, D.; Mathew, V.; et al. Fractional Flow Reserve Derived from Coronary Computed Tomography Angiography Safely Defers Invasive Coronary Angiography in Patients with Stable Coronary Artery Disease. *J. Clin. Med.* **2020**, *9*, 604. [CrossRef]
92. Yeri, A.; Shah, R.V. Comparison of Computational Fluid Dynamics and Machine Learning-Based Fractional Flow Reserve in Coronary Artery Disease. *Circ. Cardiovasc. Imaging* **2018**, *11*, e007950. [CrossRef]
93. Itu, L.; Rapaka, S.; Passerini, T.; Georgescu, B.; Schwemmer, C.; Schoebinger, M.; Flohr, T.; Sharma, P.; Comaniciu, D. A machine-learning approach for computation of fractional flow reserve from coronary computed tomography. *J. Appl. Physiol.* **2016**, *121*, 42–52. [CrossRef]
94. Han, D.; Lee, J.H.; Rizvi, A.; Gransar, H.; Baskaran, L.; Schulman-Marcus, J.; Hartaigh, B.ó.; Lin, F.Y.; Min, J.K. Incremental role of resting myocardial computed tomography perfusion for predicting physiologically significant coronary artery disease: A machine learning approach. *J. Nucl. Cardiol.* **2018**, *25*, 223–233. [CrossRef]
95. Zreik, M.; Lessmann, N.; van Hamersvelt, R.W.; Wolterink, J.M.; Voskuil, M.; Viergever, M.A.; Leiner, T.; Isgum, I. Deep learning analysis of the myocardium in coronary CT angiography for identification of patients with functionally significant coronary artery stenosis. *Med. Image Anal.* **2018**, *44*, 72–85. [CrossRef]
96. Zreik, M.; van Hamersvelt, R.W.; Khalili, N.; Wolterink, J.M.; Voskuil, M.; Viergever, M.A.; Leiner, T.; Isgum, I. Deep Learning Analysis of Coronary Arteries in Cardiac CT Angiography for Detection of Patients Requiring Invasive Coronary Angiography. *IEEE Trans. Med. Imaging* **2020**, *39*, 1545–1557. [CrossRef]
97. Coenen, A.; Kim, Y.H.; Kruk, M.; Tesche, C.; De Geer, J.; Kurata, A.; Lubbers, M.L.; Daemen, J.; Itu, L.; Rapaka, S.; et al. Diagnostic Accuracy of a Machine-Learning Approach to Coronary Computed Tomographic Angiography-Based Fractional Flow Reserve: Result From the MACHINE Consortium. *Circ. Cardiovasc. Imaging* **2018**, *11*, e007217. [CrossRef]
98. Di Jiang, M.; Zhang, X.L.; Liu, H.; Tang, C.X.; Li, J.H.; Wang, Y.N.; Xu, P.P.; Zhou, C.S.; Zhou, F.; Lu, M.J.; et al. The effect of coronary calcification on diagnostic performance of machine learning-based CT-FFR: A Chinese multicenter study. *Eur. Radiol.* **2021**, *31*, 1482–1493. [CrossRef]
99. Koo, H.J.; Kang, J.W.; Kang, S.J.; Kweon, J.; Lee, J.G.; Ahn, J.M.; Park, D.W.; Lee, S.W.; Lee, C.W.; Park, S.W.; et al. Impact of coronary calcium score and lesion characteristics on the diagnostic performance of machine-learning-based computed tomography-derived fractional flow reserve. *Eur. Heart J. Cardiovasc. Imaging* **2021**, *22*, 998–1006. [CrossRef]
100. Kumamaru, K.K.; Fujimoto, S.; Otsuka, Y.; Kawasaki, T.; Kawaguchi, Y.; Kato, E.; Takamura, K.; Aoshima, C.; Kamo, Y.; Kogure, Y.; et al. Diagnostic accuracy of 3D deep-learning-based fully automated estimation of patient-level minimum fractional flow reserve from coronary computed tomography angiography. *Eur. Heart J. Cardiovasc. Imaging* **2020**, *21*, 437–445. [CrossRef]
101. Kurata, A.; Fukuyama, N.; Hirai, K.; Kawaguchi, N.; Tanabe, Y.; Okayama, H.; Shigemi, S.; Watanabe, K.; Uetani, T.; Ikeda, S.; et al. On-Site Computed Tomography-Derived Fractional Flow Reserve Using a Machine-Learning Algorithm—Clinical Effectiveness in a Retrospective Multicenter Cohort. *Circ. J.* **2019**, *83*, 1563–1571. [CrossRef]
102. Rother, J.; Moshage, M.; Dey, D.; Schwemmer, C.; Trobs, M.; Blachutzik, F.; Achenbach, S.; Schlundt, C.; Marwan, M. Comparison of invasively measured FFR with FFR derived from coronary CT angiography for detection of lesion-specific ischemia: Results from a PC-based prototype algorithm. *J. Cardiovasc. Comput. Tomogr.* **2018**, *12*, 101–107. [CrossRef]
103. Tang, C.X.; Wang, Y.N.; Zhou, F.; Schoepf, U.J.; Assen, M.V.; Stroud, R.E.; Li, J.H.; Zhang, X.L.; Lu, M.J.; Zhou, C.S.; et al. Diagnostic performance of fractional flow reserve derived from coronary CT angiography for detection of lesion-specific ischemia: A multi-center study and meta-analysis. *Eur. J. Radiol.* **2019**, *116*, 90–97. [CrossRef]
104. Tesche, C.; Otani, K.; De Cecco, C.N.; Coenen, A.; De Geer, J.; Kruk, M.; Kim, Y.H.; Albrecht, M.H.; Baumann, S.; Renker, M.; et al. Influence of Coronary Calcium on Diagnostic Performance of Machine Learning CT-FFR: Results From MACHINE Registry. *JACC Cardiovasc. Imaging* **2020**, *13*, 760–770. [CrossRef]
105. Wang, Z.Q.; Zhou, Y.J.; Zhao, Y.X.; Shi, D.M.; Liu, Y.Y.; Liu, W.; Liu, X.L.; Li, Y.P. Diagnostic accuracy of a deep learning approach to calculate FFR from coronary CT angiography. *J. Geriatr. Cardiol.* **2019**, *16*, 42–48. [CrossRef]
106. Wardziak, L.; Kruk, M.; Pleban, W.; Demkow, M.; Ruzyllo, W.; Dzielinska, Z.; Kepka, C. Coronary CTA enhanced with CTA based FFR analysis provides higher diagnostic value than invasive coronary angiography in patients with intermediate coronary stenosis. *J. Cardiovasc. Comput. Tomogr.* **2019**, *13*, 62–67. [CrossRef]
107. Tesche, C.; De Cecco, C.N.; Baumann, S.; Renker, M.; McLaurin, T.W.; Duguay, T.M.; Bayer, R.R., 2nd; Steinberg, D.H.; Grant, K.L.; Canstein, C.; et al. Coronary CT Angiography-derived Fractional Flow Reserve: Machine Learning Algorithm versus Computational Fluid Dynamics Modeling. *Radiology* **2018**, *288*, 64–72. [CrossRef]

108. Arbab-Zadeh, A.; Miller, J.M.; Rochitte, C.E.; Dewey, M.; Niinuma, H.; Gottlieb, I.; Paul, N.; Clouse, M.E.; Shapiro, E.P.; Hoe, J.; et al. Diagnostic accuracy of computed tomography coronary angiography according to pre-test probability of coronary artery disease and severity of coronary arterial calcification. The CORE-64 (Coronary Artery Evaluation Using 64-Row Multidetector Computed Tomography Angiography) International Multicenter Study. *J. Am. Coll. Cardiol.* **2012**, *59*, 379–387. [CrossRef]
109. Chen, C.-C.; Chen, C.-C.; Hsieh, I.C.; Liu, Y.-C.; Liu, C.-Y.; Chan, T.; Wen, M.-S.; Wan, Y.-L. The effect of calcium score on the diagnostic accuracy of coronary computed tomography angiography. *Int. J. Cardiovasc. Imaging* **2011**, *27*, 37–42. [CrossRef]
110. Vavere, A.L.; Arbab-Zadeh, A.; Rochitte, C.E.; Dewey, M.; Niinuma, H.; Gottlieb, I.; Clouse, M.E.; Bush, D.E.; Hoe, J.W.M.; de Roos, A.; et al. Coronary artery stenoses: Accuracy of 64-detector row CT angiography in segments with mild, moderate, or severe calcification–a subanalysis of the CORE-64 trial. *Radiology* **2011**, *261*, 100–108. [CrossRef]
111. Arjmand Shabestari, A. Coronary artery calcium score: A review. *Iran Red. Crescent. Med. J.* **2013**, *15*, e16616. [CrossRef]
112. Agatston, A.S.; Janowitz, W.R.; Hildner, F.J.; Zusmer, N.R.; Viamonte, M., Jr.; Detrano, R. Quantification of coronary artery calcium using ultrafast computed tomography. *J. Am. Coll. Cardiol.* **1990**, *15*, 827–832. [CrossRef]
113. Yu, M.; Li, Y.; Li, W.; Lu, Z.; Wei, M.; Zhang, J. Calcification remodeling index assessed by cardiac CT predicts severe coronary stenosis in lesions with moderate to severe calcification. *J. Cardiovasc. Comput. Tomogr.* **2018**, *12*, 42–49. [CrossRef]
114. Sekimoto, T.; Akutsu, Y.; Hamazaki, Y.; Sakai, K.; Kosaki, R.; Yokota, H.; Tsujita, H.; Tsukamoto, S.; Kaneko, K.; Sakurai, M.; et al. Regional calcified plaque score evaluated by multidetector computed tomography for predicting the addition of rotational atherectomy during percutaneous coronary intervention. *J. Cardiovasc. Comput. Tomogr.* **2016**, *10*, 221–228. [CrossRef]
115. Qiao, H.Y.; Tang, C.X.; Schoepf, U.J.; Tesche, C.; Bayer, R.R., 2nd; Giovagnoli, D.A.; Todd Hudson, H., Jr.; Zhou, C.S.; Yan, J.; Lu, M.J.; et al. Impact of machine learning-based coronary computed tomography angiography fractional flow reserve on treatment decisions and clinical outcomes in patients with suspected coronary artery disease. *Eur. Radiol.* **2020**, *30*, 5841–5851. [CrossRef]
116. Liu, X.; Mo, X.; Zhang, H.; Yang, G.; Shi, C.; Hau, W.K. A 2-year investigation of the impact of the computed tomography-derived fractional flow reserve calculated using a deep learning algorithm on routine decision-making for coronary artery disease management. *Eur. Radiol.* **2021**, *31*, 7039–7046. [CrossRef]
117. Martin, S.S.; Mastrodicasa, D.; van Assen, M.; De Cecco, C.N.; Bayer, R.R.; Tesche, C.; Varga-Szemes, A.; Fischer, A.M.; Jacobs, B.E.; Sahbaee, P.; et al. Value of Machine Learning-based Coronary CT Fractional Flow Reserve Applied to Triple-Rule-Out CT Angiography in Acute Chest Pain. *Radiol. Cardiothorac. Imaging* **2020**, *2*, e190137. [CrossRef]
118. Nous, F.M.A.; Budde, R.P.J.; Lubbers, M.M.; Yamasaki, Y.; Kardys, I.; Bruning, T.A.; Akkerhuis, J.M.; Kofflard, M.J.M.; Kietselaer, B.; Galema, T.W.; et al. Impact of machine-learning CT-derived fractional flow reserve for the diagnosis and management of coronary artery disease in the randomized CRESCENT trials. *Eur. Radiol.* **2020**, *30*, 3692–3701. [CrossRef]
119. Cook, C.M.; Petraco, R.; Shun-Shin, M.J.; Ahmad, Y.; Nijjer, S.; Al-Lamee, R.; Kikuta, Y.; Shiono, Y.; Mayet, J.; Francis, D.P.; et al. Diagnostic Accuracy of Computed Tomography–Derived Fractional Flow Reserve: A Systematic Review. *JAMA Cardiol.* **2017**, *2*, 803–810. [CrossRef]
120. Gaur, S.; Ovrehus, K.A.; Dey, D.; Leipsic, J.; Botker, H.E.; Jensen, J.M.; Narula, J.; Ahmadi, A.; Achenbach, S.; Ko, B.S.; et al. Coronary plaque quantification and fractional flow reserve by coronary computed tomography angiography identify ischaemia-causing lesions. *Eur. Heart J.* **2016**, *37*, 1220–1227. [CrossRef]
121. Kawasaki, T.; Kidoh, M.; Kido, T.; Sueta, D.; Fujimoto, S.; Kumamaru, K.K.; Uetani, T.; Tanabe, Y.; Ueda, T.; Sakabe, D.; et al. Evaluation of Significant Coronary Artery Disease Based on CT Fractional Flow Reserve and Plaque Characteristics Using Random Forest Analysis in Machine Learning. *Acad. Radiol.* **2020**, *27*, 1700–1708. [CrossRef]
122. Vasquez, A.; Mistry, N.; Singh, J. Impact of Intravascular Ultrasound in Clinical Practice. *Interv. Cardiol.* **2014**, *9*, 156–163. [CrossRef]
123. Metz, J.A.; Yock, P.G.; Fitzgerald, P.J. Intravascular ultrasound: Basic interpretation. *Cardiol. Clin.* **1997**, *15*, 1–15. [CrossRef]
124. Ma, T.; Yu, M.; Li, J.; Munding, C.E.; Chen, Z.; Fei, C.; Shung, K.K.; Zhou, Q. Multi-frequency intravascular ultrasound (IVUS) imaging. *IEEE Trans. Ultrason. Ferroelectr. Freq. Control* **2015**, *62*, 97–107. [CrossRef]
125. Pu, J.; Mintz, G.S.; Biro, S.; Lee, J.-B.; Sum, S.T.; Madden, S.P.; Burke, A.P.; Zhang, P.; He, B.; Goldstein, J.A.; et al. Insights Into Echo-Attenuated Plaques, Echolucent Plaques, and Plaques With Spotty Calcification: Novel Findings From Comparisons Among Intravascular Ultrasound, Near-Infrared Spectroscopy, and Pathological Histology in 2294 Human Coronary Artery Segments. *J. Am. Coll. Cardiol.* **2014**, *63*, 2220–2233. [CrossRef]
126. Mintz, G.S.; Pichard, A.D.; Popma, J.J.; Kent, K.M.; Satler, L.F.; Bucher, T.A.; Leon, M.B. Determinants and Correlates of Target Lesion Calcium in Coronary Artery Disease: A Clinical, Angiographic and Intravascular Ultrasound Study. *J. Am. Coll. Cardiol.* **1997**, *29*, 268–274. [CrossRef]
127. Kobayashi, Y.; Okura, H.; Kume, T.; Yamada, R.; Kobayashi, Y.; Fukuhara, K.; Koyama, T.; Nezuo, S.; Neishi, Y.; Hayashida, A.; et al. Impact of target lesion coronary calcification on stent expansion. *Circ. J.* **2014**, *78*, 2209–2214. [CrossRef]
128. Nair, A.; Kuban, B.D.; Tuzcu, E.M.; Schoenhagen, P.; Nissen, S.E.; Vince, D.G. Coronary Plaque Classification With Intravascular Ultrasound Radiofrequency Data Analysis. *Circulation* **2002**, *106*, 2200–2206. [CrossRef]
129. Sonoda, S.; Hibi, K.; Okura, H.; Fujii, K.; Honda, Y.; Kobayashi, Y. Current clinical use of intravascular ultrasound imaging to guide percutaneous coronary interventions. *Cardiovasc. Interv.* **2020**, *35*, 30–36. [CrossRef]

130. Maehara, A.; Matsumura, M.; Ali, Z.A.; Mintz, G.S.; Stone, G.W. IVUS-Guided Versus OCT-Guided Coronary Stent Implantation: A Critical Appraisal. *JACC Cardiovasc. Imaging* **2017**, *10*, 1487–1503. [CrossRef]
131. Malik, A.H.; Yandrapalli, S.; Aronow, W.S.; Panza, J.A.; Cooper, H.A. Intravascular ultrasound-guided stent implantation reduces cardiovascular mortality—Updated meta-analysis of randomized controlled trials. *Int. J. Cardiol.* **2020**, *299*, 100–105. [CrossRef]
132. Chieffo, A.; Latib, A.; Caussin, C.; Presbitero, P.; Galli, S.; Menozzi, A.; Varbella, F.; Mauri, F.; Valgimigli, M.; Arampatzis, C.; et al. A prospective, randomized trial of intravascular-ultrasound guided compared to angiography guided stent implantation in complex coronary lesions: The AVIO trial. *Am. Heart J.* **2013**, *165*, 65–72. [CrossRef]
133. Sinclair, H.; Bourantas, C.; Bagnall, A.; Mintz, G.S.; Kunadian, V. OCT for the identification of vulnerable plaque in acute coronary syndrome. *JACC Cardiovasc. Imaging* **2015**, *8*, 198–209. [CrossRef]
134. Cheng, J.M.; Garcia-Garcia, H.M.; de Boer, S.P.; Kardys, I.; Heo, J.H.; Akkerhuis, K.M.; Oemrawsingh, R.M.; van Domburg, R.T.; Ligthart, J.; Witberg, K.T.; et al. In vivo detection of high-risk coronary plaques by radiofrequency intravascular ultrasound and cardiovascular outcome: Results of the ATHEROREMO-IVUS study. *Eur. Heart J.* **2014**, *35*, 639–647. [CrossRef]
135. Räber, L.; Ueki, Y. Outcomes of Intravascular Ultrasound-Guided Percutaneous Coronary Intervention in the United States. *JACC Cardiovasc. Interv.* **2020**, *13*, 1891–1893. [CrossRef]
136. Ali, Z.A.; Karimi Galougahi, K.; Maehara, A.; Shlofmitz, R.A.; Ben-Yehuda, O.; Mintz, G.S.; Stone, G.W. Intracoronary Optical Coherence Tomography 2018: Current Status and Future Directions. *JACC Cardiovasc. Interv.* **2017**, *10*, 2473–2487. [CrossRef]
137. Bae, Y.; Kang, S.J.; Kim, G.; Lee, J.G.; Min, H.S.; Cho, H.; Kang, D.Y.; Lee, P.H.; Ahn, J.M.; Park, D.W.; et al. Prediction of coronary thin-cap fibroatheroma by intravascular ultrasound-based machine learning. *Atherosclerosis* **2019**, *288*, 168–174. [CrossRef]
138. Min, H.S.; Yoo, J.H.; Kang, S.J.; Lee, J.G.; Cho, H.; Lee, P.H.; Ahn, J.M.; Park, D.W.; Lee, S.W.; Kim, Y.H.; et al. Detection of optical coherence tomography-defined thin-cap fibroatheroma in the coronary artery using deep learning. *EuroIntervention* **2020**, *16*, 404–412. [CrossRef]
139. Cho, H.; Kang, S.J.; Min, H.S.; Lee, J.G.; Kim, W.J.; Kang, S.H.; Kang, D.Y.; Lee, P.H.; Ahn, J.M.; Park, D.W.; et al. Intravascular ultrasound-based deep learning for plaque characterization in coronary artery disease. *Atherosclerosis* **2021**, *324*, 69–75. [CrossRef]
140. Hong, M.K.; Mintz, G.S.; Lee, C.W.; Park, D.W.; Choi, B.R.; Park, K.H.; Kim, Y.H.; Cheong, S.S.; Song, J.K.; Kim, J.J.; et al. Intravascular ultrasound predictors of angiographic restenosis after sirolimus-eluting stent implantation. *Eur. Heart J.* **2006**, *27*, 1305–1310. [CrossRef]
141. Song, H.G.; Kang, S.J.; Ahn, J.M.; Kim, W.J.; Lee, J.Y.; Park, D.W.; Lee, S.W.; Kim, Y.H.; Lee, C.W.; Park, S.W.; et al. Intravascular ultrasound assessment of optimal stent area to prevent in-stent restenosis after zotarolimus-, everolimus-, and sirolimus-eluting stent implantation. *Catheter. Cardiovasc. Interv.* **2014**, *83*, 873–878. [CrossRef] [PubMed]
142. Fujii, K.; Carlier, S.G.; Mintz, G.S.; Yang, Y.M.; Moussa, I.; Weisz, G.; Dangas, G.; Mehran, R.; Lansky, A.J.; Kreps, E.M.; et al. Stent underexpansion and residual reference segment stenosis are related to stent thrombosis after sirolimus-eluting stent implantation: An intravascular ultrasound study. *J. Am. Coll. Cardiol.* **2005**, *45*, 995–998. [CrossRef] [PubMed]
143. Doi, H.; Maehara, A.; Mintz, G.S.; Yu, A.; Wang, H.; Mandinov, L.; Popma, J.J.; Ellis, S.G.; Grube, E.; Dawkins, K.D.; et al. Impact of post-intervention minimal stent area on 9-month follow-up patency of paclitaxel-eluting stents: An integrated intravascular ultrasound analysis from the TAXUS IV, V, and VI and TAXUS ATLAS Workhorse, Long Lesion, and Direct Stent Trials. *JACC Cardiovasc. Interv.* **2009**, *2*, 1269–1275. [CrossRef] [PubMed]
144. Min, H.S.; Ryu, D.; Kang, S.J.; Lee, J.G.; Yoo, J.H.; Cho, H.; Kang, D.Y.; Lee, P.H.; Ahn, J.M.; Park, D.W.; et al. Prediction of Coronary Stent Underexpansion by Pre-Procedural Intravascular Ultrasound-Based Deep Learning. *JACC Cardiovasc. Interv.* **2021**, *14*, 1021–1029. [CrossRef]
145. Nishi, T.; Yamashita, R.; Imura, S.; Tateishi, K.; Kitahara, H.; Kobayashi, Y.; Yock, P.G.; Fitzgerald, P.J.; Honda, Y. Deep learning-based intravascular ultrasound segmentation for the assessment of coronary artery disease. *Int. J. Cardiol.* **2021**, *333*, 55–59. [CrossRef] [PubMed]
146. Brown, A.J.; Teng, Z.; Calvert, P.A.; Rajani, N.K.; Hennessy, O.; Nerlekar, N.; Obaid, D.R.; Costopoulos, C.; Huang, Y.; Hoole, S.P.; et al. Plaque Structural Stress Estimations Improve Prediction of Future Major Adverse Cardiovascular Events After Intracoronary Imaging. *Circ. Cardiovasc. Imaging* **2016**, *9*, e004172. [CrossRef]
147. Xie, Z.; Dong, N.; Sun, R.; Liu, X.; Gu, X.; Sun, Y.; Du, H.; Dai, J.; Liu, Y.; Hou, J.; et al. Relation between baseline plaque features and subsequent coronary artery remodeling determined by optical coherence tomography and intravascular ultrasound. *Oncotarget* **2017**, *8*, 4234–4244. [CrossRef]
148. Stone, P.H.; Saito, S.; Takahashi, S.; Makita, Y.; Nakamura, S.; Kawasaki, T.; Takahashi, A.; Katsuki, T.; Nakamura, S.; Namiki, A.; et al. Prediction of progression of coronary artery disease and clinical outcomes using vascular profiling of endothelial shear stress and arterial plaque characteristics: The PREDICTION Study. *Circulation* **2012**, *126*, 172–181. [CrossRef]
149. Calvert, P.A.; Obaid, D.R.; O'Sullivan, M.; Shapiro, L.M.; McNab, D.; Densem, C.G.; Schofield, P.M.; Braganza, D.; Clarke, S.C.; Ray, K.K.; et al. Association between IVUS findings and adverse outcomes in patients with coronary artery disease: The VIVA (VH-IVUS in Vulnerable Atherosclerosis) Study. *JACC Cardiovasc. Imaging* **2011**, *4*, 894–901. [CrossRef]
150. Zhang, L.; Wahle, A.; Chen, Z.; Lopez, J.J.; Kovarnik, T.; Sonka, M. Predicting Locations of High-Risk Plaques in Coronary Arteries in Patients Receiving Statin Therapy. *IEEE Trans. Med. Imaging* **2018**, *37*, 151–161. [CrossRef]
151. Farooq, V.; Brugaletta, S.; Serruys, P.W. The SYNTAX score and SYNTAX-based clinical risk scores. *Semin Thorac Cardiovasc Surg* **2011**, *23*, 99–105. [CrossRef] [PubMed]

152. Singh, M.; Rihal, C.S.; Lennon, R.J.; Spertus, J.; Rumsfeld, J.S.; Holmes, D.R., Jr. Bedside estimation of risk from percutaneous coronary intervention: The new Mayo Clinic risk scores. *Mayo Clin. Proc.* **2007**, *82*, 701–708. [CrossRef]
153. Chowdhary, S.; Ivanov, J.; Mackie, K.; Seidelin, P.H.; Dzavík, V. The Toronto score for in-hospital mortality after percutaneous coronary interventions. *Am. Heart J.* **2009**, *157*, 156–163. [CrossRef] [PubMed]
154. Hannan, E.L.; Farrell, L.S.; Walford, G.; Jacobs, A.K.; Berger, P.B.; Holmes, D.R., Jr.; Stamato, N.J.; Sharma, S.; King, S.B., 3rd. The New York State risk score for predicting in-hospital/30-day mortality following percutaneous coronary intervention. *JACC Cardiovasc. Interv.* **2013**, *6*, 614–622. [CrossRef] [PubMed]
155. MacKenzie, T.A.; Malenka, D.J.; Olmstead, E.M.; Piper, W.D.; Langner, C.; Ross, C.S.; O'Connor, G.T. Prediction of survival after coronary revascularization: Modeling short-term, mid-term, and long-term survival. *Ann. Thorac. Surg.* **2009**, *87*, 463–472. [CrossRef] [PubMed]
156. O'Connor, G.T.; Malenka, D.J.; Quinton, H.; Robb, J.F.; Kellett, M.A., Jr.; Shubrooks, S.; Bradley, W.A.; Hearne, M.J.; Watkins, M.W.; Wennberg, D.E.; et al. Multivariate prediction of in-hospital mortality after percutaneous coronary interventions in 1994-1996. Northern New England Cardiovascular Disease Study Group. *J. Am. Coll. Cardiol.* **1999**, *34*, 681–691. [CrossRef]
157. Rihal, C.S.; Grill, D.E.; Bell, M.R.; Berger, P.B.; Garratt, K.N.; Holmes, D.R., Jr. Prediction of death after percutaneous coronary interventional procedures. *Am. Heart J.* **2000**, *139*, 1032–1038. [CrossRef]
158. Wu, C.; Hannan, E.L.; Walford, G.; Ambrose, J.A.; Holmes, D.R., Jr.; King, S.B., 3rd; Clark, L.T.; Katz, S.; Sharma, S.; Jones, R.H. A risk score to predict in-hospital mortality for percutaneous coronary interventions. *J. Am. Coll. Cardiol.* **2006**, *47*, 654–660. [CrossRef]
159. Fanaroff, A.C.; Zakroysky, P.; Dai, D.; Wojdyla, D.; Sherwood, M.W.; Roe, M.T.; Wang, T.Y.; Peterson, E.D.; Gurm, H.S.; Cohen, M.G.; et al. Outcomes of PCI in Relation to Procedural Characteristics and Operator Volumes in the United States. *J. Am. Coll. Cardiol.* **2017**, *69*, 2913–2924. [CrossRef]
160. Iverson, A.; Stanberry, L.I.; Tajti, P.; Garberich, R.; Antos, A.; Burke, M.N.; Chavez, I.; Gössl, M.; Henry, T.D.; Lips, D.; et al. Prevalence, Trends, and Outcomes of Higher-Risk Percutaneous Coronary Interventions Among Patients without Acute Coronary Syndromes. *Cardiovasc. Revasc. Med.* **2019**, *20*, 289–292. [CrossRef]
161. Singh, M.; Lennon, R.J.; Gulati, R.; Holmes, D.R. Risk scores for 30-day mortality after percutaneous coronary intervention: New insights into causes and risk of death. *Mayo Clin. Proc.* **2014**, *89*, 631–637. [CrossRef] [PubMed]
162. Zack, C.J.; Senecal, C.; Kinar, Y.; Metzger, Y.; Bar-Sinai, Y.; Widmer, R.J.; Lennon, R.; Singh, M.; Bell, M.R.; Lerman, A.; et al. Leveraging Machine Learning Techniques to Forecast Patient Prognosis After Percutaneous Coronary Intervention. *JACC Cardiovasc. Interv.* **2019**, *12*, 1304–1311. [CrossRef] [PubMed]
163. Al'Aref, S.J.; Singh, G.; van Rosendael, A.R.; Kolli, K.K.; Ma, X.; Maliakal, G.; Pandey, M.; Lee, B.C.; Wang, J.; Xu, Z.; et al. Determinants of In-Hospital Mortality after Percutaneous Coronary Intervention: A Machine Learning Approach. *J. Am. Heart Assoc.* **2019**, *8*, e011160. [CrossRef] [PubMed]
164. Rao, S.V.; Kaul, P.R.; Liao, L.; Armstrong, P.W.; Ohman, E.M.; Granger, C.B.; Califf, R.M.; Harrington, R.A.; Eisenstein, E.L.; Mark, D.B. Association between bleeding, blood transfusion, and costs among patients with non-ST-segment elevation acute coronary syndromes. *Am. Heart J.* **2008**, *155*, 369–374. [CrossRef] [PubMed]
165. Kinnaird, T.D.; Stabile, E.; Mintz, G.S.; Lee, C.W.; Canos, D.A.; Gevorkian, N.; Pinnow, E.E.; Kent, K.M.; Pichard, A.D.; Satler, L.F.; et al. Incidence, predictors, and prognostic implications of bleeding and blood transfusion following percutaneous coronary interventions. *Am. J. Cardiol.* **2003**, *92*, 930–935. [CrossRef]
166. Rao, S.V.; McCoy, L.A.; Spertus, J.A.; Krone, R.J.; Singh, M.; Fitzgerald, S.; Peterson, E.D. An Updated Bleeding Model to Predict the Risk of Post-Procedure Bleeding Among Patients Undergoing Percutaneous Coronary Intervention: A Report Using an Expanded Bleeding Definition From the National Cardiovascular Data Registry CathPCI Registry. *JACC Cardiovasc. Interv.* **2013**, *6*, 897–904. [CrossRef]
167. Mortazavi, B.J.; Bucholz, E.M.; Desai, N.R.; Huang, C.; Curtis, J.P.; Masoudi, F.A.; Shaw, R.E.; Negahban, S.N.; Krumholz, H.M. Comparison of Machine Learning Methods With National Cardiovascular Data Registry Models for Prediction of Risk of Bleeding After Percutaneous Coronary Intervention. *JAMA Netw. Open* **2019**, *2*, e196835. [CrossRef]
168. Kim, M.S.; Dean, L.S. In-stent restenosis. *Cardiovasc. Ther.* **2011**, *29*, 190–198. [CrossRef]
169. Cassese, S.; Byrne, R.A.; Tada, T.; Pinieck, S.; Joner, M.; Ibrahim, T.; King, L.A.; Fusaro, M.; Laugwitz, K.L.; Kastrati, A. Incidence and predictors of restenosis after coronary stenting in 10 004 patients with surveillance angiography. *Heart* **2014**, *100*, 153–159. [CrossRef]
170. Singh, M.; Gersh, B.J.; McClelland, R.L.; Ho, K.K.L.; Willerson, J.T.; Penny, W.F.; Holmes, D.R. Clinical and Angiographic Predictors of Restenosis After Percutaneous Coronary Intervention. *Circulation* **2004**, *109*, 2727–2731. [CrossRef]
171. Stolker, J.M.; Kennedy, K.F.; Lindsey, J.B.; Marso, S.P.; Pencina, M.J.; Cutlip, D.E.; Mauri, L.; Kleiman, N.S.; Cohen, D.J. Predicting Restenosis of Drug-Eluting Stents Placed in Real-World Clinical Practice. *Circ. Cardiovasc. Interv.* **2010**, *3*, 327–334. [CrossRef] [PubMed]
172. Sampedro-Gomez, J.; Dorado-Diaz, P.I.; Vicente-Palacios, V.; Sanchez-Puente, A.; Jimenez-Navarro, M.; San Roman, J.A.; Galindo-Villardon, P.; Sanchez, P.L.; Fernandez-Aviles, F. Machine Learning to Predict Stent Restenosis Based on Daily Demographic, Clinical, and Angiographic Characteristics. *Can. J. Cardiol.* **2020**, *36*, 1624–1632. [CrossRef] [PubMed]

173. Steele, A.J.; Denaxas, S.C.; Shah, A.D.; Hemingway, H.; Luscombe, N.M. Machine learning models in electronic health records can outperform conventional survival models for predicting patient mortality in coronary artery disease. *PLoS ONE* **2018**, *13*, e0202344. [CrossRef]
174. Bertsimas, D.; Orfanoudaki, A.; Weiner, R.B. Personalized treatment for coronary artery disease patients: A machine learning approach. *Health Care Manag. Sci.* **2020**, *23*, 482–506. [CrossRef] [PubMed]
175. Farhadian, M.; Dehdar Karsidani, S.; Mozayanimonfared, A.; Mahjub, H. Risk factors associated with major adverse cardiac and cerebrovascular events following percutaneous coronary intervention: A 10-year follow-up comparing random survival forest and Cox proportional-hazards model. *BMC Cardiovasc. Disord.* **2021**, *21*, 38. [CrossRef]
176. Krittanawong, C.; Zhang, H.; Wang, Z.; Aydar, M.; Kitai, T. Artificial Intelligence in Precision Cardiovascular Medicine. *J. Am. Coll. Cardiol.* **2017**, *69*, 2657–2664. [CrossRef]
177. Taylor, A.J.; Bindeman, J.; Feuerstein, I.; Cao, F.; Brazaitis, M.; O'Malley, P.G. Coronary calcium independently predicts incident premature coronary heart disease over measured cardiovascular risk factors: Mean three-year outcomes in the Prospective Army Coronary Calcium (PACC) project. *J. Am. Coll. Cardiol.* **2005**, *46*, 807–814. [CrossRef]
178. Detrano, R.; Guerci, A.D.; Carr, J.J.; Bild, D.E.; Burke, G.; Folsom, A.R.; Liu, K.; Shea, S.; Szklo, M.; Bluemke, D.A.; et al. Coronary Calcium as a Predictor of Coronary Events in Four Racial or Ethnic Groups. *N. Engl. J. Med.* **2008**, *358*, 1336–1345. [CrossRef]
179. Rozanski, A.; Gransar, H.; Shaw, L.J.; Kim, J.; Miranda-Peats, L.; Wong, N.D.; Rana, J.S.; Orakzai, R.; Hayes, S.W.; Friedman, J.D.; et al. Impact of coronary artery calcium scanning on coronary risk factors and downstream testing the EISNER (Early Identification of Subclinical Atherosclerosis by Noninvasive Imaging Research) prospective randomized trial. *J. Am. Coll. Cardiol.* **2011**, *57*, 1622–1632. [CrossRef]
180. Hwang, I.-C.; Park, H.E.; Choi, S.-Y. Epicardial Adipose Tissue Contributes to the Development of Non-Calcified Coronary Plaque: A 5-Year Computed Tomography Follow-up Study. *J. Atheroscler. Thromb.* **2017**, *24*, 262–274. [CrossRef]
181. Nakanishi, R.; Rajani, R.; Cheng, V.Y.; Gransar, H.; Nakazato, R.; Shmilovich, H.; Otaki, Y.; Hayes, S.W.; Thomson, L.E.; Friedman, J.D.; et al. Increase in epicardial fat volume is associated with greater coronary artery calcification progression in subjects at intermediate risk by coronary calcium score: A serial study using non-contrast cardiac CT. *Atherosclerosis* **2011**, *218*, 363–368. [CrossRef] [PubMed]
182. Berman, D.S.; Arnson, Y.; Rozanski, A. Coronary Artery Calcium Scanning: The Agatston Score and Beyond. *JACC Cardiovasc. Imaging* **2016**, *9*, 1417–1419. [CrossRef] [PubMed]
183. Chao, H.; Shan, H.; Homayounieh, F.; Singh, R.; Khera, R.D.; Guo, H.; Su, T.; Wang, G.; Kalra, M.K.; Yan, P. Deep learning predicts cardiovascular disease risks from lung cancer screening low dose computed tomography. *Nat. Commun.* **2021**, *12*, 2963. [CrossRef] [PubMed]
184. Wolterink, J.; Leiner, T.; Takx, R.A.; Viergever, M.; Išgum, I. *An Automatic Machine Learning System for Coronary Calcium Scoring in Clinical Non-Contrast Enhanced, ECG-Triggered Cardiac CT*; SPIE: San Diego, CA, USA, 2014; Volume 9035.
185. Sandstedt, M.; Henriksson, L.; Janzon, M.; Nyberg, G.; Engvall, J.; De Geer, J.; Alfredsson, J.; Persson, A. Evaluation of an AI-based, automatic coronary artery calcium scoring software. *Eur. Radiol.* **2020**, *30*, 1671–1678. [CrossRef] [PubMed]
186. Commandeur, F.; Slomka, P.J.; Goeller, M.; Chen, X.; Cadet, S.; Razipour, A.; McElhinney, P.; Gransar, H.; Cantu, S.; Miller, R.J.H.; et al. Machine learning to predict the long-term risk of myocardial infarction and cardiac death based on clinical risk, coronary calcium, and epicardial adipose tissue: A prospective study. *Cardiovasc. Res.* **2020**, *116*, 2216–2225. [CrossRef]
187. Eisenberg, E.; McElhinney, P.A.; Commandeur, F.; Chen, X.; Cadet, S.; Goeller, M.; Razipour, A.; Gransar, H.; Cantu, S.; Miller, R.J.H.; et al. Deep Learning-Based Quantification of Epicardial Adipose Tissue Volume and Attenuation Predicts Major Adverse Cardiovascular Events in Asymptomatic Subjects. *Circ. Cardiovasc. Imaging* **2020**, *13*, e009829. [CrossRef]
188. Han, D.; Kolli, K.K.; Gransar, H.; Lee, J.H.; Choi, S.Y.; Chun, E.J.; Han, H.W.; Park, S.H.; Sung, J.; Jung, H.O.; et al. Machine learning based risk prediction model for asymptomatic individuals who underwent coronary artery calcium score: Comparison with traditional risk prediction approaches. *J. Cardiovasc. Comput. Tomogr.* **2020**, *14*, 168–176. [CrossRef]
189. Tamarappoo, B.K.; Lin, A.; Commandeur, F.; McElhinney, P.A.; Cadet, S.; Goeller, M.; Razipour, A.; Chen, X.; Gransar, H.; Cantu, S.; et al. Machine learning integration of circulating and imaging biomarkers for explainable patient-specific prediction of cardiac events: A prospective study. *Atherosclerosis* **2021**, *318*, 76–82. [CrossRef]
190. Nakanishi, R.; Slomka, P.J.; Rios, R.; Betancur, J.; Blaha, M.J.; Nasir, K.; Miedema, M.D.; Rumberger, J.A.; Gransar, H.; Shaw, L.J.; et al. Machine Learning Adds to Clinical and CAC Assessments in Predicting 10-Year CHD and CVD Deaths. *JACC Cardiovasc. Imaging* **2021**, *14*, 615–625. [CrossRef]
191. Min, J.K.; Feignoux, J.; Treutenaere, J.; Laperche, T.; Sablayrolles, J. The prognostic value of multidetector coronary CT angiography for the prediction of major adverse cardiovascular events: A multicenter observational cohort study. *Int. J. Cardiovasc. Imaging* **2010**, *26*, 721–728. [CrossRef]
192. Hadamitzky, M.; Achenbach, S.; Al-Mallah, M.; Berman, D.; Budoff, M.; Cademartiri, F.; Callister, T.; Chang, H.J.; Cheng, V.; Chinnaiyan, K.; et al. Optimized prognostic score for coronary computed tomographic angiography: Results from the CONFIRM registry (COronary CT Angiography EvaluatioN For Clinical Outcomes: An InteRnational Multicenter Registry). *J. Am. Coll. Cardiol.* **2013**, *62*, 468–476. [CrossRef] [PubMed]
193. Min, J.K.; Shaw, L.J.; Devereux, R.B.; Okin, P.M.; Weinsaft, J.W.; Russo, D.J.; Lippolis, N.J.; Berman, D.S.; Callister, T.Q. Prognostic Value of Multidetector Coronary Computed Tomographic Angiography for Prediction of All-Cause Mortality. *J. Am. Coll. Cardiol.* **2007**, *50*, 1161–1170. [CrossRef] [PubMed]

194. Johnson, K.M.; Dowe, D.A. Prognostic Implications of Coronary CT Angiography: 12-Year Follow-Up of 6892 Patients. *AJR Am. J. Roentgenol.* **2020**, *215*, 818–827. [CrossRef] [PubMed]
195. Motwani, M.; Dey, D.; Berman, D.S.; Germano, G.; Achenbach, S.; Al-Mallah, M.H.; Andreini, D.; Budoff, M.J.; Cademartiri, F.; Callister, T.Q.; et al. Machine learning for prediction of all-cause mortality in patients with suspected coronary artery disease: A 5-year multicentre prospective registry analysis. *Eur. Heart J.* **2016**, *38*, 500–507. [CrossRef] [PubMed]
196. Tesche, C.; Bauer, M.J.; Baquet, M.; Hedels, B.; Straube, F.; Hartl, S.; Gray, H.N.; Jochheim, D.; Aschauer, T.; Rogowski, S.; et al. Improved long-term prognostic value of coronary CT angiography-derived plaque measures and clinical parameters on adverse cardiac outcome using machine learning. *Eur. Radiol.* **2021**, *31*, 486–493. [CrossRef]
197. van Rosendael, A.R.; Maliakal, G.; Kolli, K.K.; Beecy, A.; Al'Aref, S.J.; Dwivedi, A.; Singh, G.; Panday, M.; Kumar, A.; Ma, X.; et al. Maximization of the usage of coronary CTA derived plaque information using a machine learning based algorithm to improve risk stratification; insights from the CONFIRM registry. *J. Cardiovasc. Comput. Tomogr.* **2018**, *12*, 204–209. [CrossRef]
198. Hoshino, M.; Zhang, J.; Sugiyama, T.; Yang, S.; Kanaji, Y.; Hamaya, R.; Yamaguchi, M.; Hada, M.; Misawa, T.; Usui, E.; et al. Prognostic value of pericoronary inflammation and unsupervised machine-learning-defined phenotypic clustering of CT angiographic findings. *Int. J. Cardiol.* **2021**, *333*, 226–232. [CrossRef]
199. Johnson, K.M.; Johnson, H.E.; Zhao, Y.; Dowe, D.A.; Staib, L.H. Scoring of Coronary Artery Disease Characteristics on Coronary CT Angiograms by Using Machine Learning. *Radiology* **2019**, *292*, 354–362. [CrossRef]
200. Antonopoulos, A.S.; Sanna, F.; Sabharwal, N.; Thomas, S.; Oikonomou, E.K.; Herdman, L.; Margaritis, M.; Shirodaria, C.; Kampoli, A.M.; Akoumianakis, I.; et al. Detecting human coronary inflammation by imaging perivascular fat. *Sci. Transl. Med.* **2017**, *9*, eaal2658. [CrossRef]
201. Oikonomou, E.K.; Marwan, M.; Desai, M.Y.; Mancio, J.; Alashi, A.; Hutt Centeno, E.; Thomas, S.; Herdman, L.; Kotanidis, C.P.; Thomas, K.E.; et al. Non-invasive detection of coronary inflammation using computed tomography and prediction of residual cardiovascular risk (the CRISP CT study): A post-hoc analysis of prospective outcome data. *Lancet* **2018**, *392*, 929–939. [CrossRef]
202. Oikonomou Evangelos, K.; Desai Milind, Y.; Marwan, M.; Kotanidis Christos, P.; Antonopoulos Alexios, S.; Schottlander, D.; Channon Keith, M.; Neubauer, S.; Achenbach, S.; Antoniades, C. Perivascular Fat Attenuation Index Stratifies Cardiac Risk Associated with High-Risk Plaques in the CRISP-CT Study. *J. Am. Coll. Cardiol.* **2020**, *76*, 755–757. [CrossRef] [PubMed]
203. Oikonomou, E.K.; Williams, M.C.; Kotanidis, C.P.; Desai, M.Y.; Marwan, M.; Antonopoulos, A.S.; Thomas, K.E.; Thomas, S.; Akoumianakis, I.; Fan, L.M.; et al. A novel machine learning-derived radiotranscriptomic signature of perivascular fat improves cardiac risk prediction using coronary CT angiography. *Eur. Heart J.* **2019**, *40*, 3529–3543. [CrossRef] [PubMed]
204. Cabitza, F.; Rasoini, R.; Gensini, G.F. Unintended Consequences of Machine Learning in Medicine. *JAMA* **2017**, *318*, 517–518. [CrossRef] [PubMed]
205. Petch, J.; Di, S.; Nelson, W. Opening the black box: The promise and limitations of explainable machine learning in cardiology. *Can. J. Cardiol.* **2021**. [CrossRef]
206. Lundberg, S.M.; Lee, S.-I. A unified approach to interpreting model predictions. In Proceedings of the 31st International Conference on Neural Information Processing Systems, Long Beach, CA, USA, 4–9 December 2017; pp. 4768–4777
207. Vokinger, K.N.; Feuerriegel, S.; Kesselheim, A.S. Mitigating bias in machine learning for medicine. *Commun. Med.* **2021**, *1*, 25. [CrossRef]
208. Tat, E.; Bhatt, D.L.; Rabbat, M.G. Addressing bias: Artificial intelligence in cardiovascular medicine. *Lancet Digit Health* **2020**, *2*, e635–e636. [CrossRef]
209. Luo, Y. Evaluating the state of the art in missing data imputation for clinical data. *Brief. Bioinform.* **2021**, *23*, bbab489. [CrossRef]
210. Dietterich, T. Overfitting and undercomputing in machine learning. *ACM Comput. Surv. (CSUR)* **1995**, *27*, 326–327. [CrossRef]
211. Murdoch, B. Privacy and artificial intelligence: Challenges for protecting health information in a new era. *BMC Med. Ethics* **2021**, *22*, 122. [CrossRef]

healthcare

MDPI

Article

A Noninvasive Risk Stratification Tool Build Using an Artificial Intelligence Approach for Colorectal Polyps Based on Annual Checkup Data

Chieh Lee [1], Tsung-Hsing Lin [2], Chen-Ju Lin [3], Chang-Fu Kuo [4], Betty Chien-Jung Pai [5], Hao-Tsai Cheng [6], Cheng-Chou Lai [7] and Tsung-Hsing Chen [8,*]

1 Department of Information Management, National Sun Yat-sen University, Kaohsiung 804, Taiwan; chiehlee850427@gmail.com
2 Department of Emergency Medicine, Kuang Tien General Hospital, Taichung City 433, Taiwan; drsixmg@gmail.com
3 Department of Industrial Engineering & Management, College of Engineering, Yuan Ze University, Chung-Li City 320, Taiwan; chenju.lin@saturn.yzu.edu.tw
4 Division of Rheumatology, Allergy, and Immunology, Linkou Chang Gung Memorial Hospital, Chang Gung University College of Medicine, Taoyuan 333, Taiwan; zandis@gmail.com
5 Craniofacial Orthodontics, Craniofacial Research Center, Chang Gung Memorial Hospital, Chang Gung University, Taoyuan 333, Taiwan; pai0072@cgmh.org.tw
6 Division of Gastroenterology and Hepatology, Department of Internal Medicine, New Taipei Municipal TuCheng Hospital, New Taipei City 236, Taiwan; hautai@cloud.cgmh.org.tw
7 Department of Colon and Rectal Surgery, Linkou Medical Center, Chang Gung Memorial Hospital, Taoyuan 333, Taiwan; lai5556@cgmh.org.tw
8 Department of Gastroenterology and Hepatology, Linkou Chang Gung Memorial Hospital, Chang Gung University College of Medicine, Taoyuan 333, Taiwan
* Correspondence: itochenyu@gmail.com

Citation: Lee, C.; Lin, T.-H.; Lin, C.-J.; Kuo, C.-F.; Pai, B.C.-J.; Cheng, H.-T.; Lai, C.-C.; Chen, T.-H. A Noninvasive Risk Stratification Tool Build Using an Artificial Intelligence Approach for Colorectal Polyps Based on Annual Checkup Data. *Healthcare* **2022**, *10*, 169. https://doi.org/10.3390/healthcare10010169

Academic Editors: Mahmudur Rahman and Francesco Faita

Received: 10 November 2021
Accepted: 12 January 2022
Published: 17 January 2022

Publisher's Note: MDPI stays neutral with regard to jurisdictional claims in published maps and institutional affiliations.

Abstract: Colorectal cancer is the leading cause of cancer-related deaths worldwide, and early detection has proven to be an effective method for reducing mortality. The machine learning method can be implemented to build a noninvasive stratifying tool that helps identify patients with potential colorectal precancerous lesions (polyps). This study aimed to develop a noninvasive risk-stratified tool for colorectal polyps in asymptomatic, healthy participants. A total of 20,129 consecutive asymptomatic patients who underwent a health checkup between January 2005 and August 2007 were recruited. Positive relationships between noninvasive risk factors, such as age, *Helicobacter pylori* infection, hypertension, gallbladder polyps/stone, and BMI and colorectal polyps were observed ($p < 0.0001$), regardless of sex, whereas significant findings were noted in men with tooth disease ($p = 0.0053$). A risk stratification tool was developed, for colorectal polyps, that considers annual checkup results from noninvasive examinations. For the noninvasive stratified tool, the area under the receiver operating characteristic curve (AUC) of obese females (males) aged <50 years was 91% (83%). In elderly patients (>50 years old), the AUCs of the stratifying tools were >85%. Our results indicate that the risk stratification tool can be built by using random forest and serve as an efficient noninvasive tool to identify patients requiring colonoscopy.

Keywords: *Helicobacter pylori* infection; colorectal polyp; teeth disease; precancerous lesions; non-invasive; risk stratifying tool; random forest

1. Introduction

Colorectal cancer (CRC) is the most common cancer worldwide and a significant public health problem in developed countries [1,2]. Most CRCs arise from polyps considered to be precancerous lesions, particularly adenomatous polyps [3–6], even though most are asymptomatic. Removal of all precancerous lesions during endoscopy has been the most effective method for preventing cancer development [6–8]. Colonoscopy is the most effective method for the search and removal of colorectal polyps. However, colonoscopy is not only time consuming and costly but also has side effects. Previous studies have reported several adverse events of colonoscopy, including perforation (0.005–0.085%) and bleeding (0.0001–0.687%) [9]. These adverse events create health hazards for patients and financial burdens for healthcare centers.

Furthermore, the increasing demand for colonoscopy drastically increases the workload of gastroenterology [10]. The increasing workload might result in undesired results such as lower adenoma detection rates per colonoscopy [11] and longer waiting times for colonoscopy [12]. As shown in [12], the median waiting time for the screening colonoscopy is 210 days with the maximum waiting time equaling 631 days in Canada. Long waiting times increases the patient's mental burden and the risk of precancerous polyps' evolvement. Therefore, healthcare centers are actively searching for a risk stratification tool that identifies patients who require colonoscopy using noninvasive examination results.

Hence, risk factors of noninvasive examination data for colorectal polyps, such as gender, age, BMI, blood pressure, gallbladder (GB) polyp/stone, *Helicobacter pylori* infection, and tooth disease (periodontal disease, chronic gingivitis, and chronic periodontitis), were collected, and a machine learning method was implemented to build a risk stratification tool for patients with colorectal polyps. Risk factors were selected based on previous studies [13–17], which reported factors exhibiting some relationship with precancerous polyps [18]. Data from 20,129 consecutive asymptomatic individuals who underwent a health checkup were collected. To date, little is known about their association. Here, we hypothesized that noninvasive risk factors may be associated with colorectal precancerous lesions. Furthermore, we hypothesized that risk factors might vary from patients groups with different demographic characteristics such as gender, age, weights, etc.

After identifying noninvasive risk factors and patient grouping criteria, a noninvasive risk stratification tool was built in order to identify patients who need colonoscopy using a machine learning method. Previous studies have investigated the possibility of identifying patients at high risk for heart disease [19] and diabetes [20] using machine learning methods. More recently, artificial intelligence approaches such as machine learning methods have been used to build a risk stratification tool for different diseases [21]. Therefore, based on the identified risk factors, a machine learning method was further employed to show that the identified risk factors can serve as predictors of precancerous lesions.

To the best of our knowledge, this is the first investigation aimed at building a noninvasive stratification tool based on risk factors from annual checkup data. This study aimed to develop a simple, noninvasive, risk factor, and noninvasive risk stratification tool for these asymptomatic populations to determine colorectal precancerous lesions.

2. Materials and Methods

2.1. Study Participants

In this retrospective study, 20,129 consecutive asymptomatic patients who underwent a health checkup between January 2005 and August 2007 at Chang Gung Memorial Hospital (approval number: 201601348B0, approved 2016/01) were recruited. This study was approved by the Ethics Committee of the Institutional Review Board of Chang Gung Memorial Hospital and conducted according to the ethical principles of the Declaration of Helsinki, as reflected in the a priori approval by the institution's human research committee. Written informed consent was obtained from all patients included in the study. Our health checkup program included physical examination, chest radiography, electrocardiography, complete blood tests, biochemical laboratory tests, urine analysis, abdominal ultrasonography, and colonoscopy. Exclusion criteria were patients who did not have colonoscopy during the course of the health checkup or had incomplete colonoscopy due to various reasons, such as poor bowel preparation or incomplete total colon inspection and BMI > 35 kg/m^2. Height and body weight, used to calculate BMI, were measured by well-trained nurses. BMI ranges were underweight, under 18.5 kg/m^2; normal weight,

18.5–25 kg/m^2; overweight, 25–30 kg/m^2; and obese, >30 kg/m^2. In our institution, the C13 urea breath test was used to detect *Helicobacter pylori* infection [22].

2.2. Colonoscopy Procedure and Abdominal Ultrasonography

For bowel preparation, patients ingested 1.5–2 L of polyethylene glycol before the procedure. All procedures were performed by experienced gastroenterologists. Endoscopic findings were classified into two subgroups: polyp and polyp-free. GB polyps on ultrasonography showed fixed, hyperechoic material attached to the lumen of the GB, without an acoustic shadow [23].

2.3. Risk Stratification Tool Building

As described in Section 2.1, all items in the annual check-up data are collected for this research. Based on previous research [13–17], we selected risk factors from the following categories: (1) patient's demographic characteristics including age, sex, weight, and height; (2) patient's medical history including hypertension, diabetes, and *Helicobacter pylori* infection; (3) colonoscopy diagnosis results including colorectal polyps, ulcerative colitis, hemorrhoids, and intestinal hemorrhage, etc.; (4) abdominal ultrasonography diagnosis including GB polyps and GB stones; (5) blood sample diagnosis results including fasting blood glucose, total cholesterol, high and low-density lipoprotein (HDL and LDL), triglycerides, etc.; (6) dental diagnosis results including periodontitis, periodontal disease, chronic periodontitis, and chronic gingivitis. All diagnosis results are binary with respect to data with 1 = positive diagnosed and 0 = otherwise. BMI is calculated based on the weight of height of the patient. Furthermore, patients' demographic data are dichotomized into binary or categorical data. Age is dichotomized as over (1)/under (0) 50 years old, and BMI is categorized as 0 (underweight (<18.5 kg/m^2)), 1 (normal (18.5–25 kg/m^2)), 2 (overweight (25–30 kg/m^2)), and 3 (obese (>30 kg/m^2)).

Our overall risk stratification tool building procedure is summarized in Figure 1 and *the Heuristic*.

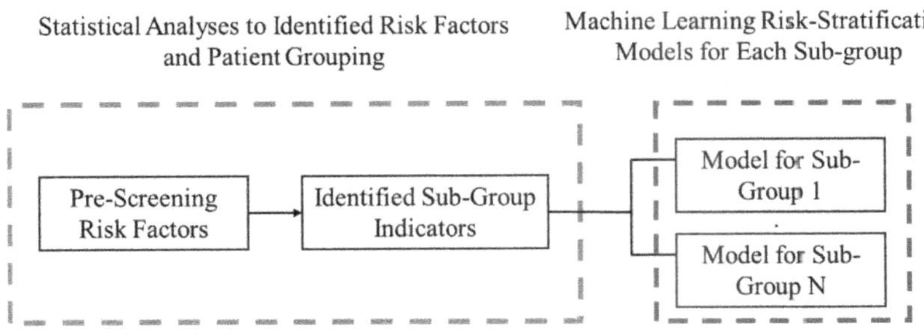

Figure 1. Diagram for proposed *Heuristic*.

The Heuristic:

Step 1: Collect data from annual health check-ups. All risk factors are indexed from $i = 1 \ldots N$, the value of the risk factor is x_i, where there are N risk factors in total.

Step 2: Pre-screen with a z-test for two sample proportions with a significance level equal to 0.05 is applied to select potential risk factors. Where the two sample proportions are calculated as For all risk factor i,

p_{hi} = the proportion of patients who has colorectal polyps for patients with risk factor $x_i = h - 1$.

That is,

p_{1i} = the proportion of patients who has colorectal polyps for patients with risk factor $x_i = 0$.

p_{2i} = the proportion of patients who has colorectal polyps for patients with risk factor $x_i = 1$.

Step 3: The null and alternative hypothesis is stated as below: Null Hypothesis: $p_{1i} = p_{2i} = \ldots . p_{hi}$

We record all risk factors which has a significantly different sample proportion between patients with and without colorectal polyps.

Step 4: Logistics regression is applied for each risk factor to calculate the discriminability for each risk factor. Based on the logistic regression, we identified the demographic risk factors which can segregate patients into different sub-groups for the machine learning process.

Step 5: Machine learning is applied to each sub-group to construct the risk stratification tool.

Step 6: We output the system of models which consisted of multiple random forest models.

Step 7: Output our four-fold-cross validation.

2.4. Statistical Analyses

Statistical analyses, including receiver operating characteristic (ROC) curve, area under ROC (AUC), multinomial logistic regression analyses, and z-test for two-sample proportions, were conducted using SAS software (version 9.4; SAS Institute, Cary, NC, USA). We use the two-sample z-test for the pre-screen tool since it is simple and efficient. Researchers might consider another pre-screen method as well. Statistical significance was set at $p < 0.05$. Simple logistic regression was applied when the independent risk factor was binary (e.g., age), and multinomial logistic regression was applied when the independent risk factor was categorical (e.g., BMI). The AUC was reported for each logistic regression. Since underweight, overweight, and obesity groups were all considered abnormal, BMI was treated as categorical instead of ordinal data. Tooth disease was identified if the patient was diagnosed with periodontal disease, chronic periodontal disease, and/or chronic gingivitis. GB equaled a score of one if GB polyps and stones were observed on abdominal ultrasonography, whereas hypertension was based on the patient's medical history and not the onsite measurement of blood pressure.

2.5. Machine Learning Algorithm

A machine learning algorithm, random forest, was adopted by using Python to build a risk stratification tool based on the risk factors identified from annual healthcare data. Discriminability was represented by AUCs. We used 75% of the data to build the model and 25% of the data to test the consistency of the model. The model building and testing process was repeated four times (four-fold validation method). Adulqader et al. [14] conducted a review on machine learning in healthcare. The authors point out the most popular classification method among all machine learning algorithms including support vector machine (SVM), random forest (RF), and Naïve Bayes. Previous studies [24–26] also use annual health check-up data to develop a risk stratification tool to serve as a screening tool for non-alcoholic fatty liver disease. Goldman et al. [25] use the decision-tree-based approach, and Fialoke et al. [26] used several other methods along with the decision-tree approach. We argue that since our risk factors are all binary data, a decision tree-based method such as RF is the most suitable method. Our machine learning algorithm is summarized as the following pseudo-code.

Machine Learning Algorithm (RF):

Step 1: Input all risk factors as vector X = <x1 ... ... xh> and the y = 1 if a patient is diagnosed with colorectal polyps, and zero otherwise. Moreover, input the demographic factors for aggregating patients into subgroups. Go to Step 2.

Step 2: Segregate all patients into subgroups. Index subgroups as k = 1 ... N for N groups in total. Let k =1 and go to Step 3.

Step 3: Input all risk factors X and y in the kth sub-group. Go to Step 4.

Step 4: Input all data in with path_name = group k, with the following specification of random foreackage in python. We selected the four-fold validation, thus 75% of data will be randomly selected for modeling building and 25% will be reserved for validation. For each run, the random forest will repeat four times for validation. Output the model and go to Step 5. Branch criterion: gini index Number of estimators (number of decision trees): 1000 Min_samples_leaf = 5 Class weight: balanced Validation: Four-fold Calculate the following statistics: Specificity = True negative/(true negative + false positive) Sensitivity = True positive/(true positive + false negative) Area Under Curve (AUC)

Step 5: Collected the outputted model and check if k = N, if not let k = k + 1 and go to Step 3, otherwise end the algorithm.

It is worth noting that all parameters are subjected to test and modified for different research topics. The parameters provided in the algorithm are the optimal parameters after our testing trials.

3. Results

3.1. Statistical Analysis

A total of 20,129 patients were enrolled, including 11,570 (57.5%) men and 8559 (42.5%) women, with a median age of 50 (range: 18–96) years, GB polyps/stones (3191, 15.85%), and tooth disease (15,346, 76.24%), as shown in Table 1. In this study, the risk factors of colorectal polyps were investigated. Each group was subdivided into two groups based on endoscopic findings: polyp and polyp-free. Logistic regression analysis was performed after adjusting for age, gender, BMI, GB polyp/stone, tooth disease, hypertension, and *Helicobacter pylori* infection to determine independent predictors of colorectal polyps. The prevalence of colorectal polyps was 27.08% (5450/20,129) and was associated with age, *Helicobacter pylori* infection, hypertension, and BMI (underweight and overweight) regardless of sex ($p < 0.0001$). Tooth disease only showed a significant difference in men ($p = 0.0053$), as shown in Table 2.

Table 1. Participants' clinical characteristics.

Total Number	*n*, %	20,129
Gender	Ratio of male to female (*n*/*n*)	11,570:8559
Polyp		
	Colorectal polyp (*n*, %)	5450, 27.08%
	Gallbladder polyps (*n*, %)	2188, 10.87%
	Gallbladder stone (*n*, %)	1106, 5.49%
Gallbladder problem		3191, 15.85%
Hypertension	(*n*, %)	1684, 8.37%
Helicobacter pylori infection	(*n*, %)	751, 3.73%
Tooth disease		15,346, 76.24%
	Periodontal disease (*n*, %)	8917, 44.30%
	Chronic gingivitis (*n*, %)	4168, 20.71%
	Chronic periodontitis (*n*, %)	11,655, 57.90%
BMI		
	Underweight (*n*, %)	805, 4%
	Normal (*n*, %)	9090, 45.16%
	Overweight (*n*, %)	6046, 30.04%
	Obesity (*n*, %)	4188, 20.81%
Age	Median (range)	50 (18–96) years
Total cholesterol		2818, 14%
HDL		2617, 13%
Triglycerides		3452, 17%

Table 2. Multinomial logistic regression analysis of variables for colorectal polyps.

		Regardless of Gender		Male		Female	
Parameters		***p*-Value**	**AUC**	***p*-Value**	**AUC**	***p*-Value**	**AUC**
Age	(>50 years = 1)	<0.0001	0.5847	<0.0001	0.5906	<0.0001	0.5900
Helicobacter pylori	(Yes = 1)	<0.0001	0.5113	<0.0001	0.5104	<0.0001	0.5092
Hypertension	(Yes = 1)	<0.0001	0.5142	0.0029	0.5084	<0.0001	0.5240
Tooth disease	Total	0.3734	0.503	0.0053	0.5118	0.1041	0.5086
Gallbladder	(Yes = 1)	<0.0001	0.514	0.002	0.5119	0.0185	0.5105
BMI							
	Underweight = 0	<0.0001		0.0012		<0.0001	
	Normal = 1	0.0055	0.5604	0.1301	0.5389	0.0341	0.5709
	Overweight = 2	<0.0001		0.0017		0.008	
	Obesity = 3						

In Table 2, we find that the risk factors differ based on gender, age, and BMI. Therefore, all patients were divided into sub-groups based on gender, age, and BMI. For each group, risk factors for GB polyps, hypertension, tooth, disease, and *Helicobacter pylori* infection were input as independent variables to predict colorectal polyps. In Table 2 we presented the AUC of risk factors with p-values of the model and AUC from the logistics equations, where the p-values are less than 0.1 for at least male or female. Results of total cholesterol, high lipoprotein cholesterol, and triglycerides are excluded since their p-values are greater than 0.1. As we can observe from Table 2, the observed significances (p-values) for risk factors are different from male to female. Thus, we separate patients with their gender for the machine learning stage. While in Table 2 we did not examine the p-value for different BMI levels, previous literature suggests BMI might significantly relate to the evolvement of colorectal polyps. For example, [27] found that overweight and underweight statuses are significantly correlated with gut microbiota and metabolism. Jain et al. [28] found that obesity significantly impacts metabolism and is accessible with colorectal cancer and polyps. Hence, we also separate patients with their status of BMI.

Figures 2 and 3 further demonstrate the significance and positive or negative impacts of each risk factor, respectively. In order to construct a risk stratification tool based on these risk factors, a random forest machine learning method was employed. In our study, age, *Helicobacter pylori* infection, and hypertension were all risk factors for colorectal polyps. A forest chart was also constructed to present estimated odds ratios for each risk factor, as shown in Figures 2 and 3. While traditional statistical methods such as logistic regression have an AUC > 0.5, discriminability is not as high as healthcare centers may wish (0.5086–0.5900). Therefore, a machine learning method is required to build a model with higher discriminability. As shown in Figures 2 and 3, abnormal body mass, age, and *Helicobacter pylori* are the most influential risk factors for colorectal polyps, regardless of the patient's gender. We also found that hypertension was a significant risk factor for colorectal polyps in male patients. Moreover, the influence of different abnormal body masses was significantly different between gender and age groups. Thus, we further divided patients according to age, gender, and body mass to obtain 16 patient subgroups ($2 \times 2 \times 4$). Since risk factors differ according to age and sex, a risk stratification model was built for each group of patients. For each subgroup, a risk stratification tool was built via a machine learning method. Building a patient-characteristic-specific risk stratification model by using the machine learning method not only enhances the discriminability of the model but also identifies a set of more precise risk factors for each patient group. Healthcare centers can utilize these risk factors to precisely diagnose patients with colorectal polyps.

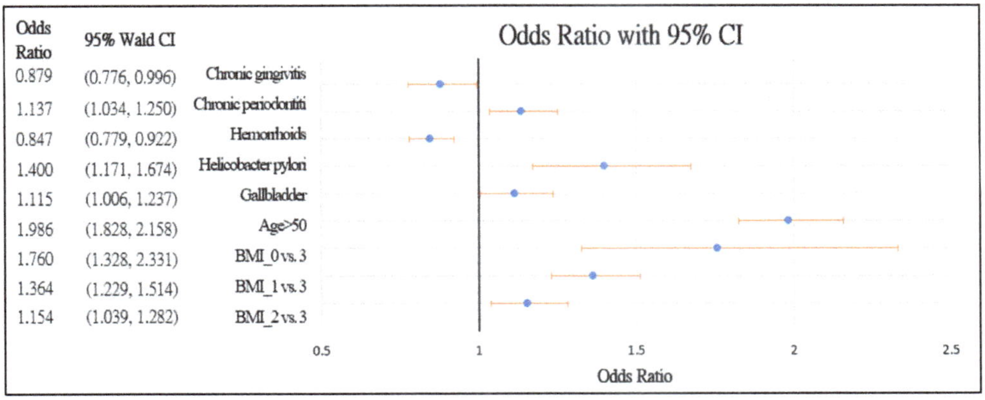

Figure 2. Forest chart of colorectal polyps' risk factors in female patients. Underweight = 0, normal = 1, overweight = 2, and obesity = 3.

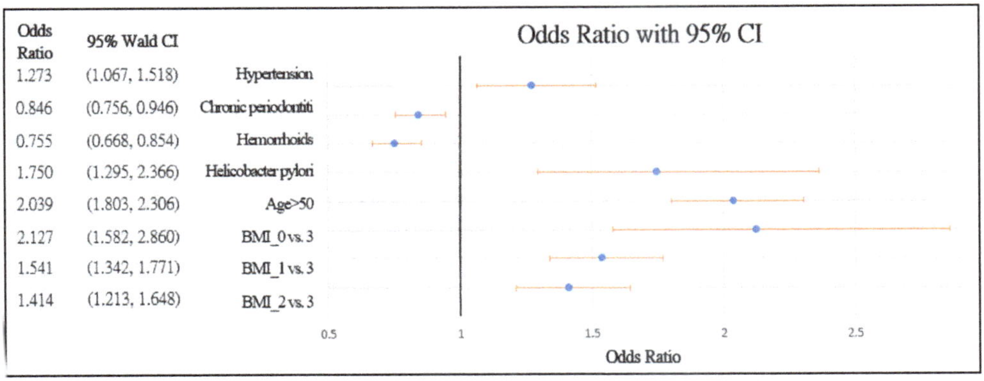

Figure 3. Forest chart of colorectal polyps' risk factors in male patients. Underweight = 0, normal = 1, overweight = 2, and obesity = 3.

3.2. Noninvasive Diagnostics Tool with Random Forests

Based on our results in Section 3.1, we separate all patients into 16 groups via their age, gender, and BMI status. The random forest algorithm in Section 2.5 is applied to each group, and validation results are summarized in Table 3. The input risk factors include hypertension, chronic periodontitis, humanoids, *Helicobacter pylori* infection, GB stones and polyps, total cholesterol, high-density lipoprotein, triglycerides, and diabetes. However, not all risk factors are significant in the final model, and the performance of the stratification model varied extensively. In women < 50 years old with a BMI > 30 kg/m^2, the random forest model's discriminability (AUC = 91%) was high compared to that in other groups. The discriminability of detecting colorectal polyps is >80% for both women and men who are obese. The noninvasive detection tool has an AUC = 80% for underweight male who is >50 years old. In general, the noninvasive colorectal polyp detection tool has a higher AUC in patients with abnormal weight.

Table 3. Noninvasive stratifying tool (random forests model).

Gender	Age	BMI	Sensitivity	Specificity	AUC
Female	<50 years old	Normal	0.22	**0.74**	0.61
		Overweight	0.09	**0.83**	**0.76**
		Obese	0.14	**0.79**	**0.91**
		Underweight	0.55	0.50	0.66
	≥50 years old	Normal	0.35	0.66	0.68
		Overweight	0.27	**0.74**	0.68
		Obese	0.34	**0.74**	**0.85**
		Underweight	0.05	0.67	**0.79**
Male	<50 years old	Normal	0.38	0.68	0.63
		Overweight	0.39	0.59	0.68
		Obese	0.29	0.67	**0.83**
		Underweight	0.11	**0.72**	**0.75**
	≥50 years old	Normal	0.56	0.47	0.67
		Overweight	0.47	0.52	**0.70**
		Obese	0.43	0.57	**0.87**
		Underweight	0.28	0.65	**0.80**

Furthermore, important risk factors identified by the random forests were examined. As shown in Table 3, in women aged >50 years and BMI > 18.5 kg/m^2, the important risk factors are hypertension, diabetes, and GB stones. In contrast, in women <50 years of age and BMI >18.5 kg/m^2, the important risk factors are GB stones and polyps. In men, for those >50 years of age and not underweight, the important risk factors are hypertension, diabetes, and high-density cholesterol. In men aged <50 years, the important risk factors are total cholesterol and high-density cholesterol. As observed, GB polyps and stones are important risk factors for predicting colorectal polyps in female patients.

4. Discussion

To the best of our knowledge, this is the first retrospective study to construct a non-invasive stratification tool for colorectal polyps based on an extensive set of risk factors identified by evaluating a possible association between colorectal polyps, GB polyps/stone, and tooth disease in healthy individuals. In this study, the participants were divided into two groups: polyp and polyp-free. Age, gender, BMI, GB polyps/stone, tooth disease (periodontal disease, chronic gingivitis, and chronic periodontitis), colorectal polyp, hypertension, and *Helicobacter pylori* infection; and triglyceride, high-density lipoprotein cholesterol, and total cholesterol were investigated. Upon disclosure, first, blood sugar status was not included since participants are required to offer their clinical data before checkup without the use of an invasive method such as "fingerstick" sampling to obtain the blood sugar level; second, the final pathological report of polyps was not illustrated because it was supposed that all polyps should be sampled for their nature to determine whether participants' potentially have colorectal polyps, which are considered to be precancerous lesions [3–6].

An association was observed between the colorectal polyp group and age, *Helicobacter pylori* infection, hypertension, and BMI regardless of gender ($p < 0.0001$). Colorectal polyps ($p = 0.0256$) and BMI (overweight, $p = 0.0111$) were significantly different among female patients. Age, *Helicobacter pylori* infection, and hypertension were common risk factors for colorectal polyps.

Regarding age, many studies have reported the association between age and colorectal polyps [29,30], suggesting that CRC screening should be performed around the age of 50–60 years in the general population owing to >80% of CRCs being diagnosed over the age of 60 years, which is consistent with our results [31–34].

Helicobacter pylori infection is highly associated with hyperplastic polyps [34–38], fundic gland polyps [34], and colorectal polyps [16,39–42]. Physiological mechanisms are still unclear, although Meira et al. [34] reported that *Helicobacter pylori* infection is associated with chronic inflammation-induced DNA damage and increased levels of serum gastrin, and *Helicobacter pylori* CagA status may be the cause of colonic neoplasm formation [43–46].

Metabolic syndrome is characterized by the presence of at least three of the following five factors—abdominal obesity, elevated triglyceride levels, decreased high-density lipoprotein cholesterol levels, hypertension, and high fasting glucose levels [47]—and contributes to various diseases, including gastric neoplasm and colorectal neoplasm [48]. In our study, hypertension and BMI were significant across genders in our analysis, and as mentioned before, noninvasive methods are available for easily obtaining factor data from individuals before endoscopy. In our study, hypertension and BMI were both significantly associated with the presence of colorectal polyps.

As discussed in [27], BMI statuses, both overweight and underweight, can alter gut metabolism, and as [28] pointed out, the change in metabolism significantly relates to colorectal cancer and polyps. We hypothesis that BMI is a significant indicator for different colorectal health; therefore, the risk factor might change from one BMI status to another. The results of AUC prove that our hypothesis is correct. For some BMI status, it is easier to identify the patient with colorectal polyps and others are not. The risk factors also differ from one BMI status to another.

The bulk of data has validated dental problems as a risk factor for colon neoplasm development [15,49]. We surmise that periodontal disease may induce chronic inflammation, resulting in immune dysregulation, and alters gut microbiota, which could be one possible pathway responsible for colorectal carcinogenesis [50–52]. It was also found that GB polyps/stones are also related to colorectal polyps, consistent with recent studies [17,53]. This may be attributed to GB polyps/stones and colorectal polyps that share some risk factors, such as obesity and metabolic syndrome [54].

In our study, there is no doubt that all aforementioned risk factors are noninvasive indicators of colorectal polyp formation [48]. Our risk stratification tool, which is built based on identified risk factors with a machine learning method, exhibits high sensitivities (70–80%) compared with that in noninvasive tools developed by previous studies (60–70%) [55]. Other decision tree-based studies [25,26] build noninvasive stratification tools using annual check-up data for non-alcoholic fatty liver obtained in AUC ranges from 85 to 87%. Compared with previous studies, the proposed model outperformed in several subgroups, such as elder obsessive individuals.

The limitations of this study were as follows: (1) its retrospective nature; (2) it was conducted at a single institution with a Taiwanese population; (3) our sigmoidoscopy is conducted under anaesthetization. Thus, our dataset excluded patients with BMI > 35 due to the protocol code of the anesthesiologist. Future researchers can build an RF model for this subgroup or collect data of non-anesthetized sigmoidoscopy diagnostics.

5. Conclusions

In this research, we proposed a new approach for building a risk stratification tool for colorectal polyps. First, we identified a set of promising risk factors using traditional statistical analysis such as z-test and logistics regression. We find that risk factors significantly differ for different genders, ages, and BMI statuses. Then, we separate patients with key demographic characteristics, which we believe each subgroup has a different set of risk factors. Then, we implement random forest to build a machine learning model to stratify patients with and without colorectal polyps. Colonoscopy verification is warranted in those

50 years of age or older, with hypertension, and infected with *Helicobacter pylori*. However, colonoscopy verification is warranted in individuals with tooth diseases and GB polyps.

For obese females, GB polyps warrant further colonoscopy verification. For males over age 50 and not underweight, hypertension is a strong indicator of possible colorectal polyps. We also find that for either underweight or obese patients, the AUC is higher than other groups. That is, abnormal weight is a strong indicator of health status, and different health statuses should be modeled differently. This is verified by our design of grouping patients with different demographic characteristics before building a machine learning model.

Our risk stratification tool can help healthcare centers identify patients who need further colonoscopy. This tool provides two major benefits: first, it helps clinicians conduct colonoscopy and discover precancerous lesions earlier to prevent cancer; second, it reduces the time and financial burden of healthcare centers in conducting unnecessary colonoscopies.

Author Contributions: C.L. conducted statistical analysis and created the machine learning algorithm and contributed to the writing of the manuscript and revised the manuscript according to reviewers' comments. C.-J.L. contributed to the implementation of the machine learning algorithm. T.-H.L. contributed to data collection, data cleaning, and manuscript writing. C.-F.K., B.C.-J.P. and H.-T.C. contributed to data cleaning, literature review, and identification of possible risk factors in this study. C.-C.L. helped with data collection. T.-H.C. provided initial ideas and research directions and finalized the manuscript. All authors contributed significantly to this study. All authors have read and agreed to the published version of the manuscript.

Funding: This work was supported by grants from Chang Gung Memorial Hospital, Taoyuan, Taiwan (CORPG3F0261).

Institutional Review Board Statement: This study was approved by the Ethics Committee of the Institutional Review Board of Chang Gung Memorial Hospital and conducted according to the ethical principles of the Declaration of Helsinki as reflected in the a priori approval by the institution's human research committee.

Informed Consent Statement: Patient consent was waived due to the retrospective nature of the study.

Data Availability Statement: The data presented in this study are available upon request from the corresponding author. The data are not publicly available due to the protection of patients' privacy and restriction from the Ethics Committee of the Institutional Review Board of Chang Gung Memorial Hospital.

Conflicts of Interest: The authors declare no conflict of interest.

References

1. Torre, L.A.; Bray, F.; Siegel, R.L.; Ferlay, J.; Lortet-Tieulent, J.; Jemal, A. Global cancer statistics, 2012. *CA Cancer J. Clin.* **2015**, *65*, 87–108. [CrossRef] [PubMed]
2. Ferlay, J.; Soerjomataram, I.; Dikshit, R.; Eser, S.; Mathers, C.; Rebelo, M.; Parkin, D.M.; Forman, D.; Bray, F. Cancer Incidence and Mortality Worldwide: Sources, methods and major patterns in GLOBOCAN 2012. *Int. J. Cancer* **2015**, *136*, E359–E386. [CrossRef] [PubMed]
3. Calderwood, A.H.; Lasser, K.E.; Roy, H.K. Colon adenoma features and their impact on risk of future advanced adenomas and colorectal cancer. *World J. Gastrointest. Oncol.* **2016**, *8*, 826–834. [CrossRef] [PubMed]
4. Schmitz, J.M.; Stolte, M. Gastric Polyps as Precancerous Lesions. *Gastrointest. Endosc. Clin. N. Am.* **1997**, *7*, 29–46. [CrossRef]
5. Zheng, E.; Ni, S.; Yu, Y.; Wang, Y.; Weng, X.; Zheng, L. Impact of gender and age on the occurrence of gastric polyps: Data analysis of 69575 southeastern Chinese patients. *Turk. J. Gastroenterol.* **2015**, *26*, 474–479. [CrossRef]
6. Islam, R.S.; Patel, N.C.; Lam-Himlin, D.; Nguyen, C.C. Gastric Polyps: A Review of Clinical, Endoscopic, and Histopathologic Features and Management Decisions. *Gastroenterol. Hepatol.* **2013**, *9*, 640–651.
7. Citarda, F.; Tomaselli, G.; Capocaccia, R.; Barcherini, S.; Crespi, M. The Italian Multicentre Study Group Efficacy in standard clinical practice of colonoscopic polypectomy in reducing colorectal cancer incidence. *Gut* **2001**, *48*, 812–815. [CrossRef]
8. Carmack, S.W.; Genta, R.M.; Graham, D.Y.; Lauwers, G.Y. Management of gastric polyps: A pathology-based guide for gastroenterologists. *Nat. Rev. Gastroenterol. Hepatol.* **2009**, *6*, 331–341. [CrossRef]
9. Kim, S.Y.; Kim, H.-S.; Park, H.J. Adverse events related to colonoscopy: Global trends and future challenges. *World J. Gastroenterol.* **2019**, *25*, 190–204. [CrossRef]

10. Greenspan, M.; Prickett, E.; Melson, J. High Clinical Patient Workload Leads to Increased Premature Adenomatous Polyp Surveillance Colonoscopy. *Am. J. Gastroenterol.* **2015**, *110*, S601. [CrossRef]
11. Almadi, M.; Sewitch, M.; Barkun, A.N.; Martel, M.; Joseph, L. Adenoma Detection Rates Decline with Increasing Procedural Hours in an Endoscopist's Workload. *Can. J. Gastroenterol. Hepatol.* **2015**, *29*, 304–308. [CrossRef]
12. Sey, M.S.L.; Gregor, J.; Adams, P.; Khanna, N.; Vinden, C.; Driman, D.; Chande, N. Wait Times for Diagnostic Colonoscopy among Outpatients with Colorectal Cancer: A Comparison with Canadian Association of Gastroenterology Targets. *Can. J. Gastroenterol.* **2012**, *26*, 894–896. [CrossRef]
13. Cappell, M.S. The pathophysiology, clinical presentation, and diagnosis of colon cancer and adenomatous polyps. *Med Clin. N. Am.* **2005**, *89*, 1–42. [CrossRef]
14. Ren, H.G.; Luu, H.N.; Cai, H.; Xiang, Y.B.; Steinwandel, M.; Gao, Y.T.; Hargreaves, M.; Zheng, W.; Blot, W.J.; Long, J.R.; et al. Oral health and risk of colorectal cancer: Results from three cohort studies and a meta-analysis. *Ann. Oncol. Off. J. Eur. Soc. Med. Oncol.* **2016**, *27*, 1329–1336. [CrossRef]
15. Momen-Heravi, F.; Babic, A.; Tworoger, S.S.; Zhang, L.; Wu, K.; Smith-Warner, S.A.; Ogino, S.; Chan, A.T.; Meyerhardt, J.; Giovannucci, E.; et al. Periodontal disease, tooth loss and colorectal cancer risk: Results from the Nurses' Health Study. *Int. J. Cancer* **2017**, *140*, 646–652. [CrossRef]
16. Brim, H.; Zahaf, M.; Laiyemo, A.O.; Nouraie, M.; Pérez-Pérez, G.I.; Smoot, D.T.; Lee, E.; Razjouyan, H.; Ashktorab, H. Gastric Helicobacter pylori infection associates with an increased risk of colorectal polyps in African Americans. *BMC Cancer* **2014**, *14*, 296. [CrossRef]
17. Liu, Y.L.; Wu, J.S.; Yang, Y.C.; Lu, F.H.; Lee, C.T.; Lin, W.J.; Chang, C.J. Gallbladder stones and gallbladder polyps associated with increased risk of colorectal adenoma in men. *J. Gastroenterol. Hepatol.* **2018**, *33*, 800–806. [CrossRef]
18. Xiao, S.; Zhou, L. Gastric cancer: Metabolic and metabolomics perspectives (Review). *Int. J. Oncol.* **2017**, *51*, 5–17. [CrossRef]
19. Ford, I.; Robertson, M.; Komajda, M.; Böhm, M.; Borer, J.S.; Tavazzi, L.; Swedberg, K. Top ten risk factors for morbidity and mortality in patients with chronic systolic heart failure and elevated heart rate: The SHIFT Risk Model. *Int. J. Cardiol.* **2015**, *184*, 163–169. [CrossRef]
20. Okada, H.; Fukui, M.; Tanaka, M.; Matsumoto, S.; Mineoka, Y.; Nakanishi, N.; Asano, M.; Yamazaki, M.; Hasegawa, G.; Nakamura, N. Visit-to-Visit Blood Pressure Variability Is a Novel Risk Factor for the Development and Progression of Diabetic Nephropathy in Patients with Type 2 Diabetes. *Diabetes Care* **2013**, *36*, 1908–1912. [CrossRef]
21. Khalilia, M.; Chakraborty, S.; Popescu, M. Predicting disease risks from highly imbalanced data using random forest. *BMC Med Inform. Decis. Mak.* **2011**, *11*, 51. [CrossRef]
22. Graham, D.Y.; Miftahussurur, M. Helicobacter pylori urease for diagnosis of Helicobacter pylori infection: A mini review. *J. Adv. Res.* **2018**, *13*, 51–57. [CrossRef]
23. Andren-Sandberg, A. Diagnosis and management of gallbladder polyps. *N. Am. J. Med. Sci.* **2012**, *4*, 203–211. [CrossRef]
24. Abdulqader, D.M.; Abdulazeez, A.M.; Zeebaree, D.Q. Machine learning supervised algorithms of gene selection: A review. *Mach. Learn.* **2020**, *62*, 233–244.
25. Goldman, O.; Ben-Assuli, O.; Rogowski, O.; Zeltser, D.; Shapira, I.; Berliner, S.; Zelber-Sagi, S.; Shenhar-Tsarfaty, S. Non-alcoholic Fatty Liver and Liver Fibrosis Predictive Analytics: Risk Prediction and Machine Learning Techniques for Improved Preventive Medicine. *J. Med. Syst.* **2021**, *45*, 22. [CrossRef]
26. Fialoke, S.; Malarstig, A.; Miller, M.R.; Dumitriu, A. Application of Machine Learning Methods to Predict Non-Alcoholic Steatohepatitis (NASH) in Non-Alcoholic Fatty Liver (NAFL) Patients. *AMIA Annu. Symp. Proc.* **2018**, *2018*, 430–439.
27. Wan, Y.; Yuan, J.; Li, J.; Li, H.; Yin, K.; Wang, F.; Li, D. Overweight and underweight status are linked to specific gut microbiota and intestinal tricarboxylic acid cycle intermediates. *Clin. Nutr.* **2020**, *39*, 3189–3198. [CrossRef]
28. Jain, R.; Pickens, C.A.; Fenton, J.I. The role of the lipidome in obesity-mediated colon cancer risk. *J. Nutr. Biochem.* **2018**, *59*, 1–9. [CrossRef]
29. Cao, W.; Hou, G.; Zhang, X.; San, H.; Zheng, J. Potential risk factors related to the development of gastric polyps. *Immunopharmacol. Immunotoxicol.* **2018**, *40*, 338–343. [CrossRef]
30. Chen, H.; Li, N.; Ren, J.; Feng, X.; Lyu, Z.; Wei, L.; Li, X.; Guo, L.; Zheng, Z.; Zou, S.; et al. Participation and yield of a population-based colorectal cancer screening programme in China. *Gut* **2018**, *68*, 1450–1457. [CrossRef] [PubMed]
31. Hussein Kamareddine, M.; Ghosn, Y.; Karam, K.; Nader, A.A.; El-Mahmoud, A.; Bou-Ayash, N.; El-Khoury, M.; Farhat, S. Adenoma Detection before and after the age of 50: A retrospective analysis of Lebanese outpatients. *BMJ Open Gastroenterol.* **2018**, *5*, 000253. [CrossRef] [PubMed]
32. Wolf, A.M.D.; Fontham, E.T.H.; Church, T.R.; Flowers, C.R.; Guerra, C.E.; LaMonte, S.J.; Etzioni, R.; McKenna, M.T.; Oeffinger, K.C.; Shih, Y.-C.T.; et al. Colorectal cancer screening for average-risk adults: 2018 guideline update from the American Cancer Society. *CA Cancer J. Clin.* **2018**, *68*, 250–281. [CrossRef] [PubMed]
33. Schreuders, E.H.; Ruco, A.; Rabeneck, L.; Schoen, R.E.; Sung, J.J.Y.; Young, G.; Kuipers, E.J. Colorectal cancer screening: A global overview of existing programmes. *Gut* **2015**, *64*, 1637–1649. [CrossRef] [PubMed]
34. Bevan, R.; Rutter, M.D. Colorectal Cancer Screening-Who, How, and When? *Clin. Endosc.* **2018**, *51*, 37–49. [CrossRef]
35. Kang, K.H.; Hwang, S.H.; Kim, N.; Kim, D.-H.; Kim, S.Y.; Hyun, J.J.; Jung, S.W.; Koo, J.S.; Jung, Y.K.; Yim, H.J.; et al. The Effect of Helicobacter pylori Infection on Recurrence of Gastric Hyperplastic Polyp after Endoscopic Removal. *Korean J. Gastroenterol.* **2018**, *71*, 213–218. [CrossRef]

36. Anjiki, H.; Mukaisho, K.-I.; Kadomoto, Y.; Doi, H.; Yoshikawa, K.; Nakayama, T.; Vo, D.T.-N.; Hattori, T.; Sugihara, H. Adenocarcinoma arising in multiple hyperplastic polyps in a patient with Helicobacter pylori infection and hypergastrinemia during long-term proton pump inhibitor therapy. *Clin. J. Gastroenterol.* **2017**, *10*, 128–136. [CrossRef]
37. Markowski, A.R.; Markowska, A.; Guzinska-Ustymowicz, K. Pathophysiological and clinical aspects of gastric hyperplastic polyps. *World J. Gastroenterol.* **2016**, *22*, 8883–8891. [CrossRef]
38. Togo, K.; Ueo, T.; Yonemasu, H.; Honda, H.; Ishida, T.; Tanabe, H.; Yao, K.; Iwashita, A.; Murakami, K. Two cases of adenocarcinoma occurring in sporadic fundic gland polyps observed by magnifying endoscopy with narrow band imaging. *World J. Gastroenterol.* **2016**, *22*, 9028–9034. [CrossRef]
39. Tongtawee, T.; Simawaranon, T.; Wattanawongdon, W. Role of screening colonoscopy for colorectal tumors in Helicobacter pylori-related chronic gastritis with MDM2 SNP309 G/G homozygous: A prospective cross-sectional study in Thailand. *Turk. J. Gastroenterol.* **2018**, *29*, 555–560. [CrossRef]
40. Kumar, A.; Kim, M.; Lukin, D.J. Helicobacter pylori is associated with increased risk of serrated colonic polyps: Analysis of serrated polyp risk factors. *Indian J. Gastroenterol.* **2018**, *37*, 235–242. [CrossRef]
41. Nam, J.H.; Hong, C.W.; Kim, B.C.; Shin, A.; Ryu, K.H.; Park, B.J.; Kim, B.; Sohn, D.K.; Han, K.S.; Kim, J.; et al. Helicobacter pylori infection is an independent risk factor for colonic adenomatous neoplasms. *Cancer Causes Control.* **2017**, *28*, 107–115. [CrossRef]
42. Meira, L.B.; Bugni, J.M.; Green, S.L.; Lee, C.-W.; Pang, B.; Borenshtein, D.; Rickman, B.H.; Rogers, A.B.; Moroski-Erkul, C.A.; McFaline, J.L.; et al. DNA damage induced by chronic inflammation contributes to colon carcinogenesis in mice. *J. Clin. Investig.* **2008**, *118*, 2516–2525. [CrossRef]
43. Thorburn, C.M.; Friedman, G.D.; Dickinson, C.J.; Vogelman, J.H.; Orentreich, N.; Parsonnet, J. Gastrin and colorectal cancer: A prospective study. *Gastroenterology* **1998**, *115*, 275–280. [CrossRef]
44. Georgopoulos, S.D.; Polymeros, D.; Triantafyllou, K.; Spiliadi, C.; Mentis, A.; Karamanolis, D.G.; Ladas, S.D. Hypergastrinemia Is Associated with Increased Risk of Distal Colon Adenomas. *Digestion* **2006**, *74*, 42–46. [CrossRef]
45. Epplein, M.; Pawlita, M.; Michel, A.; Peek, R.M.; Cai, Q.; Blot, W.J. Helicobacter pylori Protein–Specific Antibodies and Risk of Colorectal Cancer. *Cancer Epidemiol. Biomark. Prev.* **2013**, *22*, 1964–1974. [CrossRef]
46. Shmuely, H.; Passaro, D.; Figer, A.; Niv, Y.; Pitlik, S.; Samra, Z.; Koren, R.; Yahav, J. Relationship between Helicobacter pylori CagA status and colorectal cancer. *Am. J. Gastroenterol.* **2001**, *96*, 3406–3410. [CrossRef]
47. Grundy, S.M.; Brewer, H.B.; Cleeman, J.I., Jr.; Smith, S.C., Jr.; Lenfant, C. Definition of metabolic syndrome: Report of the National Heart, Lung, and Blood Institute/American Heart Association conference on scientific issues related to definition. *Circulation* **2004**, *109*, 433–438. [CrossRef]
48. Park, W.; Lee, H.; Kim, E.H.; Yoon, J.Y.; Park, J.C.; Shin, S.K.; Kil Lee, S.; Lee, Y.C.; Kim, W.H.; Noh, S.H. Metabolic syndrome is an independent risk factor for synchronous colorectal neoplasm in patients with gastric neoplasm. *J. Gastroenterol. Hepatol.* **2012**, *27*, 1490–1497. [CrossRef]
49. Chou, S.H.; Tung, Y.C.; Wu, L.S.; Chang, C.J.; Kung, S.; Chu, P.H. Severity of chronic periodontitis and risk of gastrointestinal cancers: A population-based follow-up study from Taiwan. *Medicine* **2018**, *97*, e11386. [CrossRef]
50. Lauritano, D.; Sbordone, L.; Nardone, M.; Iapichino, A.; Scapoli, L.; Carinci, F. Focus on periodontal disease and colorectal carcinoma. *Oral Implant.* **2017**, *10*, 229–233. [CrossRef]
51. Gao, Z.; Guo, B.; Gao, R.; Zhu, Q.; Qin, H. Microbiota disbiosis is associated with colorectal cancer. *Front. Microbiol.* **2015**, *6*, 20. [CrossRef] [PubMed]
52. Moutsopoulos, N.M.; Madianos, P.N. Low-Grade Inflammation in Chronic Infectious Diseases: Paradigm of Periodontal Infections. *Ann. N. Y. Acad. Sci.* **2006**, *1088*, 251–264. [CrossRef] [PubMed]
53. Stergios, K.; Damaskos, C.; Frountzas, M.; Nikiteas, N.; Lalude, O. Can gallbladder polyps predict colorectal adenoma or even neoplasia? A systematic review. *Int. J. Surg.* **2016**, *33*, 23–27. [CrossRef] [PubMed]
54. Lim, S.H.; Kim, D.H.; Park, M.J.; Kim, Y.S.; Kim, C.H.; Yim, J.Y.; Cho, K.R.; Kim, S.S.; Choi, S.H.; Kim, N.; et al. Is Metabolic Syndrome One of the Risk Factors for Gallbladder Polyps Found by Ultrasonography during Health Screening? *Gut Liver* **2007**, *1*, 138–144. [CrossRef]
55. Tanwar, S.; Vijayalakshmi, S. Comparative Analysis and Proposal of Deep Learning Based Colorectal Cancer Polyps Classification Technique. *J. Comput. Theor. Nanosci.* **2020**, *17*, 2354–2362. [CrossRef]

healthcare

MDPI

Article

Artificial Intelligence Analysis of Gene Expression Predicted the Overall Survival of Mantle Cell Lymphoma and a Large Pan-Cancer Series

Joaquim Carreras [1,*], Naoya Nakamura [1] and Rifat Hamoudi [2,3]

1 Department of Pathology, Faculty of Medicine, Tokai University School of Medicine, 143 Shimokasuya, Isehara 259-1193, Japan; naoya@is.icc.u-tokai.ac.jp
2 Department of Clinical Sciences, College of Medicine, University of Sharjah, Sharjah P.O. Box 27272, United Arab Emirates; rhamoudi@sharjah.ac.ae
3 Division of Surgery and Interventional Science, University College London, Gower Street, London WC1E 6BT, UK
* Correspondence: joaquim.carreras@tokai-u.jp; Tel.: +81-463-931-121; Fax: +81-463-911-370

Abstract: Mantle cell lymphoma (MCL) is a subtype of mature B-cell non-Hodgkin lymphoma characterized by a poor prognosis. First, we analyzed a series of 123 cases (GSE93291). An algorithm using multilayer perceptron artificial neural network, radial basis function, gene set enrichment analysis (GSEA), and conventional statistics, correlated 20,862 genes with 28 MCL prognostic genes for dimensionality reduction, to predict the patients' overall survival and highlight new markers. As a result, 58 genes predicted survival with high accuracy (area under the curve = 0.9). Further reduction identified 10 genes: *KIF18A*, *YBX3*, *PEMT*, *GCNA*, and *POGLUT3* that associated with a poor survival; and *SELENOP*, *AMOTL2*, *IGFBP7*, *KCTD12*, and *ADGRG2* with a favorable survival. Correlation with the proliferation index (Ki67) was also made. Interestingly, these genes, which were related to cell cycle, apoptosis, and metabolism, also predicted the survival of diffuse large B-cell lymphoma (GSE10846, $n = 414$), and a pan-cancer series of The Cancer Genome Atlas (TCGA, $n = 7289$), which included the most relevant cancers (lung, breast, colorectal, prostate, stomach, liver, etcetera). Secondly, survival was predicted using 10 oncology panels (transcriptome, cancer progression and pathways, metabolic pathways, immuno-oncology, and host response), and *TYMS* was highlighted. Finally, using machine learning, C5 tree and Bayesian network had the highest accuracy for prediction and correlation with the LLMPP MCL35 proliferation assay and RGS1 was made. In conclusion, artificial intelligence analysis predicted the overall survival of MCL with high accuracy, and highlighted genes that predicted the survival of a large pan-cancer series.

Keywords: mantle cell lymphoma; gene expression; MCL35 assay; artificial intelligence; machine learning; deep learning; artificial neural network; multilayer perceptron; immuno-oncology; overall survival

Citation: Carreras, J.; Nakamura, N.; Hamoudi, R. Artificial Intelligence Analysis of Gene Expression Predicted the Overall Survival of Mantle Cell Lymphoma and a Large Pan-Cancer Series. *Healthcare* **2022**, *10*, 155. https://doi.org/10.3390/healthcare10010155

Academic Editor: Mahmudur Rahman

Received: 29 October 2021
Accepted: 12 January 2022
Published: 14 January 2022

Publisher's Note: MDPI stays neutral with regard to jurisdictional claims in published maps and institutional affiliations.

1. Introduction

Mantle cell lymphoma (MCL) is a hematological neoplasia derived from B-lymphocytes, and a subtype of non-Hodgkin lymphomas (NHL) [1]. MCL represents around 7% of adult NHL, and has an incidence of four to eight cases per million people per year [2–6]. MCL affects white men, with a median age at diagnosis of 65 years. The disease frequency increases with age [7], and the incidence of this disease is on the rise in Western and developed countries [7].

MCL is a B-cell lymphoma of small and irregular cells (centrocytes) [8]. The immunophenotype of the classic variant is characterized by the expression of B-cell markers (CD19, CD20), CD5, SOX11, and cyclin D1 due to the characteristics translocation t(11; 14)(q13; q32) between *CCND1* and *IGH* locus [9–11]. MCL expresses high levels of IgM and

IgD, with a lambda light chain restriction in 80% of the cases [8,12]. At diagnosis, most of the patients present with an advanced disease, and lymphadenopathy. Primary extranodal disease is found in 20% of cases, and the gastrointestinal site in the form of lymphomatous polyposis is a characteristic location [13–15].

MCL has traditionally been considered a very aggressive and incurable lymphoma. MCL is associated with a median survival of 3–5 years, with most patients not being cured even with the newer therapeutic modalities [1,8,16]. The "leukemic" variant, which is SOX11-negative, is clinically indolent [17]. Several studies have focused on the identification of prognostic markers to identify patients with a higher probability of an aggressive disease [18–27]. Among them, the International Prognostic Index (IPI), MCL International Prognostic Index (MIPI), and proliferation index (Ki67) are extensively used [18,22]. The pathobiology of MCL comprises several pathways, mechanisms, and target genes that contribute to not only in the pathogenesis but also to aggressiveness and clinical evolution. The major oncogenic driver is *CCND1* gene of the cell cycle pathway. Other relevant genes are involved in cell cycle (*CCND2, CCND3, MYC*), response to DNA damage (*ATM, TP53*), chromatin modification (*WHSC1, MLL2, MEF2B*), apoptosis (BCL2, *BIRC3, TLR2*), and NOTCH signaling (*NOTCH1* and *NOTCH2*), NF-kB and PI3K/AKT signaling pathways, among others [8,28–31].

Neural networks are a favored analytical method for numerous predictive data mining applications because of their power, adaptability, and ease of usage. Predictive neural networks are specially valuable in applications where the underlying process is complex [32–43], such as biological systems [44]. Both the multilayer perceptron (MLP) and radial basis function (RBF) network have a feedforward architecture, because the connections in the network flow forward the input layer (predictors) to the output layer (responses). The hidden layer contains unobservable nodes or units. The value of each hidden unit is some function of the predictors. Both are supervised learning networks that perform prediction and classification. Your choice of strategy will depend on the sort of data and the level of complexity you look for to reveal; while the MLP strategy can discover more complex connections, the RBF method is faster [32,33]. We have recently shown that neural networks can predict the prognosis of diffuse large B-cell lymphoma (DLBCL) and follicular lymphoma (FL) [35,37,45], and also can predict the different subtypes of non-Hodgkin lymphomas with high accuracy [46]. In this research we focused on MCL and the workflow algorithm was improved to handle this type of lymphoma more efficiently the neural networks not only predicted the overall survival outcome and identified the most relevant genes, but the results were modulated by the inclusion of known prognostic genes and immune oncology pathways.

The main aim of the work was to use artificial neural networks (ANN) analyses and other machine learning techniques to analyze the gene expression of MCL and identify relevant prognostic markers. The principal conclusion was that ANN provided a novel analysis technique that not only confirmed known prognostic markers but also highlighted new potential pathological mechanisms.

2. Materials and Methods

2.1. Hardware

All the analyses were performed on a desktop workstation using an AMD Ryzen 7, 3700X, 8-core, processor at 2.59 GHz, 16.0 GB RAM, and a Nvidia GeForce GTX 1650 Turing architecture, 4 GB, GPU.

2.2. Software

Several software were used for data processing, preanalysis, full-analysis, and validation including EditPad Lite, Microsoft Excel, R, R Studio, IBM SPSS Statistic and Modeler, GSEA, and JMP.

The details of the software were as follows:

- EditPad Lite 8 (Just Great Software Co. Ltd., Rawai Phuket 83130, Thailand; page URL: http://www.just-great-software.com/aboutjg.html (accessed on 29 August 2021));
- Microsoft Excel 2016 [(16.0.5173.1000) MSO (16.0.5173.1000) 64-bit. Microsoft K.K., Shinagawa, Tokyo, Japan; page URL: https://www.microsoft.com/ja-jp/microsoft-365/excel (accessed on 29 August 2021)];
- R 3.6.3 (page URL: https://www.r-project.org/ (accessed on 29 August 2021) [47]);
- R Studio 1.3.959 (R Studio, Boston, MA 02210, USA; page URL: https://www.rstudio.com/products/rstudio/#rstudio-desktop (accessed on 29 August 2021));
- IBM SPSS Statistics 26 and Modeler 18 (IBM Japan Ltd., Tokyo 103-8510, Japan; page URL: https://www.ibm.com/jp-ja/analytics/spss-statistics-software (accessed on 29 August 2021));
- Gene Set Enrichment Analysis (GSEA) 4.1.0 (UC San Diego, Broad Institute, Cambridge, MA 02142, USA; page URL: http://www.gsea-msigdb.org/gsea/index.jsp (accessed on 29 August 2021) [48,49]); https://github.com/GSEA-MSigDB/gsea-desktop (accessed on 8 December 2021);
- JMP Pro 14 Statistical Discovery (SAS Institute Inc., Cary, NC 27513-2414, USA; page URL: https://www.jmp.com/ja_jp/home.html (accessed on 29 August 2021));
- Morpheus matrix visualization and analysis software (Broad Institute, Cambridge, MA 02142, USA), https://software.broadinstitute.org/morpheus) (accessed on 29 November 2021);
- String (version 11, String consortium 2020) [19]; https://string-db.org/ (accessed on 29 November 2021).

2.3. Predictive Genes and Artificial Neural Network Analysis

2.3.1. Gene Expression Series of Mantle Cell Lymphoma

The gene expression data of the MCL series GSE93291 were downloaded from the gene expression omnibus (GEO) database [50], which is located at the National Center for Biotechnology Information (NCBI) repository [page URL: https://www.ncbi.nlm.nih.gov/ (accessed on 29 August 2021)]. This database was last updated on 25 March 2019 (contact name: Professor Louis M. Staudt, National Cancer Institute, Lymphoid Malignancies Branch laboratory, Bethesda, MD 20892, USA).

The study involved retrospective gene expression profiling of samples from patients with MCL, confirmed by expert pathology consensus review. This series was created by the Lymphoma/Leukemia Molecular Profiling Project (LLMPP) [50]. These biopsies, with tumor content $\geq$ 60%, were obtained from untreated patients, with no history of previous lymphoma, who subsequently received a broad range of treatment regimens. The biopsies contributing to the set included 80 biopsies described in Rosenwald et al. [51] (classified based on established morphologic and immunophenotypic criteria, with overexpression of cyclin D1 (*CCND1*) mRNA (in most cases, immunohistochemistry demonstrated overexpression of cyclin D1 also on the protein level), 3.8 male/female ratio, median age of 62 years (range 38 to 93), multiagent treatment, and median survival 2.8 years) [51], along with additional biopsies gathered from the clinical sites of the LLMPP. The treatments of the patients was multiagent chemotherapy (R-CHOP, R-CHOP-like), six received no treatment, and no information on treatment was available for two patients.

The gene expression array used in this series was the HG-U133 plus 2 platform (GPL570, Affymetrix, Santa Clara, CA, USA). The GeneChip™ Human Genome U133 Plus 2.0 Array (#900466, ThermoFisher Scientific, Affymetrix Japan K.K., Tokyo, Japan), which is the first and most comprehensive whole human genome array. It has a complete coverage of the Human Genome U133 Set, plus 6500 additional genes for analysis of over 47,000 transcripts. The design and performance of the chip can be accessed at the following webpage: https://www.thermofisher.com/order/catalog/product/900466 (accessed on 29 December 2021).

Total RNA from MCL specimens of frozen samples from 123 patients had been extracted using the FastTrack kit from Invitrogen (Thermo Fisher Scientific Corp., Waltham,

MA 02451, USA), and biotinylated cRNA had been prepared according to the standard Affymetrix protocol from 1 microg mRNA (Expression Analysis Technical Manual, 2001, Affymetrix). The Affymetrix hybridization protocol was used: following fragmentation, 15 micrograms of cRNA were hybridized for 16 h at 45 °C on arrays from Affymetrix. Arrays were washed and stained in the Affymetrix Fluidics Station 400. The Affymetrix scanning protocol was used and the scanning had been performed by the Affymetrix 3000 scanner. The data had been analyzed with Microarray Suite version 5.0 (MA S 5.0) using Affymetrix default analysis settings and global scaling as normalization method. The trimmed mean target intensity of each array was arbitrarily set to 500. The data was normalized and log2 transformed. The original series matrix files [50] provided by the LLMPP were used for the artificial neural network analysis. The gene expression values were collapsed to symbols applying the max probe values, using the GSEA software and the gene cluster text file (*.gct) [52,53].

2.3.2. Identification of Prognostic Genes for Overall Survival

Eighty-six prognostic and pathogenic genes specific for mantle cell lymphoma (MCL) were selected from previous publications [1,8,17,22,28–31,50].

Among these 86 genes, 28 genes with prognostic value for overall survival in this GSE93291 series were selected. The selection depended on the presence of a significant p value in the Kaplan–Meier with log-rank test, after finding adequate cut-off for the stratification into low vs. high groups (Table 1).

Table 1. Prognostic and pathogenic genes of mantle cell lymphoma.

Genes (n = 86)
ADAMDEC1, ADGRG2, AKT1, AKT3, AMOTL2, ARID2, ATM, BCL2, BCL2L11, BCL6, BCOR, BIRC3, BMI1, BORCS8_MEF2B, BTK, CARD11, CASP8, CCND1, CCND2, CCND3, CD5, CD79A, CDK4, CDKN1B, CDKN2A, CDKN2C, CFLAR, CHEK1, CHEK2, CUL4A, CXCL12, CXCR4, DAZAP1, GCNA, HNRNPH1, IGFBP7, ING1, KCTD12, KIF18A, KMT2C, KMT2D, LYN, MDM2, MIR17HG, MKI67, MTOR, MYC, MYCN, NFKB1, NFKBIE, NOTCH1, NOTCH2, NSD2, PALLD, PAX5, PDGFA, PEMT, PIK3CA, PIK3CD, POGLUT3, PTEN, PTK2, RAB13, RB1, RGS1, RPGRIP1L, RRAS, SAMHD1, SELENOP, SMARCA2, SMARCA4, SMARCB1, SOX11, SYK, SYNE1, TAMM41, TERT, TET2, TMEM176B, TNFAIP3, TP53, TRAF2, UBR5, XIAP, YBX3, and *ZCCHC4*

Eighty-six genes with predictive and pathogenic role in MCL were selected from the literature. These genes were later tested for overall survival in the GSE93291 series. Only significant ones were chosen for the neural network analysis.

The cut-offs were found using SPSS software on the collapsed to symbols gene expression values dataset (i.e., each gene had only one expression value). The visual binning function created new variables based on grouping contiguous values into a limited number of distinct categories. The cutpoints were created using equal percentiles, three cutpoints and a width of 25%. After visualization of the overall survival plots with the Kaplan–Meier and log-rank test, the most adequate cut-off value was identified. Then, the Cox regression calculated the hazard-risk (contrast: indicator; reference category: first). Based on the p values (Table 2), the most relevant predictors for overall survival were *MKI67* ($p = 6.6 \times 10^{-9}$, hazard risk = 4.4), *CDK4* ($p = 3.2 \times 10^{-8}$; HR = 4.0), *CHEK1* ($p = 0.2 \times 10^{-5}$, HR = 3.0), *CCND1* ($p = 0.4 \times 10^{-5}$, HR = 3.1), and *CDKN2C* ($p = 0.8 \times 10^{-5}$, HR = 2.8). These genes belonged to the cell cycle and apoptosis pathways.

Table 2. Pathogenic genes of mantle cell lymphoma (GSE93291 series) (Method 1).

Gene	Keyword	Function	Correlation with the Overall Survival of MCL		
			beta	*p*	HR
BCL2L11	Apoptosis	B-cell apoptotic process	1.0	<0.01	2.7
BMI1	Regulation of gene expression	Component of the Polycomb group (PcG) multiprotein PRC1-like complex, negative regulation of gene expression, epigenetic	−0.5	0.042	0.6
BORCS8_MEF2B	Lysosomes	BORC complex, role in lysosomes movement and localization at the cell periphery	−1.0	<0.01	0.4
CCND1	Cell cycle	Positive regulation of G1/S transition of the mitotic cell cycle	1.1	<0.01	3.1
CCND2	Cell cycle, apoptosis	Positive regulation of G1/S transition of the mitotic cell cycle, negative regulation of apoptosis	−0.7	0.018	0.5
CDK4	Cell cycle, apoptosis	Negative regulation of G1/S transition of the mitotic cell cycle, positive regulation of apoptotic process	1.4	<0.01	4.0
CDKN2A	Cell cycle, NF-kB, apoptosis	Negative regulation of G1/S transition of the mitotic cell cycle, negative regulation of NF-kB, positive regulation of apoptotic process	1.0	<0.01	2.7
CDKN2C	Cell cycle	Negative regulation of G1/S transition of the mitotic cell cycle	1.0	<0.01	2.8
CHEK1	Cell cycle, DNA repair, apoptosis	Positive regulation of cell cycle, DNA damage checkpoint and repair, apoptosis	1.1	<0.01	3.0
CHEK2	Cell cycle, DNA repair, apoptosis	Positive regulation of cell cycle, DNA damage checkpoint and repair, apoptosis	0.8	<0.01	2.1
CXCL12	Chemotaxis, apoptosis	Cell chemotaxis, defense response, negative regulation of apoptotic process, DNA damage	−0.6	0.014	0.5
DAZAP1	Cell differentiation and proliferation	Cell differentiation, cell proliferation, positive regulation of mRNA splicing	0.8	0.016	2.3
ING1	Cell cycle	Negative regulation of cell growth, cooperates with TP53	−1.1	<0.01	0.3
MKI67	Cell proliferation	rRNA transcription	1.5	<0.01	4.4
MYC	Cell proliferation	Transcription factor that binds DNA and activates transcription of growth-related genes (positive regulation of gene expression), negative regulation of apoptotic process	0.9	<0.01	2.5
MYCN	Gene expression	Regulation of gene expression, DNA-binding	−0.5	0.052	0.6
NOTCH1	Multiple negative regulations	Affects the implementation of differentiation, proliferation, angiogenesis, and apoptotic programs. Multiple negative regulations	−0.8	<0.01	0.5
NOTCH2	Multiple regulations	Affects the implementation of differentiation, proliferation and apoptotic programs	0.6	0.020	1.8
NSD2	B-cell development	Histone methyltransferase, B-cell development (B1), and B2 activation, humoral immune response, isotype class switch recombination, germinal center formation	1.0	<0.01	2.7
PAX5	B-cell development	The commitment of lymphoid progenitors to B-lymphocyte lineage, promotes development of the mature B-cell stage.	−0.7	0.010	0.5
PIK3CA	ERBB2 signaling, apoptosis	Cell migration, ERBB2 signaling pathway, negative regulation of apoptosis,	0.5	0.042	1.7
PIK3CD	B-cell development and function	Mediates immune responses. Contributes to B-cell development, proliferation, migration, and function. Required for B-cell receptor (BCR) signaling	0.5	0.025	1.7
PTEN	Cell cycle, tumor suppressor gene	Negative regulation of G1/S transition of the mitotic cell cycle	−0.8	0.012	0.5
PTK2	Multiple regulations	Regulation of cell migration, adhesion, cell cycle progression, cell proliferation, apoptosis, MAPK/ERK1 pathway, MDM2 and TP53 recruitment	0.5	0.035	1.7
RB1	Cell cycle, tumor suppressor gene	Tumor suppressor that is a key regulator of the G1/S transition of the cell cycle	−0.5	0.043	0.6
SYNE1	Cytoskeleton	Cytoskeleton-nuclear membrane anchor activity, maintaining of subcellular spatial organization	−0.6	<0.01	0.5
TERT	Telomerase, multiple functions	Telomerase, negative regulation apoptosis, positive regulation G1/S transition of the mitotic cell cycle, negative regulation of gene expression	0.7	<0.01	2.0
XIAP	Multiple functions, regulation of caspases and apoptosis	Multi-functional protein that regulates not only caspases and apoptosis, but also modulates inflammatory signaling and immunity, copper homeostasis, mitogenic kinase signaling, cell proliferation, as well as cell invasion and metastasis	−0.8	<0.01	0.5

From an initial set of 86 genes with known pathogenic role in MCL, a final set of 28 genes were selected because their predictive value for overall survival using a Kaplan–Meier and log-rank test in the GSE93291: *P*, *p* value; HR, hazard risk. The gene information is based on UniProt [54], and Genecards [55].

2.3.3. Description of the Basic Neural Network Architecture

The multilayer perceptron (MLP) analysis was performed as previously described [35–37,45,56,57]. The architectures are shown in Figures 1–3, and the analysis outline in Figure 4. The MLP procedure produces a predictive model for one or more dependent (target) variables based on the values of the predictor variables. The MLP is a feedforward architecture, the input layer contains the predictors (our gene expression data), the hidden layer contains unobservable nodes or units, and the output layer contains the target variables. The target variables were the overall survival outcome as dead vs. alive, and the gene expression of each prognostic and pathogenic gene as a categorical variable (high vs. low expression). Figure 5, on the top right side, shows the basic neural network architecture. Of note, the basic architecture of the radial basis function (RBF) is like the MLP, but only one hidden layer characterizes it. This research used a simple type of artificial neural network, but solid enough to provide a "basic analysis unit" that conforms a more complex analysis algorithm as shown in Figure 5. A thorough description is shown in our recent publication of artificial analysis of gene expression data of diffuse large b-cell lymphoma (DLBCL) and non-Hodgkin lymphomas [46,58].

2.3.4. Parameters of the Neural Network

A thorough description of the artificial neural network procedure is described in our recent publication [58]. The predictors (covariates) were the 20,862 genes of the array. The covariates were rescaled by default to improve network training. All rescaling was performed based on the training data, even if a testing or holdout sample is defined. The method for rescaling was the standardized (subtract the mean and divide by the standard deviation (x-mean/s)). Other available methods for rescaling were the normalized ((x − min)/(max − min)), adjusted normalized ([2 × (x − min)/(max − min)] − 1), or none. The cases were randomly assigned to the training set, testing set, and holdout according to the relative number of cases, being 70%, 30%, and 0%, respectively. To avoid bias, each individual neural network underwent a random assignation of the samples into the training and testing sets.

The "best" architecture design for the analysis was searched and finally selected [58,59]. The architecture can be selected automatically (with a minimum number of units in the hidden layer of 1 and a maximum of 50) or can be a custom architecture. A custom architecture selection provides control over the hidden and output layers and can be most useful when you know in advance what architecture you want or when you need to tweak the results of the automatic architecture selection.

In a custom architecture, the number of hidden layers could be one or two. The number of units of the hidden layer could be automatically computed or custom. The activation function of the hidden layers was the hyperbolic tangent ($\gamma(c) = \tanh(c) = (e^c - e^{-c})/(e^c + e^{-c})$), or sigmoid ($\gamma(c) = 1/(1 + e^{-c})$).

The activation function of the output layer was the identity ($\gamma(c) = c$), softmax ($\gamma(c_k) = \exp(c_k)/\Sigma_j \exp(c_j)$), hyperbolic tangent, or sigmoid. Of note, the activation function chosen for the output layer determined which rescaling methods were available. The rescaling of scale dependent variables was standardized ((x − mean)/s), normalized ((x − min)/(max − min)), adjusted normalized ([2 × (x − min)/(max − min)] − 1), or none.

Several types of training were available: the batch, online, and mini-batch. The optimization algorithm included the scaled conjugate gradient, and gradient descent. The training options were the following: initial lambda (0.0000005); initial sigma (0.00005); interval center (0); and interval offset (±0.5).

The output included the network structure and network performance.

Several parameters displayed the network performance: model summary; classification results; receiver operating characteristic ROC curve; cumulative gains chart; lift chart; predicted by observed chart; and the independent variable importance analysis. ROC analysis displayed a curve for each categorical dependent variable and category and the area under each curve [35–37,45,46,56,57]. The predicting variables (predictors) were ranked according to their normalized importance for predicting the target (dependent) variable and for determining the neural network. This analysis performed a sensitivity analysis that is based on the combined training and testing samples or only on the training sample if there is no testing sample [32,33,60].

The predicted value or category and the predicted pseudo-probability for each dependent variable were saved. The synaptic weight estimates were exported to an XML file.

If it was necessary to replicate the results exactly, the same initialization value for the random number generator, data order, and variable order should be used, in addition to using the same procedure settings.

The setup of a radial basis function (RBF) is similar to the MLP. In a RBF, the activation function for hidden layer was normalized or ordinary radial basis function. Figures 1 and 2 show the general architecture for MLP and RBF [32,33,60]. Figure 3 shows the sensitivity analysis [32,33,60].

2.4. Gene Set Enrichment Analysis (GSEA)

GSEA is a method that determines whether a priori defined set of genes shows statistically concordant differences between two "biological" states (e.g., phenotypes) [48,49]. Three types of files were necessary to run the application: (1) the gene cluster text file (*.gct) with the GSE93291 gene expression dataset; (2) the phenotype data as a categorical class (e.g., dead/alive) file format (*.cls); and (3) the gene set database as a gene matrix file format (*.gmx). The GSEA parameters were the following [37]: number of permutations (1000); collapse to gene symbols; permutation type (phenotype); chip platform (GPL570, HG-U133 Plus 2); enrichment statistic (weighted); metric for ranking genes (signal2noise); gene list sorting mode (real); gene list ordering mode (descending); max size (500); and min size (15) [37].

$X^{(m)} = \left(x_1^{(m)}, \ldots, x_P^{(m)}\right)$	Input vector, pattern m, $m=1 \ldots M$.
$Y^{(m)} = \left(y_1^{(m)}, \ldots, y_R^{(m)}\right)$	Target vector, pattern m.
I	Number of layers, discounting the input layer.
J_i	Number of units in layer i. $J_0 = P$, $J_I = R$, discounting the bias unit.
Γ^c	Set of categorical outputs.
Γ	Set of scale outputs.
Γ_h	Set of subvectors of $Y^{(m)}$ containing 1-of-c coded hth categorical variable.
$a_{i:j}^m$	Unit j of layer i, pattern m, $j = 0, \ldots, J_i$; $i = 0, \ldots, I$.
$w_{i:j,k}$	Weight leading from layer $i-1$, unit j to layer i, unit k. No weights connect $a_{i-1:j}^m$ and the bias $a_{i:0}^m$; that is, there is no $w_{i:j,0}$ for any j.
$c_{i:k}^m$	$\sum_{j=0}^{J_{i-1}} w_{i:j,k} a_{i-1:j}^m$, $i=1 \ldots I$
$\gamma_i(c)$	Activation function for layer i.
w	Weight vector containing all weights $\left(w_{1:0,1}, w_{1:0,2}, \ldots, w_{I:J_{I-1},J_I}\right)$

The general architecture for MLP networks is:

Input layer: $J_0=P$ units, $a_{0:1}, \cdots, a_{0:J_0}$; with $a_{0:j} = x_j$.

ith hidden layer: J_i units, $a_{i:1}, \cdots, a_{i:J_i}$; with $a_{i:k} = \gamma_i(c_{i:k})$ and $c_{i:k} = \sum_{j=0}^{J_{i-1}} w_{i:j,k} a_{i-1:j}$ where $a_{i-1:0} = 1$

Output layer: $J_I=R$ units, $a_{I:1}, \cdots, a_{I:J_I}$; with $a_{I:k} = \gamma_I(c_{I:k})$ and $c_{I:k} = \sum_{j=0}^{J_I} w_{I:j,k} a_{i-1:j}$ where $a_{i-1:0} = 1$

Figure 1. General architecture for multilayer perceptron (MLP) networks. A neural network is a set of non-linear data modeling tools consisting of input layers plus one or two hidden layers. The multilayer perceptron procedure is a feedforward architecture. In comparison to RBF, the MLP con find more complex relationships but it is slower to compute. The MLP network is a function of one or more predictors (also called inputs or independent variables) that minimizes the prediction error of one or more target variables (also called outputs) [32,33,60].

$X^{(m)} = \left(x_1^{(m)}, \ldots, x_P^{(m)}\right)$	Input vector, pattern m, m=1,...M.
$Y^{(m)} = \left(y_1^{(m)}, \ldots, y_R^{(m)}\right)$	Target vector, pattern m.
I	Number of layers, discounting the input layer. For an RBF network, I=2.
J_i	Number of units in layer i. $J_0 = P$, $J_I = R$, discounting the bias unit. J_1 is the number of RBF units.
$\phi_j\left(X^{(m)}\right)$	jth RBF unit for input $X^{(m)}$, j=1, ...J_1.
μ_j	center of ϕ_j, it is P-dimensional.
σ_j	width of ϕ_j, it is P-dimensional.
h	the RBF overlapping factor.
$a_{i:j}^{m}$	Unit j of layer i, pattern m, $j = 0, \ldots, J_i$; $i = 0, \ldots, I$.
w_{rj}	weight connecting rth output unit and jth hidden unit of RBF layer.

There are three layers in the RBF network:

Input layer: J_0=P units, $a_{0:1}, \cdots, a_{0:J_0}$; with $a_{0:j} = x_j$.

RBF layer: J_1 units, . $a_{1:1}, \cdots, a_{1:J_1}$; with $a_{1:j} = \phi_j(X)$ and $\phi_j(X)$ described below.

Output layer: J_2=R units, $a_{I:1}, \cdots, a_{I:J_2}$; with $a_{I:r} = w_{r0} + \sum_{j=1}^{J_1} w_{rj}\phi_j(X)$.

There are many types of radial basis functions; there are two distinct types of Gaussian RBF architectures that we support:

Ordinary RBF (ORBF): This type uses the exp activation function, so the activation of the RBF unit is a Gaussian "bump" as a function of the inputs. In ORBF, the Gaussian basis function takes form

$$\phi_j(X) = \exp\left(-\sum_{p=1}^{P} \frac{1}{2\sigma_{jp}^2}(x_p - \mu_{jp})^2\right)$$

Normalized RBF (NRBF): This type uses the softmax activation function, so the activation of all the RBF units are normalized to sum to one. In NRBF networks, the basis function takes form

$$\phi_j(X) = \exp\left(-\sum_{p=1}^{P} \frac{1}{2\sigma_{jp}^2}(x_p - \mu_{jp})^2\right) \Big/ \sum_{j=1}^{J_1} \exp\left(-\sum_{p=1}^{P} \frac{1}{2\sigma_{jp}^2}(x_p - \mu_{jp})^2\right)$$

Figure 2. General architecture for radial basis function (RBF) networks. A radial basis function (RBF) network is a feed-forward, supervised learning network with only one hidden layer, called the radial basis function layer [32,33,60].

Sensitivity Analysis

For each predictor p and each input pattern m, compute:

$$d_{pm} = \max_{x_{p_1}, x_{p_2} \in S_p} \left\| \hat{Y}_{p_1}^{(m)} - \hat{Y}_{p_2}^{(m)} \right\|$$

where $\hat{Y}_{p_i}^{(m)}$ is the predicted output vector (standardized if standardization of output variable is used in training) using $\left(x_1^{(m)}, \ldots, x_{p-1}^{(m)}, x_{p_i}, x_{p+1}^{(m)}, \ldots, x_P^{(m)}\right)$ as its input, and S_p= $\left\{x_p^{\min}, x_p^{(2)}, x_p^{(3)}, x_p^{(4)}, x_p^{\max}\right\}$ for scale predictors and $\{(1,0,\ldots,0), (0,1,0,\ldots,0), \ldots, (0,0,\ldots,1)\}$ for categorical predictors.

Then compute:

$$d_p = \frac{1}{M}\sum_{m=1}^{M} d_{pm}$$

and normalize the d_ps to sum to 1, and report these normalized values as the sensitivity values for the predictors. This is the average maximum amount we can expect the output to change based on changes in the pth predictor. The greater the sensitivity, the more we expect the output to change when the predictor changes.

Figure 3. Sensitivity analysis. Independent variable importance analysis. Performs a sensitivity analysis, which computes the importance of each predictor in determining the neural network [32,33,60].

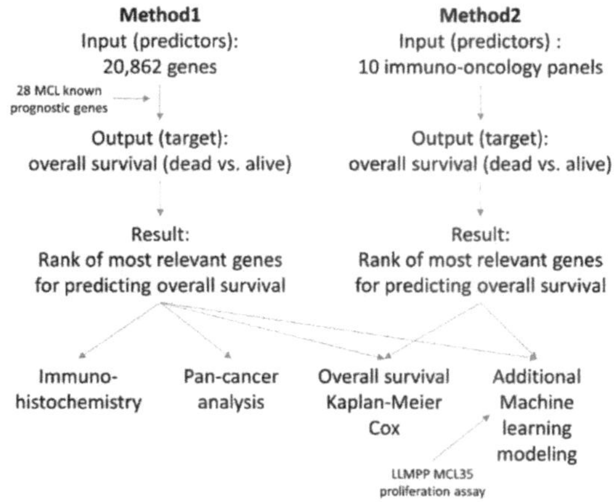

Figure 4. Summary of the analysis methodology. The analysis was comprised of two methods, one based on the analysis of 20,862 genes and a second based on 10 immuno-oncology panels. This research used artificial neural networks and several machine learning techniques to identify genes associated with the overall survival of the patients. Correlation with known MCL pathogenic genes and the LLMPP MCL35 proliferation assay was also made.

2.5. *Summary of the Research Analysis Algorithm*

The algorithms for the analysis of the gene expression data of MCL are shown in Figures 5–8.

2.5.1. Algorithm Based on the Input of 20,862 Genes (Method 1)

First, all the genes of the array were used as predictors (input layer) for the target variables (output layer) of overall survival (dead/alive) and for the 28 genes with prognostic value in MCL (high/low expression) using an artificial neural network. The neural network included both a multilayer perceptron and a radial basis function analysis for each target variable (Figure 5). In the output of each individual neural network, all the genes of the array were ranked according to their normalized importance for predicting the target variable. Then, the genes with a normalized importance above 70% were selected. In addition, the normalized importance of all the neural networks were averaged, the genes ranked according to the averaged normalized importance for prediction, and the top 1% genes were selected. As a result, the initial set of 20,862 genes was reduced to a smaller number (n = 1394).

Next, an MLP was performed using the 1394 genes as predictors (input layer) of the overall survival outcome (dead/alive, output layer); this analysis was repeated 20 times, and the top 4 MLPs with higher area under the curves were selected. The normalized importance of each 1394 were averaged between the four results and ranked from higher to lower values. Then, using multiple MLP analysis, the minimum number of genes (starting from the one with higher normalized importance) that provided the highest area under the curve was found (n = 58) (Figure 6).

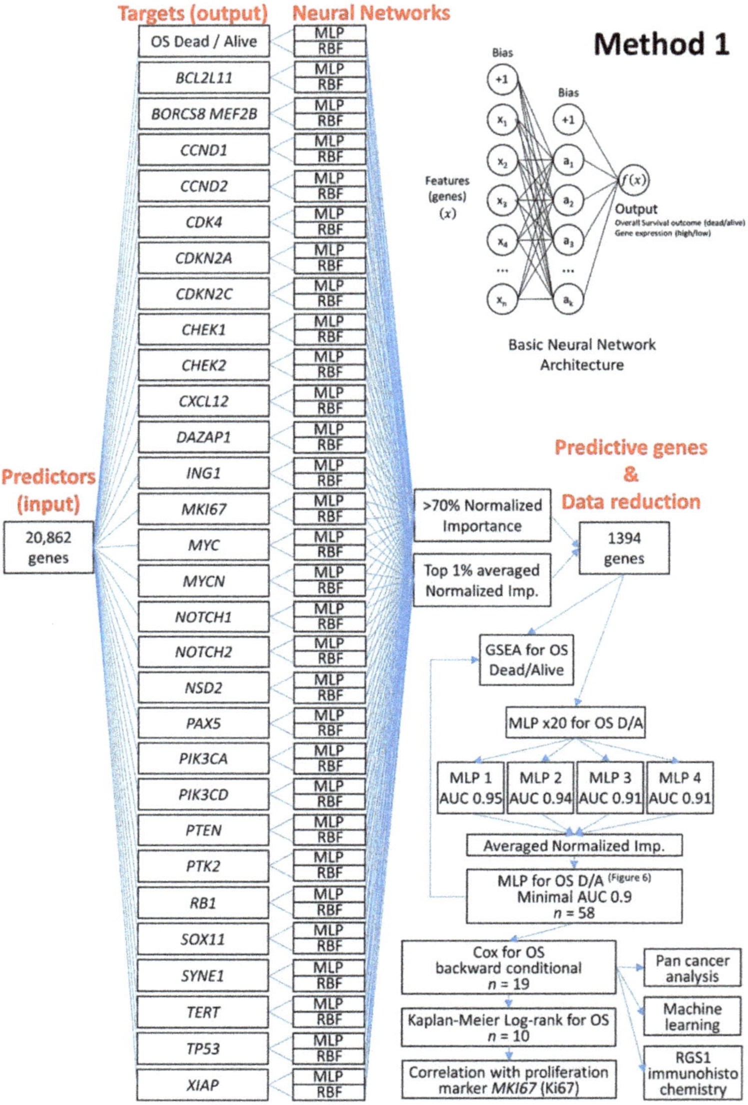

Figure 5. Artificial neural network analysis for the prediction of the overall survival of mantle cell lymphoma (Method 1). From a start point of 20,862 genes, using several neural networks, a correlation between the overall survival outcome and several mantle cell lymphoma pathogenic genes managed to reduce to a final set of 10 genes. These 10 genes correlated with the survival of the patients, but also with the proliferation index as expressed by *MKI67* gene: MLP, multilayer perceptron; RBF, radial basis function; OS, overall survival; DA, dead/alive; GSEA, gene set enrichment analysis; AUC, area under the curve.

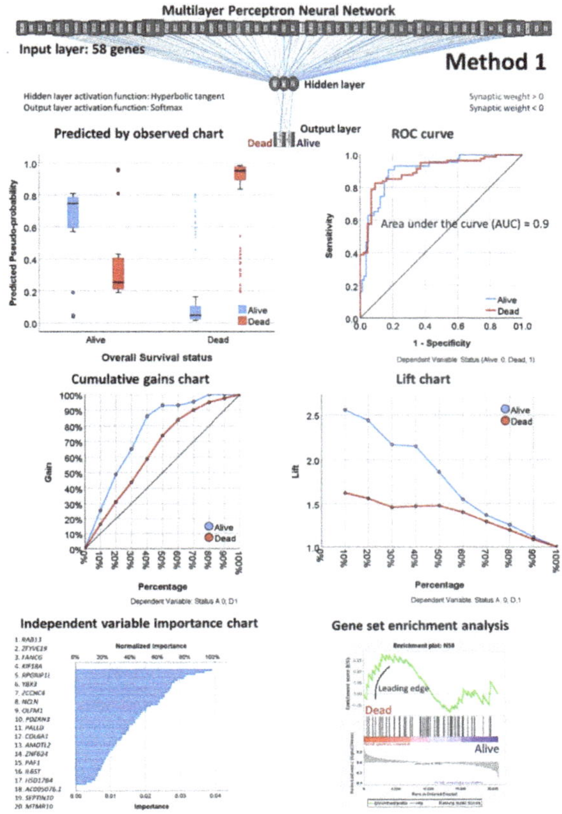

Figure 6. Multilayer perceptron analysis using the selected 58 genes (Method 1 continuation). As shown in Figure 4, the neural networks reduced the initial input of 20,862 genes to 58 predictive genes. Next, the overall survival outcome (dead/alive) was predicted using 58 genes and a neural network. Several parameters display the network performance: model summary; classification results; receiver operating characteristic ROC curve; cumulative gains chart; lift chart; predicted by observed chart; and the independent variable importance analysis. ROC analysis displays a curve for each categorical dependent variable and category and the area under each curve [34–36,44,45,55,56]. The genes were ranked according to their normalized importance for predicting the overall survival outcome as a dichotomic variables (dead vs. alive). A GSEA analysis confirmed the association toward a dead outcome. The characteristics of the network were as follows. Case processing: training $n = 93$ (76%); testing $n = 30$ (24%). Units $n = 58$. Rescaling = standardized. Hidden layer: number = 1; units = 2; activation function = hyperbolic tangent. Output layer: dependent variables = 1 (overall survival outcome dead/alive); units = 2, activation function = softmax, error function = cross-entropy. Model summary: training, cross-entropy error = 30.8, 14% of incorrect predictions; testing, cross-entropy error = 14.5, 23% of incorrect predictions. Classification: training, 86% overall correct (93.8% alive, 82% dead); testing, 77% overall correct (82% alive, 74% dead). Area under the curve = 0.9. Top 10 most relevant genes were *RAB13*, *ZFYVE19*, *FANCG*, *KIF18A*, *RPGRIP1L*, *YBX3*, *ZCCHC4*, *NCLN*, *OLFM1*, and *PDZRN3*. A complete description of the multilayer perceptron is present in our recent publication (Carreras J. et al. Artificial Neural Networks Predicted the Overall Survival and Molecular Subtypes of Diffuse Large B-Cell Lymphoma Using a Pan-cancer Immune-Oncology Panel. *Cancers* **2021**, *13*, 6384; https://doi.org/10.3390/cancers13246384) [58].

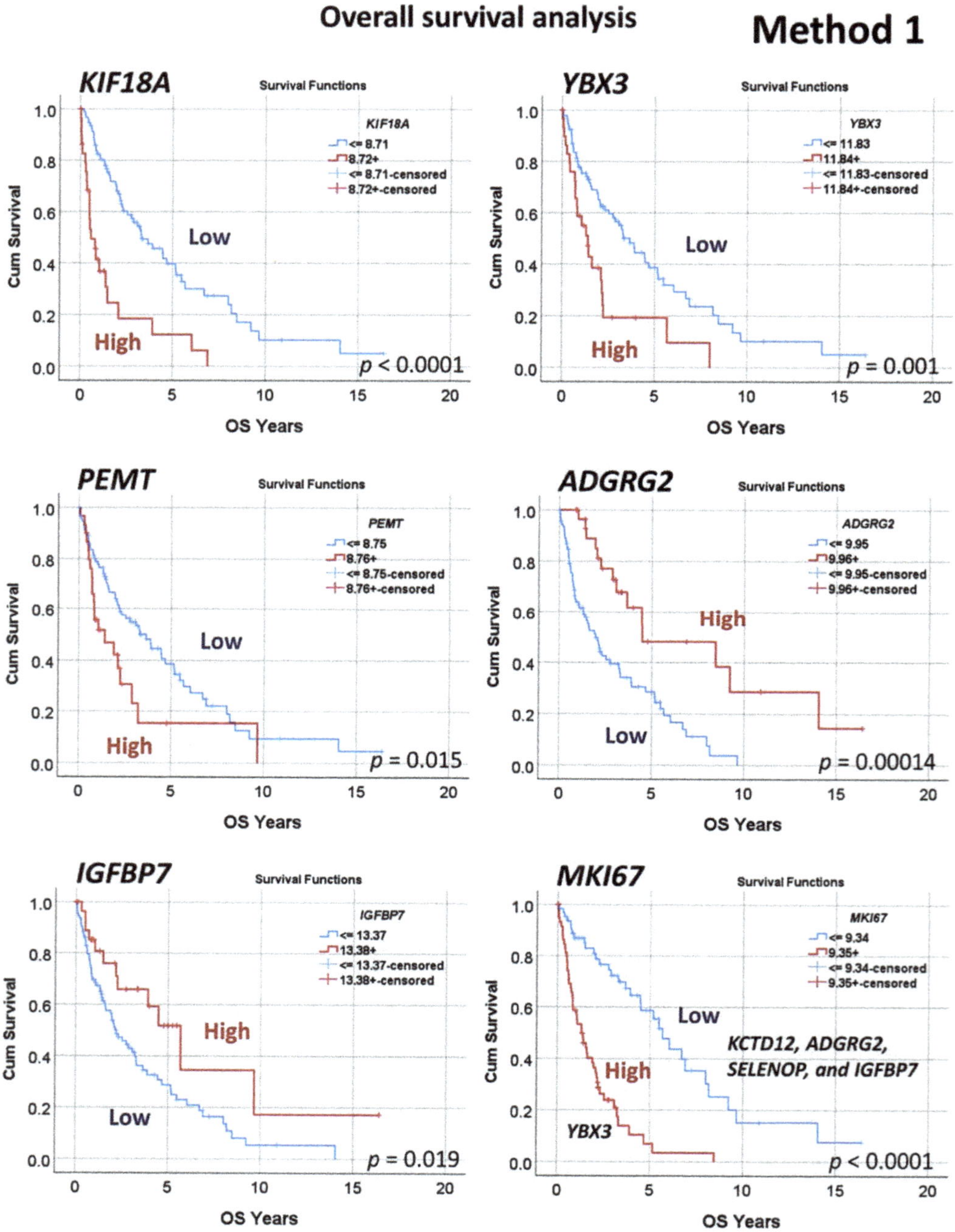

Figure 7. Overall survival analysis (Method 1 continuation). Because of the neural network analysis and dimensional reduction (Figures 4 and 5), a final set of 10 genes with overall survival relationship was highlighted. These genes not only correlated with the clinical outcome but also with the proliferation index, as expressed by *MKI67*. Of note, ki67 is a marker routinely used for prediction in mantle cell lymphoma, and the most relevant marker of the LLMPP MCL35 proliferation assay.

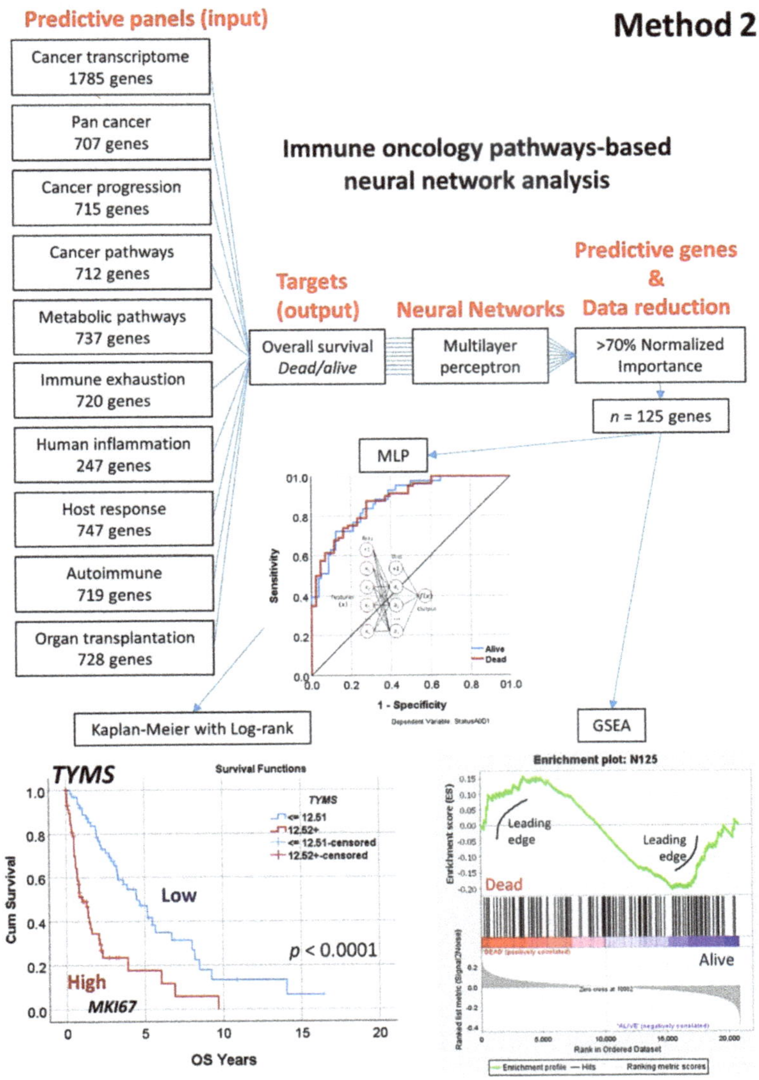

Figure 8. Artificial neural network analysis for predicting of the overall survival of mantle cell lymphoma using several immune oncology panels (Method 2). Overall survival was predicted using 10 immuno-oncology panels. After several multilayer perceptron analyses, a set of 125 genes predicted the overall survival outcome (dead/alive) with high accuracy. Among the most relevant genes, *TYMS* was highlighted. GSEA analysis had a sinusoidal-like, with some genes enriched toward dead or alive survival outcomes.

Finally, a Cox regression for overall survival (backward conditional) reduced the list to 19 genes. From these 19 genes, additional analyses included Kaplan–Meier with log-rank test for overall survival using cutoffs (Figure 7), analysis of other types of cancer ("pan-cancer analysis") (Figures 9 and 10), other machine learning (Figures 11–13), and immunohistochemistry for RGS1 (Figure 14).

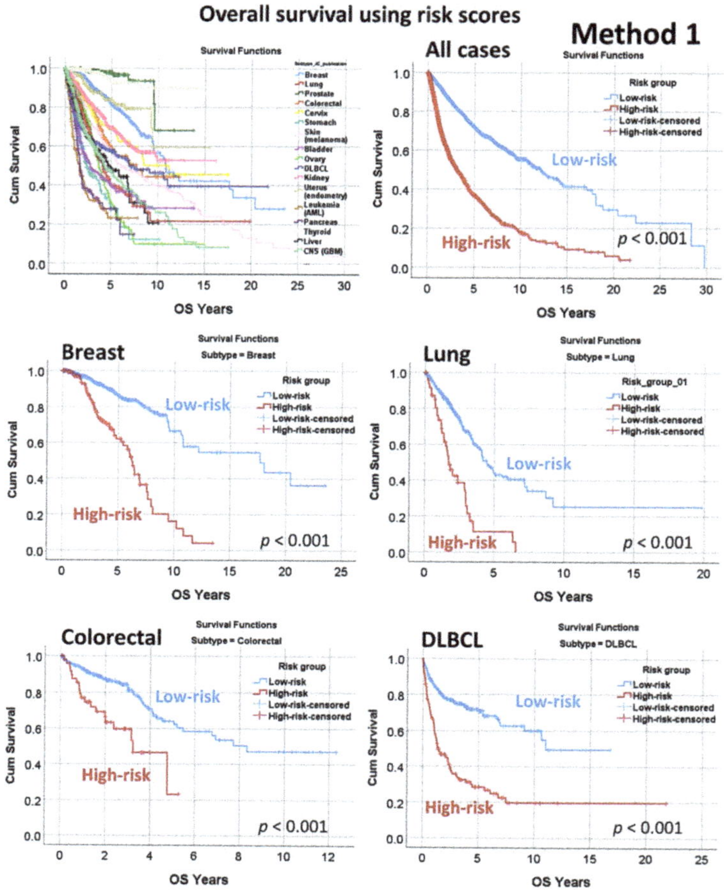

Figure 9. Overall survival in a pan-cancer series. The multilayer perceptron using the 20,862 genes identified a final set of 19 genes with prognostic value in mantle cell lymphoma. As a start point of the gene expression of the set of 19 genes and using a risk-score formula [36,46], we confirmed that these genes also contributed to the overall survival of diffuse large B-cell lymphoma (DLBCL). Additionally, these genes could also predict the overall survival of a pan-cancer series of 7289 cases from The Cancer Genome Atlas (TCGA) program that included the most frequent human cancers. Of note, the weight and direction of the overall survival association was different in each subtype of neoplasia. Risk scores were calculated by multiplying the beta values of the multivariate Cox regression analysis for overall survival of each gene with the values of the corresponding gene expressions, as previously described [58].

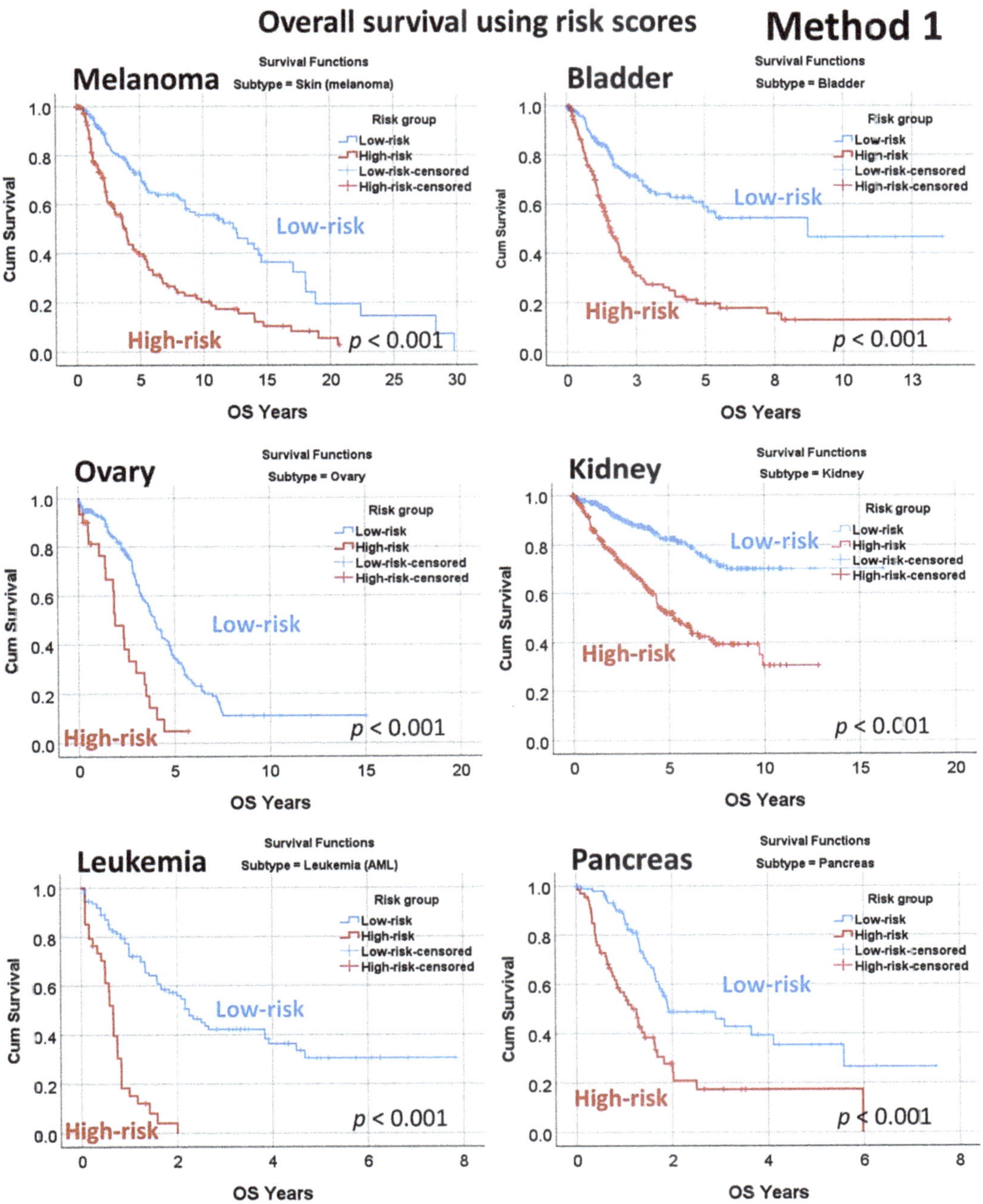

Figure 10. Overall survival in a pan cancer series.

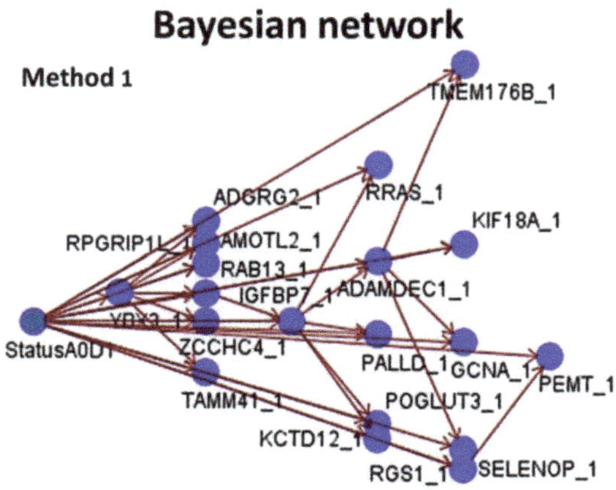

Figure 11. Bayesian network. A Bayesian network successfully modeled the overall survival outcome (dead/alive) using the 19 genes, previously identified in the neural network analysis (Figure 5, Method 1). The Bayesian network enables you to build a probability model by combining observed and recorded evidence with "common-sense" real-world knowledge to establish the likelihood of occurrences by using seemingly unlinked attributes. The node focuses on Tree Augmented Naïve Bayes (TAN) and Markov Blanket networks that are primarily used for classification. This graphical model shows the variables (nodes) and the probabilistic, or conditional, independencies between them. The links of the network (arcs) may represent causal relationships, but the links do not necessary represent direct cause and effect. This Bayesian network is used to calculate the probability of a patient of being alive or dead, given the gene expression of 19 genes, if the probabilistic independencies between the gene expression and the overall survival outcome as displayed on the graph hold true. Bayesian networks are very robust in case of missing data.

2.5.2. Algorithm Based on the Input of 10 Immune Oncology Panels (Method 2)

In comparison to the first algorithm in which the whole genes of the array were used (n = 20,862), this second algorithm used 9 different immune oncology panels as input data (7817 genes in total) (Figure 8). Nine individual MLP analysis for the prediction of overall survival outcome (dead/alive) were performed, and the genes with a normalized importance above 70% in each panel were pooled (n = 125). A GSEA analysis confirmed the association of these genes towards the dead or alive overall survival outcome (phenotype). Next, an additional MLP analysis confirmed the prediction of the overall survival outcome and ranked the 125 genes according to their normalized importance. The top genes were later tested for conventional overall survival analysis.

2.6. *Conventional Statistical Analyses*

Traditional statistics calculated the overall survival analyses. Overall survival was calculated from time of diagnosis to the last follow-up time, and recorded as alive or dead (event), following the criteria of Cheson B. D. [61,62]. Comparison between groups was performed using Kaplan–Meier analysis and the log-rank test. The Breslow and Tarone–Ware tests were also used. The Cox regression (with the method enter or backward conditional) was used to calculate the hazard-risks and the 95% confidence intervals. A p value less than 0.05 was considered statistically significant.

In case of a neural network analysis, poor prognosis/survival corresponds to the cases whose overall survival event was dead. In case of an overall survival analysis using the Kaplan–Meier test, poor prognosis corresponds to the group with lower cumulative survival proportion in the plot.

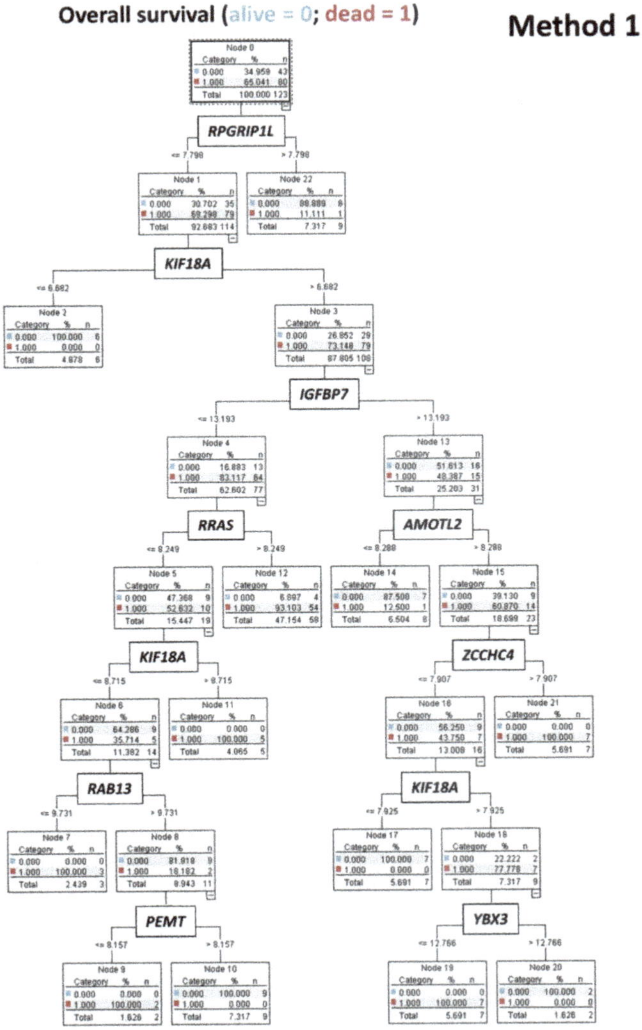

Figure 12. C5.0 decision tree model. A decision tree successfully modeled the overall survival outcome (dead/alive) using the 19 genes, previously identified in the neural network analysis (Figure 5, Method 1). This model uses the C5.0 algorithm to build either a decision tree or a rule set. A C5.0 model works by splitting the sample based on the field that provides the maximum information gain. Each subsample defined by the first split is then split again, usually based on a different field, and the process repeats until the subsamples cannot be split any further. Finally, the lowest-level splits are reexamined, and those that do not contribute significantly to the value are removed. In this model, the target field (variable) must be categorical (i.e., nominal or ordinal, such as de overall survival outcome as dead vs. alive). The input fields (predictors) can be of any type (in our analysis, the 19 genes were entered as quantitative gene expression). The C5.0 models are quite robust in the presence of problems such as missing data and large numbers of input fields. The C5.0 tree shows how using only the gene expression of 9 genes, the overall survival outcome as dead or alive can be predicted with high accuracy.

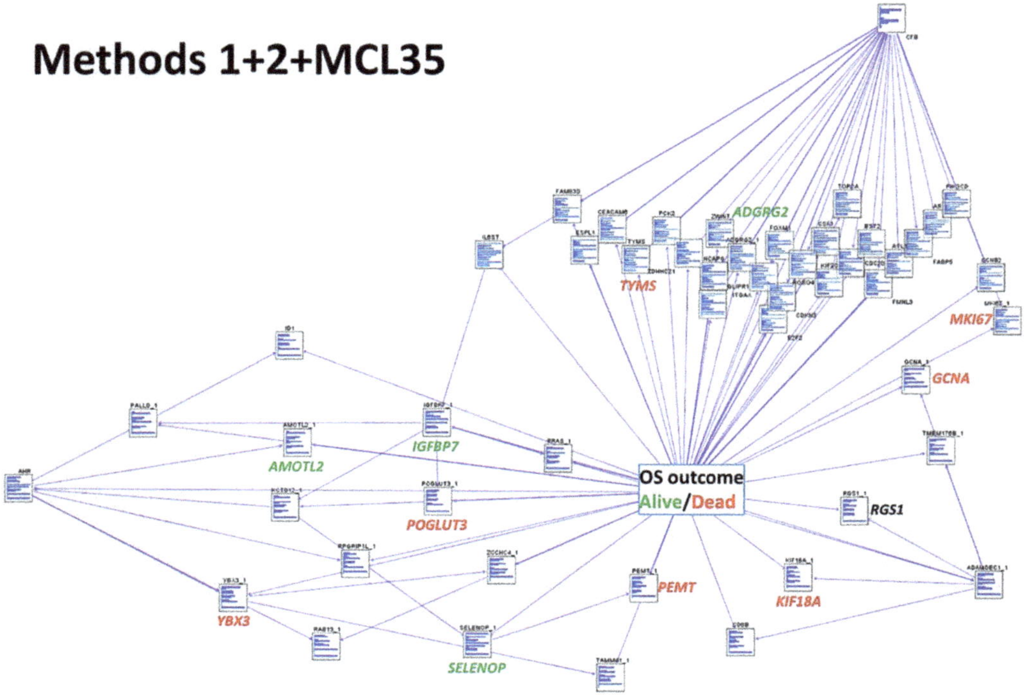

Figure 13. Addition of the MCL35 proliferation signature in a Bayesian network. A Bayesian network modeling was performed using the highlighted genes of both Methods 1 (19 genes) and Methods 2 (15) with the previously identified prognostic genes of MCL of the LLMPP, the MCL35 signature. Some of the most relevant genes are highlighted, in red for the bad, green for the good prognostic genes, and their interrelationships (arrows).

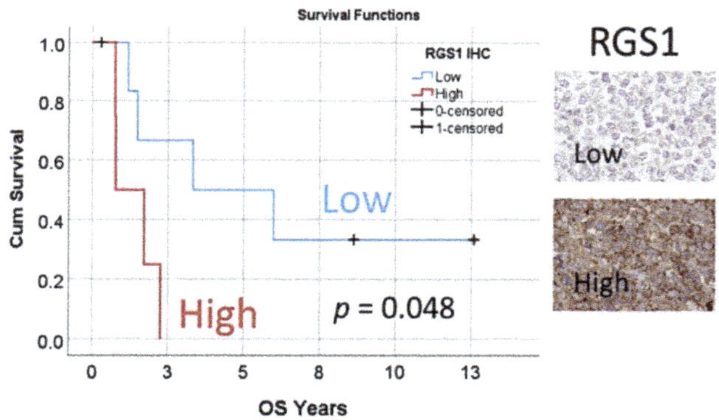

Figure 14. Overall survival according to the immunohistochemical expression of RGS1.

2.7. Immunohistochemistry

The immunohistochemistry was performed using an automated piece of equipment, Leica BOND-MAX stainer, following the manufacturer's instructions and as previously described [53,59,63–65]. The RGS1 primary antibody (rabbit polyclonal) was purchased from Thermofisher [63]. The slides were digitalized using a Hamamatsu NanoZoomer S360, scanned, and visualized using the NDP.veiw2 software.

3. Results

3.1. Highlights

- Using 20,862 genes as a start point (input layers) (Method 1), several neural network analyses correlated with the overall survival outcome and with known pathogenic genes of MCL (output layers), and a final set of 19 genes with predictive value was highlighted (Figure 5);
- This type of analysis was repeated focusing on 10 immune, cancer, and immuno-oncology panels (Method 2), and 15 genes were highlighted (Figure 8);
- Other machine learning techniques were used to predict the overall survival (Figures 11 and 12);
- The highlighted genes also predicted the overall survival of a pan-cancer series (Figures 9, 10 and A1);
- The combination of both Methods 1 (19 genes) and 2 (15 genes) with the LLMPP MCL35 assay (17) genes and analysis using several machine learning and neural networks techniques predicted the overall survival outcome (dead vs. alive) with high accuracy.

3.2. Prediction of Overall Survival Based on the 20,862 Genes of the Array (Method 1)

Dimensionality reduction refers to techniques for reducing the number of input variables in training data. Fewer input dimensions often mean correspondingly fewer parameters or a simpler architecture in the machine learning model, referred to as degrees of freedom [66]. The input layer of 20,862 predicted the overall survival of mantle cell lymphoma (MCL), using an analysis algorithm (Figure 5). The output variables (targets) were the overall survival outcome as a dichotomous variable (dead/alive), and the 28 genes (high/low expression) with prognostic relevance for the overall survival were confirmed in the same series (Table 2). Tables A1 and A2 show the complete details of the artificial neural networks. The multilayer perceptron (MLP) technique had better performance than the radial basis function (RBF): comparing area under the curve, percentage of incorrect predictions (testing set), and overall percentage of correct classification (testing set), for MLP vs. RBF, the results were 0.85 ± 0.05 vs. 0.77 ± 0.09 (p = 0.000053), 15.3% ± 5.9 vs. 26.5% ± 10.2 (p = 0.000005), and 84.7% ± 5.9 vs. 73.5% ± 10.2 (p = 0.000005), respectively. *CCND1* was the best predicted gene; in the MLP analysis *CCND1* had a percentage of incorrect predictions in the testing set of 2.8%, the lowest value among all genes (Table A1).

From the initial 20,862 genes, the list was reduced to 1394 genes, and additional multilayer perceptron analyses led to a set of 58 genes (Figure 6). The network performance of the MLP with the input of 58 genes was "good", with an area under the curve (AUC) of 0.9. The genes were ranked based on their normalized importance for prediction, and GSEA confirmed that most of these genes were associated with the death survival outcome (Figure 6); the most relevant were *KIF18A*, *FANCG*, *GCNA*, *YBX3*, *ZCCHC4*, and *DMTF1*.

Based on the 58 genes, a subsequent multivariate Cox regression analysis, backward conditional, highlighted a set of 19 genes (Table A3), and a final set of 10 genes was found after using a cut-off and a Kaplan–Meier analysis for overall survival (Table 2). *KIF18A*, *YBX3*, *PEMT*, *GCNA*, and *POGLUT3* were associated with an unfavorable overall survival, and *SELENOP*, *AMOTL2*, *IGFBP7*, *KCTD12*, and *ADGRG2* to a favorable survival (Figure 6). Finally, the 10 genes were correlated with the cell proliferation marker of *MKI67*, which is one of the most relevant genes in the pathogenesis of MCL (Table 3). The cases with low *MKI67* were associated with high *KCTD12*, *ADGRG2*, *SELENOP*, and *IGFBP7*. However, high *MKI67* associated with high *YBX3*. Table A4 shows a multivariate analysis for overall survival between *MIK67* and the 10 genes using a Cox regression.

Therefore, the dimensionality/data reduction of the Methods 1 went from 20,862 initial genes, to 1394, 58, 19, and the final 10 most relevant prognostic genes for overall survival of MCL patients.

Table 3. Kaplan–Meier analysis for prediction of overall survival outcome (Method 1).

m	Gene	Cut-Off	Log-Rank p Value	Breslow p Value	Hazard Risk	Correlation with High MKI67, Odds Ratio (OR)	OR p Value
1	*KIF18A*	8.71	<0.001	<0.001	3.5 (2.1–5.8)	1.3 (0.6–3.0)	0.499
2	*YBX3*	11.83	0.001	0.002	2.3 (1.4–3.8)	2.3 (0.9–5.3)	0.056
3	*PEMT*	8.75	0.015	0.016	1.9 (1.1–3.1)	1.1 (0.5–2.5)	0.798
4	*GCNA*	7.66	0.037	0.137	1.8 (1.0–3.3)	2.1 (0.9–4.9)	0.077
5	*POGLUT3*	8.81	0.034	0.014	1.6 (1.0–2.5)	0.9 (0.4–1.7)	0.649
6	*SELENOP*	12.81	0.028	0.048	0.6 (0.4–0.9)	0.2 (0.1–0.5)	0.001
7	*AMOTL2*	8.99	0.039	0.029	0.5 (0.3–0.9)	0.5 (0.2–1.1)	0.068
8	*IGFBP7*	13.37	0.019	0.042	0.5 (0.3–0.9)	0.2 (0.1–0.4)	<0.001
9	*KCTD12*	12.02	0.022	0.042	0.5 (0.3–0.9)	0.2 (0.1–0.5)	0.01
10	*ADGRG2*	9.95	<0.001	<0.001	0.3 (0.2–0.6)	0.2 (0.1–0.5)	0.001

This analysis is a univariate.

3.3. Prediction of Overall Survival Based on the Immuno-Oncology Panels (Method 2)

The prediction of the overall survival outcome was performed using another strategy, based on nine different immune oncology pathways, multilayer perceptron neural networks, GSEA, and Kapan–Meier analyses (Figure 8).

The characteristics and performance parameters of the neural networks are shown in Table A5. The most predictive panels (pathways) were the autoimmune (AUC = 0.98), the pan cancer human IO360 (AUC = 0.94), human inflammation (AUC = 0.89), pan cancer (AUC = 0.89), and metabolic (AUC = 0.87). Interestingly, some pathways had a more predictive power toward the dead than the alive outcome.

After selecting the genes with a normalized importance above 70% and merging, a final set of 125 was identified. A GSEA on these 125 genes had a sinusoidal-like pattern, with some genes associated toward poor (dead) and others to favorable (alive) overall survival. The genes were ranked according to their normalized importance for prediction using a multilayer perceptron analysis, and the top 15 genes were *CD8B*, *CEACAM6*, *FABP5*, *CFB*, *IL6ST*, *AHR*, *BST2*, *ROBO4*, *AR*, *ID1*, *PIK3CD*, *ITGAX*, *TYMS*, *CSF1*, and *PCK2* (normalized importance >0.68). Among them, *TYMS* was highlighted, and this gene by itself managed to predict the overall survival of the patients (Hazard risk (HR) = 3.2, 95% CI 2.0–5.0, $p = 8.9 \times 10^{-7}$). Of note, high *TYMS* also correlated with high *MIK67* expression (Fisher's exact test, $p = 0.001$).

In a multivariate Cox regression survival analysis including these top 15 genes as quantitative variables, backward conditional method, in the last step (11) the significant genes were *TYMS* ($p < 0.001$, HR = 2.6), *AR* ($p = 0.012$, HR = 1.5), and *CSF1* ($p = 0.049$, HR = 0.6).

3.4. Prediction of Overall Survival of a Pan-Cancer Series

The predictive value of the set of 19 genes, derived from neural network analysis and dimensional reduction of the initial 20,862 genes (Figure 5, Method 1), was tested for the prediction of a pan cancer series of 7289 cases from The Cancer Genome Atlas (TCGA) database and GSE10846 dataset for diffuse large B-cell lymphoma (DLBCL). Using a risk-score formula [36,46], a different overall survival of the patients was found, confirming the pathological role of these genes in cancer (Figures 9 and 10, Table A6, Figure A1). In overall high-risk versus low-risk cases, Cox regression hazard risk = 3.3 (95% CI 2.9–3.6), $p < 0.0001$.

3.5. Prediction of Overall Survival Outcome Using other Machine Learning Techniques

The predictive value of the set of 19 genes (Method 1) as quantitative variables for the overall survival outcome was modeled using other machine-learning techniques, including logistic regression, Bayesian network, discriminant analysis, KNN algorithm, LSVM, tree-AS, C5, CHAID, Quest, random, and C&R trees. Among them, the highest overall accuracy for prediction was achieved by the C5 tree (95%, 9 genes used), and Bayesian network (85%, 19 genes, Figures 11 and 12).

3.6. Combination of Method 1, Method 2, and the LLMPP MCL35 Prognostic Gene Signature

A machine learning and neural network modeling was performed using the highlighted genes of both Methods 1 (19 genes) and Methods 2 (15) with the previously identified prognostic genes of MCL of the LLMPP, the MCL35 signature [50,67–69]. All the available artificial intelligence methods were tested, and high overall accuracy for predicting was found for logistic regression (100%), Bayesian network (92%), discriminant analysis (86%), CHAID (85%), C&R tree (85%), and SVM (81%) (Table 4, Figure 13).

Table 4. Machine learning and neural network analysis of the combined Methods 1 and 2 with the MCL35 signature.

Model	Overall Accuracy for Predicting the Overall Survival	No. of Genes Used in the Final Model	Gene Names
Logistic regression	100	50	All the 50
Bayesian network	92	50	All the 50
Discriminant	86	50	All the 50
CHAID	85	6	*E2F2*, *GCNA*, *FMNL3*, *POGLUT3*, *SELENOP*, and *ZDHHC21*
C&R tree	85	21	*ADGRG2*, *CDC20*, *CEACAM6*, *ESPL1*, *FABP5*, *FAM83D*, *FMNL3*, *GCNA*, *GLIPR1*, *ID1*, *ITGAX*, *KIF2C*, *MKI67*, *RGS1*, *ROBO4*, *RPGRIP1L*, *RRAS*, *SELENOP*, *TAMM41*, *ZDHHC21*, and *ZWINT*
SVM	81	50	All the 50
KNN algorithm	78	50	All the 50
Neural network	76	50	All the 50
C5	76	3	*ESPL1*, *RPGRIP1L*, and *ZWINT*
Quest	65	50	All the 50

In this analysis, several methods were tested, including C5, logistic regression, Bayesian network, discriminant analysis, KNN algorithm, LSVM, random trees, SVM, Tree-AS, CHAID, Quest, C&R tree, and neural networks. Among them, logistic regression and Bayesian network had the best overall accuracy for predicting the overall survival (dead vs. alive). The analysis used a custom field (genes) assignment. The target variable was the overall survival as a dichotomic (binary) variable (dead vs. alive). The inputs (predictive genes) were the most relevant genes (n = 50) that were previously identified in the Methods 1 (n = 19), 2 (n = 15), and the MCL35 signature (n = 17), as follows: *ADAMDEC1, ADGRG2, AHR, AMOTL2, AR, ATL1, BST2, CCNB2, CD8B, CDC20, CDKN3, CEACAM6, CFB, CSF1, E2F2, ESPL1, FABP5, FAM83D, FMNL3, FOXM1, GCNA, GLIPR1, ID1, IGFBP7, IL6ST, ITGAX, KCTD12, KIF18A, KIF2C, MKI67, NCAPG, PALLD, PCK2, PEMT, PIK3CD, POGLUT3, RAB13, RGS1, ROBO4, RPGRIP1L, RRAS, SELENOP, TAMM41, TMEM176B, TOP2A, TYMS, YBX3, ZCCHC4, ZDHHC21*, and *ZWINT*. A total of 13 models were selected and ranked according to their overall accuracy for predicting the overall survival. In the modeling, every possible combination of options was tested, and the best models were saved. Of note, in the final models not all the genes were necessary or contributed to the model, and only the best combinations were selected (e.g., 50 genes in the Bayesian network but only 6 in the CHAID tree).

3.7. Immunohistochemical Analysis of RGS1

RGS1 was identified as an MCL prognostic gene. It was present within the set of 19 in the last step of the first analysis algorithm (Figure 5) and the Cox regression (backward conditional). The prognostic association was tested by immunohistochemistry in a series of 11 cases of MCL from Tokai University. Among the different gene candidates, *RGS1* was selected because a reliable primary antibody for immunohistochemistry was available, and we previously showed that high RGS1 protein expression correlated with poor prognosis in diffuse large b-cell lymphoma [63]. The clinicopathological characteristics of this series was the following: age (median, 72 years; range 41–82); male (9/11, 82%); lymph node and tonsil biopsy (10/11, 91%); CD3-negative (100%); CD5-positive (10/11, 91%); CD20, CD10, Cyclin D1 (*CCND1*) and BCL2-positive (100%); BCL6-positive (3/11, 27%); MUM-1(*IRF4*)-positive (9/10, 90%); proliferation index (Ki67, 10–50%).

The RGS1 protein expression was evaluated as low and high, and correlated with the overall survival of the patients ($p = 0.048$) (Figure 10). Nevertheless, no correlation was found between RGS1 and the other clinicopathological characteristics.

4. Discussion

Mantle cell lymphoma is a hematological neoplasia that belongs to the group of non-Hodgkin lymphomas (NHL) and it is derived from mature B-lymphocytes [16].

The postulated cell of origin in most of the cases is a naïve pregerminal center B-cell of the mantle zone [1,9,16,17,46], because of the absence of somatic mutations in the variable region of the heavy chain of immunoglobulin genes (*IgVH*). *IgVH* somatic mutational status is a marker of the transition of a B-lymphocyte through a follicular germinal center [70]. However, in 20–30% of the cases somatic hypermutation is found, which suggests a postgerminal origin (marginal zone) [71], and these cases are associated with a better prognosis [72]. Because of the aggressive clinical behavior of mantle cell lymphoma, it is critical to find prognostic makers that will allow identifying the patients who should receive more aggressive therapy.

Mantle cell lymphoma is characterized by increased cell division and replication, decreased response to DNA damage, and enhanced cell survival (impaired apoptosis) [16]. Some of these pathways and genes correlate with prognosis. For instance, *TP53* and *NOTCH1* mutations, overexpression of *SOX11*, and high proliferation index (Ki67 staining) associate with a poor prognosis.

This research identified new prognostic markers using gene expression data. Dimensionality reduction refers to techniques for reducing the number of input variables in training data. Fewer input dimensions often mean correspondingly fewer parameters or a simpler architecture in the machine learning model, referred to as degrees of freedom [66]. A neural network analysis correlated the 20,862 genes of the array with the overall survival outcome (dead/alive), and ranked the genes according to their normalized importance for prediction. Additionally, the analysis was enriched with the inclusion of 28 prognostic genes, which were identified from the literature and later confirmed to have prognostic relevance in this series (Table 1). Therefore, the input data of the neural network were solid and resulted in the identification of potentially relevant new prognostic markers. Additionally, the second type of neural network analysis was performed using several immune oncology pathways, which provided a more supervised training and analysis. The fact that we found a correlation of some of the highlighted genes with the expression of *MKI67*, a marker of proliferation known to be critical in mantle cell lymphoma pathogenesis, suggests that the identified new markers are also potentially relevant.

The highlighted genes influence apoptosis, angiogenesis, cell proliferation, and metabolic processes. They contribute to hematological neoplasia or cancer (Table 5). Therefore, it is expected that these genes also affect the progression of the pan cancer series.

Table 5. Function and association of the highlighted genes in neoplasia.

Gene	Function	Role in Cancer
KIF18A	Microtubule motor activity, role in mitosis	Overexpressed in various types of cancer; inhibitors are available [73]
YBX3	Translation repression, negative regulation of intrinsic apoptosis signaling	Related to myelodysplastic syndromes and acute myeloid leukemia [74]
PEMT	Negative regulation of cell proliferation, positive regulation of lipoprotein metabolic process	Critical role in breast cancer progression [75]
GCNA	Acidic repeat-containing protein, expressed in germ cells (testis)	Regulate genome stability [76,77]
POGLUT3	Protein glucosyltransferase, specifically targets extracellular EGF repeats of proteins (NOTCH1 and NOTCH3)	Related to glioblastoma multiforme tumorigenesis [78]
SELENOP	Transport of selenium, response to oxidative stress	Prostate cancer recurrence [79]
AMOTL2	Actin cytoskeleton organization, angiogenesis, cell migration, Wnt-signaling pathway	Angiogenesis in pancreatic, and proliferation in lung cancer [80,81]
IGFBP7	Cell adhesion, metabolic process (retinoic acid, cortisol), regulation of cell growth	Prognosis of acute lymphoblastic leukemia [82]
KCTD12	GABA-B receptors auxiliary subunit	Proliferation in breast cancer [83]
ADGRG2	G protein-coupled receptor signaling pathway	Tumor suppressor in endometrial cancer [84]
TYMS	Regulation of mitotic cell cycle (G1/S transition)	Association with non-Hodgkin lymphomas, prognosis of pancreatic cancer [85,86]

The gene information is based on UniProt [54], and Genecards [55]. *TYMs* was highlighted in Method 2; the rest of genes in Method 1.

It is important to point out that one could also use background information (e.g., patient age, sex, comorbidities, etc.) into the artificial neural network analyses. Incorporating such information would have a large impact on the results. In this research, the target was the prediction of the overall survival of patients based on the gene expression data as proof of concept. In future analyses, background information will be incorporated in MCL analysis, in a similar way as we have recently done in diffuse large b-cell lymphoma (DLBCL) [35].

In addition to neural networks, other machine learning techniques were tested, and the C5 tree and Bayesian networks had the best accuracy for predicting the overall survival outcome. Of note, the type of analyses used do not necessarily represent direct cause and effect, but the probabilistic or conditional independencies between the markers.

The recent advances in machine learning have led to many artificial intelligence (AI) applications, which will produce autonomous systems. However, the effectiveness of these systems is limited by the machine's current inability to explain their decision and actions to human users [87]. Therefore, explainable AI (XAI) will be essential to understand, trust, and effectively managed AI machine partners [87]. In this research, the artificial neural networks highlighted the most relevant genes according to their normalized importance for predicting the overall survival of the patients. To make the results more explainable, we performed serval additional machine learning techniques and conventional statistics to understand the results. For future work, the explanation of algorithms will be developed. Of note, in medicine, AI technologies can be clinically validated even when their function cannot be understood by their operators [88].

Future research directions will be the validation of the methodology and highlighted genes in other series of mantle cell lymphoma and non-Hodgkin lymphomas.

5. Conclusions

This research combined artificial neural networks, machine learning, and conventional statistics to model the overall survival of mantle cell lymphoma and highlight pathogenic genes. Artificial intelligence is a promising field in the understanding of hematological neoplasia, and other types of cancer.

Author Contributions: Conceptualization, J.C.; methodology, J.C.; validation, R.H.; formal analysis, J.C.; writing—original draft preparation, J.C.; writing—review and editing, J.C.; supervision, N.N.; funding acquisition, J.C. All authors have read and agreed to the published version of the manuscript.

Funding: Joaquim Carreras was funded by the Ministry of Education, Culture, Sports, Science and Technology (MEXT) and the Japan Society for the Promotion of Science, grants KAKEN 15K19061 and 18K15100, and Tokai University School of Medicine, research incentive assistant plan 2021-B04. Rifat Hamoudi was funded by Al-Jalila Foundation (grant number AJF2018090), and University of Sharjah (grant number 1901090258).

Institutional Review Board Statement: The study was conducted according to the guidelines of the Declaration of Helsinki, and approved by the Institutional Review Board and the Ethics Committee of Tokai University, School of Medicine (protocol code IRB14R-080 and IRB20-156).

Informed Consent Statement: Informed consent was obtained from all subjects involved in the study, according to a protocol approved by the National Cancer Institute institutional review board.

Data Availability Statement: The gene expression data (GEO data sets) were obtained from the publicly available database of the NCBI resources webpage, located at https://www.ncbi.nlm.nih.gov/gds (accessed on 15 August 2021).

Acknowledgments: I would like to thank all the researchers and colleagues that contributed to the generation of the GSE93291, GSE10846, and The Cancer Genome Atlas (TCGA) program.

Conflicts of Interest: The authors declare no conflict of interest.

Appendix A

Table A1. Multilayer Perceptron Neural Network Analysis of Mantle Cell Lymphoma (Method 1).

Gene	Num. Genes Top 70%	Case Processing Summary					Network Layers				Model Summary							Classification						Area under the Curve (AUC)
		Training		Testing		Input	Hidden		Output		Training			Testing				Training (% Correct)			Testing (% Correct)			
		Num.	%	Num.	%	Units	Num.	Units	Num.	Units	Cross Entropy Error	Incorrect Predictions %	Training Time	Cross Entropy Error	Incorrect Predictions %	Observed 0	Observed 1	Overall	Observed 0	Observed 1	Overall	
Dead/Alive	80	84	68.3	39	31.7	20863	1	6	1	2	38.2	21.4	01:04.9	10.4	12.8	67.6	86	78.6	88.9	86.7	87.2	0.90
SYNE1	6	90	73.2	33	26.8	20862	1	12	1	2	38.5	18.9	01:05.8	8.8	9.1	59.3	90.5	81.1	66.7	96.3	90.9	0.86
DAZAP1	80	87	70.7	36	29.3	20862	1	11	1	2	32.0	14.9	01:06.3	6.4	5.6	64	93.5	85.1	83.3	96.7	94.4	0.92
MYCN	154	85	69.1	38	30.9	20862	1	8	1	2	37.5	27.1	01:01.5	14.4	13.2	36.4	85.7	72.9	66.7	93.1	86.8	0.82
CXCL12	56	87	70.7	36	29.3	20862	1	8	1	2	40.5	19.5	00:57.4	10.1	8.3	44	95.2	80.5	83.3	93.3	91.7	0.83
NOTCH2	20	84	68.3	39	31.7	20862	1	9	1	2	29.9	20.2	00:58.2	11.8	17.9	92.3	36.8	79.8	93.1	50	82.1	0.90
CDK4	47	87	70.7	36	29.3	20862	1	11	1	2	30.4	13.8	00:51.2	13.8	22.2	91.3	66.7	86.2	100	27.3	77.8	0.89
BMI1	25	93	85.6	30	24.4	20862	1	8	1	2	53.0	26.9	00:56.3	13.2	16.7	71.7	74.5	73.1	93.8	71.4	83.3	0.81
ING1	94	76	61.8	47	38.2	20862	1	10	1	2	36.3	17.1	00:52.7	22.7	27.7	50	93.1	82.9	30.8	88.2	72.3	0.76
NSD2	38	91	74	32	26	20862	1	9	1	2	43.0	20.9	01:04.7	15.1	15.6	82.4	75	79.1	91.7	80	84.4	0.86
PTK2	6	93	75.6	30	24.4	20862	1	13	1	2	40.2	16.1	01:07.3	7.9	10	97.1	43.5	83.9	91.3	85.7	90	0.85
PIK3CA	4	76	61.8	47	38.2	20862	1	10	1	2	26.4	13.2	00:52.4	17.7	12.8	94.8	61.1	86.8	94.3	66.7	87.2	0.88
CHEK1	86	91	74	32	26	20862	1	9	1	2	45.3	27.5	00:58.7	12.9	18.8	68.8	76.7	72.5	92.9	72.2	81.3	0.85
CHEK2	8	90	73.2	33	26.8	20862	1	10	1	2	39.8	18.9	01:07.6	13.0	15.2	77.3	84.8	81.1	83.3	86.7	84.8	0.88
PIK3CD	50	82	66.7	41	33.3	20862	1	10	1	2	17.6	11.0	01:08.1	14.6	14.6	90.9	86.8	89	90.9	78.9	85.4	0.96
XIAP	22	85	69.1	38	30.9	20862	1	12	1	2	40.2	18.8	00:49.9	17.7	23.7	83.7	78.6	81.2	85.7	64.7	76.3	0.87
PAX5	23	88	71.5	35	28.5	20862	1	7	1	2	45.3	27.3	00:55.2	13.0	8.6	20	93.7	72.7	50	100	91.4	0.75
BCL2L11	12	71	57.7	52	42.3	20862	1	5	1	2	29.9	19.7	00:50.1	24.2	23.1	92.6	41.2	80.3	94.9	23.1	76.9	0.82
BORCS8_MEF2B	12	85	69.1	38	30.9	20862	1	11	1	2	39.2	21.2	00:53.3	11.6	10.5	40.9	92.1	78.8	55.6	100	89.5	0.83
PTEN	86	84	68.3	39	31.7	20862	1	10	1	2	36.0	20.2	00:57.0	12.2	7.7	92.1	42.9	79.8	93.3	88.9	92.3	0.85
MYC	10	84	68.3	39	31.7	20862	1	9	1	2	28.9	16.7	00:56.2	14.2	20.5	87.7	68.4	83.3	96.4	36.4	79.5	0.90
CCND1	23	87	70.7	36	29.3	20862	1	8	1	2	38.3	23.0	01:03.5	6.7	2.8	92.3	31.8	77	96.4	100	97.2	0.89
MKI67	2	93	75.6	30	24.4	20862	1	10	1	2	40.2	20.4	01:04.6	11.7	16.7	78	81.4	79.6	85.7	81.3	83.3	0.89
CCND2	46	76	61.8	47	38.2	20862	1	9	1	2	32.4	21.1	00:54.9	17.7	14.9	90.7	50	78.9	92.3	50	85.1	0.84
CDKN2A	112	91	74	32	26	20862	1	14	1	2	22.0	9.9	00:53.6	11.3	21.9	94.4	73.7	90.1	91.3	44.4	78.1	0.93
CDKN2C	6	90	73.2	33	26.8	20862	1	8	1	2	46.7	26.7	00:58.1	13.5	15.2	67.4	78.7	73.3	89.5	78.6	84.8	0.85
TERT	205	82	66.7	41	33.3	20862	1	9	1	2	34.6	20.7	01:00.8	14.9	19.5	93.7	31.6	79.3	93.3	45.5	80.5	0.85
NOTCH1	15	85	69.1	38	30.9	20862	1	11	1	2	32.4	17.6	00:49.1	16.3	21.1	88.2	58.8	82.4	88.5	58.3	78.9	0.85
RB1	47	88	71.5	35	28.5	20862	1	12	1	2	48.9	27.3	00:56.3	14.3	17.1	65.1	80	72.7	78.9	87.5	82.9	0.83
Combined	18	91	74	32	26	20835	1	8	29	58	1348.9	25.7	01:22.2	525.3	29.4	-	-	74.3	-	-	70.6	-
Average		85.9	70.1	37.1	30.2	20861	1	9.6	-	-	80.4	20.1	-	30.6	15.8	75.0	70.8	79.9	84.2	73.5	84.2	0.9

Input layer: standardized rescaling method for covariates. Hidden layer: hyperbolic tangent activation function. Output layer: softmax activation function, cross-entropy error function. Model summary, training, one consecutive step(s) with no decrease in error (error computations are based on the testing sample) as stopping rule.

Table A2. Radial Basis Function Neural Network Analysis of Mantle Cell Lymphoma (Method 1).

Gene	Num. Genes Top 70%	Case Processing Summary					Network Layers				Model Summary					Classification						Area under the Curve (AUC)
		Training		Testing		Input	Hidden		Output		Training			Testing		Training (% Correct)			Testing (% Correct)			
		Num.	%	Num.	%	Units	Num.	Units	Num.	Units	Sum of Squares Error	Incorrect Predictions %	Training Time	Sum of Squares Error	Incorrect Predictions %	Observed 0	Observed 1	Overall	Observed 0	Observed 1	Overall %	
Dead/Alive	37	92	74.8	31	25.2	20863	1	8	1	2	16.9	27.2	04:13.3	6.7	38.7	45.5	88.1	72.8	10.0	85.7	61.3	0.73
SYNE1	18	85	69.1	38	30.9	20862	1	8	1	2	10.4	17.6	02:46.3	7.4	23.7	40.9	96.8	82.4	27.3	96.3	76.3	0.79
DAZAP1	28	80	65	43	35	20862	1	6	1	2	8.2	16.3	02:24.1	3.1	9.3	81.8	84.5	83.8	100.0	88.2	90.7	0.93
MYCN	48	82	66.7	41	33.3	20862	1	6	1	2	11.1	20.7	02:32.2	7.4	31.7	30.0	95.2	79.3	9.1	90.0	68.3	0.78
CXCL12	50	82	66.7	41	33.3	20862	1	5	1	2	12.7	22.0	02:39.9	8.2	26.8	10.0	100.0	78.0	0.0	100.0	73.2	0.74
NOTCH2	29	92	74.8	31	25.2	20862	1	10	1	2	11.7	15.2	03:18.6	4.9	25.8	98.6	35.0	84.8	100.0	11.1	74.2	0.80
CDK4	16	82	66.7	41	33.3	20862	1	10	1	2	11.4	20.7	02:21.8	4.9	17.1	98.3	27.3	79.3	100.0	0.0	82.9	0.83
BMI1	41	90	73.2	33	26.8	20862	1	5	1	2	20.0	34.4	03:21.6	7.4	39.4	77.6	51.2	65.6	100.0	35.0	60.6	0.70
ING1	40	79	64.2	44	35.8	20862	1	4	1	2	14.8	26.6	02:14.7	7.6	22.7	0.0	100.0	73.4	0.0	100.0	77.3	0.60
NSD2	39	92	74.8	31	25.2	20862	1	10	1	2	13.6	20.7	03:11.6	4.1	9.7	85.7	72.1	79.3	85.7	94.1	90.3	0.88
PTK2	19	90	73.2	33	26.8	20862	1	3	1	2	16.2	24.4	03:15.7	5.8	24.2	100.0	0.0	75.6	100.0	0.0	75.8	0.64
PIK3CA	46	79	64.2	44	35.8	20862	1	8	1	2	12.5	24.1	02:23.1	7.7	25.0	93.3	21.1	75.9	100.0	0.0	75.0	0.74
CHEK1	51	92	74.8	31	25.2	20862	1	8	1	2	16.4	26.1	03:12.5	7.0	41.9	78.6	70.0	73.9	50.0	72.7	58.1	0.80
CHEK2	80	88	71.5	35	28.5	20862	1	9	1	2	13.5	25.0	02:57.1	5.9	22.9	59.1	90.9	75.0	66.7	88.2	77.1	0.86
PIK3CD	47	79	64.2	44	35.8	20862	1	3	1	2	12.1	20.3	02:15.3	8.0	27.3	66.7	90.7	79.7	63.3	92.9	72.9	0.83
XIAP	89	79	64.2	44	35.8	20862	1	8	1	2	10.7	17.7	02:20.4	11.0	43.2	88.4	75.0	82.3	66.7	47.8	56.8	0.80
PAX5	81	89	72.4	34	27.6	20862	1	9	1	2	14.5	24.7	02:55.3	6.0	26.5	13.0	97.0	75.3	0.0	96.2	73.5	0.71
BCL2L11	28	88	71.5	35	28.5	20862	1	8	1	2	10.9	14.8	02:51.2	4.1	14.3	100.0	43.5	85.2	96.4	42.9	85.7	0.86
BORCS8_MEF2B	41	86	69.9	37	30.1	20862	1	3	1	2	13.8	23.3	02:45.9	5.8	18.9	19.0	95.4	76.7	30.0	100.0	81.1	0.76
PTEN	23	92	74.8	31	25.2	20862	1	7	1	2	11.1	16.3	03:14.2	3.5	12.9	95.4	55.6	83.7	92.9	33.3	87.1	0.84
MYC	18	92	74.8	31	25.2	20862	1	9	1	2	9.8	16.3	03:31.2	4.1	25.8	91.8	52.6	83.7	95.0	36.4	74.2	0.90
CCND1	42	82	66.7	41	33.3	20862	1	10	1	2	11.2	19.5	02:29.4	6.0	26.8	88.3	59.1	80.5	87.9	12.5	73.2	0.81
MKI67	37	90	73.2	33	26.8	20862	1	10	1	2	12.6	21.1	03:00.8	5.0	21.2	88.0	67.5	78.9	78.6	78.9	78.8	0.89
CCND2	40	79	64.2	44	35.8	20862	1	4	1	2	12.3	24.1	02:14.5	7.6	25.0	100.0	0.0	75.9	100.0	0.0	75.0	0.74
CDKN2A	56	92	74.8	31	25.2	20862	1	6	1	2	14.1	20.7	03:02.7	5.0	25.8	97.2	15.0	79.3	100.0	0.0	74.2	0.73
CDKN2C	34	88	71.5	35	28.5	20862	1	9	1	2	17.6	21.6	02:50.9	8.9	34.3	86.8	72.0	78.4	58.3	81.8	65.7	0.78
TERT	58	79	64.2	44	35.8	20862	1	10	1	2	10.3	17.7	02:17.2	10.0	27.3	93.7	37.5	82.3	100.0	14.3	72.7	0.71
NOTCH1	71	79	64.2	44	35.8	20862	1	3	1	2	12.4	22.8	02:14.6	7.3	25.0	100.0	0.0	77.2	100.0	0.0	75.0	0.74
RB1	87	89	72.4	34	27.6	20862	1	2	1	2	22.2	47.2	02:55.3	8.7	55.9	100.0	0.0	52.8	100.0	0.0	44.1	0.49
Combined	87	93	75.6	30	24.4	20835	1	14	29	58	366.4	20.4	09:53.4	147.2	23.7	-	-	79.6	-	-	76.3	-
Average		86.0	69.9	37.0	30.1	20861	1	7.2			25.0	22.3		11.2	26.4	73.4	58.4	77.7	69.6	51.7	73.6	0.77

Input layer: standardized rescaling method for covariates. Hidden layer: softmax activation function. Output layer: identity activation function, sum of squares error function. Model summary, testing, sum of square error (the number of hidden units is determined by the testing data criterion: The "best" number of hidden units is the one that yields the smallest error in the testing data).

Table A3. Multivariate Cox regression analysis for predicting overall survival outcome (Method 1).

Num	Gene	B	SE	Wald	df	*p* Value	Hazard Risk	95.0% CI for HR	
								Lower	Upper
1	*KIF18A*	2.7	0.3	58.3	1	<0.001	14.2	7.2	28.1
2	*YBX3*	0.8	0.2	19.0	1	<0.001	2.2	1.6	3.2
3	*GCNA*	0.9	0.2	14.6	1	<0.001	2.5	1.6	4.1
4	*POGLUT3*	1.2	0.3	13.4	1	<0.001	3.2	1.7	6.0
5	*AMOTL2*	0.9	0.3	10.1	1	0.001	2.5	1.4	4.3
6	*RAB13*	1.2	0.4	9.8	1	0.002	3.3	1.6	7.0
7	*ZCCHC4*	1.1	0.3	9.5	1	0.002	2.9	1.5	5.7
8	*PEMT*	0.6	0.2	8.4	1	0.004	1.9	1.2	2.8
9	*RRAS*	0.8	0.4	4.7	1	0.029	2.2	1.1	4.4
10	*PALLD*	0.6	0.3	3.9	1	0.048	1.8	1.0	3.1
11	*ADAMDEC1*	0.7	0.4	3.5	1	0.063	1.9	1.0	3.9
12	*ADGRG2*	0.4	0.2	2.8	1	0.094	1.5	0.9	2.3
13	*IGFBP7*	−1.5	0.3	20.3	1	<0.001	0.2	0.1	0.4
14	*TMEM176B*	−1.6	0.4	18.9	1	<0.001	0.2	0.1	0.4
15	*SELENOP*	−1.0	0.2	15.6	1	<0.001	0.4	0.2	0.6
16	*RPGRIP1L*	−0.5	0.1	10.5	1	0.001	0.6	0.5	0.8
17	*TAMM41*	−0.8	0.3	7.5	1	0.006	0.4	0.3	0.8
18	*KCTD12*	−1.2	0.5	7.5	1	0.006	0.3	0.1	0.7
19	*RGS1*	−0.4	0.2	4.5	1	0.034	0.7	0.5	1.0

Cox regression, backward conditional.

Table A4. Multivariate Cox regression overall survival analysis between MKI67 and the 10 highlighted genes (Method 1).

Gene	B	SE	Wald	df	Sig.	HR	95.0% CI for HR	
							Lower	Upper
MKI67	1.3	0.3	20.5	1	0.000	3.8	2.1	6.8
YBX3	0.9	0.3	11.3	1	0.001	2.6	1.5	4.4
SELENOP	−0.5	0.3	3.0	1	0.085	0.6	0.3	1.1
POGLUT3	0.6	0.2	6.9	1	0.009	1.9	1.2	3.1
ADGRG2	−0.7	0.3	4.5	1	0.035	0.5	0.2	0.9
GCNA	0.8	0.3	5.3	1	0.021	2.2	1.1	4.2
KIF18A	1.5	0.3	26.6	1	0.000	4.3	2.5	7.6
PEMT	0.8	0.3	6.6	1	0.010	2.1	1.2	3.8

Multivariate Cox regression analysis, backward conditional. HR, hazard risk. Note: There are only 8 genes because it is a multivariate Cox regression analysis with the backward conditional method. In this method, the nonsignificant variables are eliminated.

Table A5. Multilayer perceptron analysis of the immuno-oncology pathways (Method 2).

Pathway	Num. Genes Top 70%	Case Processing Summary				Network Layers					Model Summary					Classification						Area under the Curve (AUC)
		Training		Testing		Input	Hidden		Output		Training			Testing		Training (% Correct)			Testing (% Correct)			
		Num.	%	Num.	%	Units	Num.	Units	Num.	Units	Cross Entropy Error	Incorrect Predictions %	Training Time	Cross Entropy Error	Incorrect Predictions %	Observed Alive	Observed Dead	Overall	Observed Alive	Observed Dead	Overall %	
Cancer Transcriptome	13	84	68.3	39	31.7	1785	1	6	1	2	41.1	27.4	00:03.9	17.6	23.1	58.8	82.0	72.6	55.6	83.3	76.9	0.84
Pan Cancer Human IO360	15	84	68.3	39	31.7	727	1	8	1	2	22.5	13.1	00:01.4	14.7	15.4	82.4	90.0	86.9	88.9	83.3	84.6	0.94
Pan Cancer Immune Profiling	1	84	68.3	39	31.7	707	1	5	1	2	44.9	26.2	00:01.5	15.0	12.8	64.7	80.0	73.8	88.9	86.7	87.2	0.82
Pan Cancer Progression	18	84	68.3	39	31.7	715	1	11	1	2	51.2	32.1	00:01.7	18.7	12.8	29.4	94.0	67.9	66.7	93.3	87.2	0.74
Pan Cancer Pathways	6	84	68.3	39	31.7	712	1	8	1	2	36.9	21.4	00:01.8	16.8	15.4	67.6	86.0	78.6	77.8	86.7	84.6	0.89
Metabolic Pathways	27	84	68.3	39	31.7	737	1	14	1	2	39.8	22.6	00:01.6	13.7	17.9	55.9	92.0	77.4	66.7	86.7	82.1	0.87
Immune Exhaustion	12	84	68.3	39	31.7	720	1	10	1	2	47.2	31.0	00:01.6	18.2	17.9	50.0	82.0	69.0	66.7	86.7	82.1	0.79
Human Inflammation	23	84	68.3	39	31.7	247	1	9	1	2	33.7	17.9	00:00.6	16.6	23.1	73.5	88.0	82.1	55.6	83.3	76.9	0.89
Host Response	8	84	68.3	39	31.7	747	1	9	1	2	41.1	21.4	00:01.6	18.1	20.5	67.6	86.0	78.6	66.7	83.3	79.5	0.83
Autoimmune	13	84	68.3	39	31.7	719	1	10	1	2	11.9	6.0	00:01.5	12.5	10.3	88.2	98.0	94.0	88.9	90.0	89.7	0.98
Organ Transplantation	12	84	68.3	39	31.7	728	1	11	1	2	41.5	21.4	00:01.6	15.7	10.3	64.7	88.0	78.6	88.9	90.0	89.7	0.85

Input layer: standardized rescaling method for covariates. Hidden layer: hyperbolic tangent activation function. Output layer: softmax activation function, cross-entropy error function. Model summary, training, one consecutive step(s) with no decrease in error (error computations are based on the testing sample) as stopping rule.

Table A6. Overall survival of the pan cancer series using the risk-scores.

Subtype	Overall	Low-Risk	High-Risk	K–M Log-Rank *p* Value	Cox *p* Value	Cox HR	95% CI for HR	
							Lower	Higher
Breast	962	821	141	4.0×10^{-17}	6.5×10^{-15}	4.0	2.8	5.6
Lung	475	426	49	1.0×10^{-10}	1.1×10^{-9}	3.3	2.3	4.9
Prostate	497	446	51	1.5×10^{-4}	2.0×10^{-3}	9.2	2.3	37.2
Colorectal	466	415	51	1.4×10^{-5}	3.3×10^{-5}	2.9	1.7	4.8
Cervix	191	169	22	3.4×10^{-10}	8.9×10^{-8}	7.7	3.6	16.2
Stomach	440	293	147	2.6×10^{-4}	3.1×10^{-4}	1.8	1.3	2.4
Skin (melanoma)	335	177	158	3.2×10^{-10}	1.3×10^{-9}	2.6	1.9	3.5
Bladder	389	207	182	9.2×10^{-13}	9.7×10^{-12}	3.0	2.2	4.1
Ovary	247	217	30	0.6×10^{-5}	1.5×10^{-5}	2.9	1.8	4.6
DLBCL	414	289	125	3.3×10^{-16}	1.5×10^{-14}	3.3	2.5	4.5
Kidney	792	470	322	5.9×10^{-17}	2.5×10^{-15}	3.2	2.4	4.3
Uterus (endometrium)	247	214	33	5.5×10^{-11}	2.4×10^{-8}	7.4	3.7	15.0
Leukemia (AML)	149	115	34	1.9×10^{-14}	7.0×10^{-12}	5.5	3.4	9.0
Pancreas	176	109	67	0.4×10^{-5}	9.0×10^{-6}	2.6	1.7	3.9
Thyroid	489	434	55	9.9×10^{-12}	6.4×10^{-7}	17.4	5.6	53.5
Liver	361	197	164	6.7×10^{-10}	4.0×10^{-9}	3.0	2.1	4.3
CNS (GBM)	659	209	450	2.6×10^{-17}	8.9×10^{-15}	4.5	3.1	6.6
Overall	7289	5208	2081	2.8×10^{-178}	2.5×10^{-159}	3.3	2.9	3.6

K–M, Kapan–Meier; HR, hazard risk, DLBCL, diffuse large B-cell lymphoma; AML, acute myeloid leukemia; CNS, central nervous system; GBM, glioblastoma multiforme. This analysis is univariate.

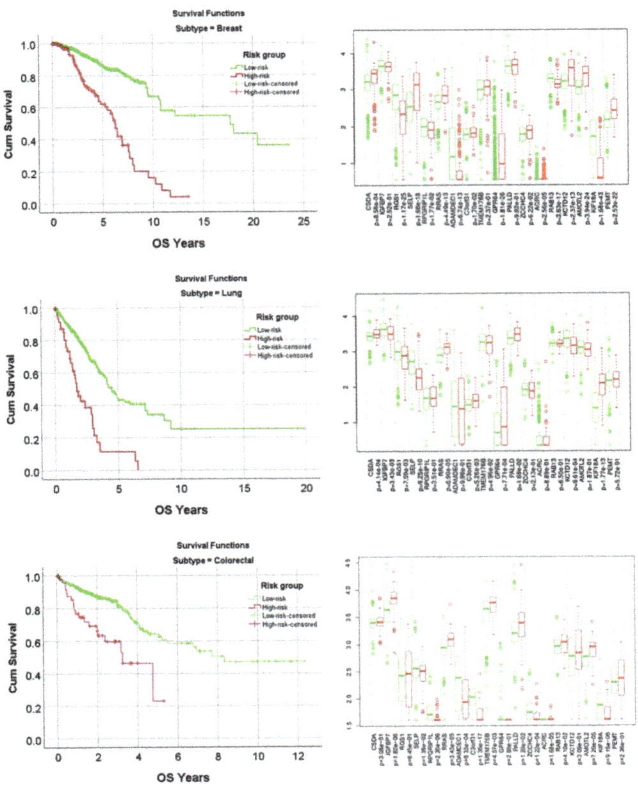

Figure A1. Differential gene expression of the set of 19 genes per cancer subtype. Based on a risk-score formula and the gene expression of 19 genes, the overall survival for each risk-group could be calculated. The contribution in the prognosis for each gene is shown on the right. This Figure is complementary to Figure 9.

References

1. Swerdlow, S.H.; Campo, E.; Pileri, S.A.; Harris, N.L.; Stein, H.; Siebert, R.; Advani, R.; Ghielmini, M.; Salles, G.A.; Zelenetz, A.D.; et al. The 2016 revision of the World Health Organization classification of lymphoid neoplasms. *Blood* **2016**, *127*, 2375–2390. [CrossRef]
2. Armitage, J.O. A clinical evaluation of the International Lymphoma Study Group classification of non-Hodgkin's lymphoma. The Non-Hodgkin's Lymphoma Classification Project. *Blood* **1997**, *89*, 3909–3918.
3. Armitage, J.O.; Weisenburger, D.D. New approach to classifying non-Hodgkin's lymphomas: Clinical features of the major histologic subtypes. Non-Hodgkin's Lymphoma Classification Project. *J. Clin. Oncol.* **1998**, *16*, 2780–2795. [CrossRef]
4. Sant, M.; Allemani, C.; Tereanu, C.; De Angelis, R.; Capocaccia, R.; Visser, O.; Marcos-Gragera, R.; Maynadie, M.; Simonetti, A.; Lutz, J.M.; et al. Incidence of hematologic malignancies in Europe by morphologic subtype: Results of the HAEMACARE project. *Blood* **2010**, *116*, 3724–3734. [CrossRef]
5. Shivdasani, R.A.; Hess, J.L.; Skarin, A.T.; Pinkus, G.S. Intermediate lymphocytic lymphoma: Clinical and pathologic features of a recently characterized subtype of non-Hodgkin's lymphoma. *J. Clin. Oncol.* **1993**, *11*, 802–811. [CrossRef] [PubMed]
6. Smith, A.; Howell, D.; Patmore, R.; Jack, A.; Roman, E. Incidence of haematological malignancy by sub-type: A report from the Haematological Malignancy Research Network. *Br. J. Cancer* **2011**, *105*, 1684–1692. [CrossRef]
7. Zhou, Y.; Wang, H.; Fang, W.; Romaguer, J.E.; Zhang, Y.; Delasalle, K.B.; Kwak, L.; Yi, Q.; Du, X.L.; Wang, M. Incidence trends of mantle cell lymphoma in the United States between 1992 and 2004. *Cancer* **2008**, *113*, 791–798. [CrossRef] [PubMed]
8. Freedman, A.S.; Aster, J.C. Clinical manifestations, pathologic features, and diagnosis of mantle cell lymphoma. In *UpToDate*; Wolters Kluwer: Waltham, MA, USA, 2021.
9. Campo, E.; Raffeld, M.; Jaffe, E.S. Mantle-cell lymphoma. *Semin. Hematol.* **1999**, *36*, 115–127.
10. Tsujimoto, Y.; Yunis, J.; Onorato-Showe, L.; Erikson, J.; Nowell, P.C.; Croce, C.M. Molecular cloning of the chromosomal breakpoint of B-cell lymphomas and leukemias with the t(11;14) chromosome translocation. *Science* **1984**, *224*, 1403–1406. [CrossRef]
11. De Wolf-Peeters, C.; Pittaluga, S. Mantle-cell lymphoma. *Ann. Oncol.* **1994**, *5* (Suppl. 1), 35–37. [CrossRef]
12. Bertoni, F.; Zucca, E.; Genini, D.; Cazzaniga, G.; Roggero, E.; Ghielmini, M.; Cavalli, F.; Biondi, A. Immunoglobulin light chain kappa deletion rearrangement as a marker of clonality in mantle cell lymphoma. *Leuk. Lymphoma* **1999**, *36*, 147–150. [CrossRef] [PubMed]
13. Argatoff, L.H.; Connors, J.M.; Klasa, R.J.; Horsman, D.E.; Gascoyne, R.D. Mantle cell lymphoma: A clinicopathologic study of 80 cases. *Blood* **1997**, *89*, 2067–2078. [CrossRef]
14. Romaguera, J.E.; Medeiros, L.J.; Hagemeister, F.B.; Fayad, L.E.; Rodriguez, M.A.; Pro, B.; Younes, A.; McLaughlin, P.; Goy, A.; Sarris, A.H.; et al. Frequency of gastrointestinal involvement and its clinical significance in mantle cell lymphoma. *Cancer* **2003**, *97*, 586–591. [CrossRef] [PubMed]
15. Ferrer, A.; Salaverria, I.; Bosch, F.; Villamor, N.; Rozman, M.; Bea, S.; Gine, E.; Lopez-Guillermo, A.; Campo, E.; Montserrat, E. Leukemic involvement is a common feature in mantle cell lymphoma. *Cancer* **2007**, *109*, 2473–2480. [CrossRef]
16. Brown, J.R.; Freedman, A.S.; Aster, J.C.; Lister, A.; Rosmarin, A. Pathobiology of mantle cell lymphoma. In *UpToDate*; Wolters Kluwer: Waltham, MA, USA, 2020.
17. Beekman, R.; Amador, V.; Campo, E. SOX11, a key oncogenic factor in mantle cell lymphoma. *Curr. Opin. Hematol.* **2018**, *25*, 299–306. [CrossRef]
18. Hoster, E.; Dreyling, M.; Klapper, W.; Gisselbrecht, C.; van Hoof, A.; Kluin-Nelemans, H.C.; Pfreundschuh, M.; Reiser, M.; Metzner, B.; Einsele, H.; et al. A new prognostic index (MIPI) for patients with advanced-stage mantle cell lymphoma. *Blood* **2008**, *111*, 558–565. [CrossRef] [PubMed]
19. Moller, M.B.; Pedersen, N.T.; Christensen, B.E. Mantle cell lymphoma: Prognostic capacity of the Follicular Lymphoma International Prognostic Index. *Br. J. Haematol.* **2006**, *133*, 43–49. [CrossRef]
20. Meusers, P.; Engelhard, M.; Bartels, H.; Binder, T.; Fulle, H.H.; Gorg, K.; Gunzer, U.; Havemann, K.; Kayser, W.; Konig, E.; et al. Multicentre randomized therapeutic trial for advanced centrocytic lymphoma: Anthracycline does not improve the prognosis. *Hematol. Oncol.* **1989**, *7*, 365–380. [CrossRef]
21. Berger, F.; Felman, P.; Sonet, A.; Salles, G.; Bastion, Y.; Bryon, P.A.; Coiffier, B. Nonfollicular small B-cell lymphomas: A heterogeneous group of patients with distinct clinical features and outcome. *Blood* **1994**, *83*, 2829–2835. [CrossRef]
22. Hartmann, E.; Fernandez, V.; Moreno, V.; Valls, J.; Hernandez, L.; Bosch, F.; Abrisqueta, P.; Klapper, W.; Dreyling, M.; Hoster, E.; et al. Five-gene model to predict survival in mantle-cell lymphoma using frozen or formalin-fixed, paraffin-embedded tissue. *J. Clin. Oncol.* **2008**, *26*, 4966–4972. [CrossRef]
23. Tiemann, M.; Schrader, C.; Klapper, W.; Dreyling, M.H.; Campo, E.; Norton, A.; Berger, F.; Kluin, P.; Ott, G.; Pileri, S.; et al. Histopathology, cell proliferation indices and clinical outcome in 304 patients with mantle cell lymphoma (MCL): A clinicopathological study from the European MCL Network. *Br. J. Haematol.* **2005**, *131*, 29–38. [CrossRef]
24. Raty, R.; Franssila, K.; Jansson, S.E.; Joensuu, H.; Wartiovaara-Kautto, U.; Elonen, E. Predictive factors for blastoid transformation in the common variant of mantle cell lymphoma. *Eur. J. Cancer* **2003**, *39*, 321–329. [CrossRef]
25. Andersen, N.S.; Jensen, M.K.; de Nully Brown, P.; Geisler, C.H. A Danish population-based analysis of 105 mantle cell lymphoma patients: Incidences, clinical features, response, survival and prognostic factors. *Eur. J. Cancer* **2002**, *38*, 401–408. [CrossRef]

26. Matutes, E.; Parry-Jones, N.; Brito-Babapulle, V.; Wotherspoon, A.; Morilla, R.; Atkinson, S.; Elnenaei, M.O.; Jain, P.; Giustolisi, G.M.; A'Hern, R.P.; et al. The leukemic presentation of mantle-cell lymphoma: Disease features and prognostic factors in 58 patients. *Leuk. Lymphoma* **2004**, *45*, 2007–2015. [CrossRef]
27. Fisher, R.I.; Dahlberg, S.; Nathwani, B.N.; Banks, P.M.; Miller, T.P.; Grogan, T.M. A clinical analysis of two indolent lymphoma entities: Mantle cell lymphoma and marginal zone lymphoma (including the mucosa-associated lymphoid tissue and monocytoid B-cell subcategories): A Southwest Oncology Group study. *Blood* **1995**, *85*, 1075–1082. [CrossRef] [PubMed]
28. Jain, P.; Wang, M. Mantle cell lymphoma: 2019 update on the diagnosis, pathogenesis, prognostication, and management. *Am. J. Hematol.* **2019**, *94*, 710–725. [CrossRef]
29. Nadeu, F.; Martin-Garcia, D.; Clot, G.; Diaz-Navarro, A.; Duran-Ferrer, M.; Navarro, A.; Vilarrasa-Blasi, R.; Kulis, M.; Royo, R.; Gutierrez-Abril, J.; et al. Genomic and epigenomic insights into the origin, pathogenesis, and clinical behavior of mantle cell lymphoma subtypes. *Blood* **2020**, *136*, 1419–1432. [CrossRef]
30. Navarro, A.; Bea, S.; Jares, P.; Campo, E. Molecular Pathogenesis of Mantle Cell Lymphoma. *Hematol. Oncol. Clin. N. Am.* **2020**, *34*, 795–807. [CrossRef]
31. Roue, G.; Sola, B. Management of Drug Resistance in Mantle Cell Lymphoma. *Cancers* **2020**, *12*, 1565. [CrossRef]
32. IBM. *IBM SPSS Neural Networks 26*; IBM: Armonk, NY, USA, 2019.
33. IBM. *IBM SPSS Neural Networks*; New tools for building predictive models; YTD03119-GBEN-01; IBM: Somers, NY, USA, 2012.
34. Banihabib, M.E.; Bandari, R.; Valipour, M. Improving Daily Peak Flow Forecasts Using Hybrid Fourier-Series Autoregressive Integrated Moving Average and Recurrent Artificial Neural Network Models. *AI* **2020**, *1*, 263–275. [CrossRef]
35. Carreras, J.; Kikuti, Y.Y.; Miyaoka, M.; Hiraiwa, S.; Tomita, S.; Ikoma, H.; Kondo, Y.; Ito, A.; Nakamura, N.; Hamoudi, R. A Combination of Multilayer Perceptron, Radial Basis Function Artificial Neural Networks and Machine Learning Image Segmentation for the Dimension Reduction and the Prognosis Assessment of Diffuse Large B-Cell Lymphoma. *AI* **2021**, *2*, 106–134. [CrossRef]
36. Carreras, J.; Kikuti, Y.Y.; Miyaoka, M.; Hiraiwa, S.; Tomita, S.; Ikoma, H.; Kondo, Y.; Ito, A.; Shiraiwa, S.; Hamoudi, R.; et al. A Single Gene Expression Set Derived from Artificial Intelligence Predicted the Prognosis of Several Lymphoma Subtypes; and High Immunohistochemical Expression of TNFAIP8 Associated with Poor Prognosis in Diffuse Large B-Cell Lymphoma. *AI* **2020**, *1*, 342–360. [CrossRef]
37. Carreras, J.; Kikuti, Y.Y.; Miyaoka, M.; Hiraiwa, S.; Tomita, S.; Ikoma, H.; Kondo, Y.; Ito, A.; Nakamura, N.; Hamoudi, R. Artificial Intelligence Analysis of the Gene Expression of Follicular Lymphoma Predicted the Overall Survival and Correlated with the Immune Microenvironment Response Signatures. *Mach. Learn. Knowl. Extr.* **2020**, *2*, 647–671. [CrossRef]
38. Lin, H.; Zheng, W.; Peng, X. Orientation-Encoding CNN for Point Cloud Classification and Segmentation. *Mach. Learn. Knowl. Extr.* **2021**, *3*, 601–614. [CrossRef]
39. Mayr, F.; Yovine, S.; Visca, R. Property Checking with Interpretable Error Characterization for Recurrent Neural Networks. *Mach. Learn. Knowl. Extr.* **2021**, *3*, 205–227. [CrossRef]
40. Pickens, A.; Sengupta, S. Benchmarking Studies Aimed at Clustering and Classification Tasks Using K-Means, Fuzzy C-Means and Evolutionary Neural Networks. *Mach. Learn. Knowl. Extr.* **2021**, *3*, 695–719. [CrossRef]
41. Shah, S.A.A.; Manzoor, M.A.; Bais, A. Canopy Height Estimation at Landsat Resolution Using Convolutional Neural Networks. *Mach. Learn. Knowl. Extr.* **2020**, *2*, 23–36. [CrossRef]
42. Silva Araújo, V.J.; Guimarães, A.J.; de Campos Souza, P.V.; Rezende, T.S.; Araújo, V.S. Using Resistin, Glucose, Age and BMI and Pruning Fuzzy Neural Network for the Construction of Expert Systems in the Prediction of Breast Cancer. *Mach. Learn. Knowl. Extr.* **2019**, *1*, 466–482. [CrossRef]
43. Škrlj, B.; Kralj, J.; Lavrač, N.; Pollak, S. Towards Robust Text Classification with Semantics-Aware Recurrent Neural Architecture. *Mach. Learn. Knowl. Extr.* **2019**, *1*, 575–589. [CrossRef]
44. Knapič, S.; Malhi, A.; Saluja, R.; Främling, K. Explainable Artificial Intelligence for Human Decision Support System in the Medical Domain. *Mach. Learn. Knowl. Extr.* **2021**, *3*, 740–770. [CrossRef]
45. Carreras, J.; Hamoudi, R.; Nakamura, N. Artificial Intelligence Analysis of Gene Expression Data Predicted the Prognosis of Patients with Diffuse Large B-Cell Lymphoma. *Tokai J. Exp. Clin. Med.* **2020**, *45*, 37–48.
46. Carreras, J.; Hamoudi, R. Artificial Neural Network Analysis of Gene Expression Data Predicted Non-Hodgkin Lymphoma Subtypes with High Accuracy. *Mach. Learn. Knowl. Extr.* **2021**, *3*, 720–739. [CrossRef]
47. Team, R.C. *R: A Language and Environment for Statistical Computing*; R Foundation for Statistical Computing: Vienna, Austria, 2013.
48. Subramanian, A.; Tamayo, P.; Mootha, V.K.; Mukherjee, S.; Ebert, B.L.; Gillette, M.A.; Paulovich, A.; Pomeroy, S.L.; Golub, T.R.; Lander, E.S.; et al. Gene set enrichment analysis: A knowledge-based approach for interpreting genome-wide expression profiles. *Proc. Natl. Acad. Sci. USA* **2005**, *102*, 15545–15550. [CrossRef] [PubMed]
49. Mootha, V.K.; Lindgren, C.M.; Eriksson, K.F.; Subramanian, A.; Sihag, S.; Lehar, J.; Puigserver, P.; Carlsson, E.; Ridderstrale, M.; Laurila, E.; et al. PGC-1alpha-responsive genes involved in oxidative phosphorylation are coordinately downregulated in human diabetes. *Nat. Genet.* **2003**, *34*, 267–273. [CrossRef] [PubMed]
50. Scott, D.W.; Abrisqueta, P.; Wright, G.W.; Slack, G.W.; Mottok, A.; Villa, D.; Jares, P.; Rauert-Wunderlich, H.; Royo, C.; Clot, G.; et al. New Molecular Assay for the Proliferation Signature in Mantle Cell Lymphoma Applicable to Formalin-Fixed Paraffin-Embedded Biopsies. *J. Clin. Oncol.* **2017**, *35*, 1668–1677. [CrossRef]

51. Rosenwald, A.; Wright, G.; Wiestner, A.; Chan, W.C.; Connors, J.M.; Campo, E.; Gascoyne, R.D.; Grogan, T.M.; Muller-Hermelink, H.K.; Smeland, E.B.; et al. The proliferation gene expression signature is a quantitative integrator of oncogenic events that predicts survival in mantle cell lymphoma. *Cancer Cell* **2003**, *3*, 185–197. [CrossRef]
52. Carreras, J.; Lopez-Guillermo, A.; Kikuti, Y.Y.; Itoh, J.; Masashi, M.; Ikoma, H.; Tomita, S.; Hiraiwa, S.; Hamoudi, R.; Rosenwald, A.; et al. High TNFRSF14 and low BTLA are associated with poor prognosis in Follicular Lymphoma and in Diffuse Large B-cell Lymphoma transformation. *J. Clin. Exp. Hematop.* **2019**, *59*, 1–16. [CrossRef]
53. Tsuda, S.; Carreras, J.; Kikuti, Y.Y.; Nakae, H.; Dekiden-Monma, M.; Imai, J.; Tsuruya, K.; Nakamura, J.; Tsukune, Y.; Uchida, T.; et al. Prediction of steroid demand in the treatment of patients with ulcerative colitis by immunohistochemical analysis of the mucosal microenvironment and immune checkpoint: Role of macrophages and regulatory markers in disease severity. *Pathol. Int.* **2019**, *69*, 260–271. [CrossRef] [PubMed]
54. UniProt, C. UniProt: The universal protein knowledgebase in 2021. *Nucleic Acids Res.* **2021**, *49*, D480–D489. [CrossRef]
55. Safran, M.; Dalah, I.; Alexander, J.; Rosen, N.; Iny Stein, T.; Shmoish, M.; Nativ, N.; Bahir, I.; Doniger, T.; Krug, H.; et al. GeneCards Version 3: The human gene integrator. *Database* **2010**, *2010*, baq020. [CrossRef]
56. Carreras, J.; Kikuti, Y.Y.; Miyaoka, M.; Roncador, G.; Garcia, J.F.; Hiraiwa, S.; Tomita, S.; Ikoma, H.; Kondo, Y.; Ito, A.; et al. Integrative Statistics, Machine Learning and Artificial Intelligence Neural Network Analysis Correlated CSF1R with the Prognosis of Diffuse Large B-Cell Lymphoma. *Hemato* **2021**, *2*, 182–206. [CrossRef]
57. Carreras, J.; Kikuti, Y.Y.; Roncador, G.; Miyaoka, M.; Hiraiwa, S.; Tomita, S.; Ikoma, H.; Kondo, Y.; Ito, A.; Shiraiwa, S.; et al. High Expression of Caspase-8 Associated with Improved Survival in Diffuse Large B-Cell Lymphoma: Machine Learning and Artificial Neural Networks Analyses. *BioMedInformatics* **2021**, *1*, 18–46. [CrossRef]
58. Carreras, J.; Hiraiwa, S.; Kikuti, Y.Y.; Miyaoka, M.; Tomita, S.; Ikoma, H.; Ito, A.; Kondo, Y.; Roncador, G.; Garcia, J.F.; et al. Artificial Neural Networks Predicted the Overall Survival and Molecular Subtypes of Diffuse Large B-Cell Lymphoma Using a Pancancer Immune-Oncology Panel. *Cancers* **2021**, *13*, 6384. [CrossRef]
59. Carreras, J.; Kikuti, Y.Y.; Hiraiwa, S.; Miyaoka, M.; Tomita, S.; Ikoma, H.; Ito, A.; Kondo, Y.; Itoh, J.; Roncador, G.; et al. High PTX3 expression is associated with a poor prognosis in diffuse large B-cell lymphoma. *Cancer Sci.* **2021**, *113*, 334–348. [CrossRef]
60. Corporation, I. *IBM SPSS Statistics Algorithms*; IBM Corporation: Armonk, NY, USA, 2017; pp. 685–686.
61. Cheson, B.D.; Horning, S.J.; Coiffier, B.; Shipp, M.A.; Fisher, R.I.; Connors, J.M.; Lister, T.A.; Vose, J.; Grillo-Lopez, A.; Hagenbeek, A.; et al. Report of an international workshop to standardize response criteria for non-Hodgkin's lymphomas. NCI Sponsored International Working Group. *J. Clin. Oncol.* **1999**, *17*, 1244. [CrossRef]
62. Cheson, B.D.; Pfistner, B.; Juweid, M.E.; Gascoyne, R.D.; Specht, L.; Horning, S.J.; Coiffier, B.; Fisher, R.I.; Hagenbeek, A.; Zucca, E.; et al. Revised response criteria for malignant lymphoma. *J. Clin. Oncol.* **2007**, *25*, 579–586. [CrossRef]
63. Carreras, J.; Kikuti, Y.Y.; Bea, S.; Miyaoka, M.; Hiraiwa, S.; Ikoma, H.; Nagao, R.; Tomita, S.; Martin-Garcia, D.; Salaverria, I.; et al. Clinicopathological characteristics and genomic profile of primary sinonasal tract diffuse large B cell lymphoma (DLBCL) reveals gain at 1q31 and RGS1 encoding protein; high RGS1 immunohistochemical expression associates with poor overall survival in DLBCL not otherwise specified (NOS). *Histopathology* **2017**, *70*, 595–621. [CrossRef]
64. Carreras, J.; Yukie Kikuti, Y.; Miyaoka, M.; Hiraiwa, S.; Tomita, S.; Ikoma, H.; Kondo, Y.; Shiraiwa, S.; Ando, K.; Sato, S.; et al. Genomic Profile and Pathologic Features of Diffuse Large B-Cell Lymphoma Subtype of Methotrexate-associated Lymphoproliferative Disorder in Rheumatoid Arthritis Patients. *Am. J. Surg. Pathol* **2018**, *42*, 936–950. [CrossRef]
65. Fujisawa, M.; Matsushima, M.; Carreras, J.; Hirabayashi, K.; Kikuti, Y.Y.; Ueda, T.; Kaneko, M.; Fujimoto, R.; Sano, M.; Teramura, E.; et al. Whole-genome copy number and immunohistochemical analyses on surgically resected intracholecystic papillary neoplasms. *Pathol. Int.* **2021**, *71*, 823–830. [CrossRef]
66. Brownlee, J. Machine Learning Mastery. Available online: https://machinelearningmastery.com/dimensionality-reduction-for-machine-learning/ (accessed on 15 October 2021).
67. Holte, H.; Beiske, K.; Boyle, M.; Troen, G.; Blaker, Y.N.; Myklebust, J.; Kvaloy, S.; Rosenwald, A.; Lingjaerde, O.C.; Rimsza, L.M.; et al. The MCL35 gene expression proliferation assay predicts high-risk MCL patients in a Norwegian cohort of younger patients given intensive first line therapy. *Br. J. Haematol.* **2018**, *183*, 225–234. [CrossRef]
68. Ramsower, C.A.; Maguire, A.; Robetorye, R.S.; Feldman, A.L.; Syrbu, S.I.; Rosenthal, A.C.; Rimsza, L.M. Clinical laboratory validation of the MCL35 assay for molecular risk stratification of mantle cell lymphoma. *J. Hematop.* **2020**, *13*, 231–238. [CrossRef]
69. Rauert-Wunderlich, H.; Mottok, A.; Scott, D.W.; Rimsza, L.M.; Ott, G.; Klapper, W.; Unterhalt, M.; Kluin-Nelemans, H.C.; Hermine, O.; Hartmann, S.; et al. Validation of the MCL35 gene expression proliferation assay in randomized trials of the European Mantle Cell Lymphoma Network. *Br. J. Haematol.* **2019**, *184*, 616–624. [CrossRef]
70. Walsh, S.H.; Thorselius, M.; Johnson, A.; Soderberg, O.; Jerkeman, M.; Bjorck, E.; Eriksson, I.; Thunberg, U.; Landgren, O.; Ehinger, M.; et al. Mutated VH genes and preferential VH3-21 use define new subsets of mantle cell lymphoma. *Blood* **2003**, *101*, 4047–4054. [CrossRef]
71. Camacho, F.I.; Algara, P.; Rodriguez, A.; Ruiz-Ballesteros, E.; Mollejo, M.; Martinez, N.; Martinez-Climent, J.A.; Gonzalez, M.; Mateo, M.; Caleo, A.; et al. Molecular heterogeneity in MCL defined by the use of specific VH genes and the frequency of somatic mutations. *Blood* **2003**, *101*, 4042–4046. [CrossRef]
72. Lai, R.; Lefresne, S.V.; Franko, B.; Hui, D.; Mirza, I.; Mansoor, A.; Amin, H.M.; Ma, Y. Immunoglobulin VH somatic hypermutation in mantle cell lymphoma: Mutated genotype correlates with better clinical outcome. *Mod. Pathol.* **2006**, *19*, 1498–1505. [CrossRef]
73. Sabnis, R.W. Novel KIF18A Inhibitors for Treating Cancer. *ACS Med. Chem. Lett.* **2020**, *11*, 2368–2369. [CrossRef]

74. Wong, J.J.; Lau, K.A.; Pinello, N.; Rasko, J.E. Epigenetic modifications of splicing factor genes in myelodysplastic syndromes and acute myeloid leukemia. *Cancer Sci.* **2014**, *105*, 1457–1463. [CrossRef]
75. Li, D.; Bi, F.F.; Chen, N.N.; Cao, J.M.; Sun, W.P.; Zhou, Y.M.; Cao, C.; Li, C.Y.; Yang, Q. Epigenetic repression of phosphatidylethanolamine N-methyltransferase (PEMT) in BRCA1-mutated breast cancer. *Oncotarget* **2014**, *5*, 1315–1325. [CrossRef]
76. Dokshin, G.A.; Davis, G.M.; Sawle, A.D.; Eldridge, M.D.; Nicholls, P.K.; Gourley, T.E.; Romer, K.A.; Molesworth, L.W.; Tatnell, H.R.; Ozturk, A.R.; et al. GCNA Interacts with Spartan and Topoisomerase II to Regulate Genome Stability. *Dev. Cell* **2020**, *52*, 53–68. [CrossRef]
77. Bjornsti, M.A.; Kaufmann, S.H. Topoisomerases and cancer chemotherapy: Recent advances and unanswered questions. *F1000Research* **2019**, *8*, 1704. [CrossRef]
78 Tsai, Y.L.; Chang, H.H.; Chen, Y.C.; Chang, Y.C.; Chen, Y.; Tsai, W.C. Molecular Mechanisms of KDELC2 on Glioblastoma Tumorigenesis and Temozolomide Resistance. *Biomedicines* **2020**, *8*, 339. [CrossRef]
79. Donadio, J.L.S.; Liu, L.; Freeman, V.L.; Ekoue, D.N.; Diamond, A.M.; Bermano, G. Interaction of NKX3.1 and SELENOP genotype with prostate cancer recurrence. *Prostate* **2019**, *79*, 462–467. [CrossRef]
80. Cui, R.; Jiang, N.; Zhang, M.; Du, S.; Ou, H.; Ge, R.; Ma, D.; Zhang, J. AMOTL2 inhibits JUN Thr239 dephosphorylation by binding PPP2R2A to suppress the proliferation in non-small cell lung cancer cells. *Biochim. Biophys. Acta Mol. Cell Res.* **2021**, *1868*, 118858. [CrossRef]
81. Guo, Z.; Wang, X.; Yang, Y.; Chen, W.; Zhang, K.; Teng, B.; Huang, C.; Zhao, Q.; Qiu, Z. Hypoxic Tumor-Derived Exosomal Long Noncoding RNA UCA1 Promotes Angiogenesis via miR-96-5p/AMOTL2 in Pancreatic Cancer. *Mol. Ther. Nucleic Acids* **2020**, *22*, 179–195. [CrossRef]
82. Silveira, V.S.; Scrideli, C.A.; Moreno, D.A.; Yunes, J.A.; Queiroz, R.G.; Toledo, S.C.; Lee, M.L.; Petrilli, A.S.; Brandalise, S.R.; Tone, L.G. Gene expression pattern contributing to prognostic factors in childhood acute lymphoblastic leukemia. *Leuk. Lymphoma* **2013**, *54*, 310–314. [CrossRef]
83. Ye, R.Y.; Kuang, X.Y.; Zeng, H.J.; Shao, N.; Lin, Y.; Wang, S.M. KCTD12 promotes G1/S transition of breast cancer cell through activating the AKT/FOXO1 signaling. *J. Clin. Lab. Anal.* **2020**, *34*, e23315. [CrossRef]
84. Ahn, J.I.; Yoo, J.Y.; Kim, T.H.; Kim, Y.I.; Broaddus, R.R.; Ahn, J.Y.; Lim, J.M.; Jeong, J.W. G-protein coupled receptor 64 (GPR64) acts as a tumor suppressor in endometrial cancer. *BMC Cancer* **2019**, *19*, 810. [CrossRef]
85. Zhou, J.Y.; Shi, R.; Yu, H.L.; Zeng, Y.; Zheng, W.L.; Ma, W.L. Association between polymorphic sites in thymidylate synthase gene and risk of non-Hodgkin lymphoma: A systematic review and pooled analysis. *Leuk. Lymphoma* **2012**, *53*, 1953–1960. [CrossRef]
86. Fu, Z.; Jiao, Y.; Li, Y.; Ji, B.; Jia, B.; Liu, B. TYMS presents a novel biomarker for diagnosis and prognosis in patients with pancreatic cancer. *Medicine* **2019**, *98*, e18487. [CrossRef]
87. Turek, M. Explainable Artificial Intelligence (XAI). Available online: https://www.darpa.mil/program/explainable-artificial-intelligence (accessed on 10 January 2022).
88. McCoy, L.G.; Brenna, C.T.A.; Chen, S.S.; Vold, K.; Das, S. Believing in black boxes: Machine learning for healthcare does not need explainability to be evidence-based. *J. Clin. Epidemiol.* 2021, *in press*. [CrossRef]

healthcare

MDPI

Article

Real-Time Tracking of Human Neck Postures and Movements

Korupalli V. Rajesh Kumar [1,*] and Susan Elias [2]

1 School of Electronics Engineering, Vellore Institute of Technology, Chennai 600127, India
2 Centre for Advanced Data Science, Vellore Institute of Technology, Chennai 600127, India; susan.elias@vit.ac.in
* Correspondence: v.rajeshkumar2016@vitstudent.ac.in

Abstract: Improper neck postures and movements are the major causes of human neck-related musculoskeletal disorders. To monitor, quantify, analyze, and detect the movements, remote and non-invasive based methods are being developed for prevention and rehabilitation. The purpose of this research is to provide a digital platform for analyzing the impact of human neck movements on the neck musculoskeletal system. The secondary objective is to design a rehabilitation monitoring system that brings accountability in the treatment prescribed, which is shown in the use-case model. To record neck movements effectively, a Smart Neckband integrated with the Inertial Measurement Unit (IMU) was designed. The initial task was to find a suitable position to locate the sensors embedded in the Smart Neckband. IMU-based real-world kinematic data were captured from eight research subjects and were used to extract kinetic data from the OpenSim simulation platform. A Random Forest algorithm was trained using the kinetic data to predict the neck movements. The results obtained correlated with the novel idea proposed in this paper of using the hyoid muscles to accurately detect neck postures and movements. The innovative approach of integrating kinematic data and kinetic data for analyzing neck postures and movements has been successfully demonstrated through the efficient application in a rehabilitation use case with about 95% accuracy. This research study presents a robust digital platform for the integration of kinematic and kinetic data that has enabled the design of a context-aware neckband for the support in the treatment of neck musculoskeletal disorders.

Keywords: inertial measurement unit; kinematic data; kinetic data; musculoskeletal disorders; neck movements; neck postures; OpenSim; random forest

Citation: Kumar, K.V.R.; Elias, S. Real-Time Tracking of Human Neck Postures and Movements. *Healthcare* **2021**, *9*, 1755. https://doi.org/10.3390/healthcare9121755

Academic Editor: Mahmudur Rahman

Received: 10 November 2021
Accepted: 16 December 2021
Published: 19 December 2021

1. Introduction

Enabling technology to monitor, measure, and manage human movements has been an active area of research with a broad spectrum of applications ranging from medical diagnostics, rehabilitation, sports, fitness, behavior analysis, and gait-based bio-metrics. The historic landscape of research publications in the field has presented the use of vision-based (video cameras), sensor-based (Inertial Measurement Units IMUs), infra-red, and RADAR-based innovations to study human movement [1–4]. Quantitative analyses have traditionally been carried out using either kinetic or kinematic data for various applications. Kinematic data are obtained using IMUs and have been used in the study of musculoskeletal disorders (MSD) [5] and gait recognition. On the other hand, kinetic data provide details of the force of the component in motion and help to analyze the activation of muscles associated with the joint in motion [6]. Kinetic data are computed from the signals obtained from a Kinesiologic Electromyography (KEMG) device and are quantitative analyses for understanding muscle force and fatigue [7–9]. The motivation for the research presented in this paper was to design a robust methodology to integrate the kinetic and kinematic features for predictive analysis of human postures and movements. IMUs are Micro-Electro-Mechanical-systems (MEMs)-based devices that are widely used to develop wearable technologies [10]. IMU data have been integrated earlier with several proprietaries and open-source-based simulation platforms for analysis and visualization of movement data [11–17]. Here, the integration of IMU with OpenSim is presented, and it is currently a focus area in top research laboratories as well. OpenSim is a free, open-source simulation and modeling tool developed at Stanford University (https://opensim.stanford.edu/) (accessed on 9 November 2021). The built-in feature-extraction functionalities and contributions by the community make OpenSim a scalable and reliable tool for analyzing human movements. Besides presenting a step-wise procedure to integrate IMU data with OpenSim, this paper presents a novel methodology of combining kinetic and kinematic data for generating an insightful analysis of human movements. OpenSim provides details of muscle activation and supports joint modeling of kinetic and kinematic parameters. In the following subsections, the detailed methodology is presented. The results and discussions that follow will highlight the significance and applications of the proposed digital platform for movement analysis.

Motivation and Proposed Work

The goal of the proposed research is to present a methodology to measure and identify the postures of the human neck for the prevention and rehabilitation of musculoskeletal disorders of the cervical region. Improper neck postures due to sedentary lifestyles are a significant cause of cervical spine dysfunction in all age groups ranging from children to the elderly [18–27]. The neck region can also be affected due to improper neck postures that are inherent in certain kinds of occupations. Other lifestyle-related activities, including sleeping in the sitting position during travel, can trigger musculoskeletal problems around the neck. The motivation for this research was to propose a novel methodology to prevent disorders of the neck through timely detection and notifications [28]. A sensor-based Smart Neckband that can precisely detect neck postures was designed to monitor and generate alert messages as a preventive measure. This neckband can also be used to take measurements of the range of motion of the neck regions during therapy and rehabilitation [28]. Research works related to the use of sensors to track movements by obtaining kinematic data have been presented extensively in the literature under the field called Actigraphy [29,30]. These approaches have also been widely used for tracking movement through commercially available wrist-worn fitness monitors. However, to identify the postures of the neck, there are several challenges and limitations in only using kinematic data obtained using sensors. Hence, we explored the possibility of integrating kinetic and kinematic data for better accuracy in the detection of neck postures. In this paper, we present a robust integrated platform for the predictive analysis of human neck postures and movements using kinetic and kinematic data. In the following sections, the methods and materials used in this research are presented.

2. Materials and Methods

There are various conventional methods for measuring or recording human neck movements. Neck Range of Motion (N ROM) measuring instruments can be designed with proximity sensors, and NROM can be calculated with local fixed points on the human face with respect to human nose as center point. From nose to ears, the distance can be calculated, and based on degrees of freedom, the NROM values can be obtained. Similarly, there is another method, using video and biomarkers; in this method, biomarkers on the human neck and simultaneous video recording can be used to track neck movements. In continuation of these methods, our proposed model will help to find human neck movements in the digital environment [31–35]. In this research, an IMU-based device is used for acquiring the kinematic data of the neck, and the OpenSim simulation modeling tool is used to generate the kinetic data of the corresponding neck movements. Predictive analysis to detect neck postures from the kinetic and kinematic data is performed using machine learning methods. Validation is shown using synthetic data.

2.1. *Kinematic Data Acquisition Using Smart Neckband*

2.1.1. IMU Neck Band

Inertial Measurement Unit (IMU) embedded in an elastic neckband captures the kinematic data required for the analysis. IMU devices are available in miniature sizes and can be used to design wearable products. For this research, Metawear CPRO, an IMU device developed at MBIENTLAB (https://mbientlab.com/metamotionc/) (accessed on 9 November 2021), has been used. It has on-chip memory, processing unit, accelerometer, gyroscope sensor, magnetometer sensor, pressure sensor, and temperature sensor. In addition to these sensors, this device has inbuilt Bluetooth support to establish communication with the associated mobile application or any Bluetooth device to transmit the device data. This IMU device can stream data for 8–24 h continuously using a 3.3 V coin-sized battery and is attached to a wearable band with an adjustable strap to fit it firmly around the neck. This neckband, when integrated with the proposed predictive analysis, can be referred to as a Smart Neckband for its context-aware functionalities. The primary use of this device is to record neck movements [28]. The technical specifications of the device and built-in sensors are given below:

- o Weight: 5.66 g;
- o Battery: 200 mAH coin battery;
- o Usage modes: 8–24 h (stream), 2–48 h (log);
- o Data Transfer: Bluetooth Low Energy Smart (BLE);
- o Flash Memory: 8 MB.
- o Built-in sensors:

Accelerometer:

Range: ±2, ±4, ±8, ±16 g;
Resolution: 16 bit;
Sampling Rate: 0.001 Hz–100 Hz stream–800 Hz log.

Gyrometer:

Range: ±125, ±250, ±500, ±1000, ±2000°/s, Resolution: 16 bit;
Sampling Rate: 0.001 Hz–100 Hz stream–800 Hz log.

Magnetometer:

Range: ±1300 μT (x,y-axis), ±2500 μT (z-axis);
Resolution: 0.3 μT;
Sampling Rate: 0.001–25 Hz.

2.1.2. Mobile Application—MetaBase

In this research, an IMU-based device is used for acquiring the kinematic data of the neck, and the OpenSim simulation modeling tool is used to generate the kinetic data of the corresponding neck movements. Predictive analysis to detect neck postures from the kinetic and kinematic data is performed using machine learning methods. In the following sections, the methods and materials used in this research are presented.

2.1.3. Sensor Data Format

The IMU device has built-in sensors to record various kinematic aspects of the movements. The format of the data captured by these sensors is given below:

- Accelerometer sensor: epoch(ms), -> time, elapsed(s), x(g), y(g), z(g);
- Gyroscope sensor: epoch(ms), -> time, elapsed(s), x(deg/s), y(deg/s), z(deg/s);
- Magnetometer: epoch (ms), -> time, elapsed(s), x(T), y(T), z(T), etc.

For the kinematic analysis of neck postures, dealt with in this research, the accelerometer sensor data were sufficient. The accelerometer sensor was operated at 100 Hz frequency with ±8 g. Figure 1 shows the kinematic data acquisition process using sensor system and MetaBase mobile application.

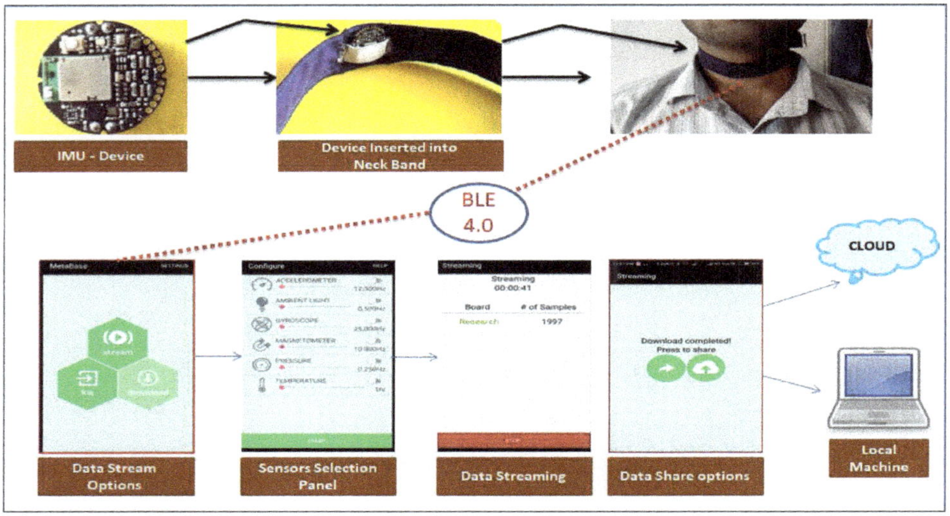

Figure 1. Kinematic data acquisition process.

2.2. *Kinetic Data Generation Using the OpenSim-Based Neck Musculoskeletal Model*

Electromyography (EMG) and Surface EMG (SEMG) are the standard methods for muscle-related data acquisition for movement analysis. The kinetic data obtained from EMG-based methods are accurate but are limited to laboratory-based studies. To carry out movement analysis using both kinematic and kinetic data, a novel methodology has been proposed in this paper. Here, we focus on the detection of postures of the neck using an innovative approach. Instead of capturing kinetic data related to muscle activation using any of the EMG-based methods, in the proposed approach, kinetic data of the corresponding neck postures are generated using a Neck Musculoskeletal Simulation Model. The Smart NeckBand captures the real-world kinematic data of the neck postures, and these data are used to generate the corresponding kinetic data relating to the muscles around the neck region using the OpenSim simulation tool.

2.2.1. OpenSim—Simulation Modeling Tool and Its Features

This is the most popular open-source tool used to create and study human musculoskeletal models and provides extensive data on kinematics and kinetics of human movement. Built-in functionalities such as Scale, Inverse Kinematics (IK), Inverse Dynamics (ID), Residual Reduction (RR), Static Optimization (STO), Computed Muscle Control (CMC), and Analyze provide support to extract all information related to the muscles, tendons, joints kinetics, and kinematics. An overview of the functionalities of OpenSim and the inbuilt tools is shown in Figure 2, and detailed information is made available by the OpenSim contributors [36–39].

Built-In Tools and Its Features

1. **Scale:** In OpenSim, Scale is a built-in tool, which is used to create user-specific musculoskeletal models, mainly used to adjust the dimensions of the skeletal system in terms of Mass. In the Scale tool, we can adjust the Scale factors and static pose weights. In this tool, we must give marker data for measurement as input. From the scale, the output is the *.osim* file, which is the main source file.
2. **Inverse Kinematics (IK):** In OpenSim, Inverse Kinematics is a built-in tool, which is used to generate the Inverse Kinematics of the musculoskeletal system concerning joint movements. The input for this tool is three-dimensional coordinate data which are in *.trc* file format (track row–column). This input file gives information about the

joint movements with respect to time in three dimensions (x,y,z). Based on this input, the IK tool will generate the motion; we can observe this in GUI—Graphical User Interface in the OpenSim tool. The output of this tool is the *.mot* file, which consists of information about the joint's motion concerning time.

3. **Inverse Dynamics (ID):** The Inverse Dynamic tool (ID) is used to determine the net forces and torques of joints, which are responsible for movement generation. IK tool output, i.e., *.mot* file motion file and ground reaction forces data in the .xml format, are fed to the ID tool as input sources. ID tool will perform the mass-dependent acceleration functions and generate the forces based on the conventional F = ma equation. The output of the ID tool is Inverse Dynamics.sto (ID-State storage file).
4. **Residual Reduction (RR):** It is a built-in tool that works like Forward Dynamics and uses a tracking controller to follow kinematics extracted from the IK tool—nothing but movements.
5. **Static Optimization (STO):** This tool helps to obtain muscle forces and activations at each instance in time. For this tool, input will be a .mot file—motion file and generates muscle forces and activations the format of .sto.
6. **Computed Muscle Control (CMC):** This tool is a major block in the OpenSim software, which computes muscle excitations, joint movements such as kinematics, and kinetics of each component present in the musculoskeletal model. This tool generates .sto files for muscles, joints, and ligaments—active and passive fibers, power, length, forces, accelerations, and positions, etc.
7. **Analyze:** This tool helps to analyze the model based on its simulation. If the duration of the simulation is long in terms of time, the Opting Analyze tool is the best option compared to the CMC tool. This tool helps to obtain accurate results in less time on the already simulated use case.

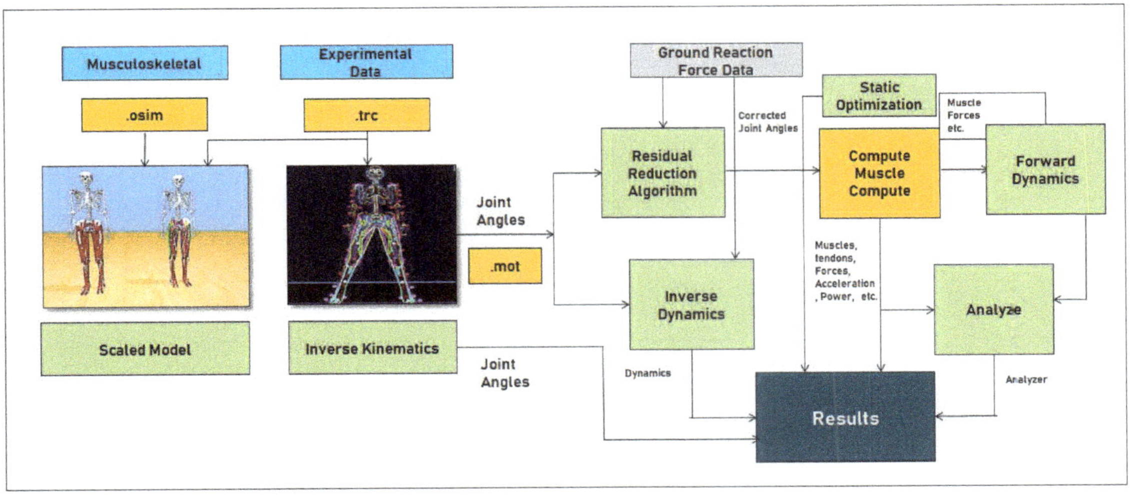

Figure 2. Overview of OpenSim functionalities.

2.2.2. Neck Musculoskeletal Model

In this research, the neck musculoskeletal model developed by [40], shown in Figure 3, was used. This is a fully flexible model for head and neck movements and versatile compared to other models [41]. The neck region consists of cervical joints (C1–C7), sixty-four muscles, and various associated tendons and ligaments. The hyoid muscles play a vital role in supporting the neck movements and hence have a vital role in the proposed predictive analysis [40,42–45]. The neck musculoskeletal model used in this research integrated the hyoid muscles, and this was a big advantage for our experimental analysis. The kinetic data provided by OpenSim includes forces, length, power, and activation levels of joints, muscles, and tendons. With the neck musculoskeletal model and research insights provided by the team [40], we simulated kinetic data of the hyoid muscles for our research analysis. The hyoid muscles ***are Digastric, Geniohyoid, Mylohyoid, Stylohyoid, Sternohyoid, Thyrohyoid, Sterno_Thyroid,*** and ***Omohyoid,*** shown in green color in Figure 3.

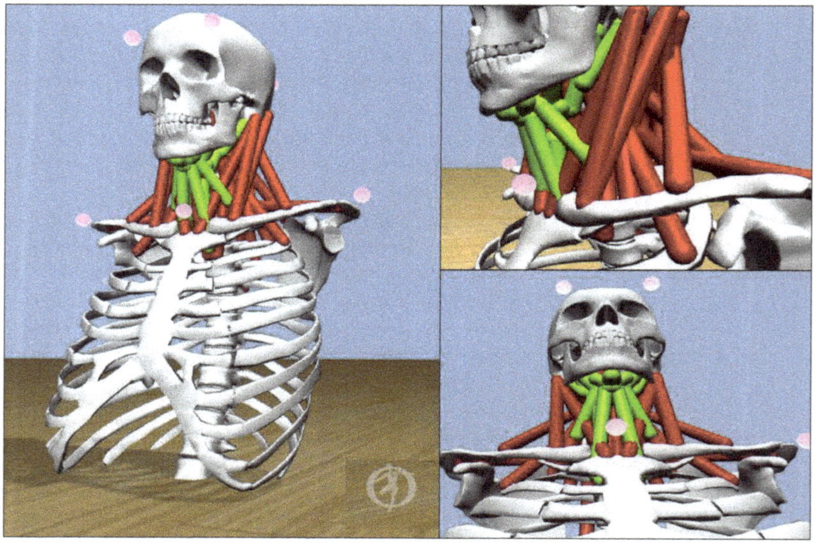

Figure 3. Hyoid muscles (in green color).

2.3. Experiments and Research Analysis

In general, the human neck has three degrees of freedom: a horizontal plane, a vertical plane, and the rolling of the head. All other asymmetric movements of the neck are variations and combinations of these three fundamental movements. At any point in time, the neck will be in one of the following nine static positions called neck postures, or it can move in any random order between these nine postures [28]. The nine static positions or neck postures are mentioned below and shown in Figure 4:

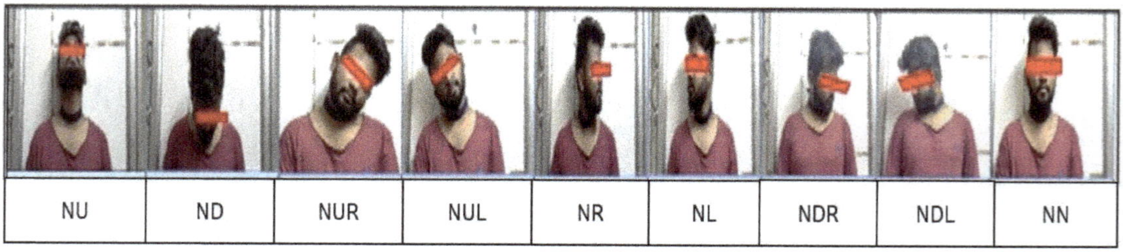

Figure 4. Neck Postures (Nine Positions)—Subject. 1. Neck at Extreme Up (**NU**), 2. Neck at Extreme Down (**ND**), 3. Neck at Extreme Right (**NR**), 4. Neck at Extreme Left (**NL**), 5. Neck at Right Up (**NRU**), 6. Neck at Right Down (**NRD**), 7. Neck at Left Up (**NLU**), 8. Neck at Left Down (**NLD**), and 9. Neck in the Middle (**NM**).

The goal of the research presented in this paper is to design a methodology to detect neck postures by training the machine learning algorithms using kinematic and kinetic data. In this research, we used Random Forest, an ensemble learning method for prediction and classification. The Smart Neckband captures the real-world data and integrates it with the OpenSim simulation platform. To effectively capture the neck kinematics, an experimental study was first carried out to determine the ideal location for the IMU device. The IMU device was fixed onto the elastic band, and the neckband could be worn in a manner that located the IMU device either in front or at the back of the neck region.

2.3.1. Experimental Study for the Location of IMU

Initially, we approached *thirty* participants for this research work. As per the Declaration of Helsinki, we guided all participants and informed them about the research procedure. In this process, we raised a query related to participants' health information, particularly about their neck/cervical problems.

Condition: Participants should participate voluntarily and should not have undergone any surgery/treatment for the neck/cervical region.

With this statement, ***eighteen*** participants were withdrawn from their participation due to their neck/cervical treatment history.

Among others, ***four*** participants were withdrawn during the research process due to personal reasons.

Finally, **eight** volunteers participated actively. Before the beginning of this research, once again, we made sure that volunteers never underwent any surgery/treatment for their neck. All the volunteers gave written consent after being fully informed about the research procedure. All the information gathered was based on the ***Declaration of Helsinki***. Participants' physical attributes were tabulated in Table 1.

Table 1. Volunteers' physical attributes.

	Volunteers = 8	
Attributes	**Means +/− SD**	**Range**
Age, year	39.5 +/− 17.6	27–52
Mass, kg	71 +/− 9.8	64–78
Height, m	1.45 +/− 0.09	1.38–1.52
Gender M/F	5/3	

The subjects wore the neckband and participated in the research study. The participants were asked to keep their necks in the nine static positions for a duration ranging from 1 min to 2 min based on their comfort level. IMU data were recorded for all participants for each of the nine neck postures with the sensor located at the FRONT side of the neck region and similarly for the BACK side of the neck. Figure 5a shows the kinematic and kinetic data extraction methods using the IMU device and OpenSim simulation model for both the front and back of neck locations. Details of the dataset are presented in Table 2.

Table 2. Dataset quantitative parameters.

IMU—Device Placement	**Total Time Duration**	**Dataset (After Pre-Processing)**
A sensor placed at the front side	1080 s for nine static positions	1080 × 5 time, acc(x,y,z), position
A sensor placed at the back side	1080 s for nine static positions	1080 × 5 time, acc (x,y,z), position

Datasets were pre-processed and modularized based on the time stamp provided by the sensor and labeled manually. The observations in the dataset were also validated using the video footage of the corresponding experimental study. The quantity of the dataset is good enough for the training and testing mechanism to predict the neck postures. From each subject, 1080-time frames of information were generated, i.e., 1080 rows of corresponding acceleration data were generated for nine positions. All together, eight subjects' data were merged into a dataset, which consisted of 8640 × 5 [rows × columns] of data. As part of the pre-processing task, NaN's (Not a Number) and NA's (Not Available) were interpreted, outliers were removed, and data were normalized.

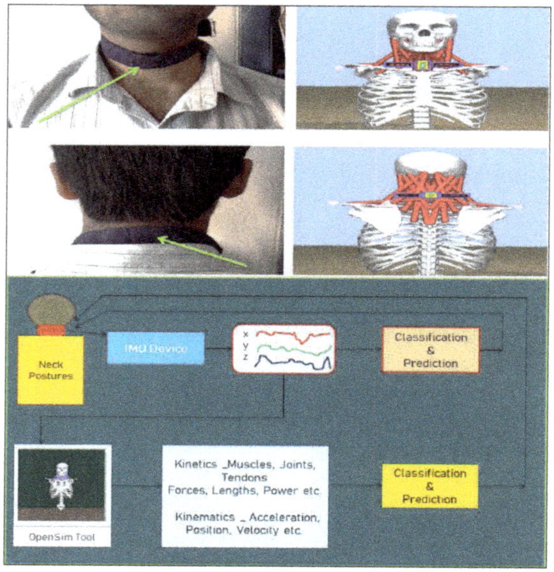

(a)

(b)

Figure 5. (**a**) IMU locations and corresponding musculoskeletal model; (**b**) IMU to OpenSim—data flow.

2.3.2. IMU Data Integration with OpenSim

- The IMU-based accelerometer sensor data format provides three-dimensional kinematic data (x,y,z).
- To export IMU kinematic data into the OpenSim simulation tool, mathematical and functional analysis is required. In OpenSim, a file with the extension .trc (track row–column) is used as an input file for the Inverse Kinematics (IK) tool, and this tool provides joint movement data as a motion file with the extension *mot'*.
- The neck-skeletal model has seven sets of markers around the skull and cervical region (four on the skull, one at the Sternum Jugular Notch, and two at the right and left acromioclavicular joints). The marker Sternum Jugular Notch (SJN) is located on the front side of the neck. The IMU-based kinematics data are mapped onto the x,y,z coordinates of SJN. The other markers are calibrated according to the functional movements. The .trc file contains the details of these markers, and it is the input file for the Inverse Kinematics (IK), and the motion file (.mot) is obtained as the output.
- The information in this .mot file is fed as input to the Computed Muscle Control (CMC) tool, which produces the data related to neck kinematics, kinetics, joints, muscles, forces, etc.
- The functional integration of IMU data and OpenSim is shown in Figure 5b. Available results were interpreted, outliers were removed, and data were normalized.

2.3.3. Smart Neckband—Comfort Level

We opted for cotton-material-based neck supportive bands, which are commercially available in the market and flexible to fit around the neck with Velcro adjustments. Then, we integrated the IMU device with the neckband. After finishing the research, we collected feedback from the participants on the comfort level of wearing the Smart Neckband. Table 3 shows the feedback given by the subjects on wearing the Smart Neckband. Based on overall feedback, we concluded that wearing this Neckband did not create any kind of disturbance for the participants, and we strongly believe that they were happy to wear it; based on their satisfaction, they had given ratings.

Table 3. Participants' feedback on wearing Smart Neckband.

Condition	Sub 01	Sub 02	Sub 03	Sub 04	Sub 05	Sub 06	Sub 07	Sub 08
Any itching around the neck?	X	X	X	X	X	X	X	X
Is it comfortable to wear?	✓	✓	✓	✓	✓	✓	✓	✓
Is it easy to adjust to your neck dimensions?	✓	✓	✓	✓	✓	✓	✓	✓
Is there any trouble/pain/discomfort from wearing this Neckband?	X	X	X	X	X	X	X	X
Please give a rating for this Smart Neckband out of 5	4	4	3.5	4.5	5	3.5	5	4.5

2.4. Predictive Analysis Using Machine-Learning Algorithms

The data collected by keeping the IMU device at the FRONTSIDE of the neck were divided into training and testing data in the ratio of 75:25 and given as input to the machine-learning algorithms to classify the nine static neck positions and to find the accuracy of the classification. Similarly, the data collected by keeping the IMU device at the BACKSIDE of the neck were processed.

2.4.1. Machine-Learning Algorithms Used in This Research

K-Nearest Neighbors (KNN): KNN is a supervised learning algorithm, which calculates the nearest distance of a similar object; this is why it is sometimes called a proximity or closeness-finding algorithm [46,47].

Decision Trees (Iterative Dichotomiser 3 (ID3)): This is a supervised learning algorithm, which uses Information Gain values to decide important contributing features to classify the data [48,49].

Random Forest Algorithm: This is a supervised learning algorithm; it is a combination of multiple Decision Trees; this ensemble algorithm works for classification and regression problems [50,51].

2.4.2. Algorithmic Responses—Result Analysis

Figure 6a shows the machine-learning algorithms' responses. Among three algorithms, the Random Forest algorithm has shown significant results in terms of classification accuracy. Further, the Random-Forest-algorithm-based confusion matrix and performance metrics were generated for the front and back location and are shown in Figure 6b,c. A total of 100% accuracy was achieved for the front-location-based classification and 99% for the back-location-based classification. From this research study, we can infer that the ideal location for the IMU device during data capture is the front side of the neck. This inference correlates with our decision to use the hyoid muscles located on the front side of the neck for the accurate classification of neck postures. In this research paper, we proposed the idea of generating kinetic data related to the hyoid muscles and using this data along with the associated kinematic data to accurately detect neck posture using classification techniques. This is presented in the following section.

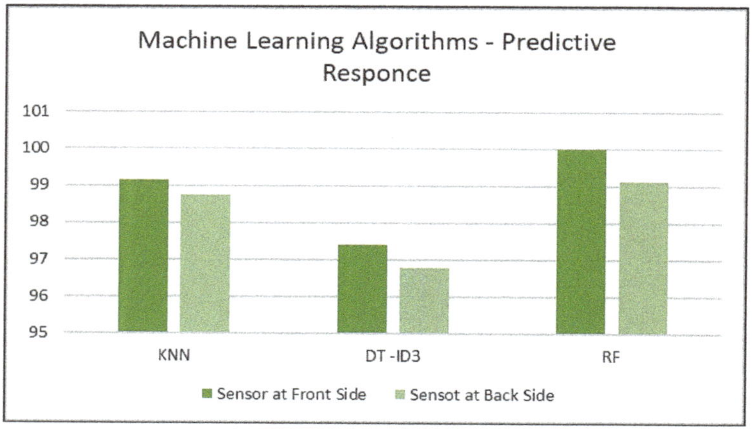

(a)

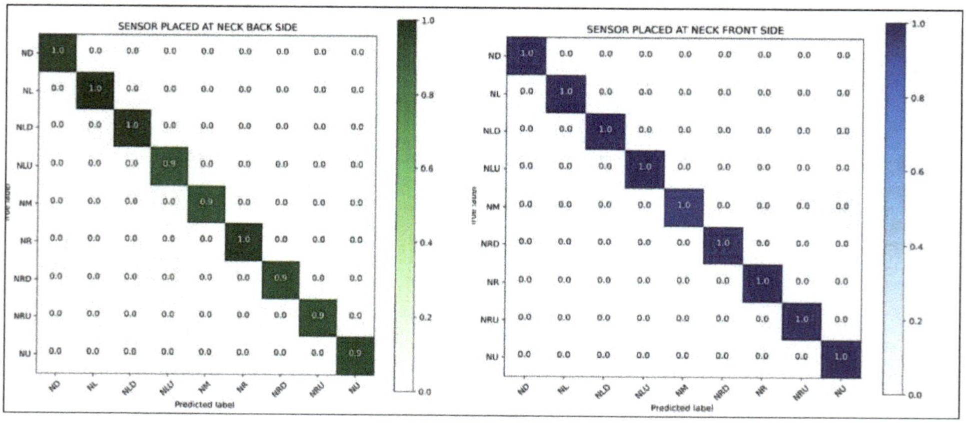

(b)

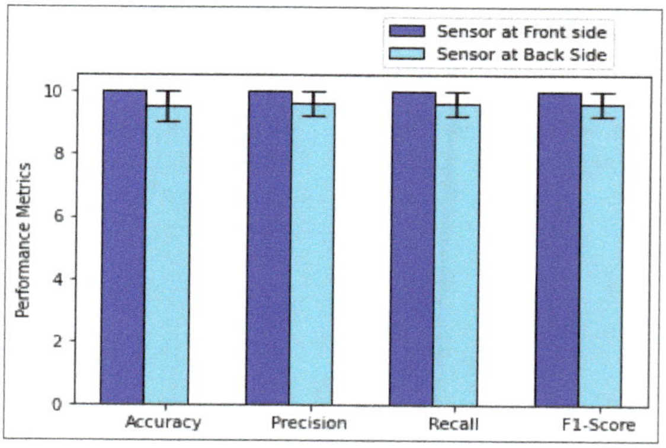

(c)

Figure 6. (**a**) Machine Learning algorithms—accuracy of the classification models; (**b**) confusion matrix—classification of neck posture based on sensor position; (**c**) performance metrics.

3. Results

3.1. Robust Integration of Kinematic and Kinetic Data

In this section, we first present the procedure to integrate kinematic and kinetic data for the experiment analysis. Integrating the IMU-based kinematic data to the OpenSim simulation modeling platform is a challenging step in the research domain [28].

3.2. Subject-Specific Neck Postures

In this research, we recorded and analyzed kinematic data for all the subjects. In this section, we present the simulation-based neck posture of ***Subject ID: 04***. The data collected from all the other subjects were used in training and testing for the classification. Figure 7 shows the neck postures of ***Subject ID: 04*** and the corresponding musculoskeletal postures in OpenSim.

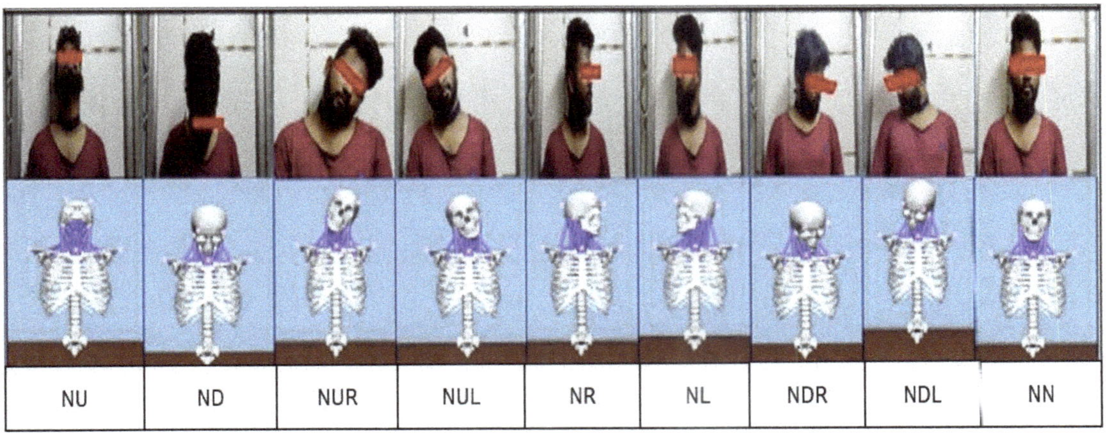

Figure 7. Subject-specific neck postures with corresponding musculoskeletal postures in OpenSim.

3.3. OpenSim—Neck-Musculoskeletal-Model-Based Kinematic and Kinetic Data Analysis

OpenSim generates neck kinematics information based on input data: acceleration, position, and velocity. We used the acceleration and position data to classify and predict the human neck posture. From Figure 7, we can observe the response of the cervical joints and other associated joints during the experimentation task.

The subjects changed the neck posture from one position to another after a time gap of about 120 sec, and with a total of nine positions in the study, 1080 s of data were captured. Figure 8a shows the variations in the neck acceleration concerning joint kinematics, and similarly, Figure 8b shows the variations in the position of the neck. OpenSim-based IK tool generates the movements (*.mot*) data, which are fed to the CMC tool as input and extracted the kinetic data as an output. From the output data, we analyzed the neck joint movements corresponding to neck positional changes. The CMC tool calculates the Body acceleration and position data.. From this data, we observed that few joints and muscles excite high for cetain neck movements, and few respond low for certain neck movements.From the corresponding figures, we can observe the changes in the neck joint's momentum.

There are eight important sets of hyoid muscles (shown in Figure 3), and many other associated hyoid muscles are attached to the hyoid bone in the neck region. These muscles help in providing free movement generation and flexibility to the neck [40,52]. The OpenSim CMC tool provides kinetic information such as forces, activation, lengths, etc. Using the CMC tool, kinetic data were extracted for the corresponding kinematic data that were captured and integrated with OpenSim. Here, the response of the tendon forces of the neck hyoid muscles was analyzed, as shown in Figure 8c.

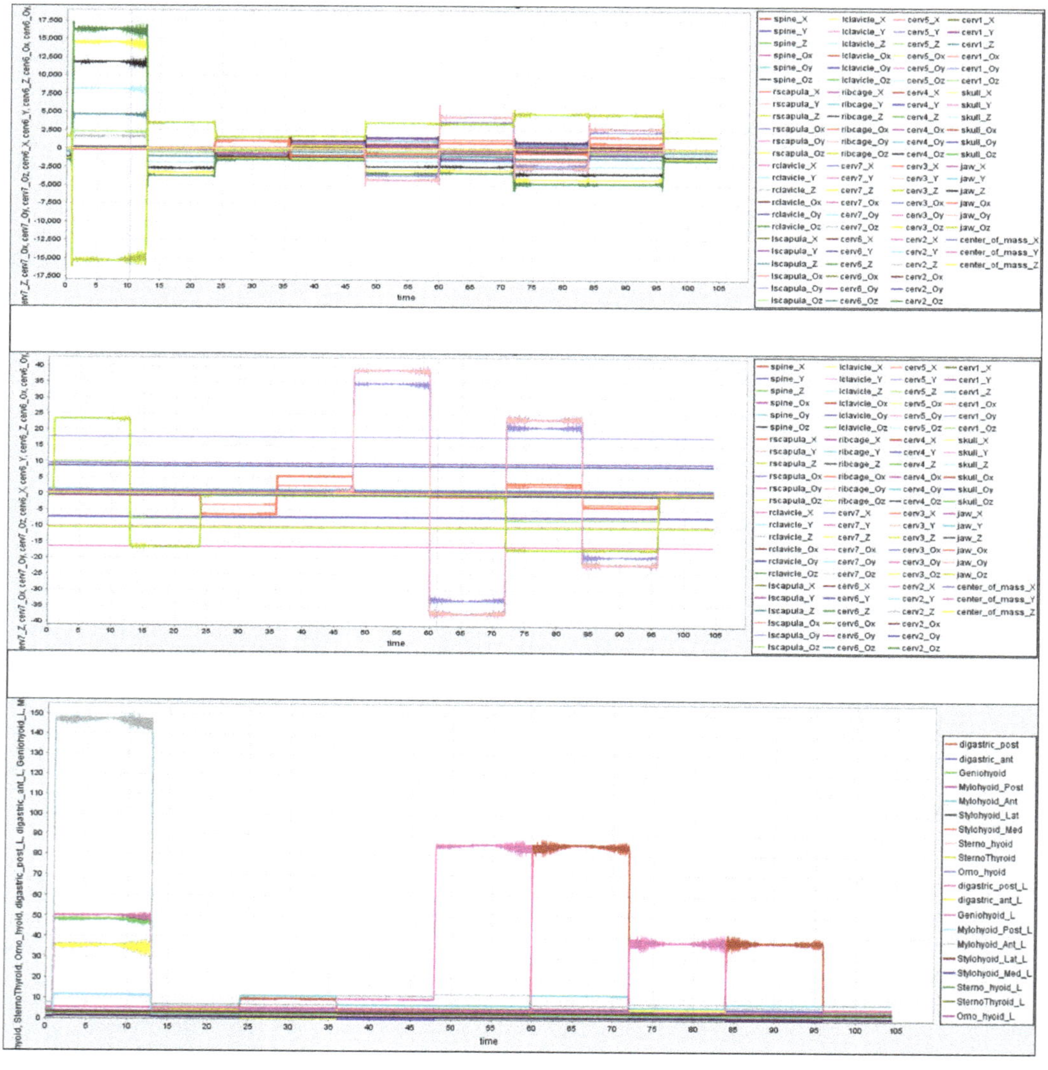

Figure 8. (**a**) Kinematics—neck acceleration; (**b**) kinematics—neck position; (**c**) kinetics—tendon forces.

3.4. Predictive Analysis—Kinematics and Kinetics

Earlier, we covered various research works carried out on posture classification methods. In this process, we did not find any potential research works similar to this research. In this respect, we observed that a few researchers performed posture prediction (hand gestures, body position, sitting, standing, squats, etc.) using Machine and Deep Learning methods. They considered sensors-, video-, and markers-based datasets for posture prediction. They achieved strong and accurate results using these methods [50,53–59]. In this research, we opted for Machine Learning algorithms for the prediction of neck postures/movements.

We have observed the performance of the Machine Learning algorithms on the datasets (Section 2.4) based on the performance metrics of the algorithms. We opted for the Random Forest (RF) algorithm for posture prediction. RF was used to classify and predict the neck posture based on the kinematics and kinetics data generated by OpenSim. The Random

Forest algorithmic approach achieved 100% accuracy in the classification of neck postures using neck acceleration and position data. These results are presented in Figure 9a,b. Similarly, the Random Forest classifier predicted nine neck postures using the response of the tendon force data of hyoid muscles and achieved 100% accuracy in the prediction. The result of the kinetic tendon force classification is shown in Figure 9c.

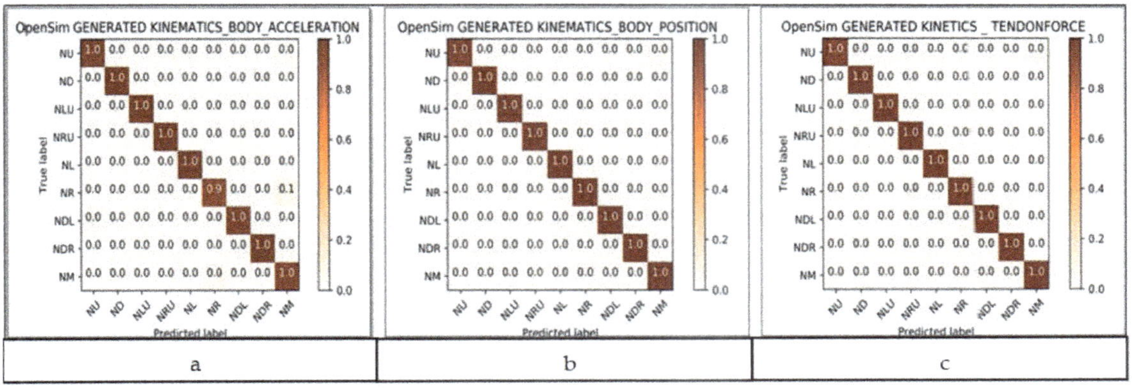

Figure 9. (**a**) Kinematics—neck acceleration; (**b**) kinematics—neck position; (**c**) kinetics—tendon forces.

4. Validation—Use Case Model: A Predictive Model for Rehabilitation Monitoring and Assessment Based on Neck Movements

4.1. Rehabilitation

Rehabilitation therapy for musculoskeletal disorders is normally prescribed by a certified physiotherapist. Based on the severity of the injury, the treatment can involve heat, cold, exercise, massage, and ultrasound methods. Exercise-based therapy is presented as a use case to demonstrate the novelty of the research proposed in this paper.

4.2. Neck Movements

For the experimental study, the subject wore the neckband and performed the neck movements as an activity in a random sequence. The sequences of movements were randomly selected neck exercises performed over 2 min of the time frame. The subject randomly moved his neck and head from one position to the other, and the following sequence is one such instance: *NM-NL-NM -NRU -NM-NLD-NM-NU-NM-NR-ND -NM-NLU-NM* (abbreviations mentioned in Section 2.3). The sequences varied for every experimental trial. In this study, neck movements were captured using the Smart Neckband, and the video recording was also performed simultaneously for sensor data segmentation and validation purposes.

4.3. Rehabilitation Monitoring System—Methodology and Results

The entire workflow of the rehabilitation monitoring system is presented in Figure 10. The complete structure of the rehabilitation monitoring system consists of two stages.

Stage 1: The workflow shows the details of how the proposed model is trained (based on Section 2.3.3).

Stage 2: The trained model is used to identify the neck postures that define the movements.

The acquired sensor dataset consists of 1238 rows and 5 features (time, accelerometer (x,y,z), movement) and is collected for 2 min. The video that was simultaneously recorded was used to trim the dataset to align with the exact neck movement data. The neck movement data were collected and saved with corresponding neck movement labels. Forty-nine samples of kinematic data points of neck postures were segregated and labeled. These labeled neck movements were exported into the OpenSim simulation tool, and corresponding kinetic data were obtained. OpenSim-based built-in tools provide options for data-capturing relevant to joint kinematics, kinetics, active and passive fiber forces of muscles and tendons, activation, lengths, velocity, power, etc. In this research, active-force data of the hyoid muscle were obtained and used for further analysis. For training the model, an integrated kinetic and kinematic dataset was used. The dataset comprised of kinetic data of various neck static postures with labels, recorded by IMU, and corresponding muscle activation forces of hyoid muscles was obtained from OpenSim.

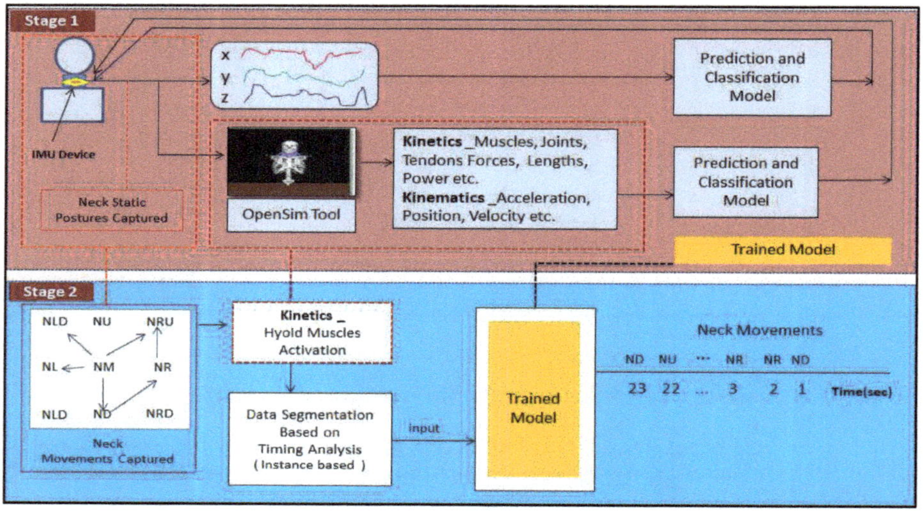

Figure 10. Block diagram of the research workflow.

4.4. Observations

The model was trained using the Random Forest algorithm [18], which was used to predict the classes of neck postures. Here, the output of neck movements is predicted in a sequence based upon time frames. Figure 11 shows that there are nine classes mapped in a circular shape. Each class represents one static neck posture point. The neck movement from one posture to another is indicated with directed arrows and is labeled in Figure 10.

Actual Movements: NM-NLU-NM-ND-NR-NM-NU-NM-NLD-NRU-NM-NL-NM.

Instance 1: Actual movements show that the neck movement started from the Neck Middle position to Neck Left Up and then came back to Neck Middle. Then, it moved from Neck Middle to Neck Down, then to Neck Right, and from there to Neck Middle. Then, it moved from Neck Up and then back to Neck Middle and then towards Neck Left Down. From there, it moved to Neck Right Up then came back to Neck Middle. Finally, it moved to Neck Left and came back to the Neck Middle position. The kinetic values of the force of hyoid muscles during the movements were used as the training data.

Instance 1: NM-NLU-NM-ND-NR-NM-NU-NM-NLD-NRU-NM-NL-NM.

Predicted 1: NM-NLU-NM-ND-NR-NM-NU-NM-NM-NRU-NM-NL-NM.

Predicted movements show that ***93.33%*** accuracy was achieved, and one neck movement was wrongly predicted.

Similarly, as a test case, we verified different neck movements for different instances. For instance, **10**, we obtained ***100%***accuracy.

Instance 10: NM-NR-NL-NM-NR-NM-ND-NM-NU-NM-NR-NU-NM-NLD-NM.

Predicted 10: NM-NR-NL-NM-NR-NM-ND-NM-NU-NM-NR-NU-NM-NLD-NM.

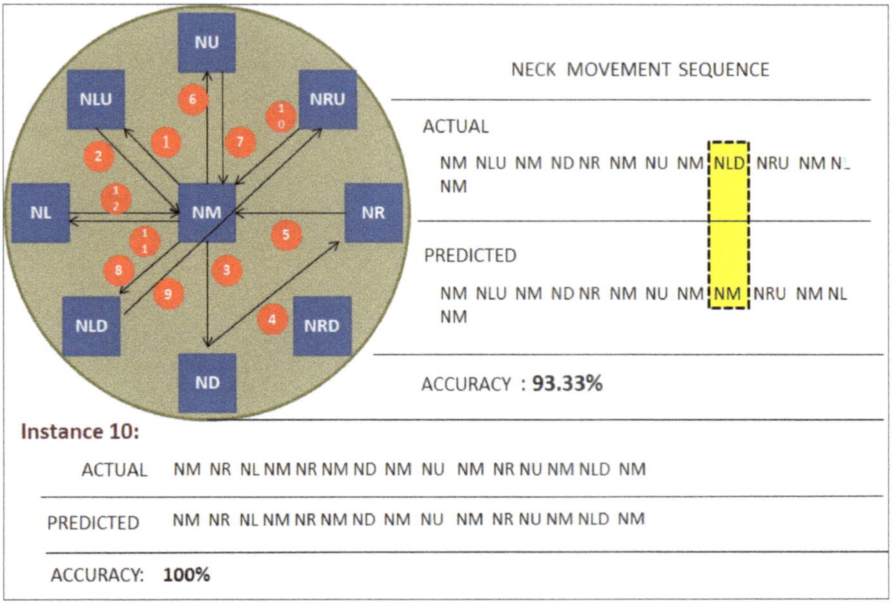

Figure 11. Rehabilitation scoring system.

4.5. Experimental Use Case—Rehabilitation Scoring System Using Synthetic Data

As a use case, a synthetic dataset was created to validate the proposed method. In this process, the dataset was created based on sample neck movements related to basic exercise patterns. These data were used to monitor and validate the rehabilitation scoring system. Figure 12 shows a 10-day neck movement monitoring and validation process.

In this process, actual neck movements were recorded as **NM-NM-NU-NU-NU-NU-NU-NU-NM-NM-ND-ND-ND-NM-NM-NU-NU** in a timeline. As indicated in Figure 11, these movements were compared with day-wise trails. In a day, two trials were conducted, and data were compared with actual movements. Based on these movements, rehabilitation scoring can be calculated. In Table 4, the rehabilitation monitoring and assessment report is summarized. Day-wise improvements are mentioned in Table 4. This result shows the effectiveness of the proposed methodology. The scale for the assessment depends on the rehabilitation scoring system shown in Figure 12.

Table 4. Rehabilitation monitoring and assessment—model analysis report.

No. of Days	No. of Trails	Rehabilitation Assessment Based on Movements
1	1	Problem: Neck Up movement needs to improve
	2	Problem: Neck Up movement needs to improve
2	1	Problem: Neck Up movement needs to improve
	2	Problem: Neck Up movement needs to improve
3	1	Problem: Neck Up movement— slightly improved compared to previous day
	2	Problem: Neck Up movement—same as the previous trial
4	1	Problem: Neck Up and Neck Down movements— needs to improve
	2	Problem: Neck Up movement needs to improve

Table 4. *Cont.*

No. of Days	No. of Trails	Rehabilitation Assessment Based on Movements
5	1	Problem: Neck Up movement. Slightly improved
	2	Problem: Neck Up movement. Good Improvement
6	1	Good Improvement
	2	Good Improvement
7	1	Good Improvement
	2	Good Improvement
8	1	Good Improvement
	2	Good Improvement
9	1	Improved
	2	Improved
10	1	Improved
	2	Improved

4.6. State of the Art: Objective vs. Research Flow

The presented research work should fill the gap between conventional physio assistive devices and technology. Recent advancements in artificial intelligence methods and hardware devices can bring a novelty in the physiotherapy processes. In this aspect, this research work can be the first approach to bring automation to the physiotherapy and assessment system.

Obj. 1: The main purpose of this research is to provide a digital platform for analyzing the impact of human neck movements on the neck musculoskeletal system.

Obj. 2: The second objective was to enable remote access to the therapist and to design a rehabilitation monitoring system that brings accountability to the treatment prescribed.

Research Flow: Figure 13 shows the entire research flow, which signifies research object 1, which highlights building a digital platform for analyzing the human neck postures and movements and their impact on the musculoskeletal system.

This research theme is offline-process-based, where we have to obtain IMU-based neck movement data and manually have to feed them into OpenSim software to extract the kinematics and kinetic data. These data are supplied to an AI engine; it predicts the postures/movements. Based on movements observations, the physiotherapist can analyze the patient's condition. In terms of automation, these movement data feed to the AI engine, which can predict the posture/ movements changes and generate the assessment status. Based on this report, physiotherapists can analyze the patient's condition and give appropriate treatment or therapy. Based on technology limitations, we have performed the entire research using an offline process. Based on advancements, we can fulfill Obj. 2 in the future.

4.7. Future Scope

Advancements in technologies such as Machine Learning, Deep Learning, and Wearable Technologies will help to bring this innovation into the limelight in physiological measuring instrumentation. As for this research, the Smart Neckband for real-time tracking of human neck movements can be helpful to assist physiotherapists in rehabilitation. Based on limitations, we have performed this research using an offline process; in the future, advancements in technology can help to build a module that works in an online mode. Based on the online working process, we can monitor patient conditions remotely. We have performed significant work on a remote monitoring system for posture/movement prediction [28]; this work can be extended to building a remote rehabilitation system that will help physiotherapists in monitoring patients in a good manner.

Days	ACTUAL MOVEMENTS (Reference) [NM NM NU NU NU NU NU NU NM NM ND ND ND NM NM NU NU]	Rehabilitation Score (%)	Subject Side View
Day 1 Trail 1	NM NM NM NM NM NM NM NM NM NM NM ND ND NM NM NM NM	47.05	
Day 1 Trail 2	NM NM NM NM NM NM NM NM NM NM ND ND ND NM NM NM NM	52.94	
Day 2 Trail 1	NM NM NM NM NM NM NM NM NM NM ND ND ND NM NM NM NM	52.94	
Day 2 Trail 2	NM NM NM NM NM NM NM NM NM NM ND ND ND NM NM NM NM	52.94	
Day 3 Trail 1	NM NM NM NM NM NU NM NM NM NM ND ND ND NM NM NM NM	58.08	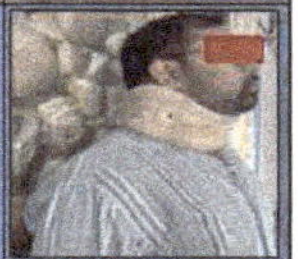
Day 3 Trail 2	NM NM NM NM NU NU NM NM NM NM ND ND ND NM NM NM NM	64.70	
Day 4 Trail 1	NM NM NM NM NM NM NM NM NM NM NM NM ND NM NM NM NM	41.17	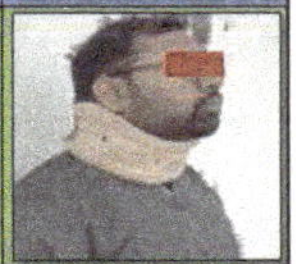
Day 4 Trail 2	NM NM NM NM NM NM NM NM NM NM NM ND ND NM NM NM NM	47.05	
Day 5 Trail 1	NM NM NM NM NM NM NU NU NM NM ND ND ND NM NM NM NU	70.58	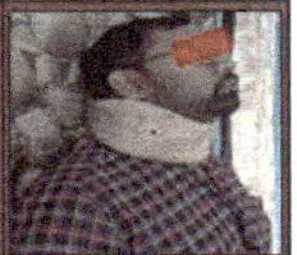
Day 5 Trail 2	NM NM NM NM NM NU NM NM NM NM ND ND ND NM NM NM NU	76.47	
Day 6 Trail 1	NM NM NM NM NM NM NU NU NM NM ND ND ND NM NM NU NU	76.47	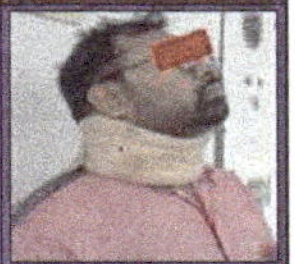
Day 6 Trail 2	NM NM NM NM NM NM NU NU NM NM ND ND NM NM NM NU NM	64.70	
Day 7 Trail 1	NM NM NU NU NU NM NU NU NM NM ND ND ND NM NM NU NM	88.23	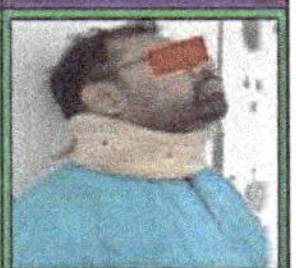
Day 7 Trail 2	NM NM NU NU NM NM NU NU NM NM ND ND ND NM NM NU NU	88.23	
Day 8 Trail 1	NM NM NU NU NU NU NU NU NM NM ND ND NM NM NM NU NU	94.11	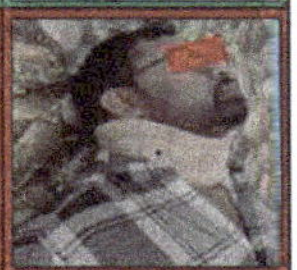
Day 8 Trail 2	NM NM NU NU NU NU NU NU NM NM ND ND ND NM NM NU NU	100	
Day 9 Trail 1	NM NM NM NU NU NU NU NU NM NM ND ND ND NM NM NU NU	94.11	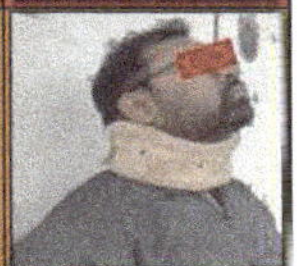
Day 9 Trail 2	NM NM NU NU NU NU NU NU NM NM NM ND ND NM NM NU NU	94.11	
Day 10 Trail 1	NM NM NU NU NU NU NU NU NM NM ND ND ND NM NM NU NU	100	
Day 10 Trail 2	NM NM NU NU NU NU NU NU NM NM ND ND ND NM NM NU NU	100	

Figure 12. Synthetic dataset-based analysis of rehabilitation system.

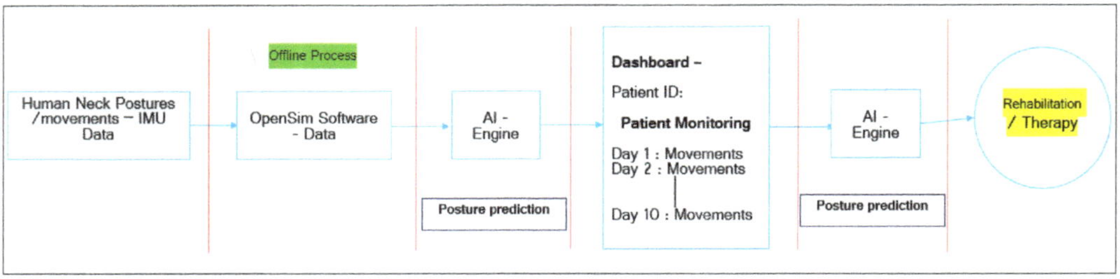

Figure 13. Research flow.

5. Conclusions

This paper presents a novel methodology to identify neck postures using kinetic and kinematic data. Improper neck postures can lead to cervical neck pain and musculoskeletal disorders. This research includes the design of a Smart Neckband consisting of an IMU device that captures kinematic data of the neck postures and movements. The OpenSim simulation tool and a neck musculoskeletal model were used to simulate the related kinetic data for the classification of neck postures. The Machine Learning algorithms achieved 100% accuracy in the prediction of neck postures. In addition to this concept, an evidence-based novel methodology is proposed for the prediction of neck movements to monitor the therapy of neuro-musculoskeletal neck disorders or injuries. Kinematic and kinetic data were integrated innovatively and used to train a model using the Random Forest algorithm. A motivating use case was presented, and this application helped to increase the potential of this innovation. The novel methodology proposed in this paper allows patients to observe their neck movements and exercise patterns to understand how specific exercises help in recovery from musculoskeletal injury. The proposed-technology-enabled system provides valuable insights to physiotherapists in understanding the progress of the patient's condition. The future scope of this research is to embed the entire research work in a single device; this can enable the therapist to have remote access and analyze the human neck movements in an online mode. It also brings in the much-needed accountability to verify if patients are following the recommended therapy. This rehabilitation monitoring mechanism can also be used for remote assessment of musculoskeletal disorders.

Author Contributions: Conceptualization, K.V.R.K.; Data curation, K.V.R.K. and S.E.; Formal analysis, S.E.; Investigation, K.V.R.K.; Methodology, K.V.R.K. and S.E.; Project administration, S.E.; Resources, K.V.R.K.; Supervision, S.E.; K.V.R.K.; Visualization, K.V.R.K.; Writing—original draft, K.V.R.K.; Writing—review & editing, S.E. All authors have read and agreed to the published version of the manuscript.

Funding: There is No Funding for this research work.

Institutional Review Board Statement: The study was conducted according to the guidelines of the Declaration of Helsinki. All the necessary documents submitted to the editor in early stage it self.

Informed Consent Statement: All the details submitted to the editor at early stage.

Data Availability Statement: Not applicable.

Conflicts of Interest: The authors declare no conflict of interest.

References

1. Ribeiro, P.C.; Santos-Victor, J.; Lisboa, P. Human activity recognition from video: Modeling, feature selection and classification architecture. In Proceedings of the International Workshop on Human Activity Recognition and Modelling (HAREM), Oxford, UK, 9 September 2005.
2. Crook, P.; Kellokumpu, V.; Zhao, G.; Pietikainen, M. Human Activity Recognition Using a Dynamic Texture Based Method. In Proceedings of the British Machine Vision Conference, Leeds, UK, 1–4 September 2008.
3. Biswas, K.K.; Basu, S.K. Gesture recognition using Microsoft Kinect. In Proceedings of the 5th International Conference on Automation, Robotics and Applications, Wellington, New Zealand, 6–8 December 2011; pp. 100–103.

4. Nandy, A.; Chakraborty, R.; Chakraborty, P. Cloth invariant gait recognition using pooled segmented statistical features. *Neurocomputing* **2016**, *191*, 117–140. [CrossRef]
5. Balakrishnan, R.; Chinnavan, E.; Feii, T. An extensive usage of hand held devices will lead to musculoskeletal disorder of upper extremity among student in AMU: A survey method. *Int. J. Phys. Educ. Sport Health* **2016**, *3*, 368–372.
6. Caron, N.; Caderby, T.; Peyrot, N.; Verkindt, C.; Dalleau, G. Validation of a method for estimating energy expenditure during walking in middle-aged adults. *Eur. J. Appl. Physiol.* **2018**, *118*, 381–388. [CrossRef] [PubMed]
7. Kadaba, M.P.; Ramakrishnan, H.K.; Wootten, M.E.; Gainey, J.; Gorton, G.; Cochran, G.V.B. Repeatability of kinematic, kinetic, and electromyographic data in normal adult gait. *J. Orthop. Res.* **1989**, *7*, 849–860. [CrossRef] [PubMed]
8. Pizzolato, S.; Tagliapietra, L.; Cognolato, M.; Reggiani, M.; Müller, H.; Atzori, M. Comparison of six electromyography acquisition setups on hand movement classification tasks. *PLoS ONE* **2017**, *12*, e0186132. [CrossRef] [PubMed]
9. Ahamed, N.U.; Taha, Z.; Alqahtani, M.; Altwijri, O.; Rahman, M.; Deboucha, A. Age Related Differences in the Surface EMG Signals on Adolescent's Muscle during Contraction. In Proceedings of the IOP Conference Series: Materials Science and Engineering Malaysia & Indonesia, Bandung, Indonesia, 16–18 November 2016.
10. Seel, T.; Raisch, J.; Schauer, T. IMU-based joint angle measurement for gait analysis. *Sensors* **2014**, *14*, 6891–6909. [CrossRef] [PubMed]
11. Shiroma, E.J.; Schepps, M.; Harezlak, J.; Chen, K.Y.; E. Matthews, C.; Koster, A.; Caserotti, P.; Glynn, N.W., Harris, T.B. Daily physical activity patterns from hip- and wrist-worn accelerometers. *Physiol. Meas.* **2016**, *37*, 1852–1861. [CrossRef] [PubMed]
12. Moncada-Torres, A.; Leuenberger, K.; Gonzenbach, R.; Luft, A.; Gassert, R. Activity classification based on inertial and barometric pressure sensors at different anatomical locations. *Physiol. Meas.* **2014**, *35*, 1245–1263. [CrossRef]
13. Nguyen, N.D.; Truong, P.H.; Jeong, G.-M. Daily wrist activity classification using a smart band. *Physiol. Meas.* **2017**, *38*, L10–L16. [CrossRef]
14. Mifsud, N.L.; Kristensen, N.H.; Villumsen, M.; Hansen, J.; Kersting, U.G. Portable inertial motion unit for continuous assessment of in-shoe foot movement. *Procedia Eng.* **2014**, *72*, 208–213. [CrossRef]
15. Crema, C.; Depari, A.; Flammini, A.; Sisinni, E.; Haslwanter, T.; Salzmann, S. IMU-based solution for automatic detection and classification of exercises in the fitness scenario. In Proceedings of the 2017 IEEE Sensors Applications Symposium (SAS), Glassboro, NJ, USA, 13–15 March 2017; pp. 1–6.
16. Georgi, M.; Amma, C.; Schultz, T. Recognizing Hand and Finger Gestures with IMU based Motion and EMG based Muscle Activity Sensing. In Proceedings of the International Conference on Bio-inspired Systems and Signal Processing (BIOSIGNALS), Lisbon, Portugal, 12–15 January 2015.
17. Chavarriaga, R.; Sagha, H.; Calatroni, A.; Digumarti, S.T.; Tröster, G.; Millan, J.D.R.; Roggen, D. The Opportunity challenge: A benchmark database for on-body sensor-based activity recognition. *Pattern Recognit. Lett.* **2013**, *34*, 2033–2042. [CrossRef]
18. Kanteshwari, K.; Sridhar, R.; Mishra, A.K.; Shirahatti, R.; Maru, R.; Bhusari, P. Correlation of awareness and practice of working postures with prevalence of musculoskeletal disorders among dental professionals. *Gen. Dent.* **2012**, *59*, 476–483.
19. Molsted, S.; Tribler, J.; Snorgaard, O. Musculoskeletal pain in patients with type 2 diabetes. *Diabetes Res. Clin. Pr.* **2012**, *96*, 135–140. [CrossRef]
20. Stefánsdóttir, R.; Gudmundsdottir, S.L. Sedentary behavior and musculoskeletal pain: A five-year longitudinal Icelandic study. *Public Health* **2017**, *149*, 71–73. [CrossRef] [PubMed]
21. Bau, J.-G.; Chia, T.; Wei, S.-H.; Li, Y.-H.; Kuo, F.-C. Correlations of Neck/Shoulder Perfusion Characteristics and Pain Symptoms of the Female Office Workers with Sedentary Lifestyle. *PLoS ONE* **2017**, *12*, e0169318. [CrossRef]
22. Baker, R.; Coenen, P.; Howie, E.; Williamson, A.; Straker, L. The Short Term Musculoskeletal and Cognitive Effects of Prolonged Sitting During Office Computer Work. *Int. J. Environ. Res. Public Health* **2018**, *15*, 1678. [CrossRef] [PubMed]
23. Citko, A.; Górski, S.; Marcinowicz, L.; Górska, A. Sedentary Lifestyle and Nonspecific Low Back Pain in Medical Personnel in North-East Poland. *BioMed. Res. Int.* **2018**, *2018*, 1–8. [CrossRef]
24. Sakakima, H.; Takada, S.; Norimatsu, K.; Otsuka, S.; Nakanishi, K.; Tani, A. Diurnal Profiles of Locomotive and Household Activities Using an Accelerometer in Community-Dwelling Older Adults with Musculoskeletal Disorders: A Cross-Sectional Survey. *Int. J. Environ. Res. Public Health* **2020**, *17*, 5337. [CrossRef] [PubMed]
25. Park, J.H.; Moon, J.H.; Kim, H.J.; Kong, M.H.; Oh, Y.H. Sedentary Lifestyle: Overview of updated evidence of potential health risks. *Korean J. Fam. Med.* **2020**, *41*, 365–373. [CrossRef]
26. David, D.; Giannini, C.; Chiarelli, F.; Mohn, A. Text neck syndrome in children and adolescents. *Int. J. Environ. Res. Public Health* **2021**, *18*, 1565. [CrossRef] [PubMed]
27. Chu, E.; Butler, K. Resolution of Gastroesophageal Reflux Disease Following Correction for Upper Cross Syndrome—A case study and brief review. *Clin. Pr.* **2021**, *11*, 322–326. [CrossRef]
28. Kumar, K.V.R.; Elias, S. Smart Neck-Band for Rehabilitation of Musculoskeletal Disorders. In Proceedings of the 2020 International Conference on COMmunication Systems & NETworkS (COMSNETS), Bengaluru, India, 7–11 January 2020.
29. Dobell, A.; Eyre, E.L.J.; Tallis, J.; Chinapaw, M.; Altenburg, T.M.; Duncan, M.J. Examining accelerometer validity for estimating physical activity in pre-schoolers during free-living activity. *Scand. J. Med. Sci. Sports* **2019**, *29*, 1618–1628. [CrossRef] [PubMed]
30. Smith, C.; Galland, B.; Taylor, R.; Meredith-Jones, K. ActiGraph GT3X+ and actical wrist and hip worn accelerometers for sleep and wake indices in young children using an automated algorithm: Validation with polysomnography. *Front. Psychiatry* **2020**, *10*, 958. [CrossRef]

31. Puerma-Castillo, M.C.; García-Ríos, M.C.; Pérez-Gómez, M.E.; Aguilar-Ferrándiz, M.E.; Peralta-Ramírez, M.I. Effectiveness of kinesio taping in addition to conventional rehabilitation treatment on pain, cervical range of motion and quality of life in patients with neck pain: A randomized controlled trial. *J. Back Musculoskelet. Rehabil.* **2018**, *31*, 453–464. [CrossRef] [PubMed]
32. Raya, R.; Garcia-Carmona, R.; Sanchez, C.; Urendes, E.; Ramirez, O.; Martin, A.; Otero, A. An inexpensive and easy to use cervical range of motion measurement solution using inertial sensors. *Sensors* **2018**, *18*, 2582. [CrossRef] [PubMed]
33. Yoon, T.-L.; Kim, H.-N.; Min, J.-H. Validity and reliability of an inertial measurement unit–based 3-dimensional angular measurement of cervical range of motion. *J. Manip. Physiol. Ther.* **2019**, *42*, 75–81. [CrossRef]
34. Moghaddas, D.; de Zoete, R.M.J.; Edwards, S.; Snodgrass, S.J. Differences in the kinematics of the cervical and thoracic spine during functional movement in individuals with or without chronic neck pain: A systematic review. *Physiotherapy* **2019**, *105*, 421–433. [CrossRef]
35. Maselli, M.; Mussi, E.; Cecchi, F.; Manti, M.; Tropea, P.; Laschi, C. A Wearable sensing device for monitoring single planes neck movements: Assessment of its performance. *IEEE Sens. J.* **2018**, *18*, 6327–6336. [CrossRef]
36. Delp, S.L.; Anderson, F.C.; Arnold, A.S.; Loan, P.; Habib, A.; John, C.T.; Guendelman, E.; Thelen, D.G. Open Sim: Open-source software to create and analyze dynamic simulations of movement. *IEEE Trans. Biomed. Eng.* **2007**, *54*, 1940–1950. [CrossRef] [PubMed]
37. Seth, A.; Sherman, M.; Reinbolt, J.A.; Delp, S.L. OpenSim: A musculoskeletal modeling and simulation framework for in silico investigations and exchange. *Procedia IUTAM* **2011**, *2*, 212–232. [CrossRef]
38. Seth, A.; Hicks, J.L.; Uchida, T.K.; Habib, A.; Dembia, C.L.; Dunne, J.J.; Ong, C.; Demers, M.S.; Rajagopal, A.; Millard, M.; et al. OpenSim: Simulating musculoskeletal dynamics and neuromuscular control to study human and animal movement. *PLoS Comput. Biol.* **2018**, *14*, e1006223. [CrossRef] [PubMed]
39. Thelen, D.G.; Anderson, F.C. Using computed muscle control to generate forward dynamic simulations of human walking from experimental data. *J. Biomech.* **2006**, *39*, 1107–1115. [CrossRef] [PubMed]
40. Mortensen, J.D.; Vasavada, A.N.; Merryweather, A.S. The inclusion of hyoid muscles improve moment generating capacity and dynamic simulations in musculoskeletal models of the head and neck. *PLoS ONE* **2018**, *13*, e0199912. [CrossRef]
41. Cazzola, D.; Holsgrove, T.; Preatoni, E.; Gill, H.; Trewartha, G. Cervical Spine Injuries: A whole-body musculoskeletal model for the analysis of spinal loading. *PLoS ONE* **2017**, *12*, e0169329. [CrossRef]
42. Falla, D.; Jull, G.; O'Leary, S.; Dall'Alba, P. Further evaluation of an EMG technique for assessment of the deep cervical flexor muscles. *J. Electromyogr. Kinesiol.* **2006**, *16*, 621–628. [CrossRef] [PubMed]
43. Wentzel, S.E.; Konow, N.; German, R.Z. Regional differences in hyoid muscle activity and length dynamics during mammalian head shaking. *J. Exp. Zoo. Part. A Ecol. Genet. Physiol.* **2010**, *315A*, 111–120. [CrossRef]
44. Zheng, L.; Jahn, J.; Vasavada, A.N. Sagittal plane kinematics of the adult hyoid bone. *J. Biomech.* **2012**, *45*, 531–536. [CrossRef] [PubMed]
45. Mao, S.; Zhang, Z.; Khalifa, Y.; Donohue, C.; Coyle, J.L.; Sejdic, E. Neck sensor-supported hyoid bone movement tracking during swallowing. *R. Soc. Open Sci.* **2019**, *6*, 181982. [CrossRef]
46. Hannan, A.; Shafiq, M.Z.; Hussain, F.; Pires, I.M. A portable smart fitness suite for real-time exercise monitoring and posture correction. *Sensors* **2021**, *21*, 6692. [CrossRef] [PubMed]
47. Siddiq, M.; Wibawa, I.P.D.; Kallista, M. Integrated Internet of Things (IoT) technology device on smart home system with human posture recognition using kNN method. In Proceedings of the IOP Conference Series: Materials Science and Engineering, the 5th Annual Applied Science and Engineering Conference (AASEC 2020), Bandung, Indonesia, 20–21 April 2020; IOP Publishing: Bandung, Indonesia, 2020; Volume 1098, p. 42065.
48. Zhang, M.; Chen, S.; Zhao, X.; Yang, Z. Research on construction workers' activity recognition based on smartphone. *Sensors* **2018**, *18*, 2667. [CrossRef]
49. Khoury, N.; Attal, F.; Amirat, Y.; Oukhellou, L.; Mohammed, S. Data-driven based approach to aid parkinson's disease diagnosis. *Sensors* **2019**, *19*, 242. [CrossRef]
50. Lee, J.; Joo, H.; Lee, J.; Chee, Y. Automatic classification of squat posture using inertial sensors: Deep learning approach. *Sensors* **2020**, *20*, 361. [CrossRef] [PubMed]
51. Jeng, P.-Y.; Wang, L.-C.; Hu, C.-J.; Wu, D. A wrist sensor sleep posture monitoring system: An automatic labeling approach. *Sensors* **2021**, *21*, 258. [CrossRef]
52. Mortensen, J.; Trkov, M.; Merryweather, A. Improved ergonomic risk factor assessment using opensim and inertial measurement units. In Proceedings of the 2018 IEEE/ACM International Conference on Connected Health: Applications, Systems and Engineering Technologies (CHASE), Washington, DC, USA, 26–28 September 2018.
53. Echeverría, V.; Avendaño, A.; Chiluiza, K.; Vásquez, A.; Ochoa, X. Presentation Skills Estimation Based on Video and Kinect Data Analysis. In Proceedings of the 2014 ACM Workshop on Multimodal Learning Analytics Workshop and Grand Challenge, Istanbul, Turkey, 12–16 November 2014; ACM Press: New York, NY, USA, 2014; pp. 53–60.
54. Kim, Y.M.; Son, Y.; Kim, W.; Jin, B.; Yun, M.H. Classification of children's sitting postures using machine learning algorithms. *Appl. Sci.* **2018**, *8*, 1280. [CrossRef]
55. Sandybekov, M.; Grabow, C.; Gaiduk, M.; Seepold, R. Posture tracking using a machine learning algorithm for a home aal environment. In *Innovation in Medicine and Healthcare*; Springer: Berlin/Heidelberg, Germany, 2019; pp. 337–347.
56. Bourahmoune, K.; Amagasa, T. *AI-Powered Posture Training: Application of Machine Learning in Sitting Posture Recognition Using the LifeChair Smart Cushion*; IJCAI: Macao, China, 16 August 2019; pp. 5808–5814.

57. Bejinariu, S.-I.; Costin, H.; Rotaru, F.; Luca, R.; Lazar, C. Deep learning based human locomotion recognition in video sequences. In Proceedings of the 2020 International Conference on e-Health and Bioengineering (EHB), Iasi, Romania, 29–30 October 2020; pp. 1–4.
58. Liaqat, S.; Dashtipour, K.; Arshad, K.; Assaleh, K.; Ramzan, N. A hybrid posture detection framework: Integrating machine learning and deep neural networks. *IEEE Sens. J.* **2021**, *21*, 9515–9522. [CrossRef]
59. Bonneau, M.; Benet, B.; Labrune, Y.; Bailly, J.; Ricard, E.; Canario, L. Predicting sow postures from video images: Comparison of convolutional neural networks and segmentation combined with support vector machines under various training and testing setups. *Biosyst. Eng.* **2021**, *212*, 19–29. [CrossRef]

healthcare

MDPI

Article

Interocular Symmetry Analysis of Corneal Elevation Using the Fellow Eye as the Reference Surface and Machine Learning

Shiva Mehravaran [1,*], Iman Dehzangi [2] and Md Mahmudur Rahman [3]

1 Department of Biology, School of Computer, Mathematical and Natural Sciences, Morgan State University, Baltimore, MD 21251, USA
2 Center for Computational and Integrative Biology, Department of Computer Science, Rutgers University, Camden, NJ 08102, USA; i.dehzangi@rutgers.edu
3 Department of Computer Science, School of Computer, Mathematical and Natural Sciences, Morgan State University, Baltimore, MD 21251, USA; md.rahman@morgan.edu
* Correspondence: shiva.mehravaran@morgan.edu

Abstract: Unilateral corneal indices and topography maps are routinely used in practice, however, although there is consensus that fellow-eye asymmetry can be clinically significant, symmetry studies are limited to local curvature and single-point thickness or elevation measures. To improve our current practices, there is a need to devise algorithms for generating symmetry colormaps, study and categorize their patterns, and develop reference ranges for new global discriminative indices for identifying abnormal corneas. In this work, we test the feasibility of using the fellow eye as the reference surface for studying elevation symmetry throughout the entire corneal surface using 9230 raw Pentacam files from a population-based cohort of 4613 middle-aged adults. The 140×140 matrix of anterior elevation data in these files were handled with Python to subtract matrices, create color-coded maps, and engineer features for machine learning. The most common pattern was a monochrome circle ("flat") denoting excellent mirror symmetry. Other discernible patterns were named "tilt", "cone", and "four-leaf". Clustering was done with different combinations of features and various algorithms using Waikato Environment for Knowledge Analysis (WEKA). Our proposed approach can identify cases that may appear normal in each eye individually but need further testing. This work will be enhanced by including data of posterior elevation, thickness, and common diagnostic indices.

Keywords: unsupervised machine learning; clustering; cornea; corneal topography; interocular symmetry; corneal elevation; keratoconus

Citation: Mehravaran, S.; Dehzangi, I.; Rahman, M.M. Interocular Symmetry Analysis of Corneal Elevation Using the Fellow Eye as the Reference Surface and Machine Learning. *Healthcare* **2021**, *9*, 1738. https://doi.org/10.3390/healthcare9121738

Academic Editor: Francesco Faita

Received: 30 November 2021
Accepted: 13 December 2021
Published: 16 December 2021

1. Introduction

The cornea is the dome-shaped layer of transparent tissue at the frontmost part of the eye globe, and its main function is to provide 75% to 80% of the refractive power of the eye [1–3]. In the frontal view, the cornea is almost circular in outline with a horizontal diameter of about 11.0–12.0 mm horizontally and 10.0–11.0 mm vertically [3]. Given the pivotal role of the cornea in vision, even small deviations from normal and subtle imperfections in the transparency and shape of the cornea can disturb the quality of the retinal image. Therefore, accurate measurement of various corneal properties such as its curvature, thickness, and elevation is an integral part of a comprehensive eye exam.

Early attempts at describing the corneal shape date back to 1619 when Scheiner used glass balls of known diameters to measure the curvature of the cornea [4]. Until quite recently, the description of the corneal shape was limited to local metrics of the corneal curvature as measured with manual keratometers and single-point measurements of the corneal thickness with ultrasound pachymeters. Technological advances in ophthalmology have provided us modern systems that perform computer analysis of photographs taken from the entire surface of the cornea, and convert the data to color-coded contour maps [5]. Today, corneal topographic categories such as "round", "oval", "bow-tie", and "irregular", that were originally described by Bogan et al. [6] in 1990 and further expanded by Rabinowitz et al. [7] in 1996 are well known to practitioners. Since their introduction, computerized imaging systems have greatly enhanced our understanding of the corneal topography in normal and disease conditions. However, identifying corneal degenerative changes in early subclinical stages remains a challenge [8,9], and there is an active area of research to develop discriminative algorithms and finetune diagnostic criteria using state-of-the-art corneal imaging systems.

The Pentacam (Oculus GmBH, Wetzlar, Germany) is a popular projection-based anterior segment imaging device that utilizes a high-resolution Scheimpflug camera that scans the anterior segment by rotating 360° around the center [10]. The system captures data from 25,000 distinct elevation points within 2 s which are used to generate a 3-dimensional virtual model of the anterior segment. Once image processing is complete, the user can choose to review various maps and displays such as the sagittal curvature, pachymetry, and elevation maps of the anterior and posterior cornea, which compose the default 4-map display. Originally, the elevation of each point on the corneal surface is measured as its distance from a reference plane tangent to the corneal apex. This is quite similar to terrain topography, where elevation is defined as the distance above sea level. However, to make subtle surface variations discernible, the displayed data is a recalculation of the raw data to express the perpendicular distance from a sphere of variable diameter and position that best fits each individual cornea.

Currently available diagnostic algorithms and classification systems are mainly based on unilateral data [8,11–13]. Since there is wide variation in the normal population that define their reference ranges, they have shown suboptimal performance in discriminating normal corneas from subclinical forms of disorders [14]. This is while measures of normal fellow corneas are strongly correlated [15–17], contralateral eyes are highly symmetric [18–22], and there is consensus that lack of symmetry should be interpreted as a red flag warranting reevaluation or further testing [20,23–27]. Nonetheless, our understanding of corneal symmetry is limited to single-point metrics (e.g., elevation at the apex, corneal thickness at the thinnest point) and local indices (e.g., simulated keratometry in the steep and flat axes); the color-coded patterns have not been classified or described, and no multi-feature or global indices have been developed yet. Some other limitations of extant literature are that studies were mostly clinic-based with small sample sizes of defined groups that are not representative of the general population, and they used relative measures of elevation displayed by the system rather than the actual elevation (height) data.

This study was designed as a proof-of-concept study for using fellow eyes as the reference surface using raw Pentacam elevation data from a large population-based sample (including normal and abnormal) with two main goals: (1) describe pancorneal symmetry patterns observed in difference colormaps, and (2) cluster the data by applying machine learning techniques. Proving the feasibility of this approach is the first step in creating a novel diagnostic index for identifying cases with subtle changes and to assess longitudinal changes in the same eye.

2. Materials and Methods

The proposal of this secondary data analysis study was reviewed and approved by the Institutional Review Board of Morgan State University. The deidentified data was obtained from the Shahroud Eye Cohort Study (ShECS) which is an observational cohort of adults between the ages of 40 and 64 years at first enrollment [28,29]. To date, three phases of the study have been completed at 5-year intervals. Of the 6311 Shahroud residents who were invited to the study in 2009, 5190 participated (82.2% participation rate), completed the interview, and had a comprehensive eye examination including anterior segment imaging with the Pentacam. For this study, we used baseline data including the deidentified cohort database (containing demographic variables including age and gender) and Pentacam elevation files directly exported from the device. The only inclusion criterion was having both the right and left eye elevation files (bilateral cases).

Data management was done in the Anaconda3 platform using various packages of Python version 3.7.4 in the Jupyter Notebook (server version 6.0.1). Waikato Environment for Knowledge Analysis (WEKA) version 3.8.4 was used for unsupervised machine learning and cluster analyses of the data and engineered features [30].

2.1. Creating Pancorneal Difference Matrices

For this step, first the IDs of right and left eyes were matched using Python's *fnmatch* function to identify cases with bilateral data. Then the 141 × 141 matrix of anterior elevation values were extracted from each Pentacam elevation file. Each data point in the 141 × 141 anterior elevation matrix corresponds to an area of 0.1 × 0.1 mm; therefore, each matrix provides a coverage of 14 × 14 mm centered on the corneal apex (x = 0, y = 0 coordinates). The process for creating the fellow-eye difference matrices were relatively similar to what has been described by Cavas-Martínez et al. [22] who assessed shape symmetry in a sample of 33 normal cases. For each matched pair, the left eye matrix was rotated 180° around its Y axis using the NumPy *flip* function to account for the mirror symmetry between fellow eyes. Then, the right eye matrix was subtracted from the flipped left eye matrix. Figure 1 provides a schematic illustration of how the contralateral eye becomes the reference surface when raw elevation data are subtracted to create a fellow-eye difference matrix.

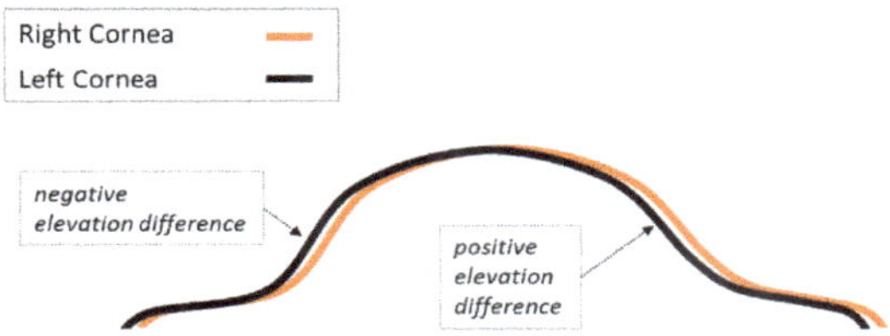

Figure 1. Schematic presentation of using the contralateral cornea as the reference surface for measuring elevation and assessing elevation symmetry between fellow eyes. Highly symmetric corneas should fit each other, and hypothetically, there will be zero distance between them. The higher the asymmetry, the bigger the area between the two surfaces.

2.2. Creating Elevation Difference Colormaps

The difference matrices created in the previous step were color-coded to 2-dimensional colormaps. Using the Matplotlib and Seaborn packages, we assigned the spectral color palette because it resembles the ones routinely used in corneal topography. As such, the scale range was set from extreme negative (plotted in dark red) to extreme positive (plotted in dark blue) and the center 0 point was plotted as bright yellow. Therefore, deviation from the middle of the scale to either side could be illustrated with ascending darker colors.

2.3. Feature Engineering

To exclude extreme outliers in the corneal periphery that could be due to the effect of the limbus, eyelids, nose shadow, pterygium, and/or data extrapolation, elevation difference matrices were masked to only keep the data in the central 6.0 mm zone of the cornea (2821 data points per case). This zone was further divided to four smaller concentric zones with diameter sizes of 2.0, 3.0, 4.0, and 5.0 mm. The data within these five zones (2-dimensional arrays) were then flattened to a single dimension using the *flatten* function of NumPy and compiled into a single data frame in which there was one row of data per participant, and the columns represented the coordinates of the 2-dimensional masked matrix. The data in each row were summarized into their descriptive statistics including skew, absolute skew, kurtosis, mean, standard deviation of the mean, absolute mean (average of absolute means), median, absolute median, minimum, maximum, absolute maximum (the larger of maximum and absolute minimum), range, and central 95% range.

The sums of negative and positive elevation difference values (Figure 1) were used to calculate the negative and positive volumes, respectively, as well as the sum of the two volumes (Total Volume) and the absolute difference between the two volumes (Volume Difference) as a measure of intraindividual asymmetry.

2.4. Cluster Analysis

In the next step, the data from difference matrices and their descriptive statistics were used as features for unsupervised machine learning analysis in WEKA. Different combinations were tested with different clustering algorithms such as the simple k-means and the simple expectation-maximization (EM) algorithms, and in some iterations, principal component analysis (PCA) was applied first for feature reduction. The outputs were inspected and compared in terms of the distribution of cases within each cluster, number of clusters, and the summary statistics of difference features.

2.5. Adding Other Indices

To make comparisons with the literature, we extracted the apical and minimum corneal thickness, maximum (simulated keratometry at the steep meridian) and mean (average of the keratometry in the steep and flat meridians) keratometry readings, and corneal astigmatism and computed the absolute interocular difference for these continuous variables. Pentacam also generates two categorical parameters, namely the quality specification (QS) and the keratoconus score (KKS) for each examined eye, which indicate the quality of the data and normality of the cornea, respectively. To examine the agreement between our clustering results and Pentacam-assigned categories, these parameters were extracted, recoded, and combined to create four bilateral categories with QS indicated as OK (Tables S1 and S2).

3. Results

A total of 9303 Pentacam elevation files were available; 4670 right eyes and 4633 left eyes. Matching the right and left data files by their study ID resulted in 4615 bilateral cases, two of which were excluded due to insufficient data points (computations returned NULL), and 4613 were included in the analysis. The mean age of this sample was 50.9 ± 6.3 years, and 41% were male.

3.1. Symmetry Patterns in Colormaps

Figure 2 illustrates four different interocular elevation difference colormaps of the same individual; the peripheral outliers were removed by masking the difference matrices to the central 6.0 mm zone, and the overall visualization was improved by setting the scale to ± 70 μm and cropping the image.

In reviewing the color-maps generated from fellow-eye difference data, we found a monochrome yellow circle to be the most common pattern showing that the interocular difference is zero or very close to zero, and the fellow corneas fit nicely with very little or no gap between them; this was named "flat" (Figure 3). Other commonly discernible patterns of interocular difference colormaps were named "tilt", "cone", "4-leaf", and "irregular". As illustrated in Figure 3, the pattern we named "tilt" demonstrated a semicircle of negative values on one side and a semicircle of positive values on the other side, separated by a yellow band (zero or near zero values). This pattern could be indicative of a difference in the imaging or visual axis between fellow eyes, and one eye is off-axis. The "cone" pattern would appear in cases where one cornea is steeper than the other, and the gap between them increases from the center to the periphery; this is the pattern one would expect to see in central keratoconus. The "4-leaf" pattern can be attributed to situations where the cornea in one eye is steeper in a certain meridian and flatter in the perpendicular meridian; these could be cases of direct symmetry especially in the presence of corneal astigmatism. Symmetry patterns that did not fit any of these categories were assigned to the "irregular" group.

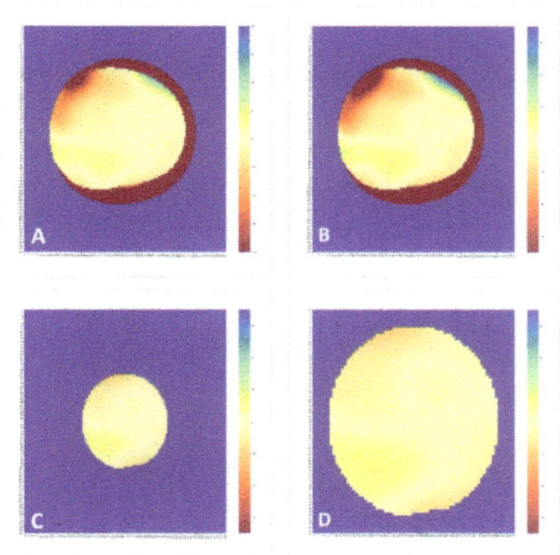

Figure 2. Fellow-eye elevation difference colormaps of a 53-year-old woman. The full 140 × 140 difference matrix using a ±100 µm scale (**A**) and a ±70 µm scale (**B**) show sharp contours in the periphery which were eliminated by masking the central 6.0 mm (**C**) and cropping out the extra data (**D**).

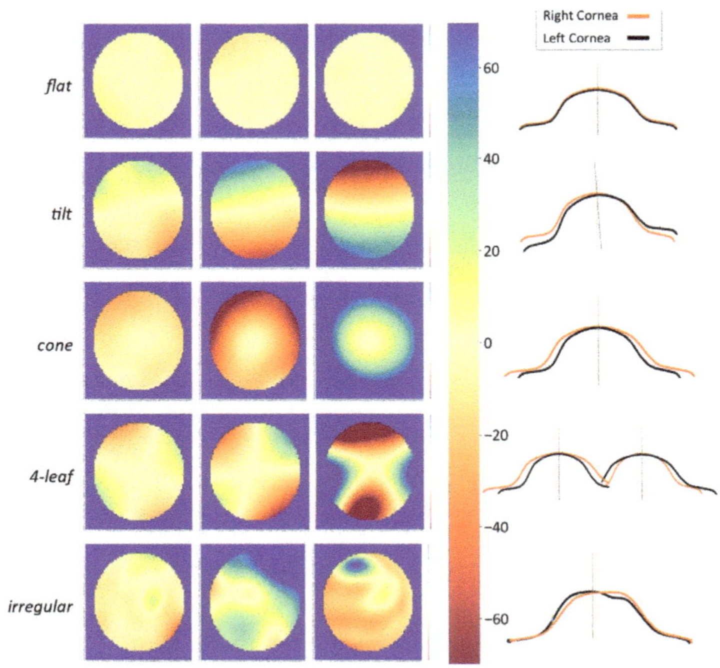

Figure 3. Sample 6.0 mm colormaps of common patterns observed in the elevation difference colormaps. The same ±70 µm scale (shown in the middle) was applied to all colormaps. The schematics on the right demonstrate how the fellow corneas fit in each category. In the flat pattern, the fellow corneas fit well, and there is minimum distance between them. In the tilt pattern, half of one cornea is below and the other half is above its fellow cornea. In the cone pattern one cornea is steeper that its fellow cornea, and the area between the two surfaces increases from the center to the periphery. In the 4-leaf pattern, one cornea is steeper in a given meridian and flatter in the perpendicular meridian compared to its fellow cornea.

3.2. Data Exploration and Feature Engineering

Figure 4 illustrates the cumulative percentage of cases in which the minimum and maximum interocular elevation difference (i.e., values of all data points) in each of the five studied zones was within the specified range. For example, all data points in the central 2.0 mm zone were within ±5.0 μm in 88.4% of the cases, within ±10.0 μm in 96.0%, and within ±15.0 μm in 97.6% of cases. In case of the central 3.0 mm zone, all data were within ±10.0 μm in 90.0% and within ±25.0 μm in % 97.6% of cases. In case of the central 6.0 mm, all data were within ±60.0 μm in 90.0%, and within ±100.0 μm in 92.7% of cases.

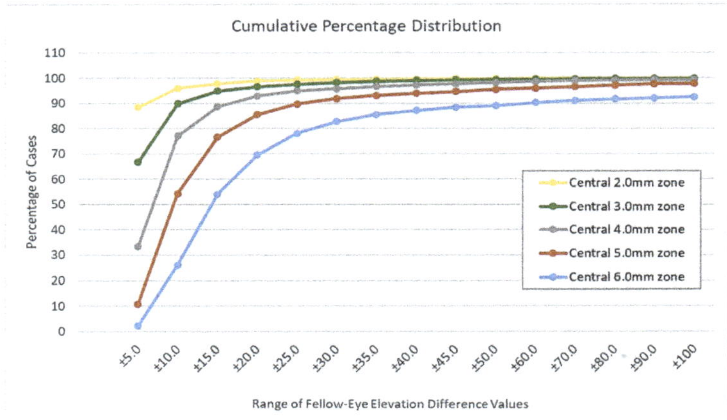

Figure 4. Cumulative percentage of cases that had all data points within a given range.

In the total sample of 4613 cases, mean elevation difference at the (0, 0) coordinates was 0.04 ± 2.0 μm (central 95% range: 7.8 μm). The descriptive statistics (measures of central tendency and variability) are summarized as their mean, standard deviation of the mean, and the central 95% range in Table S3. Both the mean and variance of the data increased in larger, more peripheral zones.

3.3. Unsupervised Machine Learning

tabref:healthcare-1512249-t001 and Figure 5 present the results of simple k-means clustering in WEKA with the following attributes: the central 95% range in the 6.0 mm zone and the absolute mean, standard deviation of the mean, and volume difference in the central 3.0 mm zone. The full sample was grouped into three clusters with 3839 (83.2%) in Cluster 1, 618 (13.4%) in Cluster 2, and 156 (3.4%) in Cluster 3; mean elevation difference at the (0, 0) coordinates was −0.0005 ± 0.32 μm, −0.016 ± 0.29 μm, and 1.12 ± 10.76 μm in Cluster 1, 2, and 3, respectively.

Table 1. WEKA output using simple k-means clustering. Attributes used in this model included the central 95% range of the 6.0 mm zone and the mean, standard deviation of the mean, and volume difference of the central 3.0 mm zone.

Attribute	Full n = 4613	Cluster 1 n = 3839	Cluster 2 n = 618	Cluster 3 n = 156
Central 95% Range/6.0 mm	20.7	14.1	42.1	97.0
Absolute Mean/3.0 mm	0.8	0.6	1.2	4.5
Standard Deviation/3.0 mm	2.0	1.3	3.7	11.4
Volume Difference/3.0 mm	5.7	4.3	8.2	30.8

Clustering Model: k-means; Number of iterations: 30; Within cluster sum of squared errors: 34.2.

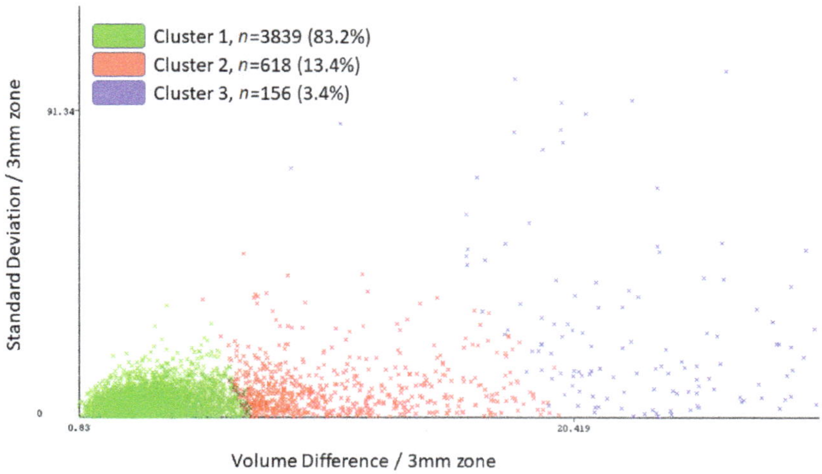

Figure 5. WEKA visualization output using simple k-means clustering. Attributes used in this model included the central 95% range of the 6.0 mm zone and the mean, standard deviation of the mean, and volume difference of the central 3.0 mm zone.

Figure 6 illustrates the central 6.0 mm interocular elevation difference maps of three random samples from each of the three clusters. In Cluster 1, the colormap pattern was "flat" in all cases; the other patterns appeared in Cluster 2 with lighter colors and in Cluster 3 with darker colors.

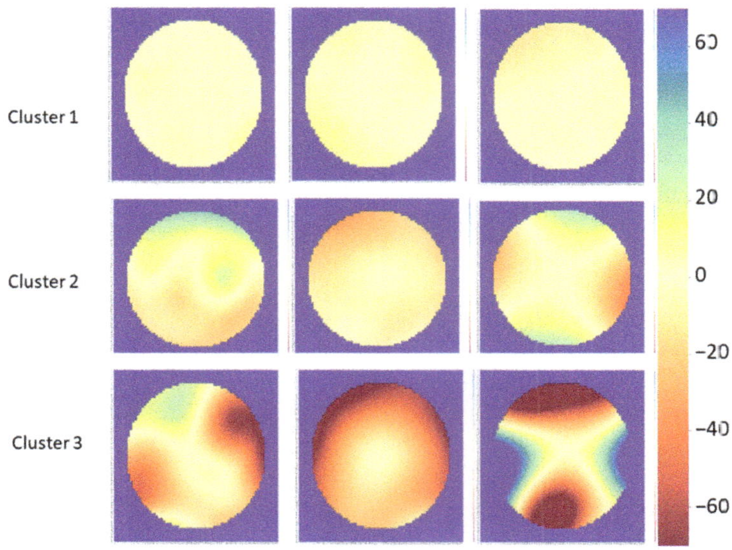

Figure 6. Sample 6.0 mm colormaps of three random cases from each of the three clusters. The scale in all colormaps is the ±70 µm scale shown on the right. These three clusters were created in using the simple k-means clusterer in WEKA and the following attributes: central 95% range of the 6.0 mm zone and the absolute mean, standard deviation of the mean, and volume difference of the central 3.0 mm zone. Cluster 1 corresponds with normal corneas and all maps showed the flat pattern. Other patterns were observed in cluster 2 and 3, although the degree of asymmetry was greater in the latter group.

Table 2 summarizes the summary statistics of the interocular anterior elevation differences in the 3 clusters within the five studied central corneal zones. Overall, both the mean and the spread of the values were smallest in Cluster 1 and highest in Cluster 3. They were also higher in larger, more peripheral corneal zones within each cluster.

Table 2. Mean and standard deviation of summary statistics of anterior elevation difference (in μm) between corresponding points on fellow eyes in the total sample and the three clusters within the five concentric central corneal zones.

Zone	Statistic	Full Sample n = 4613	Cluster 1 n = 3839	Cluster 2 n = 618	Cluster 3 n = 156
2.0 mm	Abs-Mean	0.5 ± 2.1	0.3 ± 0.3	0.6 ± 0.6	3.5 ± 10.7
	SD	1.3 ± 2.0	0.9 ± 0.4	2.2 ± 1.2	7.4 ± 8.1
	Min	−2.8 ± 3.6	−2.0 ± 1.1	−4.8 ± 2.6	−15.4 ± 11.4
	Max	2.9 ± 6.1	2.0 ± 1.2	5.0 ± 3.2	17.2 ± 28.3
	Range	5.7 ± 8.6	3.9 ± 1.6	9.8 ± 4.7	32.5 ± 34.3
	Abs-Max	3.7 ± 6.5	2.5 ± 1.1	6.2 ± 3.0	21.9 ± 28.1
3.0 mm	Abs-Mean	0.8 ± 2.3	0.6 ± 0.5	1.2 ± 1.1	5.5 ± 11.2
	SD	2.0 ± 3.1	1.3 ± 0.5	3.7 ± 1.5	12.5 ± 12.0
	Min	−4.8 ± 6.6	−3.2 ± 1.8	−8.7 ± 4.2	−28.0 ± 21.9
	Max	2.9 ± 6.1	2.0 ± 1.2	5.0 ± 3.2	17.2 ± 28.3
	Range	9.6 ± 14.3	6.4 ± 2.5	17.5 ± 6.7	57.6 ± 54.0
	Abs-Max	6.3 ± 9.8	4.2 ± 1.8	11.1 ± 4.6	38.5 ± 38.3
4.0 mm	Abs-Mean	1.3 ± 2.8	0.9 ± 0.7	1.8 ± 1.6	8.0 ± 12.6
	SD	2.9 ± 4.5	1.9 ± 0.9	5.6 ± 2.2	18.9 ± 16.3
	Min	−7.5 ± 10.5	−5.0 ± 2.9	−14.2 ± 7.0	−44.0 ± 34.2
	Max	7.8 ± 15.4	5.0 ± 6.9	15.1 ± 16.5	46.7 ± 53.4
	Range	15.3 ± 23.0	10.0 ± 7.4	29.3 ± 18.3	90.7 ± 76.1
	Abs-Max	10.2 ± 16.5	6.8 ± 6.8	19.2 ± 16.0	60.7 ± 52.1
5.0 mm	Abs-Mean	1.8 ± 3.5	1.3 ± 1.1	2.6 ± 2.5	11.0 ± 14.4
	SD	4.4 ± 6.9	2.8 ± 2.7	8.6 ± 6.0	26.9 ± 21.0
	Min	−11.1 ± 15.1	−7.3 ± 4.3	−21.7 ± 11.3	−63.3 ± 48.0
	Max	13.6 ± 35.5	9.1 ± 27.1	26.8 ± 46.7	71.8 ± 78.3
	Range	24.7 ± 43.3	16.4 ± 27.8	48.5 ± 48.4	135.1 ± 105
	Abs-Max	17.4 ± 36.4	11.7 ± 26.8	33.8 ± 45.7	92.1 ± 75.3
6.0 mm	Abs-Mean	2.6 ± 4.7	1.9 ± 2.3	4.0 ± 5.0	14.0 ± 16.9
	SD	7.3 ± 13.3	5.0 ± 9.6	13.8 ± 15.2	38.0 ± 28.2
	Min	−16.0 ± 20.9	−10.7 ± 6.4	−31.5 ± 18.1	−85.4 ± 64.6
	Max	31.3 ± 92.6	23.8 ± 83.4	53.3 ± 109.0	129.6 ± 149.8
	Range	47.3 ± 99.3	34.5 ± 84.3	84.7 ± 110.6	214.9 ± 175.5
	Abs-Max	36.8 ± 92.4	27.6 ± 82.8	64.1 ± 106.7	154.4 ± 141.8

All $p < 0.001$; ANOVA comparing the mean in the three clusters. Abs-Mean: mean of absolute mean differences; SD: standard deviation; Min: minimum; Max: maximum; Abs-Max: the larger of maximum and absolute minimum.

3.4. Assessing Clusters Using Parameters Other Than Elevation

Table 3 presents summary statistics of the studied corneal thickness and curvature parameters in the right eyes, left eyes, and the absolute interocular difference in the total sample (n = 4613) and the 3 clusters. Similar to elevation data, both the mean and spread were smallest in Cluster 1 and highest in Cluster 3.

From the total sample of 4613, 571 cases had imaging errors (QS of 1 or 2, see Table S1) in at least one eye, and they were excluded from the comparison with Pentacam normality indices. As indicated in the top section of Table 4, of the 2975 cases in the Bilateral-normal/QS-OK category (64.5% of the total sample), 2696 (90.6%) were in Cluster 1 (the cluster with the least interocular differences). However, from this same category, 22 (0.7%) were in Cluster 3 (the cluster with the highest levels of difference between fellow eyes). Central 6.0 mm fellow-eye elevation difference maps of these cases are illustrated in Tables 5 and 6. All 10 cases illustrated in Table 5 have 1.0 D or more interocular difference in corneal astigmatism; in 8 cases, the difference is 2.5 D or more. Cases #4 and #5 in Row 1 as well as case #4 in Row 2 also show considerable interocular differences in terms of corneal thickness at the apex and thinnest point. Among the remaining 12 cases (Table 6), all three cases in Row 3 have 2.5 D or more interocular difference in maximum keratometry, and case #1 in Row 3 shows more than 30 µm corneal thickness difference between fellow eyes. The four cases shown in Row 4 have 17 µm or more interocular thickness difference either at the apex, the thinnest point, or both. Finally, the five cases in Row 5 have 13.0 µm or more absolute maximum elevation difference.

Table 3. Mean ± standard deviation of the corneal thickness and curvature indices in the right end left eyes, and their absolute interocular difference in the total sample and the three clusters.

Parameter		Total	Cluster 1	Cluster 2	Cluster 3	p-Value *
		n = 4613	n = 3839	n = 618	n = 156	
Apical Thickness (µm)	OD	529.9 ± 33.7	530.9 ± 32.0	527.6 ± 34.5	515.4 ± 59.2	<0.001
	OS	530.6 ± 33.8	531.5 ± 32.2	527.9 ± 35.3	519.6 ± 55.4	<0.001
	i-dif	9.6 ± 13.2	8.0 ± 6.4	11.4 ± 11.1	39.7 ± 51.7	<0.001
Minimum Thickness (µm)	OD	524.5 ± 37.0	526.6 ± 32.1	521.2 ± 34.7	486.6 ± 93.7	<0.001
	OS	525.2 ± 35.6	527.2 ± 32.3	521.0 ± 35.9	493.5 ± 73.8	<0.001
	i-dif	10.3 ± 20.0	8.1 ± 6.4	12.0 ± 12.1	56.8 ± 89.5	<0.001
Maximum Keratometry (D)	OD	44.2 ± 1.7	44.1 ± 1.6	44.5 ± 1.9	45.5 ± 3.2	<0.001
	OS	44.2 ± 1.8	44.1 ± 1.6	44.6 ± 1.9	45.8 ± 3.8	<0.001
	i-dif	0.5 ± 0.8	0.3 ± 0.3	0.7 ± 0.6	2.4 ± 3.2	<0.001
Mean Keratometry (D)	OD	43.7 ± 1.7	43.7 ± 1.5	43.8 ± 1.8	44.0 ± 3.3	0.017
	OS	43.8 ± 1.7	43.7 ± 1.5	43.9 ± 1.8	44.3 ± 3.3	<0.001
	i-dif	0.4 ± 0.6	0.3 ± 0.2	0.6 ± 0.5	2.3 ± 2.5	<0.001
Corneal Astigmatism (D)	OD	0.9 ± 1.1	0.8 ± 0.5	1.3 ± 1.1	3.0 ± 4.2	<0.001
	OS	0.9 ± 1.1	0.8 ± 0.5	1.4 ± 1.3	3.0 ± 3.9	<0.001
	i-dif	0.5 ± 1.1	0.4 ± 0.3	1.0 ± 1.1	3.3 ± 4.7	<0.001

* ANOVA comparing the mean in the three clusters. D: diopter; OD: right eyes; OS: left eyes; i-diff: absolute difference between fellow eyes.

Table 4. Frequency distribution of the combined corneal abnormality categories in the full sample and the three clusters, and the mean (± standard deviation) interocular difference values of thickness and curvature measures.

Pentacam Category	Parameter	Total	Cluster 1	Cluster 2	Cluster 3
		(n = 4613)	(n = 3839)	(n = 618)	(n = 156)
Bilateral-normal QS-OK	n	2975 (64.5%)	2696 (90.6%)	257 (8.6%)	22 (0.7%)
	Ap-thick	8.2 ± 6.6	8.0 ± 6.2	10.0 ± 8.7	14.4 ± 10.3
	Min-thick	8.3 ± 6.5	8.1 ± 6.2	9.9 ± 8.6	12.5 ± 10.7
	MaxK	0.3 ± 0.3	0.3 ± 0.3	0.6 ± 0.6	0.9 ± 1.3
	MeanK	0.3 ± 0.3	0.3 ± 0.2	0.4 ± 0.4	1.4 ± 1.2
	Cor-ast	0.4 ± 0.5	0.3 ± 0.3	1.0 ± 1.0	1.9 ± 2.1
KS-abnormal QS-OK	n	684 (14.8%)	512 (74.9%)	144 (21.1%)	28 (4.1%)
	Ap-thick	9.5 ± 9.4	8.1 ± 6.7	11.4 ± 10.3	25.8 ± 22.1
	Min-thick	10.0 ± 11.8	8.0 ± 6.5	11.9 ± 9.7	35.3 ± 37.6
	MaxK	0.5 ± 1.0	0.4 ± 0.3	0.7 ± 0.7	2.3 ± 4.3
	MeanK	0.5 ± 0.7	0.3 ± 0.2	0.7 ± 0.6	2.0 ± 2.2
	Cor-ast	0.6 ± 1.3	0.4 ± 0.3	1.0 ± 1.1	3.3 ± 4.9
KCN-1-2 QS-OK	n	84 (1.8%)	30 (35.7%)	36 (42.9%)	18 (21.4%)
	Ap-thick	15.8 ± 14.1	8.6 ± 7.3	14.9 ± 10.5	29.6 ± 19.0
	Min-thick	16.9 ± 17.6	9.2 ± 7.2	14.4 ± 10.2	34.5 ± 27.6
	MaxK	1.2 ± 1.1	0.6 ± 0.4	1.1 ± 0.9	2.3 ± 1.6
	MeanK	0.9 ± 1.0	0.4 ± 0.3	0.9 ± 0.6	1.8 ± 1.6
	Cor-ast	1.0 ± 1.0	0.5 ± 0.4	1.0 ± 1.0	1.7 ± 1.5

Table 4. *Cont.*

Pentacam Category	Parameter	Total	Cluster 1	Cluster 2	Cluster 3
		(n = 4613)	(n = 3839)	(n = 618)	(n = 156)
KCN-3-4 QS-OK	n	10 (0.2%)	0 (0.0%)	4 (40.0%)	6 (60.0%)
	Ap-thick	33.2 ± 34.1		16.0 ± 6.8	44.7 ± 40.9
	Min-thick	29.4 ± 28.8		18.3 ± 9.8	36.8 ± 35.7
	MaxK	3.7 ± 4.4		2.1 ± 0.8	4.8 ± 5.6
	MeanK	3.2 ± 4.4		1.9 ± 0.5	4.0 ± 5.8
	Cor-ast	1.7 ± 1.3		1.9 ± 0.9	1.6 ± 1.6

Ap-thick: apical thickness (μm); Min-thick: minimum thickness (μm); MaxK: keratometry in the steep meridian (diopter); MeanK: average of keratometry in steep and flat meridians (diopter); Cor-ast: corneal astigmatism (diopter).

Table 5. Elevation difference colormaps of 10 Cluster 3 cases identified as bilaterally normal by Pentacam.

Parameter	1	2	3	4	5
Row 1	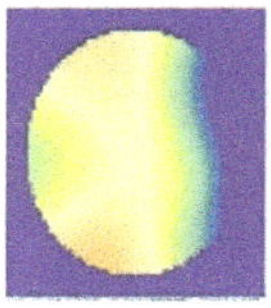	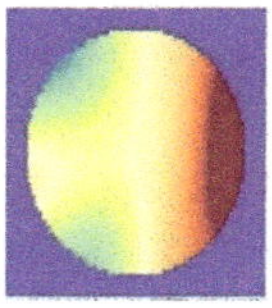		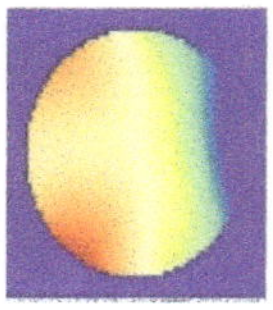	
ΔAst	6.2	5.6	5.6	4.0	3.7
ΔmaxK	0.54	0.41	0.12	0.33	0.00
ΔpAx	7.0	3.0	19.0	23.0	34.0
ΔpThin	1.0	4.0	4.0	21.0	36.0
ΔmaxEle	12.0	17.0	23.0	13.0	18.0
Row 2					
ΔAst	3.5	3.4	2.5	1.2	1.0
ΔmaxK	0.28	0.24	1.77	0.61	0.91
ΔpAx	6.0	1.0	5.0	25.0	11.0
ΔpThin	1.0	10.0	2.0	27.0	15.0
ΔmaxEle	12.0	12.0	7.0	27.0	14.0

Note: The scale in all colormaps is ±70 μm. Δ: interocular difference; Ast: anterior corneal astigmatism (diopter); maxK: maximum keratometry (diopter); pAx: pachymetry at the apex (μm); pThin: pachymetry at the thinnest point of the cornea (μm); maxEle: the larger of the maximum and absolute minimum elevation difference (μm).

Table 6. Colormaps of 12 cases in Cluster 3 that were identified as bilaterally normal by Pentacam.

Parameter	1	2	3	4	5
Row 3					
ΔAst	0.5	0.7	0.6		
ΔmaxK	5.11	3.27	2.61		
ΔpAx	35.0	3.0	22.0		
ΔpThin	31.0	2.0	25.0		
ΔmaxEle	11.0	27.0	9.0		

Table 6. *Cont.*

Parameter	1	2	3	4	5
Row 4					
ΔAst	0.2	0.1	0.1	0.9	
ΔmaxK	0.06	0.70	0.68	0.00	
ΔpAx	22.0	22.0	20.0	19.0	
ΔpThin	1.0	20.0	17.0	19.0	
ΔmaxEle	9.0	22.0	13.0	15.0	
Row 5					
ΔAst	0.4	0.2	0.3	0.2	0.4
ΔmaxK	0.54	0.70	0.06	0.34	0.06
ΔpAx	9.0	0.0	10.0	14.0	7.0
ΔpThin	8.0	1.0	10.0	12.0	7.0
ΔmaxEle	23.0	17.0	17.0	16.0	13.0

Note: ±70 μm scale. Δ: interocular difference; Ast: anterior corneal astigmatism (diopters); maxK: keratometry in the steep meridian (diopters); pAx: pachymetry at the apex (μm); pThin: pachymetry at the thinnest point of the cornea (μm); maxEle: the larger of the maximum and absolute minimum elevation difference in the central 2.0 mm zone (μm).

4. Discussion

One of the main objectives of this study was to create fellow eye anterior elevation difference colormaps and suggest descriptive names for discernible patterns. As expected, the most common pattern was "flat" showing that the interocular difference is zero or very close to zero and the fellow corneas fit nicely with very little or no gap between them (Figure 3). The "tilt" pattern could be attributed to a difference in the imaging or visual axis between fellow eyes; identifying this pattern could have implications in evaluating strabismus, prescribing corrective eyeglasses, or, as suggested by Fathy et al. [31], in screening for keratoconus. The "cone" pattern is expected in keratoconus, especially central forms. The "4-leaf" pattern can be attributed to cases of direct symmetry especially in the presence of corneal astigmatism; for these cases, creating fellow-eye difference matrices without flipping the left eye matrix could return one of the other patterns. Although the patterns of unilateral corneal topography maps have been studied and have accepted nomenclature [6,7], to the best of our knowledge, this is the first study to examine fellow-eye difference maps and give them descriptive names. Adding fellow-eye difference displays to corneal imaging systems can facilitate interocular symmetry review for eye care providers, and once they become familiar with the patterns and complete the learning curve, the approach has the potential to become an integral part of a comprehensive eye exam, especially for preoperative screening.

Recent studies of fellow-eye symmetry have looked at different corneal features and parameters including corneal biometrics [27,32], higher order aberrations [33], and corneal surface area [34]. A summary of the few studies that have examined anterior elevation symmetry is presented in Table 7 [21,25,35–39]. These studies greatly vary by methodology such as sample selection and size, the corneal topographer used for imaging, the reference surface used for measuring elevation, and the choice of elevation measure.

Table 7. Summary of fellow-eye symmetry studies reporting measures of anterior corneal elevation.

First Author [Ref #]	Studied Sample	Reference Surface	Anterior Elevation Measure	Mean Interocular Difference (μm)
Falavarjani [25]	275 normal	Float BFS with auto diameter	Maximum in the central 4.0 mm	2.2 (range, 0–21)
Durr * [21]	3835 normal	Average BFS of all eyes	Average elevation in the central 6.0 mm †	Range ± 6.0
Saad * [35]	51 normal 32 KCN	Default float BFS	Maximum/at thinnest point	Normal: 0.0 ± 0.0/0.0 ± 0.0 KCN: 0.02 ± 0.01/0.02 ± 0.01
Galletti [36]	177 normal 44 intermediate 121 KCN	No mention	At thinnest corneal location	Central 98% range Normal: 4.0 KCN: 31.0
Naderan [37]	306 normal 68 suspect 446 KCN	8.0 mm BFS	At thinnest point within the central 3.0 mm	Normal: 1.3 ± 0.7 KCS: 5.5 ± 4.8 KCN: 14.0 ± 10.4
Henriquez [38]	341 normal 50 high ammetropia 294 KCN	8.0 mm BFS	Maximum/at thinnest point	Normal: 1.4 ± 1.4/1.1 ± 1.0 KCN: 10.3 ± 11.0/8.7 ± 9.9
Eppig [39]	68 normal 350 KCN	No mention	Elevation deviation	Normal: 0.46 ± 0.39 KCN: 7.8 ± 7.4
Current Project	4615 general population	Raw data (the fellow eye)	Values and descriptive statistics of all corresponding points in the central 2.0–6.0 mm	See Tables 2 and 3

* Used Orbscan IIz; all other studies used Pentacam; † Constructed from the average of each point in the whole sample. BFS: best fit sphere; KCN: keratoconus.

As summarized in Table 7, Falavarjani et al. [25] reported a mean interocular difference of 2.2 μm (range: 0 to 21.0 μm) and suggested that a difference greater than 17.4 μm (95th percentile) should be interpreted as a potential red flag. Their results can be compared to our 4.0 mm data summarized in Table 2. The mean absolute maximum (the larger of maximum and absolute minimum) was 6.8 ± 6.8 μm, 19.2 ± 16.0 μm, and 60.7 ± 52.1 μm in Clusters 1, 2, and 3, respectively, and 10.2 ± 16.5 μm in the total sample. Therefore, the average of 2.2 μm reported in their study is even smaller than what was observed for Cluster 1 (6.8 μm) which is the group with highest symmetry in our study. This is mainly due to methodological differences; they only included healthy eyes and calculated the interocular difference at only one single point (the maximum anterior) which was measured from a spherical reference surface that may have been different between fellow eyes.

The study by Durr et al. [21] was similar to ours in that they examined a large population-based sample. Methodological differences included using Orbscan IIz, applying exclusion criteria (no history of ocular disease, ocular surgery, or recent contact lens wear), and using a reference surface based on the average best fit sphere of all right and left eyes. The average anterior elevation difference in the 6.0 mm zone in their study ranged within ±6.0 μm. Because of the methodological differences mentioned above, as well as age differences between the samples of the two studies, their result is much smaller the mean range of 27.6 μm observed for the 6.0 mm zone in Cluster 1 of our study (Table 2).

The other five reports summarized in Table 7 were clinic-based comparative studies, that enrolled two or more sample groups, one being a normal control group and one a group of keratoconus patients. Saad et al. [35] used the Orbscan IIz with the reference surface set on the default float mode. Although the intergroup differences were statistically significant (both $p < 0.001$), the mean interocular differences they observed in the maximum anterior elevation and the anterior elevation at the thinnest point of the cornea (Table 7) were very close to zero in both groups. Galletti et al. [36] and Eppig et al. [39] used the Pentacam; the reference surface is not mentioned, but perhaps the default setting was used [40]. Galletti et al. [36] included the absolute interocular difference of the anterior elevation at the thinnest point. The central 90% range for this variable was 4.0 μm in the normal comparison group and 31.0 μm in the group that was labelled as keratoconus based on Pentacam diagnostic indices. Eppig et al. [39] examined another relative measure of anterior elevation which looks at the difference in elevation values between measurements made with a standard best fit sphere and an "enhanced" best fit sphere which is calculated from the

9.0 mm central data minus the 4.0 mm around the thinnest point [41]. Naderan at al. [37] appear to have set the device to use an 8.0 mm best fit sphere. The anterior elevation measure they examined in their study is described as the "maximum at the thinnest point" of the cornea "based on the data from the 3.0 mm annular corneal diameter ring". They found a mean absolute interocular difference of 1.3 ± 0.7 μm (0.0 –7.0 μm) in the normal comparison group, 5.5 ± 4.8 μm (1.0–14.0 μm) in the keratoconus suspect group, and 14.0 ± 10.4 μm (1.0–36.0 μm) in the keratoconus group. The interocular differences observed by Henriquez et al. [37] were very similar to that reported by Naderan et al. [37]. In both cases, the interocular differences are far smaller than what was observed in our study, which again, similar to the study by Falavarjani et al. [25], could be attributed to methodological differences and the use of a variable reference surface.

This study (bottom row in Table 7) is novel in multiple ways. Firstly, the fellow cornea was used as the reference surface (Figure 1). This was based on the hypothesis that doing so would allow one to discern subtle interocular differences that may not be obvious when comparing two separate elevation maps, especially if their measurements are based on different reference surfaces. Secondly, the symmetry data was pancorneal and not limited to one or two points. The number of corresponding data points in the central 2.0 mm, 3.0 mm, 4.0 mm, and 5.0 mm of the cornea were 317, 709, 1257, and 1961, respectively, and the central 6.0 mm was represented by 2821 data points. Thirdly, from the subtraction matrix of each individual, multiple features representing their central tendency and variability (skew, mean, central 95% range, total volume, etc.) in the 2.0–6.0 mm zones of the cornea were engineered and used as attributes in machine learning and clustering algorithms. Another strength of this study is its large population-based sample (n = 4613) and inclusion of all cases.

Different combinations of a multitude of features were tested in different iterations with WEKA. To maintain simplicity and allow comparison with other studies, the next steps of the analyses were done with a 3-cluster output. As demonstrated in Table 2, both the mean and the standard deviation (spread) of the summary statistics were significantly different between the three groups; values were lowest in Cluster 1 (best symmetry) and highest in Cluster 3 (least symmetry). A similar trend was observed when the three clusters were compared in terms of corneal thickness and curvature indices (Table 3). This is because corneal features are strongly correlated. In fact, elevation-based topographers, such as the Pentacam, capture elevation data directly, while anterior and posterior corneal power data are computed from the elevation data of their corresponding surface and corneal thickness is the elevation distance between the two corneal surfaces.

A summary of interocular symmetry studies examining measures of corneal thickness and curvature is presented in Table 8 [23,24,35,37–39,42–44]. Comparison of the values shows that Cluster 1 corresponds with normal groups. As such, mean interocular differences in central and minimum corneal thickness were 8.0 μm and 8.1 μm in Cluster 1, respectively, and they ranged between 4.3 μm and 11.0 μm in the normal groups of other studies. In terms of maximum, minimum, and mean keratometry, all three values were around 0.3 D in Cluster 1, and the range reported for the normal groups summarized in Table 8 is between 0.2 D and 0.4 D. However, as demonstrated in Table 4, 11.7% of Cluster 1 cases were red-flagged by Pentacam. Since their colormap patterns were "flat", this mismatch is probably due to the fact that only anterior corneal elevation data were used for clustering, and therefore, abnormalities in the corneal thickness and posterior corneal surface were overlooked. A similar comparison shows that Cluster 2 is comparable to the keratoconus suspect group in the study by Naderan et al. [37]; other studies did not have an intermediate or suspect group. In Cluster 3, mean interocular differences were 41.1 μm and 58.3 μm for central and minimum corneal thickness, respectively, while the values in keratoconus groups of other studies are in the range of 25.9–34.0 μm and 30.2–39.8 μm for central and minimum corneal thickness, respectively. The interocular differences in maximum and mean keratometry readings are lower in Cluster 3 compared to the keratoconus groups in other studies, and minimum keratometry is in the mid-range.

Also, contrary to all other groups, the difference in minimum keratometry is higher than that of maximum keratometry. The lack of agreement between Cluster 3 and other groups is probably because the sample in our study was the general population and mirror symmetry was assumed. As such, there may be highly asymmetric cases (albeit clinically normal) due to other reasons such as anisoastigmatism, anisorule, and/or direct symmetry patterns [22,45].

Table 8. Summary of fellow-eye symmetry studies reporting measures of corneal thickness and power.

First Author [Ref #]	Group	Corneal Thickness (μm)		Simulated Keratometry (D)			
		Central	Thinnest	Steep	Flat	Mean	Diff
Myrowitz * [23]	normal	-	8.0 ± 7.0	-	-	0.5 ± 0.4	-
Khachikian [24]	normal	8.8 ± 7.2	9.0 ± 8.3	-	-	-	-
Henriquez [42]	normal	10.2 ± 7.9	11.0 ± 8.2	0.3 ± 0.3	0.3 ± 0.2	-	-
	KCN	25.9 ± 24.1	30.2 ± 29.1	3.8 ± 4.2	2.7 ± 3.3	-	-
Henriquez [38]	normal	10.3 ± 7.9	11.0 ± 8.2	-	-	-	-
	KCN	25.9 ± 24.1	30.2 ± 29.1	-	-	-	-
Dienes [43]	normal	5.6 ± 4.9	6.6 ± 5.3	0.4 ± 0.4	0.4 ± 0.4	-	-
	KCN	30.1 ± 35.8	39.7 ± 36.4	4.4 ± 5.1	2.7 ± 3.6	-	-
Kovács [44]	normal	6.3 ± 6.9	6.9 ± 7.5	0.3 ± 0.2	0.3 ± 0.2	-	-
	KCN	29.9 ± 34.3	39.8 ± 29.1	3.3 ± 2.6	2.8 ± 3.1	-	-
Naderan [37]	normal	4.3 ± 1.6	5.9 ± 2.2	0.3 ± 0.2	0.2 ± 0.2	0.2 ± 0.2	0.1 ± 0.1
	suspect	12.8 ± 10.0	13.7 ± 10.9	1.0 ± 1.2	0.6 ± 0.8	0.7 ± 0.8	1.0 ± 0.8
	KCN	29.4 ± 28.5	33.6 ± 33.2	4.3 ± 4.2	3.4 ± 3.7	3.7 ± 3.8	1.8 ± 1.5
Eppig [39]	normal	6.0 ± 5.0	6.0 ± 5.0	-	-	0.2 ± 0.2	0.4 ± 0.4
	KCN	34.0 ± 30.0	37.0 ± 32.0	-	-	3.8 ± 4.0	2.0 ± 1.7
Saad * [35]	normal	5.4 ± 4.9	6.0 ± 5.0	0.3 ± 0.3	0.4 ± 0.3	-	0.3 ± 0.3
	KCN	33.9 ± 37.0	35.7 ± 34.5	4.1 ± 2.9	2.4 ± 2.9	-	2.1 ± 2.3
Current Project †	Cluster 1	8.0 ± 6.3	8.1 ± 6.4	0.3 ± 0.3	0.3 ± 0.3	0.3 ± 0.2	0.3 ± 0.3
	Cluster 2	11.3 ± 10.5	11.9 ± 10.8	0.7 ± 0.7	0.8 ± 0.8	0.6 ± 0.5	1.0 ± 1.1
	Cluster 3	41.1 ± 53.6	58.3 ± 92.5	2.5 ± 3.3	3.1 ± 3.8	2.3 ± 2.6	3.4 ± 4.9

* Used Orbscan IIz; all other studies used Pentacam. † Excluding 571 cases with quality error > 0 in either eye. See Table 3 for full sample results.

Anisoastigmatism is defined as an interocular difference of 1.0 D or more in refractive astigmatism [46–48]. The interocular difference in anterior corneal astigmatism was 3.4 D in Cluster 3, but the range in the keratoconus groups of other studies is only 1.8–2.1 D (Table 8). Twenty-two cases in Cluster 3 were found to be bilaterally normal by Pentacam's built-in algorithms (Table 4), and 10 of them had anisoastigmatism (Table 5). The common discernable pattern in this group (Tables 5 and 6) was "tilt" (Figure 1) which can be due to interocular differences in angle kappa or how the apex, line of sight, and measurement axis line up [40] or a displaced apex [49]. One way to examine this is the interocular difference in anterior elevation at (0, 0) coordinates; the mean of this index was −0.18 μm in this subsample of 22 cases, 0.04 in the total sample, and 0.0 in Cluster 1. To control for such an effect in future research, we will apply the iterative closest point transformation algorithm described by Fathy et al. [31] before subtracting data on corresponding points.

As mentioned earlier, this study had certain limitations that need to be addressed in the follow-up work. Firstly, despite the large sample size, the age range was limited to 40–64 years who might have higher levels of corneal irregularity than younger individuals [50]. In our future work, we will use data from a younger population-based cohort [51] and/or the general sample from a clinical database. Secondly, since this was a preliminary proof of concept study, clustering algorithms were provided with anterior elevation data only; this can explain the false negative and false positive cases described above. Also, to allow for simplicity and comparability, the number of clusters was limited to three. In future work, adding features derived from posterior elevation and thickness symmetry and previously developed diagnostic indices along with a larger (or automated) number of clusters could help improve the accuracy of the algorithm and facilitate classifying symmetry patterns.

5. Patents

The concept behind this work is under patent protection by Morgan State University.

Supplementary Materials: The following are available online at https://www.mdpi.com/article/10.3390/healthcare9121738/s1, Table S1: Device-generated quality and normality indicators extracted from each CSV file and their recoding into fewer categories, Table S2: Combining the recoded quality and normality indicators into 6 bilateral categories, and Table S3: The mean ± standard deviation (central 95% range) of the descriptive statistics of the interocular elevation difference values (μm) in the central 2.0 mm–6.0 mm zones of the cornea within each individual (n = 4613).

Author Contributions: Conceptualization, S.M.; methodology, S.M., I.D., M.M.R.; software, S.M., I.D., M.M.R.; validation, S.M.; formal analysis, S.M., M.M.R.; investigation, S.M.; resources, S.M., I.D., M.M.R.; data curation, S.M.; writing—original draft preparation, S.M.; writing—review and editing, S.M., I.D., M.M.R.; visualization, S.M. All authors have read and agreed to the published version of the manuscript.

Funding: The principal investigator's (S.M.) salary is supported by the ASCEND Center for Biomedical Research (RL5GM118972) and the Center for Urban Health Disparities Research and Innovation (U54MD013376).

Institutional Review Board Statement: The study was conducted according to the guidelines of the Declaration of Helsinki, and approved by the Institutional Review Board of Morgan State University (IRB #20/10-0119, approved on 30 September 2020).

Informed Consent Statement: Not applicable.

Data Availability Statement: Restrictions apply to the availability of these data. Data was obtained from the Shahroud Eye Cohort Study.

Acknowledgments: The authors wish to thank the principal investigators of the Shahroud Eye Cohort Study Drs. M.H. Emamian, A. Fotouhi, and H. Hashemi for generously sharing the deidentified data files.

Conflicts of Interest: The authors declare no conflict of interest.

References

1. Dowling, J.E.; Dowling Jr, J.L. *Vision: How It Works and What Can Go Wrong*; MIT Press: Cambridge, MA, USA, 2016.
2. Nilsson, S.F.E.; Hoeve, J.V.; Wu, S.; Kaufman, P.L.; Alm, A. *Adler's Physiology of the Eye E-Book*; Elsevier Health Sciences: Amsterdam, The Netherlands, 2011; ISBN 978-0-323-08116-0.0.
3. Freddo, T.F.; Chaum, E. *Anatomy of the Eye and Orbit: The Clinical Essentials*; Wolters Kluwer Health: Philadelphia, PA, USA, 2017; ISBN 978-1-4698-7328-2.
4. Daxecker, F. Christoph Scheiner's Eye Studies. *Doc. Ophthalmol.* **1992**, *81*, 27–35. [CrossRef]
5. Maguire, L.J.; Singer, D.E.; Klyce, S.D. Graphic Presentation of Computer-Analyzed Keratoscope Photographs. *Arch. Ophthalmol. Chic.* **1987**, *105*, 223–230. [CrossRef] [PubMed]
6. Bogan, S.J.; Waring, G.O., III; Ibrahim, O.; Drews, C.; Curtis, L. Classification of Normal Corneal Topography Based on Computer-Assisted Videokeratography. *Arch. Ophthalmol.* **1990**, *108*, 945–949. [CrossRef] [PubMed]
7. Rabinowitz, Y.S.; Yang, H.; Brickman, Y.; Akkina, J.; Riley, C.; Rotter, J.I.; Elashoff, J. Videokeratography Database of Normal Human Corneas. *Br. J. Ophthalmol.* **1996**, *80*, 610–616. [CrossRef]
8. Motlagh, M.N.; Moshirfar, M.; Murri, M.S.; Skanchy, D.F.; Momeni-Moghaddam, H.; Ronquillo, Y.C.; Hoopes, P.C. Pentacam®Corneal Tomography for Screening of Refractive Surgery Candidates: A Review of the Literature, Part I. *Med. Hypothesis Discov. Innov. Ophthalmol.* **2019**, *8*, 177–203.
9. Zhang, X.; Munir, S.Z.; Sami Karim, S.A.; Munir, W.M. A Review of Imaging Modalities for Detecting Early Keratoconus. *Eye Lond Engl.* **2021**, *35*, 173–187. [CrossRef]
10. Pentacam Interpretation Guideline–Third Edition. Available online: https://www.pentacam.com/fileadmin/user_upload/pentacam.de/downloads/interpretations-leitfaden/interpretation_guideline_3rd_edition_0915.pdf (accessed on 13 December 2020).

11. Belin, M.W.; Kundu, G.; Shetty, N.; Gupta, K.; Mullick, R.; Thakur, P. ABCD: A New Classification for Keratoconus. *Indian J. Ophthalmol.* **2020**, *68*, 2831–2834. [CrossRef]
12. Doctor, K.; Vunnava, K.P.; Shroff, R.; Kaweri, L.; Lalgudi, V.G.; Gupta, K.; Kundu, G. Simplifying and Understanding Various Topographic Indices for Keratoconus Using Scheimpflug Based Topographers. *Indian J. Ophthalmol.* **2020**, *68*, 2732–2743. [CrossRef]
13. Hashemi, H.; Mehravaran, S. Day to Day Clinically Relevant Corneal Elevation, Thickness, and Curvature Parameters Using the Orbscan II Scanning Slit Topographer and the Pentacam Scheimpflug Imaging Device. *Middle East Afr. J. Ophthalmol.* **2010**, *17*, 44–55. [CrossRef]
14. Klyce, S.D. Chasing the Suspect: Keratoconus. *Br. J. Ophthalmol.* **2009**, *93*, 845–847. [CrossRef]
15. Cheng, C.Y.; Liu, J.H.; Chiang, S.C.; Chen, S.J.; Hsu, W.M. Statistics in Ophthalmic Research: Two Eyes, One Eye or the Mean? *Zhonghua Yi Xue Za Zhi Chin. Med. J. Free China Ed* **2000**, *63*, 885–892.
16. Armstrong, R.A. Statistical Guidelines for the Analysis of Data Obtained from One or Both Eyes. *Ophthalmic Physiol. Opt. J. Br. Coll. Ophthalmic Opt. Optom.* **2013**, *33*, 7–14. [CrossRef]
17. Zhang, H.G.; Ying, G. Statistical Approaches in Published Ophthalmic Clinical Science Papers: A Comparison to Statistical Practice Two Decades Ago. *Br. J. Ophthalmol.* **2018**, *102*, 1188–1191. [CrossRef]
18. Dingeldein, S.A.; Klyce, S.D. The Topography of Normal Corneas. *Arch. Ophthalmol. Chic.* **1989**, *107*, 512–518. [CrossRef]
19. Corbett, M.C.; O'Brart, D.P.S.; Saunders, D.C.; Rosen, E.S. The Topography of the Normal Cornea. *Eur. J. Implant Refract. Surg.* **1994**, *6*, 286–297. [CrossRef]
20. Bao, F.; Chen, H.; Yu, Y.; Yu, J.; Zhou, S.; Wang, J.; Wang, Q.; Elsheikh, A. Evaluation of the Shape Symmetry of Bilateral Normal Corneas in a Chinese Population. *PLoS ONE* **2013**, *8*, e73412. [CrossRef]
21. Durr, G.M.; Auvinet, E.; Ong, J.; Meunier, J.; Brunette, I. Corneal Shape, Volume, and Interocular Symmetry: Parameters to Optimize the Design of Biosynthetic Corneal Substitutes. *Investig. Ophthalmol. Vis. Sci.* **2015**, *56*, 4275–4282. [CrossRef]
22. Cavas-Martínez, F.; Piñero, P.P.; Fernández-Pacheco, D.G.; Mira, J.; Cañavate, F.J.F.; Alió, J.L. Assessment of Pattern and Shape Symmetry of Bilateral Normal Corneas by Scheimpflug Technology. *Symmetry* **2018**, *10*, 453. [CrossRef]
23. Myrowitz, E.H.; Kouzis, A.C.; O'Brien, T.P. High Interocular Corneal Symmetry in Average Simulated Keratometry, Central Corneal Thickness, and Posterior Elevation. *Optom. Vis. Sci. Off. Publ. Am. Acad. Optom.* **2005**, *82*, 428–431. [CrossRef]
24. Khachikian, S.S.; Belin, M.W.; Ciolino, J.B. Intrasubject Corneal Thickness Asymmetry. *J. Refract. Surg.* **2008**, *24*, 606–609. [CrossRef]
25. Falavarjani, K.G.; Modarres, M.; Joshaghani, M.; Azadi, P.; Afshar, A.E.; Hodjat, P. Interocular Differences of the Pentacam Measurements in Normal Subjects. *Clin. Exp. Optom.* **2010**, *93*, 26–30. [CrossRef]
26. Li, Y.; Bao, F.J. Interocular Symmetry Analysis of Bilateral Eyes. *J. Med. Eng. Technol.* **2014**, *38*, 179–187. [CrossRef]
27. Xu, G.; Hu, Y.; Zhu, S.; Guo, Y.; Xiong, L.; Fang, X.; Liu, J.; Zhang, Q.; Huang, N.; Zhou, J.; et al. A Multicenter Study of Interocular Symmetry of Corneal Biometrics in Chinese Myopic Patients. *Sci. Rep.* **2021**, *11*, 5536. [CrossRef] [PubMed]
28. Fotouhi, A.; Hashemi, H.; Shariati, M.; Emamian, M.H.; Yazdani, K.; Jafarzadehpur, E.; Koohian, H.; Khademi, M.R.; Hodjatjalali, K.; Kheirkhah, A.; et al. Cohort Profile: Shahroud Eye Cohort Study. *Int. J. Epidemiol.* **2013**, *42*, 1300–1308. [CrossRef]
29. About ShECS. Available online: http://en.shecs.info/ (accessed on 15 December 2020).
30. Frank, E.; Hall, M.A.; Witten, I.H. *The WEKA Workbench. Online Appendix for "Data Mining: Practical Machine Learning Tools and Techniques"*, 4th ed.; Morgan Kaufmann: San Francisco, CA, USA, 2016.
31. Fathy, A.; Lopes, B.T.; Ambrósio, R.; Wu, R.; Abass, A. The Efficiency of Using Mirror Imaged Topography in Fellow Eyes Analyses of Pentacam HR Data. *Symmetry* **2021**, *13*, 2132. [CrossRef]
32. Fraenkel, D.; Hamon, L.; Daas, L.; Flockerzi, E.; Suffo, S.; Eppig, T.; Seitz, B. Tomographically Normal Partner Eye in Very Asymmetrical Corneal Ectasia: Biomechanical Analysis. *J. Cataract Refract. Surg.* **2021**, *47*, 366–372. [CrossRef] [PubMed]
33. Alzaben, Z.; Gammoh, Y.; Freixas, M.; Zaben, A.; Zapata, M.A.; Koff, D.N. Inter-Ocular Asymmetry in Anterior Corneal Aberrations Using Placido Disk-Based Topography. *Clin. Ophthalmol. Auckl.* **2020**, *14*, 1451–1457. [CrossRef] [PubMed]
34. Crahay, F.-X.; Debellemanière, G.; Tobalem, S.; Ghazal, W.; Moran, S.; Gatinel, D. Quantitative Comparison of Corneal Surface Areas in Keratoconus and Normal Eyes. *Sci. Rep.* **2021**, *11*, 6840. [CrossRef]
35. Saad, A.; Guilbert, E.; Gatinel, D. Corneal Enantiomorphism in Normal and Keratoconic Eyes. *J. Refract. Surg.* **2014**, *30*, 542–547. [CrossRef]
36. Galletti, J.D.; Vázquez, P.R.R.; Minguez, N.; Delrivo, M.; Bonthoux, F.F.; Pförtner, T.; Galletti, J.G. Corneal Asymmetry Analysis by Pentacam Scheimpflug Tomography for Keratoconus Diagnosis. *J. Refract. Surg.* **2015**, *31*, 116–123. [CrossRef] [PubMed]
37. Naderan, M.; Rajabi, M.T.; Zarrinbakhsh, P. Intereye Asymmetry in Bilateral Keratoconus, Keratoconus Suspect and Normal Eyes and Its Relationship with Disease Severity. *Br. J. Ophthalmol.* **2017**, *101*, 1475–1482. [CrossRef] [PubMed]
38. Henriquez, M.A.; Izquierdo, L.; Belin, M.W. Intereye Asymmetry in Eyes with Keratoconus and High Ammetropia: Scheimpflug Imaging Analysis. *Cornea* **2015**, *34*, S57–S60. [CrossRef] [PubMed]
39. Eppig, T.; Langenbucher, A.; Papavasileiou, K.; Spira-Eppig, C.; Goebels, S.; Seitz, B.; El-Husseiny, M.; Lenhart, M.; Szentmáry, N. Asymmetry between Left and Right Eyes in Keratoconus Patients Increases with the Severity of the Worse Eye. *Curr. Eye Res.* **2018**, *43*, 848–855. [CrossRef]
40. Pentacam_Guideline_3rd_0218_k.Pdf. Available online: https://www.pentacam.com/fileadmin/user_upload/pentacam.de/downloads/interpretations-leitfaden/Pentacam_Guideline_3rd_0218_k.pdf (accessed on 15 December 2021).

41. Belin, M.W.; Khachikian, S.S.; Ambrósio, R.J.; Salomão, M. Keratoconus/Ectasia Detection with the Oculus Pentacam: Belin/Ambrósio Enhanced Ectasia Display. *Highlights Ophthalmol.* **2007**, *35*, 5–12.
42. Henriquez, M.A.; Izquierdo, L.J.; Mannis, M.J. Intereye Asymmetry Detected by Scheimpflug Imaging in Subjects with Normal Corneas and Keratoconus. *Cornea* **2013**, *32*, 779–782. [CrossRef]
43. Dienes, L.; Kránitz, K.; Juhász, É.; Gyenes, A.; Takács, Á.; Miháltz, K.; Nagy, Z.Z.; Kovács, I. Evaluation of Intereye Corneal Asymmetry in Patients with Keratoconus. A Scheimpflug Imaging Study. *PLoS ONE* **2014**, *9*, e108882. [CrossRef]
44 Kovács, I.; Miháltz, K.; Kránitz, K.; Juhász, É.; Takács, Á.; Dienes, L.; Gergely, R.; Nagy, Z.Z. Accuracy of Machine Learning Classifiers Using Bilateral Data from a Scheimpflug Camera for Identifying Eyes with Preclinical Signs of Keratoconus. *J. Cataract Refract. Surg.* **2016**, *42*, 275–283. [CrossRef]
45. Hashemi, H.; Asharlous, A.; Yekta, A.; Ostadimoghaddam, H.; Mohebi, M.; Aghamirsalim, M.; Khabazkhoob, M. Enantiomorphism and Rule Similarity in the Astigmatism Axes of Fellow Eyes: A Population-Based Study. *J. Optom.* **2019**, *12*, 44–54. [CrossRef]
46. Ostadimoghaddam, H.; Fotouhi, A.; Hashemi, H.; Yekta, A.A.; Heravian, J.; Hemmati, B.; Jafarzadehpur, E.; Rezvan, F.; Khabazkhoob, M. The Prevalence of Anisometropia in Population Base Study. *Strabismus* **2012**, *20*, 152–157. [CrossRef]
47. Dobson, V.; Harvey, E.M.; Miller, J.M.; Clifford-Donaldson, C.E. Anisometropia Prevalence in a Highly Astigmatic School-Aged Population. *Optom. Vis. Sci. Off. Publ. Am. Acad. Optom.* **2008**, *85*, 512–519. [CrossRef]
48. O'Donoghue, L.; McClelland, J.F.; Logan, N.S.; Rudnicka, A.R.; Owen, C.G.; Saunders, K.J. Profile of Anisometropia and Aniso-Astigmatism in Children: Prevalence and Association with Age, Ocular Biometric Measures, and Refractive Status. *Investig. Ophthalmol. Vis. Sci.* **2013**, *54*, 602–608. [CrossRef] [PubMed]
49. Belin, M.W.; Khachikian, S.S. An Introduction to Understanding Elevation-Based Topography: How Elevation Data Are Displayed—A Review. *Clin. Exp. Ophthalmol.* **2009**, *37*, 14–29. [CrossRef] [PubMed]
50. Hayashi, K.; Kawahara, S.; Manabe, S.; Hirata, A. Changes in Irregular Corneal Astigmatism with Age in Eyes with and without Cataract Surgery. *Investig. Ophthalmol. Vis. Sci.* **2015**, *56*, 7988–7998. [CrossRef] [PubMed]
51. Emamian, M.H.; Hashemi, H.; Khabazkhoob, M.; Malihi, S.; Fotouhi, A. Cohort Profile: Shahroud Schoolchildren Eye Cohort Study (SSCECS). *Int. J. Epidemiol.* **2019**, *48*, 27. [CrossRef]

healthcare

MDPI

Review

A Comprehensive Survey of Image-Based Food Recognition and Volume Estimation Methods for Dietary Assessment

Ghalib Ahmed Tahir and Chu Kiong Loo *

Department of Artificial Intelligence, Faculty of Computer Science and Information Technology, University of Malaya, Kuala Lumpur 50603, Malaysia; 12mscsgtahir@seecs.edu.pk or ghalib@siswa.um.edu.my
* Correspondence: ckloo.um@um.edu.my

Abstract: Dietary studies showed that dietary problems such as obesity are associated with other chronic diseases, including hypertension, irregular blood sugar levels, and increased risk of heart attacks. The primary cause of these problems is poor lifestyle choices and unhealthy dietary habits, which are manageable using interactive mHealth apps. However, traditional dietary monitoring systems using manual food logging suffer from imprecision, underreporting, time consumption, and low adherence. Recent dietary monitoring systems tackle these challenges by automatic assessment of dietary intake through machine learning methods. This survey discusses the best-performing methodologies that have been developed so far for automatic food recognition and volume estimation. Firstly, the paper presented the rationale of visual-based methods for food recognition. Then, the core of the study is the presentation, discussion, and evaluation of these methods based on popular food image databases. In this context, this study discusses the mobile applications that are implementing these methods for automatic food logging. Our findings indicate that around 66.7% of surveyed studies use visual features from deep neural networks for food recognition. Similarly, all surveyed studies employed a variant of convolutional neural networks (CNN) for ingredient recognition due to recent research interest. Finally, this survey ends with a discussion of potential applications of food image analysis, existing research gaps, and open issues of this research area. Learning from unlabeled image datasets in an unsupervised manner, catastrophic forgetting during continual learning, and improving model transparency using explainable AI are potential areas of interest for future studies.

Keywords: food recognition; feature extraction; automatic diet monitoring; image analysis; volume estimation; interactive segmentation; food datasets

Citation: Tahir, G.A.; Loo, C.K. A Comprehensive Survey of Image-Based Food Recognition and Volume Estimation Methods for Dietary Assessment. *Healthcare* **2021**, *9*, 1676. https://doi.org/10.3390/healthcare9121676

Academic Editor: Mahmudur Rahman

Received: 4 September 2021
Accepted: 23 November 2021
Published: 3 December 2021

1. Introduction

Despite recent advancements in medicine, the number of people affected by chronic diseases is still large [1]. This rate is primarily due to their unhealthy lifestyles and irregular eating patterns. As a result, obesity and weight issues are becoming increasingly common around the globe. Some of the more notable diseases caused by obesity include hypertension [2], blood sugar [3], cardiovascular diseases [4], and different kinds of cancers [5]. The main reported obesity issues are in developed and middle-income countries. In 2016, 1.9 billion adults 18 years and older were overweight, while 650 million were obese. With time, children are also becoming affected by obesity at an alarming rate. According to World Health Organization (WHO), over 340 million children and adolescents between 5 and 19 years were overweight or obese [6].

The prevalence of these alarming statistics poses a serious concern. However, determining the effective remedial measures depends on different factors, ranging from a person's genetics to their lifestyle choices. To cope with chronic weight problems, people often keep notes to track their dietary intake. In turn, dieticians require these records to estimate a patient's nutrient consumption. However, these methods pose a challenge for users and dieticians, especially when they have to record time and estimate nutrients of

diet intake [7]. For these reasons, recent research efforts have explored sophisticated vision-based methods to automate the process of food recognition and volume estimation [8,9]. The advancement in smartphone applications and hardware resources has made this more convenient, and present studies also show a higher retention rate of these mHealth apps than traditional methods [10]. Recent advancements in machine learning methods have further paved the way for more robust mHealth apps. Some dietary mobile applications such as DietLens [11], DietCam [12], Im2Calories [13], etc. integrate their apps with AI models for food recognition and ingredients detection to automate food logging. The Dietcam app also estimates nutrients from smartphone camera pictures.

However, automatic food recognition using a smartphone camera in the real world is considered a multi-dimensional problem, and the solution effectiveness depends upon several factors. Firstly, the model can achieve optimal classification performance by training with many food images for each class. Other than that, food recognition is a complex task that involves several domain-specific challenges. There is no spatial layout information that it can exploit like, in the case of the human body, the spatial relationship between body parts. The head is always present over the trunk of the human body [14–16] and feet towards the lower end. Similarly, the non-rigid structure of the food and intra-source variations make it even more complicated to classify food items correctly as preparation methods and cooking styles vary from region to region. Moreover, inter-class ambiguity is also a source of potential recognition problems as different food items may look very similar (e.g., soups). Moreover, in many dishes, some ingredients are concealed from view that can limit the performance of food ingredient classification models.

In addition to this, image quality from the smartphone camera is dependent on different types of cameras, lighting conditions, and orientations. As a result, the poor performance of food recognition models is highly susceptible to image distortions.

Despite these challenges, many food images possess distinctive properties to distinguish one food type from another. Firstly, the visual representations of food images are of fundamental importance as it significantly impacts classification performance. Therefore, many food-recognition methods employ handcrafted features such as shape, color, texture, and location. Recent techniques are using deep visual features for image representations. Some of these methods implement a combination of handcrafted and deep visual features for image feature representations. Secondly, for enhanced classification performance and reduced computational complexity, an appropriate selection of attributes is essential for removing redundant features from feature vectors. Finally, wisely selecting classification techniques is crucial to address food recognition challenges effectively.

Similarly, manual logging of food volume is a tedious task and involves a high rate of human error by as much as 30% [17–22]. Several solutions are proposed whose aim is to estimate food volume from smartphone camera pictures. Previous studies [23] show that using a mobile phone camera for food volume estimation increases the accuracy of the estimation of calories. Some methods involve capturing a single image, while multiple views are needed to determine accurate volume in other techniques. The food volume estimation process involves the following two steps (1) multiple images or a single image from a mobile camera is needed (2) computation of food volume from 3D construction or calibration object. Regardless of other volume estimation tasks, food volume estimation is a complex task with factors such as variations in shape and appearance due to various shapes of food and eating conditions affecting its performance.

The following research paper aims to scrutinize state-of-the-art vision-based approaches for dietary assessment to give researchers a summary of this area. Figure 1 represents the detailed scope and taxonomy of our survey study. The contribution of this survey is summarized as follows:

(1) The article briefly explores food databases for evaluating vision-based approaches and performance measures to thoroughly investigate food recognition, ingredient detection, and volume estimation methods.

(2) It presents an extensive review of food recognition techniques, including traditional methods with handcrafted features and modern deep-learning-based approaches.
(3) It provides deep insight into multi-label methods for food ingredient classification.
(4) This study surveyed most performing single-view and multi-view methods for food volume estimation.
(5) This study presents existing mobile applications that implement these approaches and other potential applications of vision-based methods in health care.
(6) The article analyzes open issues and suggests possible solutions to overcome the limitations of the existing methodologies.

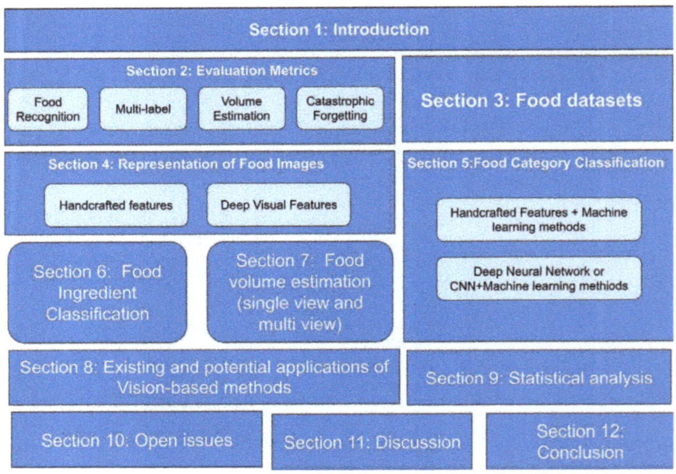

Figure 1. Scope and taxonomy of this survey paper.

It should be noted that the article is related to vision-based methods for food image analysis and their applications in the field of healthcare currently being discussed in the literature. However, the methodology of this article seeks to examine the systems more broadly by describing their important aspects similar to narrative overview [24] instead of a systematic review, some related works to the topic, or adopted search followed by a brief discussion.

Section 1 has presented the introduction of the study. The rest of the article is organized as follows. Sections 2 and 3 examine evaluation metrics and existing datasets. Section 4 examines feature extraction methods for food image representation including handcrafted and deep visual features. In Sections 5 and 6, we presented the most performing classifiers for food categorization and ingredient detection. Section 7 represents the food-volume-estimation methods. In Section 8, we provide brief information about mobile applications implementing these methods and other potential applications. Sections 9 and 10 summarize statistical analysis and open issues. To conclude, we highlight our findings and future works related to this topic.

2. Evaluation Metrics

2.1. Evaluation Metrics for Food Categorization

The performance of automatic food recognition models is highly dependent on the correct mapping of food images into their respective categories. Therefore, confusion-matrix and evaluation metrics play an essential role in determining the correctness of food recognition models. Several metrics have been discussed in the literature, and their appropriate selection depends on the requirements of specific applications. It has also been observed that a classifier may perform well under one metric but poorly under another metric. For example, in the context of an imbalanced food dataset, the data samples from

one or more classes outnumber data samples from the remaining food classes. Then a model trained on an imbalanced data set can have higher accuracy because of its good performance on the majority classes despite having bad classification performance on minority classes. Confusion matrix and other intrinsic metrics (Accuracy, Precision, Recall, and F1-score) generally used for detailed comparisons are discussed in detail below.

2.1.1. Confusion Matrix

Confusion matrices are a widely used approach to summarize the performance of a classification model in machine learning. In some cases, classification accuracy alone can be misleading, especially when there are more than two classes in a dataset or if there were an unequal number of observations present in food classes. Therefore, the confusion matrix provides a clear picture of actual and predicted classes obtained by the classification model. The confusion matrix is basically a two-dimensional matrix where each row represents an example of an actual food class and each column represents a state of the predicted food class. TP stands for true positive, TN represents the number of true negatives, FP is the number of false positives, and FN represents false negatives in the confusion matrix shown in Figure 2.

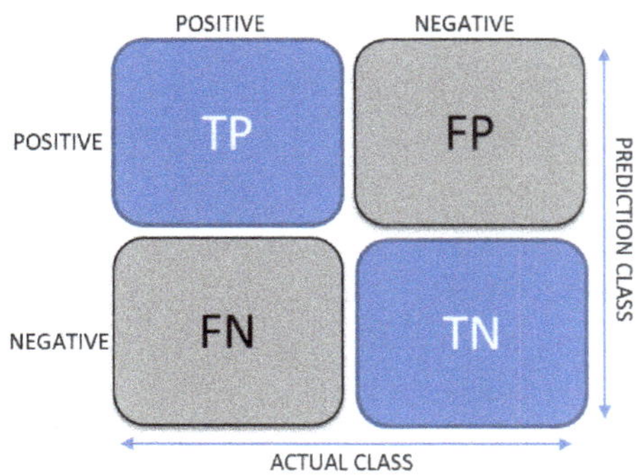

Figure 2. Confusion matrix.

2.1.2. Accuracy

The accuracy of a model determines whether the model is able to predict food classes correctly or how well a certain model can generally perform. Equation (1) represents the mathematical form of accuracy. However, accuracy cannot be used as a major performance metric, as it does not serve the purpose when there is an imbalanced dataset. Therefore, we have incorporated Precision, Recall, and F1 score to provide better insights into the results.

$$\text{Accuracy} = \frac{(TP + FN)}{(TP + FP + FN + TN)} \times 100 \tag{1}$$

Here TP refers to the true positive. True positive is an outcome where the model has correctly predicted a positive class. For example, in the case of food recognition, it refers to the food class that the model is trying to predict. TN refers to the true negatives: the prediction is correct, and the actual value is negative. In the case of food recognition, it refers to images from those food classes that the model is not trying to predict. FP refers to the false positive, and FP prediction results are wrong. For example, in the case of Food/NonFood recognition, FP refers to images that are non-food but are predicted as food. FN refers to the false negatives. It refers to those data samples which are positive but wrongly classified as negative class. For example, those food images that are classified as non-food images by model.

2.1.3. Precision

The Precision score can be defined as how often a model can correctly predict values classified as positives. In simpler words, out of all predicted positive food classes, it indicates what percentage is truly positive. This score is beneficial when the cost of false positives is high. It is calculated by Equation (2).

$$\text{Precision Score} = \frac{TP}{(TP + FP)} \tag{2}$$

2.1.4. Recall

Recall score identifies the model's ability to correctly classify food classes. It determines out of total positive food classes what percentage is predicted positives. It provides better insight when the cost of false negatives is high. It is computed by using Equation (3).

$$\text{Recall} = \frac{TP}{(TP + FN)} \tag{3}$$

2.1.5. F1 Score

F1 score represents the harmonic mean of recall and precision score. It considers both false positives and false negatives; therefore, it performs great on imbalanced datasets. It is calculated by following Equation (4).

$$\text{F1 Score} = \frac{(2 * (\text{Precision} * \text{Recall}))}{\text{Precision} + \text{Recall}} \tag{4}$$

2.2. Catastrophic Forgetting During Progressive Learning

Food datasets are open-ended due to the large variety of food dishes and different preparation styles. There are no limitations and constraints on the number of classes, and the model can progressively adapt domain variations in existing classes while learning new food classes. However, catastrophic forgetting during progressive learning causes the neural network to forget previous knowledge while learning new concepts. Catastrophic forgetting measures compute the algorithm's ability to retain previous concepts and knowledge while learning new information. Kemker et al. [25] and Chaudry et al. [26] proposed five measures of catastrophic forgetting to achieve this objective.

2.2.1. Intransigence

This refers to the difference in classification performance between the reference model trained by batch learning technique and the model trained on feature vectors using incremental learning protocol. The negative intransigence shows that incrementally learning a new set of food classes improves performance. Equation (5) denotes its mathematical form.

$$I_k = a_k^* - a_{k,k} \tag{5}$$

2.2.2. Forgetting

This refers to the difference between the highest classification performance of a particular session in previous sessions and its classification performance in the current sessions. Equation (6) computes the average forgetting of the network up to the kth session.

$$\begin{gathered} f_j^k = max_{1 \in \{1,\ldots\ldots,K-1\}}\, a_{i,j} - a_{(k,j)}, j > k \\ F_k = \tfrac{1}{k-1} \textstyle\sum_{j=1}^{k-1} f_j^k \end{gathered} \tag{6}$$

2.2.3. Base Session

This refers to the model's ability to retain the knowledge of base food classes in current sessions, as shown in Equation (7).

$$\Omega_{base} = \frac{1}{k-1}\sum\nolimits_{j=2}^{k} \frac{a_{j,1}}{a_{ideal}} \quad (7)$$

2.2.4. New Session

This is the ability of a model to recall newly learned food classes, as shown in Equation (8).

$$\Omega_{new} = \frac{1}{k-1}\sum\nolimits_{j=2}^{k} a_{j,j} \quad (8)$$

2.2.5. All Session

This refers to the retention of the previous food classes learned by the network when learning new food classes, as computed by Equation (9).

$$\Omega_{all} = \frac{1}{k-1}\sum\nolimits_{j=2}^{k} \frac{a_{j,all}}{a_{ideal}} \quad (9)$$

2.3. Evaluation Metrics for Food Ingredient Classification

Similarly, food ingredient recognition is equally important for dietary assessment applications. As food categorization is limited to the classification of generic food items present in the food images, food ingredient recognition and classification provide deep insights into the caloric content present in the food image. Therefore, food ingredient recognition applications widely incorporate multi-label classification [27]. Since food ingredient recognition is considered a multi-label problem as food images usually contain more than one ingredient. Therefore, evaluation metrics generally used for multi-label classification are different from traditional single-label classification. The following are the performance metrics are used by food ingredient recognition models.

Consider x_i, Y_i with L number of labels as training datasets. Let us assume that MLC is the training method and $Z_i = MLC(x_i)$ is the output labels (ingredients) predicted by the classification method.

2.3.1. Precision

Precision is the ratio of correctly predicted labels to the total number of actual labels, averaged across all instances. Equation (10) represents precision for food ingredient classification.

$$\text{Precision} = \frac{1}{N}\sum_{i=1}^{N}\left(\frac{MLC(x_i) \cap Y_i}{MLC(x_i)}\right) \quad (10)$$

2.3.2. Recall

Recall is computed by Equation (11). It is the ratio of correctly predicted labels to the total number of predicted labels.

$$\text{Recall} = \frac{1}{N}\sum_{i=1}^{N}\left(\frac{MLC(x_i) \cap Y_i}{MLC(Y_i)}\right) \quad (11)$$

2.3.3. F1 Score

Finally, F1 score is the harmonic mean of the precision and recall. Equation (12) represents the F1 score.

$$\text{F1 Score} = \frac{1}{N}\sum_{i=1}^{N}\left(\frac{2 * |MLC(x_i) \cap Y_i|}{|MLC(x_i)| + |Y_i|}\right) \quad (12)$$

2.4. Evaluation Metrics for Food Volume Estimation

Similarly, various studies related to food volume estimation use ground truth values to compare the accuracy of their proposed methods to determine the accurate food volume [28–39]. Unfortunately, there is no dataset available to date for accurate measurement of food volume. Nevertheless, the method proposed by [40] uses controlled experiments that require participants to click images before and after their meal to compute consumed calories, which are later compared with ground truth values. Similarly, Ref. [41] incorporated different food models to determine the true volume; however, various models failed to provide accurate information. Therefore, they implemented the water displacement method, which requires a mean of three readings to find out the true volume. Furthermore, most studies used the following equations to compute the relative error and estimate the accuracy of the method

$$e = |v - v_{approx}| \quad (13)$$

where v is the actual volume and v_{approx} is the approximate volume

$$e = \frac{1}{N}\sum_{i=1}^{n} \frac{|w_i - w_g|}{w_g} \quad (14)$$

where N is the number of food items, w_i is the estimated weight of the food item, and w_g is the ground truth value of the food.

3. Datasets Used for Food Recognition

Performance of feature extraction and classification techniques is highly dependent on the detail-oriented collection of images, which, in our case, happen to be food images. As consolidated large food image datasets, for example, UECFOOD-100, Food-101, UECFOOD-256, UNCIT-FD1200, and UNCIT-FD889 are eventually used as benchmarks to collate recognition performance of existing approaches with new classifiers. Such datasets can be distinctive in terms of characteristics, such as the total number of images in a particular dataset, cuisine type, and included food categories.

For instance, UECFOOD-100 contains 100 different sorts of food categories, and each food category has a bounding box that indicates the location of the food item in the photograph. Food categories in this dataset mainly belong to popular foods in Japan [42]. Similarly, UECFOOD-256 is another variant of UECFOOD-100. However, it differs in terms of the number of images as it contains 256 food images of different kinds [42]. Food-101 contains 101,000 real-world images that are classified into 101 food categories. It includes diverse yet visually similar food classes [43]. Similarly, the PFID food dataset is composed of 1098 food images from 61 different categories. The PFID collection currently has three instances of 101 fast foods [44]. UNCIT-FD1200 is composed of 4754 food images of 1200 types of dishes captured from actual meals. Each food plate is acquired multiple times, and the overall dataset presents both geometric and photometric variability. Similarly, UNICT-FD 889 dataset has 3583 images [45] of 889 different real food plates captured using mobile devices in uncontrolled scenarios (e.g., different backgrounds and light environmental conditions). Moreover, they capture each dish image in UNICT-FD899 multiple times to ensure geometric and photometric variability (changes in rotation, scale, and point of view) [46].

Several datasets mainly consist of various food images collected through various sources such as web crawlers and social media platforms such as Instagram, Flickr, and Facebook. Furthermore, most of these datasets contain images of foods that are specific to certain regions, such as Vireo-Food 172 [47] and ChineseFoodNet [48]. Both datasets contain Chinese dishes. Similarly, Food-50 [49], Food-85 [49], Food log [50], UECFOOD-100 [42], and UECFOOD-256 [43] contain Japanese Foods items. Turkish foods-15 [51] is limited to Turkish food items only. Furthermore, the Pakistani Food Dataset [52] accommodates Pakistani dishes, and the Indian Food Database incorporates Indian cuisines. In addition to

this, few datasets only include fruits and vegetables like VegFru [53], Fruits 360 Dataset [54], and FruitVeg-81 [55]. Furthermore, Table 1 provides a brief description about food image datasets. Figure 3 shows the system flow and Figure 4 shows the sample images from the food datasets.

Table 1. Food image datasets.

Authors	Year	Dataset	Food Category	Total # Images/Class	Image Source
S. Godwin et al. [56]	2006	Wedge Shape foods dataset	American Foods	3 categories	Controlled environment
Chen et al. [44]	2009	PFID	American Fast Foods	1038(61)	Fast food data captured in multiple restaurants
Mariappan et al. [57]	2009	TADA	Artificial And Generic Food	256(11)	Controlled environment
Yanai et al. [49]	2010	Food-50	Japanese Foods	5000(50)	Crawled from web
Hoashi et al. [49]	2010	Food-85	Japanese Foods	8500(85)	Existing food databases
Miyazaki et al. [29]	2011	Foodlog	Japanese Foods	6512(2000)	Captured by users
Marc Bosch et al. [58]	2011	FNDDS	American Foods	7000	Images of food accquired by users
Matsuda et al. [42]	2012	UECFOOD-100	Japanese Foods	14,361(100)	Captured by mobile camera
Chen et al. [48]	2012	ChineseFoodNet	Chinese dishes.	192,000(208)	Gathered from web
M.-Y. Chen et al. [48]	2012	Chen	Chinese Foods	5000/50	Crawled from the Internet
Bossard et al. [59]	2014	Food-101	American Foods	101,000(101)	Crawled from web
L. Bossard et al. [59]	2014	ETHZ Food-101	American Foods	100,000(101)	Crawled from web
Kawano et al. [43]	2014	UECFOOD-256	Japanese Foods	25,088(256)	Captured by mobile camera
T. Stutz et al. [60]	2014	Rice dataset	Generic (Rice)	1 food type	Acquired from user
Farinella et al. [46]	2014	UNCIT-FD889	Italian Foods	3583 (899)	Acquired with a smartphone
Meyers et al. [13]	2015	FOOD201-Segmented	American Foods	12625	Manually annotated dataset
Xin Wang et al. [61]	2015	UPMC Food-101	Generic	100,000(101)	Crawled from web
Cioccoa et al. [50]	2015	UNIMB 2015	Generic	2000(15)	Using a Samsung Galaxy S3 smartphone
Shaobo Fang et al. [62]	2015	TADA(19 foods)	American Foods	19 categories	Controlled environment
Xu et al. [63]	2015	Dishes	Chinese Restaurant Foods	117,504(3832)	Download from dianping
Beijbom et al. [64]	2015	Menu-Match	Generic Restaurant Food	646(41)	Captured from social media
Zhou et al. [65]	2016	Food-975	Chinese Foods	37,785(975)	Collected from restaurants
J. chen et al. [47]	2016	Vireo-Food 172	Chinese Foods	110,241(172)	Downloaded from web
Cioccoa et al. [66]	2016	UNIMB 2016	Italian Foods	1027(73)	Captured from dining tables
Hui Wu et al. [67]	2016	Food500	Generic	148,408(508)	Crawled from web
Singla et al. [68]	2016	Food-11	Generic	16,643(11)	Other food datasets
Farinella et al. [45]	2016	UNCIT-FD1200	Generic	4754(1200)	Acquired using smartphone
Jaclyn Rich et al. [69]	2016	Instagram 800k	Generic	808,964(43)	Social Media
Liang et al. [70]	2017	ECUSTFD	Generic	2978(19)	Acquired using smartphone
Güngör et al. [51]	2017	Turkish-Foods-15	Turkish Dishes	7500/15	Collected from other datasets
Pandey et al. [71]	2017	Indian Food Database	Indian Foods	5000(50)	Downloaded from web
Termritthikun et al. [72]	2017	THFood-50	Thai Foods	700/50	Downloaded from web
Ciocca et al. [73]	2017	FOOD524DB	Generic	247,636(524)	Existing food database
Hou et al. [53]	2017	VegFru	Generic (Fruit and VEG)	160,731(292)	Collected from search engine
Waltner et al. [55]	2017	FruitVeg-81	Generic (Fruit and VEG)	15,630(81)	Collected using mobile phone
Muresan et al. [54]	2018	Generic (Fruits 360 Dataset)	Fruit Dataset	71,125(103)	Camera
Qing Yu et al. [74]	2018	FLD-469	Japanese Foods	209,700(469)	Smart Phone camera
Kaur et al. [75]	2019	FoodX-251	Generic	158,000(251)	Collected from web
Ghalib et al. [52]	2020	Pakistani Food Dataset	Pakistani Dishes	4928(100)	Crawled from web
Narayanan et al. [76]		AI-Crowd	Swiss Foods	25,389	Volunteer Users
Bolaños M. et al. [77]	2016	EgocentricFood	Generic	5038(9)	Taken by a wearable egocentric vision camera
E. Aguilar et al. [78]	2019	MAFood-121	Spanish Foods	21,175	Google search engine

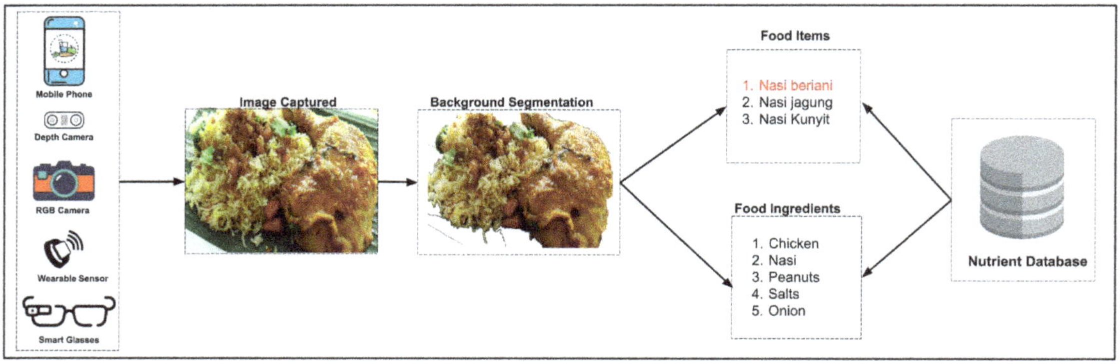

Figure 3. System Flow.

Figure 4. Sample images from few food datasets.

Therefore, it is evident from the survey that there is an immense need for broad and generic food datasets for better food recognition and enhanced performance. This necessity is because region-specific food items or datasets with fewer food categories can undermine the accuracy and performance of classification and extraction methods.

4. Representation of Food Images

Feature extraction plays a vital role in automated food recognition applications due to its noticeable impact on the recognition efficiency of an employed system. Feature extractors methods extract different food image representations. The process of feature extraction involves the identification of visual characteristics like color, shape, and texture. The main objective of feature extraction is to reduce dimensionality space [79] and extract more manageable groups from raw vectors of food images.

Moreover, selecting the right set of features ensures that relevant information is extracted from input images to perform the desired task. We categorized the feature

extraction techniques into two main types: hand-crafted and deep visual features. The term 'handcrafted' refers to identifying relevant feature vectors of appropriate objects such as shape, color, and texture. In contrast to that, the deep model provides state-of-the-art performance due to automatic feature extraction through a series of connected layers. For this reason, recent studies have adopted combinations of both hand-crafted and deep visual features for food image representation.

4.1. Handcrafted Features

The existing literature exhibits a large number of methods to employ manually designed or handcrafted features. Handcrafted features are properties obtained through algorithms using help from information available in the image. Figure 5 categorizes the handcrafted feature extraction methods. In the scenario of food image recognition, there is variation among different food types in terms of texture, shape, and color.

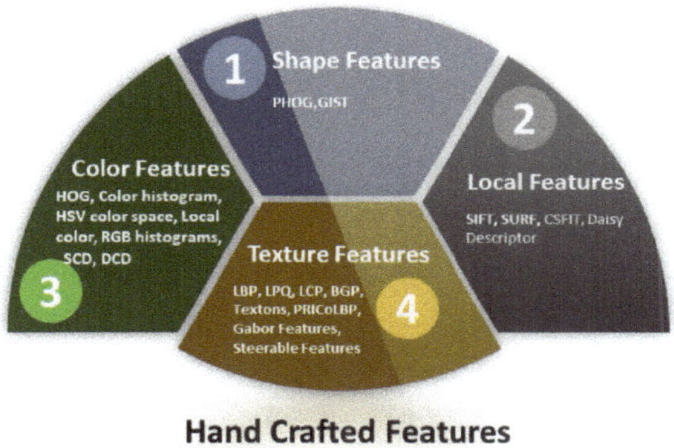

Figure 5. Handcrafted feature extraction methods.

The term 'texture' refers to homogeneous visual patterns that do not result from single colors such as sky and water [7]. Textural features usually consist of regularity, coarseness, and/or frequency. Texture-based characteristics are classified into two classes, namely statistical and transform-based models. Similarly, shape features attempt to quantify shape in ways that agree with human intuition or aid in perception based on relative proximity to well-known shapes. Based on the analysis, these shapes can be declared either perceptually similar to human perception or different. Furthermore, extracted features should remain consistent concerning rotation, location, and scaling (changing the object size) of an image. Unlike shape and texture features, color features are prevalent for image retrieval and classification because of their invariant properties concerning image translation, scaling, and rotation. The key items of the color feature-extraction process are color quantization and color space. Therefore, the resulting histogram is only discriminative when it projects the input image is to the appropriate color space. Different methods are widely employed for food classification, including hue, saturation, value (HSV); CIELab; red, green, and blue (RGB); normalized RGB; opponent color spaces; color k-means clustering; bag of color features; color patches; and color-based kernel. Although the color features from the food images distinguish between different food items, due to intra-class similarity, these features alone are not enough to accurately classify food images. For this reason, most researchers have used color features in combination with other feature extraction methods.

Hoashi et al. [49] employed bag of features, color histogram, Gabor features, and gradient histogram with multiple kernel learning for automatic food recognition of 85 different food categories. Similarly, Yang et al. [80] dealt with pairwise statistics between local features for food recognition purposes using the PFID dataset. For real-time food image recognition, Kawano and Yanai et al., 2014 [43] utilized handcrafted features such as color,

histogram of oriented gradient (HoG), and Fisher Vector (FV). Moreover, the cloud-based food recognition method proposed by Pouladzadeh et al., 2015 [81], involves features like color, texture, size, shape, and Gabor filter. They evaluated their framework on single food portions consisting of fruit and a single item of food. Furthermore, mobile food recognition systems proposed by Kawano and Yanai, 2013 [82], and Oliveira et al., 2014 [83], also used handcrafted features like color and texture. Table 2 summarizes the details of proposed methods that employ handcrafted features for food recognition.

However, identification of food involves challenges due to varying recipes and presentation styles used to prepare food all around the globe, resulting in different feature sets [84]. For instance, the shape and texture of a salad containing vegetables differ from the shape and texture of a salad containing fruits. For this reason, we should optimize the feature extraction process by extracting relevant visual information from food images. Such data are present in general information descriptors, which are a collection of visual descriptors that provide information about primary features like shape, color, texture, and so forth. Some important descriptors used in existing studies include Gabor Filter, Local Binary Patterns (LBP), Scale-invariant Feature Transform (SIFT), and color information to extract features of food images [85]. These descriptors can be applied individually or in combination with other descriptors for enhanced accuracy.

Table 2. Handcrafted features.

Reference	Year	Visual Features	Dataset	Recognition Type
Hoashi et al. [49]	2010	Bag-of-features (BoF), Color histogram, Gabor features, and gradient histogram with Multiple Kernel learning.	Used for recognition of 85 food categories	Automatic food recognition
Yang et al. [80]	2010	Deals with pair wise statistics between local features	Pittsburgh Food Image Dataset (PFID)	Food recognition
kong and Tan [86]	2011	SIFT, Guassian Region detector	Pittsburgh Food Image Dataset (PFID) and dataset consisting of food images collected from local restaurants.	Regular shaped foods recognition
Bosh et al. [85]	2011	Global feature classes: texture and color Local features: local entropy color, local color, Garbor filter, SIFT, Haar, Daisy descriptor, Steerable filters and Tamura perceptual filter	Database consisting of food images collected under controlled conditions, from nutritional studies conducted at Prudue University [58]	Food recognition and quantification
Zhang et al. [87]	2011	Color, SIFT, Shape, RGB histograms	Dataset came from online sources, which includes three types of cuisines, two dishes per cuisines were represented by 76 images	Classification of cuisines
Matsuda et al. [88]	2012	Gabor texture features, Histogram of Oriented Gradient (HoG), Bag-of-features of SIFT and CSFIT with Spatial pyramid.	Food image dataset containing 100 different food categories.	Multiple food images recognition
Kawano and Yanai [82]	2013	Bag-of-features and color histogram, HOG patch descriptor and color patch descriptor.	-	Mobile food recognition
Anthimopoulos et al. [89]	2014	Bag-of-features, SIFT and HSV color space	Visual dataset consisting of 5000 food images organized into 11 different classes	Food recognition system for diabetic patients
Tammachat and Pantuwong [90]	2014	Bag-of-features (BoF) , Texture and Color	Database consisting of 40 types of Thai food consisting of 100 images of each food type.	Food image recognition
Pouladzadeh et al. [91]	2014	Graph cut, Color and Texture	Dataset consisting of 15 different categories of fruits and food.	Food image recognition for calorie estimation
He et al. [92]	2014	Color, Texture, Dominant Color Descriptor (DCD), Scalable Color Descriptor (SCD), SIFT, Multi-scale Dense SIFT (MDSIFT), Entropy-Based Categorization and Fractal Dimension Estimation (EFD) and Gabor-Based Image Decomposition and Fractal Dimension Estimation (GFD)	Food image dataset containing 1453 images	Food image analysis

Table 2. *Cont.*

Reference	Year	Visual Features	Dataset	Recognition Type
Kawano and Yanai [43]	2014	Color, HoG and Fisher Vector	UECFOOD-256 food image dataset	Real-time food image recognition
Oliveira et al. [83]	2014	Color, Texture	Images were gathered using mobile's camera	Mobile Food Recognition
Pouladzadeh et al. [81]	2015	Color, Texture, Size, Shape, Gabor filter	System was tested on single food portions consisting of fruits and single piece of food. 100 images were chosen for training and 100 for testing purposes.	Cloud-based food recognition.
Farinella et al. [45]	2016	SIFT, Bag of Textons, PRICoLBP	UNICT-FD1200 dataset.	Food image recognition

Nonetheless, feature selection remains a complex task for food types that involve mixed and prepared foods. Such food items are difficult to identify and are not easily separable due to the proximity of ingredients in terms of color and texture features. In contrast, the evolution of deep learning methods has remarkably reduced the use of handcrafted features. This is due to their superior performance for both food categorization and ingredient detection tasks. However, handcrafted methods for feature extraction may still serve as the foundation for automated food recognition systems in the future.

4.2. Deep Visual Features

Recently, deep learning techniques have gained immense attention due to their superior performance for image recognition and classification. The deep learning approach is a sub-type of machine learning, and it trains more constructive neural networks. The vital operation of deep learning approaches includes automatic feature extraction through the sequence of connected layers leading up to a fully connected layer, which is eventually responsible for classification. Moreover, in contrast to conventional methods, deep learning techniques show outstanding performance while processing large datasets and have excellent classification potential [93,94].

Deep learning methods such as Convolutional Neural Networks (CNNs) [95], Deep Convolutional Neural Networks (DCNNs) [96], Inception-v3 [97], and Ensemble net are implemented by existing food recognition methods for feature extraction. Convolutional Neural Networks are one of the widely used deep learning techniques in the area of computer vision due to their impressive learning ability regarding visual data, and they achieve higher accuracy than other conventional techniques [98]. The DCNN technique gained popularity owing to its large-scale object recognition ability. It incorporates all major object recognition procedures such as feature extraction, coding, and learning. Therefore, DCNN is an adaptive approach for estimating adequate feature representation for datasets [99]. Similarly, Inception-v3 is also a new deep convolutional neural network technique introduced by Google. It is composed of small inception modules that are capable of producing very deep networks. As a result, this model has proved to have higher accuracy, decreased number of parameters, and computational cost in contrast to other existing models. Likewise, Ensemble Net is a deep CNN-based architecture and is a suitable method for extracting features. It is due to the outstanding performance of CNN feature descriptors as compared to handcrafted features.

Asymmetric multi-task CNN and spatial pyramid CNN [100] provides highly discriminative image representations. Jing et al. [47] proposed ARCH-D architecture for multi-class multilabel food recognition, and their model provides feature vectors for both food category and ingredient recognition. Although the feature vectors from multi-scale multi-view deep network [101] has a very high dimension, they were successful in achieving state-of-art performance. Ghalib et al. [52] proposed ARCIKELM for open-ended learning. They have employed InceptionResnetV2 for feature extraction due to their superior performance over

other deep feature extraction methods such as ResNet-50 and DenseNet201. Table 3 further provides a brief description of deep visual features.

Table 3. Deep visual features.

Reference	Year	Features	Dataset	Recognition Type
Kawano and Yanai, [102]	2014	Fisher Vector and DCNN	UECFOOD-100 and 100-class food Dataset	Food image recognition
Yanai and Kawano, [96]	2015	DCNN	UECFOOD-100 and UECFOOD- 256	Food image recognition
Christodoulidis et al. [103]	2015	CNN	Manually annotated dataset with 573 food items	Food recognition
Pouladzadeh et al. [104]	2016	Graphcut and DCNN	Database consisting of 10,000 high res images	Food recognition for calorie measurement
Hassannejad et al. [105]	2016	Inception	Food-101, UECFOOD-100 and UECFOOD-256	Food image recognition
Liu et al. [106]	2016	DCNN	Food-101, UECFOOD-256	Mobile food image recognition
Chen and Ngo, [47]	2016	Arch-D	Chinese Foods	Ingredient recognition and food categorization
Ciocca et al. [66]	2017	VGG	UNIMIB 2016	Food recognition
Termritthikun et al. [72]	2017	NU-InNet	THFOOD-50	Food recognition
Pandey et al. [71]	2017	AlexNet, GoogLeNet and ResNet	ETH Food-101 and Indian Food Image Database	Food Recognition
Liu et al. [107]	2018	GoogleNet	UECFOOD-100, UECFOOD-256 and Food-101	Food recognition for dietary assessment
McAllister et al. [108]	2018	ResNet-152, GoogLeNet	Food 5k, Food-11, RawFooT-DB and Food-101	Food recognition
Martinel et al. [109]	2018	WISeR	UECFOOD-100, UECFOOD-256 and Food-101	Food recognition
E. Aguilar et al. [110]	2018	AlexNet	UNIMIB2016	Automatic food tray analysis
S. Horiguchi et al. [111]	2018	GoogleNet	Built their own food dataset FoodLog	Food image recognition
Gianluigi Ciocca et al. [112]	2018	ResNet50	Food 475	Food image recognition and classification
B. Mandal et al. [113]	2019	SSGAN	ETH Food-101 and Indian Food Dataset	Food Recognition of Partially Labeled Data
G.Ciocca et al. [114]	2020	GoogleNet, Inception-v3, MobileNet-V2 and ResNet-50	Own dataset containing 20 different food categories of fruit and vegetables.	Food category recognition, Food state recognition
L. Jiang et al. [115]	2020	VGGNet	UECFOOD-100, UECFOOD-256 and introduced new dataset based on FOOD-101.	Food recognition and dietary assesment
C. Liu et al. [116]	2020	VGGNet, ResNet	Vireo-Food 172	Food ingredient recognition
H. Liang et al. [117]	2020		ChineseFoodNet and Vireo-Food 172	Chinese food recognition
H. Zhao et al. [118]	2020	VGGNet, ResNet and DenseNet	UECFOOD-256 and Food-101	Mobile food recognition
G. A. Tahir and C. K. Loo [52]	2020	ResNet-50, DenseNet201 and InceptionResNet-V2	Pakistani Food Dataset, UECFOOD-100, UECFOOD-256, FOOD-101 and PFID	Food recognition
C. S. Won [119]	2020	ResNet50	UECFOOD-256, Food-101 and Vireo-Food 172	Fine grained Food image recognition
Zhidong Shen et al. [120]	2020	Inception-v3, Inception-v4	Dataset was created including hundreds and thousands of images of several food categories	Food recognition and nutrition estimation

5. Food Category Classification

The primary requirement of any food recognition system is accurate identification and recognition of food components in the meal. Therefore, robust and precise food classification methods are crucial for several health-related applications such as automated dietary assessment, calorie estimation, and food journals. Image classification refers to a machine learning technique that associates a set of unspecified objects with a subset (class) learned by the classifier during the training phase. In the scenario of food image classification, food images are used as input data to train the classifier. Hence, an ideal classifier must recognize any food category explicitly included during the learning phase. The accuracy of a classifier mainly depends on the quantity and quality of images, as there are several variations in food images such as rotation, distortion, lightning distribution, and so forth. In this section, we discuss classification techniques used by traditional approaches

that use handcrafted features. Following that, we analyzed state-of-the-art deep learning models for food recognition.

5.1. Traditional Machine Learning Methods

Major classifiers used by several traditional approaches in the domain of food image recognition include Support Vector Machines (SVM) [49], Multiple Kernel Learning (MKL) [49] and K-Nearest Neighbor (KNN) [47]. It is due to their outstanding performance as compared to other classification methods.

The food recognition method proposed by [121] employs color, SIFT, and texture features to train the KNN classifier. In contrast to SVM, KNN achieved higher classification accuracy, i.e., 70%, whereas the accuracy of the SVM classifier was only 57%. Similarly, treatment of diabetic patients involves a daily insulin prandial dose to compensate for the effect of a meal, and its estimation is a complex task with carbohydrate counting being a key element. To assist patients in automating the process of counting CHO from images captured from a camera, Anthimopoulos et al. [89] applied a bag-of-features model using SIFT features. A linear SVM classifier trained on food images of 11 different food classes acquired a classification accuracy of 78%.

Chen et al. [48], employed a multi-class SVM classifier for the identification of 50 different classes of Chinese food. It includes 100 food images in each category. However, classification accuracy was only 62.7%. They further implemented a multi-class Adaboost algorithm and increased their classification accuracy up to 68.3%. Furthermore, Bejibom et al. [64] used LBP, color, SIFT, MR8, and HoG features to train an SVM image classifier. They evaluated their work on two different datasets and achieved a classification accuracy of 77.4% on the dataset presented by [48]; their classification accuracy was 51.2% when applied to the menu-matched dataset. Table 4 summarizes classifiers implemented by traditional classification methods along with their achieved classification accuracies.

Table 4. Traditional machine learning methods for food category classification.

Reference	Year	Classification Technique	Classification Accuracy	
			Top 1	**Top 5**
Hoashi et al. [49]	2010	Multiple Kernel Learning (MKL)	Own Food Dataset = 62.5%	N/A
Yang et al. [80]	2010	Support Vector Machine (SVM)	PFID = 78.0%	N/A
Kong and Tan [86]	2011	Multi-class SVM	PFID = 84%	N/A
Bosh et al. [85]	2011	Support Vector Machine (SVM)	Dataset collected = 86.1% using nutritional studies Conducted at Prudue University	N/A
Zhang et al. [87]	2011	SVM regression with RBF kernel	Own Food Dataset = 82.9%	N/A
Matsuda et al. [88]	2012	Multiple Kernel Learning (MKL) and Support Vector Machine (SVM)	Own food Dataset = 55.8%	N/A
Kawano and Yanai [82]	2013	Linear SVM and fast tookernel	N/A	81.6%
Anthimopoulos et al. [89]	2014	Linear SVM	Own Food Dataset = 78.0%	N/A
Tammachat and Pantuwong [90]	2014	Support Vector Machine (SVM)	Own Food Dataset = 70.0%	N/A
Pouladzadeh et al. [91]	2014	Support Vector Machine (SVM)	Own Food Dataset = 95%	N/A
He et al. [92]	2014	K-nearest Neighbors and Vocabulary Trees	Own Food Dataset = 64.5%	N/A
Kawano and Yanai [43]	2014	One-vs-rest	UECFOOD-256 = 50.1%	UECFOOD-256 = 74.4%
Oliveira et al. [83]	2014	Support Vector Machine (SVM)	Own Food Dataset Top 3 classification achieved between 84 and 100%	N/A
Pouladzadeh et al. [81]	2015	Cloud-based Support Vector Machine	Own Food Dataset = 94.5%	N/A
Farinella et al. [45]	2016	Support Vector Machine (SVM)	UNICT-FD1200 = 75.74%	UNICT-FD1200 = 85.68%

5.2. Deep Learning Models

Deep learning approaches have gained significant attention in the field of food recognition. This is due to their exceptional classification performance in comparison to traditional approaches [48,64]. convolutional neural network (CNN), deep convolutional neural network (DCNN), Ensemble Net, and Inception-v3 are some of the most prominent techniques used as existing methods for food image recognition purposes.

Yanai and Kawano [102] employed a deep convolutional neural network (DCNN) on three food datasets: Food-101, UECFOOD-256, and UECFOOD-100. They explored the effectiveness of pre-training and fine-tuning a DCNN model using 100 images from each food category obtained from each dataset. During evaluation, classification accuracy achieved was 78.77% for UECFOOD-100, 67.57% for UECFOOD-256, and 70.4% for Food-101. Similarly, the study presented by [105] implemented Inception-v3 deep network established by Google [97] on the same datasets, i.e., Food-101, UEC FOOD-100, and UECFOOD-256. Classification accuracy achieved using fine-tuned model V3 was greater than classification accuracy of the fine-tuned version of DCNN i.e., 88.28%, 81.45%, and 76.17% for UECFOOD-100, UECFOOD-256, and Food-101, respectively. The food recognition method proposed by [106] implemented a CNN-based approach using the Inception model on the same three datasets.

Classification accuracy achieved was 77.4%, 76.3% and 54.7% for UECFOOD-100, UECFOOD-256 and Food-101, respectively. Table 5 provides the overview of existing food recognition methods based on deep learning approaches and their classification performance.

Table 5. Deep learning models for food category classification.

Reference	Year	Classification Technique	Classification Performance	
			Top 1	Top 5
Yanai and Kawano [96]	2015	DCNN	UECFOOD-100 = 78.8% UECFOOD-256 = 67.6%	N/A
Christodoulidis et al. [103]	2015	DCNN	Own dataset = 84.9%	N/A
Chen and Ngo [47]	2016	DCNN		
Pouladzadeh et al. [104]	2016	DCNN + Graph cut	Own dataset = 99%	N/A
Hassannejad et al. [105]	2016	DCNN	ETH Food-101 = 88.3% UECFOOD-100 = 81.5% UECFOOD-256 = 76.2%	ETH Food-101 = 96.9% UECFOOD-100 = 97.3% UECFOOD-256 = 92.6%
Liu et al. [106]	2016	CNN	UECFOOD-100 = 76.3% Food-101 = 77.4%	UECFOOD-100 = 94.6% Food-101 = 93.7%
Pandey et al. [71]	2017	Ensemble Net	ETH-Food101 = 72.1% Indian Food = 73.5% Database	ETH-Food101 = 91.6% Indian Food = 94.4% Database
Ciocca et al. [66]	2017	CNN	UNIMIB 2016 = 78.3%	N/A
Termritthikun et al. [72]	2017	CNN	THFOOD-50 = 69.8%	THFOOD-50 = 92.3%
McAllister et al. [108]	2018	CNN+ANN+SVM+ Random Forest	Food-5K = 99.4% Food-11 = 91.3% RawFooT-DB = 99.3% Food-101 = 65.0%	N/A
Liu et al. [107]	2018	DCNN	UECFOOD-256 = 54.5% UECFOOD-100 = 77.5% Food 101 = 77.0%	UECFOOD-256 = 81.8% UECFOOD-100 = 95.2% Food 101 = 94.0%
Martinel et al. [109]	2018	DNN	UECFOOD-100 = 89.6% UECFOOD-256 = 83.2% Food-101 = 90.3%	UECFOOD-100 = 99.2% UECFOOD-256 = 95.5% Food-101 = 98.7%
E. Aguilar et al. [110]	2018	CNN+SVM	UNIMIB 2016 = 90.0%	N/A
Gianluigi Ciocca et al. [112]	2018	CNN	Food-475 = 81.6%	Food-475 = 95.5%

Table 5. *Cont.*

Reference	Year	Classification Technique	Classification Performance	
			Top 1	Top 5
S. Horiguchi et al. [111]	2018	Sequential Personalized Classifier (SPC) with fixed-class and incremental classification	FoodLog = 40.2% (t251-t300)	FoodLog = 56.5% (t251-t300)
B. Mandal et al. [113]	2019	Generative Adversarial Network	ETH Food-101 = 75.3% IndianFood Database = 85.3%	ETH Food-101 = 93.3% Indian Food Database = 95.6%
Aguilar-Torres et al. [122]	2019	CNN based on ResNet-50	MAFood-121 = 81.62%	N/A
Kaiz Merchant and Yash Pande [123]	2019	Inception V3	ETHZ Food-101 = 70.0%	N/A
Mezgec, S. et al. [124]	2019	Deep Learning	Own Food dataset = 93%	N/A
L. Jiang et al. [115]	2020	DCNN (Faster R-CNN)	FOOD20-with-bbx = 71.7%	FOOD20-with-bbx = 93.1%
C. Liu et al., 2020 [116]				
H. Zhao et al. [118]	2020	JDNet	UECFOOD-256 = 84.0% FOOD-101 = 91.2%	UECFOOD-256 = 96.2% FOOD-101 = 98.8%
G. A. Tahir and C. K. Loo [52]	2020	Adaptive Reduced Class Incremental Kernel Extreme Learning Machine (ARCIKELM)	Food-101 = 87.3% UECFOOD-100 = 88.7% UECFOOD-256= 76.51% PFID = 100% Pakistani Food = 74.8%	N/A
C. S. Won [119]	2020	Three-scale CNN	UECFOOD-256 = 74.1% Food 101 = 88.8% Vireo-Food 172 = 91.3%	UECFOOD-256 = 93.2% Food-101 = 98.1% Vireo-Food 172 = 98.9%
Zhidong Shen et al. [120]	2020	CNN	Own dataset = 85.0%	N/A
Jiangpeng He et al. [125]	2020	18 layer ResNet	Own dataset = 88.67%	N/A
Eduardo Aguilar et al. [126]	2020	CNN	Own dataset = 88.67%	N/A
Dario Ortega Anderez et al. [127]	2020	CNN	Own dataset = 97.10%	N/A
G. Song et al. [128]	2020	CNN	Web crawled dataset = 56.47%	Web crawled dataset = 60.33
Limei Xiao et al. [129]	2021	CNN	Own dataset = 97.42%	N/A
Lixi Deng et al. [130]	2021	ResNet-50	School lunch dataset = 95.3%	N/A

6. Food Ingredient Classification

Over the past few years, nutritional awareness among people has increased due to their intolerance towards certain types of food, mild or severe obesity problems, or simply interest in maintaining a healthy diet. This rise in nutritional awareness has also caused a shift in the technological domain, as several mobile applications facilitate people in keeping track of their diet. However, such applications hardly offer features for automated food ingredient recognition.

For this purpose, several proposed models use multi-label learning for food ingredient recognition. It can be defined [27] as the prediction of more than one output category for each input sample. Therefore, food ingredient recognition is known as a multi-label learning problem. Marc Bolanos et al. have deployed CNN as a multi-label predictor to discover recipes in terms of the list of ingredients from food images [131]. Similarly, Yunan Wang et al. [132] used multi-label learning for mixed dish recognition, as they have no distinctive boundaries among them. Therefore, labeling bounding boxes for each dish is a challenging task. Another system proposed by Amaia Salvador et al. [133] regenerates recipes from provided food images along with cooking instructions. On the other hand, Jingjing Chen and Chong-Wah Ngo [47] proposed deep architectures for food ingredient recognition and food categorization and evaluated their proposed system on a large Chinese food dataset with highly complex food images. Food ingredient recognition is often overlooked and is a challenging task, as it requires training samples under different cooking and cutting methods for robust recognition. Therefore, methods proposed by Chen et al. [134] and J. Chen et al. [135] focus on food ingredient recognition. The authors Chen et al. [134] deploy multi-relational graph convolutional network that was later evaluated

on Chinese and Japanese food datasets, resulting in 36.7% for UECFOOD-100 and 48.8% for VireoFood-172. However, Chen et al. [135] proposed DCNN based method for food ingredient recognition and achieved Top 1 accuracy up to 86.91% and Top 5 accuracy up to 97.59% for Vireo Food-251.

Furthermore, Table 6 provides brief information about accuracy scores of proposed systems along with methods and dataset used.

Table 6. Proposed methods for food ingredient classification.

Reference	Year	Dataset	Method	Recall	Precision	F1
Chen et al. [47]	2016	Vireo-Food 172	Arch-D (Multi-task)	-	-	67.17% (Micro-F1) 47.18% (Macro-F1)
		UECFOOD-100	Arch-D (Multi-task)	-	-	82.06% (Micro-F1) 95.88% (Macro-F1)
Bolaños et al. [131]	2017	Food-101	ResNet50+ Ingredients 101	73.45%	88.11%	80.11%
		Recipe 5k	ResNet50+ Recipe 5k	19.57%	38.93%	26.05%
		Recipe 5k	Inception-v3+ Recipe 5k (Simplified)	42.77%	53.43%	47.51%
Wang, Yunan, et al. [132]	2019	Economic Rice	Inception-V4 + NS (multi-scale)	71.90%	72.10%	71.40%
		Economic Behoon	Inception-V4 + NS (multi-scale)	77.60%	68.50%	69.70%
Salvador, Amaia, et al. [133]	2019	Recipe 1M	CNN Auto-Encoder	75.47%	77.13%	48.61%
J. Chen et al. [135]	2021	VireoFood-172	DCNN	-	-	75.77% (Micro-F1)

7. Food Volume Estimation

Automated food volume assessment is a convoluted task involving various challenges. Highly diverse and varying compositions of food, increasing varieties of ingredients, and different methods of preparations are only some of the factors that need to be taken into consideration. Furthermore, the quality of pictures taken for food volume estimation also impacts the accuracy. Clear pictures taken in good lighting conditions would yield different results compared to low-resolution or low-light images. Thus, far, several methods have been proposed for accurate estimation of food volume ranging from simple techniques such as pixel counting to complex methods such as 3D image reconstruction. They have been broadly categorized as either 'single image view' or 'multi-image/video view' methods in the subsequent sections. Figure 6 shows the types of food volume estimation methods.

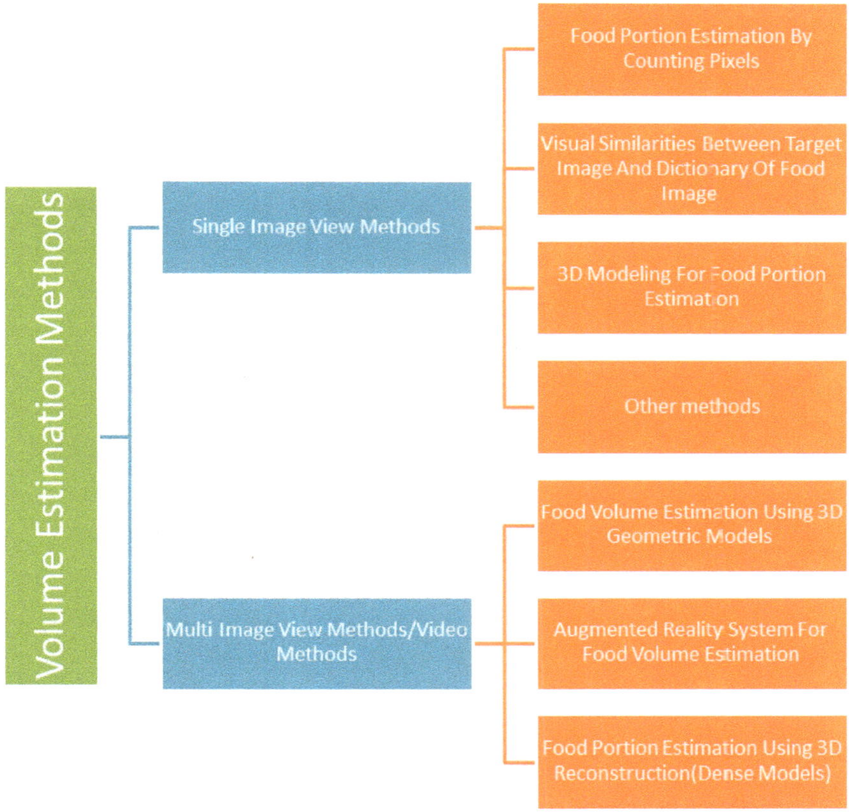

Figure 6. Food Volume Estimation Methods.

7.1. Single Image View Methods

Single-Image-View Methods for food volume estimation require only a single image for food volume estimation. These methods are relatively more user-friendly than 'multi-image view methods' because they do not require multiple images from different viewpoints. However, as a trade-off, most of the single-view methods are less accurate in contrast to multi-view methods. Table 7 summarizes single view methods for volume estimation. The following are a few common methods that use the single-view method for food portion estimation:

Table 7. Comparison of single-view methods for food volume estimation.

Reference	Year	Dataset	Results (E: Error%)	Technique
S. Fang [62]	2015	19 food items	E: <6%	3D parameters and reference objects to compute density for estimating the weight of food item
Y. He [36]	2013	1453 food images	E: 11% (beverages) 63%	"Integrated image segmentation and identification system"
T. Miyazaki [29]	2011	6512 images	E: 40%	Linear estimation
Beijbom, O [64]	2015	646 images, with 1386 tagged food items across 41 categories	E: 232 ± 7.2	Restaurant-specific food recognition considers meal as a whole entry with all of its nutrients details in DB to solve the volume estimation problem for the restaurant scenario.
Koichi Okamoto [31]	2016	20 kinds of Japanese Foods (60 test image)	E: 21.30%	Single-image-based food calorie estimation system which uses reference objects to determine food region and quadratic curve estimation from the 2D size of foods to their calories
Pettitt, C [136]	2016	Test data from N:6 participants who completed food diary during pilot sudy by wear micro camera	E: 34%	Wearable micro camera in conjunction with food dairies
Akpa Akpro Hippocrate [34]	2016	119 food images	E: 6.87%	Image processing with cutlery
Jia, W. Y [35]	2012	224 pictures	E: <10%	3D location of a circular feature from a 2D image
Yang, Y. Q [33]	2011	72 images	E: −3.55%	Single digital image, plate reference
Huang, J [39]	2015	fruits (n:6)		imaging processing
Yue, Y [41]	2012	6 food replicas	E: Length (−1.18)	A mathematical model based system involves a camera, circular object in a 3D space to compute food volume.
Zhang, W [38]	2015	15 different kinds of foods	85%	Portion estimation by counting pixels
Rob Comber [137]	2016	6 different meals	"Beef (E: −13.89 g, σ: 5.10 g), scrambled egg (E: −9.11 g, σ: 8.29 g), Jam sponge (E: −12.31 g, σ: 7.03 g) and fish pie (E: −12.59 g, σ: 5.74 g). Mean: −9.58"	Visual Assessment
S. Fang [30]	2016	10 objects		"3D geometric models and depth images."
Godwin, S. [56]	2006	Five portions of 9-inch cake, Seven portions of pizza, Pies were 9 or 10 inches	E: 25%	Estimated portion sizes using a ruler and the adjustable wedge
Hernández, Teresita [37]	2006	101 subjects, 5 foods	E: 4.8% ± 1.8%	Digital photographs printed onto a poster.
Yang et al. [138]	2021	Virtual Food Dataset and Real Food Dataset (RFD) (1500 images)	E: <9% on VFD, E: 11.6% and 20.1% on RFD.	Estimates volume by computing inner product between the probability vector from modified MobileNetV2 and the reference volume vector.
Graikos et al. [139]	2021	EPIC-KITCHENS and their own food video datasets	46.32% average MAPE on 16 test foods and 36.90% average MAPE on 6 combined meals.	Generate 3-dimensional point cloud by using depth map, segmentation mask and camera parameters. It then approximates the volume with points cloud-to-volume algorithm.
Lo, F.P.W et al. [140]	2019	Test dataset: 11 food items	E: 15.32%.	3D point cloud completion from RGB and depth images.

7.1.1. Food Portion Estimation by Counting Pixels

This method utilizes pixel count in each relevant image section to estimate food portion size. Studies [120] show that these methods are less complex than methods that rely on 3D modeling. Despite its simplicity, it gives a good estimation of portion size, thus making calculation of caloric content and nutritional facts easier.

7.1.2. Visual Similarities between Target Image and Dictionary of Food Images

This method estimates visual similarities between a given image and an existing food image dictionary. It is used by many existing systems today [29], where the caloric and nutrient contents in the food image dictionary are defined by dietary professionals to get a better approximation. The method selects first 'n' images from the dictionary and calculates the calorie content of the target image based on the average calorie content of dictionary images.

7.1.3. 3D Modeling for Food Portion Estimation

This method projects a 3D model of food portions onto 2D space or uses 3D geometric models for volume estimation. Generally, this method gives finer approximation in contrast to the other methods for single-image-view methods.

7.1.4. Other Methods

Other methods for food-portion estimation include estimating portion sizes using a ruler and adjustable wedge [56], mobile augmented reality, virtual reality [33], visual assessment [137] feature extraction, and its matching [29,64].

7.2. Multi-Image View or Video Methods

Multi-Image view or video methods require multiple images for food portion estimation. They are relatively more accurate than single-view-image methods. However, multi-image methods are less user-friendly as they require multiple images from different viewpoints in order to provide better results. Table 8 summarizes single-view methods for volume estimation. The following are a few methods that use multi-image-view techniques for food volume estimation.

Table 8. Comparison of multi-view methods for food volume estimation.

Reference	Year	Dataset	Results (E: Error%)	Technique
F. Zhu [141]	2010	3000 images	E: 1% 19 food items (97.2%)	"Camera calibration step and a 3D volume reconstruction step"
Xu Chang [141]	2013	14 to 20 images for multi-view method	E: 7.4% to 57.3%	Multi-view volume estimation using "Shape from Silhouettes" to estimate the food portion size
Kong, Fanyu [12]	2015	6 food items	84–91%	Multi-View RGB images for 3D reconstruction to estimate the volume
Trevno, Roberto [142]	2015	120 students (n = 120 meals; 57 breakfast + 63 lunch)	74% (reliability)	Digital Food Imaging Analysis (DFIA)
Jia, W. Y [143]	2014	100 food samples	E: −2.80% 30%	ebutton is used for taking pictures, and then portion size is calculated semi-automatically by using computer software
Xu, C [36]	2013		E: 10%	3D MODELLING AND POSE ESTIMATION
Rhyner, D [144]	2016	6 meals	85.10%	Multi-View RGB images, reference card and 3D model for volume estimation
T. Stutz [60]	2014	Rice, blinded servings	E: <33%	Mobile Augmented Reality System
Makhsous et al. [145]	2020	8 food items tested	40% improvement in the accuracy of volume estimation as compared to manual calculation.	Employs a mobile Structured Light System (SLS) to measure the food volume and portion size of a dietary intake.
Yuan et al. [146]	2021	Test dataset: 6 food items	E: 0.83 5.23%.	3D reconstruction from multi-view RGB images.
Lo, F.P.W et al. [140]	2019	Test dataset: 11 food items	E: 15.32%.	3D point cloud completion from RGB and depth images.

7.2.1. Food Volume Estimation Using 3D Geometric Models

This multi-image-view method uses a shape template method or 3D modeling for portion size estimation. As a single shape template is not suitable for all food types, the use of geometric models with correct food classification labels and segmentation masks in the image is important to index food labels to their respective classes of predefined geometric models. These can be used later for finding correct parameters of the selected geometric model [28,40,41,56,62].

Moreover, in 3D modeling and pose estimation, models for food are constructed in advance by using between 15 and 20 food images captured from several angles or a video sequence. Finally, food volume is estimated by registering pose from 3D models to 2D images [36].

7.2.2. Augmented Reality System for Food Volume Estimation

The use of augmented reality is also being widely used by researchers to estimate food portion size. Many systems such as Eat AR make use of it for portion size estimation [60] by developing prototypes to aid users. These prototypes generally require fiducial markers or credit-card-sized objects for overlaying 3D forms. Finally, the volume of the overlaid forms is computed using a signed volume estimation algorithm for closed 3D objects.

Similarly, the 'Serv Ar' augmented reality tool is used to provide guidance about food serving size [147]. Many of these technologies are being used with object recognition methods to identify food items and determine their caloric content. Similarly, methods that use augmented reality in combination with other portion estimation techniques have enhanced accuracy and much more interactive interfaces, resulting in a high retention rate.

7.2.3. Food Portion Estimation Using 3D Reconstruction (Dense Models)

Portion estimation by constructing dense 3D models usually requires multiple images or a video segment [139]. Joachim Dehais et al. [148] have shown the use of two views for volume estimation using 3D construction. In its first stage, the system learns about the configuration of different views, followed by the construction of a dense 3D model to extract the volume of each individual food item placed before it. Similarly, Wen Wu et al. [32] studied the use of fast food videos for caloric estimation. Most of these methods require images from different viewpoints, and for this reason, more advanced methods such as 3D construction from accidental motion can be explored for food volume estimation in the future.

7.3. Strengths and Weakness of the Food Volume Estimation Methods

Automatic food volume estimation method helps people to monitor their dietary intake suffering from chronic diseases without any expert intervention. It gives a quick result as compared to the traditional method which generally involves sending food images to the dietitian. The traditional method involves continuous involvement of dietitians, which makes it unworkable for dietitians to immediately respond to a large number of patients. Conversely, automatic food volume estimation is not standardized, as there are no existing guidelines by experts that refer to the error rate of these applications. Furthermore, different volume estimation methods vary in terms of accuracy and usability. Most of these methods are classified into two categories: single-image-view method and multiple-image-view method. Single-view-image methods are more user friendly, but their accuracy is compromised compared to multiple image view methods as it requires images from different. Therefore, standard guidelines are required for food volume estimation, which should include criteria for a balanced trade between features such as usability and accuracy, and developed applications must be verified according to the standard guidelines. Figure 7 summarizes the strengths and weaknesses of food volume estimation methods.

Automatic Food Estimation Methods	
Strengths	**Weaknesses**
Helps people to monitor dietary intake without expert's intervention.	Methods used are not standardized.
Gives quick results.	No existing guidelines from experts to depict error rate of such applications.
Immediate response to large number of patients.	Different food volume estimation methods varies in terms of accuracy and usability.
	Needs standard guidelines for a balanced trade between features like usability and accuracy

Figure 7. Strengths and weaknesses of automatic food estimation methods.

8. Existing and Potential Applications of Vision-Based Methods for Food Recognition in Healthcare

We summarized the core applications of vision-based methods for food recognition in the context of public policy and health care.

8.1. mHealth Apps for Dietary Assessment

Today, several mobile applications have been developed to monitor diet and help users to choose healthier alternatives regarding food consumption. Initially, these mobile applications were dependent on manually inputting food items by selecting from limited food databases. Therefore, such applications were not very reliable as they were prone to inaccuracies in dietary assessment, mainly extending from limited exposure to numerous food categories. With the advancement in the area of food image recognition, a large number of mHealth applications for dietary assessment use images to recognize food categories. For this purpose, existing mobile applications use different combinations of traditional and deep visual feature extraction, and classification methods for food recognition described earlier in Sections 3 and 4. Aizawa et al. [149] developed a mobile app food log, which uses traditional feature-extraction methods such as color, Bag of Features, and SIFT and uses an Adaboost classifier for classification purposes. Similarly, Ravi et al. [150] proposed the 'FoodCam' application, which uses traditional methods for feature extraction (LBP and RGB color features) and SVM for classification. Alternatively, Meyers et al. [13] employed a deep visual technique (GoogleNet CNN model) for feature extraction and classification purposes. Similarly, the Food Tracker app proposed by Jiang et al. [151] uses a deep convolutional neural network for feature extraction and classification. Furthermore, G. A. Tahir and C. K. Loo [52] utilized deep visual methods such as ResNet-50, DenseNet201, and InceptionResNet-V2 for feature extraction and Adaptive Reduced Class Incremental Kernel Extreme Learning Machine (ARCIKELM) as a classification method for their mobile application "My Diet Cam". Table 9 summarizes existing mobile applications in terms of feature extraction and classification methods used. Based on these deep visual method combinations, food recognition accuracies differ for various existing mobile applications. Therefore, apps with higher food recognition and classification accuracies gain more popularity. These apps tend to ease the dietary assessment process. Figure 8 shows the mobile application by Ravi et al. [150].

Table 9. Summary of feature extraction and classification methods used by existing mobile applications.

Reference	Year	Application Name	Food Segmentation	Feature Extraction Method	Classification Method
Aizawa et al. [149]	2013	FoodLog	No	Color, SIFT and Bag of Features	Adaboost Classifier
Oliveira et al. [83]	2014	-	Yes	Color and Texture	Support Vector Machine (SVM)
Probst et al. [152]	2015	-	-	SIFT, LBP and Color	Linear SVM
Meyers et al. [13]	2015	Im2Calories	Yes	GoogleNet CNN	GoogleNet CNN model
Ravi et al. [150]	2015	FoodCam	No	HoG, LBP and RGB Color Features	Linear SVM
Waltner et al. [55]	2017	-	Yes	RGB, HSV and LAB Color values	Random Forest Classifier
Mezgec and Seljak [153]	2017	-	-	NutriNet	NutriNet
Pouladzadeh et al. [154]	2017	-	Yes	CNN	Caffe Framework
Waltner et al. [155]	2017	-	Yes	CNN	CNN
Ming et al. [11]	2018	DietLens	-	ResNet-50 CNN	ResNet-50 CNN
Jiang et al. [151]	2018	-	Yes	Colors, Lines, Points, SIFTand Texture Features	Reverse Image Search (RIS) and Text Mining
Jianing Sun et al. [156]	2019	Food Tracker	Yes	DCNN	DCNN
G. A. Tahir and C.K. Loo [52]	2020	MyDietCam	Yes	ResNet-50, DenseNet201 and Inception ResNet-V2	Adaptive Reduced Class Incremental Kernel Extreme Learning Machine (ARCIKELM)

8.2. Harnessing Vision-Based Method to Measure Nutrient Intake during COVID-19

As the COVID-19 is a leading global challenge across the world, maintaining good nutritional status is mandatory for keeping good health to fight against the virus. Automatic vision-based methods for volume estimation and food image recognition in these nutrition tracking apps can assist patients in objectively measuring the nutrient intake of vital vitamins required for boosting the immune system.

8.3. Life's Simple 7

Life's Simple 7 health score is recently introduced based on modifiable health factors that contribute to heart health. Physical activity, non-smoking status, healthy diet, and body mass index are four modifiable health behaviors in this score. The other three modifiable factors are biological. They include blood pressure, fasting glucose, and cholesterol details. Besides cardiovascular health, Life's Simple 7 also relates to other health conditions such as venous thromboembolism, cognitive health, atherosclerosis, etc. As dietary intake plays a vital role in computing Life's Simple 7, manually measuring these factors and then calculating a Life's Simple 7 score is a very tedious process. This makes it very difficult for both middle-aged patients and elderly patients to keep track of their health. So vision-based methods can play an important role in automating the diet score. However, there are no current studies that have explored this research direction.

Figure 8. The application provides the top prediction result. This picture is taken from the study of Ghalib et al., 2020 (permission has been obtained from original author).

8.4. Enforcing Eating Ban on Public Places during COVID-19 Pandemic or Other Restricted Places

Vision-based food recognition can automate the enforcement of an eating ban at public places by automatically detecting foods from CCTV and wearable cameras to curb

the spread of the virus. Similarly, vision-based food recognition coupled with CCTV or wearable cameras and smart apps automate the enforcement of eating bans at workplaces, laboratories, etc.

8.5. Monitoring Malnutrition in Low-Income Countries

Coupling vision-based methods with wearable cameras can automatically detect foods from egocentric images with reasonable accuracy while reducing the burden of processing big data and addressing the user's privacy concerns. Egocentric images acquired from these cameras are important to study diet and lifestyle, especially in low-income countries with a high malnutrition rate. For example, Jia et al. [157] focused on gathering image data from wearable cameras and discriminating between food/non-food classes based on their tag from the CNN to study human diets. Similarly, Chen et al. [158] studied malnutrition in low- and middle-income countries by using the wearable device e-button.

8.6. Food Image Analysis from Social Media

We are in the era of social media, and food is a basic necessity of life, a great deal of content on social media platforms is related to food items. User's of these platforms frequently share new recipes, new methods of cooking, food pictures after restaurant check-in. Researchers have exploited this data on social media platforms for analyzing dietary intake. For example, Mejova et al. [159] studied food images from foursquare and Instagram to analyze the food consumption pattern in the USA. Similarly, food images on social media platforms are of different cultures. These images can be crawled and then combined together to prepare a large food database.

8.7. Food Quality Assessment

Evaluating fruit quality and freshness at the marketplace and at the user end is of increasing interest as opposed to accessing quality at the time of manufacturing. Efforts to date have focused on accessing the quality of foods using vision-based methods. For example, Ismail et al. have contributed an Apple-NDDA dataset [160] that consists of defective and non-defective apple images for food quality assessment.

9. Statistical Analysis

We provide a statistical analysis of our study based on the articles and conference proceedings gathered to write this survey paper. We surveyed research studies up to 2020 from various reputed sources: IEEE, Elsevier, ACM, and Web of Sciences. Figure 9 shows a pie chart of the distribution of surveyed food databases according to the country to which the food dishes belong. In it, generic databases are those that contain food dishes of multiple countries. We summarized the surveyed studies in two main categories: studies using handcrafted features, and studies using visual feature representation from convolutional neural networks (CNN), as shown in Figure 10. As discussed in Section 7, volume estimation methods require a single view or multiple images from different viewpoints. We presented a pie chart as shown in Figure 11 that describes the percentage of studies we surveyed according to the number of image viewpoints required to estimate food volume. For ingredient detection, all included studies used CNN due to recent interest in this extension. Similarly, for studies that have implemented mobile applications, the piechart in Figure 12 shows that 46.2% of applications implement CNN for food recognition while remaining mobile applications from surveyed studies are implementing traditional methods for feature extraction.

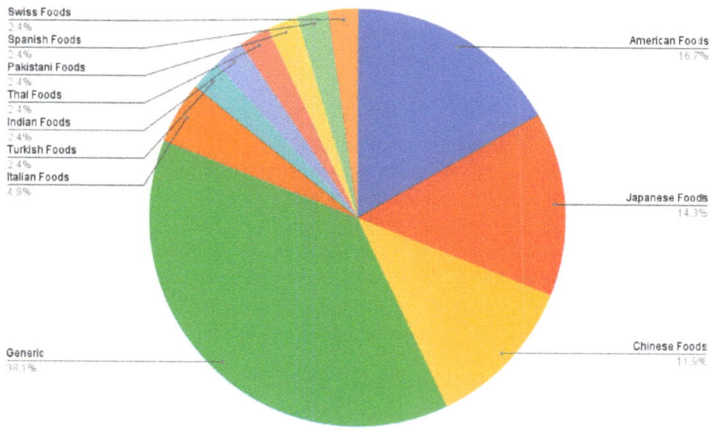

Figure 9. Percentage of datasets summarized according to the types of food. Generic refers to the multi-cultural dataset.

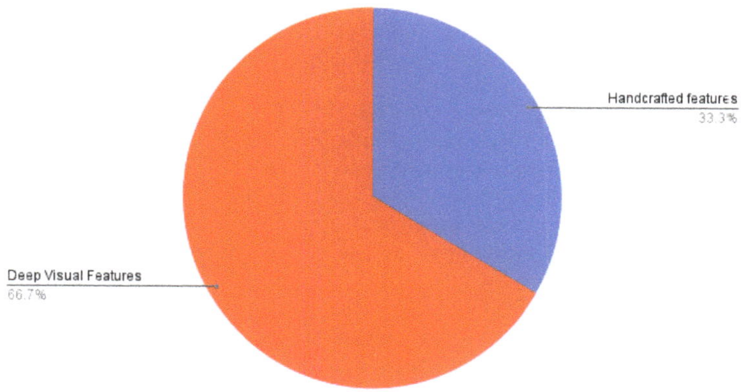

Figure 10. Percentage of studies summarized according to the type of feature extraction methods.

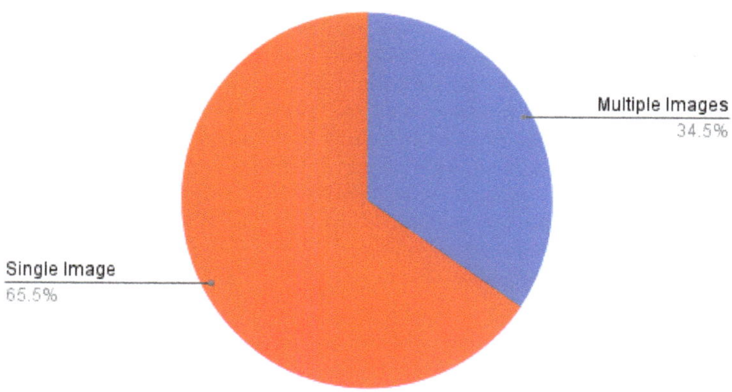

Figure 11. Volume estimation methods using single images vs. multiple images.

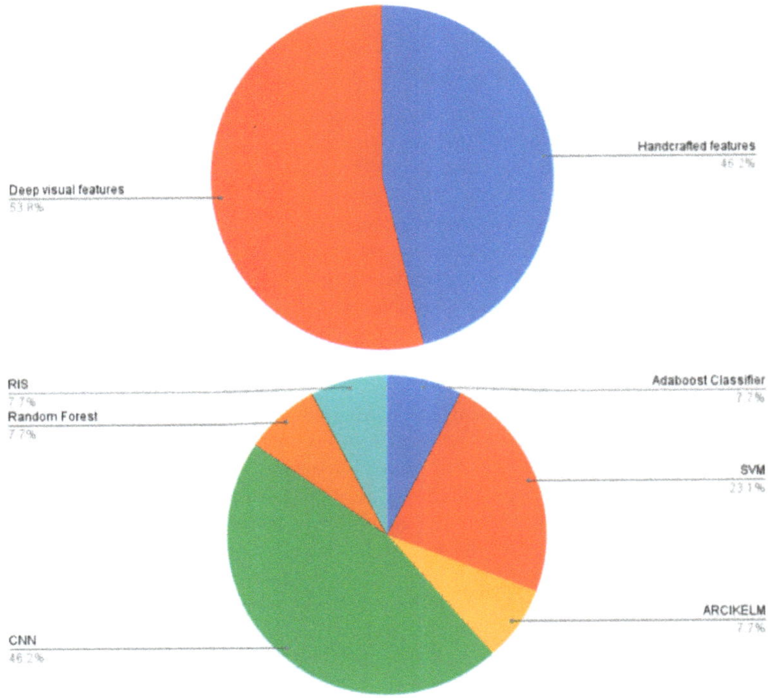

Figure 12. Percentage of studies summarized according to the type of methods employed for feature extraction from food images and the category of classifier used for food image analysis in a mobile application.

10. Open Issues

This study highlighted open issues based on the survey papers and the authors' first-hand experience with existing methodologies.

10.1. Unsupervised Learning from Unlabelled Dataset

Preparing a large comprehensive annotated data is still a challenge, as manually annotating a dataset is a difficult task with many challenges. Due to the large variety of food dishes, different styles of preparation, etc., it is difficult for an expert dietician to correctly label all the foods, especially in the preparation of a multi-culture food database. Similarly, it involves high costs and a large number of working hours to prepare such a dataset. Recent advancements in contrastive learning have opened a new research paradigm of unsupervised learning. Methods based on contrastive learning such as SimCLR [161] and SwAV [162] do not require labeled datasets and seem to be interesting potential areas of research that future works in food recognition should exploit.

10.2. Continual Learning

Food datasets are open-ended, and there is no cap on the number of dishes. So the network must adapt to continuously evolving datasets. All of these properties of food datasets have made them a strong use case for continual learning methods. One of the principal challenges in continuous learning methods is catastrophic forgetting. Catastrophic forgetting refers to completely or abruptly forgetting previously learned information while learning new classes. Many neural networks are susceptible to forgetting during continual learning. It is a prime hindrance in achieving the objective of continuously evolving

networks similarly to those of humans. Hence, researchers should also study catastrophic forgetting in the context of food databases.

10.3. Explainability

Although there have been numerous attempts, including activation methods, SHAP values [163], and distillation methods, there is still a research gap in the context of food recognition. As food recognition has many domain-specific challenges such as intraclass variations, and non-rigid structure, visualization of the reasoning behind model predictions is vital to trust its decisions. Recently, unsupervised clustering methods [164] are exploited to explain model predictions by distilling knowledge into surrogate models. They provide similar images to test images for explaining prediction results. Explaining prediction results by showing images similar to test images seems more friendly as users do not need any specific domain knowledge to understand these results.

11. Discussion

Our research provides deep insight into computer vision-based approaches for dietary assessment. It focuses on both traditional and deep learning methodologies for feature extraction and classification methods used for food image recognition and single- and multi-view methods for volume estimation. Similarly, this survey also explores and compares current food image datasets in detail, as vision-based techniques are highly dependent on a comprehensive collection of food images. In contrast to previous research work, such as work by Mohammad A. Sobhi et al. [165], Min, Weiqing, et al. [166], our survey scrutinizes traditional and current deep visual approaches for feature extraction and classification to enhance clarity in terms of their performance and feasibility. Unlike existing surveys, our survey emphasizes existing solutions developed for food ingredient recognition through multi-label learning. We also reviewed existing computer-based food volume estimation methods in detail, as they have reduced dietitians' and experts' intervention and can accurately determine the portion size of the food in contrast to the self-estimation. Finally, our research study also explores real-world applications using the prior methodologies for dietary assessment purposes.

11.1. Findings

Our findings indicate that the ultimate performance of traditional and deep visual techniques depends on the type of dataset used. This has been observed from the datasets included from the studies explored in this survey (as shown in Table 1); the three most commonly used datasets were UECFOOD-256 [43], UECFOOD-100 [42], and Food-101 [59]. UECFOOD-256 (25,088 images and 256 classes) and UECFOOD-100 (14,361 images and 100 classes of food) are Japanese food datasets consisting of Japanese food images captured by users, whereas Food-101(101,000 images and 101 classes) is an American fast food dataset containing images crawled from several websites. However, these widely used datasets are region-specific. Therefore, there is an immense need for generic food datasets for excluding regional bias from experimental results. In addition, it is also evident from this survey that deep visual techniques have replaced traditional machine learning methodologies for food image recognition. As per our survey, systems proposed after the year 2015 mainly use deep learning technologies for food classification purposes. This is due to their phenomenal classification performance. In the context of classification performance of deep visual techniques, for food–non-food classification, McAllister et al., 2018 [108] (99.4%), and Pouladzadeh et al., 2016 [104] (99%), achieved the highest top 1 classification accuracy. Pouladzadeh et al., 2016 [104], used DCNN and Graph cut on their proposed dataset, whereas McAllister et al., 2018 [108], used CNN, ANN, SVM, and random forest on the food 5k dataset. Table 5 further compares classification accuracies of proposed deep visual models. Recent advancements and exceptional performance of food image classification methods have now led researchers to explore food images from a much deeper perspective in terms of retrieval and classification of food ingredients from food images. Therefore,

we have also explored several proposed solutions for food ingredient recognition and classification. According to our survey, the system proposed by Chen et al., 2016 [47], has achieved the highest F1 score, i.e., 95.88% macro-F1 and 82.06% micro-F1, using the Arch-D method on the UECFOOD-100 dataset (as shown in Table 6). Similarly, automatic food volume estimation methods have reduced dietitians' and experts' intervention and can accurately determine the portion size of the food in contrast to the self-estimation for food volume estimation. Single-view methods involve capturing a single image, while multi-views require multiple images to determine accurate food volumes. The results in Table 8 show that multi-view methods are mostly better than single-view methods.

Finally, food category recognition, ingredient classification, and volume estimation techniques helped provide an automatic dietary assessment with reduced human intervention in mHealth apps. For this purpose, we have also surveyed several mobile applications that employ deep learning methods for dietary assessment.

11.2. Limitations and Future Research Challenges

Despite enhanced performance and classification accuracy, food image recognition and volume estimation through vision-based approaches may continue to present interesting future research challenges. This is because the performance of the methodologies used for food image identification is highly dependent on the source of images in a particular food dataset. Although a growing number of food categories are being incorporated into food image datasets such as UECFOOD-256 [43], Food 85 [49], and Food201-segmented [13], there is still an immense need for generalized, comprehensive datasets for better performance evaluation and benchmarking. Moreover, we observed that datasets with a large number of food images significantly positively impact classification accuracy. However, keeping these large image datasets updated is another challenge, especially since different types of foods are being prepared every day.

In addition to this, progressive learning during the classification phase is vital for food image datasets due to the continuous arrival of new concepts and domain variation within existing concepts. Similarly, developing frameworks interpretable by highlighting the contribution of the area of interest will improve the overall human trust level on a solution in a real-world environment.

Following food recognition, food volume estimation is a particularly complex and challenging assignment since food items have large variations in shape, texture, and appearances. Our article categorized food portion estimation methods into single-view and multi-view methods. Multi-view methods are more accurate; however, most of these methods also require calibration objects each time and images from different viewpoints, which makes the usability of these solutions tedious for elderly users.

Finally, there is a need to design and develop solutions that can respond to situations ethically. In our context, this refers to the removal of any biases concerning region-specific food preferences. It will help to ensure transparency in existing models.

12. Conclusions

In this work, we explored a broad spectrum of vision-based methods that are specifically tailored for food image recognition and volume estimation. In practice, the food recognition process incorporates four tasks: acquiring food images from the corresponding food datasets, feature extraction using handcrafted or deep visual, selection of relevant extracted features, and finally, appropriate selection of classification technique using either traditional machine learning approach or deep learning models followed by food ingredient classification to provide better insight of nutrient information. The findings of surveyed studies have shown that 38.1% of datasets are generic, which includes multicultural food dishes. Similarly, 46.2% of surveyed applications implemented CNN for food recognition, while 45.2% of mobile applications have implemented traditional methods for feature extraction. For ingredient detection, several studies used CNN due to its superior performance and recent interest. In addition, 34.5% of techniques for volume estimation

require multiple images, while the remaining methods used a single image to estimate food volume.

Despite impeccable performance exhibited by state-of-the-art approaches, there exist several limitations and challenges. There is an immense need for comprehensive datasets for benchmarking and performance evaluation of these models, as incorporating large food image datasets improves the overall performance. Consequently, when dealing with open-ended and dynamic food datasets, the classifier must be capable of open-ended continuous learning. However, existing methods have several bottlenecks, which undermine the food-recognition ability when it comes to open-ended learning, as proposed methods are prone to catastrophic forgetting. They tend to forget previous knowledge extracted from images while learning new information. Such methods work well only for fixed food image datasets. Moreover, our findings indicate that proposed techniques for food ingredient classification still struggle with performance issues when applied to prepared and mixed food items. Survey findings further indicate that CNN models employed for visual feature extraction require labeled datasets for fine-tuning and training. Preparing a labeled food dataset is a difficult task due to the large variety of food dishes. To tackle this problem, unsupervised methods based on contrastive learning seem to have good research potential.

Similarly, automatic food portion estimation methods are categorized into two major categories: single-view-image methods and multi-view-image methods. As discussed earlier, most of multi-view image methods are more accurate than single view methods, but multi-view-image methods require complex processing and images from different angles, resulting in a reduced user retention rate. Furthermore, most of the single and multi-view methods require calibration objects each time, which has made the usability of these solutions tedious for elderly patients.

Therefore, there is substantial room for innovative health care and dietary assessment applications that can integrate wearable devices with a smartphone to revolutionize this research area. Moreover, dietary assessment systems should address these challenges to provide better insights into effective health maintenance and chronic disease prevention.

Author Contributions: G.A.T. was responsible for the literature search and writing the article and approved the final version as submitted. C.K.L. contributed to the study design, reviewed the study for intellectual content, and confirmed the final version as submitted. All authors have read and agreed to the published version of the manuscript.

Funding: This research was supported by the UM Partnership Grant: Project No RK012-2019 from University of Malaya, IIRG Grant (IIRG002C-19HWB) from University of Malaya, International Collaboration Fund for project Developmental Cognitive Robot with Continual Lifelong Learning (IF0318M1006) from MESTECC, Malaysia, and ONRG grant (Project No.: ONRG-NICOP- N62909-18-1-2086)/IF017-2018 from Office of Naval and Research Global, UK.

Institutional Review Board Statement: Not applicable.

Informed Consent Statement: Not applicable.

Data Availability Statement: Not applicable.

Conflicts of Interest: The authors wish to confirm that there are no conflicts of interest.

References

1. Hajat, C.; Stein, E. The global burden of multiple chronic conditions: A narrative review. *Prev. Med. Rep.* **2018**, *12*, 284–293. [CrossRef]
2. Hall, J.E.; do Carmo, J.M.; da Silva, A.A.; Wang, Z.; Hall, M.E. Obesity-induced hypertension: Interaction of neurohumoral and renal mechanisms. *Circ. Res.* **2015**, *116*, 991–1006. [CrossRef] [PubMed]
3. Al-Goblan, A.S.; Al-Alfi, M.A.; Khan, M.Z. Mechanism linking diabetes mellitus and obesity. *Diabetes Metab Syndr. Obes.* **2014**, *7*, 587–591. [CrossRef]
4. Akil, L.; Ahmad, H.A. Relationships between obesity and cardiovascular diseases in four southern states and Colorado. *J. Health Care Poor Underserved.* **2011**, *22*, 61–72. [CrossRef] [PubMed]
5. De Pergola, G.; Silvestris, F. Obesity as a major risk factor for cancer. *J. Obes.* **2013**, *2013*, 291546. [CrossRef]

6. World Health Organization (WHO). Obesity and Overweigh. Available online: https://www.who.int/news-room/fact-sheets/detail/obesity-and-overweight (accessed on 23 August 2018).
7. Ngo, J.; Engelen, A.; Molag, M.; Roesle, J.; García-Segovia, P.; Serra-Majem, L. A review of the use of information and communication technologies for dietary assessment. *Br. J. Nutr.* **2009**, *101* (Suppl. 2), S102–S112. [CrossRef] [PubMed]
8. Mendi, E.; Ozyavuz, O.; Pekesen, E.; Bayrak, C. Food intake monitoring system for mobile devices. In Proceedings of the 5th IEEE International Workshop on Advances in Sensors and Interfaces IWASI, Bari, Italy, 13–14 June 2013; pp. 31–33. [CrossRef]
9. Haapala, I.; Barengo, N.C.; Biggs, S.; Surakka, L.; Manninen, P. Weight loss by mobile phone: A 1-year effectiveness study. *Public Health Nutr.* **2009**, *12*, 2382–2391. [CrossRef] [PubMed]
10. Chen, Y.S.; Wong, J.E.; Ayob, A.F.; Othman, N.E.; Poh, B.K. Can Malaysian young adults report dietary intake using a food diary mobile. application? A pilot study on acceptability and compliance. *Nutrients* **2017**, *9*, 62. [CrossRef]
11. Ming, Z.Y.; Chen, J.; Cao, Y.; Forde, C.; Ngo, C.W.; Chua, T.S. Food photo recognition for dietary tracking: System and experiment. In *Multimedia Modeling*; Springer: Berlin/Heidelberg, Germany, 2018; pp. 129–141.
12. Kong, F.; Tan, J. DietCam: Automatic Dietary Assessment with Mobile Camera Phones. *Pervasive Mob. Comput.* **2012**, *8*, 147–163. [CrossRef]
13. Meyers, A.; Johnston, N.; Rathod, V.; Korattikara, A.; Gorban, A.; Silberman, N.; Guadarrama, S.; Papandreou, G.; Huang, J.; Murphy, K.P. Im2Calories: Towards an Automated Mobile Vision Food Diary. In Proceedings of the 2015 IEEE International Conference on Computer Vision (ICCV), Santiago, Chile, 7–13 December 2015; pp. 1233–1241. [CrossRef]
14. Martinel, N.; Micheloni, C. Classification of local eigen-dissimilarities for person re-identification. *IEEE Signal Process. Lett.* **2015**, *22*, 455–459. [CrossRef]
15. Martinel, N.; Das, A.; Micheloni, C.; Roy-Chowdhury, A.K. Re-Identification in the function space of feature warps. *IEEE Trans. Pattern Anal. Mach. Intell.* **2015**, *37*, 1656–1669. [CrossRef] [PubMed]
16. Martinel, N.; Micheloni, C.; Foresti, G.L. Kernelized Saliency-Based Person Re-Identification Through Multiple Metric Learning. *IEEE Trans. Image Process.* **2015**, *24*, 5645–5658. [CrossRef] [PubMed]
17. Mahabir, S.; Baer, D.J.; Giffen, C.; Subar, A.; Campbell, W.; Hartman, T.J.; Clevidence, B.; Albanes, D.; Taylor, P.R. Calorie intake misreporting by diet record and food frequency questionnaire compared to doubly labeled water among postmenopausal women. *Eur. J. Clin. Nutr.* **2006**, *60*, 561–565. [CrossRef] [PubMed]
18. Bandini, L.G.; Must, A.; Cyr, H.; Anderson, S.E.; Spadano, J.L.; Dietz, W.H. Longitudinal changes in the accuracy of reported energy intake in girls 10–15 y of age. *Am. J. Clin. Nutr.* **2003**, *78*, 480–484. [CrossRef]
19. Champagne, C.M.; Baker, N.B.; DeLany, J.P.; Harsha, D.W.; Bray, G.A. Assessment of energy intake underreporting by doubly labeled water and observations on reported nutrient intakes in children. *J. Am. Diet Assoc.* **1998**, *98*, 426–433. [CrossRef]
20. Champagne, C.M.; Bray, G.A.; Kurtz, A.A.; Monteiro, J.B.; Tucker, E.; Volaufova, J.; Delany, J.P. Energy intake and energy expenditure: A controlled study comparing dietitians and non-dietitians. *J. Am. Diet Assoc.* **2002**, *102*, 1428–1432. [CrossRef]
21. Subar, A.F.; Kipnis, V.; Troiano, R.P.; Midthune, D.; Schoeller, D.A.; Bingham, S.; Sharbaugh, C.O.; Trabulsi, J.; Runswick, S.; Ballard-Barbash, R.; et al. Using intake biomarkers to evaluate the extent of dietary misreporting in a large sample of adults: The OPEN study. *Am. J. Epidemiol.* **2003**, *158*, 1–13. [CrossRef]
22. Blanton, C.A.; Moshfegh, A.J.; Baer, D.J.; Kretsch, M.J. The usda automated multiple-pass method accurately estimates group total energy and nutrient intake. *J. Nutr.* **2006**, *136*, 2594–2599. [CrossRef] [PubMed]
23. Daugherty, B.L.; Schap, T.E.; Ettienne-Gittens, R.; Zhu, F.M.; Bosch, M.; Delp, E.J.; Ebert, D.S.; Kerr, D.A.; Boushey, C.J. Novel Technologies for Assessing Dietary Intake: Evaluating the Usability of a Mobile Telephone Food Record Among Adults and Adolescents. *J. Med. Internet Res.* **2012**, *14*, e58. [CrossRef]
24. Snyder, H. Literature review as a research methodology: An overview and guidelines. *J. Bus. Res.* **2019**, *104*, 333–339. [CrossRef]
25. Ronald, K.; Marc, M.; Angelina, A.; Tyler, H.; Christopher, K. Measuring Catastrophic Forgetting in Neural Networks. *arXiv* **2017**, arXiv:1708.02072.
26. Liew, W.S.; Loo, C.K.; Gryshchuk, V.; Weber, C.; Wermter, S. Effect of Pruning on Catastrophic Forgetting in Growing Dual Memory Networks. In Proceedings of the 2019 International Joint Conference on Neural Networks (IJCNN), Budapest, Hungary, 14–19 July 2019; pp. 1–8. [CrossRef]
27. Tsoumakas, G.; Katakis, I. Multi-label classification: An overview. *Int. J. Data Warehous. Min.* **2006**, *3*, 1–3. [CrossRef]
28. He, Y.; Xu, C.; Khanna, N.; Boushey, C.J.; Delp, E.J. Food image analysis: Segmentation, identification and weight estimation. In Proceedings of the 2013 IEEE International Conference on Multimedia and Expo (ICME), San Jose, CA, USA, 15–19 July 2013; pp. 1–6. [CrossRef]
29. Miyazaki, T.; de Silva, G.C.; Aizawa, K. Image-based Calorie Content Estimation for Dietary Assessment. In Proceedings of the 2011 IEEE International Symposium on Multimedia, Dana Point, CA, USA, 5–7 December 2011; pp. 363–368. [CrossRef]
30. Fang, S.; Zhu, F.; Jiang, C.; Zhang, S.; Boushey, C.J.; Delp, E.J. A comparison of food portion size estimation using geometric models and depth images. In Proceedings of the Image Processing (ICIP), Hoenix, AZ, USA, 25–28 September 2016; Volume 2016, pp. 26–30. [CrossRef]
31. Okamoto, K.; Yanai, K. An Automatic Calorie Estimation System of Food Images on a Smartphone. In Proceedings of the Madima'16: Proceedings of the 2nd International Workshop on Multimedia Assisted Dietary Management, Amsterdam, The Netherlands, 16 October 2016; pp. 63–70. [CrossRef]

32. Wu, W.; Yang, J. Fast food recognition from videos of eating for calorie estimation. In Proceedings of the 2009 IEEE International Conference on Multimedia and Expo, New York, NY, USA, 28 June–3 July 2009; pp. 1210–1213. [CrossRef]
33. Zhang, Z.; Yang, Y.; Yue, Y.; Fernstrom, J.D.; Jia, W.; Sun, M. Food volume estimation from a single image using virtual reality technology. In Proceedings of the 2011 IEEE 37th Annual Northeast Bioengineering Conference (NEBEC), Troy, NY, USA, 1 April 2011; pp. 1–2. [CrossRef]
34. Hippocrate, E.A.A.; Suwa, H.; Arakawa, Y.; Yasumoto, K. Food Weight Estimation using Smartphone and Cutlery. In Proceedings of the First Workshop on IoT-enabled Healthcare and Wellness Technologies and Systems (IoT of Health'16), Singapore, 30 June 2016; Association for Computing Machinery: New York, NY, USA, 2016; pp. 9–14. [CrossRef]
35. Yue, Y.; Jia, W.; Sun, M. Measurement of food volume based on single 2-D image without conventional camera calibration. In Proceedings of the 2012 Annual International Conference of the IEEE Engineering in Medicine and Biology Society, San Diego, CA, USA, 29 August 2012; pp. 2166–2169. [CrossRef]
36. Xu, C.; He, Y.; Khanna, N.; Boushey, C.J.; Delp, E.J. Model-based food volume estimation using 3D pose. In Proceedings of the 2013 IEEE International Conference on Image Processing, Melbourne, VIC, Australia, 15–18 September 2013; pp. 2534–2538. [CrossRef]
37. Hernández, T.; Wilder, L.; Kuehn, D.; Rubotzky, K.; Moser-Veillon, P.; Godwin, S.; Thompson, C.; Wang, C. Portion size estimation and expectation of accuracy. *J. Food Compos. Anal.* **2006**, *19*, S14–S21. [CrossRef]
38. Zhang, W.; Yu, Q.; Siddiquie, B.; Divakaran, A.; Sawhney, H. Snap-n-Eat: Food Recognition and Nutrition Estimation on a Smartphone. *J. Diabetes Sci. Technol.* **2015**, *9*, 525–533. [CrossRef]
39. Huang, J.; Ding, H.; Mcbride, S.; Ireland, D.; Karunanithi, M. Use of Smartphones to Estimate Carbohydrates in Foods for Diabetes Management. *Stud. Health Technol. Inform.* **2015**, *214*, 121–127. [CrossRef]
40. Khanna, N.; Boushey, C.J.; Kerr, D.; Okos, M.; Ebert, D.S.; Delp, E.J. An Overview of The Technology Assisted Dietary Assessment Project at Purdue University. In Proceedings of the 2010 IEEE International Symposium on Multimedia, ISM 2010, Taichung, Taiwan, 13–15 December 2010; pp. 290–295. [CrossRef]
41. Jia, W.; Yue, Y.; Fernstrom, J.D.; Zhang, Z.; Yang, Y.; Sun, M. 3D localization of circular feature in 2D image and application to food volume estimation. In Proceedings of the 2012 Annual International Conference of the IEEE Engineering in Medicine and Biology Society, San Diego, CA, USA, 29 August 2012; pp. 4545–4548. [CrossRef]
42. Matsuda, Y.; Yanai, K. Multiple-food recognition considering co-occurrence employing manifold ranking. In Proceedings of the 21st International Conference on Pattern Recognition (ICPR2012), Tsukuba, Japan, 11–15 November 2012; pp. 2017–2020. [CrossRef]
43. Kawano, Y.; Yanai, K. FoodCam-256: A Large-scale Real-time Mobile Food RecognitionSystem employing High-Dimensional Features and Compression of Classifier Weights. In Proceedings of the 22nd ACM international conference on Multimedia (MM'14), Orlando, FL, USA, 3–7 November 2014; Association for Computing Machinery: New York, NY, USA, 2014; pp. 761–762. [CrossRef]
44. Chen, M.; Dhingra, K.; Wu, W.; Yang, L.; Sukthankar, R.; Yang, J. PFID: Pittsburgh fast-food image dataset. In Proceedings of the 2009 16th IEEE International Conference on Image Processing (ICIP), Cairo, Egypt, 7–10 November 2009; pp. 289–292. [CrossRef]
45. Farinella, G.M.; Allegra, D.; Moltisanti, M.; Stanco, F.; Battiato, S. Retrieval and classification of food images. *Comput. Biol. Med.* **2016**, *77*, 23–39. [CrossRef]
46. Farinella, G.M.; Allegra, D.; Stanco, F. A Benchmark Dataset to Study the Representation of Food Images. In Proceedings of the International Workshop on Assistive Computer Vision and Robotics (ACVR) 2014, Zurigo, Switzerland, 6–12 September 2014. [CrossRef] [PubMed]
47. Chen, J.; Ngo, C. Deep-based Ingredient Recognition for Cooking Recipe Retrieval. In Proceedings of the 24th ACM International Conference on Multimedia (MM'16), Amsterdam, The Netherlands, 15–19 October 2016; Association for Computing Machinery: New York, NY, USA, 2016; pp. 32–41. [CrossRef]
48. Chen, M.-Y.; Yang, Y.-H.; Ho, C.-J.; Wang, S.-H.; Liu, S.-M.; Chang, E.; Yeh, C.-H.; Ouhyoung, M. Automatic Chinese food identification and quantity estimation. In *SIGGRAPH Asia 2012 Technical Briefs (SA'12)*; Association for Computing Machinery: New York, NY, USA, 2012; pp. 1–4. [CrossRef]
49. Hoashi, H.; Joutou, T.; Yanai, K. Image recognition of 85 food categories by feature fusion. In Proceedings of the 2010 IEEE International Symposium on Multimedia, Taichung, Taiwan, 13–15 December 2010; pp. 296–301. [CrossRef]
50. Ciocca, G.; Napoletano, P.; Schettini, R. Food Recognition and Leftover Estimation for Daily Diet Monitoring. In Proceedings of the ICIAP 2015 International Workshops, BioFor, CTMR, RHEUMA, ISCA, MADiMa, SBMI, and QoEM, Genoa, Italy, 7–8 September 2015; Volume 9281, pp. 334–341. [CrossRef]
51. Güngör, C.; Baltacı, F.; Erdem, A.; Erdem, E. Turkish cuisine: A benchmark dataset with Turkish meals for food recognition. In Proceedings of the 2017 25th Signal Processing and Communications Applications Conference (SIU), Antalya, Turkey, 15–18 May 2017; pp. 1–4. [CrossRef]
52. Tahir, G.A.; Loo, C.K. An Open-Ended Continual Learning for Food Recognition Using Class Incremental Extreme Learning Machines. *IEEE Access* **2020**, *8*, 82328–82346. [CrossRef]
53. Hou, S.; Feng, Y.; Wang, Z. VegFru: A Domain-Specific Dataset for Fine-Grained Visual Categorization. In Proceedings of the 2017 IEEE International Conference on Computer Vision (ICCV), Venice, Italy, 22–29 October 2017; pp. 541–549. [CrossRef]

54. Mureșan, H.; Oltean, M. Fruit recognition from images using deep learning. *Acta Univ. Sapientiae Inform.* **2018**, *10*, 26–42. [CrossRef]
55. Waltner, G.; Schwarz, M.; Ladstätter, S.; Weber, A.; Luley, P.; Lindschinger, M.; Schmid, I.; Scheitz, W.; Bischof, H.; Paletta, L. Personalized dietary self-management using mobile vision-based assistance. In Proceedings of the International Conference on Image Analysis and Processing, Catania, Italy, 11–15 September 2017; Springer: Berlin/Heidelberg, Germany, 2018; pp. 385–393. [CrossRef]
56. Godwin, S.; Chambers, E.T.; Cleveland, L.; Ingwersen, L. A new portion size estimation aid for wedgeshaped. *Foods. J. Am. Diet. Assoc.* **2006**, *106*, 1246–1250. [CrossRef]
57. Mariappan, A.; Bosch, M.; Zhu, F.; Boushey, C.J.; Kerr, D.A.; Ebert, D.S.; Delp, E.J. Personal Dietary Assessment Using Mobile Devices. In Proceedings of the Computational Imaging VII. International Society for Optics and Photonics, San Jose, CA, USA, 19–20 January 2009; Volume 7246, 72460Z. [CrossRef]
58. Bosch, M.; Schap, T.; Zhu, F.; Khanna, N.; Boushey, C.J.; Delp, E.J. Integrated database system for mobile dietary assessment and analysis. In Proceedings of the 2011 IEEE International Conference on Multimedia and Expo, Barcelona, Spain, 11–15 July 2011. [CrossRef]
59. Bossard, L.; Guillaumin, M.; van Gool, L. Food-101–mining discriminative components with random forests. In Proceedings of the European Conference on Computer Vision, Zurich, Switzerland, 6–12 September 2014; pp. 446–461. [CrossRef]
60. Stütz, T.; Dinic, R.; Domhardt, M.; Ginzinger, S. Can mobile augmented reality systems assist in portion estimation? A user study. In Proceedings of the 2014 IEEE International Symposium on Mixed and Augmented Reality—Media, Art, Social Science, Humanities and Design (ISMAR-MASH'D), Munich, Germany, 10–12 September 2014; pp. 51–57. [CrossRef]
61. Wang, X.; Kumar, D.; Thome, N.; Cord, M.; Precioso, F. Recipe recognition with large multimodal food dataset. In Proceedings of the 2015 IEEE International Conference on Multimedia & Expo Workshops (ICMEW), Turin, Italy, 29 June–3 July 2015; pp. 1–6. [CrossRef]
62. Fang, S.; Liu, C.; Zhu, F.; Delp, E.J.; Boushey, C.J. Single-View Food Portion Estimation Based on Geometric Models. In Proceedings of the 2015 IEEE International Symposium on Multimedia (ISM), Miami, FL, USA, 14–16 December 2015; pp. 385–390. [CrossRef]
63. Herranz, L.; Xu, R.; Jiang, S. A probabilistic model for food image recognition in restaurants. In Proceedings of the 2015 IEEE International Conference on Multimedia and Expo (ICME), Turin, Italy, 29 June–3 July 2015; pp. 1–6. [CrossRef]
64. Beijbom, O.; Joshi, N.; Morris, D.; Saponas, S.; Khullar, S. Menu-Match: Restaurant-Specific Food Logging from Images. In Proceedings of the 2015 IEEE Winter Conference on Applications of Computer Vision, Waikoloa, HI, USA, 5–9 January 2015; pp. 844–851. [CrossRef]
65. Zhou, F.; Lin, Y. Fine-grained image classification by exploring bipartite-graph labels. In Proceedings of the IEEE Conference on Computer Vision and Pattern Recognition, Las Vegas, NV, USA, 27–30 June 2016; pp. 1124–1133. [CrossRef]
66. Ciocca, G.; Napoletano, P.; Schettini, R. Food Recognition: A New Dataset, Experiments and Results. *IEEE J. Biomed. Health Inform.* **2017**, *21*, 588–598.
67. Wu, H.; Merler, M.; Uceda-Sosa, R.; Smith, J.R. Learning to Make Better Mistakes: Semantics-aware Visual Food Recognition. In Proceedings of the 24th ACM international conference on Multimedia (MM'16), Amsterdam, The Netherlands, 15–19 October 2016; Association for Computing Machinery: New York, NY, USA, 2016; pp. 172–176. [CrossRef] [PubMed]
68. Singla, A.; Yuan, L.; Ebrahimi, T. Food/Non-food Image Classification and Food Categorization using Pre-Trained GoogLeNet Model. In Proceedings of the 2nd International Workshop on Multimedia Assisted Dietary Management (MADiMa'16), Amsterdam, The Netherlands, 16 October 2016; Association for Computing Machinery: New York, NY, USA, 2016; pp. 3–11. [CrossRef]
69. Rich, J.; Haddadi, H.; Hospedales, T.M. Towards Bottom-Up Analysis of Social Food. In Proceedings of the 6th International Conference on Digital Health Conference (DH'16), Montréal, QC, Canada, 11–13 April 2016; Association for Computing Machinery: New York, NY, USA, 2016; pp. 111–120. [CrossRef]
70. Liang, Y.; Li, J. Computer vision-based food calorie estimation: Dataset, method, and experiment. *arXiv* **2017**, arXiv:1705.07632.
71. Pandey, P.; Deepthi, A.; Mandal, B.; Puhan, N.B. FoodNet: Recognizing Foods Using Ensemble of Deep Networks. *IEEE Signal Process. Lett.* **2017**, *24*, 1758–1762. [CrossRef]
72. Termritthikun, C.; Muneesawang, P.; Kanprachar, S. NUInNet: Thai food image recognition using convolutional neural networks on smartphone. *J. Telecommun. Electron. Comput. Eng. (JTEC)* **2017**, *9*, 63–67. [CrossRef]
73. Ciocca, G.; Napoletano, P.; Schettini, R. Learning CNN-based features for retrieval of food images. In Proceedings of the New Trends in Image Analysis and Processing—ICIAP 2017, Catania, Italy, 11–15 September 2017; pp. 426–434.
74. Yu, Q.; Anzawa, M.; Amano, S.; Ogawa, M.; Aizawa, K. Food Image Recognition by Personalized Classifier. In Proceedings of the 2018 25th IEEE International Conference on Image Processing (ICIP), Athens, Greece, 7–10 October 2018; pp. 171–175. [CrossRef]
75. Kaur, P.; Sikka, K.; Wang, W.; Belongie, S.; Divakaran, A. Foodx-251: A dataset for fine-grained food classification. *arXiv* **2019**, arXiv:1907.06167.
76. Available online: https://www.aicrowd.com/challenges/food-recognition-challenge (accessed on 23 August 2021).
77. Bolaños, M.; Radeva, P. Simultaneous Food Localization and Recognition. In Proceedings of the 23rd International Conference on Pattern Recognition (ICPR) 2016 (IN PRESS), Cancun, Mexico, 4–8 December 2016.
78. Aguilar, E.; Bolaños, M.; Radeva, P. Regularized Uncertainty-based Multi-Task LearningModel for Food Analysis. *J. Vis. Commun. Image R.* **2019**, *60*, 360–370. [CrossRef]

79. Kumar, G.; Bhatia, P.K. A Detailed Review of Feature Extraction in Image Processing Systems. In Proceedings of the 2014 Fourth International Conference on Advanced Computing & Communication Technologies, Rohtak, India, 8–9 February 2014; pp. 5–12. [CrossRef]
80. Yang, S.; Chen, M.; Pomerleau, D.; Sukthankar, R. Food recognition using statistics of pairwise local features. In Proceedings of the IEEE Computer Society Conference on Computer Vision and Pattern Recognition, San Francisco, CA, USA, 13–18 June 2010; pp. 2249–2256. [CrossRef]
81. Pouladzadeh, P.; Shirmohammadi, S.; Bakirov, A.; Bulut, A.; Yassine, A. *Cloud-Based SVM for Food Categorization. Multimedia Tools and Applications*; Springer: Berlin/Heidelberg, Germany, 2014. [CrossRef]
82. Kawano, Y.; Yanai, K. Real-Time Mobile Food Recognition System. In Proceedings of the 2013 IEEE Conference on Computer Vision and Pattern Recognition Workshops, Portland, OR, USA, 23–28 June 2013; pp. 1–7. [CrossRef]
83. Oliveira, L.; Costa, V.; Neves, G.; Oliveira, T.; Jorge, E.; Lizarraga, M. A mobile, lightweight, poll-based food identification system. *Pattern Recognit.* **2014**, *47*, 1941–1952. [CrossRef]
84. Martinel, N.; Piciarelli, C.; Micheloni, C. A supervised extreme learning committee for food recognition. *Comput. Vis. Image Underst.* **2016**, *148*, 67–86. [CrossRef]
85. Bosch, M.; Zhu, F.; Khanna, N.; Boushey, C.J.; Delp, E.J. Combining global and local features for food identification in dietary assessment. In Proceedings of the 2011 18th IEEE International Conference on Image Processing, Brussels, Belgium, 11–14 September 2011; Volume 2011, pp. 1789–1792. [CrossRef]
86. Kong, F.; Tan, J. DietCam: Regular Shape Food Recognition with a Camera Phone. In Proceedings of the 2011 International Conference on Body Sensor Networks, Chicago, IL, USA, 19–22 May 2011; pp. 127–132. [CrossRef]
87. Zhang, M.M. *Identifying the Cuisine of a Plate of Food*; Tech. Report; University of California San Diego: San Diego, CA, USA, 2011. [CrossRef]
88. Matsuda, Y.; Hoashi, H.; Yanai, K. Recognition of Multiple-Food Images by Detecting Candidate Regions. In Proceedings of the 2012 IEEE International Conference on Multimedia and Expo, Melbourne, VIC, Australia, 9–13 July 2012; pp. 25–30. [CrossRef]
89. Anthimopoulos, M.M.; Gianola, L.; Scarnato, L.; Diem, P.; Mougiakakou, S.G. A food recognition system for diabetic patients based on an optimized bag-of-features model. *IEEE J. Biomed. Health Inform.* **2014**, *18*, 1261–1271. [CrossRef]
90. Tammachat, N.; Pantuwong, N. Calories analysis of food intake using image recognition. In Proceedings of the 2014 6th International Conference on Information Technology and Electrical Engineering (ICITEE), Yogyakarta, Indonesia, 7–8 October 2014; pp. 1–4. [CrossRef]
91. Pouladzadeh, P.; Shirmohammadi, S.; Yassine, A. Using graph cut segmentation for food calorie measurement. In Proceedings of the 2014 IEEE International Symposium on Medical Measurements and Applications (MeMeA), Lisbon, Portugal, 11–12 June 2014; pp. 1–6. [CrossRef]
92. He, Y.; Xu, C.; Khanna, N.; Boushey, C.J.; Delp, E.J. Analysis of food images: Features and classification. In Proceedings of the 2014 IEEE International Conference on Image Processing (ICIP), Paris, France, 27–30 October 2014; pp. 2744–2748. [CrossRef]
93. Heaton, J. Ian Goodfellow, Yoshua Bengio, and Aaron Courville: Deep learning. In *Genetic Programming and Evolvable Machines*; The MIT Press: Cambridge, MA, USA, 2016; Volume 19, 800p, ISBN 0262035618. [CrossRef]
94. Deng, L.; Yu, D. Deep learning: Methods and applications. *Found. Trends Signal Process.* **2014**, *7*, 197–387. [CrossRef]
95. LeCun, Y.; Bengio, Y.; Hinton, G. Deep learning. *Nature* **2015**, *521*, 436–444. [CrossRef]
96. Yanai, K.; Kawano, Y. Food image recognition using deep convolutional network with pre-training and fine-tuning. In Proceedings of the 2015 IEEE International Conference on Multimedia & Expo Workshops (ICMEW), Turin, Italy, 29 June–3 July 2015; pp. 1–6. [CrossRef]
97. Szegedy, C.; Vanhoucke, V.; Ioffe, S.; Shlens, J.; Wojna, Z. Rethinking the Inception Architecture for Computer Vision. In Proceedings of the 2016 IEEE Conference on Computer Vision and Pattern Recognition (CVPR), Las Vegas, NV, USA, 27–30 June 2016; pp. 2818–2826. [CrossRef]
98. Russakovsky, O.; Deng, J.; Su, H.; Krause, J.; Satheesh, S.; Ma, S.; Huang, Z.; Karpathy, A.; Khosla, A.; Bernstein, M.; et al. ImageNet Large Scale Visual Recognition Challenge. *Int. J. Comput. Vis.* **2015**, *115*, 211–252. [CrossRef]
99. Krizhevsky, A.; Sutskever, I.; Hinton, G.E. Imagenet classification with deep convolutional neural networks. *Adv. Neural Inf. Process. Syst.* **2012**, *25*, 1097–1105. [CrossRef]
100. Heravi, E.; Aghdam, H.; Puig, D. Classification of Foods Using Spatial Pyramid Convolutional Neural Network. *Artif. Intell. Res. Dev.* **2016**, *288*, 163–168. [CrossRef]
101. Jiang, S.; Min, W.; Liu, L.; Luo, Z. Multi-Scale Multi-View Deep Feature Aggregation for Food Recognition. *IEEE Trans. Image Process.* **2020**, *29*, 265–276. [CrossRef]
102. Kawano, Y.; Yanai, K. Food image recognition with deep convolutional features. In Proceedings of the 2014 ACM International Joint Conference on Pervasive and Ubiquitous Computing: Adjunct Publication (UbiComp'14 Adjunct), Seattle, WA, USA, 13–17 September 2014; Association for Computing Machinery: New York, NY, USA, 2014; pp. 589–593. [CrossRef]
103. Christodoulidis, S.; Anthimopoulos, M.; Mougiakakou, S. Food Recognition for Dietary Assessment. Using Deep. *Convolutional Neural Netw.* **2015**, *9281*, 458–465. [CrossRef]
104. Pouladzadeh, P.; Kuhad, P.; Peddi, S.V.B.; Yassine, A.; Shirmohammadi, S. Food calorie measurement using deep learning neural network. In Proceedings of the 2016 IEEE International Instrumentation and Measurement Technology Conference Proceedings, Taipei, Taiwan, 23–26 May 2016; pp. 1–6. [CrossRef]

105. Hassannejad, H.; Matrella, G.; Ciampolini, P.; de Munari, I.; Mordonini, M.; Cagnoni, S. Food Image Recognition Using Very Deep Convolutional Networks. In Proceedings of the 2nd International Workshop on Multimedia Assisted Dietary Management (MADiMa'16), Amsterdam, The Netherlands, 16 October 2016; Association for Computing Machinery: New York, NY, USA, 2016; pp. 41–49. [CrossRef]
106. Liu, C.; Cao, Y.; Luo, Y.; Chen, G.; Vokkarane, V.; Ma, Y. Deepfood: Deep learning-based food image recognition for computer-aided dietary assessment. In Proceedings of the International Conference on Smart Homes and Health Telematics, Wuhan, China, 25–27 May 2016; pp. 37–48. [CrossRef]
107. Liu, C.; Cao, Y.; Luo, Y.; Chen, G.; Vokkarane, V.; Yunsheng, M.; Chen, S.; Hou, P. A New Deep Learning-Based Food Recognition System for Dietary Assessment on An Edge Computing Service Infrastructure. *IEEE Trans. Serv. Comput.* **2018**, *11*, 249–261. [CrossRef]
108. McAllister, P.; Zheng, H.; Bond, R.; Moorhead, A. Combining deep residual neural network features with supervised machine learning algorithms to classify diverse food image datasets. *Comput. Biol. Med.* **2018**, *95*, 217–233. [CrossRef]
109. Martinel, N.; Foresti, G.L.; Micheloni, C. Wide-Slice Residual Networks for Food Recognition. In Proceedings of the 2018 IEEE Winter Conference on Applications of Computer Vision (WACV), Lake Tahoe, NV, USA, 12–15 March 2018; pp. 567–576. [CrossRef]
110. Aguilar, E.; Remeseiro, B.; Bolaños, M.; Radeva, P. Grab, Pay, and Eat: Semantic Food Detection for Smart Restaurants. *IEEE Trans. Multimed.* **2018**, *20*, 3266–3275. [CrossRef]
111. Horiguchi, S.; Amano, S.; Ogawa, M.; Aizawa, K. Personalized Classifier for Food Image Recognition. *IEEE Trans. Multimed.* **2018**, *20*, 2836–2848. [CrossRef]
112. Ciocca, G.; Napoletano, P.; Schettini, R. CNN-based features for retrieval and classification of food images. *Comput. Vis. Image Underst.* **2018**, *176–177*, 70–77. [CrossRef]
113. Mandal, B.; Puhan, N.B.; Verma, A. Deep Convolutional Generative Adversarial Network-Based Food Recognition Using Partially Labeled Data. *IEEE Sens. Lett.* **2019**, *3*, 7000104. [CrossRef]
114. Ciocca, G.; Micali, G.; Napoletano, P. State Recognition of Food Images Using Deep Features. *IEEE Access* **2020**, *8*, 32003–32017. [CrossRef]
115. Jiang, L.; Qiu, B.; Liu, X.; Huang, C.; Lin, K. DeepFood: Food Image Analysis and Dietary Assessment via Deep Model. *IEEE Access* **2020**, *8*, 47477–47489. [CrossRef]
116. Liu, C.; Liang, Y.; Xue, Y.; Qian, X.; Fu, J. Food and Ingredient Joint Learning for Fine-Grained Recognition. *IEEE Trans. Circuits Syst. Video Technol.* **2021**, *31*, 2480–2493. [CrossRef]
117. Liang, H.; Wen, G.; Hu, Y.; Luo, M.; Yang, P.; Xu, Y. MVANet: Multi-Tasks Guided Multi-View Attention Network for Chinese Food Recognition. *IEEE Trans. Multimed.* **2020**, *23*, 3551–3561. [CrossRef]
118. Zhao, H.; Yap, K.-H.; Kot, A.C.; Duan, L. JDNet: A Joint-Learning Distilled Network for Mobile Visual Food Recognition. *IEEE J. Sel. Top. Signal Process.* **2020**, *14*, 665–675. [CrossRef]
119. Won, C.S. Multi-Scale CNN for Fine-Grained Image Recognition. *IEEE Access* **2020**, *8*, 116663–116674.. [CrossRef]
120. Shen, Z.; Shehzad, A.; Chen, S.; Sun, H.; Liu, J. Machine Learning Based Approach on Food Recognition and Nutrition Estimation. *Procedia Comput. Sci.* **2020**, *174*, 448–453. [CrossRef]
121. Zhu, F.; Bosch, M.; Khanna, N.; Boushey, C.J.; Delp, E.J. Multiple Hypotheses Image Segmentation and Classification With Application to Dietary Assessment. *IEEE J. Biomed. Health Inform.* **2015**, *19*, 377–388. [CrossRef]
122. Aguilar-Torres, E.; Radeva, P. Food Recognition by Integrating Local and Flat Classifiers. In Proceedings of the 9th Iberian Conference, IbPRIA 2019, Madrid, Spain, 1–4 July 2019. [CrossRef]
123. Merchant, K.; Pande, Y. ConvFood: A CNN-Based Food Recognition Mobile Application for Obese and Diabetic Patients. In *Emerging Research in Computing, Information, Communication and Applications*; Springer: Singapore, 2019. [CrossRef]
124. Mezgec, S.; Eftimov, T.; Bucher, T.; Koroušić Seljak, B. Mixed deep learning and natural language processing method for fake-food image recognition and standardization to help automated dietary assessment. *Public Health Nutr.* **2019**, *22*, 1193–1202. [CrossRef]
125. He, J.; Shao, Z.; Wright, J.; Kerr, D.; Boushey, C.; Zhu, F. Multi-task Image-Based Dietary Assessment for Food Recognition and Portion Size Estimation. In Proceedings of the 2020 IEEE Conference on Multimedia Information Processing and Retrieval (MIPR), Shenzhen, China, 9–11 April 2020; pp. 49–54. [CrossRef]
126. Aguilar, E.; Nagarajan, B.; Khantun, R.; Bolaños, M.; Radeva, P. Uncertainty-Aware Data Augmentation for Food Recognition. In Proceedings of the 2020 25th International Conference on Pattern Recognition (ICPR), Milan, Italy, 10–15 January 2021; pp. 4017–4024. [CrossRef]
127. Ortega Anderez, D.; Lotfi, A.; Pourabdollah, A. A deep learning based wearable system for food and drink intake recognition. *J. Ambient. Intell. Hum. Comput.* **2020**, *12*, 9435–9447. [CrossRef]
128. Song, G.; Tao, Z.; Huang, X.; Cao, G.; Liu, W.; Yang, L. Hybrid Attention-Based Prototypical Network for Unfamiliar Restaurant Food Image Few-Shot Recognition. *IEEE Access* **2020**, *8*, 14893–14900. [CrossRef]
129. Xiao, L.; Lan, T.; Xu, D.; Gao, W.; Li, C. A Simplified CNNs Visual Perception Learning Network Algorithm for Foods Recognition. *Comput. Electr. Eng.* **2021**, *2*, 107152. [CrossRef]
130. Deng, L.; Chen, J.; Ngo, C.W.; Sun, Q.; Tang, S.; Zhang, Y.; Chua, T.S. Mixed Dish Recognition with Contextual Relation and Domain Alignment. *IEEE Trans. Multimed.* **2021**. [CrossRef]

131. Marc, B.; Ferrà, A.; Radeva, P. Food ingredients recognition through multi-label learning. In Proceedings of the International Conference on Image Analysis and Processing, Catania, Italy, 11–15 September 2017; Springer: Cham, Switzerland, 2017. [CrossRef]
132 Wang, Y.; Chen, J.-J.; Ngo, C.-W.; Chua, Y.-S.; Zuo, W.; Ming, Z. Mixed Dish Recognition through Multi-Label Learning. In Proceedings of the 11th Workshop on Multimedia for Cooking and Eating Activities (CEA'19), Ottawa, ON, Canada, 10 June 2019; Association for Computing Machinery: New York, NY, USA, 2019; pp. 1–8. [CrossRef]
133. Salvador, A.; Drozdzal, M.; Giro-i-Nieto, X.; Romero, A. Inverse cooking: Recipe generation from food images. In Proceedings of the IEEE/CVF Conference on Computer Vision and Pattern Recognition, Long Beach, CA, USA, 15–20 June 2019. [CrossRef]
134. Chen, J.; Pan, L.; Wei, Z.; Wang, X.; Ngo, C.-W.; Chua, T.-S. Zero-Shot Ingredient Recognition by Multi-Relational Graph Convolutional Network. In Proceedings of the AAAI Conference on Artificial Intelligence 2020, New York, NY, USA, 7–12 February 2020; Volume 34, pp. 10542–10550. [CrossRef]
135. Chen, J.; Zhu, B.; Ngo, C.-W.; Chua, T.-S.; Jiang, Y.-G. A Study of Multi-Task and Region-Wise Deep Learning for Food Ingredient Recognition. *IEEE Trans. Image Process.* **2021**, *30*, 1514–1526. [CrossRef]
136. Pettitt, C.; Liu, J.; Kwasnicki, R.; Yang, G.; Preston, T.; Frost, G. A pilot study to determine whether using a lightweight, wearable micro-camera improves dietary assessment accuracy and offers information on macronutrients and eating rate. *Br. J. Nutr.* **2016**, *115*, 160–167. [CrossRef]
137. Comber, R.; Weeden, J.; Hoare, J.; Lindsay, S.; Teal, G.; Macdonald, A.; Methven, L.; Moynihan, P.; Olivier, P. Supporting visual assessment of food and nutrient intake in a clinical care setting. In Proceedings of the Conference on Human Factors in Computing Systems, Austin, TX, USA, 5–10 May 2012; pp. 919–922. [CrossRef] [PubMed]
138. Yang, Z.; Yu, H.; Cao, S.; Xu, Q.; Yuan, D.; Zhang, H.; Jia, W.; Mao, Z.-H.; Sun, M. Human-Mimetic Estimation of Food Volume from a Single-View RGB Image Using an AI System. *Electronics* **2021**, *10*, 1556. [CrossRef]
139. Graikos, A.; Charisis, V.; Iakovakis, D.; Hadjidimitriou, S.; Hadjileontiadis, L. Single Image-Based Food Volume Estimation Using Monocular Depth-Prediction Networks. In *Universal Access in Human-Computer Interaction. Applications and Practice. HCII 2020*; Lecture Notes in Computer Science; Springer: Cham, Switzerland, 2020; Volume 12189. [CrossRef] [PubMed]
140. Lo, F.P.; Sun, Y.; Qiu, J.; Lo, B.P.L. Point2Volume: A Vision-Based Dietary Assessment Approach Using View Synthesis. *IEEE Trans. Ind. Inform.* **2020**, *16*, 577–586. [CrossRef]
141. Zhu, F.; Bosch, M.; Boushey, C.; Delp, E. An image analysis system for dietary assessment and evaluation. *Proc./ICIP Int. Conf. Image Process.* **2010**, *185*, 1853–1856. [CrossRef]
142. Trevño, R.; Ravelo, A.; Birkenfeld, E.; Murad, M.; Diaz, J. Food Weight Estimation: A Comparative Analysis of Digital Food Imaging Analysis and 24-Hour Dietary Recall. *J. Nutr. Educ. Behav.* **2015**, *47*, S105. [CrossRef]
143. Jia, W.; Chen, H.C.; Yue, Y.; Li, Z.; Fernstrom, J.; Bai, Y.; Li, C.; Sun, M. Accuracy of food portion size estimation from digital pictures acquired by a chest-worn camera. *Public Health Nutr.* **2014**, *17*, 1671–1681. [CrossRef]
144. Rhyner, D.; Loher, H.; Dehais, J.; Anthimopoulos, M.; Shevchik, S.; Botwey, R.H.; Duke, D.; Stettler, C.; Diem, P.; Mougiakakou, S. Carbohydrate Estimation by a Mobile Phone-Based System Versus Self-Estimations of Individuals With Type 1 Diabetes Mellitus: A Comparative Study. *J. Med. Internet Res.* **2016**, *18*, e101. [CrossRef] [PubMed]
145. Makhsous, S.; Mohammad, H.M.; Schenk, J.M.; Mamishev, A.V.; Kristal, A.R. A Novel Mobile Structured Light System in Food 3D Reconstruction and Volume Estimation. *Sensors* **2019**, *19*, 564. [CrossRef]
146. Yuan, D.; Hu, X.; Zhang, H.; Jia, W.; Mao, Z.; Sun, M. An automatic electronic instrument for accurate measurements of food volume and density. *Public Health Nutr.* **2021**, *24*, 1248–1255. [CrossRef]
147. Rollo, M.E.; Bucher, T.; Smith, S.P.; Collins, C.E. ServAR: An augmented reality tool to guide the serving of food. *Int. J. Behav. Nutr. Phys. Act.* **2017**, *14*, 65. [CrossRef] [PubMed]
148. Dehais, J.; Anthimopoulos, M.; Shevchik, S.; Mougiakakou, S. Two-View 3D Reconstruction for Food Volume Estimation. *IEEE Trans. Multimed.* **2017**, *19*, 1090–1099. [CrossRef]
149. Aizawa, K.; Maruyama, Y.; Li, H.; Morikawa, C. Food Balance Estimation by Using Personal Dietary Tendencies in a Multimedia Food Log. *IEEE Trans. Multimed.* **2013**, *15*, 2176–2185. [CrossRef]
150. Ravì, D.; Lo, B.; Yang, G. Real-time food intake classification and energy expenditure estimation on a mobile device. In Proceedings of the 2015 IEEE 12th International Conference on Wearable and Implantable Body Sensor Networks (BSN), Cambridge, MA, USA, 9–12 June 2015; pp. 1–6. [CrossRef]
151. Jiang, H.; Starkman, J.; Liu, M.; Huang, M. Food Nutrition Visualization on Google Glass: Design Tradeoff and Field Evaluation. *IEEE Consum. Electron. Mag.* **2018**, *7*, 21–31. [CrossRef]
152. Probst, Y.; Nguyen, D.T.; Tran, M.K.; Li, W. Dietary Assessment on a Mobile Phone Using Image Processing and Pattern Recognition Techniques: Algorithm Design and System Prototyping. *Nutrients* **2015**, *7*, 6128–6138. [CrossRef]
153. Mezgec, S.; Koroušić Seljak, B. Nutrinet: A deep learning food and drink image recognition system for dietary assessment. *Nutrients* **2017**, *9*, 657. [CrossRef]
154. Pouladzadeh, P.; Shirmohammadi, S. Mobile Multi-Food Recognition Using Deep Learning. *ACM Trans. Multimed. Comput. Commun. Appl.* **2017**, *13*, 36. [CrossRef] [PubMed]
155. Waltner, G.; Schwarz, M.; Ladstätter, S.; Weber, A.; Luley, P.; Bischof, H.; Lindschinger, M.; Schmid, I.; Paletta, L. MANGO—Mobile Augmented Reality with Functional Eating Guidance and Food Awareness. In Proceedings of the New Trends in Image Analysis and Processing—ICIAP 2015 Workshops, Genoa, Italy, 7–8 September 2015; pp. 425–432. [CrossRef]

156. Sun, J.; Radecka, K.; Zilic, Z. FoodTracker: A Real-time Food Detection Mobile Application byDeep Convolutional Neural Networks. *arXiv* **2019**, arXiv:1909.05994.
157. Jia, W.; Li, Y.; Qu, R.; Baranowski, T.; Burke, L.E.; Zhang, H.; Bai, Y.; Mancino, J.M.; Xu, G.; Mao, Z.H.; et al. Automatic food detection in egocentric images using artificial intelligence technology. *Public Health Nutr.* **2018**, *22*, 1168–1179. [CrossRef]
158. Chen, G.; Jia, W.; Zhao, Y.; Mao, Z.H.; Lo B.; Anderson, A.K.; Frost, G.; Jobarteh, M.L.; McCrory, M.A.; Sazonov, E.; Steiner-Asiedu, M. Food/Non-Food Classification of Real-Life Egocentric Images in Low- and Middle-Income Countries Based on Image Tagging Features. *Front. Artif. Intell.* **2021**, *4*, 644712. [CrossRef]
159. Mejova, Y.; Abbar, S.; Haddadi, H. Fetishizing Food in Digital Age: Foodporn Around the World. *arXiv* **2016**, arXiv:1603.00229.
160. Ismail, A.; Idris, M.Y.I.; Ayub, M.N.; Por, L.Y. Investigation of Fusion Features for Apple Classification in Smart Manufacturing. *Symmetry* **2019**, *11*, 1194. [CrossRef]
161. Chen, T.; Kornblith, S.; Norouzi, M.; Hinton, G. A Simple Framework for Contrastive Learning of Visual Representations. In Proceedings of the 37th International Conference on Machine, Learning, PMLR, Vienna, Austria, 13–18 July 2020; Volume 110, pp. 1597–1607.
162. Caron, M.; Misra, I.; Mairal, J.; Goyal, P.; Bojanowski, P.; Joulin, A. Unsupervised Learning of Visual Features by Contrasting Cluster Assignments. *arXiv* **2020**, arXiv:2006.09882.
163. Strumbelj, E.; Kononenko, I. Explaining prediction models and individual predictions with feature contributions. *Knowl. Inf. Syst.* **2014**, *41*, 647–665.
164. Arik, S.; Liu, Y.-H. Explaining Deep Neural Networks using Unsupervised Clustering. *arXiv* **2020**, arXiv:2007.07477.
165. Subhi, M.A.; Ali, S.H.; Mohammed, M.A. Vision-based approaches for automatic food recognition and dietary assessment: A survey. *IEEE Access* **2019**, *7*, 35370–35381.
166. Min, W.; Jiang, S.; Liu, L.; Rui, Y.; Jain, R. A Survey on Food Computing. *ACM Comput. Surv.* **2019**, *52*, 92. [CrossRef]

healthcare

Review

Does Artificial Intelligence Make Clinical Decision Better? A Review of Artificial Intelligence and Machine Learning in Acute Kidney Injury Prediction

Tao Han Lee [1,2], Jia-Jin Chen [1,2], Chi-Tung Cheng [3] and Chih-Hsiang Chang [1,2,*]

[1] Kidney Research Center, Department of Nephrology, Chang Gung Memorial Hospital, Linkou Branch, Taoyuan 33305, Taiwan; kate0327@hotmail.com (T.H.L.); Raymond110234@hotmail.com (J.-J.C.)
[2] Graduate Institute of Clinical Medical Science, College of Medicine, Chang Gung University, Taoyuan 33302, Taiwan
[3] Department of Trauma and Emergency Surgery, Chang Gung Memorial Hospital, Taoyuan 33305, Taiwan; atong89130@gmail.com
* Correspondence: franwisandsun@gmail.com

Abstract: Acute kidney injury (AKI) is a common complication of hospitalization that greatly and negatively affects the short-term and long-term outcomes of patients. Current guidelines use serum creatinine level and urine output rate for defining AKI and as the staging criteria of AKI. However, because they are not sensitive or specific markers of AKI, clinicians find it difficult to predict the occurrence of AKI and prescribe timely treatment. Advances in computing technology have led to the recent use of machine learning and artificial intelligence in AKI prediction, recent research reported that by using electronic health records (EHR) the AKI prediction via machine-learning models can reach AUROC over 0.80, in some studies even reach 0.93. Our review begins with the background and history of the definition of AKI, and the evolution of AKI risk factors and prediction models is also appraised. Then, we summarize the current evidence regarding the application of e-alert systems and machine-learning models in AKI prediction.

Keywords: artificial intelligence; machine learning; acute kidney injury; prediction model

Citation: Lee, T.H.; Chen, J.-J.; Cheng, C.-T.; Chang, C.-H. Does Artificial Intelligence Make Clinical Decision Better? A Review of Artificial Intelligence and Machine Learning in Acute Kidney Injury Prediction. *Healthcare* **2021**, 9, 1662. https://doi.org/10.3390/healthcare9121662

Academic Editor: Mahmudur Rahman

Received: 19 October 2021
Accepted: 26 November 2021
Published: 30 November 2021

1. Introduction

Acute kidney injury (AKI), defined as increased serum creatinine level or decreased urine output, is the most common and adverse complication of hospitalization in patients [1]. The incidence of AKI among inpatients ranges from 5% to 10%, and it ranges from 20% to 70% among patients admitted to an intensive care unit (ICU) [2–5]. AKI incidence varies by clinical condition; approximately 20% of patients with Stevens–Johnson syndrome or toxic epidermal necrolysis developed AKI, and 56% of patients with severe sepsis developed AKI. Among patients who have undergone surgery, AKI incidence varies by the type of operation, ranging from 25% for trauma surgery to as high as 50% for cardiac or aortic surgery [6,7]. Although the quality of medication data and the effectiveness of treatment have greatly improved recently, the incidence of AKI has continually increased, possibly due to the aging population and rising comorbidities, such as diabetes mellitus and hypertension.

After an initial AKI episode, the risk of chronic kidney disease (CKD), long-term dialysis and mortality are significantly increased in the affected patients [8–14]. According to a previous meta-analysis, patients with AKI had higher risks of CKD, end-stage renal disease (ESRD), and mortality than patients without AKI; the hazard ratios were 8.8, 3.1, and 2.0, respectively [10]. Among patients with AKI, those with dialysis-dependent AKI had even poorer renal outcomes than patients with non-dialysis-dependent AKI [14,15]. Although investigators had identified that patients with hypertension or diabetes mellitus, those requiring readmission for cardiovascular disease or sepsis, those receiving

cardiovascular surgery or neurosurgery, and those taking nephrotoxic agents (nonsteroidal anti-inflammatory drugs, radiocontrast, hydroxyethyl starch, and nephrotoxic antimicrobials) were prone to experience AKI [16–18]. No accurate tool has been established for identifying patients at risk of AKI and for predicting AKI occurrence. At the same time, patients only exhibit imperceptible signs of AKI or even exhibit no clinical symptoms in the early stages of AKI. Once oliguria, hematuria, or anasarca is present, patients may already have considerable parenchymal injury and require renal replacement therapy. Although research on novel biomarkers has increased in recent years, advances in clinical informatics, artificial intelligence (AI), and machine learning may enable the development of additional approaches for the prediction and estimation of AKI risk through the processing of electronic medical records (EMRs) [19]. In this article, we review the progress in the application of machine learning systems for AKI risk prediction.

1.1. AKI Definition

The definition of AKI has evolved over the past few decades, ranging from the initial Risk, Injury, Failure, Loss of kidney function, and End-stage kidney disease (RIFLE) classification and the Acute Kidney Injury Network (AKIN) criteria to the most recent Kidney Disease Improving Global Outcome (KDIGO) guidelines [1,20,21]. The KDIGO guidelines have been the most widely used definition of AKI over the past decade, and according to these guidelines, AKI is divided into stages by severity on the basis of increasing serum creatinine level and urine output rate data. However, the serum creatinine level and urine output rate are not sensitive or specific markers of AKI. The interpretation of changes in renal function is prone to error when conducted on the basis of serum creatinine level. First, because creatinine is not only glomerular-filtered but also secreted by tubules, creatinine clearance overestimates the true GFR, especially in cases of decreased renal function [22,23]. Second, serum creatinine level is influenced by muscle mass (creatinine is a product of muscle catabolism), diet (a protein-rich diet results in higher serum creatinine level), and drugs (for example, trimethoprim and cimetidine interfere with the tubular secretion of creatinine) [24,25]. Third, the production of muscular creatine is influenced by disease status; for example, it is lower and greater in severe hepatic disease and rhabdomyolysis, respectively [22,26]. Lastly, serum creatinine level is not significantly elevated until 48 h after renal injury, and delayed elevation detrimentally affects the timely identification of renal injury [27,28]. Although urine output rate may reflect renal function decline in a timelier manner, it is still affected by the patient's volemic status and is influenced by diuretic treatment.

Because both serum creatinine level and urine output rate are nonspecific and inaccurate markers of AKI, multiple novel biomarkers have been investigated for predicting or diagnosing AKI in a timely manner. The following novel biomarkers have been identified for the early detection of AKI: cystatin C, neutrophil gelatinase-associated lipocalin, kidney injury molecule 1, liver type fatty-acid binding protein, urine angiotensinogen (AGT), and calprotectin. Chen and colleagues reported that serum cystatin C, urine NGAL, and serum interleukin-18 (IL-18) played valuable roles in the early detection of AKI in a cardiac care unit (CCU) and that the areas under the receiver operating characteristic curve (AUROCs) of serum cystatin C, urine NGAL, and serum IL-18 for AKI prediction were 0.895, 0.886, and 0.841, respectively. Multiple regression analysis indicated that urine NGAL, serum IL-18, and sodium levels at CCU admission were independent risk factors for 6-month mortality. Among these factors, urine NGAL had the highest discriminatory power, and the Youden index indicated that it yielded the most accurate prediction of patient mortality [29]. Some studies have described pseudo-worsening renal failure (also termed pseudo-AKI), which is a common clinical condition in patients with cardiorenal syndrome in which increases in serum creatinine level are induced by diuretic treatment rather than by tubular necrosis or interstitial nephritis. These studies have suggested that the novel biomarker calprotectin can distinguish a true AKI episode from a pseudoepisode

of diuretics-related AKI [30,31]. Chang et al. aptly reported that calprotectin had an excellent AUROC of 0.946 for predicting intrinsic AKI [32].

Although the novel AKI biomarkers identified in recent studies have greatly improved and enabled the earlier detection of AKI, many difficulties remain in applying these biomarkers in clinical settings. Vanmassenhove and colleagues noted that the early diagnosis of AKI by using novel serum and urinary biomarkers remains cumbersome, especially in settings in which the timing and etiology of AKI are not well defined [33]. Another difficulty is that tests for novel biomarkers are not widely commercially available or can be expensive and repeat examinations may be required during the process of AKI diagnosis. Moreover, Marx et al. concluded that it is almost impossible to depend on one universal serum or urine biomarker to determine the risk, diagnosis, severity, and outcome of AKI and to discriminate between etiologies of AKI and monitor its course [34]. AKI is a nonuniform, complex condition with a wide spectrum of causes and pathophysiological mechanisms; therefore, the requirement of several biomarkers or marker panels that cover different aspects of AKI seems reasonable for standardizing diagnoses [34,35]. However, examining multiple novel biomarkers or evaluating the patient's condition by using marker panels may further increase the costs of predicting or diagnosing AKI early and accurately. Therefore, the most cost-effective method appears to be identifying which patients with AKI are at high risk before arranging a biomarker examination for them.

1.2. AKI Risk Factors and Risk Scores

Some studies that have focused on identifying significant risk factors for AKI have determined that both patient susceptibilities and exposure are crucial in AKI development. Patient susceptibilities include age, gender, race, and comorbidities. Among all comorbidities, CKD has been identified as a major risk factor for AKI due to its associated loss of autoregulation, loss of renal reserve, and susceptibility to nephrotoxic agents. Moreover, diabetes mellitus, hypertension, cardiovascular disease, hyperuricemia, obesity, and liver disease have all been reported as risk factors for AKI [19,36,37]. Exposure to sepsis, nephrotoxic agents, surgical intervention, and shock have been identified as contributors to AKI [16,17]. A multicenter international cross-sectional AKI–EPI study reported that sepsis, hypovolemia, and nephrotoxic drug exposure were the three most frequently reported etiologies of AKI in patients with a critical illness [16]. The incidence of AKI may be higher among patients with poor physical condition after certain exposure; for example, an aging patient may have a higher risk of AKI after cardiac surgery. However, AKI risk differs by the physical condition and nephrotoxic exposure; this renders accurate risk assessment challenging.

After the risk factors for AKI were identified, investigators began focusing on establishing a risk score by using a combination of independent AKI predictors, assessment of relative impact, and external validation. A precise risk prediction score must be able to identify at-risk patients and guide physicians in preventing, diagnosing, and treating the disease. Different scoring systems have been constructed for assessing the risk of AKI in specific groups of patients; these prediction models include age, gender, baseline renal function, and comorbidities, and specific predictors can be added depending on surgery type, medication, and procedure-related data.

The Mehran risk score was proposed in 2004 for analyzing the risk of AKI and the requirement of renal replacement therapy in patients with postpercutaneous coronary intervention; according to later external validation conducted in 2016, the system exhibited adequate performance for predicting contrast-induced nephropathy in patients with acute coronary syndrome who underwent coronary angiography [38,39]. Large cohort studies have revealed that surgery is a major cause of AKI, and the AKI incidence rate ranges from 25% for trauma surgery to as high as 50% for cardiac or aortic surgery [6,7,40]. Additionally, cardiac surgery is associated with the highest AKI incidence among all types of surgery, ranging from 2% to 50%, and the dialysis-dependent rate is 1% to 6% [41,42]; therefore, it is unsurprising that several prediction models have been established for AKI

risk identification in patients who plan to undergo cardiac surgery. The earliest scoring system EuroSCORE is based on European multicenter data published in 1999, and the 2010 Value of Age, Creatinine, and Ejection Fraction (ACEF) score is also based on data from European databases [43,44]. The short-term risk (Society of Thoracic Surgeons, STS) score was created in 2008 by using data from the national database of the American Society of Thoracic Surgery; this score is used to evaluate adult preoperative cardiac surgery risk, and professionals have retained and modified this prediction model [45,46]. In an externally validated study, 196 patients received mitral valve repair, and their STS and ACEF scores were compared; the STS renal failure score was the most accurate for predicting stage 2 and 3 AKI. Additionally, that study found that ACEF scores exhibited an AUROC similar to that of STS renal failure scores across all AKI predictions (ACEF and STS score AUROCs: 0.758 and 0.797, respectively), but the ACEF score includes only three prediction factors: age, creatinine, and ejection fraction; thus, the ACEF score is more convenient for clinical physicians [41]. In another study that compared the preoperative risk models of AKI in isolated coronary artery bypass grafting surgery, the EuroSCORE II, STS score, and ACEF score all performed adequately for predicting stage 3 AKI; additionally, the ACEF score exhibited satisfactory discriminatory power for predicting postoperative AKI, with an AUROC of 0.781 [47].

Besides the comorbidities and acute illness conditions, race and epidemiology factors also showed their impact on AKI incidence according to previous studies. Mathioudakis and his colleagues had reported that blacks had a 50% higher age- and sex-adjusted odds of AKI compared to whites (odds ratio: 1.51; 95% CI 1.37–1.66) based on the national databases of the U.S. This association between the black race and increased risk of AKI persisted after additional adjustment for multiple AKI-related risk factors [48]. In 2013, a meta-analysis focused on AKI incidence worldwide reported that the pooled rate of AKI according to KDIGO criteria showed a difference around the world. According to geographic regions of the world and patterns of country economies and latitude, the pooled rate of AKI appeared higher in South versus North America (29.6% versus 24.5%), Southern versus Northern Europe (31.5% versus 14.7%), and South versus Western or Eastern Asia (23.7% versus 16.7% versus 14.7%). The pooled rate of AKI appeared higher in studies from countries located south versus north of the equator (27.0% versus 22.6%), in addition, this study also revealed that the AKI incidence was high in countries that spent >10% versus ≤5% GDP on total health expenditure (25.2% versus 14.5%) [49].

Considering the influence of race and epidemiology on AKI incidence, some investigators have validated their scores against data from their country's health insurance research database to achieve high prediction performance. An example is the ADVANCIS score, which is used to predict AKI in patients who receive percutaneous coronary intervention (PCI) for coronary artery disease; the score was validated against data from Taiwan's National Health Insurance Research Database. The ADVANCIS score uses eight clinical parameters (age, diabetes mellitus, ventilator use, prior AKI, number of intervened vessels, CKD, IABP use, and cardiogenic shock), and the score ranges from 0 to 22; additionally, an ADVANCIS score of ≥6 is associated with higher in-hospital mortality risk [50]. In addition to modifying risk prediction models in accordance with epidemiological factors, researchers have included novel biomarkers as prediction factors in some modern AKI prediction score systems and have assessed the association between biomarkers and patients' clinical information. Zhou et al. established a prediction score of AKI in patients with acute decompensated heart failure by setting urine NGAL and urine AGT as risk factors [51].

Although various scoring systems have been established to address different clinical conditions, most prediction models can perform only as single-point AKI prediction models, such as predicting AKI incidence after a specific type of surgery or before the use of a contrast agent, making it difficult to reflect changes in real-time. Furthermore, some of these scoring systems cover several factors, including baseline condition, clinical data, and novel biomarkers, making them too complex for clinical use. With the development of information technology, some hospitals have integrated these prediction systems into

their medical informatics systems (MISs), and these clinical risk assessment tools have been increasingly used because they enable the automated analysis of data. Because race, genes, disease prevalence, and medication differ between countries, the combined use of an MIS and risk prediction scores potentially enable the use of data from local databases to assess the risks of AKI and the requirement of renal replacement therapy.

1.3. From Automated Electronic Alerts to AI

As MISs become more popular, systems that provide automated electronic alerts (e-alerts) have become increasingly feasible; in such a system, the electronic records and clinical information of patients are analyzed using an algorithm that predicts whether early or subclinical AKI is present [52]. These systems are expected to aid patient care by making clinical evaluation and treatment timelier. Park and his colleagues had investigated an AKI alert system with automated nephrologist consultation in which clinicians could generate automated consultations to the nephrology division while patient's serum creatinine concentration elevation of at least 1.5-fold or 0.3 mg/dL from baseline. This study reported that the early consultation with a nephrologist was greater (adjusted OR, 6.13; 95% CI, 4.80–7.82) and odds of a severe AKI event were reduced (adjusted OR, 0.75; 95% CI, 0.64–0.89) after introducing the e-alert system. However, mortality was not affected (adjusted HR, 1.07; 95% CI, 0.68–1.68) [53]. Another study used an e-alert system in ICU patients, clinician received a "pop-up" message while the e-alert system screened the serum creatinine data and detected possible AKI events following the KDIGO criteria definition. Although the sensitivity, specificity, Youden Index and accuracy of the AKI e-alert system were 99.8, 97.7, 97.5 and 98.1%, respectively, in this study, and the prevalence of diagnosis AKI and the prevalence of nephrology consultation in the e-alert group was higher than that in the non-e-alert group. There was no significant difference in the prevalence of dialysis, rehabilitation of renal function, or death in the two groups [54]. In 2017, a systemic review concluded that an e-alerts system neither reduced mortality (odds ratio [OR], 1.05; 95% CI, 0.84–1.31) nor reduced the incidence of dialysis treatment (OR, 1.20; 95% CI, 0.91–1.57) [55,56]. All six studies included in this meta-analysis used only serum creatinine change as the trigger for e-alerts, and serum creatinine change is neither a sensitive nor specific marker of kidney injury, as mentioned in the preceding paragraph. Beyond the limitation of serum creatinine as an AKI marker, e-alerts systems face challenges when used in patients without baseline renal function and those with CKD who have higher baseline creatinine levels and more significant changes in renal function following small changes in creatinine level; a wide variety of further care is provided by clinicians to patients after the receipt of e-alerts. To prescribe standardized and evidence-based clinical care after the receipt of e-alerts, a care bundle was built. The most recent guidelines prescribe no specific management options for AKI, and the treatment strategy is mainly supportive. In critically ill patients, the occurrence and severity of AKI were reduced following adherence to KDIGO guidelines detailing the management of fluids, avoidance of nephrotoxins, monitoring of serum creatinine levels and hemodynamics, and referral to a specialist. Several studies have reported a decrease in hospital-acquired AKI and AKI-associated mortality and hospitalization days when the e-alert system was combined with a care bundle, the patient's history was analyzed, the patient's urine samples were tested, a clinical diagnosis of AKI was established, the course of treatment and testing was planned, and advice was sought from a nephrologist [57–59]. Machine-learning algorithms are in high demand and require large volumes of data. With large EMR databases and powerful computing hardware, scholars have extended the application of machine learning. Recently, AI has also been applied with various machine-learning algorithms, especially deep neural networks.

2. Methods

In order to get a closer look at the investigating of machine-learning studies on AKI prediction, we searched PubMed for clinical trials and conference abstracts discussing how machine learning and AI can be used to predict AKI. Online literature searches of the PubMed database were performed, and the database search was last updated on 1 December 2020. The search strategy targeted published clinical trials, including conference abstracts that described the use of machine learning for predicting AKI in adults. The search strategy and results are detailed in Supplementary Table S1. Two investigators (T.H. Lee and J.J. Chen) independently evaluated the titles and abstracts of the retrieved studies, and articles were excluded upon initial screening if their titles or abstracts indicated that they were clearly irrelevant to the objective of the current study. Full-text reviews were then performed for the articles deemed potentially relevant to assess their eligibility for inclusion. The study inclusion criteria were as follows: (i) a study population consisting of adults and the study having a prospective or retrospective design and (ii) AKI prediction through machine learning. Case series and reports, conference abstracts, comments on other studies, and review articles were excluded.

3. Results

In total, 31 studies reported the discriminating ability of machine learning for predicting AKI (Table 1).

As shown in Table 1, the included studies predicted AKI adequately, some studies had AUROC > 0.8, and the study conducted by Koola et al. had the highest AUROC of 0.93 in logistic regression. The models outperformed diagnosis through novel biomarkers. Machine-learning models that were used to predict AKI had four to 57 covariates. These covariates were epidemiological factors, comorbidities, laboratory data, medications, and surgery types. We summarize the most commonly used covariates in these machine-learning prediction models in Figure 1. In the 31 studies, the five most commonly used covariates were creatinine, age, blood pressure, gender, and diabetes mellitus. Among these 31 studies, eight studies focused on patients' undergoing surgery (surgeries were cardiac or aortic surgeries in five studies), and the most commonly used covariates in surgical patients are illustrated in Figure 2; the five most common used covariates were gender, body mass index, age, creatinine, and surgery type.

In Table 2, we summarized the method of feature selection, data splitting and machine learning algorithm choices in enrolled studies. Different performances on predicting AKI by using different machine learning algorithms were also listed in this table. More than half of the enrolled studies used LASSO, XGBoost, or other feature selection methods to choose the covariates for machine learning, but some studies chose covariates according to clinical experience or previous reports.

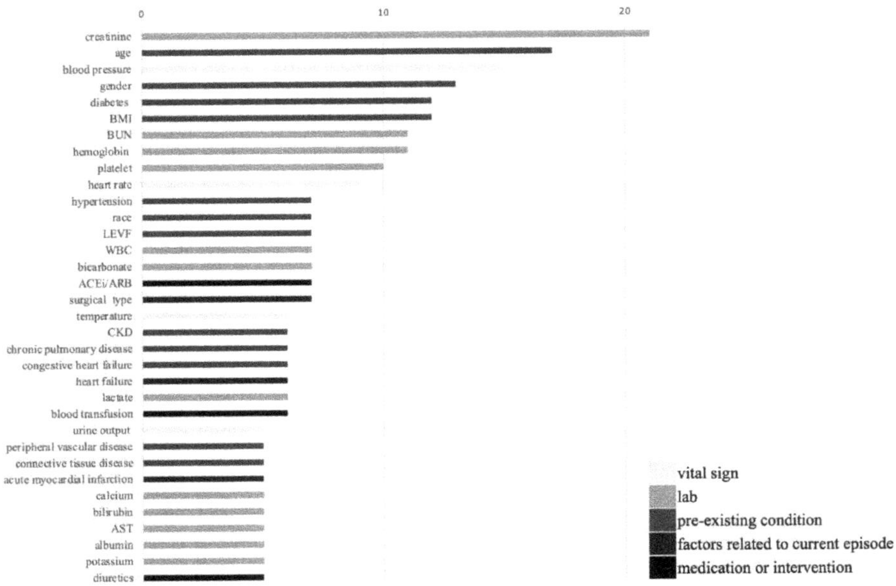

Figure 1. Covariates are most commonly used in machine-learning prediction models in the enrolled studies. The covariates are grouped by type. ACEi: angiotensin converting enzyme inhibitor; ARB: angiotensin receptor blocker; AST: aspartate aminotransferase; BMI: body mass index; BUN: blood urea nitrogen; CKD: chronic kidney disease; LVEF: left ventricle ejection fraction; WBC: white blood cell count.

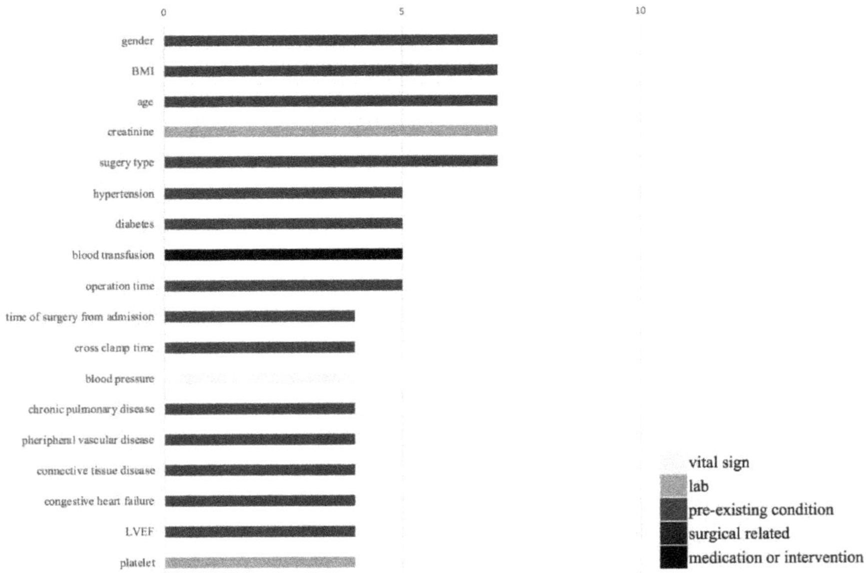

Figure 2. Covariates are most commonly used in machine-learning prediction models in enrolled surgical studies. The covariates are grouped by type. BMI: body mass index; LVEF: left ventricle ejection fraction.

Table 1. Summary of machine-learning studies on acute kidney injury (AKI) prediction.

Scheme	Year	Design	Population	AKI Definition	Timing of AKI	AKI Incidence (%)	Patient Number	External Validation	Continuous Prediction
Kate et al. [60]	2016	retrospective	medical and surgical	AKIN	during hospitalization	8.9%	25,521	no	no
Thottakkara et al. [61]	2016	retrospective	surgical	KDIGO	post operation	36.0%	50,318	no	no
Davis et al. [62]	2017	retrospective	medical and surgical	KDIGO	during hospitalization	6.8%	2003	no	no
Cheng et al. [63]	2018	retrospective	medical and surgical	KDIGO	during hospitalization	9.0%	60,534	no	no
Ibrahim et al. [64]	2018	prospective	PCI	KDIGO	pre and post intervention	4.8%	889	no	no
Koola et al. [65]	2018	retrospective	medical and surgical	KDIGO	during hospitalization	NR (41.6% HRS)	504	no	no
Koyner et al. [66]	2018	retrospective	medical and surgical	KDIGO	24 h post admission	14.4%	121,158	no	no
Huang et al. [67]	2018	retrospective	PCI	KDIGO	during hospitalization	7.4%	947,091	no	no
Lin et al. [68]	2019	retrospective	ICU	KDIGO	during hospitalization	14%	19,044	no	no
Simonov et al. [69]	2019	retrospective	medical and surgical	KDIGO	24 h post admission	11.4–19.1%	169,859	yes	no
Huang et al. [70]	2019	retrospective	PCI	AKIN	pre and post intervention	6.4%	2,076,694	no	no
Tomašev et al. [71]	2019	retrospective	medical and surgical	KDIGO	during hospitalization	13.4%	703,782	no	yes
Adhikari et al. [72]	2019	retrospective	surgical	KDIGO	post operation	46.0%	2901	no	no
Flechet et al. [73]	2019	prospective	ICU	KDIGO	during hospitalization	12% †	252	no	no
Parreco et al. [74]	2019	retrospective	medical and surgical	KDIGO	during hospitalization	5.6%	151,098	no	no
Xu et al. [75]	2019	retrospective	medical and surgical	KDIGO	during hospitalization	NR	58,976	no	no
Tran et al. [76]	2019	prospective	burn	KDIGO	during hospitalization	50.0%	50	no	no
Zhang et al. [77]	2019	retrospective	ICU	KDIGO	24 h post admission	58.1%	6682	no	no
Zimmerman et al. [78]	2019	retrospective	ICU	KDIGO	72 h post admission	16.5%	46,000	no	no
Rashidi et al. [79]	2020	retrospective and prospective	burn and trauma	KDIGO	1st week post ICU admission	50.0%	101	no	no
Zhou et al. [80]	2020	retrospective	TAAAR	NR	post operation	12.7%	212	no	no
Martinez et al. [81]	2020	retrospective	medical and surgical	KDIGO	emergency department	7.9%	59,792	no	no
Lei et al. [82]	2020	retrospective	TAAR	KDIGO	post operation	72.6%	897	no	no
Lei et al. [83]	2020	retrospective	hepatectomy	KDIGO	post operation	6.6%	1173	no	no
Qu et al. [84]	2020	retrospective	acute pancreatitis	KDIGO	during hospitalization	24.0%	334	no	no
Tseng et al. [85]	2020	retrospective	Cardiac surgery	KDIGO	post operation	24.3%	671	no	no
Sun et al. [86]	2020	retrospective	PCI	KDIGO	during hospitalization	15.1%	1495	no	no
Churpek et al. [87]	2020	retrospective	medical and surgical	KDIGO	during hospitalization	14.3%	495,971	yes	no
Hsu et al. [88]	2020	retrospective	medical and surgical	KDIGO	Community acquired AKI	8.4%	234,867	no	no
Penny-Dimri et al. [89]	2020	retrospective	Cardiac surgery	Other *	post operation	6.5%	97,964	no	no
Li et al. [90]	2020	retrospective	Cardiac surgery	KDIGO	post operation	37.5%	5533	no	no

* The AKI definition in this study was as follows: (1) new postoperative and in-hospital serum creatinine level > 200 mmol/L AND a doubling or greater increase in creatinine over the baseline preoperative value AND the patient did not require preoperative renal replacement therapy; and (2) a new inhospital requirement for renal replacement therapy. † Only reported the percentage of AKI stage 2 and stage 3. AKI: acute kidney injury; ICU: intensive care unit; PCI: percutaneous coronary intervention; TAAR: total aortic arch replacement; TAAAR: thoracoabdominal aortic aneurysm repair.

Table 2. Summary of data processing and performance of machine-learning algorithm in enrolled studies.

Study	Feature Selection Algorithm	Feature Selection Method	Data Splitting	Machine Learning Algorithm	AUROC
Kate et al. [60]	NR	NR	ten-fold cross-validation	naïve Bayes	0.654
				SVM	0.621
				decision trees	0.639
				logistic regression	0.660
Thottakkara et al. [61]	LASSO	embedded method	training data (70%); validation (30%)	naïve Bayes	0.819
				generalized additive model	0.858
				logistic regression	0.853
				support vector machine	0.857
Davis et al. [62]	according to clinical experience or previous report	NR	five-fold cross-validation	random forest	0.73
				neural network	0.72
				naïve Bayes	0.69
				logistic regression	0.78
Cheng et al. [63]	according to clinical experience or previous report	NR	ten-fold cross-validation	random forest	0.765
				AdaBoostM1	0.751
				logistic regression	0.763
Ibrahim et al. [64]	LASSO	embedded method	Monte Carlo cross-validation	logistic regression	0.79
Koola et al. [65]	LASSO	embedded method	five-fold cross-validation	logistic regression	0.93
				naïve Bayes;	0.73
				support vector machines;	0.90
				random forest;	0.91
				gradient boosting	0.88
Koyner et al. [66]	tree-based method	embedded method	ten-fold cross-validation	gradient boosting	0.9
Huang et al. [67]	XGBoost and LASSO	embedded method	training data (70%); validation (30%)	gradient boost;	0.728
				logistic regression	0.717
Lin et al. [68]	according to clinical experience or previous report	NR	five-fold cross-validation	SVM	0.86
Simonov et al. [69]	according to clinical experience or previous report	NR	training data (67%); validation (33%)	discrete-time logistic regression	0.74
Huang et al. [70]	stepwise backward selection, LASSO, premutation-based selection	embedded method	training (50%); validation (50%)	generalized additive model	0.777
Tomašev et al. [71]	L1 regularization	embedded method	training (80%); validation (5%); calibration (5%); test (10%)	recurrent neural network	0.934

Table 2. *Cont.*

Study	Feature Selection Algorithm	Feature Selection Method	Data Splitting	Machine Learning Algorithm	AUROC
Adhikari et al. [72]	F-test	filter method	five-fold cross-validation	random forest	0.86
Flechet et al. [73]	according to clinical experience or previous report	NR	NR	random forest	0.78
Parreco et al. [74]	NR	NR	NR	gradient boosting; logistic regression; deep learning	0.834 0.827 0.817
Xu et al. [75]	gradient boosting	embedded method	five-fold cross-validation	gradient boosting	0.749
Tran et al. [76]	NR	NR	Scikit-learn cross validation	k-nearest neighbor	0.92
Zhang et al. [77]	XGBoost	embedded method	bootstrap validation	gradient boosting	0.86
Zimmerman et al. [78]	logistic regression	embedded method	five-fold cross-validation	logistic regression random forest neural network	0.783 0.779 0.796
Rashidi et al. [79]	according to clinical experience or previous report	NR	Scikit-learn cross validation	recurrent neural network	0.92
Zhou et al. [80]	NR	NR	five-fold cross-validation	logistic regression linear kernel SVM Gaussian kernel SVM random forest	0.73 0.84 0.77 0.89
Martinez et al. [81]	LASSO	embedded method	ten-fold cross-validation	random forest	not provided
Lei et al. [82]	NR	NR	training data (70%); validation (30%)	Gradient boosting	0.8
Lei et al. [82]	NR	NR	training data (70%); validation (30%)	Gradient boosting Light gradient boosted machine random forest DecisionTree	0.772 0.725 0.662 0.628
Qu et al. [84]	NR	NR	ten-fold cross-validation	random forest classification and regression tree logistic regression extreme gradient boosting	0.821 0.8033 0.8728 0.9193
Tseng et al. [85]	tree-based method	embedded method	five-fold cross-validation	random forest random forest with extreme gradient boosting	0.839 0.843
Sun et al. [86]	Boruta algorithm	wrapper method	ten-fold cross-validation	random forest logistic regression;	0.82 0.69

Table 2. *Cont.*

Study	Feature Selection Algorithm	Feature Selection Method	Data Splitting	Machine Learning Algorithm	AUROC
Churpek et al. [87]	gradient boosting	embedded method	ten-fold cross-validation	gradient boosted machine	0.72
Hsu et al. [88]	XGBoost and LASSO	embedded method	five-fold cross-validation	logistic regression;	0.767
Penny-Dimri et al. [89]	tree-based method	embedded method	five-fold cross-validation	logistic regression; gradient boosted machine neural networks	0.77 0.78 0.77
Li et al. [90]	LASSO	embedded method	ten-fold cross-validation	Bayesian networks	0.736

AUROC: area under the receiver operating characteristic curve; LASSO: least absolute shrinkage and selection operator; NR: not reported; SAPS: simplified acute physiology score; SVM: support vector machine; XGB: eXtreme Gradient Boostin.

4. Discussion

Among these 31 studies, there were several studies that are worth addressing. By reviewing these studies, we found that most of these studies lacked external validation, which implies that the results cannot be extended to other populations. Two studies performed external validation. Simonov and colleagues established a real-time AKI prediction model by using an electronic health record (EHR) dataset of 169,859 hospital admissions in three hospitals. The training dataset contained the data of 60,701 patients, and the internal validation dataset contained the data of 30,599 patients from the same hospitals; external validation was performed with the data sets of 43,534 and 35,025 patients from two other hospitals. The incidence of AKI was similar in the training and external validation datasets (19.1% and 18.9%, respectively). Discrete-time logistic regression was used to train the model, a total of 35 covariates were included in the fully adjusted models, and the AUROCs for predict sustained AKI, dialysis, and death were 0.77 (95% CI, 0.76–0.78), 0.79 (95% CI, 0.73–0.85), and 0.69 (95% CI, 0.67–0.72), respectively [69,91]. This real-time prediction model was based on large cohorts including patients requiring hospitalization and those in surgical and ICU settings, and the external validation of this model was performed using the data from two other institutions, with high predictive performance found across the three diverse care settings; the subsequent prospective cohort study indicated that the clinical alert system based on this prediction model was successfully integrated into the EHR system [91]. However, this real-time prediction model still had several limitations. First, patients whose creatinine levels were $\geq$4 were excluded during the development of this prediction model, but the risk and incidence of AKI and dialysis requirements are especially high in this population. Second, this prediction model did not include urine output, one of the most sensitive markers of AKI, and thus, could delay diagnosis in patients who already had oliguria but had increased serum creatinine levels. Third, more than 30 covariates were included in this prediction model; some of these covariates are infrequently checked laboratory data, such as bicarbonate and chloride levels. Moreover, as mentioned in this report, only the model containing time-updated laboratory values had similar performance in predicting AKI, sustained AKI, dialysis, and death. Unless all of these items are regularly checked in the ICU, it is difficult to evaluate AKI risk in a timely manner. Another study that performed external validation was published by Churpek et al., the data of 48,463 admissions were included in training and internal validation datasets, and the data of 447,508 admissions were used for external validation. The AUROC for predicting development AKI within 48 h was 0.72 for the internal validation cohort and the ARUROC of the two external validation cohorts were 0.67, 0.69, whereas the AUROC for predicting the receipt of renal replacement therapy within 48 h was 0.95. However, this study had a similar limitation to that of the study by Simonov et al.; the study excluded patients with serum creatinine concentration over 3.0 mg/dL on admission [87]. Higher creatinine levels and chronic kidney disease are known risk factors for AKI. It is unfortunate that the only two studies with external validation coincidentally excluded the high-risk population from the beginning.

In addition to the lack of external validation, most of the enrolled studies only predicted AKI risk at a single time point and could not provide continual predictions. Given that patients' clinical conditions change from time to time, using laboratory, medication, and vital sign data at a single time point to perform single-point AKI risk prediction may not reflect the real-time changes of patients. One study investigated continuous risk prediction by using novel neural network algorithms. Such algorithms can process time-series data to produce time-dependent forecasts rather than forecasts that depend on summary data, as is the case in traditional methods. Tomašev et al. used the recurrent neural network to demonstrate a deep-learning approach for the continuous prediction of AKI; the approach was based on recent work on modeling adverse events from EHRs. That study was based on data provided by the United States Department of Veterans Affairs; the data were the data of 703,782 adult patients across 1243 health care facilities in the United States. By analyzing 6-hourly EHR data during hospitalization, the model predicted 55.8%

of all inpatient episodes of AKI and 90.2% of all AKIs that required subsequent dialysis. The AUROC of predicting AKI within 24-, 48-, and 72-h time windows was 0.934, 0.921 and 0.914, respectively [71]. However, the high discriminative power of this system for AKI prediction derived from a large manipulated and processed dataset; the total number of independent entries in the dataset was approximately 6 billion according to the authors, which means that data cleaning and processing were difficult and had been executed by experts in data science. External validation of this successful result may be difficult due to the differing EHR systems, clinical pathways, treatments, and examination frequencies. Therefore, it may be crucial to establish an AI-assisted prediction model on the basis of a hospital's unique clinical practices. Although real-time prediction was not performed, another study attempted to use time-series variables to improve risk prediction. Before this investigation, most postoperative AKI prediction models were based on preoperative variables. Adhikari et al. published MySurgeryRisk, a machine-learning algorithm that uses random forests to predict the postoperative AKI risk within the 3 and 7 days after surgery and the overall AKI risk. The data of 2911 patients who underwent surgery were internally validated. By combining intraoperative physiological time-series covariates with preoperative variables, machine-learning prediction models achieved an AUROC of 0.86 for predicting 7-day postoperative AKI outcomes, and AUROC was 0.84 when only the preoperative covariates of the same cohort were used. That study confirmed that postoperative AKI prediction had higher sensitivity and specificity when machine learning was applied for the dynamic incorporation of intraoperative data [72].

Most of the enrolled studies used independent cohorts; it is challenging to evaluate whether machine learning truly improved AKI risk prediction compared with the original statistics. Under this consideration, Huang et al. used the same cohort and candidate variables that were used to develop the Cath/PCI Registry AKI model as well as the data from the American College of Cardiology National Cardiovascular Data Registry collected in 1694 hospitals. That retrospective study analyzed 947,091 patients receiving PCI and concluded that the risk prediction model containing 13 variables (age, prior heart failure, cardiogenic shock within 24 h, cardiac arrest within 24 h, diabetes mellitus, coronary artery disease, heart failure within 2 weeks, preprocedure GFR and hemoglobulin, admission source, body mass index, elective or emergency PCI, and preprocedure left ventricular ejection fraction), which was validated using the generalized additive model, performed adequately, with an AUROC of 0.752 (95% CI, 0.749–0.754) and performed more highly than the original Cath/PCI Registry AKI model (AUROC, 0.711; 95% CI, 0.708–0.714). This machine-learning model also had a significantly wider predictive range than the Cath/PCI Registry AKI model did (25.3% vs. 21.6%, $p < 0.001$) and was more accurate than that model in stratifying patient risk for AKI [67].

Although machine-learning algorithms may not have matured yet and still have several limitations, they have already shown impressive performance and sensitivity in the early detection of AKI, giving clinicians useful information regarding further adverse events and long-term prognosis. By reviewing studies focused on the application of machine learning to AKI prediction, we showed that machine-learning algorithms have had a high performance for AKI prediction not only in inpatients but also in the surgical population. To date, whether the use of machine-learning algorithms for the earlier prediction of AKI risk can truly improve the prognosis of AKI remains questionable, but its ability on predicting AKI is recognized.

5. Conclusions

AKI is the most common and adverse potential complication of hospitalization, and it has a considerable negative impact on short-term and long-term patient outcomes. Although current guidelines use serum creatinine level and urine output rate for defining AKI and as the staging criteria of AKI, these markers are not sensitive or specific for AKI. With the advances in techniques, digitization of MISs and EHRs can provide more and timing information from patients' underlying disease to real-time vital sign variability

which increases the performance and sensitivity of machine-learning algorithms. Current studies reported that the AUROC of machine-learning algorithms on AKI prediction can be over 0.80. However, most of the studies were retrospective analyses and lacked external validation which implicated the results of the proposed models cannot be generalized outside the experimental population, and the variability of EHRs across hospitals may limit the widespread use of these prediction models. Besides, even though the MISs and EHRs provide continuous clinical records of patients but only one study performed continual risk prediction by using the recurrent neural network with a deep-learning approach, and only one study used time-series covariates to improve risk discrimination demonstrating that the use of machine learning to address large datasets is not popularized and continuous prediction of AKI via machine-learning algorithms still needs to be improved. Considering that the influencing factors, clinical and laboratory parameters might change over the hospitalization, the longitudinal evaluation to predict AKI continuously might be the next challenge of application of machine learning on AKI prediction. When the machine learning algorithms can provide real-time informatics of AKI prediction by dealing with complex databased of EHR, it might be worthwhile to look forward to the combination of machine-learning algorithms and e-alert systems. At that time, by using these machine-learning algorithms but not only serum creatinine level, e-alert systems will have a chance to provide more accurate and earlier alarm of AKI which might improve the prognosis of AKI after combining with the care bundle.

Supplementary Materials: The following are available online at https://www.mdpi.com/article/10.3390/healthcare9121662/s1, Table S1. Details of the search strategy source: PubMed; the search was performed on 1 December 2020.

Author Contributions: T.H.L. and J.-J.C. participated in conceptualization (create ideas and overarching research goal) and writing-original draft, C.-T.C. carried out the software programming and supporting algorithms, T.H.L. and C.-H.C. carried out the investigation and data curation; Jia-Jin Chen carried out writing-review and editing. C.-H.C. carried out the supervision and project administration. All authors have read and agreed to the published version of the manuscript.

Funding: This work was supported by the Ministry of Science and Technology (MOST), Taiwan (MOST 110-2314-B-182A-043).

Institutional Review Board Statement: Not applicable.

Informed Consent Statement: Not applicable.

Data Availability Statement: Not applicable.

Conflicts of Interest: The authors declare no conflict of interest.

References

1. Section 2: AKI Definition. *Kidney Int. Suppl.* **2012**, *2*, 19–36. [CrossRef]
2. Kuo, G.; Yang, S.Y.; Chuang, S.S.; Fan, P.C.; Chang, C.H.; Hsiao, Y.C.; Chen, Y.C. Using acute kidney injury severity and scoring systems to predict outcome in patients with burn injury. *J. Formos. Med. Assoc.* **2016**, *115*, 1046–1052. [CrossRef] [PubMed]
3. Kim, G.H.; Oh, K.H.; Yoon, J.W.; Koo, J.W.; Kim, H.J.; Chae, D.W.; Noh, J.W.; Kim, J.H.; Park, Y.K. Impact of burn size and initial serum albumin level on acute renal failure occurring in major burn. *Am. J. Nephrol.* **2003**, *23*, 55–60. [CrossRef]
4. Jenq, C.C.; Tsai, M.H.; Tian, Y.C.; Lin, C.Y.; Yang, C.; Liu, N.J.; Lien, J.M.; Chen, Y.C.; Fang, J.T.; Chen, P.C.; et al. RIFLE classification can predict short-term prognosis in critically ill cirrhotic patients. *Intensive Care Med.* **2007**, *33*, 1921–1930. [CrossRef]
5. Waikar, S.S.; Curhan, G.C.; Wald, R.; McCarthy, E.P.; Chertow, G.M. Declining mortality in patients with acute renal failure, 1988 to 2002. *J. Am. Soc. Nephrol.* **2006**, *17*, 1143–1150. [CrossRef] [PubMed]
6. Hobson, C.; Lysak, N.; Huber, M.; Scali, S.; Bihorac, A. Epidemiology, outcomes, and management of acute kidney injury in the vascular surgery patient. *J. Vasc. Surg.* **2018**, *68*, 916–928. [CrossRef]
7. Hobson, C.; Singhania, G.; Bihorac, A. Acute Kidney Injury in the Surgical Patient. *Crit. Care Clin.* **2015**, *31*, 705–723. [CrossRef]
8. Amdur, R.L.; Chawla, L.S.; Amodeo, S.; Kimmel, P.L.; Palant, C.E. Outcomes following diagnosis of acute renal failure in U.S. veterans: Focus on acute tubular necrosis. *Kidney Int.* **2009**, *76*, 1089–1097. [CrossRef]
9. Ishani, A.; Xue, J.L.; Himmelfarb, J.; Eggers, P.W.; Kimmel, P.L.; Molitoris, B.A.; Collins, A.J. Acute kidney injury increases risk of ESRD among elderly. *J. Am. Soc. Nephrol.* **2009**, *20*, 223–228. [CrossRef] [PubMed]

10. Coca, S.G.; Singanamala, S.; Parikh, C.R. Chronic kidney disease after acute kidney injury: A systematic review and meta-analysis. *Kidney Int.* **2012**, *81*, 442–448. [CrossRef]
11. Nakasone, H.; Sakugawa, H.; Fukuchi, J.; Miyagi, T.; Sugama, R.; Hokama, A.; Nakayoshi, T.; Kawakami, Y.; Yamashiro, T.; Kinjo, F.; et al. A patient with primary biliary cirrhosis associated with autoimmune hemolytic anemia. *J. Gastroenterol.* **2000**, *35*, 245–249. [CrossRef] [PubMed]
12. Lo, L.J.; Go, A.S.; Chertow, G.M.; McCulloch, C.E.; Fan, D.; Ordonez, J.D.; Hsu, C.Y. Dialysis-requiring acute renal failure increases the risk of progressive chronic kidney disease. *Kidney Int.* **2009**, *76*, 893–899. [CrossRef] [PubMed]
13. Pannu, N.; James, M.; Hemmelgarn, B.; Klarenbach, S.; Alberta Kidney Disease Network. Association between AKI, recovery of renal function, and long-term outcomes after hospital discharge. *Clin. J. Am. Soc. Nephrol.* **2013**, *8*, 194–202. [CrossRef]
14. Gammelager, H.; Christiansen, C.F.; Johansen, M.B.; Tonnesen, E.; Jespersen, B.; Sorensen, H.T. Five-year risk of end-stage renal disease among intensive care patients surviving dialysis-requiring acute kidney injury: A nationwide cohort study. *Crit. Care* **2013**, *17*, R145. [CrossRef]
15. Forni, L.G.; Darmon, M.; Ostermann, M.; Oudemans-van Straaten, H.M.; Pettila, V.; Prowle, J.R.; Schetz, M.; Joannidis, M. Renal recovery after acute kidney injury. *Intensive Care Med.* **2017**, *43*, 855–866. [CrossRef] [PubMed]
16. Hoste, E.A.; Bagshaw, S.M.; Bellomo, R.; Cely, C.M.; Colman, R.; Cruz, D.N.; Edipidis, K.; Forni, L.G.; Gomersall, C.D.; Govil, D.; et al. Epidemiology of acute kidney injury in critically ill patients: The multinational AKI-EPI study. *Intensive Care Med.* **2015**, *41*, 1411–1423. [CrossRef] [PubMed]
17. Singbartl, K.; Kellum, J.A. AKI in the ICU: Definition, epidemiology, risk stratification, and outcomes. *Kidney Int.* **2012**, *81*, 819–825. [CrossRef] [PubMed]
18. Perazella, M.A. Drug use and nephrotoxicity in the intensive care unit. *Kidney Int.* **2012**, *81*, 1172–1178. [CrossRef]
19. Gameiro, J.; Branco, T.; Lopes, J.A. Artificial Intelligence in Acute Kidney Injury Risk Prediction. *J. Clin. Med.* **2020**, *9*, 678. [CrossRef] [PubMed]
20. Bellomo, R.; Ronco, C.; Kellum, J.A.; Mehta, R.L.; Palevsky, P.; Acute Dialysis Quality Initiative Workgroup. Acute renal failure—Definition, outcome measures, animal models, fluid therapy and information technology needs: The Second International Consensus Conference of the Acute Dialysis Quality Initiative (ADQI) Group. *Crit. Care* **2004**, *8*, R204–R212. [CrossRef]
21. Mehta, R.L.; Kellum, J.A.; Shah, S.V.; Molitoris, B.A.; Ronco, C.; Warnock, D.G.; Levin, A.; Acute Kidney Injury, N. Acute Kidney Injury Network: Report of an initiative to improve outcomes in acute kidney injury. *Crit. Care* **2007**, *11*, R31. [CrossRef]
22. Perrone, R.D.; Madias, N.E.; Levey, A.S. Serum creatinine as an index of renal function: New insights into old concepts. *Clin. Chem.* **1992**, *38*, 1933–1953. [CrossRef] [PubMed]
23. Bauer, J.H.; Brooks, C.S.; Burch, R.N. Clinical appraisal of creatinine clearance as a measurement of glomerular filtration rate. *Am. J. Kidney Dis.* **1982**, *2*, 337–346. [CrossRef]
24. Van Acker, B.A.; Koomen, G.C.; Koopman, M.G.; de Waart, D.R.; Arisz, L. Creatinine clearance during cimetidine administration for measurement of glomerular filtration rate. *Lancet* **1992**, *340*, 1326–1329. [CrossRef]
25. Delanaye, P.; Cavalier, E.; Pottel, H. Serum Creatinine: Not So Simple! *Nephron* **2017**, *136*, 302–308. [CrossRef]
26. Papadakis, M.A.; Arieff, A.I. Unpredictability of clinical evaluation of renal function in cirrhosis. Prospective study. *Am. J. Med.* **1987**, *82*, 945–952. [CrossRef]
27. Green, T.P.; Mirkin, B.L. Furosemide disposition in normal and proteinuric rats: Urinary drug-protein binding as a determinant of drug excretion. *J. Pharmacol. Exp. Ther.* **1981**, *218*, 122–127. [PubMed]
28. McIlroy, D.R.; Wagener, G.; Lee, H.T. Biomarkers of acute kidney injury: An evolving domain. *Anesthesiology* **2010**, *112*, 998–1004. [CrossRef]
29. Chen, T.H.; Chang, C.H.; Lin, C.Y.; Jenq, C.C.; Chang, M.Y.; Tian, Y.C.; Hung, C.C.; Fang, J.T.; Yang, C.W.; Wen, M.S.; et al. Acute kidney injury biomarkers for patients in a coronary care unit: A prospective cohort study. *PLoS ONE* **2012**, *7*, e32328. [CrossRef] [PubMed]
30. Yui, S.; Nakatani, Y.; Mikami, M. Calprotectin (S100A8/S100A9), an inflammatory protein complex from neutrophils with a broad apoptosis-inducing activity. *Biol. Pharm. Bull.* **2003**, *26*, 753–760. [CrossRef]
31. Pepper, R.J.; Wang, H.H.; Rajakaruna, G.K.; Papakrivopoulou, E.; Vogl, T.; Pusey, C.D.; Cook, H.T.; Salama, A.D. S100A8/A9 (calprotectin) is critical for development of glomerulonephritis and promotes inflammatory leukocyte-renal cell interactions. *Am. J. Pathol.* **2015**, *185*, 1264–1274. [CrossRef]
32. Chang, C.H.; Yang, C.H.; Yang, H.Y.; Chen, T.H.; Lin, C.Y.; Chang, S.W.; Chen, Y.T.; Hung, C.C.; Fang, J.T.; Yang, C.W.; et al. Urinary Biomarkers Improve the Diagnosis of Intrinsic Acute Kidney Injury in Coronary Care Units. *Medicine* **2015**, *94*, e1703. [CrossRef] [PubMed]
33. Vanmassenhove, J.; Vanholder, R.; Nagler, E.; Van Biesen, W. Urinary and serum biomarkers for the diagnosis of acute kidney injury: An in-depth review of the literature. *Nephrol. Dial. Transplant.* **2013**, *28*, 254–273. [CrossRef] [PubMed]
34. Marx, D.; Metzger, J.; Pejchinovski, M.; Gil, R.B.; Frantzi, M.; Latosinska, A.; Belczacka, I.; Heinzmann, S.S.; Husi, H.; Zoidakis, J.; et al. Proteomics and Metabolomics for AKI Diagnosis. *Semin. Nephrol.* **2018**, *38*, 63–87. [CrossRef]
35. Kashani, K.; Cheungpasitporn, W.; Ronco, C. Biomarkers of acute kidney injury: The pathway from discovery to clinical adoption. *Clin. Chem. Lab. Med.* **2017**, *55*, 1074–1089. [CrossRef] [PubMed]
36. Lameire, N.H.; Bagga, A.; Cruz, D.; De Maeseneer, J.; Endre, Z.; Kellum, J.A.; Liu, K.D.; Mehta, R.L.; Pannu, N.; Van Biesen, W.; et al. Acute kidney injury: An increasing global concern. *Lancet* **2013**, *382*, 170–179. [CrossRef]

37. Hahn, K.; Kanbay, M.; Lanaspa, M.A.; Johnson, R.J.; Ejaz, A.A. Serum uric acid and acute kidney injury: A mini review. *J. Adv. Res.* **2017**, *8*, 529–536. [CrossRef]
38. Sgura, F.A.; Bertelli, L.; Monopoli, D.; Leuzzi, C.; Guerri, E.; Sparta, I.; Politi, L.; Aprile, A.; Amato, A.; Rossi, R.; et al. Mehran contrast-induced nephropathy risk score predicts short- and long-term clinical outcomes in patients with ST-elevation-myocardial infarction. *Circ. Cardiovasc. Interv.* **2010**, *3*, 491–498. [CrossRef]
39. Abellas-Sequeiros, R.A.; Raposeiras-Roubin, S.; Abu-Assi, E.; Gonzalez-Salvado, V.; Iglesias-Alvarez, D.; Redondo-Dieguez, A.; Gonzalez-Ferreiro, R.; Ocaranza-Sanchez, R.; Pena-Gil, C.; Garcia-Acuna, J.M.; et al. Mehran contrast nephropathy risk score: Is it still useful 10 years later? *J. Cardiol.* **2016**, *67*, 262–267. [CrossRef]
40. Uchino, S.; Kellum, J.A.; Bellomo, R.; Doig, G.S.; Morimatsu, H.; Morgera, S.; Schetz, M.; Tan, I.; Bouman, C.; Macedo, E.; et al. Acute renal failure in critically ill patients: A multinational, multicenter study. *JAMA* **2005**, *294*, 813–818. [CrossRef]
41. Chang, C.H.; Lee, C.C.; Chen, S.W.; Fan, P.C.; Chen, Y.C.; Chang, S.W.; Chen, T.H.; Wu, V.C.; Lin, P.J.; Tsai, F.C. Predicting Acute Kidney Injury Following Mitral Valve Repair. *Int. J. Med. Sci.* **2016**, *13*, 19–24. [CrossRef]
42. Wang, Y.; Bellomo, R. Cardiac surgery-associated acute kidney injury: Risk factors, pathophysiology and treatment. *Nat. Rev. Nephrol.* **2017**, *13*, 697–711. [CrossRef]
43. Nashef, S.A.; Roques, F.; Michel, P.; Gauducheau, E.; Lemeshow, S.; Salamon, R. European system for cardiac operative risk evaluation (EuroSCORE). *Eur. J. Cardiothorac. Surg.* **1999**, *16*, 9–13. [CrossRef]
44. Wykrzykowska, J.J.; Garg, S.; Onuma, Y.; de Vries, T.; Goedhart, D.; Morel, M.A.; van Es, G.A.; Buszman, P.; Linke, A.; Ischinger, T.; et al. Value of age, creatinine, and ejection fraction (ACEF score) in assessing risk in patients undergoing percutaneous coronary interventions in the 'All-Comers' LEADERS trial. *Circ. Cardiovasc. Interv.* **2011**, *4*, 47–56. [CrossRef]
45. Shahian, D.M.; Jacobs, J.P.; Badhwar, V.; Kurlansky, P.A.; Furnary, A.P.; Cleveland, J.C., Jr.; Lobdell, K.W.; Vassileva, C.; Wyler von Ballmoos, M.C.; Thourani, V.H.; et al. The Society of Thoracic Surgeons 2018 Adult Cardiac Surgery Risk Models: Part 1-Background, Design Considerations, and Model Development. *Ann. Thorac. Surg.* **2018**, *105*, 1411–1418. [CrossRef]
46. O'Brien, S.M.; Feng, L.; He, X.; Xian, Y.; Jacobs, J.P.; Badhwar, V.; Kurlansky, P.A.; Furnary, A.P.; Cleveland, J.C., Jr.; Lobdell, K.W.; et al. The Society of Thoracic Surgeons 2018 Adult Cardiac Surgery Risk Models: Part 2-Statistical Methods and Results. *Ann. Thorac. Surg.* **2018**, *105*, 1419–1428. [CrossRef]
47. Wendt, D.; Thielmann, M.; Kahlert, P.; Kastner, S.; Price, V.; Al-Rashid, F.; Patsalis, P.; Erbel, R.; Jakob, H. Comparison between different risk scoring algorithms on isolated conventional or transcatheter aortic valve replacement. *Ann. Thorac. Surg.* **2014**, *97*, 796–802. [CrossRef] [PubMed]
48. Mathioudakis, N.N.; Giles, M.; Yeh, H.C.; Haywood, C., Jr.; Greer, R.C.; Golden, S.H. Racial differences in acute kidney injury of hospitalized adults with diabetes. *J. Diabetes Complicat.* **2016**, *30*, 1129–1136. [CrossRef] [PubMed]
49. Susantitaphong, P.; Cruz, D.N.; Cerda, J.; Abulfaraj, M.; Alqahtani, F.; Koulouridis, I.; Jaber, B.L.; Acute Kidney Injury Advisory Group of the American Society of Nephrology. World incidence of AKI: A meta-analysis. *Clin. J. Am. Soc. Nephrol.* **2013**, *8*, 1482–1493. [CrossRef] [PubMed]
50. Fan, P.C.; Chen, T.H.; Lee, C.C.; Tsai, T.Y.; Chen, Y.C.; Chang, C.H. ADVANCIS Score Predicts Acute Kidney Injury After Percutaneous Coronary Intervention for Acute Coronary Syndrome. *Int. J. Med. Sci.* **2018**, *15*, 528–535. [CrossRef] [PubMed]
51. Zhou, L.Z.; Yang, X.B.; Guan, Y.; Xu, X.; Tan, M.T.; Hou, F.F.; Chen, P.Y. Development and Validation of a Risk Score for Prediction of Acute Kidney Injury in Patients With Acute Decompensated Heart Failure: A Prospective Cohort Study in China. *J. Am. Heart Assoc.* **2016**, *5*, e004035. [CrossRef] [PubMed]
52. Cheungpasitporn, W.; Kashani, K. Electronic Data Systems and Acute Kidney Injury. *Contrib. Nephrol.* **2016**, *187*, 73–83. [CrossRef] [PubMed]
53. Park, S.; Baek, S.H.; Ahn, S.; Lee, K.H.; Hwang, H.; Ryu, J.; Ahn, S.Y.; Chin, H.J.; Na, K.Y.; Chae, D.W.; et al. Impact of Electronic Acute Kidney Injury (AKI) Alerts With Automated Nephrologist Consultation on Detection and Severity of AKI: A Quality Improvement Study. *Am. J. Kidney Dis.* **2018**, *71*, 9–19. [CrossRef] [PubMed]
54. Wu, Y.; Chen, Y.; Li, S.; Dong, W.; Liang, H.; Deng, M.; Chen, Y.; Chen, S.; Liang, X. Value of electronic alerts for acute kidney injury in high-risk wards: A pilot randomized controlled trial. *Int. Urol. Nephrol.* **2018**, *50*, 1483–1488. [CrossRef]
55. Lachance, P.; Villeneuve, P.M.; Rewa, O.G.; Wilson, F.P.; Selby, N.M.; Featherstone, R.M.; Bagshaw, S.M. Association between e-alert implementation for detection of acute kidney injury and outcomes: A systematic review. *Nephrol. Dial. Transplant.* **2017**, *32*, 265–272. [CrossRef]
56. Lachance, P.; Villeneuve, P.M.; Wilson, F.P.; Selby, N.M.; Featherstone, R.; Rewa, O.; Bagshaw, S.M. Impact of e-alert for detection of acute kidney injury on processes of care and outcomes: Protocol for a systematic review and meta-analysis. *BMJ Open* **2016**, *6*, e011152. [CrossRef]
57. Kolhe, N.V.; Reilly, T.; Leung, J.; Fluck, R.J.; Swinscoe, K.E.; Selby, N.M.; Taal, M.W. A simple care bundle for use in acute kidney injury: A propensity score-matched cohort study. *Nephrol. Dial. Transplant.* **2016**, *31*, 1846–1854. [CrossRef]
58. Kolhe, N.V.; Staples, D.; Reilly, T.; Merrison, D.; McIntyre, C.W.; Fluck, R.J.; Selby, N.M.; Taal, M.W. Impact of Compliance with a Care Bundle on Acute Kidney Injury Outcomes: A Prospective Observational Study. *PLoS ONE* **2015**, *10*, e0132279. [CrossRef]
59. Hodgson, L.E.; Roderick, P.J.; Venn, R.M.; Yao, G.L.; Dimitrov, B.D.; Forni, L.G. The ICE-AKI study: Impact analysis of a Clinical prediction rule and Electronic AKI alert in general medical patients. *PLoS ONE* **2018**, *13*, e0200584. [CrossRef]
60. Kate, R.J.; Perez, R.M.; Mazumdar, D.; Pasupathy, K.S.; Nilakantan, V. Prediction and detection models for acute kidney injury in hospitalized older adults. *BMC Med. Inform. Decis. Mak.* **2016**, *16*, 39. [CrossRef] [PubMed]

61. Thottakkara, P.; Ozrazgat-Baslanti, T.; Hupf, B.B.; Rashidi, P.; Pardalos, P.; Momcilovic, P.; Bihorac, A. Application of Machine Learning Techniques to High-Dimensional Clinical Data to Forecast Postoperative Complications. *PLoS ONE* **2016**, *11*, e0155705. [CrossRef]
62. Davis, S.E.; Lasko, T.A.; Chen, G.; Siew, E.D.; Matheny, M.E. Calibration drift in regression and machine learning models for acute kidney injury. *J. Am. Med. Inform. Assoc.* **2017**, *24*, 1052–1061. [CrossRef] [PubMed]
63. Cheng, P.; Waitman, L.R.; Hu, Y.; Liu, M. Predicting Inpatient Acute Kidney Injury over Different Time Horizons: How Early and Accurate? *AMIA Annu. Symp. Proc.* **2017**, *2017*, 565–574.
64. Ibrahim, N.E.; McCarthy, C.P.; Shrestha, S.; Gaggin, H.K.; Mukai, R.; Magaret, C.A.; Rhyne, R.F.; Januzzi, J.L., Jr. A clinical, proteomics, and artificial intelligence-driven model to predict acute kidney injury in patients undergoing coronary angiography. *Clin. Cardiol.* **2019**, *42*, 292–298. [CrossRef]
65. Koola, J.D.; Davis, S.E.; Al-Nimri, O.; Parr, S.K.; Fabbri, D.; Malin, B.A.; Ho, S.B.; Matheny, M.E. Development of an automated phenotyping algorithm for hepatorenal syndrome. *J. Biomed. Inform.* **2018**, *80*, 87–95. [CrossRef] [PubMed]
66. Koyner, J.L.; Carey, K.A.; Edelson, D.P.; Churpek, M.M. The Development of a Machine Learning Inpatient Acute Kidney Injury Prediction Model. *Crit. Care Med.* **2018**, *46*, 1070–1077. [CrossRef]
67. Huang, C.; Murugiah, K.; Mahajan, S.; Li, S.X.; Dhruva, S.S.; Haimovich, J.S.; Wang, Y.; Schulz, W.L.; Testani, J.M.; Wilson, F.P.; et al. Enhancing the prediction of acute kidney injury risk after percutaneous coronary intervention using machine learning techniques: A retrospective cohort study. *PLoS Med.* **2018**, *15*, e1002703. [CrossRef] [PubMed]
68. Lin, K.; Hu, Y.; Kong, G. Predicting in-hospital mortality of patients with acute kidney injury in the ICU using random forest model. *Int. J. Med. Inform.* **2019**, *125*, 55–61. [CrossRef]
69. Simonov, M.; Ugwuowo, U.; Moreira, E.; Yamamoto, Y.; Biswas, A.; Martin, M.; Testani, J.; Wilson, F.P. A simple real-time model for predicting acute kidney injury in hospitalized patients in the US: A descriptive modeling study. *PLoS Med.* **2019**, *16*, e1002861. [CrossRef]
70. Huang, C.; Li, S.X.; Mahajan, S.; Testani, J.M.; Wilson, F.P.; Mena, C.I.; Masoudi, F.A.; Rumsfeld, J.S.; Spertus, J.A.; Mortazavi, B.J.; et al. Development and Validation of a Model for Predicting the Risk of Acute Kidney Injury Associated With Contrast Volume Levels During Percutaneous Coronary Intervention. *JAMA Netw. Open* **2019**, *2*, e1916021. [CrossRef] [PubMed]
71. Tomasev, N.; Glorot, X.; Rae, J.W.; Zielinski, M.; Askham, H.; Saraiva, A.; Mottram, A.; Meyer, C.; Ravuri, S.; Protsyuk, I.; et al. A clinically applicable approach to continuous prediction of future acute kidney injury. *Nature* **2019**, *572*, 116–119. [CrossRef] [PubMed]
72. Adhikari, L.; Ozrazgat-Baslanti, T.; Ruppert, M.; Madushani, R.; Paliwal, S.; Hashemighouchani, H.; Zheng, F.; Tao, M.; Lopes, J.M.; Li, X.; et al. Improved predictive models for acute kidney injury with IDEA: Intraoperative Data Embedded Analytics. *PLoS ONE* **2019**, *14*, e0214904. [CrossRef]
73. Flechet, M.; Falini, S.; Bonetti, C.; Guiza, F.; Schetz, M.; Van den Berghe, G.; Meyfroidt, G. Machine learning versus physicians' prediction of acute kidney injury in critically ill adults: A prospective evaluation of the AKIpredictor. *Crit. Care* **2019**, *23*, 282. [CrossRef] [PubMed]
74. Parreco, J.; Soe-Lin, H.; Parks, J.J.; Byerly, S.; Chatoor, M.; Buicko, J.L.; Namias, N.; Rattan, R. Comparing Machine Learning Algorithms for Predicting Acute Kidney Injury. *Am. Surg.* **2019**, *85*, 725–729. [CrossRef]
75. Xu, Z.; Luo, Y.; Adekkanattu, P.; Ancker, J.S.; Jiang, G.; Kiefer, R.C.; Pacheco, J.A.; Rasmussen, L.V.; Pathak, J.; Wang, F. Stratified Mortality Prediction of Patients with Acute Kidney Injury in Critical Care. *Stud. Health Technol. Inform.* **2019**, *264*, 462–466. [CrossRef] [PubMed]
76. Tran, N.K.; Sen, S.; Palmieri, T.L.; Lima, K.; Falwell, S.; Wajda, J.; Rashidi, H.H. Artificial intelligence and machine learning for predicting acute kidney injury in severely burned patients: A proof of concept. *Burns* **2019**, *45*, 1350–1358. [CrossRef] [PubMed]
77. Zhang, Z.; Ho, K.M.; Hong, Y. Machine learning for the prediction of volume responsiveness in patients with oliguric acute kidney injury in critical care. *Crit. Care* **2019**, *23*, 112. [CrossRef]
78. Zimmerman, L.P.; Reyfman, P.A.; Smith, A.D.R.; Zeng, Z.; Kho, A.; Sanchez-Pinto, L.N.; Luo, Y. Early prediction of acute kidney injury following ICU admission using a multivariate panel of physiological measurements. *BMC Med. Inform. Decis. Mak.* **2019**, *19*, 16. [CrossRef]
79. Rashidi, H.H.; Sen, S.; Palmieri, T.L.; Blackmon, T.; Wajda, J.; Tran, N.K. Early Recognition of Burn- and Trauma-Related Acute Kidney Injury: A Pilot Comparison of Machine Learning Techniques. *Sci. Rep.* **2020**, *10*, 205. [CrossRef] [PubMed]
80. Zhou, C.; Wang, R.; Jiang, W.; Zhu, J.; Liu, Y.; Zheng, J.; Wang, X.; Shang, W.; Sun, L. Machine learning for the prediction of acute kidney injury and paraplegia after thoracoabdominal aortic aneurysm repair. *J. Card. Surg.* **2020**, *35*, 89–99. [CrossRef]
81. Martinez, D.A.; Levin, S.R.; Klein, E.Y.; Parikh, C.R.; Menez, S.; Taylor, R.A.; Hinson, J.S. Early Prediction of Acute Kidney Injury in the Emergency Department With Machine-Learning Methods Applied to Electronic Health Record Data. *Ann. Emerg. Med.* **2020**, *76*, 501–514. [CrossRef]
82. Lei, G.; Wang, G.; Zhang, C.; Chen, Y.; Yang, X. Using Machine Learning to Predict Acute Kidney Injury After Aortic Arch Surgery. *J. Cardiothorac. Vasc. Anesth.* **2020**, *34*, 3321–3328. [CrossRef] [PubMed]
83. Lei, L.; Wang, Y.; Xue, Q.; Tong, J.; Zhou, C.M.; Yang, J.J. A comparative study of machine learning algorithms for predicting acute kidney injury after liver cancer resection. *PeerJ* **2020**, *8*, e8583. [CrossRef]
84. Qu, C.; Gao, L.; Yu, X.Q.; Wei, M.; Fang, G.Q.; He, J.; Cao, L.X.; Ke, L.; Tong, Z.H.; Li, W.Q. Machine Learning Models of Acute Kidney Injury Prediction in Acute Pancreatitis Patients. *Gastroenterol. Res. Pract.* **2020**, *2020*, 3431290. [CrossRef] [PubMed]

85. Tseng, P.Y.; Chen, Y.T.; Wang, C.H.; Chiu, K.M.; Peng, Y.S.; Hsu, S.P.; Chen, K.L.; Yang, C.Y.; Lee, O.K. Prediction of the development of acute kidney injury following cardiac surgery by machine learning. *Crit. Care* **2020**, *24*, 478. [CrossRef]
86. Sun, L.; Zhu, W.; Chen, X.; Jiang, J.; Ji, Y.; Liu, N.; Xu, Y.; Zhuang, Y.; Sun, Z.; Wang, Q.; et al. Machine Learning to Predict Contrast-Induced Acute Kidney Injury in Patients With Acute Myocardial Infarction. *Front. Med.* **2020**, *7*, 592007. [CrossRef]
87. Churpek, M.M.; Carey, K.A.; Edelson, D.P.; Singh, T.; Astor, B.C.; Gilbert, E.R.; Winslow, C.; Shah, N.; Afshar, M.; Koyner, J.L. Internal and External Validation of a Machine Learning Risk Score for Acute Kidney Injury. *JAMA Netw. Open* **2020**, *3*, e2012892. [CrossRef]
88. Hsu, C.N.; Liu, C.L.; Tain, Y.L.; Kuo, C.Y.; Lin, Y.C. Machine Learning Model for Risk Prediction of Community-Acquired Acute Kidney Injury Hospitalization From Electronic Health Records: Development and Validation Study. *J. Med. Internet Res.* **2020**, *22*, e16903. [CrossRef] [PubMed]
89. Penny-Dimri, J.C.; Bergmeir, C.; Reid, C.M.; Williams-Spence, J.; Cochrane, A.D.; Smith, J.A. Machine Learning Algorithms for Predicting and Risk Profiling of Cardiac Surgery-Associated Acute Kidney Injury. *Semin. Thorac. Cardiovasc. Surg.* **2020**. [CrossRef]
90. Li, Y.; Xu, J.; Wang, Y.; Zhang, Y.; Jiang, W.; Shen, B.; Ding, X. A novel machine learning algorithm, Bayesian networks model, to predict the high-risk patients with cardiac surgery-associated acute kidney injury. *Clin. Cardiol.* **2020**, *43*, 752–761. [CrossRef]
91. Ugwuowo, U.; Yamamoto, Y.; Arora, T.; Saran, I.; Partridge, C.; Biswas, A.; Martin, M.; Moledina, D.G.; Greenberg, J.H.; Simonov, M.; et al. Real-Time Prediction of Acute Kidney Injury in Hospitalized Adults: Implementation and Proof of Concept. *Am. J. Kidney Dis.* **2020**, *76*, 806–814.e801. [CrossRef] [PubMed]

healthcare

MDPI

Article

Machine Learning for Predicting the Risk for Childhood Asthma Using Prenatal, Perinatal, Postnatal and Environmental Factors

Zineb Jeddi [1], Ihsane Gryech [1,2,*], Mounir Ghogho [1,3,*], Maryame EL Hammoumi [4] and Chafiq Mahraoui [4]

1 TICLab, College of Engineering & Architecture, International University of Rabat, Rabat 11103, Morocco; zineb.jeddi@uir.ac.ma
2 ENSIAS, Mohammed V University in Rabat, Rabat 10000, Morocco
3 School of IEEE, University of Leeds, Leeds LS2 9JT, UK
4 Pediatrics Department, CHU, Rabat 10000, Morocco; drmaryamehm@gmail.com (M.E.H.); cmahraoui@gmail.com (C.M.)
* Correspondence: ihsane.gryech@uir.ac.ma (I.G.) ; mounir.ghogho@uir.ac.ma (M.G.)

Abstract: The prevalence rate for childhood asthma and its associated risk factors vary significantly across countries and regions. In the case of Morocco, the scarcity of available medical data makes scientific research on diseases such as asthma very challenging. In this paper, we build machine learning models to predict the occurrence of childhood asthma using data from a prospective study of 202 children with and without asthma. The association between different factors and asthma diagnosis is first assessed using a Chi-squared test. Then, predictive models such as logistic regression analysis, decision trees, random forest and support vector machine are used to explore the relationship between childhood asthma and the various risk factors. First, data were pre-processed using a Chi-squared feature selection, 19 out of the 36 factors were found to be significantly associated (p-value < 0.05) with childhood asthma; these include: history of atopic diseases in the family, presence of mites, cold air, strong odors and mold in the child's environment, mode of birth, breastfeeding and early life habits and exposures. For asthma prediction, random forest yielded the best predictive performance (accuracy = 84.9%), followed by logistic regression (accuracy = 82.57%), support vector machine (accuracy = 82.5%) and decision trees (accuracy = 75.19%). The decision tree model has the advantage of being easily interpreted. This study identified important maternal and prenatal risk factors for childhood asthma, the majority of which are avoidable. Appropriate steps are needed to raise awareness about the prenatal risk factors.

Keywords: asthma; machine learning; prediction; risk factors; environment; prevention; pediatrics

Citation: Jeddi, Z.; Gryech, I.; Ghogho, M.; EL Hammoumi, M.; Mahraoui, C. Machine Learning for Predicting the Risk for Childhood Asthma Using Prenatal, Perinatal, Postnatal and Environmental Factors. *Healthcare* **2021**, *9*, 1464. https://doi.org/10.3390/healthcare9111464

Academic Editor: Mahmudur Rahman

Received: 5 September 2021
Accepted: 6 October 2021
Published: 29 October 2021

1. Introduction

Asthma is the most common chronic disease among children in the world. It is a multi-factorial disease caused by a chronic inflammation of the airways. This chronic respiratory condition is characterized by several persistent symptoms, including cough, wheeze, dyspnea, and chest tightness. According to the world health organization, asthma affected 262 million people and was responsible for 461,000 deaths worldwide in 2019 [1,2]. Globally, asthma affects approximately 334 million people per year and 14% of the world's children experience asthma symptoms [3]. Even though the prevalence of childhood asthma varies between countries across the world, studies have shown that asthma prevalence is increasing at a high rate in developing countries [4], especially in densely populated areas [5]. In contrast, many developed countries have managed to slow down the increasing rate of asthma prevalence among their populations [6]. In Morocco, asthma is much more prevalent in children than in adults. The prevalence rate of asthma in children between the ages of 13 and 14 is 20%, whereas for adults, it varies between 15% and 17% [7]. Given the complex nature of this disease, several factors can be responsible for the increasing rate of

childhood asthma prevalence, including genetic predisposition factors [4], environmental factors [8], prenatal and postnatal factors as well as the other factors related to the health of the mother during pregnancy and delivery periods. Studies have shown that the mother's overall health during pregnancy in the prenatal period is significantly associated with developing asthma in the early years of childhood [9]. In fact, studies have shown that maternal diseases during pregnancy such as diabetes, atopic diseases, asthma and hypertension increase the risk of asthma for the child [10]. Moreover, other studies have also shown that forceps-assisted deliveries, maternal smoking during pregnancy, and low birth weight may also present significant risk factors for childhood asthma [10–12]. On the other hand, it was shown in [13] that frequent maternal exposure to farm animals during pregnancy can help prevent childhood asthma [14]. In the case of Morocco, the non-availability of medical data due to patients' privacy and the lack of electronic health records makes scientific research on diseases such as asthma very challenging and limited. However, because of the increasing prevalence of asthma among the pediatric population, focused efforts must be dedicated to providing a better understanding of the disease and thus elaborate better prevention and management strategies for childhood asthma. In this study, we utilize data from the Ibn Sina Hospital Center (CHUIS) to contribute to the assessment of the Asthma situation in Morocco. We first investigate perinatal, prenatal, postnatal and environmental risk factors for asthma, using patient data. We then use machine learning models to predict the occurrence of childhood asthma and to quantify the importance of the identified risk factors. It is worth pointing out that previous studies have focused on statistical methods to infer associations between asthma and risk factors.

2. Materials and Methods

In this section, we describe the process followed in our study (Figure 1). One of the main goals of this work is to lay the ground for future work on uncovering asthma risk factors in Morocco. Thus, we use a Moroccan data set.

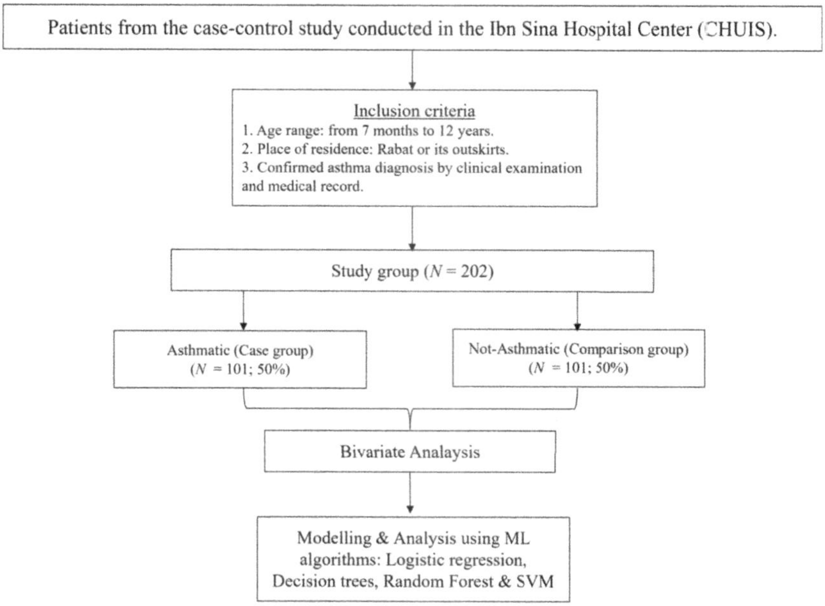

Figure 1. Flow chart of the study.

2.1. Data Collection

A case-control study of 202 children was previously conducted in the Ibn Sina Hospital Center (CHUIS). A dataset resulted from this study and was made available to us for

analysis. The study consists of children with ($N = 101$) and without ($N = 101$) asthma. The data collection was conducted over a period of 4 months, from May to September 2018. The age of the children included in the study varies from 7 months to 12 years. The data collection took place in the pneumology, allergology and infectiology service at the Children's Hospital in Rabat. The doctors participating in the study interviewed the child's mother in the local language (Moroccan dialect). The questions used for the interviews were designed by pediatricians to gather information about prenatal, perinatal and postnatal periods, as well as factors that are potentially associated with childhood asthma, including family history, environment, and other exposure features during early childhood (first two years of life). All variables were binary categorical.

2.2. Inclusion Criteria

The inclusion criteria used by the medical doctors for data collection are as follows:

- Age range: this ranges from 7 months to 12 years.
- Place of residence: only patients living in the city of Rabat or its outskirts were included in the study.
- Confirmed asthma diagnosis: the diagnosis was based on a clinical examination by a pediatrician who assessed tangible symptoms such as wheezing, chest tightness, difficulty in breathing induced by physical exercise and dry coughs, especially at night.

2.3. Data Analysis

Data were analyzed using the R software. First, we started with a primary feature selection using a Chi-squared test. This bivariate analysis allowed to assess the association between the response variable and the other variables in the dataset. Variables associations with p-value < 0.05 were considered to be significant risk factors for childhood asthma. For the modeling part, we partitioned the data into two subsets: 80% for training and 20% for testing. Second, we performed logistic regression. We used backward stepwise logistic regression to select the final model where only significant variables (p-value < 0.05) were retained in the final model. In order to identify the best model for predicting childhood asthma, we also built predictive models based on Decision Tree and Random forest techniques. Then, we used both the training and the testing data sets to compare the performance of the different models and identify the model that better predicts childhood asthma diagnosis. To evaluate the predictive ability of the different models, we used different performance metrics, namely accuracy, F1 scores, AUC-ROC, sensitivity (the false positive, Sn) and specificity (the false negative, Sp).

3. Results

Table 1 displays descriptive characteristics and the association between prenatal, perinatal, postnatal factors and childhood asthma, measured by the Chi-squared test of independence. The history of having maternal atopic tendencies and environmental factors such as cold air, strong odors, reported dust mites, pollen, mold in the child's environment and having pets (during the prenatal, perinatal and postnatal periods) were all significantly associated with childhood asthma (p-values < 0.05). Other significant factors are related to the mother's state of health, including consumption of "antibiotics/paracetamol" during pregnancy, a cesarean mode of birth, maternal overweight during pregnancy and a paternal age of more than 34 years at the child's birth. In the postnatal period and early childhood, other features were also significant predictors for asthma; these include breastfeeding, dietary diversity when the child is aged between 4 and 6 months and also when the child is aged over 6 months. Overweight and the use of antibiotics by the child in the first two years were also significantly associated with childhood asthma in the bivariate analysis.

Table 1. Descriptive characteristics and results of Chi-squared test of independence for the study sample.

Characteristics (n = 202)	Children with Asthma (N = 101, 50%)	Children without Asthma (N = 101, 50%)	Chi-Square Test (p-Value)
Factors related to family history			
Maternal atopy	28 (84.85%)	5 (15.15%)	1.263×10^{-5}
Paternal atopy	17 (65.38%)	9 (34.62%)	0.09361
History of an atopic disease in brothers or sisters	9 (56.25%)	7 (43.75%)	0.6032
Personal atopic dermatitis	13 (61.90%)	8 (0.3809524)	0.2502
Factors related to the child environment			
Reported dust mites in the child environment	31 (96.87%)	1 (3.13%)	8.089×10^{-9}
Reported pets (cats) in the child environment	7 (77.78%)	2 (22.22%)	0.08897
Reported pollen in the child environment	13 (0.7222222)	5 (27.78%)	0.04875
Reported mold in the child environment	12 (85.71%)	2 (14.20%)	0.005719
Reported cold airflow in the child environment	15 (83.34%)	3 (16.67%)	0.003115
Reported respiratory infections in family members (cold)	23 (76.67%)	7 (23.34%)	0.001589
Reported respiratory infections in family members (flu)	15 (62.50%)	9 (37.5%)	0.1931
Reported respiratory infections in family members (sinusitis)	5 (83.34%)	1 (16.67%)	0.09819
Prenatal, Perinatal and postnatal factors			
Maternal age $\leq$ 25 years	33 (76.75%)	10 (23.26%)	8.027×10^{-5}
Maternal age $\geq$ 35 years	5 (62.5%)	3 (37.50%)	0.4717
Paternal age $\leq$ 24 years	(62.50%)	(37.50%)	0.4717
Paternal age $\geq$ 34 years	22 (%)	7 (%)	0.002679
Maternal obesity during pregnancy	15 (75%)	5 (25%)	0.01878
Maternal anxiety during pregnancy	16 (69.57%)	7 (30.43%)	0.04674
Exposure to secondhand smoking during pregnancy	25 (56.82%)	19 (43.18%)	0.3076
Consumption of antibiotics/paracetamol during pregnancy	9 (90%)	1 (10%)	0.009641
Underweight child	9 (75%)	3 (25%)	0.07483
Overweight child	15 (68.18%)	7 (31.81%)	0.07149
Prematurity	5 (62.50%)	3 (37.50%)	0.4717
Cesarian mode of birth	59 (60.83%)	38 (39.17%)	0.003177
Breastfeeding	55 (38.73%)	87 (61.27%)	8.876×10^{-7}
dietary diversity for children aged between 4 and 6 months	21 (37.50%)	35 (62.50%)	0.02816
dietary diversity for children aged more than 6 months	80 (54.79%)	66 (45.20%)	0.02816
Factors related to early childhood			
Overweight during the first 2 years	11 (78.57%)	3 (21.43%)	0.02705
Consumption of antibiotics during first 2 years	32 (74.42%)	11 (25.58%)	0.0003174
Exposure to pollution in the first two years	14 (60.87%)	9 (39.13%)	0.2693

3.1. Logistic Regression

Despite its name, logistic regression (LR) is a classification model rather than a regression model. It is an efficient method for binary and linear classification. For a model with two predictors, x_1 and x_2, and one binary (Bernoulli) response variable Y, the probability for $Y = 1$, denoted as $p = P(Y = 1)$, is expressed as

$$p = \frac{1}{1 + e^{-(b_0 + b_1 x_1 + b_2 x_2)}} \quad (1)$$

where $b_0 + b_1 x_1 + b_2 x_2$ are parameters of the model. LR is the transformation of a linear regression using the Sigmoid function to restrict the value of p to be between 0 and 1.

Table 2 displays the multivariate odds ratios (OR) and the confidence intervals (2.5–97.5%) obtained by the statistical analysis of the logistic regression model. Environmental factors, including reported dust mites and the cold airflow in the child's environment were the most significant factors in predicting childhood asthma. The chances of having asthma were approximately a hundred times higher among children who were born in environments with a reported presence of mites (adjusted OR = 101.23, 95% CI = 13.39–2271.27) and 21 times higher in an environment with a persistent cold airflow (adjusted OR = 21.62, 95% CI = 2.18–335.19). Having family members with cold (adjusted OR = 5.98, 95% CI = 1.32–31.15) and flu (adjusted OR = 11.61, 95% CI = 2.31–76.33) in the environment of the child during the neonatal period also increases the chances of childhood asthma. Among mothers who reported having a history of an atopic disease, the odds of having childhood asthma were approximately nineteen-fold higher (adjusted OR = 19.04, 95% CI = 3.83–126.39). Parents age at birth was also a relevant factor to predict childhood asthma. A maternal age that is above 35 years (adjusted OR = 53.13, 95% CI = 4.24–850.82) or below 25 years (adjusted OR = 7.19, 95% CI = 1.81–33.17) as well as a paternal age that is above 34 years (adjusted OR = 13.50, 95% CI = 2.66–84.79) were found to be highly associated with childhood asthma in this model. The mode of birth was also an important factor in predicting childhood asthma, where the chances of developing asthma were almost seven-fold higher among children who were delivered via a cesarean section (adjusted OR = 6.77, 95% CI = 2.12–25.75). Breastfeeding in the first two years (adjusted OR = 0.03, 95% CI = 0.01–0.12) and diversifying the baby's diet between 4 and 6 months of age (adjusted OR = 0.35, 95% CI = 0.09–1.24) were found to be protective against childhood asthma.

Table 2. Association of prenatal factors with childhood asthma using univariate logistic regression.

Variable	OR	2.5%	97.5%
Maternal atopy	19.04	3.83	126.39
Reported dust mites in the child's environment	101.23	13.39	2271.27
Maternal age $\leq$ 25 years	7.19	1.81	33.17
Maternal age $\geq$ 35 years	53.13	4.24	850.82
Cold air in the child environment	21.62	2.18	335.19
Respiratory infections in family members (cold)	5.98	1.32	31.15
Respiratory infections in family members (flu)	11.61	2.31	76.33
Paternal age $\geq$ 34 years	13.50	2.66	84.79
Cesarean mode of birth	6.77	2.12	25.75
Breastfeeding in the first two years	0.03	0.01	0.12
Dietary diversity for children aged between 4 and 6 months	0.35	0.09	1.24

3.2. Decision Tree Model

Decision trees are one of the most popular non-parametric supervised learning methods for classification and regression. The goal of a decision tree is to create a model that predicts a targeted value by learning simple decision rules from the data features. For decision trees, internal nodes denote a test on an attribute, the branch represents an outcome of the test, and the leaf node holds a class label. In our case, we built a decision tree classifier using the features selected based on the Chi-squared test. When training the model, the metric used to perform the splits is the Gini's Diversity Index (GDI), which is a measure of the node's impurity. The size of the tree was determined by setting a minimum of 10 observations per leaf node. Each node shows respectively:

- The predicted class ('Asthma' or 'Not asthma').
- The predicted probability of asthma diagnosis.
- The percentage of observations in the node.

The decision tree in Figure 2 indicates that the most influential attribute in determining childhood asthma is the reported 'presence of dust mites in the child's environment'.

For the decision tree interpretation, the first question asked is ' are there any reported dust mites in the child's environment?'. If the answer is yes, the model verifies if the patient's mother has reported having a history of atopic diseases. If the answer is no, the model verifies if the mother had a cesarean mode of birth, if the answer is now yes, the decision tree classifies the case as non-asthmatic. Similarly, all the tree branches are interpreted in the same manner.

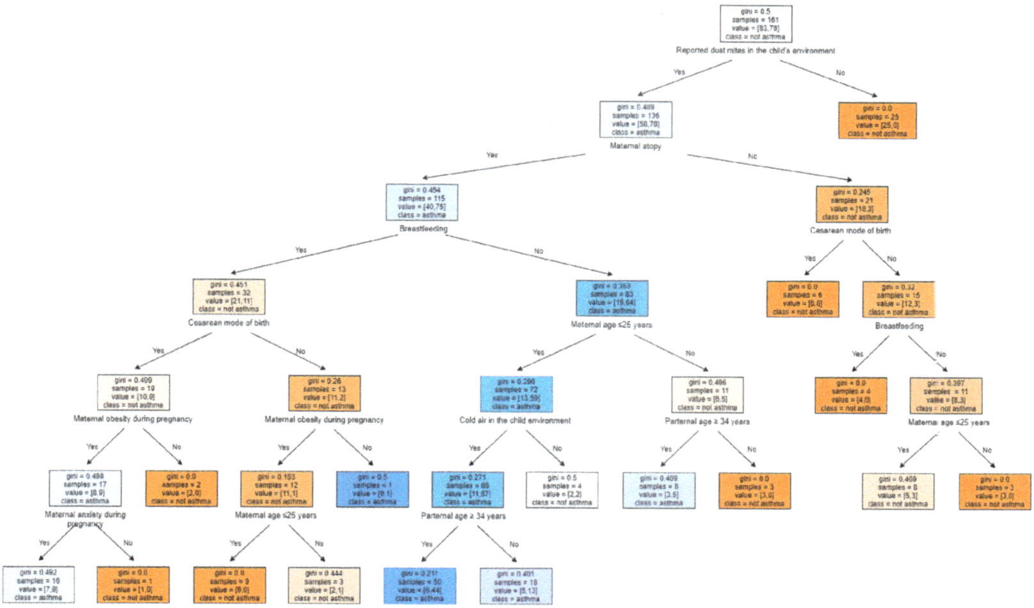

Figure 2. The obtained decision tree model-based classifier.

3.3. Random Forest Model

Random forest is a very effective ensemble learning technique that combines many classifiers to provide solutions to complex problems. After using decision trees, we decided to use random forest, which consists of many decision trees. The 'forest' of trees generated by the random forest algorithm is trained through bagging or bootstrap aggregating. Increasing the number of trees increases the precision of the outcome and reduces overfitting. In this work, we used 100 trees to 'grow' the forest (using a full feature set). The number of features randomly selected to perform each split was set to be the square root of the number of features, which is a typical choice. Since in this study, we have 36 features in total, the number of features that are randomly selected at each node is set to 6 features. The variable importance is computed using the mean decrease in the Gini index. Table 3 shows the 19 most important risk factors associated with childhood asthma.

Table 3. Variable importance using a random forest model.

Variable	Mean Decrease Gini
Breastfeeding	9.49
Reported dust mites in the child's environment	9.37
Maternal atopy	4.93
Cesarean mode of birth	4.18
Maternal age of ≤25 years	3.99
Antibiotic use during the first 2 years	3.59
Respiratory infections in family members (cold)	3.41
Paternal age of ≤25 years	2.85
Maternal obesity during pregnancy	2.05
Respiratory infections in family members (flu)	1.89
Consumption of antibiotics/paracetamol during pregnancy	1.77
Dietary diversity for children aged between 4 months and 6 months old	1.72
Dietary diversity for children aged more than 6 months	1.62
Cold airflow in the child environment	1.57
Strong odors in the child's environment	1.39
Overweight in the first 2 years	1.27
Pollen in the child environment	1.14
Mold in the child environment	0.99
Maternal age of ≥35 years	0.82

3.4. Support Vector Machine

Support vector machine is a machine learning technique that relies on kernel functions to provide the best fit to observed data [15]. It aims to map a high-dimensional feature space to the considered output. Different kernel functions can be adopted [16]. In this work, we assume a Gaussian kernel function. Hence, the prediction takes in the following form

$$\hat{Y} = \sum_{i=1}^{m} \theta_i \exp(\frac{- \parallel X - x_i \parallel^2}{\gamma}), \tag{2}$$

where $X = [X_1, \ldots, X_p]$, x_i is the value of the feature vector that corresponds to the ith observation, m is the number of observations, γ is a tuning parameter, and the θ_i's can be computed based on the cost function by evaluating the difference between the predicted values and the real values of pollutants' concentrations, to a threshold ϵ [17].

Table 4 describes the obtained results when the SVM model is adopted. It is shown that SVM did not bypass logistic regression and random forest but still yielded better results than decision trees.

3.5. Comparison of Performance of Models

In terms of predictive ability, the random forest yielded the best performance. It provided the most accurate results when predicting childhood asthma; it correctly classified 87.8% of the cases when applied to the test data set. The decision tree model has correctly classified 85.3% of the test cases. The decision tree identified "Asthma" cases with 91.30% sensitivity and "Not asthma" cases with 78% specificity. When evaluated on the test data set, the logistic regression model performed with an accuracy of 85.36%, a sensitivity of 83% and a 83% specificity (see Table 4). To settle the ambiguous results of the contest between logistic regression and decision trees. We compute a 10-fold cross validation and F1 scores, and we display an AUC-ROC for each one of our models. The average accuracy for 10-folds cross validation showed that random forest outperformed logistic regression and SVM. On the other hand, decision trees scored the lowest accuracy, but are

still helpful in terms of interpretability. Although random forest yielded the best accuracy results, it is evident from the plot in Figure 3 that the AUC for the logistic regression ROC curve is higher than that for random forest and decision trees. This means that logistic regression did a better job of classifying the positive class in the dataset. One may ask why the AUC for logistic regression is better than that of random forest, when random forest "seems" to outperform logistic regression with respect to accuracy. Our answer would be that accuracy is computed at the threshold value of 0.5. While AUC is computed by adding all the "accuracies" computed for all the possible threshold values. ROC can be seen as an average (expected value) of those accuracies when they are computed for all threshold values.

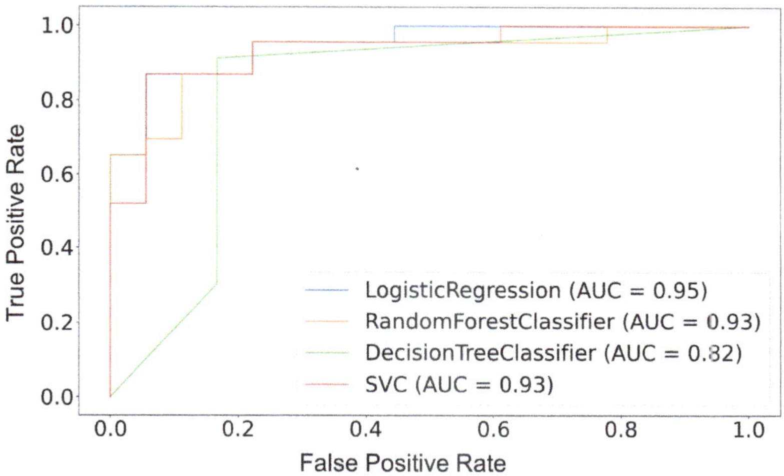

Figure 3. Models' ROC curve.

Table 4. Performance comparison of different prediction models.

Performance Metrics	Logistic Regression	Decision Tree	Random Forest	SVM
F1 score (y = Asthmatic)	0.89	0.87	0.86	0.81
F1 score (y = Not Asthmatic)	0.83	0.82	0.89	0.80
Accuracy (%)	85.36	85.3	87.8	80
Average accuracy for 10-fold cross validation (%)	82.57	75.19	84.9	82.5
Sensitivity, Sn (%)	83	91	87	67
Specificity, Sp (%)	88	78	88	94

4. Discussion

In the present study, we found that environmental factors, prenatal maternal exposures, complications during pregnancy, perinatal and postnatal personal exposures, along with other factors related to parental histories of atopy, can significantly increase the risk of asthma prevalence in pre-schooled children (children under 7 years old). As observed in previous studies [18,19], maternal histories of atopy were associated with an increased risk of childhood asthma. In this study, approximately 23.76% of the interviewed mothers reported having a history of an atopic disease. This study found that parental age at birth is significantly associated with the prevalence of asthma in 7-year-old children. Indeed, a maternal age higher than 35 years or lower than 24 were associated with high risks of childhood asthma, while a paternal age higher than 35 years was also associated with high risks of developing childhood asthma. For instance, 21.78% of asthma cases reported a paternal

age under 24 years. In previous studies, young maternal age and young paternal age were found associated with various child outcomes, including asthma prevalence in offspring; our results indicate that also maternal and paternal age of $\geq$35 years could be risk factors for childhood asthma [20–22]. In another study, using data from the Swedish Medical Birth register [23], results have shown that a decreased risk of asthma prevalence in childhood is associated with an increasing paternal age; this result was also confirmed in [22]. The difference in our results may reflect contrasting adverse factors related to behavioral, social and lifestyle agents that can characterize a middle income country such as Morocco[24]. In line with many studies [25–28], our results indicate that reported environmental factors such as cold airflow, strong odors, reported dust mites, pollen, mold and having pets in the neonatal period are significantly associated with the prevalence of childhood asthma. In this study, approximately 30.69% of asthma cases reported dust mites in their environments, 12.87% reported the presence of pollen in their surroundings, 11.88% reported the presence of mold in their surroundings and 6.93% stated an exposure to strong odors. In addition to these environmental factors, 22.77% of asthma cases reported that at least one family member had a respiratory infection (cold) in the neonatal period. Consuming antibiotics and/or paracetamol during pregnancy also was found to increase the risk of childhood asthma. Different studies provided supporting results; in [29], the authors showed that exposure to antibiotics during pregnancy was significantly associated with a small increased risk of asthma in pre-schooled children [28,29]. Different studies indicate that antibiotic use can have long-term altering effects on the vaginal bacterial flora, which may have adverse impacts on the health outcomes of the child [30,31]. Moreover, we also found that maternal obesity during pregnancy is significantly associated with asthma prevalence in children. Concerning the mode of birth, evidence for the health risks related to the perinatal period is accumulating. Children born via a cesarean delivery are at higher risks of developing autoimmune diseases such as asthma and allergies [32–35]. Our results also confirm a highly significant association between a cesarean mode of delivery and the increased risks of asthma prevalence. In our study 58.42% of children who developed asthma in their early childhood were born via a cesarean section. Furthermore, early childhood is also considered as a critical period for the occurrence of many risk factors related to environmental exposure and lifestyle habits. Although breastfeeding and delivery mode appear to modify the risk of childhood allergic outcomes, it is unclear whether they have the potential to attenuate or intensify the risk associated with developing asthma in offspring [34]. However, in our study, postnatal factors such as breastfeeding and dietary diversity between 4 months and 6 months old were found to be significantly associated with asthma prevalence among children. For instance, 45.54% of children who developed asthma did not receive maternal breastfeeding, and only 20% of patients had diverse nutrition between 4 months and 6 months old.

There are also some limitations to this study. Since the data set provided to us was obtained from a case-control study, the presence of selection bias and recall bias was a major concern. The study site, Ibn Sina childrens hospital, is an almost free of charge university hospital that cares for the local community coming from Rabat-Salé-Temara agglomeration, which is characterized by major social differences across and within areas. Although unlikely, there is also a possibility that cases and controls from places outside the hospital's service area may have come to the hospital for care hence resulting in selection bias. The outcome, i.e., child's asthma status, was determined clinically by the primary care physician, but exposure data were self-reported. Differential recall of exposures by mothers of children with asthma as compared to mothers of children without asthma could result in differential misclassification (recall) bias. Such type of bias is more common in case-control studies of children with severe medical conditions such as birth defects, and hence less likely to have occurred in our study. Furthermore, the study was designed to ask respondents about different types of exposures; thus, it is unlikely for mothers of children with asthma to remember exactly the prenatal exposures. However, to minimize interviewer bias, the researcher who interviewed study participants was blinded to the

asthma status of the child. Nonetheless, interviewer bias in face-to-face studies is difficult to eliminate completely. Prenatal exposure to pets was not measured objectively and may have resulted in misclassification errors. Obtaining access to medical data sets is very challenging due to patients' privacy and the lack of electronic health records. The current study was performed at one regional hospital, where we were only able to obtain access to data of 202 patients. Larger scale studies are needed to improve prediction performance and generalize our results beyond the regional nature of our study.

5. Conclusions

The findings of this study emphasize the potential importance of assessing prenatal, perinatal and postnatal risk factors associated with childhood asthma. In order to reduce the risks of developing childhood asthma in our population, the results from this study can provide relevant support for further use when elaborating the right prevention strategies regarding prenatal and during pregnancy care. Moreover, the risk factors identified in this study can help us predict children that are prone to develop asthma in the early stages of life and thereby allow a secure set of interventions that could prevent them from developing the disease and thus help them lead a healthy and normal childhood.

Author Contributions: Conceptualization, Z.J., I.G. and M.G.; methodology, Z.J. and I.G.; formal analysis, Z.J.; validation, M.G. and C.M.; investigation, Z.J.; resources, M.E.H. and C.M.; writing—original draft preparation, Z.J. and I.G.; writing—review and editing, Z.J., I.G. and M.G.; supervision, M.G.; project administration, M.G.; funding acquisition, M.G. All authors have read and agreed to the published version of the manuscript.

Funding: The work presented in this paper was carried out within the MoreAir project, which is partly funded by the Belgium Ministry of cooperation through the VLIR UOS program under grant MA2017TEA446A101.

Acknowledgments: We thank the pediatric department of the Ibn Sina hospital for their support and collaboration.

Conflicts of Interest: The authors declare no conflict of interest. The funders had no role in the design of the study; in the collection, analyses, or interpretation of data; in the writing of the manuscript, or in the decision to publish the results.

References

1. WHO. Asthma Fact Sheet. 2019. Available online: https://www.who.int/news-room/fact-sheets/detail/asthma (accessed on 3 May 2021).
2. Wadden, D.; Farrell, J.; Smith, M.J.; Twells, L.K.; Gao, Z. Maternal history of asthma modifies the risk of childhood persistent asthma associated with maternal age at birth: Results from a large prospective cohort in Canada. *J. Asthma* **2021**, *58*, 38–45. [CrossRef] [PubMed]
3. Flanigan, C.; Sheikh, A.; Nwaru, B.I. Prenatal maternal psychosocial stress and risk of asthma and allergy in their offspring: Protocol for a systematic review and meta-analysis. *NPJ Prim. Care Respir. Med.* **2016**, *26*, 16021. [CrossRef]
4. Yang, H.J. Impact of perinatal environmental tobacco smoke on the development of childhood allergic diseases. *Korean J. Pediatr.* **2016**, *59*, 319. [CrossRef] [PubMed]
5. Asher, M.I. Recent perspectives on global epidemiology of asthma in childhood. *Allergol. Immunopathol.* **2010**, *38*, 83–87. [CrossRef] [PubMed]
6. Martino, D.; Prescott, S. Epigenetics and prenatal influences on asthma and allergic airways disease. *Chest* **2011**, *139*, 640–647. [CrossRef] [PubMed]
7. Nafti, S.; Taright, S.; El Ftouh, M.; Yassine, N.; Benkheder, A.; Bouacha, H.; Fakhfakh, H.; Ali-Khoudja, M.; Texier, N.; El Hasnaoui, A. Prevalence of asthma in North Africa: the Asthma Insights and Reality in the Maghreb (AIRMAG) study. *Respir. Med.* **2009**, *103*, S2–S11. [CrossRef]
8. Subbarao, P.; Becker, A.; Brook, J.R.; Daley, D.; Mandhane, P.J.; Miller, E.G.; Turvey, E.S.; Sears, M.R. Epidemiology of asthma: Risk factors for development. *Expert Rev. Clin. Immunol.* **2009**, *5*, 77–95. [CrossRef] [PubMed]
9. Kashanian, M.; Mohtashami, S.S.; Bemanian, M.H.; Moosavi, S.A.J.; Moradi Lakeh, M. Evaluation of the associations between childhood asthma and prenatal and perinatal factors. International *J. Gynecol. Obstet.* **2017**, *137*, 290–294. [CrossRef] [PubMed]
10. Oliveti, J.F.; Kercsmar, C.M.; Redline, S. Pre-and perinatal risk factors for asthma in inner city African-American children. *Am. J. Epidemiol.* **1996**, *143*, 570–577. [CrossRef] [PubMed]

11. Midodzi, W.K.; Rowe, B.H.; Majaesic, C.M.; Saunders, L.D.; Senthilselvan, A. Early life factors associated with incidence of physician-diagnosed asthma in preschool children: Results from the Canadian Early Childhood Development cohort study. *J. Asthma* **2010**, *47*, 7–13. [CrossRef] [PubMed]
12. Davidson, R.; Roberts, S.E.; Wotton, C.J.; Goldacre, M.J. Influence of maternal and perinatal factors on subsequent hospitalisation for asthma in children: Evidence from the Oxford record linkage study. *BMC Pulm. Med.* **2010**, *10*, 14. [CrossRef] [PubMed]
13. Douwes, J.; Cheng, S.; Travier, N.; Cohet, C.; Niesink, A.; McKenzie, J.; Cunningham, C.; Le Gros, G.; von Mutius, E.; Pearce, N. Farm exposure in utero may protect against asthma, hay fever and eczema. *Eur. Respir. J.* **2008**, *32*, 603–611. [CrossRef] [PubMed]
14. Arif, A.A.; Veri, S.D. The association of prenatal risk factors with childhood asthma. *J. Asthma* **2019**, *56*, 1056–1061. [CrossRef] [PubMed]
15. Gryech, I.; Ghogho, M.; Elhammouti, H.; Sbihi, N.; Kobbane, A. Machine learning for air quality prediction using meteorological and traffic related features. *J. Ambient. Intell. Smart Environ. Prepr.* **2020**, *12*, 379–391. [CrossRef]
16. Vidnerová, P.; Neruda, R. Sensor Data Air Pollution Prediction by Kernel Models. In Proceedings of the 2016 16th IEEE/ACM International Symposium on Cluster, Cloud and Grid Computing (CCGrid), Cartagena, Colombia, 16–19 May 2016; pp. 666–673. [CrossRef]
17. Smola, A.J.; Schölkopf, B. A tutorial on support vector regression. *Stat. Comput.* **2004**, *14*, 199–222. [CrossRef]
18. Illi, S.; Weber, J.; Zutavern, A.; Genuneit, J.; Schierl, R.; Strunz-Lehner, C.; von Mutius, E. Perinatal influences on the development of asthma and atopy in childhood. *Ann. Allergy Asthma Immunol.* **2014**, *112*, 132–139. [CrossRef] [PubMed]
19. Jaakkola, J.J.; Nafstad, P.; Magnus, P. Environmental tobacco smoke, parental atopy, and childhood asthma. *Environ. Health Perspect.* **2001**, *109*, 579–582. [CrossRef]
20. Laerum, B.N.; Svanes, C.; Wentzel-Larsen, T.; Gulsvik, A.; Torén, K.; Norrman, E.; Gíslason, T.; Janson, C.; Omenaas, E. Young maternal age at delivery is associated with asthma in adult offspring. *Respir. Med.* **2007**, *101*, 1431–1438. [CrossRef] [PubMed]
21. Sherriff, A.; Peters, T.J.; Henderson, J.; Strachan, D. Risk factor associations with wheezing patterns in children followed longitudinally from birth to 3(1/2) years. *Int. J. Epidemiol.* **2001**, *30*, 1473–1484. [CrossRef] [PubMed]
22. Thomsen, A.M.L.; Ehrenstein, V.; Riis, A.H.; Toft, G.; Mikkelsen, E.M.; Olsen, J. The potential impact of paternal age on risk of asthma in childhood: A study within the Danish National Birth Cohort. *Respir. Med.* **2018**, *137*, 30–34. [CrossRef] [PubMed]
23. Almqvist, C.; Olsson, H.; Ullemar, V.; D'Onofrio, B.M.; Frans, E.; Lundholm, C. Association between parental age and asthma in a population-based register study. *J. Allergy Clin. Immunol.* **2015**, *136*, 1103–1105.e2. [CrossRef] [PubMed]
24. Gryech I, Ben-Aboud Y, Guermah B, Sbihi N, Ghogho M, Kobbane A. MoreAir: A Low-Cost Urban Air Pollution Monitoring System. *Sensors* **2020**, *20*, 998. [CrossRef] [PubMed]
25. Castro-Rodriguez, J.A.; Forno, E.; Rodriguez-Martinez, C.E.; Celedón, J.C. Risk and protective factors for childhood asthma: What is the evidence? *J. Allergy Clin. Immunol. Pract.* **2016**, *4*, 1111–1122. [CrossRef]
26. Segura, N.; Fraj, J.; Cubero, J.; Sobrevía, M.; Lezaun, A.; Ferrer, L.; Sebastián, A.; Colás, C. Mould and grass pollen allergy as risk factors for childhood asthma in Zaragoza, Spain. *Allergol. Immunopathol.* **2016**, *44*, 455–460. [CrossRef]
27. Celedón, J.C.; Milton, D.K.; Ramsey, C.D.; Litonjua, A.A.; Ryan, L.; Platts-Mills, T.A.; Gold, D.R. Exposure to dust mite allergen and endotoxin in early life and asthma and atopy in childhood. *J. Allergy Clin. Immunol.* **2007**, *120*, 144–149. [CrossRef]
28. Murrison, L.B.; Brandt, E.B.; Myers, J.B.; Hershey, G.K.K. Environmental exposures and mechanisms in allergy and asthma development. *J. Clin. Investig.* **2019**, *129*, 1504–1515. [CrossRef] [PubMed]
29. Mulder, B.; Pouwels, K.B.; Schuiling-Veninga, C.C.M.; Bos, H.J.; De Vries, T.W.; Jick, S.S.; Hak, E. Antibiotic use during pregnancy and asthma in preschool children: The influence of confounding. *Clin. Exp. Allergy* **2016**, *46*, 1214–1226. [CrossRef] [PubMed]
30. McKeever, T.M.; Lewis, S.A.; Smith, C.; Hubbard, R. The importance of prenatal exposures on the development of allergic disease: A birth cohort study using the West Midlands General Practice Database. *Am. J. Respir. Crit. Care Med.* **2002**, *166*, 827–832. [CrossRef] [PubMed]
31. Kuo, C.H.; Kuo, H.F.; Huang, C.H.; Yang, S.N.; Lee, M.S.; Hung, C.H. Early life exposure to antibiotics and the risk of childhood allergic diseases: An update from the perspective of the hygiene hypothesis. *J. Microbiol. Immunol. Infect.* **2013**, *46*, 320–329. [CrossRef] [PubMed]
32. Bager, P.; Wohlfahrt, J.; Westergaard, T. Caesarean delivery and risk of atopy and allergic disesase: Meta-analyses. *Clin. Exp. Allergy* **2008**, *38*, 634–642. [CrossRef] [PubMed]
33. Thavagnanam, S.; Fleming, J.; Bromley, A.; Shields, M.D.; Cardwell, C.R. A meta-analysis of the association between Caesarean section and childhood asthma. *Clin. Exp. Allergy* **2008**, *38*, 629–633. [CrossRef] [PubMed]
34. Sitarik, A.R.; Kasmikha, N.S.; Kim, H.; Wegienka, G.; Havstad, S.; Ownby, D.; Zoratti, E.; Johnson, C.C. Breast-feeding and delivery mode modify the association between maternal atopy and childhood allergic outcomes. *J. Allergy Clin. Immunol.* **2018**, *142*, 2002–2004. [CrossRef] [PubMed]
35. Pluymen, L.P.; Smit, H.A.; Wijga, A.H.; Gehring, U.; De Jongste, J.C.; Van Rossem, L. Cesarean delivery, overweight throughout childhood, and blood pressure in adolescence. *J. Pediatr.* **2016**, *179*, 111–117. [CrossRef] [PubMed]

healthcare MDPI

Review

Sensor-Based Fall Risk Assessment: A Survey

Guangyang Zhao [1], Liming Chen [2] and Huansheng Ning [1,*]

[1] School of Computer and Communication Engineering, University of Science and Technology Beijing, Beijing 100089, China; g20198892@xs.ustb.edu.cn
[2] School of Computing, University of Ulster, Newtownabbey BT37 0QB, UK; l.chen@ulster.ac.uk
* Correspondence: ninghuansheng@ustb.edu.cn

Abstract: Fall is a major problem leading to serious injuries in geriatric populations. Sensor-based fall risk assessment is one of the emerging technologies to identify people with high fall risk by sensors, so as to implement fall prevention measures. Research on this domain has recently made great progress, attracting the growing attention of researchers from medicine and engineering. However, there is a lack of studies on this topic which elaborate the state of the art. This paper presents a comprehensive survey to discuss the development and current status of various aspects of sensor-based fall risk assessment. Firstly, we present the principles of fall risk assessment. Secondly, we show knowledge of fall risk monitoring techniques, including wearable sensor based and non-wearable sensor based. After that we discuss features which are extracted from sensors in fall risk assessment. Then we review the major methods of fall risk modeling and assessment. We also discuss some challenges and promising directions in this field at last.

Keywords: fall risk assessment; fall prediction; gait monitoring; sensor

Citation: Zhao, G.; Chen, L.; Ning, H. Sensor-Based Fall Risk Assessment: A Survey. *Healthcare* **2021**, *9*, 1448. https://doi.org/10.3390/healthcare9111448

Academic Editor: Mahmudur Rahman

Received: 17 September 2021
Accepted: 21 October 2021
Published: 27 October 2021

1. Introduction

Aging population has become a common problem for major countries in the world. Compared with young people, falls are more likely to occur in the elderly. In some countries, such as the United States, falls have become a leading cause of death due to injuries in people over 65 [1]. In addition to physical damage, the elderly people with a history of falls may have greater psychological stress and a narrow scope of daily living activities, resulting in worse quality of life. Furthermore, the resulting personal and social expenditure is a large amount. Therefore, reducing the incidence of falls in the elderly is a matter of great significance. Fall risk assessment is one of the emerging promising technologies for the above goal.

In our opinion, "fall risk" refers to whether a person is prone to falling. This is how most work defines "fall risk" in our reviewed articles. They analyzed the relationship between gait characteristics and fall risk in combination with fall history in the past or fall status in the future. The items "fall detection" and "fall prediction" (fall risk assessment) are often confused. They are protective measures from different perspectives. Fall detection aims at detecting the occurrence of fall event in time so that treatment or protection during fall (e.g., air cells at waist) can be carried out right away. It is an "afterwards approach". Unlike fall detection, fall risk assessment is a "beforehand approach". It tries to identify elderly people at high fall risk. Then targeted preventive measures can be taken before the "real fall" happens. This kind of technology has a remarkable social and economic worth. It has received growing attention due to the great progress of sensing, communication and data processing technologies in recent years.

Fall risk assessment is a complicated process based on detection and analysis of factors leading to falls. There are many factors leading to falls, and they have been divided into two categories: external and internal [2]. External factors refer to environmental factors such as room layout, road conditions, to name but a few. Existing studies focus more

on internal factors. Internal factors refer to self status, including physical, cognitive and psychological. Older age has been shown to be related to falls, because aging can lead to instability in walking posture [3]. Sarcopenia is a syndrome highly relevant to falls [4]. It is a condition characterized by decreased muscle mass, muscle strength and physical performance. Sensory disturbance is another important factor related to falls [5], such as visual impairment and hearing impairment. In addition, medication, stroke, depression and postural hypotension are also internal factors. According to reviewed papers, there are scale-based and sensor-based approaches to detect and analyze internal factors.

The fall risk scale is an important tool in fall risk assessment. Scale-based assessment is suitable for most internal factors. Researchers fill out the scale through inquiry, observation and measurement. There are many scales for fall risk assessment. The most commonly used are the Berg Balance Scale [6], Tinetti Balance Scale [7], STRATIFY [8], and so on. Different scales are suitable for different situations. For example, STRATIFY was used only for elderly hospitalized patients [9]. Several reviews about scale-based fall risk assessment have been published over the years. A systematic review in 2018 by Park [9] paid attention to the quantitative analysis of the predictive validity of scales. The author pointed out that combining two assessment tools was more effective than using a single tool due to more factors were contained. Another review in 2018 by Ruggieri and colleagues [10] focused entirely on the setting, language, pathology and psychometric properties of scales. Together these papers have provided a comprehensive overview on the scales used in fall risk assessment. Given the existing works, we do not review research on the scale-based approach. It is necessary to point out that, while the scale-based approach is comprehensible and low-cost, it suffers from being subjective.

In recent years, technologies like sensing, efficient wireless communication and data processing have made significant progress. The advances and maturity of above mentioned technologies make a lot of researchers try to realize fall risk assessment by sensors. Sensor-based approaches focus on characteristics of kinematics and kinetics of a person. It tries to assess one's fall risk through motion state. Compared to the scale-based approach, the sensor-based approach is more objective, and at the same time, easy to implement. Sensor-based fall risk assessment is a complicated process that can be roughly characterized by four steps. These steps include (1) to select and fix wearable sensors to the subjects or deploy non-wearable sensors to environments to monitor the motion of the subjects, (2) to make the participants walk by rules and collect and transmit data that is obtained by the sensors, (3) to preprocess the data collected and choose or develop algorithms to establish models, (4) to use the model in last step to assess the relationship between gait status and fall risk using sensor data as input.

Compared to the number of surveys in scale-based fall risk assessment, there is a lack of comprehensive overviews on the latest development of sensor-based fall risk assessment. Considering this, a systematic survey will be of high value. It can inform the researchers of the current status and future promising directions. This survey aims at presenting a comprehensive overview on the state of the art of sensor-based fall risk assessment. It will cover the life cycle of the approach and provide descriptions and comparisons of various methods to highlight their advantages and disadvantages. In this survey, we review related works based on the order from monitoring to features to modeling and assessment. After the introduction, the organization of this article is as follows. In Section 2, we investigate the monitoring techniques used in sensor-based fall risk assessment. We then discuss about features in sensor-based fall risk assessment in Section 3. In Section 4, We review major modeling techniques and present comparison of these techniques. In Section 5, we provide insights into existing challenges of fall risk assessment. Potential future research directions of this field are listed in Section 6. The survey is concluded in Section 7.

2. Fall Risk Monitoring

A wide range of sensors, including inertial sensors like accelerometers and gyroscopes, pressure sensors, and infrared sensors, to name but a few, are used in fall risk assessment.

These various sensors have different types, functions, output signals, and technical principles. They can be classified as wearable sensors and non-wearable sensors according to the way they are deployed. Wearable sensors are the most commonly used. In the following we present the common practice in sensor-based fall risk assessment.

2.1. Wearable Sensors

Wearable sensors are sensors which are directly or indirectly fixed to human body, and they generate signals when the user moves or performs other activities. Wearable sensors can be embedded into daily objects like belts and shoes or directly fixed to the body. They can monitor movement status or physiological information when properly worn by users.

Inertial sensors and pressure sensors are the most frequently used wearable sensors in fall risk assessment. Inertial sensors mainly include accelerometers and gyroscopes. They are suitable for monitoring body motions. In Figure 1, we show a graphic example of leg flexion and extension angle during walking, which was obtained in our real application. Generally, inertial sensors are fixed to different body parts to obtain different motion features and pressure sensors are embedded into insoles. Howcroft et al. [11] placed tri-axial accelerometers on head, lower back and left and right shanks of older individuals under single-task and dual-task conditions to identify the optimal sensor combination, placement and modeling approaches for fall risk assessment. In addition, participants were required to wear pressure-sensing insoles. Accelerometer-based features used in this study were maximum, mean, and standard deviation of acceleration for different axes, cadence, stride time, fast Fourier transform (FFT) Quartile, ratio of even to odd harmonics (REOH), and maximum Lyapunov exponent (MLE). For pressure-sensing insoles, they derived features like center of pressure (CoP) path, temporal features such as stride time symmetry index between left and right limbs and impulse from the total force-time curve. The results indicated that multi-layer perceptron had a better performance than naïve Bayesian and support vector machine. In single-task fall risk classification, head sensor-based models had the best performance. Accelerometers were placed on lower limb (ankle, shank, etc.) to obtain spatiotemporal gait features like gait speed [11,12]. Doheny et al. [13] used tri-axial accelerometers on the thigh to record the process from sitting to standing during the five-times-sit-to-stand test. Weiss et al. [14] asked 107 Parkinson's patients to fix a small three-axis accelerometer to the lower back for three days respectively. Their walking quantity and quality were determined. Pressure sensors are effective for recording the changes in the plantar pressure of the human body during walking. It is an ideal assessment tool for postural stability. The study in [15] assessed fall risk of workers in the construction industry by changes of biomechanical gait stability features based on wearable insoles with pressure sensors. According to these reviewed articles, accelerometers are the most frequently used and practical wearable sensor category.

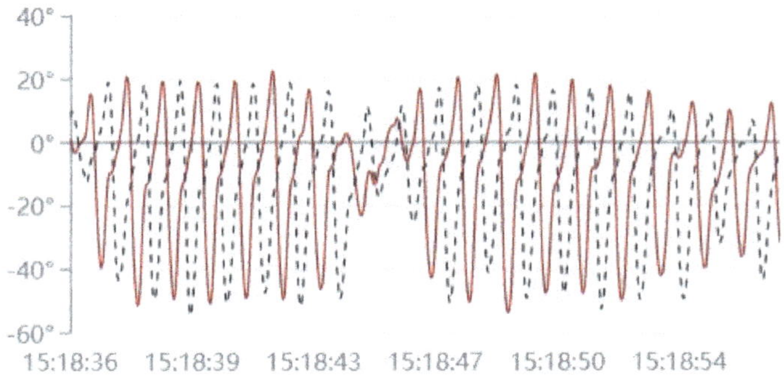

Figure 1. The angle curve of leg flexion and extension in the vertical direction during walking. The red line represents the left leg, and the black dotted line represents the right leg.

The human walking process can be regarded as several consecutive repetitive gait cycles. Generally, a gait cycle refers to the process of walking from the same foot's toe-off/heel-strike to the next toe-off/heel-strike. A gait cycle can also be divided into a swing phase and a stance phase. This means that each cycle can be recognized by detecting the flag events like heel-strike in the gait process, the gait motion can be segmented, and the features can be further extracted. Take a gait cycle of the right foot as an example. The time from the first toe off (right Toe-Off) to the first heel landing on the ground (right Heel-Strike) is the swing phase of the right foot. Then the right foot supports the weight, and the left foot enters the swing phase. It is the stance phase of the right foot until the next right Toe-Off. To detect the flag gait events, peak detection algorithms are useful [16,17]. They achieve this function by detecting repeated peaks of acceleration or angular velocity during gait.

Kinematic data of the human body is useful for knowing about gait status and fall risk assessment. It is usually obtained by inertial sensors. However, there are some types of errors in practice, which makes it challenging to obtain accurate motion information. The mounting error is an important error in applications of inertial sensors due to the misalignment from the inertial frame (sensor coordinate frame) and the global frame (body frame). Chen et al. [16] proposed a method for mounting error calibration. The method can determine the orientation of inertial frame with respect to the global frame. The results showed that it corrected the mounting error greatly. The integration drift is another kind of common errors in practice of inertial sensors. It comes from the accumulated signal noise in the process of integrating acceleration into velocity or angular velocity into angular displacement. Filters like Butterworth filter [18], Kalman filter [19] are usually used to eliminate the integration drift.

2.2. Non-Wearable Sensors

Pressure sensors can be either wearable or non-wearable. In addition to embedding to insoles or shoes, pressure sensors also can be used in treadmill or pressure platform like Wii balance board [20].

Infrared sensors and laser sensors are the most frequently used ones of non-wearable sensors. Nishiguchi et al. [21] developed a device that was based on an infrared laser sensor (laser range finder) to assess stepping performance. Further a new index "stepping response score" was created to assess fall risk of community-dwelling elderly individuals. The infrared laser sensor in this research was used to measure spatial and temporal parameters of steps by detecting position and motion of both legs.

Microsoft Kinect is often used in gait analysis and further for fall prevention. It consists of RGB cameras and infrared sensitive cameras and can produce depth images. Dubois et al. [22] proposed a system based on Microsoft Kinect camera to help preventing falls of the elderly people. They extracted three gait spatiotemporal features from the vertical displacement of the center of mass of the subject. The features are step length, step duration and gait speed. The features were compared to those obtained by the carpet. The results showed that their approach of gait analysis was effective. However, they did not further use this approach for fall risk assessment. Stone et al. [23] compared gait measurements by Kinect to those using a web-camera based system and those from a Vicon motion capture system. The results showed good agreements among them and confirmed the effectiveness of Kinect for passive fall risk assessment.

Infrared or laser sensors have the advantage of being precise. However, they suffer from many other issues. For example, the clothes worn by the subjects may affect the reflection of infrared rays, and multiple devices may be required due to limited field of view of sensors. This will raise the cost. Moreover, it usually requires participants to walk within a limited area which it can see. Compared to this kind of sensor, wearable sensors are cheaper and easier to deploy. Furthermore, wearable sensors are more flexible whether they are directly secured on body or embedded in clothes. Nevertheless, wearable sensor-

based fall risk assessment still suffers from issues like size, battery, and data transmission, to name but a few.

In the procedure of fall risk assessment, participants will be required to do assessment tasks. The most commonly used one is steady walking on the treadmill or ground. In addition, the Timed Up and Go (TUG) test [24] is often combined with sensors to assess fall risk. It measures the time it takes for the elderly to stand up and walk and then come back and sit down. The traditional TUG test is single-task. Research has found that the dual-task test results are better [25]. The five-times-sit-to-stand test (FTSS) is another test used for assessing fall risk. The time to finish this test is measured to indicate fall risk [13].

3. Sensor-Based Features for Fall Risk Characterization

In sensor-based fall risk assessment, features are extracted from sensor signals to quantify one's gait or posture characteristics. For most studies, it is necessary to identify each gait cycle by algorithms. Then the measurement values are calculated for each step. For instance, the gait speed of each step can be obtained by integrating the acceleration value obtained from accelerometers. Our survey do not focus on algorithms for identifying gait cycles or raw signals from sensors but concentrates on features extracted finally. In the following we summarize the common features according to their described gait characteristics.

3.1. Gait Intensity

Number of steps in a period is often used to reflect whether the participant is vigorous over a period of time. Cadence is the value of the number of steps divided by walking time. It reflects the intensity of gait. Too low or too high cadence during walking indicates abnormal gait patterns. Too low cadence may be related to freezing gait, and too high cadence may be related to festinating gait. These two abnormal gait patterns have been identified as highly correlated to fall incidents [26,27].

3.2. Gait Variability

Time-related features are useful in fall risk assessment. They include step time, stride time, stance time, and swing time during walking, single support time, double support time and so forth. Gait variability refers to the fluctuation in the value of a feature from one step to another [28]. These time-related features can quantify gait variability by computing average variance or standard deviation or coefficient of variation of them [29]. Hausdorff et al. [30] used the standard deviation of each participant's stride time and swing time to quantify gait variability. They placed force-sensitive insoles in participants' shoes to identify each stride. Then those required time-related features are determined for each stride by algorithms. The gait variability was further quantified by calculating the standard deviation of those features. The results showed that these two measures of gait variability were predictive of future falls, and the possibility of falling is positively correlated with the degree of gait variability. Generally, higher variability of gait means higher fall risk [31]. In addition to time-related features, trunk acceleration is also suitable for quantifying the variability of human gait [32]. The authors in [32] used standard deviation of trunk acceleration to measure gait variability under inclined conditions.

3.3. Gait Stability

Stability of gait is another important indicator for assessing fall risk. Gait stability is close to gait variability but not equal to it. Gait stability refers to the ability of maintaining gait stable when walking under small perturbations or recovering from an external perturbation. Hollman et al. [33] used variability of velocity from stride to stride to quantify gait stability. GAITRite is responsible for measuring the spatiotemporal parameters required for this study. The study was to examine whether gait stability differ in older adults compared with young adults during normal walking and walking while performing cognitive tasks. The results indicated that variability of velocity from stride to stride was greater during

dual task walking, and dual task walking had a larger impact on the older adults than young adults. Therefore, walking with cognitive tasks may increase the gait instability and risk of falls. In addition to linear statistics, non-linear measures like maximum Lyapunov exponent (λs) are also used in stability assessment [32]. Local dynamic stability (LDS) is based on λs, which can be calculated by Rosenstein's algorithm [34]. LDS is a nonlinear parameter which is derived from dynamic system theory to assess gait stability. There is a negative correlation between LDS and maximum Lyapunov exponent. When λs increases, LDS decreases and fall risk increases.

3.4. Postural Stability

Postural stability refers to the ability of maintaining body stable when standing. Melzer et al. [35] measured the postural stability of subjects in wide stance and narrow stance to find differences between fallers and non-fallers. The features were based on center of pressure (COP), which included COP path length, elliptical area, COP velocity, medio-lateral (ML) sway length, and antero-posterior sway length. The results indicated that there were no significant differences between two groups when standing in wide stance. Significant differences emerged when subjects standing in narrow stance. Fallers had significant higher values of COP path length, elliptical area, COP velocity, and ML sway length in various conditions which included eyes open, eyes closed, and eyes open while standing on the foam.

3.5. Gait Symmetry

Gait symmetry reflects the control of the lower limbs on both sides during walking. Jiang et al. [36] pointed out that gait symmetry as well as gait stability is important for fall risk assessment. There are four frequently used methods to measure gait symmetry, namely symmetry ratio, symmetry index, gait asymmetry, and symmetry angle [37]. They are showed in Table 1. Symmetry ratio index has been used in clinical measurement of gait symmetry but has a low sensitivity [38]. The symmetry index is a symmetry evaluation standard based on ground reaction forces proposed by Robinson et al. [39]. Gait asymmetry is a logarithmic transform of symmetry ratio. In [40], the authors evaluated the degree of asymmetry by comparing the swing time of the legs on both sides. Symmetry angle was proposed by Zifchock et al. [41]. Zifchock certified that symmetry angle is highly correlated with symmetry index. This suggest that symmetry angle may be a good substitute for symmetry index.

Table 1. Calculation formula of gait symmetry.

Measurement	Abbreviation	Calculation Formula
Symmetry ratio index	RI	$\left(1-\frac{x_r}{x_l}\right)*100\%$
Symmetry index	SI	$\frac{\lvert x_r-x_l\rvert}{0.5(x_r+x_l)}*100\%$
Gait asymmetry	GA	$\ln\left(\frac{x_r}{x_l}\right)*100\%$
Symmetry angle	SA	$\frac{45^\circ-\tan^{-1}\left(\frac{x_r}{x_l}\right)}{90^\circ}*100\%$

x_r and x_l: values of specific features for right and left limbs.

Among these four measures, there is no recognized standard for assessing gait symmetry. Patterson et al. [37] analyzed and compared these four measures for stroke patients and normal people. Stroke patients were at high risk of falls. They used five features including step length, swing time, stance time, double support time and ratio of swing time to stance time in the above four equations respectively. Analysis results suggested that no equation performed better in distinguishing stroke patients. On the contrary, different gait features have a more significant impact on the results. However, symmetry ratio may be

more interpretable than the others. Therefore, the authors recommended symmetry ratio as a candidate standardization.

3.6. Gait Smoothness

Gait smoothness is associated with the quality of walking control. It reflects the continuousness of walking. It is usually measured by harmonic ratio. Harmonic ratio is a frequency feature in fall risk assessment. It is the ratio between the sum of the magnitudes of the even to the odd harmonics over a single stride. A higher value of harmonic ratio indicates smoother gait when walking. Parkinson's patients and stroke patients usually perform poorly in smoothness. Low smoothness may lead to falls. Doi et al. [42] used harmonic ratio of trunk acceleration to predict falls of elderly based on prospective research method. In this study, researchers calculated the harmonic ratio of acceleration of upper and lower trunk by digital Fourier transformation in each direction. The results indicated that the harmonic ratio of upper trunk acceleration was independently associated with incidence of falls in a year. It was confirmed by ROC curve analysis that the harmonic ratio of upper trunk acceleration had high specificity for predicting potential future falls.

4. Fall Risk Modeling and Assessment Approaches

4.1. Conventional Machine Learning

The mainstream modeling approaches in sensor-based fall risk assessment are related to machine learning techniques. Conventional machine learning approaches for fall risk modeling and assessment can be classified into two categories: discriminative and generative. Discriminative models are to find a decision boundary through which samples are divided into corresponding categories. Discriminative models mainly include linear regression, logistic regression, linear discriminant analysis, Support Vector Machine (SVM), to name but a few. Generative models are to learn the boundary of each category instead of the single decision boundary. Generative modeling approaches include Naïve Bayesian classifier, k-Nearest Neighbor (KNN), Dynamic Bayesian Network (DBN) and so on. In the following, we cover the frequently used conventional machine learning approaches in sensor-based fall risk assessment.

4.1.1. Discriminative Modeling Approaches

Perhaps the most frequently used discriminative modeling approach is regression which includes linear regression and logistic regression. Linear regression is a regression analysis method that uses linear regression equations to model the relationship between independent variables and dependent variables. Liu et al. [43] used multiple linear regression to map features which derived from accelerometer data to the number of falls in the past one year. In this research, a triaxial accelerometer was mounted on participants' waist. There were 126 features extracted from the acceleration data of 68 subjects and a discriminant classifier was established according to an applied threshold value. The classifier obtained an accuracy of 97% in identifying multiple-fall fallers, and the accuracy in estimating the number of falls in the last year was 71%. This study is a retrospective analysis to explore the relationship between gait status and fall history. Therefore, its result cannot be directly used in predicting future falls. Nevertheless, it may be used as a reference in prospective studies.

Compared to linear regression, logistic regression is more commonly used. Logistic regression converges the output range from the real number domain to [0, 1] through sigmoid function. It is used in binary classification problem and more robust than linear regression. Doheny et al. [13] applied the five-times-sit-to-stand test to obtain acceleration data by an accelerometer attached to the lateral thigh to identify each sit-to-stand-to-sit phase and sit-to-stand and stand-to-sit processes. Another accelerometer was attached to the sternum to capture trunk acceleration. Participants were 39 elderly people, 19 of whom had a history of falls. Logistic regression was used to classify the participants based on their status. There were totally 70 accelerometer-derived features which were the mean and

variation of the root-mean-squared amplitude, jerk and Spectral Edge Frequency (SEF) of the acceleration during each section of the assessment. Four features were finally selected for modeling based on test-retest reliability of each feature. Their model's accuracy was 74.4%, specificity was 80.0% and sensitivity was 68.7%. It is worth mentioning that the number of participants was relatively small. It may lead to overfitting of the model.

Support Vector Machine is another discriminative modeling technique which is a kind of generalized linear classifier for binary classification. SVM finds a hyperplane to classify two categories, and the hyperplane needs to be as far away as possible from the nearest element of each category. Greene et al. [44] applied a body-worn inertial sensor (SHIMMER) which was attached to the lower back of participants and pressure data was obtained by a Tactex S4 HD pressure mat. The root mean square amplitude of the medio-lateral and the anterior-posterior acceleration was measured for quantifying postural sway of each direction. For frequency domain, the spectral edge frequency and the spectral entropy (H) were calculated for acceleration and angular velocity signals. SVM was used to distinguish between fallers and non-fallers and obtained a mean classification accuracy greater than 70%. The result was better than that of using Berg Balance Scale (BBS) as a comparison.

Discriminative approaches are used to model conditional probability and find the optimal boundary between different categories. It pays more attention to the differences between different categories than the characteristics of the sample data itself. Compared to generative modeling, it can work by less computing resource and samples and has better predicting performance in most practical cases.

4.1.2. Generative Modeling Approaches

Naïve Bayesian classifier may be the simplest generative modeling approach. It has been used with satisfying results for fall risk assessment [11,12,45,46]. Naïve Bayesian classifier is based on Bayes theorem and assumes that the features are conditional independent of each other. This assumption will lead to the decline of classification accuracy when the correlation among features is large. K-Nearest Neighbor is another generative modeling approach that can be used in fall risk assessment [46,47]. KNN method determines the category of the samples to be divided according to the category of the nearest one or several samples. The disadvantage of KNN is the large amount of calculation, because for each sample to be classified, the distance from itself to all known samples must be calculated.

Bayesian Network (BN) is a type of graphical model to describe the dependency relationship between data variables. Dynamic Bayesian Network is an extention of BN. It can represent the evolution of variables over time. Cuaya et al. [48] built two DBN models for predicting falls in next six months. One feature set was established under the guidance of experts from the Human Motion Analysis Laboratory of the INR. Another one was founded on the feature set automatically selected by forward sequential selection (FSS) algorithm. The expert-guided model showed slight advantages over the model applying FSS. However, considering the small sample size, a conclusion that expert-guided feature selection is better than FSS cannot be made.

4.2. Deep Learning

Neural networks are based on perceptron, so neural networks are sometimes called multi-layer perceptron, namely Artificial Neural Networks (ANN). Deep learning is the general name of pattern analysis methods based on ANN. The neural network layer in ANN can be divided into three categories: input layer, hidden layer and output layer. Generally speaking, all middle layers are hidden layers. Basically, deep learning approaches can be divided into two categories: supervised learning and semi-supervised/unsupervised learning. However, most studies of fall risk assessment using deep learning are based on supervised learning. There are few studies using semi-supervised or unsupervised learning approaches. Therefore, we focus on the applications of supervised learning approaches in fall risk assessment.

Deep neural networks (DNN) are composed of a large number of simple processing modules named "neurons". These "neurons" are distributed in separate layers and their common task is calculating the "activation function" of the weighted sum of their inputs. DNNs have the ability to learn directly on the raw data, so as to reduce the demand for feature engineering. Full connected DNNs are suitable for most classification tasks theoretically. However, they are rarely used in practice due to demand for large amount of data.

Nait Aicha et al. [49] used a dataset which consists of acceleration data from 296 older people. They compared Convolutional Neural Networks (CNN), Long Short-Term Memory (LSTM), a combination of them and a base model which used biomechanical features in single-task learning and multi-task learning respectively. The results indicated that deep learning methods perform better in identifying subjects and assessing fall risk with gender and age as auxiliary tasks when doing multi-task learning.

Recurrent Neural Networks (RNN) are neural networks with sequence data as input. LSTM and Bi-directional long short-term memory (BiLSTM) are the most common RNNs for studies of fall risk assessment. Meyer et al. [50] analyzed a variety of machine learning models and feature sets for fall risk assessment of patients with multiple sclerosis. For conventional machine learning methods which were logistic regression, SVM, decision tree, KNN and ensemble binary statistical classification models in this study, feature sets manually calculated from accelerometer data were used. Deep learning methods do not require features to be manually calculated. Deep leaning models can take raw data as input, extract features automatically and finish the classification task. In this study, BiLSTM which combines forward LSTM with backward LSTM was used. BiLSTM is based on Recurrent Neural Network. In discriminating patients with a fall history from those without, BiLSTM obtained an accuracy of 86% and an AUC of 0.88. It performed better than all conventional machine learning methods used in this study. It is worth mentioning that BiLSTM may be better than LSTM when considering retrospective fall status classification as an intermediate step in prospective fall risk assessment.

Recurrent Neural Networks are suitable for sequence data, which leads to frequent use. Tunca et al. [51] explored LSTM for fall risk assessment as well. LSTM is sequence-to-label classifier that can operate on sequence data directly. Considering that the existing research on fall risk assessment and gait analysis has accumulated valuable domain knowledge, this study attempted to combine the domain knowledge inherent in the spatio-temporal gait features with LSTM. Sequences of spatio-temporal gait features from a sensor-based system were used as input. Data samples consisted of four-dimensional sequences whose length are ten. Four dimensions were stride length, clearance, stance time and swing time. The length was due to 10 strides in a window which was used for data augmentation. In addition to sequence data, another LSTM model was trained by raw data to determine whether the model can implicitly learn the required features. The results showed that LSTM with sequences of gait features achieved an accuracy of 89% on a validation set and 92.1% on a separate test set.

Convolutional Neural Networks are feedforward neural networks with convolutional calculation. Although most applications of CNNs are related to pictures, they can handle most grid-like data. Time series data commonly used in fall risk assessment can be considered as one-dimensional grid-like data. Savadkoohi et al. [52] used a one-dimensional CNN with force plate time series data. Their network consists of three convolutional layers, a max pooling layer and a global average pooling layer. The authors used only two convolutional layers before the max pooling layer to minimize the information loss. Global average pooling was used to reduce the possibility of overfitting.

4.3. Knowledge-Driven Model

In addition to data-driven models mentioned in above two sections, there have been a small quantity attempts to perform fall risk assessment by knowledge-driven models. Farseeing Fall Risk Assessment Tool (FRAT-up) introduced in [53] is based on probabilistic

rules, generated automatically from a light ontology. FRAT-up is based on an assumption that the total fall risk of a person is determined by the contributions of the risk factors related to falls. The system takes the characteristics of a subject in terms of risk factors. As output FRAT-up provides an estimation of the fall risk.

Compared to data-driven models, knowledge-driven models may be not applicable to sensor-based fall risk assessment.

A comparison between different fall risk modeling approaches is showed in Table 2.

Table 2. The comparison of fall risk modeling approaches.

	Conventional Machine Learning	Deep Learning
Model type	Logistic regression, SVM, KNN, HMM	CNN, LSTM, BiLSTM
Advantage	More friendly to small sample size, cheaper, more interpretable	Higher accuracy, no need for engineering
Disadvantage	Lack of scalability	More expensive, hard to understand

5. Challenges

This survey analyzed literatures about sensor-based fall risk assessment, including sensors themselves, features extracted from sensor data and modeling approaches. There are still some challenges needed to focus on and they are listed below.

5.1. *Optimal Sensor Placement*

Different parts of human body show different motion characteristics in the process of walking. Fixing sensors on various body parts can obtain various types of data and extract various gait features. For example, sensors attached to thighs can monitor the process of sit-to-stand. Many body positions have been tested. However, researchers have not reached a consensus on the optimal position of sensors for the work of assessing fall risk.

5.2. *Better Task for Risk Assessment of Falling*

In the step of motion monitoring and data collecting of fall risk assessment, subjects are required to perform a task. It may be walking on flat ground, walking on the slope, Five-Times-Sit-to-Stand. In addition to single-task, dual-task walking has been used in some studies. However, dual-task walking does not significantly help to improve the performance of the model for fall risk assessment. It is important to find tasks that can better characterize human gait.

5.3. *Insufficient Sample Size*

In most studies we reviewed, the sample size is usually small. Many studies included no more than 100 subjects, and due to the requirement of long and intensive follow-up period, the final sample size may be smaller. In addition, the continuous tracking of a large number of subjects is also a costly work. Too little total sample size may lead to overfitting of the final model. Too few positive samples relative to total sample size may lead to distorted models.

6. Future Directions

6.1. *More Robust Feature Construction*

Feature selection is a commonly used approach of feature dimensionality reduction. It can simplify our final model and reduce the overfitting of model. Compared with automated feature selection, the extraction of raw feature set depends more on manually operating. Therefore, more effective feature selection approaches are important for feature construction before modeling. In addition, knowledge in other fields like physiatry may help find the categories of raw features which are more closely associated with falls. Most studies today focus on motion signals of human. In fact, physiological signals can also reflect human's status. Feature construction based on physiological signals may be also a future direction.

6.2. Public Database of Various Datasets

Based on the extensive literature search, we found that the studies focus much on the way sensors are placed and how to design tasks before building a model. Authors acquire data in various ways: Types and position of sensors, tasks that paticipants should do, sampling frequency, extracted features, to name but a few. Building a public database with these various datasets acquired by different ways could help compare and reuse the results.

6.3. Daily-Life Monitoring System

Most studies are carried out in laboratories or clinical environments. Professionals are in need and data acquisition is inflexible. Moreover, participants who go to lab may feel nervous when doing the task required so that they show a motion pattern which is different from the usual one. Therefore, developing an daily-life continuously gait monitoring system is important. In order to make elderly people continuously use the gait system, it cannot be intrusive and clumsy, and user-friendly design should be under consideration to make users comfortable with it. The wireless Inertial Measurement Unit (IMU) is a good choice due to its small size and convenience of data transmission. Embedding sensors into daily clothes is a new developing direction. Rosa et al. [54] developed an electric insole based system. It transferred data collected to user's smartphone in real time. And the smartphone further transferred the data to backend server to analyze. The smartphone itself also can be used for fall risk assessment due to the sensors it contains. Nishiguchi et al. [55] developed a mobile phone application to validate its capacity to quantify gait features. However, there are issues when using smartphone as the tool to assess fall risk. For example, people carry their smartphones in different places (bags, pockets, etc.), which may affect the assessment process of mobile phone.

7. Conclusions

Fall risk assessment has become a promising direction in the health industry. The development of various disciplines, e.g., machine learning, sensor networks and wireless communications, has jointly promoted the progress in this field. In this work, a survey of fall risk assessment based on sensors has been presented. We first introduced principles and methodology of this field. Then we reviewed monitoring, modeling and assessment of fall risk respectively. In particular we identified characteristics of above aspects and reviewed each individual field by category. At last, we discussed the existing challenges and promising directions in this field.

Author Contributions: Investigation and writing—original draft preparation, G.Z.; writing—review and editing, L.C.; topic selection, framework design and coordination, H.N. All authors have read and agreed to the published version of the manuscript.

Funding: This research received no external funding.

Institutional Review Board Statement: Not Applicable.

Informed Consent Statement: Not Applicable.

Data Availability Statement: Not Applicable.

Conflicts of Interest: The authors declare no conflict of interest.

References

1. Bergen, G.; Stevens, M.R.; Burns, E.R. Falls and fall injuries among adults aged $\geq$ 65 years—United States, 2014. *Morb. Mortal. Wkly. Rep.* **2016**, *65*, 993–998. [CrossRef]
2. King, M.B.; Tinetti, M.E. Falls in community-dwelling older persons. *J. Am. Geriatr. Soc.* **1995**, *43*, 1146–1154. [CrossRef]
3. Delbaere, K.; Close, J.C.; Heim, J.; Sachdev, P.S.; Brodaty, H.; Slavin, M.J.; Kochan, N.A.; Lord, S.R. A multifactorial approach to understanding fall risk in older people. *J. Am. Geriatr. Soc.* **2010**, *58*, 1679–1685. [CrossRef]
4. Landi, F.; Liperoti, R.; Russo, A.; Giovannini, S.; Tosato, M.; Capoluongo, E.; Bernabei, R.; Onder, G. Sarcopenia as a risk factor for falls in elderly individuals: Results from the ilSIRENTE study. *Clin. Nutr.* **2012**, *31*, 652–658. [CrossRef]

5. Pfortmueller, C.A.; Kunz, M.; Lindner, G.; Zisakis, A.; Puig, S.; Exadaktylos, A.K. Fall-related emergency department admission: Fall environment and settings and related injury patterns in 6357 patients with special emphasis on the elderly. *Sci. World J.* **2014**. [CrossRef] [PubMed]
6. Santos, G.M.; Souza, A.; Virtuoso, J.F.; Tavares, G.; Mazo, G.Z. Predictive values at risk of falling in physically active and no active elderly with Berg Balance Scale. *Braz. J. Phys. Ther.* **2011**, *15*, 95–101. [CrossRef]
7. Raîche, M.; Hébert, R.; Prince, F.; Corriveau, H. Screening older adults at risk of falling with the Tinetti balance scale. *Lancet* **2000**, *356*, 1001–1002. [CrossRef]
8. Oliver, D.; Britton, M.; Seed, P.; Martin, F.; Hopper, A. Development and evaluation of evidence based risk assessment tool (STRATIFY) to predict which elderly inpatients will fall: Case-control and cohort studies. *BMJ* **1997**, *315*, 1049–1053. [CrossRef] [PubMed]
9. Park, S.H. Tools for assessing fall risk in the elderly: A systematic review and meta-analysis. *Aging Clin. Exp. Res.* **2018**, *30*, 1–16. [CrossRef]
10. Ruggieri, M.; Palmisano, B.; Fratocchi, G.; Santilli, V.; Mollica, R.; Berardi, A.; Galeoto, G. Validated fall risk assessment tools for use with older adults: A systematic review. *Phys. Occup. Ther. Geriatr.* **2018**, *36*, 331–353. [CrossRef]
11. Howcroft, J.; Lemaire, E.D.; Kofman, J. Wearable-sensor-based classification models of faller status in older adults. *PLoS ONE* **2016**, *11*, e0153240. [CrossRef] [PubMed]
12. Howcroft, J.; Kofman, J.; Lemaire, E.D. Prospective fall-risk prediction models for older adults based on wearable sensors. *IEEE Trans. Neural Syst. Rehabil. Eng.* **2017**, *25*, 1812–1820. [CrossRef] [PubMed]
13. Doheny, E.P.; Walsh, C.; Foran, T.; Greene, B.R.; Fan, C.W.; Cunningham, C.; Kenny, R.A. Falls classification using tri-axial accelerometers during the five-times-sit-to-stand test. *Gait Posture* **2013**, *38*, 1021–1025. [CrossRef] [PubMed]
14. Weiss, A.; Herman, T.; Giladi, N.; Hausdorff, J.M. Objective assessment of fall risk in Parkinson's disease using a body-fixed sensor worn for 3 days. *PLoS ONE* **2014**, *9*, e96675. [CrossRef] [PubMed]
15. Antwi-Afari, M.F.; Li, H. Fall risk assessment of construction workers based on biomechanical gait stability parameters using wearable insole pressure system. *Adv. Eng. Inform.* **2018**, *38*, 683–694. [CrossRef]
16. Chen, S.; Cunningham, C.L.; Lach, J.; Bennett, B.C. Extracting spatio-temporal information from inertial body sensor networks for gait speed estimation. In Proceedings of the 2011 International Conference on Body Sensor Networks, Dallas, TX, USA, 23–25 May 2011; pp. 71–76.
17. Jiang, S.; Wang, X.; Kyrarini, M.; Gräser, A. A robust algorithm for gait cycle segmentation. In Proceedings of the 2017 25th european signal processing conference (eusipco), Kos Island, Greece, 28 August–2 September 2017; pp. 31–35.
18. Takeda, R.; Lisco, G.; Fujisawa, T.; Gastaldi, L.; Tohyama, H.; Tadano, S. Drift removal for improving the accuracy of gait parameters using wearable sensor systems. *Sensors* **2014**, *14*, 23230–23247. [CrossRef]
19. Roetenberg, D.; Slycke, P.J.; Veltink, P.H. Ambulatory position and orientation tracking fusing magnetic and inertial sensing. *IEEE Trans. Biomed. Eng.* **2007**, *54*, 883–890. [CrossRef]
20. Kwok, B.-C.; Clark, R.A.; Pua, Y.-H. Novel use of the Wii Balance Board to prospectively predict falls in community-dwelling older adults. *Clin. Biomech.* **2015**, *30*, 481–484. [CrossRef]
21. Nishiguchi, S.; Yamada, M.; Uemura, K.; Matsumura, T.; Takahashi, M.; Moriguchi, T.; Aoyama, T. A novel infrared laser device that measures multilateral parameters of stepping performance for assessment of all risk in elderly individuals. *Aging Clin. Exp. Res.* **2013**, *25*, 311–316. [CrossRef]
22. Dubois, A.; Charpillet, F. A gait analysis method based on a depth camera for fall prevention. In Proceedings of the 2014 36th Annual International Conference of the IEEE Engineering in Medicine and Biology Society, Chicago, IL, USA, 26–30 August 2014; pp. 4515–4518.
23. Stone, E.E.; Skubic, M. Evaluation of an inexpensive depth camera for passive in-home fall risk assessment. In Proceedings of the 2011 5th International Conference on Pervasive Computing Technologies for Healthcare (PervasiveHealth) and Workshops, Washington, DC, USA, 23–25 May 2011; pp. 71–77.
24. Alexandre, T.S.; Meira, D.M.; Rico, N.C.; Mizuta, S.K. Accuracy of Timed Up and Go Test for screening risk of falls among community-dwelling elderly. *Braz. J. Phys. Ther.* **2012**, *16*, 381–388. [CrossRef]
25. Yamada, M.; Aoyama, T.; Arai, H.; Nagai, K.; Tanaka, B.; Uemura, K.; Mori, S.; Ichihashi, N. Dual-task walk is a reliable predictor of falls in robust elderly adults. *J. Am. Geriatr. Soc.* **2011**, *59*, 163–164. [CrossRef] [PubMed]
26. Okuma, Y. Freezing of gait and falls in Parkinson's disease. *J. Parkinson's Dis.* **2014**, *4*, 255–260. [CrossRef] [PubMed]
27. Baker, J.M. Gait disorders. *Am. J. Med.* **2018**, *131*, 602–607. [CrossRef]
28. Callisaya, M.L.; Blizzard, L.; Schmidt, M.D.; McGinley, J.L.; Srikanth, V.K. Ageing and gait variability—A population-based study of older people. *Age Ageing* **2010**, *39*, 191–197. [CrossRef] [PubMed]
29. Brodie, M.A.; Coppens, M.J.; Ejupi, A.; Gschwind, Y.J.; Annegarn, J.; Schoene, D.; Wieching, R.; Lord, S.R.; Delbaere, K. Comparison between clinical gait and daily-life gait assessments of fall risk in older people. *Geriatr. Gerontol. Int.* **2017**, *17*, 2274–2282. [CrossRef]
30. Hausdorff, J.M.; Rios, D.A.; Edelberg, H.K. Gait variability and fall risk in community-living older adults: A 1-year prospective study. *Arch. Phys. Med. Rehabil.* **2001**, *82*, 1050–1056. [CrossRef]
31. Brach, J.S.; Berlin, J.E.; VanSwearingen, J.M.; Newman, A.B.; Studenski, S.A. Too much or too little step width variability is associated with a fall history in older persons who walk at or near normal gait speed. *J. Neuroeng. Rehabil.* **2005**, *2*, 1–8. [CrossRef]

32. Vieira, M.F.; Rodrigues, F.B.; e Souza, G.S.d.S.; Magnani, R.M.; Lehnen, G.C.; Campos, N.G.; Andrade, A.O. Gait stability, variability and complexity on inclined surfaces. *J. Biomech.* **2017**, *54*, 73–79. [CrossRef]
33. Hollman, J.H.; Kovash, F.M.; Kubik, J.J.; Linbo, R.A. Age-related differences in spatiotemporal markers of gait stability during dual task walking. *Gait Posture* **2007**, *26*, 113–119. [CrossRef]
34. Rosenstein, M.T.; Collins, J.J.; De Luca, C.J. A practical method for calculating largest Lyapunov exponents from small data sets. *Phys. D Nonlinear Phenom.* **1993**, *65*, 117–134. [CrossRef]
35. Melzer, I.; Benjuya, N.; Kaplanski, J. Postural stability in the elderly: A comparison between fallers and non-fallers. *Age Ageing* **2004**, *33*, 602–607. [CrossRef]
36. Jiang, S.; Zhang, B.; Wei, D. The elderly fall risk assessment and prediction based on gait analysis. In Proceedings of the 2011 IEEE 11th international conference on computer and information technology, Washington, DC, USA, 28 August–2 September 2011; pp. 176–180.
37. Patterson, K.K.; Gage, W.H.; Brooks, D.; Black, S.E.; McIlroy, W.E. Evaluation of gait symmetry after stroke: A comparison of current methods and recommendations for standardization. *Gait Posture* **2010**, *31*, 241–246. [CrossRef]
38. Sadeghi, H.; Allard, P.; Prince, F.; Labelle, H. Symmetry and limb dominance in able-bodied gait: A review. *Gait Posture* **2000**, *12*, 34–45. [CrossRef]
39. Robinson, R.; Herzog, W.; Nigg, B.M. Use of force platform variables to quantify the effects of chiropractic manipulation on gait symmetry. *J. Manip. Physiol. Ther.* **1987**, *10*, 172–176.
40. Błażkiewicz, M.; Wiszomirska, I.; Wit, A. Comparison of four methods of calculating the symmetry of spatial-temporal parameters of gait. *Acta Bioeng. Biomech.* **2014**, *16*, 29–35. [PubMed]
41. Zifchock, R.A.; Davis, I.; Higginson, J.; Royer, T. The symmetry angle: A novel, robust method of quantifying asymmetry. *Gait Posture* **2008**, *27*, 622–627. [CrossRef] [PubMed]
42. Doi, T.; Hirata, S.; Ono, R.; Tsutsumimoto, K.; Misu, S.; Ando, H. The harmonic ratio of trunk acceleration predicts falling among older people: Results of a 1-year prospective study. *J. Neuroeng. Rehabil.* **2013**, *10*, 1–6. [CrossRef]
43. Liu, Y.; Redmond, S.J.; Narayanan, M.R.; Lovell, N.H. Classification between non-multiple fallers and multiple fallers using a triaxial accelerometry-based system. In Proceedings of the 2011 Annual International Conference of the IEEE Engineering in Medicine and Biology Society, Berlin, Germany, 23–27 July 2011; pp. 1499–1502.
44. Greene, B.R.; McGrath, D.; Walsh, L.; Doheny, E.P.; McKeown, D.; Garattini, C.; Cunningham, C.; Crosby, L.; Caulfield, B.; Kenny, R.A. Quantitative falls risk estimation through multi-sensor assessment of standing balance. *Physiol. Meas.* **2012**, *33*, 2049. [CrossRef]
45. Colagiorgio, P.; Romano, F.; Sardi, F.; Moraschini, M.; Sozzi, A.; Bejor, M.; Ricevuti, G.; Buizza, A.; Ramat, S. Affordable, automatic quantitative fall risk assessment based on clinical balance scales and Kinect data. In Proceedings of the 2014 36th Annual International Conference of the IEEE Engineering in Medicine and Biology Society, Chicago, IL, USA, 26–30 August 2014; pp. 3500–3503.
46. Caby, B.; Kieffer, S.; de Saint Hubert, M.; Cremer, G.; Macq, B. Feature extraction and selection for objective gait analysis and fall risk assessment by accelerometry. *Biomed. Eng. Online* **2011**, *10*, 1–19. [CrossRef]
47. Similä, H.; Mäntyjärvi, J.; Merilahti, J.; Lindholm, M.; Ermes, M. Accelerometry-based berg balance scale score estimation. *IEEE J. Biomed. Health Inform.* **2013**, *18*, 1114–1121. [CrossRef]
48. Cuaya, G.; Munoz-Meléndez, A.; Carrera, L.N.; Morales, E.F.; Quinones, I.; Pérez, A.I.; Alessi, A. A dynamic Bayesian network for estimating the risk of falls from real gait data. *Med. Biol. Eng. Comput.* **2013**, *51*, 29–37. [CrossRef] [PubMed]
49. Nait Aicha, A.; Englebienne, G.; Van Schooten, K.S.; Pijnappels, M.; Kröse, B. Deep learning to predict falls in older adults based on daily-life trunk accelerometry. *Sensors* **2018**, *18*, 1654. [CrossRef]
50. Meyer, B.M.; Tulipani, L.J.; Gurchiek, R.D.; Allen, D.A.; Adamowicz, L.; Larie, D.; Solomon, A.J.; Cheney, N.; McGinnis, R.S. Wearables and deep learning classify fall risk from gait in multiple sclerosis. *IEEE J. Biomed. Health Inform.* **2020**, *25*, 1824–1831. [CrossRef]
51. Tunca, C.; Salur, G.; Ersoy, C. Deep learning for fall risk assessment with inertial sensors: Utilizing domain knowledge in spatio-temporal gait parameters. *IEEE J. Biomed. Health Inform.* **2019**, *24*, 1994–2005. [CrossRef] [PubMed]
52. Savadkoohi, M.; Oladunni, T.; Thompson, L.A. Deep Neural Networks for Human's Fall-risk Prediction using Force-Plate Time Series Signal. *Expert Syst. Appl.* **2021**, *182*, 1–19. [CrossRef]
53. Cattelani, L.; Chesani, F.; Palumbo, P.; Palmerini, L.; Bandinelli, S.; Becker, C.; Chiari, L. FRAT-up, a rule-based system evaluating fall risk in the elderly. In Proceedings of the 2014 IEEE 27th International Symposium on Computer-Based Medical Systems, New York, NY, USA, 27–29 May 2014; pp. 38–41.
54. Di Rosa, M.; Hausdorff, J.M.; Stara, V.; Rossi, L.; Glynn, L.; Casey, M.; Burkard, S.; Cherubini, A. Concurrent validation of an index to estimate fall risk in community dwelling seniors through a wireless sensor insole system: A pilot study. *Gait Posture* **2017**, *55*, 6–11. [CrossRef] [PubMed]
55. Nishiguchi, S.; Yamada, M.; Nagai, K.; Mori, S.; Kajiwara, Y.; Sonoda, T.; Yoshimura, K.; Yoshitomi, H.; Ito, H.; Okamoto, K. Reliability and validity of gait analysis by android-based smartphone. *Telemed. E-Health* **2012**, *18*, 292–296. [CrossRef]

healthcare MDPI

Article

Towards Validating the Effectiveness of Obstructive Sleep Apnea Classification from Electronic Health Records Using Machine Learning

Jayroop Ramesh *, Niha Keeran, Assim Sagahyroon and Fadi Aloul

Department of Computer Science and Engineering, American University of Sharjah, Sharjah 26666, United Arab Emirates; g00057302@aus.edu (N.K.); asagahyroon@aus.edu (A.S.); faloul@aus.edu (F.A.)
* Correspondence: jramesh@aus.edu; Tel.: +971-5-59517842

Abstract: Obstructive sleep apnea (OSA) is a common, chronic, sleep-related breathing disorder characterized by partial or complete airway obstruction in sleep. The gold standard diagnosis method is polysomnography, which estimates disease severity through the Apnea-Hypopnea Index (AHI). However, this is expensive and not widely accessible to the public. For effective screening, this work implements machine learning algorithms for classification of OSA. The model is trained with routinely acquired clinical data of 1479 records from the Wisconsin Sleep Cohort dataset. Extracted features from the electronic health records include patient demographics, laboratory blood reports, physical measurements, habitual sleep history, comorbidities, and general health questionnaire scores. For distinguishing between OSA and non-OSA patients, feature selection methods reveal the primary important predictors as waist-to-height ratio, waist circumference, neck circumference, body-mass index, lipid accumulation product, excessive daytime sleepiness, daily snoring frequency and snoring volume. Optimal hyperparameters were selected using a hybrid tuning method consisting of Bayesian Optimization and Genetic Algorithms through a five-fold cross-validation strategy. Support vector machines achieved the highest evaluation scores with accuracy: 68.06%, sensitivity: 88.76%, specificity: 40.74%, F1-score: 75.96%, PPV: 66.36% and NPV: 73.33%. We conclude that routine clinical data can be useful in prioritization of patient referral for further sleep studies.

Keywords: electronic health records; machine learning; obstructive; polysomnography; prediction; sleep apnea

Citation: Ramesh, J.; Keeran, N.; Sagahyroon, A.; Aloul, F. Towards Validating the Effectiveness of Obstructive Sleep Apnea Classification from Electronic Health Records Using Machine Learning. *Healthcare* **2021**, *9*, 1450. https://doi.org/10.3390/healthcare9111450

Academic Editor: Mahmudur Rahman

Received: 22 September 2021
Accepted: 25 October 2021
Published: 27 October 2021

1. Introduction

Sleep research is of pertinence due to its fundamental role in ensuring health and wellbeing, and as cited by the American Psychiatrist Allan Hobson "Sleep is of the brain, by the brain and for the brain" [1]. Sleep disorders are impairments of sleep architecture (consisting of sleep stages) and disrupts psycho-physical health leading to the development of a host of diseases. More than a billion adults globally between the ages of 30 to 69 years suffer from obstructive sleep apnea (OSA), the most common type of sleep-disordered breathing. 936 million of them suffer mild to moderate symptoms and 425 million suffer from moderate to severe symptoms. The highest concentration of these individuals can be found in China, followed by India, Brazil, United States of America, Pakistan, Russia, Nigeria, Germany, France and Japan [2].

OSA causes temporary lapses in breath when the upper airway at the back of the throat becomes partially or completely blocked during sleep. This can lead to fragmented sleep since the individuals need to be conscious enough to wake up and reopen their airway to resume breathing and sleep and this poor quality of sleep results in sleepiness, fatigue and considerable physiological and psychological distress. Some of the common symptoms that can help identify the disorder is disrupted breathing, excessive daytime sleepiness (EDS), morning headaches, irritability, limited attention span, snoring and dry mouth [3]. Untreated OSA has been associated with many health conditions such as obesity, cardiovascular and metabolic disorders, in addition to reduced quality of life and depression [4].

To diagnose OSA, polysomnography (PSG) conducted in a sleep laboratory is usually considered as the gold reference standard. PSG monitors and records several body functions during sleep. If there are more than 15 obstructive respiratory events per hour of sleep, then no other symptoms are needed. The PSG test defines an apnea-hypopnea index (AHI) based on the criteria above. Severity grading varies, but typically mild OSA is defined by an AHI of $5 \leq 15$, moderate OSA by AHI between $16 \leq 29$, and severe by AHI ≥ 30. This method has several limitations: (i) it is expensive and time-consuming and requires medical supervision and in addition to being confined within a hospital or clinical setting, (ii) the sleep environment will be altered and does not represent the natural sleep context of the individual, and (iii) it cannot be implemented over a long time, being limited to a span of few days. There are other tests such as the multiple sleep latency test (MSLT), maintenance of wakefulness test (MWT), CPAP titration test, all of which are conducted in a controlled environment, typically following the PSG. Home sleep tests are a limited PSG which can be taken at home allowing it to be in the patient's natural environment but it cannot determine sleep stages or other parameters which puts them at a major disadvantage. Self-assessment methods like sleep questionnaires and sleep diaries are an alternative inexpensive method which preserves the normal sleep environment but are highly subjective. Furthermore, sleep questionnaires are subject to bias due to patient reluctance in disclosing sensitive private information, or as a consequence of diminished awareness about the implications of potential sleep disorders. Sleep diaries contains more pertinent information as it is filled over a longer period of time, but has the same underlying issues as sleep questionnaires [5].

Accounting for these considerations, it is integral to develop easy-to-use and cheap accurate screening tools that can easily monitor disturbances in the population at a relatively low cost. In today's increasingly digital world, there is a large amount of health data generated by different sources such as real-time physiological data from connected wearables, electronic health records (EHR), insurance claims and social media posts. Artificial intelligence, more specifically machine learning (ML) is emerging as a powerful tool in healthcare to mine available patient data and build powerful diagnostic frameworks [6]. This paradigm is gaining momentum in the area of OSA classification with two of the aforementioned sources: physiological data and EHR.

Physiological data can be derived from electroencephalogram [7], electrocardiogram or photoplethysmogram readings acquired either during PSG or through consumer-grade wearable devices [8]. In general, the former type of data collected in sleep labs with a ground truth respiratory signal achieve noticeably better performance with any ML algorithms. While actigraphy studies are attractive owing to its applicability in community based populations, it is inherently challenging to achieve comparable OSA screening performances as those from sleep lab studies. This is a consequence of occurrences such as noise, motion artifacts or other disturbances (such as battery depletion, missing data, loose skin contact, etc.). Researchers have also developed smartphone sensor based application for sleep apnea monitoring [9] and presented contact-less sleep disorder detection using sonar techniques [10]. The physiological monitoring modalities have the common issue of requiring additional obtrusive monitoring apparatus or expert supervision, which brings to the forefront the alternative approach of using routinely acquired electronic health records to perform screening. In can be surmised that sleep physiological data such as pulse oximetry and sleep stage duration have considerable predictive ability, but are not readily available, as the expensive, time consuming and labor intensive nature of PSG limits regular monitoring and diagnosis [11,12]. Moreover, the variability in performance of such solutions over an extended period of time within a community based setting conveys a relatively low level of overall reliability.

The use of digital health records and machine learning techniques trained on Big Data publicly available can allow for the transfer the knowledge representation to generalized cases. These tests would be more accurate in identifying patients with a higher pretest probability of OSA and can rule out OSA in low-risk patients, due to the high volume, veracity, velocity, variety and value provided by the datasets [4]. There are multiple successful studies leveraging EHRs to implement effective disease prediction models in literature [13]. A study conducted using EHRs from over 1 million outpatient visits from over 500,000 patients at a major academic medical referral center in China, was used to create an AI-based diagnostic system for detection of pediatric diseases with an accuracy in the ranges of 90–95% for multiple disease categories [14]. Although traditionally predictive modelling techniques require custom datasets, with specific variables limit the scope of the applicability, especially with large feature variables, recent developments in artificial intelligence address these challenges [15]. Predictive modeling with electronic health records using the "transfer learning" approach has shown to accurately predict medical events from multiple clinics without being site specific [16]. Moreover, with the creation of flexible standardized clinical data representation formats like FHIR (Fast HealthCare Interoperability Resources), any developed models can be integrated into clinical systems [17]. One of the primary advantages of such models would be the ability to contribute to a wider population health paradigm using the routine biomarkers and patient profiles in hospitals to screen and preemptively identify at risk individuals for care. These screening methods reduce the need for patients to undergo either obtrusive tests such as PSG to even identify sleep disorders, or remote patient monitoring systems using wearables, although these approaches do have their value in screening within consumer lifestyle management applications. There is a significant cost reduction to both the clinics and patients in the deployment of clinical screening algorithms, as they would not be as expensive as PSG, and allows for consideration of patients who do not have wearable devices as well. Most literature in this intersecting area of patient health records, Big Data and deep learning focus on prediction of mortality, cardiovascular risks, diabetes and pulmonary conditions. A systematic review of recent developments in deep learning methods and their clinical outcomes with the utilization electronic health records can be observed in [18]. Their study reiterates that general conditions such as suicide risk, future disease predictions, readmission probability prediction, heart failure prediction and hospital stay duration estimation are the actively researched areas.

The experiments in [19] saw the deployment of a learning algorithm to distinguish cases of diagnosed OSA and non-cases using EHR ICD-codes across six health systems in the United States. A cohort study of adults in Canada was conducted as follows in [20], where an algorithm trained on administrative data and ICD-codes found a high degree of specificity in identifying patients with OSA. A super sparse linear integer model was developed in [21], by training the model on self-reported symptoms, self-reported medical information, demographics and comorbidities data to screen for OSA cases with considerably success. Another study [22] focused on developing a support vector machine-based prediction model using 2 to 6 features collected at clinical visits to identify patients with AHI index at 3 cut offs. The model was fivefold-cross validated and had balanced performance measures in the 70% range. It outperformed the Berlin Questionnaire, NoSAS score and Supersparse Linear Integer model for the age category for men below 65 years of age. The primary limitations between the clinical data trained models are due to oversampling of the target class (i.e., more sleep apnea cases than control group), lack of generalizability (due to limited data features), and relatively high false alarms for OSA [23]. In clinics where PSG is not possible, or there is no sleep data available, medical staff still screen using self-reported questionnaires during patient visits [24]. There is room for improvement, especially considering boosting algorithms as their ability to uncover non-linear patterns are unparalleled, even given large number of features, and make this process much easier [25].

This work presents and attempts to answer this question: "Is it possible to develop machine learning models from EHR that are as effective as those developed using sleep physiological parameters for preemptive OSA detection?". There exist no comparative studies between both approaches which empirically validates the quality of using routinely available clinical data to screen for OSA patients. The proposed work implements ensemble and traditional machine learning models to screen for OSA patients using routinely collected clinical information from the Wisconsin Sleep Cohort (WSC) dataset [26]. WSC includes overnight physiological measurements, and laboratory blood tests conducted in the following morning in a fasting state. In addition to the standard features used for OSA screening in literature, we consider an expanded range of questionnaire data, lipid profile, glucose, blood pressure, creatinine, uric acid, and clinical surrogate markers. In total, 56 continuous and categorical covariates are initially selected, the the feature dimension narrowed systematically based on multiple feature selection methods according to their relative impacts on the models' performance. Furthermore, the performance of all the implemented ML models are evaluated and compared in both the EHR and the sleep physiology experiments.

The contributions of this work are as follows:

- Implementation and evaluation of ensemble and traditional machine learning with an expanded feature set of routinely available clinical data available through EHRs.
- Comparison and subsequent validation of machine learning models trained on EHR data against physiological sleep parameters for screening of OSA in the same population.

This paper is organized as follows: Section 2 details the methodology, Section 3 presents the results, Section 4 discusses the findings, and Section 5 concludes the work with directions for future research.

2. Materials and Methods

As shown in Figure 1, the proposed methodology composes of the following five steps: (i) preprocessing, (ii) feature selection, (iii) model development, (iv) hyperparameter tuning and (v) evaluation. This process is conducted for the EHR as well as for the physiological parameters acquired from the same population in the WSC dataset.

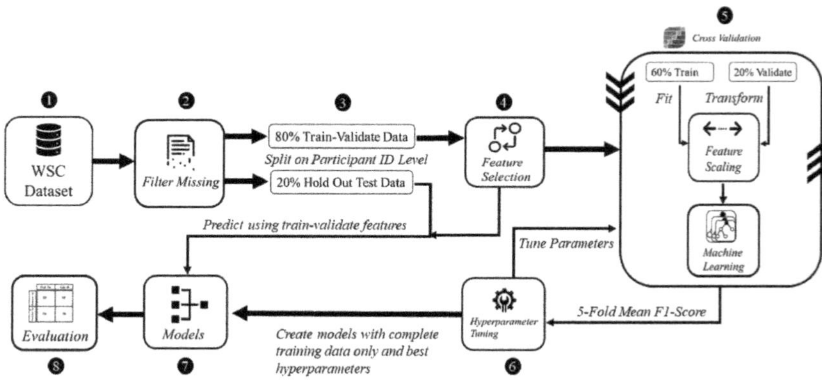

Figure 1. High level view of the proposed methodology.

OSA is a multi-factorial condition, as it can manifest alongside patients with other conditions such as metabolic, cardiovascular, and mental health disorders. Blood biomarkers can therefore be indicative of the condition or a closely associated co-morbidity, such as heart disease and metabolic dysregulation. These biomarkers include fasting plasma glucose, triglycerides, and uric acid [27]. The presence of one or the other comorbidities does not always necessarily indicate OSA, however in recent literature clinical surrogate markers reflective of particular conditions have shown considerable association with suspected OSA. Clinical surrogate markers exhibit more sensitive responses to minor changes in patient pathophysiology, and are generally more cost-effective to measure than complete laboratory analysis [28]. Thus, we derive 4 markers, Triglyceride glucose (TyG) index, Lipid Accumulation Product (LAP), Visceral Adipose Index (VAI) and the Waist-Height Ratio (WHrt), and observe their value in discriminating between OSA and non-OSA patients [29]. Ref. [30] reports LAP, VAI and TyG were reliable surrogate markers for identifying metabolic syndrome in middle-aged and elderly Chinese population. TyG was independently associated with increased OSA risk, as it is a reliable marker of insulin resistance, comprising of glucose intolerance, dyslipidemia, and hypertension [31]. This relationship is observed as insulin resistance increases due to the intermittent periods of asphyxia, hypoxia and sleep depivation caused due to OSA [32].

The Wisconsin Sleep Cohort (WSC) from University of Wisconsin-Madison is a study of 1500 participants having the causes, consequences and natural history of sleep disorders [26]. Fifty-six total features are extracted and categorized into demographics, anthropometry, blood tests, derived clinical markers, general health questionnaires, self-reported history, polysomnography derived parameters, as presented in Tables A1–A8 respectively within Appendix A. The dataset contains 2570 records of the 1500 participants assessed at four-year intervals, where each participant can have up to five records in the study. The total number of participants/patients is denoted by n_p, and the total number of health records is denoted by n_r. The demographics included age, sex, race, alcohol and smoking habits. The anthropometric features included patient height, weight, BMI, waist circumference, and neck circumference. The laboratory blood test results were obtained the morning following the overnight sleep study in a fasting state. The profiles are of fasting plasma glucose, HDL-C, LDL-C, total cholesterol, creatinine, uric acid, systolic and diastolic blood pressure. The self-reported history consisted of general health status, existing medical conditions and sleep symptoms, which were acquired through self-administered questionnaires. Finally, polysomnography derived parameters included objective information about sleep stages, sleep duration, AHI events, and oxygen saturation levels. To compare model discriminability when trained with clinical data features and PSG parameters, they are used exclusively to implement independent models.

An eighteen channel PSG system (Grass instruments model 78; Quincy, MA, USA) was used to record sleep state with electroencephalography, electrooculography, and electromyography [33]. Breathing, nasal and oral airflow, and oxyhemoglobin saturation were assessed respectively using respiratory inductance plethysmography (Respitrace; Ambulatory Monitoring, Ardsley, NY), thermocouples (ProTec, Hendersonville, TN and Validyne Engineering Corp pressure transducer, Northridge, CA) and pulse oximetry (Ohmeda Biox 3740; Englewood, CO, USA) [33]. Every 30 s of the PSG recordings were scored in terms of sleep stage and apnea and hypopnea events by trained technicians according to conventional standards [34,35]. Cessation of airflow for $\geq$10 s and discernible reduction in breathing expressed as a sum of chest and abdominal excursions with a oxyhemoglobin saturation decrease of $\geq$4% defined apnea and hypopnea events respectively [33].

The dataset was examined for missing values for deletion or imputation. Little's MCAR (Missing Completely at Random Test) confirmed the null hypothesis ($p > 0.05$) that the pattern of missing values did not have any significant relationship with the rest of the data [36]. As such, imputation would not be an effective approach, due to the large number of missing values in the records relative to the total size of the dataset itself. Thus, listwise deletion was employed to remove entire records where the clinical features of interest values were missing, or had a numeric value of 0 where domain knowledge states it is not possible (e.g., fasting plasma glucose, triglycerides). Continuous variables and categorical variables were handled separately, due to their differing mathematical characteristics. Continuous variables were scaled using the standardization technique to distribute the values around a mean with unit standard deviation. Categorical variables were converted into one-hot encoded vectors equal to the number of unique categories for each column using dummy variables.

The data records were split on a participant level into a training-validation set consisting of distinct patients ($n_p = 752$) and a hold-out testing set of ($n_p = 188$) patients.

The cleaned dataset had ($n_r = 1479$) records, where ($n_r = 853$) records exhibited OSA and ($n_r = 626$) did not have OSA. This was done as a single patient can have multiple records in the dataset, and records repeating across the both training set and testing set will introduce data leakage.For the development of both the EHR and PSG data based models, the same training-validation and hold-out sets are used. All subsequent analysis that are part of steps (i)–(iv) in the methodology is conducted using the training-validation split, and step (v) is applicable for the hold-out testing set.

The populations were split at the threshold of AHI = 5 for the total of 56 features. In all following analysis, p-values < 0.05 are the cut-off for statistical significance. We applied the Shapiro-Wilk test of normality [37] to the populations, and note deviation from Gaussian distribution. Hence, we apply the Mann Whitney U-Test [38], which is distribution agnostic, to the continuous variables. Only self-reported sleep latency, LDL-C, total cholesterol, creatinine, Horne Ostberg score, State-Trait anxiety scores, non-REM sleep duration, and percentage of sleep stage 3&4 had p-values > 0.05. The average age is above 50 for both populations, and it is more probable that some of the patients may be facing onset of age-related diseases and increasing risk of OSA [39]. However, despite the aging, the overall population appears to be healthy, without much severity in any present comorbidities.

For categorical variables, we apply Chi Square with Bonferroni-Adjusted-p-value, as post-hoc testing can reduce false positives when multiple category levels are involved. No Yates correction was employed, to yield conservatives in the obtained p-values [40]. The demographic is heavily skewed towards the Caucasian ethnicity. Other perceived differences are in distribution of sexes (more men), occurrences of previous heart attacks, hypertension issues, angina, coronary, diabetes, arthritis, congestive heart failure, existing apnea and excessive daytime sleepiness along with snore volume being relatively higher among the OSA group. In terms of lifestyle, alcohol consumption and smoking is fairly similar between the two populations.

Feature selection was conducted using only the training-validation set. To mitigate possible selection bias and reduce redundancy, consistently highly ranking common features across all feature selection methods are chosen. We run two variations of this approach to ascertain the relative importance of all features. The intersection of the top two and top twenty features from each method is taken in the two cases respectively. The lower and upper bounds for the top features experiment is decided based on the distribution of the feature importance scores. To be more specific, many features have approximately the same impact on the AHI values, and we demarcate the two points where the differences between subsequent scores are the highest.

In the feature selection process for the clinical data, biological plausibility and their effective values during correlation with OSA were considered as well [41]. Automated step-wise procedures were avoided in favor of manual feature selection to ensure that the predictions made by the model can remain interpretable by medical professionals, if needed.

Pearson's correlation coefficient estimates coefficients between the output class and each of the predictor features signifying the strength and nature of the relationship between the two [42]. The coefficient is distributed between -1 and +1, where the former is total negative correlation, and the latter is total positive correlation. 0 indicates no linear correlation between the variables. We select the continuous features with positive and negative correlation as per this method to capture linear relationships, as shown in Figure 2. The coefficient estimation does not assume normality, but does assume finite variance and finite covariance as per the central limit theorem. Kendall's Tau correlation coefficient is a non-parametric test for measuring degree of association between the output class and predictor features applicable for categorical variables [42]. It is more robust to outliers and operates on the principles of comparing concordant and discordant pairs for ordinal variables. The most impactful categorical features are selected, as shown in Figure 3. Extremely Randomized Trees Classifier is a method where a number of randomized decision trees are fitted on subsets of the dataset [43]. Each decision tree results in a different model that has been trained with a different set of features. The relative importance of each feature on the classification performance of AHI is quantified as per the Gini index, as shown by Figure 4. We apply the Mutual Information technique to ensure that all strong associations, even non-linear between the continuous and categorical features with respect to the output class of OSA have been effectively captured [44]. Information gain measures the reduction in entropy of predictor features by partitioning a dataset according to the output classes. The entropy quantifies the probability distribution of observations in the dataset belonging to positive or negative class. Higher information gain suggests higher dependency between a feature and a specific output, while 0 suggests both are independent of each other. This method accepts continuous and categorical variables, and is able to capture both linear and non-linear relationships, as shown in Figure 5.

The final feature set in the top two-features per method consisted of a total of 8 features: waist circumference, neck circumference, daily snoring frequency, snoring volume, EDS, BMI, Whrt, and weight. The final features in the top twenty-feature per method consisted of the following 11 features in addition to the previous 8 features: fasting plasma glucose, LAP, uric acid, VAI, hypertension, heart attack comorbidity, TyG, triglycerides, systolic blood pressure and age.

In the feature selection process for the PSG parameters, all the variables were continuous. Thus, Kendall's Tau was excluded, and the feature rankings from Pearson's Correlation Coefficient, Extremely Randomized Trees Classifier, and Mutual Information are shown in Figures 6–8 respectively. Unlike the clinical data features, where multiple features had relatively similar influences on the dependent AHI variable, the most important parameters from PSG are the mean desaturation percentage, and minimum level of oxygen saturation. This is expected as the apnea-hypopnea events are scored using the changes in breathing and airflow.

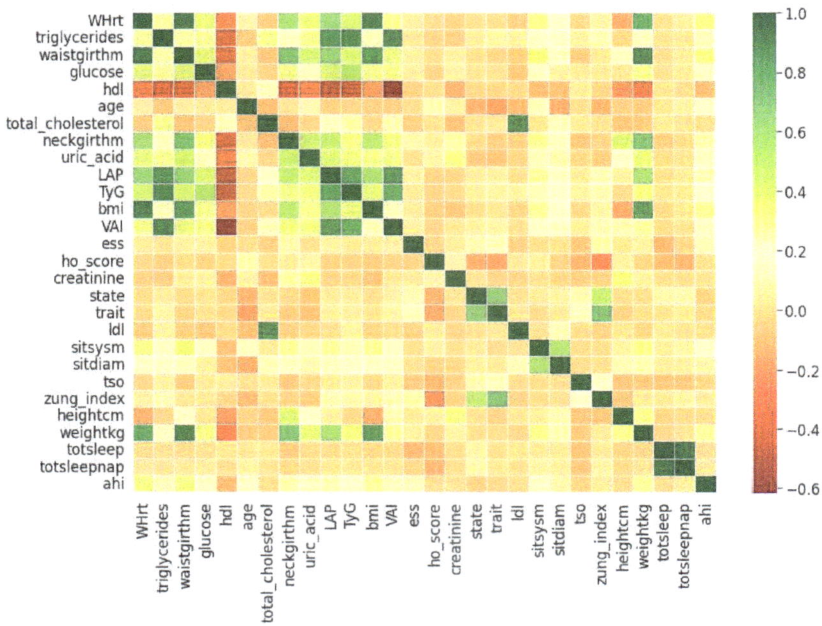

Figure 2. Clinical features ordered as per Pearson's Correlation Coefficient.

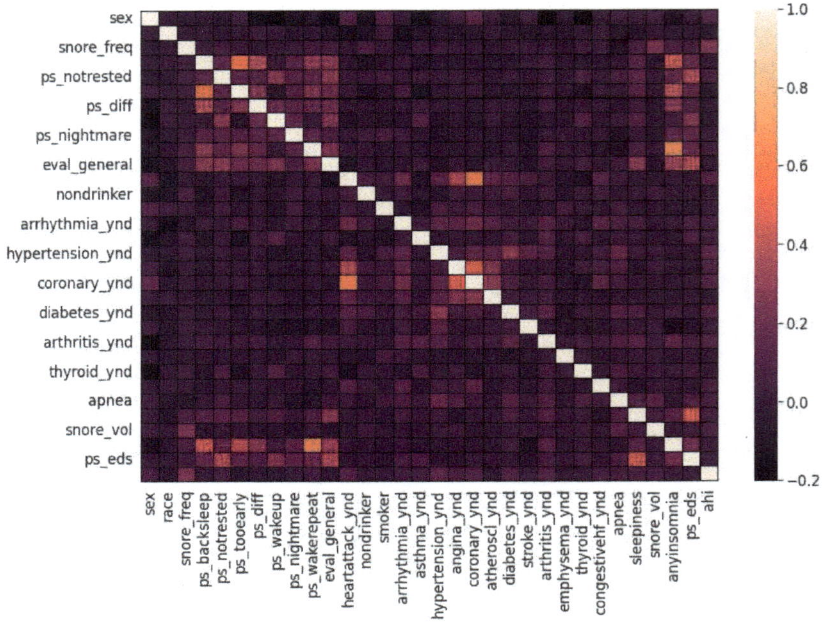

Figure 3. Clinical features ordered as per Kendall's Tau.

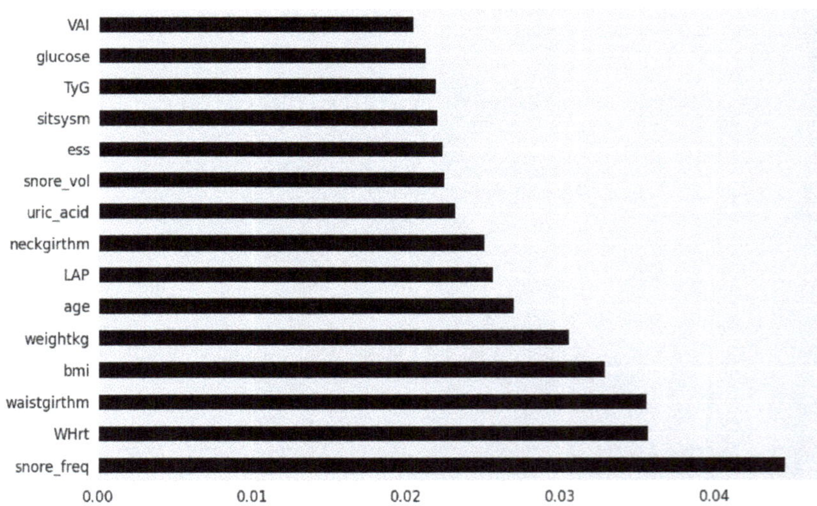

Figure 4. Clinical features ordered as per Extremely Randomized Trees.

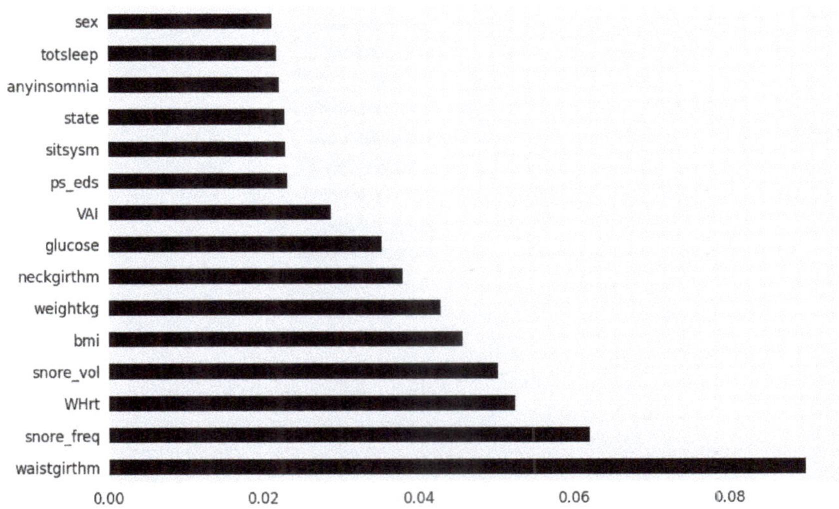

Figure 5. Clinical features ordered as per Mutual Information.

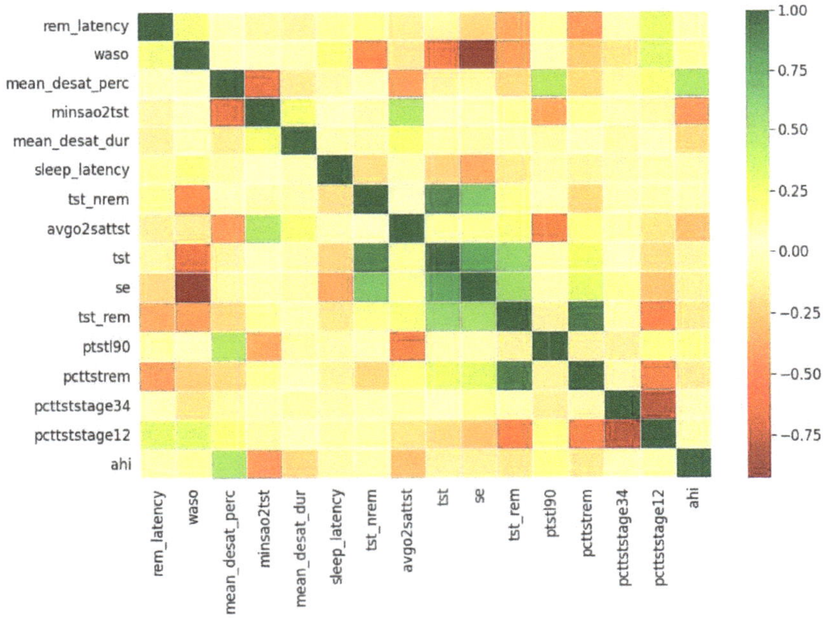

Figure 6. PSG features ordered as per Pearson's Correlation Coefficient.

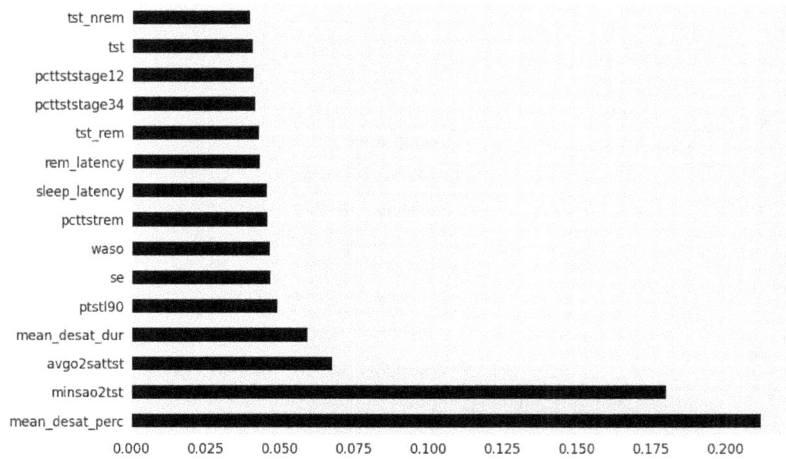

Figure 7. PSG features ordered as per Extremely Randomized Trees.

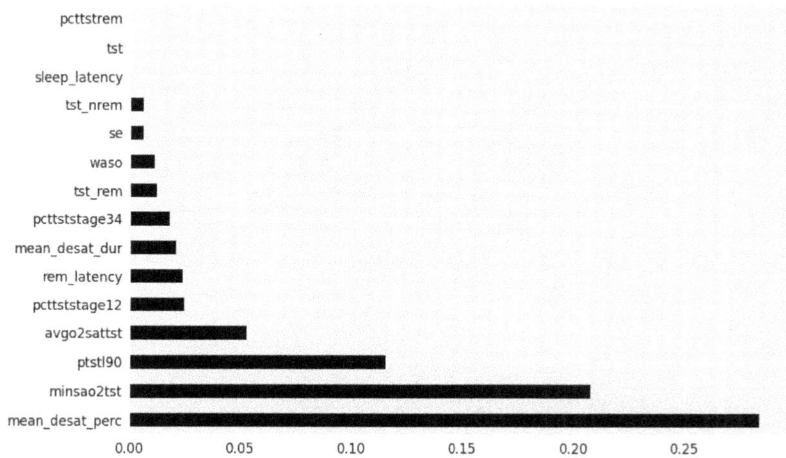

Figure 8. PSG features ordered as per Mutual Information.

The final feature set in the top two-features per method derived from oximetry consisted of a total of two features: mean oxygen desaturation percentage, and minimum level of oxygen saturation. The final feature set in the top fifteen-features per method derived from oximetry c in the top fifteen features consisted of the following 4 features in addition to the previous 2 features: sleep duration with oxygen saturation percentage below 90%, REM sleep latency, average oxygen desaturation of apnea-hypopnea event and mean oxygen desaturation duration.

Ensemble methods include "bagging" (e.g., Random Forest algorithm) and "boosting" methods (e.g., Extreme Gradient Boosting technique). Ensemble machine learning methods such as gradient boosting iteratively combines a set of weak base classification models to construct a strong learner. Gradient boosting techniques are currently being employed to attain state-of-the-art results in clinical applications [45,46]. Gradient boosting techniques sequentially minimize the residual error of preceding learners. The variation in individual base learner configuration is expected to capture different relationships in the data distribution. Its integration into a unified prediction model is similar to the concept of collecting various expert opinions on an initial prognosis, aggregating and making a final decision.

Extreme gradient boosting (XGB) [47] utilizes the gradient boosting framework, with the algorithmic enhancements of regularization, sparsity awareness, weighted quantile sketch and internal cross-validation. Light gradient booting machine (LGBM) [48] is another variant, where the key difference is in its implementation of vertical decision tree growth and gradient-based One-Side Sampling strategy. LGBM grows tree in a leaf-wise manner, as opposed to level-wise, thereby is capable of reducing delta loss more drastically. CatBoost (CB) [49] is yet another variant of gradient boosting, with the refinement strategies of symmetric tree implementation, ordered target statistics and ordered boosting to minimize prediction shift with categorical variables.

The traditional machine learning models of k-Nearest Neighbours (kNN), Support Vector (SVM) Machines and Logistic Regression (LR) are used as baseline to benchmark the performance of the ensemble techniques [50]. KNN is non-parametric learning algorithm which distributes similar instances in the same proximity defined by the Euclidean distance, and classifies new unknown instances by majority vote of their k nearest instance neighbours. SVM is an algorithm that performs prediction by optimally separating the data instances of different classes in an n dimensional space using a hyperplane and its associated support vectors. LR is an extended case of the classic linear regression method, in which one or more independent input variables predicts the probability of occurrence of a binary output variable.

We applied a hybrid hyperparameter tuning approach by combining a Bayesian Optimization variant for global search, and a genetic algorithm for local search. The methods were Tree-structured Parzen estimator (TPE) [51] and Covariance matrix adaptation evolution strategy (CMA-ES) [52] respectively. TPE constructs a probability model of the specified objective function, and identifies the ideal hyperparameters, and CMA-ES iteratively samples candidate solutions using a derivative free approach. The parameters and instantiation values for both the algorithms are based on the work presented in [53]. The optimization criteria was the aggregate cross-validation F1-score of the training-validation set in order to achieve a balanced screening system.

3. Results

All analysis were conducted using Python 3.7.12 on a workstation operating a Linux OS with 24 GB RAM, Intel Quad-Core Xeon CPU (2.3GHz), and Tesla K80 GPU (12 GB VRAM). The Python libraries used are mentioned in the subsequent paragraph.

Data was processed with numpy 1.19.5 [54] and pandas 1.1.5 [55]. Statistical methods and correlation tests were performed using scipy 1.4.1 [56]. Gradient boosting models were constructed using the standard xgboost 0.90 [47], lightgbm 2.2.3 [48] and catboost 1.0.0 [49] libraries. Baseline machine learning models were constructed using scikit-learn 1.0.0 [57]. Visualizations were made using seaborn 0.11.2 [58] and matplotlib 3.2.2 [59]. Hyperparameter tuning was performed using the Optuna 2.10.0 library [53].

The following metrics are used to ascertain the performance quality of the gradient boosting models through a 5-fold cross-validation approach: accuracy (Acc), sensitivity (Sen), specificity (Sp), positive prediction value (PPV), negative prediction value (NPV), F1-Score, and Area Under Curve (AUC). Accuracy is the proportion of correct predictions across the total test dataset. Sensitivity is the proportion of OSA patients correctly identified as positive and specificity is the proportion of non-OSA patients correctly identified as negative. Positive prediction value is the probability of positive cases correctly being OSA patients, and negative prediction value is the probability of negative cases correctly being non-OSA patients. The F1-score measures the balance between positive predictive value (cause of type-1 errors) and sensitivity (cause of type-2 errors). Area Under Curve denotes the trade-off between sensitivity and specificity, with the cut-off value identified using the Youden index.

All reported metrics of the EHR trained and oximetry trained models are obtained through evaluation on the hold-out test data in Tables 1–5. The best hyperparameters used to generate the reported results in Tables 1 and 4 are provided in Tables A9 and A10 respectively.

It is observed that the oximetry related parameters exhibit a considerably better performance for detecting OSA across all metrics with its increased impact evident particularly on specificity, as evident by Table 3. These features are capable of finding patterns whilst remaining fairly stable in small amounts of data as well, which may required for data constrained environments. Since trained specialists perform annotation of an apnea or hypopnea event based on the nature of respiration and oxygen levels, it is expected that the respective physiological parameters reflecting this are much more effective. However, in non-monitored, community-based conditions where patient apnea events are classified by automated algorithms through portable medical devices, smartphones or smart watches, the efficacy of alternate parameters needs to be examined further. Despite these observations, we can surmise that the routinely collected clinical features of waist circumference, neck circumference, BMI, and weight along with the self-reported symptoms of EDS, snoring frequency and snoring volume and derived clinical surrogate markers of lipid accumulation product and Waist-Height ratio have utility in identification of OSA. Thereby, in comparison with overnight pulse oximetry, use of electronic health records is a viable alternative, albeit for early risk screening and prioritization of OSA patients.

Table 1. Classification performance measures across ensemble and traditional models for 8 EHR features.

Model	*Acc%*	*Sen%*	*Sp%*	*F1-Score%*	*PPV%*	*NPV%*	*AUC%*
XGB	68.05	79.20	53.33	73.82	69.11	66.05	66.30
LGBM	67.41	74.15	58.52	72.13	70.21	63.20	66.33
CB	67.41	83.14	46.65	74.37	67.27	67.74	64.09
RF	68.05	77.52	55.55	73.40	69.69	65.22	66.54
kNN	67.09	77.00	54.00	72.67	68.84	64.03	65.55
LR	67.73	80.89	50.37	74.00	68.24	66.66	65.63
SVM	68.06	88.76	40.74	75.96	66.38	73.33	64.75

Table 2. Classification performance measures across ensemble and traditional models for 19 EHR features.

Model	*Acc%*	*Sen%*	*Sp%*	*F1-Score%*	*PPV%*	*NPV%*	*AUC%*
XGB	69.64	78.65	57.77	74.66	71.65	67.24	64.66
LGBM	68.37	73.60	61.48	72.57	71.58	63.84	67.53
CB	69.00	77.52	57.77	74.00	70.76	66.60	67.65
RF	65.81	73.03	56.30	70.84	68.78	61.30	64.66
kNN	63.25	69.10	55.55	68.14	67.21	57.69	62.32
LR	67.41	74.15	58.51	72.13	70.21	63.20	66.33
SVM	65.17	77.53	49.63	71.54	66.90	62.04	63.30

Table 3. Classification performance measures across ensemble and traditional models for 2 PSG features.

Model	*Acc%*	*Sen%*	*Sp%*	*F1-Score%*	*PPV%*	*NPV%*	*AUC%*
XGB	82.74	88.00	76.15	85.06	82.35	83.33	82.05
LGBM	83.04	87.42	77.48	85.20	83.08	83.00	82.97
CB	83.63	89.00	76.82	85.85	83.00	84.67	83.00
RF	83.63	87.43	78.80	85.64	84.00	83.20	83.12
kNN	82.74	88.48	75.50	85.13	82.03	83.82	82.00
LR	81.87	82.77	80.79	81.76	84.49	78.71	81.17
SVM	83.04	86.91	78.15	85.13	83.42	82.51	82.52

Table 4. Classification performance measures across ensemble and traditional models for 6 PSG features.

Model	*Acc%*	*Sen%*	*Sp%*	*F1-Score%*	*PPV%*	*NPV%*	*AUC%*
XGB	83.92	89.53	76.82	86.14	83.00	85.30	83.17
LGBM	83.33	88.50	76.82	85.56	82.84	84.05	82.65
CB	84.21	89.53	77.50	86.36	83.41	85.40	83.50
RF	84.50	89.53	78.14	86.60	83.82	85.50	86.58
kNN	83.33	88.00	77.48	85.50	83.17	83.57	82.72
LR	83.62	86.91	79.47	85.56	84.26	82.75	83.19
SVM	83.33	86.91	78.80	85.34	83.83	82.63	85.34

Table 5. A comparison of recent works developed for EHR-based screening of OSA through machine learning.

Source	Dataset	Features	Approach	*Sen%*	*Sp%*
This work	WSC ($n_p = 940$)	waist-to-height ratio, waist circumference, neck circumference, BMI, EDS, LAP, daily snoring frequency and snoring volume	SVM	83.76	40.74
[21]	Private ($n_p = 1922$)	age, hypertension, BMI and sex	SLIM	64.20	77.00
[22]	Private ($n_p = 6875$)	waist circumference and age	SVM	74.14	74.71
[60]	Private ($n_p = 279$)	waist circumference, frequency of falling asleep, subnasale to stomion length, hypertension, snoring volume, and fatigue severity score	SVM	80.33	86.96
[61]	Private ($n_p = 313$)	BMI, ESS, and number of apneas	SVM	44.7	-

4. Discussion

The primary motivation behind the application of ensemble gradient boosting algorithms in this work was an attempt to capturing higher dimensional interactions in the data, as a consequence of the multifactorial nature of OSA. The performance of the SVM, LR, and KNN baseline models are relatively similar to the performance of boosting (CatBoost, XGB and LGBM) and bagging (RF) algorithms with the top 8 features as presented in Table 1. Interestingly, the ensemble models do not fare significantly better than the traditional models in either the EHR or PSG case. For the 8 feature case, the sensitivity, F1-score and NPV of the SVM is the highest, while LGBM has higher specificity, PPV and AUC. CB has the second highest sensitivity and F1-score. For the 19-feature case, the XGB model performs the best across the metrics of accuracy, sensitivity, F1-score, PPV, and NPV while LGBM still retains the highest specificity. SVM has the second highest sensitivity but its performance across the other metrics is not as comparable. However, as the number of features increase, roughly a factor of two in this case, the overall performance begins to decrease as presented in Table 2. The F1-score, a robust metric of reliability is consistently higher for the ensemble techniques in the 19 feature case. It is possible that in the case of non-linear relationships, ensemble learning can learn more complex relations from relatively small amounts of data ($\sim$1000 samples). The intention behind selecting the most important 8 EHR features then extending to 19 EHR features, is to observe whether an increase in the number of EHR features with association to OSA can improve the specificity of detection. We note that age, triglycerides, and the existing conditions of hypertension and previous heart attack exhibit the ability to predict OSA, but it does not increase the rate of detection among the population sample available for this work. Since the focus of this work is identifying the model giving rise to the highest sensitivity for screening with the most impactful features, even at the expense of specificity, the SVM is most applicable. When we compare the EHR performance metrics to the PSG case, the disparity is evident in favor of the latter. As the number of features are increased in the PSG case, all metrics across all models exhibit a modest increase in performance. In both the 2 feature and 6 feature experiment, the CB model emerges as the best method, followed by RF. It is possible that in the EHR case that multiple features are related with each other, and there is underlying redundancy, which does not contribute towards the knowledge representation learned by the models. In contrast, the addition of more PSG features might be providing extra information, which enables the models with an improved representational understanding of the relationship between these predictors and OSA severity.

One of our contributions are in the expansion of the initial feature dimensions to 56 EHR parameters, consisting of a combination of medical history, comorbidities, clinical measurements, laboratory blood tests and self-reported symptoms. Most existing works only consider for waist circumference, neck circumference, BMI and age as the feature set, which may not completely represent the populations at risk of OSA. Risk factors underlying the decision remain poorly understood, therefore adding multiple dimensions, can potentially reduce the unnecessary referrals and account for the typically missing screening of patients with sleep apnea and minimal snoring. We additionally evaluate the role of LDL-C, HDL-C, fasting plasma glucose, uric acid and derived clinical surrogate markers of Whrt, LAP, VAI and TyG in predicting OSA, within a machine learning context. With the incorporation of additional features, we attempted to rectify the high false positive rate by increasing model specificity through holistic consideration of a complete patient medical history. Gradient boosting methods were applied with the intentions of reducing bias, improving generalization ability and reducing overfitting. Regardless, these models exhibit only marginal superiority over traditional methods such as SVM.

Waist, neck circumference and EDS have been long established as vital indicators for OSA susceptibility, and results of feature selection methods are in agreement. It is important to note that abdominal obesity is not the same as peripheral obesity. Waist circumference depends on the fatty tissues in the peritoneum, and thus, the abdominal obesity, which is known to affect upper airway functioning, a consistent symptom of OSA [62].

Frequent snoring was detected during feature selection as yet another pertinent feature for OSA prediction, and is part of the minimal feature set for the trained models. Although experts in [63] advise caution in the interpretation of snoring symptoms for assessing sleep apnea, they state it can be reliable when used in conjunction with additional clinical and physical readings, which is the case in our presented work. While the features of insomnia and daytime sleepiness (quantified by ESS) were included in feature selection, they only showed a marginal association with OSA, as opposed to the stipulations of [64,65], respectively. This can be explained by the overall minimal OSA severity levels of the dataset population used in this work.

Patient laboratory blood tests and clinical surrogate markers were introduced as auxiliary biomarker features and its value in improving the model discernibility for classification of OSA was studied. In the case where 19 features were utilized for training, fasting plasma glucose, uric acid, and LAP (dependent on on waist circumference to triglycerides ratios) showed correlation with OSA in a similar fashion to traditionally expected indicators such as EDS and BMI. Additionally, the clinical markers of systolic blood pressure, VAI, and TyG are also present. These biomarkers are associated with OSA, and is in concordance with prior literature. Although the models were not able to utilize all biomarkers relevant to OSA with equal effectiveness, the possible reasons for the findings and variations in this work are worth mentioning.

Fasting plasma glucose is arguably the strongest blood biomarker feature, ranking consistently highly behind the physical measurements and snoring features across all the feature selection methods. This is expected given its relation with sleep quality and the effect of fragmented sleep on metabolic dysregulation which causes elevated glucose levels in the body, as reported in [66]. For some patients, the presence of insulin resistance/glucose irregularity, overlaps with the OSA symptoms of upper airway narrowing and decrease reduced dilator muscle contraction. Interestingly, glucose irregularity in a sleep disordered population of males has been shown in [67] to be independent of obesity and diabetes, indicating a strong correlation with OSA severity. From the findings of [68], OSA was independently associated with decreased insulin sensitivity in a female population as well.

In this work, uric acid emerged as a viable secondary predictor for OSA. This is likely due to hyperuricemia, which is an excess of uric acid levels, has been reported to be significantly associated with OSA as well as obesity and overnight oxygen desaturation severity.

As hypothesized, it appears the Whrt and VAI and LAP indices prove to be useful indicated as well. This is expected since fat distribution, visceral fat, and body composition increases the risk of anatomical irregularities common among OSA patients, and this is stated in [69].

The VAI feature can be useful as a secondary risk factor; likely due to visceral fat being a consequence of OSA adversely influencing the systemic inflammation of the body, as observed by [70].

TyG was used as a predictor in this work, and the findings parallel the results of [71] where TyG had a noticeable independent correlation with OSA in both non-obese and non-diabetic patients.

Recent studies reveal the capability of sleep architecture, in terms of sleep stages and sleep duration, in producing effective technology enabled screening of sleep disorders. Sleep architecture is estimated by leveraging wearable sensors or smartwatches with machine learning methods and its effect on OSA screening is observed in [72,73]. Specifically, stage 1 and stage 3 sleep exhibited anomalous behavior in the case of OSA patients, as stated in [74–76]. Interestingly, the findings of our presented work does not reveal strong predictive powers when using the features of sleep stages (stage 1, stage 2, stage 3 and REM) as well as sleep duration metrics. This could be because OSA does not always reflect the same changes across all stages of sleep for all individuals, due to variations in pathophysiological factors such as airway collapsibility, muscle responsiveness, arousal thresholds, and stable ventilation. These points arise as substantial inconsistencies when conducting sleep experiments on populations with different demographical composition in terms of age, gender or ethnicity, as noted in [77,78]. This brings to light the need of extended monitoring to accurately confirm the severity of OSA in patients using sleep staging approaches as well.

The demographics in the dataset used in this work did not have many extreme cases of OSA, and the severities seem to be fairly imbalanced, in favor of mild and moderate cases. Despite the relatively older ages of the population (average age 58.02 ± 8.04), OSA outcomes and associated medical conditions were not severe. A long-term study focusing on the same population as they age to analyze the OSA predictors and symptoms can likely reveal useful insights about the impact of lifestyle, and the potential consequences of other physiological and physical features. The OSA patient distribution was skewed towards men in this dataset. It could be due to the fact that women are generally less susceptible an OSA. As mentioned in [79], female hormones increases upper airway dilator muscle tone, and reduces the risk of pharangeal collapse (upper airway collapse), a major issue among OSA patients.

The presented work builds upon the findings reported previously in [21,22,60,61], which prove the feasibility of utilizing clinical information to screen for OSA patients and prioritize them for further sleep studies. Our models were able to predict clinical cases of OSA with reasonable accuracy, sensitivity and specificity, and is competitive with the recent electronic health record based prediction studies, as shown in Table 5. Consistent limitations in previous works include relatively fewer clinical parameters, high false positive rate, and demographic constraints. We observe that our proposed SVM model achieved the highest sensitivity among the existing works, with a specificity trade-off, in order to achieve a greater screening efficiency.

We further provide evidence that routinely clinical information can be effective in classification of OSA in a population health monitoring context. From the oximetry features, it can be said that desaturation severity, which consider the duration of apnea and hypopneas and the severity of breathing cessations may be more strongly related with daytime sleepiness and other symptoms than AHI or ODI [65,80]. Results suggest that oximetry data estimated using wearables, can be leveraged in conjunction with patient EHR to improve the detection rate, decrease false positives, and identify patients with risk of OSA. To enable continuous monitoring, another method would be to integrate personal health devices such as glucometers, in addition to the wearable data, as varying levels of glucose can indicate issues with metabolic issues concurrent with OSA complications. By considering all facets of an individual's health, from associated comorbidities, to treatment and risk factors, machine learning models can reasonably indicate effective. By prioritizing patients based on symptom severity, physicians can verify which cases are urgent, and which cases are false alarms. The incorporation of specialist feedback can enable a continuous active learning process to continuously train and retrain the model for better predictability.

The limitations of these works are as follows. Low specificity when model is trained with EHR data, similar to previous works in this domain, as indicated by Table 5. Majority of the patients participating in the Wisconsin Sleep study have reported some symptoms of OSA. This leads to the prevalence being higher in this dataset than the general public, and there likely may be only minimal differences between the non-OSA and OSA populations. Furthermore, most cases in the mild severity category, where they may not be necessarily chronic, but perhaps intermittent and only exacerbated by underlying comorbidities. The Wisconsin Sleep Study was conducted over a span of 10 years with a single patient having up to five different entries, and as noted previously [39] increasing age is typically correlated with higher prevalence as well. The dataset used is saturated with the Caucasian demographic, which could hinder its applicability to other races.

5. Conclusions

Routinely available clinical information such as patient questionnaires responses and anthropometry can be used to develop screening obstructive sleep apnea (OSA) classification models. However, its relative effectiveness in comparison with models trained with physiological oximetry has not been established till this work. The purpose of this study was to incorporate additional clinical parameters such as laboratory blood tests, clinical surrogate markers and history of comorbidities for training machine learning models and empirically validate its performance against models trained on oximetry measures acquired from the same population. This study proposes a SVM for classifying OSA patients at the cut-off of apnea-hypopnea index $\geq$5 and achieved accuracy: 68.06%, sensitivity: 88.76%, specificity: 40.74%, F1-score: 75.96%, PPV: 66.36% and NPV: 73.33%, which is competitive with existing research. The findings of this study demonstrate the potential of screening models for the early detection of individuals with high pretest OSA possibility using routinely collected clinical parameters. To address the limitations of this work, a large-scale prospective study is likely needed to assess the performance of the proposed screening model on the general population.

Author Contributions: Conceptualization, A.S. and F.A.; data curation, N.K.; investigation, J.R. and A.S.; methodology, N.K.; project administration, A.S. and F.A.; resources, J.R. and N.K.; software, J.R.; supervision, A.S. and F.A.; validation, J.R.; writing—original draft, J.R.; writing—review and editing, A.S. and F.A. All authors have read and agreed to the published version of the manuscript..

Funding: This research received no external funding.

Institutional Review Board Statement: Not applicable.

Informed Consent Statement: Not applicable.

Data Availability Statement: The data presented in this study are openly available in National Sleep Research Resource at https://doi.org/10.25822/js0k-yh52, accessed on 16 September 2021.

Conflicts of Interest: The authors declare no conflict of interest.

Abbreviations

The following abbreviations are used in this manuscript:

AHI	Apnea Hypopnea Index
ACC	Accuracy
AUC	Area Under Curve
BMI	Body-Mass Index
CB	Catboost Algorithm
CMA-ES	Covariance Matrix Adaptation Evolution Strategy
EDS	Excessive Daytime Sleepiness
ESS	Epworth Sleepiness Scale
EHR	Electronic Health Records
LAP	Lipid Accumulation Product
kNN	K-Nearest Neighbours
LGBM	Light Gradient Boosting
LR	Logistic Regression
ML	Machine Learning
MSLT	Multiple Sleep Latency Test
MWT	Maintenance of Wakefulness Test
NPV	Negative Predictive Value
OSA	Obstructive Sleep Apnea
PPV	Positive Predictive Value
PSG	Polysomnography
RF	Random Forest
SEN	Sensitivity
SLIM	Supersparse Linear Integer Model
SP	Specificity
SVM	Support Vector Machines
TPE	Tree-structured Parzen Estimator
WSC	Wisconsin Sleep Cohort
VAI	Visceral Adiposity Index
XGB	Extreme Gradient Boosting

Appendix A

The complete code to reproduce this work and further details regarding the results of the statistical tests, participant IDs for training-validation and hold-out testing set split, and additional model pipeline configurations is available at https://github.com/jayrmh/EHRWSC, accessed on 8 October 2021.

Table A1. Demographic characteristics of cohort expressed as mean ± standard deviation.

Demographics	**Overall** $n_r = 1479$	**OSA (≥5)** $n_r = 853$	**No OSA (≤5)** $n_r = 626$
AHI (/h)	12.26 ± 15.21	19.7 ± 16.33	2.03 ± 1.41
Age (y/o)	58.20 ± 8.04	59.483 ± 7.78	56.67 ± 8.12
Sex (%Male)	787 (53.2)	494 (57.81)	293 (46.08)
Race (%Caucasian)	1430 (96.68)	825 (96.71)	605 (96.65)
Alcohol (%Yes)	1080 (73.00)	619 (72.56)	461 (73.64)
Smoking (%Yes)	740 (50.00)	427 (50.00)	313 (50.00)

Table A2. Anthropometric characteristics of cohort expressed as mean ± standard deviation.

Anthropometric	**Overall** $n_r = 1479$	**OSA (≥5)** $n_r = 853$	**No OSA (≤5)** $n_r = 626$
Height (cm)	169.04 ± 9.24	169 ± 9.19	168.70 ± 9.30
Weight (kg)	90.05 ± 20.50	95.27 ± 20.27	82.94 ±18.58
BMI (kg/m^2)	31.54 ± 7.05	33.33 ± 7.29	29.09 ± 5.91
Neck Circumference (cm)	38.58 ± 4.04	39.53 ± 3.83	37.30 ± 3.966
Waist Circumference (cm)	99.89 ± 16.06	104.56 ± 15.25	93.55 ± 14.93

Table A3. Blood test profile characteristics of cohort expressed as mean ± standard deviation.

Blood Tests	**Overall** $n_r = 1479$	**OSA (≥5)** $n_r = 853$	**No OSA (≤5)** $n_r = 626$
Height (cm)	169.04 ± 9.24	169 ± 9.19	168.70 ± 9.30
Weight (kg)	90.05 ± 20.50	95.27 ± 20.27	82.94 ±18.58
BMI (kg/m^2)	31.54 ± 7.05	33.33 ± 7.29	29.09 ± 5.91
Neck Circumference (cm)	38.58 ± 4.04	39.53 ± 3.83	37.30 ± 3.966
Waist Circumference (cm)	99.89 ± 16.06	104.56 ± 15.25	93.55 ± 14.93

Table A4. Clinical surrogate marker characteristics of cohort expressed as mean ± standard deviation.

Clinical Surrogate Markers	**Overall** $n_r = 1479$	**OSA (≥5)** $n_r = 853$	**No OSA (≤5)** $n_r = 626$
TyG	8.73 ± 0.60	8.82 ± 0.06	8.60 ± 0.58
LAP	340.12 ± 258.60	392.36 ± 268.20	268.92 ± 226.462
VAI	3.83 ± 3.07	4.21 ± 3.32	3.31 ± 2.66
Whrt	0.59 ± 0.09	0.61 ± 0.09	0.55 ± 0.08

Table A5. General health characteristics of cohort expressed as mean ± standard deviation.

General Health	**Overall** $n_r = 1479$	**OSA (≥5)** $n_r = 853$	**No OSA (≤5)** $n_r = 626$
Zung Depression Scale	39.73 ± 8.13	40.07 ± 8.02	39.27 ±8.25
Horne Ostberg Score	62.40 ± 9.56	62.48 ± 9.84	62.25 ± 9.18
Epworth Sleepiness Scale	8.84 ± 4.17	9.22 ± 4.20	8.31 ± 4.08
State Anxiety Score	27.20 ± 6.91	27.11 ± 6.96	27.32 ±6.84
Trait Anxiety Score	31.67 ± 8.23	31.58 ± 8.15	31.76 ±8.33

Table A6. Comorbidities characteristics of cohort expressed as mean ± standard deviation.

Comorbidities	**Overall** $n_r = 1479$	**OSA (≥5)** $n_r = 853$	**No OSA (≤5)** $n_r = 626$
Heart Attack (%yes)	61 (4.12)	50 (6.00)	11 (1.75)
Hypertension (%yes)	531 (36.00)	357 (41.80)	174 (27.79)
Arrhythmia (%yes)	203 (13.72)	126 (14.77)	77 (12.30)
Angina (%yes)	45 (3.40)	34 (4.00)	11 (1.75)
Coronary (%yes)	106 (7.16)	76 (8.90)	30 (4.79)
Atherosclerosis (%yes)	28 (1.90)	14 (1.64)	14 (2.23)
Congestive Heart Failure (%yes)	14 (0.09)	13 (0.16)	1 (1.52)
Asthma (%yes)	263 (17.70)	162 (18.99)	101 (16.13)
Emphysema (%yes)	24 (1.62)	13 (1.52)	11 (1.75)
Diabetes (%yes)	166 (11.22)	119 (13.95)	47 (7.50)
Stroke (%yes)	28 (1.90)	19 (2.22)	9 (1.43)
Thyroid (%yes)	195 (13.18)	112 (13.13)	83 (13.25)
Arthritis (%yes)	460 (31.10)	302 (35.40)	158 (25.23)
Sleep Apnea (%yes)	187 (12.64)	123 (14.42)	64 (10.22)

Table A7. Self-reported sleep characteristics of cohort expressed as mean ± standard deviation.

Sleep History	**Overall** n_r = **1479**	**OSA (≥5)** n_r = **853**	**No OSA (≤5)** n_r = **626**
Excessive Daytime Sleepiness	314 (21.23)	195 (22.86)	119 (19.00)
Sleep Latency (min)	14.78 ± 12.96	14.56 ± 11.51	15.13 ± 14.71
Trouble Falling Back to Sleep (%sometimes)	533 (36.03)	312 (36.57)	221 (35.30)
Feeling Not Rested (%rarely)	488 (33.00)	273 (32.00)	215 (34.34)
Waking Up Too Early (%rarely)	527 (35.63)	309 (36.22)	218 (24.82)
Waking Up Repeatedly (%rarely)	417 (28.19)	240 (28.13)	177 (28.27)
Difficulty Falling Asleep (%rarely)	612 (41.37)	358 (41.96)	254 (40.57)
Difficulty Waking Up (%rarely)	568 (38.40)	329 (38.56)	239 (38.17)
Frequency of Nightmares (%rarely)	666 (45.00)	393 (46.07)	273(43.61)
Frequency of Snoring (%every night)	393 (26.67)	300 (35.18)	93 (14.85)
Snoring Volume (%talkingvolume)	426 (28.80)	238 (30.03)	188 (28.00)
Sleep Satisfaction (%mostly)	1019 (68.89)	590 (69.16)	429 (68.53)

Table A8. PSG-derived oximetry characteristics of cohort expressed as mean ± standard deviation.

Oximetry	**Overall** n_r = 1479	**OSA (≥5)** n_r = 853	**No OSA (≤5)** n_r = 626
Sleep Efficiency (%)	80.64 ± 10.16	79.67 ± 10.33	81.96±9.78
Sleep Latency (min)	12.63 ± 14.77	12.15 ± 14.91	13.30 ± 14.56
Average Oxygen Saturation (%)	95.32 ± 1.56	94.88 ± 1.58	95.90 ± 1.33
Minimum Oxygen Saturation (%)	85.00 ± 7.47	82.14 ± 7.72	88.89 ± 4.93
Average Oxygen Desaturation of Apnea-Hypopnea Event (%)	4.54 ± 1.23	5.06 ± 1.35	3.83 ± 0.44
Average Duration (s) of Apnea-Hypopnea Event	35.32 ± 8.58	33.85 ± 7.37	37.32 ± 9.65
Total Sleep Duration (min)	368.32 ± 57.45	364.12 ± 58.00	374 08 ± 56.22
REM Sleep Duration (min)	61.70 ± 25.92	58.35 ± 25.12	66.27 ± 26.30
REM Sleep Percentage (%)	16.51 ± 5.91	15.79 ±5.76	17.50 ±5.98
REM Latency (min)	123.40 ± 73.82	127.51 ± 76.06	117.96 ± 70.33
NREM Sleep Duration (min)	306.62 ± 47.65	305.75 ± 48.20	307.80 ± 46.90
Stage I and II Sleep Percentage (%)	76.21 ± 9.49	77.23 ± 9.31	74.82 ± 9.56
Stage III and IV Sleep Percentage (%)	7.26 ± 7.87	6.97±7.49	7.66 ±8.35
Wake After Sleep Onset (min)	68.89 ± 40.47	72.931 ± 40.93	63.33 ± 39.21
Sleep Duration Percentage with Oxygen Saturation below 90% (%)	1.92 ± 8.16	2.94 ± 9.88	0.53 ± 4.56

Table A9. Optimal hyperparameters for all ML models attained through tuning for the 8 feature EHR experiment.

Model	**Hyperparameters**
XGB	booster: dart lambda: 8.44×10^{-5} alpha: 1.36×10^{-8} max_depth: 4 eta: 0.604 gamma: 0.630 grow_policy: depthwise sample_type: weighted normalize_type: forest rate_drop: 0.758 skip_drop: 5.32×10^{-7}

Table A9. *Cont.*

Model	Hyperparameters
LGBM	booster: gbtree lambda: 4.18×10^{-8} alpha: 0.166 max_depth: 2 *eta:* 0.005 gamma: 0.007 grow_policy: lossguide
CB	objective: logloss colsample_bylevel: 0.055 depth: 9 boosting_type: ordered bootstrap_type: MVS
RF	n_estimators: 610 max_depth: 35 min_samples_leaf: 55 min_samples_split: 56
kNN	leaf_size: 70 n_neighbors: 37
LR	C: 0.007
SVM	kernel: rbf gamma: 0.24 C: 0.148

Table A10. Optimal hyperparameters for all ML models attained through tuning for 6 feature PSG experiment.

Model	Hyperparameters
XGB	booster: dart lambda: 0.0006 alpha: 0.0003 max_depth: 4 eta: 0.009 gamma: 3.838×10^{-5} grow_policy: depthwise sample_type: weighted normalize_type: tree rate_drop: 1.2×10^{-8} skip_drop: 0.0005
LGBM	booster: gbtree lambda: 4.23×10^{-6} alpha: 3.76×10^{-7} max_depth: 2 eta: 1.14×10^{-8} gamma: 0.914 grow_policy: depthwise

Table A10. *Cont.*

Model	Hyperparameters
CB	objective: crossentropy colsample_bylevel: 0.099 depth: 4 boosting_type: ordered bootstrap_type: Bernoulli
RF	n_estimators: 350 max_depth: 79 min_samples_leaf: 7 min_samples_split: 10
kNN	leaf_size: 60 n_neighbors: 63
LR	C: 2010.58
SVM	kernel: linear gamma: 5.68 C: 1.657

References

1. Hobson, J.A. Sleep Is of the Brain, by the Brain and for the Brain. *Nature* **2005**, *437*, 1254–1256. [CrossRef] [PubMed]
2. Benjafield, A.V.; Ayas, N.T.; Eastwood, P.R.; Heinzer, R.; Ip, M.S.M.; Morrell, M.J.; Nunez, C.M.; Patel, S.R.; Penzel, T.; Pépin, J.L.; et al. Estimation of the Global Prevalence and Burden of Obstructive Sleep Apnoea: A Literature-Based Analysis. *Lancet Respir. Med.* **2019**, *7*, 687–698. [CrossRef]
3. Lévy, P.; Kohler, M.; McNicholas, W.T.; Barbé, F.; McEvoy, R.D.; Somers, V.K.; Lavie, L.; Pépin, J.L. Obstructive Sleep Apnoea Syndrome. *Nat. Rev. Dis.* **2015**, *1*, 15015. [CrossRef]
4. Semelka, M.; Wilson, J.; Floyd, R. Diagnosis and Treatment of Obstructive Sleep Apnea in Adults. *Am. Fam. Physician* **2016**, *94*, 355–360. [PubMed]
5. Ibáñez, V.; Silva, J.; Cauli, O. A Survey on Sleep Assessment Methods. *PeerJ* **2018**, *6*, e4849. [CrossRef]
6. Pépin, J.L.; Bailly, S.; Tamisier, R. Big Data in Sleep Apnoea: Opportunities and Challenges. *Respirology* **2020**, *25*, 486–494. [CrossRef]
7. Sabil, A.; Vanbuis, J.; Baffet, G.; Feuilloy, M.; Le Vaillant, M.; Meslier, N.; Gagnadoux, F. Automatic Identification of Sleep and Wakefulness Using Single-Channel EEG and Respiratory Polygraphy Signals for the Diagnosis of Obstructive Sleep Apnea. *J. Sleep Res.* **2019**, *28*, e12795. [CrossRef] [PubMed]
8. Papini, G.B.; Fonseca, P.; van Gilst, M.M.; Bergmans, J.W.M.; Vullings, R.; Overeem, S. Wearable Monitoring of Sleep-Disordered Breathing: Estimation of the Apnea—Hypopnea Index Using Wrist-Worn Reflective Photoplethysmography. *Sci. Rep.* **2020**, *10*, 13512. [CrossRef]
9. Al-Mardini, M.; Aloul, F.; Sagahyroon, A.; Al-Husseini, L. Classifying Obstructive Sleep Apnea Using Smartphones. *J. Biomed. Inform.* **2014**, *52*, 251–259. [CrossRef]
10. Nandakumar, R.; Gollakota, S.; Watson, N. Contactless Sleep Apnea Detection on Smartphones. In Proceedings of the 13th Annual International Conference on Mobile Systems, Applications, and Services. Association for Computing Machinery, MobiSys '15, Florence, Italy, 18–22 May 2015; pp. 45–57. [CrossRef]
11. Korkalainen, H.; Aakko, J.; Duce, B.; Kainulainen, S.; Leino, A.; Nikkonen, S.; Afara, I.O.; Myllymaa, S.; Töyräs, J.; Leppänen, T. Deep Learning Enables Sleep Staging from Photoplethysmogram for Patients with Suspected Sleep Apnea. *Sleep* **2020**, *43*, zsaa098. [CrossRef] [PubMed]
12. Suliman, L.; Shalabi, N.; Saad, A. Validity of Overnight Pulse Oximetry as a Screening Tool of Obstructive Sleep Apnea. *ERS* **2016**, *48*, PA2316. [CrossRef]
13. Adkins, D.E. Machine Learning and Electronic Health Records: A Paradigm Shift. *Am. J. Psychiatry* **2017**, *174*, 93–94. [CrossRef] [PubMed]
14. Liang, H.; Tsui, B.Y.; Ni, H.; Valentim, C.C.S.; Baxter, S.L.; Liu, G.; Cai, W.; Kermany, D.S.; Sun, X.; Chen, J.; et al. Evaluation and Accurate Diagnoses of Pediatric Diseases Using Artificial Intelligence. *Nat. Med.* **2019**, *25*, 433–438. [CrossRef] [PubMed]
15. Goldstein, C.A.; Berry, R.B.; Kent, D.T.; Kristo, D.A.; Seixas, A.A.; Redline, S.; Westover, M.B. Artificial Intelligence in Sleep Medicine: Background and Implications for Clinicians. *J. Clin. Sleep Med.* **2020**, *16*, 609–618. [CrossRef] [PubMed]
16. Rajkomar, A.; Oren, E.; Chen, K.; Dai, A.M.; Hajaj, N.; Hardt, M.; Liu, P.J.; Liu, X.; Marcus, J.; Sun, M.; et al. Scalable and Accurate Deep Learning with Electronic Health Records. *NPJ Digit. Med.* **2018**, *1*, 18. [CrossRef] [PubMed]

17. Mandel, J.C.; Kreda, D.A.; Mandl, K.D.; Kohane, I.S.; Ramoni, R.B. SMART on FHIR: A Standards-Based, Interoperable Apps Platform for Electronic Health Records. *J. Am. Med. Inform. Assoc.* **2016**, *23*, 899–908. [CrossRef] [PubMed]
18. Ayala Solares, J.R.; Diletta Raimondi, F.E.; Zhu, Y.; Rahimian, F.; Canoy, D.; Tran, J.; Pinho Gomes, A.C.; Payberah, A.H.; Zottoli, M.; Nazarzadeh, M.; et al. Deep Learning for Electronic Health Records: A Comparative Review of Multiple Deep Neural Architectures. *J. Biomed. Inform.* **2019**, *101*, 103337. [CrossRef] [PubMed]
19. Keenan, B.T.; Kirchner, H.L.; Veatch, O.J.; Borthwick, K.M.; Davenport, V.A.; Feemster, J.C.; Gendy, M.; Gossard, T.R.; Pack, F.M.; Sirikulvadhana, L.; et al. Multisite Validation of a Simple Electronic Health Record Algorithm for Identifying Diagnosed Obstructive Sleep Apnea. *J. Clin. Sleep Med.* **2020**, *16*, 175–183. [CrossRef]
20. Laratta, C.R.; Tsai, W.H.; Wick, J.; Pendharkar, S.R.; Johannson, K.A.; Ronksley, P.E. Validity of Administrative Data for Identification of Obstructive Sleep Apnea. *J. Sleep Res.* **2017**, *26*, 132–138. [CrossRef]
21. Ustun, B.; Westover, M.B.; Rudin, C.; Bianchi, M.T. Clinical Prediction Models for Sleep Apnea: The Importance of Medical History over Symptoms. *J. Clin. Sleep Med.* **2016**, *12*, 161–168. [CrossRef]
22. Huang, W.C.; Lee, P.L.; Liu, Y.T.; Chiang, A.A.; Lai, F. Support Vector Machine Prediction of Obstructive Sleep Apnea in a Large-Scale Chinese Clinical Sample. *Sleep* **2020**, *43*, zsz295. [CrossRef] [PubMed]
23. Caffo, B.; Diener-West, M.; Punjabi, N.M.; Samet, J. A Novel Approach to Prediction of Mild Obstructive Sleep Disordered Breathing in a Population-Based Sample: The Sleep Heart Health Study. *Sleep* **2010**, *33*, 1641–1648. [CrossRef] [PubMed]
24. Chung, F.; Abdullah, H.R.; Liao, P. STOP-Bang Questionnaire: A Practical Approach to Screen for Obstructive Sleep Apnea. *Chest* **2016**, *149*, 631–638. [CrossRef]
25. Heldt, F.S.; Vizcaychipi, M.P.; Peacock, S.; Cinelli, M.; McLachlan, L.; Andreotti, F.; Jovanović, S.; Dürichen, R.; Lipunova, N.; Fletcher, R.A.; et al. Early Risk Assessment for COVID-19 Patients from Emergency Department Data Using Machine Learning. *Sci. Rep.* **2021**, *11*, 4200. [CrossRef] [PubMed]
26. Young, T. Rationale, Design and Findings from the Wisconsin Sleep Cohort Study: Toward Understanding the Total Societal Burden of Sleep Disordered Breathing. *Sleep Med. Clin.* **2009**, *4*, 37–46. [CrossRef] [PubMed]
27. Fleming, W.E.; Holty, J.E.C.; Bogan, R.K.; Hwang, D.; Ferouz-Colborn, A.S.; Budhiraja, R.; Redline, S.; Mensah-Osman, E.; Osman, N.I.; Li, Q.; et al. Use of Blood Biomarkers to Screen for Obstructive Sleep Apnea. *Nat. Sci. Sleep* **2018**, *10*, 159–167. [CrossRef]
28. Montesi, S.B.; Bajwa, E.K.; Malhotra, A. Biomarkers of Sleep Apnea. *Chest* **2012**, *142*, 239–245. [CrossRef]
29. Wei, R.; Gao, Z.; Xu, H.; Jiang, C.; Li, X.; Liu, Y.; Zou, J.; Zhu, H.; Yi, H.; Guan, J.; et al. Body Fat Indices as Effective Predictors of Insulin Resistance in Obstructive Sleep Apnea: Evidence from a Cross-Sectional and Longitudinal Study. *Obes. Surg.* **2021**, *31*, 2219–2230. [CrossRef]
30. Li, R.; Li, Q.; Cui, M.; Yin, Z.; Li, L.; Zhong, T.; Huo, Y.; Xie, P. Clinical Surrogate Markers for Predicting Metabolic Syndrome in Middle-Aged and Elderly Chinese. *J. Diabetes Investig.* **2018**, *9*, 411–418. [CrossRef]
31. Ge, H.; Yang, Z.; Li, X.; Liu, D.; Li, Y.; Pan, Y.; Luo, D.; Wu, X. The Prevalence and Associated Factors of Metabolic Syndrome in Chinese Aging Population. *Sci. Rep.* **2020**, *10*, 20034. [CrossRef]
32. Zhou, W.; Li, C.l.; Cao, J.; Feng, J. Metabolic Syndrome Prevalence in Patients with Obstructive Sleep Apnea Syndrome and Chronic Obstructive Pulmonary Disease: Relationship with Systemic Inflammation. *Clin. Respir. J.* **2020**, *14*, 1159–1165. [CrossRef]
33. Young, T.; Finn, L.; Peppard, P.E.; Szklo-Coxe, M.; Austin, D.; Nieto, F.J.; Stubbs, R.; Hla, K.M. Sleep Disordered Breathing and Mortality: Eighteen-Year Follow-up of the Wisconsin Sleep Cohort. *Sleep* **2008**, *31*, 1071–1078. [PubMed]
34. Hori, T.; Sugita, Y.; Koga, E.; Shirakawa, S.; Inoue, K.; Uchida, S.; Kuwahara, H.; Kousaka, M.; Kobayashi, T.; Tsuji, Y.; et al. Proposed Supplements and Amendments to 'A Manual of Standardized Terminology, Techniques and Scoring System for Sleep Stages of Human Subjects', the Rechtschaffen & Kales (1968) Standard. *Psychiatry Clin. Neurosci.* **2001**, *55*, 305–310. [CrossRef]
35. Quan, S.F.; Gillin, J.C.; Littner, M.R.; Shepard, J.W. Sleep-Related Breathing Disorders in Adults: Recommendations for Syndrome Definition and Measurement Techniques in Clinical Research. *Sleep* **1999**, *22*, 667–689. [CrossRef]
36. Pedersen, A.B.; Mikkelsen, E.M.; Cronin-Fenton, D.; Kristensen, N.R.; Pham, T.M.; Pedersen, L.; Petersen, I. Missing Data and Multiple Imputation in Clinical Epidemiological Research. *Clin. Epidemiol.* **2017**, *9*, 157–166. [CrossRef]
37. Rochon, J.; Gondan, M.; Kieser, M. To Test or Not to Test: Preliminary Assessment of Normality When Comparing Two Independent Samples. *BMC Med. Res. Methodol.* **2012**, *12*, 81. [CrossRef]
38. Nachar, N. The Mann-Whitney U: A Test for Assessing Whether Two Independent Samples Come from the Same Distribution. *Tutor. Quant. Methods Psychol.* **2018**, *4*, 13–20. [CrossRef]
39. Deng, X.; Gu, W.; Li, Y.; Liu, M.; Li, Y.; Gao, X. Age-Group-Specific Associations between the Severity of Obstructive Sleep Apnea and Relevant Risk Factors in Male and Female Patients. *PLoS ONE* **2014**, *9*, e107380. [CrossRef]
40. Vickerstaff, V.; Omar, R.Z.; Ambler, G. Methods to Adjust for Multiple Comparisons in the Analysis and Sample Size Calculation of Randomised Controlled Trials with Multiple Primary Outcomes. *BMC Med. Res. Methodol.* **2019**, *19*, 129. [CrossRef]
41. Remeseiro, B.; Bolon-Canedo, V. A Review of Feature Selection Methods in Medical Applications. *Comput. Biol. Med.* **2019**, *112*, 103375. [CrossRef] [PubMed]
42. Göktaş, A.; İşçi, Ö. A Comparison of the Most Commonly Used Measures of Association for Doubly Ordered Square Contingency Tables via Simulation. *Metod. Zv.* **2011**, *8*, 17.
43. Menze, B.H.; Kelm, B.M.; Masuch, R.; Himmelreich, U.; Bachert, P.; Petrich, W.; Hamprecht, F.A. A Comparison of Random Forest and Its Gini Importance with Standard Chemometric Methods for the Feature Selection and Classification of Spectral Data. *BMC Bioinform.* **2009**, *10*, 213. [CrossRef] [PubMed]

44. Fang, L.; Zhao, H.; Wang, P.; Yu, M.; Yan, J.; Cheng, W.; Chen, P. Feature Selection Method Based on Mutual Information and Class Separability for Dimension Reduction in Multidimensional Time Series for Clinical Data. *Biomed. Signal Process. Control* **2015**, *21*, 82–89. [CrossRef]
45. Hsu, Y.C.; Weng, H.H.; Kuo, C.Y.; Chu, T.P.; Tsai, Y.H. Prediction of Fall Events during Admission Using eXtreme Gradient Boosting: A Comparative Validation Study. *Sci. Rep.* **2020**, *10*, 16777. [CrossRef] [PubMed]
46. Zhang, Z.; Zhao, Y.; Canes, A.; Steinberg, D.; Lyashevska, O. Predictive Analytics with Gradient Boosting in Clinical Medicine. *Ann. Transl. Med.* **2019**, *7*. [CrossRef] [PubMed]
47. Chen, T.; Guestrin, C. XGBoost: A Scalable Tree Boosting System. In Proceedings of the 22nd ACM Sigkdd International Conference on Knowledge Discovery and Data Mining, San Francisco, CA, USA, 13–17 August 2016; pp. 785–794. [CrossRef]
48. Ke, G.; Meng, Q.; Finley, T.; Wang, T.; Chen, W.; Ma, W.; Ye, Q.; Liu, T.Y. LightGBM: A Highly Efficient Gradient Boosting Decision Tree. *Adv. Neural Inf. Process. Syst.* **2017**, *30*, 3146–3154
49. Prokhorenkova, L.; Gusev, G.; Vorobev, A.; Dorogush, A.V.; Gulin, A. CatBoost: Unbiased Boosting with Categorical Features. In Proceedings of the 32nd International Conference on Neural Information Processing Systems, NIPS'18, Montreal, QC, Canada, 3–8 December 2018; pp. 6639–6649.
50. Jayroop Ramesh, R.A. A Remote Healthcare Monitoring Framework for Diabetes Prediction Using Machine Learning. *Healthc. Technol. Lett.* **2021**, *8*, 45. [CrossRef]
51. Bergstra, J.S.; Bardenet, R.; Bengio, Y.; Kégl, B. Algorithms for Hyper-Parameter Optimization. *Adv. Neural Inf. Process. Syst.* **2011**, *24*, 9.
52. Loshchilov, I.; Hutter, F. CMA-ES for Hyperparameter Optimization of Deep Neural Networks. *arXiv* **2016**, arXiv:1604.07269.
53. Akiba, T.; Sano, S.; Yanase, T.; Ohta, T.; Koyama, M. Optuna: A Next-Generation Hyperparameter Optimization Framework. In Proceedings of the 25th ACM SIGKDD International Conference on Knowledge Discovery & Data Mining, Anchorage, AK, USA, 4–8 August 2019; pp. 2623–2631. [CrossRef]
54. Harris, C.R.; Millman, K.J.; van der Walt, S.J.; Gommers, R.; Virtanen, P.; Cournapeau, D.; Wieser, E.; Taylor, J.; Berg, S.; Smith, N.J.; et al. Array Programming with NumPy. *Nature* **2020**, *585*, 357–362. [CrossRef]
55. McKinney, W. Data Structures for Statistical Computing in Python. In Proceedings of the Python in Science Conference, Austin, TX, USA, 28–30 June 2010; pp. 56–61. [CrossRef]
56. Virtanen, P.; Gommers, R.; Oliphant, T.E.; Haberland, M.; Reddy, T.; Cournapeau, D.; Burovski, E.; Peterson, P.; Weckesser, W.; Bright, J.; et al. SciPy 1.0: Fundamental Algorithms for Scientific Computing in Python. *Nat. Methods* **2020**, *17*, 261–272. [CrossRef]
57. Pedregosa, F.; Varoquaux, G.; Gramfort, A.; Michel, V.; Thirion, B.; Grisel, O.; Blondel, M.; Prettenhofer, P.; Weiss, R.; Dubourg, V.; et al. Scikit-Learn: Machine Learning in Python. *J. Mach. Learn. Res.* **2011**, *12*, 2825–2830.
58. Waskom, M.L. Seaborn: Statistical Data Visualization. *J. Open Source Softw.* **2021**, *6*, 3021. [CrossRef]
59. Hunter, J.D. Matplotlib: A 2D Graphics Environment. *Comput. Sci. Eng.* **2007**, *9*, 90–95. [CrossRef]
60. Kim, Y.J.; Jeon, J.S.; Cho, S.E.; Kim, K.G.; Kang, S.G. Prediction Models for Obstructive Sleep Apnea in Korean Adults Using Machine Learning Techniques. *Diagnostics* **2021**, *11*, 612. [CrossRef] [PubMed]
61. Mencar, C.; Gallo, C.; Mantero, M.; Tarsia, P.; Carpagnano, G.E.; Foschino Barbaro, M.P.; Lacedonia, D. Application of Machine Learning to Predict Obstructive Sleep Apnea Syndrome Severity. *Health Inform. J.* **2020**, *26*, 298–317. CrossRef]
62. Davidson, T.M.; Patel, M.R. Waist Circumference and Sleep Disordered Breathing. *Laryngoscope* **2008**, *118*, 339–347. [CrossRef]
63. Alakuijala, A.; Salmi, T. Predicting Obstructive Sleep Apnea with Periodic Snoring Sound Recorded at Home. *J. Clin. Sleep Med.* **2016**, *12*, 953–958. [CrossRef] [PubMed]
64. Luyster, F.S.; Buysse, D.J.; Strollo, P.J., Jr. Comorbid Insomnia and Obstructive Sleep Apnea: Challenges for Clinical Practice and Research. *J. Clin. Sleep Med.* **2010**, *06*, 196–204. [CrossRef]
65. Kainulainen, S.; Töyräs, J.; Oksenberg, A.; Korkalainen, H.; Sefa, S.; Kulkas, A.; Leppänen, T. Severity of Desaturations Reflects OSA-Related Daytime Sleepiness Better Than AHI. *J. Clin. Sleep Med.* **2019**, *15*, 1135–1142. [CrossRef]
66. Michalek-Zrabkowska, M.; Macek, P.; Martynowicz, H.; Gac, P.; Mazur, G.; Grzeda, M.; Poreba, R. Obstructive Sleep Apnea as a Risk Factor of Insulin Resistance in Nondiabetic Adults. *Life* **2021**, *11*, 50. [CrossRef] [PubMed]
67. Mullington, J.M.; Abbott, S.M.; Carroll, J.E.; Davis, C.J.; Dijk, D.J.; Dinges, D.F.; Gehrman, P.R.; Ginsburg, G.S.; Gozal, D.; Haack, M.; et al. Developing Biomarker Arrays Predicting Sleep and Circadian-Coupled Risks to Health. *Sleep* **2016**, *39*, 727–736. [CrossRef] [PubMed]
68. Kritikou, I.; Basta, M.; Vgontzas, A.N.; Pejovic, S.; Liao, D.; Tsaoussoglou, M.; Bixler, E.O.; Stefanakis, Z.; Chrousos, G.P. Sleep Apnoea, Sleepiness, Inflammation and Insulin Resistance in Middle-Aged Males and Females. *Eur. Respir. J.* **2014**, *43*, 145–155. [CrossRef] [PubMed]
69. Kim, D.H.; Kim, B.; Han, K.; Kim, S.W. The Relationship between Metabolic Syndrome and Obstructive Sleep Apnea Syndrome: A Nationwide Population-Based Study. *Sci. Rep.* **2021**, *11*, 8751. [CrossRef] [PubMed]
70. Vicente, E.; Marin, J.M.; Carrizo, S.J.; Osuna, C.S.; González, R.; Marin-Oto, M.; Forner, M.; Vicente, P.; Cubero, P.; Gil, A.V.; et al. Upper Airway and Systemic Inflammation in Obstructive Sleep Apnoea. *Eur. Respir. J.* **2016**, *48*, 1108–1117. [CrossRef] [PubMed]
71. Bikov, A.; Frent, S.M.; Meszaros, M.; Kunos, L.; Mathioudakis, A.G.; Negru, A.G.; Gaita, L.; Mihaicuta, S. Triglyceride-Glucose Index in Non-Diabetic, Non-Obese Patients with Obstructive Sleep Apnoea. *J. Clin. Med.* **2021**, *10*, 1932. [CrossRef]
72. Mostafa, S.S.; Mendonça, F.G.; Ravelo-García, A.; Morgado-Dias, F. A Systematic Review of Detecting Sleep Apnea Using Deep Learning. *Sensors* **2019**, *19*, 4934. [CrossRef]

73. Sridhar, N.; Shoeb, A.; Stephens, P.; Kharbouch, A.; Shimol, D.B.; Burkart, J.; Ghoreyshi, A.; Myers, L. Deep Learning for Automated Sleep Staging Using Instantaneous Heart Rate. *NPJ Digit. Med.* **2020**, *3*, 106. [CrossRef] [PubMed]
74. Jalilolghadr, S.; Yazdi, Z.; Mahram, M.; Babaei, F.; Esmailzadehha, N.; Nozari, H.; Saffari, F. Sleep Architecture and Obstructive Sleep Apnea in Obese Children with and without Metabolic Syndrome: A Case Control Study. *Sleep Breath.* **2016**, *20*, 845–851. [CrossRef]
75. Basunia, M.; Fahmy, S.A.; Schmidt, F.; Agu, C.; Bhattarai, B.; Oke, V.; Enriquez, D.; Quist, J. Relationship of Symptoms with Sleep-Stage Abnormalities in Obstructive Sleep Apnea-Hypopnea Syndrome. *J. Community Hosp. Intern. Med. Perspect.* **2016**, *6*, 32170. [CrossRef]
76. BaHammam, A.S.; Alshahrani, M.; Aleissi, S.A.; Olaish, A.H.; Alhassoon, M.H.; Shukr, A. Blood Pressure Dipping during REM and Non-REM Sleep in Patients with Moderate to Severe Obstructive Sleep Apnea. *Sci. Rep.* **2021**, *11*, 7990. [CrossRef] [PubMed]
77. Acosta-Castro, P.; Hirotsu, C.; Marti-Soler, H.; Marques-Vidal, P.; Tobback, N.; Andries, D.; Waeber, G.; Preisig, M.; Vollenweider, P.; Haba-Rubio, J.; et al. REM-Associated Sleep Apnoea: Prevalence and Clinical Significance in the HypnoLaus Cohort. *Eur. Respir. J.* **2018**, *52*, 1702484. [CrossRef] [PubMed]
78. Shahveisi, K.; Jalali, A.; Moloudi, M.R.; Moradi, S.; Maroufi, A.; Khazaie, H. Sleep Architecture in Patients with Primary Snoring and Obstructive Sleep Apnea. *Basic Clin. Neurosci.* **2018**, *9*, 147–156. [CrossRef] [PubMed]
79. Saaresranta, T.; Anttalainen, U.; Polo, O. Sleep Disordered Breathing: Is It Different for Females? *ERJ Open Res.* **2015**, *1*. [CrossRef]
80. Veugen, C.C.A.F.M.; Teunissen, E.M.; den Otter, L.A.S.; Kos, M.P.; Stokroos, R.J.; Copper, M.P. Prediction of Obstructive Sleep Apnea: Comparative Performance of Three Screening Instruments on the Apnea-Hypopnea Index and the Oxygen Desaturation Index. *Sleep Breath.* **2021**, *25*, 1267–1275. [CrossRef]

healthcare MDPI

Article

Patient Satisfaction and Hospital Quality of Care Evaluation in Malaysia Using SERVQUAL and Facebook

Afiq Izzudin A. Rahim [1], Mohd Ismail Ibrahim [1,*], Kamarul Imran Musa [1], Sook-Ling Chua [2] and Najib Majdi Yaacob [3]

[1] Department of Community Medicine, School of Medical Science, Universiti Sains Malaysia, Kubang Kerian, Kota Bharu 16150, Kelantan, Malaysia; drafiqrahim@student.usm.my (A.I.A.R.); drkamarul@usm.my (K.I.M.)
[2] Faculty of Computing and Informatics, Multimedia University, Persiaran Multimedia, Cyberjaya 63100, Selangor, Malaysia; slchua@mmu.edu.my
[3] Unit of Biostatistics and Research Methodology, Health Campus, School of Medical Sciences, Universiti Sains Malaysia, Kubang Kerian, Kota Bharu 16150, Kelantan, Malaysia; najibmy@usm.my
* Correspondence: ismaildr@usm.my; Tel.: +60-9767-6621; Fax: +60-9765-3370

Citation: Rahim, A.I.A.; Ibrahim, M.I.; Musa, K.I.; Chua, S.-L.; Yaacob, N.M. Patient Satisfaction and Hospital Quality of Care Evaluation in Malaysia Using SERVQUAL and Facebook. *Healthcare* **2021**, *9*, 1369. https://doi.org/10.3390/healthcare9101369

Academic Editor: Mahmudur Rahman

Received: 22 August 2021
Accepted: 12 October 2021
Published: 14 October 2021

Publisher's Note: MDPI stays neutral with regard to jurisdictional claims in published maps and institutional affiliations.

Abstract: Social media sites, dubbed patient online reviews (POR), have been proposed as new methods for assessing patient satisfaction and monitoring quality of care. However, the unstructured nature of POR data derived from social media creates a number of challenges. The objectives of this research were to identify service quality (SERVQUAL) dimensions automatically from hospital Facebook reviews using a machine learning classifier, and to examine their associations with patient dissatisfaction. From January 2017 to December 2019, empirical research was conducted in which POR were gathered from the official Facebook page of Malaysian public hospitals. To find SERVQUAL dimensions in POR, a machine learning topic classification utilising supervised learning was developed, and this study's objective was established using logistic regression analysis. It was discovered that 73.5% of patients were satisfied with the public hospital service, whereas 26.5% were dissatisfied. SERVQUAL dimensions identified were 13.2% reviews of tangible, 68.9% of reliability, 6.8% of responsiveness, 19.5% of assurance, and 64.3% of empathy. After controlling for hospital variables, all SERVQUAL dimensions except tangible and assurance were shown to be significantly related with patient dissatisfaction (reliability, $p < 0.001$; responsiveness, $p = 0.016$; and empathy, $p < 0.001$). Rural hospitals had a higher probability of patient dissatisfaction ($p < 0.001$). Therefore, POR, assisted by machine learning technologies, provided a pragmatic and feasible way for capturing patient perceptions of care quality and supplementing conventional patient satisfaction surveys. The findings offer critical information that will assist healthcare authorities in capitalising on POR by monitoring and evaluating the quality of services in real time.

Keywords: patient satisfaction; service quality; SERVQUAL; Facebook; machine learning; patient online review; Malaysia

1. Introduction

The World Health Organization (WHO) stresses that substandard care wastes significant resources and jeopardises public health by degrading human capital and decreasing productivity. Thus, in addition to providing effective coverage of essential health services and financial security in each country, delivering high-quality care or service is important in achieving the Universal Health Coverage goal [1]. At the core of delivering high-quality care is a dedication to person-centered care. Communities must be engaged in the design, implementation, and ongoing evaluation of health services to ensure that they meet local health needs. Also, striking a balance between patient expectations and quality improvement initiatives is important, since it influences patient safety, survival, and long-term health [2]. According to a systematic analysis, poor healthcare quality was the main factor leading to an increase in deaths from cardiovascular disease, neonatal trauma, and communicable illnesses [3]. As healthcare prepares for the Industrial Revolution 4.0 by becoming more patient-centered and value-driven, quality management systems must include efforts to understand and respect patients' interests, desires, and values. Because such reports can only be generated by patients, it is critical to create systems for monitoring patient experiences and to promote their use on an individual and communal level [4,5]. Patient tperception and satisfaction have been a key component of patient-centered care since the early 1990s and have been incorporated into healthcare quality of care assessment. Healthcare administrators that aim for excellence consider patient perception while creating strategies for improving treatment quality [6].

Service quality (SERVQUAL) is a commonly used technique for evaluating the quality of service in a wide variety of service environments, sectors, and nations [7]. Because the model encompasses five dimensions—tangible, reliability, responsiveness, empathy, and assurance—it efficiently measures customer service needs and perceptions [8].

SERVQUAL, Hospital Consumer Assessment of Healthcare Providers and Systems (HCAHPS), and other traditional patient satisfaction surveys are the product of years of evaluative analysis, are performed and evaluated in a methodical manner, and may evoke a wide variety of answers from patients [9,10]. However, traditional patient or public surveys used to assess the quality of healthcare services are time and resource intensive, require considerable time between hospital admission and report disclosure, frequently result in a failure to identify the underlying causes of concern, and introduce response and selection bias [11,12]. The disconnect between conventional surveys and patient perceptions and treatment quality underscored the need for developing new data sources for assessing patient perceptions and care quality [13]. Technological innovation is essential for creating new ways for rapidly assessing the quality of services at an affordable cost. Therefore, social media platforms, which are often referred to as patient online reviews (POR), have been suggested as a new way for gauging patient satisfaction and monitoring treatment quality [14,15].

There have been small number of POR studies in contrast to its exponential growth [16,17]. While it has been demonstrated that Facebook and other social media platforms can improve health outcomes through health education and information [18,19] and can be beneficial during public health crises [20,21], other studies have examined specific features of social media platforms such as reviews and ratings and their relationship to patient satisfaction and hospital quality measures [16]. For example, Facebook offers a review feature that allows users to leave narrative assessments and evaluate the performance of companies and institutions on their Facebook pages. Numerous studies have discovered a weak to moderate correlation between Facebook evaluations and traditional patient satisfaction survey metrics [22–25], while another study discovered a link between clinical quality indicators such as reduced re-admission rates and higher Facebook ratings [26]. According to recent research, hospitals with an active Facebook page had a higher number of "likes," a greater percentage of patients ready to refer the hospital, and a higher overall satisfaction score [27]. Additional study on the patient viewpoint and its relationship to hospital patients' total Facebook ratings discovered associations with a variety of issues, including wait times, treatment effectiveness, and communication [28]. With an increasing number of patients asking and freely sharing hospital evaluations on social media, feedback data may supplement conventional patient satisfaction surveys [14,27].

However, the unstructured nature of POR data collected from social media presents several difficulties, including data cleaning and processing. While this may be accomplished manually via human input, the process is lengthy, and the method's validity and reliability are often questioned [29]. A systematic evaluation of POR was proposed to accelerate the processing of large-scale online data review using sophisticated analytical techniques such as machine learning [16]. Consequently, a machine learning approach for classifying service quality themes or subjects based on unstructured social media data has the potential to significantly improve healthcare quality of care [30,31].

Additionally, the population's fondness for social media has led many healthcare institutions to use their country's most popular social media platforms for online communication and engagement with the public. According to a national survey conducted in Taiwan, Facebook has a high level of penetration and popularity in the country, which may be one of the reasons why more than half of Taiwan's hospitals have established an official Facebook profile [32]. Facebook is also a critical component of Malaysian social media use. According to a 2020 survey, 91.7 percent of Malaysian internet users utilised Facebook, and the site is projected to continue to be the country's most popular social networking site [33]. Given the popularity of Facebook in Malaysia and its expanding usage in healthcare, this study's first task was to assess the frequency of SERVQUAL dimensions in Facebook reviews of Malaysian public hospitals using a machine learning classifier and prevalence of hospital patient satisfaction. The second was to seek to establish relationships between SERVQUAL qualities and hospital patient dissatisfaction as expressed in Facebook reviews. POR analyzed using a machine learning algorithm may have value in assisting all key healthcare stakeholders in making decisions to enhance the quality of care delivered in Malaysia.

2. Related Work

2.1. Patient Satisfaction

Intellectuals have been assessing hospital patient satisfaction for years, using a range of methodologies and conceptual frameworks. An earlier study showed that patients with moderate expectations reported the highest levels of satisfaction, whereas those with excessive expectations reported the lowest levels of satisfaction [34]. When patients' expectations were met in terms of health care delivery, they reported satisfaction with such services [35]. Since those early attempts, the number of factors linked with patient satisfaction have increased dramatically and vary between research [36,37]. However, one systematic review found that two significant determinants of patient satisfaction were variables affecting the healthcare provider and patient characteristics [35]. Across studies, that study found that provider-related variables were the strongest predictor of patient satisfaction. There were nine identified determinants of healthcare services: technical care, interpersonal care, physical environment, accessibility, availability, financial resources, organisational characteristics, continuity of treatment, and care result. Research that examined the physical environment in relation to patient satisfaction ratings on social media discovered that environmental variables such as parking, cleanliness, and waiting rooms all contributed to patient satisfaction [38]. Another POR research showed that comments on the efficacy of treatment, communication, and diagnostic quality were most strongly linked with patients' overall ratings [28]. A comprehensive assessment of patient satisfaction confirmed the results, revealing that interpersonal skills and technical care features had the most positive associations with service-related factors [35].

Patient characteristics such as age, gender, education, socioeconomic status, marital status, race, religion, geographic characteristics, frequency of visits, length of stay, health status, personality, and expectations were all investigated to ascertain their associations with patient satisfaction [35]. Hospital characteristics such as location and rural regions were shown to be positively associated with patient discontent [39], even though another study found rural residents were satisfied with healthcare services [40]. Additionally, the size and type of hospital services influenced patient satisfaction [15,41]. Previously, it was believed that people would be more unhappy with a service that dealt with a greater number of patients and a bigger office. However, in a comprehensive assessment of patient satisfaction, these associations were modest and inconsistent [35]. Therefore, the research concluded that it may be worthwhile to attempt to build patient satisfaction using health care quality indicators and observe how individuals increase their satisfaction with health services. SERVQUAL and HCAHPS are two examples of systematic surveys that assess healthcare quality of care. The findings of patient satisfaction surveys may be very helpful for both healthcare professionals and patients. They aid healthcare providers in finding areas in which their services might be improved. Increased patient satisfaction with healthcare services boosts public hospital responsiveness [42]. Additionally, it enables policymakers to understand patient needs and therefore create strategic plans for more effective and high-quality services. According to studies, satisfied patients are more likely to follow their physicians' recommendations for treatment and follow-up visits, resulting in better health outcomes and hospital recommendations to others [35].

2.2. Social Media Data and Machine Learning

Social media data are often massive and present several difficulties, including data cleansing, data processing, and developing a theoretical model of social media content quality. While this may be accomplished manually via human input, the procedure is time consuming, labour intensive, and the validity and reliability of the technique are often questioned [29]. A comprehensive analysis of POR established and recommended the use of advanced analytical methods such as machine learning to accelerate the processing of huge amounts of online review data [16]. Additionally, the systematic review recommended doing an in-depth examination of the contents of online reviews rather than just comparing structured data to social media ratings. Monitoring service quality through hospital social media platforms may assist all stakeholders in detecting quality issues and minimising the need for expensive and time-consuming surveys. Despite their rarity, research on Facebook content analysis demonstrates a correlation between social media quality domains and traditional hospital quality metrics [23,28,43,44].

The word "themes" or "text classification" refers to the process of grouping together a collection of textual messages according on their content. Machine learning enables automatic topic analysis via the application of various algorithms that are classified as supervised and unsupervised learning. The existence of labels in the subset of training data distinguishes these two main categories [45]. Along with input features, supervised machine learning makes use of predefined output features. The algorithms attempt to forecast and classify the predefined feature, and their accuracy and misclassification, as well as other performance metrics, are determined by the counts of the predetermined feature that are correctly predicted or classified, or that are incorrectly predicted or classified. Manual classification is a technique that is often used in supervised learning. Numerous studies have utilised this approach to deduce the topics of contention in POR [11,12,28,46–48].

On the other hand, unsupervised learning is pattern recognition that does not need the usage of a target feature. Unsupervised algorithms identify unlabeled data's underlying groupings and then label each value. Topic modelling is a technique for automatically identifying topics within a given remark, with the most often used approach being Latent Dirichlet Allocation (LDA). Numerous studies have utilised the technique to elicit information on the themes or subjects of discussion in POR [49–54].

According to prior research, POR often addressed issues such as appointment scheduling, wait times, the efficiency of the healthcare system, and interpersonal quality [12,28,46,50]. However, other topics such as communication, technological elements, treatment effectiveness, patient safety, environment, and hospital expenses were recognised as significant concerns [13,38,52,53]. Further study of hospitals in the United States revealed that the variables most significantly linked with patients' overall ratings or satisfaction included waiting times, treatment effectiveness, communication, diagnostic quality, environmental cleanliness, and economic concerns [28]. Comparable research utilising the Consumer Assessment of Healthcare Providers and Systems (CAHPS) Dental Plan Survey [55] and Press Ganey [56] corroborated the result. Other research discovered that the issues discussed in the dissatisfaction survey mirrored the often-discussed topics of appointment access and wait time [46]. Additionally, patient discontent was often related to personnel, punctuality, and diagnostic problems, while satisfaction was significantly related to interpersonal and technical brilliance [52]. However, Yelp review research discovered that patient satisfaction was related to interpersonal quality of surgical care, while dissatisfaction was related to insurance, billing, and the cost of the hospital visit [50]. Another study examined National Health Service (NHS) tweets using the SERVQUAL model and found that the aspects of responsiveness and assurance were often addressed in negative narratives, while empathy was completely positive [53]. It is unsurprising that some subjects elicited more negative annotations than others, particularly comments about time, money, or pain, which are unlikely to be related to patient satisfaction [12].

2.3. Proposed Work

Given the exponential growth of social media in Malaysia and Southeast Asia, it is critical to use technology to improve healthcare services. Meanwhile, although Facebook is a popular social media platform, there has been very little study on machine learning and quality measures using Facebook data [28,57,58]. Given Facebook's popularity in Malaysia and its growing usage in healthcare, this research seeks to fill a void by investigating whether patient comments in Facebook Reviews can be categorised into SERVQUAL topics, and determining their association with patient satisfaction.

Additionally, this research used supervised machine learning to classify topics. Conventional patient satisfaction surveys have several disadvantages, and social media has been proposed as a potential substitute for evaluating patient satisfaction and mood in real time. According to a systematic review of the use of natural language processing (NLP) and machine learning (ML) to process and analyse patient experience data, manual classification of free text comments remains the 'gold standard' method of analysis and is currently the only way to ensure that all pertinent patient comments are coded and analysed [29]. Additionally, the analysis showed that patient inputs produced via free-text supplements to structured questionnaires such as SERVQUAL and HCAHPS were stable in nature, making them an appealing source of data for supervised learning. Numerous studies have utilised supervised machine learning to categorise POR themes [28,47,48,57,59–61]. Moreover, we suggested that SERVQUAL dimensions be used to train our machine learning topic classifier. Previous research has classified themes or subjects in POR using structured patient questionnaires such as SERVQUAL [53,62], CAHPS Dental Plan Survey [55] and HCAHPS [50]. The potential results may be compared with those obtained via traditional surveys of patient satisfaction or treatment quality.

Nevertheless, the current body of evidence is still limited, owing to a scarcity of sophisticated statistical studies linking patient satisfaction or hospital quality indicators. A systematic review suggested that more empirical research on POR be conducted using pertinent hypotheses, rigorous design, and data analytics [16]. Thus, this study should go beyond basic descriptive analysis and include the testing of theory-based hypotheses to offer additional policy implications and understanding. Previously published research has utilised analysis of variance (ANOVA) [55], various regression analytical tests [12,52,54,58], Pearson correlation [50,57] or Spearman's rank correlation [57,63]. As such, this research seeks to examine variables related with patient dissatisfaction using rigorous statistical techniques such as regression analysis.

3. Materials and Methods

This research was cross-sectional in design and took place between March 2020 and May 2021. To achieve an equilibrium between subject homogeneity and generalizability of the findings, this research comprised only government hospitals. Universal sampling was utilsed as the sample technique.

3.1. Facebook Data

WebHarvy Scraping Software (SysNucleus, Kochi, India) was used to gather data on Facebook reviews from the official Facebook pages of public hospitals in Malaysia from January 2017 to December 2019. First, via the Ministry of Health official website, any webpage link of a public hospital website was sought to be identified. Then a link to the hospital's official Facebook page inside the hospital's web page was sought. If there was no link to the hospital's official Facebook page on the hospital's website, the search was continued on the Facebook platform. When an official hospital Facebook page was discovered, the information was confirmed by utilising the hospital's official website's URL, contacting hospital officials, or using this study's operational definition for a legitimate hospital Facebook page. An 'official hospital Facebook page' was defined as one with a 'verified tick' [64] or one with the hospital's official name (RASMI in the Malay language) included in the Facebook page's name or in the description of the site. All data gathered from the official Facebook page was kept in a pro forma checklist. The Facebook accounts of hospital departments, health institutions/agencies (such as the Ministry of Health (MOH) or the Institute of Medical Research), non-governmental organizations (NGOs) and long-term care facilities were omitted. These methods of searching have also been used in previous studies [23,24,64]. Malaysia is a multilingual country with a rich variety of languages and dialects. Malay is the national language, while English is the second language. Therefore, reviews were gathered in only those languages. To guarantee that the data language was appropriate and standardised for analysis, a group of junior doctors examined and corrected any spelling and grammatical errors in online reviews written in Malay and English. Then, data in Malay language were manually translated into English for further research by junior doctors. All data were kept in a local database that was encrypted and accessible only to the research team.

3.2. Machine Learning Topics Classification

To serve as a "gold standard" for machine learning classifiers, a labeled data set was generated through manual coding. The categorisation was based on the five-dimensional SERVQUAL theoretical notion [8,65]. These categories were: (1) tangible—the appearance of physical facilities, equipment, and healthcare personnel; (2) reliability—the ability to perform the promised services accurately and reliably; (3) responsiveness—the willingness to assist the customer and provide prompt service; (4) assurance—the employee's knowledge and courtesy, as well as their ability to inspire trust and confidence; and (5) empathy—the ability to empathise with the customer. Two hospital quality managers or SERVQUAL domain experts were assigned to perform initial "open" coding on batches of three hundred Facebook reviews based on the MOH SERVQUAL patient satisfaction survey and other SERVQUAL surveys from previous studies aimed at establishing the source of the coding standard. Intercoder reliability was then determined using a randomly chosen subsample of three hundred Facebook reviews. The raters separately coded the reliability subsample. Inter-rater agreement was determined using Cohen's Kappa (k) values for each SERVQUAL dimension. The agreement between the coding of tangible (Cohen's k = 0.885, $p < 0.001$), empathy (Cohen's k = 0.875, $p < 0.001$), reliability (Cohen's k = 0.736, $p < 0.001$), and responsiveness (Cohen's k = 0.72, $p < 0.001$) was high, but the agreement for assurance (Cohen's k = 0.626, $p < 0.001$) was moderate. Cohen's k coefficient was 0.769 on average in all dimensions. The machine learning classifier was then trained on a sample of nine hundred manually labelled Facebook reviews.

The machine learning technique analysed the characteristics of the individual phrases used in the Facebook reviews, and used this data to build a topic classifier. First, the labeled dataset was pre-processed to remove URLs, numerals, punctuation marks, stop words and simplifying words using a lemmatization technique (e.g., treating as a treat). Following that, the weights of terms were calculated using the term frequency-inverse document frequency (TF-IDF) approach, which demonstrated their significance to the documents and corpus. Figure 1 explains the Natural Language Processing (NLP) techniques used in the text preprocessing phase.

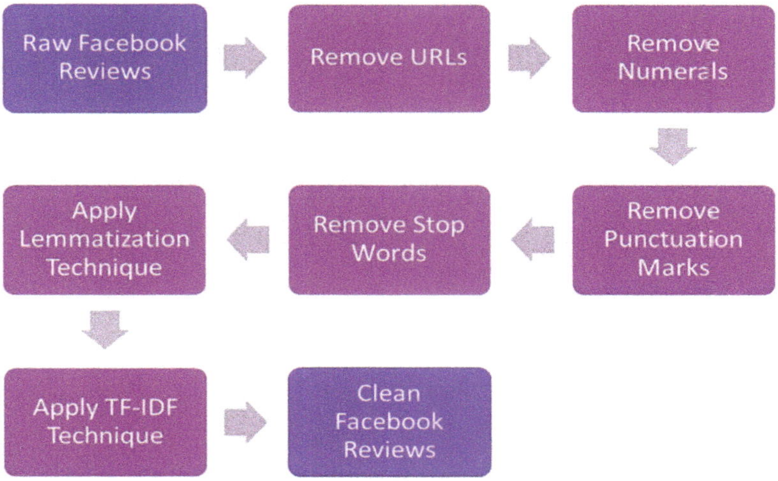

Figure 1. Text Preprocessing using natural language processing (NLP) techniques.

Iterative stratification was used to divide randomly labelled data into 80% for training and 20% for testing. Several multi-label classifier techniques were trained for topic classification, including binary relevance, label powerset, classifier chains, RAkEL (Random k-labELsets), MLkNN (multi-label k-Nearest Neighbor), and BRkNN (Binary Relevance k-NN). For each method, three main classifiers were trained: naive Bayes (NB), support vector machine (SVM), and logistic regression (LR). These classifiers are all widely used methods and have been shown to perform well on text classification tasks [29,31,66]. Multiple label classifiers were evaluated using the scikit-multilearn module in Python [67]. Finally, the various classifiers were evaluated using 5-fold cross-validation.

The 5-fold cross-validation revealed that the machine learning algorithms' F1-score performance varied between 0.69 and 0.76, suggesting that the models accurately classified the reviews. When different models and classifiers were compared, it was shown that the SVM model with classifier chains multi-label method had the highest accuracy (0.215) and F1-score (0.757). Additionally, the model had the lowest hamming loss (0.273). Hamming loss is a key performance metric in topic classification models since it measures the percentage of erroneous projected class labels. As a consequence, the machine learning classifier was trained using the chains classifier technique on the SVM model. The performance metrics for supervised machine learning with 5-fold cross-validation are summarised in Table 1. The proposed methodology general architecture is depicted in Figure 2.

Table 1. Overall ML models performance with 5-fold cross-validation.

Multilabel Classifier	Model	Accuracy	Recall	Precision	F1-Score	Hamming Loss
Binary Relevance	NB	0.147	0.761	0.701	0.730	0.315
	SVM	0.211	0.763	0.745	0.754	0.278
	LR	0.193	0.775	0.732	0.753	0.285
Label Powerset	NB	0.130	0.896	0.633	0.741	0.349
	SVM	0.166	0.799	0.679	0.734	0.323
	LR	0.158	0.825	0.669	0.739	0.326
Chains Classifier	NB	0.149	0.756	0.705	0.730	0.313
	SVM	0.215	0.761	0.753	0.757	0.273
	LR	0.191	0.770	0.727	0.748	0.290
RAkEL	NB	0.157	0.749	0.699	0.722	0.322
	SVM	0.186	0.764	0.724	0.743	0.295
	LR	0.180	0.765	0.726	0.745	0.293
MLkNN	N/A	0.140	0.737	0.697	0.715	0.327
BRkNN	N/A	0.157	0.648	0.732	0.687	0.330

NB, naive Bayes; SVM, support vector machine; LR, logistic regression.

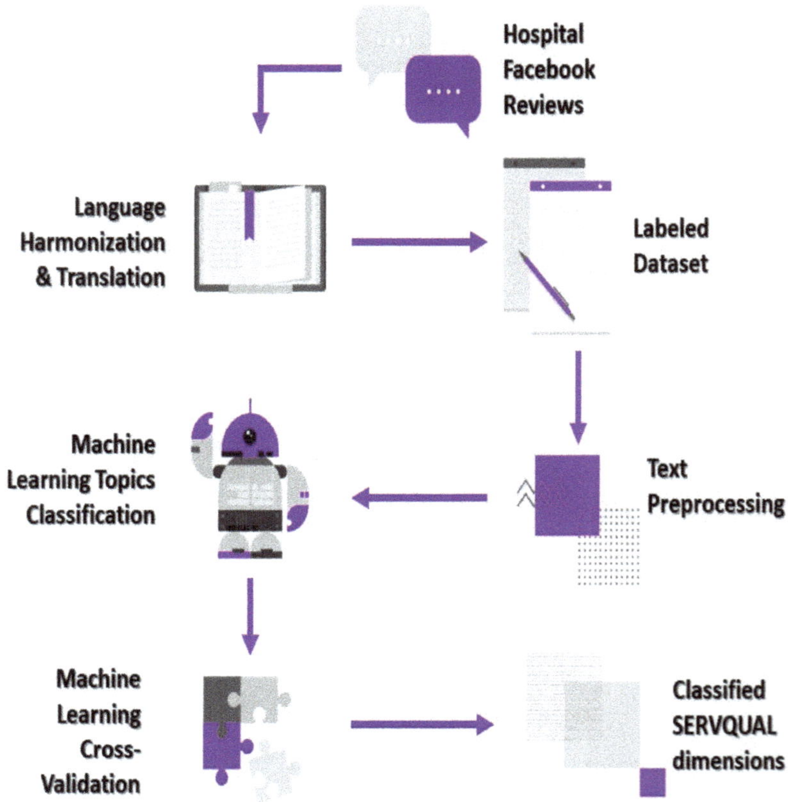

Figure 2. General architecture of proposed methodology in this study.

3.3. Outcome: Patient Dissatisfaction

Facebook review is a feature that allows people to leave narrative reviews on organisations' and companies' Facebook profiles. Since its debut in 2013, the Facebook review section has been included into the Facebook pages of many hospitals. Patients and their relatives have gradually begun to make use of it. Previously, Facebook utilised a five-star rating system until early 2018, when it switched to a binary rating system named "Recommends" or "Doesn't Recommend." This simplified the review process for users. As is the case with other social media platforms, Facebook ratings provide insight on how people feel about healthcare services. Customer recommendations were collected from hospital Facebook pages to determine patient satisfaction. Patient dissatisfaction was characterised as non-recommendation in the Facebook Review section, and patient satisfaction as recommendation. Any recommendation made outside of the Facebook review area was ignored.

3.4. Statistical Analysis

Due to the non-normal distribution of the data, medians (interquartile range [IQR]) were used for numerical data, and frequencies and percentages for categorical variables in the statistical analysis. Binary logistic regression analysis was used to evaluate the associations between patient dissatisfaction and multiple factors. Confounding variables included hospital characteristics (region, bed count, urban or rural location, and type of hospital), as well as Facebook page characteristics such as previous star ratings, acceptable hospital information on the Facebook page, and administrator reaction in the Facebook review area. These characteristics, according to previous research, were linked with patient satisfaction [12]. The data were examined to determine whether findings were statistically significant with a p value less than 0.05. All statistical tests were verified and found to be valid. Hosmer and Lemeshow tests were used to verify the model fitness, as well as the area under the receiver operating characteristic (ROC) curve. SPSS software version 26 was used to analyse the data (IBM Corp, Armonk, NY, USA).

4. Results

4.1. Hospital and Facebook Characteristics

In Malaysia, 63.7% of the 135 public hospitals have a Facebook page, with 48 of them accepting customer feedback through Facebook Review. Except for the western part of Malaysia, every region has at least 10 hospitals with a Facebook review function: 37.5% of tertiary hospitals, 8.3% of secondary hospitals, and 54.2% of primary hospitals all have Facebook review sections. The majority of these hospitals are located in cities, with an average of 730 beds. The average number of reviews on each hospital's Facebook page was 15.5 (27.5), with a previous star rating of 5.00 (1.65).

4.2. Facebook Reviews and Patient Satisfaction

A total of 3025 Facebook reviews were collected, with 1200 being used for machine learning training and the rest for association analysis. More Facebook reviews were seen at hospitals in the western (50.5%) and northern (21.5%) areas. Furthermore, urban hospitals accounted for 87.2% of all assessments, tertiary institutions for 88.8%, and the median bed count was 730. The average previous star rating on Facebook in terms of Facebook characteristics was 4.70 (1.5). The majority of Facebook reviews provided sufficient information about the hospital yet received little to no response from hospital management. Most notably, this study discovered that 73.5% were satisfied with the public hospital service, whereas 26.5% were dissatisfied. Table 2 describes hospital Facebook review characteristics.

Table 2. Hospital Facebook review characteristics (n = 1825).

Variable		n	(%)	Median	(IQR)
Hospital Features					
Region	East Coast	189	(10.4)		
	North	393	(21.5)		
	West	922	(50.5)		
	South	178	(9.8)		
	East Malaysia	143	(7.8)		
Location	Rural	234	(12.8)		
	Urban	1591	(87.2)		
Hospital Type	Primary	125	(6.8)		
	Secondary	80	(4.4)		
	Tertiary	1620	(88.8)		
Beds				730	(563)
Facebook Features					
Previous Facebook Star Ratings				4.70	(1.5)
Admin Response	No	1651	(90.5)		
	Yes	174	(9.5)		
Adequate Hospital Information	No	1651	(90.5)		
	Yes	174	(9.5)		
Patient Satisfaction	Dissatisfied	483	(26.5)		
	Satisfied	1342	(73.5)		

4.3. Classification of SERVQUAL Dimensions

Using the machine learning topics classification, there were 13.2% reviews with a tangible dimension, 68.9% reviews of reliability, 6.8% reviews of responsiveness, 19.5% reviews of assurance, and 64.3% reviews of empathy. The overall SERVQUAL dimensions are presented in Figure 3.

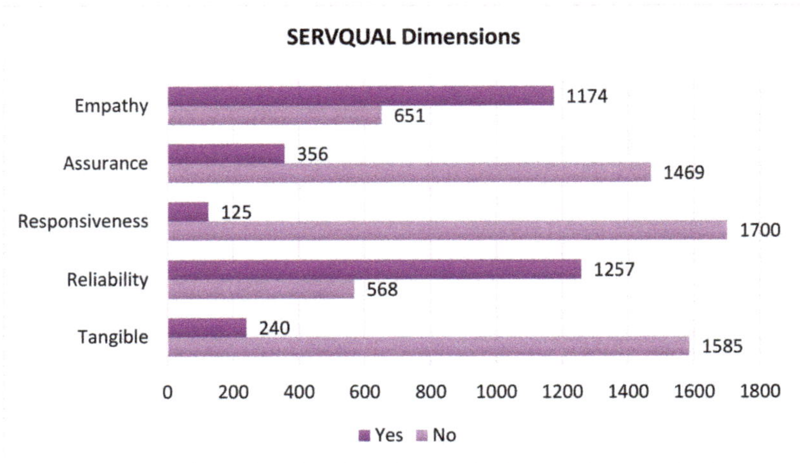

Figure 3. SERVQUAL dimensions classified by machine learning classifier (n = 1825).

4.4. Factors Associated with Patient Dissatisfaction

To assist MOH and key stakeholders in identifying areas for improvement, binary logistic regression was utilised, with patient dissatisfaction as the primary outcome. When compared with East Malaysia, a univariate study of hospital variables indicated that the three regions were related with patient dissatisfaction: West Coast (Crude OR = 2.11; 95% CI: 1.35–3.30; p = 0.001), East Coast (Crude OR = 0.63; 95% CI: 0.41–0.96; p = 0.031), and South (Crude OR = 2.38; 95% CI: 1.49–3.80; p = 0.001). In addition, patient dissatisfaction was linked to rural hospitals (Crude OR = 1.87; 95% CI: 1.40–2.49; $p < 0.001$) and tertiary hospitals (Crude OR = 0.65; 95% CI: 0.44–0.96; p = 0.030). Moreover, a relationship was discovered between previous Facebook star ratings and patient dissatisfaction (Crude OR = 0.86; 95% CI: 0.80–0.93; $p < 0.001$). Reliability (Crude OR = 1.52; 95% CI: 1.20–1.92; p = 0.001), responsiveness (Crude OR = 2.10; 95% CI: 1.45–3.04; p = 0.001), and empathy (Crude OR = 1.57; 95% CI:1.25–1.97; p = 0.001) were all significantly associated with patient dissatisfaction. The univariate study of hospital and Facebook features, as well as SERVQUAL in relation to patient dissatisfaction, is summarised in Table 3.

Table 3. Factors associated with patient dissatisfaction in univariate analysis (n = 1825).

Variables		Crude OR	95% CI	*p*-Value *
			(Lower, Upper)	
Hospital Features				
Region	East Malaysia	Ref		
	East Coast	0.63	0.41, 0.96	0.031
	North	1.08	0.75, 1.55	0.695
	West	2.11	1.35, 3.30	0.001
	South	2.38	1.49, 3.80	<0.001
Location	Urban	Ref		
	Rural	1.87	1.40, 2.49	<0.001
Hospital Type	Primary	Ref		
	Secondary	0.97	0.54, 1.76	0.924
	Tertiary	0.65	0.44, 0.96	0.030
Beds		1.00	1.00, 1.00	0.275

Table 3. *Cont.*

Variables		Crude OR	95% CI (Lower, Upper)	p-Value *
Facebook Features				
Admin Response to Review	No	Ref		
	Yes	1.24	0.88, 1.75	0.210
Adequate Hosp Info	No	Ref		
	Yes	0.80	0.53, 1.22	0.306
Facebook Star Ratings		0.86	0.80, 0.93	<0.001
SERVQUAL				
Tangible	No	Ref		
	Yes	1.25	0.93, 1.69	0.137
Reliability	No	Ref		
	Yes	1.52	1.20, 1.92	0.001
Responsiveness	No	Ref		
	Yes	2.10	1.45, 3.04	<0.001
Assurance	No	Ref		
	Yes	0.96	0.74, 1.25	0.766
Empathy	No	Ref		
	Yes	1.57	1.25, 1.97	<0.001

* Simple logistic regression.

In multivariate analysis, variables with a p-value less than 0.25 in univariate analysis were chosen throughout the model selection phase. Forward LR, backward LR, and manual selection methods were used to create a parsimonious model. The final model included hospital location and SERVQUAL dimensions other than tangible and assurance. When chosen SERVQUAL dimensions were controlled, hospitals situated in rural areas had a 100% higher likelihood of patient dissatisfaction compared with hospitals located in urban areas (95% CI:1.49–2.68; $p < 0.001$). Most importantly, when other variables were adjusted, reliability had a 113% higher likelihood of patient dissatisfaction (95% CI: 1.63–2.78; $p < 0.001$), responsiveness had a 61% higher likelihood of patient dissatisfaction (95% CI:1.09–2.38; $p = 0.016$), and empathy had a 108% higher likelihood of patient dissatisfaction (95% CI:1.63–2.69; $p < 0.001$). There was no interaction and multicollinearity in the multivariate model. The model's fitness was also satisfactory, as verified by the Hosmer and Lemeshow Test ($p = 0.875$), 73.5% of the classification table, and 61.7% of the area under the receiver operating characteristic (ROC) curve ($p < 0.001$). Table 4 details the multivariate analysis.

Table 4. Factors associated with patient dissatisfaction in multivariable analysis ($n = 1825$).

Variable		Adjusted OR	Adjusted 95% CI (Lower, Upper)	p-Value *
Location	Urban	Ref		
	Rural	2.00	1.49, 2.68	<0.001
Reliability	No	Ref		
	Yes	2.13	1.63, 2.78	<0.001
Responsive	No	Ref		
	Yes	1.61	1.09, 2.38	0.016
Empathy	No	Ref		
	Yes	2.08	1.61, 2.69	<0.001

* Multiple logistic regression, constant = −2.180, forward LR, backward LR and manual selection methods were applied, no significant interaction or multicollinearity. Hosmer and Lemeshow test = 0.875, classification table = 73.5%, area under the operating curve (ROC) = 61.7% ($p < 0.001$).

5. Discussion

POR influences patient preferences, emphasising the critical role of patient-centered health care and changing the system. The research is a critical first step in developing a strategy for utilising social media data in Malaysia, as well as a first effort to monitor public views of healthcare services using a novel data source. This is the first study to use automated computer methods to assess topics from online hospital evaluations and to characterise the content of narrative online hospital reviews in Malaysia. According to the machine learning classifier, the SERVQUAL dimension with the greatest frequency was reliability, followed by empathy. The reliability dimension was often concerned with appointment scheduling, punctuality, the healthcare system's efficacy, and the capability to keep accurate data.

Meanwhile, the problem of empathy related specifically to staff attention and helpfulness, an understanding of patient requirements, convenient hospital hours, and a commitment to the patient's best interests. These findings supported previous studies indicating that online reviews often emphasise time promise, healthcare system efficiency, and interpersonal quality [11,12,28,46,50]. However, additional topics were identified in the POR as major concerns, including communication, therapeutic effectiveness and patient safety, the environment, and hospital costs [13,38,52,53]. Moreover, most online patients reported satisfaction with the treatments provided by Malaysian hospitals. The findings supported comprehensive studies of patient online evaluations, which showed that the majority of patients were satisfied with their healthcare providers and would recommend them to family and friends [16,68].

Patient satisfaction surveys assist health care workers in identifying opportunities for service improvement. Additionally, they enable authorities to understand patient needs and create strategic plans for more effective and high-quality services [35]. This study found that hospital characteristics such as location in the western and southern regions, as well as rural locations, were associated with patient dissatisfaction. This was supported by African research [39], despite the fact that an Asian survey found rural residents to be generally satisfied with healthcare services [40]. Additionally, the size and type of hospital services had an effect on patient satisfaction [15,41]. Previously, it was believed that people would be more unhappy with a service that dealt with a greater number of patients and a bigger practice. However, this study found a negative correlation between tertiary centre and patient dissatisfaction, suggesting that patients were pleased with the service given by bigger types of hospitals, owing to the comprehensive healthcare services provided.

Interpersonal skills (empathy) were shown to be a major factor in increased patient satisfaction [35,69,70]. In this study, the empathy component was shown to be positively associated with patient dissatisfaction. The finding was confirmed by a social media study performed in China [13] and research conducted on the NHS Choices website [71], both of which revealed further negative comments regarding the doctor–patient connection, nurse service, roughness, and apathy. Moreover, a comparative study of POR in China and the United States found that the majority of complaints addressed the doctor's or hospital staff's bedside demeanour [51]. However, data from NHS Twitter showed that patients expressed a high degree of satisfaction with the empathy component of healthcare [53]. Physicians and nurses were assessed on their interactions with patients and their family or friends, including their friendliness, honesty, concern, compassion, empathy, kindness, civility, and respect for patient preferences [35,70]. Patients who were satisfied with physicians' affective behaviours were more likely to recommend them to others, according to research performed at a Scottish NHS trust [72].

Another area in which Malaysian public hospitals might improve is their reliability. A positive and statistically significant relationship was found between reliability and patient dissatisfaction in public hospitals. It is unsurprising that the majority of patient complaints or dissatisfaction voiced through POR concerned time commitment, appointment or follow-up access, and service inefficiencies [12,13,28,46,51]. Patient satisfaction was positively linked with ease of access to the hospital, convenient location, a streamlined admission and discharge procedure, and an efficient appointment system [35]. According to one study, scheduling convenience and adequate follow-up may help reduce patient dissatisfaction [54]. Additionally, local research has shown that the "lean" strategy may be effectively utilised to improve hospital reliability [73].

Responsiveness was defined as the willingness of healthcare professionals and providers to assist and give timely service to clients. A positive and statistically significant connection was found between responsiveness and patient dissatisfaction. Similar findings have been reported in earlier local research [74,75] as well as in international SERVQUAL studies [10,76]. Additionally, experimental research of the perceived SERVQUAL model using tweets from the NHS UK found that people expressed their dissatisfaction with responsiveness more than with other elements [53]. Patient satisfaction was shown to be positively linked with reduced wait times and quick treatment in a systematic study [35]. A comprehensive study showed that a wait time of more than 17 min decreased the probability of obtaining a good rating status [54].

Although this research discovered no significant connections between assurance and tangible dimensions with patient dissatisfaction, it is worth highlighting the dimensions' predictive value in POR. The quality of technical care was closely related to elements of assurance such as human competency, professionalism, and confidentiality [35]. Moreover, it pertained to the services' compliance with clinical diagnostic and treatment standards and recommendations. Numerous studies have found an association between assurance-related topics and patient satisfaction, including treatment effectiveness, diagnostic quality, and treatment side effects, utilising theme analysis of social media data [28,77]. Meanwhile, a study comparing POR in China and the United States found that both nations' citizens were dissatisfied with medical treatment [51]. Previously, it was thought that those who felt they had been treated unfairly were less satisfied with health care services. However, since some patients were unable to evaluate the technical quality of therapy due to their limited comprehension, they may have replaced their judgement for the sense of how nice and caring health professionals were toward them [35].

The physical environment was another important factor influencing patient satisfaction. Patient satisfaction was expected to be related to the pleasantness of the environment, cleanliness, noise level, food service, toilet comfort, clarity of signs and instructions, layout of equipment and facilities, and parking. Few studies have shown that patient satisfaction is influenced by attractive facilities, environmental cleanliness, and design-related factors [28,38,40,46]. However, further research showed that patients were unhappy with aspects of the hospital atmosphere based on their online assessments [46,53,61,69]. Malaysia's government has spent millions of ringgits in a series of Malaysia Plans aimed at enhancing public hospital facilities and services and building new hospitals [78]. As a result, hospital clients appreciate the upgrade and improvement of public hospital assets on social media.

These findings have a number of implications for many aspects of hospital quality of care. To begin, quality-of-care metrics and patient satisfaction can be monitored and evaluated in real time by using hospital Facebook reviews and machine learning algorithms. The method used in this study enables policymakers to make use of social media data rather than more expensive national questionnaire surveys. Moreover, there is no comparable open-standard research of patient satisfaction in Malaysia's public and private sectors. While the Ministry of Health prefers the SERVQUAL questionnaire, private hospitals may develop their own or adhere to an international standard. As a result, Facebook reviews may serve as a new barometer of patient satisfaction in each of these domains. Additionally, Facebook reviews are straightforward and accessible, reducing obstacles to obtaining information about hospital quality and helping hospitals in addressing quality-of-service problems while also alerting hospitals to possible patient safety concerns. While social media ratings are untested and unregulated, traditional patient satisfaction surveys have been validated and tested. By including additional hospital quality metrics on hospital

Facebook pages and critical information such as the official status of the Facebook site and the exact Facebook addresses, the validity of Facebook data will be increased [23].

Furthermore, this research has highlighted three SERVQUAL characteristics, namely reliability, responsiveness, and empathy, that need additional attention and improvement on the part of Malaysian healthcare authorities. Enhancing interpersonal skills training, especially for medical students, ongoing training for health professionals in the workplace, and lean model adaption will substantially enhance the quality of treatment that is currently lacking [79,80]. However, health authorities must realise that the findings are unlikely to be representative of the whole population served by hospitals. Rather than that, this study of service quality issues should be seen as a complement to more traditional data collection efforts and as an effective early warning system for hospital quality management.

Future Works and Limitations

Future study should concentrate on improving the efficacy of machine learning classifiers and collecting a bigger dataset of POR, including those from the Malaysian private sector. Second, further research is required to establish the relationship between POR and other hospital quality or clinical outcome measures, as earlier studies have done [11,12,43,63,81]. Additionally, future research may incorporate additional social media platforms (e.g., Twitter, Instagram, Tik-Tok, etc.) with specific adjustments such as a focus on the youth population (targeted audience), common public health topics discussed on social media platforms (depression, vaccination, cyberbullying, etc.), as well as identifying popular hashtags related to public health issues. The data collected from various social media platforms may offer healthcare agencies with a unique viewpoint on patients and may be utilised as a real-time public health surveillance system.

This research has a number of limitations. Due to the cross-sectional nature of the research, the possibility of a causal connection in our findings cannot be ruled out. Moreover, almost one-third of public hospitals posted feedback on Facebook. Incorporating unauthorised Facebook pages for public hospitals may have a contrasting impact. Additionally, the research dataset is considered small-scale in comparison to other POR research, due to Malaysia's small population and the relatively recent adoption of POR in the Malaysian healthcare sector. Malaysians, on the other hand, have a high rate of internet usage, which continues to grow year after year, thus a surge of POR about healthcare services may be expected over the next few years. Additionally, the main limitation was the time needed for content analysis and manual coding. Comprehensive reading and classification of datasets remains the gold standard for building machine learning-based topic classifiers and is the only way to ensure that all essential comments are coded [29]. However, it is time consuming, and in text classification, increasing the diversity of comments lowers the ability of the machine learning system to properly recognise the remark. However, if social media input becomes more prevalent, manual coding may become problematic owing to time constraints, and topic modelling may be a viable alternative. Topic modelling using Latent Dirichlet Allocation (LDA) may aid in determining how well the results fit the themes chosen by domain experts, and this unsupervised approach will allow the identification of previously undiscovered topics [82].

6. Conclusions

Patient online reviews offer healthcare authorities a practical, low-cost, and accessible way of collecting information about the quality of care they deliver. Healthcare officials have long considered how to include POR into citizen-government engagement and policymaking in order to create evidence-based reporting. Despite scholars' focus on the potential for POR data to assist in decision making, methods for realising this potential have been very restricted, often fragmentary, and non-standardised. This research suggested a systematic method for integrating POR data in order to analyse and monitor patient perceptions of the service quality at Malaysian public hospitals. Automatically classifying Facebook reviews into SERVQUAL dimensions using machine learning minimised human interference and selection bias in the study. Classification performance was verified, with an emphasis on the criticality of collecting reliable quality of care topic sets using the SERVQUAL model, and used to grasp the context of Facebook reviews. Despite the fact that the majority of POR were found to be satisfied with the hospital service, this study highlighted SERVQUAL dimensions of reliability, responsiveness, and empathy as areas for quality-of-care improvement in Malaysian public hospitals. Additionally, public hospital service in rural areas was associated with patient dissatisfaction. The results provide important insights that will aid healthcare officials and authorities in capitalising on the opportunities of POR by monitoring and assessing services' quality in order to make rapid improvements. Furthermore, the findings of traditional patient satisfaction surveys may be routinely supplemented with data from POR to continually improve and create high-quality healthcare services.

Author Contributions: Conceptualization, M.I.I. and K.I.M.; data curation, A.I.A.R.; formal analysis, A.I.A.R., S.-L.C. and N.M.Y.; funding acquisition, M.I.I.; investigation, A.I.A.R. and N.M.Y.; methodology, K.I.M., S.-L.C. and N.M.Y.; resources, K.I.M. and S.-L.C.; software, K.I.M. and S.-L.C.; supervision, M.I.I. and K.I.M.; validation, M.I.I., K.I.M., S.-L.C. and N.M.Y.; writing—original draft, A.I.A.R.; writing—review and editing, M.I.I., K.I.M., S.-L.C. and N.M.Y. All authors have read and agreed to the published version of the manuscript.

Funding: This study was funded by Fundamental Research Grant Scheme (2020), project code: FRGS/1/2020/SKK04/USM/02/3, Ministry of Higher Education, Malaysia, grant number 203/PPSP/6171293.

Institutional Review Board Statement: Ethical clearance was obtained from the Ethical and Research Committee Review of Universiti Sains Malaysia [32], code: USM/JEPeM/19120839.

Informed Consent Statement: Informed consent was not applicable for the current study because it does not involve humans.

Data Availability Statement: The Facebook data presented in this study are available on request from the corresponding author. The data are not publicly available due to privacy.

Acknowledgments: Thank you for the support given by the Ministry of Health, especially the Patient Satisfaction Unit of the Medical Development Division and Malaysian Society for Quality in Health (MSQH). The authors would like to express their gratitude to Nur Alia Binti Anuar for her important contribution to the data analysis. Additionally, the authors would like to express their gratitude to Universiti Sains Malaysia for providing a venue for this study.

Conflicts of Interest: The authors declare no conflict of interest.

References

1. WHO. *Delivering Quality Health Services: A Global Imperative for Universal Health Coverage*; World Health Organization; Organisation for Economic Co-Operation and Development; The World Bank: Geneva, Switzerland, 2018; ISBN 978-92-64-30030-9.
2. Gardner, J.W.; Linderman, K.W.; McFadden, K.L. Managing Quality Crossroads in Healthcare: An Integrative Supply Chain Perspective. *Qual. Manag. J.* **2018**, *25*, 2–17. [CrossRef]
3. Kruk, M.E.; Gage, A.D.; Joseph, N.; Danaei, G.; Garcia-Saiso, S.; Salomon, J.A. Mortality due to low-quality health systems in the universal health coverage era: A systematic analysis of amenable deaths in 137 countries. *Lancet* **2018**, *392*, 2203–2212. [CrossRef]
4. Fung, C.H.; Lim, Y.-W.; Mattke, S.; Damberg, C.; Shekelle, P.G. Systematic Review: The Evidence That Publishing Patient Care Performance Data Improves Quality of Care. *Ann. Intern. Med.* **2008**, *148*, 111–123. [CrossRef]
5. Lagu, T.; Goff, S.L.; Hannon, N.S.; Shatz, A.; Lindenauer, P.K. A Mixed-Methods Analysis of Patient Reviews of Hospital Care in England: Implications for Public Reporting of Health Care Quality Data in the United States. *Jt. Comm. J. Qual. Patient Saf.* **2013**, *39*, 7–15. [CrossRef]
6. Al-Abri, R.; Al-Balushi, A. Patient Satisfaction Survey as a Tool towards Quality Improvement. *Oman Med. J.* **2014**, *29*, 3–7. [CrossRef]
7. Ladhari, R. A review of twenty years of SERVQUAL research. *Int. J. Qual. Serv. Sci.* **2009**, *1*, 172–198. [CrossRef]
8. Parasuraman, A.; Zeithaml, V.A.; Berry, L.L. A Conceptual Model of Service Quality and Its Implications for Future Research. *J. Mark.* **1985**, *49*, 41–50. [CrossRef]
9. Alanazi, M.R.; Alamry, A.; Al-Surimi, K. Validation and adaptation of the hospital consumer assessment of healthcare providers and systems in Arabic context: Evidence from Saudi Arabia. *J. Infect. Public Health* **2017**, *10*, 861–865. [CrossRef] [PubMed]

10. Shafiq, M.; Naeem, M.A.; Munawar, Z.; Fatima, I. Service Quality Assessment of Hospitals in Asian Context: An Empirical Evidence from Pakistan. *Inq. J. Health Care Organ. Provis. Financ.* **2017**, *54*, 0046958017714664. [CrossRef] [PubMed]
11. Greaves, F.; Laverty, A.A.; Cano, D.R.; Moilanen, K.; Pulman, S.; Darzi, A.; Millett, C. Tweets about hospital quality: A mixed methods study. *BMJ Qual. Saf.* **2014**, *23*, 838–846. [CrossRef] [PubMed]
12. Hawkins, J.B.; Brownstein, J.S.; Tuli, G.; Runels, T.; Broecker, K.; Nsoesie, E.O.; McIver, D.J.; Rozenblum, R.; Wright, A.; Bourgeois, F.T.; et al. Measuring patient-perceived quality of care in US hospitals using Twitter. *BMJ Qual. Saf.* **2016**, *25*, 404–413. [CrossRef]
13. Hu, G.; Han, X.; Zhou, H.; Liu, Y. Public Perception on Healthcare Services: Evidence from Social Media Platforms in China. *Int. J. Environ. Res. Public Health* **2019**, *16*, 1273. [CrossRef] [PubMed]
14. Chakraborty, S.; Church, E.M. Social media hospital ratings and HCAHPS survey scores. *J. Health Organ. Manag.* **2020**, *34*, 162–172. [CrossRef] [PubMed]
15. Geletta, S. Measuring patient satisfaction with medical services using social media generated data. *Int. J. Health Care Qual. Assur.* **2018**, *31*, 96–105. [CrossRef]
16. Hong, Y.A.; Liang, C.; Radcliff, T.A.; Wigfall, L.T.; Street, R.L. What Do Patients Say About Doctors Online? A Systematic Review of Studies on Patient Online Reviews. *J. Med. Internet Res.* **2019**, *21*, e12521. [CrossRef] [PubMed]
17. Placona, A.M.; Rathert, C. Are Online Patient Reviews Associated with Health Care Outcomes? A Systematic Review of the Literature. *Med. Care Res. Rev.* **2021**. [CrossRef] [PubMed]
18. Farsi, D. Social Media and Health Care, Part I: Literature Review of Social Media Use by Health Care Providers. *J. Med. Internet Res.* **2021**, *23*, e23205. [CrossRef] [PubMed]
19. Giustini, D.M.; Ali, S.M.; Fraser, M.; Boulos, M.K. Effective uses of social media in public health and medicine: A systematic review of systematic reviews. *Online J. Public Health Inform.* **2018**, *10*, e215. [CrossRef] [PubMed]
20. Boon-Itt, S.; Skunkan, Y. Public Perception of the COVID-19 Pandemic on Twitter: Sentiment Analysis and Topic Modeling Study. *JMIR Public Health Surveill.* **2020**, *6*, e21978. [CrossRef]
21. Chu, W.-M.; Shieh, G.-J.; Wu, S.-L.; Sheu, W.H.-H. Use of Facebook by Academic Medical Centers in Taiwan during the COVID-19 Pandemic: Observational Study. *J. Med. Internet Res.* **2020**, *22*, e21501. [CrossRef]
22. Rahim, A.A.; Ibrahim, M.; Musa, K.; Chua, S.-L. Facebook Reviews as a Supplemental Tool for Hospital Patient Satisfaction and Its Relationship with Hospital Accreditation in Malaysia. *Int. J. Environ. Res. Public Health* **2021**, *18*, 7454. [CrossRef]
23. Bjertnaes, O.; Iversen, H.H.; Skyrud, K.D.; Danielsen, K. The value of Facebook in nation-wide hospital quality assessment: A national mixed-methods study in Norway. *BMJ Qual. Saf.* **2019**, *29*, 217–224. [CrossRef]
24. Campbell, L.; Li, Y. Are Facebook user ratings associated with hospital cost, quality and patient satisfaction? A cross-sectional analysis of hospitals in New York State. *BMJ Qual. Saf.* **2017**, *27*, 119–129. [CrossRef]
25. Hefele, J.G.; Li, Y.; Campbell, L.; Barooah, A.; Wang, J. Nursing home Facebook reviews: Who has them, and how do they relate to other measures of quality and experience? *BMJ Qual. Saf.* **2017**, *27*, 130–139. [CrossRef]
26. Lee, J.Y.; Gowen, C.R.; McFadden, K.L. An empirical study of U.S. hospital quality: Readmission rates, organizational culture, patient satisfaction, and Facebook ratings. *Qual. Manag. J.* **2018**, *25*, 158–170. [CrossRef]
27. Richter, J.P.; Kazley, A.S. Social media: How hospital facebook activity may influence patient satisfaction. *Health Mark. Q.* **2020**, *37*, 1–9. [CrossRef]
28. Zaman, N.; Goldberg, D.M.; Abrahams, A.S.; Essig, R.A. Facebook Hospital Reviews: Automated Service Quality Detection and Relationships with Patient Satisfaction. *Decis. Sci.* **2020**. [CrossRef]
29. Khanbhai, M.; Anyadi, P.; Symons, J.; Flott, K.; Darzi, A.; Mayer, E. Applying natural language processing and machine learning techniques to patient experience feedback: A systematic review. *BMJ Health Care Inform.* **2021**, *28*, e100262. [CrossRef] [PubMed]
30. Gohil, S.; Vuik, S.; Darzi, A. Sentiment Analysis of Health Care Tweets: Review of the Methods Used. *JMIR Public Health Surveill.* **2018**, *4*, e43. [CrossRef] [PubMed]
31. Zunic, A.; Corcoran, P.; Spasic, I. Sentiment Analysis in Health and Well-Being: Systematic Review. *JMIR Med. Inform.* **2020**, *8*, e16023. [CrossRef] [PubMed]
32. Yang, P.-C.; Lee, W.-C.; Liu, H.-Y.; Shih, M.-J.; Chen, T.-J.; Chou, L.-F.; Hwang, S.-J. Use of Facebook by Hospitals in Taiwan: A Nationwide Survey. *Int. J. Environ. Res. Public Health* **2018**, *15*, 1188. [CrossRef]
33. MCMC. *Internet Users Survey 2020, "IUS 2020"*; Malaysian Communications and Multimedia Commission: Cyberjaya, Malaysia, 2020; p. 160.
34. Swan, J.E.; Sawyer, J.C.; Van Matre, J.G.; McGee, G.W. Deepening the understanding of hospital patient satisfaction: Fulfillment and equity effects. *J. Health care Mark.* **1985**, *5*, 7–18.
35. Batbaatar, E.; Dorjdagva, J.; Luvsannyam, A.; Savino, M.M.; Amenta, P. Determinants of patient satisfaction: A systematic review. *Perspect. Public Health* **2016**, *137*, 89–101. [CrossRef]
36. Almasabi, M.; Yang, H.; Thomas, S. A Systematic Review of the Association between Healthcare Accreditation and Patient Satisfaction. *World Appl. Sci. J.* **2014**, *31*, 1618–1623.
37. Yunita, H.; Amal Chalik, S. Effect of Hospital Accreditation on Patient Safety Culture and Satisfaction: A Systematic Review. In Proceedings of the 6th International Conference on Public Health 2019, Solo, Indonesia, 23 October 2019; pp. 547–555.
38. Alkazemi, M.F.; Bayramzadeh, S.; Alkhubaizi, N.B.; Alayoub, A. The physical environment and patient satisfaction ratings on social media: An exploratory study. *Facilities* **2019**, *38*, 86–97. [CrossRef]

39. Yaya, S.; Bishwajit, G.; Ekholuenetale, M.; Shah, V.; Kadio, B.; Udenigwe, O. Urban-rural difference in satisfaction with primary healthcare services in Ghana. *BMC Health Serv. Res.* **2017**, *17*, 776. [CrossRef]
40. Liu, J.; Mao, Y. Patient Satisfaction with Rural Medical Services: A Cross-Sectional Survey in 11 Western Provinces in China. *Int. J. Environ. Res. Public Health* **2019**, *16*, 3968. [CrossRef]
41. Tang, L. The influences of patient's satisfaction with medical service delivery, assessment of medical service, and trust in health delivery system on patient's life satisfaction in China. *Health Qual. Life Outcomes* **2012**, *10*, 111. [CrossRef]
42. Draper, M.; Cohen, P.; Buchan, H. Seeking consumer views: What use are results of hospital patient satisfaction surveys? *Int. J. Qual. Health Care* **2001**, *13*, 463–468. [CrossRef]
43. Synan, L.T.; Eid, M.A.; Lamb, C.R.; Wong, S.L. Crowd-sourced hospital ratings are correlated with patient satisfaction but not surgical safety. *Surgery* **2021**, *170*, 764–768. [CrossRef]
44. Rahim, A.I.A.; Ibrahim, M.I.; Musa, K.I.; Chua, S.-L.; Yaacob, N.M. Assessing Patient-Perceived Hospital Service Quality and Sentiment in Malaysian Public Hospitals Using Machine Learning and Facebook Reviews. *Int. J. Environ. Res. Public Health* **2021**, *18*, 9912. [CrossRef]
45. Friedrich, S.; Groß, S.; König, I.R.; Engelhardt, S.; Bahls, M.; Heinz, J.; Huber, C.; Kaderali, L.; Kelm, M.; Leha, A.; et al. Applications of artificial intelligence/machine learning approaches in cardiovascular medicine: A systematic review with recommendations. *Eur. Hear. J.—Digit. Health* **2021**, *2*, 424–436. [CrossRef]
46. Doing-Harris, K.; Mowery, D.L.; Daniels, C.; Chapman, W.W.; Conway, M. Understanding patient satisfaction with received healthcare services: A natural language processing approach. *AMIA Annu. Symp. Proc.* **2016**, *2016*, 524–533.
47. Alemi, F.; Torii, M.; Clementz, L.; Aron, D.C. Feasibility of Real-Time Satisfaction Surveys through Automated Analysis of Patients' Unstructured Comments and Sentiments. *Qual. Manag. Health Care* **2012**, *21*, 9–19. [CrossRef]
48. Greaves, F.; Ramirez-Cano, D.; Millett, C.; Darzi, A.; Donaldson, L. Use of Sentiment Analysis for Capturing Patient Experience From Free-Text Comments Posted Online. *J. Med. Internet Res.* **2013**, *15*, e239. [CrossRef]
49. Bahja, M.; Lycett, M. Identifying patient experience from online resources via sentiment analysis and topic modelling. In Proceedings of the 3rd IEEE/ACM International Conference on Big Data Computing, Applications and Technologies, Shanghai, China, 6–9 December 2016; pp. 94–99.
50. Ranard, B.L.; Werner, R.M.; Antanavicius, T.; Schwartz, H.A.; Smith, R.J.; Meisel, Z.F.; Asch, D.; Ungar, L.H.; Merchant, R.M. Yelp Reviews Of Hospital Care Can Supplement And Inform Traditional Surveys of the Patient Experience of Care. *Health Aff.* **2016**, *35*, 697–705. [CrossRef]
51. Hao, H.; Zhang, K.; Wang, W.; Gao, G. A tale of two countries: International comparison of online doctor reviews between China and the United States. *Int. J. Med. Inform.* **2017**, *99*, 37–44. [CrossRef]
52. James, T.L.; Calderon, E.D.V.; Cook, D.F. Exploring patient perceptions of healthcare service quality through analysis of unstructured feedback. *Expert Syst. Appl.* **2017**, *71*, 479–492. [CrossRef]
53. Lee, H.J.; Lee, M.; Lee, H.; Cruz, R.A. Mining service quality feedback from social media: A computational analytics method. *Gov. Inf. Q.* **2021**, *38*, 101571. [CrossRef]
54. Ko, D.; Mai, F.; Shan, Z.; Zhang, D. Operational efficiency and patient-centered health care: A view from online physician reviews. *J. Oper. Manag.* **2019**, *65*, 353–379. [CrossRef]
55. Lin, Y.; Hong, Y.A.; Henson, B.S.; Stevenson, R.D.; Hong, S.; Lyu, T.; Liang, C. Assessing Patient Experience and Healthcare Quality of Dental Care Using Patient Online Reviews in the United States: Mixed Methods Study. *J. Med. Internet Res.* **2020**, *22*, e18652. [CrossRef]
56. Nawab, K.; Ramsey, G.; Schreiber, R. Natural Language Processing to Extract Meaningful Information from Patient Experience Feedback. *Appl. Clin. Inform.* **2020**, *11*, 242–252. [CrossRef]
57. Abirami, A.; Askarunisa, A. Sentiment analysis model to emphasize the impact of online reviews in healthcare industry. *Online Inf. Rev.* **2017**, *41*, 471–486. [CrossRef]
58. Huppertz, J.W.; Otto, P. Predicting HCAHPS scores from hospitals' social media pages: A sentiment analysis. *Health Care Manag. Rev.* **2018**, *43*, 359–367. [CrossRef]
59. Cole-Lewis, H.; Varghese, A.; Sanders, A.; Schwarz, M.; Pugatch, J.; Augustson, E. Assessing Electronic Cigarette-Related Tweets for Sentiment and Content Using Supervised Machine Learning. *J. Med. Internet Res.* **2015**, *17*, e208. [CrossRef]
60. Daniulaityte, R.; Chen, L.; Lamy, F.R.; Carlson, R.G.; Thirunarayan, K.; Sheth, A. "When 'Bad' is 'Good'": Identifying Personal Communication and Sentiment in Drug-Related Tweets. *JMIR Public Health Surveill.* **2016**, *2*, e162. [CrossRef]
61. Jung, Y.; Hur, C.; Jung, D.; Kim, M.; Ning, L.; Zimlichman, E. Identifying Key Hospital Service Quality Factors in Online Health Communities. *J. Med. Internet Res.* **2015**, *17*, e90. [CrossRef]
62. Lee, H.J.; Lee, M.; Lee, H. Tracking Social Perception on Healthcare Service Quality Using Social Media. In Proceedings of the Management Knowledeg and Learning International Conference 2018, Naples, Italy, 16–18 May 2018; p. 18.
63. Boylan, A.-M.; Turk, A.; van Velthoven, M.H.; Powell, J. Online patient feedback as a measure of quality in primary care: A multimethod study using correlation and qualitative analysis. *BMJ Open* **2020**, *10*, e031820. [CrossRef]
64. Moore, K.; Cottrell, E.; Chambers, R. Facebook in general practice: A service evaluation in one health economy. *BJGP Open* **2017**, *1*, 101181. [CrossRef]
65. Parasuraman, A.P.; Zeithaml, V.; Berry, L. SERVQUAL: A multiple- Item Scale for measuring consumer perceptions of service quality. *J. Retail.* **1988**, *64*, 12–40.

66. Bari, V.; Hirsch, J.S.; Narvaez, J.; Sardinia, R.; Bock, K.R.; Oppenheim, M.I.; Meytlis, M. An approach to predicting patient experience through machine learning and social network analysis. *J. Am. Med. Inform. Assoc.* **2020**, *27*, 1834–1843. [CrossRef] [PubMed]
67. Szymański, P.; Kajdanowicz, T. A Network Perspective on Stratification of Multi-Label Data. In Proceedings of the First International Workshop on Learning with Imbalanced Domains: Theory and Applications, Skopje, Macedonia, 22 September 2017; pp. 22–35.
68. Boylan, A.-M.; Williams, V.; Powell, J. Online patient feedback: A scoping review and stakeholder consultation to guide health policy. *J. Health Serv. Res. Policy* **2019**, *25*, 122–129. [CrossRef]
69. Ko, C.-H.; Chou, C.-M. Apply the SERVQUAL Instrument to Measure Service Quality for the Adaptation of ICT Technologies: A Case Study of Nursing Homes in Taiwan. *Healthcare* **2020**, *8*, 108. [CrossRef] [PubMed]
70. Tan, C.N.-L.; Ojo, A.; Cheah, J.-H.; Ramayah, T. Measuring the Influence of Service Quality on Patient Satisfaction in Malaysia. *Qual. Manag. J.* **2019**, *26*, 129–143. [CrossRef]
71. Brookes, G.; Baker, J.P. What does patient feedback reveal about the NHS? A mixed methods study of comments posted to the NHS Choices online service. *BMJ Open* **2017**, *7*, e013821. [CrossRef]
72. Jenkinson, C.; Coulter, A.; Bruster, S.; Richards, N.; Chandola, T. Patients' experiences and satisfaction with health care: Results of a questionnaire study of specific aspects of care. *Qual. Saf. Health Care* **2002**, *11*, 335–339. [CrossRef] [PubMed]
73. Tajudin, M.S.; Habidin, N.F. Lean Healthcare Practices Improve the Patient Performance in Public Hospitals. *Int. J. Acad. Res. Bus. Soc. Sci.* **2020**, *10*, 783–796. [CrossRef]
74. Aliman, N.K.; Mohamad, W.N. Perceptions of Service Quality and Behavioral Intentions: A Mediation Effect of Patient Satisfaction in the Private Health Care in Malaysia. *Int. J. Mark. Stud.* **2013**, *5*, 15. [CrossRef]
75. Kang, A.J. *Patients' Satisfaction towards the Healthcare Institutions Service Quality: A Comparison between Public and Private Hospitals in Klang Valley*; UTAR: Kampar, Malaysia, 2019.
76. Al-Neyadi, H.S.; Abdallah, S.; Malik, M. Measuring patient's satisfaction of healthcare services in the UAE hospitals: Using SERVQUAL. *Int. J. Health Manag.* **2018**, *11*, 96–105. [CrossRef]
77. Wagland, R.; Recio-Saucedo, A.; Simon, M.; Bracher, M.; Hunt, K.; Foster, C.; Downing, A.; Glaser, A.; Corner, J. Development and testing of a text-mining approach to analyse patients' comments on their experiences of colorectal cancer care. *BMJ Qual. Saf.* **2016**, *25*, 604–614. [CrossRef]
78. Bernama. MOH, Architects to Further Improve Hospital Infrastructure. Available online: https://www.astroawani.com/berita-malaysia/moh-architects-further-improve-hospital-infrastructure-health-dg-250636 (accessed on 7 August 2021).
79. Lawal, A.K.; Rotter, T.; Kinsman, L.; Sari, N.; Harrison, L.; Jeffery, C.; Kutz, M.; Khan, M.F.; Flynn, R. Lean management in health care: Definition, concepts, methodology and effects reported (systematic review protocol). *Syst. Rev.* **2014**, *3*, 103. [CrossRef] [PubMed]
80. Mata, Á.N.D.S.; de Azevedo, K.P.M.; Braga, L.P.; de Medeiros, G.C.B.S.; Segundo, V.H.D.O.; Bezerra, I.N.M.; Pimenta, I.D.S.F.; Nicolás, I.M.; Piuvezam, G. Training in communication skills for self-efficacy of health professionals: A systematic review. *Hum. Resour. Health* **2021**, *19*, 30. [CrossRef] [PubMed]
81. Li, Y.; Cai, X.; Wang, M. Social media ratings of nursing homes associated with experience of care and "Nursing Home Compare" quality measures. *BMC Health Serv. Res.* **2019**, *19*, 260. [CrossRef] [PubMed]
82. Kherwa, P.; Bansal, P. Topic Modeling: A Comprehensive Review. *ICST Trans. Scalable Inf. Syst.* **2018**, *7*, 159623. [CrossRef]

healthcare

MDPI

Article

The Application of Projection Word Embeddings on Medical Records Scoring System

Chin Lin [1,2,3,4,†], Yung-Tsai Lee [5,†], Feng-Jen Wu [6,†], Shing-An Lin [7], Chia-Jung Hsu [7], Chia-Cheng Lee [7,8], Dung-Jang Tsai [2,3,4,*] and Wen-Hui Fang [4,9,*]

1 School of Medicine, National Defense Medical Center, Taipei 114, Taiwan; xup6fup0629@gmail.com
2 School of Public Health, National Defense Medical Center, Taipei 114, Taiwan
3 Graduate Institute of Life Sciences, National Defense Medical Center, Taipei 114, Taiwan
4 Artificial Intelligence of Things Center, Tri-Service General Hospital, National Defense Medical Center, Taipei 114, Taiwan
5 Division of Cardiovascular Surgery, Cheng Hsin Rehabilitation and Medical Center, Taipei 112, Taiwan; andrewytlee.cvs@gmail.com
6 Department of Informatics, Taoyuan Armed Forces General Hospital, Taoyuan 325, Taiwan; army.afth@gmail.com
7 Department of Medical Informatics, Tri-Service General Hospital, National Defense Medical Center, Taipei 114, Taiwan; beeverything@hotmail.com (S.-A.L.); jayronhh@gmail.com (C.-J.H.); lcgnet@gmail.com (C.-C.L.)
8 Division of Colorectal Surgery, Department of Surgery, Tri-Service General Hospital, National Defense Medical Center, Taipei 114, Taiwan
9 Department of Family and Community Medicine, Department of Internal Medicine, Tri-Service General Hospital, National Defense Medical Center, Taipei 114, Taiwan
* Correspondence: oo800217@gmail.com (D.-J.T.); rumaf.fang@gmail.com (W.-H.F.); Tel.: +886-2-8792-3100 (ext. #18305) (D.-J.T.); +886-2-8792-3100 (ext. #12322) (W.-H.F.); Fax: +886-2-8792-3147 (D.-J.T. & W.-H.F.)
† C.L., Y.-T.L. and F.-J.W. contribute equally in the article.

Citation: Lin, C.; Lee, Y.-T.; Wu, F.-J.; Lin, S.-A.; Hsu, C.-J.; Lee, C.-C.; Tsai, D.-J.; Fang, W.-H. The Application of Projection Word Embeddings on Medical Records Scoring System. *Healthcare* **2021**, *9*, 1298. https://doi.org/10.3390/healthcare9101298

Academic Editor: Mahmudur Rahman

Received: 24 August 2021
Accepted: 28 September 2021
Published: 29 September 2021

Abstract: Medical records scoring is important in a health care system. Artificial intelligence (AI) with projection word embeddings has been validated in its performance disease coding tasks, which maintain the vocabulary diversity of open internet databases and the medical terminology understanding of electronic health records (EHRs). We considered that an AI-enhanced system might be also applied to automatically score medical records. This study aimed to develop a series of deep learning models (DLMs) and validated their performance in medical records scoring task. We also analyzed the practical value of the best model. We used the admission medical records from the Tri-Services General Hospital during January 2016 to May 2020, which were scored by our visiting staffs with different levels from different departments. The medical records were scored ranged 0 to 10. All samples were divided into a training set (n = 74,959) and testing set (n = 152,730) based on time, which were used to train and validate the DLMs, respectively. The mean absolute error (MAE) was used to evaluate each DLM performance. In original AI medical record scoring, the predicted score by BERT architecture is closer to the actual reviewer score than the projection word embedding and LSTM architecture. The original MAE is 0.84 $\pm$ 0.27 using the BERT model, and the MAE is 1.00 $\pm$ 0.32 using the LSTM model. Linear mixed model can be used to improve the model performance, and the adjusted predicted score was closer compared to the original score. However, the project word embedding with the LSTM model (0.66 $\pm$ 0.39) provided better performance compared to BERT (0.70 $\pm$ 0.33) after linear mixed model enhancement ($p < 0.001$). In addition to comparing different architectures to score the medical records, this study further uses a mixed linear model to successfully adjust the AI medical record score to make it closer to the actual physician's score.

Keywords: medical records scoring; projection word embedding; long short-term memory; bidirectional encoder representations from transformers; artificial intelligence; natural language processing; electronic health records

1. Introduction

With the increasing advancement of technology, the data amount generated by humans is growing explosively [1]. Effectively taking advantage of these growing data may bring valuable information, which many successful cases from different industries [2] have already proved. However, the majority of these data are not structured [3], which cannot be directly used by traditional analytical methods. At the same time, it is expected to employ new algorithms to use these data to allow for stronger decision-making capacity [4,5]. In recent years, with the breakthrough developments of the deep neural network in diverse fields, we are already capable of directly analyzing data in the forms of videos, texts, and voices. Hence, the focus of researches is now to develop applications to solve practical problems.

The medical system is an important field that is very suitable to develop the above-mentioned applications. Medical knowledge is accumulating quickly, making it more and more possible for doctors to have knowledge gaps [6], which may cause misdiagnoses and, thus, urgently need to be solved [7]. Computer-aided diagnosis systems have been greatly developed in recent years, aiming to solve this problem, yet unsuccessfully so far [8]. This is probably because the majority of medical data are non-structural data [9]; take cancer, for example, where about 96% of cancer diagnoses are made from pathological section reports, the data of which, however, are recorded in text descriptions and videos [10]. Thus, it is difficult for traditional models to link these original non-structural data with diagnosis information directly. With the advancement of artificial intelligence (AI) technology, the new generation of computer-aided diagnosis systems is expected to make great contributions to the intellectualization of medical systems. It can further eliminate human errors to increase the quality of medical care [11]. In 2012, AlexNet was the ILSVRC champion, leading the 3rd AI revolution [12]. Since then, more powerful deep learning models have been developed, such as VGGNet [13], Inception Net [14], ResNet [15], DenseNet [16], etc. This revolution led by deep learning has made enormous progress in image recognition tasks, driving breakthroughs in related research. Computer-aided diagnosis tools built based on deep learning technology have led to an increase in medical care quality [11]. Examples include lymph node metastasis detection [17], diabetic retinopathy detection [18], skin cancer classification [19], pneumonia detection [20], bleeding identification [21], etc. There have been over 300 studies (mostly in the last 2 years) using such technologies in medical image analysis [22]. It is worth mentioning that the most impressive capacity of deep learning technology is automatic feature extraction. With the precondition of a large database for annotation, it has been proven to reach, or even surpass, the level of human experts [15,23,24].

The current method to use a large amount of information from medical records is to code through recognition by experts and according to ICD (The International Statistical Classification of Diseases and Related Health Problems). This work is not only necessary for our national health insurance declaration system but may also be used in disease monitoring, hospital management, clinical studies, and policy planning. However, artificial classification is not only expensive but is also time-inefficient, which is the most important. For example, in disease monitoring, since the outbreak of infectious disease will cause large casualties [25], many countries have developed their disease monitoring systems specifically aiming at contagious diseases, such as the Real-time Outbreak and Disease Surveillance (RODS) system [26]. To ensure time efficiency, this system stipulates emergency physicians to input data within required time limits when identifying notifiable diseases, making it hard to be promoted to other diseases. With the advancement of data science, it has been universally expected that an automatic disease interpretation model can be developed to solve the high-cost and time-inefficient problems of artificial interpretation.

Due to the popularization of medical records electronization, a great number of studies have attempted to use this information for text mining and ICD code classification. The current technology primarily uses a bag-of-words model to standardize text medical records, then uses a support vector machine (SVM), random forest tree, and other classifiers for diagnosis classification [27–31]. However, previous studies have found that these methods were incapable of accurate diagnosis classification because of the particularity and diversity of clinical terms, where synonyms need to be properly processed before data preprocessing [10]. A complete medical dictionary integrates the currently recommended forms of clinical terms; yet, it is almost impossible due to the complexity of clinical terms. Therefore, traditional automatic classification programs can hardly make significant progress. In addition, the bag-of-words model treats different characters as different features and counts the number of features in one article. Although this makes it possible to use a dictionary to handle the synonym problem, similar characters would be considered two different features. Thus, the number of features integrated by the bag-of-words model will be strikingly huge, causing a curse of dimensionality when classified by subsequent classifiers, leading to inefficiency and slow progress of traditional algorithms.

Other than classification efficiency, the greatest challenge for traditional algorithms is new diseases. For instance, there was an H1N1 outbreak in 2009, with related cases that had never been recorded before 2008. Traditional classification algorithms are completely unable to perform proper classification of newly emerged words [27–31]. This disadvantage makes it absolutely impossible for traditional methods to reach full automation. Regarding this issue, we proposed word embedding as a technical breakthrough in disease classification. Since the 20th century, word embedding has been an important technology to allow computers to understand the semantic meaning further. Its core logic is hoping to characterize every single word into a vector in high-dimensional space and expecting similar vectors for similar characters/words to express semantic meaning [32,33]. The word2vec published by the Google team in 2013 is considered the most important breakthrough in recent word embedding studies. It has been verified to allow similar characters to have very high cosine similarity and very close Euclidean distance in vector space [34]. However, this technology has a disadvantage that, once applied, it converts an article into an unequal matrix, making it inapplicable for traditional classifiers, such as SVM and random forest trees. A general solution is to average or weighted average the word vector of all characters in an article as semanteme [35]. However, from the MultiGenre NLI (MultiNLI) Corpus competition release by the natural language research team of Stanford (https://nlp.stanford.edu/projects/snli/), we can still see that combining modern AI technology gives better efficiency to models. Language processing conducts analysis mostly based on Recurrent Neural Network (RNN) or Convolutional Neural Network (CNN). Its core principle is to use convolutional layer (does not have memory but can gradually integrate surrounding single-character information in higher-order features, requires more layers) or Long Short-Term Memory Unit (has short- and long-term memory, thus needing fewer layers) for feature extraction and is able to process information in matrix form [36]. CNN has become the primary method in all computer vision competitions. Its reason for success is a fuzzy matching technique of convolutional layer, allowing for integrating similar image features. We will be able to change the convolutional layer from recognizing similar image features to recognizing similar vocabularies through certain designs. Hence, CNN has been applied in text mining, such as semantic classification [37], short sentence searching [38], and chapter analysis [39], and has shown considerably good efficiency. In the most recent study, Bidirectional Encoder Representations from Transformers (BERT), developed by Google, has swept all kinds of natural language process competitions [40]. Yet, its core is still good work/sentence/paragraph embedding. Generally speaking, combining good embedding technology with modern deep learning neural networks is undoubtedly the best option for current natural language processing tasks.

Our team has already applied it in disease classification of discharge record summaries and proved that it compared with traditional models. AI model with combined word embedding model and CNN reduces 30% error rate in disease classification tasks, makes modeling easier by avoiding troublesome text integration preprocessing, and learns external language resources through unmonitored learning to integrate similarity among clinical clauses [41]. However, although the combination of word embedding and CNN is better in disease classification tasks than traditional methods, its accuracy still cannot be compared with humans. One of the reasons is the error in understanding the semantic meaning. Therefore, improving the word embedding model's understanding of the meaning of medical terms might increase its subsequent analytical efficiency [42]. There are two studies that have evaluated the application of word embedding models trained by different resources on biomedical NLP and found EHR-trained word embedding could better capture semantic property [43,44]. On the other hand, external data resources have a neglected advantage in that the vocabulary diversity of external internet data resources is far more than that of internal task database. This advantage will greatly affect real disease coding tasks. Hence, an embedded training process needs to be developed to maintain the vocabulary diversity of internet resources and medical terms' understanding of the internal task database. A recent word embedding comparison study showed that EHR-trained work embedding could usually better capture medical semantic meaning [43]. Even the research team of abroad Mayo Clinic uses an EHR with a large amount of data. The total number of words is only about 100,000, the vocabulary diversity of which is still far less than the external database [43,44]. This is due to the lack of some rare diseases and periodic diseases, such as the 2003 SARS outbreak and the 2009 H1N1 outbreak. Therefore, EHR-trained word embedding models are unable to include enough vocabulary. For this reason, our team developed a projection word embedding model that has the vocabulary diversity of Wikipedia/PubMed, as well as an understanding of medical terms in EHR [45].

A medical record is a historical record and also the foundation of a patient's medical care. It records the patient's conditions, reasons, results of examinations/tests, treatment methods, and results during care processes. It integrates and analyzes patients' related information, presents the executive ground of medical decisions, and even affects national health policy. The basic purpose of medical records is to remind oneself or other medical care colleagues of a patient's daily conditions and attending physician's current thoughts. When medical treatment is being performed, the medical record serves as the communication tool among physicians and means for continuous treatment. In other words, the medical record is the only text material that records a patient's conditions and focuses on all medical care personnel. A medical record is an index of medical care quality reflecting a physician's clinical thinking and diagnostic basis. It serves as the reference for learning, research, and education. Meanwhile, it also serves as the evidence for medical disputes to clarify the attribution of liabilities. The medical record is the foundation of patient care as it records the contents of patient care provided by medical personnel. Thus, all results obtained from observation or examination can be found on the medical record. Therefore, any change in the patient's condition can be found from the medical record so that the patient's current condition can be evaluated for suitable treatments. Moreover, communication with a patient should also be included in the medical record so that medical personnel can learn the patient's expectations on the treatment, resulting in a closer doctor-patient relationship. For other professionals, a detailed medical record saves a lot of communication time and avoids misunderstanding or missing the patient's previous conditions that may lead to mistreatment.

The content of medical records also has legal effects. It is the basis of insurance benefits and even affects national health policy. For example, public health studies usually need to include case information under national health insurance, and, through studying a large number of medical records, such studies can help public health researchers and medical officials to establish more suitable public health decisions and administrative rules that protect the rights and interests of both doctors and patients. Clinical decision-making guides formulated by many specialized medical associations also used information from medical records. The implicit demographic information from these medical records is also collected at the national level and published as national health demographic information to compare with other countries so as to serve as a way to communicate and learn from each other for mutual benefits.

In this study, as shown in the graphical abstract, a scoring database was established by experts performing scoring on medical records. An AI model was trained to learn experts' scoring logics so as to screen high-quality medical record summaries. In contrast, the database made up of which will have the chance to promote the establishment of other subsequent AI models, improve model accuracy, and serve as a teaching example to improve medical education efficiency.

2. Method

2.1. Data Source

In this study, inpatient medical records from Tri-Service General Hospital from 1 January 2016 to 31 December 2019 were used as the basic database, which was ethically approved by institutional review board (IRB NO. A202005104). Physicians of different levels from different departments were invited for medical records summary scoring. Scoring dimensions include different indexes, based on clinical writing standards, it contains 12 scoring items from each detailed structure of the QNOTE scale's inpatient record, including chief of complaint, history of the present illness, problem list, past medical history, medications, adverse drug reactions and allergies, social and family history, review of systems, physical findings, assessment, plan of care, and follow-up information. The completeness of each item's record, as well as the 5 structures (completeness, correctness, concordance, plausibility, and currency) of electronic medical records' examination information, are evaluated in 5 levels of the Likert scale: strongly disagree, disagree, no comment (not agree nor disagree), agree, and strongly agree. Specialists from different departments were required to review 227,689 medical records and preliminarily score them on a 10-point Likert scale based on the average of above 5 structures. These scores were then used as the training target of the AI model to represent medical record writing quality. All samples were divided into a training set (n = 74,959) and testing set (n = 152,730) based on time, and then they were evaluated by different departments. Data of the testing set was compared with the actual scores for analysis, and MAE from the Likert scale was used as the evaluation index for model performance. In the end, the aforementioned model was applied in Tri-Service General Hospital. A medical record auto-scoring system was established in the hospital so as to screen high-quality medical records for future teaching and research studies.

2.2. AI Algorithm

The collected medical records and various writing quality indicators can be used for artificial intelligence model training. The model architecture uses the word embedding and LSTM model developed by our team. The word embedding also uses the projection word embedding comparison table to perform single-character conversion mathematical vectors and uses the entire input article as the input matrix. We used projection word embedding to construct a deep convolutional network model to enable the network to integrate the transformed semantic vectors and extract written medical records based on different word combinations. First, we used the word embedding comparison table trained by Wikipedia and PubMed library, and then we used EHR to perform projection word embedding training. Next, we connected the converted text matrix in parallel so that the network can refer to two different word embedding sources simultaneously. In addition, we used different word embeddings separately as conversion sources to compare their effects on prediction performance.

2.2.1. Long Short-Term Memory (LSTM)

In RNN, the output can be given back to the network as input, thereby creating a loop structure. RNNs are trained through backpropagation. In the process of backpropagation, RNN will encounter the problem of vanishing gradient. We use the gradient to update the weight of the neural network. The problem of vanishing gradient is when the gradient shrinks as it propagates backwards in time. Therefore, the layers that obtain small gradients will not learn but will, instead, cause the network to have short-term memory.

The LSTM architecture was introduced by Hochreiter and Schmidhuber [46] to alleviate the problem of vanishing gradients. LSTMs can use a mechanism called gates to learn long-term dependencies. These gates can learn which information in the sequence is important to keep or discard. LSTMs have three gates: input, forget, and output. This is the core of the LSTM model, where pointwise addition and multiplication are performed to add or delete information from the memory. These operations are performed using the input and forget gate of the LSTM block, which also contains the output "tanh" activation function. In addition to using the original architecture and model parameters, the other settings are Epochs = 20, Batch size = 300, and Learning rate = 0.001.

2.2.2. Bidirectional Encoder Representation from Transformers (BERT)

Other than the original word embedding and LSTM architecture, BERT architecture was also used for feature extraction. BERT is a recent attention-based model with a bidirectional Transformer network that was pre-trained on a large corpus. This pre-trained model is then effectively used to solve various language tasks with fine-tuning [40,47]. In brief terms, the task-specific BERT architecture represents input text as sequential tokens. The input representation is generated with the sum of the token embeddings, the segmentation embeddings and the position embeddings [40]. For a classification task, the first word in the sequence is a unique token which is denoted with [CLS]. An encoder layer is followed with a fully-connected layer at the [CLS] position. Finally, a softmax layer is used as the aggregator for classification purposes [47]. If the NLP task has pair of sentences as in question-answer case, the sentence pairs may be separated with another special token [SEP]. BERT multilingual base model (cased) is used as transfer feature learning, and other parameters are set to Epochs = 30, Batch size = 32, and Learning rate = 0.00001.

Through these two methods, we can enable the network to learn the semantic meanings of different individual characters. We can also let the network learn from different texts, such as from Wikipedia and PubMed. Then, through EHR for Fine-tune retraining, the BERT architecture that has finished learning only needs to change from predicting its context output to predicting the categories of multiple medical record quality dimensions; then, it can be trained with medical record information.

2.3. *Linear Mixed Model Function for Medical Records Scoring Prediction*

Suppose data are collected from m independent groups of observations (called clusters or subjects in longitudinal data).

$$Y_m = X_m B_m + e_m. \tag{1}$$

Here, Y_m is an $n \times 1$ vector of the dependent variable for patient m, and X_i is an $n \times q$ matrix of all the independent variables for patient m. B_m is a $q \times 1$ unknown vector of regression coefficients, and e_m is an $n \times 1$ vector of residuals. This results in a multi-level mixed model with random effects for all samples, which is expressed as

$$Y = XB + Zu + e, \tag{2}$$

where Z is a matrix of known constants included in the information of the independent variables with random effects, and u is a matrix of random effects for all patients.

The best linear unbiased prediction (BLUP) is important for predicting the medical record score in each patient, and it can be calculated by following the steps in [48].

Y_m is an $n \times 1$ vector of the dependent variable for patient m, and X_i is an $n \times q$ matrix of all independent variables for patient m. Moreover, Z_m is an $n \times p$ matrix of independent variables with random effects for patient m. These matrices contain the observed data and are defined as

$$Y_m = \begin{bmatrix} y_{1,m} \\ y_{2,m} \\ \cdots \\ y_{n,m} \end{bmatrix}, X_m = \begin{bmatrix} 1 & x_{1,1,m} & \cdots & x_{1,q-1,m} \\ 1 & x_{2,1,m} & \cdots & x_{2,q-1,m} \\ \cdots & \cdots & \cdots & \cdots \\ x_1 & x_{n,1,m} & \cdots & x_{n,q-1,m} \end{bmatrix}, Z_m = \begin{bmatrix} 1 & x_{1,1,m} & \cdots & x_{1,p-1,m} \\ 1 & x_{2,1,m} & \cdots & x_{2,p-1,m} \\ \cdots & \cdots & \cdots & \cdots \\ x_1 & x_{n,1,m} & \cdots & x_{n,p-1,m} \end{bmatrix}. \quad (3)$$

After building the prediction tool, we have the G matrix, B vector and σ^2. G is a variance co-variance matrix of the random effects ($p \times p$), and B is the fixed effect coefficients vector ($q \times 1$). σ^2 is the variance of the residuals. We can calculate a matrix R ($n \times n$) using

$$G = \begin{bmatrix} \tau_1^2 & \tau_{12} & \cdots & \tau_{1p} \\ \tau_{12} & \tau_2^2 & \cdots & \tau_{2p} \\ \cdots & \cdots & \cdots & \cdots \\ \tau_{1p} & \tau_{2p} & \cdots & \tau_p^2 \end{bmatrix}, B = \begin{bmatrix} b_0 \\ b_1 \\ \cdots \\ b_{q-1} \end{bmatrix}, R = \sigma^2 I_{n \times n} = \begin{bmatrix} \sigma^2 & 0 & \cdots & 0 \\ 0 & \sigma^2 & \cdots & 0 \\ \cdots & \cdots & \cdots & \cdots \\ 0 & 0 & \cdots & \sigma^2 \end{bmatrix}. \quad (4)$$

If the independence assumption holds (i.e., $\begin{bmatrix} u \\ e \end{bmatrix} \sim N\left(\begin{bmatrix} 0 \\ 0 \end{bmatrix}, \begin{bmatrix} G & 0 \\ 0 & R \end{bmatrix}\right)$), then we can calculate the variance co-variance matrix (Σ_m) of Y_m using

$$\Sigma_m = Z_m G Z_m^T + R. \quad (5)$$

Finally, the BLUP of the random effect in patient m can be estimated using

$$BLUP_m = G Z_m^T \Sigma_m^{-1} (Y_m - X_m B). \quad (6)$$

We can estimate the regression coefficients (B_m) in patient m based on the above result, and B_m can be used to predict the disease progression. B_m can be calculated using

$$B_m = B + BLUP_m \quad (7)$$

Note that this calculation cannot make direct forecasts without the co-variable values. Thus, the co-variables information at the time of interest must be generated. We propose two methods for generating this information: (1) assume consistency between the last time and the time of interest and (2) predict the linear expectations. We will assess these methods in our analysis. Unquestionably, clinicians can use the most reasonable values based on their judgment to predict the co-variables at the time of interest. In summary, we can combine this method with population information to predict the medical record score.

2.4. Evaluation Criteria

We evaluated the generalization performance of each model in the training and testing samples. Mean absolute error (MAE) were used to compare the performance of the models, as follows:

$$MAE = \frac{\sum_{i=1}^{N} |y_i - \hat{y}_i|}{N}. \quad (8)$$

3. Results

The research scheme is shown in Figure 1, where a total of 227,689 medical records were scored by experts. In AI model training, the medical records were divided into the training set and testing set based on year, where 74,959 records were used to establish BERT and LSTM models, and 152,730 records were used to test record scoring. LMM was then employed to modify BERT and LSTM to establish another two models. In the end, MAE was used to compare the four models' efficiencies in predicting medical record scores.

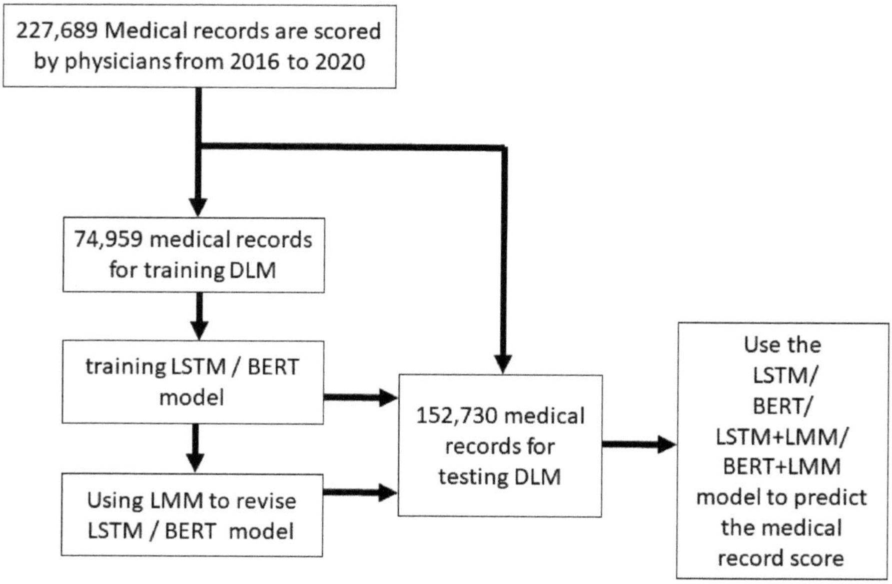

Figure 1. Training and testing sets generation. Schematic of the data set creation and analysis strategy, which was devised to assure a robust and reliable data set for training and testing of the network. Once a medical records data were placed in one of the data sets, that individual's data were used only in that set, avoiding 'cross-contamination' among the training and testing sets. The details of the flow chart and how each of the data sets was used are described in the Methods.

Table 1 shows the distribution of medical records in different departments. It can be seen that 74,959 records were included for modeling, and then 152,730 records were used for prediction. The average score from experts was 7.24 ± 1.02 for the training set and 7.67 ± 0.84 for the testing set; after BERT and LSTM modeling of medical record scoring, the average score of BERT prediction in the testing set was 7.47 ± 0.89, and 7.15 ± 1.05 for LSTM. After training through the BERT and LSTM models, the artificial intelligence model had already scored the medical records.

Table 1. Medical records distribution and scoring in the training set and testing set of different departments.

	Training Set (n = 74,959)	**Testing Set (n = 152,730)**	***p*-Value**
Department			<0.001 *
General surgery	4843 (6.5%)	10,504 (6.9%)	
Pleural surgery	1932 (2.6%)	3472 (2.3%)	
Cardiovascular surgery	3904 (5.2%)	8319 (5.4%)	
Colorectal & rectal surgery	491 (0.7%)	3479 (2.3%)	
Urology surgery	1330 (1.8%)	3313 (2.2%)	
Pediatric Surgery	99 (0.1%)	85 (0.1%)	
Plastic surgery	1748 (2.3%)	4009 (2.6%)	
Pulmonary Medicine	10,268 (13.7%)	19,065 (12.5%)	
Cardiology	2723 (3.6%)	4765 (3.1%)	
Nephrology	2473 (3.3%)	3749 (2.5%)	
Blood Oncology	9257 (12.3%)	17,110 (11.2%)	
Endocrine and metabolic	839 (1.1%)	1477 (1.0%)	
Gastroenterology	3861 (5.2%)	7372 (4.8%)	
Rheumatism, immunology and allergy	1247 (1.7%)	2624 (1.7%)	
Trauma	756 (1.0%)	940 (0.6%)	
Infection and Tropical Medicine	3701 (4.9%)	8488 (5.6%)	

Table 1. *Cont.*

	Training Set (n = 74,959)	Testing Set (n = 152,730)	p-Value
Psychiatric department	6531 (8.7%)	14,331 (9.4%)	
Neurological department	3159 (4.2%)	7374 (4.8%)	
Pediatric department	1138 (1.5%)	2474 (1.6%)	
Dental department	1223 (1.6%)	2483 (1.6%)	
Surgery department	607 (0.8%)	817 (0.5%)	
Dermatology department	5 (0.0%)	109 (0.1%)	
ENT department	2388 (3.2%)	3907 (2.6%)	
Radiology	40 (0.1%)	175 (0.1%)	
Emergency department	0 (0.0%)	300 (0.2%)	
Family and Community Medicine	188 (0.3%)	655 (0.4%)	
Nuclear Medicine Department	144 (0.2%)	153 (0.1%)	
Neurosurgery	3219 (4.3%)	6937 (4.5%)	
Orthopedic department	3482 (4.6%)	7876 (5.2%)	
Obstetrics and Gynecology	1766 (2.4%)	3222 (2.1%)	
Ophthalmology department	607 (0.8%)	903 (0.6%)	
Rehabilitation department	990 (1.3%)	2243 (1.5%)	
Experts' scores	7.24 ± 1.02	7.67 ± 0.84	<0.001 *
BERT prediction score		7.47 ± 0.89	
LSTM prediction score		7.15 ± 1.05	

*: p-value < 0.05.

Our team's projection word embedding model allowed the model to have both the vocabulary diversity of Wikipedia/PubMed and an understanding of medical terms in EHR. The concept of projection word embedding used the results of our previous studies, a concept in linear algebra that projects through matrix multiplication to allow all coordinates to convert into a new coordinate system. Such conversion changes the correlation of certain points while at the same time maintaining all current coordinates. In addition to the original projection word embedding and LSTM architecture, we attempted to use BERT architecture for feature extraction. BERT stands for Bidirectional Encoder Representations from Transformers, the elementary unit of BERT architecture is the encoder's Multi-Head Self-Attention Layer in the transformer. In contrast, the overall architecture of BERT is stacked by a bidirectional Transformer Encoder Layer. As shown in Table 2, in general, on the ground of experts' scoring, the trained scoring model BERT had a prediction score of 7.49 ± 0.28. In contrast, LSTM had 7.17 ± 0.31; after modification by the linear mixed model (LMM), BERT's and LSTM's prediction scores were 7.36 ± 0.56 and 7.33 ± 0.65, respectively. After layering different departments, such as internal medicine, surgery, obstetrics, and pediatrics, it can be learned that BERT all had higher prediction scores than LSTM, while, after LMM modification, all LSTM prediction scores increased. Through further looking into different departments, it was found that most departments' BERT prediction scores were higher than that of LSTM, and the latter increased after LMM modification.

Table 2. BERT and LSTM original prediction scores and LMM-modified scores.

	Experts' Scores	BERT Prediction Scores	LSTM Prediction Scores	LMM-Modified BERT Prediction Scores	LMM-Modified LSTM Prediction Scores
Overall	7.69 ± 0.64	7.49 ± 0.28	7.17 ± 0.31	7.36 ± 0.56	7.33 ± 0.65
Internal medicine	7.49 ± 0.66	7.37 ± 0.21	7.01 ± 0.20	7.14 ± 0.56	7.08 ± 0.65
Surgery	7.78 ± 0.55	7.49 ± 0.22	7.16 ± 0.17	7.54 ± 0.43	7.54 ± 0.51
Obstetrics and pediatrics	8.08 ± 0.69	7.68 ± 0.31	7.37 ± 0.31	7.70 ± 0.61	7.68 ± 0.79
Other departments	7.76 ± 0.60	7.57 ± 0.33	7.32 ± 0.40	7.39 ± 0.53	7.37 ± 0.61
Department					
General surgery	7.69 ± 0.74	7.48 ± 0.53	7.26 ± 0.28	7.45 ± 0.56	7.45 ± 0.57
Pleural surgery	7.87 ± 0.25	7.55 ± 0.35	7.22 ± 0.16	7.55 ± 0.43	7.64 ± 0.48
Cardiovascular surgery	7.73 ± 0.56	7.38 ± 0.37	7.01 ± 0.05	7.34 ± 0.17	7.35 ± 0.34
Colorectal & rectal surgery	7.92 ± 0.18	7.73 ± 0.37	7.22 ± 0.16	7.87 ± 0.35	7.97 ± 0.40

Table 2. *Cont.*

	Experts' Scores	BERT Prediction Scores	LSTM Prediction Scores	LMM-Modified BERT Prediction Scores	LMM-Modified LSTM Prediction Scores
Urology surgery	7.76 ± 0.18	7.48 ± 0.29	7.14 ± 0.09	7.54 ± 0.25	7.48 ± 0.37
Pediatric Surgery	6.16 ± NA	6.86 ± 0.50	7.09 ± NA	6.86 ± NA	6.65 ± NA
Plastic surgery	7.98 ± 0.08	7.58 ± 0.32	7.20 ± 0.15	7.65 ± 0.23	7.65 ± 0.29
Pulmonary Medicine	7.58 ± 0.83	7.30 ± 0.57	6.98 ± 0.19	7.26 ± 0.58	7.22 ± 0.65
Cardiology	7.19 ± 0.97	7.02 ± 0.64	6.99 ± 0.08	6.83 ± 0.68	6.75 ± 0.73
Nephrology	8.13 ± 0.69	7.54 ± 0.55	7.12 ± 0.06	7.42 ± 0.47	7.39 ± 0.60
Blood Oncology	7.21 ± 0.55	6.89 ± 0.50	6.71 ± 0.16	6.77 ± 0.52	6.71 ± 0.74
Endocrine and metabolic	7.64 ± 0.26	7.38 ± 0.35	7.17 ± 0.04	7.35 ± 0.44	7.25 ± 0.55
Gastroenterology	7.19 ± 0.25	7.15 ± 0.26	6.96 ± 0.12	7.16 ± 0.30	7.09 ± 0.33
Rheumatism, immunology and allergy	7.79 ± 0.21	7.33 ± 0.32	6.98 ± 0.14	7.29 ± 0.17	7.19 ± 0.22
Trauma	7.84 ± 1.32	7.39 ± 0.57	7.18 ± 0.02	7.21 ± 0.35	7.14 ± 0.47
Infection and Tropical Medicine	7.33 ± 0.53	7.09 ± 0.57	6.98 ± 0.07	6.94 ± 0.74	6.89 ± 0.87
Psychiatric department	8.41 ± 0.48	8.08 ± 0.47	8.00 ± 0.16	7.94 ± 0.59	7.94 ± 0.67
Neurological department	7.89 ± 0.24	7.62 ± 0.23	7.39 ± 0.06	7.60 ± 0.18	7.63 ± 0.25
Pediatric department	7.91 ± 0.85	7.51 ± 0.66	7.14 ± 0.10	7.52 ± 0.66	7.48 ± 0.93
Dental department	7.95 ± 0.25	7.05 ± 0.52	6.53 ± 0.09	6.89 ± 0.04	6.76 ± 0.04
Surgery department	7.81 ± NA	7.40 ± 0.26	7.14 ± NA	7.33 ± NA	7.25 ± NA
Dermatology department	8.58 ± NA	7.67 ± 0.64	6.83 ± NA	7.73 ± NA	7.85 ± NA
ENT department	7.37 ± 0.49	7.36 ± 0.38	7.29 ± 0.15	7.32 ± 0.47	7.37 ± 0.54
Radiology	6.85 ± NA	6.70 ± 0.17	6.67 ± NA	6.51 ± NA	6.57 ± NA
Family and Community Medicine	7.37 ± 0.41	7.19 ± 0.61	7.29 ± 0.09	6.91 ± 0.80	6.90 ± 1.15
Nuclear Medicine Department	8.76 ± NA	8.01 ± 0.45	7.54 ± NA	7.83 ± NA	8.02 ± NA
Neurosurgery	7.95 ± 0.49	7.59 ± 0.56	7.12 ± 0.07	7.78 ± 0.63	7.78 ± 0.75
Orthopedic department	7.38 ± 0.40	7.21 ± 0.34	7.09 ± 0.09	7.14 ± 0.38	7.11 ± 0.44
Obstetrics and Gynecology	8.31 ± 0.34	7.96 ± 0.41	7.67 ± 0.23	7.95 ± 0.49	7.96 ± 0.51
Ophthalmology department	7.86 ± 0.19	7.65 ± 0.26	7.56 ± 0.06	7.54 ± 0.27	7.53 ± 0.33
Rehabilitation department	8.06 ± 0.59	7.63 ± 0.41	7.29 ± 0.16	7.61 ± 0.25	7.51 ± 0.37

It can be learned from Table 3 that, when reviewer physicians' scores and AI scores were calculated using mean absolute error (MAE), both BERT and LSTM AI scores were 0.6~1.3 points lower than reviewer physicians' scores; thus, the linear mixed model (LMM) was introduced for modification, thereby reducing the score difference to 0.3~1 points, showing a significant reduction ($p < 0.001$) in score difference. The reason for the modification using LMM is that an ordinary linear regression contains only two influencing factors: fixed effect and noise. The latter is a random factor not considered in our model, while the former are those predictable factors that can also be completely divided. The AI scoring of medical records after modification by LMM is also more realistic. After department layering, it was found that, in some departments, LMM-modified MAE was not significantly reduced comparing with the original MAE. Hence, experts' scores were made into a heat map (Figure 2), where it was found that some groups of scoring physicians and scored physicians had closer scores, and were separately analyzed. In Table 4, medical record prediction scores and MAE are analyzed from Block A to H, respectively, and, except for block F, most blocks had similar record scores with previous results, and the MAE of LSTM prediction scores significantly reduced ($p < 0.05$) after LMM modification.

Table 3. The difference between the original AI/LMM-modified score and the expert score.

		Original MAE [a]	LMM-modified MAE [b]	p-Value
Overall				
	BERT	0.84 ± 0.27	0.70 ± 0.33	<0.001 *
	LSTM	1.00 ± 0.32	0.66 ± 0.39	<0.001 *
Internal medicine	BERT	0.82 ± 0.27	0.66 ± 0.37	0.007 *
	LSTM	0.96 ± 0.32	0.63 ± 0.41	<0.001 *
Surgery	BERT	0.86 ± 0.24	0.72 ± 0.25	0.011 *
	LSTM	1.04 ± 0.25	0.67 ± 0.30	<0.001 *
Obstetrics and pediatrics	BERT	1.05 ± 0.30	0.82 ± 0.32	0.069
	LSTM	1.21 ± 0.31	0.74 ± 0.44	<0.001 *

Table 3. *Cont.*

		Original MAE [a]	LMM-modified MAE [b]	p-Value
Other departments	BERT	0.79 ± 0.26	0.70 ± 0.35	0.142
	LSTM	0.96 ± 0.35	0.67 ± 0.41	<0.001 *
Department				
General surgery	BERT	0.80 ± 0.21	0.75 ± 0.15	0.645
	LSTM	1.03 ± 0.20	0.72 ± 0.12	0.003 *
Pleural surgery	BERT	0.72 ± 0.10	0.49 ± 0.26	0.200
	LSTM	0.91 ± 0.20	0.38 ± 0.27	0.100
Cardiovascular surgery	BERT	0.88 ± 0.26	0.86 ± 0.42	0.589
	LSTM	1.09 ± 0.39	0.79 ± 0.51	0.065
Colorectal & rectal surgery	BERT	0.74 ± 0.12	0.61 ± 0.25	0.686
	LSTM	0.97 ± 0.10	0.57 ± 0.34	0.057
Urology surgery	BERT	0.73 ± 0.06	0.67 ± 0.10	0.318
	LSTM	0.93 ± 0.10	0.63 ± 0.20	0.002 *
Plastic surgery	BERT	0.76 ± 0.05	0.59 ± 0.15	0.057
	LSTM	0.97 ± 0.08	0.52 ± 0.22	0.029 *
Pulmonary Medicine	BERT	0.94 ± 0.32	0.69 ± 0.29	0.040 *
	LSTM	1.14 ± 0.36	0.65 ± 0.27	0.002 *
Cardiology	BERT	1.01 ± 0.41	0.75 ± 0.33	0.136
	LSTM	1.12 ± 0.34	0.74 ± 0.34	0.024 *
Nephrology	BERT	0.89 ± 0.29	0.89 ± 0.41	0.841
	LSTM	1.22 ± 0.47	0.82 ± 0.49	0.222
Blood Oncology	BERT	0.85 ± 0.21	0.66 ± 0.23	0.130
	LSTM	0.91 ± 0.22	0.72 ± 0.28	0.195
Endocrine and metabolic	BERT	0.82 ± 0.03	0.68 ± 0.16	0.343
	LSTM	0.95 ± 0.09	0.63 ± 0.23	0.114
Gastroenterology	BERT	0.60 ± 0.11	0.42 ± 0.20	0.050 *
	LSTM	0.66 ± 0.17	0.37 ± 0.23	0.015 *
Rheumatism, immunology and allergy	BERT	0.74 ± 0.11	0.69 ± 0.13	0.548
	LSTM	1.02 ± 0.15	0.70 ± 0.16	0.032 *
Trauma	BERT	1.08 ± 0.22	0.88 ± 0.70	1.000
	LSTM	1.19 ± 0.63	0.84 ± 0.72	0.667
Infection and Tropical Medicine	BERT	0.69 ± 0.17	0.66 ± 0.81	0.028 *
	LSTM	0.78 ± 0.26	0.63 ± 0.91	0.028 *
Psychiatric department	BERT	0.73 ± 0.26	0.59 ± 0.47	0.328
	LSTM	1.03 ± 0.29	0.52 ± 0.54	0.028 *
Neurological department	BERT	0.72 ± 0.06	0.56 ± 0.06	0.002 *
	LSTM	0.82 ± 0.09	0.44 ± 0.11	0.002 *
Pediatric department	BERT	1.18 ± 0.35	0.95 ± 0.33	0.328
	LSTM	1.36 ± 0.30	0.90 ± 0.49	0.007 *
Dental department	BERT	0.96 ± 0.10	1.12 ± 0.24	0.400
	LSTM	1.52 ± 0.19	1.23 ± 0.23	0.400
ENT department	BERT	0.73 ± 0.13	0.53 ± 0.17	0.024 *
	LSTM	0.78 ± 0.15	0.46 ± 0.20	<0.001 *
Family and Community Medicine	BERT	0.75 ± 0.06	0.74 ± 0.43	0.700
	LSTM	0.80 ± 0.05	0.81 ± 0.62	0.700
Neurosurgery	BERT	1.12 ± 0.28	0.80 ± 0.10	0.002 *
	LSTM	1.21 ± 0.30	0.77 ± 0.14	0.002 *
Orthopedic department	BERT	0.78 ± 0.34	0.71 ± 0.38	0.630
	LSTM	0.92 ± 0.28	0.68 ± 0.42	0.089
Obstetrics and Gynecology	BERT	0.88 ± 0.03	0.64 ± 0.24	0.009 *
	LSTM	1.02 ± 0.19	0.53 ± 0.28	0.004 *
Ophthalmology department	BERT	0.56 ± 0.17	0.55 ± 0.26	0.690
	LSTM	0.60 ± 0.09	0.57 ± 0.30	0.222
Rehabilitation department	BERT	0.88 ± 0.12	0.77 ± 0.22	0.180
	LSTM	1.06 ± 0.38	0.77 ± 0.38	0.180

[a] Original MAE: Expert's score—BERT/LSTM prediction score. [b] LMM-modified MAE: Expert's score—LMM-modified BERT/LSTM prediction score. *: p-value < 0.05.

Table 4. Experts' scores, BERT and LSTM prediction scores, and MAE of different blocks.

Block	Experts' Score (a)	BERT Score (b)	LSTM Score (c)	*p*-Value	LMM-Modified BERT Score (d)	LMM-Modified LSTM Score (e)	*p*-Value	∣a-b∣ #	∣a-d∣ #	*p*-Value	∣a-c∣ #	∣a-e∣ #	*p*-Value
A	7.44 ± 0.66	7.35 ± 0.17	6.99 ± 0.17	<0.001 *	7.08 ± 0.56	7.02 ± 0.66	0.626	0.83 ± 0.27	0.66 ± 0.38	0.008 *	0.97 ± 0.33	0.63 ± 0.43	<0.001 *
B	7.35 ± 0.51	7.43 ± 0.06	7.32 ± 0.17	0.087	7.32 ± 0.47	7.38 ± 0.54	0.824	0.7 ± 0.13	0.51 ± 0.17	0.013 *	0.76 ± 0.16	0.45 ± 0.2	0.002 *
C	7.88 ± 0.14	7.56 ± 0.09	7.4 ± 0.1	0.016 *	7.59 ± 0.18	7.63 ± 0.24	0.740	0.69 ± 0.03	0.54 ± 0.1	0.005 *	0.77 ± 0.08	0.41 ± 0.14	<0.001 *
D	7.94 ± 1	7.43 ± 0.19	7.13 ± 0.08	0.005 *	7.57 ± 0.6	7.61 ± 0.84	0.932	1.29 ± 0.29	0.88 ± 0.31	0.042 *	1.44 ± 0.3	0.74 ± 0.35	0.004 *
E	7.74 ± 0.91	7.51 ± 0.08	6.98 ± 0.18	<0.001 *	7.19 ± 0.45	7.12 ± 0.56	0.772	1.05 ± 0.33	0.85 ± 0.4	0.227	1.25 ± 0.41	0.8 ± 0.35	0.016 *
F	7.3 ± 0.63	6.97 ± 0.24	6.61 ± 0.17	0.004 *	6.75 ± 0.54	6.69 ± 0.74	0.874	0.88 ± 0.22	0.73 ± 0.28	0.258	1 ± 0.27	0.78 ± 0.3	0.154
G	7.76 ± 0.15	7.46 ± 0.08	7.08 ± 0.11	<0.001 *	7.52 ± 0.25	7.46 ± 0.37	0.707	0.69 ± 0.08	0.63 ± 0.12	0.238	0.93 ± 0.12	0.6 ± 0.2	0.003 *
H	8.41 ± 0.47	8.1 ± 0.06	8.05 ± 0.18	0.436	7.93 ± 0.59	7.95 ± 0.67	0.962	0.71 ± 0.23	0.58 ± 0.48	0.489	1 ± 0.3	0.51 ± 0.55	0.045 *

#: The mean absolute error (MAE), the absolute value of the original score minus the predicted score. *: *p*-value < 0.05.

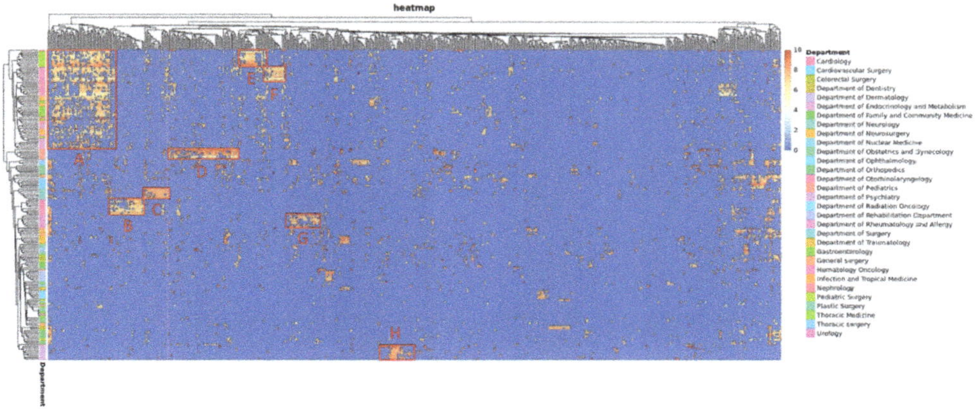

Figure 2. Heat map of medical record scores from scoring and scored physicians. *X*-axis: physicians who wrote the medical records; *Y*-axis: scoring physicians and their departments. A redder grid means record scoring physicians give a higher score to record writing physicians. There are clusters in some areas; thus, we put out some blocks and observe the block (A to H) characteristics in Table 4.

In spite of this, we were still unable to identify the reason why the MAE of certain departments had no significant reduction after LMM modification. Thus, heat map analysis was performed on LMM-modified LSTM prediction scores. Figure 3 shows that some reviewers' LMM-modified LSTM prediction scores had relatively greater MAE. After grouping using LMM modified MAE (Grade-LMM modified LSTM), experts' scores were close among groups, but BERT and LSTM prediction scores were lower than the original experts' scores. In Figure 4, We further using MAE to evaluate model efficiency, and then comparing MAE (|Grade—LMM modified BERT|, |Grade—LMM modified BERT|) of LMM-modified BERT or LSTM with the MAE (|Grade—BERT|, |Grade—LSTM|) of the original BERT or LSTM, it was found MAE was effectively reduced through LMM modification in Q1~Q3, but not in Q4. Thus, it is suspected that some scoring physicians in Q4 may have scored incorrectly.

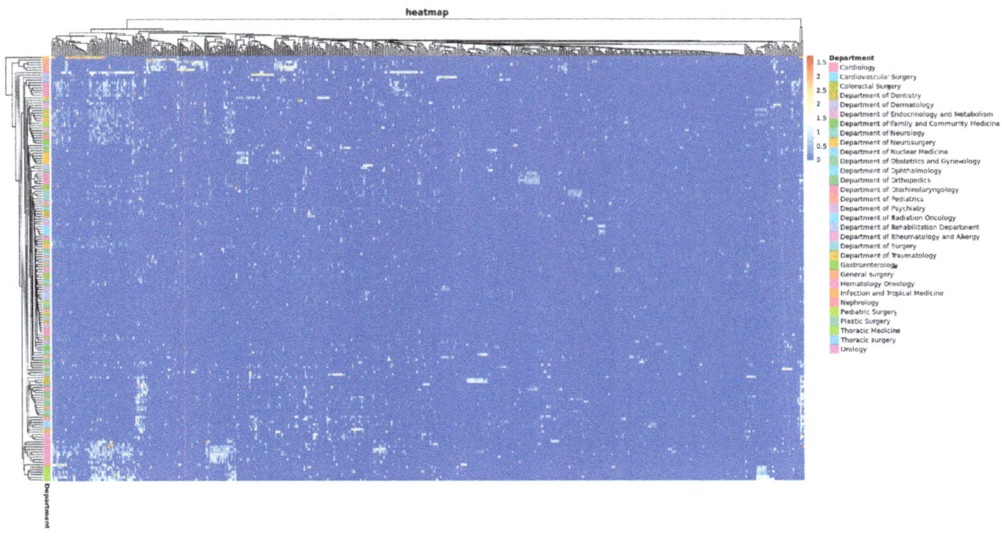

Figure 3. MAE heat map of LMM-modified LSTM prediction scores from scoring and scored physicians. *X*-axis: physicians who wrote the medical records; *Y*-axis: scoring physicians and their departments. By subtracting the MAE of the original

score from the LMM modified LSTM prediction score, and using the MAE and coring physicians to conduct a heat map analysis, it can be found that some reviewer scores are on the high side.

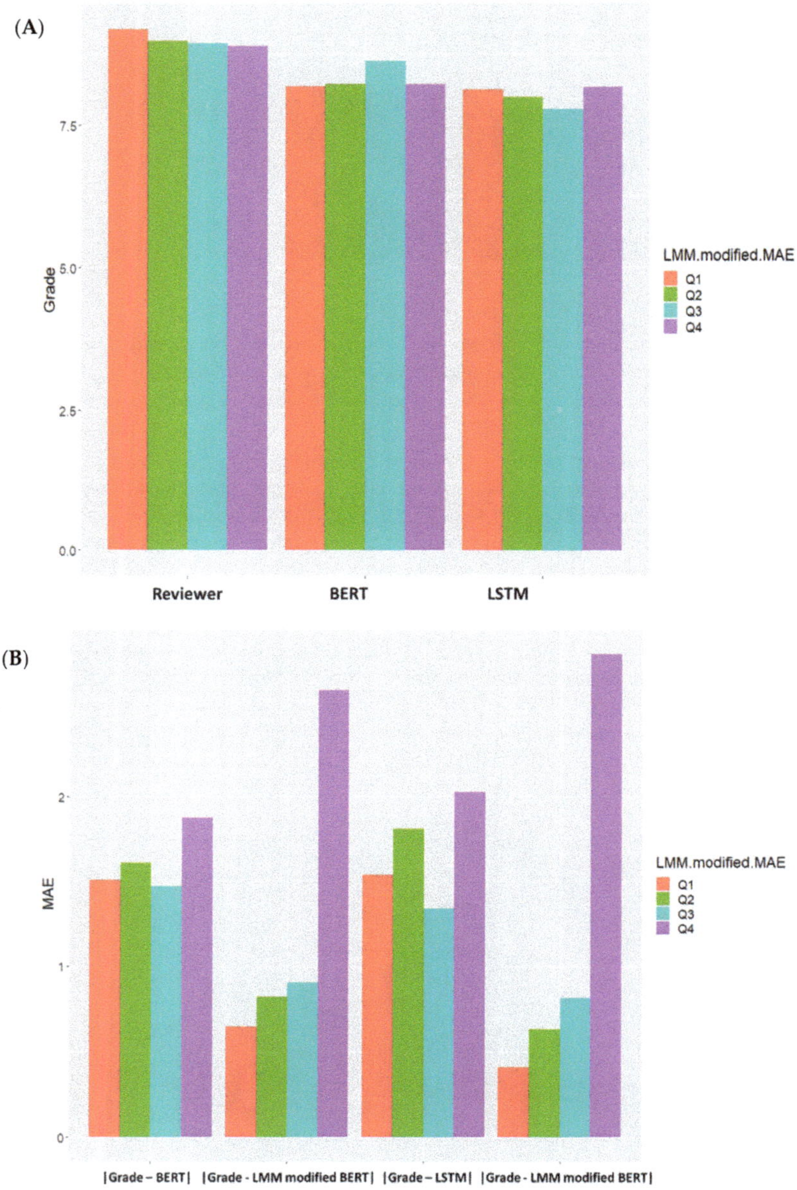

Figure 4. Using LMM modified MAE (Grade-LMM modified LSTM) for interquartile range grouping. (**A**): Compare the scores of Experts, BERT and LSTM. *Y*-axis: medical record scores, *X*-axis: Experts' score, BERT prediction score, LSTM prediction score. (**B**): Compare the original MAE with the LMM modified MAE. *Y*-axis: mean absolute error (MAE), *X*-axis: |Grade—BERT|, |Grade—LMM modified BERT|, |Grade—LSTM|, |Grade—LMM modified BERT| for model efficiency evaluation. The LMM modified MAE (Grade-LMM modified LSTM) is grouped by interquartile range and divided into Q1, Q2, Q3, and Q4.

4. Discussion

In this study, the projection word embedding model was used to develop an AI system to evaluate the writing quality of inpatient medical records. The AI system is already capable of accurate classification to level 3 ICD-10 coding, combined with results from previous studies. Since level 3 coding is already at the disease level, subsequent coding will all just be remarks (such as location), and reaching such a level will allow for the possibility of full automation of common disease classification tasks, as well as extraction of disease features from other medical descriptions, through this algorithm. In addition to the original word embedding and LSTM architecture, BERT architecture was also employed to extract disease features for medical record scoring. LMM was further used for modification to get AI scores closer to actual reviewer physicians' scores. Moreover, it was also identified that some physicians over-scored medical records. If these scoring standards can be improved in the future, a better medical writing quality could be expected.

In addition, why is the quality of medical record writing so important? Because the medical record is the historical record of the patient's health care; it is also the basis of care, and its content records the patient's condition during the care process, the reason and result of the inspection, and the treatment method and result. In recent studies, it is feasible to use electronic health records (EHR) to predict disease risk, such as atrial fibrillation (AF) [49], coronary heart disease in patients with hypertension [50], fall risk [51], multiple sclerosis disease [52], and cervical cancer [53]. Over the past two decades, the investigation of genetic variation underlying disease susceptibility has increased considerably. Most notably, genome-wide association studies (GWAS) have investigated tens of millions of single-nucleotide polymorphisms (SNPs) for associations with complex diseases. However, results from numerous GWAS have revealed that the majority of statistically significantly associated genetic variants have small effects [54] and may not be predictive of disease risks [55], and many diseases are associated with tens of thousands of genetic variants [56]. These findings have led to the resurgence of the polygenic risk score (PRS), an aggregate measure of many genetic variants weighted by their individual effects on a given phenotype. However, epidemiologic studies are expensive and complex to run, which raises the question of whether a PRS could be developed and applied in a clinical setting using genetic data that are more readily available. Recently, some scholars proposed new ideas for developing and implementing PRS predictions using biobank-linked EHR data [57].

For the medical records scoring system, this not only saves doctors the time for scoring medical records but also can get feedback immediately after the writing is completed to improve the quality of medical record writing. In the past research, clinicians spent 3.7 h per day, or 37% of their work day, on EHR [58]. There was a marked reduction in EHR time with both clinician and resident seniority. Despite this improvement, the total time spent on EHR remained exceedingly high amongst even the most experienced physicians [58]. The significance of an increasing shift towards EHR is a growing paradigm that cannot be understated, particularly in the current era of healthcare, and there is increasing scrutiny on documentation [59,60]. These increased demands can lead to EHR fatigue and physician burnout. In a survey of a general internal medicine group, 38% reported feeling burnt out, with 60% citing high documentation pressure and 50% describing too much EHR time at home [61]. Burnout has been linked to an increased risk of resident's wellbeing [62].

There are still some limitations for electronic medical records. First, this scoring system can only be used in our hospital because the medical record system of different hospitals do not talk to each other. Second, entering data into an EHR requires a doctor to spend a lot of time doing so, leading to most physicians experiencing burnout symptoms due to EMR-related workloads. Third, cyber-attacks are a perennial concern for EHRs. It is, therefore, imperative that cybersecurity is continually enhanced. Fourth, timing discrepancies occur in EHRs, and they can lead to serious clinical consequences.

In summary, combining projection word embedding and LSTM with LMM can give better prediction scores. This system can be used to assist medical record scoring so that young physicians can get immediate writing feedback, so as to improve the quality of medical record writing in my country and let the public, Medical units, and insurance units can all get better help. In the future, it may be possible to actively introduce such technologies into hospitals to achieve personalized precision medicine.

Author Contributions: Data curation, C.L., C.-J.H., and C.-C.L.; Project administration, W.-H.F.; Resources, Y.-T.L., F.-J.W., and W.-H.F.; Software, S.-A.L.; Supervision, C.L.; Visualization, D.-J.T.; Writing—original draft, Y.-T.L., F.-J.W., and D.-J.T.; Writing—review & editing, C.L. and D.-J.T. All authors have read and agreed to the published version of the manuscript.

Funding: This research received no external funding.

Institutional Review Board Statement: This study was approved by the institutional review board in Tri-Service General Hospital, Taipei, Taiwan (IRB NO. A202005104).

Informed Consent Statement: Patients' consent was waived because data were collected retrospectively and in anonymized files and encrypted from the hospital to the data controller.

Data Availability Statement: The data presented in this study are available on request from the corresponding author.

Acknowledgments: This study was supported by funding from the Ministry of Science and Technology, Taiwan (MOST109-2314-B-016-021, MOST110-2314-B-016-008 to W.H. Fang and MOST110-2314-B-016-010-MY3 to C. Lin), the National Science and Technology Development Fund Management Association, Taiwan (MOST109-3111-Y-016-002 to C. Lin and MOST110-3111-Y-016-005 to C. Lin), Cheng Hsin General Hospital (CHNDMC-109-19 to Y.T. Lee and C. Lin), and Taoyuan Armed Forces General Hospital (TYAFGH-D-109043 to F.J. Wu and C. Lin).

Conflicts of Interest: The authors declare no conflict of interest.

References

1. Hilbert, M.; Lopez, P. The world's technological capacity to store, communicate, and compute information. *Science* **2011**, *332*, 60–65. [CrossRef]
2. McAfee, A.; Brynjolfsson, E. Big data: The management revolution. *Harv. Bus. Rev.* **2012**, *90*, 60–68.
3. *Unstructured Data and the 80 Percent Rule*; Clarabridge Bridgepoints: Reston, VA, USA, 2008; Volume Q3.
4. Jeong, S.R.; Ghani, I. Semantic Computing for Big Data: Approaches, Tools, and Emerging Directions (2011–2014). *TIIS* **2014**, *8*, 2022–2042.
5. Cox, M.; Ellsworth, D. Application-controlled demand paging for out-of-core visualization. In Proceedings of the 8th Conference on Visualization'97, Phoenix, AZ, USA, 24 October 1997.
6. McDonald, C.J. Medical heuristics: The silent adjudicators of clinical practice. *Ann. Intern. Med.* **1996**, *124*, 56–62. [CrossRef]
7. National Academies of Sciences E, Medicine. *Improving Diagnosis in Health Care*; National Academies Press: Washington, DC, USA, 2016.
8. El-Kareh, R.; Hasan, O.; Schiff, G.D. Use of health information technology to reduce diagnostic errors. *BMJ Qual. Saf.* **2013**, *22* (Suppl 2), ii40–ii51. [CrossRef]
9. Murdoch, T.B.; Detsky, A.S. The inevitable application of big data to health care. *JAMA* **2013**, *309*, 1351–1352. [CrossRef]
10. Spasic, I.; Livsey, J.; Keane, J.A.; Nenadic, G. Text mining of cancer-related information: Review of current status and future directions. *Int. J. Med. Inform.* **2014**, *83*, 605–623. [CrossRef]
11. Cahan, A.; Cimino, J.J. A Learning Health Care System Using Computer-Aided Diagnosis. *J. Med. Internet* **2017**, *19*, e54. [CrossRef]
12. Krizhevsky, A.; Sutskever, I.; Hinton, G.E. Imagenet classification with deep convolutional neural networks. *Adv. Neural Inf. Process. Syst.* **2012**, *25*, 1097–1105. [CrossRef]
13. Simonyan, K.; Zisserman, A. Very deep convolutional networks for large-scale image recognition. *arXiv* **2014**, arXiv:1409.1556.
14. Szegedy, C.; Liu, W.; Jia, Y.; Sermanet, P.; Reed, S.; Anguelov, D.; Erhan, D.; Vanhoucke, V.; Rabinovich, A. Going deeper with convolutions. In Proceedings of the IEEE Conference on Computer Vision and Pattern Recognition, Boston, MA, USA, 7–12 June 2015.
15. He, K.; Zhang, X.; Ren, S.; Sun, J. Deep residual learning for image recognition. In Proceedings of the IEEE Conference on Computer Vision and Pattern Recognition, Las Vegas, NV, USA, 27–30 June 2016.
16. Huang, G.; Liu, Z.; Weinberger, K.Q.; van der Maaten, L. Densely connected convolutional networks. In Proceedings of the IEEE Conference on Computer Vision and Pattern Recognition, Honolulu, HI, USA, 21–26 July 2017.
17. Bejnordi, B.E.; Veta, M.; van Diest, P.J.; van Ginneken, B.; Karssemeijer, N.; Litjens, G.; van der Laak, J.A.W.M.; the CAMELYON16 Consortium. Diagnostic Assessment of Deep Learning Algorithms for Detection of Lymph Node Metastases in Women with Breast Cancer. *JAMA* **2017**, *318*, 2199–2210. [CrossRef]

18. Gulshan, V.; Peng, L.; Coram, M.; Stumpe, M.C.; Wu, D.; Narayanaswamy, A.; Venugopalan, S.; Widner, K.; Madams, T.; Cuadros, J.; et al. Development and Validation of a Deep Learning Algorithm for Detection of Diabetic Retinopathy in Retinal Fundus Photographs. *JAMA* **2016**, *316*, 2402–2410. [CrossRef]
19. Esteva, A.; Kuprel, B.; Novoa, R.A.; Ko, J.; Swetter, S.M.; Blau, H.M.; Thrun, S. Dermatologist-level classification of skin cancer with deep neural networks. *Nature* **2017**, *542*, 115–118. [CrossRef]
20. Rajpurkar, P.; Irvin, J.; Zhu, K.; Yang, B.; Mehta, H.; Duan, T.; Ding, D.; Bagul, A.; Langlotz, C.; Shpanskaya, K.; et al. CheXNet: Radiologist-Level Pneumonia Detection on Chest X-Rays with Deep Learning. *arXiv* **2017**, arXiv:1711.05225.
21. Grewal, M.; Srivastava, M.M.; Kumar, P.; Varadarajan, S. RADNET: Radiologist Level Accuracy using Deep Learning for HEMORRHAGE detection in CT Scans. *arXiv* **2017**, arXiv:1710.04934.
22. Litjens, G.; Kooi, T.; Bejnordi, B.E.; Setio AA, A.; Ciompi, F.; Ghafoorian, M.; van der Laak, J.A.W.M.; Ginneken, B.; Sánchez, C.I. A survey on deep learning in medical image analysis. *Med. Image Anal.* **2017**, *42*, 60–88. [CrossRef]
23. Xiong, W.; Droppo, J.; Huang, X.; Seide, F.; Seltzer, M.; Stolcke, A.; Yu, D.; Zweig, G. Achieving human parity in conversational speech recognition. *arXiv* **2016**, arXiv:1610.05256.
24. Amodei, D.; Ananthanarayanan, S.; Anubhai, R.; Bai, J.; Battenberg, E.; Case, C.; Casper, J.; Catanzaro, B.; Cheng, Q.; Chen, G.; et al. Deep speech 2: End-to-end speech recognition in english and mandarin. In Proceedings of the International Conference on Machine Learning, New York, NY, USA, 20–22 June 2016.
25. Dembek, Z.F.; Kortepeter, M.G.; Pavlin, J.A. Discernment between deliberate and natural infectious disease outbreaks. *Epidemiol. Infect.* **2007**, *135*, 353–371. [CrossRef]
26. Tsui, F.C.; Espino, J.U.; Dato, V.M.; Gesteland, P.H.; Hutman, J.; Wagner, M.M. Technical description of RODS: A real-time public health surveillance system. *J. Am. Med. Inform. Assoc. JAMIA* **2003**, *10*, 399–408. [CrossRef]
27. Koopman, B.; Zuccon, G.; Nguyen, A.; Bergheim, A.; Grayson, N. Automatic ICD-10 classification of cancers from free-text death certificates. *Int. J. Med. Inform.* **2015**, *84*, 956–965. [CrossRef]
28. Koopman, B.; Karimi, S.; Nguyen, A.; McGuire, R.; Muscatello, D.; Kemp, M.; Truran, D.; Zhang, M.; Thackway, S. Automatic classification of diseases from free-text death certificates for real-time surveillance. *BMC Med. Inform. Decis. Mak.* **2015**, *15*, 53. [CrossRef] [PubMed]
29. Koopman, B.; Zuccon, G.; Wagholikar, A.; Chu, K.; O'Dwyer, J.; Nguyen, A.; Keijzers, G. Automated Reconciliation of Radiology Reports and Discharge Summaries. In *AMIA Annual Symposium Proceedings*; American Medical Informatics Association: Bethesda, MD, USA, 2015; Volume 2015, pp. 775–784.
30. Khachidze, M.; Tsintsadze, M.; Archuadze, M. Natural Language Processing Based Instrument for Classification of Free Text Medical Records. *BioMed Res. Int.* **2016**, *2016*, 8313454. [CrossRef] [PubMed]
31. Mujtaba, G.; Shuib, L.; Raj, R.G.; Rajandram, R.; Shaikh, K.; Al-Garadi, M.A. Automatic ICD-10 multi-class classification of cause of death from plaintext autopsy reports through expert-driven feature selection. *PLoS ONE* **2017**, *12*, e0170242. [CrossRef] [PubMed]
32. Bengio, Y.; Ducharme, R.; Vincent, P.; Jauvin, C. A neural probabilistic language model. *J. Mach. Learn. Res.* **2003**, *3*, 1137–1155.
33. Yih, W.T.; Toutanova, K.; Platt, J.C.; Meek, C. Learning discriminative projections for text similarity measures. In Proceedings of the Fifteenth Conference on Computational Natural Language Learning, Portland, OR, USA, 23 June 2011.
34. Mikolov, T.; Sutskever, I.; Chen, K.; Corrado, G.S.; Dean, J. Distributed representations of words and phrases and their compositionality. In Proceedings of the Advances in Neural Information Processing Systems, Lake Tahoe, Nevada, 5–10 December 2013.
35. Arora, S.; Liang, Y.; Ma, T. A Simple but Tough-to-Beat Baseline for Sentence Embeddings. 2016. Available online: https://openreview.net/forum?id=SyK00v5xx (accessed on 27 September 2021).
36. LeCun, Y.; Bottou, L.; Bengio, Y.; Haffner, P. Gradient-based learning applied to document recognition. *Proc. IEEE.* **1998**, *86*, 2278–2324. [CrossRef]
37. Yih, W.T.; He, X.; Meek, C. Semantic Parsing for Single-Relation Question Answering. In Proceedings of the 52nd Annual Meeting of the Association for Computational Linguistics (Volume 2: Short Papers), Baltimore, MD, USA, 22–27 June 2014.
38. Shen, Y.; He, X.; Gao, J.; Deng, L.; Mesnil, G. Learning semantic representations using convolutional neural networks for web search. In Proceedings of the 23rd International Conference on World Wide Web, Seoul, Korea, 7–11 April 2014.
39. Kim, Y. Convolutional neural networks for sentence classification. *arXiv* **2014**, arXiv:1408.5882.
40. Devlin, J.; Chang, M.-W.; Lee, K.; Toutanova, K. Bert: Pre-training of deep bidirectional transformers for language understanding. *arXiv* **2018**, arXiv:1810.04805.
41. Lin, C.; Hsu, C.J.; Lou, Y.S.; Yeh, S.J.; Lee, C.C.; Su, S.L.; Chen, H.C. Artificial Intelligence Learning Semantics via External Resources for Classifying Diagnosis Codes in Discharge Notes. *J. Med. Internet Res.* **2017**, *19*, e380. [CrossRef]
42. Choi, Y.; Chiu, C.Y.; Sontag, D. Learning Low-Dimensional Representations of Medical Concepts. *AMIA Jt. Summits Transl. Sci. Proceedings. AMIA Jt. Summits Transl. Sci.* **2016**, *2016*, 41–50.
43. Wang, Y.; Liu, S.; Afzal, N.; Rastegar-Mojarad, M.; Wang, L.; Shen, F.; Kingsbury, P.; Liu, H. A comparison of word embeddings for the biomedical natural language processing. *J. Biomed. Inform.* **2018**, *87*, 12–20.
44. Pakhomov, S.V.; Finley, G.; McEwan, R.; Wang, Y.; Melton, G.B. Corpus domain effects on distributional semantic modeling of medical terms. *Bioinformatics* **2016**, *32*, 3635–3644. [CrossRef]

45. Lin, C.; Lou, Y.S.; Tsai, D.J.; Lee, C.C.; Hsu, C.J.; Wu, D.C.; Wang, M.C.; Fang, W.H. Projection Word Embedding Model With Hybrid Sampling Training for Classifying ICD-10-CM Codes: Longitudinal Observational Study. *JMIR Med. Inform.* **2019**, *7*, e14499. [CrossRef] [PubMed]
46. Hochreiter, S.; Schmidhuber, J. Long Short-term Memory. *Neural Comput.* **1997**, *9*, 1735–1780. [CrossRef] [PubMed]
47. Gao, Z.; Feng, A.; Song, X.; Wu, X. Target-dependent sentiment classification with BERT. *IEEE Access* **2019**, *7*, 154290–154299. [CrossRef]
48. Robinson, G.K. That BLUP is a good thing: The estimation of random effects. *Stat. Sci.* **1991**, 15–32. [CrossRef]
49. Hulme, O.L.; Khurshid, S.; Weng, L.C.; Anderson, C.D.; Wang, E.Y.; Ashburner, J.M.; Ko, D.; McManus, D.D.; Benjamin, E.J.; Ellinor, P.T.; et al. Development and Validation of a Prediction Model for Atrial Fibrillation Using Electronic Health Records. *JACC. Clin. Electrophysiol.* **2019**, *5*, 1331–1341. [CrossRef] [PubMed]
50. Du, Z.; Yang, Y. Accurate Prediction of Coronary Heart Disease for Patients with Hypertension from Electronic Health Records with Big Data and Machine-Learning Methods: Model Development and Performance Evaluation. *JMIR Med. Inform.* **2020**, *8*, e17257. [CrossRef] [PubMed]
51. Ye, C.; Li, J.; Hao, S.; Liu, M.; Jin, H.; Zheng, L.; Xia, M.; Jin, B.; Zhu, C.; Alfreds, S.T.; et al. Identification of elders at higher risk for fall with statewide electronic health records and a machine learning algorithm. *Int. J. Med. Inform.* **2020**, *137*, 104105. [CrossRef]
52. Ahuja, Y.; Kim, N.; Liang, L.; Cai, T.; Dahal, K.; Seyok, T.; Lin, C.; Finan, S.; Liao, K.; Savovoa, G.; et al. Leveraging electronic health records data to predict multiple sclerosis disease activity. *Ann. Clin. Transl. Neurol.* **2021**, *8*, 800–810. [CrossRef]
53. Weegar, R.; Sundström, K. Using machine learning for predicting cervical cancer from Swedish electronic health records by mining hierarchical representations. *PLoS ONE.* **2020**, *15*, e0237911. [CrossRef]
54. Manolio, T.A.; Collins, F.S.; Cox, N.J.; Goldstein, D.B.; Hindorff, L.A.; Hunter, D.J.; McCarthy, M.I.; Ramos, E.M.; Cardon, L.R.; Chakravarti, A.; et al. Finding the missing heritability of complex diseases. *Nature* **2009**, *461*, 747–753. [CrossRef] [PubMed]
55. Lo, A.; Chernoff, H.; Zheng, T.; Lo, S.-H. Why significant variables aren't automatically good predictors. *Proc. Natl. Acad. Sci. USA* **2015**, *112*, 13892–13897. [CrossRef] [PubMed]
56. Visscher, P.M.; Wray, N.R.; Zhang, Q.; Sklar, P.; McCarthy, M.I.; Brown, M.A.; Yang, J. 10 years of GWAS discovery: Biology, function, and translation. *Am. J. Hum. Genet.* **2017**, *101*, 5–22. [CrossRef] [PubMed]
57. Li, R.; Chen, Y.; Ritchie, M.D.; Moore, J.H. Electronic health records and polygenic risk scores for predicting disease risk. *Nat. Rev. Genet.* **2020**, *21*, 493–502. [CrossRef] [PubMed]
58. Verma, G.; Ivanov, A. Analyses of electronic health records utilization in a large community hospital. *PLoS ONE* **2020**, *15*, e0233004. [CrossRef]
59. Pizziferri, L.; Kittler, A.F.; Volk, L.A.; Honour, M.M.; Gupta, S.; Wang, S.; Wang, T.; Lippincott, M.; Li, Q.; Bates, D.W. Primary care physician time utilization before and after implementation of an electronic health record: A time-motion study. *J. Biomed. Inform.* **2005**, *38*, 176–188. [CrossRef] [PubMed]
60. Pizziferri, L.; Kittler, A.F.; Volk, L.A.; Shulman, L.N.; Kessler, J.; Carlson, G.; Michaelidis, T.; Bates, D.W. Impact of an Electronic Health Record on oncologists' clinic time. In Proceedings of the AMIA Annual Symposium Proceedings, Washington, DC, USA, 22–26 October 2005; p. 1083.
61. Linzer, M.; Poplau, S.; Babbott, S.; Collins, T.; Guzman-Corrales, L.; Menk, J.; Murphy, M.L.; Ovington, K. Worklife and wellness in academic general internal medicine: Results from a national survey. *J. Gen. Intern. Med.* **2016**, *31*, 1004–1010. [CrossRef] [PubMed]
62. Van der Heijden, F.; Dillingh, G.; Bakker, A.; Prins, J. Suicidal thoughts among medical residents with burnout. *Arch. Suicide Res.* **2008**, *12*, 344–346. [CrossRef]

healthcare

MDPI

Article

Forecast of the COVID-19 Epidemic Based on RF-BOA-LightGBM

Zhe Li and Dehua Hu *

School of Life Sciences, Central South University, Changsha 410083, China; zhanghang22@csu.edu.cn
* Correspondence: hudehua@csu.edu.cn

Abstract: In this paper, we utilize the Internet big data tool, namely Baidu Index, to predict the development trend of the new coronavirus pneumonia epidemic to obtain further data. By selecting appropriate keywords, we can collect the data of COVID-19 cases in China between 1 January 2020 and 1 April 2020. After preprocessing the data set, the optimal sub-data set can be obtained by using random forest feature selection method. The optimization results of the seven hyperparameters of the LightGBM model by grid search, random search and Bayesian optimization algorithms are compared. The experimental results show that applying the data set obtained from the Baidu Index to the Bayesian-optimized LightGBM model can better predict the growth of the number of patients with new coronary pneumonias, and also help people to make accurate judgments to the development trend of the new coronary pneumonia.

Keywords: COVID-19; Baidu index; random forest; bayesian optimization; LightGBM

Citation: Li, Z.; Hu, D. Forecast of the COVID-19 Epidemic Based on RF-BOA-LightGBM. *Healthcare* **2021**, *9*, 1172. https://doi.org/10.3390/healthcare9091172

Academic Editors: Mahmudur Rahman and Daniele Giansanti

Received: 21 July 2021
Accepted: 30 August 2021
Published: 6 September 2021

1. Introduction

During the outbreak of infectious diseases, social media is usually the most active platform for the exchange of information on infectious disease, and the information released is often of good real-time. Using Internet information to predict the epidemic situation of infectious diseases is one of the current research hotspots. L. Lu et al. used Baidu index and micro-index to conduct a comparative study on influenza surveillance in China [1]. J. H. Lu, School of Public Health, Sun Yat-sen University, and others studied the use of Internet search queries or social media data to monitor the temporal and spatial trends of the Avian Influenza (H7N9) in China, and the results show that the number of H7N9 cases is positively correlated with Baidu Index and Weibo Index search results in space and time [2]. J. X. Feng of the University of South Georgia and others studied the impact of Chinese social networks on the Middle East Respiratory Syndrome Coronavirus and Avian Influenza [3]. Mutual relations prove the effectiveness of using social media to predict infectious diseases. H. G. Gu et al. collected data on cases of H7N9 avian influenza in the Chinese urban population through the Internet, as well as geographic and meteorological data during the same period, and established a disease risk early warning model for human infection with H7N9 avian influenza, which can identify the high risk areas of avian influenza outbreaks and issue an early warning [4]. However, in these studies, most of the search process of network data adopts manual empirical methods to select keywords for search, and the choice of keywords often has a greater impact on search results.

At present, the focus of the world's attention is mainly on the changes in the epidemic situation of the new type of coronary pneumonia. During the four months after the outbreak of the new type of coronavirus in Wuhan, Hubei in December 2019, the epidemic information was widely disseminated on social media such as Baidu, Sina, 360, Sogou, WeChat and QQ. Google, Weibo, Zhihu, Dingxiangyuan, Twitter, Facebook, etc. also released a lot of information about the new coronavirus epidemic, especially through the Google platform to spread to the world. On 31 March 2019, Google launched a project called "COVID-19 Public Datasets" to provide a public database related to the epidemic and open it to the public for free, which means that people can freely access and analyze relevant data and information [5]. How to use this information to predict the spread of the new type of coronary pneumonia in time is an urgent research topic. Currently, X. M. Zhao and others have proposed to use big data retrospective technology to study the spreading trend and epidemic control of the new coronary pneumonia [6]. B. McCall et al. used artificial intelligence methods to predict the new type of coronary pneumonia, thereby protecting medical staff and controlling the spread of the epidemic [7]. These studies are still in the preliminary stage, and the use of network data and prediction of the new coronary pneumonia are not yet ideal.

In this article, we consider that the amount of data indexed by Baidu is large enough for us to use. Based on this, we use the first feature in the search index, namely Baidu index [8], to study the prediction of the epidemic of new coronary pneumonia. We collected data on COVID-19 cases in China from 1 January 2020 to 1 April 2020, and used the random forest feature selection method to select the optimal sub-data set, and used grid search, random search and the Bayesian optimization algorithm optimizes the 7 hyperparameters of the LightGBM (light gradient boosting machine) model. The results show that the application of the data set obtained from the Baidu index to the Bayesian-optimized LightGBM model can better predict the growth of the number of patients with new coronary pneumonia.

This paper is organized as follows. In Section 2, we introduce the data set and analysis method used in detail. Baidu index search and actual case results are compared in time and space, and the impact of keywords and selected index in Baidu index search on the results is analyzed. Model structure, data set preprocessing methods, tuning algorithm, etc. are also introduced in detail. In Section 3, the experimental results are showed and related discussions are presented. Finally, the conclusion is drawn in Section 4.

2. Materials and Methods

2.1. COVID-19 Dataset

In order to standardize prevention and treatment, on 11 February 2020, the World Health Organization named the pneumonia caused by the new coronavirus as "COVID-19" (Corona Virus Disease 2019). In this study, we first obtain the data of COVID-19 cases that occurred in China from 1 January 2020 to 1 April 2020 by searching the COVID-19 Public Datasets on the Google platform, mainly including diagnosis number and death toll, and use them as actual data. These data are released by the Centers for Disease Control (CDC), so we identify these data as CDC data , namely the CDC-Diagnosis and CDC-Death toll mentioned in this paper. Then, we can collect keywords related to COVID-19 through commonly used social networking sites, such as Baidu, Sina, 360, Sogou, WeChat, QQ, Google, Weibo, Zhihu, Dingxiangyuan, Twitter, Facebook, etc., And form a keyword library. Then use the Baidu index platform (http://index.baidu.com, (accessed on 1 April 2020)) to retrieve relevant keywords, and use the statistics of the average daily search volume of relevant Chinese keywords as social network mining data for prediction. In this article, this part of the data is identified as Baidu index data.

By searching for the name and clinical symptoms of new coronavirus pneumonia on social networking sites, we can get the following keywords: new coronavirus, fever, dry cough, fatigue, dyspnea and cough. Using the Baidu index platform to retrieve the above keywords, we can get the average daily search volume of each keyword from 1 January 2020 to 1 April 2020, that is, Baidu index data. Table 1 shows part of the data of the CDC data set and the Baidu index data set. See Appendix A for all the data.

Table 1. Partial data from CDC and Baidu Index search.

Data Source / Date	CDC-Diagnosis	Baidu-New Coronavirus	Baidu-Fever	Baidu-Dry Cough	Baidu-Fatigue	Baidu-Dyspnea	Baidu-Cough	CDC-Death Toll
1 January 2020	0	0	4001	1100	256	481	5885	0
2 January 2020	0	0	4323	1206	278	602	6448	0
3 January 2020	1	0	4212	1173	262	654	6392	0
4 January 2020	0	0	4309	1109	270	621	6570	0
5 January 2020	5	0	4327	1118	271	591	6564	0
6 January 2020	0	0	4324	1226	310	693	6404	0
7 January 2020	0	0	3920	1175	288	633	5875	0
8 January 2020	0	0	3803	1124	272	622	5354	0
9 January 2020	0	8812	3693	1131	270	579	5182	0
10 January 2020	0	2032	3700	1095	263	535	5022	0
11 January 2020	0	2879	3478	1083	237	498	5033	1
12 January 2020	0	1445	3364	1067	252	474	5011	1
13 January 2020	0	1515	3573	1118	278	494	4418	1
14 January 2020	0	4846	3479	1133	266	528	4359	1
15 January 2020	0	4191	3241	1097	245	512	4355	2

Note: CDC = Centers of Disease Control.

2.1.1. Time and Space Comparative Analysis of Baidu Index Search and Actual Cases

Based on the data obtained during the data collection phase, we have drawn the trend graph of CDC data and Baidu Index data over time, as shown in Figure 1. From Figure 1a–g, it can be seen that the keyword "dry cough" is the most commonly used keyword when Chinese netizens search for symptoms of new coronavirus pneumonia, followed by fever, dyspnea, and fatigue. We can see that in the Baidu index method, the keywords "new coronavirus" and "dry cough" are the best choices. The extracted data has the best spatio-temporal positive correlation with the actual number of cases. Through website search, we can find that these two keywords mainly appear in the columns of Baidu Baike and Baidu Health Pharmacopoeia. Therefore, it is recommended to search these two columns first when choosing keywords in the future. On the other hand, it can also be seen that the Baidu index method is used to predict the change trend of the new coronavirus pneumonia. If the keywords are not selected properly, not only will the accuracy of the prediction be low, but sometimes it may even make it impossible to predict in advance.

In addition, we can see that the CDC diagnosis number and Baidu index data have peak times, so we can compare the correlation between the Baidu index data and the CDC-Diagnosis number from the perspective of the first peak generation time and the time difference, which are shown in Table 2. From the comparative analysis of Figure 1 and Table 2, we can draw the following conclusions. The actual number of new coronavirus pneumonia cases in China reached its highest value on 12 February 2020, which was 15,152, while the Baidu Index data all reach their peak before this date, and the average value of the first peak time difference between the Baidu Index data based on the six keywords and the newly diagnosed CDC is 18 days. This is mainly because during the outbreak of the COVID-19, people like to discuss the it on social media networks. The information released on the new crown epidemic is often of good real-time. The CDC data collection comes from the national infectious disease surveillance system, where the pneumonia often requires a longer diagnosis process from onset to diagnosis, usually 7–14 days.

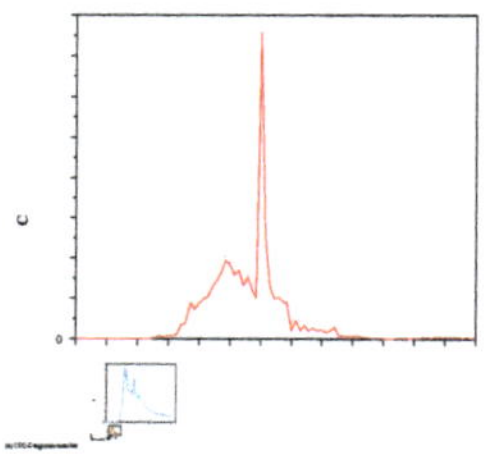

a normal state. It is further verified that the data distribution has a large skewness, and further data conversion is needed to make it conform to the normal distribution.

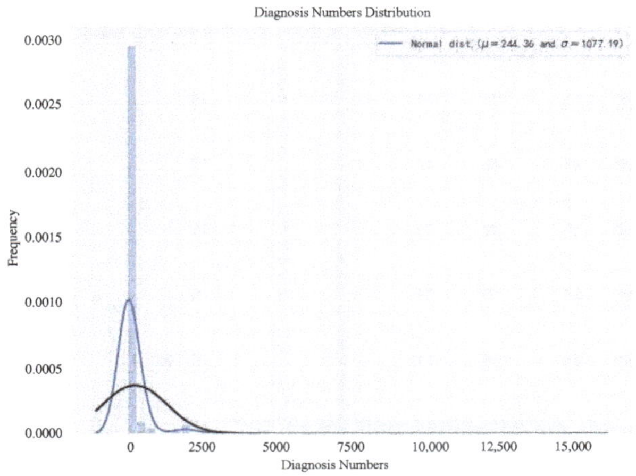

Figure 2. Original diagnosis numbers distribution diagram.

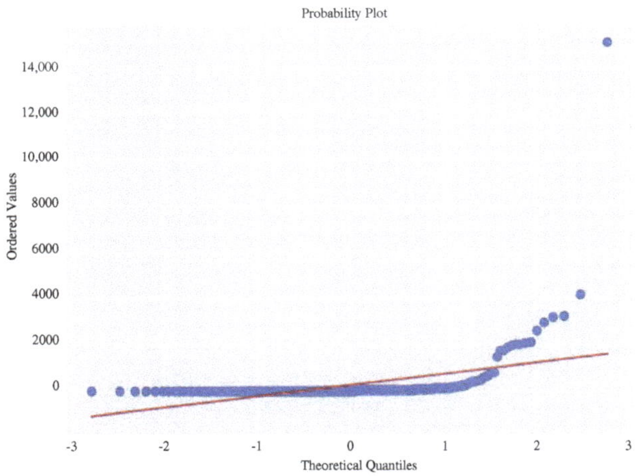

Figure 3. Original diagnosis numbers Q-Q diagram.

Figure 4 shows the relationship between Diagnosis Numbers and other attributes. It can be seen from the figure that the attributes in the data set are basically positively correlated with the attributes of Diagnosis Numbers. Figure 5 shows the relationship between all attributes, which can be represented by a heat map. The heat map uses different colors to intuitively show the relationship between different attributes, which is a very simple way of data interpretation. The values in the figure are calculated using Pearson's correlation coefficient. The calculation formula of Pearson's correlation coefficient is

$$r(X,Y) = \frac{\mathrm{Cov}(X,Y)}{\sqrt{\mathrm{Var}[X]\,\mathrm{Var}[Y]}}. \tag{1}$$

It can be seen from the heat map that the attribute of month is negatively correlated with Diagnosis Numbers. It can be seen from the above analysis that the collected data set has a certain influence on Diagnosis Numbers and can be used for the numerical prediction of Diagnosis Numbers.

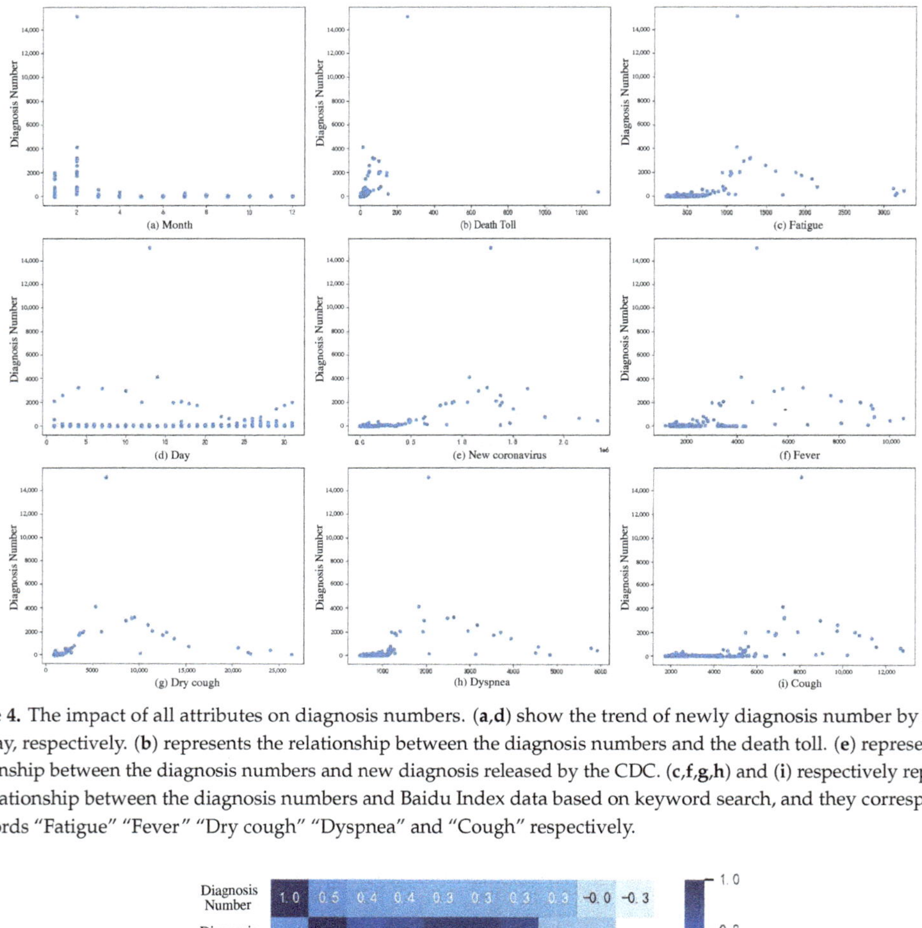

Figure 4. The impact of all attributes on diagnosis numbers. (**a**,**d**) show the trend of newly diagnosis number by month and day, respectively. (**b**) represents the relationship between the diagnosis numbers and the death toll. (**e**) represents the relationship between the diagnosis numbers and new diagnosis released by the CDC. (**c**,**f**,**g**,**h**) and (**i**) respectively represent the relationship between the diagnosis numbers and Baidu Index data based on keyword search, and they correspond to keywords "Fatigue" "Fever" "Dry cough" "Dyspnea" and "Cough" respectively.

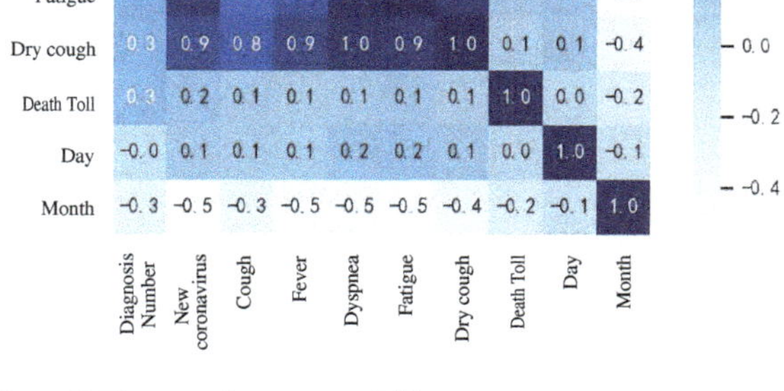

Figure 5. Heat map between variables.

2.2. RF-BOA-LightGBM

As a new cutting-edge technology, predictive models based on machine learning have been widely used in various fields of medicine. For example, Y. D. Zhang et al. proposed a new attention network model, namely ANC (attention network for COVID-19) model, which can diagnose COVID-19 more effectively and accurately [9]. X. Zhang et al. enhanced the deep learning network AlexNet to achieve a more effective classification of new coronary pneumonia [10]. Here, we consider using the RF-BOA-LightGBM (random forest-Bayesian optimization algorithm-light gradient boosting machine) model to predict the development trend of the COVID-19.

2.2.1. Model Structure

Figure 6 shows the model structure used in this article. After collecting the data, you need to perform a simple processing on the data, so that this model can "learn" the data. Then build the LightGBM model for training, but due to the many parameters of LightGBM, the effect of using the default parameters to train the data set in this article is not necessarily good, so three hyperparameter tuning algorithms are introduced here to adjust the model parameters of LightGBM Perform tuning. After finding a combination of model parameters suitable for the data set in this article, the training prediction is carried out.

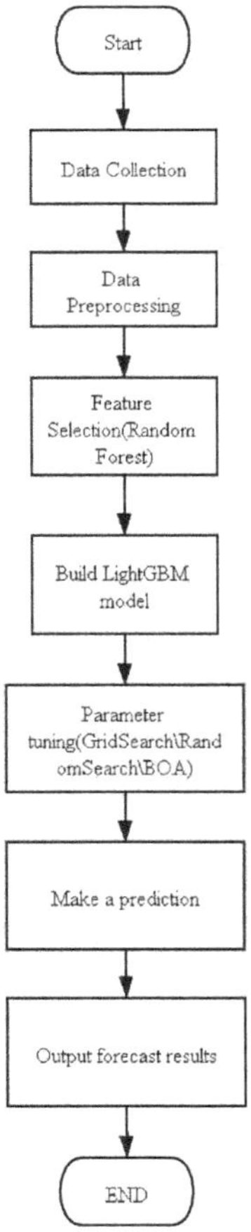

Figure 6. RF-BOA-LightGBM structure. BOA = Bayesian optimization algorithm.

2.2.2. Dataset Preprocessing

In order to enable the model to fully learn the data obtained from the Baidu Index COVID-19 vaccine, this article first made great efforts to preprocess the data. It can be seen from the foregoing that the distribution of the data in this paper presents a similar normal distribution. Therefore, this article first performs logarithmic transformation on the data to make the data satisfy the normal distribution. The data conversion formula is

$$y = \log_c(1 + \lambda_x). \quad (2)$$

Then, deal with the missing data in the data set and delete the samples with missing values (there are not many samples with missing values, which has little effect on the results). Subsequently, the date is divided into three attributes: year, month, and day, and the year attribute is deleted (the year attribute is a fixed value and has little effect on the result), which avoids the problem that the model cannot directly process the date. Finally, the maximum and minimum normalization method is used to integrate the data into (0, 1) range data, which eliminates the influence between samples of different orders of magnitude. The maximum and minimum normalization formula is as follows

$$X_{norm} = \frac{X - X_{min}}{X_{max} - X_{min}}. \quad (3)$$

The distribution graph and Q-Q graph of the processed data are shown in Figures 7 and 8 respectively. As can be seen from the figure, the data has basically satisfied the normal distribution.

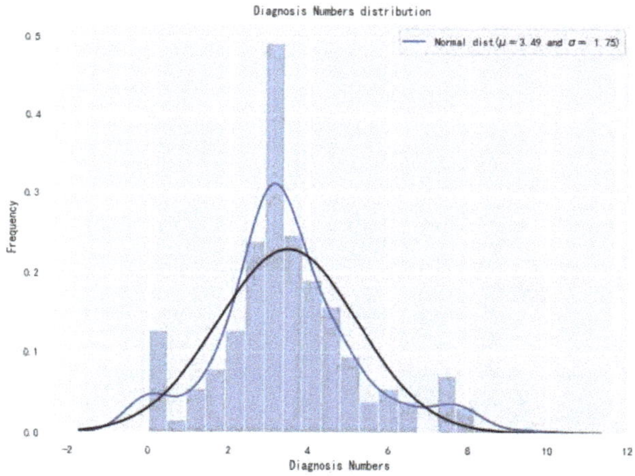

Figure 7. Distribution of diagnosis numbers after data conversion.

This data set contains feature data related to the number of new crowns, irrelevant feature data and related but redundant feature data. In the face of complex faults, it is no longer possible to accurately obtain the number of new crowns by relying only on expert experience and simple correlation analysis to perform feature selection work. Important features, so this article uses random forest (RF) out-of-bag estimation to rank the importance of new crown-related features. The random forest is used to select the features of the data set, and the features that have little influence on the prediction results are eliminated.

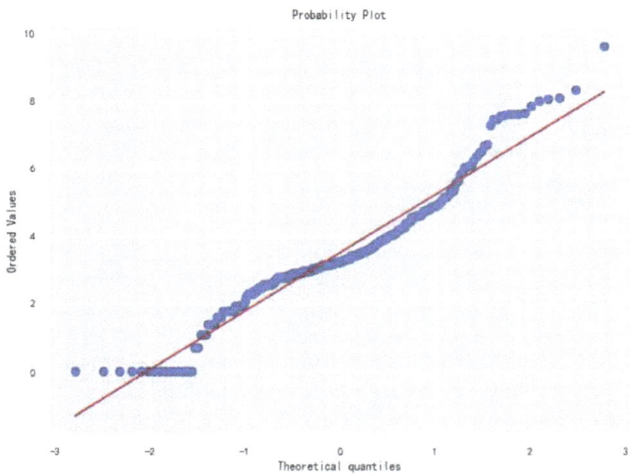

Figure 8. Diagnosis numbers Q-Q diagram after data conversion.

RF is a combined classifier based on decision trees, which can be used for feature selection [11]. RF uses the Bagging method to randomly and repeatably extract samples from the original sample set for classifier training. About 1/3 of the sample data will not be selected [12]. This data is called Out of Bag (OOB). When calculating the importance of a certain feature, use the OOB data as the base learner after the test set to test the training, and the test error rate is recorded as the out-of-bag error (errOOB). Add noise to the important features to be calculated in the OOB sample, and recalculate errOOB again. The average test error of all base learners is calculated by using the average accuracy decrease rate (MDA) as an indicator for feature importance calculation, namely

$$MDA = \frac{1}{n}\sum_{t=1}^{n}(\text{err}\,OOB_t - \text{errOOB}'_t), \quad (4)$$

where n is the number of base learners, errOOB is the out-of-bag error after adding noise.

The more the MDA index decreases, the more the corresponding feature has a greater impact on the prediction result, and the higher its importance. This feature importance calculation method is called random forest out-of-bag estimation. According to this method, the importance of fault-related features is ranked and feature selection is performed.

2.2.3. Tuning Algorithm

For the LightGBM model, there are many internal hyperparameters that affect the prediction results. However, if the value of the hyperparameter used is the default value, this hyperparameter combination may not be the optimal hyperparameter combination for the new coronavirus number prediction data set [13]. Therefore, this paper introduces three tuning algorithms, namely grid search, random search, and Bayesian optimization, to optimize some important hyperparameters of LightGBM [14]. Before adjusting the parameters of LightGBM, the optimization range of hyperparameters is generally set first. These three algorithms are briefly described below.

Grid search divides the search range into grid shapes, and adjusts the parameters according to the set step to train the model until all possible combination parameters are verified, and finally the parameter combination that gives the best result is output [15]. Because the different prediction results of the data in each group of hyperparameter combinations are also different, when the hyperparameter combination is relatively large and the search range is relatively large, the optimization speed of the grid search is very slow.

Random search is similar to grid search, but it does not verify all possible parameter combinations like grid search, but randomly combines the random value of each parameter, so the speed of random search is faster than that of Grid search [16]. However, random search may also miss the parameter combination that maximizes the prediction result.

Bayesian optimization algorithm(BOA) can quickly find the optimal parameters for the problem to be solved based on historical experience [17]. The main problem scenarios for Bayesian optimization are

$$X^* = \text{argmax}\, f(x)(x \in S), \tag{5}$$

where x is the parameter to be optimized, S is the candidate set of x variable, that is, the set of possible values of parameter x. The target selects an x from the set S such that the value of $f(x)$ is the largest or smallest. Here, the specific formula of $f(x)$ may not be known, that is, the black box function. But you can choose an x, and get the value of $f(x)$ through experiment or observation [18].

BOA has two core processes, a priori function (PF) and acquisition function (AC). The acquisition function is also called the efficiency function. Under the framework of Bayesian decision theory, many collection functions can be interpreted as evaluating the expected loss associated with f at point x, and then usually selecting the point with the lowest expected loss [19]. PF mainly uses Gaussian process regression, AC mainly uses these methods including EI (expected improvement), PI (probability of improvement) and UCB (upper confidence bound), and this article uses the EI function. The EI function can find out the global optimum without falling into the local optimum. The collection function is as follows

$$u(x) = \max(0, f' - f(x)), \tag{6}$$

where f is the collection function,and $f(x)$ is the optimized performance indicator.

The final collection function for variable x is

$$\begin{aligned} a_{\text{EI}}(x) &= \text{E}[\mathbf{u}(x) \mid x, D] = \textstyle\int_{-\infty}^{f'}(f' - f)\text{N}(f; \mathbf{u}(x), K(x,x))df \\ &= (f' - u(x))\Phi\Big(f^l; u(x), K(x,x)\Big) + K(x,x)\Big)N(f'; u(x), K(x,x)). \end{aligned} \tag{7}$$

The calculation shows that the point corresponding to the maximum value of $\mathbf{a}_{\text{EI}}$ is the best point. There are two components in Formula (7). To maximize the value of it, you need to optimize the left and right parts at the same time, that is, the left side needs to reduce the $\mu(x)$ as much as possible, and the right side needs to increase the variance (or covariance) $K(x, x)$ as small as possible. It is a typical theory on issues such as exploration and exploitation.

Upper confidence bound (UCB) can be simply understood as the upper confidence boundary. It is usually described by maximizing f instead of minimizing f. But in the case of minimization, the collection function will take the following form

$$a_{UCB}(x) = u(x) - \beta\sigma(x), \tag{8}$$

where $\beta > 0$ is a strategy parameter, and $\sigma(x) = \sqrt{K(x,x)}$ is the boundary standard deviation of $f(x)$. Similarly, UCB also includes exploitation $(u(x))$ and exploration $((x)$ modes. It can converge to the global optimal value under certain conditions.

Table 3 shows the hyperparameter combinations selected in this article and the corresponding descriptions.

Table 3. The LightGBM hyperparameters selected in this article and their functions.

Parameter	Style	Search Scope	Effect
learn_rate	float	(0.001, 0.3)	improve accuracy
max_depth	int	(3, 10)	prevent overfitting
num_leaves	int	(3, 1024)	improve accuracy
min_data_in_leaf	int	(0, 80)	prevent overfitting
feature_fraction	float	(0.2, 0.9)	accelerate
bagging_fraction	float	(0.2, 0.9)	accelerate
lambda_l1	float	(0, 10)	prevent overfitting

2.2.4. LightGBM

LightGBM is an open source decision tree-based gradient boosting framework proposed by Microsoft. As an improved version of Gradient Boosting, it has the characteristics of high accuracy, high training efficiency, support for parallelism and GPU, small memory required, and ability to handle large-scale data [20].

According to the different generation methods of the base learner, integrated learning can be divided into parallel learning and serial learning. As the most typical representative of serial learning, Boosting algorithm can be divided into Adaboost and Gradient Boosting. The main difference between them is that the former improves the model by increasing the weight of misclassified data points, while the latter improves the model by calculating negative gradients. The core idea of Gradient Boosting is to use the negative gradient of the loss function to approximate the value of the current model $f(x) = f_{j-1}(x)$ to replace the residual. Suppose the training sample is i ($i = 1, 2, \ldots, n$), the number of iterations is j ($j = 1, 2, \ldots, m$), and the loss function is $L(yi, f(xi))$, then the negative gradient r_{ij} can be expressed as

$$r_{ij} = -\left[\frac{\partial L(y_i, f(x_i))}{\partial f(x_i)}\right]_{f(x)=f_{j-1}(x)}. \tag{9}$$

Use the base learner $h_j(x)$ to fit the negative gradient r of the loss function, and find the best fit value r_j that minimizes the loss function

$$r_j = \arg\min L(y_i, f_{j-1}(x_i) + rh_j(x_i)). \tag{10}$$

Model update:

$$| \ f_j(x) = f_{j-1}(x) + r_j h_j(x). \tag{11}$$

Gradient Boosting generates a base learner in each round of iteration. Through multiple rounds of iteration, the final strong learner $F(x)$ is the base learner generated in each round and obtained by linear addition:

$$F(x) = f_m(x) \tag{12}$$

As an improved lightweight Gradient Boosting algorithm, the core ideas of LightGBM are: histogram algorithm, leaf growth strategy with depth limitation, direct support for category features, histogram feature optimization, multithreading optimization, and cache hit rate optimization. The first two features effectively control the complexity of the model and realize the lightweight of the algorithm, so this article is particularly concerned.

The histogram algorithm discretizes continuous floating-point features into L integers to construct a histogram with a width of L. When traversing the data, use the discretized value as an index to accumulate statistics in the histogram. After traversing the data once, the histogram accumulates the necessary statistics, and then find the optimal split point from the discrete values of the histogram .

The traditional leaf growth strategy can split the leaves of the same layer at the same time. In fact, the splitting gain of many leaves is low and there is no need to split, which brings a lot of unnecessary expenses. For this, LightGBM uses a more efficient leaf growth strategy: each time it searches for the leaf with the largest split gain from all the current leaves to split, and sets a maximum depth limit. While ensuring high efficiency, it also prevents the model from overfitting.

3. Results and Discussion

3.1. Performance Predictor

All models are cross-validated and the coefficient of determination (R2), mean absolute error (MAE), relative absolute error (RAE), relative square root error (RRSE), root mean square error (RMSE) are calculated, as shown below

$$R2(y,\hat{y}) = \sqrt{1 - \frac{\sum_{i=1}^{n}(y_i - \hat{y}_i)^2}{\sum_{i=1}^{n}(y_i - \bar{y})^2}}, \quad (13)$$

$$\mathrm{RMSE}(y,\hat{y}) = \sqrt{\frac{1}{n}\sum_{i=1}^{n} y_i - \hat{y}_i^2}, \quad (14)$$

$$\mathrm{MAE}(y,\hat{y}) = \frac{1}{n}\sum_{i=1}^{n}|y_i - \hat{y}_i|, \quad (15)$$

$$\mathrm{RAE}(y,\hat{y}) = \sqrt{\frac{\sum_{i=1}^{n}|y_i - \hat{y}_i|}{\sum_{i=1}^{n}|y_i - \bar{y}|}}, \quad (16)$$

$$\mathrm{RRSE}(y,\hat{y}) = \sqrt{\frac{\sum_{i=1}^{n}(y_i - \hat{y}_i)^2}{\sum_{i=1}^{n}(y_i - \bar{y})^2}}, \quad (17)$$

where y represents the true value, $\hat{y}$ represents the predicted value, $\bar{y}$ represents the average value of the true value and n is the number of test sets.

3.2. Experiment Results

Figure 9 shows the result of feature selection using random deep forest, and the features are output in descending order of importance. It can be seen from the figure that Death Toll has the greatest impact on Diagnosis Numbers, while the attribute of Month has the least impact. Finally, we selected the 7 most influential attributes for the prediction of Diagnosis Numbers.

According to the optimal parameter set of the model, the Diagnosis Numbers prediction model of COVID-19 is constructed. In this paper, LightGBM, GridSearch-LightGBM, RandomSearch-LightGBM, and BOA-LightGBM models are used for Diagnosis Numbers prediction. Table 4 shows the specific values of the optimal parameter combinations found by the three tuning algorithms.

Table 4. Specific parameter values found by three tuning algorithms.

Parameter	GridSearch	RandomSearch	BOA
learn_rate	0.632	0.828	0.355
max_depth	7	8	5
num_leaves	225	237	249
min_data_in_leaf	33	27	30
feature_fraction	0.7	0.7	0.8
bagging_fraction	0.7	0.7	0.8
lambda_l1	2.34	3.45	1.80

Note: BOA = Bayesian optimization algorithm.

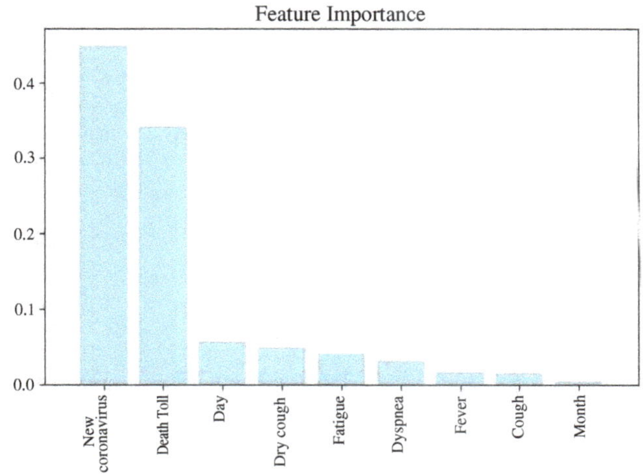

Figure 9. Feature selection results.

Table 5 shows the evaluation indicators of the prediction results of the four models. The prediction results of the model are evaluated by R2, RMSE, MAE, RAE, RRSE evaluation indicators. It can be seen from the values of the five evaluation indicators that the results of BOA-LightGBM are better than the former. RandomSearch-LightGBM and GridSearch-LightGBM have their own advantages and disadvantages. It can also be seen that the default hyperparameters of LightGBM are not suitable for the prediction of Diagnosis Numbers of COVID-19 in this article. From the approximate prediction effect, BOA-LightGBM can better analyze the relationship between historical data and can effectively predict the value of Diagnosis Numbers of COVID-19, which proves the superiority of the model.

Table 5. Model evaluation index.

Models	R2	RMSE	MAE	RAE	RRSE
LightGBM	0.820	354.945	138.939	0.535	0.424
GridSearch-LightGBM	0.865	311.918	145.266	0.548	0.368
RandomSearch-LightGBM	0.861	316.217	137.621	0.533	0.373
BOA-LightGBM	0.879	295.686	124.911	0.508	0.348

Note: GBM, gradient boosting machine; BOA, Bayesian optimization algorithm; R2, coefficient of determination; RMSE, root mean square error; MAE, mean absolute error; RAE, relative absolute error; RRSE, relative square root error.

Figure 10 is a line chart of the four algorithms to predict Diagnosis Numbers, and only part of the data is taken on the abscissa. The prediction effect of the model can be seen more intuitively from the line graph. It can be seen from the figure that in most cases, the BOA-LightGBM model can better fit the fluctuation trend of Diagnosis Numbers at some points, and the predicted value is very close to the actual value. In the figure, the points predicted by GridSearch-LightGBM are basically covered, so they are not shown in the figure, which just shows that the prediction results are not very prominent. Sometimes the prediction value of LightGBM is better than other models, but most of them are inferior to other models. So comprehensively, the BOA-LightGBM model is more in line with the changing trend of real values.

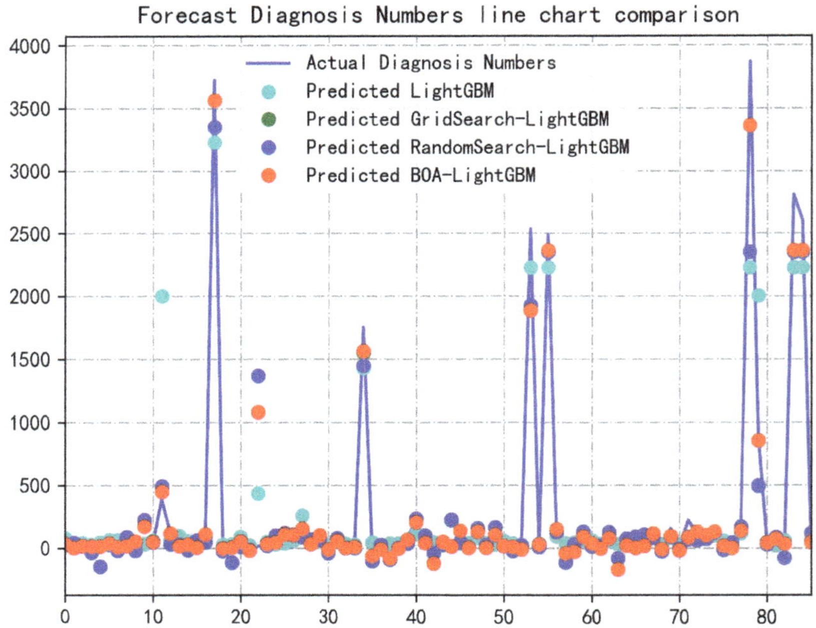

Figure 10. Comparison of predicted and true values of the four models. BOA, Bayesian optimization algorithm; GBM, gradient boosting machine;.

4. Conclusions

This study uses the Internet big data tool-Baidu Index to predict the development trend of the new coronavirus pneumonia epidemic to obtain data. By selecting appropriate keywords, data on COVID-19 cases in China from 1 January 2020 to 1 April 2020 are collected. After preprocessing the data set, the random forest feature selection method is used to obtain the optimal sub-data set. After comparing and analyzing the optimization results of the seven hyperparameters of the LightGBM model with the three optimization algorithms of grid search, random search, and Bayesian optimization. It is concluded that applying the data set obtained from the Baidu Index to the Bayesian-optimized LightGBM model can better predict the increase in the number of new coronary pneumonias, and it is a good aid to predict the new number of new coronary pneumonia in the future medical structure effect.

Author Contributions: Conceptualization, D.H.H; methodology, Z.L. and D.H.H; formal analysis, Z.L.; data curation, Z.L.; writing—original draft preparation, Z.L.; writing—review and editing, D.H.H; validation, D.H.H All authors have read and agree to the published version of the manuscript.

Funding: This research received no external funding.

Institutional Review Board Statement: Not applicable.

Informed Consent Statement: Not applicable.

Data Availability Statement: Data are available at School of Life Sciences, Central South University, China.

Conflicts of Interest: The authors declare no conflict of interest.

Appendix A. Datasets from CDC and Baidu Index Search

Table A1. Datasets from CDC and Baidu Index search.

Data Source / Date	CDC-Diagnosis	Baidu-New Coronavirus	Baidu-Fever	Baidu-Dry Cough	Baidu-Fatigue	Baidu-Dyspnea	Baidu-Cough	CDC-Death Toll
1 January 2020	0	0	4001	1100	256	481	5885	0
2 January 2020	0	0	4323	1206	278	602	6448	0
3 January 2020	1	0	4212	1173	262	654	6392	0
4 January 2020	0	0	4309	1109	270	621	6570	0
5 January 2020	5	0	4327	1118	271	591	6564	0
6 January 2020	0	0	4324	1226	310	693	6404	0
7 January 2020	0	0	3920	1175	288	633	5875	0
8 January 2020	0	0	3803	1124	272	622	5354	0
9 January 2020	0	8812	3693	1131	270	579	5182	0
10 January 2020	0	2032	3700	1095	263	535	5022	0
11 January 2020	0	2879	3478	1083	237	498	5033	1
12 January 2020	0	1445	3364	1067	252	474	5011	1
13 January 2020	0	1515	3573	1118	278	494	4418	1
14 January 2020	0	4846	3479	1133	266	528	4359	1
15 January 2020	0	4191	3241	1097	245	512	4355	2
16 January 2020	0	5174	3230	1100	267	546	4220	2
17 January 2020	4	7713	3247	1114	254	521	4008	2
18 January 2020	17	7754	3271	1060	228	492	4218	2
19 January 2020	36	29,003	3418	1182	253	548	4323	2
20 January 2020	151	266,892	4064	3684	609	1090	5324	2
21 January 2020	77	659,926	5474	10,162	1106	2073	7260	2
22 January 2020	149	852,363	6782	21,967	1711	3125	8751	3
23 January 2020	131	1,374,253	9151	26,393	3141	4840	10,229	11
24 January 2020	259	1,469,947	8108	21,718	3162	4511	9059	41
25 January 2020	688	2,330,851	10029	24,100	3253	5922	12798	56
26 January 2020	769	2,150,021	10552	20,635	3117	5779	12677	80
27 January 2020	1771	1,816,430	9406	15,323	2152	4572	11,547	106
28 January 2020	1459	2,227,942	9091	15,115	2296	4087	11,185	132
29 January 2020	1737	1,503,255	9350	13,783	2088	3940	11,351	170
30 January 2020	1982	1,372,206	9287	12,574	1943	3541	10,786	213
31 January 2020	2102	1,390,560	8855	12,974	1876	3702	10,584	259
1 February 2020	2590	1,334,127	8108	11,425	1620	2952	9741	304
2 February 2020	2829	1,374,154	7682	10,981	1491	3162	9750	361
3 February 2020	3235	1,277,132	7258	10,683	1365	2949	8517	425
4 February 2020	3887	1,244,048	6602	9504	1293	2626	7258	490
5 February 2020	3694	1,209,808	6213	8763	1349	2380	7434	563
6 February 2020	3143	1,943,197	5736	8305	1295	2179	8043	636
7 February 2020	3399	1,643,941	5789	9236	1292	2488	7261	722
8 February 2020	2656	1,185,978	5126	7287	1183	2131	6718	811
9 February 2020	3062	1,142,892	5220	8719	1187	2004	8173	908
10 February 2020	2478	1,158,302	5450	8585	1212	1946	8948	1016
11 February 2020	2015	1,061,433	4814	7421	1239	1901	8641	1113

Table A1. *Cont.*

Data / Source / Date	CDC-Diagnosis	Baidu-New Coronavirus	Baidu-Fever	Baidu-Dry Cough	Baidu-Fatigue	Baidu-Dyspnea	Baidu-Cough	CDC-Death Toll
12 February 2020	15,152	1,050,392	4590	5971	1163	1922	7908	1367
13 February 2020	5090	1,277,024	4745	6436	1125	2049	8076	1380
14 February 2020	2641	1,069,203	4140	5339	1126	1830	7197	1523
15 February 2020	2009	948,165	3295	4537	1018	1456	6452	1596
16 February 2020	2048	904,431	2994	3953	942	1205	5461	1770
17 February 2020	1886	920,373	3454	4025	1046	1406	6542	1868
18 February 2020	1749	840,490	3274	3652	1056	1278	6889	2004
19 February 2020	394	784,784	3327	3530	1038	1315	6848	2118
20 February 2020	889	800,960	3035	3071	1012	1345	6552	2236
21 February 2020	397	776,563	3003	3244	935	1269	6467	2345
22 February 2020	648	636,594	2663	3003	949	1179	5606	2442
23 February 2020	409	622,095	2777	2771	978	1172	5218	2592
24 February 2020	508	634,391	3234	2695	1025	1286	5940	2663
25 February 2020	406	550,484	3066	2550	964	1260	5462	2715
26 February 2020	433	482,726	2850	2468	896	1202	5451	2744
27 February 2020	327	478,822	2835	2403	819	1165	5354	2788
28 February 2020	427	486,394	2660	2285	845	1195	5425	2835
29 February 2020	573	496,289	2420	2213	750	1133	4655	2870
1 March 2020	202	482,280	2244	2070	686	1151	4458	2912
2 March 2020	125	441,914	2468	2123	785	1176	5326	2943
3 March 2020	119	393,118	2223	1955	755	1143	4741	2981
4 March 2020	139	441,921	2264	1970	765	1163	4967	3012
5 March 2020	143	414,142	2122	1789	680	1140	5157	3042
6 March 2020	99	376,106	2111	1658	694	1112	5186	3070
7 March 2020	44	369,780	1877	1539	723	1072	4196	3097
8 March 2020	40	368,916	1759	1480	646	1052	3993	3119
9 March 2020	19	359,426	2017	1414	687	1133	5547	3136
10 March 2020	24	335,711	1792	1288	635	1085	4164	3158
11 March 2020	15	337,491	1911	1413	633	1049	4331	3169
12 March 2020	8	353,167	1891	1575	686	1088	3967	3176
13 March 2020	11	353,857	1906	1756	641	1119	3269	3189
14 March 2020	20	332,215	1745	1358	601	1042	2788	3199
15 March 2020	16	364,033	1721	1486	657	1037	2732	3213
16 March 2020	21	324,566	1985	1555	759	1087	3845	3226
17 March 2020	13	300,185	1885	1546	673	1068	3022	3237
18 March 2020	34	295,536	1920	1491	696	1052	3198	3245
19 March 2020	39	282,990	1724	1355	663	1057	2742	3248
20 March 2020	41	300,183	1779	1227	621	1036	2705	3255
21 March 2020	46	299,291	1734	1308	641	1006	2577	3261
22 March 2020	39	285,191	1736	1102	672	1027	2829	3270
23 March 2020	78	280,841	1855	1391	704	1102	3018	3277
24 March 2020	47	278,221	1830	1457	704	1052	3215	3281
25 March 2020	67	259,091	1810	1308	656	1045	3446	3287

Table A1. *Cont.*

Data \ Source Date	CDC-Diagnosis	Baidu-New Coronavirus	Baidu-Fever	Baidu-Dry Cough	Baidu-Fatigue	Baidu-Dyspnea	Baidu-Cough	CDC-Death Toll
26 March 2020	55	261,957	1839	1094	655	1091	3114	3292
27 March 2020	54	279,082	1645	1129	592	1061	2780	3295
28 March 2020	45	264,664	1525	1065	476	998	2480	3300
29 March 2020	31	265,761	1562	1096	490	961	2364	3304
30 March 2020	48	264,442	1725	1094	601	1031	3021	3305
31 March 2020	36	239,272	1772	1071	535	1007	2676	3312
1 April 2020	35	243,582	1569	1080	565	1013	2676	3318

Note: CDC = Centers of Disease Control.

References

1. Lu, L.; Zou, Y.Q.; Peng, Y.S.; Li, K.L.; Jiang, T.J. Comparison of Baidu index and Weibo index in surveillance of influenza virus in China. *Appl. Res. Comput.* **2016**, *33*, 392–395.
2. Chen, Y.; Zhang, Y.Z.; Xu, Z.W.; Wang, X.Z.; Lu, J.H.; Hu, W.B. Avian Influenza A (H7N9) and related Internet search query data in China. *Sci. Rep.* **2019**, *9*, 10434. [CrossRef] [PubMed]
3. Fung, I.C.H.; Fu, K.W.; Ying, Y.C.; Schaible, B.; Hao, Y.; Chan, C.H.; Tse, Z.T.H. Chinese social media reaction to the MERS-CoV and avian influenza A(H7N9) outbreaks. *Infect. Dis. Poverty* **2013**, *2*, 31. [CrossRef] [PubMed]
4. Gu, H.G.; Zhang, W.J.; Xu, H.; Li, P.Y.; Wu, L.L.; Guo, P.; Hao, Y.T.; Lu, J.H.; Zhang, D.M. Predicating risk area of human infection with avian influenza A (H7N9) virus by using early warning model in China. *Chin. J. Epidemiol.* **2015**, *36*, 470–475.
5. COVID-19 Coronavirus Data. Available online: https://data.europa.eu/euodp/en/data/dataset/covid-19-coronavirus-data (accessed on 14 December 2020).
6. Zhao, X.M.; Li, X.H.; Nie, C.H. Retrospecting the spread of new coronary pneumonia based on big data and China's control of the epidemic. *Bull. Chin. Acad. Sci.* **2020**, *35*, 248–255.
7. McCall, B. COVID-19 and artificial intelligence: Protecting health-care workers and curbing the spread. *Lancet Digit. Health* **2020**, *2*, 166–167. [CrossRef]
8. Baidu Index. Available online: http://index.baidu.com/ (accessed on 1 April 2020).
9. Zhang, Y.D.; Zhang, X.; Zhu, W.G. ANC: Attention network for COVID-19 explainable diagnosis based on convolutional block attention module. *Cmes-Comp. Model. Eng.* **2021**, *127*, 1037–1058.
10. Zhang, X.; Lu, S.Y.; Wang, S.H.; Yu, X.; Wang, S.J.; Yao, L.; Pan, Y.; Zhang, Y.D. Diagnosis of COVID-19 pneumonia via a novel deep learning architecture. *J. Comput. Sci. Tech.* **2021**, *1*. [CrossRef]
11. Sylvester, E.V.A.; Bentzen, P.; Bradbury, I.R.; Clement, M.; Pearce, J.; Horne, J.; Beiko, R.G. Applications of random forest feature selection for fine-scale genetic population assignment. *Evol. Appl.* **2018**, *11*, 153–165. [CrossRef] [PubMed]
12. Li, X.K.; Chen, W.; Zhang, Q.R.; Wu, L.F. Building auto-encoder intrusion detection system based on random forest feature selection. *Comput. Secur.* **2020**, *95*, 101851. [CrossRef]
13. Al Daoud, E. Comparison between XGBoost, LightGBM and CatBoost using a home credit dataset. *Int. J. Comput. Inf. Eng.* **2019**, *13*, 6–10.
14. Frazier, P.I. A tutorial on Bayesian optimization. *arXiv* **2018**, arXiv:1807.02811.
15. Liashchynskyi, P.; Liashchynskyi, P. Grid search, random search, genetic algorithm: A big comparison for NAS. *arXiv* **2019**, arXiv:1912.06059.
16. Wang, Y.; Wang, T. Application of improved LightGBM model in blood glucose prediction. *Appl. Sci.* **2020**, *10*, 3227. [CrossRef]
17. Liang, X. Image-based post-disaster inspection of reinforced concrete bridge systems using deep learning with Bayesian optimization. *Comput.-Aided Civ. Inf.* **2019**, *34*, 415–430. [CrossRef]
18. Jones, D.R.; Schonlau, M.; Welch, W.J. Efficient global optimization of expensive black-box functions. *J. Global Optim.* **1998**, *13*, 455–492. [CrossRef]
19. Sameen, M.I.; Pradhan, B.; Lee, S. Application of convolutional neural networks featuring Bayesian optimization for landslide susceptibility assessment. *Catena* **2020**, *186*, 104249. [CrossRef]
20. Liang, W.Z.; Luo, S.Z.; Zhao, G.Y.; Wu, H. Predicting hard rock pillar stability using GBDT, XGBoost, and LightGBM algorithms. *Mathematics* **2020**, *8*, 765. [CrossRef]

healthcare MDPI

Article

Incorrect Facemask-Wearing Detection Using Convolutional Neural Networks with Transfer Learning

Jesús Tomás *, Albert Rego, Sandra Viciano-Tudela and Jaime Lloret

Instituto de Investigacion Para la Gestion Integrada de Zonas Costeras, Universitat Politecnica de Valencia, C/Paranimf 1, Grao de Gandia, 46730 Valencia, Spain; alremae@teleco.upv.es (A.R.); sandraviciano8493@gmail.com (S.V.-T.); jlloret@dcom.upv.es (J.L.)

* Correspondence: jtomas@upv.es

Abstract: The COVID-19 pandemic has been a worldwide catastrophe. Its impact, not only economically, but also socially and in terms of human lives, was unexpected. Each of the many mechanisms to fight the contagiousness of the illness has been proven to be extremely important. One of the most important mechanisms is the use of facemasks. However, the wearing the facemasks incorrectly makes this prevention method useless. Artificial Intelligence (AI) and especially facial recognition techniques can be used to detect misuses and reduce virus transmission, especially indoors. In this paper, we present an intelligent method to automatically detect when facemasks are being worn incorrectly in real-time scenarios. Our proposal uses Convolutional Neural Networks (CNN) with transfer learning to detect not only if a mask is used or not, but also other errors that are usually not taken into account but that may contribute to the virus spreading. The main problem that we have detected is that there is currently no training set for this task. It is for this reason that we have requested the participation of citizens by taking different selfies through an app and placing the mask in different positions. Thus, we have been able to solve this problem. The results show that the accuracy achieved with transfer learning slightly improves the accuracy achieved with convolutional neural networks. Finally, we have also developed an Android-app demo that validates the proposal in real scenarios.

Keywords: facemask-wearing condition; transfer learning; convolutional neural network; deep learning; facial recognition; COVID-19

Citation: Tomás, J.; Rego, A.; Viciano-Tudela, S.; Lloret, J. Incorrect Facemask-Wearing Detection Using Convolutional Neural Networks with Transfer Learning. *Healthcare* **2021**, *9*, 1050. https://doi.org/10.3390/healthcare9081050

Academic Editor: Pedram Sendi

Received: 19 July 2021
Accepted: 10 August 2021
Published: 16 August 2021

1. Introduction

1.1. Motivation

In December 2019, cases of pneumonia of unknown origin appeared in Wuhan, Hubei, China, the clinical symptoms of which were very similar to pneumonia of viral origin. In the first group of infected people, it was determined that it was a zoonotic infection, that is, a viral transmission from animals to humans [1]. Due to the taking of samples from the lower respiratory tract, employing the RT-PCR technique in real-time, genome sequencing was carried out. This fact allowed us to clarify the taxonomy of this virus. New Coronavirus 2019 (2019-nCoV) was the name that was assigned to it.

Different studies have shown the spread by aerosols and fomites of SARS-CoV-2 among humans. In addition, it has been demonstrated that the virus can be spread even before the appearance of symptoms, referring to asymptomatic patients who are carriers of the virus, which implies that they can spread it without showing any symptoms of the disease. This fact has caused the rapid evolution and expansion of the pandemic. Due to the rapid spread of the virus, the World Health Organization (WHO) declared COVID-19 a public health emergency of international concern [2]. According to the WHO, as of 2 May 2021, there have been 151,803,822 confirmed cases of COVID-19, including 3,186,538 deaths.

The first symptoms found in this new coronavirus were coughing, fevers, and respiratory distress. However, with the progress of studies, other identifying symptoms of this virus have been determined. According to an update in March 2020 by the Council of State and Territorial Epidemiologists (CSTE), the loss of smell and taste was added as one of the compatible and most characteristic symptoms of this virus [3]. Moreover, in several studies, it has been shown that the pandemic has affected us mentally, not only because of the consequences left by the virus once the patient has recovered, but also because of the restrictions to which we have been subjected to reduce the transmission of SARS-CoV-2.

Mental health is associated with both demographic and psychosocial factors that, during this pandemic, cause a higher predisposition in some people to suffer these types of problems. The psychiatric disorders associated with this pandemic are usually stress, emotional disorders, depression, anxiety, sleeping problems, panic attacks, among others [4].

In addition to the symptoms of COVID-19, its rapid transmission between humans has been the object of study. According to studies, the contagion rate is higher in closed spaces without ventilation. This fact is because viral particles are capable of traveling in microdroplets (<10 μm), also called aerosols. These aerosols are produced by the human being when we speak, sing, laugh, etc. In addition, their speed increases with the force of the flow, for example, when we run or shout. What happens is that the largest drops fall to the ground, however, its nucleus (where the virus particle is located) is suspended in the air and is capable of being inhaled by another person, producing a possible infection [5,6].

Due to the transmission of the virus by air, many countries have introduced the mask as a mandatory use for protection against possible infections. With the use of the mask, the morbidity of COVID-19 and its associated mortality has been reduced. In addition, medical care has been reduced, preventing health systems from collapsing [7]. In addition, the use of the mask together with social distancing has managed to flatten the epidemic curve. The masks have the following two main functions: on the one hand, to prevent the viral particles that travel in the aerosol from being transmitted between the general population when coughing, speaking, sneezing, etc. On the other hand, the material from which the mask is made allows the volatile particles to be filtered from the air. In addition, the use of the mask has been of great help in preventing asymptomatic patients from infecting the rest of the general population [8]. Recent studies show that surgical masks effectively reduce the emission of viral particles. Coronavirus was detected in 30 and 40% of samples collected in participants without face masks, but no virus in droplets or aerosols was detected in participants with face masks. This study was conducted on exhaled air samples of SARS-CoV and MERS-CoV from infected patients and the findings indicate that surgical masks effectively reduce the emission of viral particles [9].

The recommendations of the Centers for Disease Control and Prevention (CDC) [10,11] indicate that while wearing the surgical mask, on exhalation, the air from the nose and mouth leaves with a high velocity and is directed frontally. The particles are relatively thick, between 3 and 8 microns (1 micron = 0.001 mm), and impact directly on the inside of the mask. Even if air escapes through the edges, bacteria, or other particles, do not escape since, due to their thickness, they are not able to follow the flow lines of the air that leaves the edges as long as the adjustment is correct.

The proper use of masks requires strict adherence to general hygiene measures, among which adequate coverage of the mouth and nose, avoiding gaps between the face and the mask, stands out. A partial, incorrect, or asymmetric fit poses a high risk for the transmission of infection [12].

It is due to the great importance of the use of masks that it is necessary to control their correct use. This fact leads to an increase in the control methods of non-pharmaceutical products that allow reducing the transmission of the virus [13]. For this reason, methods based on Artificial Intelligence (AI) have taken on great relevance, which allows a more exhaustive control over mask use in public spaces or areas with large population concurrence.

1.2. The Aim of the Study

Our objective is to study the application of Machine Learning techniques to distinguish whether a person is wearing a mask properly. Therefore, this project must take into account not only the presence or absence of a mask, but also its proper use, meaning, and being able to distinguish when it is well placed and when it is not. It offers detection and warning regarding multiple possibilities of placement errors that, without this application, would be hard for a person to notice. Although one of the most obvious mistakes is the use of the mask under the nose, there are others, such as the use of the mask over glasses, which are more difficult to detect, but no less important, or the poor fitting of the mask to the face. For the development of this application, deep learning techniques have been used to recognize people's faces and proper mask use. For this aim, we need a solid training set that allows us to achieve good accuracy and reduce bias. Since there is no training set available for this task, one of the challenges of the project has been its creation.

1.3. Related Work

Due to the serious threat of the COVID-19 pandemic, novel solutions to optimize the incorrect use of prevention mechanisms are a hot topic. AI is one of the most useful techniques in adapting problems such as image classification to different situations. In this subsection, we are going to describe the most relevant works about AI and COVID-19 technical solutions.

AI and CNN solutions can solve problems by detecting patterns in images. Authors such as Chung et al., in [14], have developed an application integrated with the mobile phone capable of recognizing and classifying plants through the use of images using InceptionV3 CNN [15]. They have also implemented a prototype for identifying tree species with which real-time classification is performed remotely.

However, the current pandemic situation produced by SARS-CoV-2, in which we find ourselves, has led various authors to try to improve the control of the spread of the virus by developing applications to control the use of masks. Although, indeed, in previous years, authors such as Nieto-Rodriguez et al. had already developed this type of system to deal with other epidemics. Nieto-Rodriguez et al. in 2015 [16] developed a real-time image processing system in VGA resolution reaching 10 fps. VGA resolution allows the object to be 5 m from the camera to distinguish faces or masks. The system was developed to control the use of masks by medical personnel within operating rooms. In this way, an alarm goes off when the health personnel do not carry it because its use is mandatory.

Nevertheless, it has been in the last two years that the number of publications related to these AI systems has skyrocketed due to the critical situation we are facing.

Chen et al., in [17], have developed a mobile application that allows us to determine the service life of a facemask, indicating what period it is in, in addition to telling us what its level of effectiveness is after a period of use. To do this, they use microphotographs by extracting four characteristics of gray, employing co-occurrence matrices (GLCM) from the microphotographs of the facial mask. Using KNN, three results are obtained. The precision of these is 82.87% (macro measurements). The precision of "normal use" and "not recommended" reaches 92.00 and 92.59%.

Nonetheless, the need to control the overcrowding of people who wear or do not wear masks in public spaces has increased in importance in recent years.

Nagrath et al., in [18], have developed a design that can differentiate between the use of a facial mask or not. To perform real-time mask detection, they have used the MobileNetV2 architecture [19] as the framework for the classifier, together with the SSDMNV2 approach, which uses the single-shot multi-box detector as a face detector. By these means, they propose to use deep learning TensorFlow, Keras, and OpenCV to detect face masks. The precision of this study is 92.64% and it has an F1 score of 0.93. Mata, in [20], created a CNN model to be able to differentiate which people use a mask and which do not. It is based on a deep learning technique using an image or a video stream.

In a study carried out by Jauhari et al., in [21], the aim was to detect facial patterns to be able to detect the presence of facial masks in images. For this, it was based on Single Board Computer (SBC) Raspberry Pi. A face detection system based on the Viola Jones method was used to obtain efficient, fast, and accurate results. This method allows the adjustment of the cascade classifier to determine the area of the face in the image. The precision of this study is 90.9%. Sen et al., in [22], through the sequence of images and video, have developed a system capable of differentiating between people who wear face masks from those who do not. They use a MobileNetV2 model [19] along with python's PyTorch and OpenCV for mask detection. The model has an accuracy of 79.24%. At the same time, an entry system to public places, which differentiates people who wear a mask from those who do not, has been proposed by Balaji et al. in [23]. In addition, this system has an alarm that emits a beep with a red or green tone to alert if a person is not wearing a mask. A Raspberry-PI camera is used to capture the video and transform it into images for further processing.

Recent studies show the applicability of these types of applications. Kurlekar et al., in [24], have developed a system that can be integrated with offices, airports, and public places in general. With their application, they can detect face masks in static images as well as in real-time videos. To do this, they used Computer Vision concepts and Deep Learning, using OpenCV and Keras/TensorFlow. Sakshi et al., in [25], using Keras/TensorFlow, developed a face mask detector. The architecture on which it was based is MobileNetV2 [19]. The model has been trained with several variations to ensure that the system can identify face masks in real time through video or still images. The final objective is, through Computer Vision, to implement the model in areas of high population density, health care areas, educational institutions, etc.

In 2020, Cheng et al., in [26], proved that the detection of the use of masks was important in stopping the spread of the virus. With the use of YOLO v3-tiny, it has proven to be suitable for the real-time detection of mask use. Plus, it is small, fast, and suitable for mobile hardware deployment, as well as real-time detection. Loey et al., in [27], developed a hybrid system for the detection of face masks. They selected three data sets. The simulated masked face data set, the real-world masked face set, and the tagged faces in nature. The design of this study is composed of Resnet-50 [28], for the feature extraction component. A second component for the classification of face masks is used by this system based on Support Vector Machines (SVM) and a joint algorithm for the mask classification process. The precision of the system is 99.49%, 99.64%, and 100%, respectively, for each of the data sets studied.

In addition to the need to be able to detect the use or not of a mask, the detection of when it is used in a wrong way due to its incorrect positioning is of great relevance. This misuse significantly reduces its effectiveness against the virus. For this reason, several authors, in addition to detecting its presence or absence, have focused on detecting its correct or incorrect placement. In 2020, Rudraraju et al., in [29], developed an application based on two steps. On the one hand, it detected the use or non-use of a facial mask. After detecting a mask, it distinguished between its correct or incorrect use. To do so, it relies on fog computing. Two nodes are used to process the video sequencing. Each fog node implements two MobileNet models [19]. For face detection, Haar cascade classifiers are used. Streaming takes place locally at each fog gateway without relying on the Internet. In this way, only the mask is allowed to enter the room and only if the mask is well placed. The accuracy of this system is around 90%.

Wang et al. 2021, in [30], using hybrid machine learning techniques, proposed to detect the use of masks using a two-stage approach. In the first stage, the user wearing a face mask is detected using the Faster RCNN and InceptionV2 [15] structure model. The second step is directed to a stage of verification of real face masks implemented by a classifier through a learning system. The general accuracy for simple scenarios is 97.32%, while for more complex scenarios it is 91.13%.

Smart Screening and Disinfection Walkthrough Gate (SSDWG) was created by Hussain et al. in 2021 [31]. It is a low-cost, fast and effective virus spread detection and control system based on IoT, for all places of entry. In addition to registering body temperature through temperature sensors that do not require physical contact, the system is also capable of differentiating people who wear face masks from those who do not. For the classification, it was also added not only if they were wearing a mask but also their correct use. For this classification, VGG-16, MobileNetV2 [19], InceptionV3 [15], ResNet-50 [28], and CNN have been compared using a transfer learning approach. The use or non-use of the mask was implemented through deep learning in real time. The obtained precision was 99.81 and 99.6% using VGG-16 and MobileNetV2, respectively. In addition, the classification of the type of mask, either N-95 or surgical masks, has also been implemented. Qin et al., in [32], using super-resolution and classification networks (SRCNet), with the training from the Medical Masks database, have developed a method to identify the presence or absence of a mask. This method, in addition, is capable of identifying the most frequent error of its misuse, such as wearing the mask under the nose. The algorithm used is based on the following four steps: image pre-processing, face detection and cropping, super-resolution images, and mask detection. The precision achieved with the use of this methodology is 98.70%. Table 1 summarizes the different works presented in this section.

Table 1. Summary of the related works.

1st Author [ref]	Date	Type of Detection	Face Detector	Classification Model	Software Library	Best Accuracy
Nagrath [18]	March 2021	mask/ no mask	Single shot multibox	MobileNetV2	TensorFlow, OpenCV	92.64%
Mata [20]	April 2021	mask/ no mask	Image Data Generator	CNN	TensorFlow, OpenCV	60%
Jauhari [21]	March 2021	mask/ no mask	Cascade Viola Jones	AdaBoost	Python	90.9%
Sen [22]	February 2021	mask/ no mask	-	MobileNetV2	PyTorch, OpenCV	79.2%
Balaji [23]	2021	mask/ no mask	Viola-Jones detector	VGG-16 CNN	TensorFlow, OpenCV	96%
Kurlekar [24]	April 2021	mask/ no mask	-	CNN	TensorFlow, OpenCV, Caffe	-
Sakshi [25]	March 2021	mask/ no mask	-	MobileNetV2	TensorFlow, Keras	99%
Cheng [26]	2020	mask/ no mask	YOLO v3 -tiny	CNN + SVM	-	-
Loey [27]	January 2021	mask/ no mask	YOLO v3	Resnet50 + SVM	-	99.5%
Rudraraju [29]	September 2020	mask/ no mask/ nose out	Haar cascade classifier	MobileNet	OpenCV, Keras	90%
Wang [30]	January 2021	mask/ no mask/ nose out	Fast RCNN	InceptionV2	OpenCV, Matlab	91.1%
Hussain [31]	April 2021	mask/ no mask/ nose out	YOLO v3	VGG-16, MobileNetV2, InceptionV3, ResNet50	Keras	99.8%
Qin [32]	September 2020	mask/ no mask/ nose out	Multitask Cascaded CNN	SRCNet	Matlab	98.7%

The rest of the paper is structured as follows. In Section 2, materials and methods, we explain how we obtained the training data, how they are labeled, and the details of the intelligent system. In Section 3, the results are shown and described. Section 4 includes the discussion and, finally, Section 5, shows the conclusion along with future work.

2. Materials and Methods

This section describes the methods used in solving the problem. Our system is divided into 5 phases. Each one of the first five points in this section corresponds to these phases. In the last section, the system validation is described.

2.1. Obtaining the Data Set

The main objective of this paper is to identify mask placement errors using machine learning techniques. The drawback of machine learning is the need of a properly labeled training set.

Most of the works related to mask detection have chosen to use a synthetic corpus [33]. The idea consists in drawing, on the image of a face, the drawing of a mask. This method has been very useful in detecting if a mask is being used. However, the problem that interests us is much more complex. We want to detect small problems with the placement of the mask. The synthetic corpus would not work for our problem.

To tackle the problem, we decided to resort to citizen collaboration. We developed an application, shown in Figure 1, for mobile phones, which asked users to place the mask in different positions and take a selfie. The application was published on Google Play [34] and we went to the media to disseminate it [35]. The application was downloaded by more than 500 users during the summer of 2020 and about 3200 images were obtained, with a resolution of 360×480, half of them from the front and the other half from the side.

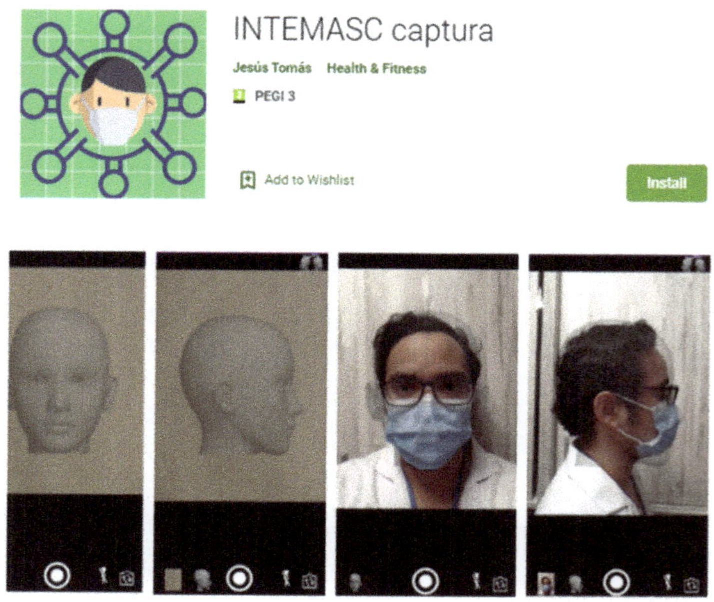

Figure 1. Application in Google Play for the acquisition of the training set.

2.2. Labeling

The labeling was carried out by a nursing group from the hospital of Ontinyent. To speed up the work, an Android application was developed that allowed us to label 12 types of problems as well as the location where the problem was evident. Figure 2 shows this application, with an image from the front and another from the side. The labels were: 1—mask incorrectly extended, 2—non-symmetrical placement, 3—incorrectly bent in the nasal part, 4—adjusted below the bridge of the nose, 5—glasses placed under the mask, 6—neck adjustment greater than 1 cm, 7—with a beard, the use of a mask is not recommended,

8—incorrectly placed rubber band, 9—lateral gap greater than 1 cm, 10—exposed nose, 11—without a mask, 12—others.

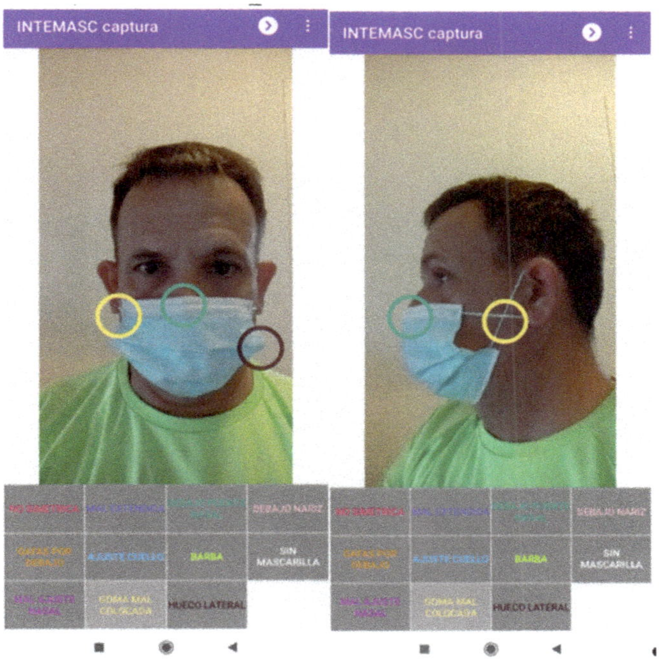

Figure 2. Application for the labeling of the training set.

Table 2 shows the statistics of the labeling process. Up to three errors could be indicated in each image. With the correct mask, 24.5% of the images were labeled, with an error of 55.9%; with two, 15.0%; and with three, 4.5%.

Table 2. Statistics of the labeling process.

	Front Image			Side Image		
	1st Error	2nd Error	3rd Error	1st Error	2nd Error	3rd Error
Correct	518	0	0	476	0	0
without a mask	350	0	0	328	0	0
adjusted below the bridge of the nose	284	74	12	261	67	12
mask incorrectly extended	247	192	58	210	104	21
exposed nose	245	1	0	218	2	0
non-symmetrical placement	108	54	15	56	10	0
lateral gap greater than 1 cm	85	21	0	92	24	1
glasses placed under the mask	82	10	1	77	9	0
incorrectly placed rubber band	81	40	6	98	47	6
incorrectly bent in the nasal part	71	11	1	68	8	0
neck adjustment greater than 1 cm	25	9	2	34	12	2
with a beard, mask is not recommended	12	0	0	11	0	0

The application asked the user to take two selfies; one from the front and the other from the side of the face. As can be seen in Table 2, some errors were better detected from the side, such as "incorrectly placed rubber band" or "lateral gap greater than 1 cm". However, it was decided that the side angle selfie would not be used given the user's great difficulty in taking them. In fact, almost 10% of the side photographs were poorly framed and had to be discarded.

In this project, we will focus on the 5 most frequent problems (see Figure 3). To achieve this, some errors have been eliminated, such as "incorrectly placed rubber band" and "neck adjustment greater than 1 cm". Others, such as "adjusted below the bridge of the nose" and "incorrect nasal bend" have been joined into a single type: "bad adjustment of nasal bridge".

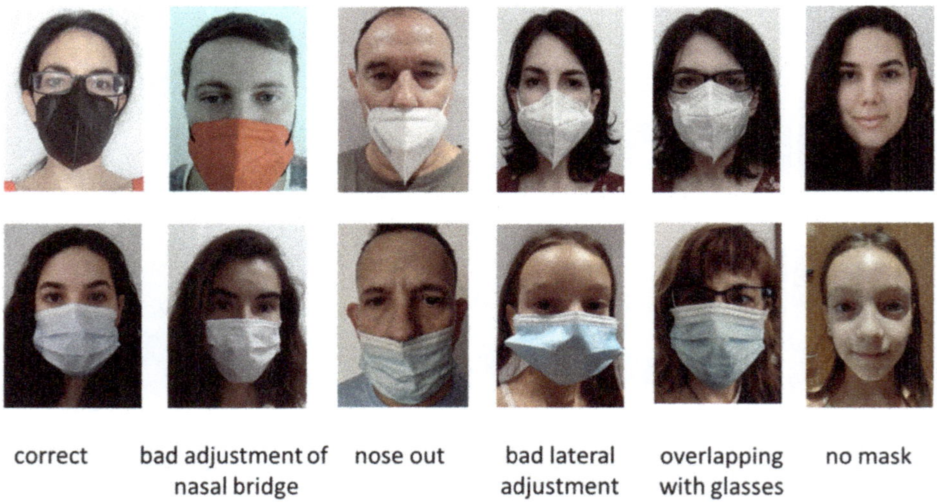

Figure 3. Example of samples of the 6 categories detected once cut.

Only the images from the front and that also only present one type of problem, or none, will be used. These images have been divided into two sets, 1000 for training and 194 for validation.

It is important to highlight that there is high subjectivity in the labeling of the data set. Determining that the mask is perfectly fitted is relative. Some tests carried out show how two experts evaluating the same set provided a difference in up to 10% of the labels.

2.3. Facial Detection and Cropping

When obtaining the data set, the volunteers were asked to place their faces on a template. However, these indications were not followed very strictly. To normalize this situation, it has been decided to perform face detection, to eliminate the edges of the image without relevant information. Rapid Object Detection Using a Boosted Cascade of Simple Features [36] was adopted for facial detection, which has been shown to perform well in obtaining facial areas.

2.4. Classification

Recently, machine learning has experienced a breakthrough thanks to the emergence of Deep Learning. More specifically, CNN are the main ones responsible for this revolution. A convolutional network is structured hierarchically. The first layers are responsible for extracting generic features from the image such as edges or textures. The following layers use these previous characteristics to search for more specific characteristics. This process is continued for several layers until it is possible to detect characteristics with a high semantic value such as the detection of eyes or nose. Finally, a conventional neural network is used to perform the classification.

Although CNNs are widely used in natural language or audio processing tasks, several studies show that their use have obtained the best results in image recognition tasks. These results make the use of CNN in our problem, a natural choice. Nevertheless, in machine learning, obtaining the data set is the most complex part. The most common is having little data.

A widely used technique to obtain the most out of the data set is Data Augmentation [37]. It consists of making small modifications to the images such as small rotations, translations, and zooming in the input images to increase the variability of the training set. After several experiments, we verified that translation and zoom operations did not improve the results. We think it may be because the face is already cropped in the images. Finally, the training dataset was randomly rotated in a range of $[-5^\circ, 5^\circ]$ and with a horizontal flip.

When few training samples are available, the Transfer Learning technique is quite useful. This method is based on using a model that was previously trained on a large data set, usually in a large-scale image classification task. This model will be used to customize this model for our task.

Transfer Learning is applied in two phases. First, we use the convolutional layers from the original model for feature extraction. The last layers, where the images are classified, are replaced to fix our problem. In the first phase the convolutional layers are fixed, only the classification layers are trained. In the second phase, known as fine tuning, all layers are unlocked, and the system is retrained. In this way, the extraction of characteristics fits our task.

We have, nowadays, a great variety of convolutional networks with dozens of layers already trained at our disposal. We can highlight MobileNet [19], Inception [15], ResNet [28], VGG [37], and Xception [38]. All these networks have been trained with the ImageNet corpus [39], a large data set with more than 14 million images where 20,000 different objects are recognized.

2.5. Decision System

This section proposes a decision system to detect, in real situations, errors in the use of masks. Depending on these errors, the system acts in the following different ways: by alerting errors, asking the user to solve the problem, etc.

The block diagram of the system is depicted in Figure 4. There are four different actors that constitute our proposal. The first one is the face detection module. This module presents a computer vision solution for face detection in real time. The second one regards classifying. The classifying module receives the input from the face detection module. Then, based on the system explained in the previous section, the classifying stage detects the errors in the facemask placement. The error and situation analysis module is included in third place. This module is an algorithm that evaluates the error given by the classifying system to select the most appropriate action. This is explained later in this section. The actuator module interacts with all the previous ones to create the warning or thanks the person for their correct use of the facemask.

Now, the only module that we need to specify is the decision system. Figure 5 shows the flow diagram of the algorithm. First, the warning level is initialized. Then, the module waits for an output, which is the probability of having a facemask placement error. When the module receives the output from the classifying module, it analyzes the predicted class with the highest probability, that is, the predicted error. Some error classes are more important than others, in which case they will be labeled as serious. For the most important errors, such as no mask detected, the system should ask the user to wear a mask, raise an alert, and block the entrance to the place if necessary. Some other errors are low-risk errors, which can be solved with a warning message to the user. When this happens, the face recognition module has to be started again. However, if there are continuous errors with the mask, due to the fact that the user does not want to wear it correctly, that would be treated as a serious error.

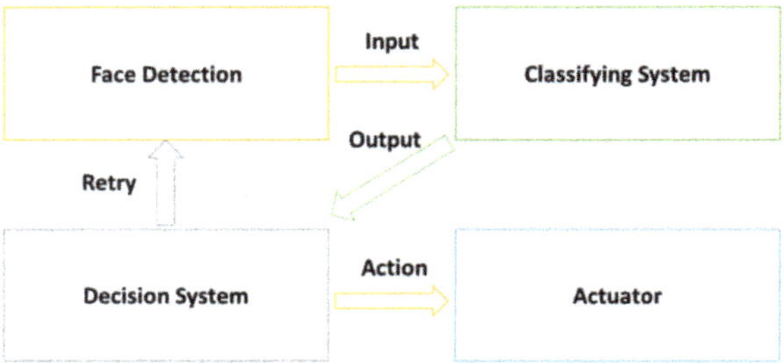

Figure 4. Block diagram of the system.

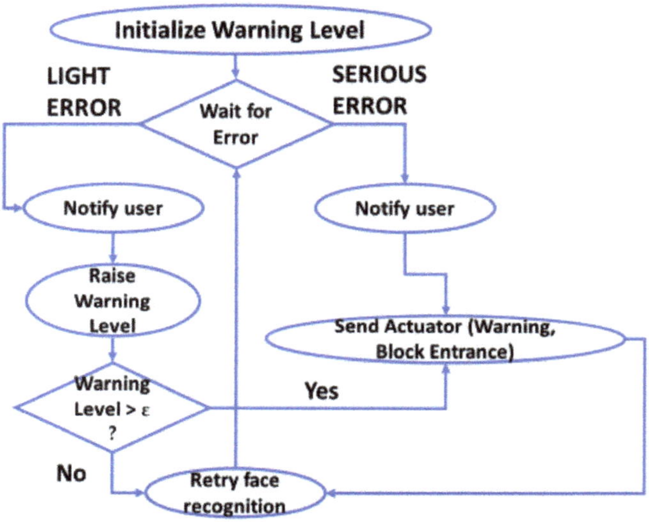

Figure 5. Flow diagram of the decision algorithm.

2.6. System Validation

To validate that the proposed method can be used in real situations, a real-time demonstration on a mobile application has been developed. The application has been developed on Android and can be downloaded from Google Play [40].

The device must be placed at the entrance of a public place, such as a hospital or educational center. Using the camera, the presence of a face will be detected, to isolate the area of interest as indicated in Section 2.3. Using a voice message, it will be indicated if a facemask-wearing problem is detected. Otherwise, the user will be thanked for its correct use.

Given the limited resources of a mobile device, it was decided to use a small and fast model. Specifically, MobileNet V2 [19] is used. As shown in the next section, very competitive results can be obtained using only 14 MB of device memory. As will also be depicted in the next section, some types of problems are not detected satisfactorily (specifically, "overlapping with glasses" and "bad lateral adjustment"). For this reason, these types of detection have not been included in the demonstration.

3. Results

To validate the proposal, we have carried out the experiments described in this section. Firstly, input images were down-sampled to 224 × 224. The Adam method was adopted as the optimizer. The network was trained for 20 epochs with an initial learning rate of 10–4 and with a learning rate dropping factor of 0.9. The batch size was 32. Transfer learning was applied for fine-tuning the same parameters, except the learning rate reduced to 10–5. The OpenCV and TensorFlow libraries were used. The link [41] shows the Python code used in each experiment.

3.1. Convolutional Neural Networks

We start by testing a traditional convolutional network. The results are shown in Figure 6. The two upper curves correspond to accuracy and the lower ones to loss. The following four convolutional layers have been used: 32 of 3 × 3, 44 of 5 × 5, 128 of 5 × 5, and 128 of 3 × 3. In each layer max-pooling (2 × 2) and ReLU activation function is used. For classification, 3 dense layers of 512, 256, and 6 neurons are used.

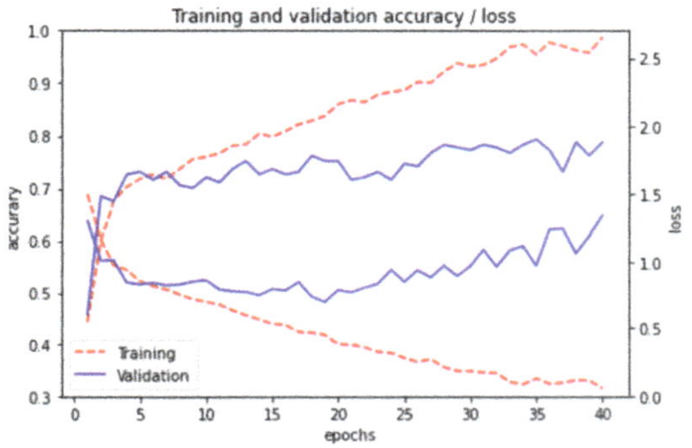

Figure 6. Accuracy and Loss for CNN model for training and validation set.

Figure 7 shows the Confusion Matrix obtained in the validation set. Each row of the matrix represents the instances in an actual class, while each column represents the instances in a predicted class. See Figure 3 for more detail on the categories. Therefore, the diagonal shows the samples that are correctly classified. For example, in this experiment, there are 25 samples labelled as "NO MASK". There are 23 samples that have been correctly classified, one as "NOSE OUT", and another as "CORRECT".

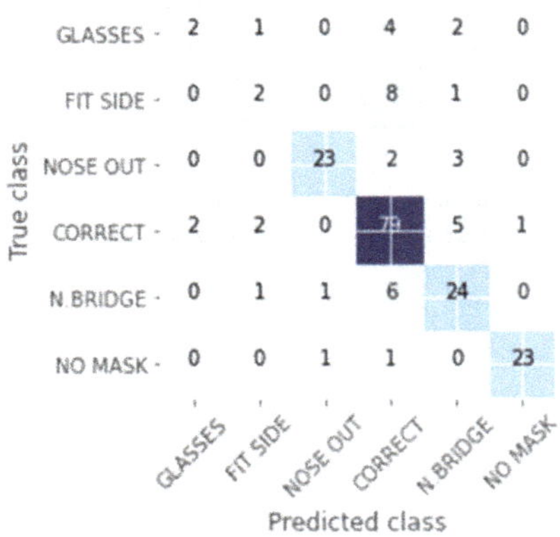

Figure 7. Confusion Matrix for CNN model in the validation set.

In this case, we can confirm that there is a high accuracy except for in the "GLASSES" and "FIT SIDE" classes. "GLASSES" corresponds to "overlapping with glasses", "N. BRIDGE" to "bad adjustment of the nasal bridge", and "FIT SIDE" to "bad lateral adjustment".

3.2. Transfer Learning

To illustrate the advantages of using Transfer Learning as the facemask-wearing condition identification network, we compare the more relevant models, including MobileNetV2 [19], Xception [38], InceptionV3 [15], ResNet-50 [28], NASNet, and VGG19 [37]. We compare features such as network size, the number of parameters, depth, and accuracy for both our task and the ImageNet task, as shown in Table 3. The first and the second rows represent the first experiment described in this section, with and without data augmentation. The rest of the rows are the different models of Transfer Learning with Data Augmentation. Depth refers to the topological depth of the network. This includes activation layers, batch normalization layers, etc. ImageNet accuracy refers to the accuracy obtained in the ImageNet task [42]. Figure 6 shows the accuracy of the transfer learning models. As can be seen in Figures 6 and 8, the precision is very noisy, varying greatly from one epoch to the next. In order to better compare the results, the accuracy shown in Table 3 corresponds to the average of the last three epochs.

Table 3. Comparison of the results obtained from the different models used.

Model	Size	Parameters	Depth	Accuracy	ImageNet Accuracy
CNN without data aug	32 MB	8.5 M	15	0.763	-
CNN	32 MB	8.5 M	15	0.797	-
MobileNet V2	14 MB	3.5 M	88	0.812	0.713
Xception	88 MB	22.9 M	126	0.802	0.790
InceptionV3	92 MB	23.9 M	159	0.819	0.779
ResNet-50	98 MB	25.6 M	-	0.742	0.749
NASNetLarge	343 MB	88,9 M	-	0.799	0.825
VGG16	528 MB	138.4 M	23	0.834	0.713

For each of the indicated models, their feature extraction layers have been used. After these, a layer of averagePooling2D is added and two dense layers of 512 and 6 neurons. The training is carried out in two phases. On epochs 1 to 20, the transfer model is locked and only the classification layers are trained. On epochs 21 to 40, fine-tuning is performed, unlocking learning of all the layers. Other configuration details are described in the Training details. Figure 8 shows the evolution of training for each model. The two upper curves correspond to accuracy and the lower ones to loss. The best results are obtained with VGG16. An improvement of 5% is observed for the results obtained with CNN.

If we analyze these results in more detail using the confusion matrix (Figure 9), we can observe how some kinds of errors such as "GLASSES" and "FIT SIDE" present lower accuracy than the others. However, the mislabeled samples have been reduced from the CNN experiment. Consequently, the accuracy is improved with this model.

Table 4 shows the classification results obtained, in the validation set, for each of the classes using the VGG16 model.

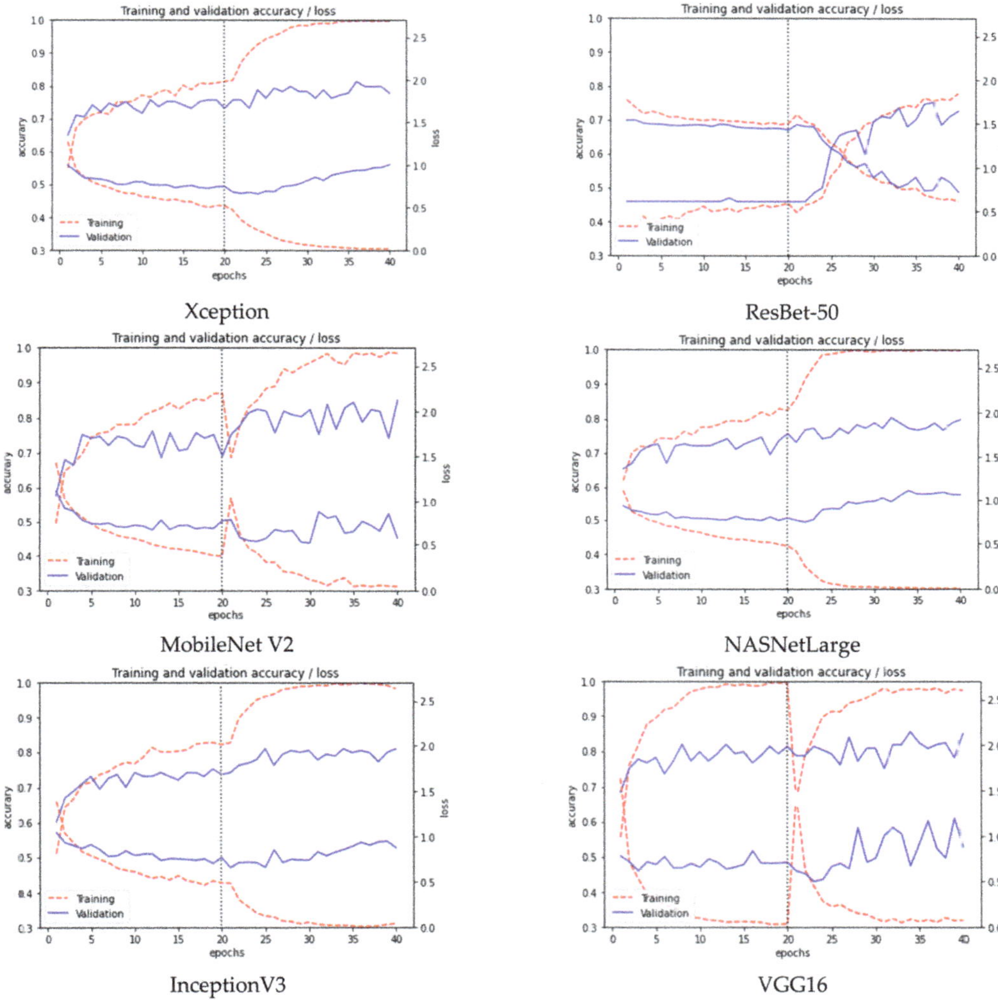

Figure 8. Accuracy for several Transfer Learning models for training and validation set.

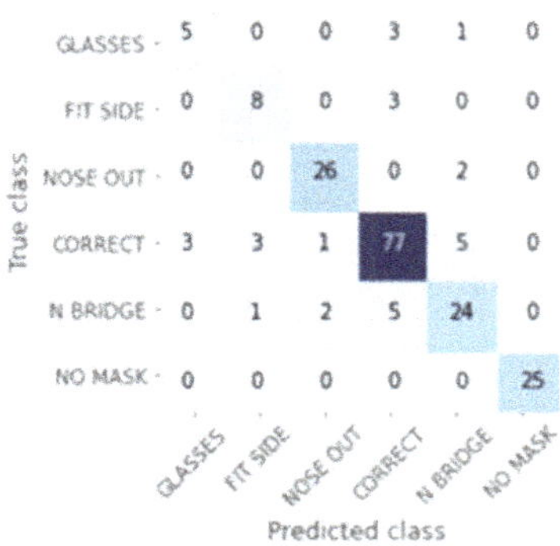

Figure 9. Confusion Matrix for VGG16 model in the validation set.

Table 4. Classification results for VGG16 model in the validation set.

	Precision	Recall	F1-Score
GLASSES	0.56	0.63	0.59
FIT SIDE	0.73	0.67	0.70
NOSE OUT	0.93	0.90	0.91
CORRECT	0.87	0.88	0.87
N.BRIDGE	0.75	0.75	0.75
NO MASK	1.00	1.00	1.00
total	0.84	0.85	0.85

4. Discussion

Due to the great general concern in society about the pandemic caused by the new virus called SARS-CoV-2 and with it, the need to use masks, many authors have developed different applications to detect the presence or absence mask as well as their proper use. In Table 1, different projects are presented with the main objective of developing applications capable of detecting masks.

While our best results' precision is, at first glance, lower than most of the related work, our system analyzes a more complex problem, and the lowered accuracy is due to this fact. By simplifying our results to "MASK" and "NO MASK", our accuracy increases to 100%, as can be seen in Figure 9, as the 25 "NO MASK" samples are classified correctly, and all the other samples are classified as one of the "MASK" classes, even if the exact errors are not always detected.

By simplifying our results to "NOSE OUT", "NO MASK", and "CORRECT" we can also use the confusion matrix shows in Figure 9. In this case, the classes "GLASSES", "FIT SIDE", and "N.BRIDGE" are unified with "CORRECT". If we obtain the new confusion matrix and perform the calculations, the new precision obtained is 97.4%. This result is comparable to those obtained in [29–32] and could be improved with specific training for these three classes.

On the one hand, references [18–27] developed a mask detection system that can differentiate between the presence or absence of a mask. On the other hand, references [29–32] are also able to detect if the nose is outside the mask. This circumstance would reveal an incorrect use of the mask, but as has been discussed in this project, it is only one of the possible incorrect uses.

In the case of our study, a system has been developed that is capable of detecting not only the presence or absence of the mask, but also different placement errors that current systems are not capable of detecting, such as the mask being placed below the nose, incorrect placement due to the use of glasses, that the nasal bridge of the mask is not correctly adjusted, or that the mask is too wide for the person, causing lateral gaps where the virus can easily enter.

To do this, different CNN-based deep learning techniques have been tested. However, the use of data augmentation does not appear to offer significant improvements, possibly due to the way the images are cropped. The transfer learning technique has been used to try to alleviate training shortages. We have tested the current most successful models. The results vary depending on the network used. The VGG16 model presents the best results (83.4% precision). This shows us that the knowledge of image processing can be used in the problem of detecting the correct use of the face mask. As library software, we have used OpenCV and TensorFlow. In addition, our system can detect with great precision other errors not usually considered in the placement of masks that have been mentioned above. By ignoring these errors, this misuse can help spread the virus.

Although in the labeling part of the corpus, more mask wearing problems were considered, in the present study we work with five types of errors. With the easiest of these to detect, such as the "no mask" or "nose out" errors, we have obtained a precision of 100 and 93%, respectively. The detection of an "incorrect adjustment of nasal bridge" error has a success rate of 75% and "incorrect lateral adjustment" a success rate of 73%. The type of error with the worst results is "overlapping with glasses" with 56%. This bad result may be due to a lack of examples in the training set.

Finally, in addition to studying the system, a mobile application has been developed. This application is accessible to all citizens and can be used to see the mistakes made regarding the placement of the mask in situ. In this case, as a classificatory model, we use MobileNetV2. This is because it demands fewer resources than others that were tested in this project and, thanks to this, it can be implemented in real time on current mobile phones, which is a requirement of our demonstration. Moreover, although the error detection precision is decreased, it is still high enough to be able to use the application to detect errors in the misplacing of a mask.

5. Conclusions

Identifying mask misuse is challenging. The limitation in the data sets is the main challenge. Data sets on mask wearing status are generally small and only identify the presence of masks. To solve the problem, we have carried out a campaign to collect images through an app, appealing to citizen participation. The samples obtained have been labeled by a group of health experts.

To our knowledge, no studies have been conducted on the identification of different misuses of masks through deep learning. The study carried out from [29–32] only detected the most obvious error, consisting of wearing the mask under the nose. Our proposal is capable of detecting, in addition to the previous mentioned issue, other types of problems, which occur very frequently. Even many of the users are unaware that they are using the mask incorrectly. However, we have not been able to detect other types of problems, such as ("incorrect lateral adjustment" and "glasses underneath"). For these cases, it is necessary to find an alternative approach or increase the number of training samples.

To validate that our proposal can be used in real situations, a real-time demonstration, an Android application, has been developed. It can be downloaded from Google Play [40]. The system is made to detect errors through the use of selfies. For this reason, for errors of bad lateral placement such as the crossing of rubbers, it would be good to teach the system to detect it with a lateral image. Although there really is a significant problem, it should be taken into account that a bad lateral adjustment indirectly causes an alteration in the frontal positioning of the mask, which can be detected with a frontal image. In this case, the system can be improved gathering more samples and teaching the system to detect, among other alterations, the crossing of the rubbers laterally.

The results support the possibility of its use in real circumstances, which makes it possible to prevent the spread of the pandemic. In future works, we want to enhance the study including other kinds of mask wearing problems and study the inclusion of other types of inputs to improve the accuracy of the "GLASSES" and "FIT SIDE" classes. On the other hand, and despite the fact that the application is capable of detecting different anomalies, it may be necessary to teach the system to differentiate between different components that currently make up the mask. This is because they have become another complement to our clothing. Many of the masks have sequins, other drawings ranging from simple squares to drawn smiles. An improvement in the system would be to verify that the application can detect these modified masks just as it does with surgical and FP2.

Furthermore, this system could be applied to different networks and scenarios. We could apply this to Smart Cities or Industrial Internet of Things environments to prevent security issues and decide when an alert should be raised.

Author Contributions: Conceptualization, J.T.; methodology, J.T. and A.R.; software, J.T.; validation, J.T. and S.V.-T.; investigation, A.R. and J.T.; data curation, J.T.; writing—original draft preparation, J.T., S.V.-T., A.R.; writing—review and editing, J.L.; supervision, J.L. All authors have read and agreed to the published version of the manuscript.

Funding: This research received no external funding.

Conflicts of Interest: The authors declare no conflict of interest.

Code Availability: The code used in the present work, including image processing and network structures, is available at [41].

References

1. Nishiura, H.; Jung, S.M.; Linton, N.M.; Kinoshita, R.; Yang, Y.; Hayashi, K.; Kobayashi, T.; Yuan, B.; Akhmetzhanov, A.R. The extent of transmission of novel coronavirus in Wuhan, China, 2020. *J. Clin. Med.* **2020**, *9*, 330. [CrossRef] [PubMed]
2. Pedersen, S.F.; Ho, Y.C. SARS-CoV-2: A storm is raging. *J. Clin. Investig.* **2020**, *130*, 2202–2205. [CrossRef]
3. Dawson, P.; Rabold, E.M.; Laws, R.L.; Conners, E.E.; Gharpure, R.; Yin, S.; Buono, S.A.; Dasu, T.; Bhattacharyya, S.; Westergaard, R.P.; et al. Loss of taste and smell as distinguishing symptoms of coronavirus disease 2019. *Clin. Infect. Dis.* **2021**, *72*, 682–685. [CrossRef] [PubMed]
4. Hossain, M.M.; Tasnim, S.; Sultana, A.; Faizah, F.; Mazumder, H.; Zou, L.; McKyer, E.L.J.; Ahmed, H.U.; Ma, P. Epidemiology of mental health problems in COVID-19: A review. *F1000Research* **2020**, *9*. [CrossRef] [PubMed]
5. Brooks, J.T.; Butler, J.C. Effectiveness of mask wearing to control community spread of SARS-CoV-2. *JAMA* **2021**, *325*, 998–999. [CrossRef]
6. Wang, J.; Pan, L.; Tang, S.; Ji, J.S.; Shi, X. Mask use during COVID-19: A risk adjusted strategy. *Environ. Pollut.* **2020**, *266*, 115099. [CrossRef]
7. Joo, H.; Miller, G.F.; Sunshine, G.; Gakh, M.; Pike, J.; Havers, F.P.; Kim, L.; Weber, R.; Dugmeoglu, S.; Waston, C.; et al. Decline in COVID-19 hospitalization growth rates associated with statewide mask mandates—10 states, March–October 2020. *Morb. Mortal. Wkly. Rep.* **2021**, *70*, 212. [CrossRef]
8. Li, T.; Liu, Y.; Li, M.; Qian, X.; Dai, S.Y. Mask or no mask for COVID-19: A public health and market study. *PLoS ONE* **2020**, *15*, e0237691. [CrossRef]
9. Leung, N.H.; Chu, D.K.; Shiu, E.Y.; Chan, K.H.; McDevitt, J.J.; Hau, B.J.; Yen, H.-L.; Li, Y.; Ip, D.K.M.; Peiris, J.S.M.; et al. Respiratory virus shedding in exhaled breath and efficacy of face masks. *Nat. Med.* **2020**, *26*, 676–680. [CrossRef]
10. U.S. Food & Administration (FDA). N95 Respiradores Y Mascarillas Quirúrgicas (Mascarillas). Available online: https://www.fda.gov/medical-devices/personal-protective-equipment-infection-control/n95-respirators-and-surgical-masks-face-masks (accessed on 11 March 2020).
11. Centers for Disease Control and Prevention. CDC 24/7. Frequently Asked Questions about Personal Protective Equipment. Available online: https://www.cdc.gov/coronavirus/2019-ncov/hcp/respirator-use-faq.html (accessed on 14 August 2021).
12. SANDETEL Junta de Andalucia. *Estado del Arte: E-Salud & E-Inclusión Estudio de Las Tecnologías de la Información y la Comunicación Aplicadas a la Salud y a la Inclusión*; Dandatel: Seville, Spain, 2011.
13. Aiello, A.E.; Perez, V.; Coulborn, R.M.; Davis, B.M.; Uddin, M.; Monto, A.S. Facemasks, Hand Hygiene, and Influenza among Young Adults: A Randomized Intervention Trial. *PLoS ONE* **2012**, *7*, e29744. [CrossRef]
14. Chung, Y.; Chou, C.A.; Li, C.Y. Central Attention and a Dual Path Convolutional Neural Network in Real-World Tree Species Recognition. *Int. J. Environ. Res. Public Health* **2021**, *18*, 961. [CrossRef]
15. Szegedy, C.; Vanhoucke, V.; Ioffe, S.; Shlens, J.; Wojna, Z. Rethinking the Inception Architecture for Computer Vision. In Proceedings of the IEEE Conference on Computer Vision and Pattern Recognition (CVPR), Las Vegas, NV, USA, 27–30 June 2016; pp. 2818–2826.
16. Nieto-Rodríguez, A.; Mucientes, M.; Brea, V.M. System for medical mask detection in the operating room through facial attributes. In *Proceedings of the Iberian Conference on Pattern Recognition and Image Analysis, Santiago de Compostela, Spain, 17–19 July 2015*; Springer: Cham, Switzerland, 2015; pp. 138–145.
17. Chen, Y.; Hu, M.; Hua, C.; Zhai, G.; Zhang, J.; Li, Q.; Yang, S.X. Face mask assistant: Detection of face mask service stage based on mobile phone. *IEEE Sens. J.* **2021**, *21*, 11084–11093. [CrossRef]
18. Nagrath, P.; Jain, R.; Madan, A.; Arora, R.; Kataria, P.; Hemanth, J. SSDMNV2: A real time DNN-based face mask detection system using single shot multibox detector and MobileNetV2. *Sustain. Cities Soc.* **2021**, *66*, 102692. [CrossRef]
19. Howard, A.G.; Zhu, M.; Chen, B.; Kalenichenko, D.; Wang, W.; Weyand, T.; Andreetto, M.; Adam, H. Mobilenets: Efficient convolutional neural networks for mobile vision applications. *arXiv Prepr.* **2017**, arXiv:1704.04861.
20. Mata, B.U. Face Mask Detection Using Convolutional Neural Network. *J. Nat. Remedies* **2021**, *21*, 14–19.
21. Jauhari, A.; Anamisa, D.R.; Negara, Y.D.P. Detection system of facial patterns with masks in new normal based on the Viola Jones method. *J. Phys. Conf. Ser.* **2021**, *1836*, 012035. [CrossRef]
22. Sen, S.; Sawant, K. Face mask detection for covid_19 pandemic using pytorch in deep learning. In *IOP Conference Series: Materials Science and Engineering*; IOP Publishing: Bristol, UK, 2021; Volume 1070, p. 012061.
23. Balaji, S.; Balamurugan, B.; Kumar, T.A.; Rajmohan, R.; Kumar, P.P. A brief Survey on AI Based Face Mask Detection System for Public Places. *Ir. Interdiscip. J. Sci. Res. IIJSR* **2021**, *5*, 108–117.
24. Kurlekar, M.S. Face Mask Detection System Using Deep Learning. *Turk. J. Comput. Math. Educ. Turcomat* **2021**, *12*, 1327–1332.
25. Sakshi, S.; Gupta, A.K.; Yadav, S.S.; Kumar, U. Face Mask Detection System using CNN. In Proceedings of the 2021 IEEE International Conference on Advance Computing and Innovative Technologies in Engineering (ICACITE), Greater Noida, India, 4–5 March 2021; pp. 212–216.
26. Cheng, G.; Li, S.; Zhang, Y.; Zhou, R. A Mask Detection System Based on Yolov3-Tiny. *Front. Soc. Sci. Technol.* **2020**, *2*, 33–41. [CrossRef]

27. Loey, M.; Manogaran, G.; Taha, M.H.N.; Khalifa, N.E.M. A hybrid deep transfer learning model with machine learning methods for face mask detection in the era of the COVID-19 pandemic. *Measurement* **2021**, *167*, 108288. [CrossRef] [PubMed]
28. He, K.; Zhang, X.; Ren, S.; Sun, J. Deep Residual Learning for Image Recognition. In Proceedings of the 2016 IEEE Conference on Computer Vision and Pattern Recognition (CVPR), Las Vegas, NV, USA, 27–30 June 2016; pp. 770–778.
29. Rudraraju, S.R.; Suryadevara, N.K.; Negi, A. Face Mask Detection at the Fog Computing Gateway. In Proceedings of the 2020 IEEE 15th Conference on Computer Science and Information Systems (FedCSIS), Sofia, Bulgaria, 6–9 September 2020; pp. 521–524.
30. Wang, B.; Zhao, Y.; Chen, C.P. Hybrid Transfer Learning and Broad Learning System for Wearing Mask Detection in the COVID-19 Era. *IEEE Trans. Instrum. Meas.* **2021**, *70*, 1–12.
31. Hussain, S.; Yu, Y.; Ayoub, M.; Khan, A.; Rehman, R.; Wahid, J.A.; Hou, W. IoT and Deep Learning Based Approach for Rapid Screening and Face Mask Detection for Infection Spread Control of COVID-19. *Appl. Sci.* **2021**, *11*, 3495. [CrossRef]
32. Qin, B.; Li, D. Identifying Facemask-Wearing Condition Using Image Super-Resolution with Classification Network to Prevent COVID-19. *Sensors* **2020**, *20*, 5236. [CrossRef]
33. Cabani, A.; Hammoudi, K.; Benhabiles, H.; Melkemi, M. MaskedFace-Net—A dataset of correctly/incorrectly masked face images in the context of COVID-19. *Smart Health* **2021**, *19*, 100144, ISSN 2352-6483. [CrossRef] [PubMed]
34. "Intemasc Captura" Application. Available online: https://play.google.com/store/apps/details?id=es.upv.mastermoviles.intemasc.captura (accessed on 13 July 2021).
35. Media Dissemination. Available online: http://mmoviles.upv.es/intemasc/stiker_difusion.mp4 (accessed on 13 July 2021).
36. Viola, P.; Jones, M. Rapid object detection using a boosted cascade of simple features. In Proceedings of the 2001 IEEE Computer Society Conference on Computer Vision and Pattern Recognition. CVPR 2001, Kauai, HI, USA, 8–14 December 2001; pp. 511–518. [CrossRef]
37. Simonyan, K.; Zisserman, A. Very Deep Convolutional Networks for Large-Scale Image Recognition. *arXiv* **2014**, arXiv:1409.1556.
38. Chollet, F. Xception: Deep Learning with Depthwise Separable Convolutions. In Proceedings of the IEEE Conference on Computer Vision and Pattern Recognition Workshops (CVPRW), Honolulu, HI, USA, 21–26 July 2017; pp. 1800–1807.
39. Russakovsky, O.; Deng, J.; Su, H.; Krause, J.; Satheesh, S.; Ma, S.; Huang, Z.; Karpathy, A.; Khosla, A.; Bernstein, M.; et al. ImageNet Large Scale Visual Recognition Challenge. *Int. J. Comput. Vis.* **2015**, *115*, 211–252. [CrossRef]
40. "Mask Detect" Application. Available online: https://play.google.com/store/apps/details?id=es.upv.mastermoviles.intemasc.rec (accessed on 14 August 2021).
41. Code Used in the Present Work. Available online: https://github.com/jesus-tomas-girones/Mask_Detect (accessed on 13 July 2021).
42. ImageNet Task. Available online: https://keras.io/api/applications (accessed on 13 July 2021).

healthcare

Review

Overview of Multi-Modal Brain Tumor MR Image Segmentation

Wenyin Zhang [1], Yong Wu [1,*], Bo Yang [2], Shunbo Hu [1], Liang Wu [3] and Sahraoui Dhelim [4]

1 School of Information Science and Engineering, Linyi University, Linyi 276000, China; zhangwenyin@lyu.edu.cn (W.Z.); hushunbo@lyu.edu.cn (S.H.)
2 Shandong Provincial Key Laboratory of Network Based Intelligent Computing, Jinan 250022, China; yangbo@ujn.edu.cn
3 School of Control Science and Engineering, Shandong University, Jinan 250061, China; wuliang@mail.sdu.edu.cn
4 School of Computer and Communication Engineering, University of Science and Technology Beijing, Beijing 100083, China; sahraoui.dhelim@xs.ustb.edu.cn
* Correspondence: wy201915120103@lyu.edu.cn; Tel.: +86-1586-208-3910

Abstract: The precise segmentation of brain tumor images is a vital step towards accurate diagnosis and effective treatment of brain tumors. Magnetic Resonance Imaging (MRI) can generate brain images without tissue damage or skull artifacts, providing important discriminant information for clinicians in the study of brain tumors and other brain diseases. In this paper, we survey the field of brain tumor MRI images segmentation. Firstly, we present the commonly used databases. Then, we summarize multi-modal brain tumor MRI image segmentation methods, which are divided into three categories: conventional segmentation methods, segmentation methods based on classical machine learning methods, and segmentation methods based on deep learning methods. The principles, structures, advantages and disadvantages of typical algorithms in each method are summarized. Finally, we analyze the challenges, and suggest a prospect for future development trends.

Keywords: image segmentation; brain tumor; magnetic resonance imaging; multi-modality

Citation: Zhang, W.; Wu, Y.; Yang, B.; Hu, S.; Wu, L.; Dhelim, S. Overview of Multi-Modal Brain Tumor MR Image Segmentation. *Healthcare* **2021**, *9*, 1051. https://doi.org/10.3390/healthcare9081051

Academic Editor: Edward J. Pavlik

Received: 23 June 2021
Accepted: 10 August 2021
Published: 16 August 2021

1. Introduction

Brain tumors can grow in cerebral vessels, nerves, brain appendages and other intracranial tissues, which seriously threaten the life and health of patients. MRI plays an important role in the diagnosis and treatment of brain tumors. It is the most widely used imaging method in brain tumor detection and clinical treatment. MRI has no radiation, no injury, and no bone artifact in the human body [1]. As a multi-parameter imaging method, MRI has high resolution in soft tissue [2]. Through the acquisition of brain image detail information, we can accurately judge the pathological and histomorphological changes to optimize the segmentation results, which is helpful to the extraction of lesions and the treatment of tumors [3]. In MRI, images of different modes can be obtained according to the difference of transverse relaxation time and longitudinal relaxation time, and images of different modes have specificity in the image information. For example, T1-weighted imaging sequence (T1) can better display the anatomical structure of various brain tissues. T1-weighted Contrast-enhanced (T1C) imaging sequence can observe the boundary information of brain tumors more clearly. T2-weighted imaging sequences (T2) enhance the lesion area and are often used to identify lesions and determine tumor type. Fluid Attenuated Inversion Recovery (FLAIR) inhibited intracranial cerebrospinal fluid and was able to better detect high signal information in the lesion area [4]. In the process of diagnosis and treatment of brain tumors, accurate segmentation of brain tumor MR images is particularly important. According to different degrees of human intervention, it can be divided into artificial segmentation, semi-automatic segmentation and automatic segmentation. Traditionally artificial segmentation has high accuracy, but it is time-consuming and laborious, and subject to the subjective judgment of doctors. In addition, this method also requires experts to have both related brain tumor image knowledge and other professional knowledge of anatomy [5]. Therefore, researchers have done tremendous work on how to improve the accuracy and efficiency of brain tumor MR image segmentation by using semi-automatic segmentation and automatic segmentation. In brain images, the amount of information that can be expressed by single-mode MR images is limited, which cannot give accurate auxiliary information to doctors. The combination of different modal images can achieve complementary information between images [6] to obtain the morphological and pathological information of brain tumors. The result of the algorithm segmentation needs to be compared with the result drawn by the doctor manually, so it is at most the same as the result of the doctor's manual segmentation. This not only saves doctors a lot of time, but also provides them with important reference information, which can assist in the diagnosis and treatment of brain tumors. Generally, the tumor segmentation process is shown in Figure 1.

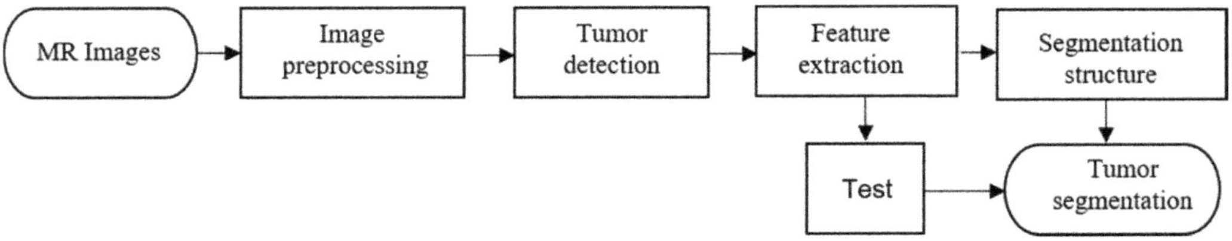

Figure 1. The flow chart of tumor segmentation.

This paper attempts to summarize the existing methods of brain tumor MR image segmentation. Firstly, this paper briefly introduces the database of brain tumor segmentation commonly used in brain tumor segmentation. Then, we introduce the basic ideas, network architecture, representative solutions, advantages and disadvantages of different methods. In addition, this paper compares the segmentation results of typical methods on BraTs database and clinical data. Finally, this paper analyzes the challenges faced by brain tumor MR image segmentation, and suggests prospects for development and direction.

2. Databases and Evaluation Measures

In the segmentation of brain tumor MR images, most research is based on public databases, and a smaller part on clinical data. After researchers obtain the results of image segmentation, they need to evaluate them. The evaluation of segmentation results can be divided into subjective evaluation and objective evaluation. Subjective evaluation needs to invest a lot of human and material resources, the evaluators rely on experience, and there is no standard answer. The subjective evaluation results of different people are generally different, and those of the same person at different times are also different. Therefore, objective evaluation measures that can be recognized by most people are particularly important in the study of brain tumor MR image segmentation.

2.1. Evaluation Measures Commonly Used

After continuous development and improvement, commonly used segmentation evaluation indicators are as follows: Dice Similarity Coefficient (*DSC*) [7], Jaccard Similarity (*JS*) [8], True Positive Rate (*TPR*) [9], Positive Predictive Value (*PPV*) [10] and Hausdorff Distance (*HD*) [11]. In order to obtain the evaluation measures introduced, we need to use the ground truths and actual segmentation results for calculation. The ground truth is an image formed by medical experts directly delineating the boundary of the relevant area of the lesion [12], and it is a standard that is unanimously recognized by researchers. The actual segmentation is the result of algorithm segmentation. Figure 2 shows the comparison between ground truth and the actual segmentation.

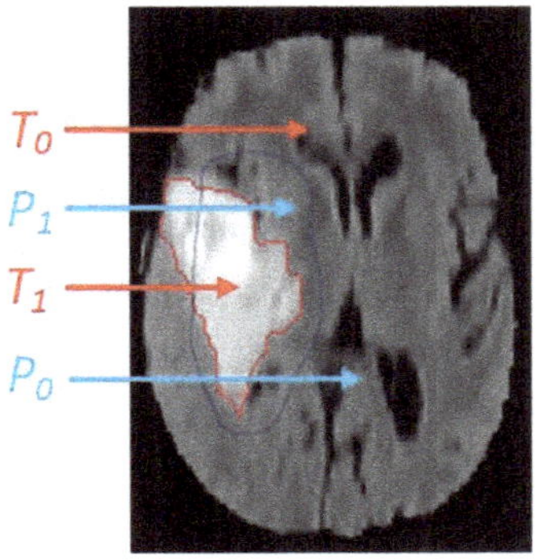

Figure 2. The comparison of actual segmentation and ground truth.

In the Figure 2, T_1 and T_0 represent the tumor area and background area of the ground truth, and P_1 and P_0 represent the tumor region and background region of the actual segmentation result.

The value of *DSC* is between [0, 1]. When *DSC* is equal to 0, the segmentation result is the worst. On the contrary, when *DSC* is equal to 1, the segmentation result is the most accurate [7], and *DSC* [13] is computed as follows:

$$DSC(P_1, T_1) = \frac{2|P_1 \cap T_1|}{|P_1| + |T_1|} \tag{1}$$

JS value is obtained by the intersection of the actual segmentation result and the ground truth and the ratio of their union, and the definition [14] is as follows:

$$JS(P_1, T_1) = \frac{|P_1 \cap T_1|}{|P_1 \cup T_1|} = \frac{|P_1 \cap T_1|}{|P_1| - |P_1 \cap T_1| + |T_1|} \tag{2}$$

TPR is obtained from the segmentation result of the algorithm and the ratio of the overlap part of ground truth to ground truth [9]. The definition of true positive rate [15] is as follows:

$$TPR(P_1, T_1) = \frac{|P_1 \cap T_1|}{|T_1|} \tag{3}$$

Positive Predictive Value (*PPV*) is also called Precision. The Positive Predictive Value is obtained by the ratio of the result correctly segmented by the algorithm to the result segmented by the algorithm. The definition [10] is as follows:

$$PPV(P_1, T_1) = \frac{|P_1 \cap T_1|}{|P_1|} \tag{4}$$

The definition of Hausdorff Distance (*HD*) is as follows:

$$HD(P_1, T_1) = MAX\{h(P_1, T_1), h(T_1, P_1)\} \tag{5}$$

$h(A, B) = \underset{a_i \in A b_j \in B}{maxmin} \|a_i - b_j\|$ can be obtained from set A and set B, $h(A, B)$ is the one-way Hausdorff distance from set A to set B, a_i means the i-th point in set A, b_j means the j-th point in set B, and $\|a_i - b_j\|$ means the distance between the point a_i and b_j [11].

2.2. Databases Commonly Used

The database commonly used for brain tumor segmentation is the BraTs (Brain Tumor Segmentation) database, and a small number of studies are based on clinical databases. This paper mainly introduces the BraTs2013, BraTs2015, BraTs2017, BraTs2018, BraTs2019,

BraTs2020 and some clinical databases. The relevant data information involved in this paper are shown in Table 1.

Table 1. Commonly used databases.

Database	Image Information	Number of Training Data	Number of Test Data	With Ground Truth		Testing Method	Data Size (mm^3)
				Training Data	Test Data		
BraTs2013	T1, T1C, T2, FLAIR	20	10	Yes	Yes	Offline	240 × 240 × 155
BraTs2015	T1, T1C, T2, FLAIR	285	110	Yes	Yes	Offline	240 × 240 × 155
BraTs2017	T1, T1C, T2, FLAIR	285	66	Yes	Yes	Offline	240 × 240 × 155
BraTs2018	T1, T1C, T2, FLAIR	285	-	Yes	No	Online	240 × 240 × 155
BraTs2019	T1, T1C, T2, FLAIR	335	-	Yes	No	Online	240 × 240 × 155
BraTs2020	T1, T1C, T2, FLAIR	369	-	Yes	No	Online	240 × 240 × 155
Clinical database	T1, T1C, T2, FLAIR	-	-	Yes	Yes	Offline	-

From Table 1, we can see that BraTs database contains four modes: T1, T1C, T2 and Flair. All image sizes are 240 mm × 240 mm × 155 mm. Before 2018, BraTs database had training data and test data, which could be tested offline. However, since 2018, there is no test data in the database and online testing is required.

2.2.1. BraTs Database

BraTs database is provided by MICCAI (Medical Image Computing and Computer Assisted Intervention) conference. This is the official database for the brain tumor MR image segmentation challenge held by the conference, and is also widely used by researchers engaged in brain tumor MR image segmentation. Since the challenge was held in 2012, the BraTs database has been updated every year. The URL of BraTs database mentioned in this paper is as follows:

BraTs2013 (from https://www.smir.ch/BRATS/Start2013, accessed on 21 May 2021),
BraTs2015 (from https://www.smir.ch/BRATS/Start2015, accessed on 21 May 2021),
BraTs2017 (from https://www.med.upenn.edu/sbia/brats2017/data.html, accessed on 21 May 2021),
BraTs2018 (from https://www.med.upenn.edu/sbia/brats2018/data.html, accessed on 21 May 2021),
BraTs2019 (from https://www.med.upenn.edu/cbica/brats2019/data.html, accessed on 21 May 2021),
BraTs2020 (from https://www.med.upenn.edu/cbica/brats2020/data.html, accessed on 21 May 2021).

In recent years, there have been a large number of studies on BraTs series databases. Table 2 shows some of these research results.

Table 2. Results of studies using BraTs series database in recent years.

Database	Method	Evaluation Measure: *DSC*		
		Whole Tumor	Core Tumor	Enhance Tumor
BRATS 2013	Tustison et al. [16]	0.871	0.781	0.741
	Pereira et al. [17]	0.83	0.78	0.73
	Havaei et al. [18]	0.86	0.77	0.73
	Shen et al. [19]	0.87	0.82	0.75
	Zhao et al. [20]	0.81	0.65	0.61
	P Bhagat et al. [21]	0.81	0.54	0.61
	Hu K et al. [22]	0.86	0.77	0.70
	Zhou Z et al. [12]	0.87	0.72	0.70

Table 2. *Cont.*

Database	Method	Evaluation Measure *DSC*		
		Whole Tumor	**Core Tumor**	**Enhance Tumor**
BRATS 2015	Casamitjana et al. [23]	0.917	0.836	0.768
	Kamnitsas et al. [24]	0.901	0.754	0.728
	Tseng et al. [25]	0.852	0.683	0.688
	Liu et al. [26]	0.87	0.62	0.68
	Iqbal et al. [27]	0.87	0.86	0.79
	Li H et al. [28]	0.890	0.733	0.726
	Hu K et al. [22]	0.87	0.76	0.75
BRATS 2017	Beers et al. [29]	0.882	0.732	0.730
	Shaikh et al. [30]	0.89	0.84	0.78
	Isensee et al. [31]	0.858	0.775	0.647
	Zhou T et al. [32]	0.885	0.846	0.734
	Po Y K et al. [33]	0.903	0.744	0.780
	Wang G et al. [34]	0.874	0.783	0.775
BRATS 2018	Wang G et al. [34]	0.908	0.869	0.807
	Subhashis B et al. [35]	0.902	0.872	0.824
	Zhou C et al. [36]	0.908	0.858	0.811
	HuA R et al. [37]	0.876	0.795	0.736
	Zhang J et al. [38]	0.876	0.810	0.773
	U Baid et al. [39]	0.848	0.769	0.668
BRATS 2019	Yogananda C et al. [40]	0.901	0.844	0.801
	Li X et al. [41]	0.886	0.813	0.771
	Wu P et al. [42]	0.891	0.817	0.757
	Zhao Y et al. [43]	0.883	0.861	0.810
	R Agravat et al. [44]	0.92	0.90	0.79
	Cheng G et al. [45]	0.905	0.820	0.764
	Ieva A et al. [46]	0.878	0.732	0.699
BRATS 2020	Lucas F et al. [47]	0.889	0.841	0.814
	Henry T et al. [48]	0.89	0.84	0.79
	Silva C et al. [49]	0.886	0.830	0.790
	Anand V et al. [50]	0.850	0.815	0.775
	Qamar S et al. [51]	0.875	0.837	0.795
	Jia H et al. [52]	0.913	0.855	0.788
	Lyu C et al. [53]	0.873	0.836	0.821

Among the work based on BraTs2013 database, Shen et al. [19] obtained good results in the segmentation of the whole tumor and its sub-regions by using the proposed structure of one subsample and three up-samples to extract stratified features. (The network diagram is shown in Figure 3a). Zhou Z et al. [12] proposed a 3D convolution pyramid module, which is a 3D dense connection architecture that can fuse multi-scale context information. This method performs well in whole tumor segmentation.

In the work based on BraTs2015 database, Iqbal et al. [27] added jump connection and interpolation operation on the basis of Segnet (the network diagram is shown in Figure 3d), which enhanced the segmentation effect of core tumor and enhanced tumor. In the whole tumor segmentation, Casamitjana et al. [23] proposed a method that uses two paths to collect low-resolution and high-resolution features from the input image, and the segmentation effect is better than with other methods.

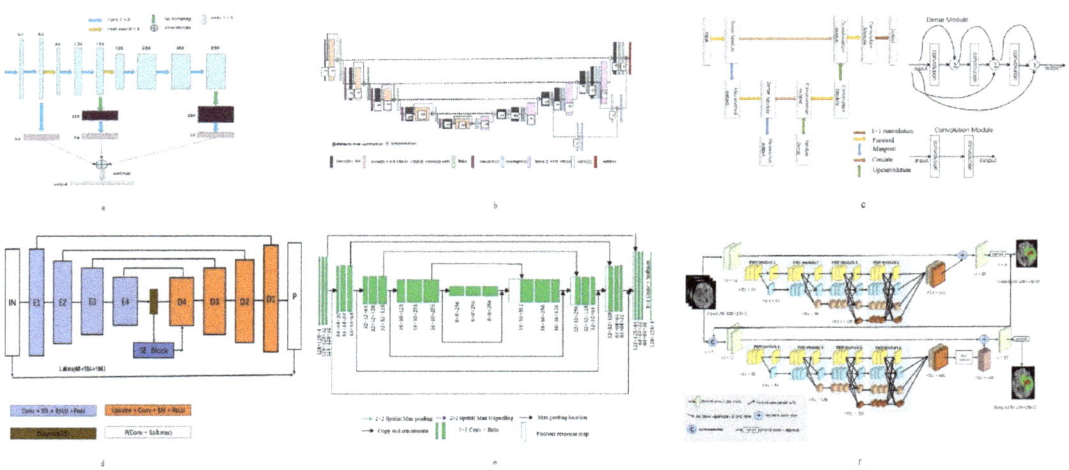

Figure 3. Network structure diagrams of better performing methods. (**a**) is the structure diagram of a sub sample and three up sampling structures proposed by Shen et al. [19]; (**b**) is the structure diagram of a 3D U-Net model combined with a priori knowledge of lesions proposed by Po et al. [26]; (**c**) is the structure diagram of the method of adding dense connections to the encoding part of the three tier codec architecture by R Agrava et al. [44]; (**d**) is the structure diagram of adding jump connection and interpolation operation on the basis of Segnet proposed by Iqbal et al. [27]; (**e**) is the structure diagram of a multi plane convolutional neural network proposed by Subhashis B et al. [35]; (**f**) is the structure diagram of hybrid high-resolution and nonlocal feature network (h2nf net) proposed by Jia h et al. [52].

In the research work based on BraTs2017 database, Po et al. [26] proposed a 3D U-Net model combined with prior knowledge of lesions. Using the original image to generate a group of patients with lesion heat map, then the generated map is employed to locate the target area. (The network diagram is shown in Figure 3b). Experiments show that this method is more effective than other methods in tumor overall segmentation and enhancement segmentation.

In the research work based on BraTs2018 database, Subhashis B et al. [35] proposed a multi-plane convolution neural network for brain tumor MR image segmentation from different anatomical planes (The network diagram is shown in Figure 3e), which showed good performance in the segmentation of whole tumor, core tumor and enhanced tumor.

In the research work based on BraTs2019 database, the method of adding dense connection to encoder part of three-layer codec architecture (The network diagram is shown in Figure 3c) proposed by R Agrava et al. [44], which has the highest precision in the segmentation of whole tumor and core tumor. In the segmentation of enhanced tumor, the deep convolution neural network improved by Zhao Y et al. [43] has the best segmentation effect.

In the research work based on BraTs2020 database, the Hybrid High-resolution and Non-local Feature Network (H2NF-Net) proposed by Jia H et al. [52] uses a single and cascaded HNF-Net to segment different brain tumor regions. Combine the prediction results as the final segmentation result. (The network diagram is shown in Figure 3f). This method works well in whole tumor and core tumor segmentation tasks. In enhanced tumor segmentation, a tumor region segmentation model that combines a two-stage codec with regularization and attention mechanism proposed by Lyu C et al. [53] works well.

2.2.2. Clinical Database

Clinical data of MR brain tumor images are collected by the hospital with the permission of the patients during their treatment. The collected MR brain images are used by doctors to judge the condition of patients and propose reasonable and effective treatment plans. Because of patient privacy and ethical issues, researchers are not allowed to use such data for research without permission from patients and hospitals. In recent years, the comparison of segmentation results based on clinical database is shown in Table 3.

Table 3. Comparison of segmentation results based on clinical databases in recent years.

Method	Data Sources	Data Volume	Evaluation Measure: *DSC*		
			Whole Tumor	Core Tumor	Enhance Tumor
Hua R et al. [37]	Local hospital	28 patients	0.864	0.804	0.722
U Baid et al. [39]	Local hospital	40 patients	0.924	0.901	0.813
Ieva A et al. [46]	Local hospital	105 patients	0.87	0.71	0.68
Shen Y et al. [54]	Local hospital	105 patients	0.894	0.790	0.653
Zhao Z et al. [55]	Local hospital	184 patients	0.785	-	-

Because the clinical data of each hospital is collected in different stages from different patients, and the equipment conditions used to collect the data are also different, it is impractical to compare the segmentation performance of these works. From the experimental results alone, the improved 3D U-Net scheme proposed by U Baid et al. [44] has higher segmentation accuracy in whole tumor, core tumor and enhanced tumor. The model consists of contraction path and expansion path. The shrinking path mainly captures the context, and the expanding path realizes the target location. The loss function, activation function and data enhancement are also considered. Therefore, each segmentation measure is increased.

3. Methods of Brain Tumor MR Image Segmentation

Segmentation methods of brain tumor MR image are mainly divided into three categories according to different segmentation principles: traditional segmentation methods, traditional machine learning-based segmentation methods and deep learning-based segmentation methods. Each category includes a variety of specific segmentation algorithms, as shown in Figure 4.

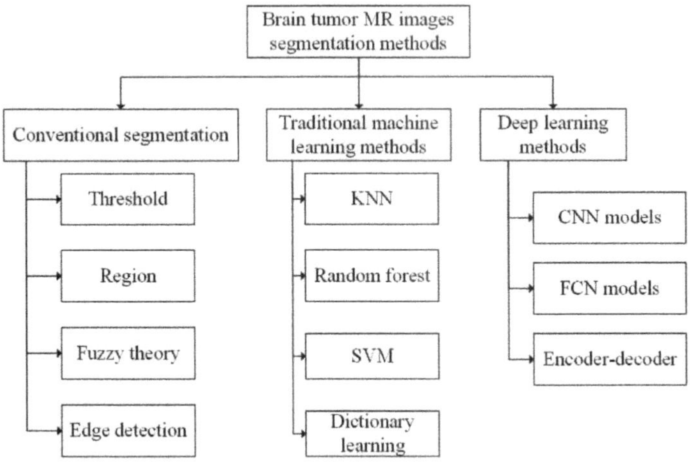

Figure 4. Brain tumor MR image segmentation methods.

3.1. Traditional Brain Tumor Segmentation Methods

According to the different theories and emphases, the traditional segmentation methods can be generally divided into four categories: threshold based segmentation, region-based segmentation, fuzzy theory based segmentation and edge detection based segmentation [56].

3.1.1. Segmentation Methods Based on Threshold

Threshold-based segmentation is the simplest method. First, it is assumed that the pixels within a range belong to the same category [57]. Brain tumor images can be divided into target region and background region by setting an appropriate threshold. Different thresholds can also be set to divide the tumor into multiple regions. After continuous research and development, the accuracy of threshold segmentation has been greatly improved. Wang Y P et al. proposed an improved threshold segmentation algorithm. The method improves the noise sensitivity in threshold segmentation by using local information of pixel neighborhood [58]. Foladivanda et al. proposed an adaptive threshold segmentation method. The method can effectively overcome the problem of uneven gray, and enhance the contrast of images, and effectively improve the DSC and JS measure of MR image segmentation of the brain tumor [59].

The segmentation method based on threshold is relatively simple, and the quality of segmentation results almost entirely depends on the size of threshold, so the selection of threshold is very important. Moreover, the threshold segmentation method can only segment simple images, and it is difficult to deal with complex images.

3.1.2. Segmentation Methods Based on Region

Common region-based segmentation methods include watershed algorithm and region-growing algorithm.

Watershed algorithm is a segmentation method based on mathematical morphology. In this algorithm, the image to be processed is compared to the terrain in geography, and the elevation of terrain is represented by the gray value of the pixel. The local minimum and its adjacent area are called the ponding basin. It is assumed that there are water permeable holes at each local minimum. With the increase of infiltration water, the ponding basin will be gradually submerged. Blocking the flow of water from a stagnant basin to a nearby basin is called a dam. When the water level reaches the peak, the infiltration process ends. These dams are called watersheds. Kaleem et al. [60] proposed a watershed segmentation method guided by setting internal or external markers to calculate the morphological gradient of the input image and internal and external markers of the original image. Then they use watershed transform to obtain the segmentation results. Rajini N et al. [61] proposed a method combining threshold segmentation and watershed. First, the image was segmented by threshold method, and then the segmented image was segmented by watershed algorithm. The experiment proved that the segmentation results obtained by this method were more accurate than those obtained by one of the two methods alone, with the average TPR measure higher than 90%.

The segmentation algorithm based on watershed can obtain a complete closed curve and provide contour information for subsequent processing, whereas the watershed algorithm is influenced by noise and easy to over segment.

The region growing algorithm draws all the pixel points conforming to the criterion into the same region via formulating a criterion, so as to achieve pixel segmentation. This kind of segmentation method has the following characteristics: (1) Each pixel must be in a certain region, and the pixels in the region must be connected, and must meet certain similar conditions; (2) different regions are disjoint, and two different regions cannot have the same property. Qusay et al. [62] proposed an automatic seed region growth method, which can automatically set the initial value of seeds, avoid the defects of manual interaction, and improve the efficiency of image segmentation.

The region-based segmentation method has the characteristics of simple calculation and high accuracy, which can extract better regional features and is more suitable for segmentation of small targets. However, it is sensitive to noise and easy to make holes in the extracted region.

3.1.3. Segmentation Methods Based on Fuzzy Theory

The segmentation methods based on fuzzy theory have also been highly valued. In brain tumor MR image segmentation, the most widely used Fuzzy theory algorithm is Fuzzy C-means clustering (FCM) [63]. Muneer K et al. [64] obtained the K-FCM method through the combination of FCM algorithm and K-means algorithm. The experiment proved that, compared with FCM, K-FCM showed higher accuracy in brain tumor MR image segmentation and could reduce the computational complexity. Guo Y et al. [65] proposed a Neutrosophic C-Means (NCM) algorithm based on fuzzy C-means and neutral set framework. The algorithm introduced distance constraint into the objective function to solve the problem of insufficient prior knowledge and achieved satisfactory segmentation results. On the basis of Super-pixel fuzzy clustering and the lattice Boltzmann method, Asieh et al. [66] proposed a level set method that can automatically segment brain tumors, which has strong robustness to image intensity and noise.

The segmentation method based on fuzzy theory can effectively solve the problem of incomplete image information, imprecision, and so on. It has strong compatibility and can be used in combination with other methods, but it is difficult to deal with large-scale data due to its large amount of computation and high time complexity.

3.1.4. Segmentation Methods Based on Edge Detection

The segmentation principle based on edge detection and target contour achieves segmentation by obtaining the edge of the target region and then obtaining the contour of the target region. Common detection operators for edge detection include Roberts operator, Sobel operator, Canny operator and Prewitt operator [67]. Jayanthi et al. [68] integrated FCM into the active contour model. The initial contour of the model is automatically selected by FCM, which reduces the human–computer interaction. Moreover, the problem of the unclear edge contour and uneven intensity in MR images was improved. The average DSC measure of segmentation by this method reached 81%.

Compared with other traditional segmentation methods, the segmentation method based on edge detection pays attention to the edge information of the image and links the edges into contours, and the anti-noise performance is stronger. But the anti-noise performance is negatively correlated with accuracy, that is, the better the anti-noise performance, the lower the accuracy. On the contrary, improved accuracy will reduce the anti-noise performance.

3.2. Segmentation Methods of Brain Tumor MR Images Based on Traditional Machine Learning

Brain tumor segmentation methods based on traditional machine learning use predefined features to train the classification model. Generally, they are divided into two levels: organizational level and pixel level. At the organizational level, the classifier needs to determine which kind of organizational structure each feature belongs to, and at the pixel level the classifier needs to determine which category each pixel belongs to. Traditional Machine Learning algorithms mainly include K-Nearest Neighbors (KNN) [69], Support Vector Machine (SVM) [70], Random Forest (RF) [71], Dictionary Learning (DL) [72], etc.

Havaei et al. [69] regarded each brain as a separate database and used the KNN algorithm for segmentation. They obtained very accurate results, and the segmentation time of each brain image is only one minute, which improves the efficiency of segmentation. Llner F et al. [70] used SVM to segment brain tumors, taking into account the changing characteristics of signal intensity and other features of brain tumor MR images. The TPR measure of this method for LGG reached 83%, and the accuracy measure for HGG reached 91%. Sher et al. [73] first segmented the image by the Otsu method and K-means clustering, then extracted the features by discrete wavelet transformation, and finally reduced the feature dimension by the PCA algorithm to obtain the best features for SVM classification. The experimental results show that the sensitivity and specificity of the scheme can reach more than 90%. Vaishnavee et al. [74] used a proximal support vector machine (PSVM). The method uses equation constraints to solve the primary linear equations, which simplifies the original problem of solving convex quadratic programming. The experiment shows that PSVM is more accurate than SVM in MR image segmentation of brain tumor. Wu et al. [75] proposed a method to first segment the image into super-voxels, then segment the tumor using MRF, estimate the likelihood function at the same time, and extract the features using a multistage wavelet filter. Nabizadeh et al. [76] proposed an automatic segmentation algorithm based on texture and contour. Firstly, the initial points were determined and the machine learning classifier was trained by the initial points. Mahmood et al. [71] proposed an automatic segmentation algorithm based on RF. This algorithm uses several important features such as image intensity, gradient and entropy to generate multiple classifiers, and classifies pixels in multispectral brain MR images by combining the results to obtain segmentation results. Selvathi et al. [77] increased the weight of the wrongly classified samples and decreased the weight of the correctly classified samples in the training process. Then the classifier gives new weights to the samples to ensure that the weights of all decision trees are positively correlated with their classification ability. Finally, the input of the improved RF consists of two parts: the image intensity feature and the original image feature extracted by curve and wavelet transformation. Experimental results show that the accuracy of the improved RF scheme is 3% higher than that of the original RF algorithm. Reza et al. [78] studied the correlation of image minimization features from the perspective of image features, effectively selected features, and finally classified features in multimodal MR images through RF. Compared with the RF algorithm alone, the proposed method can improve the DSC, PPV and TPR measure simultaneously. Meier et al. [79] trained a specific random forest classifier by semi-supervised learning. It takes image segmentation as a classification task and effectively combines the preoperative and postoperative MR image information to improve the postoperative brain tumor segmentation. The PPV and ME measure obtained by this method were 93% and 2.4%, respectively. Dictionary learning is a kind of learning method for simulating dictionary lookup. The dictionary itself is set as dictionary matrix, and the method used is sparse matrix. The process of dictionary lookup is obtained by multiplying the sparse matrix and dictionary matrix, and then the dictionary matrix and sparse matrix are optimized to minimize the error between the value searched and the original data. Chen et al. [72] transformed the super-pixel feature into a high-dimensional feature space. According to the different error values of different regions when the dictionary was modeling brain tumors, the segmentation of brain tumor MR images was realized and the segmentation accuracy was improved. Li [80] proposed a multi dictionary fuzzy learning algorithm based on dictionary learning. This algorithm effectively combines dictionary learning with fuzzy algorithm, and fully considers the differences between the target region and the background, as well as the consistency within the target region. This method can describe the gray and texture information of different regions of the image, and segment the image quickly and accurately.

The traditional machine learning algorithm is better than many traditional segmentation algorithms in algorithmic performance, but there are many shortcomings when it is used in brain tumor MR image segmentation. For example, the KNN algorithm is simple to implement, and the prediction accuracy of the brain tumor region is relatively high, but the calculation is relatively large [69]. The support vector machine has strong theory, and the final result is determined by several support vectors. The calculation is relatively simple and the generalization ability is strong, but it has higher requirements concerning the selection of parameters and kernel function [70]. Random forest can solve the problem of over-segmentation well, process multiple types of data, and has good anti-noise performance. It can parallel operation and shorten the operation time, but it has a poor effect on low-dimensional tumor data processing [71]. The algorithm based on dictionary learning is similar to the idea of dimensionality reduction, both of which reduce the computing complexity and speed up the computing speed, but also have higher requirements for tumor data [72].

3.3. Segmentation Methods of Brain Tumor MR Images Based on Deep Learning

According to different network frameworks, the brain MR image segmentation method is based on deep learning and can be divided into that based on Convolutional Neural network (Convolutional Neural Networks, CNN) of the brain MR image segmentation method, and that based on the Convolutional Neural network (Fully Convolutional Networks, FCN) MR image segmentation method of brain tumors and the brain MR image segmentation method, based on the encoder and decoder.

3.3.1. Segmentation Methods of Brain Tumor MR Images Based on CNN

Convolutional neural network belongs to the category of neural network, and its weight sharing mechanism greatly reduces the model complexity. Convolutional neural network (the network diagram is shown in Figure 5a) can directly take the image as the input, automatically extract the features, and has a high degree of invariance to the image translation, scaling and other changes. In recent years, a series of Network models based on convolutional neural Network [81], such as Network in Network [82], VGG [83], Google-Net [84], Res-Net [85], etc., have been widely used in medical image segmentation. Among them, the VGG network has a strong ability to extract features and can guarantee the convergence in the case of fewer training times. However, as the deepening of the network will cause gradient explosion and gradient disappearance, the optimization effect will start to deteriorate when the network depth exceeds a certain range.

In order to solve the problem of network degradation, He et al. [85] proposed deep Residual Network (ResNet), which achieved good results in the segmentation task [86]; Anand et al. [50] combined the 3D convolutional neural network with dense connection, pre-trained the model, and then initialized the model with the weight obtained. This method improved the DSC measure in the segmentation task of brain tumor MR images. Havaei et al. [18] constructed a cascaded dual path CNN, which took the output characteristic graph of CNN in the first stage as the additional input of CNN in the second stage. This method can effectively obtain rich background information and get better segmentation results. Lai et al. [87] reduced the tail of the original image by 98% firstly, corrected the bias field by using n4itk, then pre-segmented it by multi classification CNN, and finally obtained the final segmentation result by median filtering. The algorithm improves the DSC and PPV of segmentation significantly. Salehi et al. [6] proposed a convolutional neural network technology based on automatic context (Auto-Nets) to indirectly learn 3D image information by means of 2D convolution. This method uses 2D convolution in axial, coronal and sagittal MR images respectively to avoid complex 3D convolution operations in segmentation (The network diagram is shown in Figure 5c). Hussain et al. [88] established a correlation architecture composed of a parallel CNN layer and a linear CNN layer by adding an induction structure. This structure has achieved good results in brain tumor MR image segmentation, especially in enhancing the DSC measure to 90%. Kamnitsas et al. [24] trained 3D brain tumor images and then carried out conditional random field post-processing to obtain smoother results. Saouli et al. [89] designed a sequential CNN architecture and proposed that an end-to-end incremental network can simultaneously develop and train CNN models (the network diagram is shown in Figure 5g). The average DSC measure obtained by this method is 88%. Hu K et al. [22] proposed a more hierarchical convolution based Neural Network (Multi-Cascaded Convolutional Neural Network, MCCNN) and fully connected conditional random fields (CRFs), combined with the brain tumor segmentation method, Firstly, the brain tumor is roughly segmented by multi classification convolution neural network, and then fine segmented by fully connected random field according to the rough segmentation results, so as to achieve the effect of batch segmentation and improve the accuracy. The segmentation algorithm based on CNN can automatically extract features and process high-dimensional data, but it is easy to lose information in the process of pooling, and its interpretability is poor.

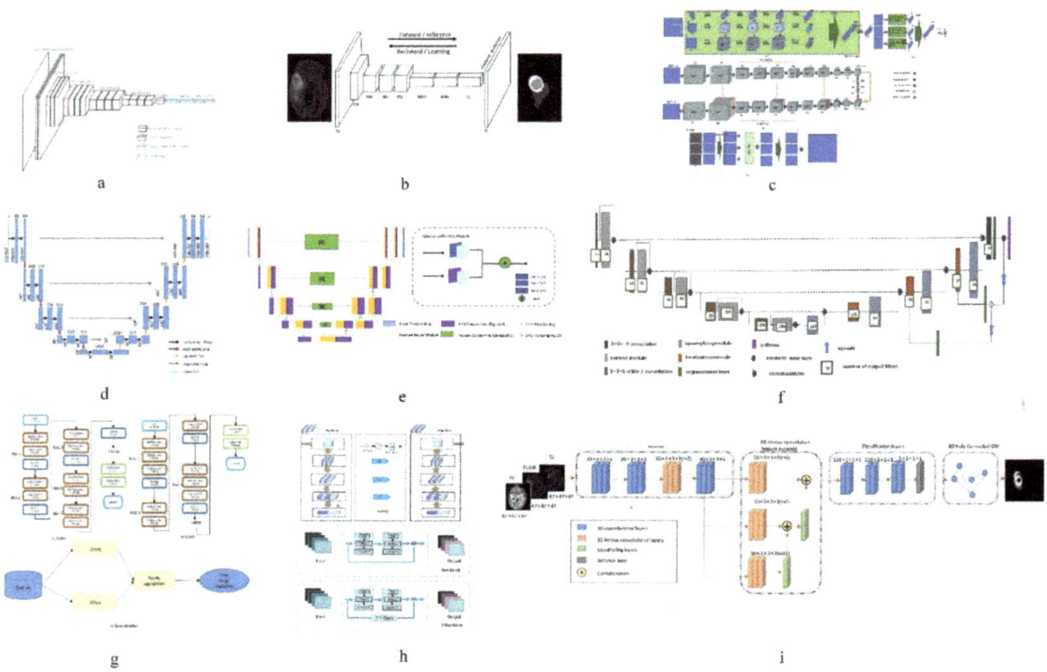

Figure 5. Network structure diagrams of some classical methods and improved methods. (**a**) is the classic CNN network model; (**b**) is the classic FCN network model; (**c**) is the structure diagram of a convolutional neural network technology based on automatic context (Auto-Nets) proposed by Salehi et al. [6]; (**d**) is the classic Encoder-Decoder network model; (**e**) is the structure diagram of a fully convolutional neural network with feature reuse module and feature integration module (f2fcn) proposed by Xue et al. [90]; (**f**) is the structure diagram of a robust neural network algorithm based on u-net proposed by isensee et al. [31]; (**g**) is the structure diagram of a sequential CNN architecture proposed by saouli et al. [89]; (**h**) is the structure diagram of attention residual U-net proposed by Zhang et al. [38]; (**i**) is a structural diagram of 3D dense connection combined with feature pyramid proposed by Zhou et al. [12].

3.3.2. Segmentation Methods of Brain Tumor MR Images Based on FCN

Compared with pixel-level classification, image-level classification and regression tasks are more suitable for using the CNN structure, because they both expect to obtain a probable value for image classification. For semantic segmentation of images, FCN works better. FCN has no requirement on the size of the input image, and there will be an up sampling process at the last convolution layer. This process can get the same result as the input image size, predicting each pixel while retaining the spatial information in the input image, so as to achieve the pixel classification. In simple terms, FCN is a method to classify and segment images at the pixel level. Therefore, the semantic segmentation model based on FCN is more in line with the requirements of medical image segmentation. Zhao et al. [20] proposed a combination of FCN with CRF for brain tumor segmentation. The method trains two-dimensional slices in axial, coronal and sagittal directions respectively, and then uses fusion strategy to combine segmented brain tumor images. Compared with the traditional segmentation methods, the segmentation speed is faster and the efficiency is higher. Xue et al. [90] proposed a fully convolutional neural network with feature reuse module and feature integration module (f2fcn). It reuses the features of different layers, and uses the feature integration module to eliminate the possible noise and enhance the fusion between different layers (the network diagram is shown in Figure 5e). The DSC and *PPV* obtained by this method are high. Zhou et al. [91] proposed a 3D atomic convolution feature pyramid to enhance the discrimination ability of the model, which is used to segment tumors of different sizes. Then, an improvement is made on the original basis [12], a 3D dense connection architecture is proposed, and a new feature pyramid module is designed by using 3D convolution (the network diagram is shown in Figure 5i). This module is used to fuse multi-scale context to improve the accuracy of segmentation. Liu et al. [26] proposed a Dilated Convolution optimization structure (DCR) based on Resnet-50, which can effectively extract local and global features, and this method can improve the segmentation PPV measure to 92%. The segmentation algorithm based on FCN can predict the category of each pixel, transform the image classification level to the semantic level, retain the position information in the original image, and obtain a result with the same size as the input image. However, the algorithm has low computational efficiency, takes up a lot of memory space, and the receptive field is relatively small.

3.3.3. Segmentation Methods of Brain Tumor MR Images Based on Encoder-Decoder Structure

The encoder-decoder structure is generally composed of an encoder and a decoder. The encoder trains and learns the input image through a neural network to obtain its characteristic map. The function of the decoder is to mark the category of each pixel after the encoder provides the feature map, so as to achieve the segmentation effect. In the segmentation tasks based on encoder-decoder structure, the structure of encoders is generally similar, mostly derived from the network structure of classification tasks, such as VGG, etc. The purpose of doing this is to obtain the weight parameters of network training through the training of a large database. Therefore, the difference of the decoder reflects the difference of the whole network to a large extent, and is also the key factor affecting the segmentation effect.

Badrinarayanan et al. [92] proposed the SegNet model. Compared with other models, this model has a deeper layer and has better performance in semantic segmentation of pixels. The encoder part of the model consists of a 13 layer vgg-16 network, and can remember the position information of the largest pixel in the encoding phase. In the decoder, the low resolution input features are up sampled to get the segmentation results. The U-Net model based on FCN is a kind of widely used brain tumor segmentation model, in which the network structure is also made up of an encoder and a decoder, and a U-Net network jump connection will code paths, used to get the characteristics of the figure to the decoding path to the corresponding position, in order to get the characteristics of the direct sampling under the coding phase into the decoding stage, thus learning more detailed characteristics. Chen et al. [93] proposed a multi-level deep network, which can obtain image multi-level information by adding auxiliary classifiers on Multi-Level Deep Medical (MLDM) and U-Net, so as to realize image segmentation. The results of DSC, PPV and TPR were 83%, 73% and 85%, respectively. In order to reduce the semantic gap between the feature mapping of encoder and decoder networks, Zhou et al. [94] proposed a variety of nested dense connection methods to connect the encoder and decoder networks. Alom et al. [95] proposed a recursive neural network and a recursive residual convolutional neural network based on U-Net. The experimental results show that the performance of the two kinds of network segmentation combined with U-Net is better than that of U-Net alone. Zhang et al. [38] introduced the attention mechanism and residual network into the traditional U-Net network and proposed an attention residual U-Net (the network diagram is shown in Figure 5h), which improved the segmentation performance of brain tumor MR images. Milletari et al. [96] proposed the V-Net model on the basis of the 3D U-Net model, which extended the original U-Net model by using a 3D convolution check. Hua et al. [37] cascaded V-Net and used the method of segmentation of the whole tumor first into sub-regions of the tumor; the accuracy of segmentation is higher than that of direct V-Net segmentation. Cicek et al. [97] proposed a 3D U-Net model to learn the features of sparse annotated volume images. On the basis of 3D U-Net, Heet et al. [98] added a Hybrid Dilated Convolution (HDC) module to increase the sensory field of neurons, overcoming the restriction that multi-scale feature extraction requires deep neural networks. Using shallow neural networks can reduce the number of model parameters and reduce the computational complexity. Tsenget et al. [25] proposed one with the depth of the layer cross-modal convolution encoder/decoder structure, in combination with MR image data of different modalities, and at the same time using the weighted and multi-stage training methods to solve the problem of unbalanced data; compared with the traditional U-Net structure, the methods of DSC, TPR and PPV measure are improved. Isensee et al. [31] improved the U-Net network model and designed a robust neural network algorithm, which prevented overfitting by expanding the amount of data (the network diagram is shown in Figure 5f). This algorithm improved the TPR measure to 91%; Haichun et al. [28] cleverly applied the improved full convolutional neural network structure to the U-Net model and proposed a novel end-to-end brain tumor segmentation method. In this method, an up-hop connection structure was designed between the encoding path and decoding path to enhance the information flow. Jia et al. [99] constructed a HNF network based on the parallel multi-scale fusion (PMF) module, and proposed a three-dimensional high-resolution and non-local feature network (HNF-NET) for multi parameter MR imaging, which can generate strong high-resolution feature representation and aggregate multi-scale context information. The expectation maximization attention (EMA) module is introduced to extract more relevant features and reduce redundant features. The DSC and HD of the whole tumor are 91.1% and 4.13%, respectively. The segmentation algorithm based on encoder-decoder can combine high-resolution and low-resolution information, and can recognize features from multiple scales, but there is only a short connection between the encoding process and the decoding process, and the connection between the two is obviously insufficient.

3.4. Summary and Analysis

This paper summarizes the existing traditional machine learning based and deep learning based brain tumor MR image segmentation methods and reviews the researchers' work in the field. It is not difficult to find that deep learning methods and techniques gradually occupy a dominant position in the field of brain tumor MR image segmentation. In the past few years, an end-to-end CNNS method and a U-Net network with codec function for brain tumor MR image segmentation have been most widely used. However, even if similar network architectures are used, the results are not identical [100,101], because data preprocessing can increase the segmentation accuracy without changing the network architecture, and can enhance the generalization ability of the network. Therefore, almost all the research has carried out data preprocessing. By comparing the segmentation performance of various methods, this paper finds that each type of method can solve some of the problems in segmentation. However, there are deficiencies in generalization. For example, brain tumor segmentation based on traditional methods is mostly simple and easy to implement, but it is difficult to process complex images, and the segmentation accuracy is generally low. Segmentation methods based on traditional machine learning are theoretically easy to understand, but it is difficult to process big data. Segmentation methods based on deep learning can extract the deep information from the image, but their interpretability is poor. The advantages and disadvantages of the brain tumor MR image segmentation method described in this paper are shown in Table 4.

Table 4. Advantages and disadvantages of various brain tumor MR image segmentation methods

	Method	Advantage	Disadvantage
Traditional segmentation methods	Segmentation method based on threshold [57,58]	Easy to implement, Fast in calculation	Low accuracy, Meaningless for small images
	Segmentation method based on Region [61,62]	Simple calculation, High accuracy, Can operate in parallel	Sensitive to noise, Easy to produce cavity, Volume effect, Easy to oversplit
	Segmentation method based on fuzzy theory [63,66]	Low image requirements, Sensitive to parameters,	Lack of theory, Imperfect system, Long time consuming
	Segmentation method based on edge monitoring [67,68]	Strong anti noise ability, Fast detection speed	Contradiction between noise, resistance and accuracy
Segmentation method based on traditional machine learning	Segmentation method based on KNN algorithm [69]	Simple, High precision, Effective noise reduction	Data correlation required, Large amount of calculation
	Segmentation method based on random forest [71,78]	Strong fitting ability, Strong anti noise ability, Fast in calculation, balance data differences	Many features are required, Easy to lose information
	Segmentation method based on support vector machine [73,74]	Easy to fit, Strong theoretical, Easy to calculate	Sensitive to kernel function, Low precision in multitasking
	Segmentation method based on dictionary learning [72,80]	Fast operation speed, good performance	High requirements for data,
Segmentation method based on deep learning	Segmentation method based on CNN [18,50]	Shared convolution kernel, Automatic feature extraction	Weak interpretability, Easily lost information, Existence of local convergence
	Segmentation method based on FCN [26,91]	Image size is not required, Classify each pixel	Efficiency is not real-time, Insensitive to details, Lack of spatial consistency
	Segmentation method based on encoder and decoder [92,93]	Multiscale feature recognition, Combined with high and low resolution information, Restore pixel position information	Insufficient contact between encoder and decoder, A large number of parameters, Slow computing speed

In recent years, there has been more and more research into brain tumor MR image segmentation. However, the DSC measure of brain tumor segmentation is only about 0.9, which due to the complexity of the brain tumor MR image and the limitation of the segmentation algorithm. In addition, there are many other challenges in the research field of brain tumor MR image segmentation, such as the generalization ability of segmentation algorithms. Most of the existing segmentation algorithms are for a single lesion, and it is difficult to generalize these to brain tumors with different conditions or even other lesions. The proportion of brain tumor background in the MR image is too large, and the proportion of tumor target region (especially the subregion of brain tumor) is too small, so it is difficult to locate accurately and effectively in the segmentation process. MR images of brain tumors are multimodal data. If the multimodal information is not handled properly, the information between images will be confused, which can lead to no improvement, or even a reduction in segmentation accuracy. Currently, many studies on brain tumor MR image segmentation are only at the theoretical stage, unable to meet the needs of medical staff and difficult to be applied in clinical practice. Deep learning has gradually become the mainstream method in brain tumor MR image segmentation. However, as a supervised learning method, deep learning relies too much on ground truth, but manual labeling is extremely difficult.

4. Future Research Directions

Through studying and summarizing the existing segmentation methods, this paper looks forward to future research directions from four aspects: data acquisition and processing, feature extraction, calculation methods and clinical application.

In recent years, with the continuous development of medical imaging, MR images of brain tumors are playing an increasingly important role in the diagnosis and treatment of brain tumors. Traditional research is mostly based on the calculation and analysis of unit point and small sample data. If the data of different institutions can be integrated and utilized, the accuracy of tumor segmentation will be greatly improved [102]. However, it is still a great challenge to find a general method to deal with all changes of brain MR images from different institutions and MRI scanners. Therefore, how to make full use of multi-site and multi center data [103] will become an area worthy of attention.

Deep learning has an ability to learn features, high efficiency in extracting features, can set the number of network layers, can be mapped to any function in theory, and can solve more complex problems. As long as there are enough brain tumor MR image data, we can obtain ideal results and good portability, which can be used in Tensorflow, Pytorch and other frameworks. Therefore, deep learning based methods will continue to be active in brain tumor MR image segmentation. However, how to improve the feature expression ability of the network is the key problem in improving the performance of the segmentation network.

With the development of artificial intelligence theories and methods, there are many efficient network architectures in the field of computer vision. How to reasonably migrate these architectures to brain tumor MR image segmentation tasks, such as using mask RCNN network [104] in image retrieval and blendmark network [105] in the instance segmentation task, to improve the detection and location ability of brain tumor and its sub regions, is a direction worth exploring.

At present, the mainstream supervised brain tumor MR image segmentation methods have limited databases and are highly dependent on ground truth, while manual labeling is extremely complex. Therefore, how to segment brain tumor MR images accurately through unsupervised learning without labels, and weakly supervised learning with a small number of labels or coarse-grained labels, or to ensure that supervised methods have unsupervised learning ability, will become a hot research direction.

With the proposal of the issue of "combining scientific research with practical problems", as well as continuous interdisciplinary collision and integration, cooperation between clinicians and computer scientists in the field of medical imaging is becoming more and more important, i.e., scientific research should meet the clinical needs of the hospital. Therefore, in the research into brain tumor MR image segmentation, how to combine clinical information, such as the deep fusion of brain tumor pathology, disease symptoms and MR image at the feature level, etc., will be an important research direction.

Author Contributions: Conceptualization, W.Z. and Y.W.; Data curation, S.H., L.W. and S.D.; Formal analysis, B.Y. and W.Z.; Funding acquisition, B.Y., W.Z. and S.H.; Investigation, Y.W. and S.D; Methodology, W.Z., Y.W. and L.W.; Project administration, B.Y. and W.Z.; Resources, W.Z.; Supervision, B.Y.; Validation, S.H.; Visualization, L.W. and S.D; Writing—original draft, Y.W. and W.Z.; Writing—review and editing, Y.W. and W.Z. All authors have read and agreed to the published version of the manuscript.

Funding: This work was supported by the National Natural Science Foundation of China (61572230, 61173078, 61771230), and the Key R & amp; D plan of Shandong (2017CXZC1206, 2019GNC106027, 2019JZZY010134), and the Youth program of Shandong Natural Science Foundation (ZR2020QF011, ZR2020QF014).

Institutional Review Board Statement: Not applicable.

Informed Consent Statement: Not applicable.

Data Availability Statement: The URL of BraTs database mentioned in this paper is as follows: BraTs2013 (from https://www.smir.ch/BRATS/Start2013, accessed on 21 May 2021), BraTs2015 (from https://www.smir.ch/BRATS/Start2015, accessed on 21 May 2021), BraTs2017 (from https://www.med.upenn.edu/sbia/brats2017/data.html, accessed on 21 May 2021), BraTs2018 (from https://www.med.upenn.edu/sbia/brats2018/data.html, accessed on 21 May 2021), BraTs2019 (from https://www.med.upenn.edu/cbica/brats2019/data.html, accessed on 21 May 2021), BraTs2020 (from https://www.med.upenn.edu/cbica/brats2020/data.html, accessed on 21 May 2021).

Conflicts of Interest: The authors declare no conflict of interest.

References

1. Lin, W. Principles of Magnetic Resonance Imaging: A Signal Processing Perspective. *IEEE Eng. Med. Biol. Mag.* **2000**, *19*, 129–130.
2. Letteboer, M.; Olsen, O.F. Segmentation of Tumors in Magnetic Resonance Brain Images Using an Interactive Multiscale Watershed Algorithm. *Acad. Radiol.* **2004**, *11*, 1125–1138. [CrossRef] [PubMed]
3. Ge, T.; Mu, N.; Li, L. A Brain Tumor Segmentation Method Based on Softmax Regression and Graph Cut. *Chin. J. Electron.* **2017**, *45*, 644–649.
4. Pham, D.L.; Xu, C.Y.; Prince, J.L. Current Methods in Medical Image Segmentation. *Annu. Rev. Biomed. Eng.* **2000**, *2*, 315–337. [CrossRef]
5. Luo, S.H.; Li, X.; Ourselin, S. *A New Deformable Model Using Dynamic Gradient Vector Flow and Adaptive Balloon Forces*; APRS Workshop on Digital Computing: Brisbane, Australia, 2003; pp. 9–14.
6. Salehi, S.; Erdogmus, D.; Gholipour, A. Auto-context Convolutional Neural Network(Auto-Net) for Brain Extraction in Magnetic Resonance Imaging. *IEEE Trans. Med. Imaging* **2017**, *36*, 2319–2330. [CrossRef] [PubMed]
7. Dice, L.R. Measures of the Amount of Ecologic Association between Species. *Ecology* **1945**, *26*, 297–302. [CrossRef]
8. Jaccard, P. The Distribution of Flora in the Alpine Zone. *New Phytol.* **1912**, *11*, 37–50. [CrossRef]
9. Storey, J.D. The Positive False Discovery Rate: A Bayesian Interpretation and the Q-Value. *Ann. Stat.* **2003**, *3*, 2013–2035. [CrossRef]
10. Fletcher Robert, H.; Suzanne, W. *Clinical Epidemiology: The Essentials*, 4th ed.; Lippincott Williams & Wilkins: Philadelphia, PA, USA, 2005; p. 45.
11. The Editor-in-Chief of Mathematics Dictionary. *Mathematics Dictionary*; Southeast University Press: Nanjing, China, 2002; Volume III.
12. Zhou, Z.X.; He, Z.S.; Shi, M.F.; Du, J.L.; Chen, D.D. 3D Dense Connectivity Network with Atrous Convolutional Feature Pyramid for Brain Tumor Segmentation in Magnetic Resonance Imaging of Human Head. *Comput. Biol. Med.* **2020**, *121*, 103766. [CrossRef] [PubMed]
13. Taha, A.; Hanbury, A. Metrics for Evaluating 3D Medical Image Segmentation: Analysis, Selection, and Tool. *BMC Med. Imaging* **2015**, *15*, 29. [CrossRef]
14. Vovk, U.; Pernus, F.; Likar, B. A Review of Methods for Correctio of Intensity Inhomogeneity in MRI. *IEEE Trans. Med. Imaging* **2007**, *26*, 405–421. [CrossRef]
15. Bland, A.D.G. Statistics Notes: Diagnostic Tests1: Sensitivity and Specificity. *BMJ* **1994**, *308*, 1552.
16. Tustison, N.J.; Shrinidhi, K.L.; Wintermark, M.; Durst, C.R.; Kandel, B.M.; Gee, J.C.; Grossman, M.C.; Avants, B.B. Optimal Symmetric Multimodal Templates and Concatenated Random Forests for Supervised Brain Tumor Segmentation (Simplified) with ANTsR. *Neuro Inform.* **2015**, *13*, 209–225. [CrossRef]
17. Pereira, S.; Pinto, A.; Correia, H.; Oliveira, J.; Rasteiro, D.M.; Silva, C.A. Brain Tumor Segmentation Based on Extremely Randomized Forest with High-level Features. In Proceedings of the 2015 37th Annual International Conference of the IEEE Engineering in Medicine and Biology Society, EMBC, Milan, Italy, 25–29 August 2015; pp. 3037–3040.
18. Havaei, M.; Davy, A.; Warde, F.D.; Biard, A.; Courville, A.; Bengio, Y.; Pal, C.; Jodoin, P.-M.; Larochelle, H. Brain Tumor Segmentation with Deep Neural Networks. *Med. Image Anal.* **2017**, *35*, 18–31. [CrossRef] [PubMed]
19. Shen, H.C.; Zhang, J.G.; Zheng, W.S. Efficient Symmetry-driven Fully Convolutional Network for Multimodal Brain Tumor Segmentation. In Proceedings of the 2017 IEEE International Conference on Image Processing, Beijing, China, 17–20 September 2017; pp. 3864–3868.
20. Zhao, X.; Wu, Y.; Song, G.; Li, Z.; Zhang, Y.; Fan, Y. A Deep Learning Model Integrating FCNNs and CRFs for Brain Tumor Segmentation. *Med. Image Anal.* **2017**, *43*, 98–111. [CrossRef]
21. Bhagat, P.K.; Choudhary, P. Multi-class Segmentation of Brain Tumor from MRI Images. *Appl. Artif. Intell. Tech. Eng. Adv. Intell. Syst. Comput.* **2019**, *698*, 543–553.
22. Hu, K.; Deng, S.H. Brain Tumor Segmentation Using Multi-Cascaded Convolutional Neural Networks and Conditional Random Field. *IEEE Access* **2019**, *7*, 2615–2629. [CrossRef]
23. Casamitjana, A.; Puch, S.; Aduriz, A.; Vilaplana, V. 3D Convolutional Neural Networks for Brain Tumor Segmentation: A Comparison of Multi-resolution Architectures. In Proceedings of the 2nd International Brain lesion: Glioma, Multiple Sclerosis, Stroke and Traumatic Brain Injuries, Athens, Greece, 17–21 October 2016; Springer: Berlin/Heidelberg, Germany, 2016; pp. 150–161.

24. Kamnitsas, K.; Ledig, C. Efficient Multi-Scale 3D CNN with Fully Connected CRF for Accurate Brain Lesion Segmentation. *Med. Image Anal.* **2017**, *36*, 61–78. [CrossRef]
25. Tseng, K.L.; Lin, Y.L.; Hsu, W.; Huang, C.Y. Joint Sequence Learning and Cross-modality Convolution for 3D Biomedical Segmentation. In Proceedings of the 2017 IEEE Conference on Computer Vision and Pattern Recognition, Honolulu, HI, USA, 21–26 July 2017; pp. 3739–3746.
26. Liu, D.; Zhang, H.; Zhao, M.; Yu, X.; Yao, S.; Zhou, W. Brain Tumor Segmentation Based on Dilated Convolution Refine Networks. In Proceedings of the 16th IEEE International Conference on Software Engineering Research, Management and Application, Kunming, China, 13–15 June 2018; pp. 113–120.
27. Iqbal, S.; Ghani, M.U.; Saba, T.; Rehman, A. Brain Tumor Segmentation in Multi-spectral MRI Using Convolutional Neural Networks(CNN). *Microsc. Res. Tech.* **2018**, *81*, 419–427. [CrossRef] [PubMed]
28. Li, H.C.; Li, A.; Wang, M.H. A Novel End-to-end Brain Tumor Segmentation Method Using Improved Fully Convolutional Networks. *Comput. Biol. Med.* **2019**, *108*, 150–160. [CrossRef]
29. Beers, A.; Chang, K.; Brown, J.; Sartor, E.; Gerstner, E.; Mammen, C.P.; Rosen, B.; Kalpathy, C.J. Sequential 3D U-Nets for Bio-logically-informed Brain Tumor Segmentation. *arXiv* **2017**, arXiv:1709.02967.
30. Shaikh, M.; Anand, G.; Acharya, G.; Amrutkar, A.; Alex, V.; Krishnamurthi, G. Brain Tumor Segmentation Using Dense Fully Convolutional Neural Network. In Proceedings of the 3rd International Brain Lesion: Glioma, Multiple Sclerosis, Stroke and Traumatic Brain Injuries, Quebec City, QC, Canada, 10–14 September 2017; Springer: Berlin/Heidelberg, Germany, 2017; pp. 309–319.
31. Isensee, F.; Wick, W.; Kickingereder, P.; Bendszus, M.; Maier, H.K. Brain Tumor Segmentation and Radio Mics Survival Prediction: Contribution to the BRATS 2017 Challenge. In Proceedings of the 3rd International Brain lesion: Glioma, Multiple Sclerosis, Stroke and Traumatic Brain Injuries, Quebec City, QC, Canada, 10–14 September 2017; Springer: Berlin/Heidelberg, Germany, 2018; pp. 287–297.
32. Zhou, T.X.; Ruan, S.; Hu, H.G.; Canu, S. Deep Learning Model Integrating Dilated Convolution and Deep Supervision for Brain Tumor Segmentation in Multi-parametric MRI. *Int. Workshop Mach. Learn. Med. Imaging* **2019**, *11861*, 574–582.
33. Po, Y.K.; Jefferson, W.; Chen, B.S. Improving 3D U-Net for Brain Tumor Segmentation by Utilizing Lesion Prior. *Comput. Sci. Comput. Vis. Pattern Recognit.* **2020**, *1907*, 00281.
34. Wang, G.; Li, W.; Ourselin, S.; Vercauteren, T. Automatic Brain Tumor Segmentation Based on Cascaded Convolutional Neural Networks with Uncertainty Estimation. *Cogn. Syst. Res.* **2020**, *59*, 304–311. [CrossRef] [PubMed]
35. Subhashis, B.; Sushmita, M. Novel Volumetric Sub-region Segmentation in Brain Tumors. *Front. Comput. Neurosci.* **2020**, *14*, 3.
36. Zhou, C.; Ding, C.; Wang, X.; Lu, Z.; Tao, D. One-Pass Multi-Task Networks with Cross-Task Guided Attention for Brain Tumor Segmentation. *IEEE Trans. Image Process.* **2020**, *29*, 4516–4529. [CrossRef] [PubMed]
37. Hua, R.; Huo, Q.; Gao, Y.; Sui, H.; Zhang, B.; Sun, Y.; Mo, S.; Shi, F. Segmenting Brain Tumor Using Cascaded V-Nets in Multimodal MR Images. *Front. Comput. Neurosci.* **2020**, *14*, 9. [CrossRef]
38. Zhang, J.; Lv, X. Ares U-Net: Attention Residual U-Net for Brain Tumor Segmentation. *Symmetry* **2020**, *12*, 721. [CrossRef]
39. Baid, U.; Talbar, S.; Rane, S.; Gupta, S.; Thakur, M.H.; Moiyadi, A.; Sable, N.; Akolkar, M.; Mahajan, A. A Novel Approach for Fully Automatic Intra-Tumor Segmentation with 3D U-Net Architecture for Gliomas. *Front. Comput. Neurosci.* **2020**, *14*, 10. [CrossRef] [PubMed]
40. Yogananda, C.; Wagner, B.; Nalawade, S.; Murugesan, G.K.; Pinho, M.C.; Fei, B.; Madhuranthakam, A.J.; Maldjian, J.A. *Fully Automated Brain Tumor Segmentation and Survival Prediction of Gliomas Using Deep Learning and MRI. Medical Image Computing and Computer Assisted Intervention, MICCAI*; Springer: Cham, Switzerland; Berlin/Heidelberg, Germany, 2020; Volume 11993, pp. 99–112.
41. Li, X.; Luo, G.; Wang, K. *Multi-step Cascaded Networks for Brain Tumor Segmentation. Medical Image Computing and Computer Assisted Intervention, MICCAI*; Springer: Cham, Switzerland; Berlin/Heidelberg, Germany, 2020; Volume 11992, pp. 163–173.
42. Wu, P.; Chang, Q. Brain Tumor Segmentation on Multimodal 3D-MRI Using Deep Learning Method. In Proceedings of the 2020 13th International Congress on Image and Signal Processing, Biomedical Engineering and Informatics, CISP-BMEI, Chengdu, China, 17–19 October 2020; pp. 635–639.
43. Zhao, Y.; Zhang, Y.; Liu, C. Bag of Tricks for 3D MRI Brain Tumor Segmentation. *Med. Image Comput. Comput. Assist. Interv.* **2020**, *11992*, 210–220.
44. Agravat, R.; Raval, M. Brain Tumor Segmentation and Survival Prediction. *arXiv* **2020**, arXiv:1909.09399V1.
45. Cheng, G.; Cheng, J.; Luo, M.; He, L.; Tian, Y.; Wang, R. Effective and Efficient Multitask Learning for Brain Tumor Segmentation. *J. Real-Time Image Process.* **2020**, *17*, 1951–1960. [CrossRef]
46. Ieva, A.; Russo, C.; Liu, S.; Jian, A.; Bai, M.Y.; Qian, Y.; Magnussen, J.S. Application of Deep Learning for Automatic Segmentation of Brain Tumors on Magnetic Resonance Imaging: A Heuristic Approach in the Clinical Scenario. *Neuroradiology* **2021**, *63*, 1253–1262. [CrossRef]
47. Lucas, F.; Sebastien, O. Generalized Wasserstein Dice Score, Distributionally Robust Deep Learning, and Ranger for Brain Tumor Segmentation: BraTs2020 Challenge. *arXiv* **2020**, arXiv:2011.01614v2.
48. Henry, T.; Carre, A.; Lerousseau, M.; Estienne, T.; Robert, C.; Paragios, N.; Deutsch, E. Brain Tumor Segmentation with Self-ensembled, Deeply-supervised 3D U-Net Neural Networks: A BraTs2020 Challenge Solution. *arXiv* **2020**, arXiv:2011.01045.
49. SilvA, C.; Pinto, A.; Pereira, S.; Lopes, A. Multi-stage Deep Layer Aggregation for Brain Tumor Segmentation. *arXiv* **2021**, arXiv:2101.00490.

50. Anand, V.K.; Grampurohit, S. Brain Tumor Segmentation and Survival Prediction Using Automatic Hard Mining in 3D CNN Architecture. *arXiv* **2021**, arXiv:2101.01546v1.
51. Qamar, S.; Ahmad, P.; Shen, L. HI-Net: Hyperdense Inception 3D U_Net for Brain Tumor Segmentation. *arXiv* **2020**, arXiv:2012.06760v1.
52. Jia, H.; Cai, W.; Huang, H.; Xia, Y. H2NF-Net for Brain Tumor Segmentation Using Multimodal MR Imaging: 2nd Place Solution to BraTs Challenge 2020 Segmentation Task. *arXiv* **2021**, arXiv:2012.15318v1.
53. Lyu, C.; Shu, H. A Two-Stage Cascade Model with Variational Auto Encoders and Attention Gates for MRI Brain Tumor Segmentation. *arXiv* **2021**, arXiv:2011.02881v2.
54. Shen, Y.; Gao, M. *Brain Tumor Segmentation on MRI with Missing Modalities. Information Processing in Medical Imaging*; Springer: Cham, Switzerland; Berlin/Heidelberg, Germany, 2019; Volume 11492, pp. 417–428.
55. Zhao, Z.R.; Zhao, Z. An Enhanced U-Net for Brain Tumor Segmentation. In Proceedings of the 2019 IEEE International Conference on Robotics and Biomimetics, ROBIO, Dali, China, 6–8 December 2019; pp. 3054–3058.
56. Tiwari, A.; Srivastava, S.; Pant, M. Brain Tumor Segmentation and Classification from Magnetic Resonance Images: Review of selected methods from 2014 to 2019. *Pattern Recognit. Lett.* **2019**, *131*, 244–260. [CrossRef]
57. Sujan, M.; Alam, N.; Abdullah, S.; Islam, M.J. A Segmentation Based Automated System for Brain Tumor Detection. *Comput. Appl.* **2016**, *153*, 41–49. [CrossRef]
58. Wang, Y.P. *Medical Image Processing*; Tsinghua University Press: Beijing, China, 2012.
59. Fooladivanda, A.; Shokouhi, S.B.; Ahmadinejad, N.; Mosavi, M.R. Automatic Segmentation of Breast and Fibro glandular Tissue in Breast MRI Using Local Adaptive Thresholding. In Proceedings of the 2014 21th Iranian Conference on Biomedical Engineering, ICBME, Tehran, Iran, 26–28 November 2014; pp. 195–200.
60. Kaleem, M.; Sanaullah, M.; Hussain, M.A.; Jaffar, M.A.; Choi, T.S. Segmentation of Brain Tumor Tissue Using Marker Controlled Watershed Transform Method. *Commun. Comput. Inf. Sci.* **2012**, *281*, 222–227.
61. Rajini, N.; Narmatha, T.; Bhavani, R. Automatic Classification of MR Brain Tumor Images Using Decision Tree. In Proceedings of the International Conference on Electronics, Communication and Information Systems, Near Madurai, Tamilnadu, India, 2–3 November 2012; pp. 10–13.
62. Qusay, A.; Isa, N. Computer-aided Segmentation System for Breast MRI Tumor Using Modified Automatic Seeded Region Growing. BMRI-MASRG. *J. Digit. Imaging* **2014**, *27*, 133–144.
63. Lei, T.; Zhang, X.; Jia, X.H. Research Progress of Image Segmentation Based on Fuzzy Clustering. *Chin. J. Electron.* **2019**, *47*, 1776–1791.
64. Muneer, K.; Joseph, K. *Performance Analysis of Combined K-mean and Fuzzy-c-Mean Segmentation of MR Brain Images Computational Vision and Bio Inspired Computing*; Springer: Cham, Switzerland; Berlin/Heidelberg, Germany, 2018; pp. 830–860.
65. Guo, Y.; Sengur, A. NCM: Neutrosophic C-Means Clustering Algorithm. *Pattern Recognit.* **2015**, *48*, 2710–2724. [CrossRef]
66. Khosravanian, A. Fast Level Set Method for Glioma Brain Tumor Segmentation Based on Super Pixel Fuzzy Clustering and Lattice Boltzmann Method. *Comput. Methods Programs Biomed.* **2020**, *198*, 105809. [CrossRef] [PubMed]
67. Canny, A. Computational Approach to Edge Detection. *IEEE Trans. Pattern Anal. Mach. Intell.* **1986**, *8*, 679–698. [CrossRef] [PubMed]
68. Jayanthi, S.; Ranganathan, H.; Palanivelan, M. Segmenting Brain Tumor Regions with Fuzzy Integrated Active Contours. *IETE J. Res.* **2019**. [CrossRef]
69. Havaei, M.; Jodoin, P.M.; Larochelle, A.H. Efficient Interactive Brain Tumor Segmentation as Within-Brain KNN Classification. In Proceedings of the International Conference on Pattern Recognition, Stockholm, Sweden, 24–28 August 2014; pp. 556–561.
70. Llner, F.; Emblem, K.; Schad, L. Support Vector Machines in DSC-based Glioma Imaging: Suggestions for Optimal Characterization. *Magn. Reson. Med.* **2010**, *64*, 1230–1236.
71. Mahmood, Q.; Basit, A. *Automatic Ischemic Stroke Lesion Segmentation in Multi-Spectral MRI Images Using Random Forests Classifier*; Springer: New York, NY, USA, 2015; pp. 266–274.
72. Chen, X.; Binh, P. Automated Brain Tumor Segmentation Using Kernel Dictionary Learning and Super Pixel-level Features. *Syst. Man Cybern.* **2016**, *10*, 1109.
73. Shil, S.; Polly, F. An Improved Brain Tumor Detection and Classification Mechanism. In Proceedings of the 2017 International Conference on Information and Communication Technology Convergence, ICTC, Jeju, Korea, 18–20 October 2017; pp. 54–57.
74. Vaishnavee, K.; Amshakala, K. An Automated MRI Brain Image Segmentation and Tumor Detection Using SOM-clustering and Proximal Support Vector Machine Classifier. In Proceedings of the 2015 IEEE International Conference on Engineering and Technology, ICETECH, Coimbatore, India, 20–20 March 2015; pp. 1–6.
75. Wu, W.; Chen, A. Brain Tumor Detection and Segmentation in A CRF (Conditional Random Fields) Framework with Pixel-pairwise Affinity and Super Pixel-level Features. *Int. J. Comput. Assist. Radiol. Surg.* **2014**, *9*, 241–253. [CrossRef]
76. Nabizadeh, N.; Kubat, M. Automatic Tumor Segmentation in Single-spectral MRI Using A Texture-based and Contour-based Algorithm. *Expert Syst. Appl.* **2017**, *77*, 1–10. [CrossRef]
77. Selvathi, D.; Selvaraj, H. Segmentation of Brain Tumor Tissues in MR Images Using Multiresolution Transforms and Random Forest Classifier with Ada Boost Technique. In Proceedings of the 2018 26th International Conference on Systems Engineering, ICSEng, Sydney, Australia, 18–20 December 2018; pp. 1–7.

78. Reza, S. Multi-fractal Texture Features for Brain Tumor and Edema Segmentation. In *Medical Imaging 2014: Computer-Aided Diagnosis*; International Society for Optics and Photonics: San Diego, CA, USA, 2014; Volume 9035, p. 903503.
79. Meier, R.; Bauer, S. Patient-specific Semi-supervised Learning for Postoperative Brain Tumor Segmentation. In Proceedings of the International Conference on Medical Image Computing and Computer-Assisted Intervention, Strasbourg, France, 27 September–1 October 2021; Springer: Cham, Switzerland; Berlin/Heidelberg, Germany, 2014; pp. 714–721.
80. Li, Y.F. A Fuzzy Method for Image Segmentation Based on Multi-dictionary Learning. *Chin. J. Electron.* **2018**, *46*, 1700–1709.
81. Chen, S.H.; Liu, W.X.; Qin, J.; Chen, L.; Bin, G.; Zhou, Y.; Huang, B. Research Progress in Computer-aided Diagnosis of Cancer Based on Deep Learning and Medical Images. *J. Biomed. Eng.* **2017**, *2*, 160–165.
82. Krizhevsky, A.; Sutskever, I.; Hinton, G.E. ImageNet Classification with Deep Convolutional Neural Networks. *Int. Conf. Neural Inf. Process. Syst.* **2012**, *60*, 1066. [CrossRef]
83. Simonyan, K.; Zisserman, A. Very Deep Convolutional Networks for Large Scale Image Recognition. *Comput. Sci.* **2014**, *6*, 1556.
84. Szegedy, C.; Liu, Y. Going Deeper with Convolutions. In Proceedings of the IEEE Conference on Computer Vision and Pattern Recognition, Boston, MA, USA, 7–12 June 2015; pp. 1–9.
85. He, K.W.; Zhang, X.; Ren, S.; Sun, J. Deep Residual Learning for Image Recognition. In Proceedings of the IEEE Conference on Computer Vision and Pattern Recognition (CVPR), Las Vegas, NV, USA, 27–30 June 2016; pp. 770–778.
86. Zhou, T.; Huo, B.Q.; Lu, H.L. Research on Residual Neural Network and Its Application in Medical Image Processing. *Chin. J. Electron.* **2020**, *48*, 1436–1447.
87. Lai, X.B.; Xu, M.S.; Xu, X.M. Multimodal MR Image Segmentation of Glioblastoma Based on Multi-class CNN. *Chin. J. Electron.* **2019**, *47*, 140–149.
88. Hussain, S.; Anwar, S.M. Segmentation of Glioma Tumors in Brain Using Deep Convolutional Neural Network. *Neuro Comput.* **2018**, *282*, 248–261. [CrossRef]
89. Saouli, R.; Akil, M.; Kachouri, R. Fully Automatic Brain Tumor Segmentation Using End-to-end Incremental Deep Neural Networks in MRI Images. *Comput. Methods Programs Biomed.* **2018**, *166*, 39–49.
90. Xue, J.; Hu, J.Y. Hypergraph Membrane System Based F2 Fully Convolutional Neural Network for Brain Tumor Segmentation. *Appl. Soft Comput. J.* **2020**, *94*, 106454. [CrossRef]
91. Zhou, Z.X.; He, Z.S.; Jia, Y.Y. AFP-Net: A 3D Fully Convolutional Neural Network with Atrous-convolution Feature Pyramid for Brain Tumor Segmentation via MRI Images. *Neuro Comput.* **2020**, *402*, 03097.
92. Badrinarayan, V.; Kendall, A. SegNet: A Deep Convolutional Encoder-Decoder Architecture for Image Segmentation. *IEEE Trans. Pattern Anal. Mach. Intell.* **2017**, *39*, 2481–2495. [CrossRef] [PubMed]
93. Chen, S.C.; Ding, C.X.; Liu, M.F. Dual-force Convolutional Neural Networks for Accurate Brain Tumor Segmentation. *Pattern Recognit.* **2019**, *88*, 90–100. [CrossRef]
94. Zhou, Z.W.; Siddiquee, M.R. *U-Net++: A Nested U-Net Architecture for Medical Image Segmentation. International Workshop on Deep Learning in Medical Image Analysis Deep Learning in Medical Image Analysis and Multimodal Learning for Clinical Decision Support*; Springer: Cham, Switzerland; Berlin/Heidelberg, Germany, 2018; pp. 3–11.
95. Alom, M.Z.; Hasan, M.; Yakopcic, C.; Taha, T.M.; Asari, K. Recurrent Residual Convolutional Neural Network Based on U-Net (R2U-Net) for Medical Image Segmentation. *Comput. Vis. Pattern Recognit.* **2018**, *5*, 06955.
96. Milletari, F.; Navab, N.; Ahmadi, S.A. V-Net: Fully Convolutional Neural Networks for Volumetric Medical Image Segmentation. In Proceedings of the International Conference on 3D Vision, Stanford, CA, USA, 25–28 October 2016; pp. 565–571.
97. Çiçek, Ö. 3D U-Net: Learning Dense Volumetric Segmentation from Sparse Annotation. In Proceedings of the International Conference on Medical Image Computing and Computer-Assisted Intervention, Cambridge, UK, 19–22 September 1999; pp. 424–432.
98. He, C.E.; Xu, H. Research on Automatic Segmentation Algorithm for Multimodal MRI Brain Tumor Images. *Acta Opt. Sin.* **2020**, *40*, 0610001.
99. Jia, H.Z.; Xia, Y. *Learning High-Resolution and Efficient Non-Local Features for Brain Glioma Segmentation in MR Images*; Medical Image Computing and Computer Assisted Intervention, MICCAI: Lima, Peru, 2020; pp. 480–490.
100. Mckinley, A.; Wiest, R.; Reyes, M. *Pooling-Free Fully Convolutional Networks with Dense Skip Connections for Semantic Segmentation, with Application to Segmentation of White Matter Lesions. Medical Image Computing and Computer Assisted Intervention*; MICCAI: Quebec City, QC, Canada, 2017; pp. 169–177.
101. Mlynarski, P.; Delingette, H. Deep Learning with Mixed Supervision for Brain Tumor Segmentation. *J. Med. Imaging* **2019**, *6*, 034002. [CrossRef] [PubMed]
102. Yuan, L.; Wei, X. Multi-center Brain Imaging Classification sing A Novel 3D CNN Approach. *IEEE Access* **2018**, *6*, 925–934. [CrossRef]
103. Glocker, B. Machine Learning with Multi-Site Imaging Data: An Empirical Study on the Impact of Scanner Effects. *arXiv* **2019**, arXiv:1910.04597.
104. He, K.; Gkioxari, G.; Dollár, P.; Girshick, R. Mask R-CNN. In Proceedings of the IEEE International Conference on Computer Vision (ICCV), Venice, Italy, 22–29 October 2017; pp. 2961–2969.
105. Chen, H.; Sun, K.Y. BlendMask: Top-Down Meets Bottom-Up for Instance Segmentation. In Proceedings of the IEEE/CVF Conference on Computer Vision and Pattern Recognition, CVPR, Seattle, WA, USA, 14–19 June 2020; pp. 8573–8581.

healthcare

MDPI

Article

Artificial Intelligence for Hospital Health Care: Application Cases and Answers to Challenges in European Hospitals

Matthias Klumpp [1,2,*], Marcus Hintze [1], Milla Immonen [3], Francisco Ródenas-Rigla [4], Francesco Pilati [5], Fernando Aparicio-Martínez [6], Dilay Çelebi [7], Thomas Liebig [8,9], Mats Jirstrand [10], Oliver Urbann [1], Marja Hedman [11], Jukka A. Lipponen [12], Silvio Bicciato [13], Anda-Petronela Radan [14], Bernardo Valdivieso [15], Wolfgang Thronicke [16], Dimitrios Gunopulos [17] and Ricard Delgado-Gonzalo [18]

1 Fraunhofer Institute for Material Flow and Logistics (IML), Josef-von-Fraunhofer-Str. 2-4, 44227 Dortmund, Germany; marcus.hintze@iml.fraunhofer.de (M.H.); oliver.urbann@iml.fraunhofer.de (O.U.)
2 Department of Business Administration, Georg-August-University of Göttingen, Platz der Göttinger Sieben 3, 37073 Göttingen, Germany
3 VTT Technical Research Centre of Finland Ltd., Kaitoväylä 1, 90571 Oulu, Finland; milla.immonen@vtt.fi
4 Polibienestar Research Institute, University of Valencia, Carrer del Serpis 29, 46022 València, Spain; francisco.rodenas@uv.es
5 Department of Industrial Engineering, University of Trento, Via Sommarive 9, 38123 Trento, Italy; francesco.pilati@unitn.it
6 NUNSYS S.L., Calle Gustave Eiffel 3, 46980 Valencia, Spain; fernando.aparicio@nunsys.com
7 Department of Management Engineering, Istanbul Technical University, Macka, Beşiktaş, 34367 İstanbul, Turkey; celebid@itu.edu.tr
8 TU Dortmund, Artificial Intelligence Unit, Otto-Hahn-Straße 12, 44221 Dortmund, Germany; thomas.liebig@tu-dortmund.de
9 Materna Information & Communications SE, Artificial Intelligence Unit, Voßkuhle 37, 44141 Dortmund, Germany
10 Fraunhofer-Chalmers Centre & Fraunhofer Center for Machine Learning, Chalmers Science Park, 41288 Gothenburg, Sweden; matsj@fcc.chalmers.se
11 Heart Center, Kuopio University Hospital and Institute of Clinical Medicine, University of Eastern Finland, Ritva Jauhiainen-Bruun, 70029 Kuopio, Finland; marja.hedman@kuh.fi
12 Department of Applied Physics, University of Eastern Finland, Yliopistonranta 1, 70210 Kuopio, Finland; jukka.lipponen@uef.fi
13 Interdepartmental Center for Stem Cells and Regenerative Medicine (CIDSTEM), Department of Life Sciences, University of Modena and Reggio Emilia, Via Gottardi 100, 41125 Modena, Italy; silvio.bicciato@unimore.it
14 Department of Obstetrics and Gynecology, University Hospital of Bern, Murtenstraße 11, 3008 Bern, Switzerland; Anda-Petronela.Radan@insel.ch
15 La Fe University Hospital Valencia, Avinguda de Fernando Abril Martorell 106, 46026 València, Spain; valdivieso_ber@gva.es
16 ATOS Information Technology GmbH, Fürstenallee 11, 33102 Paderborn, Germany; wolfgang.thronicke@atos.net
17 Department of Informatics and Telecommunications, National and Kapodistrian University of Athens, Panepistimioupolis, Ilisia, 15784 Athens, Greece; dgunopulos@gmail.com
18 Centre Suisse d'Électronique et de Microtechnique CSEM, Jaquet Droz 1, 2002 Neuchâtel, Switzerland; ricard.delgado@csem.ch
* Correspondence: matthias.klumpp@iml.fraunhofer.de

Citation: Klumpp, M.; Hintze, M.; Immonen, M.; Ródenas-Rigla, F.; Pilati, F.; Aparicio-Martínez, F.; Çelebi, D.; Liebig, T.; Jirstrand, M.; Urbann, O.; et al. Artificial Intelligence for Hospital Health Care: Application Cases and Answers to Challenges in European Hospitals. *Healthcare* **2021**, 9, 961. https://doi.org/10.3390/healthcare9080961

Academic Editor: Mahmudur Rahman

Received: 18 May 2021
Accepted: 3 July 2021
Published: 29 July 2021

Abstract: The development and implementation of artificial intelligence (AI) applications in health care contexts is a concurrent research and management question. Especially for hospitals, the expectations regarding improved efficiency and effectiveness by the introduction of novel AI applications are huge. However, experiences with real-life AI use cases are still scarce. As a first step towards structuring and comparing such experiences, this paper is presenting a comparative approach from nine European hospitals and eleven different use cases with possible application areas and benefits of hospital AI technologies. This is structured as a current review and opinion article from a diverse range of researchers and health care professionals. This contributes to important improvement options also for pandemic crises challenges, e.g., the current COVID-19 situation. The expected advantages as well as challenges regarding data protection, privacy, or human acceptance are reported. Altogether, the diversity of application cases is a core characteristic of AI applications in hospitals, and this requires a specific approach for successful implementation in the health care sector. This can include specialized solutions for hospitals regarding human–computer interaction, data management, and communication in AI implementation projects.

Keywords: COVID-19; artificial intelligence; uses cases; European hospitals; benefits

1. Introduction

Research into applications of artificial intelligence (AI) in health care and within hospitals is a crucial area of innovation [1]. Smart health care with the support of AI technologies, such as Machine Learning (ML), is needed due to specific challenges in the provision of medical support in European countries as well as in the rest of the world. It is not only the outbreak of the COVID-19 pandemic that reveals the current problems and challenges facing European hospitals. The success in the science of medicine in the last decades has had the effect of patients becoming older, frailer, and multi-morbid due to a longer lifetime expectation [2].

This is accompanied by the fact that medical care and diseases are becoming increasingly complex. Due to this medical complexity, medical personnel are becoming more and more specialized, which cannot in general be fully provided for by smaller hospitals in rural areas. Added to this is the demographic change already emerging in Europe, e.g., the population of over 80-year-olds in the EU27 will double from 6.1% in 2020 to 12.5% in 2060 [3]. Hence, more older people with their specific health problems will use the health care system. In contrast to this, the number of young well-trained medical personnel is currently decreasing and a shortage of skilled personnel, such as doctors and nurses, is already emerging in many European nations [4].

The challenges of the simultaneous increase of older and multi-morbid patients with complex diseases and the shortage of skilled personnel are also hampered by the increasing economic constraints on hospitals. An increase in chronic diseases due to aging populations and shortage of medical specialists results in resource scarcity and medical sustainability challenges. In order not to endanger the living and health standards of the European nations it will be necessary to develop applied AI-solutions to relieve the burden of increased workload as well as being instrumental to deliver efficient, effective, and high-quality health care.

Adaptability and agility at hospitals are major prerequisites in this context, and narrowing the application of AI to optimization solely does miss the point in many cases. By opening a wider range of actionable options, from personalized medical diagnosis and treatment to choices in care, sourcing, and logistics areas, AI applications will provide more important support avenues than efficiency enhancements only [5,6]. In addition, multiple benefits regarding the ongoing COVID-19 pandemic can also be expected and should be further explored, especially regarding data analysis and preventing unnecessary patient contact for health care personnel in hospitals as centres of the fight against the viral disease [7].

AI can also contribute to the fight against pandemics as COVID-19, helping hospitals focus resources on pandemic patient's treatments in the current as well as possible future situations. In this sense, most AI applications are directed at contactless analysis, diagnosis, and treatment (e.g., self-treatment and prevention), reducing the number of personal contacts and hospital visits, therefore reducing the potential spread of COVID-19 and other viral pandemics. AI in particular offers great potential for improving medical care and supporting the medical staff. The state of the art and the challenges regarding AI applications in hospitals and the health care sector are described for specific application areas in Figure 1.

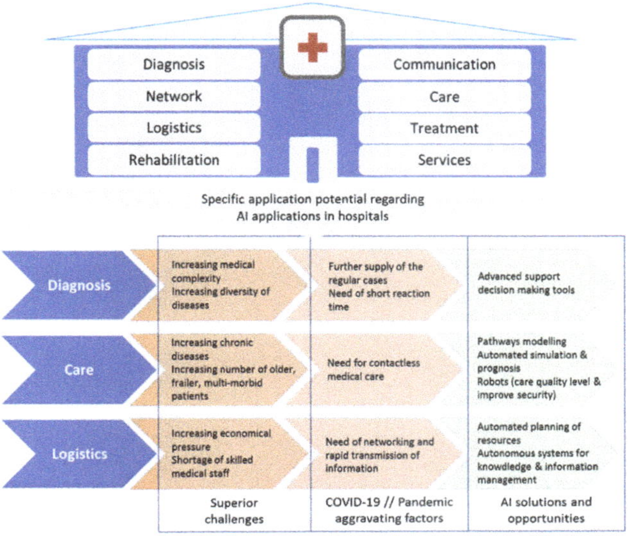

Figure 1. Interrelation structure of AI application areas for AI in hospitals.

With regards to the introduction of AI applications in hospitals, two specific questions arise, with the answers to them as the central contributions of this paper: First, what are the requirements and hospital setups for AI applications? To this end, the authors carried out a survey of different European hospitals and identified relevant projects in this field. As a result, the main fields of application of AI for hospitals are found as care, diagnosis, and logistics. The hospitals surveyed saw the greatest medical and economical potential in these three areas through the use of AI. Building on this, the paper outlines altogether 11 use cases in 9 hospitals across Europe, informing how AI can contribute to agility and efficiency in hospitals, improving health care from the resource efficiency as well as the service quality and choice side, aligned with the core hospital workflow and value adding processes. The second question is: How can a basic structure for the different AI use cases be established to avoid the mistake of developing isolated solutions that are difficult to transfer across hospitals? The authors propose three basics support areas which help to ensure a holistic approach to AI application implementation and transfer within the paper.

The paper is structured as follows: The following section is outlining the applied use case methodology for the analysis presented. The next section is describing the specific use case descriptions and expectations of hospitals towards AI applications. The following section presents a discussion regarding possible benefits and challenges as well as concept items such as human–computer interaction and medical data space concepts to overcome the challenges posed by AI applications in the hospital context. The final section provides an outlook towards future developments and challenges for AI applications in hospitals.

2. Use Case Methodology

The first step to identify the current challenges and areas of interest of European hospitals was to create a survey. The survey was carried out to obtain a differentiated view of the needs of European hospitals. Specifics were requested, such as country, type, number of patients and beds, and the main health care areas. In addition, hospital decision-makers identified specific areas of application and presented the focus and expected output of the utility of AI. The following Table 1 outlines the specific setup of these hospital characteristics for the institutions included in the survey.

Table 1. Included survey and case study hospitals in Europe.

Hospital	Type/Inpatient per Year/Outpatient per Year/Beds [1]	Main Health Area(s)	Specific Application Areas (AA)/Focus on (F)/Expected Output (EO)
University Hospital of Bern (Switzerland)	Public/6000/737,830/64 (2018) [2]	The University Clinic for Obstetrics and Gynecology of Inselspital	AA: Diagnosis F: Fetal state assessment during labor EO: AI-based decision support system for fetal state assessment during labor. The solution can assist obstetricians in accurately assessing the fetal state in clinical practice during labor.
Kuopio University Hospital (Finland)	Public/99,000/517,000/590 (2019) [3]	All branches	AA: Diagnosis F: Finding new diagnostic and treatment methods, for coronary artery disease. EO: An AI-based decision support system for selecting those patients among suspected CAD who benefit from further imaging.
Hospital of Bozen (Italy)	Public/25,064/737,830/697 (2018) [4]	All branches	AA: Care F: Rheumatological diseases and diabetes EO: An intelligent tool able to support the definition and scheduling of the different laboratory tests, medical examinations and hospitalization.
La Fe University Hospital (Spain)	Public/45,062/148,702/1004 (2019) [5]	Management of Chronicity (Integrated Care) and Active and Healthy Aging	AA: Care F: Strategic initiatives on integrated care for patients with complex chronic and/or oncological conditions EO: An intelligent tool able to improve the management of chronic patients and to characterize the use of resources throughout chronic patients' healthcare, reducing the economic burden for hospitals.
Federico II University of Naples (Italy)	Public/n.a./365,000/1000 (2019) [6]	Arterial hypertension on the cardiovascular system	AA: Care F: Arterial hypertension with particular reference to ischemia heart disease EO: Development of diagnostic and therapeutic methodologies in the field of cardiac rehabilitation; development of remote monitoring systems (telemedicine) for patients with high cardiovascular risk.

Table 1. *Cont.*

Hospital	Type/Inpatient per Year/Outpatient per Year/Beds [1]	Main Health Area(s)	Specific Application Areas (AA)/Focus on (F)/Expected Output (EO)
Orton Ltd., The Private Unit Helsinki Univ.Hospital (Finland)	Private/2000/22,000/40 [7]	Orthopedics, neurosurgery, cancer treatment, pain medicine and rehabilitation	AA: Care F: Ethical, rehabilitation and preventive care EO: Developing new tools for the treatment and rehabilitation of musculoskeletal disorders and other conditions.
Odense University Hospital (Denmark)	Public/104,229/1,104,229/1038 (2019) [8]	All branches	AA: Logistics F: Future of health care in mind, incorporating innovative clinical and logistical technologies. EO: To serve as a test bed for new medical technology, including an extensive use of robotics and AI.
Bayındır Hospital (Turkey)	Private/11,284/252,995/131 (2019) [9]	All branches	AA: Logistics F: Materials management and scheduling EO: Optimizing resource allocation and medical materials planning, reducing operational costs and patient waiting times
University Hospital Essen (Germany)	Public/50,000/195,000/1300 (2019) [10]	Genetic medicine, immunology, oncology, cardiovascular medicine and transplants	AA: Logistics//F: Care operations with materials management and supply//EO: Digitalized, patient- and employee-oriented organization. To minimize time spent for the nurses on documentation and administrative tasks to allow more time for direct patient care.

[1] Data from hospital sources. Definitions might differ due to national data regulations. [2] University Hospital of Bern: http://www.frauenheilkunde.insel.ch/de/ueber-die-klinik, accessed on 2 October 2020. [3] Kuopio University Hospital: https://www.psshp.fi/web/en/organisation/operations-and-tasks, accessed on 2 October 2020. [4] Südtiroler Sanitätsbetrieb: https://www.sabes.it/de/578.asp, accessed on 2 October 2020. [5] La Fe University Hospital: Hospital activity report, 2019. [6] Federico II University of Naples. [7] Orton Ltd. University Hospital. [8] Odense University Hospital: https://en.ouh.dk/about-ouh/key-figures, accessed on 2 October 2020. [9] Bayındır Hospital. [10] Universitätsklinikum Essen: https://www.uk-essen.de, accessed on 2 October 2020.

The framework situations for the outlined AI use cases are characterized by their specific hospital setup in a broad multitude of European hospitals. By means of surveys carried out in the hospitals participating in this analysis, different health care personnel have provided systematic answers to a structured questionnaire dealing with relevant aspects to the study. The hospitals where asked to detail current practical problems in different areas, how are they currently managing these problems, ways and mechanisms to improve in these areas by means of AI, and relevant KPIs determining qualitative and quantitative improvements related to the adoption of the AI application. As a result, after extracting the information from these surveys, use cases could be drafted for the different health institutions, based on real and actual needs and opportunities. Societies require an effective and efficient health care system and especially hospitals as nodes in a network of actors providing high-quality services, resources and serving patients. The following table summarizes the main expectations as stated by the health organizations in the survey (see Table 2).

From the expectations, a total of 11 use cases in different health areas has been envisioned. It turns out that three particular fields are of specific interest to the hospitals surveyed: diagnosis, care and logistics.

In the field of diagnosis, clinical decisions still mostly depend on the application of clinical practice guidelines, instead of being based on the use of automatic decision support tools that exploit the increasing availability of medical data from molecular assays, electronic health records, clinical and pathological images, and wearable connected sensors. Nowadays, clinicians face enormous challenges in reconciling heterogeneous clinical data and exploiting the information content to make optimal decisions when assessing a disease or its progression, and this situation has become more evident in the midst of the global COVID-19 pandemic. Thus, there is an urgent need to develop smart decision support systems, which assist clinicians in making rapid and precise diagnostic decisions through the combination of multiple data sources. AI-based methodologies for medical diagnosis and medical decision support have gained attention in the recent years as these systems hold promise to automate the diagnosis and triage processes, thus optimizing and accelerating the referral process especially in urgent and critical cases. Recently, state-of-the-art examples demonstrated that software based on AI can be used in clinical practice to improve decision-making and to achieve fast and accurate databased diagnosis of various pathologies. In particular, AI has been proven particularly helpful in areas where the diagnostic information is already digitized, such as: for detection of cancers based on molecular, genomic, and radiological data [8], making individual prognosis in psychiatry using neuroimaging [9,10] identifying strokes from computed tomography scans [11], assessing the risk of sudden cardiac death or other heart diseases based on electrocardiograms and cardiac magnetic resonance images [12,13], classifying skin lesions from skin images [14], finding indicators of diabetic retinopathy in eye images [15], and detect phenotypes that correlate with rare genetic diseases from patient facial photos [16]. The change in clinical practice through and by the means of technological innovation is today decisively enabling health care systems to face to the continuous economic, socio-demographic and epidemiological pressures [17]. However, technological innovation, although important and central, must be carefully examined and accompanied to ensure that it really corresponds to effective social innovation. As addressed by MedTech Europe, developing AI systems and algorithms for healthcare settings requires specific skillsets which are in short supply, and investment in education and training of professionals involved (e.g., data scientists, practitioners, software engineers, clinical engineers), is mandatory [18].

Table 2. Included survey and case study hospitals in Europe.

Heath Organization	Current Problems and Approach	Vision on Potential Application of AI	Expected Improvement—KPIs
University Hospital of Bern, Department of Obstetrics and Gynecology	Fetal assessment based on Cardiotocography (CTG) or electronic fetal monitoring (EFM) limitations.	Their vision is to develop a medical decision support system, which can assist obstetricians in accurately assessing the fetal state in clinical practice during labor.	Improvement of decision-making can improve fetal outcomes after delivery and avoid unnecessary medical interventions and their health implications for mother and fetus, as well as their economic implications. The KPIs of the AI application are: • Fetal outcomes, measured by clinical adaptation (APGAR score) and hypoxia (measured by arterial pH), • Invasive interventions for prematurely ending the delivery process, such as instrumental delivery or cesarean section, • Economical costs of delivery.
Kuopio University Hospital	Currently, the diagnosis of coronary heart disease has changed towards the non-invasive imaging, which has led to increasing number of patients scheduled to CCTA. Interpretation of CCTA is affected by the image quality, experience of the doctor and by other issues, which can in terms lead to unnecessary repeated or additive diagnostic imaging.	The motivation is to develop an automatic AI-based analysis system for the coronary computed tomography angiography (CCTA): To enhance diagnostic accuracy of CCTA and to guide clinical decision making. Interpretation of CCTA will be systematically guided by the standard AI-based analysis system.	The patients need only one diagnostic method and the workflow of the interpretation of CCTA become more fluent. Relevant KPIs are: • Increased number of CCTA imaging in one center, • Improved patient convenience, safety and decreased health care costs, • Improved effectiveness leads to shorter waiting times and shortened queues.
Hospital of Bozen	Limitations on healthcare resources management and chronic care pathways definition	AI tools to support the definition and scheduling of the different laboratory tests, medical examinations and hospitalization which affect STHA patients, personnel, equipment and resources inside and outside the hospital and located in multiple areas of the geographical territory of its responsibility	Ease the management of healthcare resources with a particular focus on rheumatological diseases and diabetes as chronic diseases. Relevant KPIs are: Decrease waiting time to access to scheduled medical examinations and labor tests, Average cost to provide the healthcare services to the chronic care population, Quality of the medical treatment, e.g., percentage of re-hospitalized patients.
La Fe University Hospital	Chronic diseases (CDs) represent the major cost of morbidity and mortality and lead to 86% of all deaths. In Europe, these account for more than 75% of the healthcare burden with a cost for the economy of €700 billion per year.	AI will help to: Improve the management of chronic conditions and multimorbidity in the face of aging population and its implication on public health; Contain the impact and global burden of chronic conditions, multimorbidity and frailty on individual quality of life and on healthcare systems; Strength the clinical management of complex chronic conditions and multimorbidity having a better understanding of the individual prognosis and disease evolution, and targeting personalized interventions.	Optimization of resources and the clinical flow of chronic patients at Hospital. Relevant KPIs are: Efficiency on the allocation and consumption of resources, Right assignment of chronic patient to care pathway, Decrease in turnaround time, Selection of right pathway, Avoidable episodes of care inadequate use.

Table 2. *Cont.*

Heath Organization	Current Problems and Approach	Vision on Potential Application of AI	Expected Improvement—KPIs
Federico II University of Naples	Today CVD is the leading cause of death in Europe; presently 47% of all deaths in Europe and 40% of all deaths in the European Union (EU) are attributable to CVD. This means that across Europe as a whole 4 million deaths per year currently occur due to CVD, of which 1.9 million are in the European Union	Use of AI may help clinicians in problem solving and patient's management. AI process may be used to improve process of health care management with specific regards to resource allocation, patient management.	Rapid assessment of correct management strategy. Relevant KPIs are: Improvement of timeliness in critical event treatment, Reduction of ambulatorial visits, Forecasting of avoidable critical conditions.
Odense University Hospital	Maintain high quality treatment for our patients in a demographic development scenario and increasing chronic conditions	Need to rely on AI and robots to ensure quality level and improve security in repetitive tasks, while alleviating staffing challenges.	Optimize handling of transports and logistics. Relevant KIPs are: Improve timing for transportation of patients or samples. Release of staffing resources to other tasks/areas. As well as an improved working environment for staff.

In the field of care, AI for health has shown great potential to improve healthcare efficiency, considering the relationship between health factors, including service and management, and ICT factors that include sensors, networks, data resources, platforms, applications and solutions [19]. For the hospital facilities, AI is one of the most powerful technologies from the perspectives of data, computing power and algorithms. Research in Health 4.0 has been conducted in an interdisciplinary way with a diversified set of applications and functionalities and in terms of its implementation, it has been more commonly found in hospitals' information flows, especially the ones related to healthcare treatments [20]. In this context, it is also necessary to consider and to assess the prevailing opinions and expectations among stakeholders regarding ICT health solutions, such as the improvement of factors that affect quality of life, quality of health care, patient's knowledge, monetary aspects, or data security and privacy [21]. Although the research trend in the field of chronic care is to keep a continuous monitoring of each patient (promoting continuity of health and social care), tools to identify chronic patients and analyze the use of health services (care pathways) that they perform do not exist yet, and in addition there are no AI models that facilitate the design of integrated care pathways. There is clear evidence of the relevance of organization and management of the technological issue in the health care, concept further reinforced on the light of recent COVID-19 pandemic. Assessment, supply, prioritization, appropriate usage, and exploitation are indeed not a trivial duty, and the final success of any health process is widely affected by technology management issues.

In the field of logistics, AI can be applied in the forms of optimizing ML algorithms for scheduling and transportation planning [22–24]. This has not been extended to AI-led prognosis applications at least with empirical testing. The currently existing industry standard draws on manual processes to plan and optimize resource use. Software applications are being widely used in hospitals for this problem area, such as ORBIS, Medicc or M-KIS that rely on an old architecture and non-intelligent, manual interaction with users. Even specialized software modules such as myMedis support the whole process of OR management and related resource planning but still do not use AI-based technology and thus are not able to cope with rising complexity in resource planning optimization [25–27]. It has been reported that AI adoption by key stakeholders such as doctors remains low [28], and that existing applications do not cater enough to the specific needs of human stakeholders that are supposed to interact with the systems [29]. Accordingly, a focus on human–computer interaction (HCI) spanning pre-design, design and post-design phases as well as catering to user, system, task, and interaction characteristics [30] holds the potential to increase AI adoption and user satisfaction [31]. While expertise in HCI has been developed in the fields of computer science [32,33], it has not been systematically applied to the hospital context.

3. Use Cases Descriptions and Expectations

In the field of diagnosis, we propose to advance the methods that intelligently utilize heterogeneous data from various sources and novel AI-based methods for supporting medical diagnosis and decision making inside clinics. More specifically, we propose to increase the utilization of AI-based methods in four selected use cases: diagnosing coronary artery disease (CAD), assessing fetal state during labor, diagnosing epidermolysis bullosa (a rare genetic disease) and diagnosing arrhythmias automatically. All the use cases provide heterogeneous data, which at the same time is a challenge for the medical experts to handle and on the other hand provide a possibility for the rise of novel AI-based methods in supporting diagnosis and clinical decision-making. AI-based methods also enable detection of factors in medical diagnosis that are unnoticeable for humans. Collaboration between technical and medical experts is crucial to co-create such tools to be used in clinics that are highly acceptable, highly deployed, and provide real value for patients, doctors and societies.

3.1. Use Case 1: Coronary Artery Disease Diagnosis

Among all routinely available diagnostic tests, coronary CT angiography (CCTA) has the highest sensitivity (95–99%) for detection of coronary artery disease (CAD), with a specificity of 64–83%, and it has recently set up as the first-hand diagnostic tool for stabile chest pain. However, after CCTA there are still several patients for whom the diagnosis and reason for symptoms remains unclear and further imaging studies (myocardial perfusion and/or invasive coronary angiography) are needed to decide the best way of the treatment. Training a ML algorithm to recognize those cases for whom further imaging is likely to provide essential information among the unclear cases with suspected CAD would improve the cost-efficiency and logistic of the diagnosis of chest pain patients. In other words, the aim would be to develop a tool for evaluating the risk of the patient to have prognostic CAD for customized clinical decision-making. The number of the patients with suspected CAD transmitted to hospital for diagnostic imaging is likely to grow in the future worldwide due to recently published clinical guidelines emphasizing the use of CCTA. For the study, a number of contemporary CCTA studies imaged and essential clinical data (age, sex, cardiovascular risk factors and medication) could be used to train a machine-learning algorithm such as Disease State Index (DSI), which is a method to quantify the probability to belonging to a certain disease population, originally developed to support clinicians in diagnosing Alzheimer's Disease [34].

3.2. Use Case 2: AI Based Automatic Arrhythmia Analysis

Atrial fibrillation (AF) is the most common sustained arrhythmia and is associated with significant morbidity and adverse outcomes (stroke, heart failure, death). Overall, AF is associated with five-fold greater risk of stroke. Anticoagulation therapy has been demonstrated to reduce AF-related stroke risk significantly. Paroxysmal AF (PAF) is a self-terminating recurrent form of AF. The diagnosis of PAF is often tricky since PAF episodes can be short in duration, asymptomatic and the episode incidence can be low. It is estimated that the stroke causes total costs of EUR 45 billion/year across Europe. In European countries, 1.5 million peoples are diagnosed with stroke every year, 9 million are living with stroke and it is responsible for 9% (0.4 million) of all deaths in EU [2]. Cryptogenic stroke (CS) and transient Ischemic Attack (TIA) patients and cardiac surgery patients are the three most clinically significant patient groups where PAF is often underdiagnosed. In this use case, state of the art AI-based arrhythmia analysis algorithms are developed for PAF-screening in patients with TIA or cryptogenic stroke and detection of post-operative atrial fibrillation in cardiac surgery patients. AI-based automatic arrhythmia analysis implemented in wearable sensors enables longer monitoring time with improved patient usability and still requires minimal effort from healthcare professionals. Developing novel, AI-based non-invasive methods for PAF screening, using simple wearable ECG or PPG measurement would lead to increasing rate of PAF diagnosis in cardiac surgery, CS and TIA patients. These monitoring methods will be easily exploitable and inexpensive. The timely diagnosis of PAF has an important impact since anticoagulation may save the patient's life or prevent stroke-related disabilities such as paralysis, aphasia and chronic pain. There is a high-cost saving potential, since one prevented stroke can save EUR 20,000 of direct medical costs and more than EUR 100,000 of indirect costs (disability-adjusted life years lost).

3.3. Use Case 3: Fetal State Assessment during Labour

Cardiotocography (CTG), also known as electronic fetal monitoring (EFM), is used for fetal assessment before and during labour and largely replaced the use of intermittent heart rate auscultation. Visual interpretation of CTG traces is characterized today by a great inter- and intra-observer variability with low specificity. EFM has been shown to lead to unnecessary medical interventions such as caesarean section and vaginal-operative deliveries, with the associated health consequences and economic costs. The low specificity for identifying fetal hypoxia can be partially interpreted in the context of observer variability. CTG recording is widely performed for fetal assessment during delivery and has become routine in most hospitals worldwide. A software program connected to the electrodes of the electronic fetal monitoring system (EFM) registers fetal and maternal data such as fetal heart rate and its variations, maternal heart rate, uterine contractions and fetal movements. Currently, the most specific available CTG interpretation system is the FIGO (Fédération Internationale de Gynécologie et d'Obstétrique) classification, which is most commonly used worldwide [35]. Fetal outcomes after delivery are being measured by assessing following two parameters: (1) arterial pH directly after birth (blood from the umbilical cord); (2) APGAR score assessment at 1, 5 and 10 min after delivery. This not only offers information about the fetal state, but also gives observer (obstetricians and midwives) direct feedback about previous CTG interpretation during delivery as well as prediction of fetal hypoxia/acidosis. An arterial pH under 7.15 is considered to be pathologic and is a direct indicator of fetal hypoxia. An APGAR score under 7, measured 5 min after delivery is also considered to be pathologic. APGAR as scoring system based on five fetal features—appearance, pulse, grimace, activity and respiration—providing information about the status of the new-born after delivery [36]. Considering the problematic of observer variability, four scenarios are possible when CTG interpretation is performed by obstetricians or midwives: (1) normal CTG, normal outcomes (pH/APGAR); (2) pathological CTG, normal outcomes (pH/APGAR); (3) normal CTG, pathological outcomes (pH/APGAR); (4) pathological CTG, pathological outcomes (pH/APGAR). By introducing AI interpretation, the purpose is to improve scenario 2 and 3, which will in most cases lead to avoidance of surgical interventions, since the main problem of CTG is specificity; or to performing interventions at moments where one would otherwise refrain from doing so (version 3). The AI system could provide feedback when fetal asphyxia is expected (pH < 7.15 or APGAR at 5 min < 7), as well as warnings, if applicable. The proposed AI (or ensemble of several AI instances) would help in removing the existing great inter- and intra-observer variability and would lead to a direct and positive impact on effectiveness and efficiency through: (1) decrease of unnecessary caesarean section and instrumental delivery; (2) increase of specificity for identifying fetal hypoxia; (3) decrease of unnecessary health costs derived from unnecessary surgical procedures.

3.4. Use Case 4: Diagnosis in Epidermolysis Bullosa, a Rare Genetic Disease

In Europe, a disease is considered rare when it affects less than 1 in 2000 people. There are more than 7000 rare diseases (RDs) worldwide, about 80% of them has a genetic origin and approximately 75% affect children. RDs are estimated to affect 350 million people globally [37]. In better-resourced countries, correct diagnosis of rare genetic diseases takes on average between 5.5 and 7.5 years. In Europe and United States, nearly half of the first diagnoses are only partially correct. The deployment of effective diagnostic procedures is hampered by the underestimation of the true disease frequency (owing to the lack of RDs' awareness) and by an insufficient knowledge of the disease pathophysiology and natural history combined with the paucity of validated disease-specific biomarkers. Epidermolysis bullosa (EB) is a group of inherited, genetic diseases in which the skin (and the mucous membranes) is very fragile and forms severe, chronic blisters and lesions after even minor frictions or trauma. This rare genetic disorder affects all genders, ethnic and racial groups and determines either an early death or a long-term debilitating and life-threatening condition, since the severe blistering and associated scarring and deformities result in poor quality of life and reduce life expectancy. In the world there are about 500,000 persons affected by this disease and 36,000 in the European Union (EU). EB can be classified into four major subtypes, such as dystrophic EB (DEB), junctional EB (JEB), EB simplex (EBS), and Kindler Syndrome depending on the gene mutations and the level of skin cleavage [38]. Within the subtypes, EB has different severity levels and clinical manifestations. There is an urgent need to develop efficient methods for the early diagnosis of the EB subtype, the prediction of the disease progression and, consequently, the selection of individualized, precision therapeutic strategies. In this endeavour, "omics technologies", as genomic analysis by means of next generation sequencing (NGS), have recently found applications in the diagnosis, molecular subtyping, and follow-up prediction of EB. Information retrieved from these technologies represents a substantial increase in the amount of data that can be used to support EB patients, provided that advanced computational methods are available for their integrative and combinatorial analysis. In this use case, state-of-the-art AI algorithms are developed and applied for supporting early diagnosis, sub-classification, and therapeutic stratification of EB, as an example of rare genetic disease. In particular, AI-based methods will be applied to the integrative analysis of biological (genomics, molecular, immunological, and images) and epidemiological (medical records) data with the aim to: (1) support disease and disease subtype diagnosis; (2) identify distinctive features (genomic lesions, proteins, and immunological states) associated to disease severity (biomarkers) for the prediction of disease progression; (3) detect molecular signatures for guiding patient stratification for novel means of treatment (precision therapeutics). ML algorithms can be trained to integrate phenotypic and clinical data for the prioritization of disease-related genes and mutations, for the prediction of the pathogenicity and disease clinical relevance of genetic variants, and for the identification of pathogenic variant combinations. Furthermore, AI-based methods could be used for disease comprehension and therapeutic target selection by unravelling the affected genetic and molecular players and pathways. AI and ML can be applied to detect anomalies in gene expression and to correlate transcriptional patterns with molecular mechanisms and clinical phenotypes, to learn low frequency patterns, and to deliver automated class attribution [37]. Results from these analyzes would facilitate the recommendation of optimal treatment approaches and the identification of reliable biomarkers of normal versus pathogenic states and of response to therapeutics interventions. AI methods focusing on removing the existing limitations in the correct diagnosis of EB subtypes and in the prediction of the clinical course of EB patients might achieve at least the same average accuracy as medical doctors following the latest consensus reclassification of inherited EB. The AI-based integrative analysis of biological and medical data will have a direct and positive impact on effectiveness and efficiency through: (1) decrease in the time needed for the diagnosis of the correct EB subtype and the stratification of the patient for the most effective therapeutic treatment; (2) increase in the number and efficacy of diagnostic and prognostic biomarker; (3) increase in the efficacy of selection criteria to identify patients who will benefit from ex vivo gene therapy; (4) decrease of unnecessary life-threatening conditions and health costs derived from delayed diagnosis and treatment administration.

In the field of care, AI will be applied in four other use cases: to improve the management and decision support process, specifically in the chronic care pathway and resources characterization, simulation of demand and prognosis, adverse events identification and prevention, chronic resources management support tool and monitoring of the recovery process. Novel innovative tools for simulation and prognosis would become available, projecting the demand in terms of health resources for a given characteristic population in a territory, considering temporary projections of frailty condition of population and patients. As for recovery monitoring, contactless determination of vital signs will suppose an advanced functional aspect by monitoring of all patients and not only critical cases. Patients will benefit from reduced restrictions due to cables and devices. In addition, there is a time saving for nursing staff, as they do not have to put the devices on the patient and disinfect them. Regarding prevention of adverse critical conditions, the proposed approach relies on the analysis of the entire temporal series of vital signs by means of deep neural networks and hybrid approaches.

3.5. Use Case 5: AI Chronic Management and Decision Support Engine

According to the data of the World Health Organization (WHO), respiratory diseases together with cardiovascular diseases are leading causes of death and disability in the world. Considering this premise, the use of case will focus on the analysis of data from chronic patients diagnosed with one of these four common pathologies: COPD, asthma, coronary heart disease (e.g., heart attack) and cerebrovascular disease (e.g., stroke). The objective would be to apply AI in the clinical context of chronic care to characterize the pathways and resources used, as well as anticipate the demand of resources in order to optimize the economic costs. ML could be then used to analyze data of patients related to clinical parameters (e.g., laboratory tests), use of resources (e.g., hospitalizations), sociodemographic data (e.g., age, gender), and quality of life, among others. The AI engine would be able to support two analysis processes: the chronic care pathway and resources characterization (stratify patients by degree of frailty and map pathways), and resources demand simulation and prognosis (according to each pathway/patient strata).

3.6. Use Case 6: Chronic Resources Management Support Tool

As stated by the surveyed hospitals, efficient and effective scheduling of the resources is a challenge for most hospitals. Possible resources to be scheduled are patients' beds, material, medicament and assistance kit, medical equipment (e.g., diagnostic machines) or operating theatres. The goal would be to automatically schedule the usage of the considered resources as well as to measure and improve quantitative KPIs considered relevant for the most significant hospital metrics, e.g., cost, service level, delivery time, resource utilization, etc. To achieve this objective it is necessary to carry out the following activities: (1) translating hospital needs, often presented in a medical language, in technical concepts; (2) define the scheduling problem to be tackled by the intelligent algorithm and input data; (3) development an intelligent algorithm to automatically schedule the usage of resources and to measure quantitative KPIs over time; (4) test and validation of the intelligent algorithm using real datasets with the aim to fine-tune the procedures and selection rules implemented in the algorithm; (5) continuous learning of the intelligent algorithm by its utilization, performances and evolution of the surrounding environment.

3.7. Use Case 7: Adverse Events Identification and Prevention

Clinicians require support in the identification and prevention of adverse clinical conditions (ACC), as well as in identifying the main related care pathways. The technology could support the clinician in the automatic identification of ACC, such as a reaction to a new drug assumed by the patient after a change of her/his treatment plan. The AI tools could analyze data caught by vital signs monitoring systems, such as heart rate, pressure, body temperature and other data coming from the patient, such as information inferred by dialog systems based on natural language processing that would periodically interact with the patient to identify specific symptoms. Additionally, the tools would be able to support clinical staff in case a change within the care pathway is needed due. The objective would be to identify and forecast ACC for patients with non-communicable chronic diseases, particularly referring to cardiovascular diseases, by using AI. Models and tools for the automatic identification of ACC would be preliminarily realized adopting retrospective data and classic ML algorithms using current guidelines on the management of diseases of interest. Such models and tools, however, could be continuously improved, following a continuous learning approach. Successively, the prevention of ACC could be attempted by advanced classification systems, based on a combination of deep learning and reinforcement learning approaches that will analyze time series data concerning the patient condition evolution at different stages of the care pathway.

3.8. Use Case 8: Monitoring of the Recovery Process

Monitoring of the recovery process is a key hospital process. In order to achieve a high, continuous quality, vital parameters have to be monitored constantly. Vital parameters such as the heart rate or the respiration rate are key indicators for the current health status, urgent emergencies and the recovery process. Especially, persons with chronic diseases benefit from a continuous monitoring. In areas such as operation theatres or ICUs, there is a high coverage, whereas in normal wards or floors there is little to no coverage. The objective would be to remote determination of vital parameters such as heart rate and respiration rate for an improved recovery monitoring in a patient friendly method especially for chronic diseases. This could be realized by optical sensors with remote working mode and AI algorithms such as CNNs, BNNs or adaptive optical flow. To achieve the objective it is necessary to carry out the following activities: (1) identifying of optimal positioning of optical sensors within the hospital; (2) analysis of algorithms of remote vital parameter determination in clinical environments; (3) transfer and implementation of algorithms to the clinical setting; (4) evaluation of algorithms in clinical setting by means of reference systems, which would stayed synchronized; (5) interface protocol for transmission of vital parameters to central processing unit in the hospital. It should be guaranteed that only this meta data are transferred but not the raw data, thus protecting the privacy of the patients.

In the field of logistics, AI can be implemented for example in three different use cases as described below. The main focus is the optimization of resource use. It is expected that AI will help to better predict material consumption and needs in the whole process. Besides material consumption, transport planning is a further focus point in the field of logistics.

3.9. Use Case 9: Material Consumption Recognition and Prognosis

Currently, in the University Hospital in Essen as well as many other hospitals in Europe the documentation of used materials with hospital patients is a non-digital paper-pencil process consuming a lot of human work time. Therefore, digital improvements regarding automated capture system for material consumption are a prominent request in hospitals and addressed in this use case. Together with an industry partner an innovative care trolley is developed with a camera system and the complementary AI-based software using ML to recognize the consumed objects with patient processes automatically. User interaction can be implemented according to current state-of-the-art concepts. It will provide a data recognition and prognosis tool relating actual material consumption to patient cases and therefore enabling a bottom-up planning and prognosis for optimized procurement and logistics in hospitals.

3.10. Use Case 10: Optimization of Human-Robot Teams in Hospital Logistics Operations

Odense's University Hospital (OUH) will benefit from a reactive AI-based resource management and scheduling system for material transport logistic operations. The main goal is to improve upon current task management systems with the inclusion of an AI-driven optimized scheduler that will be able to oversee all the available robots and to plan, schedule and assign tasks to the relevant hospital workforce, mainly logistic robots but also employees. The proposed task management software will have several functions and therefore will contain several different conceptual elements: (1) an automated task-generation system, based on Reinforcement Learning (RL) algorithm, that analyzes the relationship between room use and materials requirements to predict what will be needed where and when based on past experience; (2) a scheduling element that knows what transport resources are available to it, their status and where they are; and can create an optimal schedule out of transport requests generated from user input or the task generation above; (3) a reactive planning element that will rework the schedule regularly, e.g., either every hour or when new on-demand transport requests are received; (4) a transport optimizing element that analyzes the efficiency of the transport and adjusts scheduling parameters to produce maximal transport for minimal energy use and minimal task requests to humans; (5) a route generator element that creates efficient routes for the robots and sends these to robots with their new tasks, in accordance with the schedule, coupled with a route status analyzer which takes input from sensors on the robots and around the hospital to determine the location of any blockages; (6) A sensory data analyzer that can use incoming data from various infrastructure sources to inform the decision-making elements, e.g., use of elevator position to inform the route generator or use of smart cameras that can measure room occupancy for the task generator; (7) A representation of (a) task criticality, i.e., planned, urgent and critical in emergency situations, (b) the current status of the material flow, (c) the robots (name, capabilities, location, current task and status) and (d) item transport requests (also available in a form readable by humans);

(8) and a supervision element that will be utilized to identify and criticize any suboptimal decisions made by the scheduler and provide feedback that will be used as input for a reinforcement learning sub-component. Task and material flow reports collected and shared by the hospital service and logistics departments of OUH, currently exceeding 555,000 entries describing various material flow logistic cases, i.e., transfer of medication, healthcare equipment and samples, will provide a variety of types of inputs and tasks. The system could automatically obtain information from various hospital software sources, e.g., human workforce positions provided by the proposed event-based messaging system by updating and adapting the current emergency messaging solution elevator status and sensors in the hospital.

3.11. Use Case 11: Co-Development and Evaluation

Bayındır Hospital Söğütözü in Ankara is one of the three high-capacity hospitals that belongs to Bayındır Healthcare Group. Bayındır Healthcare Group have three hospitals, one medical center and seven dental clinics. All healthcare facilities material management system can be centrally monitored and controlled. This provides an additional opportunity to study the impact of planned AI implementations over multi-location inventory systems. The hospital has specific experiences and requirements regarding healthcare logistics. It has an existing barcode scanning system for collecting healthcare and inventory information that aggregates centrally for the planning the availability of medical supplies and logistics management. However, the hospital may still benefit from a new picture recognition and AI-based system in terms of time savings, reductions in human error, and an increase the safety by reducing the contact between the healthcare staff and patients. Furthermore, material management and operation room scheduling are highly interrelated in practice. Using the OR schedules to trigger the purchase of perioperative materials is expected to further reduce inventory costs and increase operational efficiency compared to independent material management systems [39]. In a comparison to standalone applications of automated inventory tracking, predictive logistics, and cognitive automation, an additional understanding of the impact of integrated AI applications on healthcare logistics operations will bring several challenges, including data storage and management, data exchange, security and privacy, and integrated decision-making.

4. Discussion: Benefits and Challenges for AI in Hospitals

The specific benefits and data as well as AI application challenges are presented and discussed in this section, based on the outlined case studies and additionally directed towards the contribution against pandemic situations, such as COVID-19.

The use cases presented in Table 3 are distinguished by specific aspects often related to the area of interest, e.g., diagnosis, care, treatment, logistics or rehabilitation, or to the targeted goals, e.g., increase the efficiency of a certain health care process, improve its quality, or increase the service level. However, the detailed description of the aforementioned case studies suggests how all the involved hospitals are affected by common challenges and potential barriers to the adoption of AI to their healthcare processes on regular basis. In particular, it is possible to define three main issues which should be properly managed to ensure an efficient and effective adoption of AI tools and techniques in the healthcare delivery processes which distinguish European hospitals. The first aspect to be considered is the human acceptance and the real adoption of AI solutions in hospitals. The resistance to automated and partially obscure tools which offer assistance in several healthcare services is a major obstacle to overcome. Leveraging such tools in traditional diagnosis, care and treatment processes is useful but often distinguished by a low level of trust, in particular by doctors and medical personnel. Furthermore, the usage of such AI solutions should not increase the complexity or time required to complete certain medical process, therefore offering an adequate and well-designed interaction with human adopters. The second challenge to be tackled to foster the adoption of AI in European hospitals is the proper management of medical data. This information is distinguished by some features which make their storage and usage much more sensitive than other data typically collected in digital environments.

However, as COVID-19 dramatically revealed, the value beyond medical data is huge. In particular, the opportunity to systematically collect data concerning the patient conditions, made diagnosis, performed treatments and defined care offer to the hospitals of the future the chance to significantly increase the efficacy and efficiency of the healthcare services delivered. The last area involved by AI structural adoption in European hospitals deals with technology selection and ethics. The former includes the complex and interrelated process of selecting a novel technology for its adoption in healthcare services, as represented by the solutions based on AI algorithms. The assessment of the most appropriate AI based technology to be adopted to ease diagnosis, treatment or care activities is a complex and distinguished by uncertain and multiple feasible outcomes with different and contrasting scenarios. The latter deals with the ethical aspects involved in the adoption of AI tools and techniques, from machine based medical decision to personalized treatments, from sharing of personal health data to acceptance of robot medical personnel. Finally, a latter aspect concerning the challenges of adopting AI in hospitals necessarily has to be mentioned, e.g., the appropriate involvement of adequate stakeholders. Indeed, this last issue is of fundamental importance to ensure the real usage of AI-based solutions in daily hospital activities by doctors, acceptance of renovated treatments and procedures by patients as well as commitment by local administrators to this modern form of health care assistance. Therefore, the process of stakeholder commitment is of paramount importance and should be adequately planned and implemented. Considering all the abovementioned challenges and potential obstacles, the following paragraphs propose possible solutions to overcome these difficulties, to ensure the adoption of AI solutions in European hospitals and maximizing the efficacy of the innovation provided. In particular, the proposed actions are grouped into three categories, human–computer interaction, medical data space, and guidebook and ethics. The linkage between these transversal activities with the application areas proposed in the manuscript is presented in the following Figure 2.

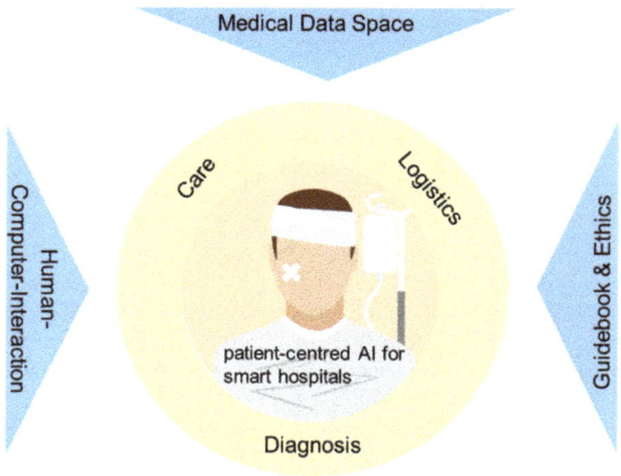

Figure 2. Linkage between transversal activities and application areas for AI adoption in European hospitals.

Table 3. AI Use Cases, AI Methods and Outcomes.

Use Case	Objectives	AI Method	Data Available	Defined Outcomes	Contributions against Pandemic Situations
Diagnosis (1) MDS for Coronary Artery Disease (CAD) diagnosis	The aim of this study is to train a ML algorithm to distinguish patients with suspected CAD to those who benefit from further imaging studies and to those who don't. In other words, to evaluate the risk of the patient to have prognostic CAD for customized clinical decision-making.	Disease State Index (DSI), which is a method to quantify the probability to belonging to a certain disease population, originally developed to support clinicians in diagnosing Alzheimer's Disease [34]. It is designed to be 'disease-agnostic', so that it can be used equally well for other diseases, provided that data are available.	For the study, a number of contemporary CCTA studies imaged in Kuopio University Hospital (KUH) as well as ECG, myocardial perfusion, invasive coronary angiography imaging and essential clinical data (age, sex and other demographic data, medical history, cardiovascular risk factors and medication) are gathered from existing clinical databases in KUH.	Algorithms and AI solutions for doctors supporting clinical decision making in CAD diagnosis.	Reduction of visits to the hospital, which increases the patient and personnel safety.
Diagnosis (2) AI based automatic arrhythmia analysis	In this use case, state-of-the-art artificial intelligence (AI) based arrhythmia analysis algorithms are developed and integrated into wearable sensors. Development of novel AI-based arrhythmia monitoring system aims to improve arrhythmia detection: Enable longer non-invasive monitoring time.	State-of-the-art AI based arrhythmia analysis algorithms are developed and utilized to atrial fibrillation (AF) screening in patients with transient ischemic attack (TIA) or cryptogenic stroke (CS) and detection of post-operative atrial fibrillation in cardiac surgery patients. Used methods: neural networks, deep learning, ML.	6000 24 h Holter recordings with arrhythmia annotations. Wearable sensor database: 700 patients (300 patients with AF episodes) with wearable sensors. New: TIA/CS database is collected: 48h home monitoring of simultaneous wearable PPG and ECG-recordings from 100 TIA/CS patients.	Developed AF-screening solution will enable long arrythmia monitoring time and increased rate of AF diagnosis. Wearable sensors offer improved patient usability and AI assisted arrythmia diagnosis requires minimal effort from healthcare professionals; AF diagnosis has important impact to patient itself, since anticoagulation may save the patient's life (prevent cardioembolic stroke). Cost saving potential: one prevented stroke can save 120,000€ to society.	Reduction of visits to the hospital, which increases the patient safety. Possibility to assess arrythmia of corona patients remotely. Increases patient and personnel safety.
Diagnosis (3) Medical decision support system for fetal assessment during labor	Improving fetal assessment with accurate prediction of fetal hypoxia and reduction of caesarean and instrumental delivery rates. Develop an AI-powered clinical decision support system.	Ensemble methods (e.g., stacking and blending) combining Explainable AI (aka XAI), neural networks (e.g., CNN and RNN), and gradient boosting techniques (e.g., XGBoost)	The maternity ward of the Department for Obstetrics and Gynecology in the University Hospital of Bern will provide a dataset of cardiotocographic (CTG) recordings. It includes physiological data such as maternal heart rate, fetal heart rate, contraction strength. The dataset is labelled by MDs.	The AI will focus on removing the existing great inter- and intra-observer variability while achieving at least the same average accuracy as medical doctors following the "Updated 2015 FIGO Intrapartum Fetal Monitoring Guidelines". The integration of our AI-powered system should lead to a direct and positive impact on effectiveness and efficiency.	Assisting personnel in diagnosis with AI in a situation where there are not enough experienced personnel available due to the tpandemic.

Table 3. *Cont.*

Use Case	Objectives	AI Method	Data Available	Defined Outcomes	Contributions against Pandemic Situations
Diagnosis (4) Diagnosis in Epidermolysis bullosa, a rare genetic disease	To support disease prediction and diagnosis through the integration of extensive biological data (images, genomics, molecular) and epidemiological (immunological, clinical, demographic, lifestyles) to identify genomic lesions, proteins and immune-logical states associated (biomarkers).	ML algorithms will be trained to integrate phenotypic and clinical data to improve accurate prediction of progress of Epidermolysis bullosa. AI-based methods will also be used for disease comprehension and therapeutic target selection by unravelling the affected genetic and molecular players and pathways.	This use case will exploit data, competencies, and facilities of the Modena EB-Hub, the center for diagnosis, research, assistance and development of innovative therapies created in January 2020 at the General Hospital of Modena.	Definition of AI-based decision support systems to expedite diagnosis, correct misdiagnosis, diagnose previously undiagnosed, and stratify EB patients for advance therapeutic intervention through the integrative analysis of clinical phenotypes and patient health records, genetic information, molecular levels, biochemical fingerprints and patient images.	Assisting doctors' in the diagnostic process during the pandemic, when the resources to be used for diagnosis is limited. Maintaining normal procedures of diagnosing other health problems during the pandemic.
Care (5) Chronic care pathway and resources characterization, simulation of demand and prognosis.	AI techniques applied to analyze the pathways of chronic care patients providing simulation and prediction capacities about the demand of use of hospital services and resources	ML techniques (neuronal networks; LSTM; statistics predictions modeling; random forest; decision trees). AI adjustment to chronic care attention, prototype testing, application evaluation (KPI).	Historical clinical records for patients with chronic diseases. Data about care plans and use of hospital services and resources (pathways) made by this group of patients based on degree of frailty. Macro parameters from population (estimate demand/prognosis)	AI agent and tool for dimensioning demand of resources, including prognosis and simulation, both at individual and population level. Intelligent assistant for redefinition/optimization of care plans	Reduction of the transmission risks by being able to re-organize the pathways according to pandemic context.
Care (6) Critical Conditions identification and prevention	Identification and prevention of critical conditions: Analysis of vital signs, automatic recognition of symptoms (e.g., skin rash, mood change) and direct interaction with patients.	Machine Learning Techniques such as DNN, Reinforcement Learning, Natural Language Processing and Statistical Methods.Adjustment chronic care, prototype, evaluation (KPI).	Test of algorithms in hospital of Bozen with either live settings or retrospective data. Retrospective data as heart rate, respiration rate, oxygen saturation and blood pressure. Moreover, general data such as age, sex, weight, height and other diseases.	AI tool for critical conditions identification and prevention along the chronic care pathway	Control of patients with COVID-19 confined to their homes, before variations in their critical conditions. Increase in patient and family safety, especially in patients with COVID-19 who live alone.
Care (7) Intelligent resources management	An intelligent algorithm is developed to efficiently manage the scheduling of hospital resources.	Evolutive, self-learning and auto-adaptive techniques focused on chronic care, prototype testing, validation through KPI.	Hospital models of processes for resource utilization. Information: processes, cost, service level, delivery time, resource utilization, medical personnel qualification.	Scheduling planning tool for optimal management of hospital care resources for patients with chronic diseases.	Reduction the transmission risks. Better planning of resources in compatibility with pandemic demand.
Care (8) Monitoring of the recovery process	Remote determination of vital parameters such as heart rate and respiration rate for an improved recovery monitoring.	Methods in the Area of computer vision and ML i.e., CNN, BNN, adaptive optical flow, SVM etc.	Recordings from lab situations available; more data will be generated within the Fraunhofer InHaus-Centre, Test of algorithms in hospital of Bozen	Software for vital parameters. Transfer to hospital environment; continuous monitoring; fast obstacle identification; safe solution; contactless	Reduction the transmission risks in professionals by reducing contact with monitored admitted patients with COVID-19.

Table 3. *Cont.*

Use Case	Objectives	AI Method	Data Available	Defined Outcomes	Contributions against Pandemic Situations
Logistics (9) Material consumption recognition and prognosis	Develop an automatic material documentation on the care wagon or the material store in the nursing ward based on computer vision. A material consumption prognoses is developed with the derived data.	ML (computer vision, CNN): Used materials are matched with patient cases and their diagnoses and treatments. Thus, it is known which and how many materials are needed by the individual patient cases.	- Material lists - Master and movement data of the materials (order history) - Demographic patient data (gender, age, weight, etc.) - Patient treatment history - Automatic stock updates for all materials on wagon	- Automatic material documentation and transport - Transparent material consumption for individual patient cases - Specified case cost calculation - Higher planning reliability for material orders - Immediate reaction to material shortage	- Improved forecasting for pandemic related uncertanities - Dynamic management of limited material (such as masks, protective visors and clothing, antiseptics, etc.) by predicting patients' disease trajectory
Logistics (10) Optimizing logistic operations	Optimize the internal logistics operations of the hospital by considering both manual and automatic transport in a resource management and scheduling framework. Generate recommendations for how to improve manual and robotic logistics, based on gathered data.	Reinforcement learning (multi-agent motion and path planning)	- Hospital maps - Data (sensor data, operational data) from robots operating at the hospital - Data from the hospitals material management system - Generating data from current hospital sensor infrastructure - Knowledge about areas that are frequented by visitors or patients probably infected with COVID-19	- Status reports for certain characteristics of automated and manual logistics operations - Recommendations for optimization of material transport - Better understanding of the events leading up to an incident report (e.g., materials arrived late, or robot stopped unexpectedly) - Facilitate future integration of robotic solutions in hospitals - Automatic avoidance of infections areas (e.g., areas frequented by visitors)	- Decreasing transmission risks to healthcare providers by minimizing the patient contact. - Optimal management of critical resources such as Intensive Care Unit (ICU) beds.
Logistics (11) Co-development and evaluation	Integration of optimization of internal logistics operations and material consumption	Predictive analytics and cognitive automation	- Material lists - Master and movement data of the materials - Demographic patient data - Patient treatment database - Availability of healthcare resources	- Adaption routines and experiences, e.g., comparison of material recognition with barcode system (already existing, comparative case) - Management of resources in multi-location setting	- Centralized planning of material consumption and shortage - Optimal assignment/scheduling of critical resources (healthcare personnel, ICU, operation rooms, etc.)

Human–Computer-Interaction: Despite progress in the field of health care data analytics, resulting in more and more prototypes and technical advancement, actual adoption by key stakeholders such as doctors remains low [28,29]. This aspect will rise in relevance when the respective systems increase in intelligence and analytical capability. Accordingly, an increased focus on human–computer interaction spanning pre-design, design and post-design phases as well as catering to user, system, task and interaction characteristics [30] holds the potential to increase AI adoption and user satisfaction in clinical practice [31].

Medical Data Space: In addition, data connections in a Medical Data Space (MDS) with distributed AI applications will help to share resources and to support specially and severely affected regions and hospitals. In additions, overall data transparency and analysis will help to fight virus outbreaks earlier through faster detection and containment options due to AI analysis. The Medical Data Space (MDS) is a specialization of the International Data Space (IDS), which provides a trustworthy, secure and cross-domain data space allowing to build an economy of data between companies of all domains and sizes. IDS was the result of R&D activities in 2015 and is now actively promoted through the Industrial Data Space Association. It is in cooperation with the OPC foundation, the FIWARE foundation and the Industrial Value Chain Initiative and the Platform Industry 4.0. The IDS and thus the MDS define an architecture of data providers and consumers, which are linked through connectors forming the data space. The architecture is defined in the IDS document describing the layers of the architecture model which in turn describe the key components necessary to realize a data space [40]. The first prototype has been presented in 2018 at the Hannover fair. The MDS concept targets the connectivity of local data spaces in hospitals for analytics and the application of AI-based algorithms for research or hospital internal use. Therefore, special services are necessary to not only store and manage the transfer of medical data securely and maintaining the sovereignty of the data owner, but it must additionally conform to requirements on anonymity and protection of personal medical data sets. Here, the element of value-added services for the data space becomes relevant enabling pseudonymization and anonymization features in the process.

Medical data of patients is a highly sensitive and therefore regulated asset which requires handling in a secure and protected environment. The Medical Data Space (MDS) builds upon the international data space to deliver a secured, controlled data storage and processing environment to build an economy of data between providers and consumers retaining sovereignty and control. The MDS extends this to address the additional medical constraints. They key concept in MDS is the trusted connector which links both parties and enforces the security and privacy policies defined. In addition to access management the MDS architecture introduces data-processing services (data-apps) which can preprocess data before or after transfer. As AI-driven smart hospitals rely basically on data targets the connectivity of local data spaces in hospitals for analytics and the application of AI-based algorithms for research or hospital internal will be used. Therefore, special services are necessary to not only store and manage the transfer of medical data securely and maintaining the sovereignty of the data owner, but it must additionally conform to requirements on anonymity and protection of personal medical data sets. Here the element of value-added services (data-apps) for the data space becomes relevant enabling specifically pseudonymization and anonymization features in the process. In future works, we plan to demonstrate that medical data space technology can provide the foundation for the development and deployment of novel AI and data management data-apps. Specifically, a pilot program for the analysis and management of in-hospital cardiac patient intervention treatment with the goal of understanding and analyzing several key factors that impact the ability and capacity of a hospital to provide treatment. The location for this future installation will be the Evaggelismos Hospital in Athens.

Guidebook and Ethics: There is clear evidence of the relevance of organization and management of the technological issue in the health care, concept further reinforced on the light of recent COVID-19 pandemic [41]. Assessment, supply, prioritization, appropriate usage and exploitation are indeed not trivial duties, and the final success of any health process is widely affected by technology management issues. In the modern re-setting of health-care delivery via technology innovation, data driven management, health technology assessment, clinical practice guidelines as well as medical leadership are the main topics that have to be addressed [42]. Knowledge management and technology innovation with their continuously growing potentiality can indeed transversally represent the answer to the demand of efficacy and efficiency of the system. Furthermore, great expectations are placed in information and communication technologies (ICT) with their contribution in the development of eHealth and closely in AI with its paramount applications in the various sectors of medical practice and public health. The change in clinical practice through and by means of the injection of technological innovation is today decisive to make the health and care systems able to face to the continuous economic, socio-demographic and epidemiological pressures [17]. However, technological innovation, although important and central, must be carefully examined and accompanied to ensure that it really corresponds to effective social innovation [43]. Furthermore, as really recently underlined by a joint report of EIT Health and McKinsey [44]. AI has indeed many potentialities for the improvement in care outcomes, patient experience and access to healthcare services. AI is thought to increase productivity and the efficiency of care delivery and allow healthcare systems to provide more and better care to more people. Finally, it can support the faster delivery of care, mainly by accelerating diagnosis time, and help healthcare systems manage population health more proactively, dynamically allocating resources to where they can have the largest impact and need. As addressed by MedTech Europe, developing AI systems and algorithms for healthcare settings requires specific skillsets which are in short supply, and investment in education and training of professionals involved (e.g., data scientists, practitioners, software engineers, clinical engineers), is mandatory [18].

Ethical issues are a major hurdle to full-scale AI application use as many cases might bring about risks such as wrong diagnosis or deviant therapy, as well as dissent among personnel due to different opinions regarding correct AI analysis and advice. Therefore, not only HCI issues but also human-human interaction and collaboration issues and ethical questions to be solved and communicated among people first of all before AI can contribute according to the full potential in health care.

5. Outlook

AI will play a significant role in future hospital health care systems. Applications such as ML will further advance the development of processes in several fields inside the hospital, of which we focus in medical diagnosis, logistics and care in this article. Important obstacles remain, such as regulations, integrations to the Electronic Health Record (EHR), standardization, medical devices certificates, training professionals, costs, updates—but this is manageable. It is important to stress that AI applications will not replace human clinicians but help them to concentrate on important human-related processes and to make correct diagnoses with less analysis and decision time. This hopefully provides them with time and focus to support patients from a specific human perspective. As a result of the developments in computational power and algorithmic advancements, combined with digitalization and improvements in data collection methods and storage technologies, the healthcare sector today is supported by AI, ML and robotics as never before in the history of medicine. Besides monitoring large-scale medical trends, these new technologies also allow measurement of individual risks based on predictions from big data analysis. AI has a key function in the healthcare management of the future. Research has already proven the game changing potential of AI in various fields of healthcare, such as those outlined in the use cases in this article. AI-based methods have been successfully developed to address several healthcare logistics problems such as appointment planning, patient and resources scheduling, resource utilization, and predicting demand for emergency departments, intensive care units, or ambulances [45]. In addition, there already exist a number of research studies which suggest that AI can perform at least as good as humans at basic healthcare functions, such as diagnosis. Today, malignant tumors are spotted more successfully by algorithms than humans [46]. As a consequence of rapid technological advancements, combined with ML's enhanced ability to transform data into insight, many of the medical tasks previously limited to humans are expected to be taken on by algorithms [47]. However, there are several reasons why it will take a long time before AI might take over comprehensive fields of activity from humans in hospitals and healthcare: recent developments show that AI systems will not replace humans on a large scale, but rather will support them in their efforts of patient care. Progressing into future times, healthcare specialists can switch to tasks and job designs focusing on unique human skills such as empathy and care. One risk within this development might be the position of healthcare providers who are unable or refuse to work in collaboration with AI applications, endangering their contributions and jobs. The most important obstacle regarding AI applications in healthcare are not the capabilities or benefits of the technologies themselves, but their applicability in medical practice. Widespread use of AI systems requires approval by regulating institutions, integration with existing systems, sufficient standardization with similar products, training of healthcare professionals, and solutions regarding issues of data privacy and security. These challenges will eventually be solved, but it will take significant time and resources [46]. The COVID-19 crisis has revealed the challenges for healthcare systems—also for future pandemic situations. This increased attention to the potential of AI in healthcare as one means of pandemic management and prevention. Major challenges in responding to COVID-19, such as managing limited healthcare resources, developing personalized treatment plans, or predicting virus spread rates, can be addressed by recent developments in AI and ML. Wynants et al. [48] have already listed 31 prediction models in a review of early studies of COVID-19. The prospective post-COVID-19 era in preparation for future pandemic events will likely feature advanced healthcare solutions in combination with operation research modeling [49]—and AI will be a crucial part of it as outlined in this paper with 11 use case studies from European hospitals. The challenges connected to such AI applications such as data management (HCI) have to be addressed soon in order to prepare hospitals for future challenges, e.g., pandemic situations [50]. This is a core challenge for health care management science and the implication for hospital practice in order to apply the full potential of AI and ML to health care systems [51].

Author Contributions: Conceptualization, M.K., M.H. (Marcus Hintze), M.I.; methodology, M.H. (Marcus Hintze), M.I., F.R.-R., F.P., F.A.-M., D.Ç.; validation, F.A.-M., D.Ç., T.L., M.J., O.U., M.H. (Marcus Hintze), J.A.L., S.B., A.-P.R.; formal analysis, M.K., B.V., W.T., D.G.; investigation, M.I., F.R.-R., F.P., F.A.-M., D.Ç., T.L., M.J., O.U., B.V., D.G., R.D.-G.; writing—original draft preparation, M.I., F.R.-R., F.P., D.Ç., T.L., M.J., O.U., M.H. (Marcus Hintze & Marja Hedman), J.A.L., S.B., B.V., W.T., R.D.-G.; writing—review and editing, M.K., M.H. (Marcus Hintze), D.G.; visualization, M.H. (Marcus Hintze), F.R.-R., T.L., M.J., A.-P.R.; supervision, M.K., M.I., F.A.-M. All authors have read and agreed to the published version of the manuscript.

Funding: This research received no external funding.

Institutional Review Board Statement: Not applicable.

Data Availability Statement: Not applicable.

Conflicts of Interest: The authors declare no conflict of interest.

References

1. Halawa, F.; Madathil, S.C.; Gittler, A.; Khasawneh, M.T. Advancing evidence-based healthcare facility design: A systematic literature review. *Heal. Care Manag. Sci.* **2020**, *23*, 453–480. [CrossRef] [PubMed]
2. McKee, M.; Merkus, S.; Edwards, N.; Nolte, E. *The Changing Role of the Hospital in European Health Systems;* Cambridge University Press: Cambridge, England, 2020.
3. Increase in the Share of the Population Aged 65 Years or Over Between 2009 and 2019. Available online: https://ec.europa.eu/eurostat/statistics-explained/index.php?title=Population_structure_and_ageing (accessed on 15 December 2020).
4. Michal, J.; Ecarnot, F. The shortage of skilled workers in Europe: Its impact on geriatric medicine. *Eur. Geriatr. Med.* **2020**, *11*, 345–347. [CrossRef]

5. Moser, E.; Narayan, G. Improving breast cancer care coordination and symptom management by using AI driven predictive toolkits. *Breast* **2020**, *50*, 25–29. [CrossRef]
6. Abràmoff, M.D.; Tobey, D.; Char, D.S. Lessons Learned About Autonomous AI: Finding a Safe, Efficacious, and Ethical Path Through the Development Process. *Am. J. Ophthalmol.* **2020**, *214*, 134–142. [CrossRef]
7. Wood, D.A.; Mahmud, E.; Thourani, V.H.; Sathananthan, J.; Virani, A.; Poppas, A.; Harrington, R.A.; Dearani, J.A.; Swaminathan, M.; Russo, A.M.; et al. Safe Reintroduction of Cardiovascular Services During the COVID-19 Pandemic. *J. Am. Coll. Cardiol.* **2020**, *75*, 3177–3183. [CrossRef]
8. McKinney, S.M.; Sieniek, M.; Godbole, V.; Godwin, J.; Antropova, N.; Ashrafian, H.; Back, T.; Chesus, M.; Corrado, G.S.; Darzi, A.; et al. International evaluation of an AI system for breast cancer screening. *Nat. Cell Biol.* **2020**, *577*, 89–94. [CrossRef]
9. Arbabshirani, M.R.; Plis, S.; Sui, J.; Calhoun, V.D. Single subject prediction of brain disorders in neuroimaging: Promises and pitfalls. *NeuroImage* **2017**, *145*, 137–165. [CrossRef] [PubMed]
10. Bzdok, D.; Meyer-Lindenberg, A. Machine Learning for Precision Psychiatry: Opportunities and Challenges. *Biol. Psychiatry: Cogn. Neurosci. Neuroimaging* **2018**, *3*, 223–230. [CrossRef] [PubMed]
11. Lee, E.-J.; Kim, Y.-H.; Kim, N.; Kang, D.-W. Deep into the Brain: Artificial Intelligence in Stroke Imaging. *J. Stroke* **2017**, *19*, 277–285. [CrossRef] [PubMed]
12. Awan, S.E.; Sohel, F.; Sanfilippo, F.M.; Bennamoun, M.; Dwivedi, G. Machine learning in heart failure: Ready for prime time. *Curr. Opin. Cardiol.* **2018**, *33*, 190–195. [CrossRef] [PubMed]
13. Hampe, N.; Wolterink, J.M.; Van Velzen, S.G.M.; Leiner, T.; Išgum, I. Machine Learning for Assessment of Coronary Artery Disease in Cardiac CT: A Survey. *Front. Cardiovasc. Med.* **2019**, *6*, 172. [CrossRef] [PubMed]
14. Esteva, A.; Kuprel, B.; Novoa, R.A.; Ko, J.; Swetter, S.M.; Blau, H.M.; Thrun, S. Dermatologist-level classification of skin cancer with deep neural networks. *Nature* **2017**, *542*, 115–118. [CrossRef] [PubMed]
15. Gulshan, V.; Peng, L.; Coram, M.; Stumpe, M.C.; Wu, D.; Narayanaswamy, A.; Venugopalan, S.; Widner, K.; Madams, T.; Cuadros, J.; et al. Development and Validation of a Deep Learning Algorithm for Detection of Diabetic Retinopathy in Retinal Fundus Photographs. *JAMA* **2016**, *316*, 2402–2410. [CrossRef]
16. Mishima, H.; Suzuki, H.; Doi, M.; Miyazaki, M.; Watanabe, S.; Matsumoto, T.; Morifuji, K.; Moriuchi, H.; Yoshiura, K.-I.; Kondoh, T.; et al. Evaluation of Face2Gene using facial images of patients with congenital dysmorphic syndromes re-cruited in Japan. *J. Hum. Genet.* **2019**, *64*, 789–794. [CrossRef]
17. Global Strategy on Human Resources for Health: Workforce 2030. Available online: https://apps.who.int/iris/bitstream/handle/10665/250368/9789241511131-eng.pdf?sequence=1 (accessed on 18 December 2020).
18. Artificial Intelligence in Medical Technology: Delivering on the Promise of Better Healthcare in Europe. Available online: https://www.medtecheurope.org/wp-content/uploads/2019/11/MTE_Nov19_AI-in-MedTech-Delivering-on-the-Promise-of-Better-Healthcare-in-Europe.pdf (accessed on 17 December 2020).
19. Xu, S.; Hu, C.; Min, D. Preparing for the AI Era Under the Digital Health Framework. In Proceedings of the 2019 ITU Kaleidoscope: ICT for Health: Networks, Standards and Innovation (ITU K), Atlanta, GA, USA, 4–6 December 2019; pp. 4–6. [CrossRef]
20. Tortorella, G.L.; Fogliatto, F.S.; Mac Cawley Vergara, A.; Vassolo, R.; Sawhney, R. Healthcare 4.0: Trends, challenges and research directions. *Prod. Plan. Control.* **2019**, *31*, 1245–1260. [CrossRef]
21. Haluza, D.; Jungwirth, D. ICT and the future of health care: Aspects of health promotion. *Int. J. Med. Inform.* **2015**, *84*, 48–57. [CrossRef]
22. Zijm, H.; Klumpp, M. Future Logistics: What to Expect, How to Adapt. In *Dynamics in Logistics. Lecture Notes in Logistics*; Freitag, M., Kotzab, H., Pannek, J., Eds.; Springer: Cham, Germany, 2017; pp. 365–379.
23. Klumpp, M. Automation and artificial intelligence in business logistics systems: Human reactions and collaboration re-quirements. *Int. J. Logist. Res. Appl.* **2018**, *21*, 224–242. [CrossRef]
24. Giusti, R.; Manerba, D.; Bruno, G.; Tadei, R. Synchromodal logistics: An overview of critical success factors, enabling tech-nologies, and open research issues. *Transp. Res. Part E Logist. Transp. Rev.* **2019**, *129*, 92–110. [CrossRef]
25. Cardoen, B.; Demeulemeester, E.; Beliën, J. Operating room planning and scheduling: A literature review. *Eur. J. Oper. Res.* **2010**, *201*, 921–932. [CrossRef]
26. Tuwatananurak, J.P.; Zadeh, S.; Xu, X.; Vacanti, J.A.; Fulton, W.R.; Ehrenfeld, J.M.; Urman, R.D. Machine Learning Can Improve Estimation of Surgical Case Duration: A Pilot Study. *J. Med Syst.* **2019**, *43*, 44. [CrossRef] [PubMed]
27. Li, F.; Gupta, D.; Potthoff, S. Improving operating room schedules. *Heal. Care Manag. Sci.* **2015**, *19*, 261–278. [CrossRef] [PubMed]
28. Kohli, R.; Tan, S.S.-L. National University of Singapore Electronic Health Records: How Can IS Researchers Contribute to Transforming Healthcare? *MIS Q.* **2016**, *40*, 553–573. [CrossRef]
29. Romanow, D.; Cho, S. Straub Editor's Comments: Riding the Wave: Past Trends and Future Directions for Health IT Research. *MIS Q.* **2012**, *36*, 8. [CrossRef]
30. Zhang, P.; Li, N. An assessment of human–computer interaction research in management information systems: Topics and methods. *Comput. Hum. Behav.* **2004**, *20*, 125–147. [CrossRef]
31. Rzepka, C.; Berger, B. User Interaction with AI-enabled Systems: A Systematic Review of IS Research. In Proceedings of the International Conference on Information Systems, San Francisco, CA, USA, 13–16 December 2018.
32. Preece, J.; Rogers, Y.; Sharp, H.; Benyon, D.; Holland, S.; Carey, T. *Human-Computer Interaction*; Addison-Wesley: Boston, MA, USA, 1994.

33. Dix, A.; Finlay, J.; Abowd, G.; Beale, R. *Human-Computer Interaction*; Pearson/Prentice-Hall: Upper Saddle River, NJ, USA, 2003.
34. Mattila, J.; Koikkalainen, J.; Virkki, A.; Simonsen, A.H.; Van Gils, M.; Waldemar, G.; Soininen, H.; Lötjönen, J.; Initiative, A.F.T.A.D.N. A Disease State Fingerprint for Evaluation of Alzheimer's Disease. *J. Alzheimer's Dis.* **2011**, *27*, 163–176. [CrossRef]
35. Ayres-de-Campos, D.; Spong, C.Y.; Chandraharan, E. FIGO consensus guidelines on intrapartum fetal monitoring: Cardiotocography. *Int. J. Gynecol. Obstet.* **2006**, *131*, 13–24. [CrossRef] [PubMed]
36. American Academy of Pediatrics, Committee on Fetus and Newborn; American College of Obstetricians and Gynecologists, Committee on Obstetric Practice. *The Apgar Score. Pediatrics* **2006**, *117*, 1444–1447.
37. Brasil, S.; Pascoal, C.; Francisco, R.; Ferreira, V.D.R.; Videira, P.A.; Valadão, A.G. Artificial Intelligence (AI) in Rare Diseases: Is the Future Brighter? *Genes* **2019**, *10*, 978. [CrossRef] [PubMed]
38. Danial, C.; Adeduntan, R.; Gorell, E.; Lucky, A.W.; Paller, A.; Bruckner, A.; Pope, E.; Morel, K.D.; Levy, M.L.; Li, S.; et al. Prevalence and Characterization of Pruritus in Epidermolysis Bullosa. *Pediatr. Dermatol.* **2014**, *32*, 53–59. [CrossRef] [PubMed]
39. Epstein, R.H.; Dexter, F. Economic analysis of linking operating room scheduling and hospital material management in-formation systems for just-in-time inventory control. *Anesth. Analg.* **2000**, *91*, 337–343. [PubMed]
40. Reference Architecture Model. IOP Publishing Internationa Data Spaces. Available online: https://www.internationaldataspaces.org/wp-content/uploads/2019/03/IDS-Reference-Architecture-Model-3.0.pdf (accessed on 18 September 2020).
41. Country & Technical Guidance—Coronavirus Disease. IOP Publishing WHO. Available online: https://www.who.int/emergencies/diseases/novel-coronavirus-2019/technical-guidance (accessed on 18 September 2020).
42. OECD. *World Health Organization Improving Healthcare Quality in Europe*; OECD: Paris, France, 2019.
43. Employment, Social Affairs & Inclusion. IOP Publishing Europa. Available online: https://ec.europa.eu/social/main.jsp?catId=1022&langId=en (accessed on 18 September 2020).
44. EIT Health, McKinsey & Company (2020) Transforming Healthcare with AI: The Impact on the Workforce and Organisations. IOP Publishing EIT Health. Available online: https://eithealth.eu/wp-content/uploads/2020/03/EIT-Health-and-McKinsey_Transforming-Healthcare-with-AI.pdf (accessed on 15 December 2020).
45. Reuter-Oppermann, M.; Kühl, N. *Artificial Intelligence for Healthcare Logistics: An Overview and Research Agenda, In Ar-tificial Intelligence and Data Mining in Healthcare*; Masmoudi, M., Jarboui, B., Siarry, P., Eds.; Springer Nature: Cham, Germany, 2021.
46. Davenport, T.; Kalakota, R. The potential for artificial intelligence in healthcare. *Futur. Heal. J.* **2019**, *6*, 94–98. [CrossRef]
47. Noorbakhsh-Sabet, N.; Zand, R.; Zhang, Y.; Abedi, V. Artificial Intelligence Transforms the Future of Health Care. *Am. J. Med.* **2019**, *132*, 795–801. [CrossRef] [PubMed]
48. Wynants, L.; van Calster, B.; Bonten, M.M.; Cllins, G.S.; Riley, R.D.; Heinze, G.; Schuit, E.; Dahly, D.L.; Damen, J.A.A.; Debray, T.P.A.; et al. Prediction models for diagnosis and prognosis of COVID-19 infection: Systematic review and critical appraisal. *BMJ* **2020**, *369*, m1328. [CrossRef] [PubMed]
49. Niessner, H.; Rauner, M.S.; Gutjahr, W.J. A dynamic simulation–optimization approach for managing mass casualty incidents. *Oper. Res. Heal. Care* **2018**, *17*, 82–100. [CrossRef]
50. Klumpp, M.; Zijm, H. Logistics Innovation and Social Sustainability: How to Prevent an Artificial Divide in Human–Computer Interaction. *J. Bus. Logist.* **2019**, *40*, 265–278. [CrossRef]
51. Bertsimas, D.; Orfanoudaki, A.; Weiner, R.B. Personalized treatment for coronary artery disease patients: A machine learning approach. *Heal. Care Manag. Sci.* **2020**, *23*, 482–506. [CrossRef] [PubMed]

healthcare

MDPI

Article

Machine Learning to Predict the Progression of Bone Mass Loss Associated with Personal Characteristics and a Metabolic Syndrome Scoring Index

Chao-Hsin Cheng [1,†], Ching-Yuan Lin [2,†], Tsung-Hsun Cho [3] and Chih-Ming Lin [4,*]

1 Division of Chest Medicine, Ten-Chan General Hospital, Chung Li, Taoyuan 320, Taiwan; starcheng2001@gmail.com
2 Department of Laboratory Medicine, Ten-Chan General Hospital, Chung Li, Taoyuan 320, Taiwan; lab02@tcmg.com.tw
3 Institute of Biomedical Informatics, National Yang-Ming-Chiao-Tung University, Taipei 112, Taiwan; eric101784@hotmail.com
4 Department of Healthcare Information and Management, Ming Chuan University, Taoyuan 333, Taiwan
* Correspondence: cmlin@mail.mcu.edu.tw; Tel.: +886-3-350-7001; Fax: +886-3-359-3880
† Chao-Hsin Cheng and Ching-Yuan Lin equally contributed as the first authors.

Abstract: A relationship exists between metabolic syndrome (MetS) and human bone health; however, whether the combination of demographic, lifestyle, and socioeconomic factors that are associated with MetS development also simultaneously affects bone density remains unclear. Using a machine learning approach, the current study aimed to estimate the usefulness of predicting bone mass loss using these potentially related factors. The present study included a sample of 23,497 adults who routinely visited a health screening center at a large health center at least once during each of three 3-year stages (i.e., 2006–2008, 2009–2011, and 2012–2014). The demographic, socioeconomic, lifestyle characteristics, body mass index (BMI), and MetS scoring index recorded during the first 3-year stage were used to predict the subsequent occurrence of osteopenia using a non-concurrence design. A concurrent prediction was also performed using the features recorded from the same 3-year stage as the predicted outcome. Machine learning algorithms, including logistic regression (LR), support vector machine (SVM), random forest (RF), and extreme gradient boosting (XGBoost), were applied to build predictive models using a unique feature set. The area under the receiver operating characteristic curve (AUC), accuracy, sensitivity, specificity, precision, and F1 score were used to evaluate the predictive performances of the models. The XGBoost model presented the best predictive performance among the non-concurrence models. This study suggests that the ensemble learning model with a MetS severity score can be used to predict the progression of osteopenia. The inclusion of an individual's features into a predictive model over time is suggested for future studies.

Keywords: osteopenia; metabolic syndrome; socioeconomic status; lifestyle; machine learning

Citation: Cheng, C.-H.; Lin, C.-Y.; Cho, T.-H.; Lin, C.-M. Machine Learning to Predict the Progression of Bone Mass Loss Associated with Personal Characteristics and a Metabolic Syndrome Scoring Index. *Healthcare* **2021**, *9*, 948. https://doi.org/10.3390/healthcare9080948

Academic Editor: Mahmudur Rahman

Received: 23 June 2021
Accepted: 25 July 2021
Published: 28 July 2021

1. Introduction

Osteoporosis is a systemic bone disease and an important public health problem because it increases the incidence and mortality of fractures and significantly increases the risk of fracture-related medical expenses [1–3]. A recent review study reported that the economic burden of osteoporosis-related fractures was significant, costing approximately USD 17.9 billion and GBP 4 billion per annum in the USA and UK, respectively [4]. In Taiwan, the prevalence of osteoporosis among the population older than 50 years increased from 17.4% in 2001 to 25.0% in 2011 [5] Approximately one-quarter of individuals older than 65 years who have been diagnosed with osteoporosis have experienced a spine or hip fracture [6].

Several physical factors have been associated with osteoporosis, including abdominal obesity, high blood pressure, dyslipidemia, and glucose metabolism abnormalities, which are all considered to be components of metabolic syndrome (MetS). Cardiovascular diseases (CVDs) have been linked to reduced bone mineral density (BMD), osteoporosis, and osteopenia [7–10]. While MetS may play a potential role in the development of osteoporosis, further research is needed to obtain hard data to support the hypothesis. Previous studies have identified similar risk factors and pathophysiological mechanisms underlying the development of both osteoporosis and atherosclerotic CVDs. There are suggestions that common underlying pathways, such as disturbed calcium homeostasis, induction of inflammatory response, and oxidative stress, are shared by the two conditions. It has been suggested that the two conditions share underlying pathways linking components of MetS as well as the coupling process of bone formation and bone reabsorption [11,12]. Evidence suggests that consideration should be given to the correction of MetS for the prevention of osteoporotic fractures [13]. Potential factors that affect MetS development have included demographic factors (including age, sex, and living area) and lifestyle behaviors (including smoking, alcohol consumption, diet, and physical activity) [14,15]. Socioeconomic status (SES) components, including income, occupation, and education, are also closely related to CVD development and metabolic indicators [15–17]. Previous analyses may have been limited by the lack of inclusion of social and lifestyle covariate factors, which may reduce the explanatory power of these analyses. To clarify the causal relationship between bone density and MetS, a prospective longitudinal study should be performed, and during the investigation, sex, age, ethnicity, lifestyle, and eating habits should not be overlooked [13,18].

For decades, artificial intelligence has been applied to the identification of risk factors or groups at risk of developing osteoporosis. The burden on health systems, the economy, and society could be lessened through the use of an artificial intelligence model to predict risk groups [19–22]. A comprehensive and low-cost method could be developed to facilitate the use of predictive models during health examinations, especially for developing countries or rural areas. However, most predictions for osteoporosis have been modeled using information for participants who have primarily been female or in specific age groups. Predictive tools should be developed to perform similarly across various populations, including greater numbers of participants across a large age range, which has not been the case for existing predictive models [22]. Additionally, few studies have performed predictive models to identify the risk for osteopenia, which represents an earlier stage of bone disorders, and to identify those at risk of osteopenia that may be useful for promoting overall bone health, especially among younger populations.

To better understand the relationship between MetS and human bone health, determining whether the underlying demographic, lifestyle, and socioeconomic causes of MetS also affect bone density is critical. Our study aimed to explore a comprehensive approach applicable to a wider population. Recent studies have reported that MetS severity scores can serve as useful indicators to assess the potential risk factors for subclinical conditions and can facilitate the development of prevention strategies during the early stages of disease development [23,24]. To our knowledge, no study has previously developed a disease prediction model that combines demographic, lifestyle, and SES with MetS score indicators. Except for the two types of research that we reported [15,21], most studies [14,16–20] conducted a cross-section (i.e., concurrence) approach to modeling the potential risks associated with bone health. A non-concurrence study examining these factors can be used to investigate the causal relationship between these factors and bone health. Using a machine learning approach, this study developed a model to predict the loss of adult bone mass among a Taiwanese population using MetS severity scores and individual risk factors.

2. Materials and Methods

2.1. Data Source

Data were obtained and analyzed from a membership-oriented private institute, a Major Health Screening Center in Taiwan. With four clinic locations around the country, the center provides periodic health examinations to approximately 600 thousand members. A detailed description of the data collection and analysis of the resulting Major Longitudinal Health-check-up-based Population Database (MJLPD) is described in detail elsewhere [25,26].

In consideration of various ethical issues within the data, the protocol of this study was evaluated and approved by the National Taiwan University Research Ethics Committee (NTU-REC 201911EM012) and the Major Health Screening Center.

2.2. Study Sample

Participants over 20 years of age who had undergone at least one standard health screening at the Center in each of three three-year stages/periods (i.e., 2006–2008, 2009–2011, and 2012–2014) from 2006 to 2014 were used to conduct the longitudinal study. All participants lacking BMD examination data or who were diagnosed with osteopenia or osteoporosis (T score < -1) at baseline (i.e., 2006–2008) were excluded from the study. For participants who had undergone multiple screenings within the three-year period, the last examination period was selected for the analysis. As a result, three questionnaires and examination measurements for each participant were collected during the nine-year period. A final total of 23,497 participants (13,012 males and 10,485 females) met the inclusion criteria and were used as our study dataset. Among the included study population, 1402 and 1805 participants were diagnosed with either osteopenia or osteoporosis during the second and third stages, respectively. Due to a relatively low positive rate, the dataset was analyzed using a random under-sampling (1:1 match) approach while applying machine learning models to mitigate the imbalance problem. A flow chart of the data collection process used to identify the study participants and define the analysis dataset is shown in Figure 1.

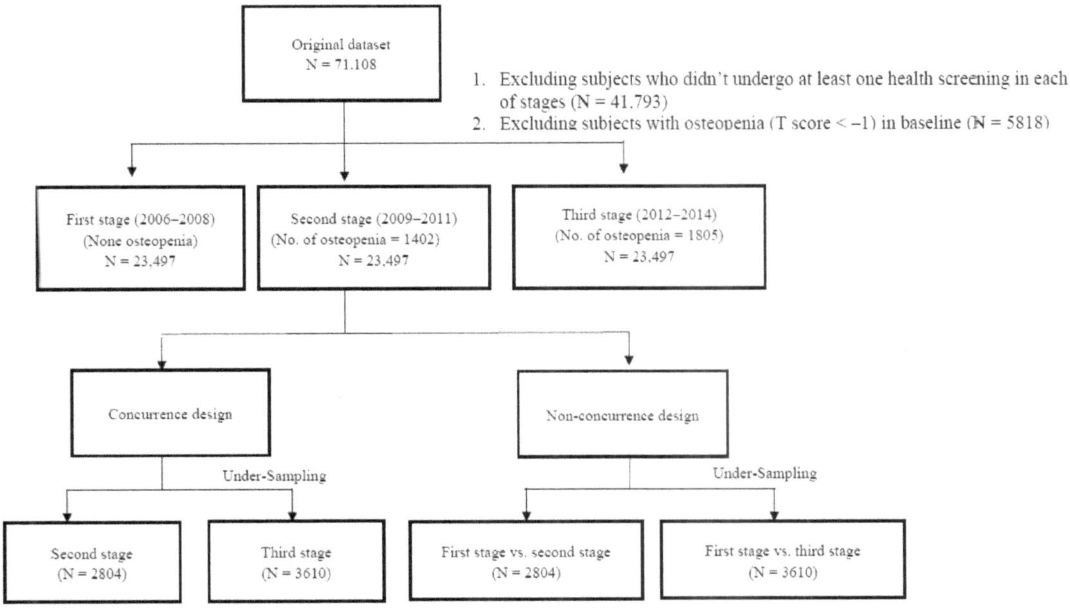

Figure 1. Flow chart of the data collection process used to identify study participants and define the analysis dataset.

2.3. Response Variables

The measurement of BMD in this study was primarily performed using a Lunar DPX-L density meter, which measures dual-energy X-ray absorption (Liberty Corp., Madison, WI, USA). Using the National Health and Nutrition Examination Survey as a reference population, gender-specific T scores were calculated, and osteoporosis was defined as a T score below −2.5 standard deviations (SDs) relative to the average population value, whereas a T score between −1.0 and −2.5 SDs was defined as low BMD (referred to as osteopenia), and a T score above −1.0 SD was defined as normal [27]. All BMD reports were independently reviewed and coded by trained research physicians. Bone measurements taken at the spine were given priority, followed by hip bones and wrist bones, and the results of all measurement sites were considered by physicians. In conducting the study of the effects of risk factors on bone health over an extended period, we collected the indicators of ongoing osteopenia or osteoporosis status among those who developed these disorders during the study period and were not diagnosed with bone disorders at baseline. Using −1.0 SD as the cutoff point in the current longitudinal study, individual BMD was treated as a dichotomous variable. The measured outcome was defined as the occurrence of bone illness, as diagnosed during the second and/or third stages for those with BMD values higher than −1.0 SD during the baseline measurement.

2.4. Explanatory Variables

Each of the study participants completed a self-administered questionnaire during screening to obtain socio-demographic characteristics and lifestyle habit information. Data collected included sex, age (classified into 20–39 years, 40–64 years, and 65 or more years), four aspects of SES (i.e., marriage, education, income, and occupation), as well as nine well-documented lifestyle habits, constituting related risk factors in past studies. Hormones, steroids, and thyroid-related treatment drugs used by patients were cataloged.

Body mass index (BMI, kg/m^2) is considered a risk factor for osteopenia and was included as a continuous variable in our analysis. Using the same databases reported in previous research [24], by back-transforming the standardized scores derived from the aforementioned equations, a covariance matrix was obtained with the MetS severity scores, calculated using waist circumference, fasting plasma glucose, systolic blood pressure, fasting triglycerides (TG), and high-density lipoprotein (HDL). First, the individual values of the five components were standardized and converted to a Z score. A confirmatory factor analysis approach was then followed to derive the score based on the five MS components, with a weighted contribution for each of the components to a latent MetS factor being determined based on both specific age ranges and genders. In the present study, a higher score denotes that a person has a more severe MetS condition, whereas lower scores indicate the lack of MetS.

Classification of the SES of participants was divided into three levels of educational attainment. Occupation was classified as unemployed, manager/owner, and non-management employee. Marriage classification was labeled as married or unmarried. Some lifestyle habits, estimated quantitatively by frequency, were classified into three levels for smoking, alcohol consumption, sugar-sweetened consumption, physical activity, and sleeping. Other habits, such as betel nut chewing, vegetarian diet, dairy intake, taking calcium supplements, and related medical treatments such as hormones, steroids, and thyroid-associated medications, were classified dichotomously.

2.5. Study Design

The physical and biochemical aspects, demographic, socioeconomic, and lifestyle characteristics of the study participants at baseline and those associated with participants who developed osteopenia or osteoporosis over the following two stages were described. In the longitudinal study, the causal relationships were analyzed using a non-concurrent design. The lifestyle characteristics, demographic, socioeconomic, and, as well as BMI and MetS scores from the first stage were used as the features recorded during the following two stages. The factors used to predict the occurrence of osteopenia or osteoporosis for the second and third stages were those identified during the initial stage. A concurrent prediction was also performed during the second and third stages using each individual's features from the stage being examined.

2.6. Feature Selection and Machine Learning

All features missing greater than 30% of values were excluded from the analysis. The missing values for the remaining features were replaced by using a multiple imputation technique. A multivariate imputation via chained equations (MICE) module was used in the R package to perform the data imputation. To identify the effects of important features on the development of osteopenia, we applied the random forest (RF) algorithm with 10 times repeated 10-fold cross-validation to select robust significant features from the training/validation (80/20) dataset, which utilized a mixture of numerical and categorical features. Both the features and the cutoff points associated with each feature were randomly chosen before each training model. Thus, the sequence of feature importance could differ during each model. Then, we averaged the ten lists of feature importance to obtain a robust selected feature list. The results demonstrated that the independent variables for forecasting the prediction included 17 of the 24 analyzed features, which were selected as a selected features dataset for further machine learning and model evaluation. The MetS score and BMI played the most important roles among the selected features (Table 1).

Table 1. Robust feature importance ranking list.

Feature	Rank	Relative Importance
MetS score	1	1.000
Body mass index	2	0.959
Age	3	0.253
Education	4	0.243
Sweetened beverage	5	0.216
Milk intake	6	0.207
Income	7	0.194
Physical activity	8	0.187
Sleep	9	0.184
Occupation	10	0.162
Cheese intake	11	0.154
Sex	12	0.151
Smoke	13	0.133
Alcohol	14	0.127
Vitamin C/E intake	15	0.105
Calcium intake	16	0.103
Marital status	17	0.102

In this study, four well-accepted machine learning algorithms, including logistic regression (LR), extreme gradient boosting (XGBoost), RF, and support vector machine (SVM), were applied to develop the concurrence and non-concurrence predictive models. A 10-fold cross-validation and grid search were used to determine the parameters of the four predictive models for the tuning of hyperparameters while training the model using the defined dataset. Using bootstrapping and 10-fold validation, the best scores were used to define the parameters for the predictive models. The test of a dataset with 80/20, which was a separated dataset from the preceding datasets, was used to avoid the development of an over-fitting model. Subsequently, 10 iterations of a receiver operating characteristic curve (ROC) analysis were employed on the randomized datasets to obtain the best area under the ROC curve (AUC). All machine learning analyses were performed using Python Software (Foundation and Python Language Reference, version 3.7.3, Beaverton, OR, USA). The libraries of Scikit-Learn 0.23.2 were implemented and used to confirm these models. The process used for feature selection and machine learning is shown in Figure 2.

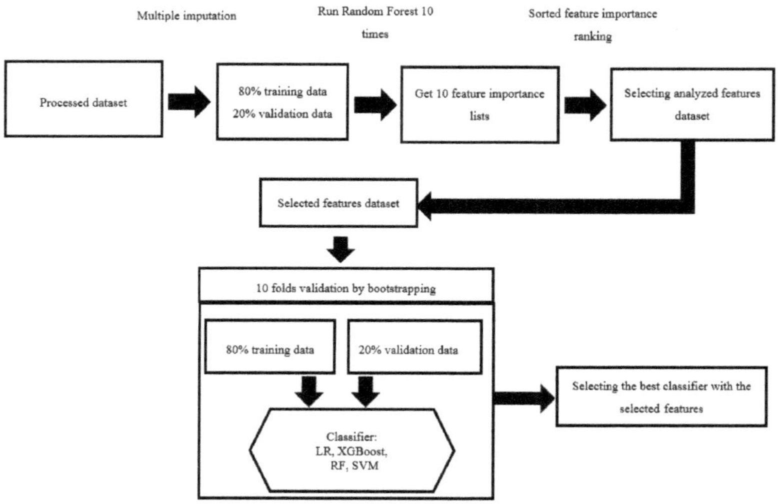

Figure 2. The process of feature selection and the application of machine learning algorithms.

2.7. Model Evaluation

The model's discrimination was measured. In this study, discrimination refers to the predictive effectiveness of the model in determining between participants with and without osteopenia. In each model, the discriminatory power was analyzed based on the AUC, while the ROC curves used were determined by plotting the true positive fraction against the false positives. For each cutoff score, the specificity (maximum subsequent sum) and sensitivity (optimal values) were calculated.

Furthermore, accuracy, precision, and F1 score evaluation indicators from the confusion matrix were used to analyze the relationship between the actual values and the predicted values for osteopenia. The precision–recall curve (PRC) was also generated to determine the tradeoff between precision and recall at different thresholds. Precision–recall is a useful measure of the success of prediction when classes are imbalanced. In the imbalanced data, the false-positive rate tends to stay at small values due to the low positive rate. Thus, ROC becomes less informative for the model performance in this situation. On the other hand, the PRC baseline is varied by the value of the positive rate, and PRC is performed by switching from false positives to precision, which provides more valuable information. A high AUC represents both high recall (i.e., sensitivity) and high precision, where high precision is associated with a low false-positive rate, and high recall relates to a low false-negative rate. F1 was calculated as 2 × precision × recall/(precision + recall), which is the harmonic mean of precision and recall. A larger F1 score indicates a more accurate model.

3. Results

Lifestyle habits, sociodemographic factors, and biochemical and physical examination items over the three study stages are presented in Table 2. Osteopenia occurrence rates during the second stage in men and women were 7.3% and 4.3%, respectively. An increase to 8.3% and 6.9% in the occurrence of osteopenia was observed during the third stage. Participants with relatively low SES, adverse habits (e.g., smoking and alcohol consumption), low sleep hours, and a vegetarian diet and who were taking related medicines during the initial baseline stage had a higher occurrence of osteopenia during the subsequent stages. Compared with the participants at baseline, the average BMI of those who developed osteopenia in subsequent stages was relatively low. The participants who went on to develop osteopenia had a lower MetS score (−0.22) in the subsequent stages compared with the average score (0.09) for the entire population at baseline.

Table 2. Explanatory variables related to osteopenia over the three study stages.

Characteristics	Participants in 2006–2008	Osteopenia in 2009–2011	Osteopenia in 2012–2014
	n (%)	n (%)	n (%)
Sex			
Male	13,012 (55.4)	953 (7.3)	1080 (8.3)
Female	10,485 (44.8)	449 (4.3)	725 (6.9)
Age (yrs)			
20–39	11,055 (47.0)	240 (2.9)	176 (3.0)
40–64	11,781 (50.1)	1029 (7.2)	1404 (8.7)
≥65	661 (2.8)	133 (14.0)	225 (16.4)
Marital status			
Unmarried	5163 (23.3)	211 (4.7)	258 (6.5)
Married	16,956 (76.7)	1100 (6.3)	1386 (7.8)
Education (yrs)			
<12	2178 (9.4)	289 (13.6)	346 (15.5)
12–15	10,529 (45.6)	635 (6.2)	777 (7.9)
≥16	10,397 (45.0)	444 (4.2)	586 (5.5)
Income (NTD/yr)			
<400,000	2676 (12.4)	226 (9.0)	297 (12.1)
400,000–799,999	5797 (26.8)	332 (6.4)	403 (8.4)
>800,000	13,174 (60.9)	699 (5.0)	899 (6.4)
Occupation			
Unemployed	3707 (17.5)	284 (7.5)	422 (10.8)
Managed	2562 (11.7)	150 (5.5)	183 (6.6)
Non-managed	15,557 (71.3)	815 (5.4)	970 (6.6)
Smoke (pack/day)			
None	18,545 (82.2)	1062 (5.6)	1503 (7.6)
≤1	3177 (14.1)	181 (6.6)	196 (7.5)
>1	839 (3.7)	71 (10.4)	57 (8.9)
Alcohol (cup/day)			
None	18,601 (83.9)	1041 (5.7)	1477 (7.6)
1	1726 (7.8)	94 (5.5)	140 (7.6)
≥2	1847 (8.3)	129 (7.2)	140 (7.8)
Chewing betel nut			
No	21,521 (93.8)	1208 (5.7)	1659 (7.5)
Yes	1428 (6.2)	93 (8.1)	102 (8.5)
Physical activity (hrs/wk)			
<1	9042 (39.6)	503 (5.4)	552 (6.7)
1–6	12,805 (56.1)	573 (6.1)	801 (7.8)
≥7	987 (4.3)	126 (7.3)	197 (10.8)
Sleep (hrs/day)			
<6	4523 (20.1)	369 (7.0)	524 (8.9)
6	16,467 (73.2)	676 (5.9)	845 (7.3)
≥7	1506 (6.7)	312 (5.1)	388 (6.9)
Vegetarian diet			
Yes	592 (2.5)	56 (8.2)	271 (7.9)
No	22,774 (97.5)	1330 (5.9)	1534 (7.6)
Sweetened beverage (cup/wk)			
None	7148 (30.8)	707 (7.2)	996 (8.8)
1–6	10,981 (47.3)	483 (5.0)	560 (6.4)
≥7	5067 (21.8)	176 (4.9)	189 (6.5)
Milk intake (cup/wk)			
None	11,545 (49.9)	701 (6.0)	871 (7.6)
1–6	9491 (41.0)	505 (5.6)	679 (7.2)
≥7	2093 (9.0)	158 (7.4)	187 (9.3)

Table 2. *Cont.*

Characteristics	Participants in 2006–2008	Osteopenia in 2009–2011	Osteopenia in 2012–2014
	n (%)	n (%)	n (%)
Cheese intake (slice/wk)			
None	13,276 (57.5)	824 (6.3)	1119 (8.4)
1–6	9390 (40.7)	503 (5.3)	581 (6.4)
≥7	430 (1.9)	33 (6.3)	32 (6.9)
Vitamin C, E intake			
Yes	4180 (17.8)	175 (4.8)	271 (7.9)
No	19,312 (82.2)	1227 (6.2)	1534 (7.6)
Calcium intake			
Yes	3990 (17.0)	326 (8.6)	403 (12.0)
No	19,502 (93.0)	1076 (5.5)	1402 (7.0)
Hypertension medicine			
Yes	1399 (6.0)	138 (7.1)	226 (9.3)
No	22,093 (94.0)	1264 (5.9)	1579 (7.5)
Diabetes medicine			
Yes	440 (1.9)	47 (7.1)	72 (8.5)
No	23,052 (98.1)	1355 (5.9)	1733 (7.7)
Thyroid medicine			
Yes	252 (1.1)	21 (6.5)	27 (7.3)
No	23,240 (98.9)	1381 (6.0)	1778 (7.7)
Lipidemia medicine			
Yes	400 (1.7)	35 (5.7)	68 (8.1)
No	23,092 (98.3)	1367 (6.0)	1737 (7.7)
Hormone medicine			
Yes	272 (1.2)	18 (7.9)	15 (7.1)
No	23,220 (98.8)	1384 (5.9)	1790 (7.7)
Body mass index (sd)	23.25 (3.41)	22.79 (3.09)	22.85 (3.07)
MetS score (sd)	0.09 (1.02)	−0.22 (0.99)	−0.22 (0.94)

The study utilized four machine learning models (i.e., LR, XGBoost, RF, and SVM) to predict osteopenia. The predictive models were corroborated using optimal parameters for each model through a grid search. The ROC and PRC curves of the generated machine learning models for the concurrence and non-concurrence designs are shown in Figures 3 and 4.

The differences between the models were more distinct when using baseline features to predict osteopenia during the second stage than when using the baseline features to predict osteopenia during the third stage. The performances of the models for predicting osteopenia occurrence during the second stage according to baseline features are shown in Table 3.

The AUC values for LR, XGBoost, RF, and SVM in the non-concurrence and concurrence models were 0.726 and 0.745, 0.753 and 0.721, 0.693 and 0.687, and 0.723 and 0.712, respectively. The F1 scores for the four algorithms in the non-concurrence and concurrence models were 0.668 and 0.689, 0.723 and 0.686, 0.656 and 0.633, and 0.688 and 0.676, respectively. Among all predictive models, the XGBoost model had the highest AUC value. Except for the LR models, most of the non-concurrence models demonstrated better predictive performance than the concomitant concurrence models. Although the concurrence LR model was associated with high AUC and PRC values of 0.745 and 0.774, respectively, the other indicators were relatively poor. The performance of the predictive models for identifying osteopenia occurrence during the third stage showed a similar pattern, with poorer performances than the models used to predict occurrence during the second stage (Table 4).

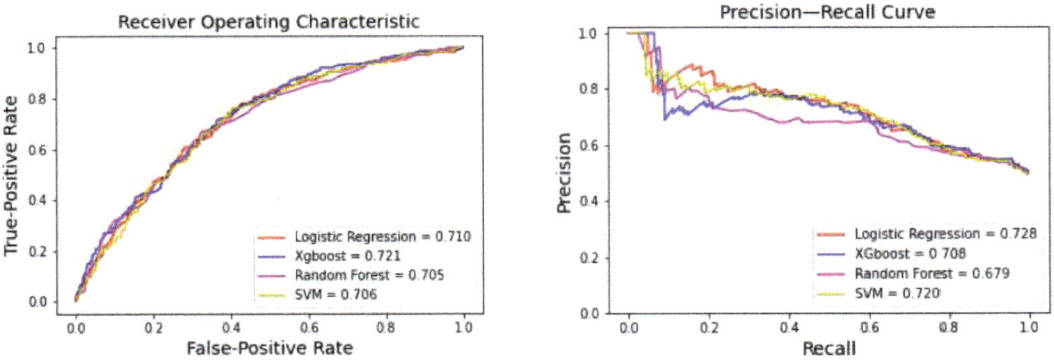

Second stage

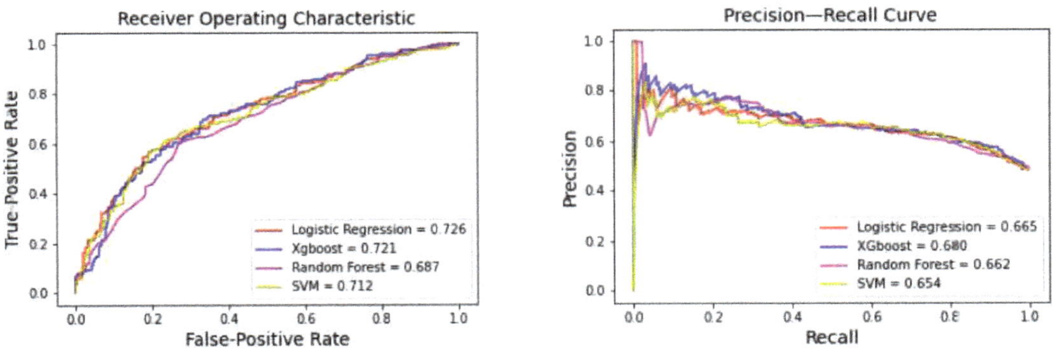

Third stage

Figure 3. ROC and PRC curves for the machine learning models with concurrence designs.

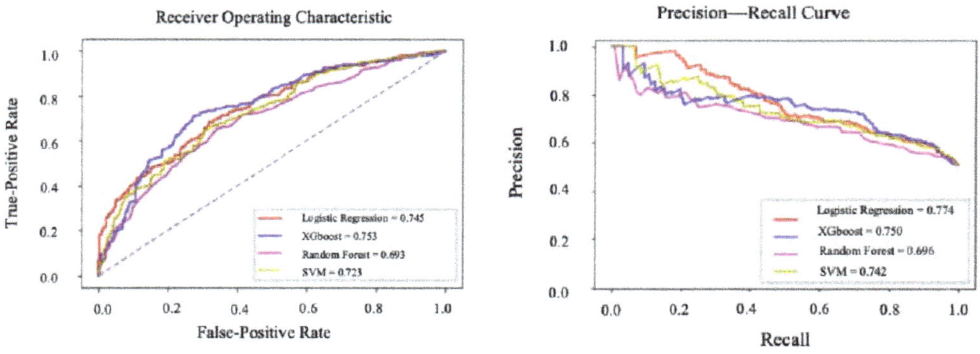

First stage vs. second stage

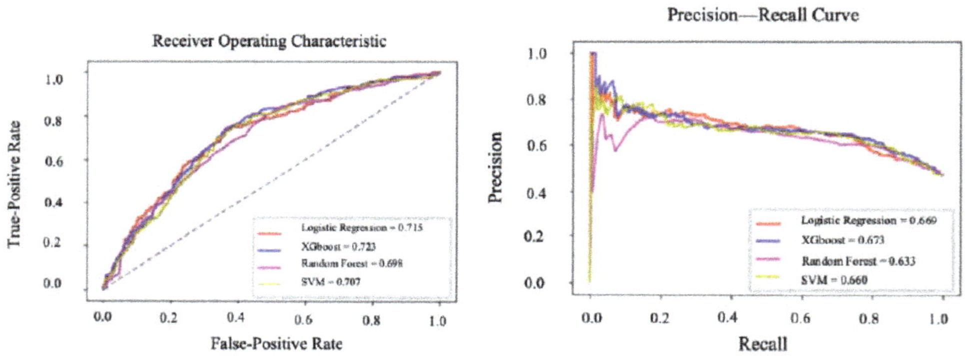

First stage vs. third stage

Figure 4. ROC and PRC curves for machine learning models with non-concurrence designs.

Table 3. Model predictions of osteopenia in the second stage (2009–2011) using concurrent and non-concurrent features.

	Logistic Regression		XGBoost		Random Forest		SVM	
	Non-Concurrent	Concurrent	Non-Concurrent	Concurrent	Non-Concurrent	Concurrent	Non-Concurrent	Concurrent
Sensitivity	0.682	0.684	0.733	0.678	0.663	0.636	0.736	0.702
Specificity	0.648	0.681	0.689	0.672	0.623	0.636	0.575	0.632
Accuracy	0.665	0.683	0.711	0.675	0.643	0.636	0.658	0.667
Precision	0.655	0.694	0.713	0.694	0.650	0.631	0.646	0.651
ROC	0.726	0.745	0.753	0.721	0.693	0.687	0.723	0.712
PRC	0.728	0.774	0.750	0.708	0.696	0.697	0.742	0.720
F1	0.668	0.689	0.723	0.686	0.656	0.633	0.688	0.676

SVM: support vector machine; XGBoost: extreme gradient boosting; ROC: receiver operating characteristic curve; PRC: precision–recall curve; Non-concurrence indicates the prediction using the individual features from the first stage (2006–2008).

Table 4. Model predictions of osteopenia in the third stage (2012–2014) using concurrent and non-concurrent features.

	Logistic Regression		XGBoost		Random Forest		SVM	
	Non-Concurrent	Concurrent	Non-Concurrent	Concurrent	Non-Concurrent	Concurrent	Non-Concurrent	Concurrent
Sensitivity	0.704	0.698	0.745	0.662	0.680	0.672	0.751	0.698
Specificity	0.646	0.620	0.633	0.657	0.622	0.660	0.600	0.627
Accuracy	0.673	0.657	0.686	0.659	0.650	0.666	0.669	0.661
Precision	0.640	0.628	0.645	0.639	0.617	0.645	0.624	0.632
ROC	0.715	0.710	0.723	0.721	0.698	0.705	0.707	0.706
PRC	0.669	0.665	0.673	0.680	0.633	0.662	0.660	0.654
F1	0.670	0.661	0.691	0.650	0.647	0.658	0.681	0.663

SVM: support vector machine; XGBoost: extreme gradient boosting; ROC: receiver operating characteristic curve; PRC: precision–recall curve; Non-concurrence indicates the prediction using the individual features from the first stage (2006–2008).

4. Discussion

Most previous studies have been conducted using a concurrence design, also known as cross-sectional design, which does not allow for the assessment of causal relationships between risk factors and BMD. By initially selecting participants without osteopenia and using a prospective dataset, the present study indicates that non-concurrent models resulted in better predictive performance and are more suitable for this empirical purpose than concurrent models while the optimal algorithm (i.e., XGBoost) is being applied. Therefore, further investigation remains necessary to verify these findings, especially for chronic disorders such as osteopenia or osteoporosis. In addition to BMI, the MetS severity score is identified as the dominant predictor of osteopenia in the present study. Though a relationship has been explored between MetS and bone health, some confusion may arise from the traditional Adult Treatment Panel criteria, such as whether individuals with two high-level MetS components have a lower CVD risk than in those with slightly elevated levels above the criteria in three or more components. Due to the limitations in the traditional MetS criteria, we instead developed the models with a MetS severity score to provide valuable evidence for healthcare societies. Additionally, the study found better predictive performance for the second stage than for the third stage, which implies that the selected features are suitable for predicting osteopenia occurrence over the short-term period of three years but may not be suitable for predictions over a longer period. It could be justified that there would be less effects of health outcomes because of even early socioeconomic or behavioral conditions since these conditions may have changed overtime due to certain personal or heath issues. In the past, risk calculators, such as the web-based Fracture Risk Assessment Tool (FRAX®) algorithm, have enabled the assessment of an individual's fracture risk using clinical risk factors, such as age and alcohol consumption [28]. A prediction of osteopenia using easily measured risk factors may alert practitioners to the condition of an individual's bone health during the early stages of bone disease and may enhance the performance of osteoporosis prevention or avoid the occurrence of future fractures. Our findings may encourage health institutes to provide prevention strategies to those who are potential osteopenia patients, which will lead to better bone health in over one thousand people or the possibility of avoiding deterioration in advance for the study participants. The results for a field with limited research provide pertinent and comprehensive information to those who seek to identify the most suitable model in bone mass loss for decision-making.

Predictive algorithms can serve as diagnostic screening tests to stratify patient populations by risk and to allow for more discrete decision-making [29]. Since screening is intended to guide interventions, high accuracy and precision testing is required. We applied four machine learning algorithms to the construction of predictive models. Generally, RF and XGBoost are ensemble learning models, and LR represents the basic machine learning model, while SVM is widely used as a predictor. As previously discussed, ensemble learning models, specifically the XGBoost model, was found to have higher prediction capabilities and lower risks of overfitting than the others, which can provide greater benefits to decision-makers who are looking for more suitable models for the prediction of healthcare demands [30]. Cruz et al. conducted a review study and summarized the different performances of various machine learning-based diagnostic models for osteoporosis among 25 studies, taking into account the artificial intelligence method applied, the number of risk factors included in the model, the number of patients evaluated, the country associated with the evaluation, and the proportions of each sex in the study population [22]. The study noted that most of the proposed systems can be very useful for the medical community, provided that analysis is not restricted to specific groups and that a spectrum of input variables is included. To the best of our knowledge, this study is the first to develop and compare various machine learning models to predict early bone mass loss that also considers socioeconomic and lifestyle conditions, in addition to MetS indicators.

Compared with linear-based models, neural networks constitute flexible nonlinear systems and may be more suitable for the prediction of outcomes when the associations between the variables are nonlinear, complex, and multidimensional, as is done when assessing the relationships among variables in complex biological systems [31]. Using neural networks, de Cos Juez et al. studied the influence of diet and lifestyle on BMD values in postmenopausal women. A questionnaire examining nutritional habits and lifestyles was used, resulting in 39 variables, such as calcium intake, protein intake, number of pregnancies, height, and BMI, for each respondent [19]. They found that these variables influenced the progression of osteoporosis. However, collecting all possible individual predictors can be difficult, and not all predictors apply to routine disease prevention. To reduce the number of input variables required to obtain an accurate predictive model, the researchers further processed the identified variables using genetic algorithms, which resulted in a model that demonstrated better performance. To test the performance of the algorithm, we performed artificial neural network models utilizing the same dataset via the machine learning module in the SAS Viya Plus package (Linux® for x64, SAS Institute Inc., Cary, NC, USA). The results showed similar performance (e.g., AUC = 0.732 for the non-concurrence model for the second stage) as that obtained for the present LR model. However, the current performance of the predictive models developed in the present study still has room for improvement. In particular, there are even lower (AUC < 0.65) performances when only the two most significant features (i.e., MetS score and BMI) are being used to predict. Other than the modelling strategy that the study used, there are several different algorithms (e.g., gradient boosting machine, decision tree, etc.), samplings, and feature selections such as data-driven approaches [32–34] that have been developed. As we performed additional analyses under varying approaches, the results showed that the synthetic minority oversampling technique could be used to optimize the performance of machine learning (Tables S1–S3). Future studies with the approach may be applicable. However, caution should be exercised to prevent adding increased uncertainty, especially in regard to a sample with a small number of examples of a minority class or a non-continuous feature space [35]. Additionally, Loke et al. studied the association between MetS and BMD and found that the correlation had a very different effect among men than among women [18]. Despite considering the effects of sex and age and using sex- and age-based MetS severity scores in the present study, subgroup analyses stratified by sex and age might provide more information in future studies.

Various issues, including patient self-selection, confounding due to various indications, and the inconsistent availability of outcome data, can result in the inadvertent introduction of bias in machine learning-based predictions [29]. The present study has some limitations that must be addressed. First, although the use of several medications was considered, information regarding some treatments related to BMD was not available, including treatments associated with chronic renal insufficiency, bone metabolic illness, chronic hepatopathy, and neoplasia. A recent study suggested that genomic data can be used to develop a predictive model for BMD using a machine learning approach [36]. However, genotypic variables were not collected in our study, which may have impacted the performance of bone mass prediction models. Additionally, the issues of information loss and selection bias raised by the under-sampling method and imputation procedure cannot be ignored. Finally, the concept of social mobility refers to the degree of SES stability or change in the trajectory of an individual's life course itself. Therefore, exposure to socioeconomic or behavioral adversities during the course of life increases an individual's long-term risk. The accumulation of risk models advocates that increased exposure, duration, and severity to adverse events during the life course increase the risk of disease development [37]. The development of a life course approach has been suggested to develop a better understanding of how a reciprocal relationship between affected factors and health changes over different life stages [38].

5. Conclusions

The study found that an individual's MetS severity score, BMI, and socioeconomic and lifestyle indicators could be used as tools to predict the progression of bone density health using an ensemble learning model. The prediction of osteopenia using easily measured risk factors may alert a physician to the precarious condition of an individual's bone health during the early stages of bone disease and may enhance the performance of preventative measures to avoid osteoporosis or further fractures, reducing the economic burdens associated with related diseases. Our findings can provide guidance for health care providers when designing health promotion measures for specific populations. However, to reflect real-world conditions, the inclusion of an individual's specific features into a predictive model, including changes that occur over time, is suggested for future studies.

Supplementary Materials: The following are available online at https://www.mdpi.com/article/10.3390/healthcare9080948/s1, Table S1: Additional model predictions of osteopenia using non-concurrent features using the other algorithms. Table S2: Additional model predictions of osteopenia using non-concurrent features selected by gradient boosting approach. Table S3. Additional model predictions of osteopenia using non-concurrent features using synthetic minority oversampling technique.

Author Contributions: Conceptualization, C.-H.C., C.-Y.L. and C.-M.L.; methodology, C.-H.C. and C.-M.L.; software, T.-H.C. and C.-M.L.; validation, C.-H.C. and C.-M.L.; formal analysis, T.-H.C. and C.-M.L.; investigation, C.-Y.L.; resources, C.-Y.L. and C.-M.L.; data curation, C.-Y.L.; writing—original draft preparation, C.-H.C., C.-Y.L. and C.-M.L.; writing—review and editing, C.-M.L.; supervision, C.-M.L.; project administration, C.-M.L.; funding acquisition, C.-H.C. and C.-M.L. All authors have read and agreed to the published version of the manuscript.

Funding: This research was funded by the Ministry of Science and Technology (MOST 109-2410-B-130-001) and the Ten-Chan General Hospital (110001).

Institutional Review Board Statement: All or part of the data used in this research were authorized by and received from the MJ Health Research Foundation (Authorization Code: MJHRF2017003A and MJHRF2018013A). The protocol of this study was evaluated and agreed to by the Research Ethics Committee, National Taiwan University (NTU-REC 201911EM012), and the Major Health Screening Center.

Informed Consent Statement: Informed consent was not required.

Data Availability Statement: Data presented in this study are not available on request from the corresponding author. Due to the General Data Protection Regulation, the data presented in this research are not publicly available.

Conflicts of Interest: The authors declare no conflict of interest. Any interpretations or conclusions described in this paper do not represent the views of the MJ Health Research Foundation.

References

1. Bliuc, D.; Nguyen, N.D.; Milch, V.E.; Nguyen, T.V.; Eisman, J.A.; Center, J.R. Mortality Risk Associated with Low-Trauma Osteoporotic Fracture and Subsequent Fracture in Men and Women. *JAMA J. Am. Med. Assoc.* **2009**, *301*, 513–521. [CrossRef] [PubMed]
2. Wright, N.C.; Looker, A.C.; Saag, K.G.; Curtis, J.R.; Delzell, E.S.; Randall, S.; Dawson-Hughes, B. The recent prevalence of osteoporosis and low bone mass in the United States based on bone mineral density at the femoral neck or lumbar spine. *J. Bone Miner. Res.* **2014**, *29*, 2520–2526. [CrossRef]
3. Muka, T.; Trajanoska, K.; Kiefte-de Jong, J.C.; Oeil, L.; Uitterlinden, A.; Hofman, A.; Dehghan, A.; Zillikens, M.C.; Franco, O.H.; Rivadeneira, F. The Association between Metabolic Syndrome, Bone Mineral Density, Hip Bone Geometry and Fracture Risk: The Rotterdam Study. *PLoS ONE* **2015**, *10*, e0129116. [CrossRef] [PubMed]
4. Clynes, M.A.; Harvey, N.C.; Curtis, E.M.; Fuggle, N.R.; Dennison, E.M.; Cooper, C. The epidemiology of osteoporosis. *Br. Med. Bull.* **2020**, *133*, 105–117. [CrossRef]
5. Chen, F.P.; Huang, T.S.; Fu, T.S.; Sun, C.C.; Chao, A.S.; Tsai, T.L. Secular trends in incidence of osteoporosis in Taiwan: A nationwide population-based study. *Biomed. J.* **2018**, *41*, 314–320. [CrossRef]
6. Hsu, W.L.; Chen, C.Y.; Tsauo, J.Y. Balance control in elderly people with osteoporosis. *J. Formos. Med. Assoc.* **2014**, *113*, 334–339. [CrossRef] [PubMed]
7. Ye, C.; Xu, M.; Wang, S.; Jiang, S.; Chen, X.; Zhou, X.; He, R. Decreased Bone Mineral Density Is an Independent Predictor for the Development of Atherosclerosis: A Systematic Review and Meta-Analysis. *PLoS ONE* **2016**, *11*, e0154740. [CrossRef] [PubMed]
8. Zhang, Y.; Feng, B. Systematic review and meta-analysis for the association of bone mineral density and osteoporosis/osteopenia with vascular calcification in women. *Int. J. Rheum. Dis.* **2016**, *20*, 154–160. [CrossRef] [PubMed]
9. Rodríguez, A.J.; Scott, D.; Hodge, A.M.; English, D.R.; Giles, G.G.; Ebeling, P.R. Associations between hip bone mineral density, aortic calcification and cardiac workload in community-dwelling older Australians. *Osteoporos. Int.* **2017**, *28*, 2239–2245. [CrossRef]
10. Zhang, Y.; He, B.; Wang, H.; Shi, J.; Liang, H. Associations between bone mineral density and coronary artery disease: A meta-analysis of cross-sectional studies. *Arch. Osteoporos.* **2020**, *15*, 24. [CrossRef] [PubMed]
11. Almeida, M.; Han, L.; Martin-Millan, M.; Plotkin, L.I.; Stewart, S.A.; Roberson, P.K.; Kousteni, S.; O'Brien, C.A.; Bellido, T.; Parfitt, A.M.; et al. Skeletal involution by age-associated oxidative stress and its acceleration by loss of sex steroids. *J. Biol. Chem.* **2007**, *282*, 27285–27297. [CrossRef] [PubMed]
12. Ding, C.; Parameswaran, V.; Udayan, R.; Burgess, J.; Jones, G. Circulating levels of inflammatory markers predict change in bone mineral density and resorption in older adults: A longitudinal study. *J. Clin. Endocrinol. Metab.* **2008**, *93*, 1952–1958. [CrossRef] [PubMed]
13. Wong, S.K.; Chin, K.Y.; Suhaimi, F.H.; Ahmad, F.; Ima-Nirwana, S. The Relationship between Metabolic Syndrome and Osteoporosis: A Review. *Nutrients* **2016**, *8*, 347. [CrossRef] [PubMed]
14. Kim, J.; Choi, Y.H. Physical activity, dietary vitamin C, and metabolic syndrome in Korean adults: The Korea National Health and Nutrition Examination Survey 2008 to 2012. *Public Health* **2016**, *135*, 30–37. [CrossRef] [PubMed]
15. Liao, C.M.; Lin, C.M. Life Course Effects of Socioeconomic and Lifestyle Factors on Metabolic Syndrome and 10-Year Risk of Cardiovascular Disease: A Longitudinal Study in Taiwan Adults. *Int. J. Environ. Res. Public Health.* **2018**, *15*, 2178. [CrossRef]
16. Yoo, S.; Cho, H.J.; Khang, Y.H. General and abdominal obesity in South Korea, 1998–2007: Gender and socioeconomic differences. *Prev. Med.* **2010**, *51*, 460–465. [CrossRef]
17. Kim, J.Y.; Kim, S.H.; Cho, Y.J. Socioeconomic status in association with metabolic syndrome and coronary heart disease risk. *Korean J. Fam. Med.* **2013**, *34*, 131–138. [CrossRef]
18. Loke, S.S.; Chang, H.W.; Li, W.C. Association between metabolic syndrome and bone mineral density in a Taiwanese elderly population. *J. Bone. Miner. Metab.* **2018**, *36*, 200–208. [CrossRef] [PubMed]
19. de Cos Juez, F.J.; Suárez-Suárez, M.A.; SánchezLasheras, F.; Murcia-Mazón, A. Application of neural networks to the study of the influence of diet and lifestyle on the value of bone mineral density in postmenopausal women. *Math. Comp. Model.* **2011**, *54*, 1665–1670. [CrossRef]
20. Liua, Q.; Cuia, X.; Chou, Y.C.; Abbodd, M.F.; Line, J.; Shieh, J.S. Ensemble artificial neural networks applied to predict the key risk factors of hip bone fracture for elders. *Biomed. Signal. Process. Control* **2015**, *21*, 146–156. [CrossRef]
21. Shioji, M.; Yamamoto, T.; Ibata, T.; Tsuda, T.; Adachi, K.; Yoshimura, N. Artificial neural networks to predict future bone mineral density and bone loss rate in Japanese postmenopausal women. *BMC Res. Notes* **2017**, *10*, 590. [CrossRef]
22. Cruz, A.S.; Lins, H.C.; Medeiros, R.V.A.; Filho, J.M.F.; da Silva, S.G. Artificial intelligence on the identification of risk groups for osteoporosis, a general review. *Biomed. Eng. Online* **2018**, *17*, 12. [CrossRef]
23. Gurka, M.J.; Ice, C.L.; Sun, S.S.; DeBoer, M.D. A confirmatory factor analysis of the metabolic syndrome in adolescents: An examination of sex and racial/ethnic differences. *Card. Diab.* **2012**, *11*, 128. [CrossRef]
24. Lin, C.M. An Application of Metabolic Syndrome Severity Scores in the Lifestyle Risk Assessment of Taiwanese Adults. *Int. J. Environ. Res. Public Health* **2020**, *17*, 3348. [CrossRef]
25. Huang, K.C.; Lee, L.T.; Chen, C.Y.; Sung, P.K. All-cause and cardiovascular disease mortality increased with metabolic syndrome in Taiwanese. *Obesity* **2008**, *1*, 1–6. [CrossRef] [PubMed]

26. Yang, X.; Tao, Q.; Sun, F.; Zhan, S. The impact of socioeconomic status on the incidence of metabolic syndrome in a Taiwanese health screening population. *Int. J. Public Health* **2012**, *57*, 551–559. [CrossRef] [PubMed]
27. Looker, A.C.; Wahner, H.W.; Dunn, W.L.; Calvo, M.S.; Harris, T.B.; Heyse, S.P.; Johnston, C.C., Jr.; Lindsay, R. Updated data on proximal femur bone mineral levels of US adults. *Osteoporos. Int.* **1998**, *8*, 468–489. [CrossRef]
28. Kanis, J.A.; Oden, A.; Johnell, O.; Johansson, H.; de Laet, C.; Brown, J.; Burckhardt, P.; Cooper, C.; Christiansen, C.; Cummings, S.; et al. The use of clinical risk factors enhances the performance of BMD in the prediction of hip and osteoporotic fractures in men and women. *Osteoporos. Int.* **2007**, *18*, 1033–1046. [CrossRef] [PubMed]
29. Chen, J.H.; Asch, S.M. Machine Learning and Prediction in Medicine Beyond the Peak of Inflated Expectations. *N. Engl. J. Med.* **2017**, *376*, 2507–2509. [CrossRef] [PubMed]
30. Shi, X.; Wong, Y.D.; Li, M.Z.F.; Palanisamy, C.; Chai, C. A feature learning approach based on XGBoost for driving assessment and risk prediction. *Accid. Anal. Prev.* **2019**, *129*, 170–179. [CrossRef]
31. Greenwood, D. An overview of neural networks. *Behav. Sci.* **1991**, *36*, 1–33. [CrossRef] [PubMed]
32. Wiener, M. Classification and Regression by randomForest. *R News* **2002**, *2*, 18–22.
33. Chawla, N.V.; Bowyer, K.W.; Hall, L.O.; Kegelmeyer, W.P. SMOTE: Synthetic minority over-sampling technique. *J. Artif. Intell. Res.* **2002**, *16*, 321–357. [CrossRef]
34. Friedman, J.H.; Hastie, T.; Tibshirani, R. Regularization paths for generalized linear models via coordinate descent. *J. Stat. Softw.* **2010**, *33*, 1–22. [CrossRef]
35. Abd Elrahman, S.M.; Abraham, A. A review of class imbalance problem. *J. Network Innov. Comput.* **2013**, *1*, 332–340.
36. Wu, Q.; Nasoz, F.; Jung, J.; Bhattarai, B.; Han, M.V.; Greenes, R.A.; Saag, K.G. Machine Learning Approaches for the Prediction Bone Mineral Density by using genomic and phenotypic data of 5130 older Men. *Sci. Rep.* **2021**, *11*, 4482. [CrossRef]
37. Camelo, L.V.; Giatti, L.; Chor, D.; Griep, R.H.; Benseñor, I.M.; Santos, I.S.; Kawachi, I.; Barreto, S.M. Associations of life course socioeconomic position and job stress with carotid intima-media thickness. The Brazilian Longitudinal Study of Adult Health (ELSA-Brasil). *Soc. Sci. Med.* **2015**, *141*, 91–99. [CrossRef] [PubMed]
38. Hoffmann, R.; Kröger, H.; Pakpahan, E. Pathways between socioeconomic status and health: Does health selection or social causation dominate in Europe? *Adv. Life Course Res.* **2018**, *36*, 23–36. [CrossRef]

healthcare MDPI

Article

Loss Weightings for Improving Imbalanced Brain Structure Segmentation Using Fully Convolutional Networks

Takaaki Sugino [1,*], Toshihiro Kawase [1], Shinya Onogi [1], Taichi Kin [2], Nobuhito Saito [2] and Yoshikazu Nakajima [1,*]

1 Department of Biomedical Information, Institute of Biomaterials and Bioengineering, Tokyo Medical and Dental University, Tokyo 101-0062, Japan; kawase.bmi@tmd.ac.jp (T.K.); onogi.bmi@tmd.ac.jp (S.O.)
2 Department of Neurosurgery, Graduate School of Medicine, The University of Tokyo, Tokyo 113-0033, Japan; tkin-tky@g.ecc.u-tokyo.ac.jp (T.K.); nsaito-nsu@m.u-tokyo.ac.jp (N.S.)
* Correspondence: sugino.bmi@tmd.ac.jp (T.S.); nakajima.bmi@tmd.ac.jp (Y.N.); Tel.: +81-3-5280-8173 (T.S. & Y.N.)

Abstract: Brain structure segmentation on magnetic resonance (MR) images is important for various clinical applications. It has been automatically performed by using fully convolutional networks. However, it suffers from the class imbalance problem. To address this problem, we investigated how loss weighting strategies work for brain structure segmentation tasks with different class imbalance situations on MR images. In this study, we adopted segmentation tasks of the cerebrum, cerebellum, brainstem, and blood vessels from MR cisternography and angiography images as the target segmentation tasks. We used a U-net architecture with cross-entropy and Dice loss functions as a baseline and evaluated the effect of the following loss weighting strategies: inverse frequency weighting, median inverse frequency weighting, focal weighting, distance map-based weighting, and distance penalty term-based weighting. In the experiments, the Dice loss function with focal weighting showed the best performance and had a high average Dice score of 92.8% in the binary-class segmentation tasks, while the cross-entropy loss functions with distance map-based weighting achieved the Dice score of up to 93.1% in the multi-class segmentation tasks. The results suggested that the distance map-based and the focal weightings could boost the performance of cross-entropy and Dice loss functions in class imbalanced segmentation tasks, respectively.

Keywords: brain structure segmentation; fully convolutional networks; class imbalance; loss weighting; magnetic resonance images

Citation: Sugino, T.; Kawase, T.; Onogi, S.; Kin, T.; Saito, N.; Nakajima, Y. Loss Weightings for Improving Imbalanced Brain Structure Segmentation Using Fully Convolutional Networks. *Healthcare* **2021**, *9*, 938. https://doi.org/10.3390/healthcare9080938

Academic Editor: Mahmudur Rahman

Received: 29 May 2021
Accepted: 22 July 2021
Published: 26 July 2021

Publisher's Note: MDPI stays neutral with regard to jurisdictional claims in published maps and institutional affiliations.

1. Introduction

Brain structure segmentation on magnetic resonance (MR) images is an essential technique for measuring, visualizing, and evaluating brain morphology. It is used for diagnosis support of psychiatric and neurodegenerative diseases, brain development analysis, and surgical planning and navigation [1,2]. It is manually performed in practice, but manual segmentation is a very laborious task and is subject to intra- and inter-operator variability [1]. Thus, it is desirable to provide an automatic accurate segmentation of brain structures. The most successful state-of-the-art approach for automated segmentation is a fully convolutional network (FCN) [3]. It enables pixel-wise segmentation in an end-to-end manner. Since it was proposed by Long et al. [3] in 2015, it has been improved for medical image segmentation [4,5] and applied to brain structure segmentation tasks [6]. However, it is often biased towards the majority (large-size) classes and suffers from low segmentation performance on the minority (small-size) classes due to a high imbalance between background and foreground classes in medical images. To address this problem, which is commonly known as the class imbalance, there are two types of approaches: data-level approaches and algorithm-level approaches [7,8].

Data-level approaches mainly alleviate the class imbalance by undersampling the majority classes [9] and oversampling the minority classes [10]. However, the majority undersampling limits the information of available data for training and the minority oversampling can lead to overfitting. On the other hand, algorithm-level approaches address the class imbalance by improving algorithms for training. The most common approach is improving loss functions. The improvement of loss functions can be carried out by using new evaluation metrics for loss function or weighting loss functions to enhance the importance of minority classes in the training process. Thus far, various types of loss functions [11–17] and loss weighting strategies [4,18–25] have been proposed to alleviate the class imbalance problem. They can be applied for any medical image segmentation tasks in a plug-and-play fashion [26]. However, it is unclear which loss function and weighting strategy should be used in different situations. Thus, it is important to reveal weighted loss functions which can enhance the capability of FCNs in brain structure segmentation tasks.

In related works, Ma et al. [26] performed a systematic study of the utility of 20 loss functions on typical segmentation tasks using public datasets and evaluated the performance of these loss functions in the imbalanced segmentation tasks. Moreover, Ma et al. [27] compared and evaluated the boundary-based loss functions, which minimize the distance between boundaries of ground-truth and predicted segmentation labels, in an empirical study. Yeung et al. [28] focused on compound loss functions, combining Dice and cross-entropy-based losses with a modulating factor of focal loss function [19] and evaluated what compound loss functions were effective to handle class imbalance problems. As shown in these related works, the effect of loss functions varies according to the situation of segmentation tasks (e.g., medical images used for segmentation, the number and size of segmentation target objects, and the degree of class imbalance). However, how the loss functions work for different segmentation targets remains undiscussed, although their accuracies were evaluated in the related works.

We test the effect of weighted loss functions in different situations of imbalanced brain structure segmentation tasks, including binary- and multi-class segmentation tasks. Especially, in this study, we focus on weighting strategies of loss functions, defined based on class frequency, predictive probability, and distance map, and aim to investigate and discuss how the loss weightings affect the performance of FCNs in brain structure segmentation tasks with different class imbalances.

2. Materials and Methods

2.1. Segmentation Target

In this study, we adopted a segmentation task of brain structures, including the cerebrum, cerebellum, brainstem, and blood vessels, on MR images. As for MR images, we used MR cisternography (MRC) and MR angiography (MRA) images (Figure 1). MRC images, i.e., heavily T2-weighted images, can clearly represent brain surface and cerebral sulci due to the high intensity of cerebrospinal fluid, whereas MRA images can highlight blood vessels. In our group, we used MRC and MRA as clinical routine MR sequences because of the ease of segmentation processing, and segmented brain parenchyma on MRC images and blood vessels on MRA images for the planning and navigation of neurosurgeries. The brain structures have different features in the MR images. The cerebrum is the largest part of the brain and has a low-level foreground–background imbalance in the MRC images. Its surface, i.e., cerebral sulci, has a bit more of a complex shape. The cerebellum is the second largest part of the brain and is located under the cerebrum. It can be considered a middle-level imbalanced target. The brainstem is a small part of the brain and is located between the cerebrum and the spinal cord. It has a high foreground–background imbalance. The brain parenchyma, i.e., the cerebrum, cerebellum, and brainstem, appears in much the same location in every MRC image volume, although its size and shape have individual differences. Its surface can be clearly visualized in MRC images due to high signal intensity of the cerebrospinal fluid around it. On the other hand, blood vessels have varying locations and shapes and appear as small white spots in MRA images. Thus, they are considered a hard-to-segment target with the high foreground–background imbalance, although they are clearly visualized in MRA images. We used the segmentation targets to fundamentally evaluate the effect of loss weightings on the FCN-based segmentation of different brain structures.

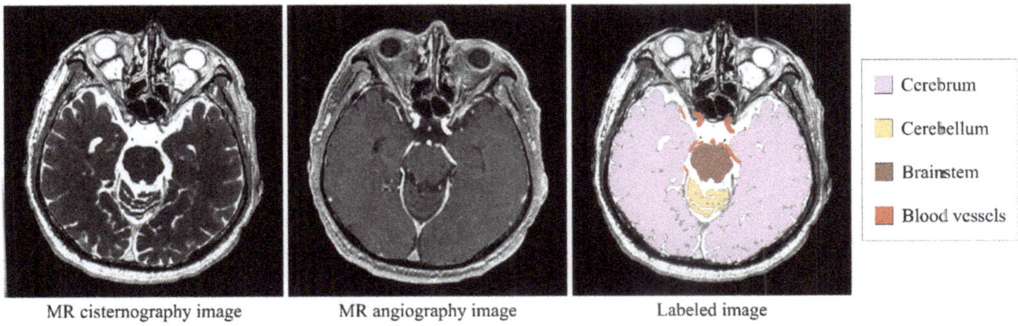

Figure 1. MR images used in this study.

2.2. Network Architecture

As an FCN architecture, we adopted a 2D U-net [4], which is one of the most popular FCN architectures for medical image segmentation. Figure 2 shows the network architecture used in this study. The U-net architecture, which consists of a symmetrical encoder–decoder architecture with skip connections, has been often adopted as a baseline FCN architecture for various medical image segmentation tasks. Many different variants of the U-net architecture have been proposed according to different medical image segmentation tasks, and moreover, a 3D U-net architecture [5] has been introduced for volumetric medical image segmentation. However, training the 3D U-net on full input MR image volumes is usually impractical due to memory limitations of the graphical processing unit (GPU). In the case of the MR image volumes used in this study, it would require at least more than 150 GB of GPU memory, which far exceeds the memory of prevalent GPUs. To overcome the memory limitation, approaches to train 3D FCNs on resized or cropped MR image volumes have been proposed. However, resizing MR image volumes to a smaller size may cause the loss of information on segmentation targets, whereas a patch-based approach [5,29] that crops MR image volumes requires the tuning of more hyperparameters (i.e., patch size), which may affect segmentation performance. Thus, in this study, we decided to use the simple 2D U-net architecture to reduce other factors affecting the results as much as possible.

2.3. Loss Functions

As shown in the related works [26–28], loss functions are an important factor for handling the class imbalance. Existing loss functions for FCN-based segmentation can be divided into four categories: distribution-based loss, region-based loss, boundary-based loss, and compound loss [26]. Distribution-based loss functions measure the dissimilarity between two distributions based on cross-entropy. Region-based loss functions quantify the mismatch or the overlap between two regions. Dice loss function [11,12] is the most common loss function in this category. Boundary-based loss functions measure the distance between two boundaries. Euclidean distance [16] or Hausdorff distance [17] metrics can be used for loss functions in this category. Compound loss functions are defined as the combinations among the distribution-, region-, and boundary-based loss functions [15,28,30–32].

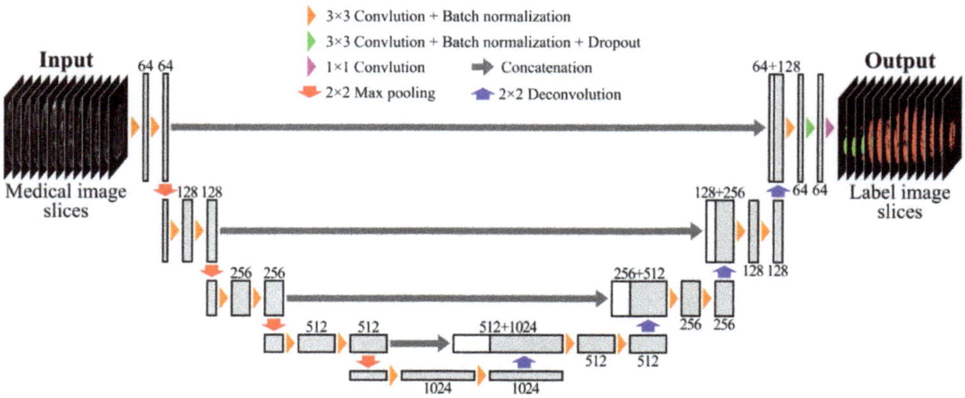

Figure 2. FCN architecture. Each box represents a set of feature maps. The number of feature maps is denoted on the top or bottom of each box.

As described in [26], most of the distribution-based and region-based loss functions can be considered as the variants of cross-entropy and Dice loss functions, respectively. Moreover, boundary-based loss functions, which are formally defined in a region-based way, have similarities to the Dice loss function. Therefore, as most of the loss functions are based on the cross-entropy and Dice loss functions, we decided to use these two loss functions in this study. The cross-entropy loss L_{CE} and the Dice loss L_{Dice} are defined as

$$L_{\mathrm{CE}} = -\frac{1}{N}\sum_{c=1}^{C}\sum_{i=1}^{N} g_{i,c}\log p_{i,c}, \tag{1}$$

$$\begin{aligned} L_{\mathrm{Dice}} &= 1 - \frac{2\sum_{c=1}^{C}\sum_{i=1}^{N} g_{i,c}p_{i,c}}{2\sum_{c=1}^{C}\sum_{i=1}^{N} g_{i,c}p_{i,c} + \sum_{c=1}^{C}\sum_{i=1}^{N}(1-g_{i,c})p_{i,c} + \sum_{c=1}^{C}\sum_{i=1}^{N} g_{i,c}(1-p_{i,c})} \\ &= 1 - \frac{2\sum_{c=1}^{C}\sum_{i=1}^{N} g_{i,c}p_{i,c}}{\sum_{c=1}^{C}\sum_{i=1}^{N} g_{i,c} + \sum_{c=1}^{C}\sum_{i=1}^{N} p_{i,c}}, \end{aligned} \tag{2}$$

where $g_{i,c}$ and $p_{i,c}$ are the ground-truth label and the predicted segmentation probability of class c at pixel i, respectively. N and C are the numbers of pixels and classes in images for a training dataset, respectively.

2.4. Loss Weighting Strategies

In highly imbalanced segmentation tasks, FCNs are likely to ignore small-size foreground classes in the training process, which results in the low segmentation accuracy of the foreground classes. This is what is called the class imbalance problem and can be alleviated by weighting the loss of small-size foreground classes. In this study, we adopted five loss weighting strategies defined based on different factors of class frequency, predictive probability, and distance map. Table 1 indicates the overview of weighted loss functions used in this study. The details of loss weightings are described below.

Table 1. Overview of the weighted loss functions.

Baseline Loss Functions	Weighting Strategies		Weighted Loss Functions
Cross-entropy loss function L_{CE}	Class frequency-based weighting	Inverse frequency weighting	$L_{\text{CE}}^{\text{Inverse}} = -\frac{1}{N}\sum_{c=1}^{C} W_c^{\text{Inverse}} \sum_{i=1}^{N} g_{i,c} \log p_{i,c}$
		Inverse median weighting	$L_{\text{CE}}^{\text{Median}} = -\frac{1}{N}\sum_{c=1}^{C} W_c^{\text{Median}} \sum_{i=1}^{N} g_{i,c} \log p_{i,c}$
	Predictive probability-based weighting	Focal weighting	$L_{\text{CE}}^{\text{Focal}} = -\frac{1}{N}\sum_{c=1}^{C}\sum_{i=1}^{N} W_{i,c}^{\text{Focal}} g_{i,c} \log p_{i,c}$
	Distance map-based weighting	Distance transform map-based weighting	$L_{\text{CE}}^{\text{DTM}} = -\frac{1}{N}\sum_{c=1}^{C}\sum_{i=1}^{N} W_c^{\text{DTM}} g_{i,c} \log p_{i,c}$
		Distance penalty term-based weighting	$L_{\text{CE}}^{\text{DPT}} = -\frac{1}{N}\sum_{c=1}^{C}\sum_{i=1}^{N} W_c^{\text{DPT}} g_{i,c} \log p_{i,c}$
Dice lossfunction L_{Dice}	Class frequency-based weighting	Inverse frequency weighting	$L_{\text{Dice}}^{\text{Inverse}} = 1 - \frac{2\sum_{c=1}^{C} W_c^{\text{Inverse}} \sum_{i=1}^{N} g_{i,c} p_{i,c}}{\sum_{c=1}^{C} W_c^{\text{Inverse}} \sum_{i=1}^{N} (g_{i,c} + p_{i,c})}$
		Inverse median weighting	$L_{\text{Dice}}^{\text{Median}} = 1 - \frac{2\sum_{c=1}^{C} W_c^{\text{Median}} \sum_{i=1}^{N} g_{i,c} p_{i,c}}{\sum_{c=1}^{C} W_c^{\text{Median}} \sum_{i=1}^{N} (g_{i,c} + p_{i,c})}$
	Predictive probability-based weighting	Focal weighting	$L_{\text{Dice}}^{\text{Focal}} = 1 - \frac{2\sum_{c=1}^{C}\sum_{i=1}^{N} W_{i,c}^{\text{Focal}} g_{i,c} p_{i,c}}{\sum_{c=1}^{C}\sum_{i=1}^{N} W_{i,c}^{\text{Focal}} (g_{i,c} + p_{i,c})}$
	Distance map-based weighting	Distance transform map-based weighting	$L_{\text{Dice}}^{\text{DTM}} = 1 - \left(2\sum_{c=1}^{C}\sum_{i=1}^{N} g_{i,c} p_{i,c}\right) / \left(2\sum_{c=1}^{C}\sum_{i=1}^{N} g_{i,c} p_{i,c} + \sum_{c=1}^{C}\sum_{i=1}^{N} W_c^{\text{DTM}} (1 - g_{i,c}) p_{i,c} + \sum_{c=1}^{C}\sum_{i=1}^{N} W_c^{\text{DTM}} g_{i,c} (1 - p_{i,c})\right)$
		Distance penalty term-based weighting	$L_{\text{Dice}}^{\text{DPT}} = 1 - \left(2\sum_{c=1}^{C}\sum_{i=1}^{N} g_{i,c} p_{i,c}\right) / \left(2\sum_{c=1}^{C}\sum_{i=1}^{N} g_{i,c} p_{i,c} + \sum_{c=1}^{C}\sum_{i=1}^{N} W_c^{\text{DPT}} (1 - g_{i,c}) p_{i,c} + \sum_{c=1}^{C}\sum_{i=1}^{N} W_c^{\text{DPT}} g_{i,c} (1 - p_{i,c})\right)$

2.4.1. Inverse Frequency Weighting

Inverse frequency weighting [24], which is one of the most common weighting strategies, is a method for weighting each class based on the class frequency. The weight is inversely proportional to the number of pixels. The smaller the size of target objects is, the higher the weight of them becomes. The inverse frequency weight W_c^{Inverse} in class c is defined by

$$W_c^{\text{Inverse}} = \frac{1}{\left(\sum_{i=1}^{N} g_{i,c}\right)^{\alpha}}, \tag{3}$$

where α is a power parameter. In this study, we used $\alpha = 1$ for the cross-entropy loss function and $\alpha = 2$ for the Dice loss function. The Dice loss function weighted by the inverse of square frequency is known as generalized Dice loss function [24].

2.4.2. Inverse Median Frequency Weighting

Inverse median frequency weighting [18] is a frequency-based weighting as with the inverse frequency weighting. The inverse median frequency weight W_c^{Median} is computed as

$$F_c = \frac{\sum_{i=1}^{N} g_{i,c}}{N}, \tag{4}$$

$$W_c^{\text{Median}} = \frac{median(F_c)}{F_c}, \tag{5}$$

where F_c is the normalized frequency of class c and $median(\cdot)$ denotes a function returning the median value of input data.

2.4.3. Focal Weighting

Focal weighting [19] is a method for putting more focus on hard-to-classify class pixels based on predictive probability. It gives a higher weight to class pixels with lower prediction confidence and reduces the loss assigned to well-classified pixels during the training process. The focal weighting $W_{i,c}^{\text{Focal}}$ is defined by

$$W_{i,c}^{\text{Focal}} = (1 - p_{i,c})^{\gamma}, \tag{6}$$

where γ is called a focusing parameter. In this study, we used $\gamma = 2$ for cross-entropy loss function as in [19] and $\gamma = 1$ for Dice loss function as in [25]. Note that for simplification, here, we did not consider the balancing factor α used in [19].

2.4.4. Distance Transform Map-Based Weighting

Distance transform map (DTM), which is computed as the Euclidean distance from the boundary of target objects, is used in the distance-based loss functions [16,17]. Figure 3b shows an example of DTM. DTM-based weighting can be performed by multiplying prediction errors by the DTM. This weighting assigns higher weights to the pixels which are more distant from the boundary of ground-truth labels. Here, we defined the DTM-based weight W_c^{DTM} as

$$DTM_c = \begin{cases} 0, & x \in \partial G_c \\ \inf\limits_{y \in \partial G_c} ||x - y||_2, & others \end{cases} \tag{7}$$

$$W_c^{\text{DTM}} = 1 + DTM_c, \tag{8}$$

where DTM_c is the distance transform map in class c, and ∂G_c denotes the boundary of ground-truth label in class c. $||x - y||_2$ denotes the Euclidean distance between pixels x and y in images.

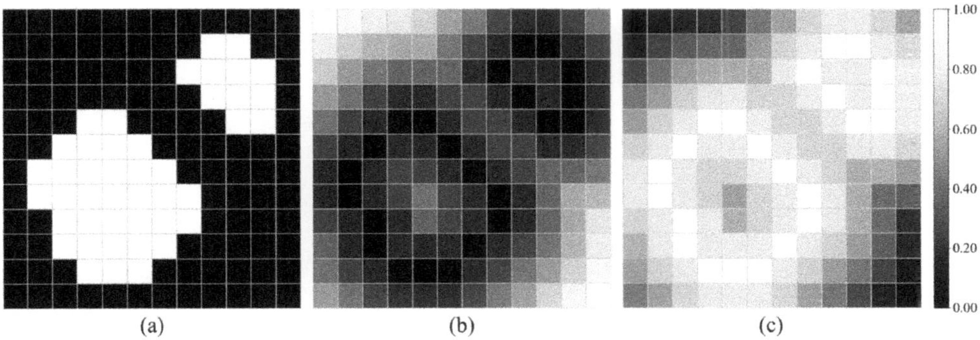

Figure 3. Distance maps for loss weighting. (**a**) Label image, (**b**) distance transform map, and (**c**) distance penalty term.

2.4.5. Distance Penalty Term-Based Weighting

Distance penalty term (DPT) is a distance map for weighting hard-to-segment boundary regions [20], in contrast to the DTM. Let DPT_c be the distance penalty term in class c. Then, DPT_c is defined as the inverse of the DTM_c, and thus, it puts higher weights on the pixels closer to the boundary of ground-truth labels in contrast with the DTM-based weighting. Figure 3c shows an example of DPT. As with the DTM-based weighting, DPT-based weighting penalizes prediction errors with the DPT. The DPT-based weight W_c^{DPT} is defined by

$$W_c^{\mathrm{DPT}} = 1 + DPT_c. \tag{9}$$

We used the cross-entropy and Dice loss functions weighted by the above five weighting strategies. Table 1 summarizes the weighted loss functions used in this study. As for the weighted Dice loss functions, $L_{\mathrm{Dice}}^{\mathrm{Inverse}}$, $L_{\mathrm{Dice}}^{\mathrm{Median}}$, and $L_{\mathrm{Dice}}^{\mathrm{Focal}}$ put their weights on both the numerator and denominator terms as in [24], while $L_{\mathrm{Dice}}^{\mathrm{DTM}}$ and $L_{\mathrm{Dice}}^{\mathrm{DPT}}$ assign their weights to the false positive (i.e., $\sum_{c=1}^{C}\sum_{i=1}^{N}(1 - g_{i,c})p_{i,c}$) and false negative (i.e., $\sum_{c=1}^{C}\sum_{i=1}^{N} g_{i,c}(1 - p_{i,c})$) terms in the denominator.

2.5. Evaluation of Loss Weighting Strategies

2.5.1. Dataset

We used the MR images of 84 patients with unruptured cerebral aneurysms, which were imaged with MRC and time-of-flight MRA sequences on a 3.0 T scanner (Signa HDxt 3.0 T, GE Healthcare, WI, USA) at the University of Tokyo Hospital, Tokyo, Japan. The MR image volumes had 144–190 slices of 512 × 512 pixels with an in-plane resolution of 0.47 × 0.47 mm^2 and a slice thickness of 1.00 mm. As a preprocessing step, the MR images were normalized to have a mean of 0 and a standard deviation of 1. The dataset consisting of 84 cases was divided into the following three subsets: training (60 cases), validation (4 cases), and test subsets (20 cases).

The ground-truth-labeled images for training and testing were manually created by using an open-source software for medical image processing (3D Slicer, Brigham and Women's Hospital, MA, USA); the cerebrum, cerebellum, and brainstem were annotated on MRC images, while blood vessels were annotated on MRA images. The manual annotation was performed by a biomedical engineer and a neurosurgeon. Table 2 indicates the frequency $\left(F_c = \sum_{i=1}^{N} g_{i,c}/N\right)$ of the foreground classes (the cerebrum, cerebellum, brainstem, and blood vessels) in the training subsets. The cerebrum was the most frequent in the foreground classes, followed by the cerebellum, brainstem, and blood vessels.

Table 2. Frequency of the foreground classes in the training subset ($n = 60$).

	Cerebrum	Cerebellum	Brainstem	Blood Vessels
Frequency	0.096	0.012	0.003	0.001

2.5.2. Segmentation Tasks

The goal of this work was to study the effect of loss weightings in different class imbalance situations. Thus, we evaluated the effect of loss weightings on both binary- and multi-class segmentation tasks. Table 3 indicates the overview of the training datasets in the binary- and multi-class segmentation tasks.

Table 3. Training datasets in binary- and multi-class segmentation tasks. BG, CR, CL, BS, and BV stand for background, cerebrum, cerebellum, brainstem, and blood vessels, respectively.

Dataset	Ratio [1]
Binary-class segmentation tasks	
Dataset 1: Cerebrum	BG : CR = 9 : 1
Dataset 2: Cerebellum	BG : CL = 86 : 1
Dataset 3: Brainstem	BG : BS = 352 : 1
Dataset 4: Blood vessels	BG : BV = 749 : 1
Multi-class segmentation tasks	
Dataset 1: Three classes	BG : CR : BV = 677 : 72 : 1
Dataset 2: Four classes	BG : CR : CL : BV = 668 : 72 : 9 : 1
Dataset 3: Five classes	BG : CR : CL : BS : BV = 666 : 72 : 9 : 2 : 1

[1] Ratio of the number of labeled voxels between foreground classes in each training dataset.

Binary-class segmentation tasks: To test how the effect of loss weightings varies according to the size of a foreground class in binary-class segmentation tasks, we evaluated the segmentation performance on the binary-class segmentation task for each of the foreground classes. Note that the binary-class segmentation tasks for the cerebrum, cerebellum, and brainstem were performed using MRC images, whereas the binary-class segmentation for blood vessels was performed using MRA images.

Multi-class segmentation tasks: To test how the effect of loss weightings varies according to the imbalance of foreground classes in multi-class segmentation tasks, we evaluated the segmentation performance on the three-, four-, and five-class segmentation tasks; the three, four, and five classes include the foreground classes of (cerebrum, blood vessels), (cerebrum, cerebellum, blood vessels), and (cerebrum, cerebellum, brainstem, blood vessels), respectively. Note that the multi-class segmentation tasks were performed using multi-modal MR images which included MRC and MRA images.

2.5.3. Network Training Procedure

In the binary- and multi-class segmentation tasks, we trained the FCN model on each training dataset using the cross-entropy and Dice loss functions with or without the loss weightings. The FCN model was trained from scratch for 30 epochs with the Adam optimization algorithm [33] (α (learning rate) $= \{1e-3,\ 1e-4,$ and $1e-5\}$, $\beta_1 = 0.9$, $\beta_2 = 0.999$, and epsilon $= 1e-7$) and a batch size of 5 in each training process. For testing, we used the best trained model in the set {learning rate, epoch} $= \{1e-3, 10\}, \{1e-3, 20\}, \{1e-3, 30\}, \{1e-4, 10\}, \{1e-4, 20\}, \{1e-4, 30\}, \{1e-5, 10\}, \{1e-5, 20\}$, and $\{1e-5, 30\}$ because the condition for good training convergence, especially learning rate and number of epochs, was different according to the loss weightings.

The FCN model with the weighted loss functions were implemented by using Keras with Tensorflow backend, and the training and prediction were performed on an Ubuntu 16.04 PC (CPU: Intel Xeon Gold 5222 3.80 GHz, RAM: 384 GB) with NVIDIA Quadro RTX8000 GPU cards for deep learning.

2.5.4. Evaluation Metrics

To quantitatively evaluate the segmentation performance, we adopted the Dice similarity coefficient (DSC), surface DSC (SDSC) [34], average symmetric surface distance (ASD), and Hausdorff distance (HD). The DSC and SDSC, overlap-based metrics, can be used for evaluating the region overlaps; the DSC measures the overlap of whole regions between ground-truth and predicted labels, whereas the SDSC measures the overlap of the two surface regions. The DSC was calculated by

$$\mathrm{DSC} = \frac{2|G \cap P|}{|G| + |P|}, \tag{10}$$

where G and P denote the regions of ground-truth and predicted labels, respectively. The SDSC was calculated by

$$\mathrm{SDSC} = \frac{\left|\partial G \cap B_{\partial P}^{(\tau)}\right| + \left|\partial P \cap B_{\partial G}^{(\tau)}\right|}{|\partial G| + |\partial P|}, \tag{11}$$

where ∂G and ∂P denote the boundaries of ground-truth and predicted labels, respectively. $B_{\partial G}^{(\tau)}, B_{\partial P}^{(\tau)} \subset \mathbb{R}^3$ are the border regions of ground-truth and predicted label surfaces at tolerance τ, which are defined as $B_{\partial G}^{(\tau)} = \{x \in \mathbb{R}^3 | \exists y \in \partial G, \ ||x - y|| \leq \tau\}$ and $B_{\partial P}^{(\tau)} = \{x \in \mathbb{R}^3 | \exists y \in \partial P, \ ||x - y|| \leq \tau\}$, respectively [26,34]. We here used $\tau = 1$ mm as in [26].

The ASD and HD, boundary distance-based metrics, can be used for evaluating the surface errors; ASD measures the average surface distance between ground-truth and predicted labels, whereas HD measures the max surface distance between them. The ASD was calculated by

$$\mathrm{ASD} = \frac{\sum_{x \in \partial G} D(x, \partial P) + \sum_{y \in \partial G} D(y, \partial G)}{|\partial G| + |\partial P|}, \tag{12}$$

where $D(a, A)$ denote the minimum Euclidean distance from a voxel a to a set of voxels A. The HD was calculated by

$$\mathrm{HD} = \max\left\{\max_{x \in \partial G} D(x, \partial P), \ \max_{y \in \partial P} D(y, \partial G)\right\}. \tag{13}$$

As for HD, in this study, 95th-percentile HD (95HD) was used, as in [27].

When the segmentation accuracy increases, the overlap-based and the boundary distance-based metrics approach 1 and 0, respectively. The evaluation metrics was implemented using the open-source code, which is available at [35].

Furthermore, we used a rank score, which was defined based on [36], to comprehensively evaluate which loss weightings worked well based on the above metrics, as in [26]. The rank score was computed according to the following steps:

Step 1. Performance assessment per case: compute metrics $m_i(loss_j, class_k, \ case_l)$ $(i = 1, \ldots, \ N_m)$ of all loss functions $loss_j$ $(j = 1, \ldots, \ 12)$ for all classes $class_k$ $(k = 1, \ldots, \ N_c)$ in all test cases $case_l$ $(l = 1, \ldots, \ 20)$, where N_m and N_c are the number of metrics and classes, respectively. Note that in this case, we used four metrics $m_i \in \{\mathrm{DSC}, \ \mathrm{SDSC}, \ \mathrm{ASD}, \ \mathrm{95HD}\}$ and a total of twelve loss functions, including cross-entropy and Dice loss functions with no weighting, Inverse, Median, Focal, DTM, and DPT weightings.

Step 2. Statistical tests: perform Wilcoxon signed-rank pairwise statistical tests between all loss functions with the values $m_i(loss_j, class_k, \ case_l) - m_i\left(loss'_j, class_k, \ case_l\right)$.

Step 3. Significance scoring: compute a significance score $s_{ik}(loss_j)$ for loss functions $loss_j$, classes $class_k$, and metrics m_i. $s_{ik}(loss_j)$ equals the number of loss functions performing significantly worse than $loss_j$ according to the statistical tests ($p < 0.05$, not adjusted for multiplicity).

Step 4. Rank score computing: compute the final rank score $R(loss_j)$ of each loss function from the mean significance score of all classes and metrics in each of the binary- and multi-class segmentation tasks by the following equation:

$$R(loss_j) = \frac{1}{N_m \times N_c} \sum_{i=1}^{N_m} \sum_{k=1}^{N_c} s_{ik}(loss_j). \quad (14)$$

3. Results

We compared the results of loss weightings (inverse frequency weighting (Inverse), inverse median frequency weighting (Median), focal weighting (Focal), distance transform map-based weighting (DTM), and distance penalty term-based weighting (DPT)) with those of no weighting (N/A). The statistical difference between N/A and each loss weighting was evaluated by the Wilcoxon signed-rank test. A p-value less than 0.05 was considered significant. Subsequently, we comprehensively evaluated the effect of loss weightings by using the rank scores.

3.1. Binary-Class Segmentation Tasks

Table 4 summarizes all the results in the binary-class segmentation tasks. Figure 4 shows the violin plots of the Dice scores. As for cross-entropy loss function, Inverse and Median provided worse results than N/A in any segmentation tasks. Focal, DTM, and DPT tended to improve the surface accuracy in the highly imbalanced segmentation tasks (i.e., segmentation of brainstem and blood vessels) although the improvement was not statistically significant. As for Dice loss function, Inverse and Median significantly improved the segmentation accuracy in the highly imbalanced segmentation tasks, compared with N/A. Focal tended to provide better results than N/A in all the binary-class segmentation tasks. The distance map-based weightings (i.e., DTM and DPT) worked well in the segmentation of brain parenchyma, but they were ineffective in the segmentation of blood vessels.

Table 4. Segmentation results of no weighting (N/A), inverse frequency weighting (Inverse), inverse median frequency weighting (Median), focal weighting (Focal), distance transform map-based weighting (DTM), and distance penalty term-based weighting (DPT) in binary-class segmentation tasks: Dice similarity coefficient (DSC), surface DSC (SDSC), average symmetric surface distance (ASD) (mm), and 95th-percentile Hausdorff distance (95HD) (mm). (a) Dataset 1: cerebrum, (b) Dataset 2: cerebellum, (c) Dataset 3: brainstem, and (d) Dataset 4: blood vessels. The results of background class are excluded in this table. Compared with the results of N/A, the significantly better and worse results are shown in bold and italic, respectively (Wilcoxon signed-rank test, $p < 0.05$, not adjusted for multiplicity).

Loss Function	Weighting	DSC	SDSC	ASD	95HD
(a) Dataset 1: Cerebrum					
Cross entropy	N/A	0.987	0.991	0.064	0.287
	Inverse	*0.970*	*0.941*	*0.424*	*3.504*
	Median	*0.981*	*0.983*	*0.135*	*0.565*
	Focal	*0.986*	*0.989*	*0.073*	*0.397*
	DTM	*0.986*	0.990	0.069	0.378
	DPT	0.987	**0.992**	0.059	0.328
Dice	N/A	0.986	0.988	0.102	0.381
	Inverse	*0.984*	0.986	*0.275*	0.495
	Median	0.985	0.990	*0.234*	0.425
	Focal	**0.988**	**0.993**	**0.054**	0.308
	DTM	0.987	**0.991**	**0.061**	0.364
	DPT	**0.987**	**0.992**	**0.066**	0.341

Table 4. *Cont.*

Loss Function	Weighting	DSC	SDSC	ASD	95HD
(b) Dataset 2: Cerebellum					
Cross entropy	N/A	0.978	0.981	0.088	0.669
	Inverse	*0.954*	*0.922*	*0.411*	*1.755*
	Median	*0.950*	*0.904*	*0.525*	*2.539*
	Focal	*0.976*	*0.976*	0.166	*2.430*
	DTM	0.978	*0.978*	0.104	0.729
	DPT	0.978	0.980	0.089	0.713
Dice	N/A	0.976	0.973	0.221	1.048
	Inverse	*0.965*	*0.940*	*1.934*	*1.975*
	Median	*0.968*	*0.950*	*2.037*	*4.568*
	Focal	0.977	**0.980**	**0.101**	0.686
	DTM	0.974	0.972	0.153	0.878
	DPT	0.976	0.975	**0.184**	2.331
(c) Dataset 3: Brainstem					
Cross entropy	N/A	0.963	0.940	0.501	4.676
	Inverse	*0.933*	*0.874*	*1.024*	*8.518*
	Median	*0.922*	*0.849*	*0.849*	*6.510*
	Focal	0.962	0.947	**0.239**	1.362
	DTM	0.965	0.951	0.280	**1.204**
	DPT	0.965	0.946	0.425	3.478
Dice	N/A	0.923	0.824	8.880	156.912
	Inverse	**0.953**	**0.921**	**0.476**	**4.770**
	Median	**0.954**	**0.926**	**0.421**	**3.365**
	Focal	**0.963**	**0.949**	**0.241**	**1.905**
	DTM	**0.961**	**0.939**	**0.332**	**4.268**
	DPT	**0.957**	**0.936**	**0.318**	**1.646**
(d) Dataset 4: Blood vessels					
Cross entropy	N/A	0.785	0.809	1.415	12.947
	Inverse	*0.642*	*0.700*	*2.008*	*16.978*
	Median	*0.647*	*0.690*	*2.222*	*18.620*
	Focal	0.783	0.812	1.351	12.353
	DTM	0.786	0.821	1.419	12.243
	DPT	0.784	0.824	1.361	12.340
Dice	N/A	0.704	0.767	1.996	16.026
	Inverse	**0.786**	**0.826**	**1.385**	**13.364**
	Median	**0.768**	**0.794**	**1.627**	14.597
	Focal	**0.785**	**0.812**	**1.518**	13.104
	DTM	0.725	0.754	*2.400*	*19.281*
	DPT	*0.648*	*0.627*	*5.999*	*40.077*

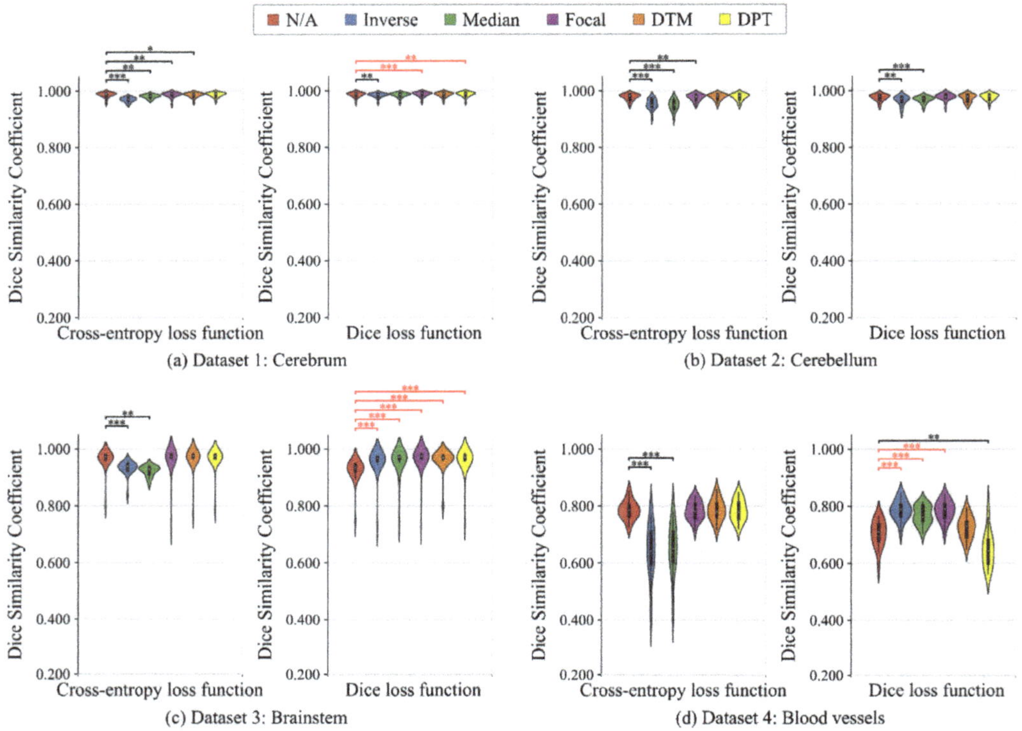

Figure 4. Violin plots of the segmentation results (Dice similarity coefficients) of no weighting (N/A), inverse frequency weighting (Inverse), inverse median frequency weighting (Median), focal weighting (Focal), distance transform map-based weighting (DTM), and distance penalty term-based weighting (DPT) in binary-class segmentation tasks. (**a**) Dataset 1: cerebrum, (**b**) Dataset 2: cerebellum, (**c**) Dataset 3: brainstem, and (**d**) Dataset 4: blood vessels. Compared with the results of N/A, the significantly worse and better results are shown in black and red, respectively (Wilcoxon signed-rank test, * $p < 0.05$, ** $p < 0.01$, and *** $p < 0.001$, not adjusted for multiplicity).

Figure 5 visualizes an example of the segmentation results of blood vessels, which are the highly imbalanced class, in the binary-class segmentation task. As for the cross-entropy loss function, N/A had difficulty in segmenting the upper blood vessels. Both Inverse and Median allowed the FCN to extract most of the upper blood vessels which N/A failed to segment, but obviously increased the overextraction. Focal provided almost the same result as N/A. Both DTM and DPT extracted the wider region of blood vessels than N/A. As for the Dice loss function, N/A had false negatives in the upper blood vessels as with the cross-entropy loss function. It also provided a few more false positives. The class frequency-based weightings, especially Inverse, improved the false positives as well as the false negatives. Focal provided better results than N/A, although it was not so much as Inverse. The results of the distance map-based weightings, especially DPT, were worse than that of N/A.

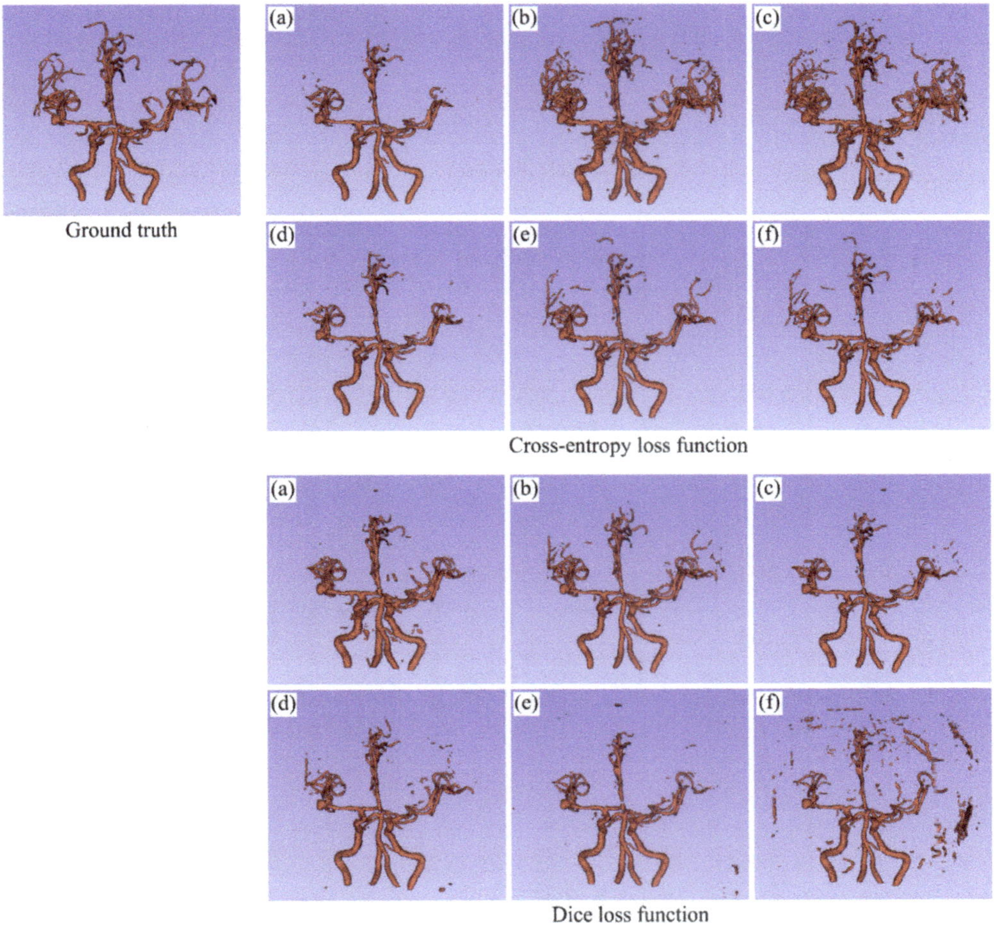

Figure 5. Visualization of the segmentation results of blood vessels in the binary-class segmentation task. (**a**) No weighting, (**b**) Inverse frequency weighting, (**c**) Inverse median frequency weighting, (**d**) Focal weighting, (**e**) Distance transform map-based weighting, and (**f**) Distance penalty term-based weighting.

3.2. Multi-Class Segmentation Tasks

Table 5 summarizes all the results in the multi-class segmentation tasks. Figure 6 shows the violin plots of the Dice scores. As for the cross-entropy loss function, Inverse and Median, as in the binary-class segmentation tasks, worsened the results in any multi-class segmentation tasks. The results of Focal, especially surface accuracies, were equivalent to or better than those of N/A in almost all the tasks. In the distance map-based weighting, DPT worked well for improvement of segmentation accuracy. As for the Dice loss function, Inverse and Median significantly improved the segmentation accuracy of blood vessels, which were a very high-level imbalanced class, in any multi-class segmentation tasks. However, Inverse also significantly worsened the segmentation accuracy of the cerebrum and cerebellum, which were relatively large-size targets. Focal provided better results than N/A for almost all the segmentation targets. The distance map-based weightings showed inconsistent results between the multi-class segmentation tasks.

Table 5. Segmentation results of no weighting (N/A), inverse frequency weighting (Inverse), inverse median frequency weighting (Median), focal weighting (Focal), distance transform map-based weighting (DTM), and distance penalty term-based weighting (DPT) in the multi-class segmentation tasks: Dice similarity coefficient (DSC), surface DSC (SDSC), average symmetric surface distance (ASD), and 95th-percentile Hausdorff distance (95HD). (a) Dataset 1: three classes, (b) Dataset 2: four classes, and (c) Dataset 3: five classes. The results of background class are excluded in this table. Compared with the results of N/A, the significantly better and worse results are shown in bold and italic, respectively (Wilcoxon signed-rank test, $p < 0.05$, not adjusted for multiplicity).

(a) Dataset 1: Three Classes

Loss Function	Weighting	Cerebrum				Blood Vessels			
		DSC	SDSC	ASD	95HD	DSC	SDSC	ASD	95HD
Cross entropy	N/A	0.979	0.965	0.507	5.635	0.778	0.810	1.926	17.142
	Inverse	*0.967*	0.956	0.265	1.256	*0.618*	*0.662*	2.448	20.272
	Median	*0.970*	0.969	0.239	1.273	*0.675*	*0.740*	1.901	17.298
	Focal	0.979	0.989	0.093	0.585	**0.796**	**0.843**	**1.195**	12.933
	DTM	0.979	0.989	0.092	0.585	0.788	**0.848**	**1.097**	**10.539**
	DPT	**0.984**	**0.992**	**0.069**	**0.492**	**0.795**	**0.836**	**1.198**	**11.321**
Dice	N/A	0.985	0.990	0.266	0.445	0.771	0.833	1.225	11.276
	Inverse	*0.896*	*0.634*	*2.290*	*17.436*	**0.800**	0.842	1.177	11.325
	Median	0.985	*0.986*	**0.109**	0.479	**0.809**	**0.848**	1.172	11.654
	Focal	0.985	*0.984*	**0.147**	0.415	0.780	*0.821*	*1.525*	*14.393*
	DTM	*0.984*	0.991	**0.068**	0.492	*0.760*	*0.817*	*1.354*	11.769
	DPT	**0.986**	**0.992**	**0.245**	0.408	0.759	0.816	1.346	12.316

(b) Dataset 2: Four classes

Loss Function	Weighting	Cerebrum				Cerebellum				Blood Vessels			
		DSC	SDSC	ASD	95HD	DSC	SDSC	ASD	95HD	DSC	SDSC	ASD	95HD
Cross entropy	N/A	0.985	0.994	0.057	0.469	0.978	0.981	0.082	0.670	0.792	0.834	1.209	11.215
	Inverse	*0.966*	*0.963*	*0.221*	*1.015*	*0.939*	*0.890*	*0.472*	*1.911*	*0.623*	*0.668*	*2.375*	*19.928*
	Median	*0.970*	*0.968*	*0.221*	*1.009*	*0.954*	*0.938*	*0.279*	*1.397*	*0.674*	*0.738*	*1.860*	*17.051*
	Focal	*0.980*	*0.990*	*0.087*	*0.575*	*0.979*	*0.982*	0.082	0.635	0.783	0.836	1.168	11.228
	DTM	**0.986**	0.994	0.059	**0.408**	*0.977*	*0.979*	0.142	2.019	*0.781*	0.827	1.247	11.639
	DPT	*0.982*	*0.992*	*0.069*	*0.505*	**0.980**	**0.986**	**0.065**	0.579	0.791	0.842	1.138	11.197
Dice	N/A	0.986	0.993	0.060	0.338	0.975	0.971	0.329	2.370	0.766	0.821	1.246	11.110
	Inverse	*0.163*	*0.066*	*18.575*	*81.644*	*0.960*	*0.949*	0.314	3.939	**0.799**	**0.840**	1.192	12.014
	Median	*0.980*	*0.984*	*0.155*	*0.524*	0.973	0.972	**0.234**	2.578	**0.780**	0.818	1.306	12.029
	Focal	0.987	0.994	**0.052**	0.352	**0.980**	**0.986**	**0.067**	**0.543**	**0.791**	0.834	1.233	11.518
	DTM	*0.971*	*0.963*	*0.198*	*1.061*	*0.956*	*0.933*	0.449	*3.654*	*0.610*	*0.630*	*5.309*	*34.425*
	DPT	0.985	*0.992*	0.064	*0.505*	0.978	0.981	**0.085**	0.593	**0.786**	0.827	1.289	12.360

Table 5. *Cont.*

(c) Dataset 3: Five classes									
Loss Function	**Weighting**	**Cerebrum**				**Cerebellum**			
		DSC	**SDSC**	**ASD**	**95HD**	**DSC**	**SDSC**	**ASD**	**95HD**
	N/A	0.981	0.991	0.083	0.552	0.977	0.980	0.127	0.855
Cross entropy	Inverse	*0.971*	*0.973*	*0.179*	*0.846*	*0.950*	*0.926*	*0.346*	*1.492*
	Median	*0.979*	*0.987*	*0.104*	*0.609*	*0.958*	*0.949*	*0.253*	*1.252*
	Focal	**0.985**	**0.993**	**0.060**	**0.469**	0.979	0.984	0.107	0.634
	DTM	*0.980*	0.990	0.085	0.552	0.979	0.982	0.093	0.898
	DPT	**0.982**	**0.993**	**0.069**	**0.502**	0.980	0.985	0.070	0.624
Dice	N/A	0.986	0.993	0.074	0.338	0.977	0.982	0.084	0.618
	Inverse	*0.000*	*0.000*	-	-	*0.955*	*0.946*	*0.221*	*1.405*
	Median	*0.984*	*0.988*	*0.107*	*0.502*	*0.974*	*0.975*	*0.171*	1.164
	Focal	**0.987**	**0.995**	**0.052**	0.291	**0.980**	**0.986**	**0.065**	0.567
	DTM	0.986	0.993	**0.068**	0.361	0.978	0.983	**0.082**	0.608
	DPT	*0.985*	*0.992*	*0.098*	*0.445*	0.974	0.977	0.095	0.747

Loss Function	**Weighting**	**Brainstem**				**Blood Vessels**			
		DSC	**SDSC**	**ASD**	**95HD**	**DSC**	**SDSC**	**ASD**	**95HD**
Cross entropy	N/A	0.961	0.942	0.266	2.083	0.790	0.846	1.084	10.471
	Inverse	*0.944*	0.937	*0.371*	1.302	*0.712*	*0.778*	*1.524*	*14.184*
	Median	*0.949*	0.928	*0.415*	1.528	*0.686*	*0.721*	*1.920*	*17.233*
	Focal	0.962	0.947	0.267	1.495	0.782	*0.830*	*1.263*	12.068
	DTM	**0.966**	0.946	0.291	2.362	*0.783*	*0.840*	*1.163*	*11.097*
	DPT	**0.964**	**0.952**	0.203	**1.343**	**0.797**	**0.855**	1.059	10.703
Dice	N/A	0.960	0.934	0.389	2.174	0.774	0.828	1.234	11.574
	Inverse	0.961	0.941	0.391	2.374	**0.801**	0.836	1.196	12.002
	Median	0.962	0.941	0.344	2.329	**0.788**	0.829	1.200	10.648
	Focal	**0.963**	**0.952**	**0.235**	**1.262**	0.783	0.828	1.300	12.835
	DTM	**0.964**	**0.944**	**0.217**	**1.288**	0.773	0.831	1.221	11.280
	DPT	0.960	0.929	0.394	3.759	*0.757*	*0.801*	*1.869*	*18.269*

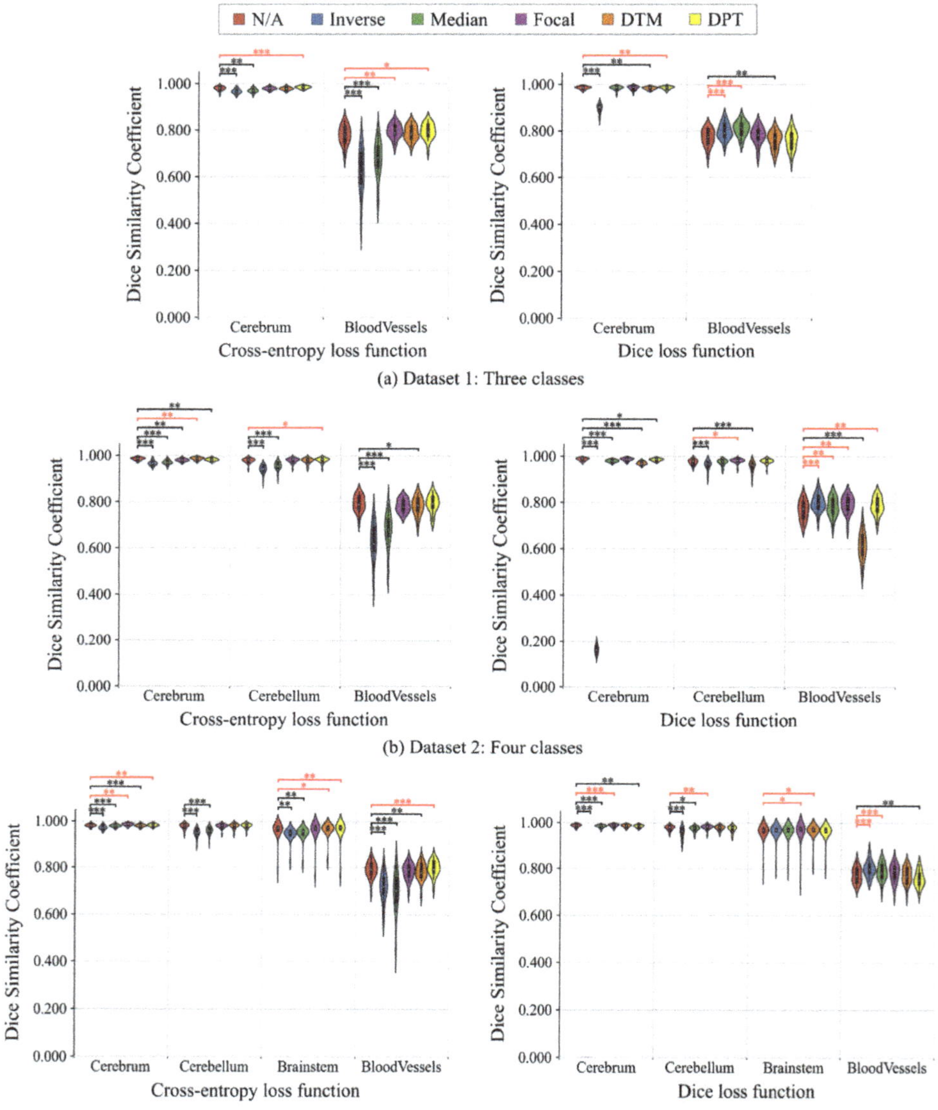

Figure 6. Violin plots of the segmentation results (Dice similarity coefficients) of no weighting (N/A), inverse frequency weighting (Inverse), inverse median frequency weighting (Median), focal weighting (Focal), distance transform map-based weighting (DTM), and distance penalty term-based weighting (DPT) in multi-class segmentation tasks. (**a**) Dataset 1: three classes, (**b**) Dataset 2: four classes, and (**c**) Dataset 3: five classes. Compared with the results of N/A, the significantly worse and better results are shown in black and red, respectively (Wilcoxon signed-rank test, * $p < 0.05$, ** $p < 0.01$, and *** $p < 0.001$, not adjusted for multiplicity).

Figure 7 visualizes an example of the segmentation results in the five-class segmentation task. It shows the false positive and false negative labels as well as the predicted labels. False positives were likely to appear around the surface of the cerebrum, cerebellum, and brainstem, while false negatives tended to appear in the upper part of blood vessels. As for the cross-entropy loss function, Inverse and Median reduced the false negatives, but more than that, they greatly increased the false positives. Focal worked well for a reduction in the false positives, although it did not reduce the false negatives. The results of the distance map-based weightings showed that DPT was a little effective in reducing the false positives and false negatives. As for Dice loss function, Inverse reduced the false negatives in blood vessels, although it failed to segment the whole cerebrum. Median worked to reduce the false negatives in blood vessels, as with Inverse. Focal slightly reduced the false positives. DTM and DPT seemed to provide almost the same results as N/A.

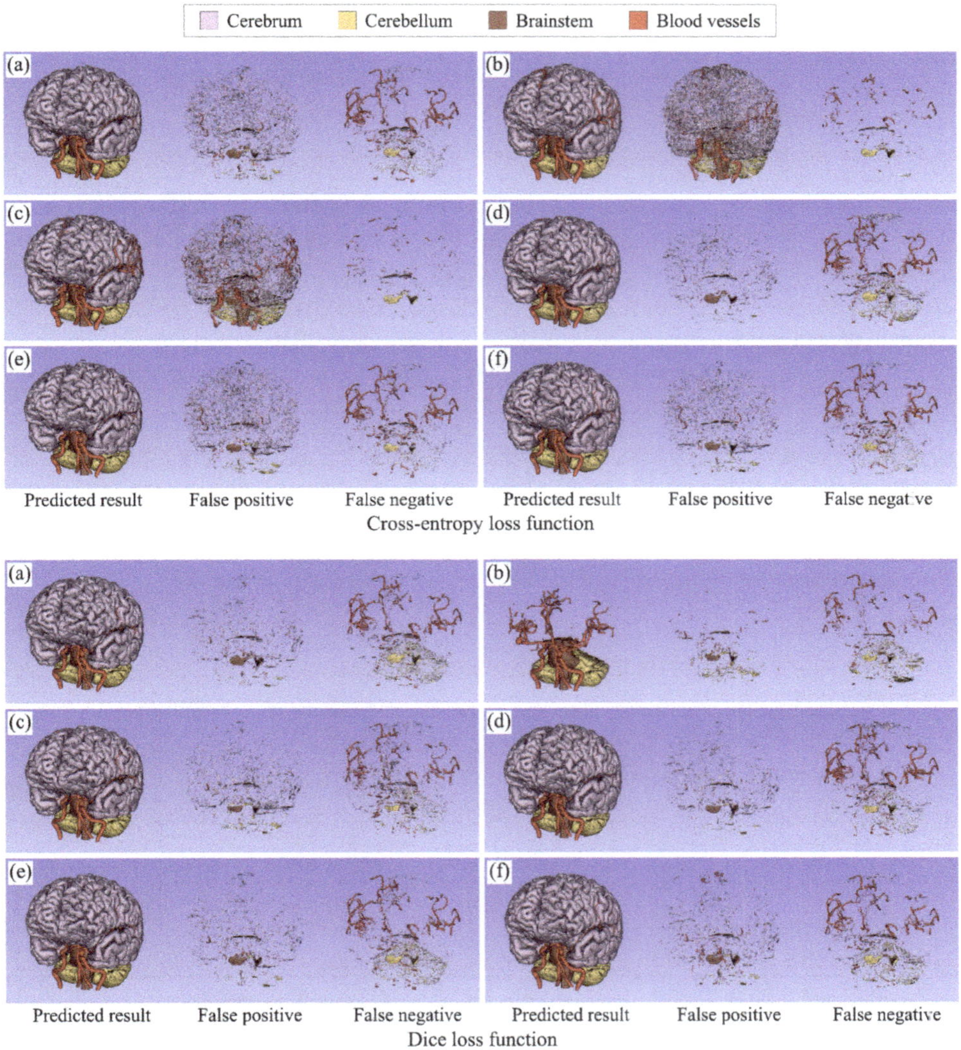

Figure 7. Visualization of the segmentation results in the five-class segmentation task. (**a**) No weighting, (**b**) inverse frequency weighting, (**c**) inverse median frequency weighting, (**d**) focal weighting, (**e**) distance transform map-based weighting, and (**f**) distance penalty term-based weighting. The segmentation results include the predicted results (left), the false positives (middle), and the false negatives (right). Note that in the result of Dice loss function with inverse frequency weighting, there are no true positive voxels in the cerebrum class and most of the background region were overestimated as the cerebrum class, but the false positives and false negatives in the cerebrum class were excluded from the figure for better visualization.

3.3. Rank Scoring

Table 6 indicates the ranking results of loss weightings in the binary- and multi-class segmentation tasks. The distance map-based weightings for cross-entropy loss function and the predictive-probability weighting for Dice loss function tended to have high rank scores in both the binary- and multi-class segmentation tasks. In the binary-class segmentation tasks, the Dice loss function with Focal showed the best ranking result. It actually obtained a high average DSC and SDSC of 92.8% and 93.3%, respectively. Compared with no weighting, it improved the DSC and SDSC values of all tasks by 0.2–8.1% and 0.5–12.5%, respectively. In the multi-class segmentation tasks, the cross-entropy loss function with DPT had the highest rank score, followed by the Dice loss function with Focal. In the five-class segmentation task, DPT achieved the highest average DSC and SDSC values of 93.1% and 94.6%, respectively.

Table 6. Ranking results of no weighting (N/A), inverse frequency weighting (Inverse), inverse median frequency weighting (Median), focal weighting (Focal), distance transform map-based weighting (DTM), and distance penalty term-based weighting (DPT) in (a) binary-class segmentation tasks and (b) multi-class segmentation tasks. The best results are shown in bold. The rank is determined based on the rank scores of segmentation results on all datasets.

(a) Binary-Class Segmentation Tasks

Loss Function	Weighting	Rank Score					Rank
		Dataset 1: Cerebrum	Dataset 2: Cerebellum	Dataset 3: Brainstem	Dataset 4: Blood Vessels	All	
Cross entropy	N/A	5.25	**7.25**	3.25	**6.00**	5.44	4
	Inverse	0.00	2.25	1.25	1.25	1.19	11
	Median	1.50	0.75	0.50	0.75	0.88	12
	Focal	3.50	4.00	6.00	**6.00**	4.88	5
	DTM	4.25	6.25	**6.50**	**6.00**	5.75	2
	DPT	5.5	6.25	4.50	**6.00**	5.56	3
Dice	N/A	2.75	4.00	0.00	2.50	2.31	10
	Inverse	1.75	1.50	3.00	5.50	2.94	8
	Median	1.75	1.00	3.50	3.75	2.50	9
	Focal	**8.5**	4.50	**6.50**	4.75	**6.06**	1
	DTM	4.5	4.25	4.25	1.75	3.69	6
	DPT	5.25	4.00	4.00	0.00	3.31	7

(b) Multi-class segmentation tasks

Loss Function	Weighting	Rank Score				Rank
		Dataset 1: Three Classes	Dataset 2: Four Classes	Dataset 3: Five Classes	All	
Cross entropy	N/A	1.50	5.75	4.13	4.08	6
	Inverse	0.63	0.83	0.81	0.78	12
	Median	1.25	1.92	0.81	1.28	11
	Focal	4.88	4.67	4.19	4.50	4
	DTM	5.63	5.25	3.69	4.64	3
	DPT	**6.75**	6.17	6.63	**6.50**	1
Dice	N/A	4.63	4.58	3.69	4.19	5
	Inverse	2.88	2.17	1.38	1.97	10
	Median	6.00	3.67	2.56	3.69	8
	Focal	3.63	**7.50**	**6.75**	6.31	2
	DTM	4.63	0.67	4.75	3.36	9
	DPT	4.88	4.67	2.44	3.72	7

4. Discussion

We evaluated the effect of loss weightings on the segmentation of the cerebrum, cerebellum, brainstem, and blood vessels from the MR images. From the segmentation results with the non-weighted loss functions, we found that the segmentation errors of the cerebrum, cerebellum, and brainstem, including false positives and false negatives, were concentrated at the edges of them, whereas the segmentation errors of blood vessels, especially false negatives, appeared in the upper part of them. This is probably because the edges of brain parenchyma or the upper blood vessels were variable according to the cases and the FCN was biased toward training image features on easier-to-segment majority regions. Thus, in order to improve the brain structure segmentation, it would be important to make the FCN focus on training image features around the edge of brain parenchyma and in the upper part of blood vessels by loss weightings. We discuss the effect of loss weightings based on the results in the binary- and multi-class segmentation tasks below. Subsequently, we also discuss the limitations of this study.

4.1. Binary-Class Segmentation Tasks

As for the cross-entropy loss function, the class frequency-based weightings (Inverse and Median) greatly increased false positives. They assign a lower uniform weight to the loss of larger-size classes, i.e., background class in the case of binary-class segmentation tasks. They gave a low uniform weight to low-confidence background pixels near the edge of the foreground, which would result in a large increase in false positives on the low-confidence background pixels, although they could also help reduce false negatives. On the other hand, the predictive probability- and the distance map-based weightings tended to improve the surface accuracy of highly imbalanced classes, i.e., the brainstem and blood vessels. Different from the class frequency-based weighting, they assign a different weight to each pixel. Using such pixel-wise weights instead of uniform weights may be appropriate for imbalanced segmentation because FCNs do not focus equally on all the pixels of the same class during training. The predictive-probability-based weighting (Focal) gives higher weights to pixels with lower prediction confidences based on the predictive probability and helps correct pixels misclassified with low prediction confidence, whereas the distance map-based weightings (DTM and DPT) define pixel-wise weights based on the distance from the edge of ground-truth labels and help correct surface segmentation errors. Thus, it is considered that these loss weightings could correct the surface error because pixels around the edge of foreground class were subject to be misclassified with low prediction confidence in the highly imbalanced segmentation tasks.

As for the Dice loss function, the class frequency-based weightings significantly improved the accuracy in the highly imbalanced segmentation tasks, although they did not work well for the cross-entropy loss function. They assigned the weight to both the denominator and numerator for the Dice loss function, which would allow the FCN to reduce false negatives without increasing false positives. The predictive probability-based weighting, which showed the best performance in Table 6, worked well for the low- and middle-level imbalanced segmentation tasks as well as the highly imbalanced segmentation tasks. This can be explained by the fact that the FCN with the Dice loss function had more pixels misclassified with low prediction confidence in the low- and middle-level imbalanced segmentation tasks, compared with that of the cross-entropy loss function. Additionally, the distance map-based weightings tended to improve the surface accuracy in the brain parenchyma segmentation. However, they were ineffective in the segmentation task of blood vessels. As shown in [16], in the case of the segmentation of objects which have variable locations and shapes, they might be able to work stably by using a scheduling strategy, i.e., gradually increasing the weight to the mismatched region with the training epochs.

4.2. Multi-Class Segmentation Tasks

The binary-class segmentation tasks included the class imbalance problem between background and foreground classes, whereas the multi-class segmentation tasks, which deal with two or more foreground classes, included the class imbalance problems not only between background and foreground classes but also among foreground classes.

However, the results in the multi-class segmentation tasks showed similar tendencies to those in the binary-class segmentation tasks, although some of them were affected by the foreground–foreground class imbalance.

The class frequency-based weightings failed to improve the segmentation performance of the FCN with the cross-entropy loss function in any multi-class segmentation tasks because they greatly increased false positives by assigning an extremely low weight to the background pixels. For the Dice loss function, they also worked negatively for the low- and middle-level imbalanced classes. Especially in the five-class segmentation task, Inverse could not segment the cerebrum at all due to the foreground–foreground class imbalance. However, it also provided the best DSC value for blood vessels. Thus, the class frequency-based weightings could work well for only objects with very high imbalance because of their extreme weighting in any segmentation tasks. The predictive probability-based weighting totally worked well for both the cross-entropy and Dice loss functions. These results suggested that despite the foreground–foreground class imbalance, it could enable FCNs to focus on the pixels misclassified with low prediction confidence, i.e., hard-to-segment pixels, by considering the predictive probability. As well, the distance map-based weightings tended to provide good segmentation results for the cross-entropy loss function. In particular, the cross-entropy loss function with DPT achieved the best performance as indicated in Table 6b. However, the distance map-based weightings provided unstable segmentation results for the Dice loss function. In this study, although we designed the Dice loss function with the distance map-based weightings by multiplying the false positive and false negative terms in the denominator by the weights, using a scheduling strategy might make the effect of the distance map-based weightings more stable, as mentioned above.

Therefore, the cross-entropy loss function with DPT and the Dice loss function with Focal achieved relatively high accuracy in any segmentation targets and tasks, but some other weightings outperformed their weightings according to segmentation targets. For example, the Dice loss function with Inverse provided better DSC and SDSC results for blood vessels than that with Focal. Therefore, in this study, we focused on the unary weighted loss functions instead of compound loss functions, but considering the difference of features in loss weightings, the combination of different weighted loss functions might lead to the further improvement of segmentation performance.

4.3. Limitations

For limitations of this work, we adopted the segmentation of brain parenchyma and blood vessels on MRC and MRA images, which is performed as a routine work in our group. However, the effect of loss weightings might depend on segmentation targets and tasks, although the results in this study reflected the features of loss weightings. Considering a wider range of applications, we should test the loss weightings in other brain structure segmentation tasks (e.g., the segmentation of white matter, gray matter, and cerebrospinal fluid on T1-weighted MR images). Second, we used the 2D U-Net architecture to investigate the effect of loss weightings with less hyperparameters. However, we would need to test 3D FCNs with the weighted loss functions, because they have been applied for volumetric brain structure segmentation. Moreover, we set default parameters for loss weightings (e.g., the focusing parameter for focal weighting) based on the previous studies, but tuning such parameters would enable the performance improvement of FCNs. Furthermore, in this study, we focused on segmenting brain structures, including blood vessels, from the MR images of patients with cerebral aneurysms, but considering the clinical practice, it would be desired to automatically detect the location of aneurysms, as in [37], in addition to the segmentation.

5. Conclusions

This paper investigated how the loss weightings work for FCN-based brain structure segmentation on MR images in different class imbalance situations. Using the 2D U-Net with cross-entropy or Dice loss functions as a baseline network, we tested the five loss weightings, which were defined based on class frequency, predictive probability, and distance map, in the binary- and multi-class brain structure segmentation on MRC and MRA images. From the experimental results, we found that the cross-entropy loss function with the distance map-based weightings, especially distance penalty term-based weighting, and the Dice loss function with the predictive probability-based weighting could stably provide good segmentation results. In the binary-class segmentation tasks, the Dice loss function with focal weighting showed the best performance and achieved a high average DSC of 92.8%, whereas in the multi-class segmentation tasks, the cross-entropy loss function with distance penalty term-based weighting provided the best performance. It achieved the highest average DSC of 93.1% in the five-class segmentation task. We also found that their weighted loss functions were relatively robust to the foreground–foreground class imbalance as well as the background–foreground class imbalance. In other words, the experimental results suggested that they could work well in the situations of both binary- and multi-class segmentation. Therefore, it may be effective to use the distance penalty term-based weighting in the cross-entropy loss function and the focal weighting in the Dice loss function. We believe that these findings would help to select weighting strategies for loss functions or design advanced loss weighting strategies.

In future work, for clinical application, we will address the detection and segmentation of a diseased area that is more highly imbalanced, such as a cerebral aneurysm, as well as its surrounding structures, by using the loss weighting strategies. Moreover, we will design compound loss functions (i.e., combination among the loss weightings) and further investigate the effect of them for different brain structure segmentation tasks.

Author Contributions: Conceptualization, T.S. and Y.N.; methodology, T.S. and Y.N.; software, T.S.; validation, all; formal analysis, T.S. and Y.N.; investigation, T.S.; resources, T.S., T.K. (Taichi Kin) and N.S.; data curation, T.S., T.K. (Taichi Kin) and N.S.; writing—original draft preparation, T.S.; writing—review and editing, T.K. (Toshihiro Kawase), S.O. and Y.N.; visualization, T.S.; supervision, N.S. and Y.N.; project administration, N.S. and Y.N.; funding acquisition, T.S., T.K. (Taichi Kin), N.S. and Y.N. All authors have read and agreed to the published version of the manuscript.

Funding: Parts of this research were supported by the Japan Agency for Medical Research and Development (AMED) (Grant Number JP21he1602001h0105) and JSPS KAKENHI (Grant Number 20K20216).

Institutional Review Board Statement: The study was conducted according to the guidelines of the Declaration of Helsinki, and approved by the Ethics Committee of Tokyo Medical and Dental University (protocol code: M2018-190 and date of approval: 29 January 2019).

Informed Consent Statement: Informed consent was obtained from all subjects involved in the study.

Data Availability Statement: Data sharing not applicable.

Conflicts of Interest: The authors declare no conflict of interest.

References

1. González-Villà, S.; Oliver, A.; Valverde, S.; Wang, L.; Zwiggelaar, R.; Lladó, X. A review on brain structures segmentation in magnetic resonance imaging. *Artif. Intell. Med.* **2016**, *73*, 45–69. [CrossRef] [PubMed]
2. Despotovic, I.; Goossens, B.; Philips, W. MRI segmentation of the human brain: Challenges, methods, and applications. *Comput. Math. Methods Med.* **2015**, *2015*, 450341. [CrossRef] [PubMed]
3. Long, J.; Shelhamer, E.; Darrell, T. Fully convolutional networks for semantic segmentation. In Proceedings of the IEEE Conference on Comput Vision and Pattern Recognition, Boston, MA, USA, 7–12 June 2015; IEEE: NW Washington, DC, USA, 2015; pp. 3431–3440.
4. Ronneberger, O.; Fischer, P.; Brox, T. U-net: Convolutional networks for biomedical image segmentation. In Proceedings of the International Conference on Medical Image Computing and Computer-Assisted Intervention, Munich, Germany, 5–9 October 2015; LNCS 9351. pp. 234–241.
5. Çiçek, Ö.; Abdulkadir, A.; Lienkamp, S.S.; Brox, T.; Ronneberger, O. 3D U-Net: Learning dense volumetric segmentation from sparse annotation. In Proceedings of the International Conference on Medical Image Computing and Computer-Assisted Intervention, Athens, Greece, 17-21 October 2016; LNCS 9351. pp. 424–432.

6. Bernal, J.; Kushibar, K.; Asfaw, D.S.; Valverde, S.; Oliver, A.; Marti, R.; Lladó, X. Deep convolutional neural networks for brain image analysis networks for brain image analysis on magnetic resonance imaging: A review. *Artif. Intell. Med.* **2019**, *95*, 64–81. [CrossRef] [PubMed]
7. Buda, M.; Maki, A.; Mazuroqski, M.A. A systematic study of the class imbalance problem in convolutional neural networks. *Neural Netw.* **2018**, *106*, 249–259. [CrossRef] [PubMed]
8. Zhou, T.; Ruan, S.; Canu, S. A review: Deep learning for medical image segmentation using multi-modality fusion. *Array* **2019**, *3*, 100004. [CrossRef]
9. Jang, J.; Eo, T.J.; Kim, M.; Choi, N.; Han, D.; Kim, D.; Hwang, D. Medical image matching using variable randomized undersampling probability pattern in data acquisition. In Proceedings of the 2014 International Conference on Electronics, Information and Communications, Kota Kinabalu, Malaysia, 15–18 January 2014; pp. 1–2.
10. Chawla, N.V.; Bowyer, K.W.; Hall, L.O.; Kegelmeyer, W.P. SMOTE: Synthetic minority over-sampling technique. *J. Artif. Intell. Res.* **2002**, *16*, 321–357. [CrossRef]
11. Milletari, F.; Navab, N.; Ahmadi, S.A. V-net: Fully convolutional neural networks for volumetric medical image segmentation. In Proceedings of the Fourth International Conference on 3D Vision, Stanford, CA, USA, 25–28 October 2016; IEEE: NW Washington, DC, USA, 2016; pp. 566–571.
12. Drozdzal, M.; Vorontsov, E.; Chartrand, G.; Kadoury, S.; Pal, C. The importance of skip connections in biomedical image segmentation. In *Deep Learning and Data Labeling for Medical Applications*; Springer: Cham, Switzerland, 2016; LNCS 10008; pp. 179–187.
13. Rahman, M.A.; Wang, Y. Optimizing intersection-over-union in deep neural networks for image segmentation. In Proceedings of the International Symposium on Visual Computing, Las Vegas, NV, USA, 12–14 December 2016; LNCS 10072. pp. 234–244.
14. Berman, M.; Triki, A.R.; Blaschko, M.B. The lovász-softmax loss: A tractable surrogate for the optimization of the intersection-over-union measure in neural networks. In Proceedings of the IEEE Conference on Computer Vision and Pattern Recognition, Salt Lake City, UT, USA, 18–22 June 2018; pp. 4413–4421.
15. Wong, K.C.L.; Moradi, M.; Tang, H.; Syeda-Mahmood, T. 3D segmentation with exponential logarithmic loss for highly unbalanced object sizes. In Proceedings of the International Conference on Medical Image Computing and Computer-Assisted Intervention, Granada, Spain, 16–20 September 2018; LNCS 11072. pp. 612–619.
16. Kervadec, H.; Bouchtiba, J.; Desrosiers, C.; Granger, E.; Dolz, J.; Ayed, I.B. Boundary loss for highly unbalanced segmentation. *Med. Image Anal.* **2019**, *67*, 101851. [CrossRef] [PubMed]
17. Karimi, D.; Salcudean, S.E. Reducing the Hausdorff distance in medical image segmentation with convolutional neural networks. *IEEE Trans. Med. Imaging* **2020**, *39*, 499–513. [CrossRef] [PubMed]
18. Eigen, D.; Fergus, R. Predicting depth, surface normal and semantic labels with a common multi-scale convolutional architecture. In Proceedings of the IEEE International Conference on Computer Vision, Santiago, Chile, 7–13 December 2015; IEEE: NW Washington, DC, USA, 2015; pp. 2650–2658.
19. Lin, T.Y.; Goyal, P.; Girshick, R.; He, K.; Dollár, P. Focal loss for dense object detection. In Proceedings of the IEEE International Conference on Computer Vision, Venice, Italy, 22–29 October 2017; IEEE: NW Washington, DC, USA, 2017; pp. 2980–2988.
20. Caliva, F.; Iriondo, C.; Martinez, A.M.; Majumdar, S.; Pedoia, V. Distance map loss penalty term for semantic segmentation. In Proceedings of the 2nd International Conference on Medical Imaging with Deep Learning, London, UK, 8–10 July 2019; pp. 1–5.
21. Salehi, S.S.M.; Erdogmus, D.; Gholipour, A. Tversky loss function for image segmentation using 3D fully convolutional deep networks. In Proceedings of the International Workshop on Machine Learning in Medical Imaging, Quebec City, QC, Canada, 10 September 2017; LNCS 10541. pp. 379–387.
22. Hashemi, S.R.; Salehi, S.S.M.; Erdogmus, D.; Prabhu, S.P.; Warfield, S.K.; Gholipour, A. Asymmetric loss functions and deep densely-connected networks for highly-imbalanced medical image segmentation: Application to multiple sclerosis lesion detection. *IEEE Access* **2018**, *7*, 1721–1735. [CrossRef] [PubMed]
23. Guerrero-Pena, F.A.; Fernandez, P.D.M.; Ren, T.I.; Yui, M.; Rothenberg, E.; Cunha, A. Multiclass weighted loss for instance segmentation of cluttered cells. In Proceedings of the 25th IEEE International Conference on Image Processing, Athens, Greece, 7–10 October 2018; pp. 2451–2455.
24. Sudre, C.H.; Li, W.; Vercauteren, T.; Ourselin, S.; Cardoso, M.J. Generalised Dice overlap as a deep learning loss function for highly unbalanced segmentations. In Proceedings of the Deep Learning in Medical Image Analysis and Multimodal Learning for Clinical Decision Support, Québec City, QC, Canada, 14 September 2017; Springer: Cham, Switzerland, 2017; LNCS 10553; pp. 240–248.
25. Li, X.; Sun, X.; Meng, Y.; Liang, J.; Wu, F.; Li, J. Dice loss for data-imbalanced NLP tasks. In Proceedings of the 58th Annual Meeting of the Association for Computational Linguistics, Online, 5–10 July 2020; Association for Computational Linguistics: Stroudsburg, PA, USA, 2020; pp. 465–476.
26. Ma, J.; Chen, J.; Ng, M.; Huang, R.; Li, Y.; Li, C.; Yang, X.; Martel, A.L. Loss odyssey in medical image segmentation. *Med. Image Anal.* **2021**, *71*, 102035. [CrossRef] [PubMed]
27. Ma, J.; Wei, Z.; Zhang, Y.; Wang, Y.; Lv, R.; Zhu, C.; Chen, G.; Liu, J.; Peng, C.; Wang, L.; et al. How distance transform maps boost segmentation CNNs: An empirical study. *Med. Imaging Deep Learn.* **2020**, *121*, 479–492.
28. Yeung, M.; Sala, E.; Schönlieb, C.B.; Rundo, L. Unified Focal loss: Generalising Dice and cross entropy-based losses to handle class imbalanced medical image segmentation. *arXiv* **2021**, arXiv:2102.04525, Preprint.

29. Huo, Y.; Xu, Z.; Xiong, Y.; Aboud, K.; Parvathaneni, P.; Bao, S.; Bermudez, C.; Resnick, S.M.; Cutting, L.E.; Landman, B.A. 3D whole brain segmentation using spatially localized atlas network tiles. *NeuroImage* **2019**, *194*, 105–119. [CrossRef] [PubMed]
30. Taghanaki, S.A.; Zheng, Y.; Zhou, S.K.; Georgescu, B.; Sharma, P.; Xu, D.; Comaniciu, D.; Hamarneh, G. Combo loss: Handling input and output imbalance in multi-organ segmentation. *Comput. Med. Imaging Graph.* **2019**, *75*, 24–33. [CrossRef] [PubMed]
31. Zhu, W.; Huang, Y.; Zeng, L.; Chen, X.; Liu, Y.; Qian, Z.; Du, N.; Fan, W.; Xie, X. AnatomyNet: Deep learning for fast and fully automated whole-volume segmentation of head and neck anatomy. *Med. Phys.* **2018**, *46*, 576–589. [CrossRef] [PubMed]
32. Xue, Y.; Tang, H.; Qiao, Z.; Gong, G.; Yin, Y.; Qian, Z.; Huang, X. Shape-aware organ segmentation by predicting signed distance maps. *AAAI Conf. Artif. Intell.* **2020**, *34*, 12565–12572. [CrossRef]
33. Kingma, D.P.; Ba, J.L. Adam: A method for stochastic optimization. *arXiv* **2014**, arXiv:1412.6980, Preprint.
34. Nikolov, S.; Blackwell, S.; Zverovitch, A.; Mendes, R.; Livne, M.; De Fauw, J.; Patel, Y.; Meyer, C.; Askham, H.; Romera-Paredes, B.; et al. Deep learning to achieve clinically applicable segmentation of head and neck anatomy for radiotherapy. *arXiv* **2018**, arXiv:1809.04430, Preprint.
35. DeepMind. Github: Library to Compute Surface Distance Based Performance Metrics for Segmentation Tasks. Available online: https://github.com/deepmind/surface-distance (accessed on 28 April 2021).
36. Antonelli, M.; Reinke, A.; Bakas, S.; Farahani, K.; Kopp-Schneider, A.; Landman, B.A.; Litjens, G.; Menze, B.; Ronneberger, O.; Summers, R.M.; et al. The Medical Segmentation Decathlon. *arXiv* **2021**, arXiv:2106.05735, Preprint.
37. Conti, V.; Militello, C.; Rundo, L.; Vitabile, S. A novel bio-inspired approach for high-performance management in service-oriented networks. *IEEE Trans. Emerg. Top. Comput.* **2020**. [CrossRef]

healthcare

MDPI

Article

Signal and Texture Features from T2 Maps for the Prediction of Mild Cognitive Impairment to Alzheimer's Disease Progression

Alejandro I. Trejo-Castro [1], Ricardo A. Caballero-Luna [2], José A. Garnica-López [2], Fernando Vega-Lara [2] and Antonio Martinez-Torteya [3,*]

1 Escuela de Ingeniería y Ciencias, Tecnológico de Monterrey, Monterrey 64849, Mexico; a00818219@itesm.mx
2 Programa de Ingeniería Biomédica, Universidad de Monterrey, San Pedro Garza García 66238, Mexico; ricardo.caballerol@udem.edu (R.A.C.-L.); jose.garnical@udem.edu (J.A.G.-L.); fernando.vega@udem.edu (F.V.-L.)
3 Departamento de Ingeniería, Universidad de Monterrey, San Pedro Garza García 66238, Mexico
* Correspondence: antonio.martinez@udem.edu; Tel.: +52-(81)-8215-1438

Abstract: Early detection of Alzheimer's disease (AD) is crucial to preserve cognitive functions and provide the opportunity for patients to enter clinical trials. In recent years, some studies have reported that features related to the signal and texture of MRI images can be an effective biomarker of AD. To test these claims, a study was conducted using T2 maps, a sequence not previously studied, of 40 patients with mild cognitive impairment (MCI) from the Alzheimer's Disease Neuroimaging Initiative database, who either progressed to AD (18) or remained stable (22). From these maps, the mean value and absolute difference of 37 signal and texture imaging features for 40 contralateral pairs of regions were measured. We used seven machine learning methods to analyze whether, by adding these imaging features to the neuropsychological studies currently used for diagnosis, we could more accurately identify patients who will progress to AD. The predictive models improved with the addition of signal and texture features. Additionally, features related to the signal and texture of the images were much more relevant than volumetric ones. Our results suggest that contralateral signal and texture features should be further investigated as potential biomarkers for the prediction of AD.

Keywords: ADNI; Alzheimer's disease; mild cognitive impairment; MRI biomarkers; signal; T2 maps; texture

Citation: Trejo-Castro, A.I.; Caballero-Luna, R.A.; Garnica-López, J.A.; Vega-Lara, F.; Martinez-Torteya, A. Signal and Texture Features from T2 Maps for the Prediction of Mild Cognitive Impairment to Alzheimer's Disease Progression. *Healthcare* **2021**, *9*, 941. https://doi.org/10.3390/healthcare9080941

Academic Editor: Mahmudur Rahman

Received: 18 June 2021
Accepted: 19 July 2021
Published: 26 July 2021

1. Introduction

Neurodegenerative diseases are a common and growing cause of mortality and morbidity, in which structural and chemical changes in the nervous system lead to the loss of neurons and progressive decline in multiple areas of functioning, including cognition, communication skills, and the ability to carry out daily activities [1]. Alzheimer's disease (AD) is the most common of these conditions, having an accumulation of amyloid-beta protein fragments outside neurons and hyperphosphorylated tau tangles within neurons as its hallmark pathology [2]. Over 110 years ago, Alois Alzheimer first described the disease that bears his name, characterizing it by deficits in memory, impairment in verbal communication, visuospatial disorders, and changes in personality such as depression [3,4]. By 2010, 35.6 million people worldwide had dementia; 60–80% of these cases were attributed to AD. However, the most alarming aspect is that a 225% increase in the number of patients with this disease is expected worldwide by mid-century, forcing countries to allocate more resources to this population and expanding the need for more caregivers [5].

Given this scenario, emphasis has been placed on predicting who will experience AD, since an early diagnosis allows patients to enroll in clinical trials, which could help to slow the progression of the disease, better preserve cognitive functions, and provide economic and emotional benefits for both caregivers and patients [6–8]. For this reason, since 1988, with Barry Reisberg's mild cognitive impairment definition (MCI), researchers have focused on distinguishing subjects with MCI who progress to AD from those who do not. MCI represents a transitional state between normal cognition and dementia, as it indicates cognitive deficits, including impairments that could be related to memory (amnestic MCI) or other cognitive abilities (non-amnestic MCI); even though not all MCI subjects progress to AD and some eventually revert to cognitive normalcy, subjects with MCI have an increased risk of developing AD [9,10].

The first criteria for diagnosing AD was created in 1984 by the National Institute of Neurological and Communicative Disorders and Stroke (NINCDS) and the Alzheimer's Disease and Related Disorders Association (ADRDA); since then, the criteria have not changed substantially [11]. Briefly, they consist of neuropsychological tests that measure cognitive decline and symptoms of the disease, such as the Mini-Mental State Examination (MMSE) to detect cognitive decline [12], Boston Naming Test (BNT) to measure language disorders [13], Geriatric Depression Scale (GDS) to identify depression [14], and the Alzheimer's Disease Assessment Scale-Cognitive Subscale (ADAS) to assess cognitive and non-cognitive function characteristics in people with AD [15]. In 2011, these criteria were revised due to the advances in the understanding of the disease. It was concluded that an AD diagnosis needed the same evidence as with the previous criteria, with the addition of one in five proposed biomarkers as potential support, one of them being an atrophy in the temporal lobes visualized by magnetic resonance imaging (MRI) [16].

Recently, signal- and texture-related features extracted from MRI scans and selected machine learning techniques have emerged as possible novel markers of AD [17]. In addition, studies of the progression of AD showed that highly asymmetrical contralateral hippocampi and amygdala may indicate an early and accelerated deterioration [18].

This work focuses on the study of the MCI to AD progression in the interest of achieving early detection of AD. Previously, we have proposed new biomarkers for AD from neuropsychological data, laboratory assays, and signal and texture features from T1-sequences, such as the magnetization-prepared rapid acquisition with gradient echo (MP-RAGE) [19]. Subsequently, we analyzed in a preliminary conference paper signal- and texture-related features from hippocampal T2 maps, finding 11 features significantly different between stable and non-stable MCI subjects. Volumetric information was non-significant, and all but one of the machine learning methods improved their accuracy for AD prediction by adding the signal- and texture-related features to the neuropsychological studies [20]. It is worth commenting that, to our knowledge, T2 maps have not been studied by other researchers for this purpose. Nevertheless, they have been used to detect other diseases such as hepatic fibrosis and acute or chronic heart failure [21,22].

The main objective in this study was to determine the predictive power of signal- and texture-related features extracted from T2 maps using all the 40 contralateral pairs available in ADNI images using both a univariate and a multivariate analysis between patients with MCI who progress to AD and those who remain stable.

2. Materials and Methods

2.1. Data

Data used in the preparation of this article were obtained from the ADNI database (adni.loni.usc.edu). The ADNI was launched in 2003 as a public–private partnership, led by Principal Investigator Michael W. Weiner, MD. The primary goal of ADNI has been to test whether serial MRI, Positron Emission Tomography (PET), other biological markers, and clinical and neuropsychological assessment can be combined to measure the progression of MCI and early AD. For up-to-date information, see www.adni-info.org.

Dual fast spin-echo images, one weighted to proton density (PD) and one to T2, and MP-RAGE images available up to April 2020 were retrieved from ADNI [23]. Additionally, segmentation maps for the MP-RAGE images generated through automatic whole-brain segmentations using multi-atlas propagation with enhanced registration were also downloaded [24].

2.2. Subject Inclusion

The experiment included only the baseline information from subjects with a baseline MCI diagnosis, between 70 and 80 years old, with available sex and years of education information and who also had the aforementioned images and segmentation map available. One subject was eliminated from the study due to poor image quality. From these, the 18 subjects who had their first AD diagnosis 2 years after their baseline visit were regarded as progressers (MCIp), while the 22 subjects who never had an AD diagnosis and participated in the study for at least 5 years were labeled as stable (MCIs). Patients who did not meet either the MCIp or the MCIs criteria were excluded from the study.

Table 1 details the demographic characteristics of the population. There was no significant difference in age and years of education between groups when tested using the Wilcoxon rank-sum test nor a significant difference in male/female proportion under a chi-squared test. MCI and AD diagnoses were determined as defined by ADNI guidelines [25].

Table 1. Demography of the population.

Group of Study	Total	MCIs	MCIp	*p*-Value
Subjects (males)	40 (32)	22 (18)	18 (14)	1.000
Years of age	75.3 ± 3.0	75.3 ± 3.2	75.2 ± 2.9	0.924
Years of education	15.7 ± 3.0	15.8 ± 3.1	15.6 ± 2.9	0.879

Mean value ± standard deviation; *p*-value of the chi-squared test (male/female proportion) or Wilcoxon rank-sum test (age and education).

2.3. MRI Processing

After the three types of images and the segmentation map were downloaded from the ADNI database for every subject, T2 maps were generated, and their 83 anatomical regions were segmented. In order to generate the T2 maps, we used the dual fast spin-echo images, namely, the PD- and T2-weighted images, each with a different echo time. As shown in (1), the T2 value for the *i*th voxel can be calculated by fitting the measured signal intensity S at each echo time TE to a mono-exponential decay function [26]:

$$S_a(i) = S_0 e^{-TE_a/T2(i)} \tag{1}$$

where S_0 is the signal intensity at zero TE. From there, and working with the signal from the PD- and T2-weighted images (S_a and S_b, respectively), we obtain (2)

$$T2(i) = \frac{TE_b - TE_a}{\ln(S_a(i)) - \ln(S_b(i))} \tag{2}$$

where T2(i) is the T2 value for the ith voxel, TE_b and TE_a represent the echo time of the T2- and PD-weighted images, respectively, and $S_a(i)$ and $S_b(i)$ represent the signal value of the ith voxel for the PD- and T2-weighted images, respectively.

To extract relevant features, it was necessary to perform a segmentation of the T2 maps. The segmentations maps downloaded from ADNI were specifically constructed for the MP-RAGE images; therefore, a registration process was required to apply these segmentation maps to the T2 maps. Spin-echo and MP-RAGE images were obtained in the same imaging session; hence, images were almost identical except for differences caused by any head movement. Using ITK [27], we performed a rigid registration between the T2 maps and their MP-RAGE counterparts using the itkVersorRigid3DTransform function with the Mattes mutual information metric, a regular step gradient descent optimizer, and a linear interpolator. Quality of the registration was confirmed visually.

2.4. Feature Extraction

Each anatomical region was measured for a set of 38 features: volume, 28 features related to signal distribution (e.g., energy, kurtosis, and skewness), and 9 texture-related features (e.g., mass scatter and compactness of the intensity projection map). Then, we proceeded to calculate the absolute difference and mean of each signal and texture measurement between contralateral regions. The final database consisted of the difference and mean features of 40 contralateral pairs, and 3 regions had no counterpart: brainstem (spans the midline), corpus callosum, and third ventricle. In total, there were 2960 features related to either the signal or texture of the T2 maps, plus the volume of the 83 regions.

2.5. Statistical Analysis

We performed a univariate and a multivariate analysis. For the former, we compared the features with the Wilcoxon rank-sum test. Then, to control the false-positive detection rate and adjust the p-values for multiple comparisons, a Benjamini–Hochberg procedure was used, and q-values were obtained [28]. A feature was determined significantly different between groups if a q-value lower than 0.05 was found.

We used FRESA.CAD Binary Classification Benchmarking, an R package that performs systematic comparisons between machine learning methods, to perform the multivariate analysis [29–31]. The methods included were: bootstrapped stage-wise model selection (BSWiMS), k-nearest neighbors (KNN) with BSWiMS features, least absolute shrinkage and selection operator (LASSO), random forest (RF), recursive partitioning and regression trees (RPART), support vector machines (SVM) with minimum-Redundancy-Maximum-Relavance (mRMR) method, and the ensemble of these methods (ENS). We performed a 100-fold cross-validation strategy, where the training sample was constructed by randomly selecting 80% of the subjects while the rest were kept for validation. For this study, we focused mainly on accuracy, sensitivity, specificity, balanced error, and the area under the receiver operating characteristic curve (ROC AUC) with a 95% confidence interval (CI).

Furthermore, in order to find the features with the highest predictive potential, we evaluated the ability of several feature-selection algorithms—integrated discrimination improvement (IDI), Kendall correlation, LASSO, mRMR, net reclassification improvement (NRI), RF, RPART, t-student test, and Wilcoxon test—in their ability to select the best set of features for several classifiers: KNN, naïve Bayes, nearest centroid with normalized root sum square distance and Spearman correlation distance, RF, and SVM. These classifiers were analyzed using the same cross-validations strategy.

2.6. Experiment Design

In order to find the predictive power of the features related to signal and texture, we performed two different experiments. The first one included the total scores from eight neuropsychology studies that are used for the diagnosis of AD, namely, MMSE, BNT, GDS, ADAS with 11 items (ADAS-11), and ADNI summary scores related with executive function, visuospatial functioning, language, and memory [32–34]. The second experiment included these 8 scores in addition to the most significant features in the univariate analysis extracted from the T2 maps.

3. Results

3.1. Univariate Analysis for Neuropsychological Studies and Volumes

The univariate analysis for the eight neuropsychological tests yielded three of them as significant: ADNI memory test (p-value = 7.765×10^{-4}), ADAS-11 (p-value = 0.004) and MMSE (p-value = 0.025). Only the first two remained significant after the Benjamini–Hochberg procedure was run with the rest of the neuropsychological tests. Regarding the volumetric information, only one feature was found to be significant under the Wilcoxon rank-sum test: the right amygdala (p-value = 0.034).

3.2. Univariate Analysis for Signal and Texture Features

Of the 2960 signal and texture features, 140 were significantly different between classes, 89 mean values and 51 absolute differences. However, after adjusting for multiple comparisons using the Benjamini–Hochberg method, none of these remained significant.

Table 2 shows the 25 features with the lowest p-values. It is worth noting that 11 of them belong to the hippocampus.

Table 2. Significant features by their q-value of β'_{1j}.

Rank	Feature	Modality	Brain Region	p-Value
1	Value at 25% [a]	Difference	Superior frontal gyrus	1.52×10^{-4}
2	Mass Scatter YY [b]	Difference	Hippocampus	5.51×10^{-4}
3	σ at 90% central value [a]	Mean	Hippocampus	0.001
4	ICV at 90% central value [a]	Mean	Hippocampus	0.002
5	Probability of value being lower than 2σ [a]	Difference	Lateral ventricle, temporal horn	0.002
6	Entropy [a]	Mean	Hippocampus	0.002
7	Energy [a]	Mean	Hippocampus	0.002
8	Value at 75% [a]	Mean	Hippocampus	0.003
9	Skewness [a]	Mean	Subcallosal area	0.004
10	Energy [a]	Mean	Subcallosal area	0.004
11	Mass Scatter YY [b]	Difference	Cerebellum	0.004
12	Value at 5% [a]	Difference	Superior frontal gyrus	0.004
13	μ signal [a]	Mean	Hippocampus	0.004
14	μ at 90% central value [a]	Mean	Hippocampus	0.005
15	Entropy [a]	Mean	Subcallosal area	0.005
16	Value at 95% [a]	Mean	Hippocampus	0.006
17	Kurtosis [a]	Mean	Hippocampus	0.006
18	Precision range [a]	Mean	Hippocampus	0.006
19	Precision range [a]	Mean	Insula	0.006
20	Value at 99.99% [a]	Difference	Anterior orbital gyrus	0.007
21	ICV at 90% central value [a]	Mean	Lateral occipitotemporal gyrus, gyrus fusiformis	0.008
22	Probability of value being greater than 3σ [a]	Mean	Cingulate gyrus, posterior part	0.008
23	Energy [a]	Mean	Cingulate gyrus, posterior part	0.008
24	Value at 25% [a]	Difference	Putamen	0.008
25	Probability of value being greater than 3σ [a]	Difference	Lateral ventricles, temporal horn	0.008

[a] Features related to the signal distribution of the image; [b] Features related to the texture of the image.

3.3. Multivariate Analysis for Neuropsychological Studies

The prediction results for the different machine learning techniques considering only neuropsychological tests are shown in Table 3. The features most frequently found in the predictive models were the total scores from the ADNI memory test and the ADAS-11. The machine learning technique with the best results was LASSO, with an accuracy of 0.675 and an ROC AUC of 0.727. However, it is worth noting that confidence intervals overlap, implying no real difference between methods.

Table 3. Results for the multivariate analysis with neuropsychological tests.

	Accuracy		ROC AUC		Specificity		Sensitivity		Balanced Error	
Technique	Mean	CI	Mean	CI	Mean	CI	Mean	CI	Mean	CI
BSWIMS	0.500	0.338–0.662	0.558	0.380–0.736	0.273	0.107–0.502	0.778	0.524–0.936	0.475	0.341–0.613
ENS	0.650	0.483–0.794	0.649	0.472–0.826	0.636	0.407–0.828	0.667	0.410–0.867	0.347	0.198–0.513
KNN	0.625	0.458–0.773	0.674	0.504–0.845	0.500	0.282–0.718	0.778	0.524–0.936	0.361	0.225–0.509
LASSO	0.675	0.509–0.814	0.727	0.567–0.888	0.636	0.407–0.828	0.722	0.465–0.903	0.321	0.177–0.469
RF	0.650	0.483–0.794	0.657	0.507–0.806	0.591	0.364–0.793	0.722	0.465–0.903	0.343	0.200–0.494
RPART	0.650	0.483–0.794	0.638	0.483–0.793	0.682	0.451–0.861	0.611	0.357–0.827	0.353	0.208–0.506
SVM	0.650	0.483–0.794	0.652	0.499–0.804	0.636	0.407–0.828	0.667	0.410–0.867	0.350	0.201–0.504

3.4. Multivariate Analysis for Neuropsychological Studies and Imaging Features

In order to include only relevant features in the selection pool to be used for each classifier, we proceeded to take the most relevant characteristics, that is, those with the lowest p-value of the univariate analyses. Twelve volumes (~15%; 12/83) and 148 features related to signal and texture (~5%; 148/2960) were considered. All eight neuropsychological tests were also included.

Table 4 shows the results obtained with each of the seven machine learning techniques for the experiment with neuropsychological information, volumetric information, and signal- and texture-related information. Comparing those results with the ones found in Table 3, it can be seen that all methods had a higher average score in accuracy and ROC AUC, except for RPART. Furthermore, we can notice that the specificity, sensitivity, and balanced error were improved. Sensitivity measures the proportion of positives that are correctly identified, and specificity measures the proportion of negatives that are correctly identified. Figure 1 shows the ROC of the most relevant machine learning methods for this experiment.

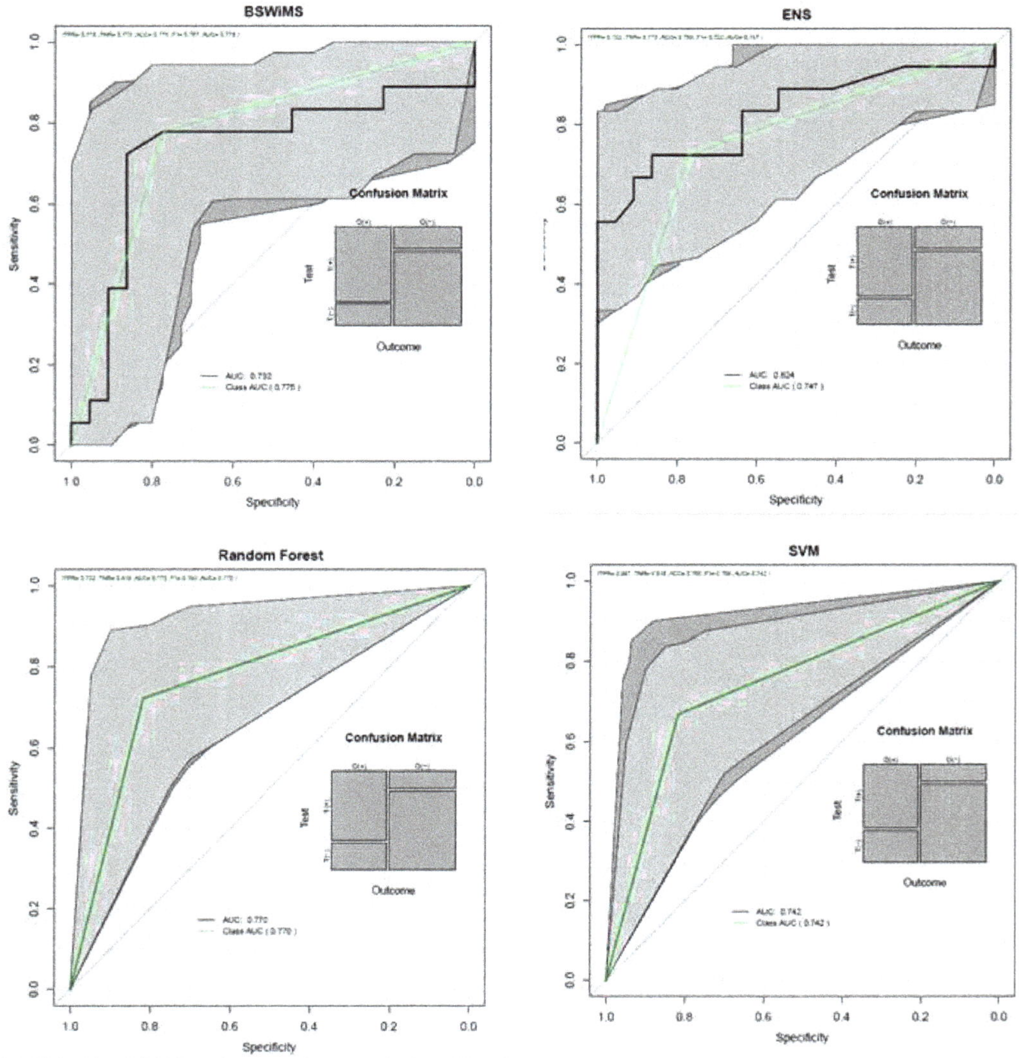

Figure 1. ROC AUC curves on the imaging features and neuropsychological test scores analysis.

Table 4. Results for neuropsychological and imaging features experiment.

Technique	Accuracy		ROC AUC		Specificity		Sensitivity		Balanced Error	
	Mean	CI	Mean	CI	Mean	CI	Mean	CI	Mean	CI
BSWIMS	0.775	0.615–0.892	0.732	0.553–0.911	0.773	0.546–0.922	0.778	0.524–0.936	0.223	0.100–0.359
ENS	0.750	0.588–0.873	0.824	0.681–0.968	0.773	0.546–0.922	0.722	0.465–0.903	0.250	0.124–0.398
KNN	0.675	0.509–0.814	0.721	0.550–0.892	0.727	0.498–0.893	0.611	0.357–0.827	0.330	0.191–0.482
LASSO	0.675	0.509–0.814	0.773	0.610–0.936	0.636	0.407–0.828	0.722	0.465–0.903	0.321	0.177–0.477
RF	0.775	0.615–0.892	0.770	0.636–0.905	0.818	0.597–0.948	0.722	0.465–0.903	0.225	0.101–0.360
RPART	0.500	0.338–0.662	0.513	0.348–0.677	0.454	0.244–0.678	0.555	0.308–0.785	0.499	0.343–0.653
SVM	0.750	0.588–0.873	0.742	0.603–0.882	0.818	0.597–0.948	0.667	0.410–0.867	0.255	0.127–0.401

As previously mentioned, we were also interested in finding out which specific features were more relevant in predicting the progression from MCI to AD. From the nine feature selection methods that were compared, the absolute difference in the Mass Scatter YY in the hippocampus, a feature related to the texture of the T2 map, was found among the six most frequent features in all of them. That is, after all feature selection methods were paired with each classifier, the frequency in which each feature was selected in the final model was computed, and this particular feature was at least the sixth most frequently selected feature every time. Similarly, the absolute difference of the value at 25% in the superior frontal gyrus, a feature related to the signal of the T2 map, was in the top-six in eight of the nine feature selection methods. Additionally, ADNI's memory test and ADAS-11 were in the top-6 in seven and six methods, respectively. Regarding volumetric information, only in the RPART method were volumes found within the 50 most frequent features. The RPART methods used on average 9.67 features per model.

4. Discussion

The present study showed that the signal and texture features extracted from T2 maps could be used in conjunction with information from neuropsychological studies for the prediction of AD. To reach this conclusion, we compared the accuracy, sensitivity, specificity, balanced error, and ROC AUC for each of the different machine learning techniques between the experiment without imaging information and the one that included it. In general, and for all metrics, there was an improvement in the different techniques by adding this information. Another important aspect to highlight is that the presence of volumes in the prediction models was inconsequential.

In a review of MRI texture analyses with machine learning techniques [17], many studies performed classification and prediction of AD. Even though these studies included a greater number of subjects, the vast majority of them focused specifically on the hippocampal region and used only one machine learning technique. Additionally, they used T1-sequences, while this study focused on T2 maps of the whole brain segmented into 40 contralateral regions. Furthermore, we were able to pinpoint specific features by performing an exhaustive feature selection analysis.

The reduction in the volume of the hippocampus was one of the first biomarkers for AD classification [35]. Later, studies have reported that the texture of the hippocampus compared to its volume predicts earlier and more effectively the progression to AD [20,36,37]. In this study, the contralateral difference in a texture-related feature measured in the hippocampi was the most frequently selected feature, and several other signal- and texture-related features from the hippocampus were found to be most relevant under the univariate analysis.

However, we were able to identify other regions of the brain as potential sources for novel biomarkers of the MCI to AD progression process. For example, the contralateral difference in a signal-related feature measured in the superior frontal gyrus was the second most frequently selected feature, and that same feature yielded the lowest p-value when the univariate analysis was run. Similarly, signal- and texture-related features measured in the lateral ventricle, the subcallosal area, and cerebellum had some of the lowest p-values from the univariate analysis and were found to be frequently selected by the different feature selection methods.

This study has several limitations; for example, we only focused on 28 features related to the signal distribution and nine to the texture of the image. However, the results we obtained motivate us to follow the recommendations of The Image Biomarker Standardization Initiative [38] in search for features that can improve our models. Additionally, the inclusion criteria forced us to work with a small population, a potential cause for the lack of significant features after the *p*-value correction in the univariate analysis; we intend to run further experiments with a larger dataset derived from more relaxed inclusion criteria. Lastly, we believe this work drives further analyses and experimentation, the most important being the inclusion of information from two different MRI sequences to enhance the models.

5. Conclusions

T2 maps segmented into 83 anatomical brain regions from 40 subjects with MCI who either progressed to AD or remained stable were analyzed and contralateral features related to the signal and texture of the maps were extracted. We identified that the contralateral difference in a texture-related feature (the absolute differences in Mass Scatter YY) extracted from the hippocampi and the contralateral difference of a signal-related feature (the signal value at 25%) extracted from the superior frontal gyrus were the most relevant features for the task of classifying between MCIp and MCIs subjects under both a univariate and a multivariate analysis. In general, signal- and texture-related features enhanced the MCI to AD predictive power of models that used information from neuropsychological tests, such as ADAS-11 and ADNI's memory test. Furthermore, we found that signal and texture information is more relevant for this task than mere volumetric information. These results suggest that contralateral signal- and texture-related information extracted from T2 maps should continue to be explored in the search for better MCI-to-AD predictive models.

Author Contributions: Conceptualization, A.I.T.-C. and A.M.-T.; methodology, A.M.-T.; software, A.I.T.-C., R.A.C.-L., J.A.G.-L. and F.V.-L.; validation, A.I.T.-C., R.A.C.-L. and J.A.G.-L.; formal analysis, A.I.T.-C. and A.M.-T.; investigation, A.I.T.-C., R.A.C.-L. and J.A.G.-L.; resources, F.V.-L. and A.M.-T.; data curation, A.I.T.-C., R.A.C.-L. and J.A.G.-L.; writing—original draft preparation, A.I.T.-C., R.A.C.-L. and J.A.G.-L.; writing—review and editing, A.M.-T.; visualization, A.I.T.-C., R.A.C.-L., J.A.G.-L. and A.M.-T.; supervision, A.M.-T.; project administration, A.M.-T.; funding acquisition, A.M.-T. All authors have read and agreed to the published version of the manuscript.

Funding: This work was partially supported by the Consejo Nacional de Ciencia y Tecnología (CONANCYT) and by Universidad de Monterrey through the Fondo de Fomentos a la Investigación Grant. The sponsors had no role in the design and conduct of the study; in the collection, analysis, and interpretation of data; in the preparation of the manuscript; or in the review or approval of the manuscript.

Institutional Review Board Statement: Not applicable.

Informed Consent Statement: Informed consent was obtained from all subjects involved in the study through the ADNI database.

Data Availability Statement: Data used in the preparation of this article were obtained from the ADNI database (adni.loni.usc.edu access date: 30 April 2020).

Acknowledgments: Data collection and sharing for this project was funded by the Alzheimer's Disease Neuroimaging Initiative (ADNI) (National Institutes of Health Grant U01 AG024904) and DOD ADNI (Department of Defense award number W81XWH-12-2-0012). ADNI is funded by the National Institute on Aging, the National Institute of Biomedical Imaging and Bioengineering, and through generous contributions from the following: AbbVie, Alzheimer's Association; Alzheimer's Drug Discovery Foundation; Araclon Biotech; BioClinica, Inc.; Biogen; Bristol-Myers Squibb Company; CereSpir, Inc.; Cogstate; Eisai Inc.; Elan Pharmaceuticals, Inc.; Eli Lilly and Company; EuroImmun; F. Hoffmann-La Roche Ltd. and its affiliated company Genentech, Inc.; Fujirebio; GE Healthcare; IXICO Ltd.; Janssen Alzheimer Immunotherapy Research & Development, LLC.; Johnson & Johnson Pharmaceutical Research & Development LLC.; Lumosity; Lundbeck; Merck & Co., Inc.; Meso Scale Diagnostics, LLC.; NeuroRx Research; Neurotrack Technologies; Novartis Pharmaceuticals

Corporation; Pfizer Inc.; Piramal Imaging; Servier; Takeda Pharmaceutical Company; and Transition Therapeutics. The Canadian Institutes of Health Research provides funds to support ADNI clinical sites in Canada. Private sector contributions are facilitated by the Foundation for the National Institutes of Health (www.fnih.org). The grantee organization is the Northern California Institute for Research and Education, and the study is coordinated by the Alzheimer's Therapeutic Research Institute at the University of Southern California. ADNI data are disseminated by the Laboratory for Neuro Imaging at the University of Southern California.

Conflicts of Interest: The authors declare no conflict of interest.

References

1. Erkkinen, M.G.; Kim, M.; Geschwind, M.D. Clinical Neurology and Epidemiology of the Major Neurodegenerative Diseases. *Cold Spring Harb. Perspect. Biol.* **2018**, *10*, a033118. [CrossRef]
2. Blennow, K.; Mattsson, N.; Schöll, M.; Hansson, O.; Zetterberg, H. Amyloid biomarkers in Alzheimer's disease. *Trends Pharmacol. Sci.* **2015**, *36*, 297–309. [CrossRef] [PubMed]
3. Zvěřová, M. Clinical aspects of Alzheimer's disease. *Clin. Biochem.* **2019**, *72*, 3–6. [CrossRef]
4. Stelzmann, R.A.; Schnitzlein, H.N.; Murtagh, F.R. An english translation of alzheimer's 1907 paper, 'über eine eigenartige erkankung der hirnrinde. *Clin. Anat.* **1995**, *8*, 429–431. [CrossRef]
5. Sosa-Ortiz, A.L.; Acosta-Castillo, I.; Prince, M.J. Epidemiology of Dementias and Alzheimer's Disease. *Arch. Med. Res.* **2012**, *43*, 600–608. [CrossRef] [PubMed]
6. Alzheimer's Association. 2019 Alzheimer's disease facts and figures. *Alzheimers Dement.* **2019**, *15*, 321–387.
7. Cadena-Hernandez, A.G.; Trejo-Castro, A.I.; Celaya-Padilla, J.M.; Tamez-Pena, J.; Martinez-Torteya, A. Longitudinal gender-specific differences in the conversion from mild cognitive impairment to Alzheimer's disease. In Proceedings of the 2018 IEEE EMBS International Conference on Biomedical & Health Informatics (BHI), Las Vegas, NV, USA, 4–7 March 2018; pp. 202–205.
8. Martinez-Torteya, A.; Trejo-Castro, A.I.; Celaya-Padill, J.M.; Tamez-Pena, J.G. Differences in the Progression from Mild Cognitive Impairment to Alzheimer's Disease between APOE4 Carriers and Non-Carriers. In Proceedings of the 2019 IEEE 19th International Conference on Bioinformatics and Bioengineering (BIBE), Athens, Greece, 28–30 October 2019; pp. 199–203.
9. Tangalos, E.G.; Petersen, R.C. Mild Cognitive Impairment in Geriatrics. *Clin. Geriatr. Med.* **2018**, *34*, 563–589. [CrossRef]
10. Petersen, R.C. Mild Cognitive Impairment. *N. Engl. J. Med.* **2011**, *364*, 2227–2234. [CrossRef]
11. McKhann, G.; Drachman, D.; Folstein, M.; Katzman, R.; Price, D.; Stadlan, E.M. Clinical diagnosis of Alzheimer's disease: Report of the NINCDS-ADRDA Work Group* under the auspices of Department of Health and Human Services Task Force on Alzheimer's Disease. *Neurology* **1984**, *34*, 939. [CrossRef]
12. Folstein, F.M.; Folstein, S.E.; McHugh, P.R. Mini-mental state. *J. Psychiatr. Res.* **1975**, *12*, 189–198. [CrossRef]
13. Kaplan, E.; Goodglass, H.; Weintrab, S. *The Boston Naming Test*; Lea & Febiger: Philadelphia, PA, USA, 1983.
14. Yesavage, J.A.; Brink, T.L.; Rose, T.L.; Lum, O.; Huang, V.; Adey, M.; Leirer, V.O. Development and validation of a geriatric depression screening scale: A preliminary report. *J. Psychiatr. Res.* **1982**, *17*, 37–49. [CrossRef]
15. Rosen, W.G.; Mohs, R.C.; Davis, K.L. A new rating scale for Alzheimer's disease. *Am. J. Psychiatry* **1984**, *141*, 1356–1364. [PubMed]
16. Jack, C.R., Jr.; Albert, M.S.; Knopman, D.S.; McKhann, G.M.; Sperling, R.A.; Carrillo, M.C.; Thies, B.; Phelps, C.H. Introduction to the recommendations from the National Institute on Aging-Alzheimer's Association workgroups on diagnostic guidelines for Alzheimer's disease. *Alzheimers Dement.* **2011**, *7*, 257–262. [CrossRef]
17. Cai, J.H.; He, Y.; Zhong, X.L.; Lei, H.; Wang, F.; Luo, G.H.; Zhao, H.; Liu, J.C. MMagnetic Resonance Texture Analysis in Alzheimer's disease. *Acad. Radiol.* **2020**, in press. [CrossRef] [PubMed]
18. Wachinger, C.; Salat, D.H.; Weiner, M.; Reuter, M. Whole-brain analysis reveals increased neuroanatomical asymmetries in dementia for hippocampus and amygdala. *Brain* **2016**, *139*, 3253–3266. [CrossRef] [PubMed]
19. Martinez-Torteya, A.; Rodriguez-Rojas, J.; Celaya-Padilla, J.M.; Galván-Tejada, J.I.; Treviño, V.; Tamez-Peña, J. Magnetization-prepared rapid acquisition with gradient echo magnetic resonance imaging signal and texture features for the prediction of mild cognitive impairment to Alzheimer's disease progression. *J. Med. Imaging* **2014**, *1*, 031005. [CrossRef]
20. Trejo-Castro, A.I. Texture and signal features from hippocampal T2 maps as biomarkers for MCI to AD progression. In Proceedings of the 2020 IEEE International Conference on Bioinformatics and Biomedicine (BIBM), Seoul, South Korea, 16–19 December 2020; pp. 772–777.
21. Yu, H.; Touret, A.S.; Li, B.; O'Brien, M.; Qureshi, M.M.; Soto, J.A.; Jara, H.; Anderson, S.W. Application of texture analysis on parametric T 1 and T 2 maps for detection of hepatic fibrosis. *J. Magn. Reson. Imaging* **2017**, *45*, 250–259. [CrossRef]
22. Baessler, B.; Luecke, C.; Lurz, J.; Klingel, K.; Das, A.; von Roeder, M.; de Waha-Thiele, S.; Besler, C.; Rommel, K.P.; Maintz, D.; et al. Cardiac MRI and Texture Analysis of Myocardial T1 and T2 Maps in Myocarditis with Acute versus Chronic Symptoms of Heart Failure. *Radiology* **2019**, *292*, 608–617. [CrossRef]
23. Jack, C.R.; Bernstein, M.A.; Borowski, B.J.; Gunter, J.L.; Fox, N.C.; Thompson, P.M.; Schuff, N.; Krueger, G.; Killiany, R.J.; Decarli, C.S.; et al. Update on the Magnetic Resonance Imaging core of the Alzheimer's Disease Neuroimaging Initiative. *Alzheimers Dement.* **2010**, *6*, 212–220. [CrossRef]

24. Heckemann, R.A.; Keihaninejad, S.; Aljabar, P.; Rueckert, D.; Hajnal, J.V.; Hammers, A. Improving intersubject image registration using tissue-class information benefits robustness and accuracy of multi-atlas based anatomical segmentation. *Neuroimage* **2010**, *51*, 221–227. [CrossRef]
25. Petersen, R.C.; Aisen, P.S.; Beckett, L.A.; Donohue, M.C.; Gamst, A.C.; Harvey, D.J.; Jack, C.R., Jr.; Jagust, W.J.; Shaw, L.M.; Toga, A.W.; et al. Alzheimer's Disease Neuroimaging Initiative (ADNI): Clinical characterization. *Neurology* **2010**, *74*, 201–209. [CrossRef] [PubMed]
26. Milford, D.; Rosbach, N.; Bendszus, M.; Heiland, S. Mono-exponential fitting in T2-relaxometry: Relevance of offset and first echo. *PLoS ONE* **2015**, *10*, e0145255. [CrossRef] [PubMed]
27. Johnson, H.J.; McCormick, M.M.; Ibanez, L. The Insight Software Consortium. In *The ITK Software Guide*, 4th ed.; Kitware, Inc.: Chapelhill, NC, USA, 2015.
28. Benjamini, Y.; Hochberg, Y. Controlling the False Discovery Rate: A Practical and Powerful Approach to Multiple Testing. *J. R. Stat. Soc. Ser. B* **1995**, *57*, 289–300. [CrossRef]
29. Martinez-Torteya, A.; Alanis, I.; Tamez-Pena, J. *FeatuRE Selection Algortihms for Computer-Aided Diagnosis: An R package. The Comprehensive R Archive Network*. 2018. Available online: https://cran.r-project.org/web/packages/FRESA.CAD/index.html (assessed on 21 July 2021).
30. Oriol, J.D.; Martinez-Torteya, A.; Trevino, V.; Alanis, I.; Vallejo, E.; Tamez-Pena, J.G. Benchmarking machine learning models for the analysis of genetic data using FRESA.CAD Binary Classification Benchmarking. *bioRxiv* **2019**. preprint.
31. Oriol, J.D.; Vallejo, E.E.; Estrada, K.; Peña, J.G.T. The Alzheimer's Disease Neuroimaging Initiative Benchmarking machine learning models for late-onset alzheimer's disease prediction from genomic data. *BMC Bioinform.* **2019**, *20*, 709.
32. Aisen, P.S.; Petersen, R.C.; Donohue, M.C.; Gamst, A.; Raman, R.; Thomas, R.G.; Walter, S.; Trojanowski, J.Q.; Shaw, L.M.; Beckett, L.A.; et al. Clinical core of the Alzheimer's disease neuroimaging initiative: Progress and plans. *Alzheimers Dement.* **2010**, *6*, 239–246. [CrossRef]
33. Crane, P.K.; Carle, A.; Gibbons, L.E.; Insel, P.; Mackin, R.S.; Gross, A.; Jones, R.N.; Mukherjee, S.; Curtis, S.M.; Harvey, D.; et al. Development and assessment of a composite score for memory in the Alzheimer's Disease Neuroimaging Initiative (ADNI). *Brain Imaging Behav.* **2012**, *6*, 502–516. [CrossRef]
34. Gibbons, L.; Carle, A.; Mackin, R.; Harvey, D. A composite score for executive functioning, validated in Alzheimer's Disease Neuroimaging Initiative (ADNI) participants with baseline mild cognitive impairment. *Brain Imaging Behav.* **2012**, *6*, 517–527. [CrossRef]
35. Jack, C.R., Jr.; Petersen, R.C.; Xu, Y.C.; O'Brien, P.C.; Smith, G.E.; Ivnik, R.J.; Boeve, B.F.; Waring, S.C.; Tangalos, E.G.; Kokmen, E. Prediction of AD with MRI-based hippocampal volume in mild cognitive impairment. *Neurology* **1999**, *52*, 1397. [CrossRef]
36. Lee, S.; Lee, H.W.; Kim, K. Magnetic resonance imaging texture predicts progression to dementia due to Alzheimer disease earlier than hippocampal volume. *J. Psychiatry Neurosci.* **2020**, *45*, 7–14. [CrossRef]
37. Sørensen, L.; Aisen, P.S.; Petersen, R.C.; Donohue, M.C.; Gamst, A.; Raman, R.; Thomas, R.G.; Walter, S.; Trojanowski, J.Q.; Shaw, L.M.; et al. Early detection of Alzheimer's disease using MRI hippocampal texture. *Hum. Brain Mapp.* **2016**, *37*, 1148–1161. [CrossRef] [PubMed]
38. Zwanenburg, A.; Leger, S.; Vallières, M.; Löck, S. The Image Biomarker Standardization Initiative: Standardized Quantitative Radiomics for High-Throughput Image-based Phenotyping. *Radiology* **2020**, *295*, 328–338. [CrossRef] [PubMed]

healthcare

MDPI

Review

A Survey on Recent Advances in Machine Learning Based Sleep Apnea Detection Systems

Anita Ramachandran [1,*] and Anupama Karuppiah [2]

1 Department of Computer Science & Information Systems, BITS, Pilani 560001, India
2 Department of Electrical & Electronics Engineering, BITS, Pilani-K K Birla Goa Campus, Near NH17B, Zuari Nagar, Sancoale 403726, India; anupkr@goa.bits-pilani.ac.in
* Correspondence: anita.ramachandran@pilani.bits-pilani.ac.in

Abstract: Sleep apnea is a sleep disorder that affects a large population. This disorder can cause or augment the exposure to cardiovascular dysfunction, stroke, diabetes, and poor productivity. The polysomnography (PSG) test, which is the gold standard for sleep apnea detection, is expensive, inconvenient, and unavailable to the population at large. This calls for more friendly and accessible solutions for diagnosing sleep apnea. In this paper, we examine how sleep apnea is detected clinically, and how a combination of advances in embedded systems and machine learning can help make its diagnosis easier, more affordable, and accessible. We present the relevance of machine learning in sleep apnea detection, and a study of the recent advances in the aforementioned area. The review covers research based on machine learning, deep learning, and sensor fusion, and focuses on the following facets of sleep apnea detection: (i) type of sensors used for data collection, (ii) feature engineering approaches applied on the data (iii) classifiers used for sleep apnea detection/classification. We also analyze the challenges in the design of sleep apnea detection systems, based on the literature survey.

Keywords: sleep apnea; machine learning; deep learning; wearable systems

Citation: Ramachandran, A.; Karuppiah, A. A Survey on Recent Advances in Machine Learning Based Sleep Apnea Detection Systems. *Healthcare* **2021**, *9*, 914. https://doi.org/10.3390/healthcare9070914

Academic Editor: Mahmudur Rahman

Received: 17 May 2021
Accepted: 13 July 2021
Published: 20 July 2021

Publisher's Note: MDPI stays neutral with regard to jurisdictional claims in published maps and institutional affiliations.

1. Introduction

Sleep apnea is a sleep disorder in which a sleeping person's breathing is disturbed. It is prevalent in adults as well as a small percentage of the juvenile population [1]. Subjects suffering from sleep apnea undergo periods of no or shallow breathing during their sleep. The former condition in which breathing stops temporarily is referred to as apnea, while the latter condition of periods of shallow breathing or airflow reduction is called hypopnea. Clinical comorbidities can result from either condition and, therefore, both are detrimental to a person's well-being [2]. The physiological symptoms of sleep apnea include snoring, gasping for air during sleep, waking up with dry mouth and, in general, low sleep quality, thereby leading to low attention, insomnia, decrease in cognitive skills, accidents, memory loss and depression. In addition to the low quality of life caused by sleep deprivation and fatigue, sleep apnea may also lead to severe issues such as diabetes, cardiovascular problems, hypertension, neurological issues, and liver problems. Due to the global prevalence of sleep apnea as well as the direct and indirect long-term problems it brings about, it is important to diagnose and treat this condition. In this paper, we review the recent state-of-the-art research in the application of machine learning for sleep apnea detection. The review covers the parameters and sensors used, and feature engineering approaches for enabling sleep apnea detection using machine learning.

There are three types of sleep apnea:

- Obstructive sleep apnea (OSA) occurs due to improper functioning of the upper respiratory tract. When the muscles of the hard palate in the back of the throat that supports that soft palate relax, the soft palate blocks the passage of air to the respiratory system. This leads to stoppage of breathing for short durations [3].

- Central sleep apnea (CSA) occurs when the brain fails to generate or transmit signals that control breathing muscles. This leads to short durations of time when the subject does not breathe at all.
- Complex sleep apnea syndrome is manifested with central apnea persisting even after obstructive events have disappeared with PAP therapy [4].

Javaheri et al. [3] describe the etiological risk factors for sleep apnea and its consequences. In this paper, we describe the recent research in the application of machine learning for sleep apnea detection. Figure 1 presents the distribution of the number of papers selected for this study from 2003 through 2021. The technical focus of this study includes the following facets of sleep apnea detection: (i) type of sensors used for data collection, (ii) feature engineering approaches applied on the data, and (iii) classifiers used for sleep apnea detection/classification.

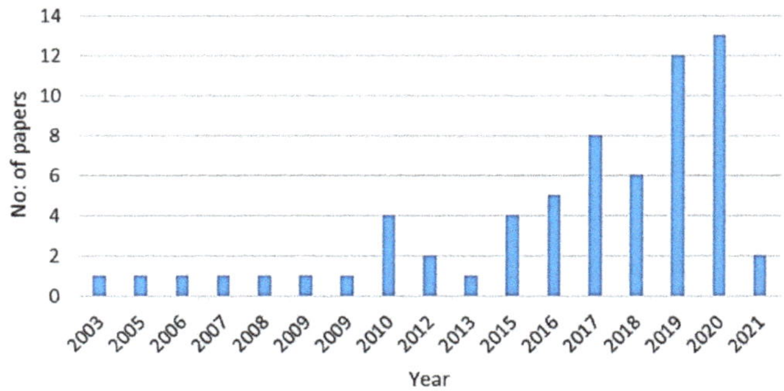

Figure 1. Year-wise Distribution of Papers.

This paper is organized as follows: In Section 2, we briefly explain how sleep apnea is diagnosed, and the biomedical parameters along with their derivatives that aid in the process. Subsequently, we examine the drawbacks of the standard tests for sleep apnea detection, and reason the need for leveraging on the advances in machine learning and wearable device technologies for the same. Section 3 details the recent studies on intelligent sleep apnea detection mechanisms using classic machine learning and deep learning based solutions, using single markers as well as sensor/feature fusion. Section 4 outlines the recent studies in sleep apnea detection using machine learning on data generated by environmental sensors and the significance of including features related health profiles, during classifier training. We conclude our paper with our observations on the various factors that influence the performance of machine learning classifiers for sleep apnea detection.

2. Background

2.1. Diagnosis of Sleep Apnea

Clinical manifestations of sleep apnea conditions include variations in oxygen saturation levels, respiratory effort, and heart rate. Gottlieb et al. [5] describes the pathophysiology, assessment and treatment of obstructive sleep apnea. The PSG test is the gold standard in the diagnosis of this condition [1]. This test is conducted in dedicated sleep labs under the supervision of trained personnel. It is time consuming, and requires subjects to be connected to instruments measuring various biomedical and physiological parameters. The test monitors upper airway flow, respiratory effort, and biomedical and physiological parameters such as electroencephalogram (EEG), electrocardiogram (ECG), and oxygen saturation (SPO2) [1]. EEG helps detect electrical activity in the brain and related disorders. This is measured using an EEG machine. ECG analyzes the rhythm of heartbeats and blood flow to the heart muscles and is measured using an ECG machine or a single lead ECG. SPO2 indicates the measure of oxygen in the blood. A pulse oximeter is used to measure

SPO2. In addition, thoracic and abdominal signals as well as acoustic signals generated by respiratory effort or snoring can also aid in the detection of sleep apnea.

Various parameters useful in the diagnosis of sleep apnea can be derived from the above-mentioned signals. Analysis of ECG yields Heart Rate Variability (HRV), ECG derived respiration (EDR), Cardiopulmonary coupling (CPC), and Ballistocardiography (BCG) parameters.

- HRV measures the variation in the time interval between consecutive heartbeats, known as the R-R interval. Previous research shows that variation in R-R interval is a symptom of apneic events, and hence can provide the physiological basis of using R-R series to detect OSA. Analyzing HRV, however, poses certain challenges. This includes special attention to signal quality and elimination of background noise, along with using a sensitive R-wave detection algorithm. Furthermore, interpretation of HRV is difficult in patients who have atrial fibrillation or those with irregular heartbeats [6].
- Instantaneous Heart Rate (IHR) is the number of times the heart would beat if successive R-R intervals were constant.
- EDR measures respiratory activity from ECG. An explanation of the relation between EDR and ECG is given in [7]. The respiratory effort causes changes in the position of the ECG electrodes, which in turn affects the amplitude of the ECG signals. EDR is the surrogate respiration signal derived from the amplitude variations of the ECG signals. There are several techniques to derive EDR from ECG [8].
- CPC quantifies the degree of coherent coupling between HRV and variations of the R-wave amplitude caused by modulation of the respiratory tidal volume. CPC can be of high or low frequency coupling (HFC, LFC); the former is indicative of stable sleep, while the latter is associated with sleep instability. A special characteristic of LFC, so-called elevated LFC, can be used to detect periods of apnea and hypopnea [9].
- Ballistocardiography (BCG) is a noninvasive method based on the measurement of body motion (body movements such as displacement, velocity, and acceleration), generated by the ejection of blood by the heart, at each cardiac cycle. This is measured using devices that can measure the body recoil force produced as a result of ejection of blood [10].
- A parameter that may be related to HRV is Pulse Rate Variability (PRV), which is measured from photoplethysmography (PPG) sensors [11]. PPG sensors use a light source and a photodetector on the skin to characterize blood circulation.

Oxygen Desaturation Index (ODI) is a metric derived from SPO2, which represents the number of times the oxygen level in blood falls for more than 10 s, divided by the number of sleep hours. ODI is defined as the number of times that oxygen desaturation was $\geq$3% per hour of sleep [12].

The above mentioned parameters are used to infer certain measures to ascertain the presence of sleep apnea, such as:

- Apnea–hypopnea index (AHI) [13] is the number of times one has apnea or hypopnea during one night, divided by the hours of sleep. In other words, AHI score is the number of apnea and hypopnea events per hour of sleep. The severity of sleep apnea is determined based on the AHI score as follows: normal (AHI < 5), mild (5 $\leq$ AHI < 15), moderate (15 $\leq$ AHI < 30), and severe (AHI $\geq$ 30).
- Respiratory Disturbance Index (RDI) factor counts the number of times respiratory difficulties disturb one's sleep. This includes, in addition to apneic and hypopneic events, respiratory effort-related arousals (RERA). RERA is the number of arousals from sleep resulting from increased respiratory effort. RDI is expressed as:
- RDI = (Number of apneas + Number of hypopneas + Number of RERAs)/sleep hours.

2.2. The Need for More Accessible Detection Mechanisms—Sensors to the Aid

While the PSG test is the gold standard in sleep apnea diagnosis, its availability, cost, requirement of trained staff, and limited capacity at sleep centers make it inaccessible to the common man, and sleep apnea is often undiagnosed or underdiagnosed, until the subject starts showing symptoms of long-term impact. Studies show that the percentage of elderly population in the world is increasing. Due to changing lifestyles, the number of elderly people living alone is also increasing. This has resulted in the emergence of geriatric healthcare homes, with round-the–the-clock staff support, albeit with high costs of maintenance. Technological advances in sensors, low power embedded systems, and machine learning have paved the way for more affordable and intelligent healthcare homes, with automatic monitoring of the subjects' vital parameters [14]. One of the possibilities of such a system is the detection of sleep apnea.

Recent advances in sensing technologies have enabled the continuous collection of various vital parameters that can lead to monitoring sleep quality in multiple ways. The use of sensors to detect sleep apnea is a widely researched area, and the application of machine learning techniques to detect apneic conditions has been found to be accurate and reliable. The parameters used to detect sleep apnea, such as ECG and SPO2, their derivatives such as HRV, BCG, ODI, thoracic and abdominal signals, pressure, and sound [15], can be obtained from biomedical sensors, environmental sensors or vision-based systems.

- Biosensors allow sensing of vital parameters. For example, ECG sensors enable the detection of HRV and R-R intervals through signal analysis. They also enable the deduction of variations in QRS (Q wave, R wave, S wave) amplitude of ECG signals and ECG derived respiration. A variant of the ECG sensor, the single lead ECG sensor, is designed to be used with wearable devices. SPO2 sensors measure oxygen saturation levels in the blood. Barometric sensors measure blood pressure.
- Environmental sensors include those that can monitor the surroundings of the subject under study. For example, sound sensors allow nocturnal sound analysis by capturing snoring via microphones. Sounds and sound patterns during inhalation and exhalation will be different from normal when the upper respiratory tract is compromised. Inertial motion unit (IMU) sensors allow deriving the position of the sleeping subject. Sensors are also placed under the bed to enable non-intrusive monitoring.
- Vision based systems allow capturing of images through image and/or video feeds. Analysis of the images and video frames enables determination of the sleeping position of the subject under study.

Leelaarporn et al. [14] provide a comprehensive review of the utilization of sensors in four different areas of smart living, including sleep monitoring. Recent research trends in the area of sleep monitoring using several types of algorithms on pulse oximetry, ECG, sounds and respiration data are described in [16]. Flemons et al. [17] studies the utility of portable monitors in diagnosing sleep apnea in adults.

3. Machine Learning in Sleep Apnea Detection Based on Biomedical Markers in Wearable Devices

Machine learning applies mathematical modelling to detect or predict anomalies or patterns, to discover new knowledge from datasets. A model trained on a given dataset is used to classify new data. Machine learning can be supervised, unsupervised, or reinforcement learning [18]. Supervised learning algorithms take a labelled dataset as input and output a hypothesis that best fits the labelled dataset. A labelled dataset provides the algorithm with an outcome variable for each record in the dataset. Unsupervised learning algorithms do not have a labelled dataset for classifier training; rather, they detect patterns in the dataset to form clusters of similar records. Reinforcement learning has a feedback ingredient that incorporates reward points for records that get correctly classified, which substantiates classifier training. While there have been studies that uses spectral/waveform analysis of signals for sleep apnea detection [19–21], the ability of machine learning classifiers to learn from input datasets and generalize for future data makes it a reliable approach in this area of research. Most studies on sleep apnea detection rely on supervised learning.

The common set of parameters that is used to detect sleep apnea was explained in a previous section. Biomedical informaticians have used various machine learning techniques to predict the accuracy of sleep apnea diagnosis using these aforementioned parameters. Of late, the effectiveness of ensemble classifiers and deep learning techniques has also been investigated. The features used for sleep apnea detection could be reported directly from sensors, or extracted from various sensor observations. There has also been extensive research into utilizing observations from one or more of these sensors using data fusion to detect sleep disorders. Studies also include the impact of extracting statistical, time and frequency domain features from the parameters, and performing dimensionality reduction to downsize the feature vectors on the classifier performance. In the following sections, we look at how classic machine learning, deep learning, and sensor fusion techniques have been applied to detect sleep apnea. Deep learning can be considered as a specialized segment of machine learning; however, the manner in which feature engineering is accomplished differs greatly from each other. A snapshot of recent research on sleep apnea detection using machine learning and deep learning with biomedical sensors is presented in Table A1.

3.1. Classic Machine Learning Based Solutions

This section presents an overview of recent research in sleep apnea detection using classic machine learning techniques. In many research papers, single biomedical markers, such as SPO2, ECG, EOG, or EEG, have been used for the detection of sleep apnea. Among these, most studies focus on using SPO2 and ECG signals because of their correlation with apneic events—research shows that heart rate and systolic blood pressure increase in response to apneic events [22]. For example, in [12], SPO2 signals are used for OSA detection. During feature engineering, ODI, total time below saturation levels (tsa), and other six features were extracted from SPO2. Various variants of decision tree (DT) classifiers were used to obtain an accuracy of 93%. In [23] too, pulse oximeter parameters are used for sleep apnea detection. PPG measurements were obtained from SPO2 sensor and analyzed to derive heart rate and breathing effort information. The best classification performance of 87% was obtained when the Linear Discriminant Analysis was used on SPO2 features and the PPG features were combined. Another study that makes use of PPG measurements extracted from SPO2 readings is [24], in which statistical and time domain SPO2 and PPG features were extracted around SPO2 drops and averaged per patient. The impact of using SPO2 and PPG features on OSA detection was analyzed here. Three SPO2 based features and two PPG features were selected for training a support vector machine (SVM) classifier. Unlike [23], it was found that the classifier based on SPO2 features along with the subjects' age yielded 77.7% accuracy, while the PPG features did not have any impact on the classifier performance. This research highlights that age is also a clear confounding parameter because of its correlation with cardiovascular health, and using age alone for OSA detection can yield a reasonable accuracy. In [25], four machine learning models are evaluated, to not just detect apnea but also ascertain its severity using only SPO2 information obtained at the patient's home. A three-step process comprising feature extraction, feature selection, and classifier evaluation was conducted. A total of 16 features were extracted from SPO2 spanning statistical, spectral, and nonlinear domains, in addition to ODI, which were input to a Fast Correlation Based Filter feature selection algorithm. An AdaBoost model built with linear discriminants as base classifiers gave the best apnea severity classification accuracy. In [13], Mostafa et al. analyzes SPO2 signals from two public datasets using Deep Belief Network (DBN). The analysis shows that while the accuracy increases with the increasing number of hidden neurons, the increase is minimal, which may not justify the trade-off between classifier performance and processing requirements. Another study that detects sleep apnea conditions employs seven features and SVM [26]. This work not only detects but also corrects apneic events via a smart pillow. The setup consists of a wearable device with a pulse oximeter, a smartphone, and an adjustable pillow. The pulse oximeter on a wearable device senses the SPO2 signal and transmits it to a smart phone. The smartphone detects the SPO2 desaturation events and issues a pillow adjustment command. The adjustable pillow adjusts its shape and height according to the command. The adjustment effect is further monitored and evaluated by the pulse oximeter, providing a closed-loop feedback system between monitoring and corrective actions. Wrist band-mobile and mobile-pillow communication is over Bluetooth. A review of approaches for detecting sleep apnea specifically using pulse oximetry data is provided in [27].

ECG is another parameter that is commonly used in the detection of sleep apnea. Hassan et al. [28] compare various machine learning classifiers on a dataset generated by a single lead ECG sensor. Statistical moment-based and empirical mode decomposition features were extracted from the raw data. Post feature extraction, Naive Bayes, k-nearest neighbor (kNN), neural network, AdaBoost, Bagging, random forest, extreme learning machine (ELM), discriminant analysis (DA) and restricted Boltzmann machine were compared for performance. ELM gave the best accuracy of 83.77%. A dataset based on single-lead ECG was used in [29] as well to detect sleep apnea. In this study, segments of ECG signals were fed into dual-tree complex wavelet transform (DTCWT) to generate frequency sub-bands. Three statistical features—variance, skewness, and kurtosis—were extracted from the DTCWT output and analyzed to determine their suitability in detecting sleep apnea. LogitBoost gave an accuracy of 84.4%. Other classifiers analyzed include DA, kNN, Artificial Neural Network (ANN), ELM, SVM, AdaBoost and Bagging. ECG signals have also been used not just for the detection of sleep apnea, but also to determine its type [20].

Previous research indicates that parameters derived from ECG such as IHR, HRV, BCG, and CPC, have also been used as markers for training classifiers to detect sleep apnea. For example, certain studies [30,31] indicate that HRV measures have a great potential to boost OSA detection. Khandoker et al. [32] highlight the effectiveness of using HRV and EDR with an SVM classifier to attain 100% accuracy in the detection of apneic events. This study also uses SVM to estimate the relative severity of OSA. In [33], kNN, quadratic discriminant analysis (QDA) and SVM were applied on statistical measures of HRV. de Chazal et al. [34] use HRV, EDR, and CPC, obtained from single lead ECG signals, for sleep apnea detection. The analysis in this study shows that CPC features along with the time-domain-based HRV parameters gave the best classification performance, with an accuracy of 89.8%. The classifier algorithm used was multiple logistic discrimination. In [35], 24 time and frequency domain features are extracted from ECG signals. This included time domain features such as mean, median, standard deviation, and mode for each NN interval series, and frequency domain features such as normalized power in various frequency ranges, and the vegetative balance index. Feature selection was performed by discarding redundant features, leading to nine features being used for training decision trees, discriminant analysis, logistic regression, support vector machines, variation of kNN, and ensemble learning classifiers. Seo et al. [36] study sleep quality and stability assessment using sleep questionnaires and ECG. Respiratory and CPC parameters were extracted from ECG signals, and results found a significant correlation between AHI and CPC. Studies related to sleep analysis using EEG signals include [37,38].

3.2. Deep Learning Based Solutions

Deep learning techniques such as Deep Neural Network (DNN), Convolutional Neural Network (CNN), Recurrent Neural Network (RNN) and Long-Short Term Memory (LSTM) are being increasingly used for diagnosing sleep apnea, both on single markers as well as with sensor/feature fusion. Feature engineering and selection is crucial to the performance of intelligent solutions, especially in the biomedical domain [39]. One of the advantages of using deep learning is that they have the capability to learn relevant features from the raw data, using neurons, convolution and pooling layers. For example, Li et al. [40] argue that that while feature engineering is essential for improving the performance of classifiers, it often depends on human expertise which can tend to be subjective. In this study, unsupervised learning algorithms with sparse auto-encoders were used to learn features from ECG signals, to decouple the dependency of subjective human expertise on crucial feature engineering aspects. Classification was carried out using SVM and ANN, and the classification performance was refined using decision fusion and Hidden Markov Model (HMM). The accuracy obtained was 85% and the sensitivity was 88.9%. Another study that performs algorithmic extraction of features is [41]. In this, a single electrooculogram (EOG) signal was used to perform automatic sleep scoring. A three-layer DBN with 500, 200, and 100 neurons was used for feature extraction and label prediction. The predicted labels and original labels were used to train an HMM model. The average accuracy of the DBN–HMM model was 83.3%. This study attempts to establish that DBN can extract features by itself without manual intervention. Novák et al. [31] study how LSTM can enable the detection of temporal dependencies in features relevant to sleep apnea detection.

Wang et al. [42] use ECG signals for sleep apnea detection. R-R intervals and R-peak amplitudes were extracted from ECG signals, and time window ANN was applied for classification. The accuracy obtained was 87.3%. Mostafa et al. [13] describe a method to detect sleep apnea using SPO2 by calculating the AHI score. The deep learning algorithm used was DBN. Performance analysis was performed on two public datasets [43,44] with SPO2 values. Pathinarupothi et al. [15] detail the use of LSTM-RNN for the detection of sleep apnea severity and explores the relation between IHR and SPO2 towards this. The research shows that OSA severity detection can be solely based on either IHR or SPO2 signals.

In [45], IHR is used as the sole marker for sleep apnea detection. This paper argues that using only IHR and its derivatives can provide 85% accuracy at best, with simple classification algorithms for classifying minute-to-minute apnea. Therefore, LSTM–RNN was employed for the identification of sleep apnea and its severity. Various configurations of LSTM–RNN, post feature extraction and selection, were used for training, which yielded 99.99% accuracy in detecting sleep apnea. Erdenebayar et al. [22] describe a comparative study of the performance of deep learning classifiers on ECG signals—the classifiers are Deep Neural Network (DNN), 1D CNN, 2D CNN, RNN, LSTM and gated-recurrent unit model (GRU). The 1D CNN and GRU models were the best performing with an accuracy and recall of 99%. Other studies include [46–48].

3.3. Sensor/Feature Fusion Techniques

Extensive study has been performed to estimate the effectiveness of sensor or feature fusion techniques to detect sleep apnea. This involves the concurrent use of two or more parameters originating from different sources and performing classification based on the values of all these parameters. For example, Memis et al. [49] apply feature-level fusion of ECG and SPO2 signals. The temporal information from the ECG and SPO2 signals was fed as input to Naïve Bayes, kNN, and SVM classifiers. SVM gave the best accuracy of 96.64%. Xie et al. [50] also explores ensembles and data fusion over ECG and SPO2 signals. When analyzed separately, the research finds that SPO2 features can detect apneic episodes better than ECG features. Various classifier combinations trained on select features from SPO2 and ECG were then analyzed for performance. Feature extraction yielded 111 ECG and 39 SPO2 features, from which 8 ECG and 31 SPO2 features were selected for classifier training. The base classifiers were combined using maximum probability, average probability, product of probability and majority voting. Garde et al. [11] extract time-domain and frequency-domain features from SPO2 and PRV, and applies Logistic Regression to detect apnea/hypopnea events.

In [51], Prabha et al. make use of HRV and Respiratory Rate Variability (RRV) from ECG and respiratory effort signals (RES), respectively. A decision making system which fuses time-domain features from HRV and RRV signals, by combining their outputs with empirically calculated weights, produced an accuracy of 100%. The weight associated with time-domain HRV features was considerably higher than that of time-domain RRV features, which indicates that HRV has a higher correlation with sleep apnea detection than RRV, although the latter may be complementing the former. This analysis concludes that the time-domain features of HRV and RRV provide sufficient information to detect OSA. Other related studies include [52,53].

4. Other Solutions

In addition to devices that measure biomedical parameters, studies show the application of environmental sensors/devices such as microphones and cameras to ascertain the presence of sleep apnea. Literature also shows the application of health profiles to detect apnea and predict the AHI values to classify the severity of apneic events. Examples of such studies are summarized below.

4.1. Using Environmental Sensors

Sleep apnea detection can be performed with externally mounted devices or ambient sensors, other than biomedical sensors. One such technique for sleep apnea detection is based on smartphones. Camcı et al. [54] use sonar waves generated by smart phones, which give information about chest movements, to detect sleep apnea. The accuracy of the system was found to be dependent on the subject's change of sleep position. Other techniques such as placing a microphone close to the subject's nose and mouth were found to be obtrusive and impacting the sleep behavior of the subjects [55,56]. Another technique relies on the use of a 3D time-of-flight camera, which records the subject's respiratory motion [57]. The signals pertaining to respiratory movement of abdominal muscles are analyzed to monitor sleep stages and detect apnea. Davidovich et al. [58] propose a novel algorithm for sleep apnea screening with a contact-free system based on a piezo-electric sensor. The setup consisted of a piezo-electric sensor, which recorded a combination of gross body motion, rib cage movements, and the cardioballistic effect. The specificity and sensitivity were found to be 89% and 88%, respectively.

Hafezi et al. [59] estimate sleep apnea severity from tracheal movements via an accelerometer attached to the participant's suprasternal notch. 7 morphological features were extracted from tracheal movements, on which a deep learning classifier using a combination of CNN and LSTM, was applied. However, this method requires wearing a patch which may be inconvenient to the subjects.

In [60], Wang et al. propose a sleep breathing monitoring mattress which utilizes the ultra-wideband (UWB) physiological sensing technique. The UWB physiological sensing is accomplished via a series of very narrow and low power pulses over wideband. If apnea is detected, the head of the mattress is lifted up to increase blood oxygen saturation and ease the apneic condition. The methodology involved dataset collection using signals recorded from the experiment using Fast Fourier Transform (FFT), feature extraction using Principal Component Analysis (PCA) and classification using kNN, AdaBoost, DT, and SVM. kNN produced better results than the rest of the classifiers.

In [56], acoustic signals placed on the ceiling above the patient's bed, were used. Subjects were classified into four sleep apnea severity groups according to their AHI. A two-stage filtering process to remove various unwanted noises and purify the sleep breathing sounds was applied. A total of 23 temporal and spectral features of the audio signal were extracted, which included the mel frequency, cepstral coefficients (MFCCs), spectral flux, and zero crossing rate. Logistic regression, SVM, DNN with 2 hidden layers were applied for classification.

In [61], machine learning models (kNN, AdaBoost, and DT) are applied on data generated by UWB sensors for sleep apnea detection. The experimental setup consists of a sleep breathing monitoring mattress which utilizes the UWB physiological sensing technique. The mattress also has a mechanism to lift up the head on detection of apneic events.

Avcı et al. [62] use abdominal, nasal, and chest respiratory signals and applied ensemble classifiers such as AdaBoost, random forest and random subspace to detect sleep apnea. Feature extraction and dimensionality reduction via PCA was performed to yield a best-case accuracy of 98.68%. Table A2 provides a snapshot of studies that apply machine learning to data generated by environmental sensors for sleep apnea detection. Ozdemir et al. In [63], a fully automatic apnea detection algorithm along with an early warning system to predict apneic events, is described. The algorithm also works on nasal respiratory airflow signals, on which feature extraction was performed. Subsequently, Randomly Select and Compute (RANSAC) algorithm was used for feature reduction on the original 39 features, and the set of features that is not significant for OSA detection is listed. SVM, kNN, and linear regression for classification are compared for learning and prediction of OSA episodes. The solution produced an accuracy of 87.6% of and sensitivity of 91.3%. Another study that makes use of airflow sensing signals for sleep apnea detection and classification of apnea severity is [55]. A total of 17 features from overnight airflow sensing samples were extracted, and fed into DNNs with various combinations of hidden layers and activation nodes per layer. The algorithm used the tanh activation function alongside the softmax classifier. Diagnosis of sleep apnea was performed using AHI threshold values of 5, 15, and 30 events/hour. The severity classification logic classified patients into four groups—no apnea, mild apnea, moderate apnea, and severe apnea. The best accuracy that DNN gave was 92.69%.

In [64], sleep data and 3D facial scans were used as features. The data collected was pre-processed for pose alignment and hole filling and analyzed using Matlab's deep learning framework. The model thus generated was tuned for performance and used for classification. The accuracy reported was 69%. However, this method requires facial images of the subject, which restricts the subject's degree of freedom while sleeping. Other studies in the area that use non-biomedical parameters include [65–67].

Non-wearable techniques for sleep apnea detection have certain advantages and disadvantages when compared with wearable devices. For example, wearable devices for sleep apnea detection have to be small in form factor and light-weight, while non-wearable techniques such as BCG-embedded beds or camera based systems do not have restrictions on their size or form factor. Another characteristic of comparison between wearable and non-wearable techniques is power consumption. Minimizing power consumption enables the wearable device to be on battery power for longer durations, which reduces the overhead of charging the devices. Power consumption of such devices occurs in three activities—sensing, processing, and communication. These three functions have to be optimized for energy saving to enable the device to be worn for long periods of time without recharging. In contrast, non-wearable devices can be connected to the main power supply, and hence need not be designed for optimized power consumption. One significant factor that affects the accuracy of sleep apnea detection in both techniques, is the placement of the sensors. Wearable devices allow round-the-clock monitoring of parameters since it does not restrict the parameter collection to a certain geographical region under study. However, non-wearable devices are sensitive to the sensing range of the devices. Environmental sensor-based systems also sometimes tend to be intrusive—for example, placing a microphone close to a subject's face while sleeping could be uncomfortable for him/her. Camera-based systems may tend to be expensive and have higher power and bandwidth requirements. Due to all these aspects, wearable devices may be conducive to at-home sleep monitoring, while non-wearable techniques may be applied in hospital environments where the mobility of the subjects is more constrained.

4.2. Health Profiles for the Detection of Sleep Apnea

There has been research that highlights the significance of including a subject's health profile in the diagnosis of sleep apnea and its severity. Mencar et al. [68] use 19 features including heart disease, diabetes, gender, BMI, age, smoking, hypertension and snoring, to explore methods to classify sleep apnea severity. Classification algorithms are applied to classify the severity of sleep apnea, and regression methods are applied to predict the AHI values. In another work, Ustun et al. [69] argue that medical information of subjects would be more suited to diagnose sleep apnea than real time sleep related symptoms. Features such as age, gender, BMI, presence of hypertension, history of heart failure, stroke, asthma, smoking, and snoring were used to train the classifiers. Seven classifiers including variants of Logistic regression, DT, and SVM were compared with a new machine learning model named SLIM (Supersparse Linear Integer Models). SLIM is a linear classification model for creating medical scoring systems, and this gave a sensitivity of 64.2% and specificity of 77%. The study supports the use of simple models with good generalization capabilities, especially for medical applications where datasets are prone to overfitting.

5. Discussion and Conclusions

In this study, we briefly summed up the causes and risks associated with sleep apnea, and the drawbacks of the related diagnostic processes. We outlined the parameters that help detect apneic events. Subsequently, we examined the application of machine learning in sleep apnea detection, with focus on wearable systems. We summarized the recent research that demonstrates feature engineering techniques and efficient use of classic machine learning, deep learning, and sensor/feature fusion algorithms to detect sleep apnea, and in some cases, classify its severity, using biomedical markers such as ECG, EEG and SPO2. The paper also briefly looked at the application of environmental sensors and information in subjects' health profiles to ascertain the presence of sleep apnea.

From our analysis, an observation is that machine learning algorithms applied to datasets in the literature survey, produce varying degrees of accuracy. This indicates that the performance of the algorithms depends on various factors such as:

(i) Data collection modalities

Factors such as type of sensors, their placement, and frequency and sensitivity of measurements, affect the training of machine learning classifiers. Among the various biomedical parameters that aid in the detection of sleep apnea, we observe that the most common of them are those from ECG, SPO2, and EEG signals. The drawback of using ECG is that the signals generated by three leads or more require a resting ECG or an ECG Holter monitor, which may be restrictive for the subject under study because of the placement of leads. Single lead ECG can be embedded within wearable devices; however, the accuracy of such devices is less than those with multilead devices. Collection of EEG data also requires the subjects to wear a headgear while sleeping, which may cause inconvenience. SPO2 sensors, such as single lead ECG sensors, can be embedded within wearable devices and, in combination with the demographic information of subjects, has been proven to provide good results in the detection of sleep apnea. Environmental sensors may constrain the subjects to a certain area under observation while sleeping (such as bed-embedded BCG sensors). Some may introduce noise in the data collection, for example, acoustic sensors are prone to errors from ambient noise.

(ii) Dataset characteristics

Characteristics of data such as its distribution and dataset features, along with the pre-processing that has been applied to it also influences the efficiency of supervised training techniques. For a classifier to be well-trained, the dataset it trains on must be balanced. In the case of sleep apnea, it has to be ensured that the number of apneic events in the dataset are comparable with that of non-apneic events. In the absence of this, the classifier gets trained for the majority classes and misclassifies the minority classes. Additionally, appropriate data pre-processing techniques and feature engineering should be performed to fine tune the classifier training.

(iii) Labelling techniques

Training machine learning models for sleep apnea detection using supervised learning techniques, requires annotation of the records in the sleep dataset. Some of the standards used in sleep stage scoring from sleep study reports are the Rechtschaffen and Kales standard (R&K) [70] and American Academy of Sleep Medicine (AASM) [71]. In practice, apneic events are annotated manually by domain experts. The process involves correlation of the subject's biomedical and physiological history with the sleep data, while adhering to the guidelines set forth by the standards. The dependency of annotation on the standards and subjective domain expertise may limit the generalization capability of the trained model.

The capability of a wearable device or an end-to-end system to store data for analysis, raise alarms on detection of abnormalities, and generate reports long-term is prudent, and especially useful in the context of geriatric care homes. Today, there are commercial devices that synchronize collected data to a smartphone periodically; however, a drawback of such a system is that at any given time, the device can be paired with only a single smartphone. The ability to support data collection and analysis at a central location would be especially beneficial in geriatric healthcare, where elderly people are saved the effort required to access and view their own reports.

Author Contributions: A.R. surveyed the existing literature and wrote the main draft of the manuscript. A.K. supervised the work, helped outline the organization of the paper and reviewed the manuscript. Both authors have read and agreed to the published version of the manuscript.

Funding: This research received no external funding.

Institutional Review Board Statement: Not applicable.

Informed Consent Statement: Not applicable.

Data Availability Statement: Not applicable.

Conflicts of Interest: The authors declare no conflict of interest.

Appendix A

Table A1. Machine/deep learning based sleep apnea detection using biomedical sensors.

Reference	Year	Subject Demographics	Signal Used	Classifiers Applied	Feature Engineering Approach	Accuracy
[22]	2019	65 male, 21 female	ECG	DNN, 1D CNN, 2D CNN, RNN, LSTM, Gated recurrent unit	Performed by deep learning algorithms. For example, while using CNN, feature map was extracted using filter kernels by the convolution layer. Dimensionality reduction was performed by the pooling layer.	Accuracy: 99.0%
[40]	2018	Apnea-ECG @	ECG	DNN, HMM SVM, ANN	Sparse auto-encoders was used to learn features via unsupervised learning.	Accuracy: 85%;
[42]	2019	Apnea-ECG @	ECG	Time window ANN	R-R intervals and R-peak amplitudes were extracted from ECG signals. Further, 6 time domain and 6 frequency domain features from R-R interval, and 6 frequency domain features from R-peak amplitudes were extracted.	Accuracy: 87.3%
[34]	2016	Apnea-ECG @	ECG	Multiple logistic discrimination	Features extracted included RR-interval, EDR, CPC and their derivatives.	Accuracy: 89.8%

Table A1. *Cont.*

Reference	Year	Subject Demographics	Signal Used	Classifiers Applied	Feature Engineering Approach	Accuracy
[31]	2008	Apnea-ECG @	ECG	LSTM, ANN and Elman Network	Feature extracted include: Time domain HRV parameters such as RMSSD (square root of mean squared differences of successive NN intervals), R-R mean (mean of R-R interval length) and NN50 (number of intervals longer than 50 ms) and frequency domain HRV features such as Low Frequency/High Frequency (LF/HF) ratio, total power for analyzed interval, Low Frequency (LF), High Frequency (HF), Very Low Frequency (VLF), normalized LF and normalized HF.	Accuracy: 82.1%
[33]	2010	5 male, 12 female, 26 years–67 years	ECG	kNN, QDA, SVM	Median, inter-quartile difference (75th and 25th percentile), and mean absolute deviations of the R-R intervals were computed for each epoch.	Accuracy: 90%
[28]	2015	Apnea-ECG @	ECG	Naive Bayes, kNN, ANN, AdaBoost, Bagging, Random Forest, ELM, DA, Restricted Boltzmann Machine	Statistical features such as mean, variance, skewness and kurtosis of the ECG signals were extracted.	Accuracy: 83.77%
[26]	2013	40 subjects	SPO2	SVM	For each detected SPO2 desaturation event, extract 7 features from a window of 150 s from the starting point of the SPO2 desaturation. Features extracted include no. of desaturation events, speed of decline in SPO2, in addition to statistical measures such as minimum and standard deviation of the SPO2 values.	Accuracy: 93.5%
[51]	2017	32 subjects, 18 to 75 years	ECG, RES	SVM	Time domain features (such as mean NN interval, standard deviation of NN interval, mean heart rate, RMSSD) and frequency domain (peak frequency, absolute power, relative power) features from HRV and RRV from ECG and RES, respectively, were computed.	Accuracy 100%
[29]	2017	Apnea-ECG @	ECG	DA, kNN, ANN, ELM, SVM, AdaBoost, Bagging, LogitBoost	Skewness, variance and kurtosis were extracted and used for classifier training.	Accuracy: 84.4%
[49]	2017	Apnea-ECG @	ECG, SPO2	Naïve Bayes, kNN, SVM	Uses concatenation of temporal information from ECG and SPO2 signals.	Accuracy: 96.64%

Table A1. *Cont.*

Reference	Year	Subject Demographics	Signal Used	Classifiers Applied	Feature Engineering Approach	Accuracy
[50]	2012	UCD #	ECG, SPO2	Adaboost, Decision Trees	Time domain features from SPO2, which measure the regularity, variability, and complexity of a time series, were extracted. From ECG, HRV and EDR-based features in both time and spectral domains were extracted.	Accuracy: 82%
[12]	2010	Apnea-ECG @	SPO2	Various variants of Decision tree classifiers	ODI indices from SPO2 were computed.	Accuracy: 93%
[15]	2017	Apnea-ECG @	ECG, SPO2	LSTM-RNN	Feature extraction was performed by LSTM.	Accuracy: 92.1%
[13]	2017	UCD #, Apnea-ECG @	SPO2	DBN	Feature extraction was by DBN.	Accuracies: 85.36% and 97.64%, respectively for the 2 datasets
[32]	2009	(1) Apnea-ECG @ (2) UCD # (3) 83 subjects; with mean +/− standard deviation age of 55.6 +/− 10.7 yrs	ECG	SVM	Feature extraction from HRV and EDR, using wavelet decomposition was performed. Feature selection was performed using a hill climbing algorithm. 14 HRV and 14 EDR were selected for classifier training.	Accuracy: 100%
[11]	2016	160 children (87 male, 59 female)	SPO2	Logistic Regression	Time and frequency domain features from SPO2 and pulse rate variability were used for classifier training.	-
[23]	2017	52 subjects	SPO2	Linear Discriminant Analysis	Features from SPO2 (such as number of desaturations > 3%, spread of SPO2, minimum and average of SPO2), PPG and PPG derived respiration were extracted for classifier training.	Accuracy: 87%
[35]	2020	Apnea-ECG @	ECG	Decision trees, DA, logistic regression, SVM, kNN, ensemble learning	24 time and frequency domain features were extracted. Feature selection by discarding redundant features, which resulted in a set of 9 features for classifier training	Accuracy: 98.7%
[52]	2020	Not specified	Respiration, SpO2, heartrate, 3-ACC signals	Gaussian Naïve-Bayes, ANN, kNN	Dataset was collected and labelled per AASM's sleep apnea judgement criteria. A train:test ratio of 8:2 was used. 5-fold cross-validation was applied. The hyper-parameters of each machine learning algorithm were set by using the average of five cross-validation data sets.	Accuracy: 95%

Table A1. *Cont.*

Reference	Year	Subject Demographics	Signal Used	Classifiers Applied	Feature Engineering Approach	Accuracy
[53]	2017	1983 subjects	Demographic information, EEG	Random Forest, XGBoost, and Light Gradient Boosting Machine	A total of 36 features were extracted from demographic information and EEG signals, including frequency and percentage of every sleep stage, time in bed, total sleep time, sleep efficiency and total number of one-step transitions overnight. Data imbalance was corrected using SMOTE analysis and feature selection was performed using statistical analysis.	Area under the curve: 0.9128
[46]	2019	SHHS (NSRR) %	Raw physiological respiratory signals	LSTM	LSTM was used to automatically learn and extract relevant features, and detect potential sleep apnea events. Direct respiration signals gave better accuracy than derived their signals such as EDR	Accuracy: 70% (approx.)
[47]	2020	Apnea-ECG @	ECG	Logistic Regression, SVM and 1D CNN	Time domain and frequency domain features of R-R interval were extracted for training logistic regression and SVM classifiers. There was no need for feature engineering with 1D CNN.	Accuracy: 88.23%
[48]	2020	MESA (NSRR) *	Respiratory signals	CNN, Markov Chain	Features learned by CNN.	Accuracy: 80.78%
[24]	2019	975 subjects	SPO2	SVM	Extracted features include simple time-domain (e.g., amplitude and length of desaturation), statistical (e.g., minimum and mean SpO2 value) and desaturation severity (e.g., area below SpO2 baseline) and quasi periodicity features (e.g., phase rectified signal averaging (PRSA)).	Accuracy: 77.7%
[41]	2015	SleepEDF Database [Expanded] ~	EOG	DBN	Feature extraction was performed by DBN.	Accuracy: 83.3%
[25]	2019	320 subjects, age 54.8 +/− 13.5 years	SPO2	Linear Discriminant Analysis, Logistic regression, Bayesian Multi Layer Perceptron, AdaBoost	Time domain features that characterize central tendency, dispersion, asymmetry, and peakedness of a given time series, frequency domain features such as PSD of SPO2 signals, ODI3 and non-linear measures were computed.	
[45]	2017	Apnea-ECG @	ECG	LSTM-RNN	Continuous time series IHR measurements were converted to a series of feature vectors, for each beat window.	Accuracy: 99.99%

@ Apnea-ECG: 70 ECG signal recordings extracted from PSG recordings with a 16-bit resolution, a sampling rate of 100 Hz [43,72]. # St Vincent's University Hospital/University College Dublin (UCD) Sleep Apnea Database: 21 males, 4 females; age 28 years—68 years [44]. % SHHS dataset Sleep Heart Health Study: The dataset consists of 5804 adults of age 40 and older. A subset consisting of 1008 female and 1092 male patients with mean age 62.5 ± 12.6 (standard deviation) years was used in the study cited [73]. * MESA (NSRR): 6814 subjects: age 45 years–84 years [74]. ~ SleepEDF Database [Expanded] 20 subjects; age 25 year–34 years [75].

Table A2. Machine/deep learning based sleep apnea detection using environmental sensors.

Reference	Year	Subject Demographics	Signal Used	Classification Approach	Methodology	Accuracy
[59]	2020	Ages 18–85 years	Tracheal movements	CNN+LSTM	Recorded tracheal movements were filtered using a bandpass filter with cut-off frequencies of 0.1 Hz and 25 Hz. Time-series sliding windowing technique with a window size of 10 s was applied. 21 morphological features were extracted from each window.	Accuracy: 84%
[64]	2018	39 male, 30 female adults	Facial images	CNN	Involved data collection, pre-processing, model generation and tuning, and classification, using Matlab based deep learning framework	Accuracy: 69%
[60]	2019	5 subjects simulating sleep apnea conditions	UWB signals	AdBoost, Decision tree, SVM, kNN		Accuracy: 98%
[54]	2017	4 subjects simulating sleep apnea conditions	Accelerometer and sonar waves from smart phones	kNN, CART	Mean, variance and range for accelerometer data and the noise level were extracted, and subsequently, no. of breaths per minute was calculated.	Accuracy: 97.7%
[58]	2016	77 male, 19 female 23–88 years	Piezo-electric sensor	Signal analysis	Gross body motion, rib cage movements, and cardioballistic effect was recorded by the piezoelectric sensor. Time and frequency domain features were extracted from motion, respiratory rate and inter-beat intervals, to calculate AHI.	-
[56]	2018	120 subjects, including 3 children and 4 adolescents	Acoustic signals from microphone placed at a distance of 1.7 m above the subject's bed	Logistic regression, SVM, DNN with 2 hidden layers	Several temporal and spectral characteristics of audio signals such as the mel frequency cepstral coefficients (MFCCs), spectral flux, and zero crossing rate were extracted.	Accuracy: 92.5%
[55]	2018	MrOS [TBD]	Airflow signals from a thermistor placed in front of the nose	SVM, AdaBoost, Regression, Deep Neural Networks	Seventeen time domain features were extracted from airflow signals, after subsampling and filtering.	Accuracy: 92.69%
[63]	2016	6 subjects	Nasal respiratory airflow signals	SVM, kNN, Linear Regression for Classfication	15 time-series features of OSA periods such as mean, variance, minimum, maximum, median values of signals were extracted. Minimum, maximum, average inspiration/expiration amplitudes and durations of nasal airflow signal were also extracted. Feature reduction was performed using RANSAC algorithm.	Accuracy: 87.6%

Table A2. *Cont.*

Reference	Year	Subject Demographics	Signal Used	Classification Approach	Methodology	Accuracy
[62]	2015	Apnea-ECG	Abdominal, chest and nasal respiratory signals	AdaBoost, Random Forest and Random Subspace	Wavelet transform based on feature extraction methods are applied on 1 min length respiration signals.	Accuracy: 98.68%
[61]	2015	3 male, 1 female Age 48 ± 6.9 years	Reflect pulses from Impulse Radio Ultra-Wide Band (IR-UWB) radar panel	Linear Discriminant	Normal and apnea epochs were extracted from the IR-UWB data. 15 statistical features were derived from these extracted epochs.	Accuracy: 73%
[65]	2020	9 subjects Age 65 years or more	Signals from pressure sensitive mat	Temporal convolutional network (TCN), bidirectional LSTM	Data pre-processing included occupancy extraction, bandpass filtering, signal combination, concatenation and normalization. TCN and bidirectional LSTM approaches were compared with SVM and threshold based approaches.	Accuracy: 95.1%
[66]	2020	4 male, 4 female Age 25 years–55 years	Respiratory signals from accelerometer and pressure transducer	CNN	Sliding window approach was used for signal processing. Proposes a system of continuous monitoring of breath, from an accelerometer-based device worn around the subject.	Accuracy: 88%
[67]	2019	20 adults	Speech signals	Random Forest	Offline detection of OSA using speech/voice analysis. This is based on the fact that speech properties of OSA patients are altered. Feature extraction was performed on audio files using Random Forest feature selection and Mann–Whitney U test ranking.	Accuracy: 87.5%

References

1. Mostafa, S.S.; Mendonça, F.; Ravelo-García, A.G.; Morgado-Dias, F. A Systematic Review of Detecting Sleep Apnea Using Deep Learning. *Sensors* **2019**, *19*, 4934. [CrossRef]
2. Spector, A.R.; Loriaux, D.; Farjat, A.E. The Clinical Significance of Apneas Versus Hypopneas: Is There Really a Difference? *Cureus* **2019**, *11*, e4560. [CrossRef]
3. Javaheri, S.; Barbé, F.; Campos-Rodriguez, F.; Dempsey, J.A.; Khayat, R.; Javaheri, S.; Malhotra, A.; Martinez-Garcia, M.A.; Mehra, R.; Pack, A.; et al. Somers, Sleep Apnea: Types, Mechanisms, and Clinical Cardiovascular Consequences. *J. Am. Coll. Cardiol.* **2017**, *69*, 841–858. [CrossRef]
4. Khan, M.T.; Franco, R.A. Complex Sleep Apnea Syndrome. *Sleep Disord.* **2014**, *2014*, 798487. [CrossRef] [PubMed]
5. Gottlieb, D.J.; Punjabi, N.M. Diagnosis and Management of Obstructive Sleep Apnea. *JAMA* **2020**, *323*, 1389–1400. [CrossRef] [PubMed]
6. Reed, M.J.; Robertson, C.E.; Addison, P.S. Heart rate variability measurements and the prediction of ventricular arrhythmias. *QJM Int. J. Med.* **2005**, *98*, 87–95. [CrossRef] [PubMed]
7. Janbakhshi, P.; Shamsollahi, M.B. ECG-derived respiration estimation from single-lead ECG using gaussian process and phase space reconstruction methods. *Biomed. Signal Process. Control.* **2018**, *45*, 80–90. [CrossRef]
8. Avci, C.; Delibasoglu, I.; Akbas, A. Sleep apnea detection using wavelet analysis of ECG derived respiratory signal. In Proceedings of the 2012 International Conference on Biomedical Engineering (ICoBE), Penang, Malaysia, 27–28 February 2012; pp. 272–275.
9. Cysarz, D.; Linhard, M.; Seifert, G.; Edelhäuser, F. Sleep Instabilities Assessed by Cardiopulmonary Coupling Analysis Increase during Childhood and Adolescence. *Front. Physiol.* **2018**, *9*, 468. [CrossRef] [PubMed]

10. Sadek, I.; Heng, T.T.S.; Seet, E.; Abdulrazak, B. A New Approach for Detecting Sleep Apnea Using a Contactless Bed Sensor: Comparison Study. *J. Med. Internet Res.* **2020**, *22*, e18297. [CrossRef]
11. Garde, A.; Dekhordi, P.; Ansermino, J.M.; Dumont, G. Identifying individual sleep apnea/hypoapnea epochs using smartphone-based pulse oximetry. In Proceedings of the 2016 38th Annual International Conference of the IEEE Engineering in Medicine and Biology Society (EMBC), Orlando, FL, USA, 16–20 August 2016; Volume 2016, pp. 3195–3198.
12. Burgos, A.; Goñi, A.; Illarramendi, A.; Bermudez, J. Real-Time Detection of Apneas on a PDA. *IEEE Trans. Inf. Technol. Biomed.* **2010**, *14*, 995–1002. [CrossRef]
13. Mostafa, S.S.; Mendonca, F.; Morgado-Dias, F.; Ravelo-Garcia, A. SpO_2 based sleep apnea detection using deep learning. In Proceedings of the 2017 IEEE 21st International Conference on Intelligent Engineering Systems (INES), Larnaca, Cyprus, 20–23 October 2017; pp. 000091–000096.
14. Leelaarporn, P.; Wachiraphan, P.; Kaewlee, T.; Udsa, T.; Chaisaen, R.; Choksatchawathi, T.; Laosirirat, R.; Lakhan, P.; Natnithikarat, P.; Thanontip, K.; et al. Sensor-Driven Achieving of Smart Living: A Review. *IEEE Sens. J.* **2021**, 1. [CrossRef]
15. Pathinarupothi, R.K.; Rangan, E.S.; Gopalakrishnan, E.A.; Vinaykumar, R.; Soman, K.P. Single Sensor Techniques for Sleep Apnea Diagnosis Using Deep Learning. In Proceedings of the 2017 IEEE International Conference on Healthcare Informatics (ICHI), Park City, UT, USA, 23–26 August 2017; pp. 524–529.
16. Mendonca, F.; Mostafa, S.S.; Ravelo-Garcia, A.G.; Morgado-Dias, F.; Penzel, T. A Review of Obstructive Sleep Apnea Detection Approaches. *IEEE J. Biomed. Health Inform.* **2018**, *23*, 825–837. [CrossRef] [PubMed]
17. Flemons, W.W.; Littner, M.R.; Rowley, J.A.; Gay, P.; Anderson, W.M.; Hudgel, D.W.; McEvoy, R.D.; Loube, D.I. Home Diagnosis of Sleep Apnea: A Systematic Review of the Literature. *Chest* **2003**, *124*, 1543–1579. [CrossRef] [PubMed]
18. Mitchell, T.M. *Machine Learning*; McGraw-Hill: New York, NY, USA, 1997; ISBN 978-0-07-042807-2.
19. Mack, D.C.; Alwan, M.; Turner, B.; Suratt, P.M.; Felder, R.A. A Passive and Portable System for Monitoring Heart Rate and Detecting Sleep Apnea and Arousals: Preliminary Validation. In Proceedings of the 1st Transdisciplinary Conference on Distributed Diagnosis and Home Healthcare, Arlington, VA, USA, 2–4 April 2006; pp. 51–54. [CrossRef]
20. Thomas, R.J.; Mietus, J.E.; Peng, C.-K.; Gilmartin, G.; Daly, R.W.; Goldberger, A.L.; Gottlieb, D.J. Differentiating Obstructive from Central and Complex Sleep Apnea Using an Automated Electrocardiogram-Based Method. *Sleep* **2007**, *30*, 1756–1769. [CrossRef] [PubMed]
21. Hsu, C.-C.; Shih, P.-T. An intelligent sleep apnea detection system. In Proceedings of the 2010 International Conference on Machine Learning and Cybernetics, Qingdao, China, 1–14 July 2010; Volume 6, pp. 3230–3233.
22. Erdenebayar, U.; Kim, Y.J.; Park, J.-U.; Joo, E.Y.; Lee, K.-J. Deep learning approaches for automatic detection of sleep apnea events from an electrocardiogram. *Comput. Methods Programs Biomed.* **2019**, *180*, 105001. [CrossRef] [PubMed]
23. Jayawardhana, M.; de Chazal, P. Enhanced detection of sleep apnoea using heart-rate, respiration effort and oxygen saturation derived from a photople-thysmography sensor. *Annu. Int. Conf. IEEE Eng. Med. Biol. Soc.* **2017**, 121–124. [CrossRef]
24. Deviaene, M.; Borzée, P.; Buyse, B.; Testelmans, D.; van Huffel, S.; Varon, C. Pulse Oximetry Markers for Cardiovascular Disease in Sleep Apnea. In Proceedings of the 2019 Computing in Cardiology (CinC), Singapore, 8–11 September 2019; pp. 1–4. [CrossRef]
25. Gutierrez-Tobal, G.C.; Alvarez, D.; Crespo, A.; Del Campo, F.; Hornero, R. Evaluation of Machine-Learning Approaches to Estimate Sleep Apnea Severity From At-Home Oximetry Recordings. *IEEE J. Biomed. Health Inform.* **2018**, *23*, 882–892. [CrossRef]
26. Zhang, J.; Zhang, Q.; Wang, Y.; Qiu, C. A real-time auto-adjustable smart pillow system for sleep apnea detection and treatment. In Proceedings of the Proceedings of the 12th International Conference on Interaction Design and Children, New York, NY, USA, 24–27 June 2013; p. 179.
27. Terrill, P.I. A review of approaches for analysing obstructive sleep apnoea-related patterns in pulse oximetry data. *Respirology* **2019**, *25*, 475–485. [CrossRef]
28. Hassan, A.R. A comparative study of various classifiers for automated sleep apnea screening based on single-lead electrocardiogram. In Proceedings of the 2015 International Conference on Electrical & Electronic Engineering (ICEEE), Rajshahi, Bangladesh, 4–6 November 2015; pp. 45–48.
29. Hassan, A.R.; Bashar, S.K.; Bhuiyan, M.I.H. Computerized obstructive sleep apnea diagnosis from single-lead ECG signals using dual-tree complex wavelet transform. In Proceedings of the 2017 IEEE Region 10 Humanitarian Technology Conference (R10-HTC), New York, NY, USA, 21–23 December 2017; pp. 43–46.
30. Gula, L.J.; Krahn, A.D.; Skanes, A.; Ferguson, K.A.; George, C.; Yee, R.; Klein, G.J. Heart Rate Variability in Obstructive Sleep Apnea: A Prospective Study and Frequency Domain Analysis. *Ann. Noninvasive Electrocardiol.* **2003**, *8*, 144–149. [CrossRef]
31. Novak, D.; Mucha, K.; Al-Ani, T. Long short-term memory for apnea detection based on Heart Rate Variability. *Annu. Int. Conf. IEEE Eng. Med. Biol. Soc.* **2008**, *2008*, 5234–5237. [CrossRef]
32. Khandoker, A.H.; Palaniswami, M.; Karmakar, C. Support Vector Machines for Automated Recognition of Obstructive Sleep Apnea Syndrome from ECG Recordings. *IEEE Trans. Inf. Technol. Biomed.* **2008**, *13*, 37–48. [CrossRef]
33. Yılmaz, B.; Asyalı, M.H.; Arıkan, E.; Yetkin, S.; Özgen, F. Sleep stage and obstructive apneaic epoch classification using single-lead ECG. *Biomed. Eng. Online* **2010**, *9*, 39. [CrossRef]
34. De Chazal, P.; Sadr, N. Sleep apnoea classification using heart rate variability, ECG derived respiration and cardiopulmonary coupling parameters. *Annu. Int. Conf. IEEE Eng. Med. Biol. Soc.* **2016**, *2016*, 3203–3206. [CrossRef] [PubMed]

35. Ivanko, K.; Ivanushkina, N.; Rykhalska, A. Identifying episodes of sleep apnea in ECG by machine learning methods. In Proceedings of the 2020 IEEE 40th International Conference on Electronics and Nanotechnology (ELNANO), Kyiv, Ukraine, 22–24 April 2020; pp. 588–593.
36. Seo, M.Y.; Hwang, S.J.; Nam, K.J.; Lee, S.H. Significance of sleep stability using cardiopulmonary coupling in sleep disordered breathing. *Laryngoscope* **2020**, *130*, 2069–2075. [CrossRef] [PubMed]
37. Gupta, V.; Pachori, R.B. FBDM based time-frequency representation for sleep stages classification using EEG signals. *Biomed. Signal Process. Control.* **2021**, *64*, 102265. [CrossRef]
38. Sharma, M.; Tiwari, J.; Acharya, U. Automatic Sleep-Stage Scoring in Healthy and Sleep Disorder Patients Using Optimal Wavelet Filter Bank Technique with EEG Signals. *Int. J. Environ. Res. Public Health* **2021**, *18*, 3087. [CrossRef] [PubMed]
39. Deviaene, M.; Testelmans, D.; Borzee, P.; Buyse, B.; Van Huffel, S.; Varon, C. Feature Selection Algorithm based on Random Forest applied to Sleep Apnea Detection. In Proceedings of the 2019 41st Annual International Conference of the IEEE Engineering in Medicine and Biology Society (EMBC), Berlin, Germany, 23–27 July 2019; Volume 2019, pp. 2580–2583.
40. Li, K.; Pan, W.; Li, Y.; Jiang, Q.; Liu, G. A method to detect sleep apnea based on deep neural network and hidden Markov model using single-lead ECG signal. *Neurocomputing* **2018**, *294*, 94–101. [CrossRef]
41. Xia, B.; Li, Q.; Jia, J.; Wang, J.; Chaudhary, U.; Ramos-Murguialday, A.; Birbaumer, N. Electrooculogram based sleep stage classification using deep belief network. In Proceedings of the 2015 International Joint Conference on Neural Networks (IJCNN), Killarney, Ireland, 12–17 July 2015; pp. 1–5.
42. Wang, T.; Lu, C.; Shen, G. Detection of Sleep Apnea from Single-Lead ECG Signal Using a Time Window Artificial Neural Network. *BioMed Res. Int.* **2019**, *2019*, 9768072. [CrossRef]
43. Goldberger, A.L.; Amaral, L.A.N.; Glass, L.; Hausdorff, J.M.; Ivanov, P.C.; Mark, R.G.; Mietus, J.E.; Moody, G.B.; Peng, C.-K.; Stanley, H.E. PhysioBank, PhysioToolkit, and PhysioNet. *Circulation* **2000**, *101*, e215–e220. [CrossRef]
44. St. Vincent's University Hospital/University College Dublin Sleep Apnea Database. Available online: https://physionet.org/content/ucddb/1.0.0/ (accessed on 9 February 2021).
45. Pathinarupothi, R.K.; Vinaykumar, R.; Rangan, E.; Gopalakrishnan, E.; Soman, K.P. Instantaneous heart rate as a robust feature for sleep apnea severity detection using deep learning. In Proceedings of the 2017 IEEE EMBS International Conference on Biomedical & Health Informatics (BHI), Orlando, FL, USA, 16–19 February 2017; pp. 293–296.
46. Van Steenkiste, T.; Groenendaal, W.; Deschrijver, D.; Dhaene, T. Automated Sleep Apnea Detection in Raw Respiratory Signals Using Long Short-Term Memory Neural Networks. *IEEE J. Biomed. Health Inform.* **2018**, *23*, 2354–2364. [CrossRef]
47. Sharan, R.V.; Berkovsky, S.; Xiong, H.; Coiera, E. ECG-Derived Heart Rate Variability Interpolation and 1-D Convolutional Neural Networks for Detecting Sleep Apnea. In Proceedings of the 2020 42nd Annual International Conference of the IEEE Engineering in Medicine & Biology Society (EMBC), Montreal, QC, Canada, 20–24 July 2020; Volume 2020, pp. 637–640.
48. Haidar, R.; Koprinska, I.; Jeffries, B. Sleep Apnea Event Prediction Using Convolutional Neural Networks and Markov Chains. In Proceedings of the 2020 International Joint Conference on Neural Networks (IJCNN), Glasgow, UK, 19–24 July 2020; pp. 1–8.
49. Memis, G.; Sert, M. Multimodal Classification of Obstructive Sleep Apnea Using Feature Level Fusion. In Proceedings of the 2017 IEEE 11th International Conference on Semantic Computing (ICSC), San Diego, CA, USA, 30 January–1 February 2017; pp. 85–88.
50. Xie, B.; Minn, H. Real-Time Sleep Apnea Detection by Classifier Combination. *IEEE Trans. Inf. Technol. Biomed.* **2012**, *16*, 469–477. [CrossRef] [PubMed]
51. Prabha, A.; Trivedi, A.; Kumar, A.A.; Kumar, C.S. Automated system for obstructive sleep apnea detection using heart rate variability and respiratory rate variability. In Proceedings of the 2017 International Conference on Advances in Computing, Communications and Informatics (ICACCI), Udupi, India, 13–16 September 2017; pp. 1303–1307.
52. Jeon, Y.; Heo, K.; Kang, S.J. Real-Time Sleep Apnea Diagnosis Method Using Wearable Device without External Sensors. In Proceedings of the 2020 IEEE International Conference on Pervasive Computing and Communications Workshops (PerCom Workshops), Austin, TX, USA, 23–27 March 2020; pp. 1–5.
53. Liu, J.; Li, Q.; Xin, Y.; Lu, X. Obstructive Sleep Apnea Detection Using Sleep Architecture. In Proceedings of the 2020 IEEE International Conference on Mechatronics and Automation (ICMA), Beijing, China, 13–16 October 2020; pp. 255–260.
54. Camci, B.; Kahveci, A.Y.; Arnrich, B.; Ersoy, C. Sleep apnea detection via smart phones. In Proceedings of the 2017 25th Signal Processing and Communications Applications Conference (SIU), Antalya, Turkey, 15–18 May 2017; pp. 1–4.
55. Lakhan, P.; Ditthapron, A.; Banluesombatkul, N.; Wilaiprasitporn, T. Deep Neural Networks with Weighted Averaged Overnight Airflow Features for Sleep Apnea-Hypopnea Severity Classification. In Proceedings of the TENCON 2018–2018 IEEE Region 10 Conference, Jeju, Korea, 28–31 October 2018; pp. 0441–0445.
56. Kim, T.; Kim, J.-W.; Lee, K. Detection of sleep disordered breathing severity using acoustic biomarker and machine learning techniques. *Biomed. Eng. Online* **2018**, *17*, 16. [CrossRef] [PubMed]
57. Falie, D.; Ichim, M. Sleep monitoring and sleep apnea event detection using a 3D camera. In Proceedings of the 2010 8th International Conference on Communications, Bucharest, Romania, 10–12 June 2010; pp. 177–180.
58. Davidovich, M.L.Y.; Karasik, R.; Tal, A.; Shinar, Z. Sleep Apnea Screening with a Contact:Free Under:the:Mattress Sensor. In Proceedings of the 2016 Computing in Cardiology Conference (CinC); Computing in Cardiology, Vancouver, BC, Canada, 11–14 September 2016; Volume 43, pp. 849–852.
59. Hafezi, M.; Montazeri, N.; Saha, S.; Zhu, K.; Gavrilovic, B.; Yadollahi, A.; Taati, B. Sleep Apnea Severity Estimation From Tracheal Movements Using a Deep Learning Model. *IEEE Access* **2020**, *8*, 22641–22649. [CrossRef]

60. Wang, C.; Chan, J.-H.; Fang, S.-H.; Cheng, H.-T.; Hsu, Y.-L. Novel Sleep Apnea Detection Based on UWB Artificial Intelligence Mattress. In Proceedings of the 2019 IEEE International Conference on Artificial Intelligence Circuits and Systems (AICAS), Hsinchu, Taiwan, 18–20 March 2019; pp. 158–159.
61. Javaid, A.Q.; Noble, C.M.; Rosenberg, R.; Weitnauer, M.A. Towards Sleep Apnea Screening with an Under-the-Mattress IR-UWB Radar Using Machine Learning. In Proceedings of the 2015 IEEE 14th International Conference on Machine Learning and Applications (ICMLA), Miami, FL, USA, 9–11 December 2015; pp. 837–842.
62. Avcı, C.; Akbaş, A. Sleep apnea classification based on respiration signals by using ensemble methods. *Bio-Med. Mater. Eng.* **2015**, *26*, S1703–S1710. [CrossRef] [PubMed]
63. Ozdemir, G.; Nasifoglu, H.; Erogul, O. A Time-Series Approach to Predict Obstructive Sleep Apnea (OSA) Episodes. In Proceedings of the 2nd World Congress on Electrical Engineering and Computer Systems and Science, Budapest, Hungary, 16–17 August 2016. [CrossRef]
64. Islam, S.M.; Mahmood, H.; Al-Jumaily, A.A.; Claxton, S. Deep Learning of Facial Depth Maps for Obstructive Sleep Apnea Prediction. In Proceedings of the 2018 International Conference on Machine Learning and Data Engineering (iCMLDE), Sydney, Australia, 3–7 December 2018; pp. 154–157.
65. Azimi, H.; Xi, P.; Bouchard, M.; Goubran, R.; Knoefel, F. Machine Learning-Based Automatic Detection of Central Sleep Apnea Events from a Pressure Sensitive Mat. *IEEE Access* **2020**, *8*, 173428–173439. [CrossRef]
66. Petrenko, A. Breathmonitor: Sleep Apnea Mobile Detector. In Proceedings of the 2020 IEEE 2nd International Conference on System Analysis & Intelligent Computing (SAIC), Kyiv, Ukraine, 5–9 October 2020; pp. 1–4.
67. Botelho, M.C.; Trancoso, I.; Abad, A.; Paiva, T. Speech as a Biomarker for Obstructive Sleep Apnea Detection. In Proceedings of the ICASSP 2019—2019 IEEE International Conference on Acoustics, Speech and Signal Processing (ICASSP), Brighton, UK, 12–17 May 2019; pp. 5851–5855.
68. Mencar, C.; Gallo, C.; Mantero, M.; Tarsia, P.; Carpagnano, G.E.; Barbaro, M.P.F.; Lacedonia, D. Application of machine learning to predict obstructive sleep apnea syndrome severity. *Health Inform. J.* **2019**, *26*, 298–317. [CrossRef]
69. Ustun, B.; Westover, M.B.; Rudin, C.; Bianchi, M.T. Clinical Prediction Models for Sleep Apnea: The Importance of Medical History over Symptoms. *J. Clin. Sleep Med.* **2016**, *12*, 161–168. [CrossRef]
70. Rechtschaffen, A.; Kales, A. *A Manual of Standardized Terminology, Techniques and Scoring System for Sleep Stages of Human Subjects*; Public Health Service, US Government Printing Office: Washington, DC, USA, 1968.
71. American Academy of Sleep Medicine. International Classification of Sleep Disorders. In *Diagnostic and Coding Mnaual*, 2nd ed.; Americal Academy of Sleep Medicine: Westchester, NY, USA, 2005.
72. Penzel, T.; Moody, G.; Mark, R.; Goldberger, A.; Peter, J. The apnea-ECG database. *Comput. Cardiol.* **2002**, *27*, 255–258. [CrossRef]
73. Quan, S.F.; Howard, B.V.; Iber, C.; Kiley, J.P.; Nieto, F.J.; O'Connor, G.T.; Rapoport, D.M.; Redline, S.; Robbins, J.; Samet, J.M.; et al. The Sleep Heart Health Study: Design, rationale, and methods. *Sleep* **1997**, *20*, 1077–1085. [CrossRef]
74. Dean, D.A.; Goldberger, A.L.; Mueller, R.; Kim, M.; Rueschman, M.; Mobley, D.; Sahoo, S.; Jayapandian, C.P.; Cui, L.; Morrical, M.G.; et al. Scaling Up Scientific Discovery in Sleep Medicine: The National Sleep Research Resource. *Sleep* **2016**, *39*, 1151–1164. [CrossRef]
75. Kemp, B.; Zwinderman, A.; Tuk, B.; Kamphuisen, H.; Oberye, J. Analysis of a sleep-dependent neuronal feedback loop: The slow-wave microcontinuity of the EEG. *IEEE Trans. Biomed. Eng.* **2000**, *47*, 1185–1194. [CrossRef] [PubMed]

healthcare

MDPI

Article

Clusters of Physical Frailty and Cognitive Impairment and Their Associated Comorbidities in Older Primary Care Patients

Sanja Bekić [1,2], František Babič [3,*], Viera Pavlišková [3], Ján Paralič [3], Thomas Wittlinger [4] and Ljiljana Trtica Majnarić [5,6]

1 General Medical Practice, 31000 Osijek, Croatia; sanja.bekic1@gmail.com
2 Faculty of Medicine, University Josip Juraj Strossmayer, 31000 Osijek, Croatia
3 Department of Cybernetics and Artificial Intelligence, Faculty of Electrical Engineering and Informatics, Technical University of Košice, 04201 Košice, Slovakia; viera.pavliskova@tuke.sk (V.P.); jan.paralic@tuke.sk (J.P.)
4 Department of Cardiology, Asklepios Hospital, 38642 Goslar, Germany; dr.wittlinger@gmx.de
5 Department of Internal Medicine, Family Medicine and the History of Medicine, Faculty of Medicine, University Josip Juraj Strossmayer, 31000 Osijek, Croatia; ljiljana.majnaric@mefos.hr
6 Department of Public Health, Faculty of Dental Medicine and Health, University Josip Juraj Strossmayer, 31000 Osijek, Croatia
* Correspondence: frantisek.babic@tuke.sk

Abstract: (1) Objectives: We aimed to identify clusters of physical frailty and cognitive impairment in a population of older primary care patients and correlate these clusters with their associated comorbidities. (2) Methods: We used a latent class analysis (LCA) as the clustering technique to separate different stages of mild cognitive impairment (MCI) and physical frailty into clusters; the differences were assessed by using a multinomial logistic regression model. (3) Results: Four clusters (latent classes) were identified: (1) highly functional (the mean and SD of the "frailty" test 0.58 ± 0.72 and the Mini-Mental State Examination (MMSE) test 27.42 ± 1.5), (2) cognitive impairment (0.97 ± 0.78 and 21.94 ± 1.95), (3) cognitive frailty (3.48 ± 1.12 and 19.14 ± 2.30), and (4) physical frailty (3.61 ± 0.77 and 24.89 ± 1.81). (4) Discussion: The comorbidity patterns distinguishing the clusters depend on the degree of development of cardiometabolic disorders in combination with advancing age. The physical frailty phenotype is likely to exist separately from the cognitive frailty phenotype and includes common musculoskeletal diseases.

Keywords: multimorbidity; primary care; physical frailty; cognitive impairment; latent cluster analysis

Citation: Bekić, S.; Babič, F.; Pavlišková, V.; Paralič, J.; Wittlinger, T.; Majnarić, L.T. Clusters of Physical Frailty and Cognitive Impairment and Their Associated Comorbidities in Older Primary Care Patients. *Healthcare* **2021**, *9*, 891. https://doi.org/10.3390/healthcare9070891

Academic Editor: Md Mahmudur Rahman

Received: 13 June 2021
Accepted: 12 July 2021
Published: 15 July 2021

1. Introduction

Population aging is a global trend in EU countries [1]. Accompanying this trend is an increase in the number of individuals with multimorbidity (a coexistence of two or more chronic diseases in the same person) and who are showing functional decline, which poses new challenges to healthcare systems, such as high requirements for utilizing healthcare services and long-term care, in particular.

Epidemiologic studies have indicated that multimorbidity increases with age and is associated with a deterioration in mental health and low physical, cognitive, and social functioning [2–4]. These observations support our current understanding of the development of common diseases of aging, such as diabetes type 2 (diabetes), cardiovascular disease (CVD), Alzheimer's dementia, and some types of cancer, as being an integrative part of the aging process [5]. Although, it has been realized that the extent to which these diseases and functional organ impairments are expressed vary between individuals, reflecting interindividual differences in rates of aging, so that the real (biological) age may fall behind or outpace the chronological age [6]. The causes and mediators of such differences are mostly unknown. According to today's prevailing theory of aging, inflammaging, a variety of stimuli operating at cellular and subcellular levels in the body, contribute to low-grade inflammation as the main driver in the acceleration of aging and the development of age-related diseases [7]. Fat tissue redistribution, which occurs with aging and is clinically visible as the abdominal type of obesity, in particular when it is combined with overnutrition and general overweight/obesity, may substantially contribute to inflammation and metabolic disorders associated with aging and the development of cardiometabolic age-related diseases, such as metabolic syndrome, diabetes, and CVD. Cerebral small-vessel disease, recognized as a pathologic mechanism underlying non-Alzheimer's cognitive disorders, is considered a part of inflammaging and reflective of the overwhelming influence of metabolic and inflammatory stimuli on pathologic changes in the brain vasculature [8].

An age-related decline in physical and cognitive capabilities can be best described by applying the concepts of physical frailty and mild cognitive impairment (MCI). Both conditions have been proven to independently increase the risk of negative health outcomes, including falls, disability, dementia, hospitalization, institutionalization, and death [9]. Frailty is considered a manifestation of reduced homeostatic reserves in many vital systems that govern neuroendocrine, energy-metabolic, and inflammation-immunologic mechanisms [10]. The transition from a prefrailty to frailty state takes place in parallel with the progression of pathophysiologic disorders, when it becomes increasingly less possible to reverse this syndrome [11,12]. The concept of MCI has been introduced to define a stage of cognitive decline between normal cognition and dementia that can be objectively measured but is still not severe enough to affect the activities of daily living [13]. Although MCI is associated with an increased risk for developing dementia, without additional complementary variables, this measure is not powerful enough to accurately predict dementia [14,15].

Emerging evidence indicates that these two disorders, physical frailty and cognitive impairment, often coexist and mutually interact, thus increasing the risk of each condition for poor health outcomes [16,17]. Although the evidence suggests that these disorders share many risk factors and mechanisms, the knowledge of common pathophysiologic pathways is still low, mainly because these disorders have been studied separately so far as independent entities [18,19].

A new entity, termed cognitive frailty, defined as the coexistence of prefrailty or frailty with MCI, has been established by the international consensus group with the aim to facilitate research on cognitive impairment that is caused by deteriorating physical health, thus distinguishing physical from neurodegenerative causes of cognitive impairment [20].

It is becoming increasingly apparent that the dynamic interplay between chronic diseases and functional impairments, which are modulated by genetic, behavioral, and environmental factors, as well as by applied treatments, directs the rates of age-related decline in physical and cognitive performance [21]. Although prospective epidemiologic studies indicate that physical frailty may be a driver of cognitive impairment, and that the opposite is less likely to occur, our knowledge concerning the exact clustering patterns of physical frailty and cognitive impairment and of their dynamics of change in the aging population is poor [17,22]. There is an increasing expert consensus that screening for cognitive impairment should be performed in all older prefrail and frail individuals with multimorbidity [17,18].

The recent shift in research on multimorbidity from disease counting to disease clustering has revealed disease patterns that could be based on common pathophysiologic pathways [23–25]. The aim of the present study was to identify clusters of physical frailty and MCI in a population of older primary care patients and to correlate the identified clusters with comorbidities and chronological age. Differences among clusters in degrees of functional declinemay reflect interindividual differences in rates of aging. Identified clusters will relate these differences to the level of the development of age-related diseases and functional organ impairments, more precisely reflecting the aging process than by using chronologic age alone [6].

2. Methods

2.1. Study Design and Participants

A cross-sectional study and retrospective analysis of the selected data used from primary care (PC) electronic health records (eHRs) were conducted in 2018 in an academic General Practice (GP) facility in the town of Osijek (currently around 60,000 inhabitants), the administrative center of Eastern Croatia. Due to the poor economic situation in this area, negative demographic trends and population aging have taken place, which has led to a high burden of chronic diseases; higher than this is the average in Croatia.

The analysis included 263 older (≥60 years) ambulatory PC patients who were enrolled at their regular visits or were invited for an interview. According to several rules of thumb, the sample size of 250 participants was sufficiently large to show statistical significance for less complicated latent class analysis (LCA) models, which was the critical analytical method used in this study [26–28].

The fact that patients were recruited from one GP facility did not hamper the representativeness of the sample, because older people living in the area have similar living conditions and are generally of a lower socioeconomic status. A good match with the general population was ensured by the fact that, in Croatia, the general population has good access to PC services, and almost all inhabitants are registered on the lists of PC physicians. The data collection from a single practice may even have some advantages by ensuring the uniformity of the diagnostic criteria and terminology that is used in communication with patients and during the diagnostic process. The fact that it was an academic GP facility ensured that the data was collected by a skilled and knowledgeable PC physician, which could guarantee the high level of data accuracy.

Of approximately 2000 patients registered in this GP facility, about a quarter were older individuals, and about a half of them entered the study. We used for analysis only community-dwelling patients to whom preventive measures, if applied, may still be beneficial and not those in home care programs or in institutions. The exclusion criteria were also acute medical conditions, exacerbations of chronic conditions, and diagnoses of psychosis or dementia. Excluded from the study were also several patients with incomplete health records. We already have two papers published using the same dataset. In the first published paper, an unsupervised learning algorithm, k-means, was applied on the data obtained from 159 patients who were enrolled first to identify clusters of numerical variables indicating mental disorders, cognitive impairment, physical frailty, and laboratory tests [29]. Information on diagnoses of chronic diseases and some functional and sensory organ impairments was used to complement the description of these clusters. When the data collection was finished, we applied the supervised latent class analysis (LCA) model on the full-sized sample of 263 patients to identify individuals with different stages of cognitive impairment and physical frailty who showed a tendency to cluster together [30]. In that paper, we presented the first part of the complex analysis, where we assessed how membership in a cluster is influenced by performances on tests of mental disorders, anxiety, and depression and by specific cognitive test tasks. In this paper, we presented the second part of the analysis, where we analyzed the differences among clusters concerning comorbidities and functional/sensory organ impairments.

2.2. Data Collection

The selection of variables that were used for analysis was based on knowledge and data availability. Data were collected from eHRs on the number and types of diagnoses of chronic diseases, the total number of prescribed medications and the number of medications with an effect on mental functions, and on laboratory tests that are routinely performed in PC to check patients' health status and which indicate metabolic disturbances and the status of inflammation and nutrition. Diagnoses of chronic diseases were recorded according to the international disease coding system (ICD-10). The laboratory test results were used from chronic disease surveillance programs and preventive check-ups and were not older than a year. The systematic way of data recording in these platforms has ensured a high level of data completeness. Only in a few cases was data missing, indicating the C-reactive protein or glomerular filtration rate, and these patients were excluded from the analysis. The laboratory test results were assessed according to appropriateness for participants' age and health status by comparing them with the laboratory reference values and recommendations from the international guidelines for managing common chronic conditions [31,32]. Information on functional/sensory organ impairments was gathered from eHRs and by patient interviews. Anthropometric measures, BMI (body mass index), a measure of general nutrition, waist circumference (a measure of abdominal obesity), and the mid-arm circumference (a measure of muscle mass loss) were performed during patient visits to add to the information on the nutritional and health status of participants who were recruited to the study [33].

To determine the level of physical frailty of participants, we used the Fried phenotypic model, which is the best-validated of available, similar measures [34]. Based on five criteria, weight loss, slow walking speed, weak grip strength (measured by the handgrip dynamometer), a subjective feeling of exhaustion, and reduced activity, this model indicates whether an individual is prefrail (1 to 2 positive criteria), frail ($\geq$3 positive criteria), or robust (no one positive criterion).

For measuring MCI, as a component of the cognitive frailty phenotype, the international consensus group recommended the Clinical Dementia Rating Scale [20]. To screen participants for MCI in this study, we used the Mini-Mental State Examination (MMSE) test, which has been broadly validated also in the elderly Croatian population [35]. This test consists of several domains, indicating either memory-related or non-memory-related (executive) functions. The MMSE cut-offs were adjusted for the participants' level of education based on the MMSE cut-off values for the Croatian population. This cut-off was 24/25 (of the maximum 30) for screening among older individuals in the general population and 26/27 for screening among those with a higher level of education (defined as $\geq$14 years of education). The MMSE test is more sensitive for diagnosing severe cognitive impairment (scores $\leq$ 17) than for distinguishing between cognitively healthy individuals and those with MCI and cannot distinguish between different types of dementia (Alzheimer's type vs. vascular type).

Information on the sociodemographic characteristics and medical history of the participants are presented in Supplementary Materials. The numerical variables are presented as the mean and the standard deviation (SD) or as the median and the interquartile range (Supplementary Materials Table S1). The categorical variables are presented with the absolute numbers and frequencies (%) (Supplementary Materials Table S2).

2.3. Statistical Analysis

The LCA method was used to identify subgroups, latent classes, as statistically distinct and clinically meaningful patterns that optimally comprehend the heterogeneity of participants in the sample regarding their achievements on the MMSE test and the Fried frailty score [30,36].

Differences in distributions of numerical variables among the clusters were analyzed using the one-way analysis of variance (ANOVA) or Kruskal–Wallis rank sum test, depending on whether numerical variables showed a normal distribution. This analysis was followed by the Games-Howell post hoc test. Differences in the categorical variables were assessed using the chi-square ($\chi2$) test and Fisher's exact test, where appropriate. Bar diagrams were used to visualize the distributions of those categorical variables for which differences among the clusters reached statistical significance.

To assess how the examined variables are associated with membership in a cluster, we used a multinomial logistic regression (MLR) model from R statistics. A cluster consisting of individuals with the best cognitive and physical performances was used as a control. We analyzed the impact of age and gender on clusters' membership in a separate MLR model. Four other models were created to show the impact on clusters' membership of (1) the level of comorbidity, presented with variables indicating the number of comorbidities and functional/sensory organ impairments and the number of prescribed medications and medications with an effect on mental functions, (2) the health-related status presented with variables indicating anthropometric measures and laboratory tests, (3) particular diagnoses of chronic diseases, and (4) functional/sensory organ impairments. Before generating these models, we checked all numerical variables in the input on collinearity, using a simple linear correlation analysis, and on multicollinearity, using the variance inflation factor (VIF) as an indicator. Variables with a high level of collinearity were not included in the models. In the third and the fourth models, only variables that were shown significant in the analysis of the differences entered the model. The AIC (Akaike Information Criterion) was used to measure the quality of the model's predictive performance [37].

3. Results

Members of the first cluster showed better cognitive and physical performance than members of the other three clusters. This cluster was therefore termed highly functional (HF). In members of the second cluster, the performance on the MMSE test was decreased, but the physical performance was good (low average frailty score); the cluster was therefore termed cognitive impairment (CI). The third cluster was termed cognitive frailty (CF), as members of this cluster showed low physical performance (increased average frailty score) and low cognitive performance (decreased average score on the MMSE test). Finally, the fourth cluster was termed physical frailty (PhyF), as its members had low physical performance (similar to members of the CF cluster) but well-preserved cognitive performance (Table 1).

Table 1. Average scores on the frailty and MMSE tests and average age across the clusters. Division of members of the clusters according to gender.

	Cluster	Number of Patients (M:F)	Average Score ± SD *	*p*-Value (Post-Hoc)	Age (Year) Average ± SD	*p*-Value (Post-Hoc)
Frailty	HF	161 (62:99)	0.58 (0.721)	<0.001 HF < CI, CF, PhyF	69.40 (5.455)	<0.001 HF < CI, CF, PhyF
	CI	63 (22:41)	0.97 (0.782)	<0.001 CI < CF, PhyF	72.33 (6.611)	<0.001 CI > HF < CF
	CF	21 (6:15)	3.48 (1.123)	<0.001 CF > CI, HF	78.62 (5.792)	<0.001 CF > CI, HF
	PhyF	18 (1:17)	3.61 (0.777)	<0.001 PhyF > CI, HF	74.72 (6.515)	<0.001 PhyF > HF
	Total	263	1.11 (1.286)		71.20 (6.434)	
MMSE	HF	161	27.42 (1.556)	<0.001 HF > CI, CF, PhyF		
	CI	63	21.94 (1.958)	<0.001 CI > CF < PhyF		
	CF	21	19.14 (2.308)	<0.001 CIF > CI, HF, PhyF		
	PhyF	18	24.89 (1.811)	<0.001 PhyF > CI, CF < HF		
	Total	263	25.27 (3.398)			

Note: HF: highly functional, CI: cognitive impairment, CF: cognitive frailty, and PhyF: physical frailty. * Higher scores on the frailty test indicate a higher level of physical frailty, whereas higher scores on the MMSE test indicate a higher level of cognitive function. The results of the post hoc test are represented by a formulation like the cluster combinations HF < CI HF < CF and HF < PhyF CI are significantly different from each other.

Table 1 also shows that individuals in the HF cluster are younger than those in other clusters, whereas those in the CF cluster are significantly older than those in CI. Individuals in the PhyF cluster are older than those in the CI cluster and younger than those in the CF cluster, but the differences are not significant. There were no significant differences in the distributions by gender (M:F) within the clusters, except for the PhyF cluster, in which women were dominant (17:1) ($\chi 2$ test, $p < 0.05$).

Table 2 shows that the PhyF cluster contains only frail individuals and that the CF cluster contains a high proportion of frail individuals and a smaller proportion of prefrail individuals. In contrast to these clusters, in the CI and HF clusters, a prevalent proportion of the individuals is prefrail (44.1% and 68.3%, respectively), and none are frail.

Table 2. Division of members of the clusters according to their frailty status and MCI diagnosis.

		Within Clusters				*p*-Value *	All
		HF	CI	CF	PhyF		
Prefrail	N	71	43	3	0	<0.001	117
	% within a cluster	44.1%	68.3%	14.3%	0.0%		44.5%
Frail	N	1	0	18	18	<0.001	37
	% within a cluster	0.6%	0.0%	85.7%	100.0%		14.1%
MCI **	N	18	47	19	7	<0.001	91
	% within a cluster	11.1%	74.6%	90.5%	38.9%		34.6%
All		161	63	21	18		263
		100.0%	100.0%	100.0%	100.0%		100.0%

Note: HF: highly functional, CI: cognitive impairment, CF: cognitive frailty, and PhyF: physical frailty. * Pearson chi-square or Fisher's Exact test, where appropriate. ** MMSE cut-offs for mild cognitive impairment (MCI) adjusted for level of education; levels in the Croatian population ≥65 years were set at ≤24 for education level <14 years and at ≤26 for education level ≥14 years.

The majority of individuals in clusters characterized by decreased cognitive function had MCI (74.6% in the CI cluster and 90.5% in the CF cluster, respectively), whereas this proportion was much smaller in the other two clusters (11.1% in the HF cluster vs. 38.9% in the PhyF cluster).

Table 3 shows that clusters in which frailty individuals are dominant (the PhyF and CF clusters), compared to clusters in which individuals are at the stage of prefrailty (the HF and CI clusters), have a higher number of chronic disease diagnoses and prescribed medications, including medications affecting mental functions. The highest number of individuals with functional/sensory organ disorders are allocated to the cluster representing the physical frailty phenotype (the PhyF cluster).

Table 3. Differences among individuals in the clusters in the level of comorbidity.

Variable	Median (Interquartile Range) Mean ± SD *				*p*-Value **	Games-Howell Post Hoc Test
	HF	CI	CF	PhyF		
Total number of diagnoses	3.00 (2.00)	3.00 (2.00)	3.84 (2.19) *	4.67 (1.88) *	0.0006	PhyF > HF PhyF > CI
Total number of prescribed medications	3.00 (3.00)	3.00 (3.00)	4.10 (1.97) *	5.17 (2.12) *	0.005	PhyF > HF PhyF > CI
Total number of medications with effect on mental functions	3.00 (2.00)	2.00 (2.00)	3.10 (1.45) *	4.17 (1.62) *	**0.01**	PhyF > HF PhyF > CI
Total number of sensory/functional disorders	2.00 (1.00)	1.00 (1.00)	2.00 (1.05) *	3.00 (1.00)	**0.009**	PhyF > HF PhyF > CI PhyF > CF

Note: *p*-values shown in bold are significant (significance level = 0.05). *—values of mean ± SD were used when Shapiro-Wilk's test confirmed the normality, **—non-parametric Kruskal-Wallis rank sum test was applied. The results of the GW post hoc test are represented by a formulation like the cluster combinations PhyF > HF and PhyF > CI are significantly different from each other.

Individuals in the CF cluster, in whom the cognitive frailty phenotype is dominant, had the lowest values for the variables indicating mid-arm circumference, HDL cholesterol, hemoglobin, erythrocyte count, and glomerular filtration rate (a marker of renal function) (Table 4).

Table 4. Differences among individuals in the clusters in the health status described with anthropometric measures and laboratory tests.

Variable	Median (Interquartile Range) Mean ± SD *				*p*-Value **	Games-Howell Post Hoc Test
	HF	CI	CF	PhyF		
BMI (kg/m^2)	29.73 (5.58)	30.35 (4.45) *	28.13 (4.83)*	28.53 (4.85)	0.19	
Waist circumference (cm)	99.0 (16.00)	101.70 (11.47)*	94.53 (11.18) *	96.61 (16.86)*	0.12	
Mid-arm circumference (cm)	32.00 (3.00)	31.59 (3.63) *	28.79 (3.12) *	30.25 (2.75)	**0.003**	CF < HF CF < CI
Fasting glucose (mmol/L)	5.50 (1.60)	5.90 (1.65)	5.30 (1.20)	5.70 (1.95)	0.18	
Total cholesterol (mmol/L)	5.76 (1.35) *	5.75 (1.24) *	5.93 (1.37) *	6.23 (1.53) *	0.21 **	
LDL cholesterol (mmol/L)	3.60 (1.40)	3.46 (1.06) *	3.46 (1.06) *	3.93 (1.32) *	0.55	
HDL cholesterol (mmol/L)	1.40 (0.40)	1.30 (0.45)	1.22 (0.31) *	1.59 (0.37) *	**0.01**	PhyF > CF
Triglycerides (mmol/L)	1.70 (0.90)	1.80 (0.95)	1.50 (0.55)	1.40 (0.60)	0.10	
Glomerular filtration rate (mL/min/1,73 m^2)	90.43 (25.50) *	86.08 (29.07) *	67.53 (20.69) *	71.72 (24.08) *	**0.0001** **	CF < HF CF < CI PhyF < HF
C-reactive protein (mg/L)	2.20 (3.20)	2.20 (3.25)	2.40 (4.85)	1.60 (2.10)	0.94	
Hemoglobin (g/L)	138.00 (15.00)	137.00 (15.00)	130.60 (12.11) *	132.20 (22.80) *	0.03	CF < HF
Erythrocyte count (x 10^{12}/L)	4.62 (0.43) *	4.56 (0.34) *	4.33 (0.40) *	4.62 (0.42)	0.03	CF < HF

Note: *p*-values shown in bold are significant (significance level = 0.05). *—values of mean ± SD were used when Shapiro-Wilk's test confirmed the normality, **—non-parametric Kruskal-Wallis rank sum test was applied. The results of the GW post hoc test are represented by a formulation like the cluster combinations CF < HF and CF < CI are significantly different from each other.

It can be seen in Table 5 that the diagnoses of chronic diseases with the most impact for distinguishing comorbidity profiles among the clusters include chronic heart disease, coronary artery disease, upper gastrointestinal tract disorders, osteoporosis, osteoarthritis, low back pain, and anxiety/depression. Of the functional/sensory organ disorders, significant differences among the clusters were shown for falls, walking difficulties, and chronic pain. The proportion of individuals with three or more diagnoses of chronic diseases (indicating the status multimorbidity) in particular clusters is as follows: 55.9% (HF), 47.6% (CI), 78.9% (CF), and 88.9% (PhyF).

Table 5. Differences among individuals in the clusters in particular diagnoses of chronic diseases and functional/sensory organ impairments.

Diagnosis	*p*-Value	Diagnosis	*p*-Value
Hypertension	0.14	Osteoporosis (confirmed)	**0.005**
Diabetes mellitus type 2	0.078	Severe osteoarthritis	0.078
Chronic obstructive pulmonary disease	1.00 *	Low back pain	**0.008**
Asthma or allergic rhinitis	0.575	Parkinson's disease	0.384 *
Chronic heart disease (failure)	**0.005** *	Urogenital diseases	0.178
Coronary artery disease	**0.039**	The thyroid gland dysfunctions	0.317
Cerebrovascular disease	0.401	Anxiety/depression	**0.011**
Periphery artery disease	0.052 *	Incontinent and other urinary bladder disorders	0.078 *
Upper gastrointestinal tract disorders	**0.030**	Significant visual loss	0.378
Chronic hepatic disorders	1.00 *	Registered hearing impairment or communication difficulties due to hearing loss	0.116
Malignant disease	0.203	Experienced falls	**<0.001**
Chronic pain complaints	**0.008**	Walking with support or visible impaired	**<0.001**
		≥3 Dg of chronic diseases	**<0.001**

Note: Fisher's exact test (marked with *, when participants in a cluster < 5% or $N \leq 10$). *p*-values shown in bold are significant (significance level = 0.05).

Figure 1 shows that participants with the diagnosis of chronic heart disease are mostly allocated to the third (CF) cluster, then to the second (CI) cluster and the fourth (PhyF) cluster. Participants diagnosed with coronary artery disease are mostly allocated to cluster 3 (CF) and then to cluster 4 (PhyF). The diagnosis of upper gastrointestinal tract disorders is more prevalent in clusters that are marked by physical frailty (the CF and PhyF clusters) than in the other two clusters (the HF and CI clusters). In fact, upper gastrointestinal tract disorders are mostly present in cluster 4 (PhyF). The diagnoses of osteoporosis and low back pain (syndroma lumbale) are mostly present in cluster 4 (PhyF), whereas the diagnosis of osteoarthritis is more prevalent in clusters CF and PhyF than in the other two clusters (HF and CI). The frequency of the diagnosis of anxiety/depression is relatively high in all clusters, but the highest frequency for this diagnosis is found in cluster 4 (PhyF).

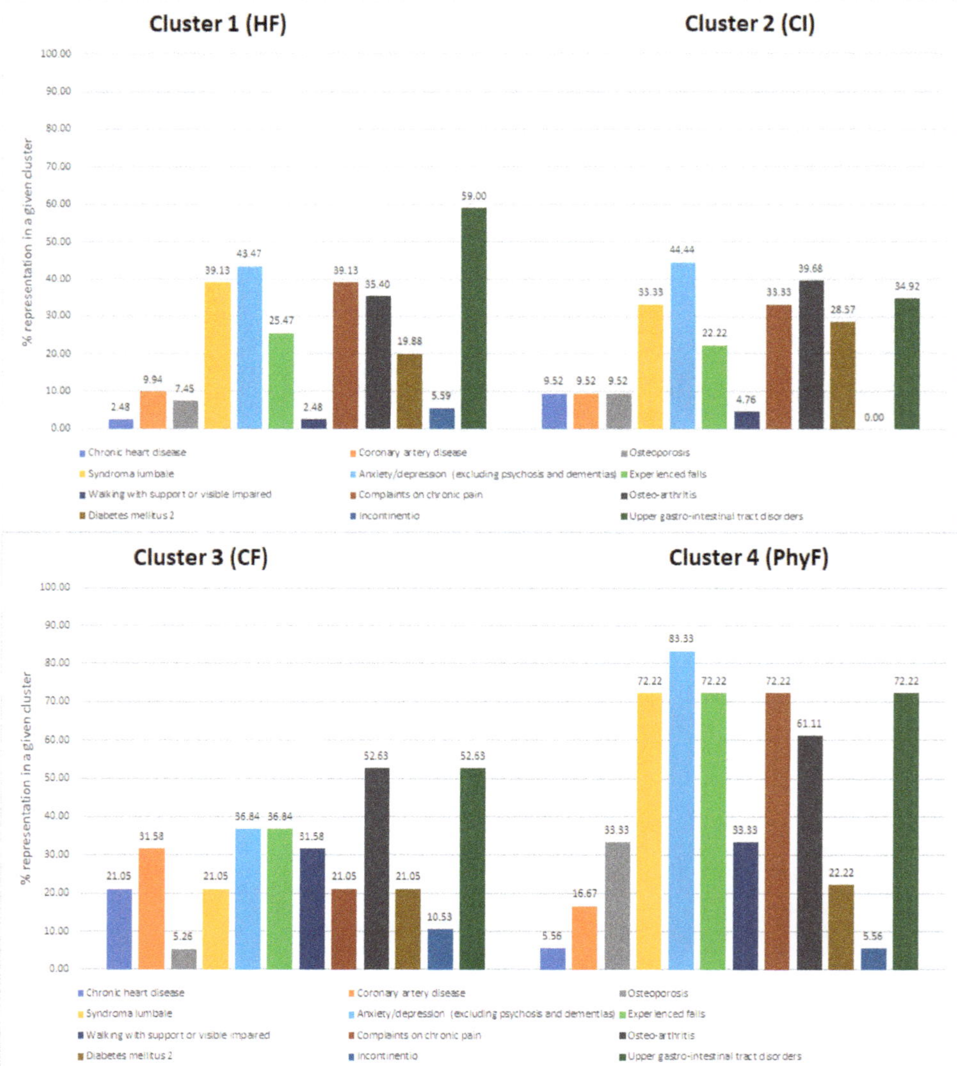

Figure 1. Graphical presentation of the differences among clusters in the diagnoses of chronic diseases and functional/sensory organ impairments.

Most participants who experienced falls were allocated to clusters 3 (CF) and 4 (PhyF). Those in cluster 4 (PhyF), more often than those in the cluster 3 (CF), experienced falls with bone fractures. Subjective walking difficulties were expressed mostly by participants in clusters 3 (CF) and 4 (PhyF). Chronic pain was a hallmark of cluster 4 (PhyF).

The MLR model presented in Table 6 shows that increased age had an impact on memberships to all three pathologic clusters (CI, CF, and PhyF). The gender imbalance only had an impact on the PhyF cluster.

Table 6. A multinomial logistic regression model indicating differences in the clinical profiles of individuals in the clusters according to their age and gender.

	Cluster CI		Cluster CF		Cluster PhyF	
	z-Value	OR	z-Value	OR	z-Value	OR
Gender = male					−2.29	0.17 (0.05–0.60)
Age	3.46	1.09 (1.05–1.15)	4.96	1.29 (1.19–1.40)	3.75	1.19 (1.10–1.28)

As shown in Tables 7–10, the variables that best characterize cluster 2 (CI) as compared to the control cluster 1 (HF) included increased fasting blood glucose, increased hemoglobin, and the diagnosis of chronic heart disease.

Table 7. A multinomial logistic regression model indicating the differences in the levels of comorbidities among individuals in the clusters.

	Cluster CI		Cluster CF		Cluster PhyF	
	z-Value	OR	z-Value	OR	z-Value	OR
Total number of sensory/functional disorders			−1.92	0.72 (0.64–1.51)	2.92	2.34 (1.45–3.78)

Table 8. A multinomial logistic regression model indicating the differences in their health status, defined by anthropometric measures and laboratory tests, among individuals in the clusters.

	Cluster CI		Cluster CF		Cluster PhyF	
	z-Value	OR	z-Value	OR	z-Value	OR
HDL cholesterol			−2.11	0.12 (0.02–0.63)		
Fasting glucose	2.64	1.23 (1.10–1.40)				
Mid arm circumference			−2.15	0.84 (0.73–0.96)		
Glomerular filtration rate			−2.35	0.97 (0.96–0.99)	−2.37	0.97 (0.95–0.99)
Haemoglobin	3.04	1.06 (1.03–1.09)				

Table 9. A multinomial logistic regression model indicating the differences in their health status, defined by particular diagnoses of chronic diseases, among individuals in the clusters.

	Cluster CI		Cluster CF		Cluster PhyF	
	z-Value	OR	z-Value	OR	z-Value	OR
Chronic heart disease (failure) = yes	2.20	4.78 (1.49–15.35)	1.89	5.05 (1.23–20.77)		
Dg of osteoporosis (confirmed) = yes					2.24	4.30 (1.47–12.54)
Dg of anxiety/depression (excluding psychosis and dementias) = yes					2.10	4.17 (1.36–12.73)

Table 10. A multinomial logistic regression model indicating the differences in their health status, defined by particular functional/sensory organ impairments, among individuals in the clusters.

	Cluster CI		Cluster CF		Cluster PhyF	
	z-Value	OR	z-Value	OR	z-Value	OR
Complaints on chronic pain = yes					2.19	3.57 (1.37–9.30)
Experienced falls = yes					2.70	5.03 (1.88–13.47)
Walking with support or visible impaired = yes			3.89	18.89 (5.45–65.53)	2.83	8.93 (2.50–31.92)

Having decreased values for HDL cholesterol, mid-arm circumference, and glomerular filtration rate (a measure of decreased renal function), together with walking difficulties and the diagnosis of chronic heart disease, increased the probability of belonging to cluster 3 (CF).

The most prominent clinical characteristics of cluster 4 (PhyF) included decreased renal function, as indicated by the variable glomerular filtration rate, and the highest rate of functional/sensory organ impairments—in particular, including chronic pain, walking difficulties, and falls. Of comorbidities specifically associated with cluster 4 (PhyF) were diagnoses of osteoporosis and anxious–depressive disorders.

4. Discussion

The identified clusters (latent classes) represent patterns of two main age-related functional disorders, physical frailty and cognitive impairment, that most optimally describe the functional heterogeneity of older, ambulatory PC patients. Indeed, trajectories for rates of aging have not yet been identified; therefore, dividing an older population into such clusters can show individuals who share similar levels of risk for some negative health outcomes [5]. An assessment of the possible at-risk individuals in the clusters were described by many sociodemographic and health-related characteristics. As there is no adequate research framework for investigating multimorbidity, this method can be applied to manage older patients with multimorbidity in a more integral manner than is currently possible when chronic diseases are considered as independent entities [38,39].

Overall, it can be said that HF and CI clusters represent the early stages of frailty (all individuals are robust or prefrail) and CF and PhyF clusters, in which cognitive frailty and physical frailty phenotypes are dominant and represent the final pathways in the development of frailty. Accordingly, individuals in the two latter clusters are generally older, present with more chronic conditions, and use more medications than individuals in the two former clusters. These results support evidence suggesting that the accumulation of comorbidities with age, together with the effect of polypharmacy that accompanies it, governs the transitions of an individual's health status to states of greater frailty and disability [22,40,41]. It is important to know how older persons in a population are distributed into these clusters, because only when multimorbidity is combined with frailty does it significantly increase the vulnerability of older persons for different stressors, predisposing them to increased mortality [41].

Differences between the CI and CF clusters in the rates of cognitive performance, together with the switch in participation from the dominant participation of prefrail to the dominant participation of frail individuals, between these two clusters support the evidence indicating that cognitive performance progressively declines across the frailty states and that, in prefrail individuals, cognitive impairment is an early sign of comorbidity-related cerebral involvement [22]. Moreover, this result supports the knowledge indicating that the likelihood of adopting the cognitive frailty phenotype strongly depends on an advancement in age [20].

While the effect of age is obviously important for functional decline to develop, the effect of pre-existing health conditions and behavioral coping strategies may direct the course of the pathophysiology disorders, either towards the development of the physical frailty phenotype or the cognitive frailty phenotype. A better understanding of these external influences could improve our capabilities to cope with the modifiable factors that accelerate aging. To reveal if there is a well-functioning group among very old (80+) individuals corresponding with the course of aging, termed as successful aging, only a large-scale study could give an answer [5]. These statements can better come to one's senses if analyzing clusters separately from each other or in comparison to each other. Thus, individuals in the HF cluster are the youngest and healthiest. However, they are not wholly free from chronic medical conditions. The moderate presence of prefrail individuals in this cluster (44.1%) can be considered as what Strandberg called "primary frailty" (a vicious cycle in which mild frailty precedes and potentiates the development of most comorbidities) [42]. This stage of frailty, in individuals in the HF cluster, can be explained by increased BMI and waist circumference values, indicating overweightness in combination with the abdominal type of obesity. Important to know in these terms is that the global pandemic of obesity, also affecting the older part of the population, can modify the expression of frailty, which was originally viewed as a state of the body shrinking [34,43]. The abdominal type of obesity represents a sign of ectopic fat accumulation and is detrimental to the development of age-related diseases by contributing to increased systemic inflammation [7]. In muscles, this ectopic fat storage is associated with muscle wasting and weakness, reducing the physical performance in obese individuals [43]. Obesity can further contribute to the development of frailty by acting through obesity-related comorbidities, such as anxious–depressive disorders and chronic lumbar pain, as also indicated by our results, by mechanisms such as reduced mobility and motivation for activities [44].

A comparison of the HF and CI clusters has shown that these clusters share many clinical characteristics, such as the number of chronic diseases and prescribed medications, fairly justified anthropometric indices of obesity, and relatively good renal function. It is expected, as these two clusters represent the early stages of frailty. Yet, they differ from each other in that the individuals in the CI cluster are significantly older and have worse CV profiles, the characteristics of which include higher rates of diabetes and chronic heart disease, longer diabetes duration, and worse diabetes control, as indicated by higher fasting serum glucose (chronic hyperglycemia). Based on its worse cardiometabolic profile, the CI cluster exhibits higher rates of prefrail individuals than the HF cluster (68.3% vs. 44.1%). It is due to the fact that both diabetes and CVD are considered a part of inflammaging and are also closely associated with frailty [45,46]. By putting these results into a broader context of inflammaging, then a wide range of comorbidities with a common pathophysiologic background, that may precede or overlap with CVD, also contribute to the close association of CVD with frailty [47,48].

According to the inflammaging theory, metabolic and inflammatory factors, by acting over time in a vicious cycle, may intensify cardiometabolic comorbidities [5,7]. Differences between the CI cluster and the HF cluster in expressing cognitive impairment (74.6% vs. 11.1% of the members with MCI and a decreased average MMSE score in the CI cluster but not in the HF cluster) can also be viewed in this context. In this case, the cerebral small-vessel disease is thought to be that structural correlate of the brain that makes a link between the intensification of cardiometabolic disorders and worsening of cognitive function [8].

Further, in the same context, our results indicate that individuals in the CF cluster, who are significantly older than those in the CI cluster, also have more CVD. This worse CV profile can explain the higher frailty rate and worse MMSE score of individuals in the CF cluster. A prominent feature of this profile is the markedly decreased renal function, which, in the CF cluster but not in the CI cluster, reached the level of chronic kidney disease (glomerular filtration rates <60 mL/min/1.73 m2) [31]. Impaired renal function is a common and concomitant disorder of CVD and cardiometabolic conditions and associated with increased inflammation and the risk of developing malnutrition, sarcopenia (muscle wasting), and frailty [49,50]. All these conditions were found to overlap in the geriatric population, and sarcopenia is increasingly being considered a marker of frailty [51,52].

What else matters when considering the effect of chronic renal impairment on developing the cognitive frailty phenotype, as our results and evidence indicate is the level of this impairment and of the burden of associated disorders [53]. In this regard, individuals in the CF cluster have significantly lower renal function and a higher burden of CVD than individuals in the CI cluster and are also characterized with higher levels of inflammation/malnutrition, as indicated with low HDL cholesterol and a higher level of muscle loss (sarcopenia), as indicated by the lower mid-arm circumference [33,54]. This pathophysiologic background, associated with inflammation–malnutrition and sarcopenia, may underlie the fully developed frailty state in individuals in the CF cluster.

As underscored by our results, individuals in the CF cluster do not differ significantly from those in the PhyF cluster in age, the level of comorbidity and medicalization, and the degree of renal function decline, but they are, nevertheless, characterized by significant cognitive impairment, while individuals in the cluster PhyF are not. Therefore, chronic kidney disease must coexist with some higher degree of CVD expression for cognitive impairment to occur. This conclusion arises from the results indicating that there is a difference between these two clusters in the level of the expression of CVD (including diagnoses of chronic heart disease and coronary artery disease), this level being higher in the CF than in the HF cluster. At higher levels of CV comorbidities, we would expect that the level of inflammation also increases, governing the development of clinically significant malnutrition and muscle loss [55]. According to the inflammaging theory, accelerated cerebral small-vessel disease is a result of the action of intensive cardiometabolic factors and increased inflammation on the cerebral vasculature or of the long duration of these factors [8].

We could not show that there are variations in the levels of inflammation among the clusters, but the real reason could be the limited scope of the laboratory tests used in the study. It is becoming increasingly clear that the commonly used inflammatory marker, CRP, also used in this study, is not suitable for all clinical situations and that only a set of variables, indicating related and overlapping disorders, would be effective for detecting variations in the levels of inflammation in older population groups [56]. In case our hypothesis is true, the total burden of cardiometabolic disorders and the level of inflammation/malnutrition and muscle loss would be a better correlate of decreased cognitive function and the presence of a cognitive frailty phenotype than just decreased renal function. Evidence that the coexistence of kidney and heart diseases, relative to the stages of progression of these disorders, contributes to the development of malnutrition, inflammation, frailty, and cognitive impairment is scarce, however, since these disorders have only been examined separately as two independent disorders [57,58].

What else would be important, is an interplay between different disorders and the dynamics of progression, which only could be assessed by longitudinal examinations. According to some observations, if frailty develops before cognitive impairment, dementia will not develop, in contrast to what happens when cognitive impairment continues to progress in parallel with an advancement in the frailty status [59].

This is likely to be indicated, although indirectly, by our results showing that a disease pattern that specifically marks the PhyF cluster and that can be used to help explain the physical frailty phenotype, a hallmark of this cluster, is markedly different from the comorbidity patterns of clusters that are characterized with significant cognitive impairment (the CI and CF clusters). A disease pattern that typically marks the PhyF cluster includes the highest expression of functional/sensory organ impairments, especially concerning those that are known to accompany musculoskeletal diseases, such as walking difficulties, chronic pain, and falls; common musculoskeletal diseases (in particular, osteoporosis and lower back pain); and mental disorders (anxiety and depression). In addition, only in this cluster does gender imbalance play a significant role in cluster membership, with women being dominant.

Like our findings, other evidence also suggests that musculoskeletal diseases are more prevalent in women than in men and, in particular, in older women with multimorbidity [60]. Musculoskeletal diseases have been recognized as a leading cause of physical disability and associated with chronic pain and frailty [61,62]. When integrating several pieces of this evidence, there is indication of a close association between chronic pain and mental disorders, anxiety, and depression and that mental disorders usually coexist with musculoskeletal diseases in comorbidity patterns [60,63]. Older persons with chronic pain experience and anxiety and depression use opioid analgesics or psychotropic medications more often than others [64]. These medications might contribute to the higher expression of functional organ impairments and the development of the physical frailty phenotype in individuals in the PhyF cluster [65].

Although individuals in the CF and PhyF clusters share similar levels of comorbidity of chronic diseases and medications prescription and, in particular, the diagnosis of osteoarthritis was recorded at higher rates in these clusters than in clusters where frailty is in early stages (the HF and CI clusters), which could contribute to higher levels of inflammation in the CF and PhyF clusters, the hallmark of the PhyF cluster, as indicated by our results, includes a combination of musculoskeletal disorders—in particular, osteoporosis and low back pain syndrome, with anxiety and depression—associated with the domination of female gender [66]. In this regard, the description of the clinical profile of individuals in the PhyF cluster integrates several pieces of evidence, indicating that women are more prone than men to anxiety/depression, multimorbidity, musculoskeletal diseases, and frailty [62,67,68]. There are opinions that a higher psychological vulnerability is the common proxy for developing musculoskeletal diseases and frailty in older women, with inadequate coping mechanisms and obesity having mediating roles [68].

As indicated by our results, objectively visible or subjectively experienced walking difficulties may serve as a simple sign for recognizing older, frail individuals, regardless of their cognitive function status.

5. Strengths and Limitations

In this study, we presented an innovative approach of how to select older people in the population based on the levels of physical frailty and cognitive impairment, considered together, as clusters, which approach, to some extent, reflects differences in the rates of aging and could be important from a prognostic perspective. However, this study also had some limitations that did not allow generalization of the results. These limitations included the low number of participants, especially the small number of individuals in the clusters indicating physical frailty and cognitive frailty phenotypes, which represent the final pathways in the development of frailty. The small size of these clusters may partly be a consequence of the participants' recruitment bias, due to the fact that many frail and immobile persons were not covered by the study. In addition, very old persons (old 80 years and more) were mostly uninvolved, which may have impaired the real picture of the distribution of functional impairments among older people in the community. Further limitations included a low sensitivity of the applied MMSE test for detecting early signs of cognitive impairment and the lack of some variables indicating inflammation, because they were not recorded in the electronic health records.

6. Conclusions

This study aimed to identify clusters of physical frailty and cognitive impairment in a population of older ($\geq$60), ambulatory, primary care patients. A comorbidity pattern that may distinguish the clusters depends on the degree of development of cardiometabolic disorders in combination with advancing age. The physical frailty phenotype is likely to exist separately from the cognitive frailty phenotype. A distinction between the two is likely to be related to variations in the expression of CVD and musculoskeletal diseases and to a gender-related predisposition for chronic diseases.

Supplementary Materials: The following are available online at https://www.mdpi.com/article/10.3390/healthcare9070891/s1, Table S1. Descriptive statistics of numerical variables; Table S2. Descriptive statistics of categorical variables.

Author Contributions: Conceptualization, L.T.M. and T.W.; methodology, L.T.M. and J.P.; validation, S.B. and L.T.M.; investigation, F.B., V.P. and J.P.; data curation, L.T.M. and S.B.; writing—original draft preparation, S.B., F.B., L.T.M.; writing—review and editing, J.P. and T.W.; visualization, F.B. and V.P.; funding acquisition, J.P. All authors have read and agreed to the published version of the manuscript.

Funding: This work was partially supported by the Slovak Grant Agency of the Ministry of Education and Academy of Science of the Slovak Republic under grant No. 1/0685/21 and The Slovak Research and Development Agency under grants No. APVV-16-0213 and APVV-17-0550.

Institutional Review Board Statement: The study was conducted according to the guidelines of the Declaration of Helsinki and approved by the Ethics Committee of the Faculty of Medicine of the Josip Juraj Strossmayer University of Osijek (No. 641-01/18-01/01).

Informed Consent Statement: Informed consent was obtained from all subjects involved in the study.

Data Availability Statement: The data presented in this study are available on request from the corresponding author. The data are not publicly available due to ongoing research.

Conflicts of Interest: The authors declare no conflict of interest.

References

1. European Commission. Ageing Report. Policy Challenges for Aaging Societies. News, 25 May 2018, Brussels. 2018. Available online: https://ec.europa.eu/info/news/economyfinance/policy-implications-ageing-examined-new-report-2018-may-25_en (accessed on 13 July 2021).
2. Barnett, K.; Mercer, S.W.; Norbury, M.; Watt, G.; Wyke, S.; Guthrie, B. Epidemiology of multimorbidity and implications for health care, research, and medical education: A cross-sectional study. *Lancet* **2012**, *380*, 37–43. [CrossRef]
3. Hanlon, P.; Nicholl, B.I.; Dinesh, J.B. Frailty and pre-frailty in middle-aged and older adults and its association with multimorbidity and mortality: A prospective analyses of 493 737 UK biobank participants. *Lancet Public Health* **2018**, *3*, e323–e332. [CrossRef]
4. Nicholson, K.; Griffith, L.E.; Sohel, N.; Raina, P. Examining early and late onset of multimorbidity in the Canadian Longitudinal Study on Aging. *J. Am. Geriatr. Soc.* **2021**. [CrossRef]
5. Franceschi, C.; Garagnani, P.; Morsiani, C.; Conte, M.; Santoro, A.; Grignolio, A.; Moni, D.; Capri, M.; Salvioloi, S. The Continuum of Aging and Age-Related Diseases: Common Mechanisms but Different Rates. *Front. Med.* **2018**, *12*, 61. [CrossRef]
6. Tuttle, C.S.L.; Maier, A.B. Towards a biological geriatric assessment. *Exp. Gerontol.* **2017**, *107*, 102–107. [CrossRef] [PubMed]
7. Franceschi, C.; Garagnani, P.; Parini, P.; Giuliani, C.; Santoro, A. Inflammaging: A new immune-metabolic viewpoint for age-related diseases. *Nat. Rev. Endocrinol.* **2018**, *14*, 576–590. [CrossRef]
8. Li, T.; Huang, Y.; Cai, W.; Chen, X.; Men, X.; Lu, T. Age-related cerebral small vessel disease and inflammaging. *Cell Death Dis.* **2020**, *11*, 932. [CrossRef]
9. Sargent, L.; Nalls, M.; Starkweather, A.; Hobgood, S.; Thompson, H.; Amella, E.J. Shared biological pathways for frailty and cognitive impairment: A systematic review. *Ageing Res. Rev.* **2018**, *47*, 149–158. [CrossRef] [PubMed]
10. Fried, L.P.; Qian-Li, X.; Cappola, A.R.; Ferrucci, L.; Chaves, P.; Varadham, R. Nonlinear multisystem physiological dysregulation associated with frailty in older women: Implications for etiology and treatment. *J. Gerontol. A Biol. Sci. Med. Sci.* **2009**, *64*, 1049–1057. [CrossRef] [PubMed]
11. Kuzuya, M. Process of physical disability among older adults—Contribution of frailty in the super-aged society. *Nagoya J. Med. Sci.* **2012**, *74*, 31–37. [PubMed]
12. O'Caoimh, R.; Galluzzo, L.; Rodríguez-Laso, Á.; van der Heyden, J.; Lamprini-Koula, M.; Ciutzan, M.; Lopez-Samaniego, L.; Liew, A. Transitions and trajectories in frailty states over time: A systematic review of the European Joint Action ADVANTAGE. *Annali dell'Istituto Superiore di Sanita* **2018**, *54*, 246–252.
13. Albert, M.S.; DeKosky, S.T.; Dickson, D.W. The diagnosis of mild cognitive impairment due to Alzheimer's disease: Recommendations from the National Institute on Aging-Alzheimer's Association workgroups on diagnostic guidelines for Alzheimer's disease. *Alzheimer Dement.* **2011**, *7*, 270–279. [CrossRef] [PubMed]
14. Tabert, M.H.; Manly, J.J.; Liu, X.; Pelton, G.H.; Rosenblum, S.; Jacobs, M.; Zamora, D.; Goodkind, M.; Bell, K.; Stern, Y.; et al. Neuropsychological prediction of conversion to Alzheimer disease in patients with mild cognitive impairment. *Arch. Gen. Psychiatry* **2006**, *63*, 916–924. [CrossRef] [PubMed]
15. Babič, F.; Pusztová, L.; Majnarić, L.T. Mild Cognitive Impairment Detection Using Association Rules Mining. *Acta Inform. Pragensia* **2020**, *9*, 92–107. [CrossRef]
16. Ávila-Funes, J.A.; Amieva, H.; Bargerger-Gateau, P.; Le Goff, M.; Raoux, N.; Ritchie, K.; Carriere, I.; Tavernier, B.; Tzourio, C.; Gutierrez-Robledo, L.M.; et al. Cognitive Impairment Improves the Predictive Validity of the Phenotype of Frailty for Adverse Health Outcomes: The Three-City Study. *J. Am. Geriatr. Soc.* **2009**, *57*, 453–456. [CrossRef] [PubMed]
17. Godin, J.; Armstrong, J.J.; Rockwood, K.; Andrew, M.K. Dynamics of Frailty and Cognition After Age 50: Why It Matters that Cognitive Decline is Mostly Seen in Old Age. *J. Alzheimers Dis.* **2017**, *58*, 231–232. [CrossRef]

18. Canevelli, M.; Cesari, M. Cognitive frailty: Far from clinical and research adoption. *J. Am. Med Dir. Assoc* **2017**, *18*, 816–818. [CrossRef]
19. Ma, L.; Chan, P. Understanding the Physiological Links Between Physical Frailty and Cognitive Decline. *Aging Dis.* **2020**, *11*, 405–418. [CrossRef] [PubMed]
20. Kelaiditi, E.; Cesari, M.; Canevelli, M.; van Kan, G.A.; Oussel, P.J.; Gillette-Guyonnet, S.; Ritz, P.; Duveau, F.; Soto, M.E.; Provencher, V.; et al. Cognitive frailty: Rational and definition from an (I.A.N.A./I.A.G.G.) international consensus group. *J. Nutr. Health Aging* **2013**, *17*, 726–734. [CrossRef]
21. Calderón-Larrañaga, A.; Vetrano, D.L.; Ferrucci, L.; Mercer, S.W.; Marengoni, A.; Onder, G.; Eriksdotter, M.; Fratiglioni, L. Multimorbidity and functional impairment- bidirectional interplay, synergistic effects and common pathways. *J. Intern. Med.* **2019**, *285*, 255–271. [CrossRef]
22. Rosado-Artalejo, C.; Carnicero, J.A.; Losa-Reyna, J.; Guadalupe-Grau, A.; Guiterrez-Avila, G.; Alfaro-Acha, A.; Rodriguez-Artalejo, F.; Rodriguez-Manas, L.; Garcia-Garcia, F.J. Cognitive performance across 3 frailty phenotypes: Toledo study for healthy aging. *J. Am. Med Dir. Assoc.* **2017**, *18*, 785–790. [CrossRef]
23. Breinholt, L.F.; Hauge Pedersen, M.; Friis, K.; Glümer, C.; Lasgaard, M. A latent class analysis of multimorbidity and the relationship to socio-demographic factors and health related quality of life. A national population-based study of 162.283 Danish adults. *PLoS ONE* **2017**, *12*, e0169426.
24. Violán, C.; Foguet-Boreu, Q.; Fernández-Bertolín, S.; Guisado-Clavero, M.; Cabrera-Bean, M.; Formiga, F.; Carbrera-Bean, M.; Formiga, F.; Valderas, J.M.; Roso-Llorach, A. Soft clustering using real-world data for the identification of multimorbidity patterns in an elderly population: Cross-sectional study in a Mediterranean population. *BMJ Open* **2019**, *9*, e029594. [CrossRef] [PubMed]
25. Marengoni, A.; Roso-Llorach, A.; Vetrano, D.L.; Fernández-Bertolín, S.; Guisado-Clavero, M.; Violán, C.; Calderon-Larranaga, A. Patterns of multimorbidity in a population-based cohort of older people, sociodemographic, lifestyle, clinical, and functional differences. *J. Gerontol. Ser. A Boil. Sci. Med Sci.* **2020**, *75*, 798–805. [CrossRef] [PubMed]
26. Nylund, K.L.; Asparouhov, T.; Muthén, B. Deciding on the number of classes in latent class analysis and growth mixture modeling: A Monte Carlo simulation study. *Struct. Equ. Modeling A Multidiscip. J.* **2013**, *14*, 535–569. [CrossRef]
27. Iacobucci, D. Structural equations modeling: Fit indices, sample size and advanced topics. *J. Consum. Psychol.* **2010**, *20*, 90–98. [CrossRef]
28. Austin, P.C. Estimating multilevel logistic regression models when the number of clusters is low: A comparison of different statistical software procedures. *Int. J. Biostat.* **2010**, *6*, 16. [CrossRef]
29. Bekić, S.; Babič, F.; Filipčić, I.; Trtica Majnarić, L. Clustering of Mental and Physical Comorbidity and the Risk of Frailty in Patients Aged 60 Years or More in Primary Care. *Med. Sci. Monit.* **2019**, *25*, 6820–6835. [CrossRef]
30. Trtica Majnarić, L.; Bekić, S.; Babič, F.; Pusztová, L'.; Paralič, J. Cluster Analysis of the Associations among Physical Frailty, Cognitive Impairment and Mental Disorders. *Med. Sci. Monit.* **2020**, *26*, e924281-1–e924281-12.
31. Levey, A.S. A decade after the KDOQI CDK guidelines. *Am. J. Kidney Dis.* **2012**, *60*, 683–685. [CrossRef]
32. Mach, F.; Baigent, C.; Catapano, A.L.; Koskinas, K.C.; Casula, M.; Badimon, L.; Chapman, M.J.; De Backer, G.G.; Delgado, V.; Ference, B.A.; et al. 2019 ESC/EAS Guidelines for the Management of Dyslipidaemias: Lipid Modification to Reduce Cardiovascular Risk: The Task Force for the Management of Dyslipidaemias of the European Society of Cardiology (ESC) and European Atherosclerosis Society (EAS). *Eur. Heart J.* **2020**, *41*, 111–188. Available online: https://academic.oup.com/eurheartj/article/41/1/111/5556353?login=true (accessed on 13 July 2021). [CrossRef]
33. Landi, F.; Russo, A.; Liperoti, R.; Pahor, M.; Tosato, M.; Capoluongo, E.; Bernabei, R.; Onder, G. Midarm muscle circumference, physical performance and mortality: Results from the aging and longevity study in the Sirente geographic area ilSIRENTE study. *Clin. Nutr.* **2010**, *29*, 441–447. [CrossRef]
34. Fried, L.P.; Tangen, C.M.; Walston, J. Cardiovascular Health Study Collaborative Research Group. Frailty in older adults: Evidence for a phenotype. *J. Gerontol. A Biol. Sci. Med. Sci.* **2001**, *56*, M146–M156. [CrossRef]
35. Boban, M.; Malojčić, B.; Mimica, N.; Vocovic, S.; Zrillic, I.; Hof, P.R.; Simic, G. The reliability and validity of the Mini Mental State Examination in the elderly Croatian population. *Dement. Geriatr. Cogn. Disord.* **2012**, *33*, 385–392. [CrossRef] [PubMed]
36. Hageenars, J.A.; McCutcheon, A.L. *Applied Latent Class Analysis*; Cambridge University Press: Cambridge, UK, 2002; p. e924281.
37. Akaike, H. Information Theory and an Extension of the Maximum Likelihood Principle. In *Proceedings of the 2nd International Symposium on Information Theory, Tsahkadsor, Armenia, 2–8 September 1971*; Petrov, B.N., Csaki, F., Eds.; Akademia Kiado: Budapest, Hungary, 1973.
38. Rijken, M.; Hujala, A.; van Ginneken, E.; Melchiorre, M.G.; Groenewegen, P.; Schellevis, F. Managing multimorbidity: Profiles of integrated care approaches targeting people with multiple chronic conditions in Europe. *Health Policy* **2018**, *122*, 44–52. [CrossRef] [PubMed]
39. Majnarić, L.T.; Babič, F.; O'Sullivan, S.; Holzinger, A. AI and Big Data in Healthcare: Towards a More Comprehensive Research Framework for Multimorbidity. *J. Clin. Med.* **2021**, *10*, 766. [CrossRef] [PubMed]
40. Gill, T.M.; Gahbauer, E.A.; Allore, H.G.; Ling Han, L. Transitions between frailty states among community-living older person. *Arch. Intern. Med.* **2006**, *166*, 418–423. [CrossRef] [PubMed]
41. Strandberg, T.E.; Lindström, L.; Jyväkorpi, S.; Urtamo, A.; Pitkälä, K.H.; Kivimäki, M. Phenotypic frailty and multimorbidity are independent 18-year mortality risk indicators in older men: The Helsinki Businessmen Study (HBS). *Eur. Geriatr. Med.* **2021**. [CrossRef]

42. Strandberg, T.E.; Pitkälä, K.H. Frailty in elderly people. *Lancet* **2007**, *369*, 1328–1329. [CrossRef]
43. Porter, S.K.N.; McDonald, S.R.; Bales, C.W. Obesity and physical frailty in older adults: A scoping review of lifestyle intervention trials. *J. Am. Med. Dir. Assoc.* **2014**, *15*, 240–250. [CrossRef] [PubMed]
44. Chou, L.; Brady, S.R.; Urquhart, D.M.; Teichtahl, A.J.; Cicuttini, F.M.; Pasco, J.A.; Brennan-Olsen, S.L.; Wluka, A.E. The Association Between Obesity and Low Back Pain and Disability Is Affected by Mood Disorders: A Population-Based, Cross-Sectional Study of Men. *Medicine* **2016**, *95*, e3367. [CrossRef] [PubMed]
45. Sinclair, A.J.L. Diabetes and frailty: Two converging conditions? *Can. J. Diabetes* **2016**, *40*, 77–83. [CrossRef]
46. Kleipool, E.E.F.; Hoogendijk, E.O.; Trappenburg, M.C.; Handoko, M.L.; Huisman, M.; Peters, M.J.; Muller, M. Frailty in older adults with cardiovascular disease: Cause, effect or both? *Aging Dis.* **2018**, *9*, 489–497. [CrossRef]
47. Ferrucci, L.; Fabbri, E. Inflammageing: Chronic inflammation in ageing, cardiovascular disease, and frailty. *Nat. Rev. Cardiol.* **2018**, *15*, 505–522. [CrossRef] [PubMed]
48. Forman, D.E.; Maurer, M.S.; Boyd, C.; Brindis, R.; Salive, M.E.; Horne, F.M.; Bell, S.P.; Fulmer, T.; Reuben, D.B.; Zieman, S. Multimorbidity in Older Adults with Cardiovascular Disease. *J. Am. Coll. Cardiol.* **2018**, *71*, 2149–2161. [CrossRef]
49. Said, S.; Hernandez, G.T. The link between chronic kidney disease and cardiovascular disease. *J. Nephropathol.* **2014**, *3*, 99–104. [CrossRef]
50. Foley, R.N.; Wang, C.; Ishani, A.; Collins, A.J.; Murray, A.M. Kidney function and sarcopenia in the United States general population: NHANES III. *Am. J. Nephrol.* **2007**, *27*, 279–286. [CrossRef]
51. Cesari, M.; Landi, F.; Vellas, B.; Bernabei, R.; Marzetti, E. Sarcopenia and physical frailty: Two sides of the coin. *Front. Aging Neurosci.* **2014**, *6*, 192. [CrossRef] [PubMed]
52. Gingrich, A.; Volkert, D.; Kiesswetter, E.; Thomanek, M.; Bach, S.; Sieber, C.C.; Zopf, Y. Prevalence and overlap of sarcopenia, frailty, cachexia and malnutrition in older medical inpatients. *BMC Geriatr.* **2019**, *19*, 120. [CrossRef] [PubMed]
53. Wu, P.Y.; Chao, C.T.; Chan, D.C.; Huang, J.W.; Hung, K.Y. Contributors, risk associates, and complications of frailty in patients with chronic kidney disease: A scoping review. *Ther. Adv. Chronic Dis.* **2019**, *5*, 2040622319880382. [CrossRef]
54. Rysz, J.; Gluba-Brzózka, A.; Rysz-Górzyńska, M.; Franczyk, B. The Role and Function of HDL in Patients with Chronic Kidney Disease and the Risk of Cardiovascular Disease. *Int. J. Mol. Sci.* **2020**, *21*, 601. [CrossRef] [PubMed]
55. Hou, P.; Xue, H.P.; Mao, X.E.; Li, Y.N.; Wu, L.F.; Liu, Y.B. Inflammation markers are associated with frailty in elderly patients with coronary heart disease. *Aging* **2018**, *10*, 2636–2645. [CrossRef] [PubMed]
56. Calder, P.C.; Ahluwalia, N.; Albers, R.; Bosco, N.; Bourdet-Sicard, R.; Haller, D.; Holgate, S.T.; Jönsson, L.S.; Latulippe, M.E.; Marcos, A.; et al. A consideration of biomarkers to be used for evaluation of inflammation in human nutritional studies. *Br. J. Nutr.* **2013**, *109*, S1–S34. [CrossRef] [PubMed]
57. Berl, T.; Henrich, W. Kidney-Heart Interactions: Epidemiology, Pathogenesis, and Treatment. *Clin. J. Am. Soc. Nephrol.* **2008**, *1*, 8–18. [CrossRef]
58. Cao, C.; Hu, J.X.; Dong, Y.F.; Zhan, R.; Li, P.; Su, H.; Peng, Q.; Wu, T.; Huang, X.; Sun, W.H.; et al. Association of Endothelial and Mild Renal Dysfunction with the Severity of Left Ventricular Hypertrophy in Hypertensive Patients. *Am. J. Hypertens* **2016**, *29*, 501–508. [CrossRef]
59. Chu, N.M.; Bandeen-Roche, K.; Tian, J.; Kasper, J.D.; Gross, A.L.; Carlson, M.C.; Xue, Q.L. Hierarchial development of frailty and cognitive impairment: Clues into etiological pathways. *J. Gerontol. Series A* **2019**, *74*, 1761–1770. [CrossRef]
60. Duffield, S.J.; Ellis, B.M.; Goodson, N.; Walker-Bone, K.; Conaghan, P.G.; Margham, T.; Loftis, T. The contribution of musculoskeletal disorders in multimorbidity: Implications for practice and policy. *Best Pract. Res. Clin. Rheumatol.* **2017**, *31*, 129–134. [CrossRef] [PubMed]
61. Ensrud, K.E.; Ewing, S.K.; Taylor, B.C.; Finh, H.A.; Stone, K.L.; Cauley, J.A.; Tracey, J.K.; Hochberg, M.C.; Rodondi, N.; Cawthon, P.M. Study of osteoporotic fracture research group. Frailty and risk of falls, fracture, and mortality in older women: The study of osteoporotic fractures. *J. Gerontol. A Biol. Sci. Med. Sci.* **2007**, *62*, 744–751. [CrossRef] [PubMed]
62. Otones Reyes, P.; Garcia Perea, E.; Pedroz Marcos, A. Chronic pain and frailty in community-dwelling older adults: A systematic review. *Pain Manag. Nurs.* **2019**, *20*, 309–315. [CrossRef] [PubMed]
63. Marshall, P.W.M.; Schabrun, S.; Knox, M.F. Physical activity and the mediating effect of fear, depression, anxiety, and catastrophizing on pain related disability in people with chronic low back pain. *PLoS ONE* **2017**, *12*, e0180788. [CrossRef]
64. Majnarić, L.T.; Wittlinger, T.; Stolnik, D.; Babič, F.; Bosnić, Z.; Rudan, S. Prescribing Analgesics to Older People: A Challenge for GPs. *Int. J. Environ. Res. Public Health* **2020**, *17*, 4017. [CrossRef]
65. Saraf, A.A.; Peterson, A.W.; Simmons, S.F.; Schnelle, J.F.; Bell, S.P.; Kripalani, S.; Meyers, A.P.; Mixon, A.S.; Long, E.A.; Jakobsen, J.M.L.; et al. Medications associated with geriatric syndromes (MAGS) and their prevalence in older hospitalized adults discharged to skilled nursing facilities. *J. Hosp. Med.* **2016**, *11*, 694–700. [CrossRef]
66. Hall, A.J.; Stubbs, B.; Mamas, M.A.; Myint, P.K.; Smith, T.O. Association between osteoarthritis and cardiovascular disease: Systematic review and meta-analysis. *Eur. J. Prev. Cardiol.* **2016**, *23*, 938–946. [CrossRef]
67. El-Gabalawy, R.; Mackenzie, C.S.; Shooshtari, S.; Saren, J. Comorbid physical health conditions and anxiety disorders: A population-based exploration of prevalence and health outcomes among older adults. *Gen. Hosp. Psychiatry* **2011**, *33*, 556–564. [CrossRef]
68. Lohmann, M.; Dumenci, L.; Mezuk, B. Depression and Frailty in Late Life: Evidence for a Common Vulnerability. *J. Gerontol. Ser. B* **2016**, *71*, 630–640. [CrossRef]

healthcare MDPI

Article

Visual Algorithm of VR E-Sports for Online Health Care

Sang-Guk Lim [1], Se-Hoon Jung [2,*] and Jun-Ho Huh [3,*]

1 School of Culture Contents, Youngsan University, Busan 48015, Korea; gooki7@ysu.ac.kr
2 School of Creative Convergence, Andong National University, Andong 36729, Korea
3 Department of Data Science, (National) Korea Maritime and Ocean University, Busan 49112, Korea
* Correspondence: jungsh@anu.ac.kr (S.-H.J.); 72networks@kmou.ac.kr (J.-H.H.)

Abstract: The need for non-face-to-face online health care has emerged through the era of "untact". However, there is a lack of standardization work and research cases on the exercise effect of immersive content. In this study, the possibility of the exercise effect of VR e-sports among e-sports cases were presented through a visual algorithm analysis. In addition, the evaluation criteria were established. The research method compares and analyzes e-sports cases and VR e-sports cases by applying existing evaluation research cases. It also sets up a new evaluation standard. As for the analysis result, the device immersion method and interaction range were set through an algorithm analysis; FOV and frame immersion were set through typification; the user recognition method and interaction method were set through the visual diagram. Then, each derived result value was quantified and a new evaluation criterion was proposed.

Keywords: VR; virtual reality; VR e-sports; AI; immersive content; visual algorithm; visual system; online health care; game

Citation: Lim, S.-G.; Jung, S.-H.; Huh, J.-H. Visual Algorithm of VR E-Sports for Online Health Care. *Healthcare* **2021**, 9, 824. https://doi.org/10.3390/healthcare9070824

Academic Editor: Mahmudur Rahman

Received: 18 May 2021
Accepted: 15 June 2021
Published: 29 June 2021

1. Introduction

The COVID-19 pandemic has not only brought confusion and control to reciprocal exchanges, as well as the entire society and economy, but also is causing mentally and physically serious illnesses. In particular, one of the causes is the absence of exercise, along with limited physical activities. With the prolongation of the pandemic, however, these have increasingly emerged as serious issues, whose resolution will require different non-contact approaches to activity and alternative environments [1]. A new coinage, "ontact," was made amidst these changes [2]. It is a concept of adding "on," which represents connecting to the outside world online, to "untact," which means non-contact. That is, it is an online in-person approach, emerging as a new flow, following the extended spread of the COVID-19 pandemic in 2020. Living in an era when people live to be 100 years old, modern people concern themselves with health and make efforts to keep their health in their busy daily life. Health care has now become an essential element in such a life [3]. After the outbreak of COVID-19, our daily lives have halted, and our physical activities restricted. In addition, functional restoration of the body for respiratory improvement after eradication of the virus is important, and in that respect, rehabilitation plays an important role in acute COVID-19 management [4]. As a way to overcome this crisis, interest in personal health care-type immersive contents based on physical exercise is increasing. Games or e-sports [5] are essential to the culture of indulging in leisure and pleasure, even briefly, to keep physical health and release mental stress. In recent years, e-sports have settled down as part of play culture among most of the young generation and formed a global fandom culture [6].

E-sports is a method of online communication over the Internet through screen devices, generally in the form of games. Recently, it has also established itself as a sport through the competitive structure, and it can be expanded from a culture of enjoying alone to a "health care"-type sport that experiences and communicates through an online network and achieves goals based on physical exercise.

Therefore health care-type immersive content has huge expandability as indoor sports fit for the "ontact" trend amidst the COVID-19 pandemic. There are various types of games according to devices, including mobile games, representative RPG games on a PC, home console games, and VR games. VR e-sports as an extended concept are part of the newly emerging extended reality (XR) culture.

In other words, VR e-sports are not only a fun game, but also has a higher screen scalability than existing e-sports and an excellent sense of immersion through interaction with the human body and the device. In addition, VR e-sports, which have been expanded to AR and MR, have sufficient potential for development as personal health care-type immersive content based on physical exercise. Therefore, it seems that VR game-type health care that is revitalized to meet the needs of the times is necessary. However, most of the currently distributed health care products are monitor-based, one-way products, and the standard for the exercise effect has not been established. In particular, from a technical point of view, it is a method of following the actions shown through a small frame monitor and checking the user's actions using sense. Most of them are focused on the operation method rather than the fun, so we think the immersion is also significantly lower. From an academic point of view, there are no academic research results or empirical evaluation standards for its effectiveness, and it is true that public awareness of how to use it is still lacking. In addition, in the case of virtual reality devices that have been continuously developed in recent years, the standardization process for the resulting values has not been established, and the operating system through "VISION" has not been clearly established.

Recently, the importance of physical activity has been highlighted in preparation for the post-COVID-19 era. Therefore, it suggests the possibility of safe, diverse and fun online e-sports health care in terms of prevention and rehabilitation. In the process, we intend to expand the fun elements of e-sports and the possibility of substituting physical activity through VR e-sports. In other words, by proposing a new evaluation tool that can measure the impact of VR e-sports on physical activity as a role of prevention and rehabilitation during a specific pandemic, the direction of realistic health care to be produced in the future is presented and the evaluation necessary to verify its effectiveness. The purpose of this study is to present a tool.

Findings about various forms of health care-type immersive content have been recently published overseas in the fields of medicine and fitness, for example the "Superpower Glass Intervention" project based on Google Glass, whose VR-based effects were tested for the behavioral analysis and treatment of autistic children [7]. A case study applied VR to the treatment of lumbar pain [8]. In another case, VR gaming technologies were applied and proven effective for the relief of pain in physical therapy [9]. Additionally, a study case on the rehabilitation efficacy of VR through comparison of virtual reality rehabilitation and conventional rehabilitation in Parkinson's disease [10].

In addition, studies on osteoporosis in patients with Down's syndrome [11] and studies on various health-related side effects caused by sarcopenia [12] remind us once again how important health and rehabilitation are in our lives.

A Korean smart health care company, Omni C&S, incorporated VR technologies into its smart health care solution OMNIFITMindcare, which helps to manage mental health [13]. Another South Korean company, Kakao VX, made a home exercise equipment called "VR Smart Home Training" and developed a variety of content, including OhShape [14]. Samsung announced a VR health care solution to help users exercise at home by themselves by connecting their avatars to artificial intelligence. Health-related research and devices are launched in various forms.

These are, however, at the experimentation stage. Research is being conducted on clear standardized methods to measure exercise or its effects with a lack of public awareness of them. Basically, how do you get users to engage? Under what conditions can exercise continue? How to build an evaluation of the effectiveness of home health care without environmental risk factors? etc., are presented as research questions. In this study, we intend to propose a new evaluation method to verify the experience method of general health care products. In addition, there is a need for methodology to examine their sustainability, since they are for health care practices by oneself at home. The present study, thus, proposed an analysis framework to test exercise effects, based on a methodology of applying and keeping gaming methods to such health care-type immersive content through the comparison analysis of e-sports and VR e-sports, as well as the result values of visualization of algorithms established in the research process. The findings of the study may serve as a guide for developers to apply to the UI/UX of immersive devices and as a humanistic guidebook for users to understand a method of experiencing and enjoying extended reality. The investigator employed such research methods as examining various research cases and visualizing a cognitive system built by immersive devices and physical activities in an algorithm form. Results values were used to propose an analysis framework to test exercise effects. An experiment was conducted to quantify differences in physical activities between the game rules of e-sports and those of VR e-sports, and to examine their efficiency as exercise effects through qualitative assessment.

After visualizing an operating algorithm between an immersive device and its user, the investigator used the result values as an analysis framework to compare and assess e-sports and VR e-sports in exercise effects. By introducing "ontact" cases in preparation of the post-COVID-19 era, the present study proposed a way of managing personal health, as well as finding joy in daily life. It will serve as a safety net to maintain our daily life in a never-ending war against another virus.

2. Theoretical Background

2.1. Case Study

Recently, many research cases related to health care have been published due to COVID-19 and are being commercialized. Most products deliver information visually through a 2D-based monitor. It is a way to compete and immerse yourself by recognizing the user's motion and interacting with the character in the screen to verify and score the exercise effect. In this context, the operating method of e-sports is very similar. Sit at a table and use a 2D monitor to control and immerse in the movement of the character you control and compete with your opponent online. The interest in e-sports is evident from the recent various humanistic research cases, and academic research efforts are emerging as well as technological developments. Given the periodic characteristics of today in a mix of virtualism and reality, there is a need to investigate the structural understanding and visual perception process of "vision" or looking at an object. The meanings of vision are expanding with mobility added to the basic screen forms, including VR, AR, MR, and physical computing in a wearable [15] method, which allows one to wear a device on the body and secure a direct view. New spaces of vision mediated in this way are post-Cartesian, post-perspective, and post-physical, but still remain within the limits of frames on a screen [16]. Visual illusionism represents the history of reproduction around presence from Giotto, before the law of perspective, to da Vinci that began to use the law of perspective in full scale [17]. Making an algorithm for the stages of visual information processing, including sense, perception, and cognition, in a visual system fit for the digital era involves the ability [18] to recognize and distinguish visual stimuli and understanding stimuli from the connection of previous experiences and perceptions, rather than responding to various human senses. It is also to sublimate it as a role and value of culture, through the academic interpretation of technological development.

Vision requests the interpretation and insight of a subject that is "cognitive," rather than a simple physical act of "seeing" by the dictionary definition [19]. When an observer sees an object through his or her eyes, it is necessary to obtain information from the visual characteristics of the object, categorize it, classify it, and select it. [20]. The figure illustrates that the human visual structure is recognized through various devices and leads to physical activities. Analyzing such a process and turning it into an algorithm through e-sports are required in the development of various devices, in addition to immersive content. As a recent study on e-sports, there have been studies on e-sports user behavior and development of e-sports measures [21]. There was various research, including the one [22] on the effects of physical activities in virtual reality, in "Experience on Demand" by Jeremy Bailenson. Health care-related research on neurological disorders and strokes examined cases of overcoming lumbar pain through virtual reality and reported that it reduced pain [23]. Another research used VR therapy to treat arachnophobia. Hoffman and Peterson published a research paper in the medical journal, PAIN in 2000, reporting that virtual reality reduced pain more than general games with the distraction technique in "Spider World" [24]. Other researchers analyzed the visual system. Hal Foster (2012) presented his study on the modern visual system in his "Vision and Visuality" [25]. Jeong Jeong-ju (2014) published a "Study on the Expansion of Communication in Media Art with the Window Metaphor" [26]. Lim Sang Guk and Kim Chee Yong (2018) proposed a digital visual system fit for the 21st century in their "Study on Changes to Digital Visuality in the 21st Century," based on Lacan's "notion of the real gaze" [27]. Still others conducted research on algorithms related to the visual system. In his "Study on the Visualization Methods of Poetry with Algorithm-Based Modeling," Kim Ju-seop (2013) used "poetry" to turn images into algorithms [28]. Kim Min-seok, Choi Woo-seong, and Jeong Sun-yeong (2018) published "Design and Implementation of an Algorithm Visualization-Based Cluster Analysis Learning System" [29]. Lim Sang Guk (2020) built "An HMB-Based Interactive Immersive Media Algorithm with L-System" [30].

Developed by Aristid Lindenmayer in 1968, the parallel rewriting system, "L-System" was introduced into computer graphics by Alvy Ray Smith in 1984. Today, it serves useful purposes in the procedural modeling of plant growth, among other things [31]. Using L-System, Kim Ju-seop (2013) proposed a method of recreating each poem in the form of an organic tree in nature, reflecting their unique characteristics in the digital space by limiting the text scope to the literary genre of "poetry," using algorithm-based modeling (procedural modeling). Kim Ju-seop (2013) offered a special explanation that L-System consisted of symbols and rules that replace symbols. Park Jin-wan visualized and presented Korean genealogy in "Visual Genealogy" to create a new story, rather than functionality or aesthetics [32]. Lim Sang Guk (2020) visualized cases of immersive content devices recently used across various fields through the analysis of their visual systems based on image categorization and text listing. He was able to understand the characteristics of media and methods of seeing by the period and offer a guide for UI/UX analysis in media development, based on comparison results.

2.2. *Study Case of Visual Algorithm Realization*

The investigator had to visualize or categorize a visual system to build a visual algorithm needed in the present study, and further, how to recognize and utilize visual information into images in the process of visual information processing, including sense, perception, and cognition in relations between diverse devices and users. Therefore, a method for algorithmizing the visual system was established, and the cases were investigated. To date, there is no evaluation tool that can analyze and verify the digital visual system. However, in this study, we would like to propose a tool for visual system analysis through various cases priority, this paper intends to utilize the following four types of representative papers. In one of such research cases, Jeong Jeong-ju (2014) proposed a traditional Cartesian visual system [33] that divided the traditional visual system into three types, including perspective, camera obscura, and panorama, in "Study on the Expansion of Communication in Media Art with the Window Metaphor." In his "Vision and Visuality," Hal Foster (2012) introduced a case of the notion of the real gaze for a visual system about Lacan's "gaze" under the traditional Cartesian visual system. Crary insisted on a need for observers recognizing vision that had nothing to do with an "act of looking" [34]. Lacan maintained that the "Cartesian visual model" should inevitably be replaced with a new visual model capable of containing a sense mechanism via the nerves [35]. Based on these arguments, Lim Sang Guk (2017) in his "Study on the Characteristics of Visuality Changes and the Expansion of Digital Frames in the 21st Century," in addition to Lim Sang Guk and

Kim Chee Yong (2018) in their "Study on Changes to Digital Visuality in the 21st Century based on Lacan's Notion of the Real Gaze," reorganized the human visual system into a digital visual system of the 21st century and visualized it into a "notion of the real gaze" by using Lacan's "notion of the real gaze". An example of the research case of "Construction of HMD-based interactive immersive media algorithm using L-System" by Im Sang-guk (2020) is given. Among them, let us look at the research results of Im Sang-guk (2017), Im Sang-guk, and Kim Kim-yong (2018), which are research cases on changes in the visual system. The result is shown in Figure 1. As seen in Figure 1, the above-mentioned research cases defined the modern visual system of four elements, including perspective, camera obscura, gaze, and panorama, based on the "notion of the real gaze".

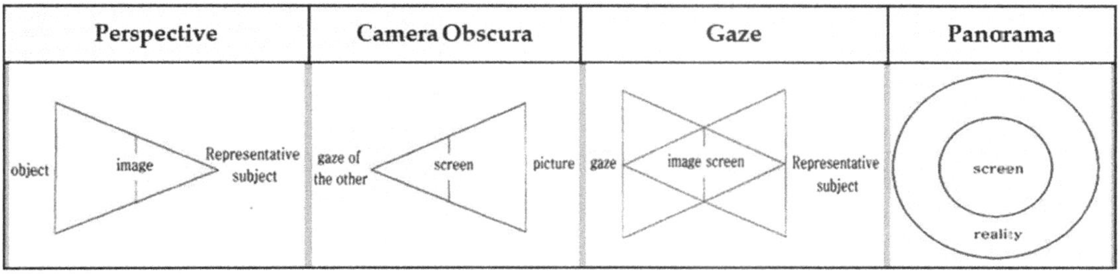

Figure 1. Lacan's "notion of the real gaze".

In Figure 1, the notion of the real gaze is a visualized image of a dual visual system in a digital device. The "window" at the center is an "image screen" showing images. That is, screens in perspective are considered as "windows" to figure out an object by viewers as the origin of the term perspective, which means "seeing through" other spaces beyond the screen, suggests [36]. Viewers look at the apex (vanishing point) on the right through a window from the left side. However, in Lacan's viewpoint, "gaze" represents another eye to look at viewers. On the other hand, on a digital visual system, viewers look at moving images (character) on a liquid crystal display, rather than a vanishing point in the traditional perspective. Such images are virtual images reproduced by the computer. This relationship leads to the formation of interactions between viewers and devices. As a study related to text visualization of images, Kim Ju-seop (2013) reported that character strings in L-System worked as a series of orders to draw Figure 2, based on algorithms. That is, it is "F -> F [+F] F". In character strings of current rules, all "F" symbols are replaced with "F [+F] F". As this rule is applied twice, it expands into the next character string.

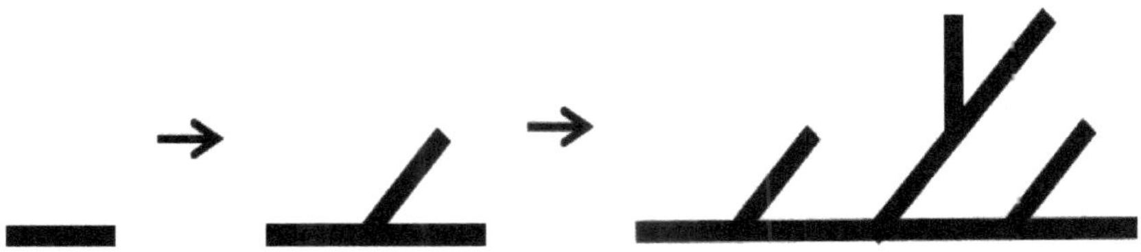

Figure 2. Example of branch generation using L-System.

Following "F -> F [+F] F -> F [+F] F [+F [+F] F] F [+F] F," geometric meanings were granted to each symbol:

- F: drawing '_' clockwise at the current position;
- +: changing the direction at 45 degrees counterclockwise;
- [: saving the current position;
-]: returning to the position saved last.

In this context, as if implementing an image through a string, we intend to apply the method of reconstructing text algorithmically through an image to this study. Figure 3 presents a case of applying users' HMD-based cognitive process to L-System and turning the process into an algorithm in the rewriting method of character strings with a VR device in Lim Sang Guk (2020)'s "An HMB-Based Interactive Immersive Media Algorithm with L-System." The process of users looking at an HMD device is sequenced in texts, and problems with the process, including dizziness and difficult cognition, are checked out in the regeneration process of images to figure out the process of viewers communicating with information images. As seen in these two research cases, cognition relations between devices and viewers were turned into a visual algorithm through the interpretation of images, based on a system of visualizing texts or rewriting character strings.

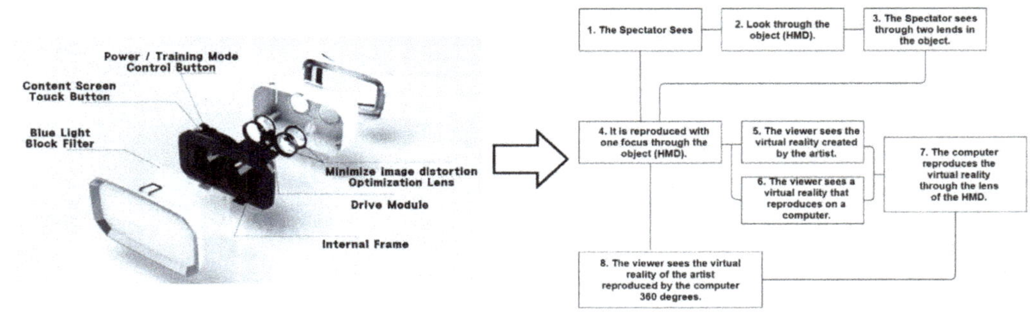

Figure 3. HMD visual system algorithm using L-System.

The next case categorized the visual systems of various immersive devices and visualized them through the "notion of the real gaze". As seen in Table 1, recently launched immersive devices were listed according to the degree of immersion in the monitoring method of single frames generally used.

Table 1. Image tangible and visual illustration of realistic content cases.

Single Frame	Extension Frame	VR	AR	MR	Hologram

			Image Typing		
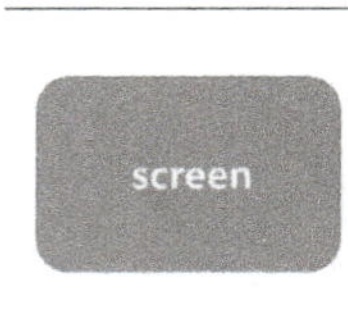 screen	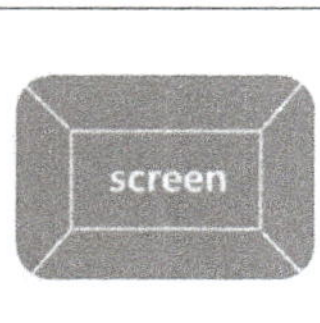 screen	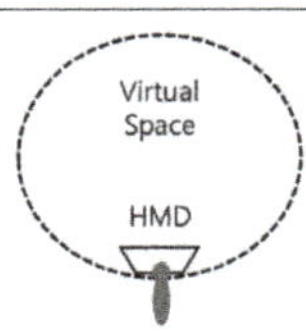 Virtual Space HMD	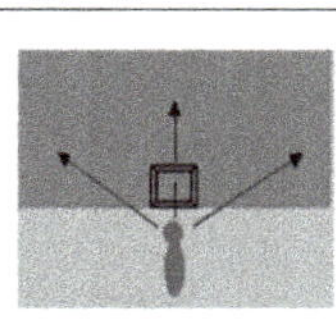	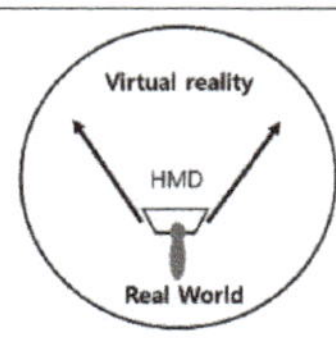 Virtual reality HMD Real World	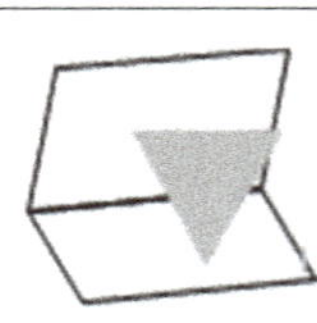
			FOV		
90°′	180°~270°	360°	360°	360°	360°
			Algorithm Visualization		
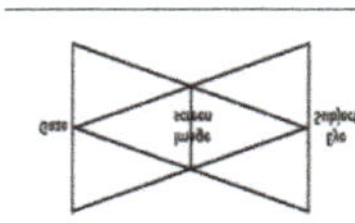	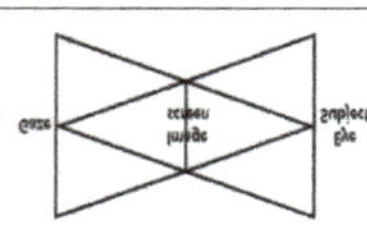	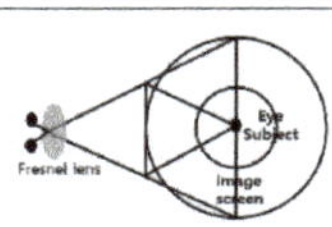	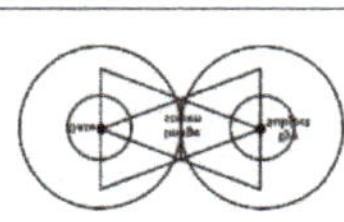	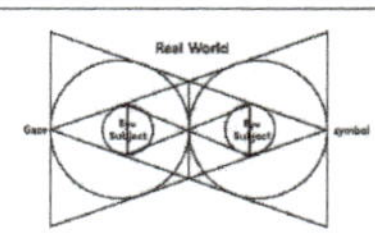	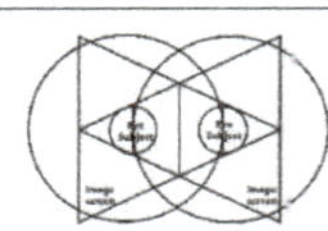

In the enjoyment of e-sports, the basic key is the duration of exercise to generate personal exercise ability effects. You need to exercise consistently for a long period in order to experience the effects. In this sense, immersion is an essential element in exercise effects. In relations between devices and users, a frame is a basic component of vision and important element to enhance immersion. The physical size of a frame works to increase immersion, expand the scope of vision, and maximize movement in the users' enjoyment of e-sports. In other words, there are clear differences in effectiveness between single frames on the monitor screen for immersion, movement of gaze, and activity of the body, and 360-degree spaces for complete immersion and expandability of gaze.

A VR device, HMD, creates a 360-degree space by blocking a gaze in a complete real space. Although it excels in immersion, it can have limits in physical activities. Users have to enjoy a game on the original spot, since they have no visual field secured in a real space. Of device cases proposed as solutions, AR and MR have the greatest advantage of allowing users to enjoy virtual images together in a real space. The expandability of frames represents a physical field of view. As its scope expands, immersion enhances and exercise effects are maximized. In this context, the cases in Table 1 were examined to figure out changes to users' visual frames. A field of view of approximately 90° is created for a single frame, and one of approximately 180~270° is created for an extended frame. Body movements are shown in a limited manner, according to the scope of field of view. VR, on the other hand, maximizes immersion with a full 360° space. Users are, however, restricted for their physical activities, due to the blockage of real spaces, which points to a disadvantage that they have to stand still or sit down to play a game. Unlike VR, AR allows users to play a game, while looking at a real space in a 360° space. In AR, exercise effects are maximized, as users are allowed to move their hands and move around easily. Furthermore, it can function as a media-based tool. In AR, however, augmented images are narrow within the limited media frame size of a field of view, thus lowering immersion. MR supplements the disadvantages of VR and AR and highlights their advantages. Users can secure a field of view of 360° and move their bodies freely, which suggests that MR can serve as a media-based exercise machine.

2.3. Interactive Visual Algorithm Visualization Research Case

One of the characteristics of the vision perspective in the digital era of the 21st century is a dual visual system. As seen in Table 1 above, the human vision system moves toward a three-dimensional system beyond a two-dimensional one. In this digital convergence era, various realities coexist together including actual reality, virtual reality, and augmented reality. In other words, the recent converged media is characterized by mutual communication and interactions between users and their devices, through optical seek-through and display-based interactive methods. As illustrated in Figure 4, the most important part in this mutual interactive method is the perception of the body through physical computing or the kinetic and tracking principle. Following the development of media, viewers' bodies have been a central research subject in various aspects. The keyword of viewer participation in works through the body is the biggest interest of contemporary artists and a characteristic of digital art in the field of digital art. The bodies of viewers represent their interactive natural ego in works and are considered as subjects of perception in the traditional concept of cognition. Now, people have to see and feel with their bodies in the sense of vision with the simple perspective of visual frames replaced by the body perception perspective of frames.

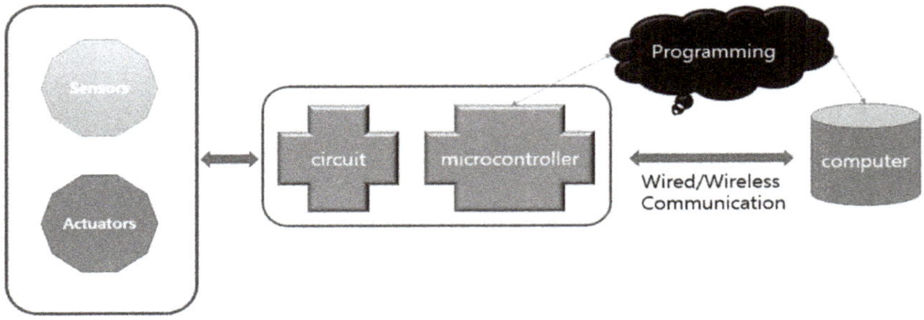

Figure 4. Principles of physical computing.

In his study "An HMB-Based Interactive Immersive Media Algorithm with L-System," Lim Sang Guk (2020) classified interactive immersive media into three types in Table 2. They are "inter-media people," "communion," and "in-media people" types. In the first "inter-media people" type, viewers look at a monitor and act accordingly, and their acts are recognized through kinetic sensors and trigger the reactions of images on the screen. Viewers interact with the monitor through camera sensors detecting their physical movements, recognize it visually, and move along with it. In their movements, they form ties with characters on the screen and interact with them through their visual body cognition. In the next "communion" type, viewers touch the screen themselves. Unlike the objective "inter-media people" type, the "communion" type involves tactile interactions through direct body touches. Viewers touch the monitor screen directly and move accordingly, feeling reactions on the touch screen themselves. There are huge differences in the amount of exercise, according to frame sizes and degree of relationship with the screen. This type enhances viewers' immersion further. In the last "in-media people" type, viewers interact with a work by moving their bodies in it. That is, they become a part of the work. The scope of frames builds a three-dimensional space, and viewers increase their activity level through their body movements and maximize their immersion.

Table 2. Three cases of interactive type.

(1) Hive "Iris" 2012	(2) Air-Screen Interactive 7.5 m^2 2013	(3) Rodrigo Carvalho "Break Down" 2014
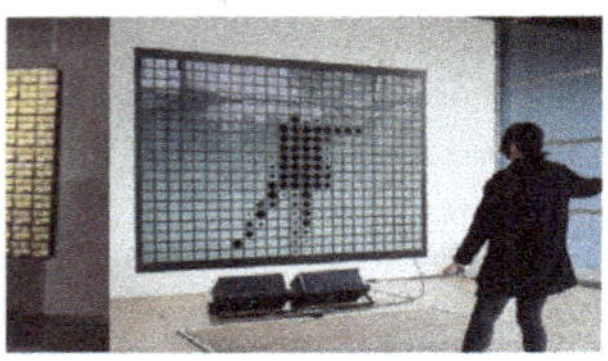		

These research cases show that the digital visual systems of the 21st century are based on moving images and viewers' participation. With the involvement of a medium called the computer, based on interactions, an interactive dual visual system is built. Viewers identify with beings (characters) in the media by moving themselves at a position facing a work. Then, they develop their tactile sense by touching the media (touch screen), being divided into two subjects (viewers in reality and virtual viewers in the media) and becoming immersed at the boundary (screen). Lastly, they walk into a work and become the work (subject) itself. Table 3 shows outcomes of reproducing interactive case analysis results in Lacan's visual plate.

Table 3. Interactive visual plate.

	Cross-Human Humanoid	**Sympathetic**	**Humanoid in the Media**
Interactive Case			
Interactive type	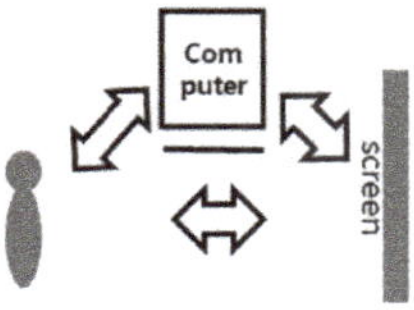	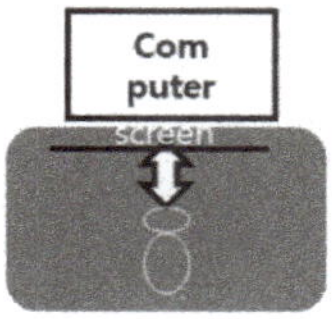	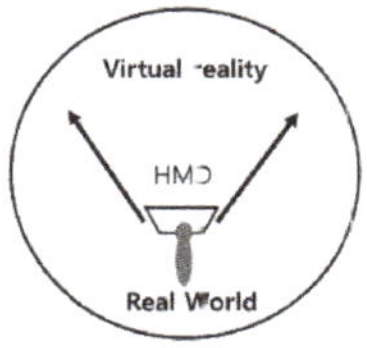
Visual plate	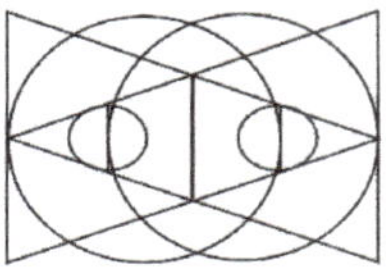	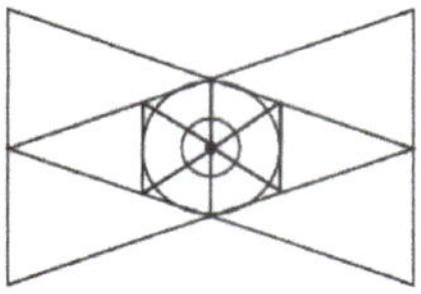	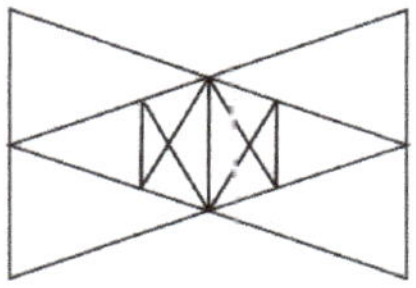

These findings indicate that the scope of frames was limited in viewers' relationships with a screen facing them, given the amount of exercise according to the fields of view, immersion, and activity scope of viewers facing a work in the "inter-media people" type. In the "in-media people type," the amount of exercise increased, according to the expanding frames, widening fields of view, and rising utilization rate of spaces.

3. Analysis Method

3.1. Analysis Targets and Applicable Devices

The present study proposed and applied an analysis method based on these various research cases. It needed objects of analysis and devices to be applied to test the exercise effects of e-sports and VR e-sports, and to propose plans for their vitalization. In other words, the study needed to undergo a process of textualizing and visualizing connections in the cognition method of "vision-body" between various devices and users. This was followed by quantifying results values and testing through qualitative evaluation. Based on the research cases above, the study then proposed objects and frameworks of analysis. First, the objects of analysis in the study included three of the most popular games in the e-sports industry of South Korea and three of the most popular games in VR e-sports. As seen in Table 4, the top three popular games in e-sports in the nation were "League of Legends," "Battleground," and "StarCraft: Remastered." The top three popular games in VR e-sports were "VR Beat Saber," "VR Dragon Flight", and the MR e-sports game, "HADO".

The cases of e-sports in the nation are basically divided between mobile games, based on a smartphone, and RPG games, based on a PC. The present study focused on PC games with great frame expandability. The cases of VR e-sports selected in the study included the HMD-based VR game "Beat Saber," the special force VR game " Dragon Flight" version, and the MR-based game "HADO".

Second, devices related to the objects of analysis were applied based on the cases of immersive devices in Table 5. That is, the characteristics of five representative immersive devices in their instructions were categorized in relations between users and their devices.

Table 4. Analysis target.

E-sports	League of Legends	Battleground	StarCraft: Remastered

VR e-sports	VR Beat Saber	VR Dragon Flight	VR HADO

Table 5. Immersive device case.

Single Frame	Extension Frame	VR	AR	MR

Third, relationships in the cognition method of "vision-body" between users and their devices in the game management method were applied to L-System, based on the objects of analysis and immersive devices to propose an analysis framework. This process was described in the rewriting method of character strings. Based on the algorithm analysis frameworks in Kim Ju-seop's (2013) "Study on the Visualization Methods of Poetry through Algorithm-Based Modeling" and Lim Sang Guk's (2020) "An HMB-Based Interactive Immersive Media Algorithm with L-System," the visual cognition processes of objects of analysis, devices, and users were turned into algorithms through texts.

Fourth, the texts that were turned into algorithms were categorized into visual images and the "notion of the real gaze" based on the "notion of the real gaze" proposed in "Study on Changes to Digital Visuality in the 21st Century, based on Lacan's Notion of the Real Gaze" of Lim Sang Guk and Kim Chee Yong (2018). Result values of fields of view were obtained to test immersion based on the scope of the image frames.

Fifth, the study digitized body movements seen through interactive relations between users and their devices and tested users' exercise effects in their utilization of objects of analysis and devices. That is, it digitized the utilization scope of spaces, based on the movement degree and travel scope of users' eyes, hands, feet, and bodies in the gaming method.

3.2. Proposed Experiment Method and Analysis Tool

The study applied the analysis frameworks for the experimentation methods according to the analysis methods in Figure 5.

As seen in Table 5, the analysis framework used in the experimentation method underwent the process of A, B, C, and D, which involved selecting objects of analysis, applying them to L-System, and turning them into algorithms, based on texts in the rewriting method of character strings, categorizing them into images through algorithm analysis, and turning them into "notion of the real gaze". This process generated the result values of user experiences and exercise effects in e-sports and VR e-sports.

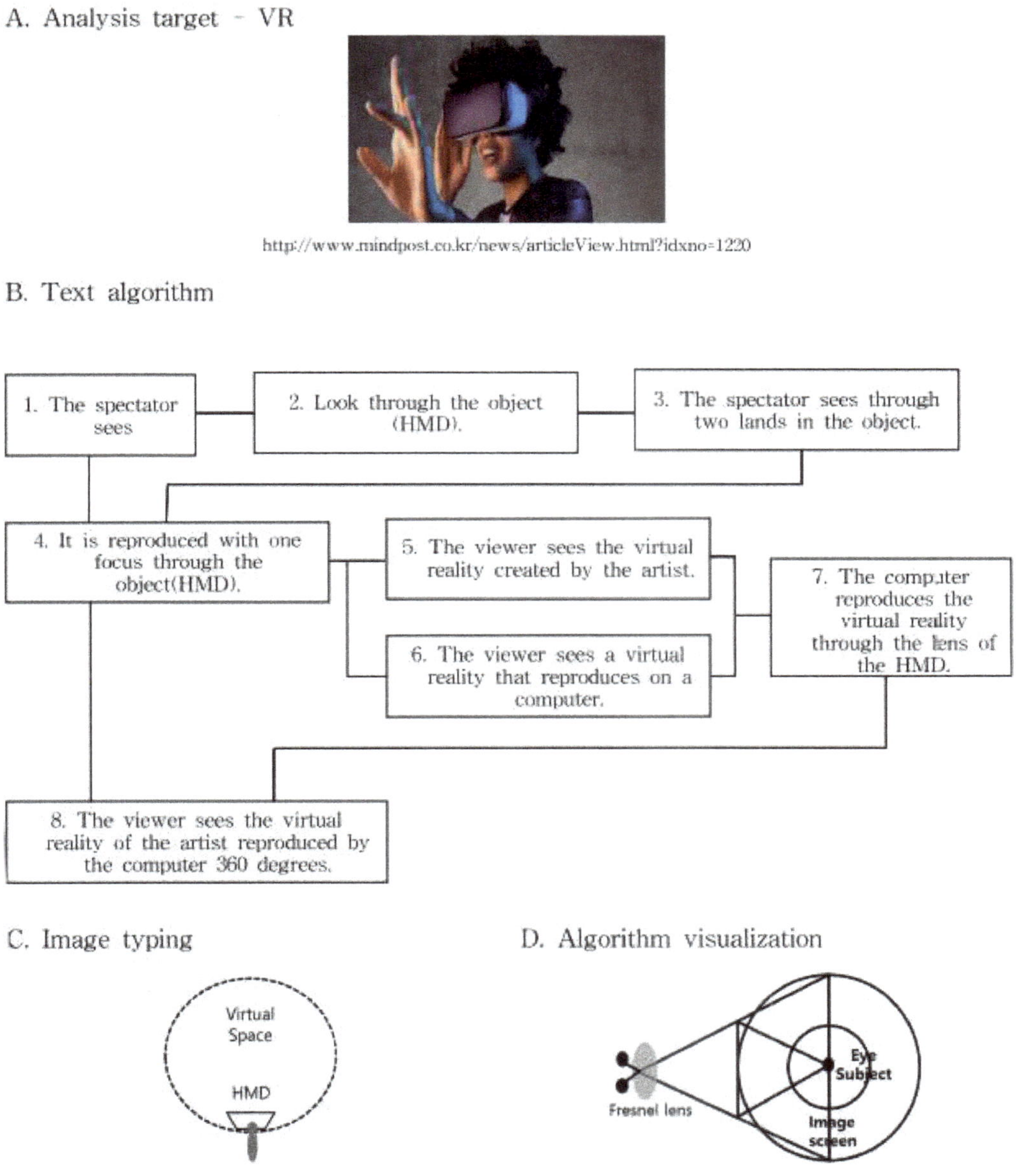

Figure 5. Experiment method model.

Based on the analysis frameworks of A, B, C, and D above, the study proposed analysis criteria for result values to be generated. The criteria would cover result values under each analysis framework and tests of exercise immersion, scope, and effects in e-sports and VR e-sports. Under the framework of "A" for objects of analysis, killer content and devices were selected to be used in e-sports and VR e-sports, based on cases needed to measure fun and exercise effects for exercise persistence, as discussed above. The framework of "B" for the textualization of algorithms found rules of various devices in users' cognition process and grounds for users' immersion and fun, based on their characteristics. The framework of "C" for categorization expressed relationships between users and their devices in images and measured their immersion and scope of activities, based on fields of view through their frames. Users' amounts of exercise were also measured, according to the scope of their space utilization and methods of content management, based on the categorization of characteristics of killer content and devices into images. The final framework of "D" for "notion of the real gaze" offered some grounds to predict UI/UX according to users' immersion and interactive management with their devices, as well as the cognition methods of their bodies. Based on the outcomes, the study proposed a guide to predict killer content to be created and methods for users to utilize their devices. The analysis criteria can be found in Table 6.

Table 6. Analysis tools and standards for experimentation.

<table>
<tr><th>Analysis Tool</th><th>A. Analysis Object</th><th colspan="4">B. Algorithm</th><th colspan="4">C. Typification</th><th colspan="4">D. Diagram of Gaze</th></tr>
<tr><td rowspan="4">Analysis standard</td><td rowspan="4">A. Immersive Device</td><td colspan="4">B-1. Device immersion method</td><td colspan="4">C-1. FOV</td><td colspan="4">D-1. User recognition method</td></tr>
<tr><td>Visual</td><td>Auditory</td><td colspan="2">Tactile</td><td>90</td><td>120</td><td>180</td><td>360</td><td>Eyes</td><td>Hands</td><td>Foot</td><td>Body</td></tr>
<tr><td colspan="4">B-2. Interaction range</td><td colspan="4">C-2. Frame immersion</td><td colspan="4">D-2. Interaction method</td></tr>
<tr><td>x</td><td>Y</td><td>z</td><td>xyz</td><td colspan="4">C-2a. Cross-human humanoid
C-2b. Sympathetic
C-2c. Humanoid in the media</td><td>Eyes</td><td>Hands</td><td>Foot</td><td>Body</td></tr>
</table>

Seeing is a process of an observer looking at an object with his or her eyes, obtaining information from its visual characteristics, categorizing and classifying it, and making a choice [37]. Wong (1994) reported that shapes, sizes, positions, and colors accounted for the most important parts in the visualization of conceptual elements through one's eyes [38]. According to Stephen, sizes, shapes, spaces, and colors of visual elements are very important comparison elements [39]. In this context, it is possible to check which sensory organs are used by users in their devices for immersion under "B" of turning relationships between users and their devices into algorithms, based on texts, according to the criteria of experimental evaluation in Table 6. It is also possible to measure their scope of activities by tracing their body movements moving from the x-axis to the y-axis or traveling along the x-, y-, and z-axes in their interactions with their devices. Under "C," it is possible to categorize instructions between users and their devices into visual images and visualize the degree of their frame utilization in checking their fields of view and uses of their devices. In VR HMD, a field of view (FOV) is important because it plays a big part in increasing the sense of reality in virtual reality. After the visual system of camera obscura, panoramas expanded the physical sensory experiences of subjects in realistic and verified spaces established by the law of principle based on the effects of technological reproduction [40].

Human eyeballs have an average of 110 FOV. An experiment can help to obtain FOV, secure FOV, and promote communion with a device to check user's immersion. Under "D" of the "notion of the real gaze" between users and their devices, it is possible to visualize which process is used by users to communicate and commune with which sensory organs through which body parts. The result values can be used to predict the efficiency and measurement criteria of exercise effects. The evaluation criteria in Table 6 were used as analysis frameworks in the experiment of the study. Additionally, the scientific numerical range for measuring the effects of e-sports and VR e-sports exercise is defined as shown in Table 7.

Table 7. Numericalization method for measuring exercise effect.

Numericalization Tool	Digitization Method	Numericalization Range
B-1. Device immersion method	Visual	**Depending on the level of physical use 1~5**
	Auditory	
	Tactile	
B-2. Interaction range	X-Y	**Depending on the scope of the interaction 1~10**
	X-Y-Z	
C-1. FOV	90°	**According to the line of sight 90°~360°**
	120°	
	180°	
	360°	
C-2. Frame immersion	C-2a. Cross-human humanoid	**Steps 1 to 3 depending on the device and the degree of immersion of the human body**
	C-2b. Sympathetic	
	C-2c. Humanoid in the media	
D-1. User recognition method	1. Eyes	**According to the range of body use 1~4**
	2. Hands	
	3. Feet	
	4. Body	
D-2. Interaction method	1. Eyes	
	2. Hands	
	3. Feet	
	4. Body	

3.3. Experiment Method

In the experiment, an experiment model was built based on these analysis frameworks and applied to the e-sports and VR e-sports to be analyzed. A database was built with the result values to propose an automation system to apply various devices and provide their result values.

Based on the database, the study demonstrated that VR e-sports had practical effects on exercise abilities and proposed a text algorithm to detect FOV and the scope of physical activities in the process of categorizing images and turning them into "notions of the real gaze". Figure 6 shows the overall flow chart of an algorithm to detect exercise abilities in the proposed immersive content. As seen in this flow chart, the analysis framework to be applied to the experiment was divided into "B-1" of the device immersion methods (visual, auditory, and tactile) and "B-2" of the scope of interactions (X, Y, Z, and XYZ) in "B" of algorithms. In "C" of categorization, it was classified into "C-1" of FOV (90, 120, 180, and 360°) and "C-2" of frame immersion (cross-human humanoid, sympathetic, and humanoid in the media). In "D" of the "notion of the real gaze," it was tested with "D-1" of users' cognition methods (eyes, hands, feet, and bodies) and "D-2" of interaction methods (eyes, hands, feet, and bodies). These testing methods can help to predict the persistence and efficiency of exercise over time, based on result values.

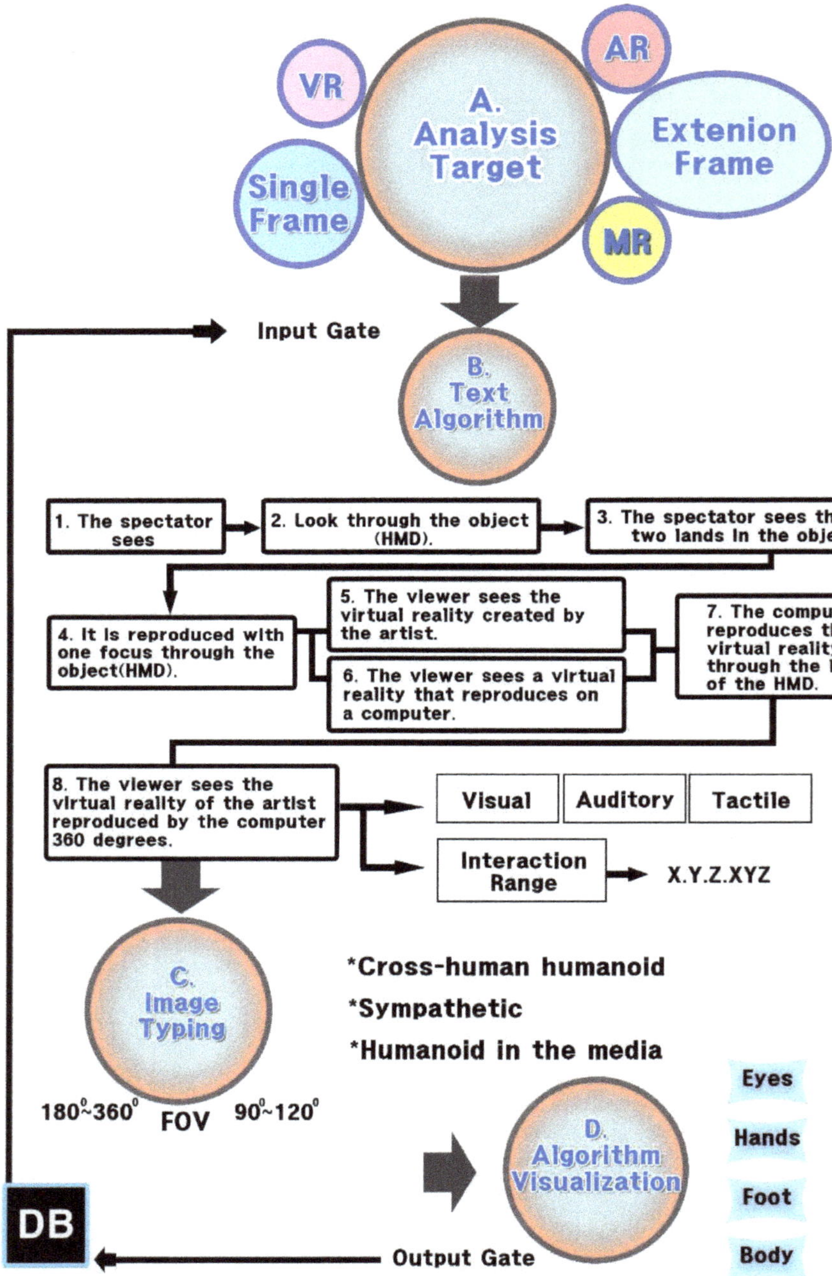

Figure 6. Flow chart of sensory content exercise ability detection algorithm.

4. Experimental Results and Discussion

4.1. B. Text Algorithm Analysis

4.1.1. E-Sports Field

Among the subjects of the experiment analysis, "Battleground" in the field of e-sports was analyzed with a text algorithm from the viewpoint of the user's visual experience. The results are shown in Figure 7. As shown in the experimental results, the user establishes a communication relationship with the character displayed on the monitor screen and thinks that he and the character are the same. In addition, the sense of immersion is enhanced through mouse control. Due to the nature of the e-sports field, athletes immerse themselves in the characters moving on the monitor screen through a visual method. It is also connected to the tactile sense through mouse control. Additionally, the sound is transmitted through the headset as an auditory sensory experience. These results reveal the processes of how the user is immersed through the relationship with the device, and the result is to be subdivided into numerical values.

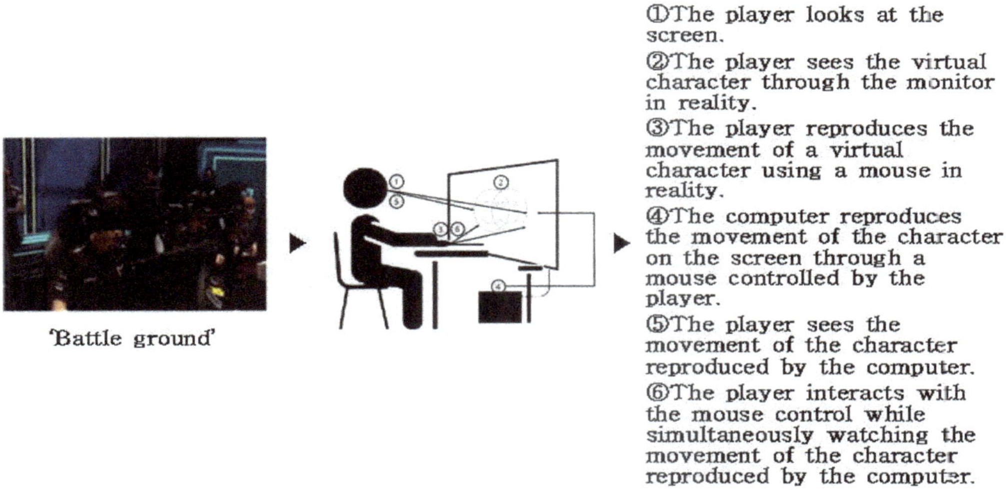

Figure 7. Example of experimental result of "text algorithm analysis" in the e-sports field.

In the above results, we looked at the experimental results of "Battleground" in the field of e-sports, used as an experiment tool. Based on the results, the following case analysis was conducted. The result is shown in the following Table 8 text algorithm analysis result in e-sports field.

Based on the above experiment results, let us analyze the results of (B-1) the device immersion method about "Battleground" in the e-sports field. The feeling of immersion is formed from the eyes of users ① and ⑤ and the movement of the character in ②. In addition, immersion is formed in the movements of the ③ and ⑥ mouse controls and the ② character. Therefore, if the device immersion method is subdivided into visual, auditory, and tactile senses, it can be seen that the visual parts of ①, ⑤, and ② and the tactile parts of ③, ⑥, and ② form a sense of immersion. In addition, the user's headset creates an immersive feeling in the auditory part. In order to quantify these results, the degree of immersion in sight, hearing, and tactile sense based on ⑤ is expressed as a number, and the results are shown in Figure 8. The rest of the cases were also tested in the same way.

Table 8. Text algorithm analysis result in e-sports field.

Analysis Target E-Sports	Text Algorithm
 a. League of Legends (LoL)	①. The player looks at the screen. ②. The player sees the virtual character through the monitor in reality. ③. The player reproduces the movement of a virtual character using a mouse in reality. ④. The computer reproduces the movement of the character on the screen through a mouse controlled by the player. ⑤. The player sees the movement of the character reproduced by the computer. ⑥. The player interacts with his mouse control, while watching the movement of the character reproduced by the computer.
 b. Battleground	①. The player looks at the screen. ②. The player sees the virtual character through the monitor in reality. ③. The player reproduces the movement of a virtual character using a mouse in reality. ④. The computer reproduces the movement of the character on the screen through a mouse controlled by the player. ⑤. The player sees the movement of the character reproduced by the computer. ⑥. The player interacts with the mouse control, while simultaneously watching the movement of the character reproduced by the computer.
 c. StarCraft: Remastered	①. The player looks at the screen. ②. The player sees the virtual character through the monitor in reality. ③. The player reproduces the movement of a virtual character using a mouse in reality. ④. The computer reproduces the movement of the character on the screen through a mouse controlled by the player. ⑤. The player sees the movement of the character reproduced by the computer. ⑥. The player interacts with the mouse control, while simultaneously watching the movement of the character reproduced by the computer.

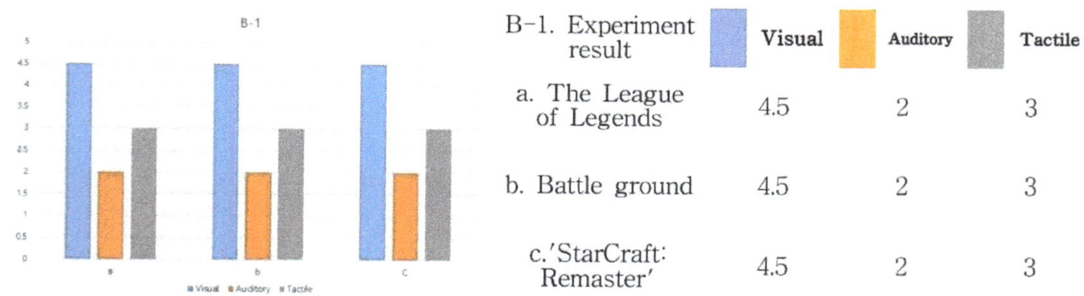

Figure 8. B-1 device immersion method test result.

The analysis targets A. League of Legends (LoL), B. Battleground, and C. StarCraft: Remastered all experience through the same operating system, and the results are the same. In other words, vision is the largest, and next, a sense of immersion is formed through a controller using mouse manipulation. Next is B-2; let us look at the range of interactions in Figure 7. Basically, it is an experience method shown in "Battleground" in the field of e-sports. First of all, the user is seated. Therefore, since the monitor screen to be viewed is flat, the user's gaze forms an interactive range following the movement of the character from the x-axis to the y-axis. In addition, it can be seen that the interaction of the mouse movement occurs only in a fixed position of the controller device. Therefore, based on 10, it can be seen that the range of interaction between the visual and mouse controller is mostly limited to the X-axis and Y-axis. In other words, it can be seen that the visual perception method on the monitor screen and the sense of immersion in mouse operation are flat, in that it is a 2D space. In addition, the same experimental results were found in League of

Legends (LoL) and "StarCraft: Remastered" in the e-sports field. The results are shown in Figure 9. Other cases were also tested in the same way.

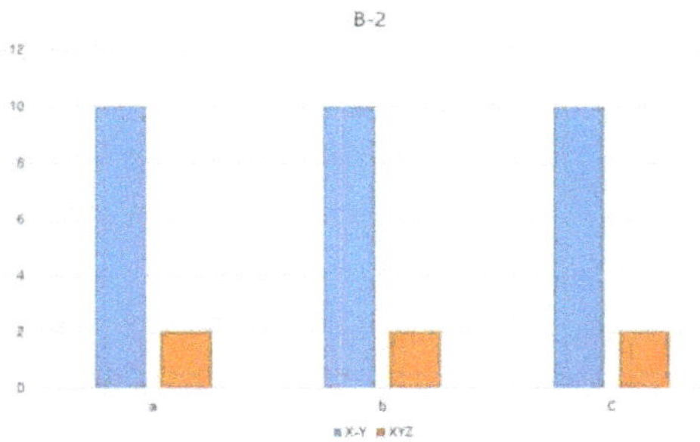

B-2. Experiment result	X-Y	X-Y-Z
a. The League of Legends	10	2
b. Battle ground	10	2
c.'StarCraft: Remaster'	10	2

Figure 9. B-2 interaction range experiment results.

4.1.2. VR E-Sports Field

The next experiment analyzes the "Dragon Fly" in the field of VR E-sports, among the analysis targets with a text algorithm from the viewpoint of the user's visual experience. The results are shown in Figure 10 below. As can be seen from the experimental results, the user establishes a consensus between the character in the game and the user through the HMD and considers himself and the character as one. In addition, the sense of immersion is accelerated through game devices (weapons). Due to the nature of the VR e-sports field, players actually experience their physical movements in a 360-degree screen. The characters in the game are immersed in a complete virtual space through a visual method and connected with a tactile sense through a game device (weapon). The headset sound also accelerates your auditory immersion.

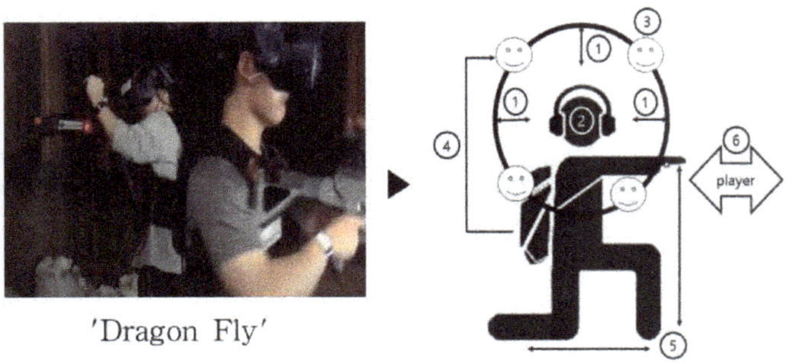

① The player looks at the screen.
② The player sees the virtual character through the media (HMD).
③ The player sees the movement of the character in the virtual through the media (HMD).
④ The player sees the movement of the character in the virtual reproduced by the computer.
⑤ The player interacts and moves with the player's body according to the movement of the character in the virtual reproduced by the computer.
⑥ The player interacts with the movement of the character reproduced by the computer and communicates with other players through movement.

Figure 10. Example of experimental result of "text algorithm analysis" in the VR e-sports field.

Let us analyze the final result, based on the experiment result of "Dragonfly" in the field of VR e-sports that was used as an experiment tool earlier. The following Table 9 shows the result of text algorithm analysis in the field of VR e-sports.

Table 9. VR e-sports text algorithm analysis result.

Analysis Target VR E-Sports	Text Algorithm
 a. Beat Saber	①. The player looks at the screen. ②. The player sees the virtual character through the media (HMD). ③. The player sees the movement of the character in the virtual through the media (HMD). ④. The player sees the movement of the character in the virtual reproduced by the computer. ⑤. The player interacts and moves with the player's body according to the movement of the character in the virtual reproduced by the computer.
 b. Dragonfly	①. The player looks at the screen. ②. The player sees the virtual character through the media (HMD). ③. The player sees the movement of the character in the virtual through the media (HMD). ④. The player sees the movement of the character in the virtual reproduced by the computer. ⑤. The player interacts and moves with the player's body according to the movement of the character in the virtual reproduced by the computer. ⑥. The player interacts with the movement of the character reproduced by the computer and communicates with other players through movement.
 c. HADO	①. The player looks at the screen. ②. The player sees the virtual image (character) and reality at the same time through the media (HMD). ③. The player sees the motion of the virtual image (character) reproduced by the computer and the opponent in reality at the same time. ④. The movement of the player is recognized by the computer and reproduced as an image (character) in the virtual. It also shows the opponent's movement in reality at the same time. ⑤. The player interacts with the player's body according to the motion of the virtual image (character) reproduced by the computer and moves with the opponent's movement in reality. ⑥. Athletes interact with fellow (team) players, while interacting with the movement of images (characters) reproduced by a computer. Communicate through movement.

Based on the above experiment results, B-1, "save bits" in the VR e-sports field, analyzing the results of the device immersion method creates a sense of immersion in the visual parts of the user in ① and ③ and the movement of the character in ② and ③. In addition, immersion is formed through interaction between the movement of user ⑤ and movement of character ③. Therefore, subdividing into visual, auditory, and tactile senses, it can be seen that the sense of immersion is formed through the visual parts of ①, ③, and ② and the movement of the user's body in ⑤. In other words, the tactile sense of immersion is formed through the movement of the body. In addition, a sense of immersion is formed in the auditory part through the user's headset.

In addition, the auditory part plays a large role in the game operation characteristics of Beat Saber, which has a strong musical element. Next, let us look at the analysis of "Dragon Fly". Immersion is formed from the user's gaze in ① and ③ and the movement of the character in ②. In addition, immersion is formed through interaction between the movement of user ⑤ and movement of character ③. In addition, with number ⑥, immersion is also formed through communication with other users. Therefore, let us look at it by subdividing it into sight, hearing, and touch. It can be seen that immersion is formed through communication with the character through the visual parts of ①, ③, and ②, the body movement in ⑤, and the relationship with other users in ⑥. In other words, the characteristic of "Dragon Fly" appeared that the tactile sense of immersion was formed through the movement of the body.

In addition, immersion is being formed in the auditory part through the user's headset. Finally, looking at the analysis of HADO, immersion is formed from the user's perspective in ① and ③ and the movement of the character in ②. The characteristic part is that in HADO; the other user is seen along with the character image in ③ and ④. In other words, a virtual image and another user in reality are simultaneously visible through visual communication. In addition, in step ⑤, communication through the movement of the character and the movement of the other user plays a large role. In addition, in step ⑥, visual and tactile communication relationships are formed through communication with fellow teams, along with the movement of the virtual character and the other user. In order to quantify these results, the degree of immersion in visual, auditory, and tactile sensations, based on ⑤, was expressed as a number. The results are shown in Figure 11.

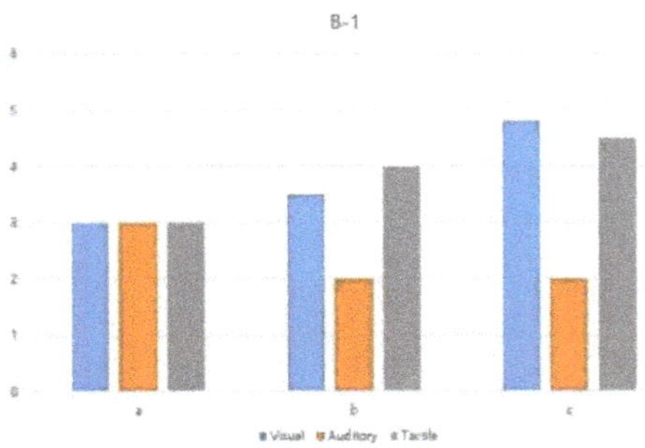

B-1. Experiment result	Visual	Auditory	Tactile
a. Beat save	3	3	3
b. Dragonfly	3.5	2	4
c. HADO	4.8	2	4.5

Figure 11. B-1 device immersion method test result.

Next, let us look at the "B-2" interaction range in Figure 11. First, in the experience method shown in Beat Saber, the monitor screen viewed by the user is completely immersive in a 360-degree three-dimensional space through the HMD, but the user is fixed in place to experience it. Therefore, the interaction range of the user's gaze appears narrowly along the movement of the character from the x-axis to the y-axis. However, it can be seen that the user's body moves up and down, left and right freely, and through interaction with the controller, there is a lot of movement of the user's whole body, such as the hands and feet. Next, in the experience method shown in Dragonfly, the monitor screen viewed by the user is 360 degrees through the HMD, and a complete immersion is formed. However, unlike Beat Saber, users can move their body. Through communication with other users, the interaction of the body movement and device immersion is better than Beat Saber. However, in that the real space is blocked, due to the nature of the HMD, the inconvenience of moving appears as a disadvantage to the interaction.

Finally, in HADO, there is no restriction on the movement of users, in that they can see both real and virtual images using the AR method. In addition, it can be seen that not only the interaction between the device and the user, but also the communication through the team and the competitive relationship with the other team expands the sense of immersion. Therefore, on the basis of 10, the range of visual interaction, device interaction, and user interaction could be analyzed by subdividing into X-axis, Y-axis and X-axis, Y-axis, Z-axis. In other words, in the interaction between the user and the device, there was a lot of body movement, due to the characteristics of the music and user interaction method of Beat Saber. However, when using the HMD, movement was formed only in a fixed position by blocking the eyes of the real space. In Dragonfly, interaction was additionally formed through communication between users. However, the movement of the body was limited, due to the blocking of reality by the HMD. However, in HADO, it can be seen that interaction is maximized through a game method that utilizes the relationship between users and the characteristic that real space and virtual images coexist. Therefore, the numerical values of the results are shown in Figure 12 below.

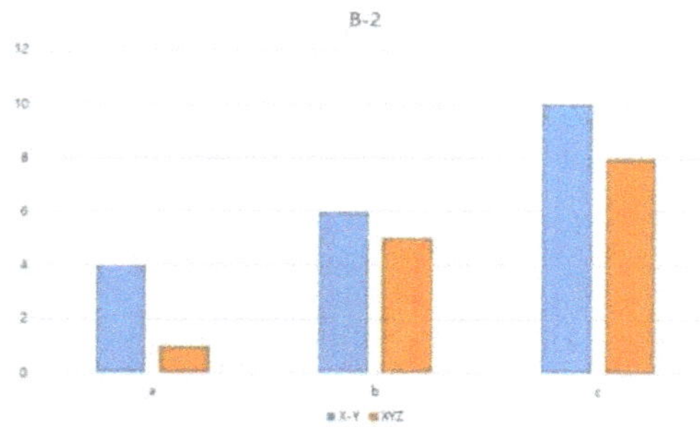

B-2. Experiment result	X-Y	X-Y-Z
a. Beat save	4	1
b. Dragonfly	6	5
c. HADO	10	8

Figure 12. B-2 interaction range experiment results.

4.2. C. Typing Analysis

The result of categorization based on the features of the frame appearing in the field of e-sports and VR e-sports is shown in Table 10 below.

Table 10. Visualization of the analysis target frame.

Analysis Target	Frame Format		
E-sport	 League of Legends (LoL) 	 Battleground 	 StarCraft: Remastered 
VR e-sport	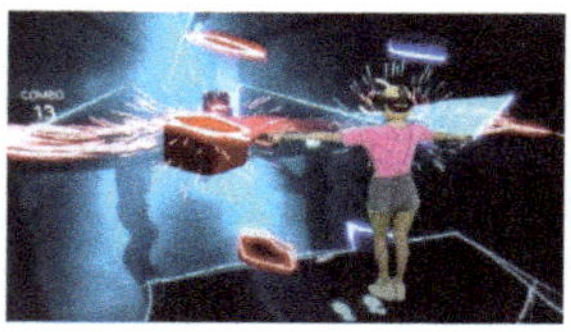 Beat Saber 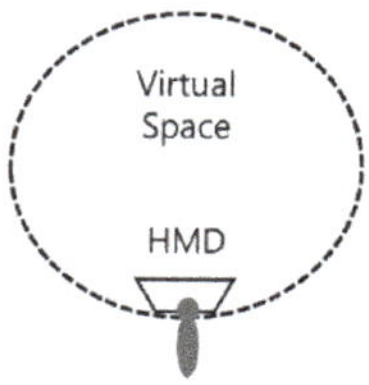	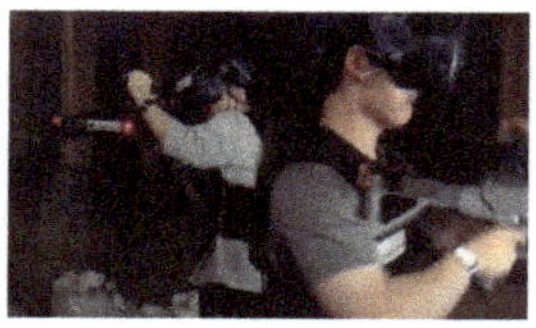 Dragonfly 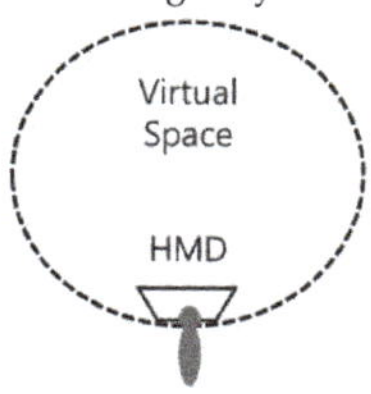	 HADO 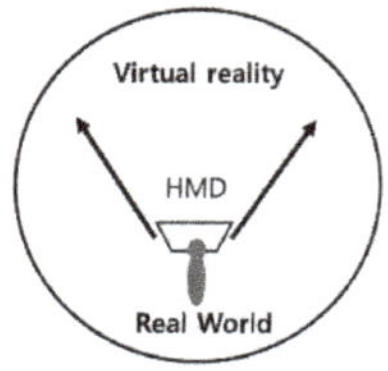

According to David Bodwell's discussion, the frame was interpreted as a boundary concept of a domain called "a rectangular border that influences the degree to which the size of the situation in the screen is controlled and understood" [41]. This means the concept of a boundary between a formal and physical concept and a spatial category as an object in which a certain method or image is represented [42]. As can be seen from the above results, "League of Legends (LoL)", "Battleground", and "StarCraft: Remastered" in e-sports all appear in the same typology. The reason for this is basically that e-sports are played in front of a monitor. Therefore, players play games from a two-dimensional perspective of a monitor device and experience in reality through virtual images and visual interactions. According to Habert Zettle, the expansion of the aspect ratio as a horizontal aspect ratio from 4:3 to 16:9 is intended to satisfy human visual needs. It has been said that this may be because the human field of view is longer horizontally than vertically, and human life is made more horizontally than vertically [43]. Therefore, the sense of immersion may vary, depending on the size of the frame, but the method of operating in front of the monitor frame is the same. On the other hand, in the field of VR e-sports, "Beat Saber" and "Dragonfly" showed the same typification, and in "HADO", different types of typification appeared. The reason is that "Beat Saber" and "Dragonfly" use HMD-based devices to completely block reality and experience a 360-degree virtual space in the visual. However, the disadvantage is that the user's visual reality is blocked, and the body movement is not free. However, there is a difference in that "HADO" uses a device that uses AR. That is, by augmenting a virtual image in a real space and visually showing it, the reality and the virtual can be seen at the same time. In this respect, the advantage is that users can move freely and be more immersed in the game. As a result, the feeling of immersion increases when the monitor frame is enlarged or the viewing angle is visually expanded, rather than the feeling of immersion felt in the way of using a single monitor. Therefore, through these features, it can be seen that the user's sense of immersion or interaction may vary, depending on which device is used.

Table 10 shows the "C-1" FOV range experiment and "C-2". Let us look at the frame immersion experiment results. The FOV, or field of view, plays an important role in enhancing the user's sense of immersion. Therefore, the larger the viewing angle, the higher the sense of immersion. This is also the case in "C-2". In the frame immersion experiment, the feeling of immersion changed according to the interaction method between the device and the user. The way users interact with each other by touching them with their hands or entering a virtual space creates a greater sense of immersion than users who only visually look at the monitor. Therefore, through the experimental results, the sense of immersion between the analysis cases was quantified. Results can be seen in Figures 13 and 14.

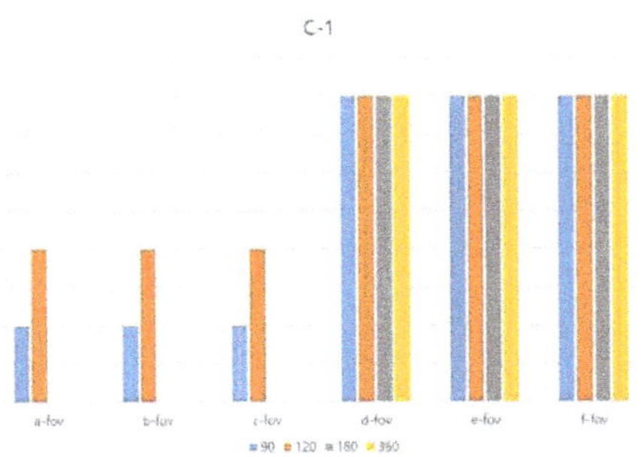

C-1. Experiment result	90°	120°	180°	360°
a. The League of Legends	90°	120°		
b. Battle ground	90°	120°		
c. 'StarCraft: Remaster'	90°	120°		
d. Beat save	90°	120°	180°	360°
e. Dragonfly	90°	120°	180°	360°
f. HADO	90°	120°	180°	360°

Figure 13. C-1 FOV range test result.

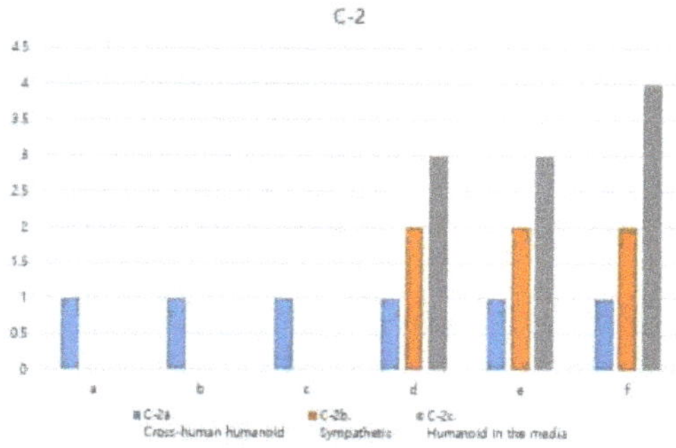

C-2. Experiment result	C-2a	C-2b	C-2c
a. The League of Legends	1		
b. Battle ground	1		
c.'StarCraft: Remaster'	1		
d. Beat save	1	2	3
e. Dragonfly	1	2	3
f. HADO	1	2	4

Figure 14. C-2 frame immersion test result.

As can be seen from the results, perception of an object is from grasping the features of the external structure to recognize the morphological features of the proposed structure. Therefore, the expansion of the frame is the creation of a perceptual range in which the eye and the object to be seen are recognized as the processes of change appearing in the physical size, and the visual border as the visible limit area is determined. As shown in the above result, League of Legends, Battleground, and StarCraft: Remastered are formed from a 90 to a 120° viewing angle. However, in Beat Saber, Dragonfly, and HADO, the viewing angle was formed from 90 to 360°. Therefore, the result is that the sense of immersion is higher in d, e, and f. That is, it can be said that the user's body movement is proportional to the visual range and the frame range of the device. Therefore, it can be said that the greater the range of physical activity of the user or the distance of visual movement, the greater the exercise effect. On the other hand, Beat Saber and Dragonfly visually block reality, and movement may be limited because the user can only move in a virtual space. However, HADO can see the real space and the virtual space at the same time, so it can be said that the user's physical activity and immersion are the most among the six cases.

In Figure 14, C-2 frame immersion can also be seen in the experimental results. In step 1, "C-2a," the cross-human humanoid method, the user only looks at the monitor device. In step 2, "C-2b," the sympathetic method is a form in which the user interacts with the monitor device. Finally, in step 3, "C-2c," the humanoid in the media method is a form in which the user interacts with the device through physical communication, so it can be said that the most immersive and physical activities are experienced. Looking at the results, in the first stage including League of Legends, Battleground, and StarCraft: Remastered, a cross-human humanoid method was shown. In Beat Saber, Dragonfly, and HADO, not only the 1st stage but also the 2nd and 3rd stages are visible. In particular, HADO shows the highest level of immersion and user activity among the three features.

4.3. D. Image of Gaze

Table 11 shows the gaze image of the target based on the features of the frame appearing in the field of e-sports and VR e-sports.

Table 11. Image of the target's gaze.

a. League of Legends	**b. Battleground**	**c. StarCraft: Remastered**	**d. Beat Saber**	**e. Dragonfly**	**f. HADO**

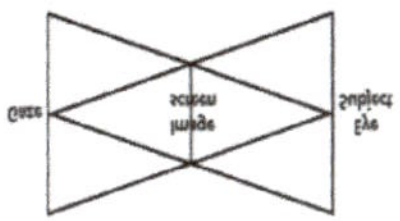	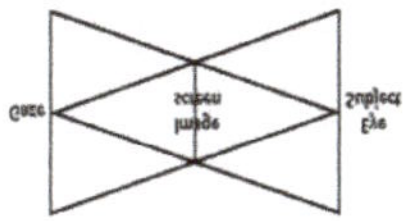	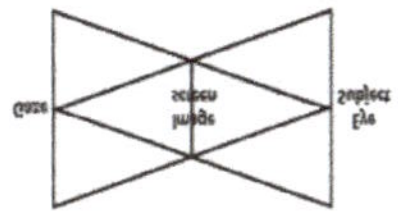	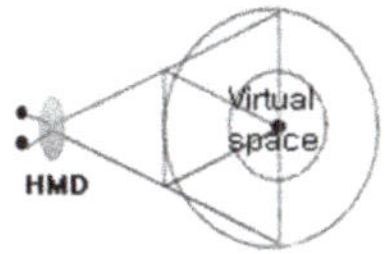	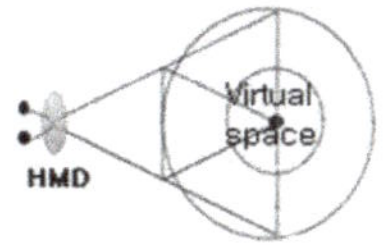	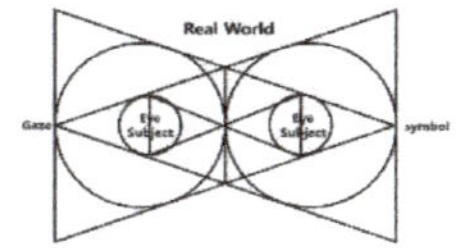

The process of visualizing a target's gaze is a visual representation of a text algorithm in a form that graphically describes the visual system between the device and the user. Therefore, the simpler the image structure, the simpler the visual system, and the more complex the visual system, the more processes the images overlap. Looking at the results above, in League of Legends, Battle ground, and StarCraft: Remastered, the image of the gaze is the basic visual structure of the "gaze," and the relationship between the user and the device are facing each other.

On the other hand, in Beat Saber and Dragonfly, the visual system forms a dual visual system in the virtual space, and it can be seen that the center of the user's gaze exists in the virtual space. In addition, in HADO, it can be seen that not only the visual system, but also the user's body is in the virtual space. Therefore, it can be seen that the user's gaze and physical activity are immersed in the virtual space as it goes from a. to f. Based on this conclusion, the "D-1" user recognition method, and "D2," let us look at the results of the interaction method. The results can be seen in Figure 15.

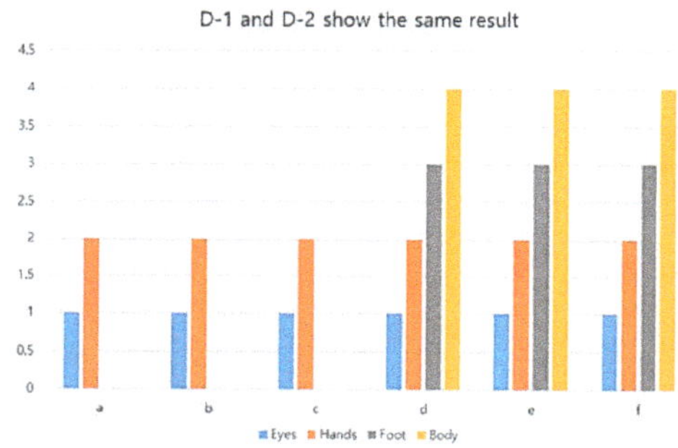

Experiment result	D-1 Recognition method				D-2 Interaction method			
	Eyes	Hands	Foot	Body	Eyes	Hands	Foot	Body
a. The League of Legends	1	2			1	2		
b. Battle Ground	1	2			1	2		
c. 'StarCraft: Remaster'	1	2			1	2		
d. Beat save	1	2	3	4	1	2	3	4
e. Dragonfly	1	2	3	4	1	2	3	4
f. HADO	1	2	3	4	1	2	3	4

Figure 15. D-1. user recognition method and D-2. interaction method experimental results.

In Figure 15, "D-1" user recognition method and "D-2" interaction method can be found through the experimental results. In the cognitive method, apply a number from 1 to 4 in the order of eyes, hands, feet, and body. Considering that the eye is 1, it is judged that the degree of immersion has increased when both the eye and the hand are recognized at the same time. Therefore, 6 experimental cases are classified into "D-1" and "D-2," and the result of the experiment increases from a to f, as shown above. The results of "D-1" and "D-2" are the same.

5. Conclusions

Following the outbreak of COVID-19, modern people have developed a need for a variety of health care content that they can enjoy safely at home. Their expectations and needs for immersive content are especially growing. A prominent need is found for killer content to satisfy both fun and health in today's reality where people suffer a lack of physical activity. There have been diverse efforts to develop such content devices, but empirical analysis framework and tests of exercise effects are still in shortage and far from standardization. Thus, the present study set out to propose an analysis framework for e-sports equipped with both entertainment and sports elements and demonstrate the possibilities of VR e-sports.

The experiment results of the study show that analysis data of immersion and exercise abilities was generated between users and their devices in six research cases of e-sports and VR e-sports. Close relationships between users and their devices were found through text algorithm analysis under "B," and the results were used to digitize "B-1" of device immersion methods and "B-2" of scope of interactions. Under "C" of categorization, the study measured and digitized "C-1" of FOV and "C-2" of frame immersion and obtained the independent result values of each device. Under "D" of "notion of the real gaze," the study segmented "D-1" of users' cognition methods and "D2" of interaction methods by the body part, identified physical characteristics according to the degree of immersion, and digitized the results. The final results were put in a diagram in Table 12.

Table 12. Final experiment result.

			a. League of Legends	b. Battle-ground	c. StarCraft: Remastered	d. Beat Saber	e. Dragonfly	f. HADO
A. Analysis Target								
B. Text algorithm analysis	B-1. Device immersion	Visual	4.5	4.5	4.5	3	3.5	4.8
		Auditory	2	2	2	3	2	2
		Tactile	3	3	3	3	4	4.5
	B-2. Interaction range	X-Y	10	10	10	4	6	10
		X-Y-Z	2	2	2	1	5	8
C. Typification	C-1. FOV	90°	90°	90°	90°	90°	90°	90°
		120°	120°	120°	120°	120°	120°	120°
		180°	-	-	-	180°	180°	180°
		360°	-	-	-	360°	360°	360°
	C-2. Frame immersion	C-2a. Cross-human humanoid	1	1	1	1	1	1
		C-2b. Sympathetic	-	-	-	2	2	2
		C-2c. Humanoid in the media	-	-	-	3	3	3
D. Diagram of gaze	D-1. User recognition method	1. Eyes	1	1	1	1	1	1
		2. Hands	2	2	2	2	2	2
		3. Feet	-	-	-	3	3	3
		4. Body	-	-	-	4	4	4
	D-2. Interaction method	1. Eyes	1	1	1	1	1	1
		2. Hands	2	2	2	2	2	2
		3. Feet	-	-	-	3	3	3
		4. Body	-	-	-	4	4	4

These findings show that under "A" of objects of analysis, a variety of devices were used, based on game content in League of Legends, Battleground, StarCraft: Remastered, Beat Saber, Dragon Flight, and HADO. That is, while a, b, and c were managed in a way of visualizing planar frames on the monitor, d, e, and f used immersive content devices that were recently attracting attention and featured three-dimensional visual frames in 360 degrees. In particular, d and e offer excellent visual immersion by blocking a real space completely, but a lot of inconvenience is created, due to the reality that has been blocked. On the other hand, f can be managed in AR or MR to supplement this disadvantage, having great potential for utilization in the future. In the experiment results, the visual system between users and their devices was turned into an algorithm, based on texts under "B" of algorithm analysis. The results were used to digitize the results values of "B-1" of device immersion methods (visual, auditory, and tactile), according to the characteristics of each device. The result values indicate that immersion was great in visual parts overall. Moving from a to f, users increased their utilization of devices through their movements, based on their tactile part as much as their visual part, growing their physical activity level. Under "B-2" of scope of interactions, the study compared the X-, Y- and X-, Y-, and Z-axes in the scope of movements in physical activities, based on devices. Users used a device in a fixed position and managed their movements on the X, Y-axis in most cases in a, b, and c. However, in d, e, and f, they moved along the X-, Y-, and Z-axes by standing up to move around spaces and moving their arms and legs. In f, featuring a mixed reality between the real and virtual worlds, users were very active with their movements and had no limits in their scope of activities.

Under "C" of categorization, the study digitized "C-1" of FOV values in a process of turning text algorithms between users and their devices into images to figure them out more easily. Broader FOV meant higher immersion and expanded scope of physical activities. The result values show that users' maximum FOV was within a scope of 90∼120° in a, b, and c, and that an FOV of 360° was basically created in d, e, and f. Allowing users to perceive reality freely, f especially maximized the scope of FOV. Under "C-2" of frame immersion, analysis was conducted in further segments of "C-2a," cross-human humanoid, "C-2b," sympathetic, and "C-2c," humanoids in the media. Result values would vary according to whether there were physical contacts in relationships between users and their devices. Users would experience greater immersion when entering a virtual space and communicating with their devices by touching the screen, than when simply looking at the monitor screen. The outcome was in the form of "C-2a" in a, b, and c, in which users had no physical contact at all with the monitor screen offering virtual images. Both "C-2b" and "C-2c" were found in d, e, and f that were managed through users touching the screen and moving their bodies. Moving from d through e to f, users engaged in more active communication with their devices and experienced greater immersion and physical activity competence according to their movements and device operation methods.

Under "D" of "notion of the real gaze," the study turned the characteristics of the experimentation process into visual algorithms. Any devices can be visualized through "notion of the real gaze" in the visualization process of information. The result values were used to digitize "D-1" of users' cognition methods. They showed in which body parts the users had the greatest immersion in their devices through their eyes, hands, feet, and bodies, and in which parts they exhibited the highest activity. In a, b, and c, they were mostly managed through eyes and hands with physical movements immensely focused on visual parts. However, in d, e, and f, users were able to move their bodies through the movement of their feet, as well as their eyes and hands. In f, their body activity performed with the greatest excellence and induced decisive immersion in users. Moving from a to f, users recorded increasingly higher physical utilization and had immersion through the movement of their entire bodies, instead of physical parts. In "D2" of interaction methods, a, b, and c enabled communication through eyes and hands, whereas d, e, and f allowed users to interact with their devices through the movement of their feet and bodies, as well as their eyes and hands.

As a result, the results of using the analysis tool among A. analysis targets a. League of Legends, b. Battle Ground, c. StarCraft: Remastered, d. Beat Save, e. Dragon Fly, and f. HADO. It was found that the most effective example of the value was f. HADO. Based on these results, the form of realistic health care to be produced in the future requires a structure that has high frame scalability and can compete or cooperate with others online. In addition, it can be conveniently used at home, and exercise prevention and rehabilitation effect through safety, space expandability, and excellent immersion are considered to be great.

The present study mentioned a need for immersive content and research in the field of health care during the COVID-19 era, analyzing various immersive content in related research and identified an analysis framework to measure exercise abilities. The findings are presented in Table 13. The study put in various devices, conducted an experiment with an analysis framework, and tested their effects through digitization in the process of A-B-C-D.

Table 13. Realistic content exercise effect verification tool.

Analysis Method	Analysis Tool	Numericalization Process
A. Analysis target		
B. Text algorithm analysis	B-1. Device immersion method	Visual Auditory Tactile
	B-2. Interaction range	X-Y X-Y-Z
C. Typification	C-1. FOV	90° 120° 180° 360°
	C-2. Frame immersion	C-2a. Cross-human humanoid C-2b. Sympathetic C-2c. Humanoids in the media
D. Image of gaze	D-1. User recognition method	1. Eyes 2. Hands 3. Feet 4. Body
	D-2. Interaction method	1. Eyes 2. Hands 3. Feet 4. Body

The present study offered a set of criteria to analyze and understand immersive content that was further diversified and advanced. It is difficult to evaluate various devices with a simple technical approach. Because the human body connectivity of recent immersive devices is very high. Therefore, evaluation criteria suitable for the new digital visual system and criteria for verifying various devices are presented in this research paper. Until now, there is no clear health care product through VR technology in the health care field.

However, we are living in an era in which modern life patterns are changing and global epidemics appear. Recently, research on safe immersive devices to be used and revived in anticipation of a new virtual space called metaverse has already begun. Therefore, it is expected that the psychological and physical evaluation criteria for future health care products will be newly presented through this study. It is expanded from the current 2D-based visual system to a 360-degree 3D-based visual system. That is, an extended evaluation standard from the UI/UX perspective between humans and media is formed. In addition, various health care products using VR and AR formats can be produced. Finally, even if a pandemic such as COVID-19 recurs, it is expected that anyone from the elderly to children can enjoy exercise and prepare for prevention and rehabilitation through a safe, convenient, and fun way at home. In addition, I hope that this research paper will serve as a guide in producing products with both fun and exercise effects in the process of producing home health care products in the future. The study also raised a need for additional researches with many practical experiments and clearer and more effective researches, based on collaboration with medical and health care professionals. Everyone hopes that this will be the last virus crisis. If another one happens in the future, they will hopefully come up with wiser measures than the current situation and convey messages of overcoming and hope, rather than frustration and fear to the lives of modern people.

Author Contributions: Conceptualization, S.-G.L., S.-H.J. and J.-H.H.; data curation, S.-G.L.; formal analysis, S.-G.L., S.-H.J. and J.-H.H.; funding acquisition, S.-G.L.; investigation, S.-G.L., S.-H.J. and J.-H.H.; methodology, S.-G.L., S.-H.J. and J.-H.H.; project administration, S.-G.L., S.-H.J. and J.-H.H.; resources, S.-G.L. and J.-H.H.; software, S.-G.L. and S.-H.J.; supervision, S.-H.J. and J.-H.H.; visualization, S.-G.L. and J.-H.H.; writing—original draft, S.-G.L., S.-H.J. and J.-H.H.; writing—

review and editing, S.-H.J. and J.-H.H. All authors have read and agreed to the published version of the manuscript.

Funding: This work was supported by Youngsan University Research Fund of 2020.

Institutional Review Board Statement: Not applicable.

Informed Consent Statement: Not applicable.

Data Availability Statement: Not applicable.

Conflicts of Interest: The authors declare no conflict of interest.

References

1. Lee, S.H. E-Sports Research. *J. Korea e-Sports Soc.* **2019**, *1*, 1–27.
2. Kim, T.J.; Huh, J.H.; Kim, J.M. Bi-directional education contents using VR equipments and augmented reality. *Multimed. Tools Appl.* **2018**, *77*, 30089–30104. [CrossRef]
3. Neozensoft. Available online: https://m.blog.naver.com/neozensoft/222078050509 (accessed on 11 May 2021).
4. Coraci, D.; Fusco, A.; Frizziero, A.; Giovannini, S.; Biscotti, L.; Padua, L. Global approaches for global challenges: The possible support of rehabilitation in the management of COVID-19. *J. Med. Virol.* **2020**, *92*, 1739–1740. [CrossRef] [PubMed]
5. Available online: https://quaresma.tistory.com/267 (accessed on 11 May 2021).
6. Kang, S.K.; Chae, H.S. A Research on e-Sports Fandom as a Cultural Performance: A Depth Interview Study. *Korean Women's Assoc. Commun. Stud.* **2011**, *18*, 5–39.
7. CNBC, Google Glass Can Help Children with Autism Understand Emotion. Available online: https://www.youtube.com/watch?v=CpCC6okoVHI (accessed on 11 May 2021).
8. National Institute of Neurological Disorders and Stroke. Available online: http://www.ninds.nih.gov/disorders/backpain/detail_backpain.htm (accessed on 11 May 2021).
9. Hoffman, H.G.; Richards, T.L.; Coda, B.; Bills, A.R.; Blough, D.; Richards, A.L.; Sharar, S.R. Modulation of thermal pain-related brain activity with virtual reality: Evidence from fMRI. *Neuroreport* **2004**, *15*, 1245–1248. [CrossRef] [PubMed]
10. Pazzaglia, C.; Imbimbo, I.; Tranchita, E.; Minganti, C.; Ricciardi, D.; Monaco, R.L.; Padua, L. Comparison of virtual reality rehabilitation and conventional rehabilitation in Parkinson's disease: A randomised controlled trial. *Physiotherapy* **2020**, *106*, 36–42. [CrossRef] [PubMed]
11. Carfì, A.; Liperoti, R.; Fusco, D.; Giovannini, S.; Brandi, V.; Vetrano, D.L.; Onder, G. Bone mineral density in adults with Down syndrome. *Osteoporos. Int.* **2017**, *28*, 2929–2934. [CrossRef] [PubMed]
12. Lorenzi, M.; Bonassi, S.; Lorenzi, T.; Giovannini, S.; Bernabei, R.; Onder, G. A review of telomere length in sarcopenia and frailty. *Biogerontology* **2018**, *19*, 209–222. [CrossRef] [PubMed]
13. MIK Hot Spot. Available online: https://www.youtube.com/watch?v=Cm4JoKBK3Zo (accessed on 11 May 2021).
14. VR Ohshape, Odders Lab. Available online: https://www.youtube.com/watch?v=dpomy6XGAbQ (accessed on 11 May 2021).
15. Kim, D.G. *Wearable Device Trends and Implications*; Korea Information Society Development Institute: Seoul, Korea, 2013; Volume 25.
16. Friedberg, A. *The Virtual Window: From Alberti to Microsoft*; MIT Press: Cambridge, MA, USA, 2006; pp. 1–6.
17. Gombrich, E.H. *Art & Illusion*; Cha, M.R., Translator; Yeolhwadang: Paju, Korea, 2004; p. 282.
18. Ham, H.Y. Study on Visual Perception Effect of Motion Graphic Motion. Master's Thesis, Chung-Ang University, Seoul, Korea, 2004; p. 25.
19. John, B. *Image Visual and Media, Dongmun Line*; Dongmun: Seoul, Korea, 2005; p. 242.
20. Arnheim, R. *Visual Thinking*; Kim, J.O., Translator; Mijinsa: Seoul, Korea, 2004; pp. 223–275.
21. Oh, S.S.; Kim, D.H. Analysis of research trends on e-sports. *J. Korean Soc. Wellness* **2012**, *7*, 113–121.
22. Jeremy, B. *Experience on Demand*; W. W. Norton & Company: New York, NY, USA, 2019.
23. "Virtual Reality Pain Reduction", HITLab. Available online: https://www.hitl.washington.edu/projects/vrpain/ (accessed on 11 May 2021).
24. "VR Therapy for Spider Phobia", HITLab. Available online: https://www.hitl.washington.edu/projects/exposure/ (accessed on 11 May 2021).
25. Hal, F. *Vision and Visuality*; Busan Kyungsung University: Busan, Korea, 2004; pp. 1–21.
26. Jung, J.J. A Study on the Expansion of Communication in Media Art Seen as a Creative Metaphor. Ph.D. Thesis, Kookmin University, Seoul, Korea, 2014.
27. Lim, S.G.; Kim, C.Y. A Study on the Change of Digital Visuality in the 21st Century through Lacanian Perspective: From the perspective of digital frame expansion. *J. Multimed. Soc.* **2018**, *21*, 638–647.
28. Kim, J.S.; de Fremery, W. A Study on Visualization Method Using Algorithm-based Modeling. *Digit. Des. Res.* **2013**, *13*, 61–70.
29. Kim, M.S.; Choi, W.S.; Joung, S.Y. Design and Implementation of Cluster Analysis Learning System Based on Algorithm Visualization. In Proceedings of the Korean Information Science Society, Pyeongchang, Korea, 17–20 October 2018; pp. 933–935.
30. Lim, S.G. Construction of HMD-based Interactive Realistic Media Algorithm Using L-System. *J. Korea Multimed. Soc.* **2020**, *23*, 58–66.

31. Ebert, D.S.; Musgrave, F.K.; Peachey, D.; Perlin, K.; Worley, S. *Texturing & Modeling: A Procedural Approach*; Morgan Kaufmann Publisher: San Francisco, CA, USA, 2013.
32. Kim, H.Y.; Bag, J.Y. A Review on Expressive Materials and Approaches to Text Visualization. *J. Korea Contents Assoc.* **2013**, *13*, 64–72. [CrossRef]
33. Crary, J. *Techiques of the Observer*; MIT Press: Cambridge, MA, USA, 1999; p. 88.
34. Jacques, L. *Jacques Lacan Seminar 11*; Maeng, J.H.; Lee, S.R., Translators; Saemulgyeol: Seoul, Korea, 2008; p. 129.
35. Itou, T. *Photography and Painting*; Kim, K.Y., Translator; Visual and Language: Seoul, Korea, 1987; p. 13.
36. Wong, U. *Design and Morphology*; Choi, G.R., Translator; Gugje Book Publishing: Seoul, Korea, 1994; p. 43.
37. Stephen, H. *Appearance & Reality: A Visual Handbook for Artists, Designer, and Makers*; Cambium Press: Danbury, CT, USA, 2000; pp. 27–31.
38. Grau, O. *Virtual Art: From Illusion to Immersion*; MIT Press: Cambridge, MA, USA, 2003; p. 5.
39. Bordwell, D. *Film Art; Theory and Practice*; Joo, J.S., Translator; Busan Kyungsung University: Busan, Korea, 1993; p. 167.
40. Aumont, J. *L'image*; Oh, J.M., Translator; Dongmunseon: Seoul, Korea, 2006; p. 193.
41. Zettle, H. *Aesthetic Principles and Methods of Video Production*; Communication Book; Calameo: Seoul, Korea, 2005; p. 127.
42. Arnheim, R. *Art and Visual Perception*; Kim, C.I., Translator; Mijinsa: Seoul, Korea, 2000; p. 52.
43. Aumont, J. *L'image*; Oh, J.M., Translator; Dongmunseon: Seoul, Korea, 2006; p. 32.

healthcare

MDPI

Article

Post-Analysis of Predictive Modeling with an Epidemiological Example

Christina Brester [1,*], Ari Voutilainen [2], Tomi-Pekka Tuomainen [2], Jussi Kauhanen [2] and Mikko Kolehmainen [1]

[1] Department of Environmental and Biological Sciences, University of Eastern Finland, Yliopistonranta 1 E, P.O. Box 1627, FI-70211 Kuopio, Finland; mikko.kolehmainen@uef.fi

[2] Institute of Public Health and Clinical Nutrition, University of Eastern Finland, Yliopistonranta 1 C, P.O. Box 1627, FI-70211 Kuopio, Finland; ari.voutilainen@uef.fi (A.V.); tomi-pekka.tuomainen@uef.fi (T.-P.T.); jussi.kauhanen@uef.fi (J.K.)

* Correspondence: kristina.brester@uef.fi or christina.brester@gmail.com

Abstract: Post-analysis of predictive models fosters their application in practice, as domain experts want to understand the logic behind them. In epidemiology, methods explaining sophisticated models facilitate the usage of up-to-date tools, especially in the high-dimensional predictor space. Investigating how model performance varies for subjects with different conditions is one of the important parts of post-analysis. This paper presents a model-independent approach for post-analysis, aiming to reveal those subjects' conditions that lead to low or high model performance, compared to the average level on the whole sample. Conditions of interest are presented in the form of rules generated by a multi-objective evolutionary algorithm (MOGA). In this study, Lasso logistic regression (LLR) was trained to predict cardiovascular death by 2016 using the data from the 1984–1989 examination within the Kuopio Ischemic Heart Disease Risk Factor Study (KIHD), which contained 2682 subjects and 950 preselected predictors. After 50 independent runs of five-fold cross-validation, the model performance collected for each subject was used to generate rules describing "easy" and "difficult" cases. LLR with 61 selected predictors, on average, achieved 72.53% accuracy on the whole sample. However, during post-analysis, three categories of subjects were discovered: "Easy" cases with an LLR accuracy of 95.84%, "difficult" cases with an LLR accuracy of 48.11%, and the remaining cases with an LLR accuracy of 71.00%. Moreover, the rule analysis showed that medication was one of the main confusing factors that led to lower model performance. The proposed approach provides insightful information about subjects' conditions that complicate predictive modeling.

Keywords: post-analysis of data-driven models; rule design; multi-objective optimization; model performance; prediction of cardiovascular death

Citation: Brester, C.; Voutilainen, A.; Tuomainen, T.-P.; Kauhanen, J.; Kolehmainen, M. Post-Analysis of Predictive Modeling with an Epidemiological Example. *Healthcare* **2021**, *9*, 792. https://doi.org/10.3390/healthcare9070792

Academic Editor: Mahmudur Rahman

Received: 21 May 2021
Accepted: 22 June 2021
Published: 24 June 2021

1. Introduction

The increasing volume of data collected and the expanding computational resources dictate current trends in data-driven modeling [1]. It is no longer surprising that models outperform human experts in many areas [2]. Yet this high performance goes hand-in-hand with significant growth in model complexity. This tendency results in greater intricacy of the use of data-driven models in the medical domain, where model interpretability is of primary importance [3]. While predicting diseases, one may even choose to use less accurate "white box" models (easily interpretable with a simple structure), such as decision trees or rules, rather than to unpack "black box" models (hardly interpretable with a complex structure) [4,5].

To address the growing complexity of data-driven models and to understand the non-trivial logic behind their decisions, a number of methods for post-analysis have been proposed recently [6]. Some of them apply local approximations with simpler but interpretable models [7], while others estimate feature importance using permutation techniques [8]; additionally, they search for influential samples that greatly affect model parameters [9]. Alternatively, there are methods applied while generating a model that aim to find a trade-off between model accuracy and complexity [10]. At this stage, optimal sampling techniques or model structures that lead to higher accuracy and lower complexity might be determined [11,12].

In this study, we addressed a particular question of model post-analysis, i.e., the conditions in the sample that lead to higher or lower model performance. To answer this question in the low-dimensional predictor space, error heatmaps can be used, since they allow the visualization and identification of subpopulations with high and low error rates considering one or two predictors [13]. Then, decision trees inherently perform error analysis, as their terminal nodes contain subpopulations—whose characteristics are discovered by moving up the tree—and model performance is easily estimated for each terminal node [14]. Moreover, stratification is widely used when comparing the model performance of different subpopulations (men vs. women, young vs. old, with vs. without a particular condition), which may positively affect model predictive ability compared to training on combined samples [15]. However, these approaches are not suitable for multivariate error analysis in the high-dimensional predictor space. Alternatively, an "unreliability" score, proposed by Myers et al. [16], is individually applied for each subject, and its high values indicate decreased model accuracy. This approach requires additional analysis of cases with high unreliability to extract knowledge about "difficult" subjects.

Taking into account the limitations of the existing studies, in the presented epidemiological example we revealed "easy" and "difficult" cases for the model when it operated in the high-dimensional predictor space. Particularly, we trained Lasso logistic regression (LLR) to predict cardiovascular death using data from the Kuopio Ischemic Heart Disease Risk Factor Study (KIHD) [17]. After validating the model, we generated a set of compact rules that covered samples with extremely high or low model performance. Generally speaking, this paper presents the model-independent approach for post-analysis and shows that this approach not only reveals when the model predictions are less reliable, but also finds out from which perspectives the model requires improvement and what kind of new samples might be collected to bring more information into "difficult" regions of the predictor space. Although we did not have a specific hypothesis to test, this exploratory post-analysis yielded interesting findings.

2. Materials and Methods

2.1. KIHD: Baseline Cohort

The data utilized in the current study were collected in 1984–1989 in the city of Kuopio and the surrounding area in Eastern Finland, whose population was recorded to have one of the highest rates of coronary heart disease [17,18]. The KIHD study is an ongoing project, where the study outcomes are derived from the national registers annually.

The baseline examinations (1984–1989) comprised 2682 randomly selected middle-aged men (42–60 years old), whose health state was carefully described with thousands of physiological, clinical, biochemical, psychological, and socioeconomic measurements. As a starting point, 950 predictors were preselected by the domain expert to perform predictive modeling. In our experiments, the outcome variable was "death from a cardiovascular disease by 2016" referring to codes I00–I99 of the 10th International Classification of Diseases (ICD 10) [19]. Before training a predictive model, preprocessing was applied (Figure 1). Every time we trained the model, the predictors were normalized to the interval (0, 1) using the scaler fitted on the training data.

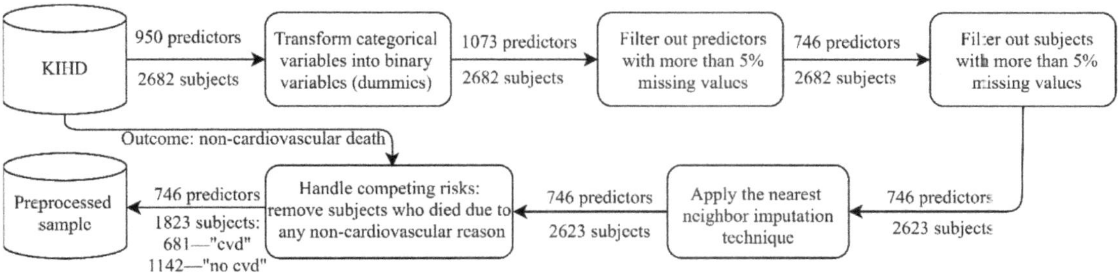

Figure 1. Preprocessing the sample from the Kuopio Ischemic Heart Disease Risk Factor Study (KIHD): "cvd" corresponds to "cardiovascular death by 2016", while "no cvd" corresponds to "no cardiovascular death (alive) by 2016". The nearest neighbor imputation technique implemented by Troyanskaya et al. was utilized [20].

2.2. Multi-Objective Rule Design for the Model Post-Analysis

Let us consider a dataset of n subjects described with m predictors each: $X = \left\{x^{(1)}, x^{(2)}, \ldots, x^{(n)}\right\}$, where $x^{(i)} \in \mathbb{R}^m$, $i = \overline{1,n}$, and an outcome variable: $y = \left\{y^{(1)}, y^{(2)}, \ldots, y^{(n)}\right\}$, where $y^{(i)} \in \{0,\ 1\}$. In the context of epidemiological data, $y^{(i)} = 0$ and $y^{(i)} = 1$ designate "negative" and "positive" regarding a diagnosis. The sample $(X,\ y) = \left\{\left(x^{(1)}, y^{(1)}\right), \left(x^{(2)}, y^{(2)}\right), \ldots, \left(x^{(n)}, y^{(n)}\right)\right\}$ is used to train a model that performs mapping $f : x^{(i)} \longmapsto y^{(i)}$, i.e., predicts a value of $y^{(i)}$ for a given vector of predictors $x^{(i)}$.

In this study, we applied an LLR model, since in our previous experiments, it demonstrated the highest performance on the KIHD data [21]. LLR describes a relationship between a linear combination of predictors and the probability of having a disease in the form of a sigmoid function, $P(y = 1|x) = 1/1 + \exp(-(\omega_0 + \omega_1 x_1 + \ldots + \omega_m x_m))$, as a traditional logistic regression model does [22]. However, LLR is a penalized regression, whose cost function includes an L1-regularization term $\parallel \omega \parallel_1$ in addition to the cross-entropy error $\sum_{i=1}^{n} \log\left(1 + \exp\left(-y^{(i)} \cdot \left\langle \omega, x^{(i)} \right\rangle\right)\right)$ [23]:

$$\min_{\omega} \parallel \omega \parallel_1 + C \cdot \sum_{i=1}^{n} \log\left(1 + \exp\left(-y^{(i)} \cdot \left\langle \omega, x^{(i)} \right\rangle\right)\right) \tag{1}$$

where C defines a shrinkage parameter (the regularization amount) equal to 0.15 in this study.

If $P(y^{(i)} = 1|x^{(i)}) \geq \alpha$, then, $y^{(i)} = 1$; otherwise, $y^{(i)} = 0$. For the imbalanced sample, a cutoff value α requires adjustment (this typically equals 0.5 for the balanced sample); therefore, α is defined as a proportion of "positive" cases in the training sample.

To estimate the model performance on the test data, we executed 50 independent runs of a k-fold cross-validation with stratification, where $k = 5$. LLR was implemented using the scikit-learn library [24]. In each run, we collected the true values of $y^{(i)}$ and the model outcomes on the test data to calculate the number of times the predictions were correct and wrong for each subject. Then, for "positive" cases, we produced the amount of true positive (TP) and false negative (FN) predictions: TP_i and FN_i; for "negative" cases, we calculated the number of true negative (TN) and false positive (FP) predictions: TN_i and FP_i. This model evaluation served as a basis for further analysis, aiming to reveal groups of "easy" and "difficult" cases.

Usually, the model performance is described with statistics averaged over the whole sample. These aggregated values are helpful when comparing models, but they do not reflect the distribution of errors in the space of predictors. In other words, the model performance varies for different subjects and we might be more or less confident in the correctness of predictions depending on the subject's characteristics. Therefore, we propose an approach to automatically generate a set of rules that represent the conditions leading to higher or lower model performance compared to the average level.

Let *rule* denote a set of concurrent conditions "x_j equals a_j":

$$rule = (x_1 = a_1) \text{ and } (x_2 = a_2) \ldots \text{and } (x_j = a_j) \text{ and} \ldots = \bigcap_{j \in J} (x_j = a_j), \quad (2)$$

where a_j is a particular level (category) of a predictor x_j and J contains indices of predictors included in the rule. To define levels a_j, we first need to introduce categories for each predictor x_j based on its distribution. For dichotomous variables, as well as for ordinal or continuous variables with fewer than eight different values in the sample, i.e., $|x_j| < 8$: $a_j \in \{0, 1, \ldots, \max(x_j)\}$. For ordinal or continuous variables with eight and more possible values, i.e., $|x_j| \geq 8$, we introduce N_{bins} intervals using the k-means clustering method, $N_{bins} = 4$. Each interval corresponds to a particular level of x_j: $\left[x_j^{lower_0}, x_j^{upper_0}\right] \to 0, \ldots, \left[x_j^{lower_{(N_{bins}-1)}}, x_j^{upper_{(N_{bins}-1)}}\right] \to (N_{bins} - 1)$.

Then, to generate rules of interest, i.e., to select predictors x_j and to define their levels a_j, we solve two three-objective optimization problems:

Problem 1 (3): To define rules that describe subgroups of subjects with the maximum true positive rate (TPR) and true negative rate (TNR) (i.e., "easy cases"):

$$\begin{cases} TPR(rule) \to \max \\ TNR(rule) \to \max \\ N_{subjects}(rule) \to \max \end{cases} \quad (3)$$

Problem 2 (4): To define rules that describe subgroups of subjects with the minimum TPR and TNR (i.e., "difficult cases"):

$$\begin{cases} TPR(rule) \to \min \\ TNR(rule) \to \min \\ N_{subjects}(rule) \to \max \end{cases} \quad (4)$$

In both problems, the third criterion is used to maximize the number of subjects covered by the rule, which aims to find as general a pattern as possible.

To solve the multi-objective Problems (3) and (4), we apply the Non-dominated Sorting Genetic Algorithm III (NSGA III), which is a stochastic optimization algorithm operating with a population of solutions whose quality improves during the search [25]. NSGA III is based on a Pareto-dominance idea and returns a set of nondominated solutions (in our case, a set of rules), of which one cannot be preferred over another. Each solution in the population is coded with a binary string, a so called "chromosome", and genetic operators such as selection, crossover, and mutation are applied to binary strings so that new solutions with better values of objective criteria are produced. Figure 2 explains how a binary coding is used to represent rules (2):

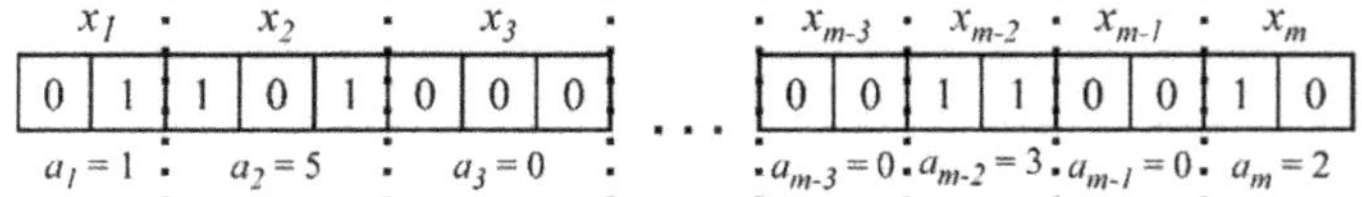

Figure 2. The binary representation of a rule. Parts of a binary string consecutively code the predictors' levels or their absence in the rule.

The number of bits n_j^{bits} to code a_j depends on the amount of levels n_j^{levels} introduced for x_j, plus one additional level, meaning the absence of x_j in the rule: $n_j^{bits} = \left\lceil \log_2\left(n_j^{levels} + 1\right) \right\rceil$. If, after decoding, $a_j > \left(n_j^{levels} + 1\right)$, this also means the absence of x_j in the rule.

In this study, we used NSGA III implemented in the Platypus package [26]. Table 1 contains the algorithm settings.

Table 1. The Non-dominated Sorting Genetic Algorithm III (NSGA III) settings used in the experiment.

Setting (Parameter) Names	Setting (Parameter) Values
Selection	Tournament selection with a tournament size of 2
Crossover	Half-uniform crossover
Mutation	Bin-flip mutation
Solution representation	Binary code → Gray code
M-objective problem	3
Outer divisions, p_{out}	20
Inner divisions, p_{in}	0
Reference points, H	$H = \binom{M + p_{out} - 1}{p_{out}} + \binom{M + p_{in} - 1}{p_{in}} = 231$
Population size	The smallest multiple of four greater than H, i.e., 232
Generations	200
Probability distribution for initializing solutions in the starting population	$P\left(x_j \text{ is not included in the rule}\right) = 0.95$ $P\left(x_j \text{ is included in the rule}\right) = 0.05$

Given the stochastic nature of NSGA III, we ran the algorithm 25 times for each problem, i.e., Problem 1 and 2 (3, 4), and then combined the final populations from all of the runs. Figure 3 summarizes the pipeline described; the source code might be found on GitHub [27].

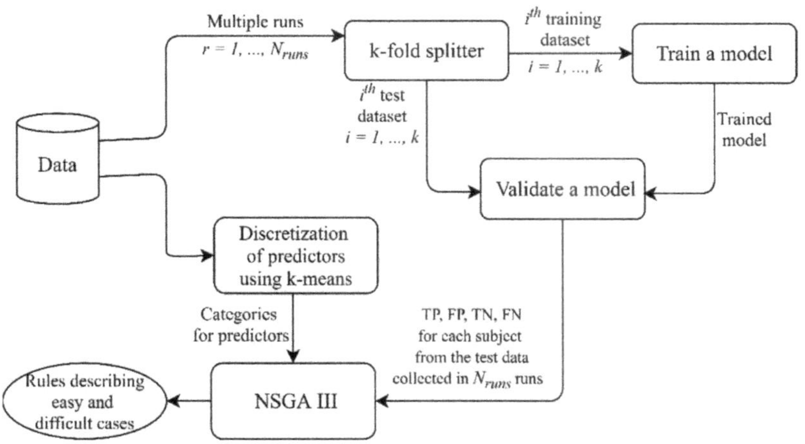

Figure 3. The pipeline implemented in this study. In several independent runs of cross-validation, the model performance was estimated using the test data and the results were collected for further analysis. Discretization was applied to continuous predictors to define their levels (categories) used for generating rules. NSGA III found combinations of predictors and their values describing "easy" cases, i.e., subgroups of subjects with a large number of true positive (TP) and true negative (TN) predictions, and "difficult" cases, i.e., subgroups of subjects with a large number of false positive (FP) and false negative (FN) predictions.

In general, any multi-objective evolutionary algorithm (MOGA) might be applied in the proposed approach (Figure 3); therefore, the overall time complexity depends on the optimization algorithm used. Time complexities of different MOGAs are discussed in the article by Curry and Dagli [28].

Lastly, from all generated rules that excessively describe subgroups of different sizes with various levels of TPR and TNR, we selected the final set of rules meeting the following criteria:

1. There are at least $Supp$ subjects covered by the rule (the minimum rule support): $Supp = 30$.
2. The difference between TPR and TNR does not exceed $\gamma \cdot \max(TPR(rule), TNR(rule)))$: $\gamma = 0.1$ (the rule is equally valid for "positive" and "negative" cases).
3. The average model accuracy for subjects covered by the rule is either lower than α_{diff} or higher than α_{easy} (to define "difficult" and "easy" cases correspondingly): $\alpha_{diff} = 50\%$, $\alpha_{easy} = 95\%$.

From a practical point of view, the final set of rules allowed us to reveal the conditions (subjects' features) that lead to higher or lower model performance, compared to the average level. First, this knowledge is helpful for revising the sample and collecting new data. Second, using these rules, we can introduce four categories of subjects: "easy" cases that are covered only by the rules with $Accuracy(rule) \geq \alpha_{easy}$; "difficult" cases that are covered only by the rules with $Accuracy(rule) \leq \alpha_{diff}$; "ambiguous" cases that are covered by the rules with $Accuracy(rule) \geq \alpha_{easy}$ and the rules with $Accuracy(rule) \leq \alpha_{diff}$; "not covered cases" that are covered by none of the rules. Applying the final set of rules to unseen data and categorizing samples in such a way provides us with additional information about the probability of wrong predictions and allows us to be more or less confident in the model outcome.

3. Results

First, we trained the LLR model predicting cardiovascular death in 50 runs of five-fold cross-validation and estimated the model performance on the test data: accuracy = 72.527%, TPR = 72.485%, and TNR = 72.552%. Despite splitting the data into the training and test samples randomly, for some subjects the correctness of the model predictions did not vary across the multiple runs: the model outcome was always (or in most of the runs) either right or wrong. To generate rules that would reveal the conditions of "easy" and "difficult" cases, we used a preprocessed set of predictors: we filtered out predictors that were not selected by LLR in at least one run. Thus, in the further analysis, 191 predictors were involved.

Next, after 25 independent runs of NSGA III, we ended up with 4355 rules for Problem 1 (3) and 4583 rules for Problem 2 (4). Figure S1 in the Supplementary Materials presents these initial sets of rules in the criterion space TPR–TNR.

Then, we selected the final set of rules using criteria 1–3 from Section 2.2 and, as a result, we obtained 43 rules representing "easy" cases, for which the model accuracy was higher than $\alpha_{easy} = 95\%$, and 39 rules that represent "difficult" cases, for which the model accuracy was lower than $\alpha_{diff} = 50\%$. These particular threshold values were chosen to reveal the subjects' characteristics that lead to the extremely high and extremely low model accuracy, provided that the accuracy for the remaining "ambiguous" and "non-covered" cases is close to the model accuracy on the whole sample. We also aimed to have the number of "non-covered" cases along with "ambiguous" cases at least lower than half of the sample. Thus, the selected 82 rules divided the KIHD sample into four categories, whose characteristics, with respect to the model accuracy and the number of subjects, are given in Table 2. Moreover, to support our choice, a description of subjects' categories obtained for other α_{easy} and α_{diff} is presented in the Supplementary Materials (Figures S2 and S3).

Table 2. Four categories of subjects in the KIHD sample after applying the final set of rules. The "Accuracy, %" columns represent the mean accuracy estimated in 50 runs of five-fold cross-validation. "The number of cases" columns contain the absolute number of subjects, with the percentage of the whole sample in parentheses.

	Accuracy, %			The Number of Cases			
	Easy	Difficult	Ambiguous and Non-Covered	Easy	Difficult	Ambiguous	Non-Covered
No cvd	95.73	46.76	66.55	414	264	91	373
Cvd	96.28	50.17	76.05	100	172	31	378
Overall	95.84	48.11	71.00	514 (28.20%)	436 (23.92%)	122 (6.69%)	751 (41.20%)

To visualize the groups of subjects covered by the final set of rules, we introduced an 82-dimentional rule space, where each dimension defined whether a subject was covered by the rule or not. These binary vectors were used by t-SNE (t-Distributed Stochastic Neighbor Embedding) to project subjects into the two-dimensional space [29]. Figure 4 illustrates the "easy", "difficult", and "ambiguous" cases and how the model accuracy changes for different clusters of subjects. These clusters mean that, within two large groups of "easy" and "difficult" cases, there are different conditions leading to an increase or decrease in model performance.

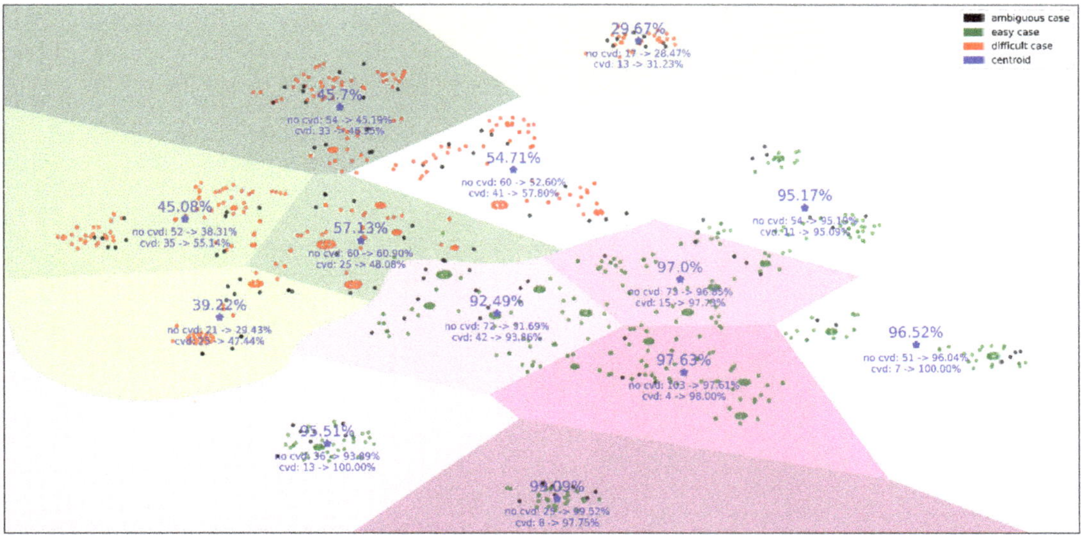

Figure 4. The KIHD subjects mapped from the 82-dimentional rule space onto the plane. This figure contains the subjects covered by the final set of rules. A separating line between "easy" and "difficult" cases was drawn by a support vector machine [30], and then borders between clusters were defined using the k-means method [31]. The overall model accuracy, true positive rate (TPR), and true negative rate (TNR) within clusters averaged over 50 independent runs of five-fold cross-validation, as well as the number of subjects who died by 2016 ("cvd") and stayed alive by 2016 ("no cvd"), are given for each cluster.

In general, the results can be analyzed from two perspectives. First, we may pay attention to the rules that describe a particular subject and make conclusions at the subject level about the predictors that make it "easy" or "difficult" for the model. Second, we may carry out analysis at the rule level by reviewing the combinations of predictors and their values that commonly lead to "easy" or "difficult" cases. For example, the two following rules describe "difficult" cases:

$$\begin{gathered}(\text{AMIHIST} = \text{no})\ \textit{and}\ (\text{BLPRESMED} = \text{yes})\ \textit{and}\ (\text{DIUR} = \text{yes})\ \textit{and}\ (\text{ISCHAEMIA} = \text{no}),\\ N_{subjects} = 38,\ TPR = 45.18\%\ \text{and}\ TNR = 42.95\%,\end{gathered} \tag{5}$$

where AMIHIST is "Myocardial infarction in the past", BLPRESMED is "Drug for blood pressure in last 7 days", DIUR is "Diuretics", and ISCHAEMIA is "Ischemia in exercise stress test".

$$\begin{gathered}(\text{SMOHIST} = \text{no})\ \text{and}\ (\text{PRESCRIP} = \text{yes})\ \text{and}\ (\text{WAGE} = \text{yes})\ \text{and}\ (54 \leq \text{AGE} \leq 55),\\ N_{subjects} = 86,\ TPR = 48.00\%\ \text{and}\ TNR = 45.45\%,\end{gathered} \tag{6}$$

where SMOHIST is "Smoking history, ever," PRESCRIP is "Drugs prescribed by a doctor in last 7 days", WAGE is "Salary: wages (if retired, then no)", and AGE is "Age, years".

What is interesting is that, in both groups, the subjects took medication. However, the first group had neither myocardial infarction in the past nor ischemia in exercise stress test, i.e., rule (5), and the second group had never smoked, i.e., rule (6). This implies that, in these rules, medication works as a factor that confuses the model for both the "cvd" and "no cvd" groups.

Lastly, to generalize the analysis of predictors and to extract the most important ones from the whole final set of rules, for each predictor, we calculated the number of unique "easy", "difficult", and "ambiguous" subjects covered by the rules that contain this predictor. Figure 5 presents the 50 most important predictors and their levels from the final set of rules. As can be noted, the same predictors with the same values might be involved in the rules that describe "easy" and "difficult" cases. For example, "ISCHAEMIA = no" in most of the cases corresponds to "easy" subjects (11.41% of the whole sample), but it also relates to "difficult" subjects (3.68% of the whole sample). On the contrary, the predictors "PRESCRIP = no" and "PGLINHIB = no" refer to "easy" subjects, whereas "PRESCRIP = yes" and "PGLINHIB = yes" correspond to "difficult" subjects. Moreover, "SELFMED = no" and "PGLIMED = no" only describe "easy" subjects.

Thus, these results support our previous conclusion about the role of medication in model performance: Its presence adversely affects both TPR and TNR [32].

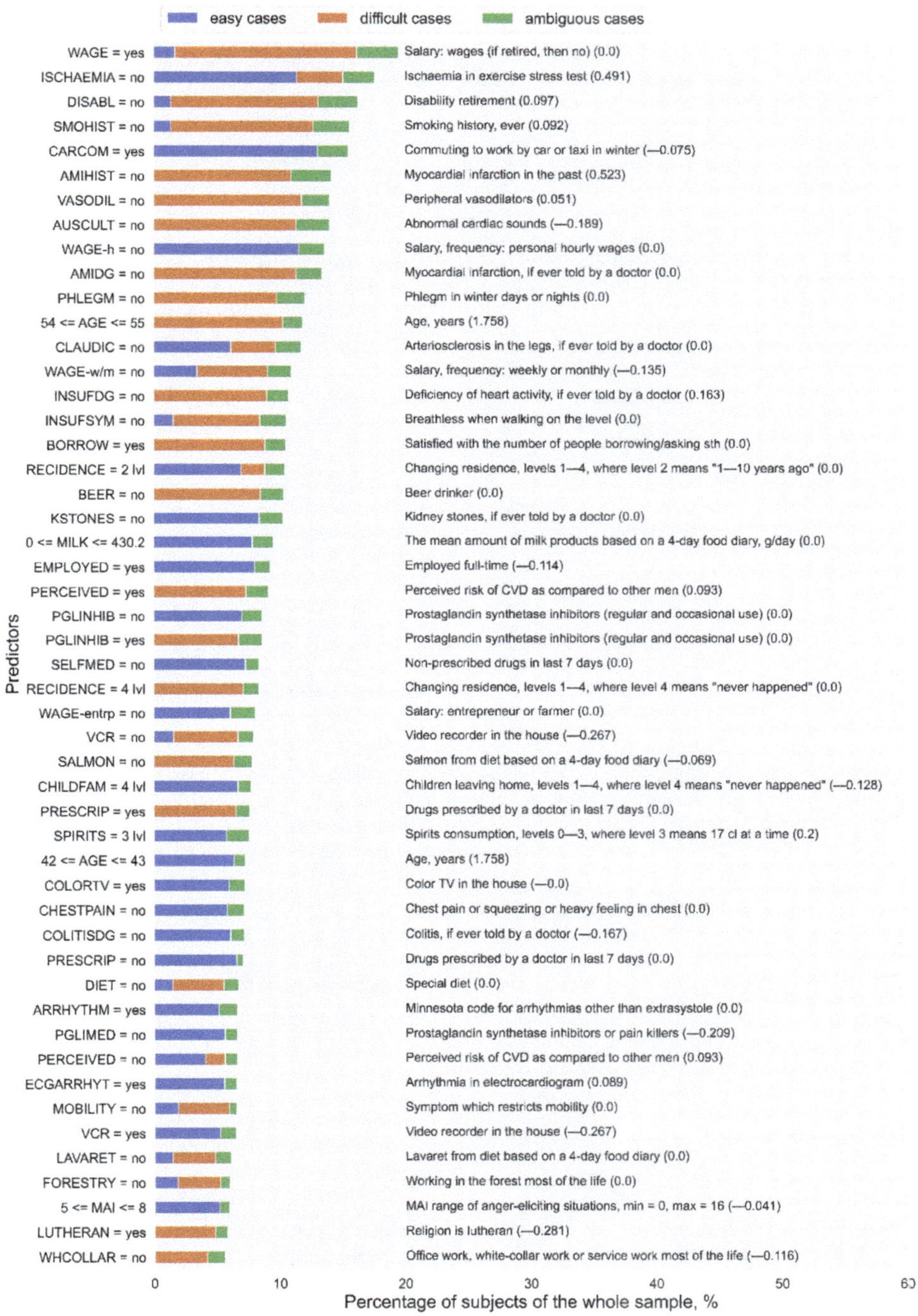

Figure 5. The 50 most important predictors and their values extracted from the final set of rules for the Lasso logistic regression (LLR) model using the KIHD sample. The numbers in parentheses are the weight coefficients in the LLR model for the corresponding predictors: Some of these variables were not selected as important for LLR, but they were important for analyzing "easy" and "difficult" cases.

4. Discussion

Typically, model performance is evaluated using the whole sample, which gives its average estimate. Yet it remains unclear for which samples the model is prone to making right predictions and for which samples its predictions are likely to be wrong. The high-dimensional predictor space commonly makes the analysis of results even more complicated. Therefore, in this study, we proposed an approach that enabled us to generate a set of rules that explain which samples were "easy" (predictions were more accurate) or "difficult" (predictions were less accurate) for the LLR model, trained and tested on the high-dimensional epidemiological data.

Since the average level of accuracy achieved by LLR was only 72.5% when predicting cardiovascular death for subjects from the KIHD cohort, the additional post-analysis was aimed at revealing those subjects' features that lead to high or low model performance. First, the knowledge about "difficult" subjects might be helpful for revising the sample, as they are the first candidates for double-checking. Moreover, collecting new samples, if it takes place, with characteristics corresponding to "difficult" cases might increase the model performance by bringing more information to the poorly modeled areas of the predictor space. Then, applying the set of rules to unseen data provides more confidence in the model predictions if new samples are categorized as "easy" cases. Knowing the weak spots of the model is of the utmost importance for clinical applications, where additional tests should be performed for "difficult" cases to avoid wrong predictions.

Moreover, as a well-interpretable tool, rules explicitly report the logic behind decision making while analyzing each subject, which is especially valuable for medicine. At the same time, when extracting the most important predictors from the final set of rules, we generalized the results at the whole sample level and obtained common patterns. Thus, in the KIHD sample, taking no medication (prescribed or non-prescribed drugs, drugs for back or joint pain, drugs for blood pressure, prostaglandin synthetase inhibitors or pain killers, diuretics, and beta-blockers), absence of other diseases (kidney stones, colitis, chronic bronchitis, gallbladder disease, migraine, and restricted mobility), and a high standard of living defined by socioeconomic predictors (having a color television, a video recorder, and/or a dishwasher and using a car or a taxi in winter) are common traits of "easy" cases for the LLR model when predicting cardiovascular death. Conversely, "difficult" cases included subjects who had taken medication (prescribed drugs, drugs for back or joint pain, drugs for blood pressure, drugs for hypertension, prostaglandin synthetase inhibitors or pain killers, and diuretics) [32].

Although in this study, the approach proposed was applied for LLR post-analysis, it is model-independent and might be useful for better understanding of any model behavior, which is especially helpful when "black box" models are trained on high-dimensional data. Such post-analysis is also useful for "white box" models, because variables that are important for making predictions differ from variables that are informative for describing "easy" and "difficult" cases.

The main limitation of this study is that all of our findings are applicable to the KIHD cohort, but they cannot be extrapolated to other populations. To make more general conclusions, e.g., about the role of medication in predicting cardiovascular death, we need to perform the same analysis for other populations, and then check if there are rules and variables similar for different populations. Additionally, since this post-analysis was applied to a particular data-driven model with a certain prediction horizon, for other models or different prediction horizons the analysis should be re-run.

Despite these limitations, the approach proposed is useful for many epidemiological studies, as the model accuracy has been reported to be far from 100% [33–36]. "Difficult" cases should be identified so that clinicians can revise the model predictions and use their expertise when the model is likely to make a mistake. This approach, i.e., when the model is augmented with the human expertise in some cases, corresponds to the philosophy of responsible machine learning (and responsible artificial intelligence, in general), which is being discussed by researchers nowadays [37]. Thus, such post-analysis is extremely important from a practical perspective, as it supports model deployment in a "responsible" way.

Supplementary Materials: The following are available online at https://www.mdpi.com/article/10.3390/healthcare9070792/s1, Figure S1: The initial sets of rules generated when solving Problems 1 and 2. Each point corresponds to one rule in the criterion space TPR-TNR, wherein color indicates the number of subjects covered by the rule. Figure S2: The overall accuracy for "easy", "difficult", "ambiguous" and "non-covered" subjects averaged over 50 runs of 5-fold cross-validation for different values of the thresholds α_easy and α_diff. Figure S3: The percentage of "easy", "difficult", "ambiguous", and "non-covered" subjects of the whole KIHD sample.

Author Contributions: Conceptualization, C.B.; data curation, C.B. and A.V.; formal analysis, C.B.; funding acquisition, M.K.; Methodology, C.B., A.V., T.-P.T. and M.K.; project administration, J.K.; software, C.B.; supervision, T.-P.T. and M.K.; validation, A.V. and T.-P.T.; visualization, C.B.; writing—original draft, C.B.; writing—review and editing, A.V., T.-P.T. and M.K. All authors have read and agreed to the published version of the manuscript.

Funding: The authors wish to acknowledge the University of Eastern Finland's Doctoral School and the Faculty of Science and Forestry for the financial support (to Christina Brester).

Institutional Review Board Statement: The study was conducted according to the guidelines of the Declaration of Helsinki, and approved by the Institutional Review Board of the University of Kuopio (study name "Sepelvaltimotaudin vaaratekijät") on 1 December 1983.

Informed Consent Statement: Informed consent was obtained from all subjects involved in the study.

Data Availability Statement: The data presented in this study are available on request from the Institute of Public Health and Clinical Nutrition, University of Eastern Finland (A.V., T.-P.T., J.K.) The data are not publicly available due to its sensitive nature.

Conflicts of Interest: The authors declare no conflict of interest.

References

1. Kagiyama, N.; Shrestha, S.; Farjo, P.D.; Sengupta, P.P. Artificial intelligence: Practical primer for clinical research in cardiovascular disease. *J. Am. Heart Assoc.* **2019**, *8*, e012788. [CrossRef] [PubMed]
2. Gulshan, V.; Peng, L.; Coram, M.; Stumpe, M.C.; Wu, D.; Narayanaswamy, A.; Venugopalan, S.; Widner, K.; Madams, T.; Cuadros, J.; et al. Development and validation of a deep learning algorithm for detection of diabetic retinopathy in retinal fundus photographs. *JAMA* **2016**, *316*, 2402–2410. [CrossRef] [PubMed]
3. Verghese, A.; Shah, N.H.; Harrington, R.A. What this computer needs is a physician: Humanism and artificial intelligence. *JAMA* **2018**, *319*, 19–20. [CrossRef] [PubMed]
4. Stead, W.W. Clinical implications and challenges of artificial intelligence and deep learning. *JAMA* **2018**, *320*, 1107–1108. [CrossRef] [PubMed]
5. Stiglic, G.; Kocbek, P.; Fijacko, N.; Zitnik, M.; Verbert, K.; Cilar, L. Interpretability of machine learning-based prediction models in healthcare. *WIREs Data Min. Knowl. Discov.* **2020**, e1379. [CrossRef]
6. Interpretable Machine Learning. A Guide for Making Black Box Models Explainable. Available online: https://christophm.github.io/interpretable-ml-book/ (accessed on 14 December 2020).
7. Ribeiro, M.T.; Singh, S.; Guestrin, C. Model-agnostic interpretability of machine learning. In Proceedings of the 2016 ICML Workshop on Human Interpretability in Machine Learning (WHI 2016), New York, NY, USA, 23 June 2016.
8. Cava, W.; Bauer, C.; Moore, J.H.; Pendergrass, S.A. Interpretation of machine learning predictions for patient outcomes in electronic health records. *AMIA Annu. Symp. Proc.* **2020**, *2019*, 572–581.
9. Koh, P.W.; Liang, P. Understanding black-box predictions via influence functions. In Proceedings of the 34th International Conference on Machine Learning, Sydney, Australia, 6–11 August 2017.
10. Aguilera-Rueda, V.J.; Cruz-Ramírez, N.; Mezura-Montes, E. Data-driven Bayesian Network learning: A bi-objective approach to address the bias-variance decomposition. *Math. Comput. Appl.* **2020**, *25*, 37. [CrossRef]
11. Ghose, A.; Ravindran, B. Interpretability with accurate small models. *Front. Artif. Intell.* **2020**, *3*, 3. [CrossRef]
12. Veiga, R.V.; Barbosa, H.J.C.; Bernardino, H.S.; Freitas, J.M.; Feitosa, C.A.; Matos, S.M.A.; Alcântara-Neves, N.M.; Barreto, M.L. Multiobjective grammar-based genetic programming applied to the study of asthma and allergy epidemiology. *BMC Bioinform.* **2018**, *19*, 245. [CrossRef]
13. Responsible-AI-Widgets. Available online: https://github.com/microsoft/responsible-ai-widgets/ (accessed on 11 June 2021).
14. Singla, S.; Nushi, B.; Shah, S.; Kamar, E.; Horvitz, E. Understanding Failures of Deep Networks via Robust Feature Extraction. *arXiv* **2021**, arXiv:2012.01750v2. Available online: https://arxiv.org/abs/2012.01750v2 (accessed on 22 June 2021).

15. Sajeev, S.; Champion, S.; Beleigoli, A.; Chew, D.; Reed, R.L.; Magliano, D.J.; Shaw, J.E.; Milne, R.L.; Appleton, S.; Gill, T.K.; et al. Predicting Australian adults at high risk of cardiovascular disease mortality using standard risk factors and machine learning. *Int. J. Environ. Res. Public Health* **2021**, *18*, 3187. [CrossRef] [PubMed]
16. Myers, P.D.; Ng, K.; Severson, K.; Kartoun, U.; Dai, W.; Huang, W.; Anderson, F.A.; Stultz, C.M. Identifying unreliable predictions in clinical risk models. *NPJ Digit. Med.* **2020**, *3*, 8. [CrossRef] [PubMed]
17. Salonen, J.T. Is there a continuing need for longitudinal epidemiologic research? The Kuopio Ischaemic Heart Disease Risk Factor Study. *Ann. Clin. Res.* **1988**, *20*, 46–50. [PubMed]
18. Kauhanen, J. Kuopio Ischemic Heart Disease Risk Factor Study. In *Encyclopedia of Behavioral Medicine*; Gellman, M.D., Turner, J.R., Eds.; Springer: New York, NY, USA, 2013. [CrossRef]
19. International Statistical Classification of Diseases and Related Health Problems. 10th Revision (ICD-10). Available online: https://icd.who.int/browse10/2016/en#/IX (accessed on 25 October 2020).
20. Troyanskaya, O.; Cantor, M.; Sherlock, G.; Brown, P.; Hastie, T.; Tibshirani, R.; Botstein, D.; Altman, R.B. Missing value estimation methods for DNA microarrays. *Bioinformatics* **2001**, *17*, 520–525. [CrossRef]
21. Brester, C.; Voutilainen, A.; Tuomainen, T.P.; Kauhanen, J.; Kolehmainen, M. Epidemiological predictive modeling: Lessons learned from the Kuopio Ischemic Heart Disease Risk Factor Study. *Inform. Health Soc. Care.* under review.
22. Hosmer, D.W.; Lemeshow, S. *Applied Logistic Regression*, 2nd ed.; John Wiley & Sons, Inc.: New York, NY, USA, 2000.
23. Tibshirani, R. Regression shrinkage and selection via the lasso. *J. R. Stat. Soc. Ser. B* **1996**, *58*, 267–288. [CrossRef]
24. Pedregosa, F.; Varoquaux, G.; Gramfort, A.; Michel, V.; Thirion, B.; Grisel, O.; Blondel, M.; Prettenhofer, P.; Weiss, R.; Dubourg, V.; et al. Scikit-learn: Machine learning in Python. *JMLR* **2011**, *12*, 2825–2830.
25. Deb, K.; Jain, H. An evolutionary many-objective optimization algorithm using reference-point-based nondominated sorting approach, part I: Solving problems with box constraints. *IEEE Trans. Evol. Comput.* **2014**, *18*, 577–601. [CrossRef]
26. Platypus. A Free and Open Source PYTHON library for Multiobjective Optimization. Available online: https://github.com/Project-Platypus/Platypus (accessed on 25 October 2020).
27. Intelligent System for Model Design (isMODE) in Personalized Medicine. Available online: https://github.com/christinabrester/isMode (accessed on 14 December 2020).
28. Curry, D.M.; Dagli, C.H. Computational complexity measures for many-objective optimization problems. *Procedia Comput. Sci.* **2014**, *36*, 185–191. [CrossRef]
29. van der Maaten, L.J.P.; Hinton, G.E. Visualizing data using t-SNE. *JMLR* **2008**, *9*, 2579–2605.
30. Chang, C.C.; Lin, C.J. LIBSVM: A library for support vector machines. *ACM Trans. Intell. Syst. Technol.* **2011**, *2*, 1–27. [CrossRef]
31. Hastie, T.; Tibshirani, R.; Friedman, J.H. *The Elements of Statistical Learning: Data Mining, Inference, and Prediction*, 2nd ed.; Springer: New York, NY, USA, 2009.
32. Pajouheshnia, R.; Damen, J.; Groenwold, R.; Moons, K.; Peelen, L. Treatment use in prognostic model research: A systematic review of cardiovascular prognostic studies. *Diagn. Progn. Res.* **2017**, *1*, 15. [CrossRef] [PubMed]
33. Yang, L.; Wu, H.; Jin, X.; Zheng, P.; Hu, S.; Xu, X.; Yu, W.; Yan, J. Study of cardiovascular disease prediction model based on random forest in eastern China. *Sci. Rep.* **2020**, *10*, 5245. [CrossRef] [PubMed]
34. Huang, Y.-C.; Li, S.-J.; Chen, M.; Lee, T.-S.; Chien, Y.-N. Machine-Learning Techniques for Feature Selection and Prediction of Mortality in Elderly CABG Patients. *Healthcare* **2021**, *9*, 547. [CrossRef] [PubMed]
35. Alaa, A.M.; Bolton, T.; Di Angelantonio, E.; Rudd, J.; van der Schaar, M. Cardiovascular disease risk prediction using automated machine learning: A prospective study of 423,604 UK Biobank participants. *PLoS ONE* **2019**, *14*, e0213653. [CrossRef] [PubMed]
36. Rezaee, M.; Putrenko, I.; Takeh, A.; Ganna, A.; Ingelsson, E. Development and validation of risk prediction models for multiple cardiovascular diseases and Type 2 diabetes. *PLoS ONE* **2020**, *15*, e0235758. [CrossRef]
37. Responsible Machine Learning with Error Analysis. Available online: https://towardsdatascience.com/responsible-machine-learning-with-error-analysis-a7553f649915 (accessed on 11 June 2021).

healthcare

MDPI

Systematic Review

The Role of Neural Network for the Detection of Parkinson's Disease: A Scoping Review

Mahmood Saleh Alzubaidi [1,*], Uzair Shah [1], Haider Dhia Zubaydi [2], Khalid Dolaat [1], Alaa A. Abd-Alrazaq [1], Arfan Ahmed [1] and Mowafa Househ [1,*]

1 College of Science and Engineering, Hamad Bin Khalifa University, Doha 53, Qatar; uzsh31989@hbku.edu.qa (U.S.); khdo31645@hbku.edu.qa (K.D.); aabdalrazaq@hbku.edu.qa (A.A.A.-A.); arahmed@hbku.edu.qa (A.A.)

2 National Advanced IPv6 Centre, Universiti Sains Malaysia, Gelugor 11800, Malaysia; haidardhia@yahoo.com

* Correspondence: maal28902@hbku.edu.qa (M.S.A.); mhouseh@hbku.edu.qa (M.H.)

Abstract: *Background:* Parkinson's Disease (PD) is a chronic neurodegenerative disorder that has been ranked second after Alzheimer's disease worldwide. Early diagnosis of PD is crucial to combat against PD to allow patients to deal with it properly. However, there is no medical test(s) available to diagnose PD conclusively. Therefore, computer-aided diagnosis (CAD) systems offered a better solution to make the necessary data-driven decisions and assist the physician. Numerous studies were conducted to propose CAD to diagnose PD in the early stages. No comprehensive reviews have been conducted to summarize the role of AI tools to combat PD. *Objective:* The study aimed to explore and summarize the applications of neural networks to diagnose PD. *Methods:* PRISMA Extension for Scoping Reviews (PRISMA-ScR) was followed to conduct this scoping review. To identify the relevant studies, both medical databases (e.g., PubMed) and technical databases (IEEE) were searched. Three reviewers carried out the study selection and extracted the data from the included studies independently. Then, the narrative approach was adopted to synthesis the extracted data. *Results:* Out of 1061 studies, 91 studies satisfied the eligibility criteria in this review. About half of the included studies have implemented artificial neural networks to diagnose PD. Numerous studies included focused on the freezing of gait (FoG). Biomedical voice and signal datasets were the most commonly used data types to develop and validate these models. However, MRI- and CT-scan images were also utilized in the included studies. *Conclusion*: Neural networks play an integral and substantial role in combating PD. Many possible applications of neural networks were identified in this review, however, most of them are limited up to research purposes.

Keywords: Parkinson's disease; neural network; deep learning; classification

Citation: Alzubaidi, M.S.; Shah, U.; Dhia Zubaydi, H.; Dolaat, K.; Abd-Alrazaq, A.A.; Ahmed, A.; Househ, M. The Role of Neural Network for the Detection of Parkinson's Disease: A Scoping Review. *Healthcare* **2021**, *9*, 740. https://doi.org/10.3390/healthcare9060740

Academic Editor: Mahmudur Rahman

Received: 25 April 2021
Accepted: 26 May 2021
Published: 16 June 2021

1. Introduction

1.1. Background

The human brain is the primary controller part of the human body. Any minor damage to any of its parts will severely affect other organs—one of its adverse effects is Parkinson's disease (PD) [1]. "PD is a chronic and progressive neurodegenerative disease" [2], and it occurs mainly in people over 50 years old [3]. Its symptoms start slowly and increase over time. PD symptoms are characterized such as motor and nonmotor [4]. Motor symptoms include movement disorders, shaking, walking issues [5], stiffness, and postural instability [6], while nonmotor symptoms including cognitive dysfunction, mood disorder [7], depression, and anxiety [8].

Parkinson's is the second worse neurodegenerative disease worldwide after Alzheimer's disease. In 2019, its incident rate ranged from 40.37 to 53.89 per 100,000 population per year in the US alone [9]. Diagnosis of PD in an early stage is an important issue to mitigate its complications. However, no medical test is available to diagnose it in the early stages conclusively. In a traditional clinical setup, the physician will ask the patient to perform some mental and physical tasks (e.g., moving and walking around) [10] or take the magnetic resonance imaging (MRI) and/or Positron emission tomography–computed tomography (PET/CT) scan of the brain. However, it is challenging to differentiate PD from other neurological disorders, and it depends on the radiologist's experience to distinguish and identify it precisely. Therefore, a computer-aided diagnosis (CAD) system helps the radiologist interpret MRI scans. In 2003, the authors of [7] made a CAD system to monitor body acceleration to detect the freezing of gait in PD patients.

Several studies were conducted to implement machine learning approaches to detect PD and differentiate it from other common neurological diseases. Feature engineering is the difficult part of deploying such systems, and it is expensive to identify the relevant features in the data. When automatic feature extraction methods and techniques (CNN, RNN) were proposed, most researchers used deep learning and neural network to detect PD due to automatic feature extraction, learning more complex patterns, and high accuracy. Therefore, this scoping review aims to explore and summarize the applications of deep learning and neural network in PD diagnosis.

1.2. Research Problem and Objectives

The scope of this paper is limited to the detection of Parkinson's disease (PD) in the early stage using neural networks. The patient dataset such as electronic health record (EHR) and medical image can be analyzed using neural network (NN) features; in particular, patient's data can undergo many processes; analysis, segmentation, augmentation, scaling, normalization, sampling, aggregation, and sifting, in order to obtain accurate prediction that assists healthcare ecosystem and stakeholders in the healthcare domain. Many studies have been recently conducted to address and propose a solution to mitigate and prevent neurodegenerative disorders such as PD. However, most of these studies and research are dispersed. Therefore, summarizing NN technologies' involvement in resolving challenges related to PD is needed; an appropriate summarization allows new researchers to understand the current role of neural networks against PD. It will open new opportunities for researchers to have the necessary base that allows them to build on instead of starting from ground zero.

Many studies have been carried out to cover AI techniques that have been used to mitigate and prevent PD [11–14]. These approaches are conducted in reviews or surveys that generally focus on artificial intelligence (AI) applications such as patient diagnosis, epidemiological monitoring, and drug and vaccine discovery [15]. Nevertheless, a massive number of research papers are constantly being published, which has overwhelmed electronic databases. Therefore, it is necessary to carry out an updated review that focuses on the uses of neural networks in PD prevention.

This review aims to identify and illustrate neural network technology's role in detecting PD early, based on the following aspects: (1) identifying the role of neural networks in PD detection, (2) highlighting the recent algorithms applied on PD datasets, (3) observing dataset types, (4) categorizing the type of PD based on symptoms, (5) investigating the best results achieved by the research community, and (6) providing a recommendation for researchers and healthcare individuals. The outcome can be used in the healthcare sector as guidance for developers who consider neural network's utilization to improve the public health capability as a response to PD.

2. Methodology

We carried out a scoping review to explore the evidence on neural network's application in diagnosing Parkinson's disease in a structured manner. In this section, we listed the details of the adopted methodology to conduct this review. For this purpose, PRISMA Extension for Scoping Reviews (PRISMA-ScR) [16] was used for this scoping review.

2.1. Search Strategy

2.1.1. Search Sources

We selected five bibliographic databases (PubMed, IEEE, ACM, ScienceDirect, and Google Scholar) to retrieve the research studies relevant to the topic. We scanned only 100 articles from Google Scholar; these articles were chosen after scanning based on their relevance to fit this paper. The backward and forward reference checking lists were not performed due to the sufficient number of included studies. The search process was performed from 24 February to 1 March 2021.

2.1.2. Search Terms

In the present review, we considered two different search terms based on population and intervention. Given the population of "Parkinson's disease" and intervention of "deep learning", the search strategy was conducted as follows: (("Parkinson's disease" OR "Parkinson*" OR "Parkinsonism" OR "paralysis agitans" OR "shaking palsy") AND (" artificial intelligence* " OR " machine learning" OR " neural network*" OR " deep learning" OR "natural language processing" OR "neural network*" OR "supervised learning" OR "unsupervised learning" OR "ensemble learning" OR "reinforcement learning")) total retrieved studies in (Appendix A).

2.2. Study Eligibility Criteria

This study aims to summarize and review the application/use of deep learning, particularly in diagnosing Parkinson's disease. Therefore, only the following studies were eligible to satisfy the below criteria: a deep learning approach or technique introduced or developed that primarily focused on diagnosing Parkinson's disease. Further, some constraints on the types of publication and the language of the studies were made. Only studies published in English between 2018 and 2021 are selected, and only peer-reviewed articles, conference proceedings, reports, theses, dissertations were admitted. Reviews, conference abstracts, commentaries, proposals, editorials were excluded. The details of exclusion and inclusion for study selection are listed in Table 1.

Table 1. Inclusion and exclusion criteria.

Criteria	Specified Criteria
Inclusion	• Studies that aim to diagnose Parkinson's using deep learning technique or approach • Studies that published from 2018 onwards • Empirical studies only • Only written in English
Exclusion	• Abstract • Review including an overview, scoping review, etc. • Non-English studies • Non-peer-reviewed articles

2.3. Study Selection

The study selection process was conducted in two stages (screening title and abstracts of retrieved studies and screening full text of the studies selected in the first stage). In the first stage, the first reviewer, MA, independently screened all the retrieved studies' titles and abstracts; due to time constraints, the second reviewer, US, and the third reviewer, KD, reviewed the first half and second half of the complete set of articles, respectively. The Rayyan software, a web-based systematic review tool, was employed for screening title and abstract [17]. In the second stage, the first reviewer, MA, performed the first stage's full-text screening of the identified studies. Any disagreement between reviewers was resolved through consensus and discussion.

2.4. Data Extraction and Data Synthesis

To extract the study-specific information and data, an extraction form was created and tested by eight included studies (Appendix B). MA and US undertook the data extraction, and the data were extracted to the excel sheet to summarize the following: general characteristic of included studies (e.g., country, types, and year of publication), aim/purpose of the study, type of Parkinson's disease, branch/type neural network, type of validation, performance metrics, the dataset used to train and test the model, number of Parkinson's and healthy samples, type of dataset, size of the dataset, data collection device or sensor, and dataset source. We used the narrative approach to synthesis the extracted data.

3. Results

3.1. Search Results

In total, 1061 studies were retrieved by searching through 5 recognized E-Databases. Then, 190 (17.90%) were removed due to duplication, while 871 (82.09%) went through title and abstract screening; in this screening, we excluded 598 (56.36%) studies due to various reasons, as shown in Figure 1. The remaining 273 (25.73%) studies went through the full-text screening, and 181 (17.05%) studies were excluded, as detailed in Figure 1. In total, 91(8.67%) studies were included in this review.

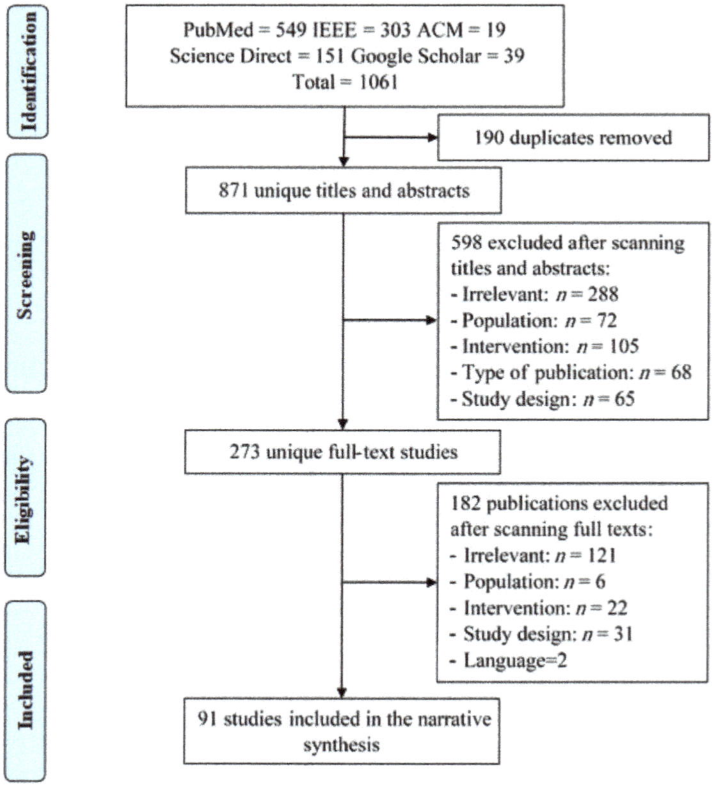

Figure 1. PRISMA chart.

3.2. General Description of the Included Studies

As shown in Table 2, the included citations were published in more than 30 different countries, as shown in Figure 2, about 13 studies from the US (14.13%), followed by 9 studies from China and India (9.78%) (Figure 3). This shows that numerous papers were published in the last 3 years; for instance, 30 papers (32.60%) were published in 2019 and 2020. More than half (56.2%) of the included studies were conference papers. However, most conference papers (n = 18) were published in 2018, and 2020, respectively, and only (n = 16) conferences article were reported in 2019. In addition, (n = 39) journal articles were published in last few years: (n = 10) in 2018; (n = 14) in 2019; (n = 12) in 2020; and (n = 3) in 2021.

Table 2. General characteristics of the included studies (n = 91).

Characteristics	Studies, n (%)	Ref.
Year of publication	2021: 4 (4.34)	[6,18–20]
	2020: 30 (32.60)	[21–51]
	2019: 30 (32.60)	[4,52–85]
	2018: 28 (30.43)	[3,86–105]
Country	US: 13 (14.13)	[18,27,29,57,61,65,74,82,87,88,92,95,104]
	China: 9 (9.78)	[33,40,52,53,66,67,85,89,90]
	India: 9 (9.78)	[3,31,37,50,51,55,60,63,105]
	Canada: 6 (6.52)	[35,38,45,46,83,93]
	UK: 4 (5.43)	[48,58,62,103]
	Korea: 4 (4.34)	[30,41,56,98]
	Turkey: 4 (4.34)	[4,36,77,101]
	Brazil: 3 (3.26)	[75,97,102]
	Australia: 3 (3.26)	[20,42,100]
	Italy: 3 (3.26)	[21,49,96]
	Spain: 3 (3.26)	[76,91,94]
	Greece: 2(2.17)	[54,99]
	Bangladesh: 2 (2.17)	[44,59]
	Japan: 2 (2.17)	[6,72]
	Lebanon: 2 (2.17)	[68,69]
	Malaysia: 2 (2.17)	[39,84]
	Germany: 2 (2.17)	[71,79]
	Morocco: 2 (2.17)	[23,25]
	Saudi Arabia: 2 (2.17)	[28,80]
	Singapore: 2 (2.17)	[32,81]
	Belgium: 1 (1.08)	[43]
	Colombia: 1 (1.08)	[70]
	France: 1 (1.08)	[47]
	Lithuania: 1 (1.08)	[22]
	Netherlands: 1 (1.08)	[78]
	Pakistan: 1 (1.08)	[86]
	Palestine: 1 (1.08)	[73]
	Portugal: 1 (1.08)	[64]
	Russia: 1(1.08)	[26]
	Slovakia: 1 (1.08)	[19]
	Romania: 1 (1.08)	[24]
	Egypt: 1 (1.08)	[34]
Type of publication	Conference: 52 (56.52)	[3,18,21–27,29–31,36,44–51,53–61,63–66,68–71,74,77,79,81–89,92,95,96,99–105]
	Journal article: 39(42.39)	[4,6,19,20,28,32–35,37–43,52,62,67,72,73,75,76,78,80,90,91,93,94,97,98]

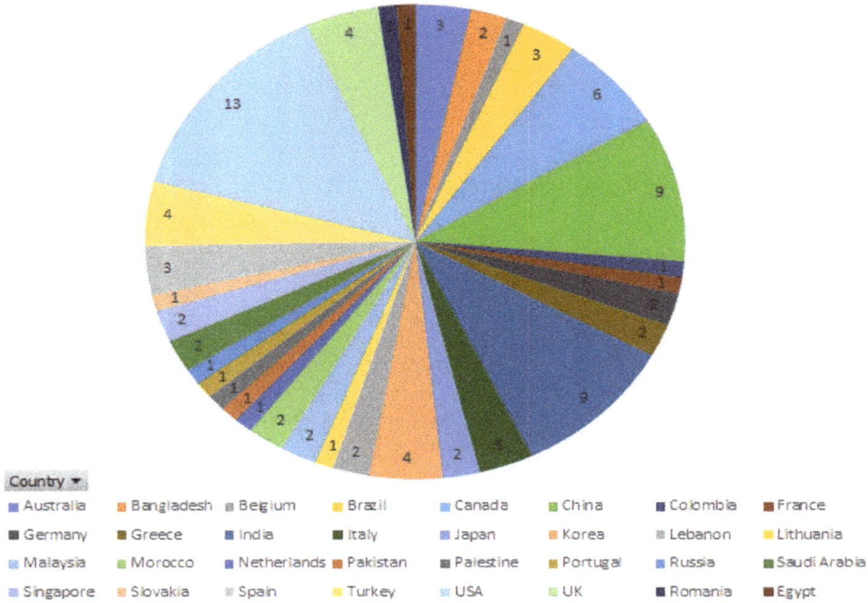

Figure 2. Number of publications for each country.

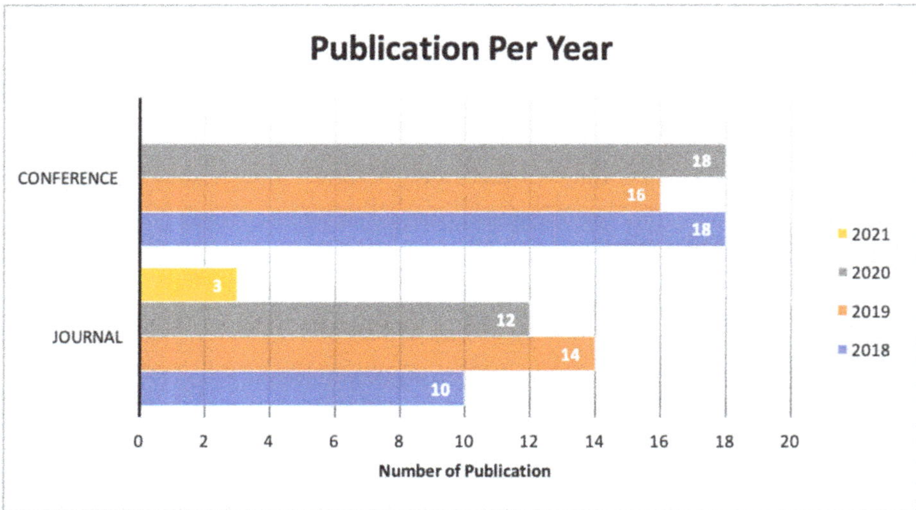

Figure 3. Type of publication and year.

3.3. Description of Detection Techniques

The study's primary aim is to investigate the role of neural networks in the diagnosis of PD. We classified neural networks into five well-known algorithms used in the included studies: CNNs, RNNs, FNNs, ANNs, and other NNs. Around half of the included studies used convolution neural networks (n = 37); afterward, other neural networks (n = 31) were implemented in the included studies, followed by artificial neural networks (ANNs) (n = 10), recurrent neural networks (RNNs) (n = 9), and fuzzy neural networks (FNNs), as shown in Table 3. In the end, the most imitated neural network architecture in the included studies was LSTM (n = 11) [6,34,36,38,40,65,70,74,77,80,83], VGG (n = 3) [18,27,58], and DNN (n = 6) [34,35,60,91,92,103]. Recently, with the developments of new techniques such as convolutional neural network [101] and transfer learning [63], deep learning gained significant advances in the computer vision tasks, e.g., ImageNet [77]. Therefore, most of the studies used different imaging data to diagnose PD, such as MRI (n = 12) [41,47,54,56,58,66,72,78,82,86,90,95] and handwritten images (n = 9) [3,19,25,30,69, 75,101,102], as well as PET and CT imaging (n = 6) [28,59,67,71,88,90] and DaTscan imaging (n = 4) [54,76,99,103]. However, CNN and transfer learning techniques were not limited to imaging data; they also learn complex features from voices and signal data [29]. Numerous studies used the biomedical voice (n = 21) [4,6,22,23,29,33,44,48,50,52,53,55,60,61,73,74, 84,93,100,104,105] and biometric signal (n = 14) [26,31,34,36,45,46,57,62,64,65,68,89,96,98]; a few of the included studies used EEG and EMG signals (n = 5) [32,39,51,83,85].

As shown in Figure 4, some studies target specific symptoms of PD, such as freezing of gait, vocal impairment, and tremor disorder. A more limited number of included studies proposed a deep learning approach to detect tremor disorder (n = 5) and vocal impairment (n = 13). However, various studies used the deep learning technique to diagnosis PD (n = 50), in general, and freezing of gait (FoG) (n = 23), in particular.

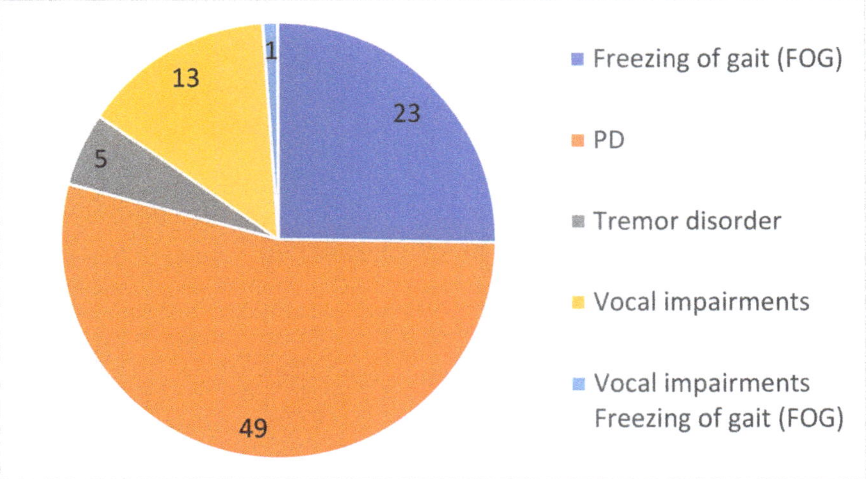

Figure 4. Different symptoms of Parkinson's disease in the included studies.

As reported in Table 3, the neural network is divided into five main branches (CNN, RNN, ANN, FNN, NN); all types of subclassification techniques are listed as backbone model; moreover, we noticed that LSTM was heavily used in a different study (n = 11), followed by none deep learning classifier SVM (n = 8); however, we have reported SVM in this review because many studies used neural networks to perform data extraction, but the classification was handled by the machine learning classifier such as SVM; hence, DNN was used and reported in (n = 6), and a predefined model such as VGG was used in (n = 3); other types of algorithms that were used rarely depended on each of the studies' design or achieved a remarkable result.

Table 3. Description of PD detection techniques ($n = 91$).

Characteristics	Studies, *n*								
Type of PD symptoms	**PD: 49**		**FoG: 23**		Vocal impairments:13		Tremor disorder: 5		Vocal impairments and FoG:1
Dataset Source	**Public:57**				Private:31		NA: 3		
Type of Dataset	MRI: 12 DaTscan: 4 PET&CT images:6 Handwriting Images: 9		Biomedical Voice:21		Biometric signal: 14		EEG and EMG: 5		VGRF time series: 4 Video: 4
Neural Network					CNN: 37 RNN:9 ANN: 10 FNN:4 Other NN: 31				
Model Backbone	LSTM: 11 SVM: 8	DNN: 4 VGG: 4	Autoencoders (AE): 2 DCNN: 2 MLPs: 2	Inception v3: 1 AlexNet: = 1 RestNet: 1 U-Net:1	WGAN: 1 ASE: 1 SSAE: 1 LSVRC: 1 DNMLDM: 1 DPRNN: 1	LRNN: 1 MTL: 1 GCN: 1 GS-RNN: 1 NR-LBP: 1 TCN: 1	OPF: 1 FRP: 1 FCNN: 1 EFMMOneR: 1	Encoder-Decoder DBN: 1 MOGA: 1 BiLSTM: 1	SSM-PCA: 1 SNN: 1
Training dataset Volume	≥80%: 20		≥70%: 19		≥60%: 5	≥50%: 3	≥40%: 1		NA: 43
Testing dataset Volume	≥50%: 3		≥40%: 1		≥30%: 6	≥20%: 18	≥10%:8	≥5%: 2	NA: 53
Validation Method	10-FCV: 29		5-FCV: 12		LOSO: 3	LOPO: 2 LOOCV: 2 3-FCV: 2	4-FCV: 1 6-FCV: 1 7-FCV: 1 8-FCV: 1	Holdout: 1	NA: 36
Evaluation Metrics	Accuracy: 56		Recall/Sensitivity: 35		Specificity: 24		Precision: 16	F1-Score: 7	AUC: 8
Developed software					Diagnosis dashboard: 1				

In most of the studies, the dataset was divided into three parts training, testing, and validation due to the limited number of studies that divided the datasets only into the training set and validation set, as presented in Table 3. We reported only the training and testing datasets. Furthermore, most of the experiments (*n* = 21) used ≥80%) volume of the training dataset, and (*n* = 9) used (≥70%). However, only few experiments provided less volume of the training dataset, as seen in (*n* = 5) used (≥60%) and (*n* = 3), (*n* = 1) used (≥50%), (≥40%), respectively. However, (*n* = 43) of the studies did not mention the volume of the training dataset. In addition, the volume of the testing dataset is not clarified in most of the studies; we noticed that (*n* = 53) did not specify the volume of the testing dataset that was used during the experiment; however, the volume of (≥20%) was mostly used in (*n* = 18), followed by (≥10%) that were mentioned in (*n* = 9), and the volume of (≥30%) was observed in (*n* = 6). The testing dataset is usually used in low volume, compared to the training dataset; however, we noticed that half of the dataset (≥50%) was used only in (*n* = 3). In addition, low volumes of testing dataset, i.e., (≥5%) and (≥40%), are reported in (*n* = 2) and (*n* = 1), respectively.

The validation method is highly considered in this review; we have reported all the studies' validation mechanisms. The most common K-fold cross validation (K-FCV) methods used are the tenfold cross-validation, which was used in (*n* = 30), followed by fivefold cross-validation in (*n* = 12), whereas fewer K-FCV methods were reported as threefold cross-validation, fourfold cross-validation, sixfold cross-validation, sevenfold cross-validation, and eightfold cross-validation in (*n* = 2), (*n* = 1), (*n* = 1), (*n* = 1), and (*n* = 1), respectively. Furthermore, other validation methods such as LOSO, LOPO, LOOCV, and holdout were rarely used, and are reported in (*n* = 3), (*n* = 2), (*n* = 2), and (*n* = 1), respectively. However, (*n* = 36) did not mention any type of validation method within their experiments.

Various evaluation metrics used to check each model's performance and accuracy are the most commonly used metrics to calculate the model's efficiency in predicting the result based on the testing dataset. In (*n* = 57), the accuracy of the models was reported. On the other hand, along with the accuracy, other evaluation methods were used, such as recall/sensitivity that was reported in (*n* = 36), followed by specificity in (*n* = 24) and precision (*n* = 17); however, few studies (*n* = 8) used area under the curve (AUC) as an evaluation metric.

During summarization of all (*n* = 91) results, unfortunately, we did not come across any empirical validation/real-life implementation in any hospital. Moreover, from the (*n* = 91) studies, we only found one study that developed diagnosis software that identified any neurological disorders such as PD and that can be employed in the medical center [51].

3.4. Dataset Description

3.4.1. Public Dataset

As discussed, an earlier total number of the public dataset (*n* = 57), Table 4, summarized the most used (*n* = 36) public available dataset sources and repositories (*n* = 36), e.g., Parkinson Progression Markers Initiative database (PPMI), UCI database repo, and PhysioNet; these were the most used datasets to develop and validate the AI models. Other public dataset sources used by the included studies were as follows: Kaggle, HandPD, DaphNet, the NTUA Parkinson Dataset, Neurovoz corpus, PC-GITA database, etc.

Table 4 only provides a sample of the public datasets used within the included studies. As seen, the number of males in the PD sample is higher than the number of females, and the number of males in healthy control is higher than the number of females in most cases. Furthermore, different types of hardware devices were used to collect the dataset; we have noticed that most of the data are in the form of images collected with different devices, starting from hospital imaging device including MRI, CT, DaTscan and ending with smartphone images that were used to capture handwriting or drawing of the PD samples (*n* = 28) and (*n* = 4) for recording video.

Table 4. Public dataset descriptions.

Dataset	Source/Host	Used Device/Sensor	Number of PD Patient		Number of Healthy Control		Ref.
			Male	Female	Male	Female	
Public	PhysioNet (*n* = 4)	16 sensors s under each foot 8 per foot	59	34	40	32	[21,49,55,81]
	The University of California, Irvine Machine Learning repository UCI (*n* = 10)	NA	84	40	23	41	[3,4,23,33,44,53,55,60, 84,105]
	Neurovoz corpus	NA	32	20	27	29	[74]
	PPMI (Parkinson Progression Markers Initiative) database (*n* = 14)	MRI Machine	129		57		[20,28,41,47,59,66,67, 76,82,86,88,90,94,95]
	The NTUA Parkinson Dataset (*n* = 1)	DaTscan and MRI Machine	55		23		[99]
	PC-GITA database (*n* = 1)	Professional audio card	25	25	25	25	[50]
	Department of Neurology in Cerrahpasa Faculty of Medicine, Istanbul University (*n* = 1)	Wacom Cintiq 12WX graphics tablet	57		15		[101]
	HandPD dataset Botucatu Medical School, São Paulo State University (*n* = 2)	Smartphone Camera	59	15	6	12	[19,102]
	Daphnet Dataset University of California, Irvine Machine Learning repository (*n* = 2)	sensor was attached to a belt and above the ankle and above the knee	7	3	NA	NA	[62,65]
	Parkinsons drawing spirals and waves Kaggle	Tablet for capture the drawing	27		28		[30,101]

Biometric signal and time-sensor-based dataset were collected using the digital keyboard or sensor/accelerometer (*n* = 16) attached to the PD and healthy control sample or placed at a different angle to measure the severity of the freezing gait or the tremor. Moreover, devices such as a high-quality standalone microphone or smartphone were used to collect the biomedical voice dataset, and (*n* = 15) reported a public vocal dataset. Moreover, in the public dataset, only (*n* = 11) reported the gender of PD and healthy control sample, and only (*n* = 5) studies identified each sample's mean age.

3.4.2. Private Dataset

As mentioned, the earlier total number of private datasets (*n* = 31) is shown in Table 5. We summarized the dataset that was clearly explained within studies (*n* = 5). This dataset was collected and labeled in different entities such as hospitals, universities, and research centers. The number of PD and healthy control samples are reported, including gender. Table 5 only provides a sample of the private datasets used within the included studies. The number of males in the PD sample is higher than the number of females, whereas the number of females in health control is higher than the number of males. Furthermore, different types of hardware devices were used to collect the dataset; we have noticed that most of the data were in the form of images collected with different devices, starting from hospital imaging device including MRI, CT, DaTscan and ending with smartphone images that were used to capture handwriting or drawing of the PD samples (*n* = 11).

Table 5. Private dataset descriptions.

Dataset	Source/Host	Used Device/Sensor	Number of PD Patient		Number of Healthy Control		Ref.
			Male	Female	Male	Female	
Private	Wearable Bio mechatronics Laboratory at Western University	wearable assistive devices for suppressing tremor	13		NA	NA	[46]
	Pacific Parkinsons Research Centre (PPRC)"	wearable headset with 27 electrodes to capture the EEG signals.	10	10	11	9	[83]
	Hospital at Sun Yat-sen University	64-electrode Geodesic Sensor Net (Electrical Geodesics Inc.)	25	15	18	12	[85]
	RMIT University, Melbourne, Australia	Apple iPhone 6S plus®	41		40		[100]
	n/A	Digital software keyboard	18		15		[79]

Biometric signal and time-serious-based dataset were collected using the digital keyboard or sensor/accelerometer (n = 14) attached to the PD and healthy control sample or placed at a different angle to measure the severity of the freezing gait or the tremor. Moreover, devices such as a high-quality standalone microphone or smartphone were used to collect the biomedical voice dataset, and (n = 6) reported a private vocal dataset. Moreover, in the private dataset, only (n = 4) reported the gender of PD and healthy control sample, and only (n = 4) studies identified each sample's mean age.

4. Discussion

4.1. Principal Findings

Although this study focuses on identifying and addressing deep learning and neural network application to detect Parkinson's disease in the early stage, we found some proposed models show promising results and can be employed in hospitals. This review provides recommendations for professional healthcare and researchers based on the included studies' outcomes. Moreover, we noticed that five studies [21,37,49,55,81] used the Vertical Ground Reaction (VGRF) dataset, which was obtained from PhysioNet hub to train the classification models including fuzzy neural networks (FNNs), stacked 2D CNNs, deep neural networks (DNNs), artificial neural networks (ANNs), and neighborhood representation local binary pattern (NR-LBP). However, DNN in [49] surprisingly achieved outstanding results for early detection of PD using the VGRF dataset, compared to the other studies.

Furthermore, for imaging dataset including MRI, PET CT, and DaTSCAN were mainly obtained from Parkinson Progression Markers Initiative (PPMI) to train classifier, as seen in [20,28,41,47,59,66,67,76,82,86,88,90,94,95]; hence, among all studies, CNN in [20] and FNN in [28] achieved an outstanding result for image classification.

We found that most of the biomedical voice measurements dataset was obtained from the University of California (UCI) Irvine Machine Learning repository; in [53,84] and [23], the same dataset is used; however, 19 achieved outstanding result using the sequential model in a deep neural network for detection PD based on voice measurement. In [33,44], and [4], the same voice measurement datasets with 756 instances and 754 attributes were used to identify PD, and the autoencoder neural network in [33] achieved better results than other studies.

Electroencephalograph (EEG) dataset was obtained from a different source and used in five studies [32,38,51,83,85]. In [38,83], we found that long short-term memory (LSTM) achieved outstanding results, indicating the best option to deal with EEG data. On the other hand, seven studies [3,19,25,27,40,69,101,102] focused on the classification of handwriting image to identify PD in the early stage, and we found that outstanding results were achieved in ANN + SVM in [3], dual-path RNN (DPRNN) in [40], and CNN + Optimum-Path Forest (OPF) in [102], respectively.

As mentioned earlier, the detection of PD using a neural network is not an easier task than other types of diseases because PD symptoms (vocal disorder, tremor disorder, freezing gait disorder) are inconsistent, and it is difficult to collect data concerning the type of the device. Therefore, many public repositories mainly focus on collecting and process certain types of datasets. Moreover, based on our findings, we can conclude that the sequential model in DNN and autoencoder neural network proved to be suitable models for PD detection from speech. Moreover, DNN is recommended to identify PD from VGFR data. Additionally, CNN is still on top for medical image classification such as MRI, PET/CT, and DaTSCAN. Moreover, the FNN shows significant results in classifying a medical image. On the other hand, in regard to images of handwritings, we found that ANN with machine learning classifier SVM had a remarkable result for the identification of PD from handwriting.

Based on the findings of this review, we can highlight the most used repositories that contain PD public datasets for the research community as follows: (1) UCI Repository of Machine Learning Database, University of California; (2) PhysioNet Laboratory for Computational Physiology, Massachusetts Institute of Technology; (3) Parkinson's Progression Markers Initiative (PPMI); (4) Pacific Parkinson's Research Institute; and (5) Botucatu Medical School, São Paulo State University, Brazil.

4.2. Strengths and Limitations

4.2.1. Strengths

This review covered deep learning neural network techniques used for PD detection regardless of the characteristics, country, and study design. We claim that this review is a comprehensive study of neural network approaches used for PD detection. It will help researchers to understand how neural network is used efficiently for detecting PD in early stages. Compared with other reviews [106–108] that do not focus on PD disease, this review is unique in its field because it describes and summarizes features of the identified neural network models, datasets, available repository, type of PD evaluation, validation, and research implication. Moreover, this review is different from the previously mentioned reviews by following the latest version of PRISMA-ScR [16]. Unlike other reviews, we retrieved the studies from the most popular computer science and healthcare database to determine the most relevant studies possible.

4.2.2. Limitations

In the beginning, we carried out a primary search from 2015 to 2021 through the five selected databases, and we retrieved a massive number of studies. Therefore, we limited our search to the period between 2018 to 2021. Due to that, we may have missed some significant studies. Due to many studies that we included (n = 92), backward and forward reference checking was not performed in this review. PD is an extensive topic and divided into many types of diseases, including various symptoms. Therefore, we may have missed categorizing some diseases from a clinical perspective.

4.3. Practical and Research Implications

Although this review investigates the neural networks used to detect Parkinson's disease (PD), some applications could significantly mitigate this neurodegenerative disorder. Nowadays, computer-aided diagnosis systems are essential because they are less time consuming and more user friendly. For example, the authors of [51] designed a GUI system that physicians may use for fast diagnosis of Parkinson's disease in its early stages. Researchers can also use the system to continue their future research on disease diagnosis, especially neurodegenerative disorders. The system will show the patient's disease progression and help clinicians monitor the disease in its early stages.

Furthermore, the system can differentiate between PD patients and healthy subjects and compare various parameters (EEG, EMG, MRI/PET scan). In both PD and control subjects, the model can detect the region of dopamine output in the substantia nigra. As a result, the proposed model would be a novel solution containing all of the PD detection parameters in a single window, which would be extremely useful for disease monitoring.

In the included studies [6,18,19,30,61,75,87,91,92,96,98,101], clinicians could obtain PD Patient data in telemonitoring using devices such as tablets and smartphones. It is a promising solution because they can increase monitoring frequency without putting a strain on professional resources during the COVID-19 pandemic. However, the cost of training and testing the detection algorithm on a smartphone was too high; thus, the results were measured on a remote server and then transferred to the computer.

Clinical studies can refer to a video recorded for the patient while performing physical activities such as a PD bed test. As mentioned, in [18,43,70,87], a neural network was able to identify the symptoms of PD through a video sample of the patient. In the future, the clinical studies may analyze any video recorded in the hospital for other patients, for example, during therapy sessions, and predict if this patient is suspected of having PD in the future.

5. Conclusions

This scoping review summarized studies by investigating the use of neural networks, specifically deep learning algorithms, for early diagnosis of PD based on various data collected from different public and private sources (91 studies), including medical image, biomedical voice, and sensor signal, for both PD and healthy control samples. Included studies were categorized into different groups based on the neural network model, type of PD symptoms, and type of dataset. Additionally, the most used dataset and best performance model were highlighted based on the detection of particular symptoms of PD in this review. All technical experiment methods were reported, including submodel, dataset volume, training, testing, evaluation metrics, and validation type. We indicated any real-time implementation used in each hospital or university setting, and based on this review, we recommended particular suggestions for healthcare professionals. Future work could be a meta-analysis to examine each study and provide a comprehensive comparison between them in terms of quality.

Author Contributions: Conceptualization, formal analysis, methodology, and writing—original draft, M.S.A., U.S. and K.D.; Data curation and investigation, A.A.A.-A. and A.A.; Writing—review & editing, H.D.Z.; Supervision and validation, M.H. All authors have read and agreed to the published version of this manuscript.

Funding: This research received no external funding.

Institutional Review Board Statement: Not applicable.

Data Availability Statement: Not applicable.

Conflicts of Interest: The authors declare no conflict of interest.

Abbreviations

PD	Parkinson's disease
FoG	Freezing of gait
NA	Not Available
MRI	Magnetic Resonance Imaging
PET	Positron emission tomography
CT	Computerized tomography
EEG	Electroencephalogram
EMG	Electromyography
VGRF	Vertical Ground Reaction Force
CNN	Convolutional Neural Network
RNN	Recurrent Neural Network
ANN	Artificial Neural Network
FNN	Fuzzy Neural Network
NN	Neural Network
DBN	Deep belief network

MOGA	Multi-Objective Genetic Algorithm
BiLSTM	Bidirectional Long short-term memory
LSTM	Long short-term memory
OPF	Optimum-Path Forest
FRP	Fuzzy Recurrence Plot
DCNN	Deep Convolutional Neural Network
FCNN	Fully Connected-Neural Network
DNN	Deep Neural Network
EFMMOneR	Fuzzy Minmax Neural Network with The One R Attribute Evaluator
LRNN	Layer Recurrent Neural Network
MTL	Multi-Task Learning
GCN	Graph Convolutional Network
GS-RNN	Gradient Stabilized Recurrent Neural Network
NR-LBP	Neighborhood Representation Local Binary Pattern
TCN	Temporal Convolutional Neural network
WGAN	Wasserstein Generative Adversarial Networks
SAE	Stacked Auto Encoder
SSAE	Stacked Sparse Auto-Encoder
LSVRC	Large Scale Visual Recognition Challenge
DNMLDM	Deep Neural Mapping Large Margin Distribution Machine
MLP	Multiple Layer Perceptron
SVM	Support Vector Machine
SSM-PCA	Scaled Subprofile Modeling Using Principal Component Analysis
SNN	Siamese Neural Network
FCV	Fold-Cross Validation
LOOCV	Leave-One-Out Cross-Validation
LOSO	Leave One Subject Out
AUC	Area Under Curve

Appendix A

Table A1. Used search terms and total number of retrieved studies per database.

Database Name	Used Research Terms	Number of Retrieved Studies
PubMed	("Parkinson's Disease" OR "Parkinson*" OR "Parkinsonism" OR "paralysis agitans" OR "shaking palsy") AND ("artificial intelligence " OR " machine learning" OR " neural network*" OR " Deep learning" OR "natural language processing" OR "Neural network*" OR "supervised learning" OR "unsupervised learning" OR "ensemble learning" OR "reinforcement learning")	549
IEEE	"Parkinson's Disease" OR "Parkinson*" OR "Parkinsonism" OR "paralysis agitans" OR "shaking palsy" AND "artificial intelligence" OR "machine learning" OR "neural network*" OR "Deep learning" OR "natural language processing" OR "Neural network*" OR "supervised learning" OR "unsupervised learning" OR "ensemble learning" OR "reinforcement learning"	303
ACM	("Parkinson's Disease" OR "Parkinson*" OR "Parkinsonism" OR "paralysis agitans" OR "shaking palsy") AND ("artificial intelligence" OR "machine learning" OR "neural network*" OR "Deep learning" OR "natural language processing" OR "Neural network*" OR "supervised learning" OR "unsupervised learning" OR "ensemble learning" OR "reinforcement learning")	19
Science Direct	("Parkinson's Disease" OR "Parkinson" OR "Parkinsonism" OR "paralysis agitans" OR "shaking palsy") AND ("artificial intelligence" OR "machine learning" OR "neural network" OR "Deep learning")	151
Google Scholar	("Parkinson's Disease" OR "Parkinson*" OR "Parkinsonism" OR "paralysis agitans" OR "shaking palsy") AND (" artificial intelligence " OR " machine learning" OR " neural network *" OR " Deep learning" OR "natural language processing" OR "Neural network*" OR "supervised learning" OR "unsupervised learning" OR "ensemble learning" OR "reinforcement learning")	39
	Total studies 2018–2021	**1061**

Appendix B

Table A2. Data extraction form.

Concept	Definition
Study Characteristics	
Author	The first author of the study.
Year Submission	The year in which the study was submitted.
Country of publication	The country where the study was published.
Publication type	The paper type (i.e., peer-reviewed, conference or preprint).
AI technique characteristics	
Purpose/use of AI	What are the applications or uses of AI in diagnosis of Parkinson (e.g., diagnosis, classification, and detection)?
AI branches	The branches/areas that were used (e.g., traditional machine learning, deep learning, natural language processing).
AI models/algorithms	The specific AI models or algorithms that were used (e.g., Decision tree, Random forest, Convolutional neural network).
Dataset Characteristics	
Data sources	Source of data that were used for the development and validation of AI models/algorithms (e.g., public databases, clinical settings, government sources).
Data types	Type of data that were used for the development and validation of AI models/algorithms (e.g., radiology images, biological data, laboratory data).
Dataset size	The total number of data that were used for the development and validation of AI models/algorithms.
Type of validation	How the dataset was split/used to develop and test the proposed models/algorithms (e.g., Train-test split, K-fold cross-validation, External validation).
Proportion of training set	Percentage of the training set of the total dataset.
Proportion of validation set	Percentage of validation set of the total dataset.
Proportion of test set	Percentage of the test set of the total dataset.
Type of device	The device used to collect the data (e.g., accelerometer, smartphone, etc.)
At-risk group	The number of Parkinson's participants included in the study.
Control group	The number of healthy participants included in the study

References

1. Alissa, M. Parkinson's Disease Diagnosis Using Deep Learning. *arXiv* **2021**, arXiv:2101.05631.
2. Burke, R.E.; O'Malley, K. Axon degeneration in Parkinson's disease. *Exp. Neurol.* **2013**, *246*, 72–83. [CrossRef]
3. Ranjan, A.; Swetapadma, A. An Intelligent Computing Based Approach for Parkinson Disease Detection. In Proceedings of the Proceedings of 2018 2nd International Conference on Advances in Electronics, Computers and Communications, ICAECC 2018, Bangalore, India, 9–10 February 2018. [CrossRef]
4. Gunduz, H. Deep Learning-Based Parkinson's Disease Classification Using Vocal Feature Sets. *IEEE Access* **2019**, *7*, 115540–115551. [CrossRef]
5. "GBD Compare" Data Visualizations. Available online: https://vizhub.healthdata.org/gbd-compare/ (accessed on 26 May 2021).
6. Quan, C.; Ren, K.; Luo, Z. A Deep Learning Based Method for Parkinson's Disease Detection Using Dynamic Features of Speech. *IEEE Access* **2021**, *9*, 10239–10252. [CrossRef]
7. Rana, A.Q.; Ahmed, U.S.; Chaudry, Z.M.; Vasan, S. Parkinson's disease: A review of non-motor symptoms. *Expert Rev. Neurother.* **2015**, *15*, 549–562. [CrossRef]
8. Sveinbjornsdottir, S. The clinical symptoms of Parkinson's disease. *J. Neurochem.* **2016**, *139*, 318–324. [CrossRef]
9. "Diagnosis Parkinson's Disease" NHS Choices. Available online: https://www.nhs.uk/conditions/parkinsons-disease/diagnosis/ (accessed on 26 May 2021).
10. Parkinson's Disease Information Page. *National Institute of Neurological Disorders and Stroke*; U.S. Department of Health and Human Services: Washington, DC, USA, 2020.

11. Belić, M.; Bobić, V.; Badža, M.; Šolaja, N.; Đurić-Jovičić, M.; Kostić, V.S. Artificial intelligence for assisting diagnostics and assessment of Parkinson's disease—A review. *Clin. Neurol. Neurosurg.* **2019**, *184*, 105442. [CrossRef]
12. Sibley, K.G.; Girges, C.; Hoque, E.; Foltynie, T. Video-Based Analyses of Parkinson's Disease Severity: A Brief Review. *J. Parkinsons. Dis.* **2021**, 1–11. [CrossRef]
13. Varrecchia, T.; Castiglia, S.F.; Ranavolo, A.; Conte, C.; Tatarelli, A.; Coppola, G.; Di Lorenzo, C.; Draicchio, F.; Pierelli, F.; Serrao, M. An artificial neural network approach to detect presence and severity of Parkinson's disease via gait parameters. *PLoS ONE* **2021**, *16*, e0244396. [CrossRef]
14. Khachnaoui, H.; Mabrouk, R.; Khlifa, N. Machine learning and deep learning for clinical data and PET/SPECT imaging in Parkinson's disease: A review. *IET Image Process.* **2020**, *14*, 4013–4026. [CrossRef]
15. Maclagan, L.C.; Visanji, N.P.; Cheng, Y.; Tadrous, M.; Lacoste, A.M.; Kalia, L.V.; Marras, C. Identifying drugs with disease-modifying potential in Parkinson's disease using artificial intelligence and pharmacoepidemiology. *Wiley Online Libr.* **2020**, *29*, 864–872. [CrossRef] [PubMed]
16. Tricco, A.C.; Lillie, E.; Zarin, W.; O'Brien, K.K.; Colquhoun, H.; Levac, D.; Moher, D.; Peters, M.D.J.; Horsley, T.; Weeks, L.; et al. PRISMA extension for scoping reviews (PRISMA-ScR): Checklist and explanation. *Ann. Intern. Med.* **2018**, *169*, 467–473. [CrossRef] [PubMed]
17. Ouzzani, M.; Hammady, H.; Fedorowicz, Z.; Elmagarmid, A. Rayyan-a web and mobile app for systematic reviews. *Syst. Rev.* **2016**, *5*, 1–10. [CrossRef]
18. Ali, M.R.; Hernandez, J.; Dorsey, E.R.; Hoque, E.; McDuff, D. Spatio-Temporal Attention and Magnification for Classification of Parkinson's Disease from Videos Collected via the Internet. In Proceedings of the 2020 15th IEEE International Conference on Automatic Face and Gesture Recognition (FG 2020); IEEE: New York, NY, USA, 2020; pp. 207–214. [CrossRef]
19. Gazda, M.; Hires, M.; Drotar, P. Multiple-Fine-Tuned Convolutional Neural Networks for Parkinson's Disease Diagnosis from Offline Handwriting. *IEEE Trans. Syst. Man, Cybern. Syst.* **2021**, 1–12. [CrossRef]
20. Mohammed, F.; He, X.; Lin, Y. An easy-to-use deep-learning model for highly accurate diagnosis of Parkinson's disease using SPECT images. *Comput. Med. Imaging Graph.* **2021**, *87*, 101810. [CrossRef]
21. Aversano, L.; Bernardi, M.L.; Cimitile, M.; Pecori, R. Fuzzy neural networks to detect parkinson disease. In Proceedings of the 2020 IEEE International Conference on Fuzzy Systems (FUZZ-IEEE), Glasgow, UK, 19–24 July 2020; pp. 1–8. [CrossRef]
22. Abayomi-Alli, O.O.; Damasevicius, R.; Maskeliunas, R.; Abayomi-Alli, A. BiLSTM with Data Augmentation using Interpolation Methods to Improve Early Detection of Parkinson Disease. In Proceedings of the 2020 Federated Conference on Computer Science and Information Systems, FedCSIS 2020, Sofia, Bulgaria, 6–9 September 2020; Polish Information Processing Society PTI: Warsaw, Poland, 2020; Volume 21, pp. 371–380. [CrossRef]
23. Asmae, O.; Abdelhadi, R.; Bouchaib, C.; Sara, S.; Tajeddine, K. Parkinson's Disease Identification using KNN and ANN Algorithms based on Voice Disorder. In Proceedings of the 2020 1st International Conference on Innovative Research in Applied Science, Engineering and Technology, IRASET 2020, Meknes, Morocco, 16–19 April 2020; IEEE: New York, NY, USA, 2020. [CrossRef]
24. Tuan, A.M.; Andrei, A.G.; Ionescu, B. Freezing of gait detection for parkinson's disease patients using accelerometer data: Case study. In Proceedings of the 2020 8th E-Health and Bioengineering Conference, EHB 2020, Iasi, Romania, 29–30 October 2020; IEEE: New York, NY, USA, 2020. [CrossRef]
25. Aghzal, M.; Mourhir, A. Early Diagnosis of Parkinson's Disease based on Handwritten Patterns using Deep Learning. In Proceedings of the 4th International Conference on Intelligent Computing in Data Sciences, ICDS 2020, Fez, Morocco, 21–23 October 2020; IEEE: New York, NY, USA, 2020. [CrossRef]
26. Moshkova, A.; Samorodov, A.; Ivanova, E.; Fedotova, E. High Accuracy Discrimination of Parkinson's Disease from Healthy Controls by Hand Movements Analysis Using LeapMotion Sensor and 1D Convolutional Neural Network. In Proceedings of the Proceedings—2020 Ural Symposium on Biomedical Engineering, Radioelectronics and Information Technology, USBEREIT 2020, Yekaterinburg, Russia, 14–15 May 2020; IEEE: New York, NY, USA, 2020; pp. 62–65. [CrossRef]
27. Shaban, M. Deep Convolutional Neural Network for Parkinson's Disease Based Handwriting Screening. In Proceedings of the ISBI Workshops 2020—International Symposium on Biomedical Imaging Workshops, Iowa City, IA, USA, 4 April 2020; IEEE: New York, NY, USA, 2020. [CrossRef]
28. Wang, W.; Lee, J.; Harrou, F.; Sun, Y. Early Detection of Parkinson's Disease Using Deep Learning and Machine Learning. *IEEE Access* **2020**, *8*, 147635–147646. [CrossRef]
29. Suhas, B.N.; Mallela, J.; Illa, A.; Yamini, B.K.; Atchayaram, N.; Yadav, R.; Gope, D.; Ghosh, P.K. Speech task based automatic classification of ALS and Parkinson's Disease and their severity using log Mel spectrograms. In Proceedings of the SPCOM 2020—International Conference on Signal Processing and Communications, Bangalore, India, 19–24 July 2020; IEEE: New York, NY, USA, 2020. [CrossRef]
30. Chakraborty, S.; Aich, S.; Sim, J.-S.; Han, E.; Park, J.; Kim, H.C. Parkinson's Disease Detection from Spiral and Wave Drawings using Convolutional Neural Networks: A Multistage Classifier Approach. In *Proceedings of the International Conference on Advanced Communication Technology, ICACT, Phoenix Park, Korea, 16–19 February 2020*; IEEE: New York, NY, USA, 2020; Volume 2020, pp. 298–303. [CrossRef]
31. Shreya Prabhu, K.; Joy Martis, R. Diagnosis of Parkinson's Disease using Computer Aided Tool based on EEG. In Proceedings of the 2020 IEEE 17th India Council International Conference, INDICON 2020, New Delhi, India, 10–13 December 2020; IEEE: New York, NY, USA, 2020; pp. 1–4. [CrossRef]

32. Oh, S.L.; Hagiwara, Y.; Raghavendra, U.; Yuvaraj, R.; Arunkumar, N.; Murugappan, M.; Acharya, U.R. A deep learning approach for Parkinson's disease diagnosis from EEG signals. *Neural Comput. Appl.* **2020**, *32*, 10927–10933. [CrossRef]
33. Xiong, Y.; Lu, Y. Deep Feature Extraction from the Vocal Vectors Using Sparse Autoencoders for Parkinson's Classification. *IEEE Access* **2020**, *8*, 27821–27830. [CrossRef]
34. Ashour, A.S.; El-Attar, A.; Dey, N.; El-Kader, H.A.; El-Naby, M.M.A. Long short term memory based patient-dependent model for FOG detection in Parkinson's disease. *Pattern Recognit. Lett.* **2020**, *131*, 23–29. [CrossRef]
35. El Maachi, I.; Bilodeau, G.-A.; Bouachir, W. Deep 1D-Convnet for accurate Parkinson disease detection and severity prediction from gait. *Expert Syst. Appl.* **2020**, *143*, 113075. [CrossRef]
36. Oktay, A.B.; Kocer, A. Differential diagnosis of Parkinson and essential tremor with convolutional LSTM networks. *Biomed. Signal Process. Control.* **2020**, *56*, 101683. [CrossRef]
37. Yurdakul, O.C.; Subathra, M.; George, S.T. Detection of Parkinson's Disease from gait using Neighborhood Representation Local Binary Patterns. *Biomed. Signal Process. Control.* **2020**, *62*, 102070. [CrossRef]
38. Shah, S.A.A.; Zhang, L.; Bais, A. Dynamical system based compact deep hybrid network for classification of Parkinson disease related EEG signals. *Neural Networks* **2020**, *130*, 75–84. [CrossRef] [PubMed]
39. Veeraragavan, S.; Gopalai, A.A.; Gouwanda, D.; Ahmad, S.A. Parkinson's Disease Diagnosis and Severity Assessment Using Ground Reaction Forces and Neural Networks. *Front. Physiol.* **2020**, *11*. [CrossRef] [PubMed]
40. Xu, S.; Wang, Z.; Sun, J.; Zhang, Z.; Wu, Z.; Yang, T.; Xue, G.; Cheng, C. Using a deep recurrent neural network with EEG signal to detect Parkinson's disease. *Ann. Transl. Med.* **2020**, *8*, 874. [CrossRef] [PubMed]
41. Chakraborty, S.; Aich, S.; Kim, H.-C. Detection of Parkinson's Disease from 3T T1 Weighted MRI Scans Using 3D Convolutional Neural Network. *Diagn.* **2020**, *10*, 402. [CrossRef]
42. Hu, K.; Wang, Z.; Wang, W.; Martens, K.A.E.; Wang, L.; Tan, T.; Lewis, S.J.G.; Feng, D.D. Graph Sequence Recurrent Neural Network for Vision-Based Freezing of Gait Detection. *IEEE Trans. Image Process.* **2019**, *29*, 1890–1901. [CrossRef] [PubMed]
43. Filtjens, B.; Nieuwboer, A.; D'Cruz, N.; Spildooren, J.; Slaets, P.; Vanrumste, B. A data-driven approach for detecting gait events during turning in people with Parkinson's disease and freezing of gait. *Gait Posture* **2020**, *80*, 130–136. [CrossRef]
44. Sarker, Y.; Mondal, N.I.; Fahim, S.R.; Shahriar, S.; Sarker, S.K.; Das, S.K. A Novel Diagnosis System Using Regularized Encoder-Decoder Based Generative Probabilistic Network for Parkinson's Disease. In *2020 IEEE Region 10 Symposium (TENSYMP)*; IEEE: New York, NY, USA, 2020; pp. 1444–1447. [CrossRef]
45. Ibrahim, A.; Zhou, Y.; Jenkins, M.E.; Trejos, A.L.; Naish, M.D. The Design of a Parkinson's Tremor Predictor and Estimator Using a Hybrid Convolutional-Multilayer Perceptron Neural Network. In Proceedings of the 2020 42nd Annual International Conference of the IEEE Engineering in Medicine & Biology Society (EMBC), Montreal, QC, Canada, 20–24 July 2020; IEEE: New York, NY, USA, 2020; Volume 2020, pp. 5996–6000. [CrossRef]
46. Ibrahim, A.; Zhou, Y.; Jenkins, M.E.; Naish, M.D.; Trejos, A.L. Parkinson's Tremor Onset Detection and Active Tremor Classification Using a Multilayer Perceptron. In Proceedings of the 2020 IEEE Canadian Conference on Electrical and Computer Engineering (CCECE), London, ON, Canada, 30 August–2 September 2020; IEEE: New York, NY, USA, 2020; pp. 1–4. [CrossRef]
47. Ramirez, V.M.; Kmetzsch, V.; Forbes, F.; Dojat, M. Deep Learning Models to Study the Early Stages of Parkinson's Disease. In Proceedings of the 2020 IEEE 17th International Symposium on Biomedical Imaging (ISBI), Iowa City, IA, USA, 3–7 April 2020; IEEE: New York, NY, USA, 2020; pp. 1534–1537. [CrossRef]
48. Bielby, J.; Kuhn, S.; Colreavy-Donnelly, S.; Caraffini, F.; O'Connor, S.; Anastassi, Z.A. Identifying Parkinson's Disease Through the Classification of Audio Recording Data. In Proceedings of the 2020 IEEE Congress on Evolutionary Computation (CEC), Glasgow, UK, 19–24 July 2020; IEEE: New York, NY, USA, 2020; pp. 1–7. [CrossRef]
49. Aversano, L.; Bernardi, M.L.; Cimitile, M.; Pecori, R. Early Detection of Parkinson Disease using Deep Neural Networks on Gait Dynamics. In Proceedings of the 2020 International Joint Conference on Neural Networks (IJCNN), Glasgow, UK, 19–24 July 2020; IEEE: New York, NY, USA, 2020; pp. 1–8. [CrossRef]
50. Karan, B.; Sahu, S.S.; Mahto, K. Stacked auto-encoder based Time- frequency features of Speech signal for Parkinson disease prediction. In Proceedings of the 2020 International Conference on Artificial Intelligence and Signal Processing (AISP), Amaravati, India, 10–12 January 2020; IEEE: New York, NY, USA, 2020; pp. 1–4. [CrossRef]
51. Saikia, A.; Majhi, V.; Hussain, M.; Barua, A.R.; Paul, S.; Verma, J.K. Machine Learning based Diagnostic System for Early Detection of Parkinson's Disease. In Proceedings of the 2020 International Conference on Computational Performance Evaluation (ComPE), Shillong, India, 2–4 July 2020; IEEE: New York, NY, USA, 2020; pp. 275–279. [CrossRef]
52. Ali, L.; Zhu, C.; Zhang, Z.; Liu, Y. Automated Detection of Parkinson's Disease Based on Multiple Types of Sustained Phonations Using Linear Discriminant Analysis and Genetically Optimized Neural Network. *IEEE J. Transl. Eng. Heal. Med.* **2019**, *7*, 1–10. [CrossRef]
53. Haq, A.U.; Li, J.; Memon, M.H.; Khan, J.; Din, S.U.; Ahad, I.; Sun, R.; Lai, Z. Comparative Analysis of the Classification Performance of Machine Learning Classifiers and Deep Neural Network Classifier for Prediction of Parkinson Disease. In Proceedings of the 2018 15th International Computer Conference on Wavelet Active Media Technology and Information Processing (ICCWAMTIP), Chengdu, China, 14–16 December 2018; IEEE: New York, NY, USA, 2018; pp. 101–106. [CrossRef]
54. Kollia, I.; Stafylopatis, A.-G.; Kollias, S. Predicting Parkinson's Disease using Latent Information extracted from Deep Neural Networks. In Proceedings of the 2019 International Joint Conference on Neural Networks (IJCNN), Budapest, Hungary, 14–19 July 2019; IEEE: New York, NY, USA, 2019; pp. 1–8. [CrossRef]

55. Shivangi; Johri, A.; Tripathi, A. Parkinson Disease Detection Using Deep Neural Networks. In Proceedings of the 2019 Twelfth International Conference on Contemporary Computing (IC3), Noida, India, 8–10 August 2019; IEEE: New York, NY, USA, 2019; pp. 1–4. [CrossRef]
56. Giuliano, M.; Garcia-Lopez, A.; Perez, S.; Perez, F.D.; Spositto, O.; Bossero, J. Selection of voice parameters for Parkinson´s disease prediction from collected mobile data. In Proceedings of the 2019 XXII Symposium on Image, Signal Processing and Artificial Vision (STSIVA), Bucaramanga, Colombia, 24–26 April 2019; IEEE: New York, NY, USA, 2019; pp. 1–3. [CrossRef]
57. Tahafchi, P.; Judy, J.W. Freezing-of-Gait Detection Using Wearable-Sensor Technology and Neural-Network Classifier. In Proceedings of the 2019 IEEE Sensors Applications Symposium (SAS), Montreal, QC, Canada, 27–30 October 2019; IEEE: New York, NY, USA, 2019; pp. 1–4. [CrossRef]
58. Yagis, E.; De Herrera, A.G.S.; Citi, L. Generalization Performance of Deep Learning Models in Neurodegenerative Disease Classification. In Proceedings of the 2019 IEEE International Conference on Bioinformatics and Biomedicine (BIBM), San Diego, CA, USA, 18–21 November 2019; IEEE: New York, NY, USA, 2019; pp. 1692–1698. [CrossRef]
59. Rumman, M.; Tasneem, A.N.; Farzana, S.; Pavel, M.I.; Alam, A. Early detection of Parkinson's disease using image processing and artificial neural network. In Proceedings of the 2018 Joint 7th International Conference on Informatics, Electronics & Vision (ICIEV) and 2018 2nd International Conference on Imaging, Vision & Pattern Recognition (icIVPR), Kitakyushu, Japan, 25–29 June 2018; IEEE: New York, NY, USA, 2018; pp. 256–261. [CrossRef]
60. Anand, A.; Haque, A.; Alex, J.S.R.; Venkatesan, N. Evaluation of Machine learning and Deep learning algorithms combined with dimentionality reduction techniques for classification of Parkinson's Disease. In Proceedings of the 2018 IEEE International Symposium on Signal Processing and Information Technology (ISSPIT), Louisville, KY, USA, 6–8 December 2018; IEEE: New York, NY, USA, 2018; pp. 342–347. [CrossRef]
61. Wroge, T.J.; Ozkanca, Y.; Demiroglu, C.; Si, D.; Atkins, D.C.; Ghomi, R.H. Parkinson's Disease Diagnosis Using Machine Learning and Voice. In Proceedings of the 2018 IEEE Signal Processing in Medicine and Biology Symposium (SPMB), Philadelphia, PA, USA, 1 December 2018; IEEE: New York, NY, USA, 2018; pp. 1–7. [CrossRef]
62. Arami, A.; Poulakakis-Daktylidis, A.; Tai, Y.F.; Burdet, E. Prediction of Gait Freezing in Parkinsonian Patients: A Binary Classification Augmented With Time Series Prediction. *IEEE Trans. Neural Syst. Rehabilitation Eng.* **2019**, *27*, 1909–1919. [CrossRef]
63. Pandit, T.; Nahane, H.; Lade, D.; Rao, V. Abnormal Gait Detection by Classifying Inertial Sensor Data using Transfer Learning. In Proceedings of the 2019 18th IEEE International Conference On Machine Learning And Applications (ICMLA), Boca Raton, FL, USA, 16–19 December 2019; IEEE: New York, NY, USA, 2019; pp. 1444–1447. [CrossRef]
64. Fernandes, C.; Fonseca, L.; Ferreira, F.; Gago, M.; Costa, L.; Sousa, N.; Ferreira, C.; Gama, J.; Erlhagen, W.; Bicho, E. Artificial Neural Networks Classification of Patients with Parkinsonism based on Gait. In Proceedings of the 2018 IEEE International Conference on Bioinformatics and Biomedicine (BIBM), Madrid, Spain, 3–6 December 2018; IEEE: New York, NY, USA, 2018; pp. 2024–2030. [CrossRef]
65. Torvi, V.G.; Bhattacharya, A.; Chakraborty, S. Deep Domain Adaptation to Predict Freezing of Gait in Patients with Parkinson's Disease. In Proceedings of the 2018 17th IEEE International Conference on Machine Learning and Applications (ICMLA), Orlando, FL, USA, 17–20 December 2018; IEEE: New York, NY, USA, 2018; pp. 1001–1006. [CrossRef]
66. Zhang, X.; Yang, Y.; Wang, H.; Ning, S.; Wang, H. Deep Neural Networks with Broad Views for Parkinson's Disease Screening. In Proceedings of the 2019 IEEE International Conference on Bioinformatics and Biomedicine (BIBM), San Diego, CA, USA, 18–21 November 2019; IEEE: New York, NY, USA, 2019; pp. 1018–1022. [CrossRef]
67. Dai, Y.; Tang, Z.; Wang, Y.; Xu, Z. Data Driven Intelligent Diagnostics for Parkinson's Disease. *IEEE Access* **2019**, *7*, 106941–106950. [CrossRef]
68. Abdallah, M.; Saad, A.; Ayache, M. Freezing of Gait Detection: Deep Learning Approach. In Proceedings of the 2019 International Arab Conference on Information Technology (ACIT), Al Ain, United Arab Emirates, 3–5 December 2019; IEEE: New York, NY, USA, 2019; pp. 259–261. [CrossRef]
69. Taleb, C.; Khachab, M.; Mokbel, C.; Likforman-Sulem, L. Visual Representation of Online Handwriting Time Series for Deep Learning Parkinson's Disease Detection. In Proceedings of the 2019 International Conference on Document Analysis and Recognition Workshops (ICDARW), Sydney, NSW, Australia, 22–25 September 2019; IEEE: New York, NY, USA, 2019; Volume 6, pp. 25–30. [CrossRef]
70. Reyes, J.F.; Montealegre, J.S.; Castano, Y.J.; Urcuqui, C.; Navarro, A. LSTM and Convolution Networks exploration for Parkinson's Diagnosis. In Proceedings of the 2019 IEEE Colombian Conference on Communications and Computing (COLCOM), Barranquilla, Colombia, 5–7 June 2019; IEEE: New York, NY, USA, 2019; pp. 1–4. [CrossRef]
71. Zhao, Y.; Cumming, P.; Rominger, A.; Zuo, C.; Shi, K.; Wu, P.; Wang, J.; Li, H.; Navab, N.; Yakushev, I.; et al. A 3D Deep Residual Convolutional Neural Network for Differential Diagnosis of Parkinsonian Syndromes on 18F-FDG PET Images. In Proceedings of the 2019 41st Annual International Conference of the IEEE Engineering in Medicine and Biology Society (EMBC), Berlin, Germany, 23–27 July 2019; IEEE: New York, NY, USA, 2019; Volume 2019, pp. 3531–3534. [CrossRef]
72. Kiryu, S.; Yasaka, K.; Akai, H.; Nakata, Y.; Sugomori, Y.; Hara, S.; Seo, M.; Abe, O.; Ohtomo, K. Deep learning to differentiate parkinsonian disorders separately using single midsagittal MR imaging: A proof of concept study. *Eur. Radiol.* **2019**, *29*, 6891–6899. [CrossRef] [PubMed]

73. Sadek, R.M.; Mohammed, S.A.; Abunbehan, A.R.K.; Ghattas, A.K.H.A.; Badawi, M.R.; Mortaja, M.N.; Abu-Naser, S.S. Parkinson's Disease Prediction Using Artificial Neural Network. *Comput. Sci.* **2019**, *3*, 1–8. Available online: http://dstore.alazhar.edu.ps/xmlui/handle/123456789/302 (accessed on 26 May 2021).
74. Bhati, S.; Velazquez, L.M.; Villalba, J.; Dehak, N. LSTM Siamese Network for Parkinson's Disease Detection from Speech. In Proceedings of the 2019 IEEE Global Conference on Signal and Information Processing (GlobalSIP), Ottawa, ON, Canada, 11–14 November 2019; IEEE: New York, NY, USA, 2019; pp. 1–5. [CrossRef]
75. Afonso, L.C.; Rosa, G.H.; Pereira, C.R.; Weber, S.A.; Hook, C.; Albuquerque, V.H.C.; Papa, J.P. A recurrence plot-based approach for Parkinson's disease identification. *Futur. Gener. Comput. Syst.* **2019**, *94*, 282–292. [CrossRef]
76. Ortiz, A.; Munilla, J.; Martínez-Ibañez, M.; Górriz, J.M.; Ramírez, J.; Salas-Gonzalez, D. Parkinson's Disease Detection Using Isosurfaces-Based Features and Convolutional Neural Networks. *Front. Aging Neurosci.* **2019**, *13*, 48. [CrossRef] [PubMed]
77. Wodzinski, M.; Skalski, A.; Hemmerling, D.; Orozco-Arroyave, J.R.; Noth, E. Deep Learning Approach to Parkinson's Disease Detection Using Voice Recordings and Convolutional Neural Network Dedicated to Image Classification. In Proceedings of the 2019 41st Annual International Conference of the IEEE Engineering in Medicine and Biology Society (EMBC), Berlin, Germany, 23–27 July 2019; IEEE: New York, NY, USA, 2019; Volume 2019, pp. 717–720. [CrossRef]
78. Manzanera, O.M.; Meles, S.K.; Leenders, K.L.; Renken, R.J.; Pagani, M.; Arnaldi, D.; Nobili, F.; Obeso, J.; Oroz, M.R.; Morbelli, S.; et al. Scaled Subprofile Modeling and Convolutional Neural Networks for the Identification of Parkinson's Disease in 3D Nuclear Imaging Data. *Int. J. Neural Syst.* **2019**, *29*, 1950010. [CrossRef] [PubMed]
79. Iakovakis, D.; Diniz, J.A.; Trivedi, D.; Chaudhuri, R.K.; Hadjileontiadis, L.J.; Hadjidimitriou, S.; Charisis, V.; Bostanjopoulou, S.; Katsarou, Z.; Klingelhoefer, L.; et al. Early Parkinson's Disease Detection via Touchscreen Typing Analysis using Convolutional Neural Networks. In Proceedings of the 2019 41st Annual International Conference of the IEEE Engineering in Medicine and Biology Society (EMBC), Berlin, Germany, 23–27 July 2019; IEEE: New York, NY, USA, 2019; Volume 2019, pp. 3535–3538. [CrossRef]
80. Pham, T.D.; Wardell, K.; Eklund, A.; Salerud, G. Classification of short time series in early Parkinsons disease with deep learning of fuzzy recurrence plots. *IEEE/CAA J. Autom. Sin.* **2019**, *6*, 1306–1317. [CrossRef]
81. Hoang, N.S.; Cai, Y.; Lee, C.-W.; Yang, Y.O.; Chui, C.-K.; Chua, M.C.H. Gait classification for Parkinson's Disease using Stacked 2D and 1D Convolutional Neural Network. In Proceedings of the 2019 International Conference on Advanced Technologies for Communications (ATC), Hanoi, Vietnam, 17–19 October 2019; IEEE: New York, NY, USA, 2019; pp. 44–49. [CrossRef]
82. Li, S.; Lei, H.; Zhou, F.; Gardezi, J.; Lei, B. Longitudinal and Multi-modal Data Learning for Parkinson's Disease Diagnosis via Stacked Sparse Auto-encoder. In Proceedings of the 2019 IEEE 16th International Symposium on Biomedical Imaging (ISBI 2019), Venice, Italy, 8–11 April 2019; IEEE: New York, NY, USA, 2019; pp. 384–387. [CrossRef]
83. Lee, S.; Hussein, R.; McKeown, M.J. A Deep Convolutional-Recurrent Neural Network Architecture for Parkinson's Disease EEG Classification. In Proceedings of the 2019 IEEE Global Conference on Signal and Information Processing (GlobalSIP), Ottawa, ON, Canada, 11–14 November 2019; IEEE: New York, NY, USA, 2019; pp. 1–4. [CrossRef]
84. Sayaydeh, O.; Mohammad, M.F. Diagnosis of the Parkinson Disease Using Enhanced Fuzzy Min-Max Neural Network and OneR Attribute Evaluation Method. In Proceedings of the 2019 International Conference on Advanced Science and Engineering (ICOASE), Zakho - Duhok, Iraq, 2–4 April 2019; IEEE: New York, NY, USA, 2019; pp. 64–69. [CrossRef]
85. Shi, X.; Wang, T.; Wang, L.; Liu, H.; Yan, N. Hybrid Convolutional Recurrent Neural Networks Outperform CNN and RNN in Task-state EEG Detection for Parkinson's Disease. In Proceedings of the 2019 Asia-Pacific Signal and Information Processing Association Annual Summit and Conference (APSIPA ASC), Lanzhou, China, 18–21 November 2019; IEEE: New York, NY, USA, 2019; pp. 939–944. [CrossRef]
86. Shah, P.M.; Zeb, A.; Shafi, U.; Alam Zaidi, S.F.; Shah, M.A. Detection of Parkinson Disease in Brain MRI using Convolutional Neural Network. In Proceedings of the 2018 24th International Conference on Automation and Computing (ICAC), Newcastle Upon Tyne, UK, 6–7 September 2018; IEEE: New York, NY, USA, 2018. [CrossRef]
87. Ajay, J.; Song, C.; Wang, A.; Langan, J.; Li, Z.; Xu, W. A pervasive and sensor-free Deep Learning system for Parkinsonian gait analysis. In Proceedings of the 2018 IEEE EMBS International Conference on Biomedical & Health Informatics (BHI), Las Vegas, NV, USA, 4–7 March 2018; IEEE: New York, NY, USA, 2018; pp. 108–111. [CrossRef]
88. Adams, M.P.; Yang, B.; Rahmim, A.; Tang, J. Prediction of outcome in Parkinson's disease patients from DAT SPECT images using a convolutional neural network. In Proceedings of the 2018 IEEE Nuclear Science Symposium and Medical Imaging Conference Proceedings (NSS/MIC), Sydney, NSW, Australia, 10–17 November 2018; IEEE: New York, NY, USA, 2018; pp. 1–4. [CrossRef]
89. Xia, Y.; Zhang, J.; Ye, Q.; Cheng, N.; Lu, Y.; Zhang, D. Evaluation of deep convolutional neural networks for detection of freezing of gait in Parkinson's disease patients. *Biomed. Signal Process. Control.* **2018**, *46*, 221–230. [CrossRef]
90. Gong, B.; Shi, J.; Ying, S.; Dai, Y.; Zhang, Q.; Dong, Y.; An, H.; Zhang, Y. Neuroimaging-based diagnosis of Parkinson's disease with deep neural mapping large margin distribution machine. *Neurocomputing* **2018**, *320*, 141–149. [CrossRef]
91. Camps, J.; Samà, A.; Martín, M.; Rodríguez-Martín, D.; Pérez-López, C.; Arostegui, J.M.M.; Cabestany, J.; Català, A.; Alcaine, S.; Mestre, B.; et al. Deep learning for freezing of gait detection in Parkinson's disease patients in their homes using a waist-worn inertial measurement unit. *Knowl. Based Syst.* **2018**, *139*, 119–131. [CrossRef]
92. Prince, J.; De Vos, M. A Deep Learning Framework for the Remote Detection of Parkinson'S Disease Using Smart-Phone Sensor Data. In Proceedings of the 2018 40th Annual International Conference of the IEEE Engineering in Medicine and Biology Society (EMBC), Honolulu, HI, USA, 18–21 July 2018; IEEE: New York, NY, USA, 2018; Volume 2018, pp. 3144–3147. [CrossRef]

93. Lahmiri, S.; Dawson, D.A.; Shmuel, A. Performance of machine learning methods in diagnosing Parkinson's disease based on dysphonia measures. *Biomed. Eng. Lett.* **2017**, *8*, 29–39. [CrossRef] [PubMed]
94. Martínez-Murcia, F.J.; Górriz, J.M.; Ramírez, J.; Ortiz, A. Convolutional Neural Networks for Neuroimaging in Parkinson's Disease: Is Preprocessing Needed? *Int. J. Neural Syst.* **2018**, *28*, 1850035. [CrossRef]
95. Zhang, X.; He, L.; Chen, K.; Luo, Y.; Zhou, J.; Wang, F. Multi-View Graph Convolutional Network and Its Applications on Neuroimage Analysis for Parkinson's Disease. *AMIA Annu. Symp. Proc.* **2018**, *2018*, 1147–1156. Available online: https://www.ncbi.nlm.nih.gov/pubmed/30815157 (accessed on 1 June 2021).
96. Loconsole, C.; Cascarano, G.D.; Lattarulo, A.; Brunetti, A.; Trotta, G.F.; Buongiorno, D.; Bortone, I.; De Feudis, I.; Losavio, G.; Bevilacqua, V.; et al. A comparison between ANN and SVM classifiers for Parkinson's disease by using a model-free computer-assisted handwriting analysis based on biometric signals. In Proceedings of the 2018 International Joint Conference on Neural Networks (IJCNN), Rio de Janeiro, Brazil, 8–13 July 2018; IEEE: New York, NY, USA, 2018; pp. 1–8. [CrossRef]
97. Pereira, C.R.; Pereira, D.R.; Rosa, G.H.; Albuquerque, V.H.; Weber, S.A.; Hook, C.; Papa, J.P. Handwritten dynamics assessment through convolutional neural networks: An application to Parkinson's disease identification. *Artif. Intell. Med.* **2018**, *87*, 67–77. [CrossRef] [PubMed]
98. Kim, H.B.; Lee, H.J.; Lee, W.W.; Kim, S.K.; Jeon, H.S.; Park, H.Y.; Shin, C.W.; Yi, W.J.; Jeon, B.; Park, K.S. Validation of Freezing-of-Gait Monitoring Using Smartphone. *Telemed. e-Health* **2018**, *24*, 899–907. [CrossRef]
99. Vlachostergiou, A.; Tagaris, A.; Stafylopatis, A.; Kollias, S. Multi-Task Learning for Predicting Parkinson's Disease Based on Medical Imaging Information. In Proceedings of the 2018 25th IEEE International Conference on Image Processing (ICIP), Athens, Greece, 7–10 October 2018; IEEE: New York, NY, USA, 2018; pp. 2052–2056. [CrossRef]
100. Khojasteh, P.; Viswanathan, R.; Aliahmad, B.; Ragnav, S.; Zham, P.; Kumar, D.K. Parkinson's Disease Diagnosis Based on Multivariate Deep Features of Speech Signal. In Proceedings of the 2018 IEEE Life Sciences Conference (LSC), Montreal, QC, Canada, 28–30 October 2018; IEEE: New York, NY, USA, 2018; pp. 187–190. [CrossRef]
101. Khatamino, P.; Canturk, I.; Ozyilmaz, L. A Deep Learning-CNN Based System for Medical Diagnosis: An Application on Parkinson's Disease Handwriting Drawings. In Proceedings of the 2018 6th International Conference on Control Engineering & Information Technology (CEIT), Istanbul, Turkey, 25–27 October 2018; IEEE: New York, NY, USA, 2018; pp. 1–6. [CrossRef]
102. Passos, L.A.; Pereira, C.R.; Rezende, E.R.S.; Carvalho, T.J.; Weber, S.A.T.; Hook, C.; Papa, J.P. Parkinson Disease Identification Using Residual Networks and Optimum-Path Forest. In Proceedings of the 2018 IEEE 12th International Symposium on Applied Computational Intelligence and Informatics (SACI), Timisoara, Romania, 17–19 May 2018; IEEE: New York, NY, USA, 2018; pp. 325–330. [CrossRef]
103. Vlachostergiou, A.; Tagaris, A.; Stafylopatis, A.; Kollias, S. Investigating the Best Performing Task Conditions of a Multi-Tasking Learning Model in Healthcare Using Convolutional Neural Networks: Evidence from a Parkinson's Disease Database. In Proceedings of the 2018 25th IEEE International Conference on Image Processing (ICIP), Athens, Greece, 7–10 October 2018; IEEE: New York, NY, USA, 2018; pp. 2047–2051. [CrossRef]
104. Zhang, H.; Wang, A.; Li, D.; Xu, W. DeepVoice: A voiceprint-based mobile health framework for Parkinson's disease identification. In Proceedings of the 2018 IEEE EMBS International Conference on Biomedical & Health Informatics (BHI), Las Vegas, NV, USA, 4–7 March 2018; IEEE: New York, NY, USA, 2018; pp. 214–217. [CrossRef]
105. Marar, S.; Swain, D.; Hiwarkar, V.; Motwani, N.; Awari, A. Predicting the occurrence of Parkinson's Disease using various Classification Models. In Proceedings of the 2018 International Conference on Advanced Computation and Telecommunication (ICACAT), Bhopal, India, 28–29 December 2018; IEEE: New York, NY, USA, 2018; pp. 1–5. [CrossRef]
106. Moon, S.; Ahmadnezhad, P.; Song, H.-J.; Thompson, J.; Kipp, K.; Akinwuntan, A.E.; Devos, H. Artificial neural networks in neurorehabilitation: A scoping review. *Neurorehabilitation* **2020**, *46*, 259–269. [CrossRef] [PubMed]
107. Shahid, N.; Rappon, T.; Berta, W. Applications of artificial neural networks in health care organizational decision-making: A scoping review. *PLoS ONE* **2019**, *14*, e0212356. [CrossRef]
108. Shatte, A.B.R.; Hutchinson, D.M.; Teague, S.J. Machine learning in mental health: A scoping review of methods and applications. *Psychol. Med.* **2019**, *49*, 1426–1448. [CrossRef] [PubMed]

healthcare

MDPI

Article

The Prediction Model of Medical Expenditure Appling Machine Learning Algorithm in CABG Patients

Yen-Chun Huang [1,2], Shao-Jung Li [3,4,5,6], Mingchih Chen [1,2,*] and Tian-Shyug Lee [1,2,*]

1 Graduate Institute of Business Administration, College of Management, Fu Jen Catholic University, New Taipei City 24205, Taiwan; hivicky92@gmail.com
2 Artificial Intelligence Development Center, Fu Jen Catholic University, New Taipei City 242062, Taiwan
3 Cardiovascular Research Center, Wan Fang Hospital, Taipei Medical University, Taipei City 116, Taiwan; leeshaojung@gmail.com
4 Taipei Heart Institute, Taipei Medical University, New Taipei City 231, Taiwan
5 Department of Surgery, School of Medicine, College of Medicine, Taipei Medical University, Taipei City 116, Taiwan
6 Division of Cardiovascular Surgery, Department of Surgery, Wan Fang Hospital, Taipei Medical University, Taipei City 116, Taiwan
* Correspondence: 081438@mail.fju.edu.tw (M.C.); 036665@mail.fju.edu.tw (T.-S.L.)

Abstract: Most patients face expensive healthcare management after coronary artery bypass grafting (CABG) surgery, which brings a substantial financial burden to the government. The National Health Insurance Research Database (NHIRD) is a complete database containing over 99% of individuals' medical information in Taiwan. Our research used the latest data that selected patients who accepted their first CABG surgery between January 2014 and December 2017 (n = 12,945) to predict which factors will affect medical expenses, and built the prediction model using different machine learning algorithms. After analysis, our result showed that the surgical expenditure (X4) and 1-year medical expenditure before the CABG operation (X14), and the number of hemodialysis (X15), were the key factors affecting the 1-year medical expenses of CABG patients after discharge. Furthermore, the XGBoost and SVR methods are both the best predictive models. Thus, our research suggests enhancing the healthcare management for patients with kidney-related diseases to avoid costly complications. We provide helpful information for medical management, which may decrease health insurance burdens in the future.

Keywords: National Health Insurance Research Database; NHIRD; CABG; machine learning; medical expenditure predict; feature selection

Citation: Huang, Y.-C.; Li, S.-J.; Chen, M.; Lee, T.-S. The Prediction Model of Medical Expenditure Appling Machine Learning Algorithm in CABG Patients. *Healthcare* **2021**, *9*, 710. https://doi.org/10.3390/healthcare9060710

Academic Editor: Mahmudur Rahman

Received: 24 May 2021
Accepted: 8 June 2021
Published: 10 June 2021

1. Introduction

Coronary artery bypass grafting (CABG) is the most common cardiac surgery to treat patients with severe coronary artery circulation blockages. After CABG, the patient will have the following two different situations: one is gradual recovery, the other is due to the complications that lead the patient to rehospitalization again [1]. Therefore, readmission is an essential outcome of CABG surgery, and it has a high incidence in 30 and 90 days [2–4]. Furthermore, it is a severe problem because it is directly related to the medical expenses that patients and hospitals must incur, substantially increasing healthcare costs and bringing a vast economic budget. However, the expenditure after CABG surgery remains poorly predicted. The various studies point out preoperative comorbidities, multiple complications, and medical expenses are essential variables that can affect the survival of CABG surgery patients [5–7].

This research used the National Health Insurance Research Database (NHIRD) to delineate this issue. It has been used widely and diversely in many academic studies [8]. Thus, the research results of NHIRD gradually become an indicator for clinical decisions, no matter in physicians or the government.

There are three aims in this research. First, we would use feature selection to identify the essential variables that affect postoperative expenditures. Secondly, we would use different feature selection methods to rank the essential variables. Last, we use different machine learning methods to build an appropriate medical expenditure prediction model for patients who underwent CABG. The information could effectively reduce medical expenditures, improve the quality of healthcare institutions, and provide essential references for medical management policy advice.

2. Materials and Methods

2.1. Data Source

Taiwan's NHIRD has been built since 1995 and the coverage rate is about nearly 99%. NHIRD provides Taiwanese personal medical information, including primary demographic data and previous diseases. In addition, the NHIRD also covered all actual and most extensive healthcare data, including patients' original outpatient, inpatient record, treatment, expenditure, diagnosis code, and admission dates. The codes were based on the International Classification of Disease, 9th Revision, Clinical Modification (ICD-9-CM); the 10th Revision was added to the database on 1 January 2016. This study was designed as a population-based study on 23 million national health insurance beneficiaries enrolled in Taiwan [9]. NHIRD provides a comprehensive long-term follow-up of all claimed records for the benefit of the NHI program. All personal information was anonymized and deidentified in NHIRD. Thus, Fu-Jen University's ethics institutional review board in Taiwan was exempted from ethical review (C108121), and the requirement to obtain informed consent was waived.

2.2. Study Population

This research selected the patients who had accepted CABG surgery (procedure codes 68023A, 68023B, 68024A, 68024B, 68025A, 68025B) between 1 January 2014 and 31 December 2017, from the Taiwan NHIRD (n = 13,078). The date of newly CABG surgery is the index date. There were 133 patients that were not eligible for the study. To ensure that this study was only included the cases that received CABG operation for the first time, patients who had CABG surgery before the initial surgery year (n = 81) were excluded, and we also excluded the patients who were under 18 years old (n = 21) and missing information (n = 31) in this research. After excluding those unqualified patients for this study, 12,945 latest CABG surgery patients were included in our research from 1 January 2014 to 31 December 2017, and all followed up until 31 December 2018 (Figure 1).

2.3. Comorbidities and Risk Factors

The baseline characteristic variables in this study included sex (male/female), age, Charlson comorbidity index (CCI), and CHA2DS2-VAS scores [10,11]. Each patient's comorbidities could be traced to the date before the CABG surgery (2002–2013). The comorbidities included diabetes mellitus (DM), hypertension, hyperlipidemia, myocardial infarction (MI), liver cirrhosis, congestive heart failure (CHF), coronary artery disease (CAD), peripheral vascular disease (PVD), acute pancreatitis, malignant dysrhythmia, atrial fibrillation (AF), transient ischemic attack (TIA), chronic kidney disease (CKD), acute coronary syndrome (ACS), chronic obstructive pulmonary disease (COPD), stroke, cancer, acute kidney failure (AKF), major bleeding, intracranial bleeding, end-stage renal disease (ESRD), and renal disease. Hospital reginal characteristics were as follows: hospital area type, hospital accreditation (medical center/non-medical center), and hospital ownership (public/private). The vessel numbers of percutaneous coronary intervention (PCI), hemodialysis, peritoneal dialysis, blood transfusion (94001C, 94002C, 4013C, 94015C, 94003C), and mechanical ventilation uses (57001B, 57002B, 57003B) in one year before surgery and during the corresponding surgery are also the risk factors in this study.

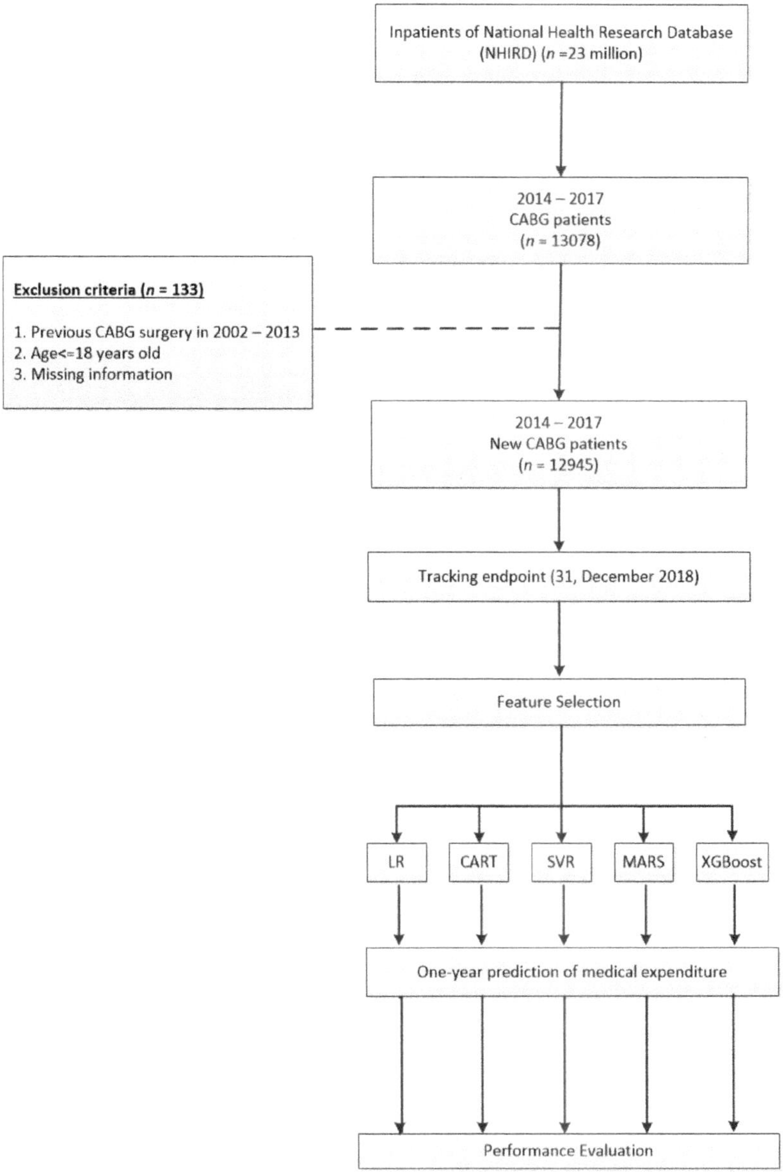

Figure 1. Flowchart of the patients who underwent the first CABG surgery between 2014 and 2017.

2.4. Variable and Outcome Definitions

This study used the total surgical expenditures of the corresponding CABG surgery and the patient's medical expenditures in the previous year as predictive variables. The total surgical expenditures of each patient were calculated by the claimed records, including examination, anesthesia, treatment, drug, operation-related expenses, and other medical services during CABG hospitalization.

To define the primary outcome, this research used one-year cumulative expenditures after discharge to reflect the medical expenditures as primary outcomes. Therefore, we added up the total expense on outpatients and hospitalization after discharged for one year, and this variable is a prediction variable (Y) (Figure 2). All expenses are identified in New Taiwan dollars (NT$).

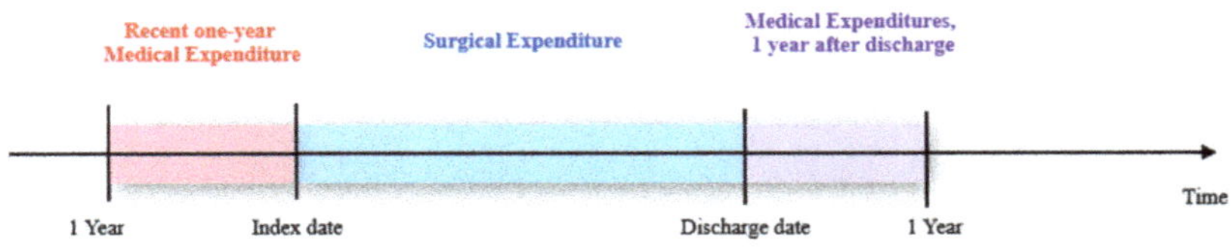

Figure 2. The definition of three different medical expenses associated with CABG surgery.

2.5. Feature Selection and Prediction Models Implementation

When doctors make clinical decisions, they must review the patient's past medical records and current examination results one by one. This not only consumes time for searching, but also slows down the speed to make precise decisions immediately. Thus, feature selection (FS) is an essential preprocessing step before model prediction. By calculating different machine learning algorithms, removing irrelevant factors, we could reduce errors in clinical decisions and improve accuracy [12,13].

Medical expense is a continuous variable. Therefore, linear regression (LR) is often used for continuous numerical estimation, a model established by finding the relationship between the independent and dependent variables. In the training set, this research used five kinds of machine learning, including LR, classification and regression tree (CART), support vector regression (SVR), multi-variate adaptive regression splines (MARS), and XGBoost (extreme gradient boosting) to train by selecting the relevant features for medical expense prediction. In order to avoid overfitting, in the training process, we used five-fold cross-validation.

In more detail, we partitioned the training data into five stratified subsets, 80% of training data were used for training, and 20% of training data were used for validation. Subsequently, we repeated the above processes five times, each subset was used once as a validation dataset. After that, we obtained the average estimated results and used five different indicators to evaluate each prediction model.

2.5.1. Linear Regression (LR)

Linear regression is the association between the dependent variable and one or more independent variables. Through the establishment of the regression model, the variable (y) can be predicted. Before building a prediction model, data must be a normally distributed.

2.5.2. Classification and Regression Tree (CART)

CART can solve the regression and classification problem of multi-dimensional output. It is a kind of flow diagram tree structure; each node was the attribute variable. The branch is a test outcome, and the tree leaves present classification [14]. The method of CART for selection criteria is to use the Gini index. The Gini index is a measure of inequality, and it is usually used to measure income imbalance and can be used to measure any uneven distribution. A number between 0 and 1. 0 is entirely equal, and 1 is entirely unequal.

2.5.3. Support Vector Regression (SVR)

The main algorithm of SVM is the "kernel". When data cannot be linearly divided into lower dimensions, the kernel can transfer them to a higher dimensional divided linearly. SVR is an extension of SVM. In order to solve the problem of nonlinear, SVR is the model for considering the risk of structural, minimizing the generalization error, and maximizing hyper-plane margin to reduce the tolerated error [15,16].

2.5.4. Multi-variate Adaptive Regression Splines (MARS)

Friedman proposed the MARS method in 1991 [17]. MARS is a non-parametric regression and flexible model, and it has consisted of the weighted sum of the basis splines piecewise polynomial functions. The optimal variable is hidden in the high-dimensional data. Through variable interactions, MARS can find the best variable easier [18].

2.5.5. XGBoost (Extreme Gradient Boosting)

XGBoost methods were proposed by Chen et al. in 2016 [19]. It is an ensemble method based on decision tree methods. The framework in this method is gradient boosting, and model builds are sequential. Therefore, it can minimize errors, maximize models' performance, and reduce tree construction time. The central idea in XGBoost is to make a new model to correct the errors in the previous training model, then make the prediction [20].

2.6. Validation Index

This study used different machine learning methods for the prediction of one-year medical expenses after discharge. The validation index of the model was the reference data for determining the quality and accuracy of the model, which depended on the model attributes.

In order to evaluate the performance of the model, this study used five different indicators to measure the prediction result, which was widely and easily understood. These five performance metrics represented the following three different types: absolute error, scaled error, and percentage error. The absolute error group contained the mean absolute error (MAE) and root mean squared error (RMSE), mean square error (MSE), mean absolute scaled error (MASE), and the group of percentage error includes mean absolute percentage error (MAPE) [21,22].

The mathematical formula of these statistical validation metrics for evaluating the models was demonstrated as follows in Table 1.

Table 1. Error measures for the performance metrics equations.

Type of Error	Metrics		Equations
Absolute error	MAE	Mean absolute error	$\frac{1}{n}\sum_{i=1}^{n}\left|\left(a^i-b^i\right)\right|$
	RMSE	Root mean square error	$\sqrt{\frac{1}{n}\sum_{i=1}^{n}\left(a^i-b^i\right)^2}$
Scaled error	MSE	Mean square error	$\frac{1}{n}\sum_{i=1}^{n}\left(a^i-b^i\right)^2\Big|$
	MASE	Mean absolute scaled error	$\frac{1}{n}\sum_{i=1}^{n}\frac{\left|a^i-b^i\right|}{\frac{1}{n-1}\sum_{i=2}^{n}\left|a^i-b^i\right|}$
Percentage error	MAPE	Mean absolute percentage error	$\frac{1}{n}\sum_{i=1}^{n}\left(\left|\frac{a^i-b^i}{b^i}\right|\right)\times 100$

The indicators were frequently and widely used as a performance index among different prediction models [23]. The lower the deviation, the better the accuracy of the prediction model.

MAPE is one of the most popular indicators to use. If MAPE < 0.1, model has high accurate discrimination; $0.11 \leq$ MAPE < 0.2, model has good discrimination; $0.21 \leq$ MAPE < 0.50, model has acceptable discrimination; MAPE > 0.51, model is an inaccurate [23–25].

The above indicators were used to measure the prediction error in each model. Where n was the total amount of patients, b presented the actual medical expense, a represented the predicted medical expense.

2.7. Statistical Analysis

This research selected new CABG patients between 2014 and 2017, which was based on the disease's demographic characteristics and history. All results were expressed as the number and percentages, N (%), for categorical variables. Means with standard deviation were presented as mean ± SD for continuous variables.

2.7.1. Hardware Equipment

MOHW provides an environment for data analysis, the main analyzed computer CPU is intel i7-8700, the main host memory is 128 GB, the brand of system disk type is Western Digital (WD10EZEX) 1T.

Research data were provided from NHIRD, which is the largest volume of data in Taiwan. All analysis data will be stored in the other replacement hard disks (disk type: WD (DC HC310) 6T), which will be kept by the Health and Welfare Data Science Center (HWDC).

2.7.2. Software

Patient data extraction was implemented in SAS version 9.4 (SAS Institute INC., Cary, NC, USA). Variable selection and model establishment is based on the relevant R statistical software (250 Northern Ave, Boston, MA 02210, R studio 3.6.1; https://www.rstudio.com/products/rstudio/). We used R package "stats", "e1071", "earth", "rpart", "XGBoost" to construct the prediction models LR, CART, SVR, MARS, and XGBoost, respectively.

3. Results

3.1. Demographic Characteristics of Study Population

A total of 12,945 new CABG surgery patients was selected from 1 January 2014 to 31 December 2017. The patient's demographic characteristics and comorbidities are shown in Table 2. We analyzed 44 variables that possibly affected one-year medical expenses after discharge (Y). In the baseline factors, the patients' age (X1) was 63.72 ± 10.65 years, the distribution in gender (X40) was 9,917 (76.61%) and 3,028 (23.39%) for males and females, respectively. CHA2DS (X2) was 3.29 ± 1.95 points, the score of CCI (X3) was 4.23 ± 2.82, and whether the patient had a significant illness (X41) was 16.28%. The factors during the CABG surgery (surgical variables) contained the following: surgical expenditure (X4) was 547,037 ± 436,611 (thousand NTD$), length of stay (X5) was 20.30 ± 12.02 days, blood transfusion (X6) was 7.94 ± 9.29 bags, mechanical ventilation use (X7) was 4.67 ± 15.55 days, the average of anastomosis was 2.40 ± 0.80 vessels (X8) and the average of PCI vessels (X9) was 1.19 ± 0.44.

Table 2. Demographic data of new CABG patients in NHIRD from 2014 to 2017.

	Variables	Mean ± SD
Y	One-year medical expenditure after discharge (thousand NTD$)	906,693 ± 710,020
Baseline		
X1	Age	63.72 ± 10.65
X2	CHA2DS score	3.29 ± 1.95
X3	CCI score	4.23 ± 2.82
Surgical variables		
X4	Surgical expenditure(thousand NTD$)	547,037 ± 436,611
X5	Length of stay (LOS)	20.30 ± 12.02
X6	Blood transfusion, (Bag)	7.94 ± 9.29
X7	Mechanical ventilation, (Day)	4.67 ± 15.55
X8	Anastomosis vessels	2.40 ± 0.80
X9	The number of PCI vessels	1.19 ± 0.44

Table 2. *Cont.*

	Variables	Mean ± SD
One year before surgery		
X10	Hospitalization	1.02 ± 1.31
X11	ED visits	1.27 ± 0.67
X12	Blood transfusion, (Bag)	4.08 ± 3.89
X13	Mechanical ventilation	4.74 ± 11.44
X14	Medical expenditure (thousand NTD$)	169,699 ± 247,396
X15	The number of HD Dialysis	11.96 ± 5.01
X16	The number of PD Dialysis	10.65 ± 2.91
X17	The number of PCI vessels	1.73 ± 1.13
Comorbidities		**N (%)**
X18	Diabetes mellitus	8142 (62.9)
X19	Hypertension	6370 (49.21)
X20	Hyperlipidemia	10,273 (79.36)
X21	Myocardial infarct	5132 (39.64)
X22	Liver cirrhosis	367 (2.84)
X23	Congestive heart failure	6687 (51.66)
X24	Coronary artery disease	12,047 (93.06)
X25	Peripheral vascular disease	2977 (23)
X26	Acute pancreatitis	432 (3.34)
X27	Malignant dysrhythmia	763 (5.89)
X28	Atrial fibrillation	1366 (10.55)
X29	Transient ischemic attack	4139 (31.97)
X30	Chronic kidney disease	3812 (29.45)
X31	Acute coronary syndrome	7384 (57.04)
X32	Chronic obstructive pulmonary disease	5036 (38.9)
X33	Stroke	4125 (31.87)
X34	Cancer	838 (6.47)
X35	Acute kidney failure	1514 (11.7)
X36	Major bleeding	3019 (23.32)
X37	Intracranial bleeding	357 (2.76)
X38	End stage renal disease	830 (6.41)
X39	Renal disease	3731 (23.82)
Baseline		
X40	**Gender**	
	Male	9917 (76.61)
	Female	3028 (23.39)
X41	Major illness	2108 (16.28)
Hospital Variables		
X42	**Hospital Area Type**	
	Central	1958 (15.13)
	Northern	8039 (62.10)
	Southern	2659 (20.54)
	Eastern	289 (2.23)

Table 2. *Cont.*

	Variables	Mean ± SD
X43	**Hospital ownership**	
	Public	4558 (35.21)
	Private	8387 (64.79)
X44	**Hospital accreditation**	
	Medical center	8012 (61.89)
	Non-medcial center	4933 (38.11)

Abbreviations: CCIS: Charlson comorbidity index score; SD: standard deviation; ED: emergency department; MI: myocardial infarct; CHF: congestive heart failure; CAD: coronary artery disease; PVD: peripheral vascular disease; AF: atrial fibrillation; TIA: transient ischemic attack; CKD: chronic kidney disease; ACS: acute coronary syndrome; COPD: chronic obstructive pulmonary disease; AKF: acute kidney failure; DM: diabetes mellitus; ESRD: end-stage renal disease.

The variables about one year before surgery were the average of hospitalization (X10), emergency department visits (X11; 1.27 ± 0.67), blood transfusion (X12; 4.08 ± 3.89 bags), mechanical ventilation (X13; 4.74 ± 11.44 days), medical expenditure (X14; 169,699 ± 247,396 thousand NTD$), hemodialysis (X15; 11.96 ± 5.01), peritoneal dialysis (X16; 10.65 ± 2.91), and 1.73 ± 1.13 PCI vessels (X17).

The comorbidities variables included the following: X18 diabetes mellitus (DM; 62.9%), X19 hypertension (49.21%), X20 hyperlipidemia (79.36%), X21 myocardial infarct (MI; 39.64%), X22 liver cirrhosis (2.84%), X23 congestive heart failure (CHF; 51.66%), X24 coronary artery disease (CAD; 93.06%), X25 peripheral vascular disease (PVD; 23%), X26 acute pancreatitis (3.34%), X27 malignant dysrhythmia (5.89%), X28 atrial fibrillation (10.55%), X29 transient ischemic attack (TIA; 31.97%), X30 chronic kidney disease (CKD; 29.45%), X31 acute coronary syndrome (ACS; 57.04%), X32 chronic obstructive pulmonary disease (COPD; 38.9%), X33 stroke (31.87%), X34 cancer (6.47%), X35 acute kidney failure (AKF; 11.7%), X36 major bleeding (23.32%), X37 intracranial bleeding (2.76%), X38 end-stage renal disease (ESRD; 6.41%) and X39 renal disease (28.82%).

X42 to X44 were hospital variables. The hospital area type (X42) was 15.13%, 62.10%, 20.54%, and 2.23% in central, northern, southern, and eastern, respectively. X43, different hospital ownership was 35.21% in public and 64.79 in private hospitals. Hospital accreditation (X44) was 61.89% in a medical center, and the non-medical center was 38.11%.

3.2. The Ranking Number of Feature Selection on CABG

After feature selection, we ranked the importance of each variable among different machine learning models that can provide helpful information for model building. Every algorithm has a different calculation. Thus, the variables selected were also different. For example, to determine the relative risk factors about the one-year medical expense after discharge, each important variable could provide helpful information through different feature selection methods. Huang et al. [5] point out that using fewer features was more efficient in model building.

This research used 44 variables [4,5,7,26–31], which depended on the physician's clinical experience and literature review. Moreover, it used five different machine learning methods to predict after filtering factors, the highest score (10 points) was the most crucial factor, which will be the first on the rank; on the other hand, the lowest predictor was ranked the last (1 point). We listed the ranking degree and average in each variable in the following Table 3.

Table 3. Importance ranking for each predictor of medical expense, by using five different machine learning methods.

	Variables	LR	SVR	CART	MARS	XGBoost	Average
X1	Age	1	0	0	0	5	1.2
X2	CHA2DS score	0	1	2	0	1	0.8
X3	CCI score	0	6	4	0	0	2
X30	Chronic kidney disease	0	7	6	0	8	4.2
X35	AKF	0	2	0	0	0	0.4
X38	ESRD	2	3	1	0	3	1.8
X39	Renal Disease	0	5	5	0	0	2
X44	Major illness	3	4	0	0	0	1.4
	Surgical variables						
X4	Surgical expenditure	10	10	10	9	10	9.8
X6	Blood transfusion	0	0	3	0	2	1
X7	Mechanical ventilation	0	0	7	0	6	2.6
	One year before surgery						
X12	Blood transfusion	4	0	0	0	4	1.6
X13	Mechanical ventilation	5	0	0	0	0	1
X14	Medical expenditure	8	9	9	10	9	9
X15	The number of HD Dialysis	9	8	8	8	7	8
X16	The number of PD Dialysis	6	0	0	0	0	1.2
X17	The number of PCI vessels	7	0	0	0	0	1.4

After screenings and analyses, the variable with a higher score was selected as the predicted value in this research. Through the calculation of different machine learning algorithms, each variable will have a different relative importance rank.

In the LR model, the most crucial variable was the surgical expenditure (X4). The other two variables, HD dialysis (X15) and medical expenditure (X14), were both from one year before surgery. Therefore, the top three essential variables of SVR, CART, and MARS are the same as LR. However, for XGBoost, the top two essential variables are still X4 and X14, and the third most important variable was CKD. Therefore, the essential variable in the LR, CART, SVR, MARS, and XGBoost models was surgical expenditure (X4; average point 9.8 points) and one-year medical expenditure before surgery (X14; average point: 9 points), and the number of HD (X15; average point: 8 points).

In general, we knew these three variables (X4, X14, X15) could affect one-year medical expenditures after discharge in CABG patients.

In order to clarify and simplify the predictors, we averaged the scores in each important variable for more equality, as shown in Figure 3. The result depicts the variables that possibly affect one-year medical expenditure after discharge.

The top five critical variables were surgical expenditure (X4), the one-year factors before surgery, medical expenditure (X14), the number of HD, CKD (X30), and the mechanical ventilation use during the CABG surgery (X7).

3.3. *Performance of 5 Different Prediction Models*

After the feature selection, we performed LR, CART, SVR, MARS, XGBoost prediction models. Then, to identify the lowest value in each indicator, we evaluated the following metrics: MAE, MSE, MASE, MAPE, and MAPE. For example, from the overall results in Table 4, after feature selection by CART and after XGBoost was used to make a prediction, MSE (0.0490) and RMSE (0.2214) were the lowest values.

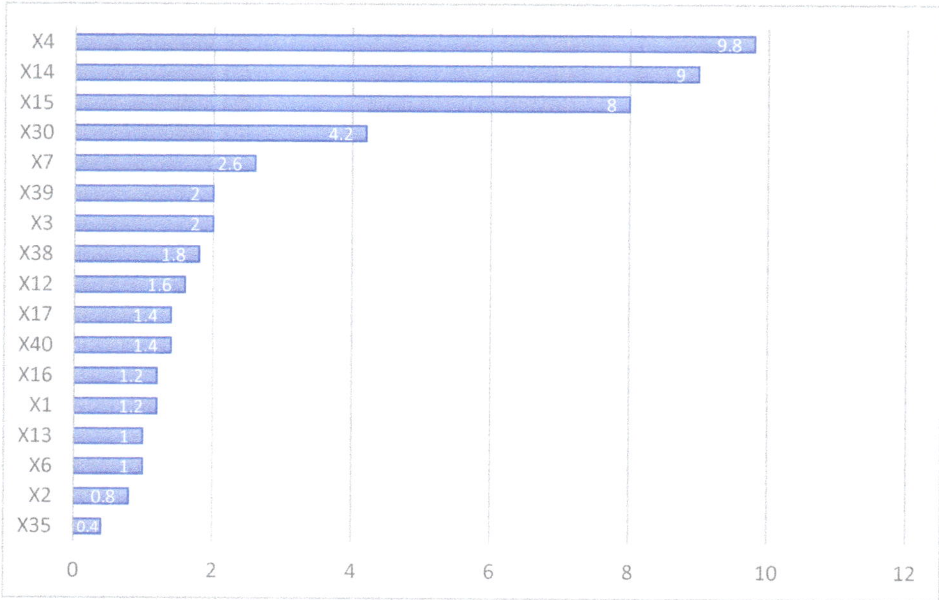

Figure 3. The average score after feature selection using five methods.

Table 4. Performance evaluation of prediction models after feature selection.

FS Methods	ML Method	MAE	MSE	MASE	RMSE	MAPE
LR (10 variables)	LR	0.1965	0.0813	0.3120	0.2851	0.0143
	SVR	0.1381	0.0580	0.2192	0.2407	0.0100
	MARS	0.1663	0.0591	0.2640	0.2431	0.0121
	CART	0.2024	0.0815	0.3214	0.2855	0.0148
	XGBoost	0.1458	0.0491	0.2315	0.2216	0.0106
SVR (10 variables)	LR	0.1987	0.0743	0.3155	0.2725	0.0145
	SVR	0.1345	0.0542	0.2136	0.2328	0.0097
	MARS	0.1652	0.0587	0.2623	0.2422	0.0120
	CART	0.2024	0.0815	0.3214	0.2855	0.0148
	XGBoost	0.1449	0.0491	0.2300	0.2216	0.0105
CART (10 variables)	LR	0.2002	0.0749	0.3178	0.2738	0.0146
	SVR	0.1354	0.0544	0.2149	0.2331	0.0098
	MARS	0.1652	0.0587	0.2623	0.2422	0.0120
	CART	0.2024	0.0815	0.3214	0.2855	0.0148
	XGBoost	0.1433	0.0490	0.2275	0.2214	0.0104
MARS (3variables)	LR	0.2070	0.0794	0.3287	0.2818	0.0151
	SVR	0.1302	0.0532	0.2067	0.2307	0.0094
	MARS	0.1667	0.0593	0.2647	0.2436	0.0121
	CART	0.2024	0.0815	0.3214	0.2855	0.0148
	XGBoost	0.1466	0.0499	0.2328	0.2233	0.0107

Table 4. *Cont.*

FS Methods	ML Method	MAE	MSE	MASE	RMSE	MAPE
XGBoost (10 variables)	LR	0.1985	0.0739	0.3151	0.2719	0.0145
	SVR	0.1344	0.0540	0.2134	0.2324	0.0097
	MARS	0.1652	0.0586	0.2622	0.2420	0.0120
	CART	0.2024	0.0815	0.3214	0.2855	0.0148
	XGBoost	0.1443	0.0492	0.2292	0.2218	0.0105

Abbreviations: LR: linear regression; SVR: support vector regression; CART: classification and regression tree; MARS: multi-variate adaptive regression splines; AUC: area under the curve; XGBoost: extreme gradient boosting; FS: feature selection; ML: machine learning.

We used the variables that were selected by MARS and SVR to build the prediction model. There were three indicators to show the lowest value, namely, MAE (0.1302), MASE (0.2067), and MAPE (0.0094). Thus, MARS only selected three variables and used SVR to make the best predictive model in this research compared to other combined methods.

4. Discussion

NHIRD provides a lot of medical information, and each patient could be traced for a long follow-up time. Therefore, we used NHIRD to make the medical expense prediction. The latest year of the NHIRD database is 2018. Therefore, we selected new CABG surgery patients between 2014 and 2017. The primary purpose of our study was to evaluate which factors could predict the one-year medical expenses after discharge of CABG patients, and build an expense prediction model. Most research discusses mortality, readmission, and the relationship between diseases and surgery [1,4,5,15,28,32–34]. However, only a few studies explored medical expenses, even forecasting. For example, Mehaffey et al. in 2018 [29], analyzed that each additional complication would cause an exponential cost increase. Baciewicz et al. in 2018 [28] referred that because sicker patients needed a high blood transfusion, it led to the increased expense. From the above results, we could know that the baseline variables, including age (X1), CHA2DS score (X2), CCI score (X3), CKD (X30), AKF (X35), ESRD (X38), renal disease (X39), major illness (X44), the variables one year before surgery (total medical expense (X14), blood transfusion (X12), mechanical ventilation use (X13), the number of HD (X15), PD (X16), and PCI vessels (X17)), the surgical variables (surgical expenditure (X4), blood transfusion (X6) and mechanical ventilation use (X7)), all positively influenced one-year medical expense after discharge.

In this study, we used multiple stages to analyze and predict the one-year medical expense after discharge. First, we used the feature selection method to find the essential variables that affect the medical expense. Secondly, after finding out the important variables, we selected five different machine learning models to build a prediction model and evaluate the performance. Besides, through feature selection, we found the folowing several exciting variables: CKD (X30), AKF (X35), ESRD (X38), and renal disease (X39). Although they are all associated with the renal condition, those variables do not have an exceptionally high ranking that is easy to be overlooked, they are topics worthy of further study. For example, Chou et al. [35] in 2014 evaluated that dialysis patients who underwent CABG surgery had better survival than PCI surgery; Chen et al. [36] analyzed that dialysis is associated with higher risk and mortality with CABG patients. Furthermore, Liao et al. [7] found that ESRD patients have a higher medical expense after CABG surgery. From the above results, it could be known that for kidney disease patients who accepted their first CABG surgery, a one-year expense after discharge would be relatively high.

The medical expenditure in preoperative one-year (X4), surgical expense (X14), and the number of HD were the most critical medical expense predictors. Furthermore, after the predictions model was built, we could use the 3 or 10 variables selected by MARS or CART, respectively, to apply SVR and XGBoost methods and achieve a better medical expense prediction model.

5. Conclusions

Our study developed a multiple-stage model to evaluate the one-year medical expense after discharge for those first-time CABG patients. Our model could find that the corresponding operation variables could predict one-year medical expenditure after CABG. Furthermore, postoperative complications will increase the medical expense [28]. In our results, we found that patients with kidney problems, including previous HD, PD, ESRD, renal disease, and CAD, all have a high connection with the forecast medical expenses after CABG surgery. Therefore, hospitals should enhance healthcare management on specific disease prevention, especially the CABG patients with kidney-related diseases.

Our study suggests that the SVR and XGBoost models are an adequate tool to make a medical expense prediction model, through MARS and CART feature selection. The research can bring the benefits of providing the references for medical management with specific diseases that could reduce the expense through effective control, and the government's burdens could also be decreased.

Author Contributions: Conceptualization, Y.-C.H. and S.-J.L.; data curation, M.C.; formal analysis, Y.-C.H., M.C. and T.-S.L.; funding acquisition, T.-S.L.; investigation, Y.-C.H.; methodology, Y.-C.H. and M.C.; project administration, M.C. and T.-S.L.; software, Y.-C.H.; supervision, S.-J.L., M.C. and T.-S.L.; validation, S.-J.L.; writing–original draft, Y.-C.H.; writing–review and editing, Y.-C.H. and S.-J.L. All authors have read and agreed to the published version of the manuscript.

Funding: The Ministry of Science and Technology (MOST-107-2221-E-030-011-).

Institutional Review Board Statement: The study was conducted according to the guidelines of the Declaration of Helsinki and approved by the institutional review board of Fu Jen Catholic University (protocol code C108121 and date of approval 5 March 2020).

Informed Consent Statement: Informed consent was obtained from all subjects involved in the study.

Data Availability Statement: The data presented in this study are not available on request from the corresponding author. Due to the General Data Protection Regulation, the data presented in this research are not publicly available.

Acknowledgments: The authors would like to sincerely thank the editor and reviewers for their kind comments, and appreciate the Ministry of Health and Welfare who provided NHIRD.

Conflicts of Interest: The authors declare no conflict of interest.

References

1. Febriani, V.; Lestari, D.; Mardiyati, S.; Devila, S. Predicting readmission risk after coronary artery bypass graft surgery using logistic regression model. *J. Phys. Conf. Ser.* **2021**, *1725*, 012083. [CrossRef]
2. Hannan, E.L.; Racz, M.J.; Walford, G.; Ryan, T.J.; Isom, O.W.; Bennett, E.; Jones, R.H. Predictors of Readmission for Complications of Coronary Artery Bypass Graft Surgery. *JAMA* **2003**, *290*, 773–780. [CrossRef]
3. Shah, R.M.; Zhang, Q.; Chatterjee, S.; Cheema, F.; Loor, G.; Lemaire, S.A.; Wall, M.J.; Coselli, J.S.; Rosengart, T.K.; Ghanta, R.K. Incidence, Cost, and Risk Factors for Readmission After Coronary Artery Bypass Grafting. *Ann. Thorac. Surg.* **2019**, *107*, 1782–1789. [CrossRef] [PubMed]
4. Zea-Vera, R.; Zhang, Q.; Amin, A.; Shah, R.M.; Chatterjee, S.; Wall, M.J.; Rosengart, T.K.; Ghanta, R.K. Development of a Risk Score to Predict 90-Day Readmission After Coronary Artery Bypass Graft. *Ann. Thorac. Surg.* **2021**, *111*, 488–494. [CrossRef]
5. Huang, Y.-C.; Li, S.-J.; Chen, M.; Lee, T.-S.; Chien, Y.-N. Machine-Learning Techniques for Feature Selection and Prediction of Mortality in Elderly CABG Patients. *Health* **2021**, *9*, 547. [CrossRef]
6. Raza, S.; Sabik, J.F.; Ainkaran, P.; Blackstone, E.H. Coronary artery bypass grafting in diabetics: A growing health care cost crisis. *J. Thorac. Cardiovasc. Surg.* **2015**, *150*, 304–312.e2. [CrossRef]
7. Liao, K.-M.; Kuo, L.-T.; Lu, H.-Y. Hospital costs and prognosis in end-stage renal disease patients receiving coronary artery bypass grafting. *BMC Nephrol.* **2020**, *21*, 333. [CrossRef] [PubMed]
8. Chen, Y.-C.; Yeh, H.-Y.; Wu, J.-C.; Haschler, I.; Chen, T.-J.; Wetter, T. Taiwan's National Health Insurance Research Database: Administrative health care database as study object in bibliometrics. *Science* **2011**, *86*, 365–380. [CrossRef]
9. Tsai, M.-Y.; Hu, W.-L.; Chiang, J.-H.; Huang, Y.-C.; Chen, S.-Y.; Hung, Y.-C.; Chen, Y.-H. Improved medical expenditure and survival with integration of traditional Chinese medicine treatment in patients with heart failure: A nationwide population-based cohort study. *Oncotarget* **2017**, *8*, 90465–90476. [CrossRef] [PubMed]

10. Tian, Y.; Yang, C.; Liu, H. CHA2DS2-VASc score as predictor of ischemic stroke in patients undergoing coronary artery bypass grafting and percutaneous coronary intervention. *Sci. Rep.* **2017**, *7*, 11404. [CrossRef]
11. Yin, L.; Ling, X.; Zhang, Y.; Shen, H.; Min, J.; Xi, W.; Wang, J.; Wang, Z. CHADS2 and CHA2DS2-VASc Scoring Systems for Predicting Atrial Fibrillation following Cardiac Valve Surgery. *PLoS ONE* **2015**, *10*, e0123858. [CrossRef]
12. Yu, L.; Liu, H. Feature Selection for High-Dimensional Data: A Fast Correlation-Based Filter Solution. In Proceedings of the 20th International Conference on Machine Learning (ICML-03), Washington, DC, USA, 21–24 August 2003; pp. 856–863.
13. Chandrashekar, G.; Sahin, F. A survey on feature selection methods. *Comput. Electr. Eng.* **2014**, *40*, 16–28. [CrossRef]
14. Kuo, C.-Y.; Yu, L.-C.; Chen, H.-C.; Chan, C.-L. Comparison of Models for the Prediction of Medical Costs of Spinal Fusion in Taiwan Diagnosis-Related Groups by Machine Learning Algorithms. *Healthc. Informatics Res.* **2018**, *24*, 29–37. [CrossRef]
15. Liu, G.; Zhang, Y.; Zhang, W.; Hu, L.; Lv, T.; Cheng, H.; Hu, Y.; Huang, J. A Risk Prediction Model of Readmission after coronary artery bypass grafting (CABG) in China. *Res. Sq.* **2020**. [CrossRef]
16. Hamdi, T.; Ben Ali, J.; Di Costanzo, V.; Fnaiech, F.; Moreau, E.; Ginoux, J.-M. Accurate prediction of continuous blood glucose based on support vector regression and differential evolution algorithm. *Biocybern. Biomed. Eng.* **2018**, *38*, 362–372. [CrossRef]
17. Friedman, J.H. Multivariate Adaptive Regression Splines. *Ann. Stat.* **1991**, *19*, 1–67. [CrossRef]
18. Lee, T.S.; Dai, W.; Huang, B.L.; Lu, C.J. Data mining techniques for forecasting the medical resource consumption of patients with diabetic nephropathy. *Int. J. Manag. Econ. Soc. Sci.* **2017**, *6*, 293–306.
19. Chen, T.; Guestrin, C. XGBoost: A Scalable Tree Boosting System. In Proceedings of the 22nd ACM SIGKDD International Conference on Knowledge Discovery and Data Mining, San Francisco, CA, USA, 13–17 August 2016; pp. 785–794.
20. Dovgan, E.; Gradišek, A.; Luštrek, M.; Uddin, M.; Nursetyo, A.A.; Annavarajula, S.K.; Li, Y.-C.; Syed-Abdul, S. Using machine learning models to predict the initiation of renal replacement therapy among chronic kidney disease patients. *PLoS ONE* **2020**, *15*, e0233976. [CrossRef] [PubMed]
21. Hudaverdi, T.; Akyildiz, O. Investigation of the site-specific character of blast vibration prediction. *Environ. Earth Sci.* **2017**, *76*, 138. [CrossRef]
22. Popoola, S.I.; Adetiba, E.; Atayero, A.A.; Faruk, N.; Calafate, C.T. Optimal model for path loss predictions using feed-forward neural networks. *Cogent Eng.* **2018**, *5*, 5. [CrossRef]
23. Rodea-Montero, E.R.; Guardado-Mendoza, R.; Rodríguez-Alcántar, B.J.; Rodríguez-Nuñez, J.R.; Núñez-Colín, C.A.; Palacio-Mejía, L.S. Trends, structural changes, and assessment of time series models for forecasting hospital discharge due to death at a Mexican tertiary care hospital. *PLoS ONE* **2021**, *16*, e0248277. [CrossRef] [PubMed]
24. Chen, W.-J.; Jhou, M.-J.; Lee, T.-S.; Lu, C.-J. Hybrid Basketball Game Outcome Prediction Model by Integrating Data Mining Methods for the National Basketball Association. *Entropy* **2021**, *23*, 477. [CrossRef] [PubMed]
25. Juang, W.-C.; Huang, S.-J.; Huang, F.-D.; Cheng, P.-W.; Wann, S.-R. Application of time series analysis in modelling and forecasting emergency department visits in a medical centre in Southern Taiwan. *BMJ Open* **2017**, *7*, e018628. [CrossRef]
26. Lee, T.S.; Li, S.J.; Jiang, Y.; Shia, B.C.; Chen, M. Cost analysis of coronary artery bypass grafting surgery under single-payer reimbursement in Taiwan. *Int. J. Appl. Sci. Eng.* **2020**, *17*, 419–428. [CrossRef]
27. Hyer, J.M.; White, S.; Cloyd, J.; Dillhoff, M.; Tsung, A.; Pawlik, T.M.; Ejaz, A. Can We Improve Prediction of Adverse Surgical Outcomes? Development of a Surgical Complexity Score Using a Novel Machine Learning Technique. *J. Am. Coll. Surg.* **2020**, *230*, 43–52. [CrossRef] [PubMed]
28. Baciewicz, F.A. Show me the money (cost). *J. Thorac. Cardiovasc. Surg.* **2018**, *155*, 883–884. [CrossRef]
29. Mehaffey, J.H.; Hawkins, R.; Byler, M.; Charles, E.J.; Fonner, C.; Kron, I.; Quader, M.; Speir, A.; Rich, J.; Ailawadi, G. Cost of individual complications following coronary artery bypass grafting. *J. Thorac. Cardiovasc. Surg.* **2018**, *155*, 875–882. [CrossRef]
30. Yount, K.W.; Isbell, J.M.; Lichtendahl, C.; Dietch, Z.; Ailawadi, G.; Kron, I.L.; Kern, J.A.; Lau, C.L. Bundled Payments in Cardiac Surgery: Is Risk Adjustment Sufficient to Make It Feasible? *Ann. Thorac. Surg.* **2015**, *100*, 1646–1652. [CrossRef]
31. Riordan, C.J.; Engoren, M.; Zacharias, A.; Schwann, T.A.; Parenteau, G.L.; Durham, S.J.; Habib, R.H. Resource Utilization in Coronary Artery Bypass Operation: Does Surgical Risk Predict Cost? *Ann. Thorac. Surg.* **2000**, *69*, 1092–1097. [CrossRef]
32. Benuzillo, J.; Caine, W.; Ms, R.S.E.; Roberts, C.; Lappe, D.; Doty, J. Predicting readmission risk shortly after admission for CABG surgery. *J. Card. Surg.* **2018**, *33*, 163–170. [CrossRef]
33. Cheng, Y.-T.; Chen, D.-Y.; Wu, V.C.-C.; Chou, A.-H.; Chang, S.-H.; Chu, P.-H.; Chen, S.-W. Effect of Previous Coronary Stenting on Subsequent Coronary Artery Bypass Grafting Outcomes. *J. Thorac. Cardiovasc. Surg.* **2020**. [CrossRef]
34. Alghafees, M.A.; Alsubaie, N.A.; Alsadoon, L.K.; Aljafari, S.A.; Alshehri, E.A.; Suliman, I.F. Thirty-day readmission rates and associated risk factors after coronary artery bypass grafting. *J. Taibah Univ. Med Sci.* **2020**, *15*, 292–297. [CrossRef]
35. Chou, C.-L.; Hsieh, T.-C.; Wang, C.-H.; Hung, T.-H.; Lai, Y.-H.; Chen, Y.-Y.; Lin, Y.-L.; Kuo, C.-H.; Wu, Y.-J.; Fang, T.-C. Long-term Outcomes of Dialysis Patients After Coronary Revascularization: A Population-based Cohort Study in Taiwan. *Arch. Med Res.* **2014**, *45*, 188–194. [CrossRef]
36. Chen, S.-W.; Chang, J.C.-H.; Lin, Y.-S.; Wu, V.C.-C.; Chen, D.-Y.; Tsai, F.-C.; Hung, M.-J.; Chu, P.-H.; Lin, P.-J.; Chen, T.-H. Effect of dialysis dependence and duration on post-coronary artery bypass grafting outcomes in patients with chronic kidney disease: A nationwide cohort study in Asia. *Int. J. Cardiol.* **2016**, *223*, 65–71. [CrossRef]

healthcare

MDPI

Article

Machine-Learning Techniques for Feature Selection and Prediction of Mortality in Elderly CABG Patients

Yen-Chun Huang [1,2], Shao-Jung Li [3,4,5,6,†], Mingchih Chen [1,2,*], Tian-Shyug Lee [1,2,*] and Yu-Ning Chien [2,7]

1. Graduate Institute of Business Administration, College of Management, Fu Jen Catholic University, New Taipei City 24205, Taiwan; hivicky92@gmail.com
2. Artificial Intelligence Development Center, Fu Jen Catholic University, New Taipei City 242062, Taiwan; 151294@mail.fju.edu.tw
3. Cardiovascular Research Center, Wan Fang Hospital, Taipei Medical University, Taipei 242, Taiwan; leeshaojung@gmail.com
4. Taipei Heart Institute, Taipei Medical University, Taipei 242, Taiwan
5. Department of Surgery, School of Medicine, College of Medicine, Taipei Medical University, Taipei 242, Taiwan
6. Division of Cardiovascular Surgery, Department of Surgery, Wan Fang Hospital, Taipei Medical University, Taipei 242, Taiwan
7. Master Program of Big Data Analysis in Biomedicine, College of Medicine, Fu Jen Catholic University, New Taipei City 242062, Taiwan

* Correspondence: 081438@mail.fju.edu.tw (M.C.); 036665@mail.fju.edu.tw (T.-S.L.)

† The author has contributed equally to this work and share first authorship.

Abstract: Coronary artery bypass surgery grafting (CABG) is a commonly efficient treatment for coronary artery disease patients. Even if we know the underlying disease, and advancing age is related to survival, there is no research using the one year before surgery and operation-associated factors as predicting elements. This research used different machine-learning methods to select the features and predict older adults' survival (more than 65 years old). This nationwide population-based cohort study used the National Health Insurance Research Database (NHIRD), the largest and most complete dataset in Taiwan. We extracted the data of older patients who had received their first CABG surgery criteria between January 2008 and December 2009 (n = 3728), and we used five different machine-learning methods to select the features and predict survival rates. The results show that, without variable selection, XGBoost had the best predictive ability. Upon selecting XGBoost and adding the CHA2DS score, acute pancreatitis, and acute kidney failure for further predictive analysis, MARS had the best prediction performance, and it only needed 10 variables. This study's advantages are that it is innovative and useful for clinical decision making, and machine learning could achieve better prediction with fewer variables. If we could predict patients' survival risk before a CABG operation, early prevention and disease management would be possible.

Keywords: National Health Insurance Research Database; NHIRD; older adults; CABG; machine learning; overall survival prediction; feature selection

Citation: Huang, Y.-C.; Li, S.-J.; Chen, M.; Lee, T.-S.; Chien, Y.-N. Machine-Learning Techniques for Feature Selection and Prediction of Mortality in Elderly CABG Patients. *Healthcare* **2021**, *9*, 547. https://doi.org/10.3390/healthcare9050547

Academic Editor: Mahmudur Rahman

Received: 21 March 2021
Accepted: 26 April 2021
Published: 7 May 2021

1. Introduction

Advancing age leads to markedly increasing coronary artery disease (CAD), a common heart disease and the leading global cause of mortality [1], significantly increasing the global healthcare burden [2]. Coronary artery bypass grafting (CABG) is an efficient treatment for patients with CAD in myocardial revascularization [3]. The risk of CABG surgery is approximately 1–3%. CABG is also high-cost surgery [4]. In recent years, various studies evaluated CABG risk on survival rate, medical cost, and follow-up of different CAD treatment strategies [3–8].

However, there is no complete research using an extensive database to build an integral machine-learning model for predicting and evaluating which risk factors could preoperatively affect older adults' survival rate. Thus, this research used the National Health Insurance Research Database (NHIRD), with a sufficiently large data sample of Taiwan, which provided all real and large healthcare data, including patients' original clinical records, treatments, inhospital expenditures, and diagnosis codes. In addition to the patients' basic characteristics and disease history, we used variables before one year and during the operation as predictive indicators. Therefore, if we could predict patients' mortality risk before a CABG operation, take early prevention and disease management for those high-risk patients would be possible. Our studies used multistage selection, which contains feature-searching methods and prediction-model development based on logistic regression (LGR), random forest (RF), classification regression tree (CART), extreme gradient boosting (XGBoost), and multivariate adaptive regression splines (MARS). The model receives as input several preoperative medical factors and their characteristics. To find the correct factors that affect the outcomes and reduce distortion, model performance relies on feature selection (Nguyen, 2010).

There were three purposes of this retrospective population-based study. The first research object was to analyze older adults' survival rate after CABG surgery within a 10-year follow-up. Second, we used different feature-selection methods to investigate which risk factors were crucial variables that could affect survival. Lastly, we aimed to determine the best prediction survival model for older adults receiving CABG procedures, and to identify the associated factors in the prediction model that determine surgery risk factors.

2. Materials and Methods

2.1. Data Source

There are around 23 million people in Taiwan. The National Health Insurance Research Database (NHIRD) enrolls nearly 99% of Taiwanese enrollees in the National Health Insurance (NHI) program [9]. NHIRD contains the personal information of patients who participate in the NHI program, including outpatient and inpatient information, and surgical procedure codes, and it enables the continuous tracking of all claimed records from each patient. The diagnosed codes were International Classification of Diseases, Ninth Revision; Clinical Modification (ICD-9-CM); the Tenth Revision (ICD-10-CM) in Taiwan was fully adopted from 1 January 2016. According to the abovementioned advantages, the NHIRD provides complete and comprehensive long-term follow-up for each patient. Demographic ID information in NHIRD was anonymized and deidentified. This study was exempted from a full ethical review by the Fu Jen Catholic University ethics institutional review board in Taiwan (C108121), and the requirement to obtain informed consent was waived.

2.2. Study Population

To understand the important factors that affect older patients' survival rate after CABG surgery, this retrospective cohort study enrolled patients over 65 years old from 1 January 2008 to 31 December 2009, from the NHIRD, Taiwan. We selected patients who had first undergone CABG operation (the operation code of only one anastomosis vessel is 68023A and 68023B, 68024A and 68024B are 2 vessels, and 68025A and 68025B are 3 diseased vessels). CABG's initial surgery date was used as the index date to ensure that this study focused on older individuals; patients under 65 years old (n = 3533) were excluded. We also excluded those who had had CABG surgery before the index year (between 2002 and 2007; n = 39), had died in the hospital (n = 434), and those with missing information (n = 5). According to these criteria, a total of 4162 patients undergoing CABG surgery were divided into two groups, dead and alive patients $\geq$65 years old, between 1 January 2008 and 31 December 2009 (Figure 1).

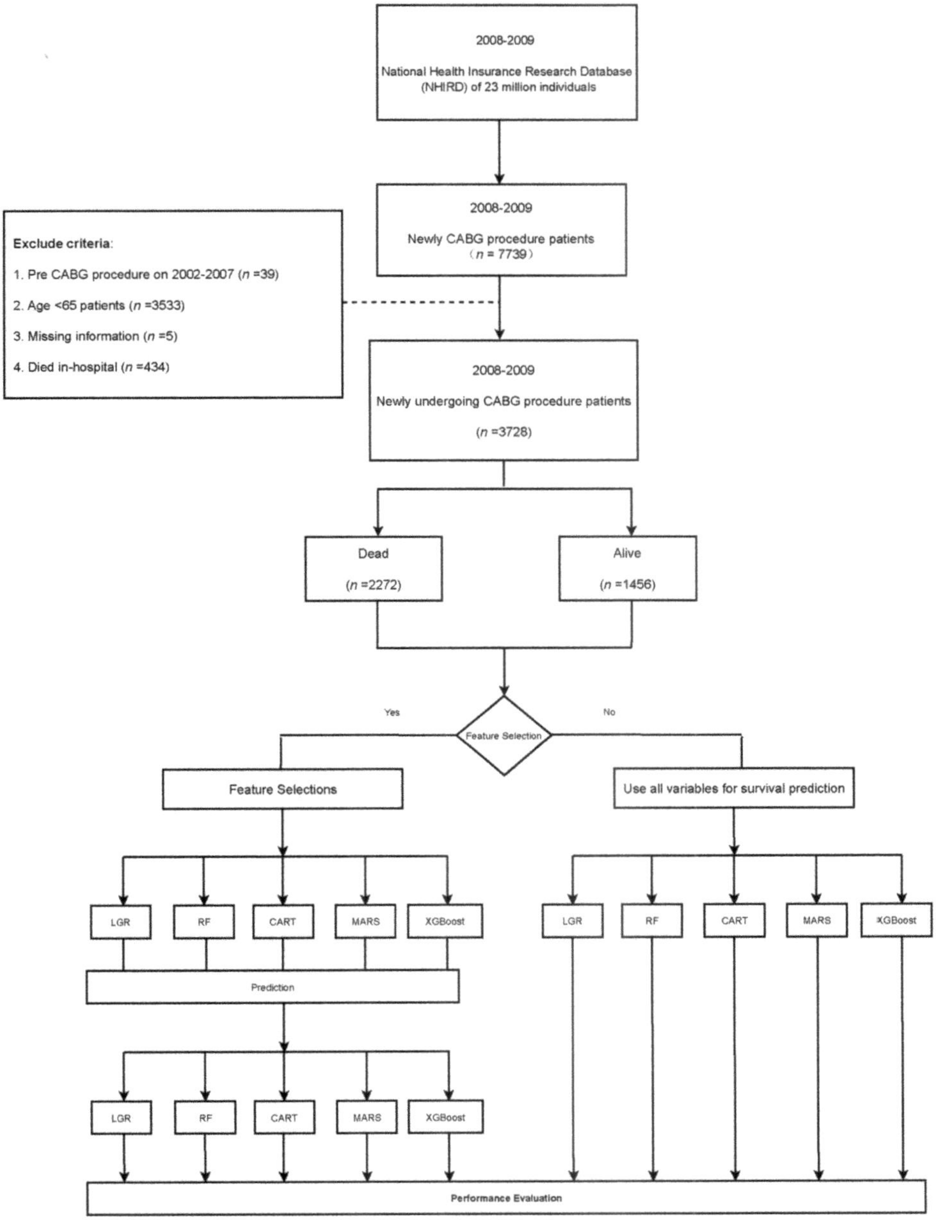

Figure 1. Patient selection and further analysis of 3728 older adult patients who had undergone first-time coronary artery bypass surgery grafting (CABG) between 2008 and 2009.

2.3. Comorbidities and Variable Definitions

In this research, the baseline characteristic variables were sex, Charlson comorbidity index (CCI) score, number of anastomosis vessels, and patient comorbidities (Supplementary Materials) including: hypertension, hyperlipidemia, diabetes mellitus (DM), congestive heart failure (CHF), peripheral vascular disease (PVD), coronary artery disease (CAD), chronic obstructive pulmonary disease (COPD), myocardial infarction (M), chronic kidney disease (CKD), end-stage renal disease (ESRD), and stroke. Blood transfusion (94001C, 94002C, 4013C, 94015C, 94003C), mechanical ventilation (57001B, 57002B, 57003B) in the preoperative one year, and CHA2DS2-VASc score [10,11] were also included. CHA2DS2-VAS was calculated for each research patient using a history of hypertension, diabetes mellitus, congestive heart failure, and vascular disease. Age between 65 and 74 years old, and female gender were 1 point. Two points were assigned for a history of ischemic stroke and transient ischemic attack (ICD-9-CM codes: 433–438; ICD-10-CM: I63.0–9, G45.9) or age $\geq$ 75 years old.

The date of comorbidities was defined as the date before the index date, which could be traced back to 2002–2007. Primary outcomes were overall survival rate of older adults after the CABG procedure, and cause of death was provided by the NHIRD death registry data. Patients in this study were all followed up from the index date until the date of death or the end of the research (31 December 2018).

2.4. Feature-Selection and Machine-Learning Prediction Models

The hospital must update each patient's information every day. After long-term accumulation, much medical information is accumulated. We also used the NHIRD to determine key factors that affect the survival of older adults from the first CABG surgery. The medical records contained numerous items. Therefore, before making predictions, features were reduced through feature selection (FS), an essential preprocessing step [12].

However, models have different abilities to predict survival. Some studies used machine methods for an early diagnosis of bipolar disorder, prostate-cancer-specific survival, erectile dysfunction, CKD, and medical cost [13–17]. This research used multiple-stage selection methods to uncover potential collinearity among variable subsets and evaluate the response variable's predictive performance. After that, we used a fivefold cross-validation process to verify the model of LGR, RF, CART, XGBoost, and MARS (for classification or continuous variables) to compare the predicted performance with all variables and evaluate the classification results after feature selection per classification method [18,19]. The classification model's performance indicators were mean accuracy, kappa, sensitivity, specificity, and area under the ROC curve (AUC). The evaluation performance of the AUC value was defined by Hosmer et al. [17]: AUC $\geq$ 0.9, outstanding discrimination; 0.8 $\leq$ AUC < 0.9, good discrimination; 0.7 $\leq$ AUC < 0.8, acceptable/fair discrimination; 0.6 $\leq$ AUC < 0.7, poor discrimination; and AUC < 0.6, no discrimination [13]. The greater the accuracy, sensitivity, specificity, and kappa values are, the better the model is.

In this research, we used five different machine-learning methods to construct predictive models and conducted the best feature selection for evaluating the mortality of the CABG patients.

2.4.1. LGR

Logistic regression is a classical prediction method suitable for predicting general binary classification problems. The central concept of LGR is the natural logarithm of an odds ratio by logit [20]. It is used to analyze the relationship between dependent and independent variables. The predicted variable Y has only two possibilities: yes (1) and no (0).

2.4.2. RF

Random forest (RF) is an ensemble method, and the classifier in the original RF algorithm is a classification and regression tree (CART) that is based on the bagging algorithm and bootstrap aggregation. It randomly selects variables to split when the CART tree grows [21]. The out-of-bag (OOB) error of random forest is the average error of each weak sample using an approximate test error to measure performance [22]. Lastly, each tree was based on node impurity to improve the amplitude of the random forest and find out the importance of variables.

2.4.3. MARS

MARS is a nonparametric statistical method developed by physicist Friedman et al. (1991) [23]. It is flexible regression processing that can automatically create a criterion model and separate linear-regression slopes to process multiple complex data and establish prediction models.

Approximated nonlinearity is adopted using separate linear-regression slopes in different intervals of the independent variable space. For the best MARS model, the first stage uses a forward algorithm to construct many possible basic functions and corresponding knots to initially overfit the data. We used the generalized cross-validation criterion (GCV) to generate the best combination in the second stage [22].

MARS can also use dummy variables to deal with missing values, and it does not need to assume the distribution of demand functions and errors.

2.4.4. CART

Breiman et al. developed the classification and regression-tree algorithm in 1984 [24]. In the process of the CART algorithm, a series of rules are generated through recursion. First, CART builds a maximal tree to divide the two subsets into left and right through binary splits, and calculates the impurity by using the Gini index under each attribute segmentation. Nodes and leaf nodes start from the root during analysis. The smallest Gini index is used to determine segmented attributes and values. Then, the parent node can divide two exclusive children from each node, and iteratively calculate until the whole decision tree stops growing and is constructed [22].

2.4.5. XGBoost

The algorithm applied by XGBoost is a gradient-boosting decision tree (GBDT) that can be used for both classification and regression problems [25]. The greedy method optimizes the maximal gain of the objective function during the construction of each tree layer. The idea of the algorithm is to continuously add trees and perform feature splitting to grow a tree. Each time a tree is added, it learns a new function to fit the residual of the last prediction.

Lastly, multiple learners are added together to make the final prediction, and the accuracy rate is higher than that of a single one. To solve overfitting, XGBoost controls the complexity of the model by using regularization terms, and objective function optimization uses the second derivative of the Taylor expansion loss function to compute pseudoresiduals [22].

2.5. Statistical Analysis

Both cohorts were stratified into two groups (dead and alive) and compared using Pearson's chi-squared tests for categorical variables. Demographic data at baseline presented numbers and percentages as *n* (%). Independent sample t-tests assessed continuous variables as means and standard deviations (mean ± SD) to compare the difference. All significance thresholds were associated with 2-tailed p values < 0.05. Data extraction was performed using SAS version 9.4 (SAS Institute Inc., Cary, NC, USA). Variable selection and model establishment was carried out with R statistical software (R studio 3.5.1; http://www.r-project.org (accessed on 12 January 2021)).

3. Results

3.1. Demographic Characteristics of Study Population

The demographic data and comorbidities of the patients who accepted their first CABG surgery are listed in Table 1. We included ≥65 year-old adults who had fulfilled the criteria from 1 January 2008, to 31 December 2009, in the Taiwan NHIRD. The dead group was 2272 (69.98%), and the alive group was 1456 (71.09%). In comparison, male patients had higher mortality than that of female patients.

Table 1. Demographic features of older CABG adults in Taiwan from 2008 to 2009.

Variables		≥65 Dead (*n* = 2272)		≥65 Alive (*n* = 1456)		*p*-Value
		n	%	*n*	%	
Sex	Female	682	30.02	421	28.91	0.471
	Male	1590	69.98	1035	71.09	
Age, mean (SD), y		74.30 (5.60)		71.27 (4.78)		<0.001
Follow up years, Mean (SD)		4.42(3.14)		10.05 (0.57)		<0.001
Follow up years, Median		4.22		10.02		-
CHA2DS score, mean (SD)		4.21 (1.67)		3.30 (1.57)		<0.001
Comorbidities						
DM		1477	65.01	739	50.76	<0.0001
Hypertension		624	27.46	379	26.03	0.335
Hyperlipidemia		1522	66.99	1056	72.53	<0.001
MI		1182	52.02	560	38.46	<0.001
Liver cirrhosis		50	2.2	10	0.69	<0.001
CHF		1385	60.96	563	38.67	<0.001
CAD		2222	97.8	1435	98.56	0.098
PVD		541	23.81	248	17.03	<0.0001
Acute pancreatitis		43	1.89	21	1.44	0.301
Malignant dysrhythmia		104	4.58	58	3.98	0.385
Intracranial bleeding		53	2.33	14	0.96	0.002
AF		348	15.32	159	10.92	<0.001
TIA		951	41.86	424	29.12	<0.0001
CKD		572	25.18	129	8.86	<0.0001
ACS		1490	65.58	810	55.63	<0.0001
COPD		1043	45.91	558	38.32	<0.0001
Stroke		947	41.68	423	29.05	<0.0001
Cancer		164	7.22	66	4.53	<0.001
CCIS scores	0	75	3.3	139	9.55	<0.0001
	1	269	11.84	330	22.66	
	2	383	16.86	362	24.86	
	3	424	18.66	239	16.41	
	4	341	15.01	165	11.33	
	5	275	12.1	115	7.9	
	6+	505	22.23	106	7.28	
Mean (SD)		3.86 (2.40)		2.59 (1.93)		<0.0001
Surgical Variables						
Anastomosis vessels, mean (SD)		2.64 (0.72)		2.79 (0.77)		<0.001
Length of stay (LOS), mean (SD)		25.59 (14.77)		18.29 (9.15)		<0.001
Blood transfusion, (Bag), mean (SD)		10.89 (14.68)		7.23 (5.31)		<0.001
Mechanical ventilation, (Day), mean (SD)		7.16 (13.90)		2.76 (3.09)		<0.001
Surgical cost		611,701 (488,753)		394,843 (165,389)		<0.001
One Year Before Surgery						
Outpatient visits, mean (SD)		37.70 (23.34)		32.36 (20.13)		<0.001
Hospitalization, mean (SD)		1.91 (1.34)		1.45 (0.82)		<0.001
ED visits, mean (SD)		58	2.55	14	0.96	<0.001
Blood transfusion, (Bag), mean (SD)		3.83 (3.69)		4.09 (4.87)		0.636
Mechanical ventilation, (Day), mean (SD)		5.55 (13.48)		3.93 (4.05)		0.373
Medical cost (related cardiology department), mean (SD) (thousand NT$)		81,957 (107,098)		60,969 (80,674)		<0.0001
Medical cost (thousand NT$)		155,186 (197087)		91,439 (98,235)		<0.0001

CCIS = Charlson comorbidity index score; SD: standard deviation; ED: Emergency departmen; MI: Myocardial infarct; CHF: Congestive heart failure; CAD: Coronary artery disease; PVD: Peripheral vascular disease; AF: Atrial fibrillation; TIA: Transient ischemic attack; CKD: Chronic kidney disease; ACS: Acute coronary syndrome; COPD: Chronic obstructive pulmonary disease ; AKF: Acute kidney failure ; DM: Diabetes mellitus.

Statistically significant results were demonstrated for the dead and alive groups. The mean follow-up periods were 4.42 ± 3.14 and 10.05 ± 0.57 years ($p < 0.001$), respectively, and the other data were as follows, as described in the brackets: CHA2DS score (4.21 ± 1.67 vs. 3.30 ± 1.57, $p < 0.001$), diabetes (65.01 vs. 50.76, $p < 0.001$), myocardial infarction (52.02 vs. 38.46, $p < 0.001$), liver cirrhosis (2.2 vs. 0.69, $p < 0.001$), peripheral vascular disease (PVD; 23.81 vs. 17.03, $p < 0.001$), congestive heart failure (CHF; 60.96 vs. 38.57, $p < 0.001$), intracranial bleeding (2.33 vs. 0.96, $p = 0.002$), atrial fibrillation (AF; 15.32 vs. 10.92, $p < 0.001$), transient ischemic attack (TIA; 41.86 vs. 29.12, $p \leq 0.001$), chronic kidney disease (CKD; 25.18 vs. 8.86, $p \leq 0.001$), acute coronary syndrome (ACS; 65.58 vs. 55.63, $p < 0.001$), chronic obstructive pulmonary disease (COPD; 45.91 vs. 38.32, $p < 0.001$), stroke (41.68 vs. 29.05, $p < 0.001$), cancer (7.22 vs. 4.53, $p < 0.001$) and CCI scores (3.86 ± 2.40 vs. 2.59 ± 1.93, $p < 0.001$).

The surgical variables were significantly different in terms of cost (TWD 611,701 ± 488,753 vs. TWD 394,843 ± 165,389, $p < 0.001$), the average diameter of anastomosis vessels (2.64 ± 0.72 vs. 2.79 ± 0.77, $p < 0.0001$), the length of stay (25.59 ± 14.77 vs. 18.29 ± 9.15, $p < 0.001$), blood transfusion (10.89 ± 14.68 vs. 7.23 ± 5.31, $p < 0.001$), and mechanical ventilation (7.16 ± 13.90 vs. 2.76 ± 3.09, $p < 0.001$). In addition, variables of 1 year before surgery, such as the mean number of outpatient department visits (37.70 ± 23.34 vs. 32.36 ± 20.13, $p < 0.001$), emergency department visits (2.55 vs. 0 96, $p = 0.0006$), hospitalization visits (1.91 ± 1.34 vs. 1.45 ± 0.82, $p < 0.0001$), the mean bag of blood transfusion (13.34 vs. 4.60, $p = 0.0006$), the length of mechanical ventilation (11.09 vs. 3.85, $p < 0.001$), and medical cost (155,186 ± 197,087 vs. 91,439 ± 98,235, $p < 0.001$), were also statistically significantly different between the dead and alive groups of older adults who had undergone first CABG surgery.

3.2. Results of Feature Selection on CABG

To determine which risk factors could predict survival among older CABG patients, we used different feature-selection methods to determine them. Ranking first was the most important. A total of 72 variables were included in this study, and each variable had its ranking in 5 different methods after filtering (Table 2)—the studied characteristics included surgical, recent 1-year variables, and the patient's baseline. LGR selected 17 variables. RF selected a total of 11 variables. CART chose nine variables. XGBoost and MARS both selected seven variables. Among those methods, LOS, CHA2DS2 score, and CKD were only selected by CART. CART, XGBoost, and MARS all selected the risk factors of surgical cost, patient's age, renal disease, and CCI score as essential variables.

Table 2. Ranking of essential variables of older CABG adults.

Variables	LGR (17 Variables)	RF (11 Variables)	CART (9 Variables)	MARS (7 Variables)	XGBoost (7 Variables)
Surgical Variables					
Blood transfusion, (Bag), mean	1				
Length of stay (LOS), mean			4		
Surgical cost			3	1	1
One Year Before Surgery					
ED visits, mean	4	6			
Outpatient visits, mean	15				
Hospitalization, mean		3			
Mechanical ventilation, (Day), mean	16	7			7
Blood transfusion, (Bag), mean		1			
Medical cost			8		6

Table 2. *Cont.*

Variables	LGR (17 Variables)	RF (11 Variables)	CART (9 Variables)	MARS (7 Variables)	XGBoost (7 Variables)
		Baseline			
Age		11	5	3	2
CHF	7	4		6	5
CKD			7		
ACS	12				
CAD	2				
CCI score			9	2	3
COPD	11				
PVD	14				
Diabetes mellitus		5		5	
Renal disease			1	4	4
Major illness	8				
Ischemic stroke	3				
CHA2DS2 scores			2		
Ulcer disease	17			7	
Hypertension	6				
Hyperlipidemia		2			
AKF	13				
Acute pancreatitis	10				
Connective tissue disease	9	8			
Moderate or severe renal disease	5	9	6		
Moderate or severe liver disease		10			

Through different variable-selection algorithm methods, we could make predictions with these variable combinations.

3.3. Performance of Different Prediction Models

Lastly, we used the results of different feature-selection methods and nonfeature selection to produce five different prediction models: LGR, RF, CART, MARS, and XGBoost. In order to predict survival, the ability of each model was an independent validation dataset. The results showed that, without variable selection (72 variables), the predictive ability of XGBoost was the best (accuracy: 0.7225) among the five models (as shown in Table 3). LGR, RF, and CART individually used 17,119 variables. XGBoost had the best predictive ability (accuracy: 0.7131) and only required seven variables. The best forecasting ability among these five methods was logistic regression (accuracy: 0.7184). We also added three risk factors to the variable selections of XGBoost and MARS—CHA2DS score, acute pancreatitis, and AKF—for further predictive analysis. Adding these three variables can improve the ability of prediction models. Overall, the feature-selection method opted for XGBoost, with surgical cost, CCI scores, age, renal disease, diabetes, CHF, ulcer disease, and three risk factors (AKF, acute pancreatitis, and CHA2DS2-VAS score). The average accuracy for MARS was 0.7225; MARS was ranked as the best and only needed ten variables.

Table 3. Performance evaluation of prediction models on nonselection and after feature selection.

	Method	Accuracy	Kappa	Sensitivity	Specificity	AUC
Overall (72 variables)	LGR	0.7198	0.4427	0.6711	0.7939	0.7926
	RF	0.7077	0.3965	0.7355	0.6655	0.7784
	MARS	0.7104	0.4294	0.6444	0.8108	0.7890
	CART	0.6930	0.3360	0.8111	0.5135	0.7031
	XGBoost	**0.7225**	0.4394	0.7044	0.7500	0.7934

Table 3. *Cont.*

	Method	Accuracy	Kappa	Sensitivity	Specificity	AUC
LGR selection (17 variables)	LGR	0.6179	0.2752	0.4888	0.8141	0.6981
	RF	**0.6260**	0.2829	0.5177	0.7905	0.6912
	MARS	0.6219	0.2771	0.5088	0.7939	0.6917
	CART	0.5911	0.2292	0.4533	0.8006	0.6576
	XGBoost	0.6246	0.2845	0.5044	0.8074	0.6977
RF selection (11 variables)	LGR	0.6876	0.3960	0.5866	0.8412	0.7784
	RF	0.6916	0.3937	0.6244	0.7939	0.7637
	MARS	0.6890	0.3817	0.6444	0.7567	0.7675
	CART	0.6930	0.3360	0.8111	0.5135	0.7031
	XGBoost	**0.6983**	0.4161	0.5977	0.8513	0.7790
CART selection (9 variables)	LGR	0.7091	0.4009	0.7311	0.6756	0.7624
	RF	0.6554	0.3464	0.5200	0.8614	0.7557
	MARS	0.7091	0.3954	0.7488	0.6486	0.7653
	CART	0.6930	0.3360	0.8111	0.5135	0.7031
	XGBoost	**0.7131**	0.4062	0.7444	0.6655	0.7652
MARS selection (7 variables)	LGR	0.6876	0.3960	0.5866	0.8412	0.7784
	RF	0.6916	0.3937	0.6244	0.7939	0.7637
	MARS	0.6890	0.3817	0.6444	0.7567	0.7675
	CART	0.6930	0.3360	0.8111	0.5135	0.7031
	XGBoost	**0.6983**	0.4161	0.5977	0.8513	0.7790
XGBoost selection (7 variables)	LGR	**0.7184**	0.4186	0.7444	0.6790	0.7739
	RF	0.6903	0.3800	0.6600	0.7354	0.7453
	MARS	0.7131	0.4096	0.7333	0.6824	0.7683
	CART	0.6930	0.3360	0.8111	0.5135	0.7031
	XGBoost	0.7104	0.4212	0.6733	0.7658	0.7763
XGBoost selection and 3 risk factors (10 variables)	LGR	0.6890	0.3937	0.6044	0.8175	0.7807
	RF	0.7037	0.4008	0.6911	0.7229	0.7727
	MARS	**0.7225**	0.4233	0.7600	0.6655	0.7831
	CART	0.6930	0.3360	0.8111	0.5135	0.7031
	XGBoost	0.6970	0.4069	0.6200	0.8141	0.7845
MARS selection and 3 risk factors (10 variables)	LGR	0.6916	0.3964	0.6155	0.8074	0.7780
	RF	0.6836	0.3806	0.6088	0.7972	0.7629
	MARS	0.7024	0.3998	0.6844	0.7297	0.7722
	CART	0.6930	0.3360	0.8111	0.5135	0.7031
	XGBoost	**0.7077**	0.4190	0.6600	0.7804	0.7806

Abbreviations: LGR: logistic regression; RF: random forest; CART: classification and regression tree; MARS: multivariate adaptive regression splines; AUC: area under the curve; XGBoost: extreme gradient boosting.

4. Discussion

This population-based cohort study was based on NHIRD, which is the largest observational database from Taiwan. The strengths of using NHIRD are as follows: (1) it included various individual medical information; (2) each patient could be tracked for a long-term follow-up; (3) it could show current diagnostic and therapeutic modes in the real world. The purpose of the research was to find the risk factors that could predict survival rates with different combinations of feature-selection methods and prediction models. We evaluated the survival to discharge and risks factors of older adults after the first CABG from 2008 to 2009 and followed up to 10 years. Our study showed that, without variable selection, XGBoost had the best predictive ability. By selecting XGBoost and adding the CHA2DS score, acute pancreatitis, and acute kidney failure for further predictive analysis, MARS had the best prediction performance and only needed 10 variables.

Previously, most studies focused on chronic or vascular diseases that had been acquired before the CABG surgery [26]. No known study investigated using preoperative and perioperative variables as predictor factors for long-term survival probability. A previous history of DM and CKD is a decisive risk factor for cardiovascular diseases, such as CAD and CHF. In part, most are contributed from aging [5,27,28], MI, AF, chronic renal failure, abnormal renal function, and renal failure have higher mortality after CABG [6,26,29–31]. Liu et al. found that $\geq$65 age, the female sex, diabetes, congenital heart disease, hypertension on Levels 2 and 3, and using private insurance contributed to a higher risk of readmission [1]. The score of CHA2DS2-VASc was employed as a risk-measurement tool; it was recorded in treatment guidelines for stroke prevention and is a factor for predicting stroke. Tian et al. suggest that CHA2DS2-VASc score should be on the clinical application [10]. This study demonstrated two significant findings: first, preoperative 1-year and perioperative variables are significant predictors. Second, after applying machine-learning variable screening and prediction methods, it is clearer to identify which variables could affect survival. Furthermore, we could also use fewer factors to achieve good predictive ability. Our study's limitations are the lack of clinical lab data, such as family history, and detailed health-check values.

5. Conclusions

On the basis of our research, we developed multiple-stage frameworks to build a survival model for predicting the mortality of older adults who had undergone their first CABG. The advantages of this study are that it is innovative and practical in clinical research. Furthermore, we could achieve better prediction with only 10 variables. This could help clinicians make decisions more quickly and encourage patients towards earlier healthcare management.

Supplementary Materials: The following are available online at https://www.mdpi.com/article/10.3390/healthcare9050547/s1, ICD-9-CM and ICD-10-CM codes used for diagnosis in this study.

Author Contributions: Conceptualization, T.-S.L., S.-J.L. and M.C.; data curation, Y.-C.H.; formal analysis, Y.-C.H.; methodology, Y.-C.H. and M.C.; project administration, T.-S.L., S.-J.L. and M.C.; software, Y.-C.H.; supervision, T.-S.L.; validation, T.-S.L., S.-J.L. and M.C.; writing—original draft, Y.-C.H.; writing—review and editing, Y.-C.H., S.-J.L. and Y.-N.C. All authors have read and agreed to the published version of the manuscript.

Funding: This research received no external funding.

Institutional Review Board Statement: The study was conducted according to the guidelines of the Declaration of Helsinki, and approved by the Institutional Review Board of Fu Jen Catholic University (protocol code C108121; date of approval, 5 March 2020).

Informed Consent Statement: Informed consent was obtained from all subjects involved in the study.

Data Availability Statement: Data presented in this study are not available on request from the corresponding author. Due to the General Data Protection Regulation, the data presented in this research are not publicly available.

Acknowledgments: The authors would like to thank the editor and the reviewers for their valuable comments. The authors sincerely appreciate NHIRD, which was provided by the Ministry of Health and Welfare.

Conflicts of Interest: The authors declare no conflict of interest.

References

1. Liu, G.; Zhang, Y.; Zhang, W.; Hu, L.; Lv, T.; Cheng, H.; Hu, Y.; Huang, J. Risk Prediction Model of Readmission after Coronary Artery Bypass Grafting (CABG) in China. *Res. Sq.* **2020**. [CrossRef]
2. Malmberg, M.; Gunn, J.; Rautava, P.; Sipilä, J.; Kytö, V. Outcome of Acute Myocardial Infarction Versus Stable Coronary Artery Disease Patients Treated with Coronary Bypass Surgery. *Ann. Med.* **2021**, *53*, 70–77. [CrossRef] [PubMed]
3. Chang, Y.-C.; Chiang, J.-H.; Lay, I.-S.; Lee, Y.-C. Increased Risk of Coronary Artery Disease in People with a Previous Diagnosis of Carpal Tunnel Syndrome: A Nationwide Retrospective Population-Based Case-Control Study. *BioMed Res. Int.* **2019**, *2019*, 1–8. [CrossRef] [PubMed]
4. Lee, T.-S.; Li, S.-J.; Jiang, Y.; Shia, B.-C.; Chen, M. Cost Analysis of Coronary Artery Bypass Grafting Surgery under Single-Payer Reimbursement in Taiwan. *Int. J. Appl. Sci. Eng.* **2020**, *17*, 419–428. [CrossRef]

5. Chen, S.-W.; Chang, C.-H.; Lin, Y.-S.; Wu, V.C.-C.; Chen, D.-Y.; Tsai, F.-C.; Hung, M.-J.; Chu, P.-H.; Lin, P.-J.; Chen, T.-H. Effect of Dialysis Dependence and Duration on Post-Coronary Artery Bypass Grafting Outcomes in Patients with Chronic Kidney Disease: A Nationwide Cohort Study in Asia. *Int. J. Cardiol.* **2016**, *223*, 65–71. [CrossRef] [PubMed]
6. Chou, C.-L.; Hsieh, T.-C.; Wang, C.-H.; Hung, T.-H.; Lai, Y.-H.; Chen, Y.-Y.; Lin, Y.-L.; Kuo, C.-H.; Wu, Y.-J.; Fang, T.-C. Long-term Outcomes of Dialysis Patients After Coronary Revascularization: A Population-based Cohort Study in Taiwan. *Arch. Med. Res.* **2014**, *45*, 188–194. [CrossRef]
7. Milojevic, M.; Head, S.J.; Parasca, C.A.; Serruys, P.W.; Mohr, F.W.; Morice, M.-C.; Mack, M.J.; Ståhle, E.; Feldman, T.E.; Dawkins, K.D.; et al. Causes of Death Following PCI Versus CABG in Complex CAD. *J. Am. Coll. Cardiol.* **2016**, *67*, 42–55. [CrossRef]
8. Zhang, Z.; Kolm, P.; Grau-Sepulveda, M.V.; Ponirakis, A.; O'Brien, S.M.; Klein, L.W.; Shaw, R.E.; McKay, C.; Shahian, D.M.; Grover, F.L.; et al. Cost-Effectiveness of Revascularization Strategies. *J. Am. Coll. Cardiol.* **2015**, *65*, 1–11. [CrossRef]
9. Kuo, C.-S.; Lu, C.-W.; Chang, Y.-K.; Yang, K.-C.; Hung, S.-H.; Yang, M.-C.; Chang, H.-H.; Huang, C.-T.; Hsu, C.-C.; Huang, K.-C. Effectiveness of 23-Valent Pneumococcal Polysaccharide Vaccine on Diabetic Elderly. *Medicine* **2016**, *95*, e4064. [CrossRef]
10. Tian, Y.; Yang, C.; Liu, H. CHA2DS2-VASc Score as Predictor of Ischemic Stroke in Patients Undergoing Coronary Artery Bypass Grafting and Percutaneous Coronary Intervention. *Sci. Rep.* **2017**, *7*, 1–7. [CrossRef]
11. Yin, L.; Ling, X.; Zhang, Y.; Shen, H.; Min, J.; Xi, W.; Wang, J.; Wang, Z. CHADS2 and CHA2DS2-VASc Scoring Systems for Predicting Atrial Fibrillation following Cardiac Valve Surgery. *PLoS ONE* **2015**, *10*, e0123858. [CrossRef]
12. Nguyen, H.T.; Petrović, S.; Franke, K. A Comparison of Feature-Selection Methods For intrusion Detection. In Proceedings of the International Conference on Mathematical Methods, Models, and Architectures for Computer Network Security, St. Petersburg, Russia, 8–10 September 2010; pp. 242–255.
13. Hu, Y.-H.; Chen, K.; Chang, I.-C.; Shen, C.-C. Critical Predictors for the Early Detection of Conversion from Unipolar Major Depressive Disorder to Bipolar Disorder: Nationwide Population-Based Retrospective Cohort Study. *JMIR Med. Inform.* **2020**, *8*, e14278. [CrossRef]
14. Lin, Y.-T.; Lee, M.T.-S.; Huang, Y.-C.; Liu, C.-K.; Li, Y.-T.; Chen, M. Prediction of Recurrence-Associated Death from Localized Prostate Cancer with a Charlson Comorbidity Index–Reinforced Machine Learning Model. *Open Med.* **2019**, *14*, 593–606. [CrossRef]
15. Chen, Y.-F.; Lin, C.-S.; Hong, C.-F.; Lee, D.-J.; Sun, C.; Lin, H.-H. Design of a Clinical Decision Support System for Predicting Erectile Dysfunction in Men Using NHIRD Dataset. *IEEE J. Biomed. Health Inform.* **2018**, *23*, 2127–2137. [CrossRef]
16. Krishnamurthy, S.; Kapeleshh, K.S.; Dovgan, E.; Luštrek, M.; Gradišek Piletič, B.; Srinivasan, K.; Li, Y.-C.; Gradišek, A.; Syed-Abdul, S. Machine Learning Prediction Models for Chronic Kidney Disease using National Health Insurance Claim Data in Taiwan. *medRxiv* **2020**. [CrossRef]
17. Hosmer, J.D.W.; Lemeshow, S.; Sturdivant, R.X. *Applied Logistic Regression*; John Wiley & Sons: Hoboken, NJ, USA, 2013.
18. Almustafa, K.M. Prediction of Heart Disease and Classifiers' Sensitivity Analysis. *BMC Bioinform.* **2020**, *21*, 1–18. [CrossRef]
19. Austin, P.C.; Ghali, W.A.; Tu, J.V. A Comparison of Several Regression Models for Analysing Cost of CABG Surgery. *Stat. Med.* **2003**, *22*, 2799–2815. [CrossRef]
20. Peng, C.-Y.J.; Lee, K.L.; Ingersoll, G.M. An Introduction to Logistic Regression Analysis and Reporting. *J. Educ. Res.* **2002**, *96*, 3–14. [CrossRef]
21. Breiman, L. Random Forests. *Mach. Learn.* **2001**, *45*, 5–32. [CrossRef]
22. Wu, T.-E.; Chen, H.-A.; Jhou, M.-J.; Chen, Y.-N.; Chang, T.-J.; Lu, C.-J. Evaluating the Effect of Topical Atropine Use for Myopia Control on Intraocular Pressure by Using Machine Learning. *J. Clin. Med.* **2020**, *10*, 111. [CrossRef]
23. Friedman, J.H. Multivariate Adaptive Regression Splines. *Ann. Stat.* **1991**, *19*, 1–67. [CrossRef]
24. Breiman, L.F.; Friedman, J.H.; Olshen, R.A.; Stone, C.J. *Classification and Regression Trees*; Chapman and Hall Pacific Grove, CA, USA, 1984.
25. Chen, T.; Guestrin, C. XGBoost: A Scalable Tree Boosting System. In Proceedings of the 22nd ACM SIGKDD International Conference on Knowledge Discovery and Data Mining, San Francisco, CA, USA,, 13–17 August 2016; pp. 785–794.
26. Carr, B.M.; Romeiser, J.; Ruan, J.; Gupta, S.; Seifert, F.C.; Zhu, W.; Shroyer, A.L. Long-Term Post-CABG Survival: Performance of Clinical Risk Models Versus Actuarial Predictions. *J. Card. Surg.* **2015**, *31*, 23–30. [CrossRef] [PubMed]
27. Feng, W.-H.; Chu, C.-Y.; Hsu, P.-C.; Lee, W.-H.; Su, H.-M.; Lin, T.-H.; Yen, H.-W.; Voon, W.-C.; Lai, W.-T.; Sheu, S.-H. The Effects of Secondary Prevention after Coronary Revascularization in Taiwan. *PLoS ONE* **2019**, *14*, e0215811. [CrossRef]
28. Raza, S.; Sabik, J.F.; Ainkaran, P.; Blackstone, E.H. Coronary Artery Bypass Grafting in Diabetics: A Growing Health Care Cost Crisis. *J. Thorac. Cardiovasc. Surg.* **2015**, *150*, 304–312. [CrossRef]
29. Liao, K.-M.; Kuo, L.-T.; Lu, H.-Y. Hospital Costs and Prognosis in End-Stage Renal Disease Patients Receiving Coronary Artery Bypass Grafting. *BMC Nephrol.* **2020**, *21*, 1–9. [CrossRef]
30. Fengsrud, E.; Englund, A.; Ahlsson, A. Pre- and Postoperative Atrial Fibrillation in CABG Patients have Similar Prognostic Impact. *Scand. Cardiovasc. J.* **2016**, *51*, 21–27. [CrossRef]
31. Pollock, B.D.; Filardo, G.; Da Graca, B.; Phan, T.K.; Ailawadi, G.; Thourani, V.; Damiano, J.R.J.; Edgerton, J.R. Predicting New-Onset Post-Coronary Artery Bypass Graft Atrial Fibrillation with Existing Risk Scores. *Ann. Thorac. Surg.* **2018**, *105*, 115–121. [CrossRef]

MDPI
St. Alban-Anlage 66
4052 Basel
Switzerland
Tel. +41 61 683 77 34
Fax +41 61 302 89 18
www.mdpi.com

Healthcare Editorial Office
E-mail: healthcare@mdpi.com
www.mdpi.com/journal/healthcare

www.ingramcontent.com/pod-product-compliance
Lightning Source LLC
Chambersburg PA
CBHW061452220926
51850CB00004B/70
* 9 7 8 3 0 3 6 5 3 7 4 2 9 *